Lecture Notes in Computer Science 5303

David Forsyth Philip Torr
Andrew Zisserman (Eds.)

Computer Vision – ECCV 2008

10th European Conference on Computer Vision
Marseille, France, October 12-18, 2008
Proceedings, Part II

Springer

Volume Editors

David Forsyth
University of Illinois at Urbana Champaign, Computer Science Department
3310 Siebel Hall, Urbana, IL 61801, USA
E-mail: daf@cs.uiuc.edu

Philip Torr
Oxford Brookes University, Department of Computing
Wheatley, Oxford OX33 1HX, UK
E-mail: philiptorr@brookes.ac.uk

Andrew Zisserman
University of Oxford, Department of Engineering Science
Parks Road, Oxford OX1 3PJ, UK
E-mail: az@robots.ox.ac.uk

Library of Congress Control Number: Applied for

CR Subject Classification (1998): I.4, I.2.10, I.5.4, I.5, I.7.5

LNCS Sublibrary: SL 6 – Image Processing, Computer Vision, Pattern Recognition, and Graphics

ISSN 0302-9743
ISBN-10 3-540-88685-0 Springer Berlin Heidelberg New York
ISBN-13 978-3-540-88685-3 Springer Berlin Heidelberg New York

Springer is a part of Springer Science+Business Media

springer.com

Typesetting: Camera-ready by author, data conversion by Scientific Publishing Services, Chennai, India
Printed on acid-free paper SPIN: 12553570 06/3180 5 4 3 2 1 0

Preface

Welcome to the 2008 European Conference on Computer Vision. These proceedings are the result of a great deal of hard work by many people. To produce them, a total of 871 papers were reviewed. Forty were selected for oral presentation and 203 were selected for poster presentation, yielding acceptance rates of 4.6% for oral, 23.3% for poster, and 27.9% in total.

We applied three principles. First, since we had a strong group of Area Chairs, the final decisions to accept or reject a paper rested with the Area Chair, who would be informed by reviews and could act only in consensus with another Area Chair. Second, we felt that authors were entitled to a summary that explained how the Area Chair reached a decision for a paper. Third, we were very careful to avoid conflicts of interest.

Each paper was assigned to an Area Chair by the Program Chairs, and each Area Chair received a pool of about 25 papers. The Area Chairs then identified and ranked appropriate reviewers for each paper in their pool, and a constrained optimization allocated three reviewers to each paper. We are very proud that every paper received at least three reviews.

At this point, authors were able to respond to reviews. The Area Chairs then needed to reach a decision. We used a series of procedures to ensure careful review and to avoid conflicts of interest. Program Chairs did not submit papers. The Area Chairs were divided into three groups so that no Area Chair in the group was in conflict with any paper assigned to any Area Chair in the group. Each Area Chair had a "buddy" in their group. Before the Area Chairs met, they read papers and reviews, contacted reviewers to get reactions to submissions and occasionally asked for improved or additional reviews, and prepared a rough summary statement for each of the papers in their pool.

At the Area Chair meeting, groups met separately so that Area Chairs could reach a consensus with their buddies, and make initial oral/poster decisions. We met jointly so that we could review the rough program, and made final oral/poster decisions in groups. In the separate meetings, there were no conflicts. In the joint meeting, any Area Chairs with conflicts left the room when relevant papers were discussed. Decisions were published on the last day of the Area Chair meeting.

There are three more somber topics to report. First, the Program Chairs had to deal with several double submissions. Referees or Area Chairs identified potential double submissions, we checked to see if these papers met the criteria published in the call for papers, and if they did, we rejected the papers and did not make reviews available. Second, two submissions to ECCV 2008 contained open plagiarism of published works. We will pass details of these attempts to journal editors and conference chairs to make further plagiarism by the responsible parties more difficult. Third, by analysis of server logs we discovered that there had been a successful attempt to download all submissions shortly after the deadline. We warned all authors that this had happened to ward off dangers to intellectual property rights, and to minimize the chances that an attempt at plagiarism would be successful. We were able to identify the responsible party, discussed this matter with their institutional management, and believe we resolved the issue as well as we could have. Still, it is important to be aware that no security or software system is completely safe, and papers can leak from conference submission.

We felt the review process worked well, and recommend it to the community. The process would not have worked without the efforts of many people. We thank Lyndsey Pickup, who managed the software system, author queries, Area Chair queries and general correspondence (most people associated with the conference will have exchanged e-mails with her at some point). We thank Simon Baker, Ramin Zabih and especially Jiří Matas for their wise advice on how to organize and run these meetings; the process we have described is largely their model from CVPR 2007. We thank Jiří Matas and Dan Večerka, for extensive help with, and support of, the software system. We thank C. J. Taylor for the 3-from-5 optimization code. We thank the reviewers for their hard work. We thank the Area Chairs for their very hard work, and for the time and attention each gave to reading papers, reviews and summaries, and writing summaries.

We thank the Organization Chairs Peter Sturm and Edmond Boyer, and the General Chair, Jean Ponce, for their help and support and their sharing of the load. Finally, we thank Nathalie Abiola, Nasser Bacha, Jacques Beigbeder, Jerome Bertsch, Joëlle Isnard and Ludovic Ricardou of ENS for administrative support during the Area Chair meeting, and Danièle Herzog and Laetitia Libralato of INRIA Rhône-Alpes for administrative support after the meeting.

August 2008

Andrew Zisserman
David Forsyth
Philip Torr

Organization

Conference Chair

Jean Ponce	Ecole Normale Supérieure, France

Honorary Chair

Jan Koenderink	EEMCS, Delft University of Technology, The Netherlands

Program Chairs

David Forsyth	University of Illinois, USA
Philip Torr	Oxford Brookes University, UK
Andrew Zisserman	University of Oxford, UK

Organization Chairs

Edmond Boyer	LJK/UJF/INRIA Grenoble–Rhône-Alpes, France
Peter Sturm	INRIA Grenoble–Rhône-Alpes, France

Specialized Chairs

Frédéric Jurie	Workshops	Université de Caen, France
Frédéric Devernay	Demos	INRIA Grenoble–Rhône-Alpes, France
Edmond Boyer	Video Proc.	LJK/UJF/INRIA Grenoble–Rhône-Alpes, France
James Crowley	Video Proc.	INPG, France
Nikos Paragios	Tutorials	Ecole Centrale, France
Emmanuel Prados	Tutorials	INRIA Grenoble–Rhône-Alpes, France
Christophe Garcia	Industrial Liaison	France Telecom Research, France
Théo Papadopoulo	Industrial Liaison	INRIA Sophia, France
Jiří Matas	Conference Software	CTU Prague, Czech Republic
Dan Večerka	Conference Software	CTU Prague, Czech Republic

Program Chair Support

Lyndsey Pickup	University of Oxford, UK

Administration

Danile Herzog	INRIA Grenoble–Rhône-Alpes, France
Laetitia Libralato	INRIA Grenoble–Rhône-Alpes, France

Conference Website

Elisabeth Beaujard	INRIA Grenoble–Rhône-Alpes, France
Amaël Delaunoy	INRIA Grenoble–Rhône-Alpes, France
Mauricio Diaz	INRIA Grenoble–Rhône-Alpes, France
Benjamin Petit	INRIA Grenoble–Rhône-Alpes, France

Printed Materials

Ingrid Mattioni	INRIA Grenoble–Rhône-Alpes, France
Vanessa Peregrin	INRIA Grenoble–Rhône-Alpes, France
Isabelle Rey	INRIA Grenoble–Rhône-Alpes, France

Area Chairs

Horst Bischof	Graz University of Technology, Austria
Michael Black	Brown University, USA
Andrew Blake	Microsoft Research Cambridge, UK
Stefan Carlsson	NADA/KTH, Sweden
Tim Cootes	University of Manchester, UK
Alyosha Efros	CMU, USA
Jan-Olof Eklund	KTH, Sweden
Mark Everingham	University of Leeds, UK
Pedro Felzenszwalb	University of Chicago, USA
Richard Hartley	Australian National University, Australia
Martial Hebert	CMU, USA
Aaron Hertzmann	University of Toronto, Canada
Dan Huttenlocher	Cornell University, USA
Michael Isard	Microsoft Research Silicon Valley, USA
Aleš Leonardis	University of Ljubljana, Slovenia
David Lowe	University of British Columbia, Canada
Jiří Matas	CTU Prague, Czech Republic
Joe Mundy	Brown University, USA
David Nistér	Microsoft Live Labs/Microsoft Research, USA
Tomáš Pajdla	CTU Prague, Czech Republic
Patrick Pérez	IRISA/INRIA Rennes, France
Marc Pollefeys	ETH Zürich, Switzerland
Ian Reid	University of Oxford, UK
Cordelia Schmid	INRIA Grenoble–Rhône-Alpes, France
Bernt Schiele	Darmstadt University of Technology, Germany
Christoph Schnörr	University of Mannheim, Germany
Steve Seitz	University of Washington, USA

Richard Szeliski	Microsoft Research, USA
Antonio Torralba	MIT, USA
Bill Triggs	CNRS/Laboratoire Jean Kuntzmann, France
Tinne Tuytelaars	Katholieke Universiteit Leuven, Belgium
Luc Van Gool	Katholieke Universiteit Leuven, Belgium
Yair Weiss	The Hebrew University of Jerusalem, Israel
Chris Williams	University of Edinburgh, UK
Ramin Zabih	Cornell University, USA

Conference Board

Horst Bischof	Graz University of Technology, Austria
Hans Burkhardt	University of Freiburg, Germany
Bernard Buxton	University College London, UK
Roberto Cipolla	University of Cambridge,UK
Jan-Olof Eklundh	Royal Institute of Technology, Sweden
Olivier Faugeras	INRIA, Sophia Antipolis, France
Anders Heyden	Lund University, Sweden
Aleš Leonardis	University of Ljubljana, Slovenia
Bernd Neumann	University of Hamburg, Germany
Mads Nielsen	IT University of Copenhagen, Denmark
Tomáš Pajdla	CTU Prague, Czech Republic
Giulio Sandini	University of Genoa, Italy
David Vernon	Trinity College, Ireland

Program Committee

Sameer Agarwal
Aseem Agarwala
Jörgen Ahlberg
Narendra Ahuja
Yiannis Aloimonos
Tal Arbel
Kalle Åström
Peter Auer
Jonas August
Shai Avidan
Simon Baker
Kobus Barnard
Adrien Bartoli
Benedicte Bascle
Csaba Beleznai
Peter Belhumeur
Serge Belongie
Moshe Ben-Ezra
Alexander Berg
Tamara Berg
James Bergen
Marcelo Bertalmio
Bir Bhanu
Stan Bileschi
Stan Birchfield
Volker Blanz
Aaron Bobick
Endre Boros
Terrance Boult
Richard Bowden
Edmond Boyer
Yuri Boykov
Gary Bradski
Chris Bregler
Thomas Breuel
Gabriel Brostow
Matthew Brown
Michael Brown
Thomas Brox
Andrés Bruhn
Antoni Buades
Joachim Buhmann
Hans Burkhardt
Andrew Calway
Rodrigo Carceroni
Gustavo Carneiro
M. Carreira-Perpinan
Tat-Jen Cham
Rama Chellappa
German Cheung
Ondřej Chum
James Clark
Isaac Cohen
Laurent Cohen
Michael Cohen
Robert Collins
Dorin Comaniciu

James Coughlan
David Crandall
Daniel Cremers
Antonio Criminisi
David Cristinacce
Gabriela Csurka
Navneet Dalal
Kristin Dana
Kostas Daniilidis
Larry Davis
Andrew Davison
Nando de Freitas
Daniel DeMenthon
David Demirdjian
Joachim Denzler
Michel Dhome
Sven Dickinson
Gianfranco Doretto
Gyuri Dorko
Pinar Duygulu Sahin
Charles Dyer
James Elder
Irfan Essa
Andras Ferencz
Rob Fergus
Vittorio Ferrari
Sanja Fidler
Mario Figueiredo
Graham Finlayson
Robert Fisher
François Fleuret
Wolfgang Förstner
Charless Fowlkes
Jan-Michael Frahm
Friedrich Fraundorfer
Bill Freeman
Brendan Frey
Andrea Frome
Pascal Fua
Yasutaka Furukawa
Daniel Gatica-Perez
Dariu Gavrila
James Gee
Guido Gerig
Theo Gevers
Christopher Geyer
Michael Goesele
Dan Goldman
Shaogang Gong
Leo Grady
Kristen Grauman
Eric Grimson
Fred Hamprecht
Edwin Hancock
Allen Hanson
James Hays
Carlos Hernández
Anders Heyden
Adrian Hilton
David Hogg
Derek Hoiem
Alex Holub
Anthony Hoogs
Daniel Huber
Alexander Ihler
Michal Irani
Hiroshi Ishikawa
David Jacobs
Bernd Jähne
Hervé Jégou
Ian Jermyn
Nebojsa Jojic
Michael Jones
Frédéric Jurie
Timor Kadir
Fredrik Kahl
Amit Kale
Kenichi Kanatani
Sing Bing Kang
Robert Kaucic
Qifa Ke
Renaud Keriven
Charles Kervrann
Ron Kikinis
Benjamin Kimia
Ron Kimmel
Josef Kittler
Hedvig Kjellström
Leif Kobbelt
Pushmeet Kohli
Esther Koller-Meier
Vladimir Kolmogorov
Nikos Komodakis
Kurt Konolige
Jana Košecká
Zuzana Kukelova
Sanjiv Kumar
Kyros Kutulakos
Ivan Laptev
Longin Jan Latecki
Svetlana Lazebnik
Erik Learned-Miller
Yann Lecun
Bastian Leibe
Vincent Lepetit
Thomas Leung
Anat Levin
Fei-Fei Li
Hongdong Li
Stephen Lin
Jim Little
Ce Liu
Yanxi Liu
Brian Lovell
Simon Lucey
John Maccormick
Petros Maragos
Aleix Martinez
Iain Matthews
Wojciech Matusik
Bruce Maxwell
Stephen Maybank
Stephen McKenna
Peter Meer
Etienne Mémin
Dimitris Metaxas
Branislav Mičušík
Krystian Mikolajczyk
Anurag Mittal
Theo Moons
Greg Mori
Pawan Mudigonda
David Murray
Srinivasa Narasimhan
Randal Nelson

Ram Nevatia
Jean-Marc Odobez
Björn Ommer
Nikos Paragios
Vladimir Pavlovic
Shmuel Peleg
Marcello Pelillo
Pietro Perona
Maria Petrou
Vladimir Petrovic
Jonathon Phillips
Matti Pietikäinen
Axel Pinz
Robert Pless
Tom Pock
Fatih Porikli
Simon Prince
Long Quan
Ravi Ramamoorthi
Deva Ramanan
Anand Rangarajan
Ramesh Raskar
Xiaofeng Ren
Jens Rittscher
Rómer Rosales
Bodo Rosenhahn
Peter Roth
Stefan Roth
Volker Roth
Carsten Rother
Fred Rothganger
Daniel Rueckert
Dimitris Samaras
Radim Šára
Eric Saund
Silvio Savarese
Daniel Scharstein
Yoav Schechner
Kondrad Schindler
Stan Sclaroff
Mubarak Shah
Gregory Shakhnarovich
Eli Shechtman
Jianbo Shi
Kaleem Siddiqi
Leonid Sigal
Sudipta Sinha
Josef Sivic
Cristian Sminchişescu
Anuj Srivastava
Drew Steedly
Gideon Stein
Björn Stenger
Christoph Strecha
Erik Sudderth
Josephine Sullivan
David Suter
Tomáš Svoboda
Hai Tao
Marshall Tappen
Demetri Terzopoulos
Carlo Tomasi
Fernando Torre
Lorenzo Torresani
Emanuele Trucco
David Tschumperlé
John Tsotsos
Peter Tu
Matthew Turk
Oncel Tuzel
Carole Twining
Ranjith Unnikrishnan
Raquel Urtasun
Joost Van de Weijer
Manik Varma
Nuno Vasconcelos
Olga Veksler
Jakob Verbeek
Luminita Vese
Thomas Vetter
René Vidal
George Vogiatzis
Daphna Weinshall
Michael Werman
Tomáš Werner
Richard Wildes
Lior Wolf
Ying Wu
Eric Xing
Yaser Yacoob
Ruigang Yang
Stella Yu
Lihi Zelnik-Manor
Richard Zemel
Li Zhang
S. Zhou
Song-Chun Zhu
Todd Zickler
Lawrence Zitnick

Additional Reviewers

Lourdes Agapito
Daniel Alexander
Elli Angelopoulou
Alexandru Balan
Adrian Barbu
Nick Barnes
João Barreto
Marian Bartlett
Herbert Bay
Ross Beveridge
V. Bhagavatula
Edwin Bonilla
Aeron Buchanan
Michael Burl
Tiberio Caetano
Octavia Camps
Sharat Chandran
François Chaumette
Yixin Chen
Dmitry Chetverikov
Sharat Chikkerur
Albert Chung
Nicholas Costen
Gabriela Oana Cula
Goksel Dedeoglu
Hervé Delingette
Michael Donoser

Mark Drew
Zoran Duric
Wolfgang Einhauser
Aly Farag
Beat Fasel
Raanan Fattal
Paolo Favaro
Rogerio Feris
Cornelia Fermüller
James Ferryman
David Forsyth
Jean-Sébastien Franco
Mario Fritz
Andrea Fusiello
Meirav Galun
Bogdan Georgescu
A. Georghiades
Georgy Gimel'farb
Roland Goecke
Toon Goedeme
Jacob Goldberger
Luis Goncalves
Venu Govindaraju
Helmut Grabner
Michael Grabner
Hayit Greenspan
Etienne Grossmann
Richard Harvey
Sam Hasinoff
Horst Haussecker
Jesse Hoey
Slobodan Ilic
Omar Javed
Qiang Ji
Jiaya Jia
Hailin Jin
Ioannis Kakadiaris
Joni-K. Kämäräinen
George Kamberov
Yan Ke
Andreas Klaus
Georg Klein
Reinhard Koch
Mathias Kolsch
Andreas Koschan
Christoph Lampert
Mike Langer
Georg Langs
Neil Lawrence
Sang Lee
Boudewijn Lelieveldt
Marc Levoy
Michael Lindenbaum
Chengjun Liu
Qingshan Liu
Manolis Lourakis
Ameesh Makadia
Ezio Malis
R. Manmatha
David Martin
Daniel Martinec
Yasuyuki Matsushita
Helmut Mayer
Christopher Mei
Paulo Mendonça
Majid Mirmehdi
Philippos Mordohai
Pierre Moreels
P.J. Narayanan
Nassir Navab
Jan Neumann
Juan Carlos Niebles
Ko Nishino
Thomas O'Donnell
Takayuki Okatani
Kenji Okuma
Margarita Osadchy
Mustafa Ozuysal
Sharath Pankanti
Sylvain Paris
James Philbin
Jean-Philippe Pons
Emmanuel Prados
Zhen Qian
Ariadna Quattoni
Ali Rahimi
Ashish Raj
Visvanathan Ramesh
Christopher Rasmussen
Tammy Riklin-Raviv
Charles Rosenberg
Arun Ross
Michael Ross
Szymon Rusinkiewicz
Bryan Russell
Sudeep Sarkar
Yoichi Sato
Ashutosh Saxena
Florian Schroff
Stephen Se
Nicu Sebe
Hans-Peter Seidel
Steve Seitz
Thomas Serre
Alexander Shekhovtsov
Ilan Shimshoni
Michal Sofka
Jan Solem
Gerald Sommer
Jian Sun
Rahul Swaminathan
Hugues Talbot
Chi-Keung Tang
Xiaoou Tang
C.J. Taylor
Jean-Philippe Thiran
David Tolliver
Yanghai Tsin
Zhuowen Tu
Vaibhav Vaish
Anton van den Hengel
Bram Van Ginneken
Dirk Vandermeulen
Alessandro Verri
Hongcheng Wang
Jue Wang
Yizhou Wang
Gregory Welch
Ming-Hsuan Yang
Caspi Yaron
Jieping Ye
Alper Yilmaz
Christopher Zach
Hongyuan Zha
Cha Zhang
Jerry Zhu
Lilla Zollei

Sponsoring Institutions

Table of Contents – Part II

People

Faces

Poster Session II

Tracking

Floor Fields for Tracking in High Density Crowd Scenes

Saad Ali and Mubarak Shah

Computer Vision Lab, University of Central Florida, Orlando, USA
{sali,shah}@eecs.ucf.edu

Abstract. This paper presents an algorithm for tracking individual targets in high density crowd scenes containing hundreds of people. Tracking in such a scene is extremely challenging due to the small number of pixels on the target, appearance ambiguity resulting from the dense packing, and severe inter-object occlusions. The novel tracking algorithm, which is outlined in this paper, will overcome these challenges using a *scene structure based force model*. In this force model an individual, when moving in a particular scene, is subjected to global and local forces that are functions of the layout of that scene and the locomotive behavior of other individuals in the scene. The key ingredients of the force model are three floor fields, which are inspired by the research in the field of evacuation dynamics, namely *Static Floor Field* (SFF), *Dynamic Floor Field* (DFF), and *Boundary Floor Field* (BFF). These fields determine the probability of move from one location to another by converting the long-range forces into local ones. The SFF specifies regions of the scene which are attractive in nature (e.g. an exit location). The DFF specifies the immediate behavior of the crowd in the vicinity of the individual being tracked. The BFF specifies influences exhibited by the barriers in the scene (e.g. walls, no-go areas). By combining cues from all three fields with the available appearance information, we track individual targets in high density crowds.

1 Introduction

Tracking individuals in a high density crowd scene is challenging for a number of reasons: 1) the number of pixels on an object decreases with the increasing density of the objects; 2) constant interaction among the individuals in a crowd makes it hard to discern individuals from one another; 3) occlusions caused by inter-object interactions result in the loss of observation of the target object; 4) the mechanics of a human crowd is complex as it exhibits goal-directed dynamics and psychological characteristics which in turn influence how an individual person will behave in a crowd. Examples of high density crowd scenes are shown in Fig. 1.

To overcome some of these challenges, we have developed an algorithm for tracking individual targets in *high density crowd scenes* containing hundreds of people at a time. The proposed algorithm is based on the observation that

D. Forsyth, P. Torr, and A. Zisserman (Eds.): ECCV 2008, Part II, LNCS 5303, pp. 1–14, 2008.

Fig. 1. Examples of high density crowd scenes. (a)-(c) Hundreds of people participating in marathons. (d) A scene from a densely packed railway station in India. (e) A group of people moving in opposite directions.

the locomotive behavior of an individual in a crowded scene is a function of collective patterns evolving from the space-time interactions of individuals among themselves and with the layout of the scene. These collective behavioral patterns, therefore, can be included as an auxiliary source of information to constrain the likely locations or paths that can be taken by the target object in the scene. In other words, natural crowd flow and scene constraints influencing the behavior of a person in a dense crowd can be used as priors to impose high-level direction for tracking purposes. Our novel model called the 'scene structure force model', is proposed to directly incorporate such prior knowledge and influences.

In our tracking algorithm, the crowd is treated as a collection of mutually interacting particles. This is a reasonable model, because when people are densely packed individual movement is restricted and members of the crowd can be considered granular particles. In order to track a specific individual in the crowd, we model the instantaneous movement of that person (or particle) with a matrix of preferences containing the probabilities of a move in a certain preferred direction. The probabilities take into consideration multiple sources of information and constraints arising from the appearance of the target individual, flow of the crowd and the structure of the scene. The flow of the crowd and scene structure incorporated by introducing the concept of *floor fields*, which model the interactions between individuals and their preferred direction of movement by transforming long ranged forces into local ones. For instance, a long range force that compels an individual in a crowd to move towards the exit door, can be converted into a local force such that it increases the instantaneous probability of a move in that direction. The transition probability of a tracked person then depends on the strength of the floor field in his/her neighborhood. The concept of a floor field is inspired by the field of evacuation dynamics [1,2], where floor fields are *manually* designed to simulate behaviors of pedestrians in panic situations. We compute three floor fields *automatically* from the visual data: a 'Static Floor Field' (SFF), a 'Boundary Floor Field' (BFF), and a 'Dynamic Floor Field' (DFF).

2 Related Work

Tracking is a popular research area in computer vision [6,16]. In this section, we briefly review tracking algorithms that were designed specifically for crowded scenes. Interested readers are referred to the recent survey by Yilmaz *et al.* [10]

for an in-depth review of the tracking literature. For tracking in crowded environments, the method proposed by Zhao *et al.* [14] was one of the first algorithms. Their algorithm used articulated ellipsoids to model the human shape, color histograms to model appearance, and a Gaussian distribution to model the background for segmentation. The initial detection of objects was based on the their segmentation scheme ([13]) proposed earlier. However, their method is not suitable for the high density crowd situations that we are dealing with, because in such a situation the whole human body is rarely visible, which makes it impossible to fit elaborate ellipsoid models of human body shapes. Brostow *et al.* [12] presented a probabilistic framework for the clustering of feature point trajectories for detection and tracking of individual pedestrians in crowds. They hypothesized that pairs of points that appear to move together were likely to be part of the same individual and as such could be used for detection and tracking.

There are other interesting tracking methods which were developed for tracking sparse crowds of ants [6], hockey players [9], crowds of clumped people [8], a dense flock of bats [4], and biological cells [5]. In general, these tracking methods are object centric and do not exploit any high level or global knowledge that may aid the tracking algorithm. This is one of the major difference between our tracking algorithm and these approaches; we have integrated high level constraints resulting from crowd flow and scene structure into the tracking algorithm.

3 Tracking Framework

The crowd flow in the scene is treated as a collection of mutually interacting particles. Therefore, given a video $E = [f_1, f_2, \ldots, f_N]$, where N is the total number of frames, the image space is discretized into cells where each cell is occupied by a single particle $o_{\mathbf{x}i}$. Here, $\mathbf{x}i = (xi, yi)$ is the coordinate of the i-th pixel at which the particle is located. The relationship between cells, pixels and particles is as follows: Cells form a grid over the spatial extent of the image and each cell is always associated with a single particle, although a cell can contain more than one pixel depending upon the resolution of the grid. A particle represents all the pixels in the cell. For tracking, the target individual is represented by a set of particles $\mathcal{P} = [\ldots, o_{\mathbf{x}i}, \ldots]$ (red particles in Fig. 2(a)) as an individual can span multiple cells in the image. An appearance template, H, of the target is then computed using all the pixels corresponding to the underlying pixels represented by particles $o \in \mathcal{P}$. The target moves from one cell to the next at discrete time steps, $t \rightarrow t + 1$, according to a transition probability that determines the likely direction of the motion. This transition probability is associated with the centroid (yellow particle in Fig. 2(a)) although its computation uses information from all the particles in the set $\mathcal{P}$. The transition probability is determined by two factors: 1) the similarity between the appearance templates at the current location and the next; 2) the influence generated by the floor fields. Formally, if the target individual is currently at cell i (the cell containing the yellow particle in Fig. 2(c)), then the probability of moving to a neighboring cell j (cells

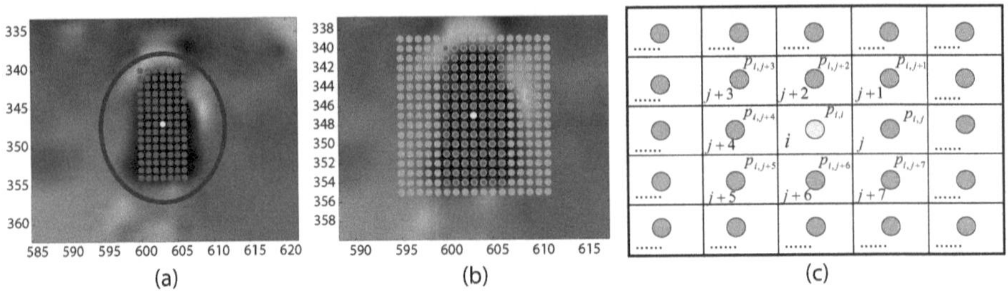

Fig. 2. (a) The particles $o \in \mathcal{P}$ belonging to the individual we want to track. Each particle occupies a single cell with the yellow particle representing the centroid of the object. In order to track the target, we track the yellow particle through the scene. (b) The green particles represent the search area for the possible next location of the yellow particle. (c) The matrix of preferred walking directions around the yellow particle. Each value in the matrix is the probability of the yellow particle moving from the center cell i to the surrounding cell. The transition probability p_{ij} is computed using 1.

containing green particles in Fig. 2(c)) is:

$$p_{ij} = Ce^{k_D D_{ij}} e^{k_S S_{ij}} e^{k_B B_{ij}} R_{ij}, \tag{1}$$

where D_{ij}, S_{ij}, and B_{ij} are the influences of the DFF, SFF, and BFF, respectively. While k_D, k_S, and k_B are the coupling strength of the tracked object to the DFF, SFF, and BFF, respectively. R_{ij} is the similarity measure between the initial appearance template H and the new appearance template of the target computed at location j. C is a normalization constant.

The benefit of using the above equation is that it employs the crowd flow and scene layout constraints to weight the appearance similarity information when tracking the person. The crowd flow constraints are captured by the DFF and SFF, while the scene layout constraints are captured by the BFF. The weighting can be understood in the following manner: if the crowd follows a path say from the cell i to j, the SFF will have a high value for S_{ij} and in turn will favor that direction of motion. This will increase the likelihood of matching the current appearance template of the target at location i with the one at location j at next time instance. Similarly, if immediate crowd behavior is such that it is moving from the cell i to $j+1$, the DFF will have a higher value for $D_{i,j+1}$ and will favor that direction of motion. This will increase the likelihood of matching the appearance template at location i with the one at location $j+1$. Thus we can use the crowd flow information to gain more confidence about our appearance matching scores. Next, we describe the algorithm for computing S_{ij}, D_{ij}, and B_{ij} from their respective floor fields.

3.1 Static Floor Field - S_{ij}

The SFF is aimed at capturing attractive and constant properties of the scene. These properties include preferred areas, such as dominant pathes often taken by the crowd as it moves through the scene, and preferred exit locations. In our

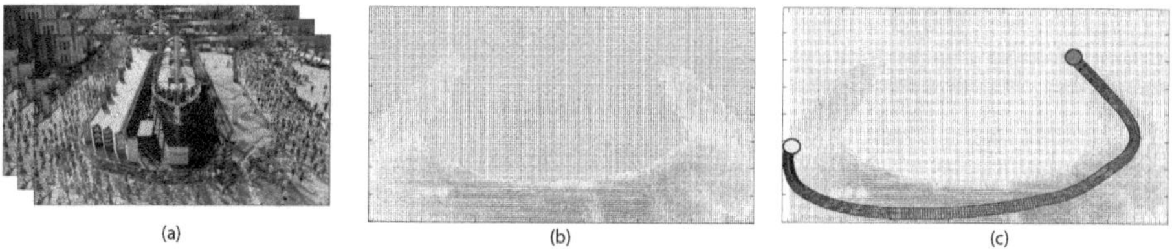

Fig. 3. (a) The dense optical flow for frames, $[f_1, f_2, \ldots, f_M]$, of the video. (b) The computed point flow field. (c) The sink seeking process: the yellow circle represents the initial location, while the red circle shows the corresponding sink. The black windows represent the area used to weight the local velocity and propagate the sink seeking process. The red trajectory represents the 'sink seeking path', while the number of black windows represents the corresponding number of sink steps.

framework, the SFF is computed only once for a given scene during the learning period, which spans initial $M << N$ frames. The SFF computation steps are: i) Computation of a point flow field; ii) Sink Seeking.

Point Flow Field. A 'point flow field' represents the instantaneous changes of motions present in the video. Each vector in this field is a 4-dimensional vector obtained by augmenting the local flow vector with the position information. The new vector is referred to as a 'point flow vector', hence the name 'point flow field'. Using the first M frames of an input video, $E = [f_1, f_2, \ldots, f_M]$, a dense optical flow can be computed between consecutive frames using the method of [15]. Then, for each cell (or pixel) i, a point flow vector, $Z_i = (X_i, V_i)$, is computed, which includes both the location $X_i = (x_i, y_i)$ and the optical flow vector $V_i = (v_{x_i}, v_{y_i})$. Note that V_i is the mean of $(M - 1)$ optical flow vectors computed at pixel i from the first M frames of the video. All flow vectors averaged over M frames of the video then constitute the 'Point Flow Field', which represents the smoothed out motion information of the video in that interval. This smoothed motion information assists in computing the dominant properties (paths, exits) of the scene, which is the primary objective of the SFF. Fig. 3(a) shows flow vectors generated for a marathon video using the dense optical flow computation [15]. The resulting point flow field is given in Fig. 3(b).

Sink Seeking Process. The point-flow field is then used to discover the regions in the scene called 'sinks'. The idea behind the sink seeking process is that the behavior of large crowds can be described as goal directed and rational, i.e. the members of the crowd have clear knowledge of what and where their goals (destinations) lie in the scene [3]. Therefore, if we know the locations of the sinks, which are the desired goals, then for any given point in the scene we can compute a local force representing the tendency of the individual at that point to move towards the nearest sink. This local force will be a function of the shortest distance to the sink in terms of the appropriate distance metric.

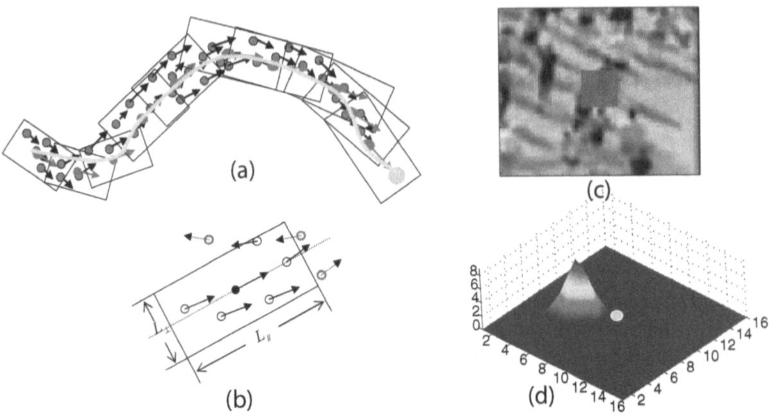

Fig. 4. (a) Sink seeking (red: the states of the point flow in the sink seeking process, orange: the sink, rectangles: sliding windows, yellow: the sink path); (b) Sliding window (solid circle: the point flow under consideration; rectangle: sliding window; hollow circles: neighboring points; red circles: non-neighboring points).(c) The region where we want to compute the DFF. (d) The computed DFF where the yellow circle represents the cell i. In this case, the DFF is capturing the dynamic relationship between the cell i and the neighboring cells.

In order to compute the sinks and their shortest distances, we initialize a grid of particles over the point flow field of the scene. Then, a particle dropped at a non-zero velocity location has the tendency to move to a new position under the influence of the neighboring point flow vectors. It then moves from that new position to the next one and continues this process. In order to optimally combine the influence of the neighboring point flow vectors, the velocity at each new position is re-estimated as the weighted sum of its neighboring velocities (Fig. 4(a)). The weights are computed using a kernel density method. If all the weights are below a threshold, it implies that the new velocity is not significant enough to drive the particle to the next position. Therefore, the particle will stop and the process of pursuing a new location is discontinued. We call this process the *sink seeking process* and the last state (stopping state) of the process is called the *sink*. The corresponding path taken by the particle to reach the sink is called the *sink path*(Figures 3(c) and 4(a)). The length of the sink path is the minimum number of steps required to reach the closest exit location in the scene. The number of steps taken during the sink-seeking process to reach the sink are called the *seek steps*. This is also the distance metric used for representing the shortest distance. Note that the sink seeking process is carried out for each point in the point-flow field, thus generating one sink path per point. Formally, let $\{Z_1, Z_2, \cdots, Z_n\}$ is the point flow field of the video, where the state of the point i is defined as: $\widetilde{Z}_{i,t} = (\widetilde{X}_{i,t}, \widetilde{V}_{i,t}), t = 1, 2, ...,$ and computed as:

$$\widetilde{Z}_{i,1} = Z_i \ , \quad \widetilde{X}_{i,t+1} = \widetilde{X}_{i,t} + \widetilde{V}_{i,t}, \tag{2}$$

$$\widetilde{V}_{i,t} = \frac{\sum_{n \in Neighbor(\widetilde{X}_{i,t})} V_n W_{t,n}}{\sum_{n \in Neighbor(\widetilde{X}_{i,t})} W_{t,n}} \ , \ W_{t,n} = \exp\left(-\left\|\frac{\widetilde{V}_{t-1} - V_n}{h_{t-1}}\right\|^2\right), \tag{3}$$

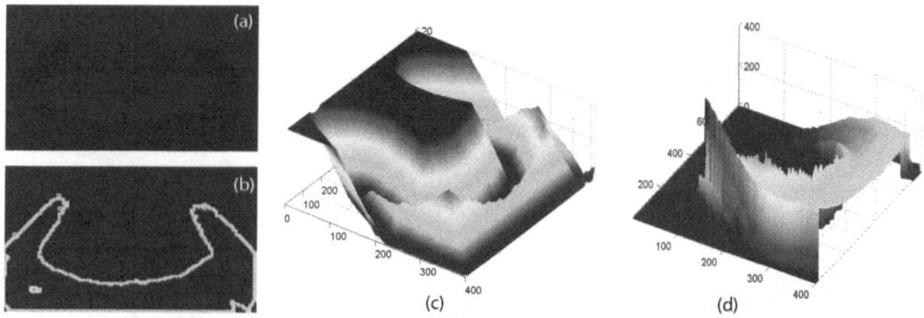

Fig. 5. (a) Crowd flow segmentation obtained by the method described in [11]. (b) The edge map obtained from the segmentation. (c) The boundary floor field for the sequence shown in Fig. 1c. The higher values in the field represent the decreasing effect of the repulsive potential generated by the barriers. In this case, the barrier effect vanishes for distances greater than 20 pixels. (d) The static floor field computed by our algorithm for the sequence shown in Fig. 1(c).

In the previous equations, it is clear that the new position of a point only depends on the location and velocity at the previous state. However the new velocity $\widetilde{V}_{i,t+1}$ depends not only on the previous velocity, but also on the observed velocities of its neighbors, which represent the motion trend of a local group. In this work, we employ the kernel based estimation which is similar to the mean shift approach [16]. However, there is one important difference. In mean shift tracking, the *appearance* of pixels within a small neighborhood of the object is used to determine the location of the object in the next frame. In our approach, we use *the location and the velocity* of the neighboring points in the point flow field to determine the next location.

SFF Generation. The SFF is generated by using the sink steps for each sink path. We find the location (x, y) (in the image space) at which each sink path starts and place the value of corresponding 'sink step' at that location. Fig. 5(d) shows the computed SFF for the sequence in Fig. 1(c). It is interesting to note that the shape of the SFF emphasizes the notion that if you place a particle at any location, it will roll down towards the exit. This is precisely what the goal oriented dynamics of the crowd in this scene represent. In the tracking algorithm, the shape of the SFF translates into a force in the direction that requires the least number of steps to reach the nearest exit or sink. That is, the difference between the values in cell i and j in this field is the measure of the S_{ij} parameter of 1. Other SFFs are shown in Fig. 6.

3.2 Boundary Floor Field - B_{ij}

The purpose of the BFF is to capture influences generated by barriers or walls in the scene. This influence is usually repulsive in nature. The computation of the BFF requires the localization of physical and virtual barriers in the scene. The virtual barriers arise from the presence of dynamically distinct crowd flows

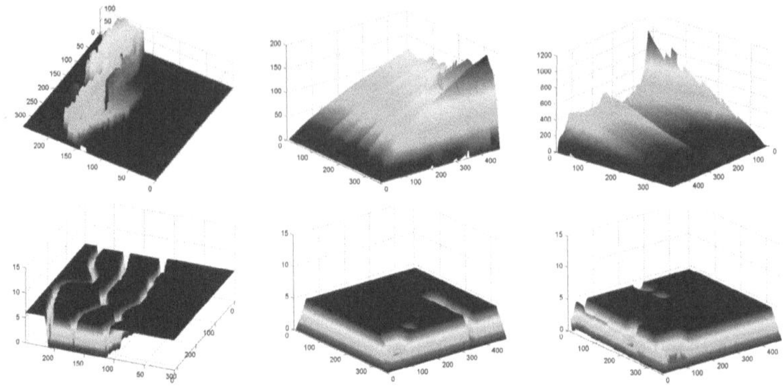

Fig. 6. The SFFs (top) and BFFs (bottom) of various sequences. Left: For the sequence in Fig. 1(e). Center: For the marathon sequence in Fig. 1(a). Right: For the marathon sequence in Fig. 1(b).

in the scene. The computation of the BFF is performed after a set time interval ΔT_B, and works on a group of frames defined by the parameter N_B. That is, computation of BFF at time t uses frames $[f_t, f_{t+1}, \ldots, f_{t+N_B}]$.

The computation of BFF is based on the crowd flow segmentation algorithm proposed earlier by the authors [11]. In that algorithm, physical and virtual barriers in the scene were represented by the ridges of the Finite Time Lyapunov Exponent (FTLE) Field. The FTLE field is then used to compute a segmentation map in which different labels represented different crowd flow segments. In order to generate the BFF, we use this segmentation map and compute an edge map by retaining only the boundary pixels of each segment. Next, the shortest distance between the wall/barrier and each pixel is determined by computing the distance transform of this edge map (see Figs. 5a-c). Note that when a distance is larger than a certain threshold, the barrier effect vanishes completely. This vanishing effect is represented by the flattening of the surface (the red region) in Fig. 5(c). The difference between the values in cell j and i represent the measure of B_{ij} in 1. Examples of BFFs are presented in Fig. 6.

3.3 Dynamic Floor Field - D_{ij}

The objective of the DFF is to determine the behavior of the crowd around the individual being tracked. The instantaneous information about the motion of the crowd is an important cue for constraining likely future locations. In our framework, the instantaneous interaction among the members of the crowd is extracted by using a particle based representation. For a given scene, the DFF is computed at each time period by using a sliding window of N_D frames. That is, for computing the DFF at time t, we use frames $F_D = [f_t, f_{t+1}, \ldots, f_{t+N_D}]$. We first compute the optical flow between consecutive frames in F_D and stack them together to generate a 3D volume of optical flow fields. Next, a grid of particles is overlaid on the first flow field of the volume and numerically advected [11].

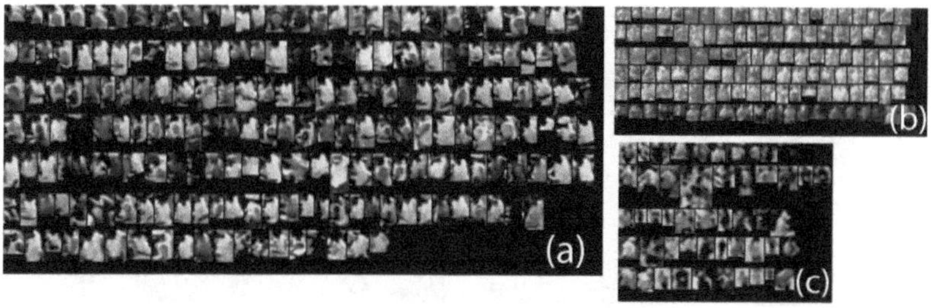

Fig. 7. Chips used for tracking. (a) Marathon-1. (b) Marathon-2. (c) Marathon-3.

During the advection, whenever a particle jumps from a cell i to one of the neighboring cells j the value of interaction between these cells increases by one. The DFF can only have non-negative integer values and there is one DFF per cell where each DFF captures the dynamic interaction between the target cell i and remaining cells in the scene. A visualization of the DFF is shown in Fig. 4 (c)-(d). Since the DFF is meant to capture the local interaction of particles around the tracked individual, Fig. 4(c)-(d) represents the shape of the DFF, but only in that local neighborhood. The peak in Fig. 4(d) represents the location where most particles end up if they pass through the cell containing the yellow particle.

4 Experiments and Discussion

A detailed experimental analysis was performed on the three marathon sequences shown in Figure 6. In addition, qualitative results are shown for a busy train-station sequence. In all the experiments, tracking began by selecting a rectangular region around the target object and using it to compute the gray-level appearance template. At each time instant, the next position of the target was chosen according to Equation 1, where the matrix of preferences around the current target location were twice the size of the selected rectangular region. We do not adapt the size of the window during the tracking. The appearance similarity was computed using normalized cross correlation and the template was progressively updated at each time instant. We set the values of k_S, k_D, and k_B equal to 0.02 for all experiments. The tracking results were stable for small changes in the values of these coupling factors. We used the first 50 frames of each sequence to construct the SFF. To compute the BFF and DFF, the values of $N_B = 20$ and $N_D = 5$ were used.

Marathon-1. This sequence (Fig. 1a) captures participants in a marathon from an overhead camera. It is a difficult sequence due to the severe occlusion among the participants, and the similar looking outfits worn by most of the athletes. The sequence has 492 frames, but each athlete remains in the field of view, on average, for 120 frames. We manually selected 199 individuals, shown in Fig. 7(a), from various frames for tracking. The average size of the selected chip was 14×22 pixels.

Fig. 8. Displays trajectories of individuals who were successfully tracked by our method. (a) Marathon-1. (b) Marathon-2. (c) Marathon-3. (d) Train Station.

A set of trajectories generated by our tracing algorithm is shown in Fig. 8(a). In total, we were able to successfully track 143 out of 199 individuals, i.e. the correct label was maintained throughout the time duration for which the athlete was visible in the FOV. Quantitative analysis of the tracking was performed by generating ground-truth trajectories for 50 athletes, which were selected randomly from the initial set of 199 athletes. The ground-truth is generated by manually tracking the centroid of each selected athlete. The ground-truth shows that these 50 athletes were visible for an average of 77 frames, and our algorithm tracked them for an average of 72 frames. This is summarized by the first 50 bars in the graph of Fig. 9. The average tracking error is summarized by the first 50 green bars in the graph of Fig. 12(a). The tracking error in a given frame is defined as the distance in pixels between the centroid of the object in the ground-truth and the centroid returned by our tracking algorithm. The tracking failure on this sequence (10(a)) occurred in situations where the target was completely occluded by either another athlete or the street-light in the scene. Since we did not use any prediction mechanism, we could not recover from full occlusion. However, partial occlusions were handled by our tracker.

Marathon-2. This sequence (Fig. 1(b)) also involves a marathon. However, the camera in this sequence is installed on a high-rise building to increase the FOV. As a result, the number of pixels on each individual is fewer. In addition, there are drastic illumination changes when athletes move into the shadow of the neighboring buildings. This sequence has 333 frames. We manually selected 120 individuals (Fig. 7(b)) from various frames for tracking. The average size of the selected chip was 13×16 pixels.

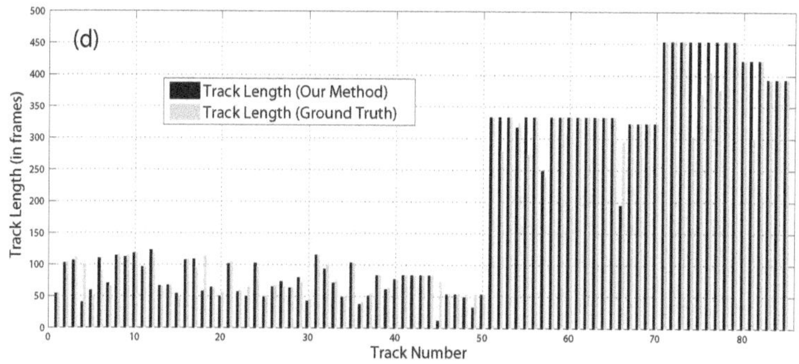

Fig. 9. A comparison of track lengths against the ground-truth: 1 to 50 Marathon-1; 51-70 Marathon-2; 71-85 Marathon-3

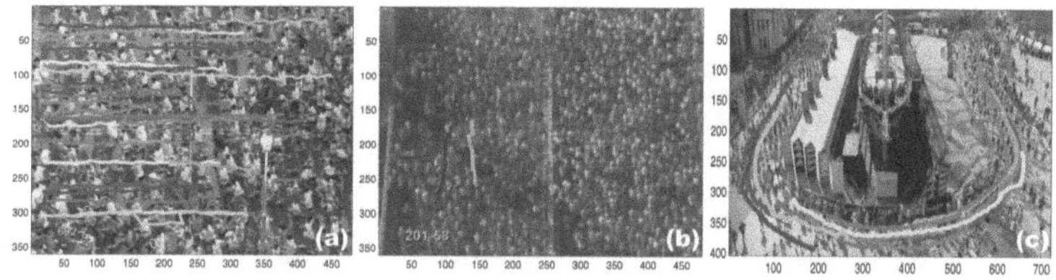

Fig. 10. The failure cases. (a) Marathon-1. (b) Marathon-2. (c) Marathon-3.

A set of trajectories generated by our tracing algorithm is shown in Fig. 8(b). In total, we were able to track 117 of the 120 (97.5%) individuals. A quantitative analysis was performed by generating ground truths for 20 of the athletes. The length of ground-truth trajectories and trajectories generated by our tracker is summarized by bars 51-70 in Fig. 9. The average tracking error is summarized by the green bars (51-70) in the graph of Fig. 12(b). It can be observed that our tracking was very accurate, in most cases, and able to overcome the illumination changes with the aid of the DFF and SFF. Some of the tracking failures on this sequence are shown in Fig. 10(b). These tracking failures were the result of severe illumination variation causing change in the appearance of the target.

Marathon-3. The third sequence (Fig. 1(c)) is extremely challenging due to two factors: 1) appearance drastically changes due to the U-shape of the path; 2) the number of pixels on target varies due to the perspective effect. The fewer pixels make it more difficult to resolve even partial occlusions. The sequence is 453 frames long. We manually selected 50 individuals (Fig. 7(c)) for tracking. The average size of the selected chip was 14×17 pixels. In total, we were able to track 38 of the 50 (76%) individuals without any tracking error (Fig. 8(c)). A comparison of track lengths returned by our algorithm with 15 ground-truth trajectories is given in Fig. 9. The average tracking error is summarized by bars 71-85 in Fig. 12(a). In addition, we performed tracking on a busy train-station sequence shown in Fig. 1(d). There, we tracked 20 individuals and some qualitative results are shown in Fig. 8(d).

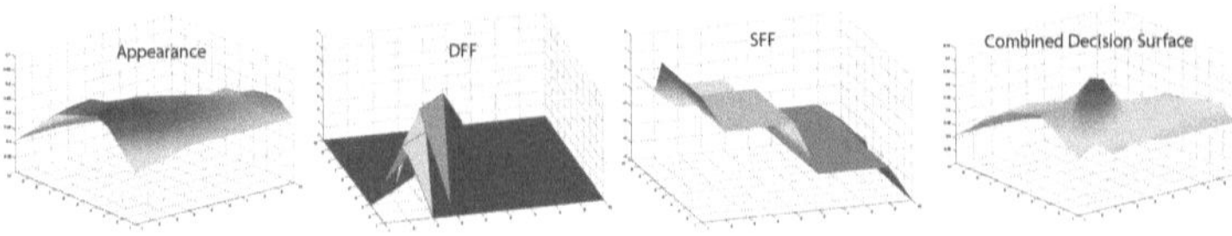

Fig. 11. (Left to Right)The appearance similarity surface, local DFF, local SFF, and the final decision surface obtained by merging appearance, the DFF, and the SFF according to 1

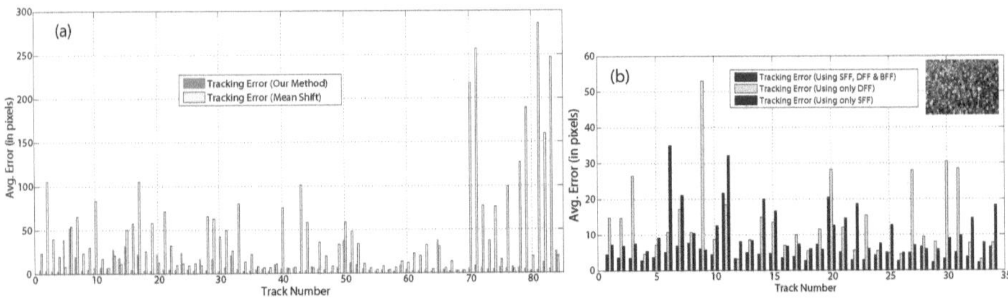

Fig. 12. (a) Comparison of the tracking error of our method against the mean-shift tracker. The bars represent the average error over the entire track. The length of the track is given in Fig. 9(1 to 50 Marathon-1; 51-70 Marathon-2; 71-85 Marathon-3). (b) Contribution made by different floor fields towards the tracking accuracy.

Analysis. One of our experiments is now discussed in detail to provide an intuitive insight on how the floor fields help in improving tracking. For this purpose, we picked a track from Marathon-3, where the athlete was wearing a black shirt and running away from the camera (Fig. 2(a)). During the course of tracking, the athlete's appearance became ambiguous as neighboring athletes were also wearing black shirts. Fig. 11(a) (top-left) shows the surface of appearance similarity measure, which was obtained by matching the template in a 16×16 neighborhood, for one of those instances. The surface was relatively flat, showing a lack of a good match for the tracked person in the current frame. If we were to use only this surface, there was a high probability that the tracker would jump onto one of the neighboring athletes wearing the similar clothing. However, floor fields helped resolve this ambiguity, which is apparent in the final decision surface (Fig. 11(a) bottom-right). The DFF shown in Fig. 11(a)(top-right) guided the tracker by emphasizing the direction taken by most particles from the current location of the target. Similarly, the SFF shown Fig. 11(a)(bottom-left) allowed the tracker to consider the direction that the objects take to the exit the scene. This way, the DFF and SFF helped resolve the appearance ambiguity.

Mean-Shift Comparison. We performed a quantitative analysis by comparing the results with a color based mean-shift tracker. The comparison was performed using the ground-truth generated for the three marathon sequences. The mean-shift was initialized using the same regions and appearance was updated during the course of tracking. The tracking error was computed as the distance between the centroid of the target returned by the tracker and the corresponding centroid in the ground-truth. The error is then averaged over the entire video. The results are summarized in Fig. 12(a). The green bars in the graph correspond to the average error of our tracking algorithm, while the yellow bars correspond to the average error committed by the mean-shift tracker. The tracking error of the mean-shift is especially higher for the trajectories of Marathon-3 sequence (bars 71-85). The reason was the gradual change in the appearance of the athlete as he/she runs along the curved portion of the track. The mean-shift algorithm that only uses the appearance information was not able to cope with it and often drifted away very quickly. While, using the crowd flow and scene constraints in the form of floor fields, our algorithm was able to stay on the target for much longer time durations. In general, it can be observed from the graph (Fig. 12(a)) that our method works much better than the mean-shift tracking method. This verifies our initial observation that when tracking in videos with high density crowds appearance alone is not a reliable cue, and, other sources of information present in the scene should be exploited.

Contribution of Floor Fields. Next an experiment was performed to test the contribution of floor fields towards the accuracy of tracking. The experiment was performed using those ground-truth trajectories from Marathon-1 for which we obtained successful tracking results using all three floor fields. We picked these trajectories as it would allow us to establish whether all three floor fields contributed towards the success or not. There were 35 such trajectories in total in Marathon-1. To measure the contribution of different floor fields, we ran our tracker first using only the SFF and then using only the DFF. The error was computed in a manner similar to that of the mean-shift experiment. The graph in Fig. 12(b) shows the comparison. It can be observed that we obtained the minimum error by using all three fields. However, the error committed by the DFF and SFF is not consistent across all the tracks. By carefully observing these 35 trajectories, we noted that SFF based tracking commits less error when the tracked athlete was running in more or less straight path with no side-ways movement, i.e. when he was perfectly obeying the geometry of the learned SFF (Fig. 6:Center) which is true for most athletes in this scene. The DFF based tracking committed relatively higher error in situations when the neighboring athletes of the target were trying to go sideways to over-take each other, and far less error in cases when the target athlete and the neighboring athletes were running in tandem.

5 Conclusion

We have developed an algorithm for tracking targets in high density crowd scenes. The algorithm utilizes crowd flow and scene layout constraints to predict the future locations of the target. These constraints are captured by learning three floor fields, namely the DFF, the SFF, and the BFF. These floor fields together with the appearance information is then used to track individuals in complex scenarios. The experimental results on complex videos verified that using high level knowledge about the scene in the form of floor fields is helpful in improving the tracking accuracy. The present work can take a number of future research directions: 1) the individual target-tracking can be extended to multi-target tracking by treating the multi-target configuration as a multi-particle system and adding particle-particle interaction terms in the tracking framework; 2) the computation of floor fields can be improved to handle multi-modality of motion at sources, sinks, and paths in between. This will be important for handling the crowded situations where people can move in multiple directions.

Acknowledgements. This research was funded by the US Government VACE program. We would like to thank Min Hu for her valuable contribution towards this research.

References

1. Burstedde, C., Klauck, K., Schadschneider, A., Zittartz, J.: Simulation of Pedestrian Dynamics Using a Two-dimensional Cellular Automation. Physica A 295(3) (2001)
2. Kirchner, A., et al.: Simulation of Evacuation Processes Using a Bionics-Inspired Cellular Automaton Model for Pedestrian Dynamics, vol. 312(1-2) (2002)
3. Lee, S.C., Hughes, R.L.: Exploring Trampling and Crushing in a Crowd. J. Transp. Engrg. 131(8) (2005)
4. Betke, M., et al.: Tracking Large Variable Numbers of Objects in Clutter. In: IEEE CVPR (2007)
5. Li, K., Kanade, T.: Cell Population Tracking and Lineage Construction Using Multiple-Model Dynamics Filters and Spatiotemporal Optimization. In: International Workshop on Microscopic Image Analysis with Applications in Biology (2007)
6. Khan, Z., Balch, T.R., Dellaert, F.: An MCMC-based Particle Filter for Tracking Multiple Interacting Targets. In: Pajdla, T., Matas, J(G.) (eds.) ECCV 2004. LNCS, vol. 3024, pp. 279–290. Springer, Heidelberg (2004)
7. Stauffer, C.: Estimating Tracking Sources and Sinks. In: IEEE Workshop on Event Mining (2003)
8. Gennari, G., Hager, G.D.: Probablistic Data Association Methods in Visual Tracking of Groups. In: IEEE CVPR (2004)
9. Cai, Y., et al.: Robust Visual Tracking of Multiple Targets. In: Leonardis, A., Bischof, H., Pinz, A. (eds.) ECCV 2006. LNCS, vol. 3954. Springer, Heidelberg (2006)
10. Yilmaz, A., et al.: Object Tracking: A Survey. ACM Journal of Computing Surveys 38(4) (2006)
11. Ali, S., Shah, M.: A Lagrangian Particle Dynamics Approach for Crowd Flow Segmentation and Stability Analysis. IEEE CVPR (2007)
12. Brostow, G., Cipolla, R.: Unsupervised Bayesian Detection of Independent Motion in Crowds. In: IEEE CVPR (2006)
13. Zhao, T., Nevatia, R.: Bayesian Human Segmentation in Crowded Situations. In: IEEE CVPR (2003)
14. Zhao, T., Nevatia, R.: Tracking Multiple Humans in Crowded Environment. In: IEEE CVPR (2004)
15. Gurka, R., et al.: Computation of Pressure Distribution Using PIV Velocity Data. In: Proc. of the 3rd International Workshop on Particle Image Velocimetry (1999)
16. Comaniciu, D., Meer, P.: Mean Shift: A Robust Approach Toward Feature Space Analysis. In: IEEE TPAMI (2002)

The Naked Truth: Estimating Body Shape Under Clothing

Alexandru O. Bălan and Michael J. Black

Department of Computer Science, Brown University, Providence, RI 02912, USA
{alb,black}@cs.brown.edu

Abstract. We propose a method to estimate the detailed 3D shape of a person from images of that person wearing clothing. The approach exploits a model of human body shapes that is learned from a database of over 2000 range scans. We show that the parameters of this shape model can be recovered independently of body pose. We further propose a generalization of the visual hull to account for the fact that observed silhouettes of clothed people do not provide a tight bound on the true 3D shape. With clothed subjects, different poses provide different constraints on the possible underlying 3D body shape. We consequently combine constraints across pose to more accurately estimate 3D body shape in the presence of occluding clothing. Finally we use the recovered 3D shape to estimate the gender of subjects and then employ gender-specific body models to refine our shape estimates. Results on a novel database of thousands of images of clothed and "naked" subjects, as well as sequences from the HumanEva dataset, suggest the method may be accurate enough for biometric shape analysis in video.

1 Introduction

We address the problem of reliably estimating a person's body shape from images of that person wearing clothing. Estimation of body shape has numerous applications particularly in the areas of tracking, graphics, surveillance and forensic video analysis. To be practical any method for estimating body shape must recover a representation that is invariant to changes in body pose. To that end, we exploit a parametric 3D body model and show that the 3D body shape parameters are largely invariant to body pose. Additionally, such a model must be robust to clothing which obscures the true body shape. Here we build on the concept of the visual hull, which represents a bound on the underlying shape. In the case of a clothed person, the visual hull may only provide a loose bound on body shape. To gain tighter bounds we exploit the pose-invariance of the shape model to combine evidence from multiple poses. As a person moves, the constraints provided by the visual hull change as the clothing becomes looser or tighter on different body parts. We combine these constraints to estimate a *maximal silhouette-consistent parametric 3D shape*. Using a unique dataset of subjects with both minimal and normal clothing we demonstrate that a person's body shape can be recovered from several images of them wearing clothes.

D. Forsyth, P. Torr, and A. Zisserman (Eds.): ECCV 2008, Part II, LNCS 5303, pp. 15–29, 2008.

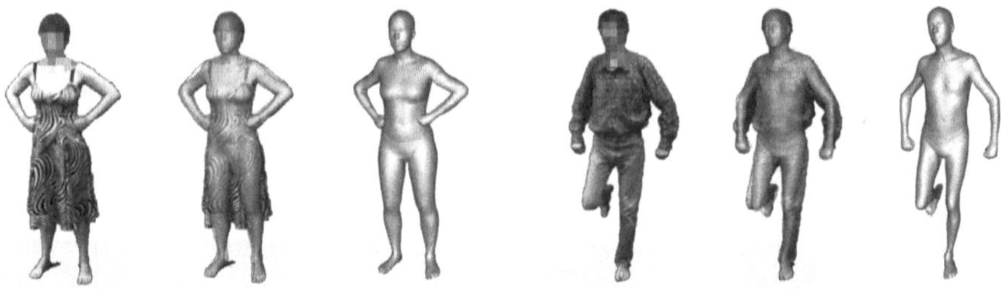

Fig. 1. Shape under clothing. Body shape recovery for two clothed subjects. For each subject we show (left to right) one of four input images, the 3D body model superimposed on the image (giving the impression of "X-ray" vision), and the estimated body model.

To our knowledge this is the first work to attempt to recover a detailed estimate of a person's 3D body shape from natural (ie. standard CCD) images when the body is obscured by clothing. The approach is illustrated in Figure 1. This work is based on a previously proposed parametric 3D body model called SCAPE [1]. The model characterizes human body shape using a low-dimensional shape subspace learned from a database of over 2000 range scans of humans. Recent work has shown that the parameters of this model can be directly estimated from image silhouettes [2,3] or from visual hulls [4] but has been restricted to people wearing tight-fitting clothing.

Our method rests on two key hypotheses. First: human body shape can be recovered independently of body pose. We test this hypothesis using a unique dataset of "naked" subjects in several poses captured with 4 calibrated cameras. We estimate their body shape in each pose both independently and in a batch fashion that combines information from multiple poses. We find that the variability in shape parameters across pose is small relative to the variability across subjects. We exploit this relative pose-independence of shape to combine multiple poses to more accurately estimate a single 3D body shape.

The second hypothesis is that images of the human body in clothing provide sufficient constraints to infer the likely 3D body shape. Of course a garment or costume could be worn which completely obscures or provides false information about the body shape. In normal "street clothes" however, we argue that many constraints exist that can be combined with a learned model of 3D body shape to infer the true underlying shape. To formalize this, we define the notion of a maximal silhouette-consistent parametric shape (MSCPS) that generalizes the notion of a visual hull. A visual hull has two properties [5,6]. First, the true 3D object lies completely within the visual hull (and its projection into images lies within the image silhouettes). Second, each facet of the visual hull touches the surface of the object. In the case of clothing, property 1 holds but property 2 does not. The object itself is obscured such that the silhouette contours may, or may not, correspond to true object boundaries. Rather, the silhouettes provide a bound on the possible shape which may or may not be a tight bound. Note also that, with clothing, in some poses the bound may be tight in some places and loose in others and these locations may change with pose.

In place of the visual hull, we define the MSCPS that optimizes the following weak constraints: 1) the shape lies inside the visual hull; 2) the volume of the shape is maximal; and 3) the shape belongs to a parametric family. In our case this family is the family of 3D human body shapes. Constraint 2 is required to avoid the trivial solution where the estimated shape is made arbitrarily small. In general, each of these constraints can be viewed as a weak constraint with the last one being a statistical prior over 3D shapes. We go a step beyond previous work to deal with time-varying constraints and non-rigid, articulated objects, by requiring constraint 1 hold over multiple poses. We also use the fact that portions of the body may actually provide tight constraints on shape. For example, when a person wears short sleeves, their bare arms provide cues not only about the arm shape, but also about the overall weight of the person. Consequently we automatically detect skin regions and exploit tight constraints in these regions.

Central to our solution is a learned human body model. We go beyond previous work to use three different models: one for men, one for women, and one gender-neutral model combining both men and women. We find that gender can be reliably inferred in most cases by fitting both gender-specific models to the image data and selecting the one that best satisfies all the constraints. Given this estimated gender, we then use a gender-specific model to produce a refined shape estimate. To our knowledge this is the first method to estimate human gender directly from images using a parametric model of body shape.

In summary, the key contributions described here include: a shape optimization method that exploits shape constancy across pose; a generalization of visual hulls to deal with clothing; a method for gender classification from body shape; and a complete system for estimating the shape of the human body under clothing. The method is evaluated on thousands of images of multiple subjects.

2 Previous Work

There are various sensing/scanning technologies that allow fairly direct access to body shape under clothing including backscatter X-ray, infra-red cameras and radio waves. While our body fitting techniques could be applied to these data, for many applications, such as forensic video analysis, body shape must be extracted from standard video images. This problem is relatively unexplored. Rosenhahn *et al.* [7] proposed a method to track lower limbs for a person wearing a skirt or shorts. Their approach uses a generative model to explicitly estimate parameters of the occluding clothing such as the cloth thickness and dynamics. In their work, they assume the shape of the body and cloth measurements are known *a priori* and do not estimate them from image evidence. There has been recent interest in generative models of cloth [8,9] but the huge variability in clothing appearance makes the use of such models today challenging.

Most human shape estimation methods attempt to estimate the shape *with* the clothing and many of these techniques are based on visual hulls [10]. Visual hull methods (including voxel-based and geometric methods) attempt to reconstruct the *observed* 3D shape with the silhouette boundary providing an outer bound on that shape. A detailed review is beyond the scope of this paper. We focus instead on those methods that have tried to restrict the shape lying inside the visual hull. Several authors have noted that, with small numbers of images, the visual hull provides a crude bound on object shape. To address this in the case of people, Stark and Hilton [11] combine silhouettes with internal structure and stereo to refine the 3D surface. They still assume the true surface projects to match the image silhouette features.

More generally, Franco *et al.* [12] impose weak assumptions on the underlying shape. They define a notion of a set of visual shapes that are consistent with the observed silhouettes (silhouette-consistent). The key contribution of their work is the idea of adding an assumption of shape smoothness which regularizes the set of possible 3D shape solutions. The observed silhouette is always considered as providing a tight bound on the surface with the priors compensating for an impoverished set of views. Note, however, in our problem, the visual hull is not the goal. Our case is different in that the object we care about (the human body) is obscured (by clothing) meaning that observed silhouette boundaries often do not provide tight constraints. We build on the notion of a visual shape set to define a person-specific prior model of the underlying shape.

In related work Corazza *et al.* [4] fit a SCAPE body model to visual hulls extracted using eight or more cameras. They do this in a single pose and assume tight-fitting clothing. We use a more detailed body model than they did and do not explicitly reconstruct the visual hull. Instead, we fit directly to image data and this allows us to use a smaller number of cameras (4 in our case).

Most visual hull reconstruction methods assume rigid objects. With non-rigid clothing we find it important to integrate information over time to constrain the underlying 3D shape. In related work, Cheung *et al.* [13] combine information over time by performing rigid alignment of visual hulls at different time instants. Knowing a rigid alignment over time effectively provides additional views. They also extend this idea to articulated body parts but focus only on recovering the bounding volume. Grauman *et al.* [14,15] estimate a 3D shape consistent with a temporal sequence of silhouettes using assumptions on the smoothness and shape transitions. They apply this method to silhouettes of humans and recover a visual hull using an example-based non-parametric model of body shapes. They do not use a parametric body model or explicitly attempt to infer the shape under clothing.

Most previous work on gender classification from images has focused on faces (e.g. [16]), but in many situations the face may be too small for reliable classification. The other large body of work is on estimating gender from gait (e.g. [17]). Surprisingly, this work typically takes silhouettes and extracts information about gait while throwing away the body shape information that can provide direct evidence about gender. We believe ours is the first method to *infer* a parametric 3D human body shape from images of clothed people and to use it for gender classification.

Fig. 2. Dataset. Example images from the clothing dataset shown here after background subtraction. *(top)* All subjects in the "naked condition" (NC); *(middle)* single subject in the "clothed condition" (CC); *(bottom)* a variety of the 11 different poses.

3 Methods

Datasets. To test our ability to infer shape under clothing we collected a dataset of 6 subjects (3 male and 3 female); a small sample of images from the dataset is shown in Figure 2. Images of the subjects were acquired in two conditions: 1) a "naked condition" (NC) where the subjects wore minimal tight fitting clothing (Figure 2 (top)) and 2) a "clothed condition" (CC) in which they wore a variety of different "street" clothes (Figure 2 (middle)). Each subject was captured in each condition in a fixed set of 11 postures, several of which are shown in Figure 2 (bottom). All postures were performed with 6 - 10 different sets of "street" clothing (trials) provided by the subjects. Overall, the dataset contains 53 trials with a total of 583 unique combinations of people, clothing and pose (a total of 2332 images). For each of these, images were acquired with four hardware synchronized color cameras with a resolution of 656 x 490 (Basler A602fc, Basler Vision Technologies). A full green-screen environment was used to remove any variability due to imprecise foreground segmentation. The cameras as well as the ground plane were calibrated using the Camera Calibration Toolbox for Matlab [18] and the images were radially undistorted. Foreground silhouette masks were obtained using a standard background subtraction method, performed in the HSV color space to account for background brightness variations induced by the presence of the foreground in the scene (e.g. shadows).

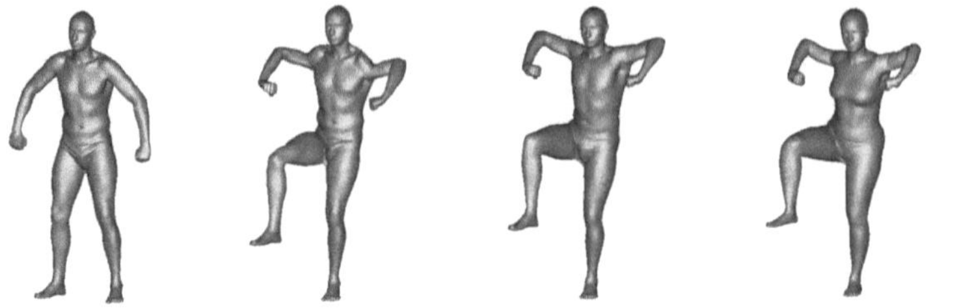

Fig. 3. SCAPE deformation process. A series of transformations are applied in succession to the reference mesh, x, including articulated rotations, R, non-rigid deformations, Q, and identity shape changes, D.

To test the generality of our methods in less than ideal circumstances, we also performed experiments on the generic HumanEva dataset [19] where the poses were "natural" and background subtraction was imperfect. This dataset only contains 2 subjects, but we tested our approach on approximately 200 frames in a wide variety of postures and with various levels of real silhouette corruption.

Parametric Body Model: SCAPE. We employ a parametric body model called SCAPE [1] that is able to capture both variability of body shapes between people, as well as articulated and non-rigid pose deformations; for more detail, the reader is referred to the original SCAPE paper. The model is derived from a large training set of human laser scans, which have been brought in full correspondence with respect to a reference mesh. Given a mesh x we deform it into a mesh y that has a different body shape and different pose. The deformation process (Figure 3) is defined as a set of transformations applied in succession to edges Δx of each triangle of the reference mesh, followed by a least-squares optimization that reconstructs a consistent mesh y:

$$y(\theta, \beta) = \arg\min_{y} \sum ||R^p(\theta) D^t_{U,\mu}(\beta) Q^t_{\alpha}(\theta) \Delta x - \Delta y||^2 \ . \tag{1}$$

The mesh is divided into 15 body parts, with the orientations specified by 3×3 part rotations $R^p(\theta)$ computed from a reduced set of joint angles θ of a kinematic skeleton that has 32 degrees of freedom. Non-rigid pose-dependent deformations $Q^t_{\alpha}(\theta)$ are linearly predicted for each triangle t from neighboring joint angles, with the linear coefficients α learned from training data of a single subject scanned in 70 different poses. Changes in 3D body shape between people are captured by shape deformations D^t which are 3×3 transformations that are applied to each triangle t in the reference mesh to deform it into an instance mesh (i.e. 3D range scan). We construct a training set of such deformations between the instance mesh and over 2000 body scans of North American adults with roughly equal gender representation (Civilian American and European Surface Anthropometry Resource (CAESAR), SAE International). We learn a low-dimensional

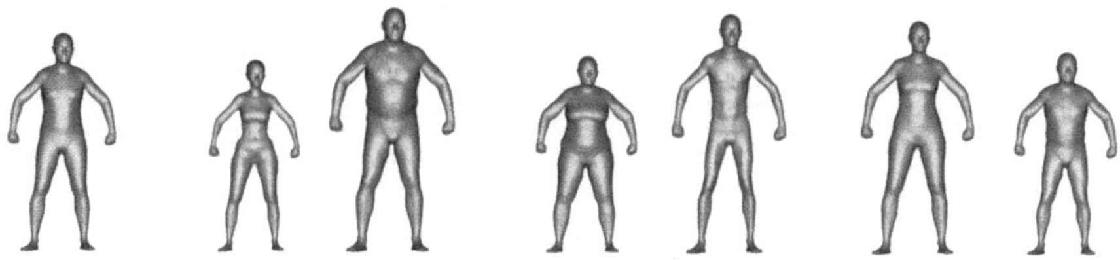

Fig. 4. 3D Body Shape Model. The mean gender-neutral body shape is shown *(left)* followed by deviations from the mean along the first three principal components (± 3 standard deviations).

linear model of human shape variability using incremental principal component analysis (iPCA) [20]. For each individual, the D^t matrices are concatenated into a single column vector that is approximated as $D_{U,\mu}(\beta) = U\beta + \mu$ where μ is the mean body shape, U are the first n eigenvectors given by iPCA and β is a vector of linear coefficients that characterizes a given shape. The variance of each shape coefficient β_j is given by the eigenvalues $\sigma^2_{\beta,j}$.

We learn separate models for male and female subjects, as well as a gender-neutral model with all the subjects. For our experiments we use the first $n = 20$ principal components which account for roughly 70% of the variance in the gender-neutral case and 65% of the variance in the gender specific cases. Figure 4 shows the mean gender-neutral shape and deviations from the mean captured by the first three principal components.

Objective Function. Our model is parameterized by a set of joint angles θ, including global position and orientation, and shape parameters β. We adopt a generative approach in which the parameters define a 3D body pose and shape model which is then projected into the camera views to produce predicted image silhouettes $S^e_{k,\beta,\theta}$. These are then compared with observed silhouettes, S^o_k, obtained by foreground segmentation in each camera view k. Specifically we define the asymmetric distance between silhouettes S and T as

$$\tilde{d}(S,T) = \sum_{i,j} (S_{ij} \cdot C_{ij}(T)) / (\sum_{i,j} S_{ij})^{3/2} , \qquad (2)$$

where S_{ij} are the pixels inside silhouette S and $C_{ij}(T))$ is a distance function which is zero if pixel (i,j) is inside T and is the Euclidean distance to the closest point on the boundary of T for points outside. The denominator is a normalization term that gives invariance to the size of the silhouette.

We first define the objective function for the NC using the bi-directional objective used by Bălan *et al.* [2]. Later we will extend this to deal with clothing. The objective function uses a symmetric distance to match the estimated and observed silhouettes over the K camera views:

$$E(\beta,\theta) = \sum_{k=1}^{K} \tilde{d}(S^e_{k,\beta,\theta}, S^o_k) + \tilde{d}(S^o_k, S^e_{k,\beta,\theta}) . \qquad (3)$$

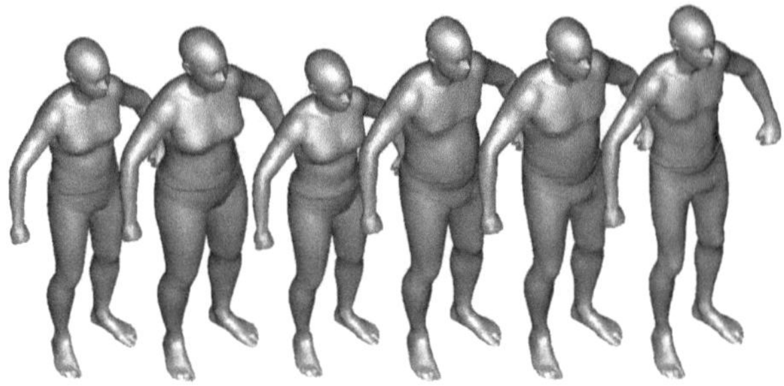

Fig. 5. Quantitative Evaluation. Ground truth body shapes are shown for the six subjects. For each subject the height, waist size and chest size are used for quantitative evaluation. The waist and chest are shown by red lines.

Optimization Method. Unlike previous work we minimize $E(\beta, \theta)$ using a gradient-free direct search simplex method. For the clothing dataset, 11 canonical pose parameters θ^0 were hand defined and used for initialization of each pose. To help avoid local minima we alternate between optimizing pose and shape in an incremental fashion: after initializing with the predefined pose and mean shape model, we begin by optimizing the global position and a few shape coefficients. We then continue optimizing the joint angles of individual joints along the kinematic chain (hips, shoulders, elbows), and then jointly optimize all joint angles together with a few more shape coefficients. In the last phase we optimize the full set of 20 principal shape components.

Evaluation Measures. We quantitatively evaluate the accuracy of our 3D body models using a variety of derived biometric measurements such as height, waist size, chest size, etc. These body measurements were collected directly from the subjects, from laser range scans of four of the subjects, and from the results of fitting the body in the NC. We verified the accuracy of the body fit in the NC using the direct measurements and then took this to be the ground truth for the remainder of the paper. Figure 5 shows the recovered 3D body shape models used for ground truth. Displayed on the models in red are the locations used to compute derived measurements for chest and waist size.

4 Shape Constancy

Previous work [2] only optimized the body shape at a particular time instant. Here we take a different approach and integrate information about body shape over multiple poses. Our first hypothesis is that the SCAPE model provides a representation of body shape that is invariant to body pose. To test this hypothesis we optimize body shape and pose for each posture independently in the NC. Figure 6 (left) shows three examples of the 3D body shape recovered for one of the subjects in this fashion. Figure 6 (right) plots the variance in the recovered shape coefficients across pose for all subjects (yellow) versus the

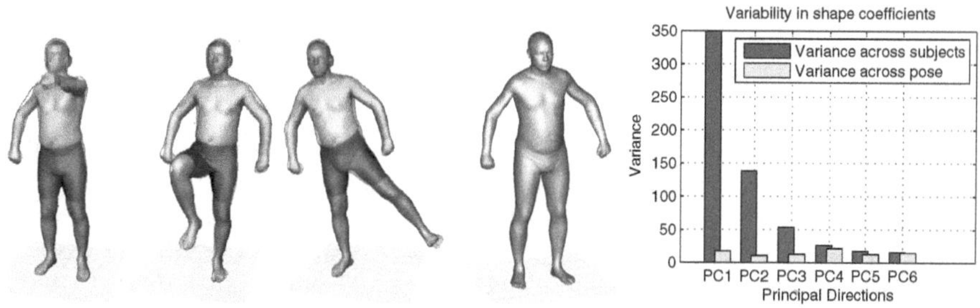

Fig. 6. Invariance of body shape to pose. *(left)* Reconstructions of 3D body shape for three individual poses using four camera views. *(middle)* Model recovered by combining across 11 different poses. *(right)* Variance in shape coefficients across subjects and poses.

variability in these shape coefficients across subjects. The variation of the major 3D shape parameters with pose is small relative to the variation across people.

To exploit this constancy we define a "batch" optimization that extends the objective function to include P different poses:

$$E(\beta, \theta_1, \ldots, \theta_P) = \sum_{p=1}^{P} \sum_{k=1}^{K} \tilde{d}(S^e_{k,\beta,\theta_p}, S^o_{k,p}) + \tilde{d}(S^o_{k,p}, S^e_{k,\beta,\theta_p}) \ . \tag{4}$$

We alternate between optimizing a single set of shape parameters β applicable to all postures, and optimizing the pose parameters θ_p independently for each posture. Figure 6 (middle) shows the body shape recovered by integrating across pose. Note that establishing this property of shape invariance with respect to pose is useful for tracking applications and biometric shape analysis.

5 Clothing

Above we established the basic model, its optimization and its application to shape estimation in the absence of loose clothing. We now turn to our main contribution which is a method for body shape recovery with clothing.

Maximal Silhouette-Consistent Parametric Shape. Laurentini [6] introduced the concept of *visual hull* to represent the maximal 3D object shape consistent with observed image silhouettes. Our assumption is that the true object shape belongs to a parametric family and lies completely within the visual hull. We also generalize the concept to cases where the visual hull only provides a loose bound to the object's surface. Since the problem becomes ill-posed in this case, we regularize the problem by making use of prior knowledge of the object shape and integrate information from time-series data. Specifically, we introduce the concept of a *maximal silhouette-consistent parametric shape* that weakly satisfies three constraints: 1. the projected model falls completely inside the foreground silhouettes; 2. the volume of the 3D model is maximal; and 3. the shape of the object belongs to a parametric family of shapes (in our case human bodies).

We satisfy the first constraint by penalizing the regions of the projected model silhouette, S^e, that fall outside the observed foreground silhouette S^o. That is, we use exactly the same distance function as defined above: $E_{inside}(\beta, \theta) = \tilde{d}(S^e_{k,\beta,\theta}, S^o_k)$. For the second constraint, we would like the projected model to explain as much of the foreground silhouette as possible; if the subject were not wearing clothing this would just be the second term from above $\tilde{d}(S^o_k, S^e_{k,\beta,\theta})$. In the more general setting where people wear clothing or interact with objects, the observed foreground silhouettes will be too large producing a bias in the shape estimates. To cope with this, we employ two strategies. The first is to down-weight the contribution of the second constraint, meaning it is more important for the estimated shape to project inside the image silhouette than to fully explain it. The second is to use features in the image that we are more confident accurately conform to the underlying shape. In particular, we detect skin-colored regions and, for these regions, we give the second constraint full weight. We denote by S^s the detected skin regions and by $S^o \backslash S^s$ non-skin regions of the observed foreground silhouette. The "expansion" constraint is then written as

$$E_{expand}(\beta, \theta) = \tilde{d}(S^s_k, S^e_{k,\beta,\theta}) + \lambda \tilde{d}(S^o_k \backslash S^s_k, S^e_{k,\beta,\theta}) \tag{5}$$

with $\lambda \ll 1$.

Finally to enforce domain knowledge, we define priors on shape parameters, as well as on joint angles. We use very weak priors designed to prevent wildly unnatural shapes but which do not bias the estimates for "normal" body shapes. Specifically we add a penalty on the shape coefficients β_i of the form

$$E_{shape}(\beta) = \sum_j \max\left(0, \frac{|\beta_j|}{\sigma_{\beta,j}} - \sigma_{thresh}\right)^2 . \tag{6}$$

No penalty is paid for small values of β and we only start penalizing shape coefficients when they exceed $\sigma_{thresh} = 3$ standard deviations from the mean. We also enforce a prior, $E_{pose}(\theta)$, on body pose which is uniform within joint angle limits and only penalizes poses beyond these limits.

Finally, the objective function we minimize is

$$E_{clothes}(\beta, \Theta) = \sum_{p=1}^{P} \sum_{k=1}^{K} \frac{E_{inside}(\beta, \theta_p)}{\sigma_o^2} + \frac{E_{expand}(\beta, \theta_p)}{\sigma_o^2} + E_{shape}(\beta) + E_{pose}(\theta_p)$$

where $\Theta = \theta_1, \ldots, \theta_P$ represents the different body poses. We assume a normal distribution over the silhouette distance metric $\tilde{d}$ with variance σ_o^2 that accounts for noisy image observations.

One advantage of our method is that being robust to clothing also provides some robustness to silhouettes that are too large (e.g. due to shadows). Also, in future work, we will learn a statistical model for E_{expand} that captures the deviations of clothing from the naked form observed in training data.

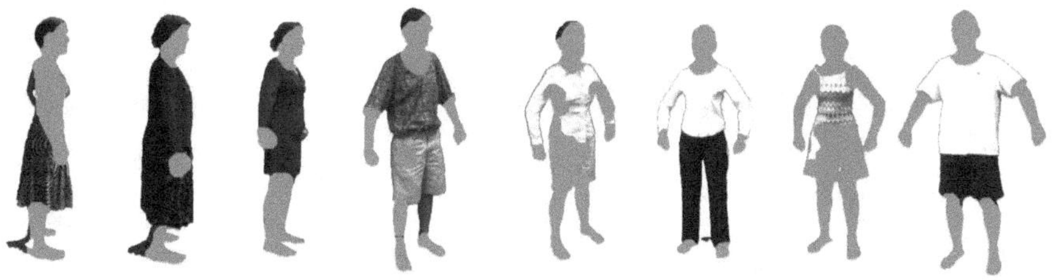

Fig. 7. Skin segmentation. Examples of segmented people and the regions identified as skin, shown in orange. Not that this segmentation need not be perfect to provide useful constraints on the shape fitting.

Skin Detection and Segmentation. In order to detect skin colored regions in an image, multiple skin detectors are trained. We use a leave-one-out cross-validation method and train one classifier for each person using all the other people in the database. Hence the skin of the left-out subject is segmented using a skin model that excludes his/her data. The skin detectors are built from training data using a simple non-parametric model of skin pixels in hue and saturation space. Figure 7 shows several examples of people in different clothing and the identified skin regions.

6 Results

Gender classification. For gender classification, we estimated the pose and the first 6 shape parameters in each *test instance* using the gender-neutral shape model. After convergence, we kept the pose parameters fixed and re-estimated the shape parameters with both gender-specific shape models. The best fitting model according to the objective function corresponded to the true gender 86% of the time when the optimization was performed on individual poses. Taking the majority classification across all poses in a trial increased the classification accuracy to 90.6%. Finally we estimated shape parameters in batch fashion over all poses in a trial, and the gender classification improved to 94.3% accuracy on the dataset of 53 trials with natural clothing.

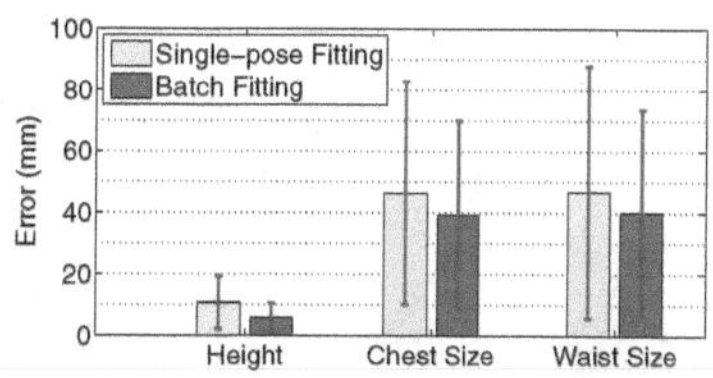

Fig. 8. Quantitative evaluation of shape. Accuracy of estimated body measurements (height, waist, chest) relative to ground truth. Batch estimation across pose decreases the errors.

Shape under Clothing. We recovered the body pose and shape for all 583 independent poses and the 53 batch trials in the CC. Fitting takes approximately 40min for a single model on a 2GHz processor. A few representative results[1] are shown in Figure 9. We quantitatively evaluate the accuracy with respect

[1] See http://www.cs.brown.edu/research/vision/scapeClothing for additional results.

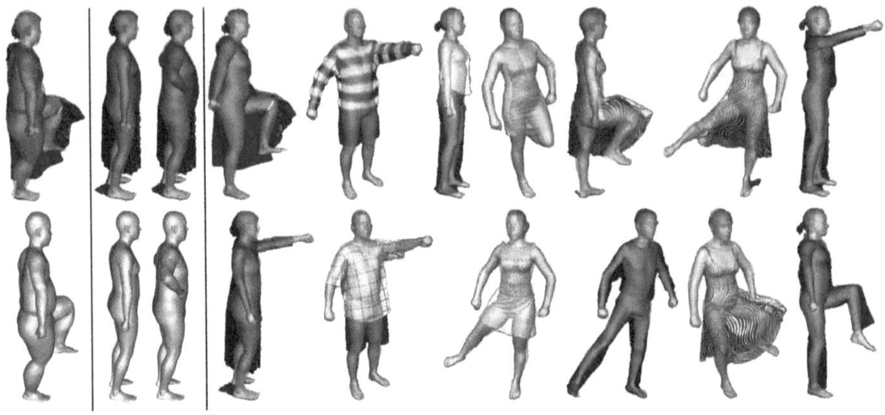

Fig. 9. Example shapes estimated by 3 different methods. *(left)* using the clothing-oblivious method of [2] for a single pose and without enforcing priors on shape leads to unrealistic overestimates of size/shape; *(middle)* estimated shapes using the proposed clothing-robust objective function formulation (7) applied only to individual poses; *(right)* shapes estimated using the multi-pose shape consistency constraint.

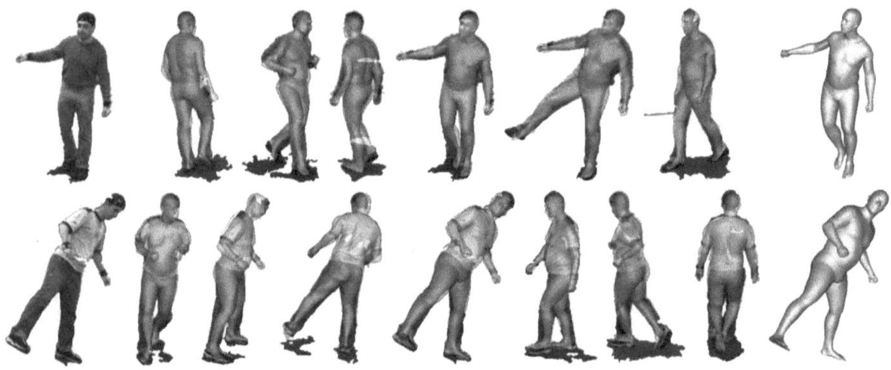

Fig. 10. Example body shapes from the HumanEva-II dataset. (*left*) One segmented frame; (*middle*) several frames with the estimated model overlaid; (*right*) estimated body shape.

to biometric measurements derived from the ground truth shape computed in batch fashion from the NC data. Figure 8 shows how errors in height, waist and chest size decrease by combining information across pose.

We also tested our method on the publicly available HumanEva-II dataset which consists of two sequences with the subjects S2 and S4 walking and jogging in a circle, followed by a leg-balancing action. A kinematic tree tracking algorithm using a coarse cylindrical body model was used to obtain rough initial pose estimates at each frame which were subsequently refined during shape estimation using our framework. We used a subset of the frames spanning a wide variety of postures to estimate the body shape (Figure 10); these results suggest the method is relatively robust to errors in foreground segmentation. The optimization converged in all test cases we considered.

7 Conclusions

We defined a new problem of inferring 3D human shape under clothing and presented a solution that leverages a learned model of body shape. The method estimates the body shape that is consistent with an extended definition of the visual hull that recognizes that shape bounds provided by the visual hull may not be tight. Specifically, the recovered shape must come from the parametric family of human shapes, it should lie completely within the visual hull, and it should explain as much of the image evidence as possible. We observed that by watching people move, we could obtain more constraints on the underlying 3D body shape than in a single pose. Consequently, we exploited the relative pose-independence of our body shape model to integrate constraints from multiple poses by solving for the body pose at each time instant and a single 3D shape across all time instants. We integrated a skin detector to provide tight constraints on 3D shape when parts of the body are seen unclothed. We also showed that gender could be reliably classified based on body shape and defined gender-specific shape models to provide stronger priors on the unobserved shape. The method was tested on a laboratory dataset of people in a fixed set of poses and two clothing conditions: with and "without" clothes. The latter gave us "ground truth" with which to evaluate the method. We also demonstrated the method for more natural sequences of clothed people from the HumanEva dataset [19].

Privacy concerns must be addressed for any technology that purports to "see" what someone looks like under their clothes. Unlike backscatter X-ray and infrared sensors, our approach does not *see through* clothing. It does not have any information about the person's body that is not available essentially to the naked eye; in this sense it is not intrusive. The unwanted production of a likeness or facsimile of a person's unclothed shape might still be considered a violation of privacy. It is important to maintain a distinction between body shape measurements and the graphical representation of those measurements as a realistic 3D model; it is the latter which has the greatest potential for concern but this may not be needed for many vision applications.

There are many applications and future directions for this new line of research. For human tracking in video it may be useful to estimate limb lengths, body shape parameters and body mass as these could be used in the inference of dynamics. Body shape parameters could be used in visual tracking applications to identify and re-acquire subjects who come in and out of the field view. For forensic video applications, the extraction of body shape parameters could be useful in identifying suspects. There are also many applications of these methods in personal fitness, retail apparel and computer games. In future work we will extend these methods to extract body shape from monocular image sequences.

Acknowledgments. This work was supported in part by the Office of Naval Research (N00014-07-1-0803), NSF (IIS-0535075), by an award from the Rhode Island Economic Development Corporation (STAC), and by a gift from Intel Corp. The authors thank L. Reiss for assistance with image segmentation and biometric measurements.

References

1. Anguelov, D., Srinivasan, P., Koller, D., Thrun, S., Rodgers, J., Davis, J.: SCAPE: Shape completion and animation of people. SIGGRAPH 24, 408–416 (2005)
2. Bălan, A.O., Sigal, L., Black, M.J., Davis, J.E., Haussecker, H.W.: Detailed human shape and pose from images. In: CVPR (2007)
3. Bălan, A.O., Black, M.J., Sigal, L., Haussecker, H.W.: Shining a light on human pose: On shadows, shading and the estimation of pose and shape. In: ICCV (2007)
4. Mündermann, L., Corazza, S., Andriacchi, T.: Accurately measuring human movement using articulated ICP with soft-joint constraints and a repository of articulated models. In: CVPR (2007)
5. Cheung, K.M., Baker, S., Kanade, T.: Visual hull alignment and refinement across time: A 3D reconstruction algorithm combining shape-from-silhouette with stereo. In: CVPR, vol. 2, pp. 375–382 (2003)
6. Laurentini, A.: The visual hull concept for silhouette-based image understanding. PAMI 16, 150–162 (1994)
7. Rosenhahn, B., Kersting, U., Powell, K., Klette, R., Klette, G., Seidel, H.P.: A system for articulated tracking incorporating a clothing model. Machine Vision and Applications (MVA) 18, 25–40 (2007)
8. Salzmann, M., Pilet, J., Ilic, S., Fua, P.: Surface deformation models for nonrigid 3D shape recovery. PAMI 29, 1481–1487 (2007)
9. White, R., Crane, K., Forsyth, D.: Capturing and animating occluded cloth. In: ACM Transactions on Graphics (TOG), SIGGRAPH (2007)
10. de Aguiar, E., Theobalt, C., Stoll, C., Seidel, H.P.: Marker-less deformable mesh tracking for human shape and motion capture. In: CVPR, pp. 2502–2509 (2007)
11. Stark, J., Hilton, A.: Surface capture for performance-based animation. IEEE Comp. Graphics and Applications 27, 21–31 (2007)
12. Franco, J.S., Lapierre, M., Boyer, E.: Visual shapes of silhouette sets. In: Proc. 3rd Int. Symp. on 3D Data Processing, Visualization and Transmission (2006)
13. Cheung, K.M., Baker, S., Kanade, T.: Shape-from-silhouette across time: Part II: Applications to human modeling and markerless motion tracking. IJCV 63, 225–245 (2005)
14. Grauman, K., Shakhnarovich, G., Darrell, T.: A Bayesian approach to image-based visual hull reconstruction. In: CVPR, vol. 1, pp. 187–194 (2003)
15. Grauman, K., Shakhnarovich, G., Darrell, T.: Inferring 3D structure with a statistical image-based shape model. In: ICCV, pp. 641–648 (2003)
16. Moghaddam, B., Yang, M.: Learning gender with support faces. PAMI 24, 707–711 (2002)
17. Li, X., Maybank, S., Yan, S., Tao, D., Xu, D.: Gait components and their application to gender recognition. IEEE Transactions on Systems, Man, and Cybernetics, Part C: Applications and Reviews 38, 145–155 (2008)

18. Bouguet, J.Y.: Camera calibration toolbox for Matlab. Technical report (CalTech)
19. Sigal, L., Black, M.J.: HumanEva: Synchronized video and motion capture dataset for evaluation of articulated human motion. Technical Report CS-06-08, Brown University (2006)
20. Brand, M.: Incremental singular value decomposition of uncertain data with missing values. In: Heyden, A., Sparr, G., Nielsen, M., Johansen, P. (eds.) ECCV 2002. LNCS, vol. 2350, pp. 707–720. Springer, Heidelberg (2002)

Temporal Surface Tracking Using Mesh Evolution

Kiran Varanasi, Andrei Zaharescu, Edmond Boyer, and Radu Horaud

LJK - INRIA Rhône-Alpes, France

Abstract. In this paper, we address the problem of surface tracking in multiple camera environments and over time sequences. In order to fully track a surface undergoing significant deformations, we cast the problem as a mesh evolution over time. Such an evolution is driven by 3D displacement fields estimated between meshes recovered independently at different time frames. Geometric and photometric information is used to identify a robust set of matching vertices. This provides a sparse displacement field that is densified over the mesh by Laplacian diffusion. In contrast to existing approaches that evolve meshes, we do not assume a known model or a fixed topology. The contribution is a novel mesh evolution based framework that allows to fully track, over long sequences, an unknown surface encountering deformations, including topological changes. Results on very challenging and publicly available image based 3D mesh sequences demonstrate the ability of our framework to efficiently recover surface motions .

1 Introduction

Tracking the surface of moving objects is of central importance when modeling dynamic scenes using multiple videos. This key step in the modeling pipeline yields temporal correspondences which are necessary when considering motion related applications such as motion capture. Furthermore, it allows recovery of improved and consistent descriptions of object shapes and appearances.

In this work we address the problem of capturing the evolution of a moving and deforming surface, in particular moving human bodies, given multiple videos. A large variety of directions can be followed, depending on the *a priori* knowledge of the observed shape, on the representation chosen for surfaces and on the information taken into account for deformations. *Model-based approaches* assume a known model of the observed surface, which is tracked over time sequences, hence solving for time correspondences. This model can be locally rigid, e.g [1–3], or deformable, e.g. [4,5]. Unfortunately, exact models need to be available, which is seldom the case in general situations. In particular, the topology of the surface can evolve over time as shown in Figure 1. As a consequence, approaches in this category are restricted to specific scenarios.

In contrast, non model-based approaches try to find displacement fields between 2 different instants in the sequence. In this category, *scene flow approaches* consider dense vector fields with various representations including voxels [6,7], implicit representations [8] or meshes [9]. However, the associated differential methods are limited to small displacements between successive frames. Alternatively, *feature-based*

D. Forsyth, P. Torr, and A. Zisserman (Eds.): ECCV 2008, Part II, LNCS 5303, pp. 30–43, 2008.

(a) (b) (c)

Fig. 1. An example of a surface for which the topology can hardly be known in advance. The belt of the dress forms a new protrusion that appears and disappears (a)-(b). (c) Dense point trajectories computed from (a) to (b).

approaches [10–12] consider meshes and allow for larger motions by casting the problem as a labeling between 2 meshes using local geometric or photometric information. This labeling solves for partial correspondences between 2 frames only and might lose efficiency when applied over long sequences, in particular as topological changes occur. Our objective is different but complementary and targeted at providing full mesh evolution over time sequences.

Our approach is grounded on the observation that natural surfaces are usually arbitrary shaped and difficult to model *a priori*. In addition, shapes can significantly evolve over a time sequence. For instance, human bodies are usually covered by clothes whose topologies can change. To handle such deformations, we use meshes which are morphed from one frame to another. Like feature-based approaches, we use photometric cues provided by images and geometric cues provided by the recovered meshes. However, instead of looking for a dense match between the vertices of the 2 meshes, we use a sparse, but robust set of matches and its associated displacement vector field to drive a full consistent mesh evolution, with possible topological changes. This approach provides both a consistent surface evolution over time and dense point trajectories on the surface.

The framework we propose assumes little about the observed surface, thus relaxing the constraints for markers, known models or limited deformations and displacements. It allows for the recovery of trajectories of points, as shown in Figure 1, on a surface undergoing significant deformations including topological changes. Instead of using traditional *Eulerian methods* e.g. level sets [13], a major innovation is to cast the problem within a mesh evolution framework that performs mesh morphing, thereby avoiding *Eulerian* limitations such as complexity and inappropriateness for tracking interface properties, e.g. vector displacements.

The remainder of this paper is organized as follows. Related works are reviewed in section 2. The proposed approach is outlined in section 3. The recovery of displacement vector fields is described in section 4 and 5. The mesh deformation is then explained in section 6. Experimental results obtained with publicly available sequences are shown in section 7, before concluding in section 8.

2 Related Work

Surfaces observed in multiple views can be fully tracked through a deformable model, which is fitted to image related observations, see for instance [14] for a review on 3D deformable models. This operation appears to be difficult, unless a precise model is available. This is particularly true with recent works [9, 15] that propose to use a laser-scanned model of the surface prior to tracking. Unfortunately, precise models will not be available in general situation. Moreover the fixed topology assumption significantly limits the application domain.

An alternative is to directly estimate surface motions between temporal frames and a significant effort has been put in that direction over the past years. *Scene flow approaches* recover dense motion fields using derivatives of the image signal [6, 7, 16]. In [8] this is used within a variational framework to fully track surfaces using level sets. As noticed in [12], flow-based approaches are nevertheless limited to small displacements, as a consequence of finite difference approximations of derivatives.

Another class of approaches solve for shape matching. Assume that shape models, e.g. meshes, can be recovered from images independently over time sequences, using for instance [17, 18, 8]. Then temporal correspondences can be obtained through vertex mapping between successive meshes. While providing displacement fields between frames, temporal correspondences yet only partially solve the problem of surface tracking since the transformation that maps a surface onto another remains unknown. Nevertheless, this can be seen as a first step towards full surface tracking. The associated labeling problem can be solved in various ways. Point based approaches, e.g. [19, 20], register sets of points but do not account for shape information, i.e. mesh connectivity. More closely related to our framework, numerous mesh based approaches have been proposed. Some solve for correspondences indirectly through embeddings, e.g. [21–23], with the price of an often difficult intermediate step. Other approaches bypass this step and directly seek correspondences between meshes. For instance [10, 11] successfully match 2 different poses of a *proper* mesh, e.g. a range-scanned model, using geometric features only. However, these approaches do not easily extend to real objects' surfaces recovered from images since the associated meshes can vary drastically between successive frames and furthermore, their topologies can change. In that case, photometric cues are advantageously added to geometric features to make the labeling feasible as in [12]. While allowing for large motions, labeling approaches seek dense correspondences, which are difficult to obtain on a regular basis over long sequences.

To the best of our knowledge, no previous work has attempted to perform the full tracking of an unknown mesh undergoing deformations with possibly topological changes, using multiple videos. Our method bridges the gap between model-based and non model-based approaches since we evolve a mesh using temporal correspondences. To this purpose, we build on some of the ideas already used in the approaches mentioned above. We combine the interest of both photometric and geometric cues for robust matching [12] with Laplacian diffusion [9] to get a dense displacement field. The resulting vector field is used to initialize a consistent mesh evolution between successive frames. To demonstrate the robustness of our scheme, we have used challenging real data sets with large motion and topological changes.

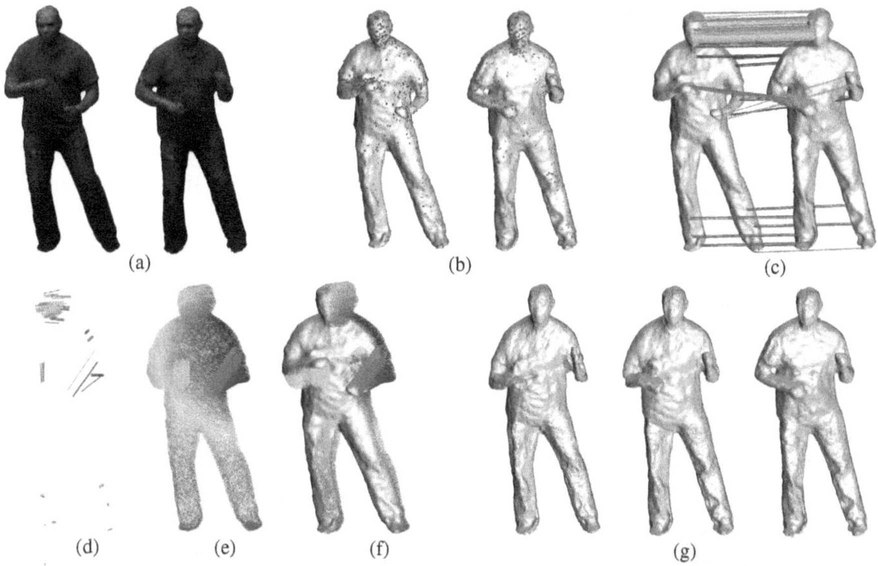

Fig. 2. The different consecutive steps of the proposed framework: (a) original meshes $\mathcal{S}^t$ and $\mathcal{M}^{t+1}$; (b) features extracted; (c) feature matching; (d) the associated sparse-displacement; (e)-(f) the dense displacement field after Laplacian diffusion; (g) mesh-morphing (observe the topological change that takes place around the right arm of the model - the right elbow de-attaches from the body, creating a genus change)

3 Approach Outline

We consider multiple camera environments and we assume that multiple calibrated videos of an object with closed surfaces are available. We also assume that 3D mesh models $\mathcal{M}^{t\in[1..n]}$ of the object at different time instances $[1..n]$ estimated using multi-view 3D modeling approaches, e.g. [17, 18, 8], are available. These meshes $\mathcal{M}^{t\in[1..n]}$ correspond to discrete values of the time continuous mesh $\mathcal{S}^t$. In order to recover $\mathcal{S}^t$, the mapping of $\mathcal{S}^t$ onto $\mathcal{M}^{t+1}$ is iteratively estimated using the following 3 consecutive steps, starting with $\mathcal{S}^1 = \mathcal{M}^1$:

1. Sparse match: photometric and geometric cues are used to match a set of points between $\mathcal{S}^t$ and $\mathcal{M}^{t+1}$ (see Figure 2-c). Unlike previous approaches, only a sparse set of correspondences is expected.
2. Motion diffusion: the identified correspondences define a sparse displacement field over $\mathcal{S}^t$ (cf. figure 2-d). This field is propagated over all vertices by Laplacian diffusion hence preserving local shape details [24] (cf. Figure 2-e).
3. Mesh evolution: The dense displacement field is applied to the vertices of $\mathcal{S}^t$ yielding a new mesh. The resulting mesh $\hat{\mathcal{S}}^t$ is then morphed to $\mathcal{M}^{t+1}$ by minimizing the signed distance to $\mathcal{M}^{t+1}$ (cf. Figure 2-fg). Mesh consistency within the optimization is enforced using [25]. The final optimized mesh defines $\mathcal{S}^{t+1}$.

The Laplacian diffusion allows a partial vertex matching only and yields to a good estimation of the motion at all vertex locations. Nevertheless, an additional step is required to guarantee that the resulting mesh fits the observations $\mathcal{M}^{t\in[1..n]}$ and also to guaranty its correctness, e.g. manifoldness. This is in contrast with the work [9] which also uses Laplacian diffusion to evolve a reference mesh to the observed posture, but without refinement and therefore without guaranties. The 3 above steps are detailed in the following sections.

4 Feature Matching

The primary step of our approach is to obtain a set of good feature matches across the two frames. In contrast to the labeling approaches mentioned previously, we do not intend here to produce a dense match over mesh vertices, but only a robust selection. To this end, we first detect a set of interest points and provide them with a variety of distinctive features (photometric and geometric). These sets of feature vectors are then matched across in an exhaustive manner, to compute a preliminary set of potential matches. The error of the matching is defined in terms of difference between the different feature vectors. We employ a two-step minimization procedure (a coarse step will guide a finer step) in order to avoid local minima. We detail each of these steps in the following subsections.

4.1 Feature Extraction

For each frame, we are provided with a 3D mesh representation S^t coupled with a set of images I_i^t depicting camera views of the object from different angles.

Image Features. We use corners as image features. If silhouettes are available, we use them to constrain the features (in practice, we erode the silhouettes by $\alpha = 3$ pixels to eliminate the features close to the boundary). The feature points are computed as maxima of the determinant of the image Hessian matrix. We have chosen Speeded-Up Robust Features (SURF) [26] as an image descriptors, because of their robustness in wide-baseline stereo and because of their increased speed of computation due to integral images. We back-project the detected interest points onto the 3D mesh and assign the corresponding SURF feature, together with color features, i.e. hue, saturation and value (HSV). The color for the 3-D point is calculated as the median color in the visible cameras. Figure 3-a illustrates the distribution of the feature points over a sample 3D mesh.

Mesh Features. Geodesic distances between mesh points offer crucial information in matching non rigid shapes. We use a feature called the *normalized geodesic integral* [27], which is defined as:

$$\mu(V) = \int_{P\in S} G(V,P)dS \quad , \quad \mu_n(V) = \frac{\mu(V) - Min_{P\in S}\mu(P)}{Max_{P\in S}\mu(P)}$$

where $G(V, P)$ denotes the geodesic distance between the points P and V and $\mu(V)$ is defined as the sum of the geodesic distances from V to all points on S. After normalization, $\mu_n(V)$ provides a continuous function whose value indicates the apparent nearness of a point to the center of the object. Its maxima will correspond to the extremities of the object. Figure 3-b illustrates the distribution of this function over the sample 3D mesh.

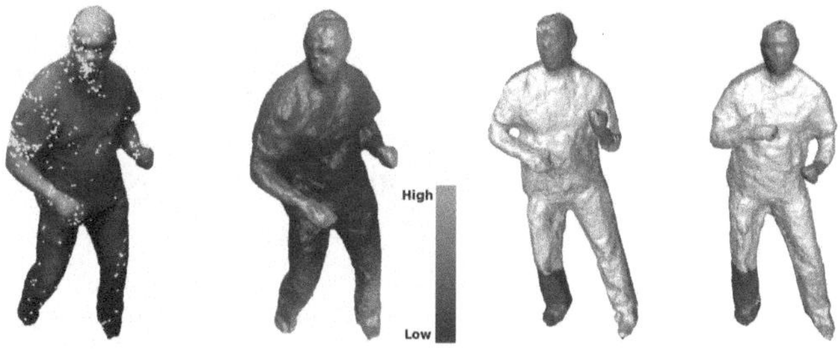

(a) (b) (c)

Fig. 3. (a) SURF feature points back-projected onto the 3D mesh (b) The geodesic integral function over the 3D mesh (c) Surface Protrusions detected as the maxima of the geodesic integral - the left figure has a collapsed protrusion

4.2 Coarse Matching by Surface Protrusions

We proceed to identify the extrema of the geodesic integral $\mu_n(v)$. This is done by simply imposing a threshold on the value of $\mu_n(v)$ and selecting the points on the mesh that lie above this threshold. Such points lie in compact clusters, typically corresponding to the different protrusions of the object. As shown in Figure 3-c, between time-frames, some of these protrusions collapse onto the surface inducing changes in the topology of the mesh. We will devise an algorithm whose goal is to correctly detect the topological changes and match the extrema accordingly.

We select the local extremum of the function $\mu_n(v)$ as representative for each cluster. The extent of each protrusion is defined by a local geodesic neighborhood from the representative point. We assign a feature to the protrusion based on color distribution in this region. This is a highly distinctive feature defined on a large neighborhood. These features can be visualized as in Figure (4- a). We now proceed to match each of these protrusions uniquely across the two frames. This problem is formulated as an error minimization problem with the following error:

$$E = \sum_i \Psi(X_i^t, X_i^{t+1}) + \sum_{i,j} |G(X_i^t, X_j^t) - G(X_i^{t+1}, X_j^{t+1})|$$

where X_i^{t+1} denotes the match of a protrusion X_i^t, $\Psi(X_i^t, X_i^{t+1})$ denotes the error computed through color features and $G(X_i, X_j)$ denotes the geodesic distance between

(a) (b) (c)

Fig. 4. (a) Color features computed on surface protrusions (b) The surface clustered into regions based on color (c) Surface protrusions matched with each other. Note that even the collapsed protrusion is matched correctly.

the protrusions X_i and X_j. Since the number of detected protrusions is typically very small, we proceed to do an exhaustive search to solve the matching problem.

In the particular case of tracking adjacent frames, we add an additional term $||X_i^t - X_i^{t+1}||$ in the error, which denotes the Euclidean distance between the two matched protrusions. This means that the two matched protrusions are not too far from each other - a strong but valid assumption in our case. This helps the algorithm in resolving issues raised by symmetry (such as between the two hands, or between the two legs).

For a human body in a normal condition, the number of protrusions are 5 (head + 2 hands + 2 legs). When there is a collapse (hand/leg touching a part of the body), the number of detected protrusions will be 4 or lesser. There will be an implicit mismatch in the number of protrusions detected, and this will be detected trivially. The correspondences of such collapsed protrusions will be left without a match in the previous minimization step. It should be noted that such collapses will not damage the geodesic distances between protrusions which did not collapse themselves. In this way, we efficiently use the geodesic distances only where they are meaningful.

To handle collapsed protrusions, we cluster the target surface region based on color, and select the most appropriate cluster based on the error $\Psi(X_i^t, X_i^{t+1})$ (or $\Psi(X_i^t, X_i^{t+1}) + ||X_i^t - X_i^{t+1}||$) for adjacent frames. The surface regions clustered according to color are shown in Figure 4-b. The result of the matching is shown in Figure 4-c.

4.3 Fine Matching by Feature Points

The previous step provides a good initialization to perform feature matching at a finer level. We intend to produce a selection of feature matches that are representative of the surface and that are mutually consistent with each other. We define an error function based on color, SURF features and the array of geodesic distances of the feature points from each of the protrusions which are matched successfully in the above step (without collapse). We select a set of best features P_i^t, P_i^{t+1} based on this error, and prune this further to impose mutual spatial consistency.

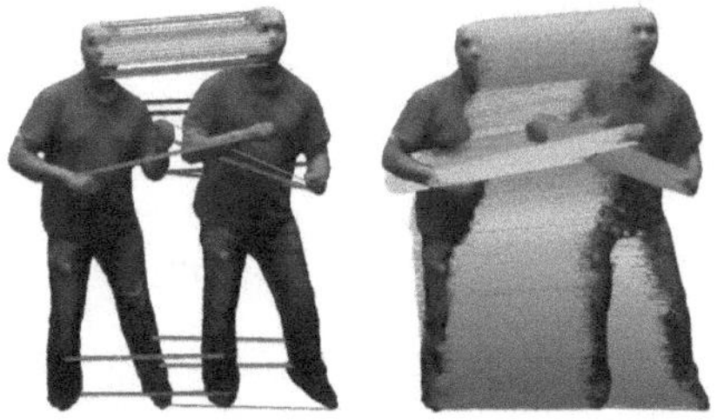

(a) (b)

Fig. 5. (a) The set of identified feature matches (b) The dense motion field computed by Laplacian diffusion

If two feature points P_1^t, P_2^t are found to be geodesically near to each other, we connect them by a link which encodes the Euclidean length $|\overrightarrow{P_1^t P_2^t}|$ of the line joining the two points, and the angles it makes with the normals $\hat{P_1^t}, \hat{P_2^t}$ at both the ends. Then the two pairs of matches (P_1^t, P_1^{t+1}) and (P_2, P_2^{t+1}) are checked for mutual spatial consistency in terms of the elastic stretch (γ_s) and twist (γ_{t1}, γ_{t2}) of the link, defined as:

$$\begin{aligned}
\gamma_s &= \Gamma_s(|\overrightarrow{P_1^t P_2^t}| - |\overrightarrow{P_1^{t+1} P_2^{t+1}}|)\\
\gamma_{t1} &= \Gamma_t(\theta(\overrightarrow{P_1^t P_2^t}, \hat{P_1^t}) - \theta(\overrightarrow{P_1^{t+1} P_2^{t+1}}, \hat{P_1^{t+1}}))\\
\gamma_{t2} &= \Gamma_t(\theta(\overrightarrow{P_1^t P_2^t}, \hat{P_2^t}) - \theta(\overrightarrow{P_1^{t+1} P_2^{t+1}}, \hat{P_2^{t+1}}))
\end{aligned}$$

where $\theta(\boldsymbol{v1}, \boldsymbol{v2})$ denotes the angle between two vectors, and Γ_s, Γ_t denote two Gaussian penalty functions.

Furthermore, pairs are checked for parity in order (ρ_1) and orientation (ρ_2) with respect to the nearest matched protrusion (X^t, X^{t+1}), and defined as:

$$\begin{aligned}
&\rho_1 : Sign(|\overrightarrow{P_1^t X^t}| - |\overrightarrow{P_2^t X^t}|) = Sign(|\overrightarrow{P_1^{t+1} X^{t+1}}| - |\overrightarrow{P_2^{t+1} X^{t+1}}|)\\
&\rho_2 : \theta(\overrightarrow{P_1^t X^t} \times \overrightarrow{P_2^t X^t}, \overrightarrow{P_1^{t+1} X^{t+1}} \times \overrightarrow{P_2^{t+1} X^{t+1}}) < 180^\circ
\end{aligned}$$

An example of the set of identified feature matches is shown in Figure 5-a.

5 Motion Diffusion

The feature matches computed as above provide the initialization for the transformation of the mesh. We propagate them across the entire mesh by Laplacian diffusion.

5.1 Approach

The geometric Laplacian operator is a way of encoding the local curvature of the mesh. This has proven to be useful in a variety of mesh applications [24], such as interactive mesh editing. This operator provides an efficient approach to deform the mesh while preserving the local shape information. If N_V is the number of vertices, the Laplacian matrix L of size $N_V \times N_V$ is defined by these equations

$$\begin{aligned} L(i,j) &= wt(i,j) \quad \forall_{j \in N(i)} \\ L(i,j) &= 0 \quad \forall_{j \notin N(i)} \\ L(i,i) &= -1 * \textstyle\sum_{j \in N(i)} wt(i,j) \end{aligned}$$

where $N(i)$ is the set of vertices sharing an edge with the vertex i, and $wt(i,j)$ is the weight of the edge as defined by mean-value coordinates [24].

The feature matches computed earlier as initialization, now define 3 matrices L_x, L_y and L_z, corresponding to the three dimensions X, Y and Z. If the number of feature matches is N_F, these matrices shall be of order $(N_V + N_F) \times N_V$. The first N_V rows shall be identical to the L matrix. The later rows are defined by constraints $(\forall i \in \{features\})$ (similarly for L_y, L_z):

$$\begin{aligned} L_x(i,j) &= 0 \quad \forall j \neq i \\ L_x(i,i) &= \lambda \end{aligned}$$

where λ is a weighting factor we set to 4000.

The 3D coordinates of the mesh vertices are stacked into three vectors X^t, Y^t and Z^t. Similar to the matrix L_x, we append the vector X^t by adding N_F new elements $\{\lambda * X_F^{t+1}\}$ where X_F^{t+1} are the X-coordinates of the feature matches in the frame $t+1$. The diffusion of the matches is done as a matrix inversion.

$$X^{t+1} = (L_x^\top L_x)^{-1} L_x^\top * X_t$$

The matrix L_x being extremely sparse, this inversion can be efficiently implemented using Cholesky factorization.

Thus we propagate the mesh S^t via Laplacian diffusion to $\hat{S}^t$. An example of the dense motion field obtained from a sparse set of feature matches is shown in Figure 5-b. In [9] the Laplacian operator is also used to diffuse motion information over meshes. However, motion is limited to rotation since flow information is considered. In contrast, we propagate full displacement vectors as obtained by matching feature points between $\hat{S}^t$ and the observed mesh $\mathcal{M}^{t+1}$.

6 Mesh Deformation

The matching and diffusion steps presented in the previous sections provide us with a dense displacement field over the mesh $\mathcal{S}^t$. As mentioned before, such motion field is a good estimate of the true motion field between time t and $t+1$. However, it will not guarantee the exact overlap with the mesh observed at $t+1$, i.e. $\mathcal{M}^{t+1}$, nor the correctness of the resulting mesh. Therefore, a final step is needed in order to ensure both convergence to the observations and correctness. Our approach is motivated by the fact that the solution mesh $\mathcal{M}^{t+1}$ and the propagated Laplacian mesh $\hat{\mathcal{S}}^t$ are different, but nearby. Thus a solution is to perform surface-morphing, that is starting from the source surface, i.e. $\hat{\mathcal{S}}^t$, and evolving it towards the destination surface $\mathcal{M}^{t+1}$. To this purpose, we have used [25] as an explicit surface evolution approach which handles self-intersections and topological changes and guarantees correctness. To drive the surface evolution, we adopt here a simple morphing scheme introduced in [28] and described below.

6.1 Approach

Consider an open set $O_A \subset \mathbb{R}^3$ representing the source object, enclosed by surface $S_A = \partial O_A$, and similarly the open set $O_B \subset \mathbb{R}^3$ representing the target object, enclosed by $S_B = \partial O_B$. Consider the signed distance u_B of S_B, as defined by:

$$u_B(x) = \begin{cases} -d(x, S_B) & \forall x \subset O_B, \\ d(x, S_B) & \text{otherwise}, \end{cases} \tag{1}$$

where $d(x, y)$ is the Euclidean distance between x and y in $\mathbb{R}^3$. Following [28], the surface motion that maximizes the overlap between the morphed object and S_B is defined by:

$$\frac{\partial S}{\partial t} = -u_B(x)\mathbf{N}(x), \tag{2}$$

where x is a point on the surface S and $\mathbf{N}(x)$ is the normal to S at x. The strategy described above will converge to the desired solution if the surface of departure S_A and the destination surface S_B overlap.

6.2 Discussion

In a few cases, certain tracks are temporarily lost, due to the non-overlap of certain parts of the surface between the propagated Laplacian and the next frame. Such an example can be observed in Figure 6-f, where the left hand had the fist properly propagated (due to the protrusion region matching), but not the forearm (due to the lack of features). This caused the signed distance function based evolution to collapse a sub-part of the forearm and regrow it from the upper-arm and the fist. These rare cases can be addressed by interpolating the trajectories from the neighboring vertices which are tracked correctly.

If we are not satisfied with the propagated Laplacian, we can also try to increase the number of the matches by exploring the neighborhoods of the sparse matches detected by our method. These increased matches are then diffused in a similar fashion using the mesh Laplacian. At a minor additional computational cost, this process produces a better initialization for the mesh deformation step.

Another remark is that instead of surface morphing, one could also consider other functions. One such choice is multi-view stereo photo-consistency. We have experimented with such a distance function [8], observing that the optimizer could not easily handle situations where the source mesh is relatively far from the destination mesh. This is in part due to its coarse to fine nature. Another benefit of the mesh morphing approach is that, assuming there is some overlap between the source $\hat{S}^t$ and the destination $\mathcal{M}^{t+1}$ meshes, it is guaranteed that the approach will converge, with potential topological changes. In addition, every vertex will reach the destination mesh. Once reached, it will neither move nor oscillate.

7 Results

7.1 Qualitative Evaluation

For our evaluation we have been using sequences from two sources: the *dance-1* (used to exemplify the method) and the *dance-2* sequences are available publicly on our website[1]. The *pop2lock* and *flashkick* sequences were made available to us by the Surface Motion Capture project at the University of Surrey [29].

The *pop2lock* sequence provides us with full 3-D reconstruction results, together with the camera calibration, input images and silhouettes, using 8 cameras (1920x1080). The *dance-2* sequence is captured using 8 cameras (780x582). For this last sequence we decided to test the limits of the algorithm. We have used rougher 3-D surfaces approximation obtained via a fast visual hull reconstruction (exclusively based on silhouettes). Despite their coarse nature and topological changes, we still obtain consistent point trajectories. We ensured proper mesh sampling via edge collapses and edge swaps, such that each edge is around 3 pixels when projected onto the image (*pop2lock* meshes - 12,000 vertices; *dance-2* - 3,000 vertices). Coarser meshes were used for computing geodesics (1,500 vertices).

Our results are presented in Figure 6, with a close-up of a topological change illustrated in Figure 7. The tracks are color-coded, where cooler colors represent earlier frames. Additional convincing results are provided as a video [2], the natural way of displaying temporal information.

We were able to successfully track without problems *long* sequences of over 100 frames with *large* inter-frame shifts. The running times are satisfactory, depending a lot

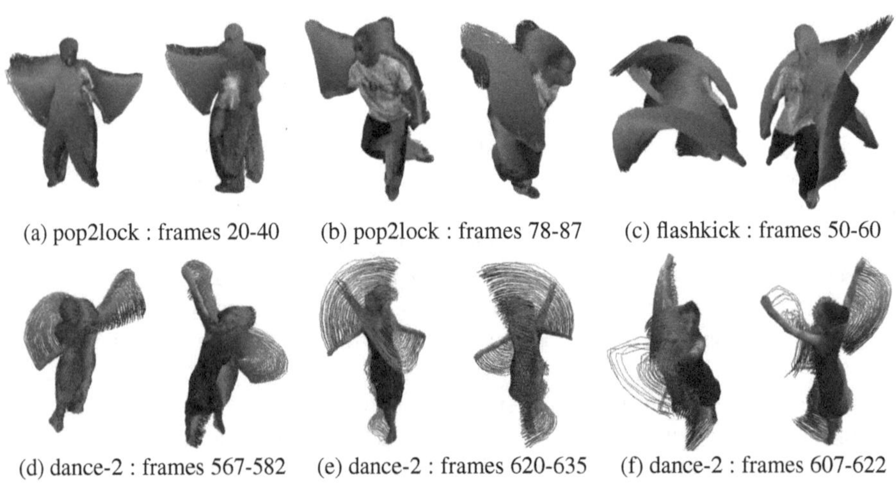

(a) pop2lock : frames 20-40 (b) pop2lock : frames 78-87 (c) flashkick : frames 50-60

(d) dance-2 : frames 567-582 (e) dance-2 : frames 620-635 (f) dance-2 : frames 607-622

Fig. 6. The tracked trajectories are presented in a color coded scheme where cooler colors represent earlier frames. On top the *pop2lock* and *flashkick* sequences (Univ. of Surrey) and bottom the *dance-2* sequence (INRIA).

[1] https://charibdis.inrialpes.fr/html/sequences.php

[2] https://perception.inrialpes.fr/Publications/2008/VZBH08/ECCV08.mp4

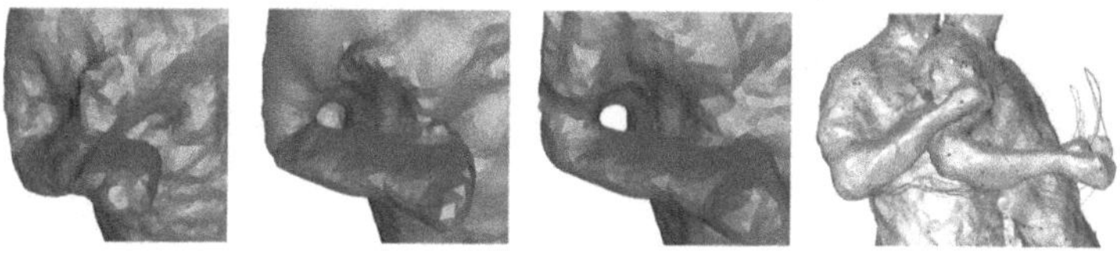

Fig. 7. Mesh deformation over topological change (from left to right): initialization; intermediate step; final step; Overall algorithm behavior (sparse set of matches shown only for ease of visualization purposes)

on the mesh density and the number of images used within each frame. As an example, in the *dance-2* sequence, an inter-frame surface tracking is produced in about 30 seconds, whereas for the *pop2lock* dataset, it takes about 2.5 minutes.

7.2 Numerical Evaluation

Lack of proper ground truth makes quantitative assessment of 3D tracking algorithms difficult. A manual labeling could be inconsistent because the accuracy of the tracks needs to be measured with high precision. Due to the absence of real world test data, we evaluated the trajectories of our algorithm against known deformations of a 3D graphical model. We used an artificially textured female humanoid model (figure 8-a), and the multi-view video is captured using a 16 camera setup. An example trajectory computed by our algorithm is visualized in figure 8-b.

We evaluated the error in point trajectories with respect to the average edge length, which defines the resolution for temporal correspondences. Figure 8-c shows such errors for 600 points randomly distributed over the mesh and as obtained with independent estimations of the surface evolutions between frames. We observe that the error is less than half the average edge length after 50 frames. In the same duration, the average true

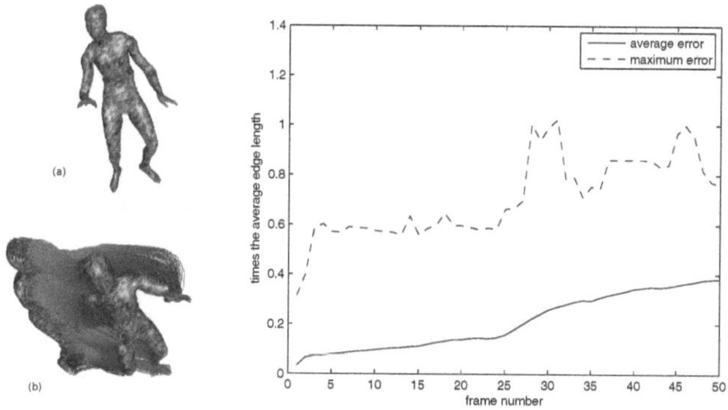

(c)

Fig. 8. Numerical Evaluation : (a) Example mesh with texture (b) Computed trajectories (c) Error over the sequence

deformation encountered by each point is about 10 times the average edge length. Thus we stay within reasonable limits of accuracy in producing our tracks.

8 Conclusion

In conclusion, we have presented a robust algorithm for temporal mesh tracking that incorporates the following key ingredients: it uses both geometric and photometric information in a coarse to fine fashion in order to efficiently solve for a sparse set of matches; it uses Laplacian propagation to obtain a dense match set; it ensures proper evolution using a mesh-morphing approach that is capable of dealing with topological changes. Thus, we are able to perform surface tracking with *large displacements* of surfaces with *topological changes* over *long sequences* in the context of multiple camera environments. In addition, our algorithm performs gracefully even when provided with inexact surfaces. For future work, we are considering real time motion capture systems, improvements of surface recovery, and reducing the drift in trajectories by time integration.

References

1. Gavrila, D., Davis, L.: 3-D model-based tracking of humans in action: a multi-view approach. In: Proceedings of IEEE Conference on Computer Vision and Pattern Recognition, San Francisco, USA (1996)
2. Kakadiaris, I., Metaxas, D.: Model-based estimation of 3d human motion. IEEE Transactions on PAMI 22, 1453–1459 (2000)
3. Carranza, J., Theobalt, C., Magnor, M., Seidel, H.P.: Free-viewpoint video of human actors. In: Proc. ACM Siggraph 2003, San Diego, USA, pp. 569–577 (2003)
4. DeCarlo, D., Metaxas, D.: Optical flow constraints on deformable models with applications to face tracking. International Journal of Computer Vision 38(2), 99–127 (2000)
5. Salzmann, M., Pilet, J., Ilic, S., Fua, P.: Surface deformation models for non-rigid 3–d shape recovery. IEEE Transactions on PAMI 29, 1481–1487 (2007)
6. Vedula, S., Rander, P., Collins, R., Kanade, T.: Three-Dimensional Scene Flow. IEEE Transactions on PAMI 27(3), 474–480 (2005)
7. Neumann, J., Aloimonos, Y.: Spatio-Temporal Stereo Using Multi-Resolution Subdivision Surfaces. International Journal of Computer Vision 47, 181–193 (2002)
8. Pons, J.P., Keriven, R., Faugeras, O.: Multi-view stereo reconstruction and scene flow estimation with a global image-based matching score. International Journal of Computer Vision 72(2), 179–193 (2007)
9. de Aguiar, E., Theobalt, C., Stoll, C., Seidel, H.: Marker-less Deformable Mesh Tracking for Human Shape and Motion Capture. In: Proceedings of IEEE Conference on Computer Vision and Pattern Recognition, Minneapolis, USA (2007)
10. Anguelov, D., Srinivasan, P., Pang, H.C., Koller, D., Thrun, S., Davis, J.: The correlated correspondence algorithm for unsupervised registration of nonrigid surfaces. In: Proceedings of Conference on Neural Information Processing Systems, Cambridge, USA (2004)
11. Bronstein, A., Bronstein, M., Kimmel, R.: Calculus of non-rigid surfaces for geometry and texture manipulation. IEEE Transaction on Visualization and Computer Graphics 13(5), 902–913 (2007)

12. Starck, J., Hilton, A.: Correspondence labelling for wide-time free-form surface matching. In: Proceedings of the 11th International Conference on Computer Vision, Rio de Janeiro, Brazil (2007)
13. Osher, S., Fedkiw, R.: Level Set Methods and Dynamic Implicit Surfaces. Springer, Heidelberg (2003)
14. Montagnat, J., Delingette, H., Scapel, N., Ayache, N.: Representation, shape, topology and evolution of deformable surfaces. application to 3d medical image segmentation. Technical Report 3954, INRIA (2000)
15. Bickel, B., Botsch, M., Angst, R., Matusik, W., Otaduy, M., Pfister, H., Gross, M.: Multiscale capture of facial geometry and motion. In: ACM Computer Graphics (Proceedings SIGGRAPH) (2007)
16. Carceroni, R., Kutulakos, K.: Multi-View Scene Capture by Surfel Sampling: From Video Streams to Non-Rigid 3D Motion, Shape and Reflectance. International Journal of Computer Vision 49(2-3), 175–214 (2002)
17. Hernandez, C., Schmitt, F.: Silhouette and stereo fusion for 3D object modeling. Computer Vision and Image Understanding 96, 367–392 (2004)
18. Furukawa, Y., Ponce, J.: Carved Visual Hulls for Image-Based Modeling. In: Proceedings of the 9th European Conference on Computer Vision, Graz, Austria (2006)
19. Besl, P., McKay, N.: A method for registration of 3-d shapes. IEEE Transactions on PAMI 14(2), 239–256 (1992)
20. Chui, H., Rangarajan, A.: A new point matching algorithm for non-rigid registration. Computer Vision and Image Understanding 89(2-3), 114–141 (2003)
21. Zhang, D., Hebert, M.: Harmonic Maps and Their Applications in Surface Matching. In: Proceedings of IEEE Conference on Computer Vision and Pattern Recognition, Fort Collins, USA (1999)
22. Zigelman, G., Kimmel, R., Kiryati, N.: Texture mapping using surface flattening via multidimensional scaling. IEEE Transactions on Visualization and Computer Graphics 8(2), 198–207 (2002)
23. Starck, J., Hilton, A.: Spherical Matching for Temporal Correspondence of Non-Rigid Surfaces. In: Proceedings of the 10th International Conference on Computer Vision, Beijing, China (2005)
24. Sorkine, O.: Laplacian mesh processing. In: Eurographics Conference (2005)
25. Zaharescu, A., Boyer, E., Horaud, R.: Transformesh: a topology-adaptive mesh-based approach to surface evolution. In: Proceedings of the 8th Asian Conference on Computer Vision, Tokyo, Japan (2007)
26. Bay, H., Tuytelaars, T., van Gool, L.: Surf: Speeded up robust features. In: Proceedings of the 9th European Conference on Computer Vision, Graz, Austria (2006)
27. Hilaga, M., Shinagawa, Y., Kohmura, T., Kunii, T.: Topology matching for fully automatic similarity estimation of 3d shapes. In: ACM Computer Graphics (Proceedings SIGGRAPH) (2001)
28. Breen, D.E., Whitaker, R.T.: A level-set approach for the metamorphosis of solid models. IEEE Transaction on Visualization and Computer Graphics 7, 173–192 (2001)
29. Starck, J., Hilton, A.: Surface capture for performance based animation. IEEE Computer Graphics and Applications 27(3), 21–31 (2007)

Grassmann Registration Manifolds for Face Recognition*

Yui Man Lui and J. Ross Beveridge

Department of Computer Science, Colorado State University,
Fort Collins, CO 80523, USA
{lui,ross}@cs.colostate.edu

Abstract. Motivated by image perturbation and the geometry of manifolds, we present a novel method combining these two elements. First, we form a tangent space from a set of perturbed images and observe that the tangent space admits a vector space structure. Second, we embed the approximated tangent spaces on a Grassmann manifold and employ a chordal distance as the means for comparing subspaces. The matching process is accelerated using a coarse to fine strategy. Experiments on the FERET database suggest that the proposed method yields excellent results using both holistic and local features. Specifically, on the FERET *Dup*2 data set, our proposed method achieves 83.8% rank 1 recognition: to our knowledge the currently the best result among all non-trained methods. Evidence is also presented that peak recognition performance is achieved using roughly 100 distinct perturbed images.

1 Introduction

The use of manifolds has received great attention in recent years. This is because manifolds capture the image variability and provide more information than a single image. There are two main schools of thought for making use of manifolds. The first is to learn the structure of manifolds whereas the other is to model the manifold directly.

Many manifold learning techniques like ISOMAP [1] and LLE [2] attempt to unfold the curved manifold onto a flat space. These manifold learning techniques need a large amount of training data and dense sampling on a manifold. Such rich training data may not be available in some real-world applications.

Another school of thought is to model the image manifold directly. Image perturbation that synthesizes a single image to a set of registration images has been employed to model nonlinear image manifolds [3] [4] [5] [6]. The manifold distances are derived by matching spanning sets, for example the tangent distance [3] and the joint manifold distance [4]. However, these methods only consider the best match between two spanning sets and ignore the curved geometry of manifolds, and the distance metric remains in Euclidean spaces.

* This material is based in part upon work supported by the National Science Foundation under Grant No. 0413284 and No. 0434351.

D. Forsyth, P. Torr, and A. Zisserman (Eds.): ECCV 2008, Part II, LNCS 5303, pp. 44–57, 2008.

This paper formally introduces the concept of the Grassmann registration manifold and demonstrates its utility in the context of face recognition. We construct a tangent space from a set of perturbed images. The structure of the approximated tangent space is essentially a vector space. Therefore, rather than proceeding on an unknown image manifold, we embed the approximated tangent space on a Grassmann manifold where each element on this manifold represents a subspace. The key aspect of this embedding is the use of a geodesic distance which is well defined on Grassmann manifolds, thus the underlying geometry is exercised.

Our proposed method assumes local linearity and selects k nearest neighbors from a registration manifold to form a tangent space. The tangent space is then approximated and embedded on a Grassmann manifold. A chordal distance is used to compute the distance between vector spaces. To ease the illumination effect, histogram equalization followed by a Gabor filter is applied. We employ a holistic image as a primary mode as well as an ensemble of classifiers using both holistic and local features.

Comparisons to seven well-known and recent algorithms on the FERET database suggest our algorithm is competitive with the best, achieving a rank 1 recognition rate of 83.8% on the the most challenging probe set (*Dup*2). Only one of the seven algorithms does better at 85.0%, and unlike our method, this algorithm is a trained method. While of course training can be valuable, it introduces a set of concerns about generalization that non-trained methods avoid.

The rest of this paper is organized as follows: Related work is discussed in Section 2. Grassmann manifolds and subspace distances are reviewed in Section 3. The Grassmann registration manifolds are presented in Section 4. The image features and image preprocessing method are revealed in Section 5. The outline of our proposed algorithm is given in Section 6. The coarse and fine matching strategy is introduced in Section 7. Finally, the experimental results and conclusions are provided in Section 8 and Section 9 respectively.

2 Related Work

Image perturbation that expands a single image to an image-set has been exercised for more than a decade and proven to be be useful and effective. In this section, we summarize some of the related methods.

Simard et al. [3] considered using synthesized images to model a nonlinear manifold. Tangent planes are approximated by a first order Taylor expansion. The distance between tangent planes, called tangent distance, is then used to measure the distance between manifolds. However, the tangent distance is still defined in the Euclidean space, and the geometry of manifolds has not been examined.

Fitzgibbon and Zisserman [4] computed a joint manifold distance to cluster appearances. The manifold is captured by a mean and a set of basis vectors. Affine transformation is added to a set of spanning images to overcome geometric deformation. This makes the measure affine resilient for appearance clustering. Although the span of a linear space is computed as a subspace distance, only the infimum between points in two subspaces is considered. As a result, the underlying geometry of manifolds remains imperceptible.

Nakayama and Kumakura [5] extracted three facial expressions (smiles, anger, and screams) using an eigenspace method. Then synthesized facial expressions are produced and added into a model as additional training data. Support vector machines were used to perform the classification.

Martínez [7] employed image perturbation to model localization errors using a single training image per class (i.e. subject). Perturbed images are utilized for learning all possible localization errors. These localization errors are modeled as a subspace by means of a mixture of Gaussians distribution for all training images such that imprecise localization can be tolerated.

Arandjelovic and Cipolla [6] modeled a face manifold as a probability distribution in video sequences. The face manifold is repopulated by a set of randomly drawn affine synthesized images. These synthesized images are augmented in the face manifold. The RANSAC algorithm is applied to remove outliers. The image data are then projected on a kernel PCA space and a symmetry version of the Kullback-Leibler divergence is exploited to compute the distance. Essentially, this method models face images on an unknown manifold.

Lui et al. [8] embedded a set of images on Grassmann manifolds and computed a geodesic distance for a subset of FRGC, version 2, experiment 4. This method achieves 15% and 40% improvements for face identification and face verification over MSM by switching to a geodesic distance. Moreover, the verification rates can be further boosted up by 50% when cohort normalization is employed in conjunction with image-set matching.

3 Grassmann Manifolds and Subspace Distances

To facilitate our discussion of Grassmann registration manifolds, we briefly describe the properties of Grassmann manifolds, canonical angles, and subspace distances. Detail descriptions on these subjects can be found in [9].

3.1 Grassmann Manifolds

A Grassmann manifold $\mathbb{G}_{n,p}$ is a set of p-dimensional linear subspaces of $\mathbb{R}^n$ (p-planes in $\mathbb{R}^n$) for $0 < p \leq n$. This Grassmann manifold has a natural quotient representation $\mathbb{G}_{n,p} = \mathbb{V}_{n,p} \;/\; \mathbb{O}_p$, where $\mathbb{V}_{n,p}$ is a Stiefel manifold (a set of $n \times p$ orthonormal matrices) and $\mathbb{O}_p$ is the orthogonal group. This representation states that two matrices belong to the same equivalence class if their columns span the same p dimensional subspace. Hence, the entire equivalence class can be represented as the subspace spanned by the columns of a given matrix $\mathcal{Y}$.

$$[\mathcal{Y}] = \{\mathcal{Y}Q_p : Q_p \in \mathbb{O}_p\} \tag{1}$$

In other words, a point on the Grassmann manifold is a linear subspace which may be specified by any arbitrary orthogonal basis.

3.2 Canonical Angles

Given two vectors x and $y \in \mathbb{R}^n$, the angle between these two vectors is defined as

$$\angle\{x, y\} = \arccos \frac{\langle x, y \rangle}{\| x \| \| y \|} \tag{2}$$

For subspaces, one can recursively define a set of angles between them which are called canonical angles, also known as principal angles [10]. Let $[\mathcal{X}]$ and $[\mathcal{Y}]$ be two subspaces in $\mathbb{R}^n$ then the canonical angles, $\angle_k\{[\mathcal{X}], [\mathcal{Y}]\}$, can be defined recursively as

$$\cos(\theta_k) = \max_{u \in [\mathcal{X}]} \max_{v \in [\mathcal{Y}]} u^T v = u_k^T v_k \tag{3}$$

subject to

$$\| u \| = \| v \| = 1, \quad u^T u_i = 0, \quad v^T v_i = 0, \quad i = 1, \ldots, k-1$$

Clearly, $\angle_k\{[\mathcal{X}], [\mathcal{Y}]\} \in [\ 0,\ \frac{\pi}{2}\ ]$, and Θ is a vector of all canonical angles.

3.3 Subspace Distances

Since the Grassmannian space is curved, the shortest distance between points on this space is geodesic. Wong [11] shows that the geodesic distance is defined using a set of canonical angles shown as

$$d_g(\mathcal{X}, \mathcal{Y}) = \| \Theta \|_2 \tag{4}$$

However, this distance is not differentiable everywhere. An alternative measure of geodesic distance is a chordal distance [12] defined as

$$d_c(\mathcal{X}, \mathcal{Y}) = \| \sin \theta \|_2 \tag{5}$$

The chordal distance approximates the geodesic distance when the planes are close and is differentiable everywhere. We employ $d_c(\mathcal{X}, \mathcal{Y})$ as our subspace distance metric.

4 Grassmann Registration Manifolds

4.1 Registration Manifolds Formation

Sampling and characterizing a registration manifold is the key step in our proposed approach. Given a pair of eye coordinates, we determine a set of affine parameters for geometric normalization. The affine transformation maps the

(x, y) coordinate from a source image to the (u, v) coordinate of a normalized image. The transformation can be written as follows:

$$\begin{pmatrix} u \\ v \\ w \end{pmatrix} = \begin{pmatrix} \cos(\theta) & -\sin(\theta) & dx \\ \sin(\theta) & \cos(\theta) & dy \\ 0 & 0 & 1 \end{pmatrix} \begin{pmatrix} 1 & q & 0 \\ 0 & 1 & 0 \\ 0 & 0 & 1 \end{pmatrix} \begin{pmatrix} 1+s_x & 0 & 0 \\ 0 & 1+s_y & 0 \\ 0 & 0 & 1 \end{pmatrix} \begin{pmatrix} x \\ y \\ z \end{pmatrix} \tag{6}$$

where the first matrix describes rotation and translation, the second matrix represents skew, and the third matrix denotes scaling. These transformed coordinates can be re-written more compactly as:

$$\begin{pmatrix} u \\ v \end{pmatrix} = \begin{pmatrix} p_1 & p_3 & p_5 \\ p_2 & p_4 & p_6 \end{pmatrix} \begin{pmatrix} x \\ y \\ 1 \end{pmatrix} \tag{7}$$

Equation 7 reveals that there are six control parameters for the affine transformation. In this paper, a set of registration images are sampled by perturbing these six affine parameters as shown in Equation 8.

$$\begin{pmatrix} u \\ v \end{pmatrix} = \begin{pmatrix} p_1 + \Delta p_1 & p_3 + \Delta p_3 & p_5 + \Delta p_5 \\ p_2 + \Delta p_2 & p_4 + \Delta p_4 & p_6 + \Delta p_6 \end{pmatrix} \begin{pmatrix} x \\ y \\ 1 \end{pmatrix} \tag{8}$$

Specifically, we perturb the initial affine parameters with Δp in a $\pm$ range, such that we synthesize 3^6 (729) perturbed images. These 729 images reside on an affine registration manifold $\mathcal{M}$. In our experiments, we employ bilinear interpolation for sampling the registration manifold, and set Δp_1, Δp_2, Δp_3, and Δp_4 as {-0.03, 0, 0.03}, and Δp_5 and Δp_6 as {-3, 0, 3}. An example of registration images is given in Fig. 1.

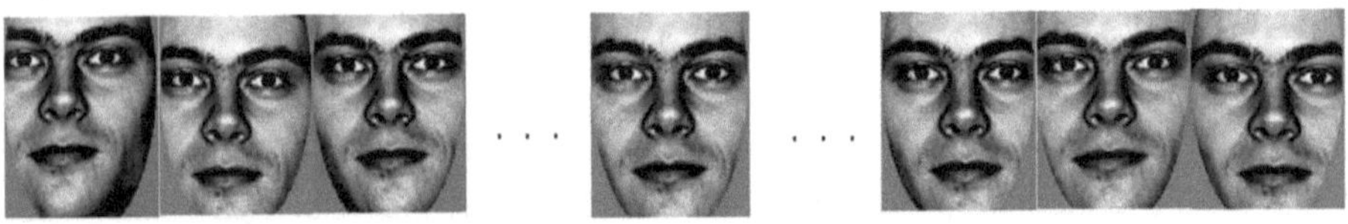

Fig. 1. Examples of registration images

4.2 Grassmann Registration Manifolds Formation

The affine registration manifold has a nonlinear structure [3]. One way to utilize the sampled registration manifold is to assume local linearity and explore its tangent space. By the mean value theorem [13], we can approximate a tangent vector using a nearby secant. A collection of these approximated tangent vectors at a point $x \in \mathbb{R}^n$ forms a tangent space. The logical choice for the base point x

is the canonical sampled image ($\Delta p_i = 0$ in Equation 8) illustrated by the center image in Fig. 1.

The selected base point x is assumed to be an interior point that has a neighborhood homeomorphic to an open ball. Then, given k nearest neighbors $\{x_1^*, x_2^*, \ldots, x_k^*\}$ where $x_i^* \in \mathcal{M}$, a tangent space centered at x on a registration manifold $\mathcal{M}$ may be represented as:

$$T_x\mathcal{M}: \quad x + \text{span}\{x - x_1^*, x - x_2^*, \ldots, x - x_k^*\} \tag{9}$$

where $\{x - x_1^*, x - x_2^*, \ldots, x - x_k^*\}$ are the approximated tangent vectors around x. This tangent space $T_x\mathcal{M}$ has a vector space structure and any point on this tangent space can be reconstructed as:

$$x + \sum_{i=1}^{k} \alpha_i (x - x_i^*) \tag{10}$$

The k nearest neighbors define the tangent basis. Because we assume local linearity about x, the k nearest neighbors are defined using an Euclidean distance:

$$x_k^* = x_{k-1}^* \bigcup \underset{x_j \notin x_{k-1}^*}{\text{argmin}} \parallel x - x_j \parallel_2^2 \tag{11}$$

where $x_0^* = \{\}$. Although we apply the same image perturbation to generate each sampled image, the k nearest neighbors for different face images are usually different. This is because changing the face image may change the relative Euclidean distances between different perturbed images.

Recall that a Grassmann manifold $\mathbb{G}_{n,p}$ is a set of p-dimensional linear subspaces of $\mathbb{R}^n$. Because a tangent space admits a vector space structure, and the approximated tangent vectors are the bases spanning the subspace, we use these tangent vectors to embed a linear subspace on a Grassmann manifold. The benefits of embedding a subspace on a Grassmann manifold are that its properties and distance metrics are well studied. Since we embed the registration manifold on a Grassmann manifold, we call the resulting manifold the Grassmann Registration Manifold (GRM).

Selecting the number of nearest neighbors to approximate the tangent space has significant implications. At the extreme low end, $k = 1$, minimizing the chordal distance equates to maximizing the correlation between pairs of images (Equation 2). At the other extreme, taking hundreds or thousands of samples is not only computationally burdensome; at some point the local linearity assumption is stretched beyond the breaking point. The effect of different choices for k on the FERET *Dup*2 data set is shown in Fig. 2. As Fig. 2 depicts, the rank 1 recognition rate starts to decrease when k is larger than 128. In our experiments, we choose $k = 100$. As Fig. 2 suggests, any value around 100 can be expected to perform well.

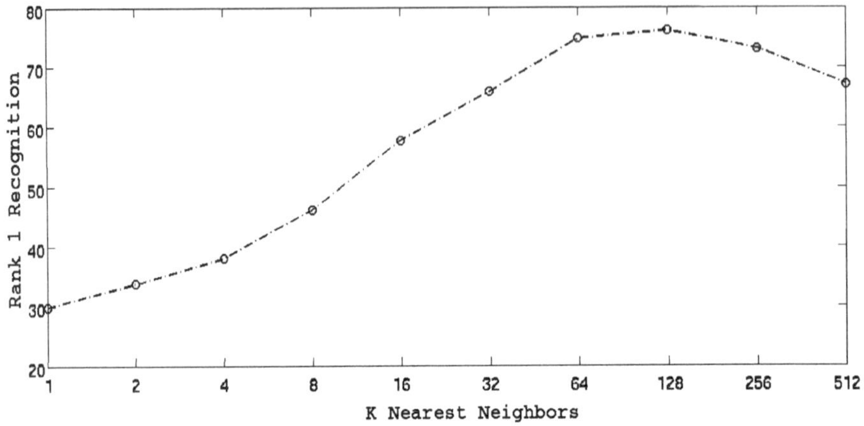

Fig. 2. The effect of K nearest neighbors

5 Image Features and Image Preprocessing

A holistic image is regarded as the whole face whereas local features can be divided into local components and local regions [14]. Local components are typically an interesting part of the face, for example an eye. Local regions are derived from regular sampling patterns, for example a grid laid over the face. In this paper, we employ a holistic image, local components, and local regions for face recognition. We choose the local components as the upper face, lower face, eyes, left eye and right eye. Local regions are 3×3 and 5×5 facial windows.[1] Examples of a holistic face, local components, and local regions are given in Fig. 3.

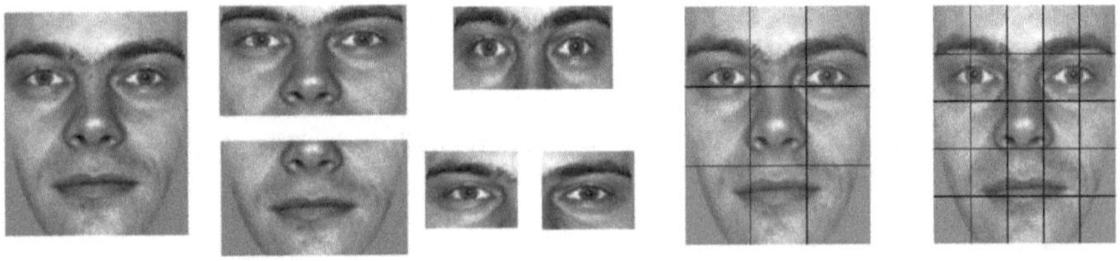

Fig. 3. Examples of a holistic face, local components, and local regions

For each probe and gallery image, the 729 registration images are sampled using eye coordinates provided with the data along with the perturbation process

[1] Since most of the lower left and lower right regions are covered by a mask in the 5×5 facial window, we eliminate these two regions from our feature set.

described previously. An elliptical mask is applied to remove the background. At this point the local regions are left alone while histogram equalization and a Gabor filter [15] are applied to the holistic image and local components. Treating local regions differently from local components makes classifiers based upon local regions and local components more independent. The specific DC-free Gabor filter is defined as follows:

$$\psi_{u,v}(z) = \frac{\| k_{u,v} \|^2}{\sigma^2} e^{(-\|k_{u,v}\|^2 \|z\|^2 / 2\sigma^2)} \left[e^{ik_{u,v}z} - e^{-\sigma^2/2} \right], \quad k_{u,v} = \frac{k_c}{f^v} e^{i\phi_u} \tag{12}$$

where u and v are the control parameters for orientations and scales, respectively, and z is the position. In our experiments, we set the $f = \sqrt{2}$, $k_c = \frac{4\pi}{5}$, $\sigma = \frac{3\pi}{2}$, $v = 0$, and $\phi_u = 0$.

6 The Grassmann Registration Manifold Algorithm

Putting all things together, the outline of the proposed Grassmann Registration Manifold (GRM) algorithm is given as follows:

- Use eye coordinates to determine the initial affine registration parameters $p_1, \ldots, p_6$ for each image. (Equation 7)
- Sample the affine registration manifold by perturbing the affine parameters (Equation 8)
- Compute the k nearest neighbors $\{x_1^*, x_2^*, \ldots, x_k^*\}$ from the registration manifold (Equation 11)
- (Optional) Apply histogram equalization and a Gabor filter (Equation 12)
- Construct the tangent space (Equation 9)
- Embed the approximated tangent space and compute canonical angles (Equation 3)
- Compute the subspace distance (Equation 5)

7 Coarse to Fine Matching Strategy

Initial studies of cumulative match curves indicate that 95% of the correct identifications are included in the top 10% of gallery candidates using just holistic image matching and $k = 16$, i.e. 16 registration manifold samples. Consequently, considerable time can be saved with a coarse to fine matching strategy as depicted in Fig. 4. The approach is coarse to fine not in the traditional sense of reduced image resolution, but instead in the sense of a reduced number of gallery candidates.

Note that each probe image has its own gallery candidates. Once the 16 sample matching is completed, the top 10% gallery candidates are passed to the 32 sample level. In the 32 sample level, only these 10% gallery candidates will be searched. Similarly, the top 5% of the gallery candidates are propagated to the 100 sample level and are examined. We can further propagate the top 5% gallery candidates to all local features. This coarse to fine strategy speeds up the matching process by roughly a factor of 800, (20×40).[2]

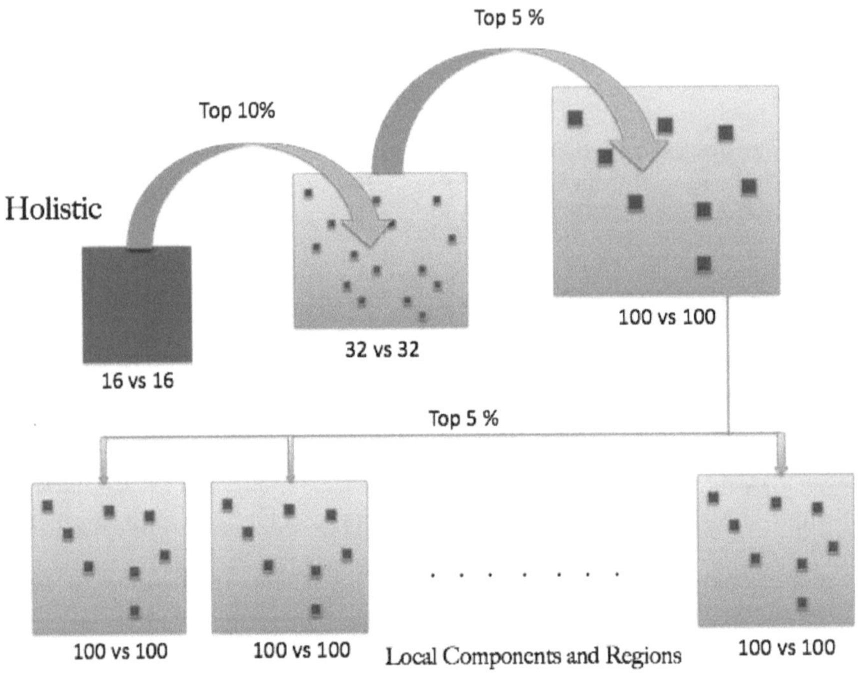

Fig. 4. The proposed coarse to fine matching strategy

8 Experiments

8.1 Data Collection

The performance of our proposed algorithm is compared to well-known algorithms on the FERET database [16]. The frontal view imagery of the FERET database is divided into 5 categories: Fa, Fb, Fc, $Dup1$, and $Dup2$, containing 1,196, 1,195, 194, 722, and 234 faces, respectively. Both Fa and Fb are taken in the same day with the same illumination condition but with different facial expressions. Fc is taken at the same day as Fa but with different illumination condition. $Dup1$ is acquired on different days from Fa. $Dup2$ is acquired at least one year apart from Fa. Following the FERET protocol, Fa is always the gallery and Fb, Fc, $Dup1$, and $Dup2$ are used as probe sets.

[2] 5% candidates and 40 features.

8.2 Prior Art on the FERET Database

The Elastic Bunch Graph Matching (EBGM) algorithm [17] was one of the top algorithms for the FERET database for more than half of a decade until the Local Binary Pattern (LBP) [18] was introduced. Since then, many LBP-like algorithms have been reported and continue to push the performance envelope for FERET. Zhang et al. [19] applied Gabor filters to extract local patterns and encoded them as phase-quadrant and XOR patterns. Tan and Triggs [20] proposed to combine Gabor and LBP features, and projected it on a discriminant space using kernel discriminative common vectors. Zou et al. [14] used a large set, 4, 172, of Gabor jets and achieved excellent results.

Liu et al. [21] create 9 images by horizontally and vertically shifting samples (left/right and up/down). The authors view these images as the basis for linear spatial filters because the 9 translated images make up a filter mask. Subsequently, the subspace distance between local patches is computed and the aggregated score is used for final classification. We include this algorithm in our comparison because it involves subspaces defined by different registration samples, and is thus related to our own work. However, Liu et al. only present their approach in the context of linear spatial filters, whereas we consider the underlying geometrical interpretation. In such, we make use of local linearity and do not choose the same image set all the times.

Table 1 summarizes the results of the above algorithms and two variants of our proposed method for the FERET database. The rightmost column indicates whether an algorithm requires training.

Table 1. Rank 1 recognition rate on the FERET database

Methods	Fb	Fc	Dup1	Dup2	Trained
EBGM [17]	95.0	82.0	59.0	52.0	Yes
Gabor-LBP-KDCV [20]	98.0	98.0	90.0	85.0	Yes
Weighted LBP [18]	97.0	79.0	66.0	64.0	Yes
Non-Weighted LBP [18]	93.0	51.0	61.0	50.0	No
HGPP [19]	97.6	98.9	77.7	76.1	No
GaborJets [14]	99.5	99.5	85.0	79.5	No
SIS [21]	91.0	90.0	68.0	68.0	No
GRM-Holistic	94.1	92.8	70.8	76.9	No
GRM-Local	97.5	97.9	79.5	83.8	No

8.3 Results with the Holistic Feature

Using the entire face image, our **GRM-Holistic** achieves 94.1%, 92.8%, 70.8% and 76.9% rank 1 recognition for *Fb*, *Fc*, *Dup*1, and *Dup*2, respectively. These results already outperform half of the top algorithms shown in Table 1. Specifically, the proposed **GRM-Holistic** ranks second among all non-trained algorithms and third among all algorithms for *Dup*2, the probe set for which most

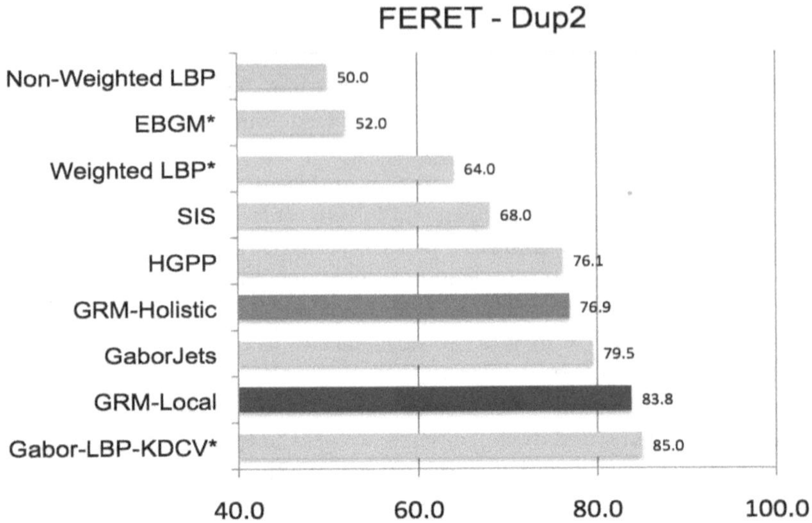

Fig. 5. Rank 1 recognition rate on the FERET database for the selected algorithms (Non-Weighted LBP [18], EBGM [17], Weighted LBP [18], SIS [21], HGPP [19], Gabor-Jets [14], and Gabor-LBP-KDCV [20]) where * indicates trained methods

algorithms have the greatest difficulty. To better visualize the ranking for *Dup*2, the results are shown from worst to best in Fig. 5.

It is worth to mention the variability associated with these rank 1 recognition rates. Assuming a binomial model [22] for rank 1 identification success/failure, a 95% confidence interval for the rank 1 recognition rate is about ± 0.05 for the FERET *Dup*2 results.

8.4 Results with Holistic + Local Features

While there is general agreement that peak performance is typically achieved using local features, Section 5 described the local features we have chosen. Fig. 6 shows the rank 1 recognition rates achieved when a GRM algorithm is constructed using each individual feature.

Neither space allows us to identify each feature, nor is it actually that important. What is important is first to note that the left most column in Fig. 6 is the holistic image as a feature, and it is consistently one of the best individual features. Hence, we can dismiss any thought that a single local feature consistently does much better than the whole image. Second, there is a fair amount of apparently random variation between features and between probe sets.

More specifically, a careful study of Fig. 6 reveals that no single local feature consistently ranks in the top 3 for all data sets. Consequently, the strength of using local features comes from the combination of independent decisions boosting recognition performance [23]. A good example is our *Dup*1 results shown in

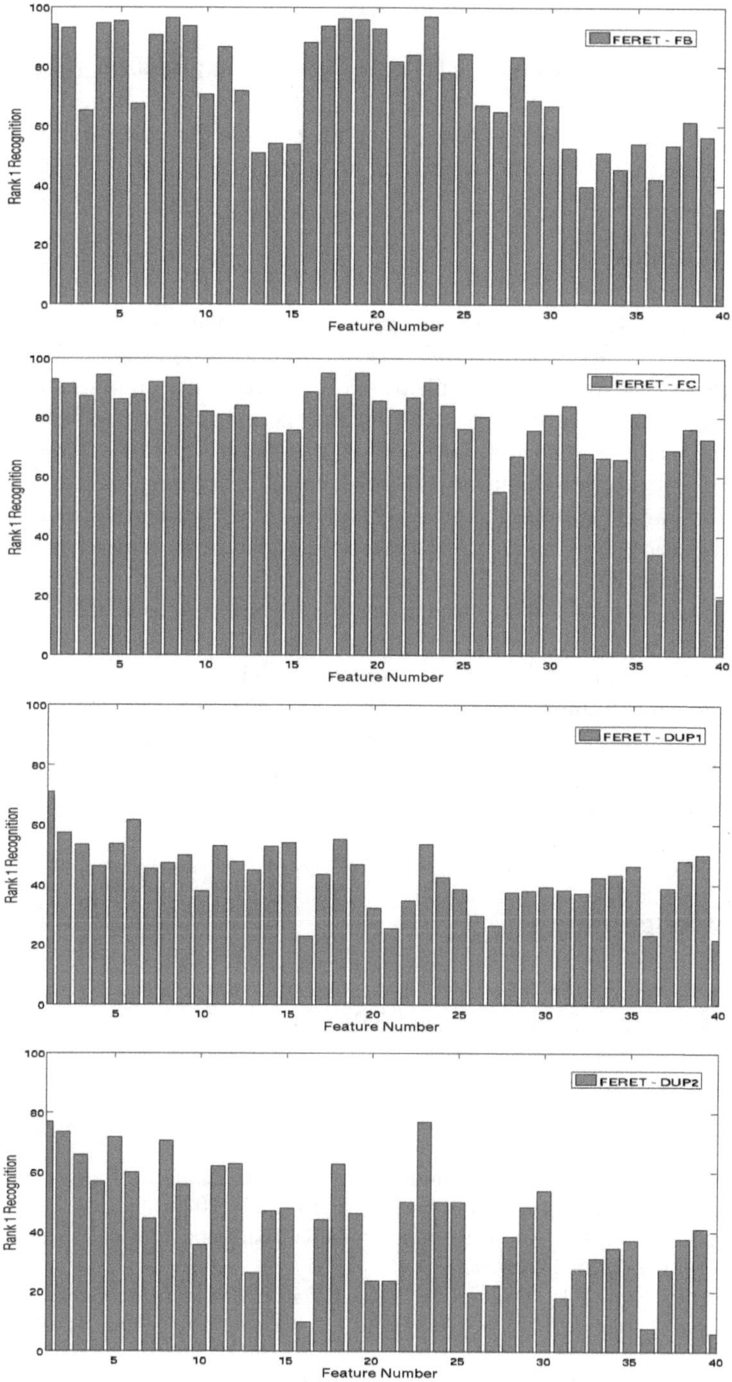

Fig. 6. Recognition rate for individual feature: The first row is Fb, the second row is Fc, the third row is Dup1 and the last row is Dup2

Fig. 6. In this paper, we use the majority voting rule [23] to combine all the feature outcomes.

Using the whole image plus local features, the **GRM-Local** algorithm achieves 97.5%, 97.9%, 79.5% and 83.8% rank 1 recognition for *Fb*, *Fc*, *Dup*1, and *Dup*2, respectively. The 83.8% rank 1 recognition for *Dup*2 is the best result for all non-trained algorithms and second among all algorithms depicted in Fig. 5.

9 Conclusions

A novel method (GRM) that embeds registration images on a Grassmann manifold is presented. We demonstrate that the underlying geometry is an important characteristic and proper utilization of this attribute enhances recognition results in the context of face recognition. Unlike many manifold learning algorithms, our proposed method does not require dense samples or training data. Therefore, our proposed method is more generic. Empirical evidence suggests that approximately 100 local samples from the affine registration manifold optimizes recognition performance. A coarse to fine matching strategy is introduced for fast computation.

A relatively simple holistic image version of our proposed GRM algorithm does very well on FERET tests, and introduction of local features further boosts recognition performance. Specifically, among all the non-trained algorithms considered, we achieve the best results of the FERET *Dup*2 data set.

Introducing the Grassmann registration manifold and associated algorithm further promotes a general trend of research into image manifolds. Our future work will focus in further exploring the question of how best to locally sample the affine registration manifold.

Acknowledgements

The authors thank Mr. David S. Bolme from Colorado State University for calculating the error bounds for the FERET data sets.

References

1. Tenenbaum, J., de Silva, V., Langford, J.: A global geometric framework for nonlinear dimensionality reduction. Science, 2319–2323 (2000)
2. Roweis, S., Saul, L.: Nonlinear dimensionality reduction by locally linear embedding. Science, 2323–2326 (2000)
3. Simard, P., Cun, Y.L., Denker, J.: Efficient pattern recognition using a new transformation distance. Advances in Neural Information Processing System 5, 50–58 (1992)
4. Fitzgibbon, A., Zisserman, A.: Joint manifold distance: a new approach to appearance based clustering. In: IEEE Proc. Computer Vision and Pattern Recognition, Wisconsin (2003)

5. Nakayama, M., Kumakura, T.: Face identification performance using facial expressions as perturbation. In: Duch, W., Kacprzyk, J., Oja, E., Zadrożny, S. (eds.) ICANN 2005. LNCS, vol. 3696, pp. 557–562. Springer, Heidelberg (2005)
6. Arandjelovic, O., Cipolla, R.: An information-theoretic approach to face recognition from face motion manifolds. Image and Vision Computing 24(6), 639–647 (2006)
7. Martinez, A.: Recognizing imprecisely localized, partially occluded, and expression variant faces from a single sample per class. IEEE Transactions on Pattern Analysis and Machine Intelligence 24(6), 748–763 (2002)
8. Lui, Y.M., Beveridge, J.R., Draper, B.A., Kirby, M.: Image-set matching using a geodesic distance and cohort normalization. In: IEEE International Conference on Automatic Face and Gesture Recognition, Amsterdam, The Netherlands (2008)
9. Edelman, A., Arias, R., Smith, S.: The geometry of algorithms with orthogonal constraints. SIAM J. Matrix Anal. Appl. (2), 303–353 (1999)
10. Hotelling, H.: Relations between two sets of variants. Biometrika 28, 321–377 (1936)
11. Wong, Y.C.: Differential geometry of grassmann manifolds. Proc. Nat. Acad. Sci. 47, 589–594 (1967)
12. Conway, J., Hardin, R., Sloane, N.: Packing lines, planes, etc.: Packings in grassmannian spaces. Experimental Mathematics 5(2), 139–159 (1996)
13. Simon, A.: Calculus with analytic geometry. Scott, Foresman and Company (1982)
14. Zou, J., Ji, Q., Nagy, G.: A comparative study of local matching approach for face recognition. IEEE Transactions on Image Processing 16, 2617–2628 (2007)
15. Lades, M., Vorbruggen, J.C., Buhmann, J., Lange, J., von der Malsburg, C., Wurtz, R.P., Konen, W.: Distortion invariant object recognition in the dynamic link architecture. IEEE Transactions on Computers 42(3), 300–311 (1993)
16. Phillips, P., Moon, H., Rizvi, S., Rauss, P.: The feret evaluation methodology for face recognition algorithms. IEEE Transactions on Pattern Analysis and Machine Intelligence 22(10), 1090–1104 (2000)
17. Wiskott, L., Fellous, J.M., Kruger, N., Malsburg, C.: Face recognition by elastic bunch graph matching. IEEE Transactions on Pattern Analysis and Machine Intelligence 19, 775–779 (1997)
18. Ahonen, T., Hadid, A., Pietikinen, M.: Face recognition with local binary patterns. In: European Conference on Computer Vision, Czech Republic, pp. 469–481 (2004)
19. Zhang, B., Shan, S., Chen, X., Gao, W.: Histogram of gabor phase patterns (hgpp): A novel object representation approach for face recognition. IEEE Transactions on Image Processing 16, 57–68 (2007)
20. Tan, X., Triggs, B.: Fusing gabor and lbp feature sets for kernel-based face recognition. In: IEEE International Workshop on Analysis and Modeling of Faces and Gestures, pp. 235–249 (2007)
21. Liu, J., Chen, S., Zhou, Z.H., Tan, X.: Single image subspace for face recognition. In: IEEE International Workshop on Analysis and Modeling of Faces and Gestures, pp. 205–219 (2007)
22. McNemar, Q.: Note on the sampling error of the difference between correlated proportions or percentages. Psychometrika, 153–157 (1947)
23. Lam, L., Suen, C.Y.: Application of majority voting to pattern recognition: An analysis of its behavior and performance. IEEE Transactions on Systems, Man and Cybernetics, Part A 27(5), 553–568 (1997)

Facial Expression Recognition Based on 3D Dynamic Range Model Sequences

Yi Sun and Lijun Yin

Department of Computer Science, State University of New York at Binghamton
Binghamton, New York, 13902 USA

Abstract. Traditionally, facial expression recognition (FER) issues have been studied mostly based on modalities of 2D images, 2D videos, and 3D static models. In this paper, we propose a spatio-temporal expression analysis approach based on a new modality, 3D dynamic geometric facial model sequences, to tackle the FER problems. Our approach integrates a 3D facial surface descriptor and Hidden Markov Models (HMM) to recognize facial expressions. To study the dynamics of 3D dynamic models for FER, we investigated three types of HMMs: temporal 1D-HMM, pseudo 2D-HMM (a combination of a spatial HMM and a temporal HMM), and real 2D-HMM. We also created a new dynamic 3D facial expression database for the research community. The results show that our approach achieves a 90.44% person-independent recognition rate for distinguishing six prototypic facial expressions. The advantage of our method is demonstrated as compared to methods based on 2D texture images, 2D/3D Motion Units, and 3D static range models. Further experimental evaluations also verify the benefits of our approach with respect to partial facial surface occlusion, expression intensity changes, and 3D model resolution variations.

1 Introduction

Research on FER has been based primarily on findings from Psychology and particularly on the Facial Action Coding System [1]. Many successful approaches have utilized Action Units (AU) recognition [2,3,4,5,6,7,8] or Motion Units (MU) detection [9,10,11]. Other well-developed approaches concentrate on facial region features, such as manifold features [12] and facial texture features [13,14]. Ultimately, however, all of above methods focus on most commonly used modality: 2D static images or 2D videos.

Recently, the use of 3D facial data for FER has attracted attention as the 3D data provides fine geometric information invariant to pose and illumination changes. There is some existing work for FER using 3D models created from 2D images [15] or from 3D stereo range imaging systems [16,17]. However, the 3D models that have been used are all *static*. The most recent technological advances in 3D imaging allow for real-time 3D facial shape acquisition [18,19] and analysis [20]. Such 3D sequential data captures the dynamics of time-varying facial surfaces, thus allowing us to use 3D dynamic surface features or 3D motion units (rather than 2D motion units) to scrutinize facial behaviors at a detailed level. Wang *et al* [18] have successfully developed a hierarchical framework for tracking high-density 3D facial sequences. The recent work in [20] utilized dynamic 3D models of six subjects for facial analysis and editing based on the generalized facial manifold of a standard model.

D. Forsyth, P. Torr, and A. Zisserman (Eds.): ECCV 2008, Part II, LNCS 5303, pp. 58–71, 2008.

Motivated by the recent work of 3D facial expression recognition reported by Yin *et al* [16] based on a static 3D facial expression database [21], we extend the facial expression analysis to a dynamic 3D space. In this paper, we propose a spatio-temporal 3D facial expression analysis approach for FER using our newly-created 3D dynamic facial expression database. This database contains 606 3D facial video sequences with 101 subjects: each subject has six 3D sequences corresponding to six prototypic facial expressions. Our approach uses 3D labeled surface type to represent the human facial surface and transforms the feature to an optimal compact space using linear discriminative analysis. Such a 3D surface feature representation is relatively robust to changes of pose and expression intensities. To explore the dynamics of 3D facial surfaces, we investigated a 1D temporal HMM structure and extended it to a pseudo-2D HMM and a real 2D HMM. There have been existing HMM-based approaches for FER using 2D videos [7,9,22], by which either a 1D HMM or multi-stage 1D-HMM was developed. However, no true 2D-HMM structure was applied to address the FER problem. Our comparison study shows that the proposed real 2D-HMM structure is better than the 1D-HMM and pseudo 2D-HMM in describing the 3D spatio-temporal facial properties.

In this paper, we conducted comparative experiments using our spatio-temporal 3D model-based approach with approaches based on 2D/3D motion units, 2D textures, and 3D static models. The experimental results show that our approach achieves a 90.44% person-independent recognition rate in distinguishing the six prototypic expressions, which outperforms the other compared approaches. Finally, the performance of our proposed approach was evaluated on its robustness dealing with 1) partial facial surface occlusion, 2) expression intensity changes, and 3) 3D model resolution variations. The paper is organized as follows: we first introduce our new 3D dynamic facial expression database in Section 2. We then describe our 3D facial surface descriptor in Section 3 and the HMM classifiers in Section 4. The experimental results and analysis are reported in Section 5, followed by the conclusion in Section 6.

2 Dynamic 3D Face Database

There are some existing public 3D *static* face databases, such as FRGC 2.0 [23], BU-3DFE [21], etc. However, to the best of our knowledge, there is no *3D dynamic* facial expression database publicly available. To investigate the usability and performance of the 3D dynamic facial models for FER, we created a dynamic 3D facial expression database [24] using the Dimensional Imaging's 3D dynamic capturing system [19]. The system captures a sequence of stereo images and produces the range models using a passive stereo-photogrammetry approach. At the same time, 2D texture videos of the dynamic 3D models are also recorded. Figure 1 shows the dynamic 3D face capture system with three cameras. Each subject was requested to perform the six prototypic expressions (*i.e., anger, disgust, fear, smile, sad, and surprise*) separately. Each 3D video sequence captured one expression at a rate of 25 frames per second and each 3D video clip lasts approximately 4 seconds with about 35,000 vertices per model. Our database currently consists of 101 subjects including 606 3D model sequences with 6 prototypic expressions and a variety of ethnic/racial ancestries. An example of a 3D facial sequence is shown in Figure 1. More details can be found in [24].

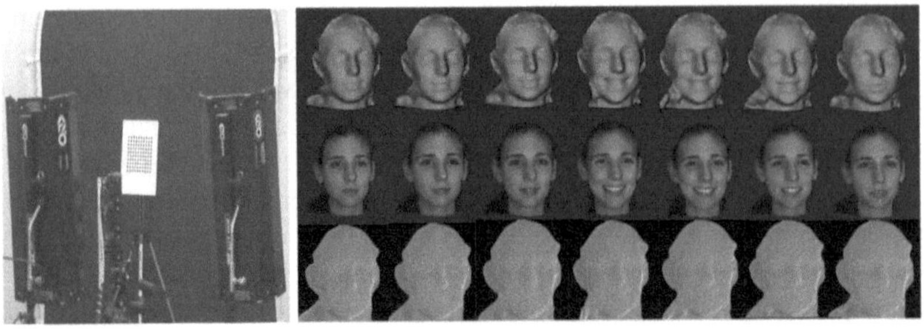

Fig. 1. Left:Dynamic 3D face capturing system setup. Right: sample videos of a subject with smile expression(from top to bottom: shaded models, textured models, and wire-frame models with 83 tracked control points).

3 3D Dynamic Facial Surface Descriptor

The dynamic 3D face data provides both facial surface and motion information. Considering the representation of facial surface and the dynamic property of facial expressions, we propose to integrate a facial surface descriptor and Hidden Markov Models to analyze the spatio-temporal facial dynamics. It is worth noting that we aim at verifying the usefulness and merits of such 3D dynamic data for FER in contrast to the 2D static/dynamic data or 3D static data. Therefore, we do not focus on developing a fully automatic system for FER in this paper. Our system is outlined in Figure 2, which consists of model pre-processing, HMM-based training, and recognition.

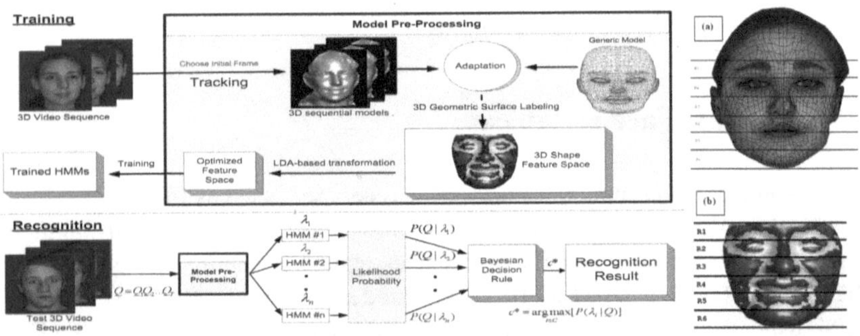

Fig. 2. Left: Framework of the FER system. Right: sub-regions defined on an adapted model (a) and a labeled model (b).

In the first stage, we adapt a generic model (*i.e., tracking model*) to each range model of a 3D model sequence. The adaptation is controlled by a set of 83 pre-defined key points (colored points on the generic model in Figure 2). After adaptation, the correspondence of the points across the 3D range model sequence is established. We apply a surface labeling approach [25] to assign each vertex one of eight primitive shape types. Thus, each range model in the sequence is represented by a "label map", G, as shown in the 3D shape feature space of Figure 2, where different colors represent different labeled shape types. We use Linear Discriminative Analysis (LDA) to transform the label map to an optimal compact space to better separate different expressions. Given the optimized features, the second stage is to learn one HMM for each expression. In recognition, the temporal/spatial dynamics of a test video is analyzed by the trained HMMs. As a result, the probability scores of the test video to each HMM are evaluated by the Bayesian decision rule to determine the expression type of the test video.

3.1 Facial Model Tracking and Adaptation

As the high-resolution range models vary in the number of vertices across 3D video frames, we must establish the vertices' correspondences and construct a common feature vector. To do so, we applied a generic model adaptation approach to "sample" the range models. This process consists of two steps: control points tracking and generic model adaptation. A set of 83 pre-defined key points is tracked using an active appearance model based approach on 2D video sequences [26,19], where the key points in the initial frame are manually picked. To reduce the tracking error, a post-processing procedure was applied by manually correcting some inaccurately tracked points. Since the 2D texture and the 3D range model of each frame are matched accurately from the system, the key points tracked in the 2D video can be exactly mapped to the 3D range surface. This semi-automatic approach allows us to obtain accurate control points on the sequential models. Figure 1 (bottom row) shows an example of a tracked sequence. The adaptation procedure is as follows: Given the N (=83) control points $\boldsymbol{U_i} = (u_{i,x}, u_{i,y}, u_{i,z})^T \in R^3$ on the generic model and the corresponding tracked points $\boldsymbol{V_i} \in R^3$ on each range model, we use the radial basis function (RBF) to adapt the generic model on the range face model. The interpolation function is formulated as:

$$f(\boldsymbol{p}) = \boldsymbol{c}_1 + [\boldsymbol{c}_2\boldsymbol{c}_3\boldsymbol{c}_4] \times \boldsymbol{p} + \sum_{i=1}^{N} \lambda_i \varphi_i (|\boldsymbol{p} - \boldsymbol{U}_i|) \tag{1}$$

where $\boldsymbol{p_i}$ is a non-control vertex on the generic model and φ_i is the RBF for $\boldsymbol{U_i}$. All coefficients $\boldsymbol{c_k}$(k=1,..,4) are determined by solving the equation: $f(\boldsymbol{U_i}) = \boldsymbol{V_i}, i = 1...N$, where $\sum_{i=1}^{N} \lambda_i = 0$ and $\sum_{i=1}^{N} \boldsymbol{U_i}\lambda_i = (0,0,0)^T$. Then, the non-control vertex $\boldsymbol{p_i}$ is mapped to $f(\boldsymbol{p_i})$. The result of adaptation provides sufficient geometric information for subsequent labeling. Figure 2(a) shows an example of an adapted model.

3.2 Geometric Surface Labeling

3D facial range models can be characterized by eight primitive surface features: convex peak, concave pit, convex cylinder, convex saddle, concave saddle, minimal surface, concave cylinder, and planar [25]. After the tracking model is adapted to the range model, each vertex of the adapted model is labeled as one of the eight primitive features. This surface labeling algorithm is similar to the approach described in [16]. The difference is that eight primitive features rather than twelve features are used for our expression representation because we apply a local coordinate system for feature calculation. Let $p = (x, y, z)$ be a point on a surface S, N_p be the unit normal to S at point p, and X_{uv} be a local parameterization of surface S at p. A polynomial patch

$z(x,y) = \frac{1}{2}Ax^2 + Bxy + \frac{1}{2}Cy^2 + Dx^3 + Ex^2y + Fxy^2 + Gy^3$is used to approximate the local surface around p by using X_u , X_v and N_p as a local orthogonal system. We then obtain the principal curvatures by computing the eigenvalues of the Weingarten matrix: $W = (A, B; B, C)$. After obtaining the curvature values of each vertex, we apply the classification method described in [25] to label each vertex of the adapted model. Thus, each range model is represented by a label map $G = [g_1, g_2, ..., g_n]$, composed of all vertices' labels on the facial region. Here, g_i is label types and n is the number of vertices in the facial region of the adapted model.

3.3 Optimal Feature Space Transformation

We now represent each face model by its label map *G*. We use LDA to project *G* to an optimal feature space O_G that is relatively insensitive to different subjects while preserving the discriminative expression information. LDA defines the within-class matrix S_w and the between-class matrix S_b. It transforms a *n*-dimensional feature to an optimized *d*-dimensional feature O_G by $O_G = {D_O}^T \cdot G$, where $d < n$, $D_O = arg\left(max_D |\left(D^T S_b D\right) / \left(D^T S_w D\right)\right)$ and *D*, projection matrix. For our experiments, the discriminative classes are 6 expressions, thus the reduced dimension *d* is 5.

4 HMM Based Classifiers

Facial expression is a spatio-temporal behavior. To better characterize this property, we used Hidden Markov Models to learn the temporal dynamics and the spatial relationships of facial regions. In this section, we describe the Temporal-HMM (T-HMM), Pseudo Spatio-Temporal HMM (P2D-HMM), and real 2D HMM (R2D-HMM), progressively. P2D-HMM is extended from T-HMM, and in turn, R2D HMM is extended from P2D-HMM. As we will discuss, R2D-HMM is the most appropriate method for learning dynamic 3D face models to recognize expressions.

4.1 Temporal HMM

Each prototypic expression is modeled as an HMM. Let $\lambda = [A, B, \pi]$ denote an HMM to be trained and N be the number of hidden states in the model, we denote the states as $S = \{S_1, S_2, ..., S_N\}$ and the state at t is q_t (see top row of Figure 3). $A = \{a_{ij}\}$is the state transition probability distribution, where $a_{ij} = P[q_{t+1} = S_j | q_t = S_i]$, $1 \leq i, j \leq N$. $B = \{b_j(k)\}$is the observation probability distribution in state j, k is an observation . We use Gaussian distributions to estimate each $B = \{b_j(k)\}$, where $b_j(k) = P[k|q_t = S_j] \sim N(\mu_j, \Sigma_j), 1 \leq j \leq N$. Let $\pi = \{\pi_i\}$ be the initial state distribution, where $\pi_i = P[q_0 = S_i], 1 \leq i \leq N$. Then, given an observation sequence, $O = O_1O_2...O_T$, where O_i denote an observation at time i, the training procedure is: *Step 1:* Take the optimized feature representation O_G of each observed 3D range face model as an observation. *Step 2:* Initialize the HMM model λ. Each observed model of a sequence corresponds to one state and is used to estimate the parameters in the observation matrix B . Set the initial values of A and π based on observations. *Step 3:* Use the forward-backward algorithm [27] to derive an estimation of the model parameter $\lambda = [A, B, \pi]$ when $P(O|\lambda)$ is maximized. Finally, we derived 6 HMMs; each represents one of the six prototypic expressions.

Given a query model sequence, we follow the Step 1 of the training procedure to represent it as $Q = Q_1 Q_2 ... Q_T$, where the optimized representation of each frame is one observation, denoted as $O_G = (O_{G,1}, O_{G,2}, O_{G,3}, O_{G,4}, O_{G,5})$. Using the forward-backward method, we compute the probability of the observation sequence given a trained HMM_i as $P(Q|\lambda_i)$. We use the Bayesian decision rule to classify the query sequence $c^* = argmax[P(\lambda_i|Q)], i \in C$, where $P(\lambda_i|Q) = \frac{P(Q|\lambda_i)P(\lambda_i)}{\sum_{j=1}^{C} P(Q|\lambda_j)P(\lambda_j)}$ and C is the number of the trained HMM models. Since this method trace the temporal dynamics of facial sequences, we denote it as a Temporal HMM(T-HMM). The top row of Figure 3 shows the structure of a 6-state T-HMM. The decision to classify a query sequence to an expression using the T-HMM is denoted as $Decision^T$.

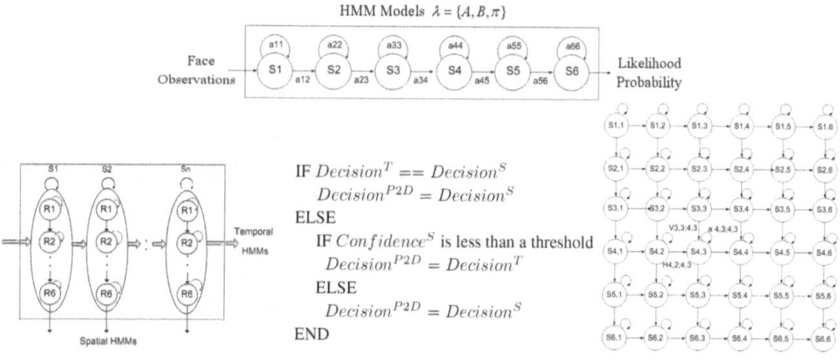

Fig. 3. Top:T-HMM; Bottom-left and middle:P2D-HMM and its decision rule; Bottom-right:R2D-HMM

4.2 Pseudo 2D Spatio-temporal HMM

Facial characteristics are not only represented by temporal dynamics (inter-frame) but also by spatial relationships (intra-frame). To model these properties of 3D faces, we investigated the structure of HMMs in the spatial domain combined with the temporal domain, a structure called P2D-HMM.

Spatial HMM (S-HMM): Based on the feature points tracked on the facial surface (e.g., contours of eyebrows, eyes, nose, mouth, and chin), we subdivide each 3D frame model of a sequence into six regions, as shown in Figure 2(b) ($R_1, R_2, ..., R_6$). We then build a 6-state 1D HMM, corresponding the six regions, as shown in each column of P2D-HMM in Figure 3. Similar to the case of entire face regions in the previous section, we transform the labeled map of each sub-region of a frame to an optimized feature space using LDA, denoted as $O_{Gi} = (O_{Gi,1}, O_{Gi,2}, O_{Gi,3}, O_{Gi,4}, O_{Gi,5}), (i = 1..6)$, where i is the region index of a frame model. We trained one HMM for each expression. Given a query face sequence with a length N, we compute the likelihood score of each frame and use the Bayesian decision rule to decide the frame's expression type. We make a final decision $Decision^S$ using majority voting. Thus, the query model sequence is recognized as an expression if this expression is the majority result among N frames. As this method tracks spatial dynamics of a facial surface, we call it a spatial HMM (S-HMM).

Combination of Spatial and Temporal HMMs: To model both spatial and temporal information of 3D face sequences, we combine the S-HMM and the T-HMM to construct a pseudo 2D HMM (P2D-HMM) (see Figure 3). The final decision $Decision^{P2D}$ is based on both $Decision^{S}$ and $Decision^{T}$. The decision rule of the P2D-HMM is also described Figure 3. Here, we define $Confidence^{S}$ as the ratio of the number of majority votes versus the total number of frames in the query model sequence. In our experiment, we took 6 frames as a sequence, and chose the threshold for this ratio as 0.67. As a consequence, if at least 4 frames of a query sequence are recognized as expression A by the S-HMM, we determine the query sequence is A. Otherwise, the result comes from the $Decision^{T}$. Essentially, P2D-HMM uses the learned facial temporal characteristics to compensate for the learned facial spatial characteristics.

4.3 Real 2D Spatio-temporal HMM

The aforementioned HMM-based approaches are essentially 1-D or pseudo-2D approaches. However, the dynamic 3D facial models are four dimensional (i.e., 3D plus time). Considering the complexity of high-dimensional HMMs and motivated by the work of Othman *et al* [28] for 2D face recognition, we developed a real 2D HMM (R2D-HMM) architecture to learn the 3D facial dynamics over time. As shown in Figure 3 (bottom-right), this architecture allows for both spatial (vertical) and temporal (horizontal) transition to each state simultaneously. The number of states along spatial (vertical) or temporal (horizontal) axes are all six. Simply put, each 3D sequence contains 6 temporal states, and each frame contains 6 spatial states from top to bottom. The transition from region R_i of the previous frames to another region R_j of the current frame can be learned from the R2D-HMM. In Figure 3, $H_{4,2;4,3}$ and $V_{3,3;4,3}$ are the horizontal and vertical transition probabilities from the state $S_{4,2}$ and the state $S_{3,3}$ to the current state $S_{4,3}$ respectively, and $a_{4,3;4,3}$ is the self-transition probability of the state $S_{4,3}$. Let $O_{r,s}$ be the observation vector of the rth region of the sth frame in a 3D video sequence, the corresponding set of feature vectors is defined as $O_{\{m,n\}} = \{O_{r,s} : 1 \leq r \leq m, 1 \leq s \leq n\}$. The feature vector set of the past observation blocks $O_{<m,n>}$ is derived by excluding the current observation block $O_{m,n}$, where $O_{<m,n>} = O_{\{m,n\}} - O_{m,n}$. Note that the joint probability of the current state and the observations up to the current observation $P\left(q_{m,n} = S_{a,b}, O_{\{m,n\}}\right)$ can be predicted based on past observation blocks in a recursive form:

$$
\begin{aligned}
P\left(q_{m,n} = S_{a,b}, O_{\{m,n\}}\right) &= P\left(O_{m,n} | q_{m,n} = S_{a,b}\right) \\
&\cdot \left| \sum_{i,j=1,1}^{M,N} P\left(q_{m,n} = S_{a,b} | q_{m-1,n} = S_{i,j}\right) P\left(q_{m-1,n} = S_{i,j}, O_{\{m-1,n\}}\right) \right. \\
&\cdot \left. \sum_{k,l=1,1}^{M,N} P\left(q_{m,n} = S_{a,b} | q_{m,n-1} = S_{k,l}\right) P\left(q_{m,n-1} = S_{k,l}, O_{\{m,n-1\}}\right) \right|^{1/2}
\end{aligned} \tag{2}
$$

Similar to the standard 1-D HMM, approach, the state matrix is denoted as $\delta_{m,n}(a,b) = \max_{q_{m-1,m}, q_{m,n-1}} P\left[q_{m,n} = S_{a,b}, O_{1,1}, ...O_{m,n}|\lambda\right]$. The observation probability distribution $B_{a,b}(O_{m,n})$ is given by

$$B_{a,b}(O_{m,n}) = \frac{1}{\left[2\pi\right]^{v/2} \Sigma^{1/2}} \cdot e^{\frac{\left(O_{m,n}-\mu_{a,b}\right)\Sigma_{a,b}^{-1}\left(O_{m,n}-\mu_{a,b}\right)^{T}}{2}} \tag{3}$$

Using the Viterbi algorithm, we estimate the model parameter λ as $P(O, Q^*|\lambda)$ is maximized, where $P(O, Q^*|\lambda) = \max_{a,b}\left[\delta_{M,N}(a,b)\right]$, and Q^* is the optimal state sequence. This structure assumes the state transitions to be left-to-right horizontally and top-to-bottom structure vertically. We set the transition matrix in the diagonal direction to be zeros using the same calculation as described in [28]. The expected complexity of the R2D-HMM method is only two times that of the 1D T-HMM structure with the same number of states. In our experiment, given a six-frame sequence, the observation vector is defined by a 6×6 matrix O, in which each cell is an observation block denoted as $O_{r,s} = (O_{r,s,1}, O_{r,s,2}, O_{r,s,3}, O_{r,s,4}, O_{r,s,5})$ $(r, s = 1...6)$, where s is the frame index, r is the region index of the frame s, and $O_{r,s}(r, s = 1...6)$ is the optimized feature after the label map of the region r of the frame s is transformed using LDA.

5 Experiments and Analysis

We conducted person-independent experiments on 60 subjects selected from our database. To construct the training set and the testing set, we generated a set of 6-frame subsequences from each expression sequence. To do so, for each expression sequence of a subject, we chose the first six frames as the first subsequence. Then, we chose 6-consecutive frames starting from the second frame as the second subsequence. The process is repeated by shifting the starting index of the sequence every one frame till the end of the sequence. The rationale for this shifting is that a subject could come to the recognition system at any time, thus the recognition process could start from any frame. As a result, 30780 ($= 95 \times 6 \times 54$) subsequences of 54 subjects were derived for training, and 3420 ($= 95 \times 6 \times 6$) subsequences of the other 6 subjects were derived for testing. Following a ten-fold cross-validation, we report the average recognition rates of the ten trials as the final result. Our database contains not only the 3D dynamic model sequences but also the associated 2D texture videos. This allows us to compare the results using both 3D data and 2D data of same subjects simultaneously. In the following section, we report the results of our proposed approaches using the 3D dynamic data and their comparative results of the existing approaches using 2D data and 3D static data. All the experiments were conducted in a person-independent fashion.

5.1 Comparison Experiments

Dynamic 3D region-based approaches: We conducted experiments using the Temporal 1D-HMM (T-HMM), Pseudo-2D HMM (P2D-HMM), and Real 2D HMM (R2D-HMM) based on the 3D dynamic surface descriptor. As previously discussed, our facial feature descriptor is constructed from vertices' labels of either entire face region or local facial regions, and we dubbed these methods as "3D *region-based*" approaches. The experimental results are reported in the bottom three rows of the right of Table 1.

Static 2D/3D region-based approaches: *(1) 2D static texture baseline:* We used the Gabor-wavelet based approach [14] as a 2D static baseline method. We used 40 Gabor kernels including 5 scales and 8 orientations and applied them to the 83 key points on the 2D texture frames of all video sequences. *(2) 3D static models baseline:* The LLE based [29], PCA-based, and LDA-based approaches [30] were implemented as the 3D static baseline methods for comparison. The input vector for these three approaches is feature G as explained in section 3.2. For the LLE-based method, we first transform the label map G of each range model to the LLE space and select key frames using k-means clustering. Then, all selected key frame models are used as the gallery models for classification. We use majority voting to classify each 3D query model in the test set. The PCA-based approach and LDA-based approach take the labeled feature G as input vector and apply the PCA and LDA for the recognition. *(3) 3D static models using surface histograms*: We implemented the algorithm reported in [16] as the 3D static baseline method for comparison. We treat each frame of the 3D model sequences as a 3D static model. Based on [16], a so-called primitive surface feature distribution (PSFD) face descriptor is implemented and applied for six-expression classification using LDA. As seen from Table 1, our dynamic 3D model-based HMM approaches outperforms the above static 2D/3D-based approaches. The performance of the PSFD approach is relatively low when it is tested on our 3D dynamic database because its feature representation is based on the static model's surface feature distribution (i.e., histogram). Such a representation may not detect local surface changes in the presence of low-intensity expressions.

Dynamic 2D/3D MU-based approaches: To verify the usefulness of 3D motion units (MU) derived from our dynamic 3D facial models, and to compare it with the 2D MU-based approaches and our dynamic 3D region-based approaches, we implemented the approach reported by Cohen *et al* [9] as the MU-2D baseline method.

(1) *MU-2D based*: According to [9], 12 motion units (MUs) are defined (as the 12 motion vectors) in areas of eyebrows, eye lids, lips, mouth corner and cheeks (see the left three images of Figure 4). Since we have tracked 83 key points on both 2D videos and 3D models as described in Section 3.1, the 12 MU points can be obtained from the tracking result. Note that although more MU points can be used from the tracking (as studied by Pantic *et al* in [8]), to be a fair comparison to the baseline approach, we only used the same 12 MU points as the ones defined in [9]. To compensate for the global rigid motion, we align current frame with the first frame using the estimated head orientation and movement from our adapted 3D tracking model. As such, the displacement vector of a MU point in frame *i* is obtained by $Disp(i) = F_i - F_{ne}$, where F_{ne} is the position of the MU point in the first frame (with a neutral expression) and F_i is the position of the MU point in the frame i. Figure 4 (left three images) is an example of the 2D MUs derived from a video sequence. In our experiment, we used the 12 MUs, derived from the 2D videos, as the input to the baseline HMM [9] to classify the six prototypic expressions. (2) *MU-3D based*: This is an extension of MU-2D

Fig. 4. An example of MUs. Left three: 2D-MUs on the initial frame, motion vectors of MUs from the initial frame to the current frame, and MUs on the current frame of a 2D sequence. Right four: 3D-MUs on the initial frame, 3D motion vectors of MUs with two different views, and MUs on the current frame of a 3D sequence.

method. It derives 3D displacement vectors of the 12 MUs from the dynamic 3D facial videos. Similarly, the 3D model of the current frame is also aligned to the 3D model of the first frame. The compensated 3D motion vectors are then used for HMM classification. Note that although the motion vectors of 2D and 3D models look alike in frontal view, they are actually different since 3D MUs also have motions perpendicular to the frontal view plane, as illustrated in the 2nd image from right of Figure 4. From Table 1, the MU-2D approach achieves a comparable result to that reported in [9] in the case of person-independent recognition. The MU-3D approach outperforms the MU-2D approach because 3D models provides more motion information for FER. Nevertheless, it is not superior to our 3D label-based spatio-temporal approaches because the MU-based approaches do not take advantage of entire facial surface features and rely on very few feature points for classification, and thus are relatively sensitive to the influence of the inaccurate feature detection. The experiment also shows that our 3D label-based R2D-HMM method achieves the best recognition result (90.44%). However, the confusion matrix (Table 2) shows that *sad, disgust*, and *fear* expressions are likely to be mis-classified as *anger*. Our R2D-HMM based approach does not rely on a few features. On the contrary, it takes advantage of entire 3D facial features as well as their 3D dynamics, and thus is more closely matched to the 3D dynamic data and more tolerant to individual feature errors than other compared approaches are.

Table 1. Facial expression recognition results summary

Model property	Method	Recognition rate	Model property	Method	Recognition rate
static 2D	Gabor-wavelet based	63.72%	dynamic 2D	MU-2D	66.95%
static 3D	LLE-based method	61.11%	dynamic 3D	MU-3D	70.31%
static 3D	PCA-based method	70.79%	dynamic 3D	T-HMM based	80.04%
static 3D	LDA-based method	77.04%	dynamic 3D	P2D-HMM based	82.19%
static 3D	PSFD method	53.24%	dynamic 3D	R2D-HMM based	90.44%

Table 2. Confusion matrix using R2D-HMM method

In/out	Anger	Disgust	Fear	Smile	Sad	Surprise
Anger	92.44%	3.68%	1.94%	1.32%	0.00%	1.42%
Disgust	8.28%	87.58%	1.27%	1.27%	0.96%	0.64%
Fear	7.45%	3.42%	85.40%	0.62%	0.00%	3.11%
Smile	0.44%	0.22%	0.66%	97.81%	0.00%	0.87%
Sad	13.12%	1.56%	0.63%	4.06%	80.32%	0.31%
Surprise	0.33%	0.00%	0.00%	0.33%	0.00%	99.34%

5.2 Performance Evaluation Using R2D-HMM

To further evaluate our spatio-temporal based approaches for 3D dynamic facial expression recognition, we conducted experiments to test the robustness of our R2D-HMM method with respect to three aspects: partial facial surface occlusion, expression intensity variation, and 3D model resolution variations.

Partial facial surface occlusion: Limited by views used in our current face imaging system, the facial surface may be partially missing due to the pose variation. To test the robustness of our proposed 3D facial descriptor and the dynamic HMM based classifier, we simulated the situation by changing the yaw and pitch angles of the facial models and generating a set of partially visible surfaces under different views. Ideally, we shall use the ground-true data collected systematically from a variety of views. However, it is hard (as well as expensive) to have such collection due to the difficulty to control the exact degree of pose during the subjects' motion. As such, in this paper we adopt the simulation approach for this study. Such a simulation allows us to study the performance of our proposed expression descriptor in the condition of partial surface invisible with a controllable degree of rotation. For the set of visible surfaces at different orientations, we report the recognition rate separately. Figure 5 shows the facial expression recognition rates with different yaw and pitch angles. The recognition results are based on our proposed dynamic-3D R2D-HMM based approach and the static-3D LDA-based approach. Generally, it shows that our dynamic approach outperforms the static approach in any situation since the motion information compensates for the loss of spatial information.

As shown in the the bottom row of Figure 5, our approach achieves a relatively high recognition rate (over 80%) even when the yaw and pitch angles change to 60 degrees, which demonstrates its robustness to the data loss due to the partial data invisible. The first row of Figure 5 shows the FER rate when the pose changes in only one dimension (yaw/pitch). Out of the useful range (i.e., either pitch or yaw angle changes exceed 150 degrees from the frontal view), the FER rate degrades to zero dramatically because of the paucity of useful information for recognition. The recognition curve of yaw's rotation within the useful range (Top-Left of Figure 5) is approximately symmetric with respect to the zero yaw angle. The recognition rate does not decrease too much even when the yaw angle is close to 90-degree (corresponds to half face visible). This is because either the left part or the right part of a face compensates for the other in the 3D space due to the approximate symmetric appearance of the face along the nose profile. However, the recognition curve of tilts rotation within the useful range is a little asymmetric as shown in the Top-Right of Figure 5. When the face is tilted up, the recognition rate is degraded not as much as when the face is tilted down in the same degree. This asymmetric property implies that the lower part of the face may provide more useful information than the upper part for expression recognition.

Variation of expression intensity: Our approach can also deal with variations of expression intensity since it not only includes different levels of intensities but also considers their dynamic changes. Based on our observation, we simply separated each 3D video sequence into two parts: a low intensity sequence (e.g., subsequences close to the starting or ending frames showing near-neutral expressions) and a high intensity

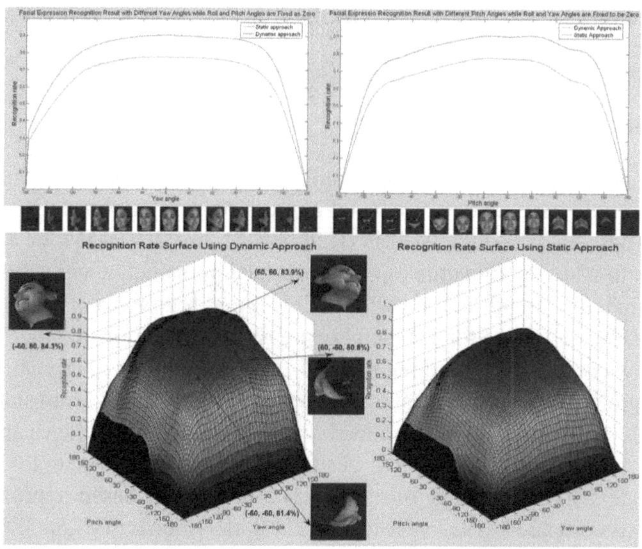

Fig. 5. FER results with simulated partial data missing scenario. Top: FER rate curves with respect to yaw rotation only and pitch rotation only; Bottom: FER rates surface with both yaw and pitch rotations. The facial pictures in the bottom illustrate the visible parts of a face when the yaw and pitch angles change to +/- 60 degrees. The recognition rates are also denoted besides the pictures.

sequence (subsequences excluding the low-intensity sequences). We performed the test on the low-intensity and high-intensity expressions individually using the R2D-HMM approach and the static PSFD approach [16]. Our training set includes both levels of intensities. The results show that the R2D-HMM method can detect both weak and strong expressions well. It achieves a 88.26% recognition rate of low intensity expressions and 91.58% recognition rate of high intensity expressions. However, the PSFD method has 71.72% recognition rate of high intensity expressions. It has less than 50% recognition rate for low intensity expressions. The main reason is that the static surface histogram descriptor may not be able to capture small variations of facial features as our 3D surface label descriptor does. In addition, the high performance of our approach is also attributed to the applied R2D-HMM classifier, which learns temporal transitions of dynamic facial surfaces effectively for both low-intensity and high-intensity expressions.

Variation of facial model resolutions: We down-sampled the test models to a low-resolution version with around 18,000 vertices, which is almost half the resolution of the original facial models (35,000 vertices) used for training. We then conducted the experiment to see whether the proposed approach works well for facial models with different resolutions. Based our R2D-HMM approach, the recognition rate for the low resolution models is 89.78%, which is comparable to the result of high resolution models (90.44%). This demonstrates that our approach has certain robustness to different resolutions, despite the fact that different resolutions could blur or sharpen the shape of facial surface. This result is supported by the psychological finding: blurring the shape information has little effect on the recognition performance as long as the motion channel is presented [31].

6 Discussions and Conclusions

In this paper, we proposed a spatio-temporal approach to study the viability of using dynamic 3D facial range models for facial expression recognition. Integrating the 3D facial surface descriptor and the HMMs (R2D-HMM, or P2D-HMM, or T-HMM), our system is able to learn the dynamic 3D facial surface information and achieves 90.44% person-independent recognition rate with both low and high intensities. In general, the HMM has been widely used for 2D facial expression recognition and face recognition. However, the way that we applied the real 2D-HMM to address 3D dynamic facial expression recognition is novel. We have extended the work of FER from static 3D range data to 3D videos. Many previous studies showed that sequential images are better than static images for FER [9,7]. We have verified that this statement holds true for 3D geometric models. The advantage of our 3D dynamic model based approach has been demonstrated as compared to several existing 2D static/video based and 3D static model based approaches using our new 3D dynamic facial expression database. This database will be made public to the research community. Ultimately, however, our focus was to study the usefulness of the new dynamic 3D facial range models for facial expression recognition rather than develop a fully automatic FER system. Our current work requires a semi-automatic process to select feature points at the initial stage. A fully automatic system with a robust 3D feature tracking will be our next stage of the development. To investigate the recognition performance in terms of large pose variations, we will design a new approach to measure the exact pose degree during the capture of ground-true spontaneous expressions. In addition, we will also investigate an approach to detect 3D action units and integrate the motion vector information with our surface label descriptor in order to improve the current FER performance.

Acknowledgement

This material is based upon the work supported in part by the National Science Foundation under grants IIS-0541044, IIS-0414029, and the NYSTAR's James D. Watson Investigator Program.

References

1. Ekman, P., Friesen, W.: The Facial Action Coding System. Consulting Psychologists Press, San Francisco (1978)
2. Tong, Y., Liao, W., Ji, Q.: Facial action unit recognition by exploiting their dynamic and semantic relationships. IEEE Trans. on PAMI 10, 1683–1699 (2007)
3. Pantic, M., Rothkrantz, L.: Automatic analysis of facial expressions: the state of the art. IEEE Trans. PAMI (2000)
4. Yang, P., Liu, Q., Metaxas, D.: Boosting coded dynamic features for facial action units and facial expression recognition. In: CVPR 2007 (2007)
5. Bartlett, M., et al.: Fully automatic facial action recognition in spontaneous behavior. In: FGR 2006, pp. 223–228 (2006)

6. Donato, G., Bartlett, M., Hager, J., Ekman, P., Sejnowski, T.: Classifying facial actions. IEEE Trans. PAMI 21(10), 974–989 (1999)
7. Lien, J., et al.: Subtly different facial expression recognition and expression intensity estimation. In: CVPR 1998 (1998)
8. Pantic, M., Patras, I.: Detecting facial actions and their temporal segments in nearly frontal-view face image sequences. In: IEEE Int'l Conf. on Systems, Man and Cybernetics 2005, pp. 3358–3363 (2005)
9. Cohen, I., Sebe, N., Garg, A., Chen, L., Huang, T.: Facial expression recognition from video sequences: temporal and static modeling. Journal of CVIU 91 (2003)
10. Sebe, N., et al.: Authentic facial expression analysis. Image Vision Computing 12, 1856–1863 (2007)
11. Zeng, Z., et al.: Spontaneous emotional facial expression detection. Journal of Multimedia 5, 1–8 (2006)
12. Chang, Y., Hu, C., Turk, M.: Probabilistic expression analysis on manifolds. In: IEEE Inter. Conf. on CVPR 2004 (2004)
13. Zhao, G., Pietikainen, M.: Dynamic texture recognition using local binary patterns with an application to facial expressions. IEEE Trans. on PAMI (2007)
14. Lyons, M., et al.: Automatic classification of single facial images. IEEE Trans. PAMI 21, 1357–1362 (1999)
15. Zalewski, L., Gong, S.: Synthesis and recognition of facial expressions in virtual 3d views. In: FGR 2004 (2004)
16. Wang, J., Yin, L., Wei, X., Sun, Y.: 3d facial expression recognition based on prmitive surface feature distribution. In: IEEE CVPR (2006)
17. Wang, P., Verma, R., et al.: Quantifying facial expression abnormality in schizophrenia by combining 2d and 3d features. In: CVPR 2007 (2007)
18. Wang, Y., Huang, X., Lee, C., et al.: High resolution acquisition, learning and transfer of dynamic 3d facial expressions. In: EUROGRAPHICS 2004 (2004)
19. Di3D, I. (2006), `http://www.di3d.com`
20. Chang, Y., Vieira, M., Turk, M., Velho, L.: Automatic 3d facial expression analysis in videos. In: IEEE ICCV 2005 Workshop on Analysis and Modeling of Faces and Gestures (2005)
21. Yin, L., Wei, X., Sun, Y., Wang, J., Rosato, M.: A 3d facial expression database for facial behavior research. In: IEEE FGR 2006 (2006)
22. Yeasin, M., et al.: From facial expression to level of interest: a spatio-temporal approach. In: CVPR 2004 (2004)
23. Phillips, P., Flynn, P., et al.: Overview of the face recognition grand challenge. In: CVPR 2005 (2005)
24. Yin, L., Chen, X., Sun, Y., Worm, T., Reale, M.: A high resolution 3d dynamic facial expression database. In: IEEE FGR 2008 (2008)
25. Sun, Y., Yin, L.: 3d face recognition using two views face modeling and labeling. In: IEEE CVPR 2005 Workshop on A3DISS (2005)
26. Cootes, T., Edwards, G., Taylor, C.: Active appearance models. IEEE Trans. PAMI 23 (2001)
27. Rabiner, L.: A tutorial on hidden markov models and selected applications in speech recognition. Proceedings of IEEE, 77(2) (1989)
28. Othman, H., Aoulnasr, T.: A separable low complex 2d hmm with application to face recognition. IEEE PAMI 25 (2003)
29. Saul, L., Roweis, S.: Think globally, fit locally: Unsupervised learning of low dimensional manifolds. Journal of Machine Learning Research 4, 119–155 (2003)
30. Martinez, A., Kak, A.: Pca versus lda. IEEE Trans. on PAMI 23, 228–233 (2003)
31. Wallraven, C., et al.: Psychophysical evaluation of animated facial expressions. In: Proc. of the 2nd Symposium on Applied Perception in Graphics and Visualization (2005)

Face Alignment Via Component-Based Discriminative Search

Lin Liang, Rong Xiao, Fang Wen, and Jian Sun

Microsoft Research Asia
Beijing, China
{lliang,rxiao,fangwen,jiansun}@microsoft.com

Abstract. In this paper, we propose a component-based discriminative approach for face alignment without requiring initialization[1]. Unlike many approaches which locally optimize in a small range, our approach searches the face shape in a large range at the component level by a *discriminative search* algorithm. Specifically, a set of *direction classifiers* guide the search of the configurations of facial components among multiple detected modes of facial components. The direction classifiers are learned using a large number of aligned local patches and misaligned local patches from the training data. The discriminative search is extremely effective and able to find very good alignment results only in a few (2$\sim$3) search iterations. As the new approach gives excellent alignment results on the commonly used datasets (e.g., AR [18], FERET [21]) created under-controlled conditions, we evaluate our approach on a more challenging dataset containing over 1,700 well-labeled facial images with a large range of variations in pose, lighting, expression, and background. The experimental results show the superiority of our approach on both accuracy and efficiency.

1 Introduction

Fitting a shape or template to a given image has been studied many years in computer vision [27,2]. The goal of face alignment is to match a 2D face shape or appearance with a facial image. Applications of face alignment range from face recognition [26], face tracking, facial expression analysis, 3D face modeling, to face cartoon animation [3]. Today, with the explosively increasing of face photos on the web and the desktop, new applications such as face search and annotation [10] have raised new requirements for the face alignment: fully automatic, efficient, and robust to facial images in under-controlled conditions.

Active Shape Model (ASM) [6] and Active Appearance Model (AAM) [4] are two most representative face alignment models. In ASM [6], a Point Distribution model captures the shape variants and gradient distributions of a set of landmark points describe the local appearance. The shape parameters are iteratively updated by locally finding the best nearby match for each landmark point. In AAM [4], the appearance is modeled globally by PCA on the mean shape coordinates (also called "shape-normalized

[1] We assume that a very rough face location is given by a face detector.

D. Forsyth, P. Torr, and A. Zisserman (Eds.): ECCV 2008, Part II, LNCS 5303, pp. 72–85, 2008.

frame"). The shape parameters are locally searched using a linear regression function on the texture residual. Many variants of ASM or AAM have been proposed, for example, Bayesian Tangent Shape Model (BTSM) [28], Mixture or non-linear shape model [7,22], View-based AAM [8], Direct Appearance Model (DAM) [12], Hierarchical DDMCMC [16], Constrained Markov Network [15], Inverse Compositional updating [19], 3D face alignment [11], Boosted Appearance Model (BAM) [17], and Boosted regression [9].

However, the models described above require good initialization because most of them are based on gradient descent or local search from a rough initial shape. The objective functions of these models might have a large number of local minimums in the high dimensional solution space. Even using the coarse-to-fine strategy, these approaches often get stuck into a local minimum if the initial shape is far from the correct solution, especially on the facial images in real applications.

A natural solution for the initialization problem is sampling - starting from multiple initial shapes. For instance, hierarchical CONDENSATION [24] uses 50 samples in each iteration and three-level pyramid to make the result less dependent on the initialization. The sampling method reduces the sensitivity to the initialization but is computational expensive. Moreover, it is unclear that how many samples are sufficient to cover the correct solution.

In this paper, we propose a component-based discriminative approach to address the efficiency and robustness problems. Our approach searches the face shape at the facial component level. We first detect a number of candidate *modes* for facial components (e. g., eyes, nose, mouth, and profiles) and construct an initial shape. Then, a *discriminative search* algorithm searches a new position for each facial component. The searching direction is determined by learned *direction classifiers*, and the searching step is adaptively set according to the candidate modes. For each facial component, nine direction classifiers (eight directions + one of "no move") are trained on aligned and misaligned local patches from the training data in the mean shape coordinates. Using our discriminative search among candidate modes, better configurations of facial components can be effectively and rapidly discovered in very large searching range. The shape is further refined using the components as constraints. Our experiments demonstrate that we are able to visit very good fitted shape only within a few (usually 2-3) iterations.

Our approach is unique in several aspects: we optimize the shape at the component level among candidate modes so that we have the ability to jump out the local minimums; we learn how to search facial components from the training data. The direction classifier is trained on a large number of positive examples and negative examples; it is fully automatic without requiring initialization; it is a very efficient deterministic algorithm which significantly outperforms the sampling based approach on the accuracy, and is much faster on the speed.

The discriminative approach has been applied to in the face alignment [17,9] and visual tracking [1,20]. Our proposal differs from these works in that: 1) we train the classifier to predict the searching direction. It is more robust than directly predicting the exact position; 2) we design an searching algorithm which can effectively utilize the multi-modes, noisy, and incomplete observations (component detection results).

2 Problem Formulation

Notation. Given an image I, the task of face alignment is to locate a number of N facial feature points $\mathbf{S_f} = \{s_i^f\}_{i=1}^N$ on the 2D image. In this work, we also define C facial component points $\mathbf{S_c} = \{s_i^c\}_{i=1}^C$. Figure 2 shows 11 facial components (eyes, brows, nose, mouth, upper/lower lips, left/right/lower profiles) we used.

Applying PCA analysis on a set of training shapes in the tangent shape space [28], the face shape $\mathbf{S} = [\mathbf{S_f}, \mathbf{S_c}]$ can be expressed as:

$$\mathbf{S}(\mathbf{p}) = T(\overline{\mathbf{x}} + \mathbf{Q}\mathbf{b}; \mathbf{t}), \tag{1}$$

where $\overline{\mathbf{x}}$ is the mean shape vector, $\mathbf{Q}$ is a shape basis matrix, and $\mathbf{b}$ is a vector of tangent shape parameters, and $T(\cdot; \mathbf{t})$ is a global similarity transformation with parameters $\mathbf{t}$. The shape parameters are denoted as $\mathbf{p} = [\mathbf{t}, \mathbf{b}]$.

To find the shape parameters $\mathbf{p}$ for a given image I, an iterative searching algorithm is usually used in ASM [6] and its variants: find an observed shape $\mathbf{y}$ by locally finding the best nearby match for each point; optimize the shape parameters $\mathbf{p}$ based on new matched points. In the second step, we optimize the following energy function:

$$E(\mathbf{p}|\mathbf{y}) = \Delta_y^T \mathbf{\Sigma}_\mathbf{y}^{-1} \Delta_y^T + \lambda \cdot \mathbf{b}^T \mathbf{\Sigma}_\mathbf{b}^{-1} \mathbf{b}, \tag{2}$$

where $\Delta_y = \mathbf{S}(\mathbf{p}) - \mathbf{y}$ is the difference between estimated shape $\mathbf{S}(\mathbf{p})$ and the observed shape $\mathbf{y}$, $\mathbf{\Sigma}_\mathbf{y}^{-1}$ is the image noise covariance matrix, $\mathbf{\Sigma_b}$ is the covariance matrix of the shape, and λ is a regularization factor. The shape parameters $\mathbf{p}$ can be optimized using EM algorithm [28].

As we mentioned before, the above model is sensitive to the initialization and may not always to converge the correction solution since the search is only performed in a small range in each iteration. A natural solution is to provide prior estimates of the position to constrain some of the shape points. In [5], Cootes and Taylor show that extra user-specified constraints can greatly improve the AAM model. In this work, we also follow this general idea: we detect facial components and use multiple detected modes as constrains.

Component-constrained optimization. Suppose we have detected positions $\widehat{\mathbf{S}}_\mathbf{c} = \{\widehat{\mathbf{s}}_\mathbf{i}^\mathbf{c}\}_{\mathbf{i}=1}^\mathbf{C}$ of facial components and corresponding variances $\{\sigma_i\}_{i=1}^C$, we can enforce the component constraints to the energy (2):

$$E'(\mathbf{p}|\mathbf{y}) = E(\mathbf{p}|\mathbf{y}) + \lambda_c \Delta_c^T \mathbf{\Sigma}_\mathbf{c}^{-1} \Delta_c^T, \tag{3}$$

where $\Delta_c = \mathbf{S_c}(\mathbf{p}) - \widehat{\mathbf{S}}_\mathbf{c}$ is the difference of estimated component positions $\mathbf{S_c}(\mathbf{p})$ with the constrained positions $\widehat{\mathbf{S}}_\mathbf{c}$, covariance matrix $\mathbf{\Sigma_c} = \mathrm{diag}(\sigma_1^2, ..., \sigma_C^2)$ models the uncertainty of detected positions, and λ_c is a weighting parameter.

However, the component detectors are usually error-prone. Because the local patch alone is not discriminative enough, multiple modes might be detected or no mode can be detected for one component. Even worse, the detected modes are rather noisy and contain localization errors. Thus, the challenge is how to effectively and efficiently use these multi-modes, noisy, and incomplete observations.

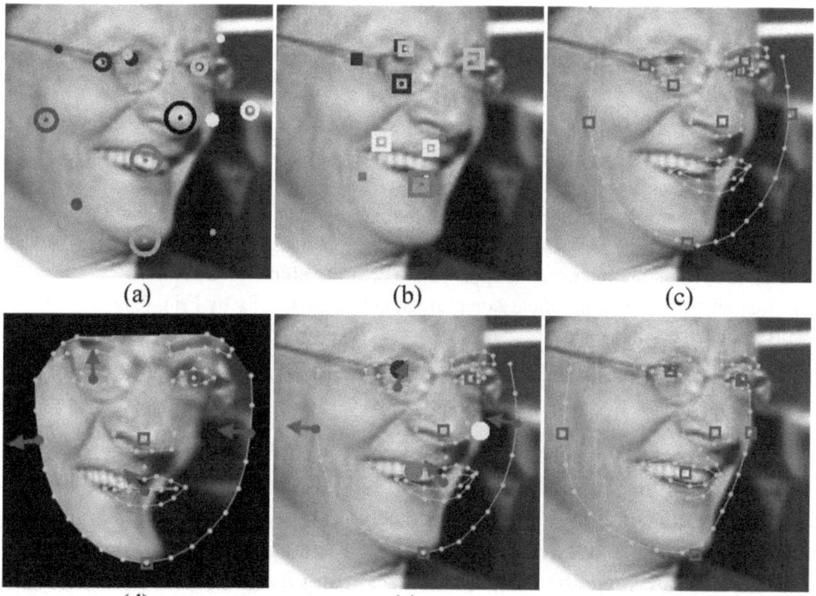

Fig. 1. An example. (a-b) detected modes of facial components. We use the color and shape to distinguish the modes from different components. The size of each mode represents its weight. (c) fitted initial shape $\mathbf{S}_0$ using the best mode (marked as small red square) with maximal weight of each component. (d) warped appearance to the mean shape coordinate. In this coordinate, we apply direction classifier to determine the searching direction of each component. The red arrows indicate the decided directions and the red rectangles represent "no move". (e) solid circles are searched new modes. (f) the fitted shape after one step searching. **We encourage the reader to see this figure in the electronic version.**

A simple solution is to sample each component's modes based on their confidence. But, the combinations of all modes of whole components could be large, which result in an inefficient algorithm. More essentially, the sampling approach is difficult to handle the case that no correct mode is detected.

To address this problem, in the next section, we present a component-based discriminative searching algorithm which can efficiently find very good configurations of components only in a few iterations.

3 Component-Based Discriminative Search

3.1 Algorithm Overview

Before we describe the details, we first give an overview of our algorithm. At first, we detect multiple possible positions, *modes*, of each facial component by a component detector (section 3.2). Using the best mode (with largest wight) of each component as constraint, we search an initial shape $\mathbf{S}_0$ by minimizing the energy (3). Then, starting

Table 1. The flowchart of our algorithm

- Detect the face
- Detect the components' modes in the face region
- Constrain each component at the mode with maximal weight, search an initial shape $\mathbf{S}_0$ using Eqn. (3).
- for $k = 1$ to K
  (1) For each component, perform discriminative search
    - Determine the searching direction by direction classifiers
    - Move the component to an appropriate position along the direction

  (2) Search with new component constraints, get the shape $\mathbf{S}_k$ by Eqn. (3)
  (3) Evaluate the shape score $f(\mathbf{S}_k)$
- The output shape $\mathbf{S}^* = \max_{\mathbf{S}_k} f(\mathbf{S}_k)$

from the initial shape, we search the new mode for each component by a discriminative search algorithm (section 3.4). The searching direction is determined by boosted direction classifiers (section 3.3). Using the new searched modes as constraints, we re-optimize the face shape. We repeat this step for K iterations. Finally, we calculate the score $f(\mathbf{S}_k)$ of all visited shapes $\{\mathbf{S}_0, ..., \mathbf{S}_K\}$ using the Boosted Appearance Model (BAM) [17], which is a discriminative approach for evaluating the goodness of a fitted shape. We pick the best one as the final output. The algorithm flowchart is shown in Table 1.

Figure 1 shows a simple example for the illustration purpose. We draw the component detection results on Figure 1 (a) (eyes, nose, mouth, and three profiles) and Figure 1 (b) (brows, upper/lower lips). Figure 1 (c) is the fitted initial shape and Figure 1 (d) is warped image from the initial shape to the mean shape coordinates. In this image, we perform the first step of the discriminative search - applying direction classifiers to decide the searching direction of each component. Then, in Figure 1 (e), we inverse-warp the searching directions back to the original image coordinates and perform the second step of the discriminative search - moving the components to appropriate positions along the searching directions. Note that some components decide not to move. Figure 1 (f) shows the re-optimized shape after one step searching.

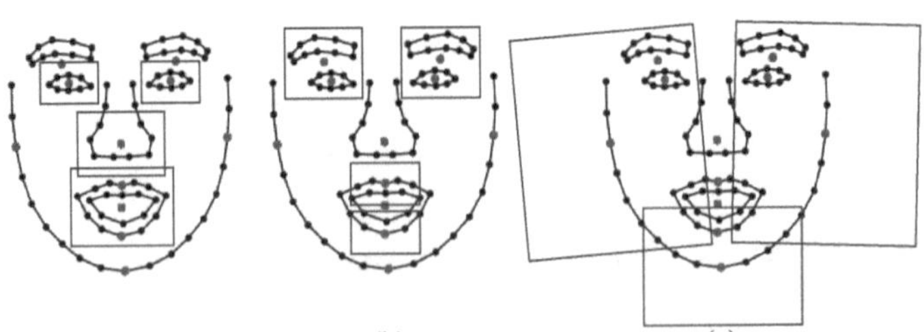

(a) (b) (c)

Fig. 2. Component definition and image patches for the detection. Red points are the positions of components. Blue rectangles are extracted image patch for the training of the component detection: (a) eyes, nose, and mouth. (b) brows, upper/lower lips. (c) three profiles.

3.2 Component Detection

For each component, we train a component detector using Harr wavelet features by Adaboost [25]. The patch definitions of components are shown in Figure 2. For three profiles, we experimentally find that the large patches are better than the small patches. We collect 4000 positive examples for each component. The negative examples are generated by "boostrap" in each stage of the cascade training. In the detection phase, we scan all patches at multiple scales with the face region. For the efficiency consideration, we only run the detectors on a regular grid with 3 pixels interval.

The confidence output of the detector is defined as:

$$c = 1/(1 + \phi_h \exp(-r)) \tag{4}$$

where r is the output score of the last strong classifier (at h-th layer) it passed, and ϕ_h is the ratio of negative examples to positive examples for the h-th layer (obtained during training process) [14]. The final confidence value $c(x, y)$ at the position (x, y) is the sum of the confidences by all detectors at multiple scales. Thus, we obtain a set of points with confidence values.

To reduce the number of candidate point, we fit a Mixture of Gaussian (MoG) on these points. We first find local maximums to initial the MoG fitting algorithm. Note that we use the confidence value as the weight of each point. We also merge very closed clusters after fitting. Finally, we use MoG's components as our detected *modes*: $\{w(\mathbf{m}_i), \mathbf{m}_i\}$, where $w(\mathbf{m}_i)$ and $\mathbf{m}_i$ are weight and center of one cluster in MoG.

In the component-constraint optimization Equation (3), a covariance matrix $\mathbf{\Sigma_c}$ = diag(σ_i^2) models the uncertainty of detected mode. In our implementation, we define the variance as $\sigma_i = \frac{1}{w(\mathbf{m}_i)+\epsilon}$, where ϵ is set to 0.01;

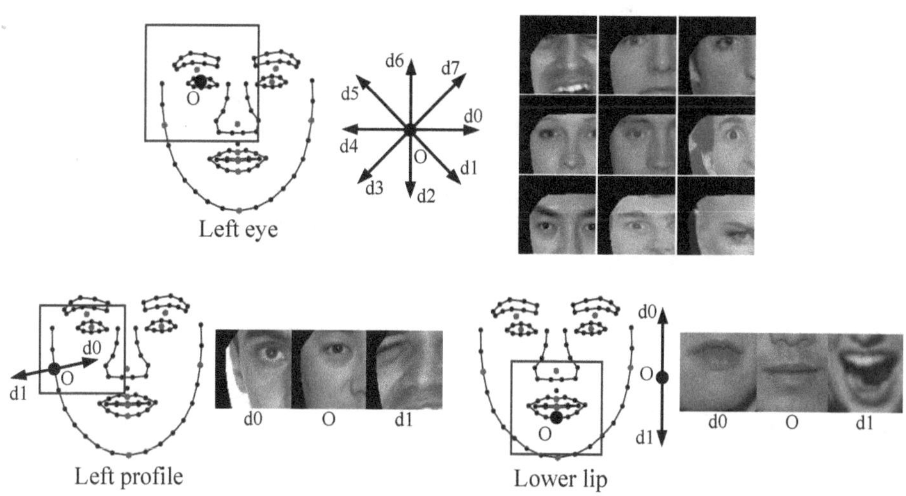

Fig. 3. Direction classifiers. Top: nine kinds of patches for left eye. O is "no move" and $d_0, d_1, ..., d_8$ are for eight directions. Bottom-left: three kinds of patches for left profile. Bottom-right: three kinds of patches for lower lip.

3.3 Learning Direction Classifiers

The goal of direction classifier is: given a patch in the facial image, we want to know whether the patch is well-aligned with a specific facial component or not; if not, what the relative direction between the patch and the facial component is. To achieve this goal, we take a discriminative approach and learn from the data.

We generate training examples from 4,000 well-labeled facial images in four ways: 1) 4,000 well-aligned "true" shapes ; 2) 60,000 randomly perturbed shapes from the true shapes by translation, rotation and scaling; 3) 30,000 randomly perturbed shapes then locally optimized using equation (2); 4) 30,000 randomly perturbed shapes then locally optimized using randomly sampled the modes as constraints. We warp all shapes and images to the mean shape coordinates. The mean image is normalized to 263×263.

For each component, we extract total 120,000, 40×40 image patches. The size of the local patch is designed to include other components that may be helpful as the context. We divide those patches into three classes based on the distance to the "true" position: well-aligned (< 5 pixels distance), mis-aligned (> 20 pixels distance), and the others. We discard the last class since they are ambiguous. We further divide the mis-aligned patches into eight sets based on eight discrete directions with respect to the 'true" position. Finally, we have nine sets of patches with known relative direction w.r.t the true position, as shown in the top row of Figure 3.

Using these training patches, we train nine classifiers for different directions (the direction of the well-aligned set is "no move"). For an individual direction, we use the corresponding patches as positive samples and all the other pathes as negative samples. The classifier is trained by Adaboost on Gabor feature (8 directions and 5 scales filter responses). The Gabor feature is a $40 \times 40 \times 8 \times 5$ vector. We use the stump decision as the weak learner on each dimension. The number of weak learner is around 200-300 on average.

For left and right profiles, it is reasonable to only consider three directions: left, right, and no move. For brows and upper/lower lips, we treat them as the child components of eyes and mouth respectively. Therefore, we only consider up, down, and no move since we will move them after the movements of eyes and mouth in the discriminative search. The example patches for these components are shown in the bottom row of Figure 3.

Equipped with the direction classifiers, we will introduce a discriminative search algorithm to effectively search new positions of facial components in the next subsection.

3.4 Discriminative Search

Given the current shape, we first warp the image to the mean shape coordinates. For each component, we calculate the raw scores $[s(O), s(d_0), ..., s(d_8)]$ of nine direction classifiers and select the optimal direction with maximal score. Then the optimal direction is transformed back to the image coordinate. We denote the optimal direction and position of the component under the image coordinate as $\mathbf{d}^* = [d_x^*, d_y^*]$ and $\mathbf{s}$, the associated score as $s(\mathbf{d}^*)$. We also define a set of modes $\mathcal{M}$ which are within the searching range of $\mathbf{d}^*$, viz

$$\mathcal{M} = \{\mathbf{m}_k | \langle \mathbf{m}_k - \mathbf{s}, \mathbf{d}^* \rangle \geq 0.7\}. \qquad (5)$$

Table 2. The pseudo code of discriminative search for an individual component

```
Input: component's position s and weight w
Output: component's new position s' and weight w'
• Find the optimal direction d* with maximal classifier score s(d*)
• if the direction is "no move"
    s' = s
    if s(d*) > 0
        w' = w,
    else
        w' = 0
• else (other directions)
    construct the mode set M = {m_k | <m_k - s, d*> >= 0.7}
    if M != Ø
        select best mode m_k* = arg max_{m_k} {<(m_k - s), d*> · w(m_k)}
        s' = m_k*
        w' = w(m_k*)
    else
        s' = s + 0.1 · FaceWidth · d*
        w' = 1
```

Based on these information, we search each component using strategies as follow:

1. $s(\mathbf{d}^*) > 0$ and $\mathcal{M} \neq \emptyset$
   The score of the direction is high and $\mathcal{M}$ is not empty. We move the component to a mode in $\mathcal{M}$. We select the mode $\mathbf{m}_k^*$ in $\mathcal{M}$ by considering both direction and mode's weight $w(\mathbf{m}_k)$:
$$\mathbf{m}_k^* = \arg\max_{\mathbf{m}_k}\{\langle(\mathbf{m}_k - \mathbf{s}), \mathbf{d}^*\rangle \cdot w(\mathbf{m}_k)\}. \tag{6}$$
   The new mode $\mathbf{m}_k^*$ is used as a constraint for the shape searching.
2. $s(\mathbf{d}^*) > 0$ and $\mathcal{M} = \emptyset$
   The score of the direction is high but $\mathcal{M}$ is empty. In this case, we move the component a constant step along the direction $\mathbf{d}^*$. The step size is set as $0.1 \times$ face width. The new position constrains the shape searching. This strategy is designed to handle the situation that no correct mode is detected.
3. $s(\mathbf{d}^*) \leq 0$
   The score of the direction is low, which means that the direction classifier is not sure about the proposed searching direction. In this case, we do not move the component and do not use component's position as the constraint.

Notice that for the direction "no move", we only apply the last strategy. The pseudo code of discriminative search for a single component is shown in Table 2. Due to the hierarchical nature of the facial components, we use a two level searching scheme. We first search eyes, nose, mouth and profiles at the top level. For brows and upper/lower lips, we only search them when their parent components (brows are children of eyes, and lips are children of mouth) are judged as "no move".

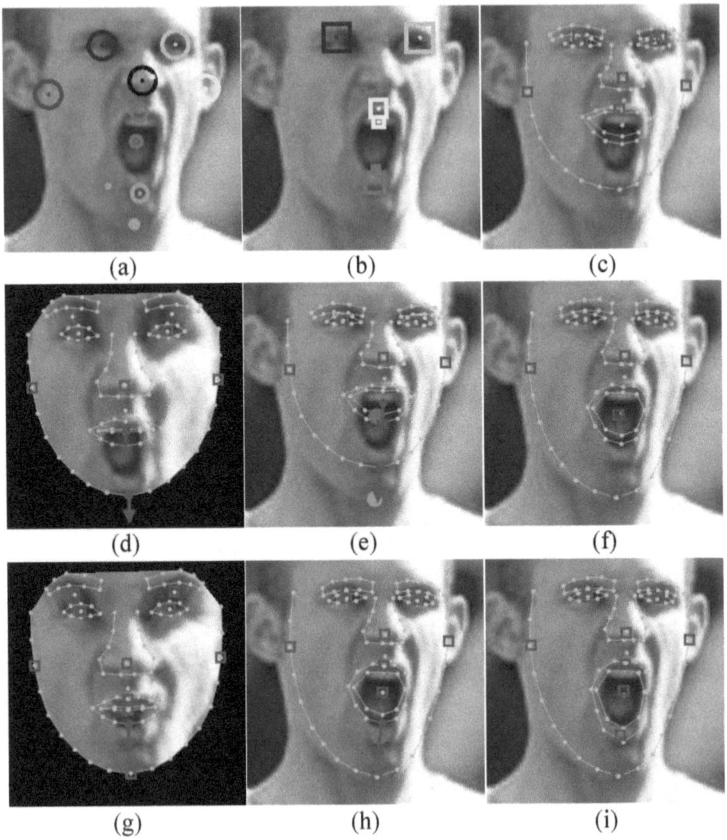

Fig. 4. Hierarchical search. (a-c) detected modes and initial shape. (d-f) top level discriminative search (eyes, nose, mouth, and profiles). (g-i) bottom level discriminative searching (brows and lips). Upper and lower lips are more accurately located.

Figure 4 shows an example. In the first iteration, mouth and lower profile are moved toward "down" (Figure 4 (d-e)). The fitted shape is shown in Figure 4 (f). In this step, we do not use upper and lower lips as constraints. In the next iteration, lower lip is moved toward "down" (Figure 4 (d-e)). The final fitted shape is obtained using upper and lower lips as constraints (Figure 4 (i)).

The discriminative search also has the capability to handle the case that no correct mode is detected for partial facial components. Figure 4 is an example.

4 Experiments

Datasets. To evaluate our algorithm, we collect 6,471 images from multiple databases, including AR database [18], FERET database [21], PIE database [23], and Labeled Faces in the Wild (LFW) [13]. We randomly select 4,002 images for training and 2,469 images for testing. To analysis the performance on different data sets, we separate the

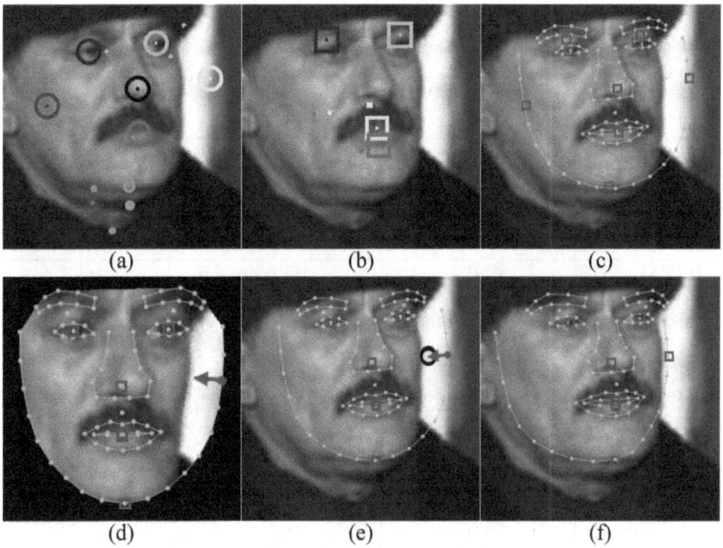

Fig. 5. Handling incorrect mode. (a-b) are detected modes of components. No correct mode of right profile is detected. (c) initial shape. (d) warped image in the mean shape coordinates. The direction classifier decides to move the right profile to "left" by a constant step. (e) direction in the image coordinates. (f) fitted shape after one step search.

Table 3. The accuracy of direction classifiers

Component	Accuracy	Component	Accuracy
left eye	98.4%	right eye	98.6%
left brow	97.8%	right brow	97.4%
nose	98.9%	mouth	97.1%
upper lip	97.6%	lower lip	97.0%
left profile	98.2%	right profile	98.0%

testing image into two sets A and B. The set A contains 753 images that are from AR, FERET, and PIE. These images are collected under controlled conditions. The set B contains 1,716 images that are all from LFW which are collected from the web and show much larger variations in pose, expression, lighting, focus, and background. See [13] for the detailed description of this database.

Performance of direction classifiers. First, we evaluate the performance of our direction classifiers. We generate 14,897 aligned and 48,663 misaligned shapes on all 2,469 testing images and extract the image patches of all facial components. The directions classifiers are applied on these patches to estimate the searching directions. Table 3 shows the accuracy (percentage of correctly predicted directions w.r.t the "true" directions) of the estimated searching detection of each component. Even using the simple maximal score rule, the average accuracy rate is very high - above 97%. It demonstrates the ability of the learned direction classifiers on the prediction of directions.

Table 4. Comparison on the set A. Each column is the percentage of images with RMSE less than a given threshold.

	RMSE< 5 pixels	RMSE< 7.5 pixels	RMSE< 10 pixels
Our approach	94.0%	98.9%	99.3%
SMP (50 samples)	91.3%	97.8%	99.1%
SMP (20 samples)	91.2%	97.6%	99.1%
SMP (5 samples)	89.6%	96.3%	98.9%
CBTSM	88.7%	95.8%	98.5%
BTSM	89.0%	95.6%	98.1%

Table 5. Comparison on the set B

	RMSE< 5 pixels	RMSE< 7.5 pixels	RMSE< 10 pixels
Our approach	74.7%	93.5%	97.8%
SMP (50 samples)	63.3%	86.8%	95.1%
SMP (20 samples)	63.1%	86.6%	94.9%
SMP (5 samples)	58.2%	83.0%	92.7%
CBTSM	57.2%	82.8%	92.1%
BTSM	51.4%	72.5%	85.2%

Comparison. We compare our approach with Bayesian Tangent Shape Model (BTSM) [28], constrained BTSM (CBTSM) (using the most confident mode of each component as constraint), and sampling-based approach (SMP). The SMP samples the mode of each component based on its weight and uses the sampled modes as the constraints. We select the best fitted sample at the output using the same shape score function [17]. For all the algorithms except for SMP, the initial shape is set as a mean shape transformed by the optimal similarity transformation using the most confident modes. All algorithms uses four level pyramid for the coarse-to-fine search. In our approach, we performs 3 iterations in the discriminative search.

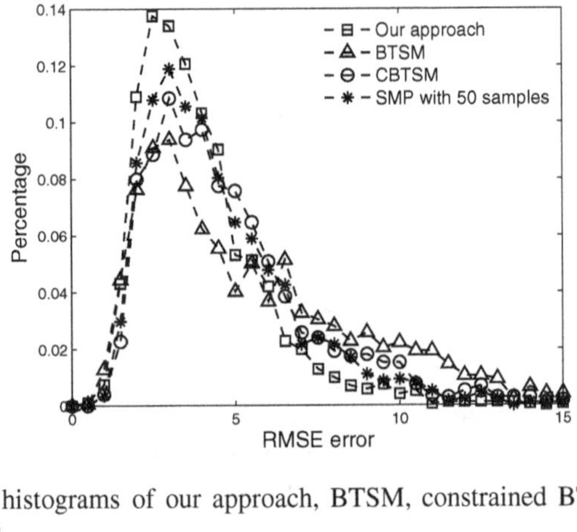

Fig. 6. The RMSE histograms of our approach, BTSM, constrained BTSM and the sampling method on the set B

Table 6. The effect of iteration number in our algorithm. The second row is the percentage of correctly aligned samples at each iteration.

Iteration 0	Iteration 1	Iteration 2	Iteration 3	Iteration 4	Iteration 5
82.8%	92.9%	93.4%	93.5%	93.5%	93.5%

Fig. 7. Aligned examples by our approach. The images in the top row are from the set A. The other images are from the set B which is a more challenging dataset.

Set A. Table 4 shows the percentage of images with RMSE (Root Mean Squared Error w.r.t the manually labeled "true" shape) less than given thresholds. All images are resized to 200-300 pixels. As we can see, with the help of component detectors, all algorithms perform very well on this data set and our result is slightly better. It indicates the commonly used AR, FERET and PIE databases created in controlled conditions are largely saturated.

Set B. Table 5 are the percentages of images with RMSE less than given thresholds and Figure 6 shows corresponding RMSE histograms. The RMSE histograms give more accurate distribution of images in given RMSE intervals. As we can see, our result is significantly better on this more challenging data set. Due to the large variation of this data set, the performance of component detectors degrade. Consequently, for many cases, the initial shape might be far from the correct shape. The performances of BTSM and CBTSM drop a lot because the most confident modes may not be correct. Using the sampling strategy, SMP is better. But the the improvement is marginal when we increase the sample number from 20 to 50. A possible reason is that the correct position is not within the detected modes. In our approach, the discriminative search algorithm can move a component along the estimated searching direction even if the component detector has failed. Thus, our algorithm can still find a better shape. As shown in Table 5, for the threshold of 7.5 pixels, our result is 93.5% and the best next one is 86.8%.

Iteration number. We also test the effect of iteration number in our algorithm on the set B. Table 6 shows the percentage of correctly aligned samples (within the threshold of 7.5 pixels) vis the number of iterations. It indicates that our algorithm can correctly align most of the images within 3 iterations. Note that the computation cost of the discriminative search is neglectable since the direction classifier is extremely fast ($<$1ms). Therefore, our whole approach is about 3 times slower than BTSM and CBTSM but 7 times faster than SMP with 20 samples.

Figure 7 shows example alignment results by our approach on the images with non-frontal view, large expression, poor lighting, and complex background.

5 Conclusion

In this paper, we have presented a discriminative approach for face alignment. By learning from the data, a new discriminative search algorithm can very effectively find a good configuration of face shape at the component level. The algorithm is very efficient and able to handle the initialization problem.

The performance of our approach is greatly benefit from the discriminative learning: detecting components' modes, searching components, and evaluating the shape. Therefore, it can be improved using a larger training data with high variability to make the face alignment more robust in the uncontrolled, real applications. To obtain more accurate results, we are going to train the direction classifiers at the finer level and apply them after the optimization at the component level.

References

1. Avidan, S.: Support vector tracking. IEEE Trans. on PAMI 26(8), 1064–1072 (2004)
2. Blake, A., Isard, M.: Active Contours. Springer, London (1998)
3. Chen, H., Xu, Y.-Q., Shum, H.-Y., Zhu, S.-C., Zheng, N.-N.: Example-based facial sketch generation with non-parametric sampling, vol. 2, pp. 433–438 (2001)

4. Cootes, T., Edwards, G., Taylor, C.: Active appearance models. In: Burkhardt, H., Neumann, B. (eds.) ECCV 1998. LNCS, vol. 1407. Springer, Heidelberg (1998)
5. Cootes, T.F., Taylor, C.J.: Constrained active appearance models. In: ICCV, vol. I, pp. 748–754 (2001)
6. Cootes, T.F., Taylor, C.J., Graham, J.: Active shape models – their training and application. Computer Vision and Image Understanding 61, 38–59 (1995)
7. Cootes, T.F., Taylor, C.J.: A mixture model for representing shape variation. Image and Vision Computing 17(8), 567–574 (1999)
8. Cootes, T.F., Walker, K., Taylor, C.J.: View-based active appearance models. Automatic Face and Gesture Recognition, 227–232 (2000)
9. Cristinacce, D., Cootes, T.F.: Boosted regression active shape models, vol. 2, pp. 880–889 (2007)
10. Cui, J., Wen, F., Xiao, R., Tang, X.: Easyalbum: An interactive photo annotation system based on face clustering and re-ranking. In: SIGCHI (2007)
11. Gu, L., Kanade, T.: 3d alignment of face in a single image, vol. 1, pp. 1305–1312 (2006)
12. Hou, X., Li, S.Z., Zhang, H., Cheng, Q.: Direct appearance models, vol. I, pp. 828–833 (2001)
13. Huang, G.B., Ramesh, M., Berg, T., Miller, E.L.: Labeled faces in the wild: A database for studying face recognition in unconstrained environments. Technical Report, 25(12):07–49 (October 2007)
14. Li, Y., Ai, H., Yamashita, T., Lao, S., Kawade, M.: Tracking in low frame rate video: A cascade particle filter with discriminative observers of different lifespans. In: CVPR, pp. 1–8 (2007)
15. Liang, L., Wen, F., Xu, Y., Tang, X., Shum, H.-Y.: Accurate face alignment using shape constrained markov network. In: CVPR 2006 (2006)
16. Liu, C., Shum, H.-Y., Zhang, C.: Hierarchical shape modeling for automatic face localization. In: ECCV, pp. 687–703 (2002)
17. Liu, X.: Generic face alignment using boosted appearance model. In: CVPR 2007 (2007)
18. Martinez, A.M., Benavente, R.: The ar face database. CVC Technical Report (24) (1998)
19. Matthews, I., Baker, S.: Active appearance models revisited. IJCV 60(2), 135–164 (2004)
20. Blake, A., Williams, O., Cipolla, R.: Sparse bayesian learning for efficient visual tracking. IEEE Trans. on PAMI 27(8), 1292–1304 (2005)
21. Phillips, P.J., Moon, H., Rauss, P.J., Rizvi, S.: The feret evaluation methodology for face recognition algorithms. IEEE Trans. on PAMI 22(10), 1090–1104 (2000)
22. Romdhani, S., Cong, S., Psarrou, A.: A multi-view non-linear active shape model using kernel pca. In: BMVC 1999 (1999)
23. Sim, T., Baker, S., Bsat, M.: The cmu pose, illumination, and expression database. IEEE Trans. on PAMI 25(12), 1615–1618 (2003)
24. Tu, J., Zhang, Z., Zeng, Z., Huang, T.: Face localization via hierarchical condensation with fisher boosting feature selection. In: CVPR, vol. 2, pp. 719–724 (2004)
25. Viola, P., Jones, M.J.: Robust real-time face detection. In: IJCV, vol. 57(2), pp. 137–154 (2004)
26. Wiskott, L., Fellous, J.M., Krüger, N., von der Malsburg, C.: Face recognition by elastic bunch graph matching. In: ICIP 1997 (1997)
27. Yuille, A.L., Cohen, D.S., Hallinan, P.W.: Feature extraction from faces using deformable templates. In: CVPR, pp. 104–109.
28. Zhou, Y., Gu, L., Zhang, H.-J.: Bayesian tangent shape model: Estimating shape and pose parameters via bayesian inference. In: CVPR, vol. 1, pp. 109–116 (2003)

Improving People Search Using Query Expansions
How Friends Help to Find People

Thomas Mensink and Jakob Verbeek

LEAR - INRIA Rhône Alpes - Grenoble, France
{thomas.mensink, jakob.verbeek}@inria.fr

Abstract. In this paper we are interested in finding images of people on the web, and more specifically within large databases of captioned news images. It has recently been shown that visual analysis of the faces in images returned on a text-based query over captions can significantly improve search results. The underlying idea to improve the text-based results is that although this initial result is imperfect, it will render the queried person to be relatively frequent as compared to other people, so we can search for a large group of highly similar faces. The performance of such methods depends strongly on this assumption: for people whose face appears in less than about 40% of the initial text-based result, the performance may be very poor. The contribution of this paper is to improve search results by exploiting faces of other people that co-occur frequently with the queried person. We refer to this process as 'query expansion'. In the face analysis we use the query expansion to provide a query-specific relevant set of 'negative' examples which should be separated from the potentially positive examples in the text-based result set. We apply this idea to a recently-proposed method which filters the initial result set using a Gaussian mixture model, and apply the same idea using a logistic discriminant model. We experimentally evaluate the methods using a set of 23 queries on a database of 15.000 captioned news stories from *Yahoo! News*. The results show that (i) query expansion improves both methods, (ii) that our discriminative models outperform the generative ones, and (iii) our best results surpass the state-of-the-art results by 10% precision on average.

1 Introduction

Over the last decade we have witnessed an explosive growth of image and video data available both on-line and off-line. This has lead to the need for methods to index, search, and manipulate such data in a semantically meaningful manner; bridging the apparent gap between low-level features and semantics [1]. Much research has addressed this problem using so called 'supervised' techniques that require explicit manual annotations to establish correspondences between low-level features and semantics, these correspondences are then captured in models that generalize the correspondence to other images. For example, image categorization has made significant progress using this approach [2]. Learning semantic relations from weaker forms of supervision is currently an active and broad line of research. Work along these lines includes learning correspondence between keywords and image regions [3,4], and learning image retrieval and auto-annotation with keywords [5,6]. In this work images were labeled with

D. Forsyth, P. Torr, and A. Zisserman (Eds.): ECCV 2008, Part II, LNCS 5303, pp. 86–99, 2008.

United Nations Secretary General ***Kofi Annan*** stands with U.N. Security Council President and U.S. Ambassador to the U.N. **John D. Negroponte** as Annan ...

North Korean leader ***Kim Jong Il***, and Russian President **Vladimir Putin** walk after talks in Vladivostok, Friday, Aug. 23, 2002. North Korean leader Kim ...

Fig. 1. Two example images with captions. The queried person is marked in italic, detected named entities in bold, and detected faces are marked by yellow rectangles.

multiple keywords per image, requiring resolution of correspondences between image content and semantic categories. Supervision from even weaker forms of annotation are also explored, e.g. based on images and accompanying text [7,8], and video with scripts and subtitles [9,10].

In this paper we aim to improve search for people in databases of captioned news photographs; see Figure 1 for two examples taken from *Yahoo! News*. Automatic analysis of news streams is important as they are major sources in the information need of people, and news articles are published at a high frequency. When searching for images of a certain person, a simple system could (i) query the database for captions containing the the name, and (ii) rank or filter the result images by the confidence level of a face detector. An example of such a system is Google Portrait [11], which queries Google with the name and then applies a fast face detector. Recently such a system has also been integrated in Google's advanced image search. Although such a system correctly rejects images without faces (or at least those without face detections), the performance of such a system is clearly limited by the fact that it returns all images with detected faces, not only those depicting the queried person.

Identification of faces in news photographs is a challenging task, significantly more so than recognition in the usual controlled setting of face recognition: we have to deal with imperfect face detection and alignment procedures, and also with great changes in pose, expression, and lighting conditions, and poor image resolution and quality. Perhaps even more importantly, labeled data sets for learning classifiers are not generally available, and tedious to produce. However, these difficulties are partly compensated by the information contained in captions. Recently it was shown that initial text-based results can be significantly improved by filtering faces on the basis of visual features [12,13]. While using the caption alone leads to a disappointing precision of 44% (fraction of faces belonging to the queried person among all returned faces, averaged over queries for 23 people), adding face analysis increases average precision to 71%, at a recall of 85%. Others [14] have considered resolving all face-name associations in databases of captioned news images. The potential advantage of solving name-face associations for multiple names at once is that the faces associated with one name may resolve ambiguities for other names.

We explore the middle ground between solving the complete name-face association problem, and analysis of only the initial result set. We do so by starting with a set of news stories found by querying the captions with a name, which we refer to as the query set. We then extend the query set by querying the database for names that appear frequently together with the queried person; we refer to these people as 'friends' of the queried person. We use the 'query expansion' —the set of faces found in images with friends appearing in the caption— to obtain a notion of whom we are *not* looking for.

We apply this idea to a generative mixture model to filter the text-based results, as well as to a linear discriminant method to filter the initial results. We find that query expansion gives dramatic improvements in the failure mode of existing work: cases where the queried person accounts for less than 40% of the query set. We find that both the generative and discriminative method benefit from query expansion, and that the discriminative model performs best albeit being computationally more demanding.

In the next section we discuss related work and the idea of query expansion in more detail. In section Section 3 we then describe the baseline methods; their counterparts using query expansion follow in Section 4. We present our experimental results in Section 5, and our conclusions in Section 6.

2 Related Work on Finding People in News Images

Previous work on finding faces of specific people analyzed only the faces returned from a text-based query over the captions. The assumption underlying these methods is that the query set consists of a large group of highly similar faces of the queried person, plus faces of many other people appearing each just a few times. The goal is thus to find a single coherent compact cluster in a space that also contains many outliers. A graph-based method was proposed in [12]: nodes represent faces in the query set, and edges encode similarity between faces. The faces in the subset of nodes with maximum density are returned as the faces representing the queried person. The density of a subset of nodes is defined as the number of edges that do not exit the subset divided by the number of nodes in the subset. In [13] we compared this method to one based on a Gaussian mixture model. One Gaussian is fitted on all the faces in the query set, while a second Gaussian is fitted to a subset of faces that are believed to represent the queried person. To deal with the fact that it is not known which faces belonging to the queried person, the EM algorithm is used to fit the model, under the constraint that at most one face per image can depict the queried person.

We found performance of these methods to deteriorate strongly as the frequency of the queried person in the result set drops below about 40%, contradicting their underlying assumption. In this case the faces of the queried person are obscured by many faces of other people, some of which will probably also occur quite often, as we expect strong correlations on which people co-occur on news images.

Methods that aim at solving the complete name-face association problem are motivated by the idea that finding name-face associations for all names jointly can resolve

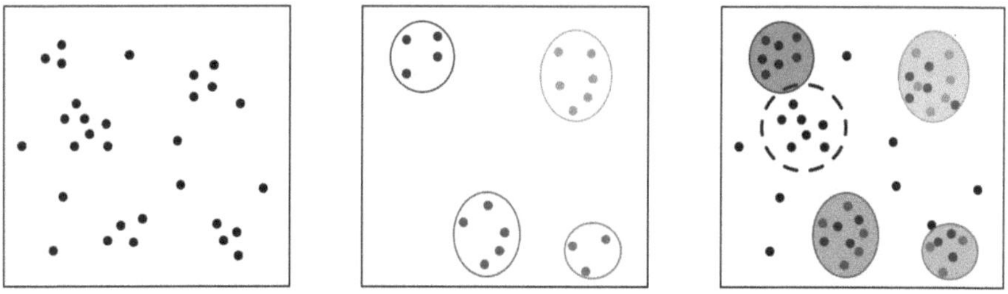

Fig. 2. Schematic illustration of how friends help to find people. The distribution of face features in the query-set (left), the query expansion with color coded faces of four people that co-occur with the queried person (middle), and how models of these people help to identify which faces in the query set are not the queried person (right).

ambiguities that would otherwise arise [14]. They work by applying a named entity detector on the captions, and a face detector on the images, and then finding the associations. The associations are constrained to assign at most one name to each face, and vice-versa. Thus if one face matches well with the model associated with a particular name, this provides evidence that the other faces should be associated with other names. In this manner documents that do not contain the name of the queried person in the caption also help to solve the query, as they provide data to learn the models for other names that co-occur with the queried person.

However, solving the complete name-face association problem is complex: captions may contain multiple names, and images multiple faces, which together have a combinatorial number of possible associations. Allowing at most one name to be associated with each face and vice-versa, a document with 6 faces and 7 names already yields 37633 possible associations. In our implementation of such a system, that relies solely on detected faces and names and does not use an additional language model, we find it to produce results that are worse than those obtained with a system that only tries to find the queried person: 54% precision with 61% recall, compared to 69% and 77% precision with 75% recall reported in [12] and [13] respectively. The worse performance of the full name-face associations is due to local maxima during learning and insufficient training data to make reliable estimates of the models for people appearing infrequently.

In this paper we aim to combine the best aspects of both approaches: (i) We *do* extend the initial query set, but only with a limited set of *relevant* stories, and model only people that co-appear relatively often. (ii) We *do not* solve the complex full name-face association problem –not even over the query expansion– but only determine in each image which (if any) face represents the queried person. By querying the database for names that frequently appear together with the queried name we collect faces that help understand whom we are *not* looking for. For example, suppose that in captions for the query *Tony Blair* the names *George Bush* and *Gordon Brown* occur often. By querying for faces of *George Bush* and *Gordon Brown* we can then rule out faces in the query set that are very similar to many faces returned for *George Bush* or *Gordon Brown*. See Figure 2 for a schematic illustration of the idea.

Our use of query expansion differs from other work using query expansion for document and image retrieval [15,16], where query expansion is used to re-query the database to obtain more similar documents or images. On the contrary, we use query expansion to obtain faces of related, but different, people to help us identify the queried person by contrasting him against these.

3 Basic Methods to Filter Text-Based Query Results

Below we describe our two baseline methods to filter the query set obatined using text-based search. Both methods allow at most one face from each image to be identified as the queried person. While sometimes violated, several cases exist where a person is photographed against a background that contains the face of the person on eg. a poster, this constraint improves the results considerably.

3.1 Face Filtering Using a Gaussian Mixture Model

A Gaussian mixture model has been previously used to solve the complete name-face association task by framing the problem as a constrained clustering problem [14]. Each cluster represents a name and the facial features of the person are modeled by a single Gaussian; an additional Gaussian models faces not assigned to any name. In the case of single-person querying, this approach results in a Gaussian mixture with just two mixture components [13]: one foreground model representing the queried person, and one generic face model. Below, we present the model in detail.

For each document in the query set we introduce an (unknown) assignment variable γ to represent which, if any, face in the image belongs to the queried person. Clearly, for a document with F face detections in the image, the number of assignments is just $(F{+}1)$: selecting one of the faces, or none ($\gamma = 0$). We then define a mixture model over the features of the detected faces $\mathcal{F} = \{f_1, \ldots, f_F\}$, marginalizing over the assignment variable γ. We use a prior over γ which is uniform over all non-zero assignments, i.e. $p(\gamma = 0) = \pi$ and $p(\gamma) = (1 - \pi)/F$ for $i \neq 0$.

$$p(\mathcal{F}) = \sum_{\gamma=0}^{F} p(\gamma)p(\mathcal{F}|\gamma), \tag{1}$$

$$p(\mathcal{F}|\gamma) = \prod_{i=1}^{F} p(f_i|\gamma), \tag{2}$$

$$p(f_i|\gamma) = \begin{cases} p_{\mathrm{BG}}(f_i) = \mathcal{N}(f_i; \mu_{\mathrm{BG}}, \Sigma_{\mathrm{BG}}) & \text{if } \gamma \neq i \\ p_{\mathrm{FG}}(f_i) = \mathcal{N}(f_i; \mu_{\mathrm{FG}}, \Sigma_{\mathrm{FG}}) & \text{if } \gamma = i \end{cases} \tag{3}$$

The parameters of the generic face model p_{BG} are set to the mean and variance of the faces in the query set, and we use the EM algorithm to obtain a maximum likelihood estimate the remaining parameters $\{\pi, \mu_{\mathrm{FG}}, \Sigma_{\mathrm{FG}}\}$. We initialize the EM algorithm in the E-step by using uniform responsibilities over the assignments: in this way faces in documents with few faces are relatively more important than faces in documents with many faces. After parameter optimization, we use the assignment maximizing $p(\gamma|\mathcal{F})$ to determine which, if any, face represents the queried person.

3.2 Face Filtering Using Linear Discriminant Analysis

Our motivation for the linear discriminant method, is to improve over the Gaussian mixture method, without resorting to a method based on pairwise similarities as in [12,13], which would be computationally costly when the query set contains many faces. We chose to use sparse multinomial logistic regression (SMLR) [17] since we are using high-dimensional face features (1664 dimensional, see Section 5). SMLR is a multi-class classifier, which is useful when considering query expansions in the next section. SMLR has been reported to perform equal or better than Support Vector Machines and Relevance Vector Machines on different benchmark classification problems [17,18].

Let f denote features, and $y \in \{1, \ldots, C\}$ to denote a class label, then the conditional probability of y given f is defined as as a soft-max over linear score functions:

$$p(y = c|f) = \frac{\exp(w_c^\top f)}{\sum_{c'=1}^{C} \exp(w_{c'}^\top f)}. \tag{4}$$

The likelihood is combined with a sparsity promoting Laplace prior over the parameters: $p(w) \propto \exp(-\lambda \|w\|_1)$, where $\|\cdot\|_1$ denotes the L_1 norm, and λ is set by cross-validation. Note that we can fix $w_1 = 0$ without loss of generality.

To learn the weight vectors we explore two strategies to use the set of noisy positive examples in the query set. The first is to simply treat all faces in the query set as positive examples ($y = 2$), and to use a random sample of faces from the database as negative examples ($y = 1$). The second strategy takes into account that each image in the query may contain at most one face of the queried person. To do this we learn the classifier iteratively, starting with all faces in query set as positive examples, and at each stage transferring the faces that are least likely to be the queried person from the positive to the negative set. At each iteration we transfer a fixed number of faces, which could involve several faces from a document as long as there remains at least one face from each document in the positive set. The last condition is necessary to avoid that a trivial classifier will be learned that classifies all faces as negative.

Once the classifier weights have been learned, we score the $(F+1)$ assignments with the log-probability of the corresponding classifier responses, e.g. for $\gamma = 1$ the score would be $\ln p(y_1 = 2|f_1) + \sum_{i=2}^{F} \ln p(y_i = 1|f_i)$.

4 Using Query Expansion to Enhance Basic Filtering Methods

In this section we consider how query expansion can improve the results of the methods presented in the previous section. The original query is expanded as follows. First, we apply a named entity detector to the captions of the documents found using the initial text-based query, and count the number of occurrences of all names. The 15 most frequently detected names, we refer to these names as 'friends', will be used to query the database again. For each friend, we use only the documents in which the queried person does not appear in the caption, and if there are less than five such documents we discard the documents for the friend all together. This condition ensures a minimum amount of data from which we learn the model for each friend. We consider two ways to exploit the query expansion: (i) using the faces from the query expansion as a batch, and (ii) using the described basic methods to process the faces returned on the query for each friend.

4.1 Query Expansion for Gaussian Mixture Filtering

The basic Gaussian mixture method uses only the faces in the query set. The first way to use the query expansion keeps the model as it is, but fits the 'background' Gaussian to the query expansion instead of the query set. In this manner the background Gaussian will be biased towards the friends of the queried person, and the foreground Gaussian is less likely to lock into one of the friends.

To use the query expansion in a more precise manner, we first apply the basic model to the query results of each friend. We then combine the foreground Gaussian of each friend with a generic background Gaussian in a mixture to form a more detailed query-specific background model. Thus, the difference with the basic model is that we set

$$p_{\mathrm{BG}}(f) = \frac{1}{N+1} \sum_{n=0}^{N} \mathcal{N}\left(f; \mu_n, \Sigma_n\right), \tag{5}$$

where $n = 0$ refers to the generic background model, and $n = 1, \ldots, N$ refer to the foreground models learned for friends 1 up to N. As before, we then fix p_{BG} and run the EM algorithm to find p_{FG} and the most likely assignment γ in each document.

4.2 Query Expansion for Linear Discriminant Filtering

The basic linear discriminant method uses a random sample from the database as negative examples to discriminate from the (noisy) positive examples in the query set. Our first way to use the query expansion is to replace this random sample with faces found when querying for friends. When there are not enough faces in the expansion (we require at least as many faces as the dimensionality to avoid trivial separation of the classes), we use additional randomly selected faces.

To use query expansion in a more precise manner we first use the basic linear discriminant method to learn a classifier to separate each friend $n = 1, \ldots, N$ from randomly selected faces, and similarly for the queried person. Let w_n denote the weight vectors learned for the friends, w_{FG} the weight vector learned for the queried person, and $w_{\mathrm{BG}} = 0$ the weight vector for the negative class. We then combine these weight vectors into a new multi-class classifier. Using this new multi-class classifier the probability that a face corresponds to the queried person is given by

$$p(y = \mathrm{FG}|f) = \frac{\exp(w_{\mathrm{FG}}^{\top} f)}{\exp(w_{\mathrm{FG}}^{\top} f) + \exp(w_{\mathrm{BG}}^{\top} f) + \sum_{n=1}^{N} \exp(w_n^{\top} f)}, \tag{6}$$

and similarly for the friends ($y \in \{1, \ldots, N\}$), and the background label ($y = \mathrm{BG}$). Note that in this manner the likelihood ratio between the background label and the queried person $p(y = \mathrm{BG}|f)/p(y = \mathrm{FG}|f)$ remains unchanged, and the same holds for the likelihood ratio of the background versus a friend. But, as desired, faces that are likely to be a friend get a lower probability to represent the queried person.

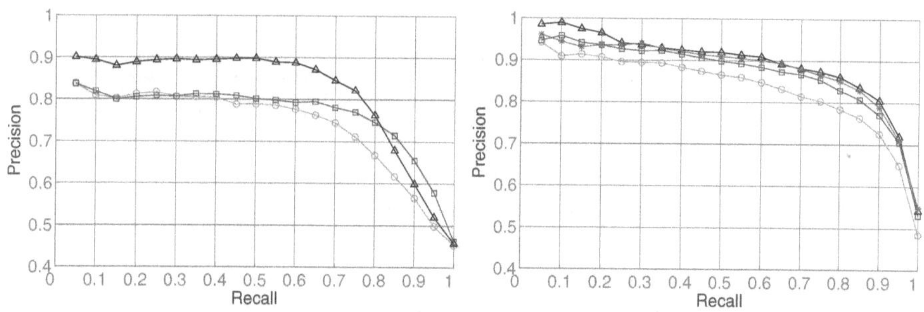

Fig. 3. Left panel: generative model using the query set to fit the background model (green, circles), the query expansion to fit the background (red, squares), and modeling each friend (blue, triangles). Right Panel: SMLR using the query set as positive examples (green circles), with iterative clean-up of the query set (red squares), using query expansion in the latter (blue triangles), and with the multi-class model (purple stars).

5 Experimental Results

We present results of two experiments. In the first experiment we evaluate performance of our different methods with and without query expansion using a database of captioned images downloaded from *Yahoo! News*. In the second experiment we consider whether the learned models are also useful to classify faces in the absence of captions.

5.1 Performance Evaluation of People Search Based on Captions

Data Set and Pre-processing Pipeline. We evaluate the results of searches performed for the same 23 people also used in [12,13] over a collection of about 15.000 captioned news photographs downloaded from *Yahoo! News* in 2002–2003.[1] For each query we manually labeled all faces detected in the images returned on the text-based query.

On the image side, we used an off-the-shelf face detector [19], and facial feature detector [9]. The facial feature detector locates nine points on the face, and another four are determined from them. Before extracting features, each face image is filtered to suppress noise and to compensate for low-frequency lighting variations using a difference-of-Gaussian filter that has been reported to yield significantly better results for a collection of features [20]. For each of the 13 points on the face we calculate a 128 dimensional SIFT descriptor [21], yielding a $13 \times 128 = 1664$ feature vector for each face. On the caption side, we use also use an of-the-shelf named entity detector [22] to find the 'friends' for the query extension. The initial text-based query is solved by a simple string matching against all captions.

Experimental Setup. We test the two basic methods described in Section 3 in three settings: (i) not using query expansion, (ii) using the query expansion as a batch, and (iii) using the query expansion while including processing of it using the basic method. We

[1] We would like to thank Tamara Berg for sharing the data set to us.

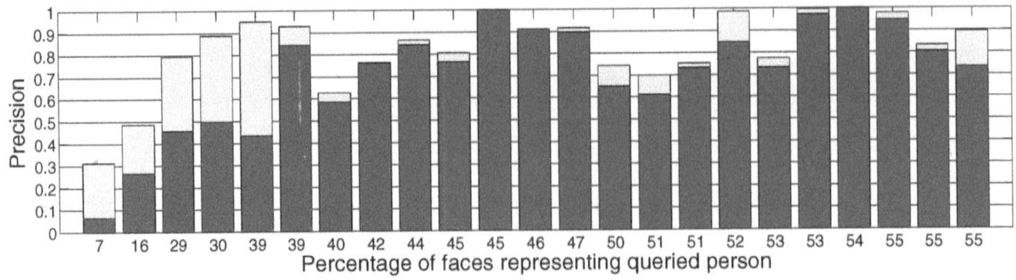

Fig. 4. Precision for the 23 queries obtained with the generative model, when using the query to fit the background model (green bars), and precision increase when using query expansion with a Gaussian for each friend (yellow bars). Queries are sorted by the percentage of faces in the query set that represent the queried person (see hor. axis).

evaluate performance in terms of precision and recall measured over the faces detected in images returned on the text-based query. Precision-recall curves are averages over the curves obtained for individual queries: at a given level of recall we average over the 23 precisions.

Experimental Results. The precision-recall curves obtained using the generative model are presented in the left panel of Figure 3. When using a single Gaussian to model the background, fitting it to a random set of faces from the database or to the faces in the query expansion leads to very similar results. For clarity we omitted the result using random faces from the figure. An overview of precisions obtained for different people at 75% recall on average is is given in Figure 4. See the figure captions for more detail.

From the results we can draw the following conclusions. First, if we use a single Gaussian to model the background, then using the query set is a suboptimal choice: better performance is obtained when it is fitted to a random set of faces or to the query expansion. Second, the query expansion leads to substantial performance increases when we fit a Gaussian to each friend using the basic method: we observe three cases where precision is increased by more than 30%. In accordance with our goal, improvements are largest for people that have a low frequency among the faces in the query set.

The right panel of Figure 3 shows results obtained using the linear discriminant method, see the figure caption for details. Experimentally, we found values of $\lambda \in [5, 15]$ to give good results. This leads to sparse classifiers with on average 102 non-zero values in each weight vector (out of 1664) in the multi-class model learned on the expansions. The results lead to the following conclusions. First, iterative transfer of faces from the positive to the negative set leads to substantially better results. Second, query expansion improves results further by up to 4%. Third, the multi-class use of the query expansion does not lead to further improvements, in fact for many levels of recall it leads to a small drop in performance.

In Figure 5 we show the result obtained using the multi-class model for a query for Vladimir Putin. For a collection of faces from the query set the break-down over different classes is shown. Note that not all expansion classes are pure, in the sense that only faces of one 'friend' are associated with them. However it is not critical for good retrieval performance how faces of other people are distributed over the BG class, and

Fig. 5. Faces in the query set for Vladimir Putin (78% precision, 91% recall), assigned to the FG class (left), the BG class (middle), and classes associated with 'friends': Silvio Berlusconi, Hu Jintao, Sergei Ivanov, Jacques Chirac (from top to bottom, right)

classes associated with other names: the only requirement is to make a good separation between the queried person and other people, not to recognize those related people.

We see that for both methods best results are obtained when using query expansion. We find the number of friends in the expansion not to have a significant impact on results once more than five are used.

In Figure 6 we show a summary of the best results obtained with the proposed methods, as well as the results obtained with the graph based methods of [12,13]. From the latter we consider the method using k-Nearest Neighbour graphs based on the number of matches between the same facial feature detections used by our generative and discriminative methods. We see that our discriminative method clearly outperforms the graph-based methods, and that our generative method with query expansion performs better than the graph-based methods up to about 85% recall. The results of the graph-based methods are slightly different from those reported in [13] because before running these methods again, we removed several duplicate images from the data set and corrected some errors in the ground truth labeling.

Comparing the results obtained with the generative and discriminative method we note an interesting difference in the condition under which query expansion becomes most useful. For the former we need to model each friend in the expansion separately,

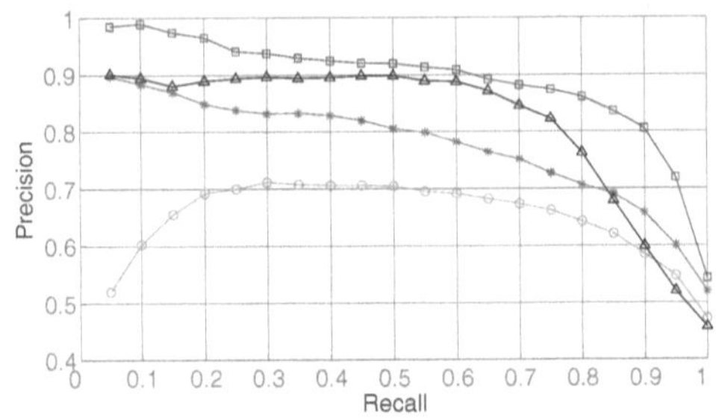

Fig. 6. An overview of the best results obtained with our discriminative model (red squares), our generative model (blue triangles), and the graph-based methods from Guillaumin et al. [13] (purple stars), and Ozkan and Duygulu [12] (green circles)

and little is gained by fitting one simple Gaussian model to all the expansion data. For the latter the situation is reversed: using all faces in the query expansion as a bulk leads to a substantial gain, while modelling each friend separately does lead to further improvement. This difference can be understood as follows: a single Gaussian is unlikely to be a precise generative model as the expansion will contain faces of many different people, and it is unlikely that this will separate the queried person better than using a generic background model. Although the discriminative model is also limited, in this case we explicitly search for a linear separation of the query set and the expansion, rather than hope that the generative model will lead to good separation.

A second interesting observation is that without query expansion the discriminative model performs substantially better than the generative one (increases in the range $10-20\%$). However, when query expansion is used this difference becomes much smaller: less than 5% precision for recall in the range $25 - 75\%$. The generative model may therefore be preferred from a practical point of view when operating at these recall levels, as it is much faster to train using an efficient EM algorithm.

5.2 Classifying Faces without Accompanying Captions

Once a model has been learned from a database of captioned news images, we can use the model also to find the person in other images even if they do not have a caption. To test this we have learned a linear discriminant model for ten people from the *Yahoo! News* database, and then apply the model to images of the *Labeled Faces in the Wild* database [23]. This database contains about 13.233 images of people cropped to head-and-shoulders, the faces are labeled with names, but captions are not available. The database is also gathered from material of Berg et.al. [14], but includes a ground-truth labeling of all faces, and allows us to test our models on faces that were not in the result of a text-based query. To allow unambiguous use of the labels, we selected the 11.948 images for which our face detector found exactly one face; in 289 images no face was detected, and in 996 more than one face was detected.

We use trained discriminative models to classify all faces, and then calculated precision at three levels of recall, averaged over the ten persons. As before, the precisions

Fig. 7. For each query the first ten images returned from *Labeled Faces in the Wild*

of the multi-class and two-class models were found to be very similar: here differences were smaller than 1%. The precision among the first 10 faces was 99% (1 error), degrading to 92% at 25 faces, and to 67% at 100% recall. These precisions are comparable to those measured on the query set from the *Yahoo! News* database. This is encouraging because in the current setting many more negative images are present in the test set. Figure 7 shows the first ten images for these ten queries.

6 Conclusions

We have shown how query expansion leads to improved results when searching for people in captioned news images. Although queries for which text-based search in the caption leads to a low fraction of relevant faces remain difficult, we have made significant progress in these cases, boosting precision by 20% up to 50% for the generative model in the five most difficult cases.

Query expansion is particularly useful for the Gaussian mixture approach to filter the initial query result. Without query expansion the discriminative method clearly outperformed the generative one ($\sim$ 10% difference in precision), but these differences become smaller using query expansion ($\sim$ 5% differences in precision).

We achieve performance levels that are significantly higher than those obtained in [12,13] using similarity-based methods. We obtain our results with methods that do not require calculation of pairwise similarities and are therefore much faster when many faces are processed. Our best method (SMLR + expansion) obtains a precision of 87.4% (83.6%) for a recall of 75% (85%), while the best previously reported result on these queries only reaches 77.6% (73.0%) for the same recall values. It is not obvious how query expansions can be used in a similarity-based approach, as the graph-density score takes into account only the similarities between faces selected for the queried person, and not those between selected and non-selected faces.

In our final experiment we have shown that the learned models also perform well when classifying faces without captions. This suggests that once a model has been learned from caption-based supervision, we can also identify these people in images where they are not named in the caption.

A demo of the generative model with query expansion on the *Yahoo! News* data set is available at `http://lear.inrialpes.fr/~verbeek/facefinder`.

References

1. Smeulders, A., Worring, M., Santini, S., Gupta, A., Jain, R.: Content-based image retrieval at the end of the early years. IEEE Transactions on Pattern Analysis and Machine Intelligence 22(12), 1349–1380 (2000)
2. Ponce, J., Berg, T., Everingham, M., Forsyth, D., Hebert, M., Lazebnik, S., Marszalek, M., Schmid, C., Russell, B., Torralba, A., Williams, C., Zhang, J., Zisserman, A.: Selected Proceedings of the first PASCAL Challenges Workshop. In: The 2005 PASCAL Visual Object Classes Challenge. LNCS (LNAI). Springer, Heidelberg (2006)
3. Lazebnik, S., Schmid, C., Ponce, J.: Affine-invariant local descriptors and neighborhood statistics for texture recognition. In: Proceedings of the IEEE International Conference on Computer Vision, pp. 649–655 (2003)
4. Verbeek, J., Triggs, B.: Region classification with Markov field aspect models. In: Proceedings of the IEEE Conference on Computer Vision and Pattern Recognition (2007)
5. Barnard, K., Duygulu, P., Forsyth, D., de Freitas, N., Blei, D., Jordan, M.: Matching words and pictures. Journal of Machine Learning Research 3, 1107–1135 (2003)
6. Grangier, D., Monay, F., Bengio, S.: A discriminative approach for the retrieval of images from text queries. In: Proceedings of the European Conference on Machine Learning, pp. 162–173 (2006)
7. Bressan, M., Csurka, G., Hoppenot, Y., Renders, J.M.: Travel blog assistant system. In: Proceedings of the International Conference on Computer Vision Theory and Applications (2008)
8. Jain, V., Learned-Miller, E., McCallum, A.: People-LDA: Anchoring topics to people using face recognition. In: Proceedings of the IEEE International Conference on Computer Vision (2007)

9. Everingham, M., Sivic, J., Zisserman, A.: Hello! My name is... Buffy - automatic naming of characters in TV video. In: Proceedings of the British Machine Vision Conference, pp. 889–908 (2006)
10. Laptev, I., Marszałek, M., Schmid, C., Rozenfeld, B.: Learning realistic human actions from movies. In: Proceedings of the IEEE Conference on Computer Vision and Pattern Recognition (2008)
11. Marcel, S., Abbet, P., Guillemot, M.: Google portrait. Technical Report IDIAP-COM-07-07, IDIAP (2007)
12. Ozkan, D., Duygulu, P.: A graph based approach for naming faces in news photos. In: Proceedings of the IEEE Conference on Computer Vision and Pattern Recognition, pp. 1477–1482 (2006)
13. Guillaumin, M., Mensink, T., Verbeek, J., Schmid, C.: Automatic face naming with caption-based supervision. In: Proceedings of the IEEE Conference on Computer Vision and Pattern Recognition (2008)
14. Berg, T., Berg, A., Edwards, J., Maire, M., White, R., Teh, Y., Learned-Miller, E., Forsyth, D.: Names and faces in the news. In: Proceedings of the IEEE Conference on Computer Vision and Pattern Recognition, pp. 848–854 (2004)
15. Buckley, C., Salton, G., Allan, J., Singhal, A.: Automatic query expansion using SMART: TREC 3. In: Proceedings of the Text Retrieval Conference, pp. 69–80 (1995)
16. Chum, O., Philbin, J., Sivic, J., Isard, M., Zisserman, A.: Total recall: Automatic query expansion with a generative feature model for object retrieval. In: Proceedings of the IEEE International Conference on Computer Vision (2007)
17. Krishnapuram, B., Carin, L., Figueiredo, M., Hartemink, A.: Sparse multinomial logistic regression: fast algorithms and generalization bounds. IEEE Transactions on Pattern Analysis and Machine Intelligence 27(6), 957–968 (2005)
18. Perronnin, F., Dance, C.: Fisher kernels on visual vocabularies for image categorization. In: Proceedings of the IEEE Conference on Computer Vision and Pattern Recognition (2007)
19. Mikolajczyk, K., Schmid, C., Zisserman, A.: Human detection based on a probabilistic assembly of robust part detectors. In: Proceedings of the European Conference on Computer Vision, pp. 69–81 (2004)
20. Tan, X., Triggs, B.: Enhanced local texture feature sets for face recognition under difficult lighting conditions. In: Zhou, S.K., Zhao, W., Tang, X., Gong, S. (eds.) AMFG 2007. LNCS, vol. 4778, pp. 168–182. Springer, Heidelberg (2007)
21. Lowe, D.: Distinctive image features from scale-invariant keypoints. International Journal of Computer Vision 60(2), 91–110 (2004)
22. Deschacht, K., Moens, M.: Efficient hierarchical entity classification using conditional random fields. In: Proceedings of Workshop on Ontology Learning and Population (2006)
23. Huang, G., Ramesh, M., Berg, T., Learned-Miller, E.: Labeled faces in the wild: a database for studying face recognition in unconstrained environments. Technical Report 07-49, University of Massachusetts, Amherst (2007)

Fast Automatic Single-View 3-d Reconstruction of Urban Scenes

Olga Barinova[1],Vadim Konushin[2], Anton Yakubenko[1], KeeChang Lee[3], Hwasup Lim[3], and Anton Konushin[1]

[1] Moscow State University, Department of Computational Mathematics and Cybernetics, Graphics & Media Lab
{obarinova,toh,ktosh}@graphics.cs.msu.ru
[2] The Keldysh Institute of Applied Mathematic Russian Academy of Sciences
vadim@graphics.cs.msu.ru
[3] Samsung Advanced Institute of Technology
{lkc.lee,hwasup.lim}@samsung.com

Abstract. We consider the problem of estimating 3-d structure from a single still image of an outdoor urban scene. Our goal is to efficiently create 3-d models which are visually pleasant. We chose an appropriate 3-d model structure and formulate the task of 3-d reconstruction as model fitting problem. Our 3-d models are composed of a number of vertical walls and a ground plane, where ground-vertical boundary is a continuous polyline. We achieve computational efficiency by special preprocessing together with stepwise search of 3-d model parameters dividing the problem into two smaller sub-problems on chain graphs. The use of Conditional Random Field models for both problems allows to various cues. We infer orientation of vertical walls of 3-d model vanishing points.

1 Introduction

Inferring the 3-d structure of pictured scene from single image is a challenging problem for current computer vision systems. Whereas understanding geometry of the scene for a human is quite easy, this problem remains open for a computer due to intrinsic ambiguity introduced by perspective projection. However, for man-made environments such as urban scenes where certain structural regularities are present, introducing additional constraints on the scene geometry might help to disambiguate the problem.

Most methods for estimation of 3-d models from a single image make strong assumptions about the scene geometry. For example, Delage, Lee and Ng [3][4] considered indoor images and classified the image into ground and vertical walls to produce a simple visually pleasing 3-d model. In [1][2] Hoiem, Efros and Hebert proposed to use similar constraints to model outdoor scenes. They classified image regions into geometric classes and then used classification results to create coarse pop-up type scene model composed of ground, walls and sky. Criminisi, Reid and Zisserman [5] determined affine structure of the image using vanishing points with application in 3-d graphical modeling.

D. Forsyth, P. Torr, and A. Zisserman (Eds.): ECCV 2008, Part II, LNCS 5303, pp. 100–113, 2008.

Fig. 1. Outline of the proposed algorithm (a) Source image (b) Preprocessed image: tilt correction, horizon detection and vanishing point estimation (c) Different positions of ground-vertical border along the vertical axis (d) Resulting model

In contrast, Saxena, Sun and Ng [7][8][9] made no explicit assumptions about the structure of the scene. They inferred both the 3-d orientation and location of the small planar regions in the image using a Markov Random Field by learning the relation between the image features and the location/orientation. MAP estimate in their model is efficiently performed by solving a linear program.

Our goal is to automatically create visually pleasant 3-d models. We choose a subset of the images (urban scenes) and fix an appropriate structure of 3-d model that allows to formulate the task of 3-d reconsruction as model fitting problem.

The structure of our 3-d models is as follows. We assume that the environment is composed of a flat ground with vertical walls where ground-vertical boundary is a continuous polyline (see Fig. 2). Each vertical wall of 3-d model corresponds to the building wall in the scene. Structural regularities of man-made environments allow tilt-correction of an image, horizon and vanishing point estimation, that provides information aboout scene geometry. Geometric structure of our 3-d models resembles the one used in [1][2] and [3][4], but in contrast to these works our 3-d model structure is rigidly fixed and can be inferred explicitly.

Assumption about continuity of ground-vertical boundary being realistic for most of urban scenes makes resulting estimates more stable and reduces the dimension of the search space. Configuration of ground-vertical boundary determines geometry of 3-d model. Given an image of urban scene, our algorithm automatically infers parameters of polyline and produces 3-d model. The question of detecting ground/vertical "occlusion boundary" automatically has also been looked at in [6]. In contrast to this work we search for ground-vertical boundary in explicit form.

Another focus of this work is the efficiency of 3-d reconstruction algorithm, which becomes particularly important when processing images of reasonable resolution. Using probabilistic framework we suggest stepwise search of ground-vertical boundary parameters. Model fitting problem is divided into two smaller sub-problems on chain graphs. We construct Conditional Random Field models incorporating various types of information (appearance, geometric properties

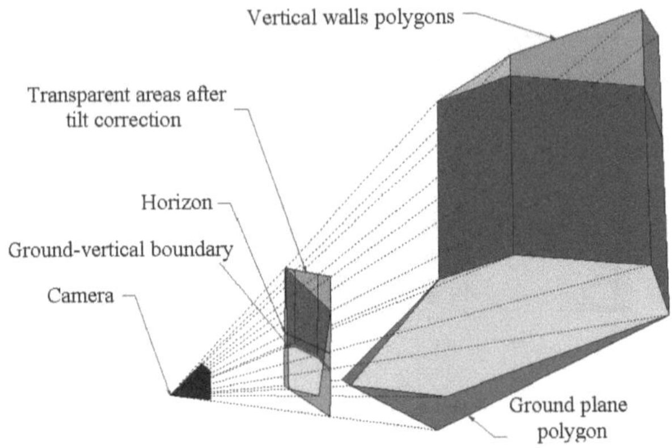

Fig. 2. 3-d model structure 3-d model is composed of a number of vertical walls and a ground plane

and context) for both problems. Inference in each model is highly efficient due to simplicity of graph structure. An outline of our algorithm is presented in Fig. 1.

The paper is structured as follows. In the next section we describe special preprocessing of an image which includes tilt-correction and vanishing points estimation. Section 3 explains how the high-level model for building-ground boundary is divided into two Conditional Random Field models. Sections 4 and 5 explain each model in details. Modeling procedure is described in Section 6. Experiments and conclusions are given in the final two sections.

2 Image Preprocessing

In this work image preprocessing is an essential part of the system. The outline of preprocessing is as follows. First, we apply Canny edge detection [10] to an image and then extract straight lines using a method described in [11]. Extracted lines are further used for tilt correction, horizon level estimation and vanishing points estimation. Preprocessing steps are demonstrated in Fig. 3.

2.1 Tilt Correction

In order to capture a building entirely, in one photo, a camera is usually pitched up. This leads to so called keystone effect (see Fig. 4). In keystoned images rectangles, like window frames, become trapezoids that are wider at the bottom (in case of pitching up). The vertical (with respect to the ground plane) lines become inclined. Non-zero camera roll angle additionally leads to the inclined horizon line.

The structure of our 3-d model requires building side borders to be vertical in the images. Given a tilted image, we need to compute a corresponding virtual

(a) (b) (c) (d)

Fig. 3. Preprocessing scheme (a) Source image (b) Tilt corrected image (c) Extracted lines after filtering (d) Horizon estimation and vanishing point estimation. Lines, which refer to the same vanishing point, are shown with the same colors

view with zero pitch and roll angles. For this task we use automatic tilt correction algorithm similar to the one, described in [12]. It uses the extracted line segments that correspond to a vertical vanishing point and searches for the best rotational homography, which transforms them into vertical lines. Due to the abundant presence of vertical lines in man-made environments this algorithm is quite robust. An example of tilt correction result is demonstrated in Fig. 3(b). The same homography transform is then applied to all straight line segments.

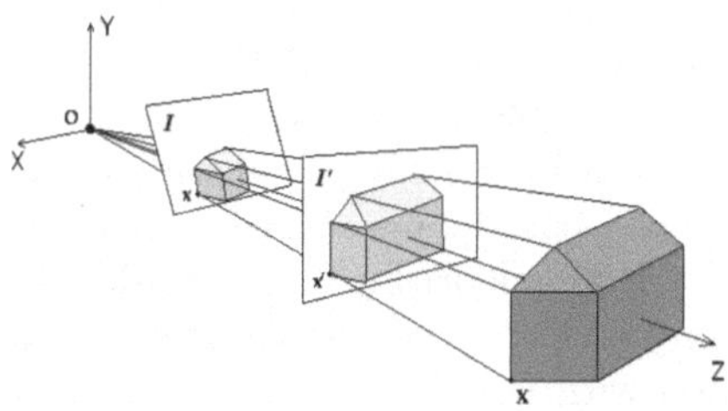

Fig. 4. Camera pitch and roll correction I' - a virtual view corresponding to a source image plane I

2.2 Horizon Level and Vanishing Points Estimation

Horizon level estimation. Due to perspective rules all vanishing points lie at the same straight line, which is called horizon. Our algorithm for horizon level estimation is based on the idea that horizon is transformed into level line after tilt correction and all vanishing points are assumed to have the same y-coordinate. We can take y-coordinate of any vanishing point for horizon level. In this work we choose the one with maximum support.

This vanishing point is estimated in a way similar to [5] using locally optimized RANSAC [14]. A pair of lines is chosen at random and their intersection is taken for vanishing point. Then all the lines that if proceeded lie close to this point are considered as inliers. After that vanishing point is recalculated as a closest point to all the inliers using least squares fit. Then set of inliers is changed according to estimated coordinates of vanishing point. This process is repeated several times, and at the end we choose a vanishing point with maximum support.

Lines filtering. Due to presence of various non-building objects in urban scenes a considerable amount of detected straight lines are not horizontal and do not refer to vanishing points. These lines are harmful for vanishing point estimation and should better be pruned out. In this work those lines are filtered out using machine learning like in [13].

Extracted line segments are classified by boosted trees trained on a set of manually labeled images into two categories: horizontal lines and clutter. We use color cues, geometric cues and gradient cues for classification. Geometric cues include coordinates of segment end points, its direction and length, and the fraction of segments that are almost parallel with it. Color cues include means and variances of color components and intensity calculated over the area around each line segment. We also use mean value of gradient along the line segment which characterizes edge strength. Fig. 3(c) shows lines left after filtering.

Vanishing points estimation. In this work vanishing point estimation is performed in two steps. First we perform lines grouping and then estimate vanishing point for each group of lines.

Lines grouping procedure is based on the fact that all lines which refer to the same vanishing point cross horizon level close to each other. Consider distribution of x-coordinates of crossings between horizon and line segments. Groups of parallel lines form the modes of this distribution. We seek for the modes of distribution using mean shift clustering [15].

Since y-coordinates of vanishing points are already known, we need to estimate only their x-coordinates. X-coordinate of one vanishing point for each group of lines is estimated with an algorithm quite similar to the one for horizon estimation. The difference is that we choose single line instead of a pair each time and take it's intersection with horizon level as a vanishing point. Lines marked as outliers by RANSAC are filtered out and are not used further for estimation orientation of the walls. Vanishing points and horizon level estimation are shown in Fig. 3(d).

3 Stagewise Inference of 3-d Model Parameters

In this work we assume that ground-vertical boundary is specified by a continuous polyline. Due to perspective constraints, each polyline segment passes through a vanishing point. Let's fix the left end of polyline (p_x^1, p_y^1), then a single polyline segment connects (p_x^1, p_y^1) with corresponding vanishing point (v_x^1, h). Let's denote the crossing of the first segment with the border between first and second walls by (p_x^2, p_y^2). Then single line segment passes through (p_x^2, p_y^2) and (v_x^2, h) etc. This implies that if x-coordinates of all polyline fractures and corresponding vanishing points are fixed then only a single polyline passes through (p_x^1, p_y^1) (see Fig. 5). Polyline is completely determined by a set af $2n + 1$ parameters $\{p_x^1, ..., p_x^n; v_x^1, ..., v_x^n; h; p_y^1\}$.

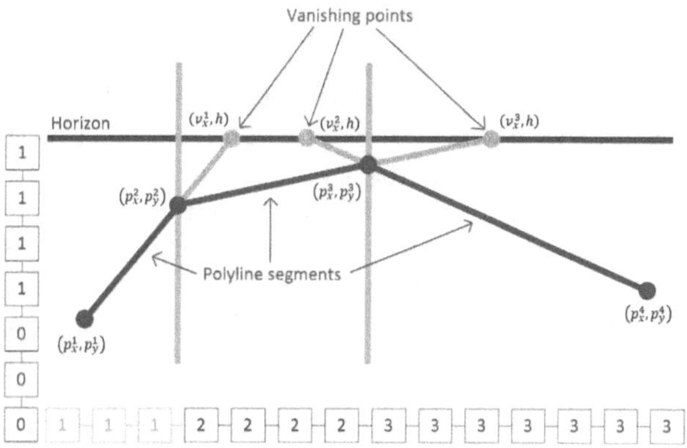

Fig. 5. Ground-vertical boundary Labeled chains represent results of polyline fractures estimation and polyline vertical positioning

Conditional probability formula allows dividing search for model parameters into two subproblems:

$P(p_x^1, ..., p_x^n; v_x^1, ..., v_x^n; h; p_y^1|I) =$

$$P(p_y^1|p_x^1, ..., p_x^n; v_x^1, ..., v_x^n; h; I)P(p_x^1, ..., p_x^n; v_x^1, ..., v_x^n; h|I) \quad (1)$$

The first subproblem includes estimation of x-coordinates of polyline's fractures $p_x^2, ..., p_x^n$ and vanishing points $v_x^1, ...v_x^n$ corresponding to each line segment. As far as vanishing points coordinates have been calculated at the stage of preprocessing, we just need to come up with correspondence between vanishing points and model walls. In this work for each wall that lies in between p_x^i and p_x^{i+1} we choose corresponding vanishing point that most of horizontal lines between p_x^i and p_x^{i+1} refer to. So the first problem evolves into estimation of $n-1$ x-coordinates of the polyline's fractures.

The second problem involves vertical positioning of the polyline with fixed $p_x^2, ..., p_x^{n-1}; v_x^1, ...v_x^n; h$ parameters. If any point of polyline is known to lie below (above) the horizon level then polyline is guaranteed to lie entirely below (above) the horizon level. This finding involves a constraint $p_y^1 < h$ that reduces a search space for p_y^1 to a segment below horizon level.

4 A Conditional Random Field Model for Fractures of Ground-Vertical Boundary

Partition of picture into vertical walls is represented by a labeled chain graph (see Fig. 5). Each node of the graph corresponds to a vertical band in the image. Label l of a graph node means that corresponding vertical band in a image belongs to model wall with index l. Total number of labels K is chosen beforehand. As long as we allow different walls to share the same vanishing point, adjacent walls can actually lie on the same plane. Thus parameter K can be viewed as maximum possible number of vertical walls in 3-d model.

We introduce additional constrains that gurantee resulting model walls to be connected. For each pair of adjacent graph nodes the label of the right node can not be less than the label of the left one, or $p(l_{i+1} < l_i) = 0$, label of right node can not differ from the label of the left node by more than 1, or $p(l_{i+1} - l_i > 1) = 0$, label of the left end of the chain equals 1, $p(l_1 \neq 1) = 0$, label of the right end of the chain equals K, $p(l_w \neq n) = 0$, where w is the total number of graph nodes and K is maximum number of vertical walls.

The entire set of such graph labelings can be described in feature space with the parameter set θ. Desired output consists of the parameter set θ, plus a labeling $\mathbf{l}$.

The use of a Conditional Random Field allows incorporation of appearance, geometric and context cues in a single unified model. The intuition behind the model used in this work is the following. Horizontal lines which lie on the same building wall all meet at a same vanishing point, while adjacent building walls are not parallel and refer to different vanishing points. Due to lighting conditions color of each building wall in the image has low variance while different building walls have different colors. The borders between adjacent building walls are characterized with local discontinuities in color and orientation of lines.

CRF model for vertical walls takes the following form:

$$P(\mathbf{l}|I,\theta) = \prod_{i=1}^{w} p(l_i|\theta) \times \prod_{i=1}^{w-1} p(l_i, l_{i+1}), \tag{2}$$

where w stands for a total number of graph nodes, unary potential $p(l_i, \theta)$ captures distribution of features inside every vertical wall, pairwise potential $p(l_i, l_{i+1})$ captures the "edginess" between two adjacent graph nodes; θ is a vector of internal parameters of CRF model.

We use Expectation Maximization algorithm (EM) for CRF likelihood maximization. The first step (the "E" step) that includes searching for the best labeling $\mathbf{l}$ given fixed cluster parameters θ can be effectively solved using dynamic programming. The second step (the "M" step) is to fix the labeling $\mathbf{l}$ and find the best clusters parameters.

We adopt a nonparametric approach [20] for calculating unary potential. Suppose we are given a kernel function $K(\mathbf{f}_i; \mathbf{f}_j)$ which measures the affinity between features $\mathbf{f}_i$ of graph node i and features $\mathbf{f}_j$ of graph node j. Value of $K(\mathbf{f}_i; \mathbf{f}_j)$ shows how much we believe the two points originated from the same process when all we know is their coordinates. Suppose that we have an approximation of an optimal labeling of the chain. Then a nonparametric density estimator for probability that graph node i belongs to a label l given set of N points drawn from the distribution of features inside l-th wall is calculated as:

$$P(l_i = l|\mathbf{f}_{l1}, .., \mathbf{f}_{lN}) = \frac{1}{N} \sum_{j=1}^{N} K(\mathbf{f}_i, \mathbf{f}_{lj}), \tag{3}$$

(a) (b) (c) (d)

Fig. 6. Training examples (a),(b) borders of 3-d model walls are manually marked for training the first model (c),(d) optimal positions of polyline along the vertical axis are manually marked for training second model, fixed polyline fractures are inferred from the first model

where $\mathbf{f}_{l1},..,\mathbf{f}_{lN}$ are features of reference graph node that belong to l-th wall. Vector $\theta = \{\mathbf{f}_{11},..,\mathbf{f}_{1N},...,\mathbf{f}_{K1},..,\mathbf{f}_{KN}\}$ contains features of graph nodes that belong to each label after each E-step of EM. Reference graph nodes in this work are chosen equally spaced inside each wall.

4.1 CRF Model Training

In this work we used a method based on piecewise training [17]. Piecewise training involves dividing the CRF model into pieces, each trained independently. So we need to find $K(\mathbf{f}_i; \mathbf{f}_j)$ and $p(l_i, l_{i+1})$ and use hand-labeled images of urban scenes for this purpose. The ground truth for learning included pictures where borders of vertical planes were marked manually (see Fig. 6(a)(b))).

Both $K(\mathbf{f}_i; \mathbf{f}_j)$ and $p(l_i, l_{i+1})$ are searched in exponential form:

$$K(\mathbf{f}_i; \mathbf{f}_j) = [1 + \exp(\sum_{m=1}^{M} \lambda^m |\mathbf{f}_i^{(m)} - \mathbf{f}_j^{(m)}|)]^{-1} \quad (4)$$

$$p(l_i, l_{i+1}) = [1 + \exp(-\sum_{m=1}^{M} \mu^m |\mathbf{f}_i^{(m)} - \mathbf{f}_{i+1}^{(m)}|)]^{-1}, \quad (5)$$

where $\mathbf{f}_i^{(m)}$ is the m-th component of feature vector, $\lambda = (\lambda^1, ..., \lambda^m)$ and $\mu = (\mu^1, ..., \mu^m)$ - coefficients of exponential models.

We use logistic regression for searching λ over a training set. Training set for learning $K(\mathbf{f}_i; \mathbf{f}_j)$ consists of pairs of features labeled with -1s if both corresponding graph nodes belong to the same wall and $+1$s if graph nodes belong to different walls. CRF pairwise term is learned from only positive examples, because of high class imbalance. Training set for learning $p(l_i, l_{i+1})$ consists of pairs of features which correspond to a pair of graph nodes around a marked edge. We use the first principal component obtained by PCA as coefficients in exponential model.

Due to requirements of computational efficiency of the algorithm, we use quite simple features $\mathbf{f}$ for estimation of ground-vertical boundary fractures. Consider

ith graph node, that corresponds to a vertical band of pixels between a_i and b_i. All features $\mathbf{f}_i$ are calculated over R rectangular regions limited by horizon level at the bottom with left borders $a_i - \delta_r, r = 1, ..., R$ and right borders $b_i + \delta_r, r = 1, ..., R$ respectively.

Feature vector includes simple color statistics and frequencies of lines referring to every vanishing point. To simplify calculation of these features we exploit integral images [22].

5 Vertical Positioning of Polyline

We formulate problem of polyline vertical positioning on a chain graph (see Fig. 4). Each graph node correspond to a fixed value of p_y^1 and in consequence a single boundary position along the vertical axis. The goal is to assign 1s to graph nodes that refer to vertical walls of 3-d model and 0s to graph nodes that refer to ground plane.

We again introduce Conditional Random Field, which incorporates different characteristics of ground-vertical boundary. The model takes into accout local color discontinuities around the boundary, similarity of ground color on different pictures, differences in color between each building and the ground below it, distinctive angles between of polyline's segments corresponding to visually-pleasant 3-d models.

Our CRF model for this problem has the following form:

$$P(\mathbf{l}|I) = \prod_{i=1}^{h} p(l_i) \times \prod_{i=1}^{h-1} p(l_i, l_{i+1}) \tag{6}$$

where h is a number of graph nodes, unary potential $p(l_i)$ captures color distribution inside region between polyline and horizon level and a region below polyline, $p(l_i, l_{i+1})$ captures local discontinuities around polyline position and angles between segments of polyline.

We again use piecewise training of CRF. In our experiments training set consists of pictures, where users manually marked ground truth positions of polyline which lead to visually pleasant models (see Fig. 6 (c)(d)). We exploit calibrated probabilistic outputs from boosting [19] for both unary and pairwise potentials. For example, unary potential takes a form:

$$p(l_i) = [1 + \exp(A_1 \cdot F(\tilde{\mathbf{f}}_i + B_1)]^{-1}, \tag{7}$$

where $\tilde{\mathbf{f}}_i$ is a feature vector of ith graph node, $F(.)$ is boosting output, A_1 and B_1 are Platt scaling parameters [19]. To avoid issues with imbalanced classes we use random undersampling [21].

Features $\tilde{\mathbf{f}}_i$ for unary potential include mean values of color and intensity around polyline. Features $\tilde{\tilde{\mathbf{f}}}_i$ for pairwise potential include mean and minimum angles between polyline segments and horizon, mean and minimum angles between adjacent polyline segments, mean, maximum and minimum y-cordinate of polyline, differences in color between regions below and above the polyline.

Again all color features are calculated over different rectangular regions around polyline segments and integral images are exploited for speed-up of calculation.

6 Modeling

Given the continuous polyline, horizon level in pixels and camera focal length in pixels it is possible to construct the 3-d scene. Our modeling scheme resembles the one, proposed in [23]. Scene is modeled as a horizontal polygon for the ground plane and a set of connected vertical trapeziform polygons for the rest of the scene. One vertical polygon is reconstructed for each straight segment of the polyline. The texturing is performed by straightforward projective mapping of tilt-corrected images onto reconstructed geometry. The idea of the algorithm is illustrated in Fig. 2.

Each vertex of the polyline in the image and bottom corners of the image should be projected onto the horizontal ground plane of the 3-d scene in order to determine the ground plane polygon. After that vertical planes with infinite heights can be also reconstructed. Top line in the image should be projected onto the reconstructed vertical planes to determine height of each vertical polygon. Since image is tilt corrected we can consider virtual views with zero pitch and roll angles, which mean that camera view vector is parallel to the ground plane and image center Y coordinate is equal to horizon level in the image. Z coordinate of the camera center affects only the scale of the resulting model and if it is specified exactly, we get a metric reconstruction as a result. Focal length in pixels can be calculated from EXIF data of JPEG file.

7 Experiments

Our image database is composed of 200 photos taken in a number of districts of Moscow and Seoul. Our test database also includes some photos taken in Moscow, Seoul and some images from Make3D and Automatic Photo Pop-up image bases [24][25]. We aim at outdoor urban scenes, so many of images of landscapes, indoor scenes, etc. from those databases are not included in our test set. These images are resized so that each image resolution is approximately 1.5 megapixel.

We have tested our algorithm, Make3D and Automatic Photo Pop-up on our test image database. Some of the test images and screenshots of the resulting models are shown in Fig. 7.

We did not use any heuristic measures for the quality of the models. The only criteria was the subjective opinion on the visual quality of the 3-d model. Resulting models were demonstrated to 4 users. Users were asked to answer whether the model had fine or unsatisfactory visual quality. The results of this visual comparison are shown in Table 1.

Due to our building-specific model, results of the proposed algorithm are visually more plausible, than the results of the compared algorithms. Hoiem el. al. and Make3D reported about 33% and 65% of visually pleasing models

Fig. 7. Comparison of the proposed algorithm with Automatic Photo Pop-up and Make3D

correspondingly, but due to the specific structure of the images used for testing the compared algorithms performed worse.

Another advantage of the proposed algorithm is its low computational time and consumed memory. In our experiments with 1.5 megapixel images, using PC

Table 1. Comparison of our algorithm with Automatic Photo Pop-up and Make3D algorthms on our Seoul test database

	Our algorithm	Automatic Photo Pop-up	Make3D
Fine results (%)	43%	36%	25%
Best among 3 algorithms (%)	52%	27%	21%

(a) (b) (c) (d)

Fig. 8. Bad results (a)-original image, (b)-resulting ground-vertical boundary, (c),(d)-screenshots of 3-d model. The first example demonstrates error in polyline vertical positioning, second example shows errors in estimation of polyline fractures, third example shows the case when ou assumption about continuity of ground-vertical boundary does not suit to scene geometry.

Pentium 2.8 GHz with 1 GB RAM, average modeling time was about 8 seconds, while required memory didnt exceed 30 MBs. Hoiem et. al. reported about 1.5 minutes processing time for 800x600 image on 2.13 GHz Athlon using unoptimized MATLAB code. It was not mentioned about memory consumption, but in our experiments Auto Pop Up algorithm required up to 120 MBs. Algorithm proposed in [7][8][9] is slower and consumes more memory, than Automatic Photo Pop-up.

In some cases the proposed algorithm does not produce good looking resulting 3-d models. The most often reasons are incorrectly estimated fractures of the polyline or errors in its estimated position along the vertical axis. Another source of error is in the assumption on the scene layout and the constraint of the continious polyline of the ground-vertical boundary. Examples of such errors are demonstrated in Fig. 8.

Our algorithm, trained on images taken in a small number of districts, showed promising results. We believe that having a larger and more diverse dataset of training images would improve the algorithm.

8 Conclusions

We presented an algorithm for inferring rough 3-d structure from a single image of urban scene. Unlike Saxena et al we impose constraints on the geometry assuming that scene is composed of ground and a number of vertical walls, which is the case for urban scenes. In contrast to Hoiem et al we search for ground-vertical boundary, which completely defines the model in explict form.

Employing greedy search of 3-d model parameters we divide the problem into two smaller sub-problems. For each subproblem we incorporate appearance, geometric properties and context into CRF model trained with supervised learning and then solve for geometry via MAP inference. We compared our method with those proposed by Saxena et al and Hoiem et al on our test set composed of pictures of Seoul, Moscow and pictures downloaded from internet. This comparison have shown that our method provides 3-d model with comparable or superior visual quality at significantly less computational expense.

Acknowledgements. Special thanks to Alexander Sobolev, Anna Kuzmishkina and Olga Senyukova for valuable comments and help during all stages of this research. Many thanks to Victor Gaganov for priceless advises and help in improving the quality of the paper text.

References

1. Hoiem, D., Efros, A., Hebert, M.: Automatic photo pop-up. In: Proc. of ACM SIGGRAPH (2005)
2. Hoiem, D., Efros, A., Hebert, M.: Geometric context from a single image. In: Proc. of ICCV (2005)
3. Delage, E., Lee, H., Ng, A.: Automatic single-image 3d reconstructions of indoor manhattan world scenes. In: Proc. of ISRR (2005)
4. Delage, E., Lee, H., Ng, A.: A dynamic bayesian network model for autonomous 3d reconstruction from a single indoor image. In: Proc. of CVPR (2006)
5. Criminisi, A., Reid, I., Zisserman, A.: Single view metrology. IJCV 40 (2000)
6. Hoiem, D., Sten, A.N., Efros, A.A., Hebert, M.: Recovering Occlusion Boundaries from a Single Image. In: Proc. of ICCV (2007)
7. Saxena, A., Sun, M., Ng, A.: Learning 3-D Scene Structure from a Single Still Image. In: Proc. of ICCV workshop on 3D representation for Recognition (2007)
8. Saxena, A., Chung, S., Ng, A.: Depth Reconstruction from a Single Still Image. In: IJCV (2007)
9. Saxena, A., Chung, S., Ng, A.: Learning Depth from Single Monocular Images. In: Proc. of NIPS, vol. 18 (2005)
10. Canny, J.: A Computational Approach To Edge Detection. PAMI 8 (1986)
11. Kosecka, J., Zhang, W.: Video Compass. In: Heyden, A., Sparr, G., Nielsen, M., Johansen, P. (eds.) ECCV 2002. LNCS, vol. 2353, pp. 476–491. Springer, Heidelberg (2002)
12. Vezhnevets, V., Konushin, A., Ignatenko, A.: Interactive image-based urban modeling. In: Proc. of PIA, pp. 63–68 (2007)

13. Barinova, O., Kuzmishkina, A., Vezhnevets, A., Vezhnevets, V.: Learning class specific edges for vanishing point estimation. In: Proc. of Graphicon, pp. 162–165 (2007)
14. Chum, O., Matas, J., Kittler, J.: Locally Optimized RANSAC. In: DAGM Symposium on Pattern Recognition, vol. 25, pp. 236–243 (2003)
15. Comaniciu, D., Meer, P.: Mean shift: A robust approach toward feature space analysis. IEEE Trans. on Pattern Analysis and Machine Intelligence 24, 5 (2002)
16. Lafferty, J., McCallum, A., Pereira, F.: Conditional random fields: Probabilistic models for segmenting and labeling sequence data. In: Proc. of ICML (2001)
17. Sutton, C., McCallum, A.: Piecewise training of undirected models. In: Proc. of UAI, vol. 21 (2005)
18. Shotton, J., Winn, J., Rother, C., Criminisi, A.: TextonBoost for Image Understanding: Multi-Class Object Recognition and Segmentation by Jointly Modeling Texture, Layout, and Context. In: IJCV (2007)
19. Niculescu-Mizil, A., Caruana, R.: Obtaining Calibrated Probabilities from Boosting. In: Proc. of UAI, pp. 413–420 (2005)
20. Wasserman, L.: All of Nonparametric Statistics. Springer, Heidelberg (2006)
21. Hulse, J., Khoshgoftaar, T., Napolitano, A.: Experimental perspectives on learning from imbalanced data. In: Proc. of ICML, pp. 935–942 (2007)
22. Viola, P., Jones, M.: Robust Real-time Object Detection. In: IJCV (2001)
23. Kang, H., Pyo, S., Anjyo, K., Shin, S.: Tour into the picture using a vanishing line and its extension to panoramic images. In: Proc. Eurographics, pp. 132–141 (2001)
24. `http://www.cs.uiuc.edu/homes/dhoiem/projects/data.html`
25. `http://make3d.stanford.edu/`

Fourier Analysis of the 2D Screened Poisson Equation for Gradient Domain Problems

Pravin Bhat[1], Brian Curless[1], Michael Cohen[1,2], and C. Lawrence Zitnick[1]

[1] University of Washington
[2] Microsoft Research

Abstract. We analyze the problem of reconstructing a 2D function that approximates a set of desired gradients and a data term. The combined data and gradient terms enable operations like modifying the gradients of an image while staying close to the original image. Starting with a variational formulation, we arrive at the "screened Poisson equation" known in physics. Analysis of this equation in the Fourier domain leads to a direct, exact, and efficient solution to the problem. Further analysis reveals the structure of the spatial filters that solve the 2D screened Poisson equation and shows gradient scaling to be a well-defined sharpen filter that generalizes Laplacian sharpening, which itself can be mapped to gradient domain filtering. Results using a DCT-based screened Poisson solver are demonstrated on several applications including image blending for panoramas, image sharpening, and de-blocking of compressed images.

1 Introduction

Accurately and efficiently recovering a depth map or an image from gradients has become a common problem in computer vision and computer graphics. In photometric stereo, for instance, one measures gradients (normals) to the surface and then "integrates" them to recover a depth map. In gradient domain compositing applications, one combines the gradients of multiple images and then solves for the underlying image most compatible with those gradients. In both cases, the problem is over-constrained; in general, no function exists whose gradients match the input gradients. The goal then is to project to the nearest function whose gradients approximate the inputs. A common approach is to employ a least squares metric and integrate over the domain, leading to a Poisson equation. This equation may be solved using, e.g., multigrid methods [1], fast marching methods [2], or so-called fast Poisson solvers [3].

In many applications the use of fast Poisson solvers based on the Fast Fourier Transform (FFT) are overlooked. This is due in part to the fact that fast Poisson solvers are restricted in the class of problems they can handle; e.g., spatially varying weights are not supported. Further, more complex approaches such as multigrid methods are perceived to be much faster.

In this paper, we expand the set of gradient domain problems that may be solved directly and exactly in the Fourier domain by including a data term that the reconstructed function must also approximate. This extra term enables

D. Forsyth, P. Torr, and A. Zisserman (Eds.): ECCV 2008, Part II, LNCS 5303, pp. 114–128, 2008.

operations like modifying the gradients of an image while staying close to the original image or combining independently measured depth maps and normals. We pose the problem in a variational framework and arrive at a modification to the Poisson equation, a result that corresponds to a 2D version of the "screened Poisson equation" known in physics [4].

Additional analysis reveals the structure of the spatial filters used to solve the 2D screened Poisson equation. Further, we show that uniformly scaling the gradients in an image, which intuitively ought to sharpen it, does in fact precisely correspond to a linear sharpen filter. This filter generalizes the standard Laplacian subtraction filter, which we show can also be interpreted in the same variational framework. In that framework, we show that, unlike Laplacian subtraction, gradient scaling includes a penalty term for large gradients.

We demonstrate results using an FFT-based screened Poisson solver for a set of image operations including image blending for panoramas, image sharpening, and de-blocking of compressed images. The FFT approach is direct and exact, unlike efficient least squares solvers which are typically iterated to within some error tolerance [5,6]. Though not quite as fast as the fastest solvers, the performance does scale well, and the implementation is very simple given commonly available libraries.

The paper is organized as follows: Section 2 describes previous work. Next in Section 3 we formulate the problem using a variational framework. In Section 4 we map our variational approach to the Fourier domain followed by a mapping to the spatial domain in Section 5; in these sections, we also analyze gradient-based sharpen filters. A description of our FFT-based screened Poisson solver is presented in Section 6 with results in Section 7. We conclude our paper with a discussion in Section 8.

2 Related Work

Gradient domain problems that map to Poisson equations can arise in numerous scenarios in vision and graphics. Simchony et al. [7] describe a number of such scenarios in vision including shape-from-shading, the lightness problem, and optical flow. The problem of integrability of normals, in addition to shape-from-shading, is important in photometric stereo [8] and Helmholtz stereo [9]. In computer graphics, gradient domain methods have become an essential tool in the image processing toolbox. Examples include tone-mapping of high dynamic range images [10], Poisson image editing [11], and digital photomontage [12].

The introduction of a data function term is less common, though it is a natural extension and is getting increasing attention. An early example is given by Horn for shape-from-shading, in the form of a term that minimizes the difference between the reflectance map and image irradiance, along with the surface gradient [13]. Several papers combine depth information with normals to improve depth map reconstruction [14,15,16]. Lischinski et al. [17] apply strokes as data constraints for gradient-based scattered data interpolation. More recently, Bhat et al. [18] have introduced a variety of image operations based on gradients and data images to enable new image and video processing filters like saliency sharpening, suppressing block artifacts in compressed images, and non-photorealistic rendering. All of these methods take a linear (weighted or unweighted) least squares approach to representing and solving their problems. Agrawal et al. [19] explore a wide space of formulations (some of them non-linear) for surface reconstruction that includes the possibility of a data term. Agrawal's thesis [20] is an excellent survey of surface reconstruction techniques from gradients. In this paper, we analyze the gradients plus data term using a formulation that corresponds to unweighted least squares, but analyze it in the Fourier domain and develop a corresponding fast solver.

Linear least squares techniques are among the most common approaches to solving problems described above. Szeliski [21] provides a nice summary of these approaches. A particular advantage of these approaches is their flexibility in modeling irregularly shaped domains and spatially varying weights. Agarwala [5] develops a quad-tree based solution for gradient domain compositing with uniform weights. A number of problems have a regular domain and uniform weights. Additional solvers include the fast marching method for integrating surface normals, introduced by Ho et al. [2] and the streaming multigrid method of Kazhdan and Hoppe [6]. Frankot and Chellapa [22] approached the same problem using a Fourier basis, and others have adopted cosine basis for their particular boundary conditions [22][23]. These latter approaches fall into the category of fast Poisson solvers [3]. We follow this Fourier approach to the problem of gradient domain integration that includes a data term.

We also note that Weiss [24] provides a method for computing a discrete spatial filter that minimizes squared differences between filtered versions of an image and corresponding inputs. Our analysis focuses on the continuous problem, specifically variational and Fourier analysis of the screened Poisson equation.

3 Variational Formulation

In this section, we describe the standard gradient integration problem and its Poisson solution and then expand this result to include a data function term.

The problem of computing a function $f(x, y)$ whose gradient $\nabla f(x, y)$ is as close as possible to a given gradient field $\mathbf{g}(x, y)$ is commonly solved by minimizing the following objective:

$$\iint \|\nabla f - \mathbf{g}\|^2 \, dx \, dy. \tag{1}$$

Note that $\mathbf{g}$ is a vector-valued function that is generally not a gradient derived from another function. (If $\mathbf{g}$ were derived from another function, then the optimal f would be that other function, up to an unknown constant offset.)

It is well-known that, by applying the Euler-Lagrange equation, the optimal f satisfies the following Poisson equation:

$$\nabla^2 f = \nabla \cdot \mathbf{g}, \tag{2}$$

which can be expanded as $f_{xx} + f_{yy} = g^x_x + g^y_y$, where $\mathbf{g} = (g^x, g^y)$. Subscripts in x and y correspond to partial derivatives with respect to those variables. We have superscripted g^x and g^y to denote the elements of $\mathbf{g}$ rather than subscript them, which would incorrectly suggest they are partial derivatives of the same function.

We now expand the objective beyond the standard formulation. In particular, we additionally require $f(x, y)$ to be as close as possible to some data function $u(x, y)$. The objective function to minimize now becomes:

$$\iint \lambda_d (f - u)^2 + \|\nabla f - \mathbf{g}\|^2 \, dx \, dy, \tag{3}$$

where λ_d is a constant that controls the trade-off between the fidelity of f to the data function versus the input gradient field.

To solve for the function f that minimizes this integral, we first isolate the integrand:

$$L = \lambda_d (f - u)^2 + \|\nabla f - \mathbf{g}\|^2 = \lambda_d (f - u)^2 + (f_x - g^x)^2 + (f_y - g^y)^2. \tag{4}$$

The function f that minimizes this integral satisfies the Euler-Lagrange equation:

$$\frac{\partial L}{\partial f} - \frac{\partial}{\partial x}\frac{\partial L}{\partial f_x} - \frac{\partial}{\partial y}\frac{\partial L}{\partial f_y} = 0. \tag{5}$$

Substituting and differentiating, we then have:

$$2\lambda_d (f - u) - 2(f_{xx} - g^x_x) - 2(f_{yy} - g^y_y) = 0. \tag{6}$$

Rearranging gives us:

$$\lambda_d f - (f_{xx} + f_{yy}) = \lambda_d u - (g^x_x + g^y_y) \tag{7}$$

or equivalently:

$$\lambda_d f - \nabla^2 f = \lambda_d u - \nabla \cdot \mathbf{g}. \tag{8}$$

The left-hand side of this equation is a screened Poisson equation, typically studied in three dimensions in physics [4]. Our analysis will be in 2D. As expected, setting $\lambda_d = 0$ nullifies the data term and gives us the Poisson equation.

4 Fourier Solution

In this section we analyze the 2D screened Poisson equation the Fourier domain. As with fast Poisson solvers, we can solve the screened Poisson equation (Equation 8) by taking its Fourier transform. First, we adopt the (s_x, s_y) spatial frequency notation of Bracewell [25] and recall that for a given function h and its Fourier transform, $\mathcal{F}\{h\} = H$, we have $\mathcal{F}\{h_x\} = i2\pi s_x H$, $\mathcal{F}\{h_y\} = i2\pi s_y H$, $\mathcal{F}\{h_{xx}\} = -4\pi^2 s_x^2 H$, and $\mathcal{F}\{h_{yy}\} = -4\pi^2 s_y^2 H$.

Simply transforming the left and right sides of Equation 7 gives us:

$$\lambda_d F + 4\pi^2 s_x^2 F + 4\pi^2 s_y^2 F = \lambda_d U - i2\pi s_x G^x - i2\pi s_y G^y, \tag{9}$$

where F, U, G^x, and G^y are the Fourier transforms of f, u, g_x, and g_y respectively. Solving for F, we find:

$$F = \frac{\lambda_d U - i2\pi s_x G^x - i2\pi s_y G^y}{\lambda_d + 4\pi^2 \left(s_x^2 + s_y^2\right)}. \tag{10}$$

We can interpret this equation as combining, in the numerator, the data function with "sharpened" versions (after taking derivatives) of the gradient field components, and then low-pass filtering the result with the denominator, which tends to dampen high frequencies.

Note that when $\lambda_d = 0$, Equation 10 simplifies to the well-known fast Poisson solver result:

$$F = \frac{-i2\pi s_x G^x - i2\pi s_y G^y}{4\pi^2 \left(s_x^2 + s_y^2\right)}. \tag{11}$$

This solution, however, is undefined at $s_x = s_y = 0$, corresponding to an unknown DC term (constant offset) which must be supplied. Thus, there exists a null space of solutions, and the operation is not strictly invertible. This observation is evident by examining the objective in Equation 1, which is invariant to constant offsets to f. This situation does not arise, however, when a data term is present $\lambda_d > 0$, in which case we find $F(0,0) = U(0,0)$.

4.1 Image Sharpening through Gradient Amplification

The previous section showed how gradient domain functionals that can be written in the form given by Equation 3 can be solved in the Fourier domain. It turns out that many interesting image processing filters can be written quite intuitively in this form, as will be discussed in Section 7.

In this section, we explore one intuitive example: sharpening an image by scaling up its gradients. Consider taking an image u and sharpening it by boosting its gradients ∇u by a constant factor c_s. Clearly, if one simply required an output image with scaled gradients, a simple (and optimal) solution would be to scale the image intensities by c_s. But, in addition to possibly pushing the intensities out of displayable range, this output image has drifted substantially from the intensities of the input image.

Instead, we can formulate an objective that trades off fidelity to the image against fidelity to the amplified gradients:

$$\iint \lambda_d (f - u)^2 + \|\nabla f - c_s \nabla u\|^2 \, dx \, dy. \tag{12}$$

This objective function is equivalent to the one in Equation 3, where we now have $\mathbf{g} = c_s \nabla u$. The Euler-Lagrange solution is then:

$$\lambda_d f - \nabla^2 f = \lambda_d u - c_s \nabla^2 u. \tag{13}$$

Similarly, the Fourier domain versions of gradient functions become: $G^x = i2\pi s_x U$ and $G^y = i2\pi s_y U$. Substituting into Equation 10, we obtain:

$$F = \left[\frac{\lambda_d + 4\pi^2 c_s \left(s_x^2 + s_y^2\right)}{\lambda_d + 4\pi^2 \left(s_x^2 + s_y^2\right)}\right] U = \left[\frac{1 + 4\pi^2 (c_s/\lambda_d) \left(s_x^2 + s_y^2\right)}{1 + 4\pi^2 (1/\lambda_d) \left(s_x^2 + s_y^2\right)}\right] U. \tag{14}$$

For $c_s > 1$, this equation results in boosting the high frequencies of u, and thus f is a sharpened version of u. As $(s_x^2 + s_y^2) \to \infty$, we can see that the frequency amplification levels off to a constant, i.e., to c_s, as illustrated in Figure 1, in which A_{GA} corresponds to the frequency filter in brackets in Equation 14.

5 Spatial Solution

Given our Fourier analysis we can map our problem back into the spatial domain. We can analyze the denominator in Equation 10 and treat it as a filter being applied to the numerator. Consider the denominator:

$$\frac{1}{\lambda_d + 4\pi^2 \left(s_x^2 + s_y^2\right)}. \tag{15}$$

Since this filter is radially symmetric, we can compute its Fourier transform using the Hankel transform, with radial frequency $\rho = \sqrt{s_x^2 + s_y^2}$ and spatial radius $r = \sqrt{x^2 + y^2}$. The following transform is known:

$$\mathcal{F}_\rho \left\{\frac{1}{a^2 + \rho^2}\right\} = 2\pi K_0(2\pi a r), \tag{16}$$

where $\mathcal{F}_\rho\{\}$ is the Hankel transform from the frequency to the spatial domain, and K_0 is the zeroth order modified Bessel function of the second kind. This function has the approximate shape of a rotationally symmetric exponential function over the radius e^{-r}/r and can be seen in Figure 1. (In fact, in three dimensions, the exponential form e^{-r}/r is known to be the exact solution for the screened Poisson equation [4].)

With some simple algebraic manipulations, we arrive at:

$$\mathcal{F}_\rho \left\{\frac{1}{\lambda_d + 4\pi^2 \left(s_x^2 + s_y^2\right)}\right\} = \frac{1}{2\pi} K_0(2\pi\sqrt{\lambda_d} r) = \frac{1}{2\pi} K_0(2\pi\sqrt{\lambda_d (x^2 + y^2)}). \tag{17}$$

The numerator in Equation 10 corresponds to $\lambda_d u - g_x^x - g_y^y$ in the spatial domain, and so the final result is the following convolution:

$$f = \frac{1}{2\pi} K_0(2\pi\sqrt{\lambda_d (x^2 + y^2)}) * (\lambda_d u - g_x^x - g_y^y). \tag{18}$$

Thus, we see that f can be obtained by subtracting the divergence of the gradient field from the input image u, and then blurring the result with the K_0 filter. Note that as λ_d increases, the support of this filter becomes smaller; i.e., a stronger data term shrinks the support of the blurring filter.

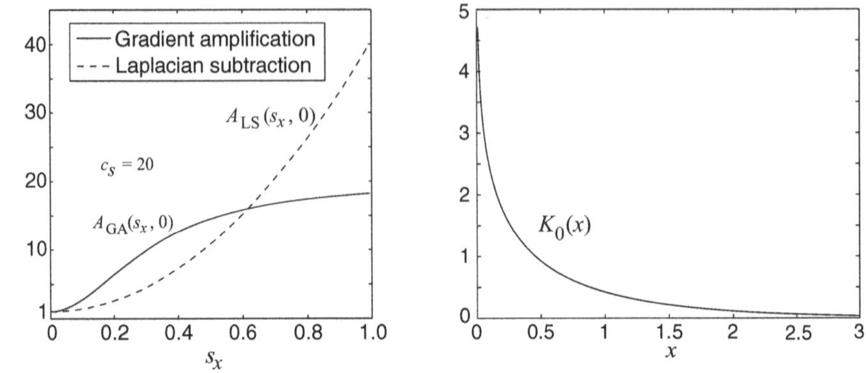

Fig. 1. Left: The frequency domain filters for gradient amplification (A_{GA}) and Laplacian subtraction (A_{LS}). Here, we set the parameters as follows: $\lambda_s = 1$, $\lambda_d = 4$, and $c_s = 20$. **Right:** The zeroth order modified Bessel function of the second kind, $K_0(x)$. In 2D, $K_0(r)$ is rotationally symmetric, and, while infinite at $r = 0$, is integrable.

5.1 Spatial Domain Sharpening

Here we determine the spatial domain filter associated with gradient amplification. Starting with Equation 14, we can decompose the frequency filter as follows:

$$\frac{\lambda_d + 4\pi^2 c_s \left(s_x^2 + s_y^2\right)}{\lambda_d + 4\pi^2 \left(s_x^2 + s_y^2\right)} = c_s - \frac{\lambda_d(c_s - 1)}{\lambda_d + 4\pi^2 \left(s_x^2 + s_y^2\right)}. \tag{19}$$

The inverse Fourier transform of the first term is a scaled Dirac delta function, and the second term follows from Equation 17, giving us:

$$\mathcal{F}_\rho \left\{ \frac{\lambda_d + 4\pi^2 c_s \left(s_x^2 + s_y^2\right)}{\lambda_d + 4\pi^2 \left(s_x^2 + s_y^2\right)} \right\} = c_s \delta(x, y) - \frac{\lambda_d(c_s - 1)}{2\pi} K_0(2\pi\sqrt{\lambda_d(x^2 + y^2)}). \tag{20}$$

Convolving this with the data function then gives:

$$f(x, y) = c_s u(x, y) - \frac{\lambda_d(c_s - 1)}{2\pi} K_0(2\pi\sqrt{\lambda_d(x^2 + y^2)}) * u(x, y). \tag{21}$$

Note that when amplifying gradients, $c_s > 1$, so the operation amounts to blurring the image with the $K_0()$ filter and subtracting it from the original image.

5.2 Relationship to Laplacian Subtraction Sharpening

In this section, we relate sharpening by gradient amplification to the more conventional sharpening by Laplacian subtraction. First, we return to the Fourier domain, define another constant $\lambda_s = c_s/\lambda_d$, and re-write Equation 14 as:

$$F = \left[\frac{1 + 4\pi^2 \lambda_s \left(s_x^2 + s_y^2\right)}{1 + 4\pi^2 (1/\lambda_d) \left(s_x^2 + s_y^2\right)} \right] U = A_{GA} U. \tag{22}$$

If we then let $\lambda_d \to \infty$ while holding λ_s constant, we arrive at:

$$F = \left[1 + 4\pi^2\lambda_s \left(s_x^2 + s_y^2\right)\right] U = A_{\mathrm{LS}} U. \tag{23}$$

This equation is precisely the Fourier transform of a commonly known sharpen filter (the Laplacian subtraction filter):

$$f = u - \lambda_s \nabla^2 u. \tag{24}$$

Thus, the gradient domain sharpen filter subsumes another common sharpen filter. At the same time, unlike this common sharpen filter, our new filter has an additional parameter to control the amount of high frequency gain. As mentioned above, the parameter c_s controls the maximum frequency amplification.

It is interesting to note, as justified in Appendix A, that the Laplacian subtraction filter can also be interpreted in a variational framework. In particular, it minimizes the following:

$$\iint (f-u)^2 - 2\lambda_s \nabla f \cdot \nabla u \, dx \, dy. \tag{25}$$

In words, the desired function f must trade off being close to the input image u against maximizing the dot product between ∇f and ∇u. The second part of this objective favors f with large gradients aligned with the input image gradients.

Furthermore, one can take the objective for gradient sharpening (Equation 12) and show that minimizing it is equivalent to minimizing the following objective:

$$\iint (f-u)^2 - 2\lambda_s \nabla f \cdot \nabla u + \frac{1}{\lambda_d} \|\nabla f\|^2 \, dx \, dy. \tag{26}$$

This objective has almost the same form as the Laplacian subtraction objective, augmented with a penalty on gradient magnitudes. This observation gives some intuition for the difference between the two sharpen filters in the frequency domain; while the Laplacian subtraction filter amplifies high frequencies without bound, the gradient amplification filter rolls off to a constant due to the additional penalty on large gradients (Figure 1).

6 Discrete Solution

The Fourier method provides a direct solution to the screened Poisson equation (Equation 8). In practice, we operate on sampled representations and solve the problem with discrete derivatives and the discrete Fourier transform (DFT).

In particular, we can first re-write Equation 7 as:

$$\lambda_d \hat{f} - (\hat{d}_x * \hat{d}_x * \hat{f} + \hat{d}_y * \hat{d}_y * \hat{f}) = \lambda_d \hat{u} - (\hat{d}_x * \hat{g}^x + \hat{d}_y * \hat{g}^y), \tag{27}$$

where $\hat{d}_x$ and $\hat{d}_y$ are discrete derivative filters in the x and y directions respectively and $\hat{f}$, $\hat{u}$, $\hat{g}^x$, and $\hat{g}^y$ are sampled versions of their continuous counterparts from Section 4. Typical choices for these discrete derivatives are forward, backward, or central differences. Taking the DFT of this equation gives us:

$$\lambda_d \hat{F} - \hat{D}_x^2 \hat{F} - \hat{D}_y^2 \hat{F} = \lambda_d \hat{U} - \hat{D}_x \hat{G}^x - \hat{D}_y \hat{G}^y, \tag{28}$$

where capitalization indicates having taken the DFT of a given discrete function. Rearranging, we arrive at:

$$\hat{F} = \frac{\lambda_d \hat{U} - \hat{D}_x \hat{G}^x - \hat{D}_y \hat{G}^y}{\lambda_d - \hat{D}_x^2 - \hat{D}_y^2} = \frac{\hat{H}}{\lambda_d - \hat{D}_x^2 - \hat{D}_y^2}, \tag{29}$$

where, for efficiency, $\hat{H}$ is computed as the DFT of $\hat{h} = \lambda_d \hat{u} - \hat{d}_x * \hat{g}^x - \hat{d}_y * \hat{g}^y$.

In practice, using the DFT is problematic, as it implicitly assumes the input sequence is periodic, placing the left boundary next to the right, and top next to bottom, before filtering. Instead, we employ the discrete cosine transform (DCT), which implicitly performs reflections across boundaries before tiling the plane periodically. We note that switching to a cosine basis has been explored by other researchers [22][23] and corresponds to Neumann boundary conditions.

A few implementation details follow. We compute the input gradients in $\hat{g}^x$ and $\hat{g}^y$ using backward differences, while the derivative filters $\hat{d}_x$ and $\hat{d}_y$ are implemented as forward differences when computing $\hat{h}$. Using backward then forward differences for derivatives also means that the derivative terms in the denominator of Equation 29 correspond to the DCT of the standard discrete 2D Laplacian. Because this filter has extremely local support in the spatial domain, we can compute its transform efficiently using the brute force DCT and without explicitly storing it.

We employ type-I DCTs using the FFTW library [26]. A single forward transform is needed to transform $\hat{h}$ to $\hat{H}$. A single reverse transform is needed to compute $\hat{f}$ from $\hat{F}$. For color images, this process is repeated for each color channel independently. The FFTW library can compute transforms in place in memory with single precision floating point arithmetic; therefore the maximum amount of memory required by our solver at any given time is four bytes for each pixel in the image. The FFTW library computes each transform in $O(n \log n)$ time, where n is the number of pixels in the image.

7 Applications and Results

In this section, we demonstrate the practical utility of our Fourier domain analysis of gradient domain problems. We start with several examples in image processing and then compare performance to several state-of-the-art methods currently used to solve one of these applications.

7.1 Gradient Domain Filters and Applications

The *de-blocking* filter proposed by Bhat et al. [18] suppresses blocking artifacts seen in highly compressed images by selectively attenuating gradients that lie

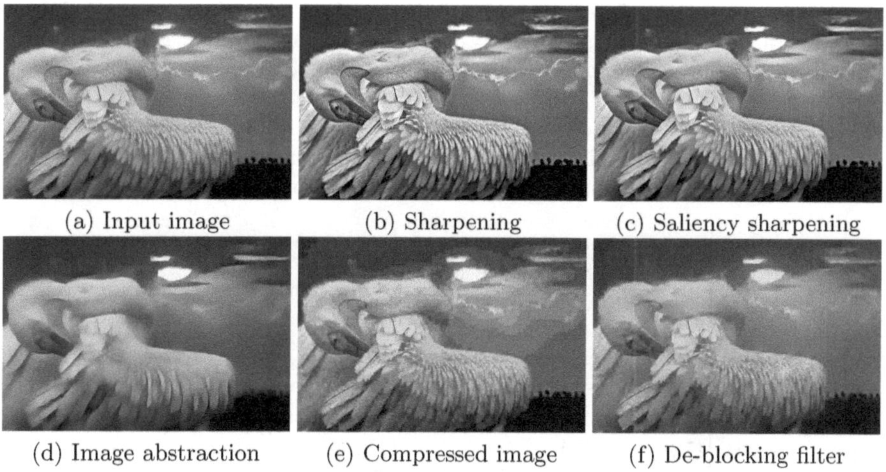

Fig. 2. Image processing with data and gradient terms using the FFT solver. Sub-figures (b,c,d) show the application of various image enhancement filters on the input image (a). Note that the blockiness in (b) arises from sharpening JPEG compression artifacts present in the original image. Sub-figure (f) shows the de-blocking filter applied to a compressed image (e).

across macro-block boundaries. The modified gradients are integrated to obtain a de-blocked image, while a data term ensures the result does not drift too far from the compressed image. See Figures 2e,2f for an example.

The *saliency sharpening filter* [18] builds on the gradient amplification sharpening filter presented in Section 4.1 in order to sharpen large scale features in an image without amplifying the noise or background clutter. This filter works by using an edge detector to detect large scale edges in the image and then selectively amplifying only those gradients that lie across large edges. Figures 2b,2c compare uniform gradient scaling to saliency sharpening.

The *image abstraction* filter defined by Bhat et al. [18] operates by suppressing small scale textures in an image. Similar to the saliency sharpening filtering, this filter uses an edge detector to determine which gradients give rise to salient features in an image. Then it selectively attenuates the non-salient gradients, thus abstracting away fine texture and noise from the input image (Figure 2d).

In *gradient domain compositing* applications [12], one combines the gradients of multiple images and then solves for the underlying image most compatible with those gradients. When no data function is used these applications can also be solved using regular Poisson solvers. Our method supports both types of gradient domain compositing applications – those with or without a data function.

7.2 Comparison to Gradient-Only Methods

We now compare the performance of our method on a gradient domain compositing application (no data term) to several state-of-the-art methods. The

Table 1. A comparison of memory and run-time performance of our method (FS) to other state-of-the-art methods (QT [5], out-of-core SM [6], HB [21], LHAB) [27]. We obtained the performance data for the non-Fourier methods from the QT and SM papers; the in-core speed of SM is better than shown here, but at very significant memory cost. Our method was evaluated on a different but a comparable speed processor (Intel Pentium 4 2.66GHz). Note also that the image dimensions in these datasets are not tuned for FFT, which explains some of the variation in timing with size.

	Size	Time (s)					Memory (MB)				
Dataset	(MP)	FS	QT	SM	HB	LHAB	FS	QT	SM	HB	LHAB
St. Emilion	10	31	9	NA	3639	160	40	24	NA	362	1044
Beynac	12	22	8	17	3357	177	48	16	190	435	1252
Rainier	23	79	14	33	6446	268	92	27	110	620	1790
Sedona	36	85	29	NA	NA	NA	144	52	NA	NA	NA
Edinburgh	51	187	122	79	NA	NA	204	123	123	NA	NA
Crag	68	172	78	NA	NA	NA	272	96	112	NA	NA
RedRock	88	270	118	118	NA	NA	352	112	NA	NA	NA

application here concerns seamless image stitching to create large panoramic images. Agarwala [5] presented a method specialized for this application by relying on the solution to the Poisson equation being smooth in most regions of the domain. He compared the performance of his method (QT) to that of a preconditioned conjugate gradient solver using two different preconditioners - hierarchical basis (HB) preconditioning [21] and locally adaptive hierarchical basis (LAHB) preconditioning [27]. More recently (contemporaneous with our own work), Kazhdan and Hoppe [6] developed a streaming multigrid (SM) Poisson solver. We ran our Fourier solver (FS) on the evaluation dataset presented in the QT paper [5]. Table 1 compares the performance of all these methods.

As can be seen from the evaluation presented in Table 1, our method is somewhat slower than QT and SM, and substantially better than HB and LAHB. It should be noted that our method and HB have very low code complexity; each taking about 100 lines of C++ code to implement using freely available libraries (FFTW [26] in our case). In contrast, QT, SM, and LAHB are relatively complex to implement. Also, QT is tuned to the image stitching problem and SM is tuned to gradient domain integration, while our method solves a more general class of problems. Of course, the HB and LAHB preconditioners can be used with linear solvers to handle a still more general class of problems.

The FS timings include a small amount of overhead introduced by FFTW to determine or "plan" an efficient DCT algorithm given the image dimensions. We chose the lowest overhead planner. However, if the image dimensions are common or many transformations are to be applied (e.g., for interactive processing or for video), then a slower planner can be employed to give more efficient transforms, and the plan can be stored for later use. In addition, FFTW supports multiple processors and cores, and in our experiments with up to 8 processors and a good plan, we have observed roughly linear speed-ups with the number of processors. The choice of planner is simply a flag passed to the FFTW library, and executing on multiple processors/cores requires adding a single line of code.

8 Discussion

In this paper, we have shown how gradient domain variational problems that include a data term can be mapped to a 2D screened Poisson equation. This problem can be studied in both the Fourier and spatial domain. This analysis gives insights into how gradient amplification corresponds to a linear sharpen filter and can be related to a standard Laplacian subtraction filter. Moreover, we show that this screened Poisson equation can be solved directly and efficiently using DCT's. By handling a data term, we are able to demonstrate a number of useful applications in image processing. We also note that the DCT formulation is extremely simple to implement with standard, optimized FFT libraries with multi-processor support.

The primary limitation of this approach is the inability to handle spatially varying weights on the gradient and data term constraints. Analysis shows that inserting such weighting into the variational formulation results in product terms that become convolutions in the frequency domain. In addition, we require complete, regular domains. Still, a number of applications can operate with constant weights over regular domains, or may potentially be initialized with an unweighted solution over a regular domain using our DCT solver to speed convergence of a more general solver.

Acknowledgments

We thank Sameer Agarwal and Eli Shechtman for helpful discussions that generalized our initial results on gradient domain filtering. Sameer also identified the connection to the screened Poisson equation. This work was supported by Microsoft, Adobe, the University of Washington Animation Research Labs, and an NVIDIA fellowship.

A Variational Formulation of Laplacian Subtraction

In Section 4.1, we showed that gradient amplification has a variational formulation that maps to a sharpen filter. Here we show that a more common sharpen filter – Laplacian subtraction (Equation 24) – can arise from another variational formulation. Consider the following objective to minimize:

$$\iint (f-u)^2 - 2\lambda_s \nabla f \cdot \nabla u \, dx\, dy. \tag{30}$$

We can isolate and expand the integrand:

$$L_{\mathrm{LS}} = (f-u)^2 - 2\lambda_s f_x u_x - 2\lambda_s f_y u_y. \tag{31}$$

Applying the Euler-Lagrange equation, the minimizing function must satisfy:

$$2(f-u) - 2\lambda_s u_{xx} - 2\lambda_s u_{yy} = 0. \tag{32}$$

Rearranging gives us:

$$f = u - \lambda_s(u_{xx} + u_{yy}) = u - \lambda_s \nabla \cdot u. \tag{33}$$

which is exactly the sharpen filter based on Laplacian subtraction.

Note that we do not claim that the form of the objective function (Equation 30) is unique. Indeed, adding other functions that do not include f or its derivatives to the integrand L_{LS} will yield the same result. Rather, it illustrates one objective function that has intuitive meeting and does lead to the Laplacian subtraction filter.

It is also instructive to relate the integrand L_{LS} to the integrand in gradient amplification:

$$L_{\text{GA}} = \lambda_d(f - u)^2 + \|\nabla f - c_s \nabla u\|^2 \tag{34}$$

$$= \lambda_d(f - u)^2 + (f_x - c_s u_x)^2 + (f_y - c_s u_y)^2. \tag{35}$$

We can divide this integrand by λ_d, because it does not affect the minimum solution. Doing so and expanding gives us:

$$L_{\text{GA}} = (f - u)^2 + \frac{1}{\lambda_d} f_x^2 - 2\frac{c_s}{\lambda_d} f_x u_x + \frac{c_s^2}{\lambda_d} u_x^2 + \frac{1}{\lambda_d} f_y^2 - 2\frac{c_s}{\lambda_d} f_y u_y + \frac{c_s^2}{\lambda_d} u_y^2. \tag{36}$$

We can drop the terms that do not depend on f or its derivatives, since again these do not affect the solution. Thus, we can omit $\frac{c_s^2}{\lambda_d} u_x^2$ and $\frac{c_s^2}{\lambda_d} u_y^2$. After also substituting $\lambda_s = c_s/\lambda_d$, we arrive at:

$$L_{\text{GA}} = (f - u)^2 + \frac{1}{\lambda_d} f_x^2 - 2\lambda_s f_x u_x + \frac{1}{\lambda_d} f_y^2 - 2\lambda_s f_y u_y \tag{37}$$

$$= (f - u)^2 - 2\lambda_s \nabla f \cdot \nabla u + \frac{1}{\lambda_d} \|\nabla f\|^2. \tag{38}$$

We can now see that the gradient amplification integrand can be related to the Laplacian subtraction integrand very simply:

$$L_{\text{GA}} = L_{\text{LS}} + \frac{1}{\lambda_d} \|\nabla f\|^2. \tag{39}$$

Again, as $\lambda_d \to \infty$ while holding λ_s constant, the gradient amplification integrand becomes the Laplacian subtraction integrand.

References

1. Trottenberg, U., Oosterlee, C.W., Schüller, A.: Multigrid. Academic Press, London (2000)
2. Ho, J., Lim, J., Yang, M., Kriegman, D.: Integrating surface normal vectors using fast marching method. In: European Conference on Computer Vision, vol. III, pp. 239–250 (2006)
3. Strang, G.: Introduction to Applied Mathematics. Wellesley-Cambridge Press (1986)

4. Fetter, A.L., Walecka, J.D.: Theoretical Mechanics of Particles and Continua. Courier Dover (2003)
5. Agarwala, A.: Efficient gradient-domain compositing using quadtrees. ACM Trans. Graph. 26, 94:1–94:5 (2007)
6. Kazhdan, M., Hoppe, H.: Streaming multigrid for gradient-domain operations on large images. In: ACM Transactions on Graphics (Proc. of ACM SIGGRAPH 2008) (to appear, 2008)
7. Simchony, T., Chellappa, R., Shao, M.: Direct analytical methods for solving poisson equations in computer vision problems. IEEE Transactions on Pattern Analysis and Machine Intelligence 12, 435–446 (1990)
8. Horn, B.: Robot Vision. MIT Press, Cambridge (1986)
9. Zickler, T.E., Belhumeur, P.N., Kriegman, D.J.: Helmholtz stereopsis: Exploiting reciprocity for surface reconstruction. Int. J. Comput. Vision 49, 215–227 (2002)
10. Fattal, R., Lischinski, D., Werman, M.: Gradient domain high dynamic range compression. In: SIGGRAPH 2002: Proceedings of the 29th annual conference on Computer graphics and interactive techniques, pp. 249–256. ACM Press, New York (2002)
11. Pérez, P., Gangnet, M., Blake, A.: Poisson image editing. In: SIGGRAPH 2003: ACM SIGGRAPH 2003 Papers, pp. 313–318. ACM Press, New York (2003)
12. Agarwala, A., Dontcheva, M., Agrawala, M., Drucker, S., Colburn, A., Curless, B., Salesin, D., Cohen, M.: Interactive digital photomontage. ACM Trans. Graph. 23, 294–302 (2004)
13. Horn, B.K.P.: Height and gradient from shading. Int. J. Comput. Vision 5, 37–75 (1990)
14. Horovitz, I., Kiryati, N.: Depth from gradient fields and control points: bias correction in photometric stereo. Image Vision Comput. 22, 681–694 (2004)
15. Nehab, D., Rusinkiewicz, S., Davis, J., Ramamoorthi, R.: Efficiently combining positions and normals for precise 3d geometry. ACM Trans. Graph. 24, 536–543 (2005)
16. Ng, H., Wu, T., Tang, C.: Surface-from-gradients with incomplete data for single view modeling. In: International Conference on Computer Vision, pp. 1–8 (2007)
17. Lischinski, D., Farbman, Z., Uyttendaele, M., Szeliski, R.: Interactive local adjustment of tonal values. In: SIGGRAPH 2006: ACM SIGGRAPH 2006 Papers, pp. 646–653. ACM Press, New York (2006)
18. Bhat, P., Zitnick, L., Cohen, M., Curless, B.: A perceptually-motivated optimization-framework for image and video processing. Technical Report UW-CSE-08-06-02, University of Washington (2008)
19. Agrawal, A.K., Raskar, R., Chellappa, R.: What is the range of surface reconstructions from a gradient field? In: Leonardis, A., Bischof, H., Pinz, A. (eds.) ECCV 2006. LNCS, vol. 3951, pp. 578–591. Springer, Heidelberg (2006)
20. Agrawal, A.: Scene Analysis under Variable Illumination using Gradient Domain Methods. Ph.D thesis, University of Maryland (2006)
21. Szeliski, R.: Fast surface interpolation using hierarchical basis functions. IEEE Trans. Pattern Anal. Mach. Intell. 12, 513–528 (1990)
22. Frankot, R.T., Chellappa, R.: A method for enforcing integrability in shape from shading algorithms. IEEE Trans. Pattern Anal. Mach. Intell. 10, 439–451 (1988)
23. Georghiades, A.S., Belhumeur, P.N., Kriegman, D.J.: From few to many: Illumination cone models for face recognition under variable lighting and pose. IEEE Transactions on Pattern Analysis and Machine Intelligence 23, 643–660 (2001)
24. Weiss, Y.: Deriving intrinsic images from image sequences. In: International Conference on Computer Vision, vol. II, pp. 68–75 (2001)

25. Bracewell, R.N.: The Fourier Transform and Its Applications, 2nd edn. McGraw-Hill, New York (1986)
26. Frigo, M., Johnson, S.G.: FFTW for version 3.0 (2003)
27. Szeliski, R.: Locally adapted hierarchical basis preconditioning. In: SIGGRAPH 2006: ACM SIGGRAPH 2006 Papers, pp. 1135–1143. ACM Press, New York (2006)

Anisotropic Geodesics for Perceptual Grouping and Domain Meshing*

Sébastien Bougleux, Gabriel Peyré, and Laurent Cohen

Université Paris-Dauphine, CEREMADE, 75775 Paris Cedex 16, France
{bougleux,peyre,cohen}@ceremade.dauphine.fr

Abstract. This paper shows how Voronoi diagrams and their dual Delaunay complexes, defined with geodesic distances over 2D Reimannian manifolds, can be used to solve two important problems encountered in computer vision and graphics. The first problem studied is perceptual grouping which is a curve reconstruction problem where one should complete in a meaningful way a sparse set of noisy curves. From this latter curves, our grouping algorithm first designs an anisotropic tensor field that corresponds to a Reimannian metric. Then, according to this metric, the Delaunay graph is constructed and pruned in order to correctly link together salient features. The second problem studied is planar domain meshing, where one should build a good quality triangulation of a given domain. Our meshing algorithm is a geodesic Delaunay refinement method that exploits an anisotropic tensor field in order to locally impose the orientation and aspect ratio of the created triangles.

1 Introduction and Related Concepts

Various important problems in computer vision and computer graphics require the integration of a local anisotropy over a complicated planar domain. This local anisotropy is dictated by directional features such as curves or textures that should be exploited to perform sampling, segmentation, grouping or meshing tasks. In this paper, we focus on two representative problems in these fields: perceptual grouping of salient features and meshing of planar domains.

The proposed approach encodes the local anisotropy within a tensor field that corresponds to a Riemannian metric. This local information is integrated into global constraints thanks to the geodesic distance over this Riemannian domain. Based on this geodesic distance, one can define the Riemannian Voronoi diagram of a point set [1] and its dual Delaunay complex to perform grouping and meshing tasks. Such geometrical structures are not easy to compute, and several approximate solutions have been proposed (see for instance [2], [3]). Here, we adopt an approach based on the Fast Marching algorithm to compute geodesic distances and Voronoi diagrams.

This section reviews basis facts about Riemannian manifolds and explains how geodesic distances can be computed efficiently with Fast Marching methods.

* This work was partially supported by ANR Grant SURF-NT05-2_45825.

D. Forsyth, P. Torr, and A. Zisserman (Eds.): ECCV 2008, Part II, LNCS 5303, pp. 129–142, 2008.

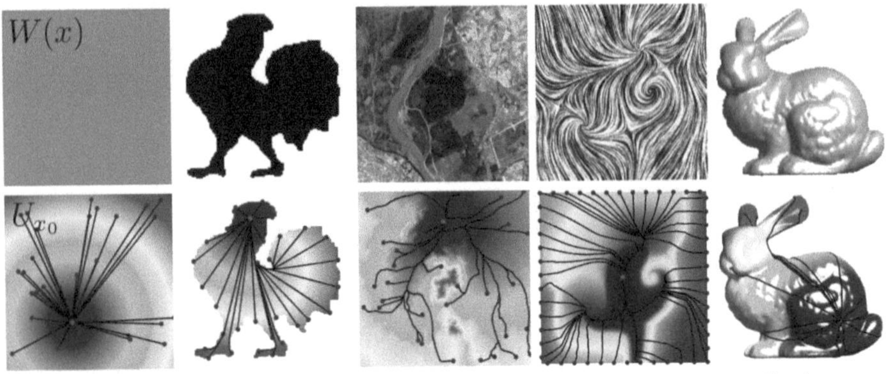

Fig. 1. Examples of Riemannian metrics (top row) and geodesic distances and curves (bottom row). The colormap indicates the geodesic distance to the starting point. From left to right: Euclidean ($H(x) = \mathrm{Id}_2$) in $\mathbb{R}^2$ and then restricted to a planar domain, isotropic ($H(x) = W(x)\mathrm{Id}_2$ with W computed from the image), Riemannian manifold ($H(x)$ is the structure tensor of the image, see (8)) and 3D surface ($H(x)$ corresponds to the first fundamental form).

Riemannian metric, geodesic distances and shortest paths. This paper considers 2D Riemannian manifolds that are defined over a compact planar domain $\Omega \subset \mathbb{R}^2$. The boundary $\partial\Omega$ of Ω is assumed to be a set of closed smooth curves. At each point $x \in \Omega$, one has a tensor $H(x) \in \mathbb{R}^{2\times 2}$ which is a positive symmetric matrix. This tensor field defines a local metric that allows to measure the length of a piecewise smooth curve $\gamma : [0,1] \to \Omega$ as follows

$$L(\gamma) \overset{\text{def.}}{=} \int_0^1 \sqrt{\gamma'(t)^{\mathrm{T}} H(\gamma(t))\gamma'(t)}\mathrm{d}t.$$

In image processing and computer vision, the manifold is often the image domain $\Omega = [0,1]^2$ equipped with a metric H derived from the image to process. In computer graphics and numerical analysis, one often deals with complicated planar domains with holes and corners. Figure 1 shows several Riemannian metrics. An important issue is thus to design, for a specific application, a tensor field H that encodes the important information about the problem to solve.

At each location $x \in \Omega$, the Riemannian tensor can be diagonalized as

$$H(x) = \lambda_1(x)e_1(x)e_1(x)^{\mathrm{T}} + \lambda_2(x)e_2(x)e_2(x)^{\mathrm{T}}, \tag{1}$$

where $\lambda_1 \geqslant \lambda_2 > 0$ are the eigenvalue fields, and e_1, e_2 are the corresponding orthogonal (un-oriented) eigenvector fields. A curve γ passing at location $\gamma(t) = x$ with speed $\gamma'(t)$ has a shorter local length if $\gamma'(t)$ is collinear to $e_1(x)$ rather than to another direction. Hence shortest paths tend to be tangent to the field e_1.

The anisotropy of a metric $H(x)$ is defined as

$$\alpha(x) \overset{\text{def.}}{=} \frac{\lambda_1 - \lambda_2}{\lambda_1 + \lambda_2} = \frac{\sqrt{(a-b)^2 + 4c^2}}{a+b} \in [0,1], \quad \text{for } H(x) = \begin{pmatrix} a & c \\ c & b \end{pmatrix}. \tag{2}$$

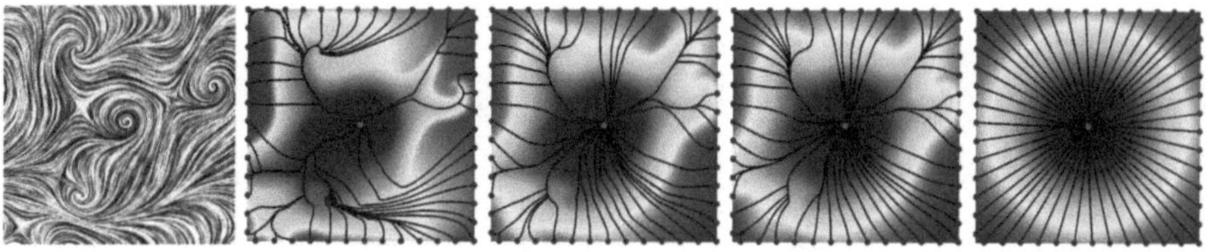

Fig. 2. Distance maps and geodesics for a tensor field with a decreasing anisotropy α. The tensor field is computed from the structure tensor of f as defined in (8).

A metric with $\alpha(x)$ close to 1 is highly directional near x, whereas a metric with $\alpha(x)$ close to 0 is locally isotropic near x.

The geodesic distance between two points of Ω is defined as

$$d(x,y) \stackrel{\text{def.}}{=} \min_{\gamma \in \mathcal{P}(x,y)} L(\gamma), \quad \forall\, (x,y) \in \Omega^2, \tag{3}$$

where $\mathcal{P}(x,y) \stackrel{\text{def.}}{=} \{\gamma \setminus \gamma(0){=}x \wedge \gamma(1){=}y\}$ denotes the set of piecewise smooth curves joining x and y. The distance d turns the domain Ω into a metric space.

A shortest path γ between two points $(x,y) \in \Omega^2$, according to the Riemannian metric, is called a geodesic. It satisfies $L(\gamma) = d(x,y)$.

Distance map. In order to perform the numerical computation of geodesic curves and distances, we fix a set of starting points $\mathcal{S} = (x_k)_k \subset \Omega$ and we define the distance map to this set

$$U_{\mathcal{S}}(x) \stackrel{\text{def.}}{=} \min_k d(x, x_k), \quad \forall x \in \Omega.$$

An important theoretical result is that if $x \mapsto H(x)$ is continuous, $U_{\mathcal{S}}$ is the unique viscosity solution of the following Hamilton-Jacobi non-linear partial differential equation (PDE)

$$\|\nabla_x U_{\mathcal{S}}\|_{H(x)^{-1}} = 1, \quad \text{with} \;\; U_{\mathcal{S}}(x_k) = 0, \; \forall k, \tag{4}$$

where $\|v\|_A = \sqrt{v^T A v}$. This classical result (see for instance [4]) allows to replace the optimization problem (3) that defines geodesic distances by a PDE.

Once the distance map $U_{\mathcal{S}}$ has been computed by solving (4), one can extract a geodesic γ between a point x and its closest point $x_k \in \mathcal{S}$ by the following gradient descent

$$\gamma'(t) = -\frac{H(\gamma(t))^{-1}\nabla_{\gamma(t)} U_{\mathcal{S}}}{\|H(\gamma(t))^{-1}\nabla_{\gamma(t)} U_{\mathcal{S}}\|}, \quad \text{with} \;\; \gamma(0) = x. \tag{5}$$

The geodesic curve γ extracted using this gradient descent is parametrized with unit speed since $\|\gamma'\| = 1$. In particular, if $\mathcal{S} = \{x_1\}$ is restricted to one point, then one can compute the geodesic curve between x_1 and any point in Ω.

Fig. 3. Anisotropic front propagation from 9 starting points. The colormap indicates the values of the distance map at a given iteration of the FM algorithm. The metric is computed using the structure tensor (see (8)), of the background image.

Figure 2 shows examples of geodesic curves computed from a single starting point $\mathcal{S} = \{x_1\}$ located at the center of the domain $\Omega = [0,1]^2$, and a set of points on the boundary of Ω. The geodesics are computed for a metric H, derived from the image $f : \Omega \to \mathbb{R}$, and whose anisotropy α (defined in (2)) is decreasing, thus making the Riemannian space progressively closer to the Euclidean space.

For the particular case of an isotropic metric $H(x) = W(x)\mathrm{Id}_x$, the geodesic distance and the shortest path satisfy

$$||\nabla_x U_{\mathcal{S}}|| = W(x) \quad \text{and} \quad \gamma'(t) = -\frac{\nabla_x U_{\mathcal{S}}}{||\nabla_x U_{\mathcal{S}}||}. \tag{6}$$

This corresponds to the Eikonal equation, that has been used to compute minimal paths weighted by W [5].

Numerical computations of geodesic distances. In order to make all the previous definitions effective in practical situations, one needs a fast algorithm to compute the geodesic distance map $U_{\mathcal{S}}$. The Fast Marching (FM) algorithm, introduced by Sethian [6], is a numerical procedure to efficiently solve in $O(n\log(n))$ operations the discretization of (6) in the isotropic case. Several extensions of the FM algorithm have been proposed in order to solve (4) for a generic metric, see for instance Kimmel and Sethian [7] for triangulated meshes and Spira and Kimel [8], Bronstein et al. [9] for parametric manifolds.

In our numerical experimentations, we use the $O(n\log(n))$ FM algorithm developed by Prados et al. [10]. As any FM method, it computes the distance map $U_{\mathcal{S}}$ by progressively propagating a front, starting from the points in $\mathcal{S}$. Figure 3 shows an example of FM computation with an anisotropic metric. The front propagates faster along the direction of the texture. This is due to the alignment of the principal direction field e_1 (of the metric) with the textures.

2 Riemannian Voronoi Diagrams and Dual Complexes

This section shows how several concepts from computational geometry can be extended to the setting of Riemannian metrics. Given a 2D Riemannian manifold as defined in Section 1, and a set of points $\mathcal{S} = \{x_i\}_{i=1}^m \subset \Omega$, one can define combinatorial and geometrical structures that organize the domain Ω and the set $\mathcal{S}$ with respect to the geometry of the manifold.

Voronoi diagrams. Riemannian Voronoi diagrams [1] considered in this paper have the same definition as the Euclidean Voronoi diagram and its extensions. Given a metric and a set of points $\mathcal{S}$, the partition $\mathcal{V}_\mathcal{S}$ of the domain Ω in Voronoi cells $\mathcal{C}_i$ is given by

$$\mathcal{V}_\mathcal{S} \stackrel{\text{def.}}{=} \bigcup_{x_i \in \mathcal{S}} \mathcal{C}_i = \Omega \quad \text{with} \quad \mathcal{C}_i \stackrel{\text{def.}}{=} \{x \in \Omega \setminus \forall j \neq i,\ d(x_i, x) \leqslant d(x_j, x)\}\,. \tag{7}$$

Using the anisotropic geodesic distance (3) for d, $\mathcal{V}_\mathcal{S}$ is a Riemannian Voronoi diagram of the set S, restricted to the domain Ω. This anisotropic geodesic Voronoi diagram shares several properties with the Euclidean Voronoi diagram. In particular, anisotropic geodesic Voronoi cells have the advantage to be connected. This is not necessarily the case for the so-called anisotropic Voronoi diagram [3], defined with a simplified anisotropic distance $\tilde{d}(x,y) = ((x-y)^{\mathrm{T}} H(x)(x-y))^{1/2}$ that is not based on shortest paths.

Two Voronoi cells $\mathcal{C}_i$ and $\mathcal{C}_j$ are adjacent if $\mathcal{C}_i \cap \mathcal{C}_j \neq \emptyset$. This intersection defines a Voronoi edge that is the set of points of Ω located at equal distance from the points x_i and x_j of $\mathcal{S}$. With the anisotropic geodesic distance (3), such an edge is an open or closed curve. A Voronoi vertex is the intersection point of at least three adjacent Voronoi cells. When the set $\mathcal{S}$ is in general position, a Voronoi vertex is the intersection point of exactly three Voronoi cells

$$\mathcal{V}_{i,j,k} \stackrel{\text{def.}}{=} \mathcal{C}_i \cap \mathcal{C}_j \cap \mathcal{C}_k.$$

In the sequel, the set $\mathcal{S}$ is assumed to be in general position, and the set of Voronoi vertices is noted $\Sigma_\mathcal{S}$.

Delaunay complex. The Delaunay complex $\mathcal{D}_\mathcal{S}$ of the set of points $\mathcal{S}$ is the simplicial complex dual to the anisotropic geodesic Voronoi diagram. It is the set of triangles, edges and vertices that can be defined as follows:

- the set of vertices of $\mathcal{D}_\mathcal{S}$ is the set of points $\mathcal{S}$,
- two points of $\mathcal{S}$ are connected by an edge of $\mathcal{D}_\mathcal{S}$ if their respective Voronoi cells are adjacent in $\mathcal{V}_\mathcal{S}$

$$(x_i, x_j) \in \mathcal{D}_\mathcal{S} \iff \mathcal{C}_i \cap \mathcal{C}_j \neq \emptyset.$$

- three points of $\mathcal{S}$ form a triangle of $\mathcal{D}_\mathcal{S}$ if their respective Voronoi cells have a non-empty intersection, that is a Voronoi vertex

$$(x_i, x_j, x_k) \in \mathcal{D}_\mathcal{S} \iff \mathcal{C}_i \cap \mathcal{C}_j \cap \mathcal{C}_k \neq \emptyset.$$

To each Delaunay edge (x_i, x_j) corresponds a double point $w_{i,j}$, defined as the closest point to x_i and x_j, that is on the Voronoi edge $\mathcal{C}_i \cap \mathcal{C}_j$

$$w_{i,j} \stackrel{\text{def.}}{=} \underset{x \in \mathcal{C}_i \cap \mathcal{C}_j}{\operatorname{argmin}}\ d(x, x_i) \stackrel{(3)}{=} \underset{x \in \mathcal{C}_i \cap \mathcal{C}_j}{\operatorname{argmin}}\ d(x, x_j).$$

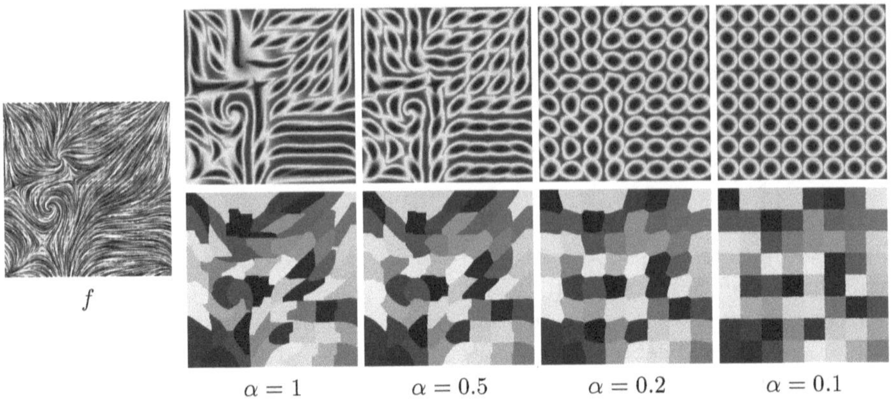

Fig. 4. Anisotropic distances (top row) and Voronoi diagrams (bottom row) with a decreasing anisotropy α. The metric tensor is computed using the structure tensor (8).

Contrary to the Euclidean case, the Delaunay complex $\mathcal{D}_\mathcal{S}$ defined above does not always define a valid planar triangulation. Some points of $\mathcal{S}$ can be isolated, i.e. incident to only one edge (their corresponding Voronoi cells are adjacent to only one other cell). Also, some triangles can overlap. If the sampling $\mathcal{S}$ is in general position and dense enough, then one can prove that $\mathcal{D}_\mathcal{S}$ is a valid planar triangulation, see [1] for more details.

While the set of edges of $\mathcal{D}_\mathcal{S}$ forms a planar graph (straight line edges), its geometric realization connects points of $\mathcal{S}$ with geodesic paths. The geometric realization $\gamma_{i,j} \in \mathcal{P}(x_i, x_j)$, of a Delaunay edge $(x_i, x_j) \in \mathcal{D}_\mathcal{S}$, is defined as the union of the two geodesics joining the double point $w_{i,j}$ to x_i and x_j.

Numerical Computation. The Voronoi diagram (7) can be computed in parallel to the computation of the geodesic distance map $U_\mathcal{S}$. This requires to track the index of the closest point in $\mathcal{S}$ of any point in the front during the Fast Marching propagation. This can be done with any algorithm mentioned in Section 1. During this propagation, one can also keep track of double points and Voronoi vertices. For each Delaunay edge $(x_i, x_j) \in \mathcal{D}_\mathcal{S}$, its double point $w_{i,j}$ corresponds to the first meeting location of the two fronts emanating from both x_i and x_j. After the propagation, the geometric realization of each Delaunay edge (x_i, x_j) is extracted by solving two gradient descents, according to (5), to compute the two geodesics joining $w_{i,j}$ to x_i and to x_j.

All the geometric tools defined above can thus be extracted using a single front propagation, which requires $O(n \log(n))$ operations. Furthermore, when they have been computed for a set of points $\mathcal{S}$, they can be extended to $\mathcal{S} \cup \{x_{m+1}\}$ by a local propagation restricted to the cell $\mathcal{C}_{m+1}$, thus requiring on average $O(n \log(n)/m)$ operations.

Figure 4 shows examples of Voronoi diagrams for Riemannian metrics with a decreasing anisotropy. One can see how the Voronoi cells $\mathcal{C}_i$ are stretched along the main tensor direction e_1 for highly anisotropic metrics.

3 Application to Perceptual Grouping

Perceptual grouping is a curve reconstruction problem where one wants to extract a curve from an image containing a sparse set of curves, eventually embedded in noise. This problem is relevant both to model good continuation perception laws and to develop efficient edge detection methods (see for instance [11]). In this paper we restrict ourselves to the detection of a set of non-intersecting open or closed curves, although other kinds of topological or regularity constraints could be enforced.

Our algorithm extends the isotropic geodesic grouping method of Cohen [12] by designing a Riemannian metric that propagates the anisotropy of the sparse curves to the whole domain. This metric helps to disambiguate difficult situations where some curves are close from one to each other. This allows a better reconstruction with less user intervention. The idea of using anisotropic information to perform perceptual grouping is introduced in [13] where the completed contours are local minimizer of a saliency field. Many variational definitions of perceptual contours have been proposed using local regularity assumption, for instance with the elastica model of Mumford [14]. In contrast, our completed contours are anisotropic shortest paths that connect Riemannian Voronoi cells, thus being the global minimum of a length criterion.

Design of an anisotropic tensor field. The Riemannian metric H needs to be computed from the noisy input image $f:[0,1]^2 \rightarrow \mathbb{R}$. In order to compute robustly the local direction of the features, we use a local pooling of the gradient information that constitutes a sparse set of tensors with a confidence measure. This sparse tensor field is then integrated by diffusion into a dense field.

The local orientation of a feature around a pixel x is given by the vector orthogonal to the gradient $v(x) = (\nabla_x f)^{\perp}$, which is computed numerically with finite differences (using maybe some little smoothing to cancel noise). This local direction information can be stored in a rank-1 tensor $T_0(x) = v(x)v(x)^{\mathrm{T}}$. In order to evaluate the local anisotropy of the image, one needs to average this tensor

$$T(x) = T_0 * G_\sigma(x), \tag{8}$$

where the four entries of the tensor are smoothed against a Gaussian kernel G_σ of width $\sigma > 0$. The metric H corresponds to the so-called structure tensor, see for instance [15]. This local tensor T is able to extract both the local direction of edges and the local direction of textural patterns (see Figure 4, left).

The structure tensor field T defined by (8) gives a robust estimation of the local anisotropy only close to image features where the gradient is large. In homogeneous areas (typically outside the salient features), the tensor is nearly isotropic with small eigenvalues. In order to have a dense anisotropic field, one needs to extend the anisotropy over the whole domain using some kind of interpolation. This notion of interpolation of local orientations is similar to the computation of good continuation field, as studied for instance in stochastic completion fields [16] or tensor voting [17].

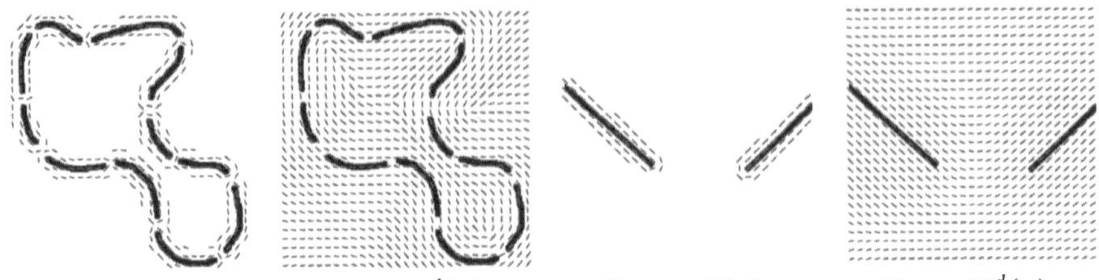

Fig. 5. Computation of a dense tensor field $T^d(x)$ from a sparse structure tensor $T(x)$

In this paper, we propose a simple interpolation method that computes a dense tensor field with a linear diffusion outside a region of high confidence. The region of high saliency is computed by thresholding the anisotropy map α of T defined in (2)

$$\Omega_\alpha \overset{\text{def.}}{=} \{x \in \Omega \setminus \alpha(x) > 1 - \varepsilon\},$$

where ε is a small constant. In our numerical examples we choose $\varepsilon = 0.05$. The orientation of the tensors in Ω_α are computed with a high confidence, and in order to compute a dense tensor field, we compute the following steady state of a heat diffusion

$$\forall\, x \notin \Omega_\alpha, \quad \Delta T^d_{i,j}(x) = 0 \quad \text{and} \quad \forall\, x \in \Omega_\alpha, \quad T^d_{i,j}(x) = T_{i,j}(x),$$

where Δ is the Laplacian with Neumann reflecting conditions on the boundary $\partial\Omega$ and where $(T_{i,j})_{i,j=1,2}$ are the components of the structure tensor.

In order to turn the dense tensor field T^d into a Riemannian metric, we apply a non-linear mapping to its eigenvalues,

$$T^d(x) = \mu_1 e_1 {e_1}^{\mathrm{T}} + \mu_2 e_2 {e_2}^{\mathrm{T}} \quad \Longrightarrow \quad H(x) = \psi_1(\mu_1) e_1 {e_2}^{\mathrm{T}} + \psi_2(\mu_2) e_2 {e_2}^{\mathrm{T}}. \quad (9)$$

where ψ_i is a decreasing function. In this paper, we use $\psi_i(x) = (\eta + |x|)^{-1}$ for a small value of η. Figure 5 shows examples of the proposed method.

Anisotropic Perceptual Grouping. Our anisotropic grouping algorithm proceeds by computing a perceptual graph $\tilde{\mathcal{D}}_\mathcal{S}$ of a set of points $\mathcal{S}$ provided by the user. This perceptual graph is a sub-graph of the set of edges of $\mathcal{D}_\mathcal{S}$.

The use of the anisotropic metric H defined in (9) helps to reduce the user intervention by grouping curves that obey a good continuation property with respect to the anisotropic tensor field. This is performed by ordering the edges $(x_i, x_j) \in \mathcal{D}_\mathcal{S}$ with respect to their respective geodesic distance $d(x_i, x_j)$. The edges are progressively inserted as long as the reconstructed curves are closed or open but non-intersecting. This topological constraint is enforced by monitoring the current degree δ_i of each point, which should be smaller or equal to 2. The algorithm is detailed in Table 1.

Figure 6 compares the results of perceptual grouping using an isotropic metric (which is equivalent to the algorithm developed in [12]) to our algorithm that

Table 1. Anisotropic perceptual grouping algorithm

1. *Initialization:* set $\tilde{\mathcal{D}}_S \leftarrow \emptyset$, and $\Pi \leftarrow \mathcal{D}_S$.
2. *Select edge:* set $(x_i, x_j) \leftarrow \underset{(x,y)\in\Pi}{\text{argmin}}\, d(x,y)$. Remove it: $\Pi \leftarrow \Pi - \{(x_i, x_j)\}$.
3. *Check topology:* if $\delta_i < 2$ and $\delta_j < 2$, then update $\tilde{\mathcal{D}}_S \leftarrow \tilde{\mathcal{D}}_S \cup \{(x_i, x_j)\}$ and set $\delta_j \leftarrow \delta_j + 1$ and $\delta_i \leftarrow \delta_j + 1$.
4. *Stop:* while $\Pi \neq \emptyset$, go back to 2.

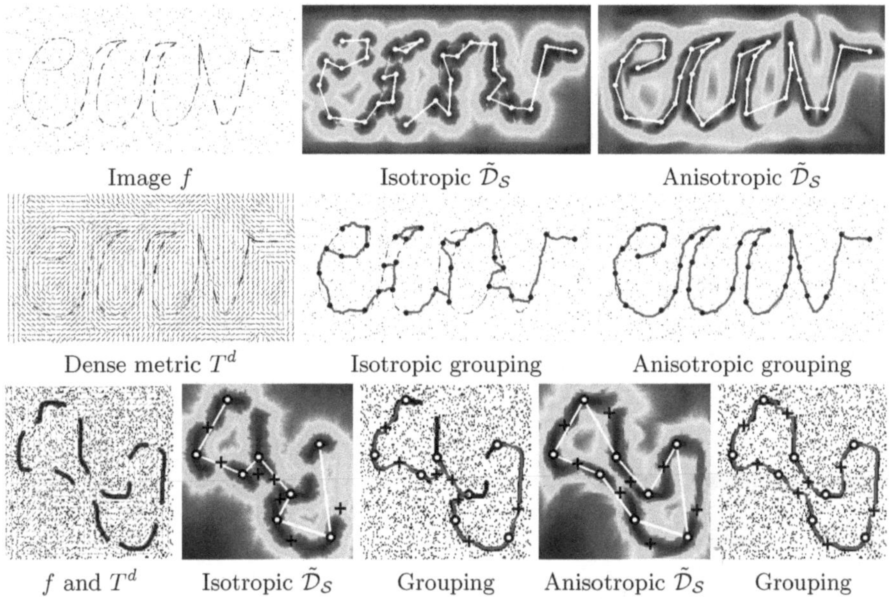

Fig. 6. Comparison of isotropic and anisotropic Riemannian metrics for perceptual grouping

uses the dense tensor field T^d. Contrary to the isotropic method, our anisotropic metric enables a correct grouping that obeys a good continuation property.

4 Application to Meshing of Planar Domains

Planar domain meshing requires to build a good quality triangulation of a given domain. Triangles with anisotropic shape and varying sizes are highly desirable because of their capability to represent efficiently functions with directional singularities that one encounters for instance in parabolic PDE's near shocks. This section proposes a new anisotropic meshing algorithm that conforms the shape and size of the triangles to the tensor field of a Riemannian metric.

Classical planar domain meshing algorithms are based on Euclidean Delaunay triangulation. They proceed by progressively inserting Voronoi vertices, see for instance [18,19]. These points are inserted in order to split triangles that are poorly shaped, and also to ensure a minimum size of the triangles.

Anisotropic meshes can be built using a local modification of the metric [20] or anisotropic elastic forces [21], bubble packing [22], simplified anisotropic metric [3]. This latter approach is used in conjunction with Ruppert's Delaunay refinement algorithm to provide anisotropic triangulation with guaranties on the aspect ratio of the triangles. This algorithm is extended to domains with curves by [23], and an alternative construction of the anisotropic Voronoi diagram is proposed in [24]. The simplified distance has been applied to image sampling [25], optimal samples placement with centroidal tessellations [26] and surface remeshing [27].

n this paper, we propose a new Delaunay refinement algorithm that extends the algorithm of Ruppert [18], and the one of Labelle and Shewchuck [3], with anisotropic geodesic distances. It also extends the isotropic farthest point seeding strategy of [28] with anisotropic metrics and domains with arbitrary boundaries.

As these algorithms, our anisotropic meshing algorithm proceeds by iteratively inserting points to an already computed set of points $\mathcal{S}$. The inserted points are the Voronoi vertices of the anisotropic geodesic Voronoi diagram $\mathcal{V}_\mathcal{S}$. The resulting mesh is the Delaunay complex $\mathcal{D}_\mathcal{S}$ dual to $\mathcal{V}_\mathcal{S}$. The main difficulty is that $\mathcal{D}_\mathcal{S}$ is not necessarily a valid planar triangulation if $\mathcal{S}$ is not dense enough. As described in Section 2, this is due to the presence of some isolated points or some overlapping triangles. As in [3] and [24], the algorithm automatically add points to overcome these problems and to guaranty a valid planar triangulation.

At least one point of $\mathcal{S}$ is located on each boundary curve, and these boundary points segment $\partial\Omega$ as a set of sub-curves

$$\partial\Omega = \bigcup_{i,j} \theta_{i,j} \quad \text{with} \quad \begin{cases} \theta_{i,j} \in \mathcal{P}(x_i, x_j), \\ \theta_{i,j} \cap \mathcal{S} = \{x_i, x_j\}, \end{cases}$$

(one can have $x_i = x_j$ if there is only one point on a curve).

In order to compute an anisotropic triangular mesh with triangles of high quality with respect to the local metric, one needs to iteratively insert the Voronoi vertex $\mathcal{V}_{i,j,k}$ dual to the Delaunay triangle (x_i, x_j, x_k) with the smallest circumradius to shortest edge ratio

$$\rho(\mathcal{V}_{i,j,k}) = \frac{U_\mathcal{S}(\mathcal{V}_{i,j,k})}{\min(d(x_i, x_j), d(x_j, x_k), d(x_k, x_i))},$$

which is a quantity computed for each Voronoi vertex in parallel to the FM propagation. In the Euclidean domain, a triangle (x_i, x_j, x_k) with a high value of $\rho(\mathcal{V}_{i,j,k})$ is badly shaped since its smallest angle is close to 0. As explained in [3], this property extends to an anisotropic metric H if angles are measured using the inner product defined by H.

One can also consider the farthest point strategy to refine the mesh, e.g. the Voronoi vertex $\mathcal{V}_{i,j,k}$ with the greatest distance $U_\mathcal{S}(\mathcal{V}_{i,j,k})$ is iteratively inserted.

Table 2. Anisotropic planar domain meshing algorithm

1. *Initialization:* set $\mathcal{S}$ with at least one point on each curve of Ω, compute $U_\mathcal{S}$ with a Fast Marching.
2. *Boundary enforcement:* while it exists $\theta_{i,j} \subset \partial\Omega$ encroached by some $x_k \in \mathcal{S}$, subdivide: $\mathcal{S} \leftarrow \mathcal{S} \cup \{y_{i,j}\}$ (see text). Update $U_\mathcal{S}$ with a local Fast Marching.
3. *Triangulation enforcement:* while it exists $(x_i, x_j) \in \mathcal{D}_\mathcal{S}$ with x_i or x_j isolated, insert $\underset{x \in \mathcal{C}_i \cap \mathcal{C}_j}{\operatorname{argmax}} U_\mathcal{S}(x)$. Update $U_\mathcal{S}$ with a local Fast Marching.
4. *Select point:* $s^\star \leftarrow \underset{s \in \Sigma_\mathcal{S} - \partial\Omega}{\operatorname{argmin}} \rho(s)$ (or $s^\star \leftarrow \underset{s \in \Sigma_\mathcal{S} - \partial\Omega}{\operatorname{argmax}} U_\mathcal{S}(s)$).
   - If in $\mathcal{S} \cup \{s^\star\}$, $s^\star$ encroaches some $\theta_{i,j} \subset \partial\Omega$, subdivide: $\mathcal{S} \leftarrow \mathcal{S} \cup \{y_{i,j}\}$.
   - Otherwise, add it: $\mathcal{S} \leftarrow \mathcal{S} \cup s^\star$.

   Update $U_\mathcal{S}$ with a local Fast Marching.
5. *Stop:* while $\rho(s^\star) > \rho^\star$ or $U_\mathcal{S}(s^\star) > U^\star$, go back to 2.
6. *Triangulation enforcement:* if it exists two adjacent triangles that overlap, add the dual Voronoi vertex of one of them to $\mathcal{S}$, update $U_\mathcal{S}$ with a local Fast Marching, and go back to 2.

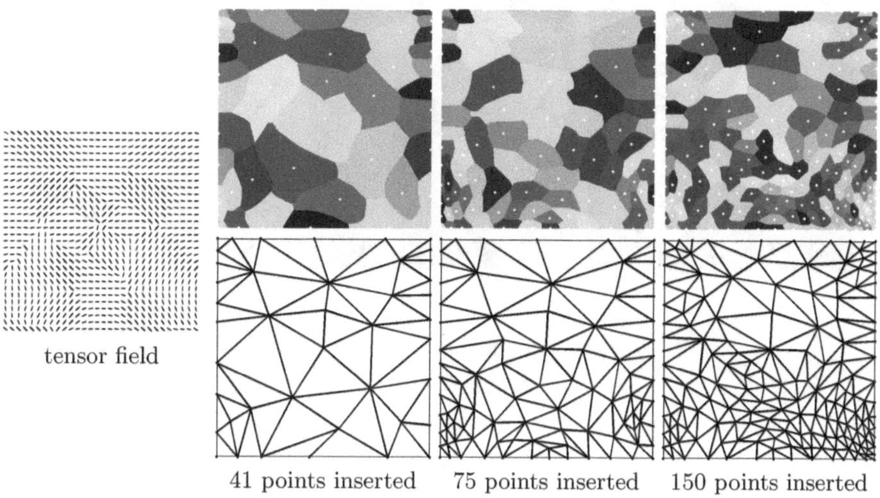

Fig. 7. Meshing of a square with an increasing number of points, which is controlled by the parameter $\rho^\star$. The initial point set $\mathcal{S}$ is the four corner points.

The major obstacle to these insertions (points to get a planar triangulation and Voronoi vertices) is that the boundary $\partial\Omega$ should always be represented in the triangulation. This can be broken if a point x_k is located too close to the boundary. A boundary sub-curve $\theta_{i,j} \in \partial\Omega$ is said to be encroached by a point $x_k \in \mathcal{S}$, $x_k \notin \partial\Omega$, if the Voronoi cell $\mathcal{C}_k$ of x_k satisfies $\mathcal{C}_k \cap \partial\Omega \neq \emptyset$. Thus, the edge (x_i, x_j) cannot be part of the Delaunay triangulation, and it is automatically split by the algorithm by inserting the point $y_{i,j} \in \theta_{i,j} \cap \mathcal{C}_k$ as close to the midpoint

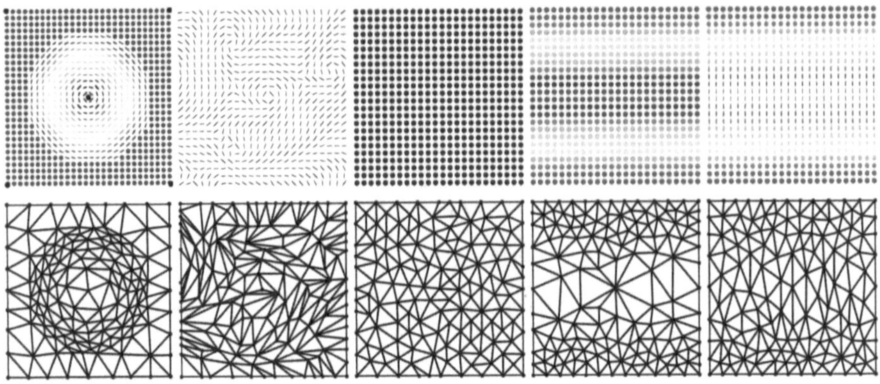

Fig. 8. Meshing of a square with the farthest point strategy, which is controlled by the parameter U^*

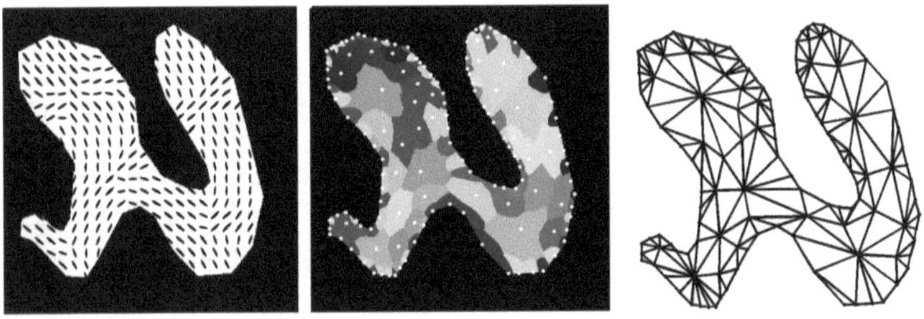

Fig. 9. Meshing of a complex domain

of $\theta_{i,j}$ as possible. Similarly, Voronoi vertices are not added if they encroach any boundary sub-curve (the sub-curve is subdivided instead).

Table 2 details this algorithm. A bound $\rho^\star$ on ρ enforces the refinement to reach some quality criterion, while a bound $U^\star$ enforces a uniform refinement to match some desired triangle density.

Figure 7 shows examples of meshing with a decreasing value of the parameter $\rho^\star$. The number of points is more and more important while $\rho^\star$ is decreasing. Figure 8 shows results of meshing with the farthest point strategy (parameter U^*) for several tensor fields.

The tensor field can be given by the user (top row) or computed from some background image (using the structure tensor, equation (8)), which can be interesting to perform image approximation by linear splines on triangles.

Figure 9 shows that our algorithm can handle domain Ω or arbitrary complexity. Here the tensor field is computed to follow the medial axis of the domain, thus stretching the triangles in an appropriate direction.

5 Conclusion

This paper has detailed how several notions from computational geometry extend seamlessly to the geodesic setting. This allows to incorporate important knowledge about the directionality of the features to solve more efficiently problems in computer vision and graphics. We explore two particular applications, perceptual grouping and domain meshing, where this anisotropy allows to overcome several limitations of previous approaches.

References

1. Leibon, G., Letscher, D.: Delaunay triangulations and Voronoi diagrams for Riemannian manifolds. In: Symp. on Computational Geometry, pp. 341–349. ACM, New York (2000)
2. Frey, P.J., George, P.-L.: Mesh generation. Application to finite elements, 2nd edn. Wiley, Chichester (2008)
3. Labelle, F., Shewchuk, J.R.: Anisotropic voronoi diagrams and guaranteed-quality anisotropic mesh generation. In: Symp. on Computational Geometry, pp. 191–200 (2003)
4. Crandall, M.G., Ishii, H., Lions, P.L.: User's guide to viscosity solutions of second order partial differential equations. Bull. Amer. Math. Soc. 27, 1–67 (1992)
5. Cohen, L.D., Kimmel, R.: Global Minimum for Active Contour models: A Minimal Path Approach. International Journal of Computer Vision 24(1), 57–78 (1997)
6. Sethian, J.A.: Level Sets Methods and Fast Marching Methods, 2nd edn. Cambridge University Press, Cambridge (1999)
7. Kimmel, R., Sethian, J.: Computing Geodesic Paths on Manifolds. Proc. Natl. Acad. Sci. 95(15), 8431–8435 (1998)
8. Spira, A., Kimmel, R.: An efficient solution to the eikonal equation on parametric manifolds. Interfaces and Free Boundaries 6(3), 315–327 (2004)
9. Bronstein, A.M., Bronstein, M.M., Kimmel, R.: Weighted distance maps computation on parametric three-dimensional manifolds. Journal of Computational Physics (accepted, 2007)
10. Prados, E., Lenglet, C., Pons, J.P., Wotawa, N., Deriche, R., Faugeras, O.D., Soatto, S.: Control theory and fast marching methods for brain connectivity mapping. In: Proc. CVPR 2006 (2006)
11. Ren, X., Fowlkes, C.C., Malik, J.: Learning probabilistic models for contour completion in natural images. Int. Journal of Computer Vision 77, 47–63 (2008)
12. Cohen, L.D.: Multiple contour finding and perceptual grouping using minimal paths. Journal of Mathematical Imaging and Vision 14(3), 225–236 (2001)
13. Guy, G., Medioni, G.: Inferring global perceptual contours from local features. International Journal of Computer Vision 20(1-2), 113–133 (1996)
14. Mumford, D.: Elastica and computer vision. In: Algebraic geometry and its applications, pp. 491–506
15. Kothe, U.: Edge and junction detection with an improved structure tensor. In: Michaelis, B., Krell, G. (eds.) DAGM 2003. LNCS, vol. 2781, pp. 25–32. Springer, Heidelberg (2003)
16. Williams, L., Jacobs, D.W.: Stochastic completion fields: a neural model of illusory contour shape and salience. Neural Comput. 9(4), 837–858 (1997)

17. Medioni, G., Lee, M.S., Tang, C.K.: A Computational Framework for Segmentation and Grouping. Elsevier, Amsterdam (2000)
18. Ruppert, J.: A Delaunay refinement algorithm for quality 2-dimensional mesh generation. J. Algorithms 18(3), 548–585 (1995)
19. Chew, L.P.: Guaranteed-quality mesh generation for curved surfaces. In: Proc. of SCG 1993, pp. 274–280. ACM, New York (1993)
20. Borouchaki, H., George, P., Mohammadi, B.: Delaunay mesh generation governed by metric specifications. Part I. algorithms. Finite Elem. Anal. Des. 25(1-2), 61–83 (1997)
21. Bossen, F.J., Heckbert, P.S.: A pliant method for anisotropic mesh generation. In: 5th Intl. Meshing Roundtable, October 1996, pp. 63–74 (1996)
22. Yamakawa, S., Shimada, K.: High quality anisotropic tetrahedral mesh generation via ellipsoidal bubble packing. In: IMR, pp. 263–274 (2000)
23. Yokosuka, Y., Imai, K.: Guaranteed-quality anisotropic mesh generation for domains with curves. In: Proc. of EWCG 2006 (2006)
24. Boissonnat, J.D., Wormser, C., Yvinec, M.: Anisotropic diagrams: Labelle shewchuk approach revisited. In: CCCG, pp. 266–269 (2005)
25. Feng, Z., Hotz, I., Hamann, B., Joy, K.I.: Anisotropic noise samples. IEEE Transactions on Visualization and Computer Graphics 14(2), 342–354 (2008)
26. Du, Q., Wang, D.: Anisotropic centroidal Voronoi tessellations and their applications. SIAM Journal on Scientific Computing 26(3), 737–761 (2005)
27. Valette, S., Chassery, J., Prost, R.: Generic remeshing of 3d triangular meshes with metric-dependent discrete Voronoi diagrams. IEEE Trans. Visu. Comp. Graph. 14(2), 369–381 (2008)
28. Peyré, G., Cohen, L.D.: Geodesic remeshing using front propagation. Int. J. Comput. Vision 69(1), 145–156 (2006)

Regularized Partial Matching of Rigid Shapes

Alexander M. Bronstein and Michael M. Bronstein

Technion – Israel Institute of Technology, Haifa 32000, Israel
{bron,mbron}@cs.technion.ac.il

Abstract. Matching of rigid shapes is an important problem in numerous applications across the boundary of computer vision, pattern recognition and computer graphics communities. A particularly challenging setting of this problem is partial matching, where the two shapes are dissimilar in general, but have significant similar parts. In this paper, we show a rigorous approach allowing to find matching parts of rigid shapes with controllable size and regularity. The regularity term we use is similar to the spirit of the Mumford-Shah functional, extended to non-Euclidean spaces. Numerical experiments show that the regularized partial matching produces better results compared to the non-regularized one.

1 Introduction

Shape matching is one of the cornerstone problems arising in many applications across the boundaries of computer vision, shape recognition and computer graphics communities. By saying *matching*, we actually refer here to two separate problems: *similarity* and *correspondence*. The first problem consists of computing a "distance" between the shapes and is usually dealt with in computer vision and pattern recognition applications. The purpose of the second is finding the matching points between the shapes; it arises in computer graphics community in shape synthesis applications. Usually, the two problems are inter-related and one can be solved as a byproduct of the other.

Many matching algorithms exist in the literature depending on the nature of shapes and their representation [17]. A more complicated setting of shape matching is encountered when the shapes have similar parts but are dissimilar as a whole. For example, the shapes of a centaur and a human fall into this category [9]. A semantically correct matching between such shapes would be the upper part of the body, which both the centaur and the human share. We call this type of matching *partial*. Partial matching plays a crucial role in many practical problems, in which the data to be matched are not available entirely due to acquisition imperfections. For example, if one wishes to match two instances of the same shape acquired using a three-dimensional scanner, the shapes may be different due to scanning artifacts (noise, holes, etc.) It is often important to have control over the selected parts, e.g. controlling their size and regularity.

In computer graphics applications, rigid shape matching is usually performed using different flavors of the classical *iterative closest point* (ICP) algorithm [6,2,12,8]. ICP algorithms try to optimally align the shapes by finding a rigid transformation minimizing a surface-to-surface distance between them. Partial matching can be performed using ICP with the rejection or down-weighting of points with a "bad" correspondence. This approach allows to find best matching parts of shapes, but does not allow direct

D. Forsyth, P. Torr, and A. Zisserman (Eds.): ECCV 2008, Part II, LNCS 5303, pp. 143–154, 2008.

control of their size and regularity. Another common family of approaches is based on finding a set of features on the shapes, followed by computation of invariant local descriptor for each feature. The reader is referred to [17] for a comprehensive review of these methods. A recent study on using local descriptor-based approaches for partial shape similarity is presented in [7].

In the computer vision community, the problem of partial matching of objects is sometimes approached using the "recognition by parts" idea [15]: segmenting the shape into significant parts and matching pairs of parts as whole shapes [10,16,3]. The main difficulty of this approach is that the definition of "significant parts" is semantic rather than geometric, and therefore, automatically finding such parts is not a well-defined problem [1]. Trying to avoid this problem, Latecki *et al.* [11] proposed a partial similarity criterion that removed the arbitrarity of shape partition. A meaningful part is defined as the most similar common part of two shapes, and is practically found by simplifying the shapes until they look the most similar. Bronstein *et al.* [4] extended this idea, proposing a Paretian framework for the computation of partial matching of *nonrigid* shapes. Their approach consists of solving a multicriterion optimization problem trying to simultaneously maximize the similarity and the size of the parts and allows control the size of the selected parts. The main disadvantage of this approach is that it ignores the shape of the selected parts, which theoretically can be irregular.

In this paper, we show that the method of [4] can also be used for partial matching of rigid shapes and improve it by taking into account the regularity of parts besides their size. The regularity term proposed here is similar to the spirit of the Mumford-Shah [13] and Chan-Vese [5] functionals. The rest of the paper is organized as follows. In Section 2, we present the problem of partial rigid shape matching and iterative closest point (ICP) algorithms with rejection of points attempting to solve this problem. We point out the main drawback of this method: the part size and regularity are not directly controllable. In Section 3, we present our approach for regularized partial shape matching with controllable part size and regularity and in Section 4 describe its numerical computation. Section 5 shows experimental results. Finally, Section 6 concludes the paper.

2 Problem Definition

Let X and Y be two rigid shapes, considered here as a subsets of the Euclidean space $\mathbb{R}^3$. The shapes are said to be *congruent* if applying a *rigid motion* to one of the shape we can obtain the other, or in other words, $X = RY + t$ (here R is a rotation matrix and t is a translation vector). Using congruence as an equivalence relation, we say that two rigid shapes are equivalent if they are congruent.

In practice, shapes rarely happen to be precisely congruent. We can still use the degree of incongruence of two shapes in order to quantify their dissimilarity. It is common to use the *Hausdorff distance*

$$d_{\mathrm{H}}(X,Y) = \max\left\{ \max_{x\in X}\min_{y\in Y} \|x-y\|\,,\ \max_{y\in Y}\min_{x\in X} \|x-y\| \right\} \quad (1)$$

to measure the distance between to subsets of $\mathbb{R}^3$. Trying to minimize d_{H} over all the possible rigid motions, we obtain a criterion of rigid shape dissimilarity,

$$d(X,Y) = \min_{R,t} d_{\mathrm{H}}(X, RY+t) \tag{2}$$

X and Y are congruent if and only if $d(X,Y)=0$.

Defining $y^* : X \to Y$ to be the function mapping a point in x on X to the closest point $y^*(x) = \arg\min_{y\in Y} \|x-y\|^2$ on Y, and, analogously, $x^*(y) = \arg\min_{x\in X} \|x-y\|^2$, to be the closest point on X, the Hausdorff distance can be rewritten as

$$d_{\mathrm{H}}(X,Y) = \max\left\{ \max_{x\in X} \|x - y^*(x)\|, \max_{y\in Y} \|y - x^*(y)\| \right\}. \tag{3}$$

Using this formulation, the distance in (2) can be computed using an iterative two-stage process: first, find the correspondence between X and Y. Second, the optimal rigid motion parameters R, t are found to minimize the distance (3) between the closest points on X and Y. The process is repeated until convergence. This class of method is commonly known as *iterative closest point* (ICP) algorithms [6,2]. The outcome of these algorithms is two-fold: the correspondences $x^*(Y)$ and $y^*(X)$ and the dissimilarity $d(X,Y)$ between the shapes X and Y. In practice, the L_∞ distance (3) is usually replaced by an L_2 version,

$$d_{\mathrm{H}}(X,Y) = \int_X \|x - y^*(x)\|^2 dx + \int_Y \|y - x^*(y)\|^2 dy, \tag{4}$$

or its non-symmetric variant, involving only the first term.

2.1 Partially Similar Shapes

In some cases, it may happen that while the shapes X and Y are dissimilar as whole, their parts are similar. Using straightforwardly rigid matching methods will produce meaningless results, failing to capture the partial similarity of shapes. In order to find a partial matching between X and Y, we need to determine similar parts $X' \subseteq X$ and $Y' \subseteq Y$. A common method used for this purpose is a modification of the ICP algorithm with rejection or down-weighting of dissimilar point pairs. Using this approach, the Hausdorff distance in the ICP algorithm is replaced for example by

$$d_{\mathrm{H}}(X,Y) = \int_X \|x - y^*(x)\|^2 w(x) dx, \tag{5}$$

where $w(x)$ is a weighting function rejecting points with bad correspondence. The "quality" of correspondence can be determined by the distance between x and $y^*(x)$ or the mismatch between the normals to X and Y at the points x and $y^*(x)$, or a combination of both. For example, $w(x)$ can be defined as

$$w(x) = \begin{cases} 1 : \langle n_X(x), n_Y(y^*(x))\rangle \geq \tau_{\mathrm{n}} \text{ and } \|x - y^*(x)\| \leq \tau_{\mathrm{d}} \\ 0 : \text{else} \end{cases} \tag{6}$$

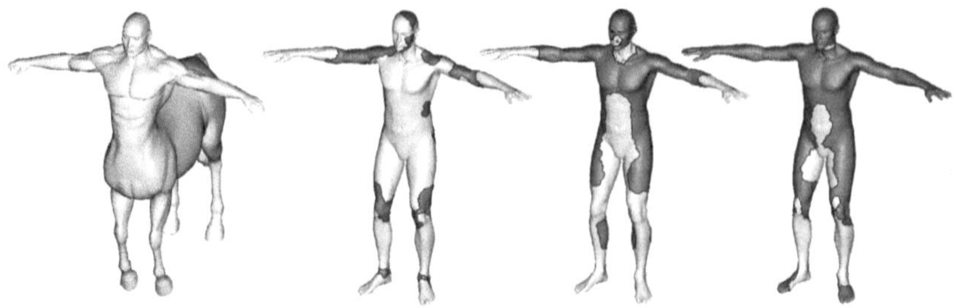

Fig. 1. Matching between a centaur (left) and a man (three rightmost shapes) using ICP with different degree of rejection of points, corresponding approximately to 25%, 50%, and 75% of the shape area (left-to-right). Note that the obtained parts suffer from irregularity.

where τ_n and τ_d are some fixed thresholds. Points for which $w = 0$ will be rejected, i.e., will not contribute to $d_\mathrm{H}(X,Y)$. The matching parts are defined as $X' = \{x \in X : w(x) = 1\}$ and $Y' = y^*(X')$.

By changing the rejection thresholds τ_n and τ_d, it is possible to control the size (area) of the parts (Figure 1). However, this control is indirect, as there is no explicit relation between the surface-to-surface distance and the area of the parts – the size of the parts is known only *post factum*. Another drawback of this method is evident from Figure 1: the matching parts are irregular and contain multiple disconnected components and holes. The influence of the rejection thresholds on the part size and regularity is indirect and depends on the specific shapes, and thus is not explicitly controllable.

3 Regularized Partial Matching of Shapes

The main contribution of this paper is a framework for partial matching of shapes with controllable size and regularity of parts. We call this approach *regularized partial matching*. Given two partially similar shapes X and Y, our goal is to select two parts $X' \subseteq X$ and $Y' \subseteq Y$, such that they are as regular, similar and large as possible. It is clear that the three criteria are conflicting. For example, increasing the size of the parts makes them less similar, and in order to make the parts regular we may sacrifice both size and similarity.

More formally, we have a multicriterion optimization problem with three objectives: dissimilarity $d_\mathrm{H}(X',Y')$, *partiality* $p(X')+p(Y') = \mathrm{area}(X) - \mathrm{area}(X') + \mathrm{area}(Y) - \mathrm{area}(Y')$ and *irregularity* $r(X') + r(Y')$, which we wish to simultaneously minimize over all the possible parts,

$$\min_{X' \subseteq X, Y' \subseteq Y} (d(X',Y'), p(X') + p(Y'), r(X') + r(Y')). \tag{7}$$

The solution of the multicriterion optimization problem (7) is the set of parts (X^*, Y^*) achieving an optimal tradeoff between the criteria, in the sense that there exists no other pair of parts (X', Y') with $d_\mathrm{H}(X',Y') < d(X^*,Y^*)$, $p(X')+p(Y') < p(X^*)+p(Y^*)$ and $r(X')+r(Y') < r(X^*)+r(Y^*)$ holding simultaneously. Such a solution is called

Pareto optimal and is not unique. One of the ways to solve the above problem is by fixing some partiality p_0 and minimizing

$$\min_{X' \subseteq X, Y' \subseteq Y} d(X', Y') + \mu(r(X') + r(Y')) \text{ s.t. } p(X') + p(Y') \leq p_0, \tag{8}$$

where λ is a non-negative coefficient, determining the relative importance of the similarity and regularity of the parts.

As the criterion of part irregularity, the simplest choice is the length of the part boundary $\partial X'$,

$$r(X') = \int_{\partial X'} d\ell. \tag{9}$$

Using this definition, the minimization problem of the part irregularity $r(X')$ subject to fixed partiality $p(X') = p_0$ can be regarded as an *isoperimetric problem*. In case of two-dimensional shapes (subsets of the Euclidean plane) the minimum is achieved by a circle, which in many applications is undoubtfully the most regular shape. Unfortunately, there exists no known extension of this result to curved surfaces, on which, in general, we may find two parts with the same area and boundary length having an arbitrarily large number of disconnected components. In the case when the topological regularity of the part is important, $r(X')$ can be defined as the *genus* of X', which according to the Gauss-Bonnet theorem can be expressed as

$$r(X') = 1 - \frac{1}{2}\int_{X'} K(x)dx - \frac{1}{2}\int_{\partial X'} \kappa_g(x)d\ell, \tag{10}$$

where K and κ_g are the Gaussian and geodesic curvatures, respectively. Such a definition will penalize parts having holes or multiple disconnected components (and, thus, larger genus), and provide the desired topological regularity. In general, different irregularity criteria can be defined to suit the specific application needs. In the sequel, we will stick to the boundary length irregularity (9) for simplicity.

3.1 Fuzzy Formulation

The solution of (8) involves minimization over the set of all pairs of parts of X and Y, which can be thought of as minimization over all pairs of binary *membership functions* $u : X \to \{0, 1\}$ and $v : Y \to \{0, 1\}$, specifying for each point in X and Y whether it belongs to the part or not. Such a discrete minimization problem is clearly computationally intractable. As a remedy, we replace the binary membership functions u and v by a *fuzzy* approximation $u : X \to [0, 1]$ and $v : Y \to [0, 1]$, bringing the problem back to a tractable continuous formulation. We will denote by $\tilde{X} = (X, u)$ and $\tilde{Y} = (Y, v)$ the *fuzzy parts* of X and Y, respectively. In this formulation, the dissimilarity between two parts is given by

$$d(\tilde{X}, \tilde{Y}) = \min_{R,t} d_{\mathrm{H}}(\tilde{X}, R\tilde{Y} + t), \tag{11}$$

where $R\tilde{Y} + t = (RY + t, v)$, and

$$d_{\mathrm{H}}(\tilde{X}, \tilde{Y}) = \int_X \|x - y^*(x)\|^2 u(x)dx + \int_Y \|y - x^*(y)\|^2 v(y)dy. \tag{12}$$

The sets of the closest points $x^*(Y)$ and $y^*(X)$ are defined as before, and are functions of X and Y. Denoting by $e_{X,Y}(x) = \|x - y^*(x)\|^2$ and $e_{Y,X}(y) = \|y - x^*(y)\|^2$ the local measures of surface misalignment, we can formulate the dissimilarity as

$$d(\tilde{X}, \tilde{Y}) = \min_{R,t} \int_X e_{X,RY+t}(x)u(x)dx + \int_Y e_{RY+t,X}(y)v(y)dy. \tag{13}$$

Here, for the sake of simplicity, we will stick to these basic definitions of surface misalignment. In practice, other measures involving, for example, the misalignment of the normals of X and Y can be used as well.

To conclude our discussion on fuzzy approximation, we need to "fuzzify" the partiality and the irregularity terms. The partiality of a fuzzy part $\tilde{X}$ straightforwardly becomes

$$p(\tilde{X}) = \int_X (1 - u(x))dx. \tag{14}$$

The irregularity term is slightly more elaborate, since in the fuzzy part there is no more boundary in the strict sense. However, adopting the Mumford-Shah spirit, we can replace integration along the boundary by integration of the band in which the membership function changes from small to large values [13,5],

$$r(\tilde{X}) = \int_X h(u(x)) \|\nabla_X u(x)\| \, dx \tag{15}$$

where $h(t) \approx \delta(t - 0.5)$ is an approximation of the Dirac delta function, and $\nabla_X u(x)$ is the intrinsic gradient of u at the point x. The quantity $\|\nabla_X u(x)\|$ can be thought of as the length of the extrinsic gradient vector $\nabla_{\mathbb{R}^3} u$ projected on the tangent space of X at a point x.

3.2 Alternating Minimization Algorithm

Substituting the fuzzy approximation of dissimilarity, partiality and regularity into the minimization problem (8), we obtain

$$\begin{aligned} \min_{R,t,u,v} \; & \int_X (e_{X,RY+t}(x)u(x) + \mu h(u(x)) \|\nabla_X u(x)\|)dx + \\ & \int_Y (e_{RY+t,X}(y)v(y) + \mu h(v(y)) \|\nabla_Y v(y)\|)dy \\ \text{s.t.} \; & \int_X u(x)dx + \int_Y v(y)dy \geq (1 - p_0)(\text{area}(X) + \text{area}(Y)). \end{aligned} \tag{16}$$

Here, for convenience, we normalized the partiality parameter p_0 by the total area of X and Y. By fixing the membership functions u and v, the minimization over R and t becomes

$$\min_{R,t} \int_X e_{X,RY+t}(x)u(x)dx + \int_Y e_{RY+t,X}(y)v(y)dy, \tag{17}$$

which can be solved using a weighted ICP algorithm. On the other hand, fixing R and t yields the linearly constrained nonlinear minimization problem

$$\begin{aligned}\min_{u,v} & \int_X (e_{X,Y}(x)u(x) + \mu h(u(x)) \|\nabla_X u(x)\|)dx + \\ & \int_Y (e_{Y,X}(y)v(y) + \mu h(v(y)) \|\nabla_Y v(y)\|)dy \\ \text{s.t.} & \int_X u(x)dx + \int_Y v(y)dy \geq (1 - p_0)(\text{area}(X) + \text{area}(Y)). \end{aligned} \quad (18)$$

We can therefore solve (16) by alternatingly solving (17) and (18), which is expressed in the following framework algorithm:

1. *Weighted rigid alignment:* Fix u and v, and compute R and t minimizing (17).
2. Update $Y = RY + t$, and compute the misalignment fields $e_{X,Y}$ and $e_{Y,X}$.
3. *Optimal part selection:* Fix $e_{X,Y}$ and $e_{Y,X}$, and find u and v minimizing (18).

In what follows, we are going to present a discretization and a numerical scheme for this algorithm.

4 Numerical Framework

We assume the shape X to be given as a triangular mesh with N vertices $\{\mathbf{x}_1, ..., \mathbf{x}_N\} \subset \mathbb{R}^3$ represented as an $N \times 3$ matrix $\mathbf{X}$, and T faces, represented as a $T \times 3$ matrix $\mathbf{T}$ of vertex indices. For brevity, we will continue writing X referring to the mesh $(\mathbf{X}, \mathbf{T})$. We denote by $\mathbf{P}$ be the sparse $T \times N$ matrix, whose elements p_{ij} are set to $\frac{1}{3}$ if triangle i shares the vertex $\mathbf{x}_j$, and 0 otherwise. The matrix $\mathbf{P}$ can be thought of as a projection operator, converting a function defined on the mesh vertices into a function defined on the mesh faces. We denote by $\mathbf{a}'$ the $T \times 1$ vector, whose elements are areas of the mesh triangles. The $N \times 1$ vector and by $\mathbf{a} = \mathbf{P}\mathbf{a}'$ discretizes the area elements at each vertex on the mesh X. The membership function u is represented as the $N \times 1$ vector $\mathbf{u} = (u_1, ..., u_N)^{\mathrm{T}}$. The shape Y is represented similarly as a mesh $Y = (\mathbf{Y}, \mathbf{S})$, containing M vertices $\{\mathbf{y}_1, ..., \mathbf{y}_M\}$ and S faces. We denote by $\mathbf{Y}$, $\mathbf{S}$, $\mathbf{Q}$, $\mathbf{b}$, and $\mathbf{v}$ the counterparts of $\mathbf{X}$, $\mathbf{T}$, $\mathbf{P}$, $\mathbf{a}$, and $\mathbf{u}$, respectively.

4.1 Weighted Rigid Alignment

The solution of (17) in Step 1 of our framework algorithm is carried out using a symmetric weighted ICP algorithm. We fix $\mathbf{u}$ and $\mathbf{v}$, and iterate the following steps until convergence:

1. Compute the closest points

$$\mathbf{x}_i^* = \arg \min_{\mathbf{x} \in \{\mathbf{x}_1, ..., \mathbf{x}_N\}} \|\mathbf{y}_i - \mathbf{x}\|; \qquad \mathbf{y}_i^* = \arg \min_{\mathbf{y} \in \{\mathbf{y}_1, ..., \mathbf{y}_M\}} \|\mathbf{x}_i - \mathbf{y}\|, \quad (19)$$

and construct the $3 \times M$ matrix $\mathbf{X}^*$ and the $3 \times N$ matrix $\mathbf{Y}^*$, whose rows are the $\mathbf{x}_i^*$, and the $\mathbf{y}_i^*$, respectively.

2. Compute the weighted centroids of the matrices $(\mathbf{X}, \mathbf{X}^*)$ and $(\mathbf{Y}^*, \mathbf{Y})$,

$$\overline{\mathbf{x}} = \sum_{i=1}^{N} u_i \mathbf{x}_i + \sum_{i=1}^{M} v_i \mathbf{x}_i^*, \qquad \overline{\mathbf{y}} = \sum_{i=1}^{N} u_i \mathbf{y}_i^* + \sum_{i=1}^{M} v_i \mathbf{y}_i, \tag{20}$$

and construct the centered matrices $\overline{\mathbf{X}} = (\mathbf{x}_1 - \overline{\mathbf{x}}, ..., \mathbf{x}_M - \overline{\mathbf{x}}, \mathbf{x}_1^* - \overline{\mathbf{x}}, ..., \mathbf{x}_N^* - \overline{\mathbf{x}})$ and $\overline{\mathbf{Y}} = (\mathbf{y}_1^* - \overline{\mathbf{y}}, ..., \mathbf{y}_M^* - \overline{\mathbf{y}}, \mathbf{y}_1 - \overline{\mathbf{y}}, ..., \mathbf{y}_N - \overline{\mathbf{y}})$.
3. Compute the weighted covariance matrix $\mathbf{H} = \overline{\mathbf{X}}\mathbf{W}\overline{\mathbf{Y}}^{\mathrm{T}}$, where $\mathbf{W} = \mathrm{diag}(\mathbf{u}, \mathbf{v})$ is an $(N+M) \times (N+M)$ diagonal matrix with the elements of $\mathbf{u}$ and $\mathbf{v}$ on the diagonal. Compute the singular value decomposition $\mathbf{H} = \mathbf{U}\boldsymbol{\Lambda}\mathbf{V}^{\mathrm{T}}$, and set $\mathbf{R} = \mathbf{U}\mathbf{V}^{\mathrm{T}}$, and $\mathbf{t} = \overline{\mathbf{x}} - \mathbf{R}\overline{\mathbf{y}}$.
4. Replace the columns $\mathbf{y}_i$ of $\mathbf{Y}$ with their transformed versions $\mathbf{R}\mathbf{y}_i + \mathbf{t}$.

As a stopping condition, a norm of the change of $\mathbf{Y}$, e.g. $\sum \|\mathbf{y}_i - \mathbf{R}\mathbf{y}_i - \mathbf{t}\|^2 \leq \tau$ can be used for some small threshold τ. Once the ICP algorithm is terminated, we compute the local surface misalignment as

$$\begin{aligned} \mathbf{e}_{X,Y} &= (\|\mathbf{x}_1 - \mathbf{y}_1^*\|^2, ..., \|\mathbf{x}_N - \mathbf{y}_N^*\|^2)^{\mathrm{T}}, \\ \mathbf{e}_{Y,X} &= (\|\mathbf{y}_1 - \mathbf{x}_1^*\|^2, ..., \|\mathbf{y}_M - \mathbf{x}_M^*\|^2)^{\mathrm{T}}. \end{aligned} \tag{21}$$

Note that like in the classical ICP, when $e_{X,Y}(x) = \|x - y^*(x)\|^2$ and $e_{Y,X}(y) = \|y - x^*(y)\|^2$ are used, the best alignment between $\tilde{X}$ and $\tilde{Y}$ for a given correspondence is given analytically using the SVD of the covariance matrix $\mathbf{H}$.

4.2 Optimal Part Selection

The solution of (18) in Step 3 of our framework algorithm consists of finding the best parts $\mathbf{u}$ and $\mathbf{v}$ given fixed $\mathbf{e}_{X,Y}$ and $\mathbf{e}_{Y,X}$. We discretize the dissimilarity term $d(\tilde{X}, \tilde{Y})$ as $d(\mathbf{u}, \mathbf{v}) = \mathbf{e}_{X,Y}^{\mathrm{T}}\mathbf{A}\mathbf{u} + \mathbf{e}_{Y,X}^{\mathrm{T}}\mathbf{B}\mathbf{v}$, where $\mathbf{A} = \mathrm{diag}(\mathbf{a})$ and $\mathbf{B} = \mathrm{diag}(\mathbf{b})$ are, respectively, the $N \times N$ and $M \times M$ diagonal matrices with the elements of $\mathbf{a}$ and $\mathbf{b}$ on the diagonal. Arranging $\mathbf{e}_{X,Y}$ and $\mathbf{e}_{Y,X}$ into an $(M+N) \times 1$ vector $\mathbf{e}^{\mathrm{T}} = (\mathbf{e}_{X,Y}^{\mathrm{T}}, \mathbf{e}_{Y,X}^{\mathrm{T}})$, and $\mathbf{u}$ and $\mathbf{v}$ into the vector $\mathbf{w}^{\mathrm{T}} = (\mathbf{u}^{\mathrm{T}}, \mathbf{v}^{\mathrm{T}})$, we can write $d(\mathbf{w}) = \mathbf{e}^{\mathrm{T}}\mathbf{C}\mathbf{w}$, where $\mathbf{C} = \mathrm{diag}(\mathbf{a}, \mathbf{b})$ is the $(M+N) \times (M+N)$ diagonal matrix, containing the elements of $\mathbf{a}$ and $\mathbf{b}$ on the diagonal.

In order to discretize the irregularity term $r(\tilde{X})$, we first need to approximate the norm of the intrinsic gradient $\nabla_X u$ on the mesh X. Assuming a first-order approximation of u, the gradient $\nabla_X u$ is constant one each face of the mesh. Given a triangle i formed by the vertices $\mathbf{x}_{t_{i,1}}, \mathbf{x}_{t_{i,2}}, \mathbf{x}_{t_{i,3}}$, the gradient norm can be expressed as $g_i = \sqrt{\boldsymbol{\delta}^{\mathrm{T}}(\mathbf{X}_i^{\mathrm{T}}\mathbf{X}_i)^{-1}\boldsymbol{\delta}}$, where $\mathbf{X}_i = (\mathbf{x}_{t_{i,2}} - \mathbf{x}_{t_{i,1}}, \mathbf{x}_{t_{i,3}} - \mathbf{x}_{t_{i,1}})$ is a 3×2 matrix with the local system of coordinates of triangle t, and $\boldsymbol{\delta} = (u_{t_{i,2}} - u_{t_{i,1}}, u_{t_{i,3}} - u_{t_{i,1}})$ is the vector of the membership function differences. Arranging the g_i's as the elements of the $T \times 1$ vector $\mathbf{g}_X(\mathbf{u})$, we can express the irregularity term as $r(\mathbf{u}) = h(\mathbf{u})^{\mathrm{T}}\mathbf{A}\mathbf{P}\mathbf{g}_X(\mathbf{u})$, where $h(\mathbf{u})^{\mathrm{T}} = (h(u_1), ..., h(u_N))$ is the non-linearity h applied element-wise to the vector $\mathbf{u}$. The second irregularity term $r(\mathbf{v})$ is discretized in a similar way, yielding $r(\mathbf{v}) = h(\mathbf{v})^{\mathrm{T}}\mathbf{B}\mathbf{Q}\mathbf{g}_Y(\mathbf{v})$. Since the gradients $\mathbf{g}_X(\mathbf{u})$ and $\mathbf{g}_Y(\mathbf{v})$ are intrinsic, they are invariant to rigid motion of X and Y and depend only on $\mathbf{u}$ and $\mathbf{v}$. In terms of the combined vector $\mathbf{w}$, the regularity $r(\tilde{X}) + r(\tilde{Y})$ can be written as

$$r(\mathbf{w}) = (h(\mathbf{u})^{\mathrm{T}}, h(\mathbf{v})^{\mathrm{T}}) \begin{pmatrix} \mathbf{AP} & \\ & \mathbf{BQ} \end{pmatrix} \begin{pmatrix} \mathbf{g}_X(\mathbf{u}) \\ \mathbf{g}_Y(\mathbf{v}) \end{pmatrix} = h(\mathbf{w})^{\mathrm{T}} \mathbf{D} \mathbf{g}(\mathbf{w}), \quad (22)$$

where $\mathbf{D} = \mathbf{C}\,\mathrm{diag}(\mathbf{P}, \mathbf{Q})$ and $\mathbf{g}(\mathbf{w})^{\mathrm{T}} = (\mathbf{g}_X(\mathbf{u})^{\mathrm{T}}, \mathbf{g}_Y(\mathbf{v})^{\mathrm{T}})$.

The partiality term $p(\tilde{X})$ is discretized as the inner product $p(\mathbf{u}) = \mathbf{a}^{\mathrm{T}}(\mathbf{1} - \mathbf{u})$, where $\mathbf{1}$ is an $N \times 1$ vector of ones. The second partiality term is discretized in a similar manner as $p(\mathbf{v}) = \mathbf{b}^{\mathrm{T}}(\mathbf{1} - \mathbf{v})$. Using these expressions, the inequality constraint in minimization problem (17) can be rewritten as the linear constraint $\mathbf{a}^{\mathrm{T}}\mathbf{u} + \mathbf{b}^{\mathrm{T}}\mathbf{v} \geq (1 - p_0)\alpha$, where $\alpha = \mathbf{a}^{\mathrm{T}}\mathbf{1} + \mathbf{b}^{\mathrm{T}}\mathbf{1}$ is the sum of the areas of the meshes X and Y. Using the combined vector $\mathbf{w}$, we can write $\mathbf{c}^{\mathrm{T}}\mathbf{w} \geq (1 - p_0)\alpha$ with $\mathbf{c}^{\mathrm{T}} = (\mathbf{a}^{\mathrm{T}}, \mathbf{b}^{\mathrm{T}})$.

Plugging the discretized dissimilarity, partiality and irregularity terms into problem (18), we obtain the

$$\min_{\mathbf{w}} \mathbf{e}^{\mathrm{T}}\mathbf{C}\mathbf{w} + \mu\, h(\mathbf{w})^{\mathrm{T}} \mathbf{D}\mathbf{g}(\mathbf{w}) \text{ s.t. } \begin{cases} \mathbf{c}^{\mathrm{T}}\mathbf{w} \geq (1 - p_0)\alpha \\ \mathbf{0} \leq \mathbf{w} \leq \mathbf{1} \end{cases} \quad (23)$$

Since the objective function has a well-structured sparse Hessian, the use of second-order minimization algorithms is appealing. In our implementation, the problem is solved using a reduced Newton descent algorithm [14], where the projection enforces the bound constraints $\mathbf{0} \leq \mathbf{w} \leq \mathbf{1}$. The inequality constraint $\mathbf{c}^{\mathrm{T}}\mathbf{w} \geq (1 - p_0)\alpha$ is introduced by adding an augmented Lagrangian term to the objective function. The reduced Newton system is solved using the modified Cholesky factorization, guaranteeing a descent direction at each iteration. Since part regularization has a low-pass filtering effect, the solution of (18) at low mesh resolution usually provides a sufficiently good initialization for the following resolution levels. Our practice shows that about four resolution levels improve convergence by about one order of magnitude.

The solution of the regularized matching problem yields the vector $\mathbf{w}$, representing the fuzzy part membership functions, and the transformation $\mathbf{R}, \mathbf{t}$ bringing those parts into best alignment. Since generally $\mathbf{w}$ will contain a continuous range of values from 0 to 1, the final step of the algorithm requires to "defuzzify" the fuzzy parts, i.e. convert $\mathbf{u}$ and $\mathbf{v}$ into binary membership functions. This is done by finding a threshold τ satisfying

$$\tau = \arg\min_{\tau} \tau \quad \text{s.t.} \quad \mathbf{c}^{\mathrm{T}} \chi(\mathbf{w} \geq \tau) \geq (1 - p_0)\alpha, \quad (24)$$

where $\chi(\mathbf{w} \geq \tau) = (\chi_1, ..., \chi_{M+N})^{\mathrm{T}}$ is the indicator vector with the elements $\chi_i = 1$ for $w_i \geq \tau$, and $\chi_i = 0$ otherwise. The crisp parts are then given by $\mathbf{u} = \chi(\mathbf{u} \geq \tau)$ and $\mathbf{v} = \chi(\mathbf{v} \geq \tau)$.

5 Results

In order to visualize the advantage of the proposed regularized matching framework, we compared it to unregularized matching performed by means of ICP with rejection of points based on distance and angle between corresponding normal vectors. Figures 2

Fig. 2. Best partial matching of four rigid objects without regularization. Partiality p_0 is set to 0.4, 0.4, 0.65 and 0.65 (left-to-right). Two topmost rows present the best matching parts; bottom row shows their alignment. Highlighted are the matching parts.

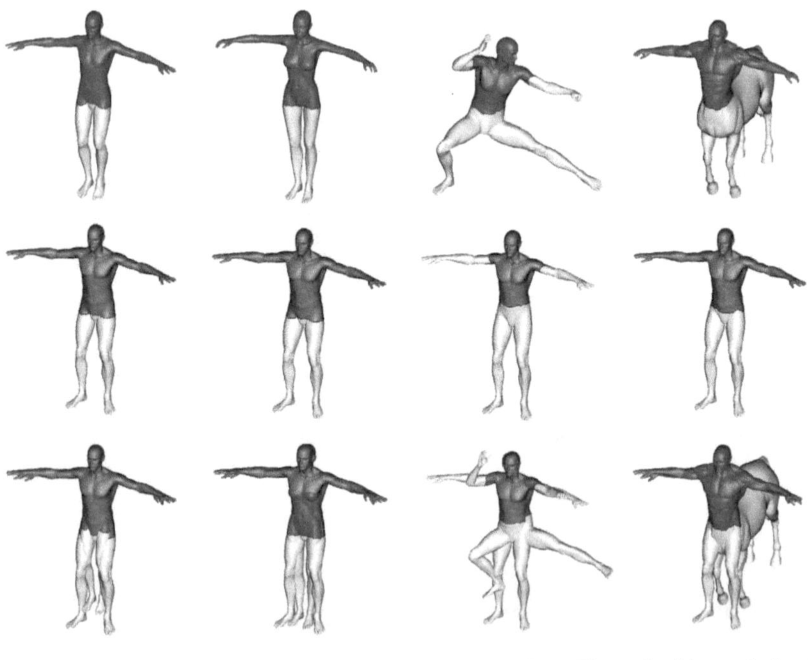

Fig. 3. Best partial matching of four rigid objects from Figure 2 with regularization

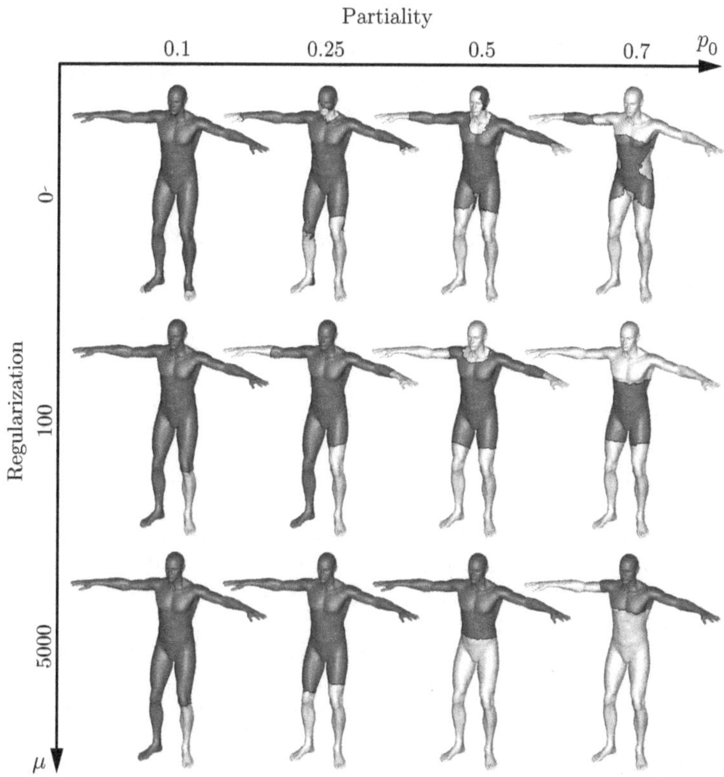

Fig. 4. Best partial matching of the male shape with the female shape from Figure 2 for different values of p_0 and μ. Note how for a fixed value of p_0, the area of the part remains the same, while its boundary length decreases with the increase of μ, resulting in a more regular part (see e.g. the third column).

and 3 show partial matching of four different rigid shapes with and without regularization, respectively. The use of regularization produces meaningful parts, unlike the unregularized case, where the parts contain many spurious disconnected components and holes. By changing the parameters p_0 and μ, we can control the size and the regularity of the selected parts.

Figure 4 shows the optimal part for different values of partiality p_0 and regularization parameter μ. Fixing p_0 and increasing μ gradually decreases the boundary length of the part, making it more regular.

6 Conclusions

In this paper, we considered the Paretian approach for partial matching of shapes proposed in [4] and based on a multicriterion problem of simultaneous maximization of similarity and size of the parts. We extended this approach, proposing a different definition of part significance, taking into account the regularity of parts besides their size. We showed an efficient computation scheme based on fuzzy approximation, which allowed formulating the regularization in the spirit of the Mumford-Shah functional [13]. Though down-weighting of outliers is common in data fitting (including ICP algorithms), we believe that our paper proposes for the first time a rigorous method for selecting the weights optimally, in a controllable manner, and driven by the shape geometry, rather than resorting to heuristics. In our future studies, we intend to explore other definitions of topological and geometric regularity, and extensions of the proposed framework to nonrigid shapes.

References

1. Basri, R., Costa, L., Geiger, D., Jacobs, D.: Determining the similarity of deformable shapes. Vision Research 38, 2365–2385 (1998)
2. Besl, P.J., McKay, N.D.: A method for registration of 3D shapes. Trans. PAMI 14, 239–256 (1992)
3. Boiman, O., Irani, M.: Similarity by composition. In: Proc. NIPS (2006)
4. Bronstein, A.M., Bronstein, M.M., Bruckstein, A.M., Kimmel, R.: Paretian similarity of non-rigid objects. In: Sgallari, F., Murli, A., Paragios, N. (eds.) SSVM 2007. LNCS, vol. 4485, pp. 264–275. Springer, Heidelberg (2007)
5. Chan, T.F., Vese, L.A.: Active contours without edges. Trans. Image Processing 10(2), 266–277 (2001)
6. Chen, Y., Medioni, G.: Object modeling by registration of multiple range images. In: Proc. Conf. Robotics and Automation (1991)
7. Funkhouser, T., Shilane, P.: Partial matching of 3D shapes with priority-driven search. In: Proc. SGP (2006)
8. Gelfand, N., Mitra, N.J., Guibas, L., Pottmann, H.: Robust global registration. In: Proc. Symp. Geometry Processing (SGP) (2005)
9. Jacobs, D., Weinshall, D., Gdalyahu, Y.: Class representation and image retrieval with non-metric distances. Trans. PAMI 22, 583–600 (2000)
10. Kupeev, K., Wolfson, H.: On shape similarity. In: Proc. ICPR, pp. 227–237 (1994)
11. Latecki, L.J., Lakaemper, R., Wolter, D.: Optimal Partial Shape Similarity. Image and Vision Computing 23, 227–236 (2005)
12. Mitra, N.J., Gelfand, N., Pottmann, H., Guibas, L.: Registration of point cloud data from a geometric optimization perspective. In: Proc. Eurographics Symposium on Geometry Processing, pp. 23–32 (2004)
13. Mumford, D., Shah, J.: Boundary Detection by Minimizing Functionals. Image Understanding (1990)
14. Nocedal, J., Wright, S.: Numerical optimization. Springer, New York (1999)
15. Pentland, A.: Recognition by parts. In: Proc. ICCV, pp. 612–620 (1987)
16. Siddiqi, K., Kimia, B.: A shock grammar for recognition. In: Proc. CVPR, pp. 507–513 (1996)
17. Tangelder, J.W.H., Veltkamp, R.C.: A survey of content based 3D shape retrieval methods. In: Proc. Shape Modeling and Applications, pp. 145–156 (2004)

Compressive Sensing for Background Subtraction

Volkan Cevher[1], Aswin Sankaranarayanan[2], Marco F. Duarte[1], Dikpal Reddy[2], Richard G. Baraniuk[1], and Rama Chellappa[2]

[1] Rice University, ECE, Houston TX 77005
[2] University of Maryland, UMIACS, College Park, MD 20947

Abstract. Compressive sensing (CS) is an emerging field that provides a framework for image recovery using sub-Nyquist sampling rates. The CS theory shows that a signal can be reconstructed from a small set of random projections, provided that the signal is sparse in some basis, e.g., wavelets. In this paper, we describe a method to directly recover background subtracted images using CS and discuss its applications in some communication constrained multi-camera computer vision problems. We show how to apply the CS theory to recover object silhouettes (binary background subtracted images) when the objects of interest occupy a small portion of the camera view, i.e., when they are sparse in the spatial domain. We cast the background subtraction as a sparse approximation problem and provide different solutions based on convex optimization and total variation. In our method, as opposed to learning the background, we learn and adapt a low dimensional compressed representation of it, which is sufficient to determine spatial innovations; object silhouettes are then estimated directly using the compressive samples without any auxiliary image reconstruction. We also discuss simultaneous appearance recovery of the objects using compressive measurements. In this case, we show that it may be necessary to reconstruct one auxiliary image. To demonstrate the performance of the proposed algorithm, we provide results on data captured using a compressive single-pixel camera. We also illustrate that our approach is suitable for image coding in communication constrained problems by using data captured by multiple conventional cameras to provide 2D tracking and 3D shape reconstruction results with compressive measurements.

1 Introduction

Background subtraction is fundamental in automatically detecting and tracking moving objects with applications in surveillance, teleconferencing [1,2] and even 3D modeling [3]. Usually, the foreground or *the innovation* of interest occupies a sparse spatial support, as compared to the background and may be caused by the motion and the appearance change of objects within the scene. By obtaining the object silhouettes on a single image plane or multiple image planes, a background subtraction algorithm can be performed.

In all applications that require background subtraction, the background and the test images are typically fully sampled using a conventional camera. After the foreground estimation, the remaining background images are either discarded or embedded back into the background model as part of a learning scheme [2]. This sampling process is

D. Forsyth, P. Torr, and A. Zisserman (Eds.): ECCV 2008, Part II, LNCS 5303, pp. 155–168, 2008.

inexpensive for imaging at the visible wavelengths as the conventional devices are built from silicon, which is sensitive to these wavelengths; however, if sampling at other optical wavelengths is desired, it becomes quite expensive to obtain estimates at the same pixel resolution as new imaging materials are needed. For example, a camera with an array of infrared sensors can provide night vision capability but can also cost significantly more than the same resolution CCD or CMOS cameras.

Recently, a prototype single pixel camera (SPC) was proposed based on the new mathematical theory of *compressive sensing* (CS) [4]. The CS theory states that a signal can be perfectly reconstructed, or can be robustly approximated in the presence of noise, with sub-Nyquist data sampling rates, provided that it is *sparse* in some linear transform domain [5,6]. That is, it has K nonzero transform coefficients with $K \ll N$, where N is the dimension of the transform space. For computer vision applications, it is known that natural images can be sparsely represented in the wavelet domain [7]. Then, according to the CS theory, by taking random projections of a scene onto a set of test functions that are incoherent with the wavelet basis vectors, it is possible to recover the scene by solving a convex optimization problem. Moreover, the resulting *compressive measurements* are robust against packet drops over communication channels with graceful degradation in reconstruction accuracy, as the image information is fully distributed.

Compared to conventional camera architectures, the SPC hardware is specifically designed to exploit the CS framework for imaging. An SPC fundamentally differs from a conventional camera by (i) reconstructing an image using only a single optical photodiode (infrared, hyperspectral, etc.) along with a digital micromirror device (DMD), and (ii) combining the sampling and compression into a single nonadaptive linear measurement process. An SPC can directly scale from the visual spectra to hyperspectral imaging with only a change of the single optical sensor. Moreover, enabled by the CS theory, an SPC can robustly reconstruct the scene from much fewer measurements than the number of reconstructed pixels which define the resolution, given that the image of the scene is compressible by an algorithm such as the wavelet-based JPEG 2000.

Conventional cameras can also benefit by processing in the compressive sensing domain if their data is being sent to a central processing location. The naïve approach is to transmit the raw images to the central location. This exacerbates the communication bandwidth requirements. In more sophisticated approaches, the cameras transmit the information within the background subtracted image, which requires an even smaller communication bandwidth than the compressive samples. However, the embedded systems needed to perform reliable background subtraction are power hungry and expensive. In contrast, the compressive measurement process only requires cheaper embedded hardware to calculate inner products with a previously determined set of test functions. In this way, the compressive measurements require comparable bandwidth to transform coding of the raw data. They trade off expensive embedded intelligence for more computational power at the central location, which reconstructs the images and is assumed to have unlimited resources.

The communication bandwidth and camera hardware limitations make it desirable to directly reconstruct the sparse foreground innovations within a scene without any intermediate image reconstruction. The main idea is that the background subtracted images can be represented sparsely in the spatial image domain and hence the CS reconstruction theory should be applicable for directly recovering the foreground. For natural images, we use wavelets as the transform domain. Pseudo-random matrices provide an incoherent set of test functions to recover the foreground image. Then, the following questions surface (i) how can we detect targets without reconstructing an image? and (ii) how can we directly reconstruct the foreground without reconstructing auxiliary images?

In this paper, we describe a method based on CS theory to directly recover the sparse innovations (foreground) of a scene. We first show that the object silhouettes (binary background subtracted images) can be recovered as a solution of a convex optimization or an orthogonal matching pursuit problem. In our method, the object silhouettes are learned directly using the compressive samples without any auxiliary image reconstruction. We then discuss simultaneous appearance recovery of objects using the compressive measurements. In this case, we show that it may be necessary to reconstruct one auxiliary image. To demonstrate the performance of the proposed algorithm, we use field data captured by a compressive camera and provide background subtraction results. We also show results on field data captured by conventional CCD cameras to simulate multiple distributed single-pixel cameras and provide 2D tracking and 3D shape reconstruction results.

While the idea of performing background subtraction on compressed images is not novel, there exist no cameras that record MPEG video directly. Both Aggarwal et al. [8] and Lamarre and Clark [9] perform background subtraction on a MPEG-compressed video using the DC-DCT coefficients of I frames, limiting the resolution of the BS images by 64. Our technique is tailored for CS imaging, and not compressed video files. Lamarre et al. [9] and Wang et al. [10] use DCT coefficients from JPEG pictures and MPEG videos, respectively, for representation. Toreyin et al. [11] similarly operate on the wavelet representation. These methods implicitly perform decompression by working on every DCT/wavelet coefficient of every image. We never have to go to the high dimensional images or representations during background subtraction, making our approach particularly attractive for embedded systems and demanding communication bandwidths. Compared to the eigenbackground work of Oliver et al. [12], random projections are universal so there is no need to update bases - the only basis needed is the sparsity basis for difference images, hence no training is required. The very recent work of Uttam, Goodman and Neifeld [13] considers background subtraction from adaptive compressive measurements, with the assumption that the background-subtracted images lie in a low-dimensional subspace. While this assumption is acceptable when image tiling is performed, background-subtracted images are sparse in an appropriate domain, spanning a union of low-dimensional subspaces rather than a single subspace.

Our specific contributions are as follows:

1. We cast the background subtraction problem as a sparse signal recovery problem where convex optimization and greedy methods can be applied. We employ Basis Pursuit Denoising methods [14] as well as total variation minimization [5] as convex objectives to process field data.
2. We show that it is possible to recover the silhouettes of foreground objects by learning a low-dimensional compressed representation of the background image. Hence, we show that it is not necessary to learn the background itself to sense the

innovations or the foreground objects. We also explain how to adapt this representation so that our approach is robust against variations of the background such as illumination changes.
3. We develop an object detector directly on the compressive samples. Hence, no foreground reconstruction is done until a detection is made to save computation.

2 The Compressive Sensing Theory

2.1 Sparse Representations

Suppose that we have an image $\boldsymbol{X}$ of size $N_1 \times N_2$ and we vectorize it into a column vector $\boldsymbol{x}$ of size $N \times 1$ $(N = N_1 N_2)$ by concatenating the individual columns of $\boldsymbol{X}$ in order. The nth element of the image vector $\boldsymbol{x}$ is referred to as $x(n)$, where $n = 1, \ldots, N$. Let us assume that the basis $\boldsymbol{\Psi} = [\psi_1, \ldots, \psi_N]$ provides a K-sparse representation of $\boldsymbol{x}$:

$$\boldsymbol{x} = \sum_{n=1}^{N} \theta(n)\psi_n = \sum_{l=1}^{K} \theta(n_l)\psi_{n_l}, \tag{1}$$

where $\theta(n)$ is the coefficient of the nth basis vector ψ_n (ψ_n: $N \times 1$) and the coefficients indexed by n_l are the K-nonzero entries of the basis decomposition. Equation (1) can be more compactly expressed as follows

$$\boldsymbol{x} = \boldsymbol{\Psi}\boldsymbol{\theta}, \tag{2}$$

where $\boldsymbol{\theta}$ is an $N \times 1$ column vector with K-nonzero elements. Using $\|\cdot\|_p$ to denote the ℓ_p norm where the ℓ_0 norm simply counts the nonzero elements of $\boldsymbol{\theta}$, we call an image $\boldsymbol{X}$ as K-sparse if $\|\boldsymbol{\theta}\|_0 = K$.

Many different basis expansions can achieve sparse approximations of natural images, including wavelets, Gabor frames, and curvelets [5,7]. In other words, a natural image does not result in an exactly K-sparse representation; instead, its transform coefficients decay exponentially to zero. The discussion below also applies to such images, denoted as compressible images, as they can be well-approximated using the K largest terms of $\boldsymbol{\theta}$.

2.2 Random/Incoherent Projections

In the CS framework, it is assumed that the K-largest $\theta(n)$ are not measured directly. Rather, $M < N$ linear projections of the image vector $\boldsymbol{x}$ onto another set of vectors $\boldsymbol{\Phi} = [\phi'_1, \ldots, \phi'_M]'$ are measured:

$$\boldsymbol{y} = \boldsymbol{\Phi}\boldsymbol{x} = \boldsymbol{\Phi}\boldsymbol{\Psi}\boldsymbol{\theta}, \tag{3}$$

where the vector $\boldsymbol{y}$ $(M \times 1)$ constitutes the compressive samples and the matrix $\boldsymbol{\Phi}$ $(M \times N)$ is called the *measurement matrix*. Since $M < N$, recovery of the image $\boldsymbol{x}$ from the compressive samples $\boldsymbol{y}$ is underdetermined; however, as we discuss below, the additional *sparsity* assumption makes recovery possible.

The CS theory states that when (i) the columns of the sparsity basis $\boldsymbol{\Psi}$ cannot sparsely represent the rows of the measurement matrix $\boldsymbol{\Phi}$ and (ii) the number of measurements M is greater than $\mathcal{O}\left(K \log\left(\frac{N}{K}\right)\right)$, then it is possible to recover the set of nonzero entries of $\boldsymbol{\theta}$ from $\boldsymbol{y}$ [5,6]. Then, the image $\boldsymbol{x}$ can be obtained by the linear transformation of $\boldsymbol{\theta}$ in (1). The first condition is called the incoherence of the two bases and it holds for many pairs of bases, e.g., delta spikes and the sine waves of the Fourier basis. Surprisingly, incoherence also holds with high probability between an arbitrary basis and a randomly generated one, e.g., i.i.d. Gaussian or Bernoulli/Rademacher ± 1 vectors.

2.3 Signal Recovery Via ℓ_1 Optimization

There exists a computationally efficient recovery method based on the following ℓ_1-optimization problem [5,6]:

$$\widehat{\boldsymbol{\theta}} = \arg\min \|\boldsymbol{\theta}\|_1 \quad \text{s. t. } \boldsymbol{y} = \boldsymbol{\Phi}\boldsymbol{\Psi}\boldsymbol{\theta}. \tag{4}$$

This optimization problem, also known as *Basis Pursuit* [6], can be efficiently solved using polynomial time algorithms.

Other formulations are used for recovery from noisy measurements such as Lasso, Basis Pursuit with quadratic constraint [5]. In this paper, we use Basis Pursuit Denoising (BPDN) for recovery:

$$\widehat{\boldsymbol{\theta}} = \arg\min \|\boldsymbol{\theta}\|_1 + \frac{1}{2}\beta\|\boldsymbol{y} - \boldsymbol{\Phi}\boldsymbol{\Psi}\boldsymbol{\theta}\|_2^2, \tag{5}$$

where $0 < \beta < \infty$ [14]. When the images of interest are smooth, a strategy based on minimizing the total variation of the image works equally well [5].

3 CS for Background Subtraction

With background subtraction, our objective is to recover the location, shape and (sometimes) appearance of the objects given a test image over a known background. Let us denote the background, test, and difference images as $\boldsymbol{x}_b$, $\boldsymbol{x}_t$, and $\boldsymbol{x}_d$, respectively. The difference image is obtained by pixel-wise subtraction of the background image from the test image. Note that the support of $\boldsymbol{x}_d$, denoted as $\mathcal{S}_d = \{n | n = 1, \ldots, N; |\boldsymbol{x}_d(n)| \neq 0\}$, gives us the location and the silhouettes of the objects of interest, but not their appearance (see Fig. 1).

3.1 Sparsity of Background Subtracted Images

Suppose that $\boldsymbol{x}_b$ and $\boldsymbol{x}_t$ are typical real-world images in the sense that when wavelets are used as the sparsity basis for $\boldsymbol{x}_b$, $\boldsymbol{x}_t$, and $\boldsymbol{x}_d$, these images can be well approximated with the largest K coefficients with hard thresholding [15], where K is the corresponding sparsity proportional to the cardinality of the image support. The images $\boldsymbol{x}_b$ and $\boldsymbol{x}_t$ differ only on the support of the foreground, which has a cardinality of $P = |\mathcal{S}_d|$ pixels with $P \ll N$. Moreover, we assume that images have uniform complexity in space. We model the sparsity of the real world images as a function of their

Fig. 1. (*Left*) Example background image. (*center*) Test image. (*Right*) Difference image. Note that the vehicle appearance also shows the curb in the background, which it occludes. The images and are from the PETS 2001 database.

size: $K_{\text{scene}} = K_b = K_t = (\lambda_0 \log N + \lambda_1)N$, where $(\lambda_0, \lambda_1) \in \mathbb{R}^2$. We assume that the difference image is also a real-world image on a restricted support (see Fig. 1(c)), and similarly we approximate its sparsity as $K_d = (\lambda_0 \log P + \lambda_1)P$.

The number of compressive samples M necessary to reconstruct $\boldsymbol{x}_b$, $\boldsymbol{x}_t$, and $\boldsymbol{x}_d$ in N dimensions are then given by $M_{\text{scene}} = M_b = M_t \approx K_{\text{scene}} \log(N/K_{\text{scene}})$ and $M_d \approx K_d \log(N/K_d)$. When $M_d < M_{\text{scene}}$, a smaller number of samples is needed to reconstruct the difference image than the background or foreground images. We empirically show in Section 5 that this condition is almost always satisfied when the sizes of the difference images are smaller than original image sizes for natural images.

3.2 The Background Constraint

Let us assume that we have multiple compressive measurements $\boldsymbol{y}_{bi}$ ($M \times 1$, $i = 1, \ldots, B$) of training background images $\boldsymbol{x}_{bi}$, where $\boldsymbol{x}_b$ is their mean. Each compressive measurement is a random projection of the whole image, whose distribution we approximate as an i.i.d. Gaussian distribution with a constant variance $\boldsymbol{y}_{bi} \sim \mathcal{N}(\boldsymbol{y}_b, \sigma^2 \boldsymbol{I})$, where the mean value is $\boldsymbol{y}_b = \boldsymbol{\Phi}\boldsymbol{x}_b$. When the scene changes to include an object which was not part of the background model and we take the compressive measurements, we obtain a test vector $\boldsymbol{y}_t = \boldsymbol{\Phi}\boldsymbol{x}_t$, where $\boldsymbol{x}_d = \boldsymbol{x}_t - \boldsymbol{x}_b$ is sparse in the spatial domain.

In general, the sizes of the foreground objects are relatively smaller than the size of the background image; hence, we model the distribution of the *literally* background subtracted vector as $\boldsymbol{y}_d = \boldsymbol{y}_t - \boldsymbol{y}_b \sim \mathcal{N}(\boldsymbol{\mu}_d, \sigma^2 \boldsymbol{I})$ $(M \times 1)$, where $\boldsymbol{\mu}_d$ is the mean. Note that the appearance of the objects constructed from the samples $\boldsymbol{y}_d$ would correspond to the literal subtraction of the test frame and the background; however, their silhouette is preserved (Fig. 1(c)).

The number of samples M in $\boldsymbol{y}_b$ is greater than M_d as discussed in Sect. 3.1, but is not necessarily greater than or equal to M_b or M_t; hence, it may not be sufficient to reconstruct the background. However, the background image $\boldsymbol{x}_b$ still satisfies the constraint $\boldsymbol{y}_b = \boldsymbol{\Phi}\boldsymbol{x}_b$. To be robust against small variations in the background and noise, we consider the distribution of the ℓ_2 distances of the background frames around their mean $\boldsymbol{y}_b$:

$$\|\boldsymbol{y}_{bi} - \boldsymbol{y}_b\|_2^2 = \sigma^2 \sum_{n=1}^{M} \left(\frac{y_{bi}(n) - y_b(n)}{\sigma} \right)^2 . \quad (6)$$

When M is greater than 30, this sum can be well approximated by a Gaussian distribution due to the central limit theorem. Then, it is straightforward to show that we have $\|\boldsymbol{y}_{bi} - \boldsymbol{y}_b\|_2^2 \sim \mathcal{N}\left(M\sigma^2, 2M\sigma^4\right)$. When we have a test frame with a foreground object, the same distribution becomes $\|\boldsymbol{y}_t - \boldsymbol{y}_b\|_2^2 \sim \mathcal{N}\left(M\sigma^2 + \|\boldsymbol{\mu}_d\|_2^2, 2M\sigma^4 + 4\sigma^2\|\boldsymbol{\mu}_d\|_2^2\right)$.

Since σ^2 scales the whole distribution and $1/M \ll 1$, the logarithm of the ℓ_2 distances in (6) can be approximated quite accurately with a Gaussian distribution. That is, since $u \ll 1$ implies $1 + u \approx e^u$, we have $\mathcal{N}\left(M\sigma^2, 2M\sigma^4\right) = M\sigma^2\mathcal{N}\left(1, \frac{2}{M}\right) = M\sigma^2\left(1 + \sqrt{\frac{2}{M}}\mathcal{N}(0,1)\right) \approx M\sigma^2 \exp\left\{\sqrt{\frac{2}{M}}\mathcal{N}(0,1)\right\}$. This derivation can also motivated by the fact that the square-root of the Chi-squared distribution can be well approximated by a Gaussian [16].

Hence, (6) can be used to approximate

$$\log \|\boldsymbol{y}_{bi} - \boldsymbol{y}_b\|_2^2 \sim \mathcal{N}\left(\mu_{bg}, \sigma_{bg}^2\right), \tag{7}$$

where μ_{bg} is the mean and σ_{bg}^2 is the variance term, which does not depend on the additive noise in pixel measurements. Equation (7) allows some variability around the constraint $\boldsymbol{y}_b = \boldsymbol{\Phi}\boldsymbol{x}_b$ that the background image needs to satisfy in order to cope with the small variations of the background and the measurement noise. However, the samples $\boldsymbol{y}_d = \boldsymbol{y}_t - \boldsymbol{y}_b$ can be used to recover the foreground objects. We learn the log-Normal parameters in (7) from the data using maximum likelihood techniques.

3.3 Object Detector Based on CS

Before we attempt any reconstruction, it is a good idea to determine if the test image has any differences from the background. Using the results from Sect. 3.2, the ℓ_2 distance of $\boldsymbol{y}_t$ from $\boldsymbol{y}_b$ can be subsequently approximated by

$$\log \|\boldsymbol{y}_t - \boldsymbol{y}_b\|_2^2 \sim \mathcal{N}\left(\mu_t, \sigma_t^2\right). \tag{8}$$

When the object is small, σ_t^2 should be on the same order size of σ_{bg}^2, while μ_t is different from μ_{bg} in (7). Then, to test the hypothesis of whether there is a new object, the optimal detector would be a simple threshold test since we would be comparing two Gaussian distributions with similar variances. When σ_t^2 is significantly different from σ_{bg}^2, the optimal test can be a two sided threshold test [17]. For our case, we simply use a constant times the standard deviation of the background as a threshold and declare that there is a new object if $\left|\log \|\boldsymbol{y}_t - \boldsymbol{y}_b\|_2^2 - \mu_{bg}\right| \geq c\sigma_{bg}$.

3.4 Foreground Reconstruction

For foreground reconstruction, we use BPDN with a fixed point continuation method [18] and total variation (TV) optimization with an interior point method [5] on the background subtracted compressive measurements. The BPDN solver is the fastest among the proposed algorithms because it solves an unconstrained optimization problem. During the reconstruction, we lose the actual appearance of the objects as the obtained measurements also contain information about the background. Although it is known that the subtracted image is a sum of two components that exclusively appear in $\boldsymbol{x}_b$ and $\boldsymbol{x}_t$, it is difficult, if not impossible, to unmix them without taking enough measurements to recover $\boldsymbol{x}_b$ or $\boldsymbol{x}_t$. Hence, if the appearances of the objects are needed, a straightforward way to obtain them would be to either reconstruct the test image by taking enough compressive samples and then use the binary foreground image as a mask, or reconstruct and mask the background image and then add the result to the foreground estimate.

3.5 Adaptation of the Background Constraint

We define two types of changes in a background: drifts and shifts. A background drift consists of gradual changes that occur in the background such as illumination changes in the scene and may result in immediate unwanted foreground estimates. A background shift is a major and sudden change in the definition of the background, such as a new vehicle parked within the scene. Adapting to background shifts at the sensing level is quite difficult because high level logical operations are required, such as detecting the new object and deciding that it is uninteresting. However, adapting to background drifts is essential for a robust background subtraction system as it has immediate impacts on the foreground recovery.

The background constraint $\boldsymbol{y}_b$ needs to be updated continuously if the background subtraction system is to be robust against the background drifts. Otherwise, the drifts may accumulate and trigger unwanted detections. In the compressive sensing framework, this can be done as follows. Once we obtain an estimate of the difference image $\widehat{\boldsymbol{x}}_d$ with one of the reconstruction algorithms discussed in the previous section, we determine the compressive samples that should be generated by it: $\widehat{\boldsymbol{y}}_d = \boldsymbol{\Phi}\widehat{\boldsymbol{x}}_d$. Since we already have $\boldsymbol{y}_d = \boldsymbol{y}_t - \boldsymbol{y}_b$, we can substitute the de-noised difference estimate to obtain the background estimate of the current frame: $\widehat{\boldsymbol{y}}_b = \boldsymbol{y}_t - \widehat{\boldsymbol{y}}_d$. Then, a running average can be used to update the background with a learning rate of $\alpha \in (0,1)$ as follows:

$$\boldsymbol{y}_b^{\{j+1\}} = \alpha\left(\boldsymbol{y}_t^{\{j\}} - \widehat{\boldsymbol{y}}_d^{\{j\}}\right) + (1-\alpha)\boldsymbol{y}_b^{\{j\}}, \tag{9}$$

where j is the time index.

Unfortunately, this update rule does not suffice for compensating background shifts, such as new stationary targets. Consider a pixel whose intensity value changes because of a background shift. This pixel will then be identified as an outlier in the background model. The corresponding pixel in the background model will not be updated in (9). Hence, for all future frames, the pixel will continue to be classified as part of the foreground. This problem can be handled by allowing for a second moving average of the frames, which updates all pixels within the image as in [19].

Hence, we use the following updates:

$$\begin{aligned}\boldsymbol{y}_{\mathrm{ma}}^{\{j+1\}} &= \gamma\boldsymbol{y}_t^{\{j\}} + (1-\gamma)\boldsymbol{y}_{\mathrm{ma}}^{\{j\}},\\ \boldsymbol{y}_b^{\{j+1\}} &= \alpha\left(\boldsymbol{y}_t^{\{j\}} - \widehat{\boldsymbol{y}}_e^{\{j\}}\right) + (1-\alpha)\boldsymbol{y}_b^{\{j\}},\end{aligned} \tag{10}$$

where $\boldsymbol{y}_{\mathrm{ma}}$ is the simple moving average, $\gamma \in (0,1)$ is the moving average learning rate, and $\widehat{\boldsymbol{y}}_e = \boldsymbol{\Phi}\widehat{\boldsymbol{x}}_{\mathrm{ma}}$. Consider a global illumination change. The moving average update integrates the pixel's illumination change over time, whose speed depends on γ. In

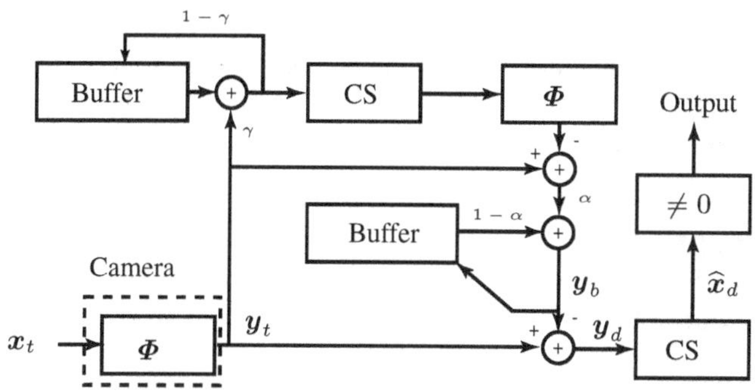

Fig. 2. Block diagram of the proposed method

subsequent frames, the value of the moving average will approach the intensity value observed at the pixel. This implies that when used as a detection image, the moving average will stop detecting the pixel as foreground. Once this happens, the pixel will be updated in the background update, making the background model adaptive to global changes in illumination. A disadvantage of this approach is that if the targets stay stationary for extended periods of time, they become part of the background. However, if they move again, they can be detected. Figure 2 illustrates the outline of the proposed background subtraction method.

4 Limitations

In this section, we discuss some of the limitations of the specific compressive sensing approach to the background subtraction presented in this paper. Some of these limitations can be caused by the hardware architecture, whereas others are due to our image models. Note that our formulation is general enough that we do not require an SPC for operation. CS can be used for rateless coding of BS images. If a centralized vision system is used with no background subtraction at the camera, then our methods can be used at conventional cameras for processing in the compressive domain to reduce communication bandwidth and be robust against packet drops.

The SPC architecture uses a DMD to generate a random sampling pattern and sends the resulting inner product of the incident light field from the scene with the random pattern to the optical sensor to create a compressive measurement. By changing the random pattern in time, a set of M consecutive measurements can be made about the scene using the same optical sensor, which form the measurement vector $\boldsymbol{y}$. The current DMD arrays can change their geometric configuration approximately 10 to 40K times per second. For example, with a rate of 30K times per second, we can construct at most a 300×300 resolution background subtracted image with 1% compression ratios at 30fps. Although the resolution may not be sufficient for some applications, it will improve as the capabilities of the DMD arrays increase.

In our background modeling, we assume that the background and foreground images exhibit sparsity. We argued that the background subtracted image has a lower sparsity and hence can be reconstructed with fewer samples that is necessary to reconstruct the background or the foreground images. When the images of interest do not show sparsity (e.g., they are white noise), our approach can still be applied. That is, the difference image $\boldsymbol{x}_d$ is always sparse regardless of the sparsities of $\boldsymbol{x}_b$ and $\boldsymbol{x}_t$ if its support cardinality P is much smaller than N.

5 Experiments

5.1 Background Subtraction with an SPC

We performed background subtraction experiments with an SPC; in our test, the background $\boldsymbol{x}_b$ consists of the standard test *Mandrill* image, with the foreground $\boldsymbol{x}_t$ consisting of a white rectangular patch as shown in Fig. 3. Both the background and the foreground were acquired using pseudorandom compressive measurements ($\boldsymbol{y}_b$ and $\boldsymbol{y}_t$, respectively) generated by a Mersenne Twister algorithm with a 64×64 pixel resolution [20]. We obtain measurements for the subtraction image as $\boldsymbol{y}_d = \boldsymbol{y}_t - \boldsymbol{y}_b$. We reconstructed both the background, test, and difference images, using TV minimization. The reconstruction is performed using several measurement rates ranging from 0.5% to 50%. In each case, we compare the subtraction image reconstruction with the difference between the reconstructed test and background images. The resulting images are shown in Fig. 3, and show that for low rates the background and test images are not recovered accurately, and therefore the subtraction performs poorly; however, the sparser foreground innovation is still recovered correctly from the difference of the measurements, with rates as low as 1% being able to recover the foreground at this low resolution.

5.2 The Sparsity Assumption

In our formulation, we assumed that the sparsity of natural images has the following form: $K = (\lambda_0 \log N + \lambda_1)N$. To test this assumption, we used the Berkeley Segmentation Data Set (BSDS) as a natural image database [21] and obtained wavelet approximations of various block sizes varying from 2×2 to 256×256 pixels. To approximate the sparsity K of any given tile size, we determined the minimum number of wavelet coefficients that results in a compression with -40dB distortion with respect to the image itself. Figure 4 shows that our sparsity assumption is justified for natural images, and illustrates that the necessary number of compressive samples is monotonic with the tile size. Therefore, if the innovations in the image are smaller than the image, it takes fewer compressive samples to recover them. In fact, the total number of samples necessary to reconstruct is rather close to linear: $M \approx \kappa N^{1-\delta}$ where $\delta \ll 1$. In general, the λ parameters are scene specific (Fig. 4(*Right*)). Hence, the exact number of compressive measurements needed may vary.

5.3 Multi-view Ground Plane Tracking

Background subtraction forms an important pre-processing component for many vision applications. In this regard, it is important to see if the imagery generated using compressive measurements can be used in such applications. In this section, we demonstrate

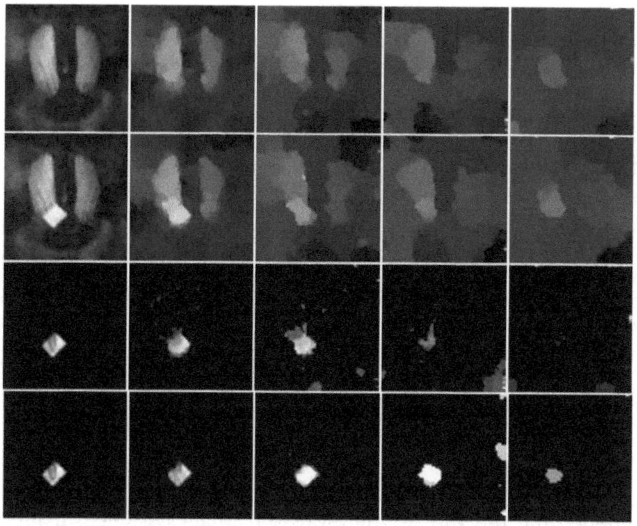

Fig. 3. Background subtraction experimental results using an SPC. Reconstruction of background image (top row) and test image (second row) from compressive measurements. Third row: conventional subtraction using the above images. Fourth row: reconstruction of difference image directly from compressive measurements. The columns correspond to measurement rates M/N of 50%, 5%, 2%, 1% and 0.5%, from left to right. Background subtraction from compressive measurements is feasible at lower measurement rates than standard background subtraction.

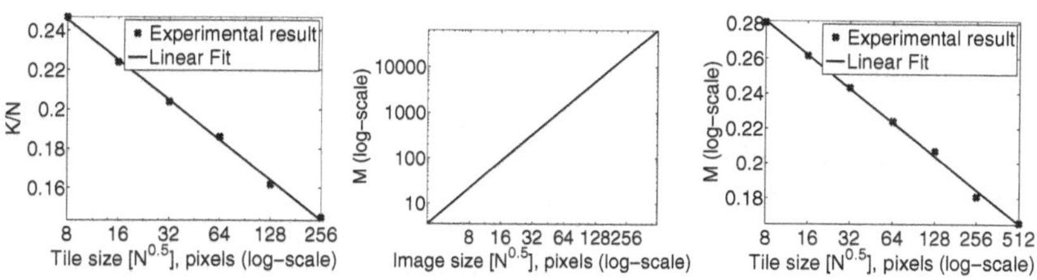

Fig. 4. (*Left*) Average sparsity over N as a function of the tile size for the images in BSDS. (center) Number of compressive measurements needed to reconstruct an image of different sizes from BSDS. (*Right*) Average sparsity over N as a function of the tile size for the images in PETS 2001 data set.

a multi-view tracking application where accurate background subtraction is key in determining overall system performance.

In Figure 5, we show results on a multi-view ground plane tracking algorithm over a sequence of 300 frames with 20% compression ratio. We first obtain the object silhouettes using the compressive samples at each view. We use wavelets as the sparsifying basis Ψ. At each time instant, the silhouettes are mapped on to the ground planes and averaged. Objects on the ground plane (e.g., the feet) combine in synergy while those off the plane are in parallax and do not support each other. We then threshold to

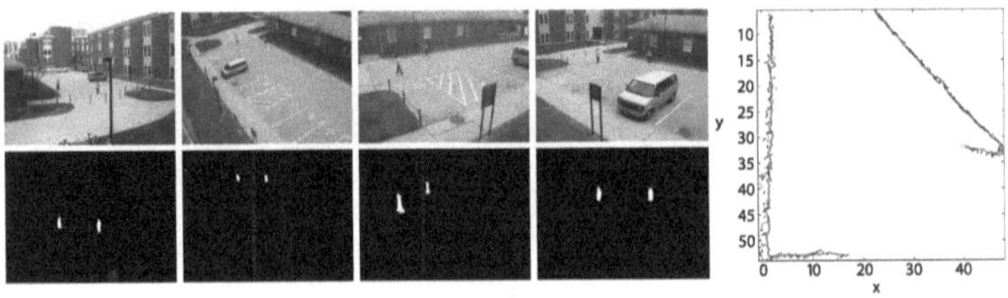

Fig. 5. Tracking results on a video sequence of 300 frames. (*Left*) The first two rows show sample images and background subtraction results using the compressive measurements, respectively. The background subtracted blobs are used to detect target location on the ground plane. The right figure shows the detected points using CS (blue dots) as well as the detected points using full images (black). The distances are in meters.

obtain potential target locations as in [22]. The outputs indicate the background subtracted images are sufficient to generate detections that compare well against the detections generated using the full non-compressed images. Hence, using our method, the communication bandwidth of a multi camera localization system can be reduced to one-fifth if the estimation is done at a central location.

5.4 Adaptation to Illumination Changes

To compare the performance of the background constraint adaptations (9) (drift adaptive) and (10) (shift adaptive), we test them on a sequence where there is a global illumination change due to sunlight. To emphasize the differences, we use the delta basis (0/1 in spatial domain) as the sparsifying basis $\boldsymbol{\Psi}$. This basis creates much noisier background subtraction images than wavelets, but it is quite illustrative for the purposes of this comparison.

Figure 6 shows the results of the comparison. The images on top are the original images. The middle row corresponds to the update in (10) whereas the bottom row images correspond to the update in (9). The update in (10) allows the background constraint to keep track of the sudden change in illumination. Hence, the resulting images are cleaner and continue to improve. This results in much lower false alarm rates for the same detection probability (see Fig. 6(*Right*)). For the receiver operating characteristics (ROC) curves, we use the full images, run the background subtraction algorithm proposed in [19], and obtain baseline background subtracted images. We then compare the pixels on the resulting target from different updates to calculate the detection rate. We also compare the spurious detections in the rest of the images to generate the ROC curve.

5.5 Silhouettes vs. Difference Images

We have used a multi camera set up for a 3D voxel reconstruction using the compressive measurements. Figure 7(*Left*) shows the ground truth and the difference image reconstructed using CS, which incorporates elements from the background, such as the camera setup behind the subject, affecting the final reconstruction. Hence, the difference

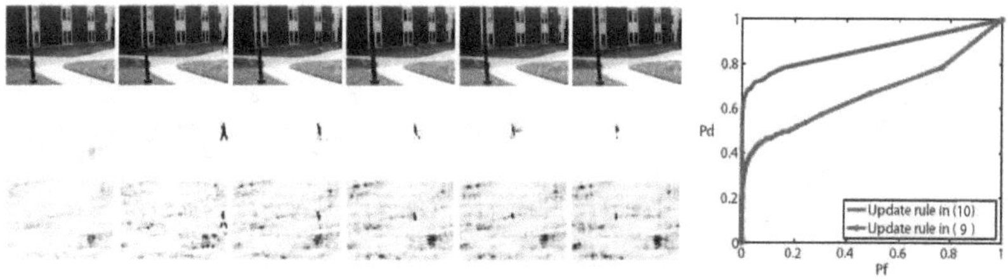

Fig. 6. Background subtraction results on a sequence with changing illumination using (9) and (10) for background constraint updates. Outputs are shown with identical parameters used for both models. Note that for the same detection output, the update rule (10) produces much less false alarm. However, (10) has twice the computational cost as (9).

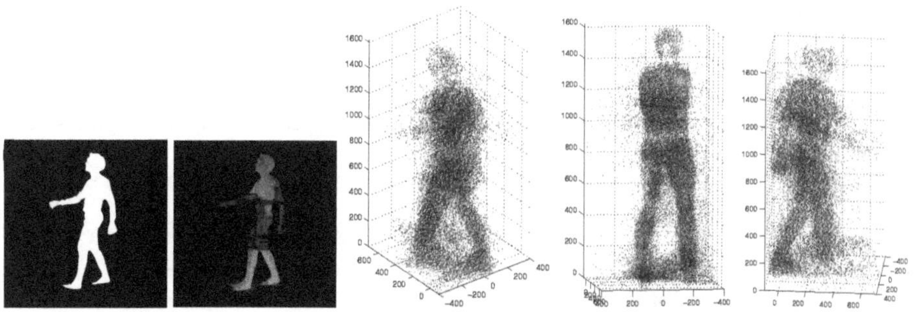

Fig. 7. (*Left*) Ground truth detections marked in white and unthresholded background difference image reconstruction using compressive samples with 40% compression. (*Right*) Reconstructed 3D point clouds of the target.

images do not always result in the desired silhouettes. Figure 7(*Right*) shows the voxel reconstruction with four cameras with 40% compression, which is visually satisfactory despite the artifacts in the difference images.

6 Conclusions

We demonstrated that the CS framework can be used to directly reconstruct sparse innovations on a background scene with a significantly fewer data samples than the conventional methods. As opposed to acquiring the minimum amount of measurements to recover a background and the test image, we can exploit the sparsity of the foreground to perform background subtraction by using even fewer measurements (M_d measurements as opposed to M_b). We illustrated that due to the linear nature of the measurements, it is still possible to adapt to the changes in the background directly in the compressive domain. In addition, it is possible to formulate an object detector. By exploiting sparsity in background subtracted images in multi-view tracking and 3D reconstruction problems, we can reduce sampling costs and alleviate communication and storage burdens while obtaining comparable estimation performance.

References

1. Elgammal, A., Harwood, D., Davis, L.: Non-parametric model for background subtraction. In: IEEE FRAME-RATE Workshop. Springer, Heidelberg (1999)
2. Piccardi, M.: Background subtraction techniques: a review. In: IEEE International Conference on Systems, Man and Cybernetics, vol. 4 (2004)
3. Cheung, G.K.M., Kanade, T., Bouguet, J.Y., Holler, M.: Real time system for robust 3 D voxel reconstruction of human motions. In: CVPR, pp. 714–720 (2000)
4. Wakin, M.B., Laska, J.N., Duarte, M.F., Baron, D., Sarvotham, S., Takhar, D., Kelly, K.F., Baraniuk, R.G.: An architecture for compressive imaging. In: ICIP, Atlanta, GA, October 2006, pp. 1273–1276 (2006)
5. Candes, E.: Compressive sampling. In: Proceedings of the International Congress of Mathematicians (2006)
6. Donoho, D.L.: Compressed Sensing. IEEE Trans. Info. Theory 52(4), 1289–1306 (2006)
7. Mallat, S., Zhang, S.: Matching pursuits with time-frequency dictionaries. Matching pursuits with time-frequency dictionaries 41(12), 3397–3415 (1993)
8. Aggarwal, A., Biswas, S., Singh, S., Sural, S., Majumdar, A.K.: Object Tracking Using Background Subtraction and Motion Estimation in MPEG Videos. In: ACCV, pp. 121–130. Springer, Heidelberg (2006)
9. Lamarre, M., Clark, J.J.: Background subtraction using competing models in the block-DCT domain. In: ICPR (2002)
10. Wang, W., Chen, D., Gao, W., Yang, J.: Modeling background from compressed video. In: IEEE Int. Workshop on VSPE of TS, pp. 161–168 (2005)
11. Töreyin, B.U., Çetin, A.E., Aksay, A., Akhan, M.B.: Moving object detection in wavelet compressed video. Signal Processing: Image Communication 20(3), 255–264 (2005)
12. Oliver, N., Rosario, B., Pentland, A.: A Bayesian Computer Vision System for Modeling Human Interactions. In: ICVS. Springer, Heidelberg (1999)
13. Uttam, S., Goodman, N.A., Neifeld, M.A.: Direct reconstruction of difference images from optimal spatial-domain projections. In: Proc. SPIE, San Diego, CA, August 2008, vol. 7096 (2008)
14. Chen, S.S., Donoho, D.L., Saunders, M.A.: Atomic Decomposition by Basis Pursuit. SIAM Journal on Scientific Computing 20, 33 (1998)
15. Mallat, S.: A Wavelet Tour of Signal Processing. Academic Press, London (1999)
16. Cevher, V., Chellappa, R., McClellan, J.H.: Gaussian approximations for energy-based detection and localization in sensor networks. In: IEEE Statistical Signal Processing Workshop, Madison, WI, August 26–29 (2007)
17. Van Trees, H.L.: Detection, Estimation, and Modulation Theory, Part I. John Wiley & Sons, Inc., Chichester (1968)
18. Hale, E.T., Yin, W., Zhang, Y.: A fixed-point continuation method for ℓ_1-regularized minimization with applications to compressed sensing. Technical Report TR07-07, Rice University Department of Computational and Applied Mathematics, Houston, TX (2007)
19. Joo, S., Zheng, Q.: A Temporal Variance-Based Moving Target Detector. In: Proc. IEEE Int. Workshop on Performance Evaluation of Tracking and Surveillance (PETS) (2005)
20. Matsumoto, M., Nishimura, T.: Mersenne Twister: A 623-Dimensionally Equidistributed Uniform Pseudo-Random Number Generator. ACM Transactions on Modeling and Computer Simulation 8(1), 3–30 (1998)
21. Martin, D., Fowlkes, C., Tal, D., Malik, J.: A database of human segmented natural images and its application to evaluating segmentation algorithms and measuring ecological statistics. In: Proc. 8th Int'l Conf. Computer Vision, July 2001, vol. 2, pp. 416–423 (2001)
22. Khan, S.M., Shah, M.: A multi-view approach to tracking people in crowded scenes using a planar homography constraint. In: Leonardis, A., Bischof, H., Pinz, A. (eds.) ECCV 2006. LNCS, vol. 3954, pp. 133–146. Springer, Heidelberg (2006)

Robust 3D Pose Estimation and Efficient 2D Region-Based Segmentation from a 3D Shape Prior

Samuel Dambreville, Romeil Sandhu, Anthony Yezzi, and Allen Tannenbaum

Georgia Institute of Technology

Abstract. In this work, we present an approach to jointly segment a rigid object in a 2D image and estimate its 3D pose, using the knowledge of a 3D model. We naturally couple the two processes together into a unique energy functional that is minimized through a variational approach. Our methodology differs from the standard monocular 3D pose estimation algorithms since it does not rely on local image features. Instead, we use global image statistics to drive the pose estimation process. This confers a satisfying level of robustness to noise and initialization for our algorithm, and bypasses the need to establish correspondences between image and object features. Moreover, our methodology possesses the typical qualities of region-based active contour techniques with shape priors, such as robustness to occlusions or missing information, without the need to evolve an infinite dimensional curve. Another novelty of the proposed contribution is to use a unique 3D model surface of the object, instead of learning a large collection of 2D shapes to accommodate for the diverse aspects that a 3D object can take when imaged by a camera. Experimental results on both synthetic and real images are provided, which highlight the robust performance of the technique on challenging tracking and segmentation applications.

1 Motivation and Related Work

2D image segmentation and 2D-3D pose estimation are ubiquitous tasks in computer vision applications and have received a great deal of attention in the past few years. These two fundamental techniques are usually studied separately in the literature. In this work, we combine both approaches in a variational framework. To appreciate the contribution of this work, we recall some of the results and specifics of both fields.

2D-3D pose estimation aims at determining the pose of a 3D object relative to a calibrated camera from one unique or a collection of 2D images. By knowing the mapping between the world coordinates and image coordinates from the camera calibration matrix, and after establishing correspondences between 2D features in the image and their 3D counterparts on the model, it is then possible to solve the pose transformation (from a set of equations that express these correspondences). The literature concerned with 3D pose estimation is very large and a complete survey is beyond the scope of this paper. However, most methods can be distinguished by the type of *local* image features used to establish correspondences, such as points [1], lines or segments [2,3], multi-part curve segments [4], or complete contours [5,6].

Segmentation consists of separating an object from the background in an image. The geometric active contour (GAC) framework, in which a curve is evolved continuously to

D. Forsyth, P. Torr, and A. Zisserman (Eds.): ECCV 2008, Part II, LNCS 5303, pp. 169–182, 2008.

capture the boundaries of an object, has proven to be quite successful at performing this task. Originally, the method focused on extracting local image features such as edges to perform segmentation; see [7,8] and the references therein. However, edge-based techniques can suffer from the typical drawbacks that arise from using local image features: high sensitivity to noise or missing information, and a multitude of local minima that result in poor segmentations. Region-based approaches, which use global image statistics inside and outside the contour, were shown to drastically improve the robustness of segmentation results [9,10,11,12]. Region-based techniques are able to deal with various statistics of the object and background such as distinct mean intensities [10], Gaussian distributions [11,12] or intensity histograms [13,14] as well as a wide variety of photometric descriptors such as grayscale values, color or texture [15]. Further improvement of the GAC approach consists of learning the shape of objects and constrain the contour evolution to adopt familiar shapes, to make up for poor segmentation results obtained in the presence of noise, clutter, occlusion or when the statistics of the object and background are difficult to distinguish (see e.g., [16,17,18,19]).

Motivation/Contribution. Our goal is to combine the strengths of both techniques and to avoid some of their typical weaknesses, to robustly both segment 2D images and estimate the pose of an arbitrary 3D object which shape is known.

In particular, we use a region-based approach to continuously drive the pose estimation process. This global approach avoids using local image features and, hence, addresses two shortcomings that typically arise from doing so in many 2D-3D pose estimation algorithms: Firstly, finding the correspondence between local features in the image and on the model is a non-trivial task, due for instance to their viewpoint dependency - no local correspondences need to be found in our global approach. Secondly, local image features may not even exist or can be difficult to detect in a reliable and robust fashion in the presence of noise clutter or occlusion. Furthermore, simplifying assumptions usually need to be made on the class of shapes that a 2D-3D pose estimation technique can handle. Many approaches are limited to simple shapes that can be described using geometric primitives such as corners, lines, circles or cylinders. Recent work focused on free-form objects, which admit a manageable parametric description as in [5]. However, even this type of algebraic approaches can become unmanageable for objects of arbitrary and complex shape. Our approach can deal with rigid object of *arbitrary* shape, represented by a 3D level set [20] or a 3D cloud of points (Figure 1).

Conversely, a shortcoming of the GAC framework using shape priors is that 2D shapes are usually learned to segment 2D images. Hence, a large collection of 2D shapes needs to be learned to represent the wide variation in aspect that most natural 3D objects take, when projected onto the 2D image plane. Our region-based approach benefits from the knowledge of the object shape that is compactly described by a *unique* 3D model. Acquisition of 3D models can be readily accomplished using range scans [21] or structure from motion approaches [22], notably. In addition, and in contrast to the GAC framework, the proposed method does not involve the evolution of an infinite dimensional contour to perform segmentation, but only solves for the finite dimensional pose parameters (as is common for 2D-3D pose estimation approaches).

This results in a much simplified framework that avoids dealing with problems such as infinite dimensional curve representation, evolution and regularization .

Relation to Previous Work. In this paper, we exploit many ideas from recent variational approaches that address the problem of structure from motion and stereo reconstruction from multiple cameras ([22,23] or [24]). Originally, the authors in [22,23] presented a method to reconstruct the 3D shape of an object from multiple 2D views obtained from calibrated cameras. The present contribution aims at performing a somewhat opposite task: given the 3D model of an object, perform the segmentation of 2D images and recover the 3D pose of the object relative to a *unique* camera. This is the first time that the framework in [22,23] is adapted and employed in the specific context of segmenting 2D images from a unique camera, using the knowledge of a 3D model. The framework in [22] has also recently been extended in [25] to address the problem of multiple camera calibration. In the present work, the camera is assumed to be calibrated. However, this assumption could easily be dropped by also solving for the optimal camera calibration parameters as presented in [25].

We note that, although the use of 3D shape knowledge to perform the 2D segmentation of regions presents obvious advantages, the literature dealing with this type of approaches is strikingly thin. The pieces of work closest to the proposed contribution are probably [26] and [27]. In [26], the authors evolve an (infinite dimensional) active contour as well as 3D pose parameters to minimize a joint energy functional encoding both image information and 3D shape knowledge. Our method differs from the aforementioned approach in many crucial aspects: We optimize a *unique* energy functional, which allows us to circumvent the need to determine ICP-like correspondences and to perform costly back-projections between the segmenting contour and the shape model at each iteration. Also, we perform optimization *only* in the finite dimensional space of the Euclidean pose parameters. In addition to being computationally efficient, this allows our technique to be less likely to be trapped in local minima, resulting in robust performances as demonstrated in the experimental part. In [27], the method of [26] is successfully simplified by performing energy minimization only in the space of 3D pose parameters. Thus, the method of [27] and our contribution present some similarities. However, the energy minimization method and resulting algorithms are radically different: In [27], an algebraic approach is used that involves establishing correspondences and back-projections between the 3D and 2D world, as well as linearizing the resulting system of equations. Consequently, important information about the geometry of the 3D model is lost through the algebraic approach. In contrast, our approach relies on surface differential geometry (see e.g., [28]) to link geometric properties of the model surface and its projection in the image domain. This allows us to derive the partial differential equations necessary to perform energy optimization, as well as exploit the knowledge of the 3D object to its full extent.

Our technique uses a 3D shape prior in a region-based framework, and can thereby be expected to be robust to noise or occlusion. Hence, an obvious application of the proposed approach is the robust tracking of 3D rigid objects in 2D image sequences. Our approach is, thus, also related to a wealth of methods concerned with the problem of model-based monocular tracking (see [29] for a recent survey).

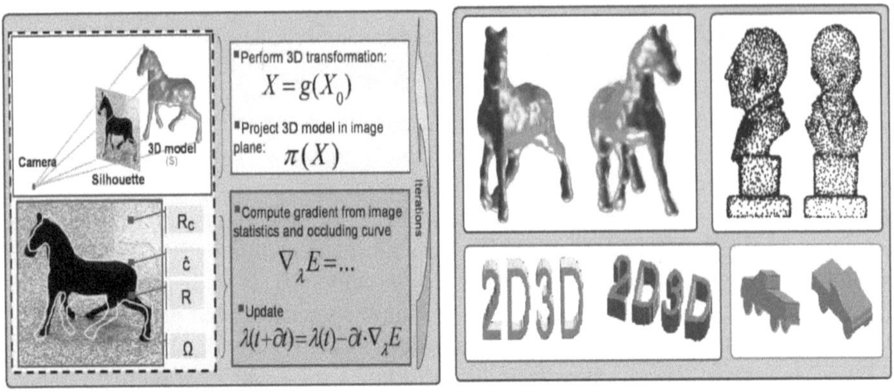

Fig. 1. *Left:* Schema summarizing our segmentation/pose estimation approach from a 3D model, in 4 steps. *Right:* Different views of the 3D models used (rendered surfaces or cloud of points).

2 Proposed Approach

We suppose we have at our disposal the 3D surface model of an object. Our goal is to find the (Euclidean) transformation that needs to be applied to the model so that it coincides with the object of interest in the referential attached to a calibrated camera. We now describe the proposed approach in details, starting with our choice of notation. An overview of the method can be found in Figure 1.

2.1 Notation

Let $\mathbf{X} = [X, Y, Z]^T$ denote the coordinates of a point in $\mathbb{R}^3$, measured with respect to a referential attached to the imaging camera. We denote by I the image, by $\Omega \subset \mathbb{R}^2$ the image domain, and by $d\Omega$ its area element. We assume the camera is modeled as an ideal perspective projection[1]: $\pi : \mathbb{R}^3 \mapsto \Omega; \mathbf{X} \mapsto \mathbf{x}$, where $\mathbf{x} = [x, y]^T = [X/Z,\ Y/Z]^T$ denotes coordinates in Ω.

Let S be the smooth surface in $\mathbb{R}^3$ defining the shape of the object of interest. The (outward) unit normal to S at each point $\mathbf{X} \in S$ will be denoted by $\mathbf{N} = [N_1, N_2, N_3]^T$. To determine the pose of S with respect to the camera, we define the identical reference surface S_0, whose pose is known.[2] Denoting by X_0 the coordinates of points on S_0, one can locate S in the camera referential via the transformation $g \in SE(3)$, such that $S = g(S_0)$, or written point-wise $\mathbf{X} = g(\mathbf{X}_0) = \mathbf{R}\mathbf{X}_0 + \mathbf{T}$, with $\mathbf{R} \in SO(3)$ and $\mathbf{T} \in \mathbb{R}^3$. The parameters of the rigid motion g will be denoted by $\lambda = [\lambda_1, \ldots, \lambda_6]^T = [t_x, t_y, t_z, \omega_1, \omega_2, \omega_3]^T$ (rotations are represented in exponential coordinates [32]).

Let $R = \pi(S) \subset \Omega$, be the region of the image on which the surface S projects (i.e., the region of Ω corresponding to imaging S). Let $R^c = \Omega \setminus R$ and $\hat{c} = \partial R$ denote the

[1] More general models of cameras (see [30,31]) can straightforwardly be handled. We make this assumption here to simplify the presentation.

[2] One can assume that the center of gravity of S_0 coincides with the camera center and the rotation is known.

complement and the boundary of R, respectively (Figure 1). The curve $\hat{c} \subset \Omega$ is the projection of the curve $C \subset S$ that delineates the visible part of S from the camera: $\hat{c} = \pi(C)$. The 2D curve $\hat{c}$ and 3D curve C will be respectively referred to as the "silhouette" and the "occluding curve".

2.2 Energy Functional

In [22], the authors used an image formation approach to define a cost functional measuring the discrepancy between the photometric properties of the surface S (as well as the 3D-background), and the pixel intensities of multiple images. The resulting energy involved back-projections to the surface S to guarantee the coherence between the measurements obtained from multiple cameras. In this present work, we are interested in segmenting a *unique* image and adopt a somewhat different approach directly inspired from region-based active contours techniques [10,11,12,13]. Most region-based approaches assume that the pixels corresponding to the object of interest or the background are distinct with respect to a certain grouping criterion. To perform segmentation, a closed curve is evolved to increase the discrepancy between the statistics of the pixels located in its interior and exterior. Accordingly, we define an energy of the form:

$$E = \int_{\mathrm{R}} r_O\big(I(\mathbf{x}), \hat{c}\big) d\Omega + \int_{\mathrm{R^c}} r_B\big(I(\mathbf{x}), \hat{c}\big) d\Omega, \tag{1}$$

where $r_O : \mathcal{Z}, \Omega \mapsto \mathbb{R}$ and $r_B : \mathcal{Z}, \Omega \mapsto \mathbb{R}$ are two monotonically decreasing functions measuring the matching quality of the image pixels with a statistical model over the regions R and R^c, respectively. The space $\mathcal{Z}$ corresponds to the photometric variable (such as gray-scale intensity, color or texture vector) chosen to perform segmentation. Hence, depending on the choices for r_O, r_B, and $\mathcal{Z}$ a larger class of images than the ones fitting the specific hypotheses made in [22] can be dealt with. The energy E measures the discrepancy between the statistical properties of the pixels located inside and outside the curve $\hat{c}$, and does not involve any back-projections.

It is expected that E is minimal when R and R^c correspond to the object and background in I, respectively. Most region-based approaches evolve an infinite dimensional curve, which amounts to exploring unconstrained shapes of the segmenting contour. Since we assume that the 3D shape of the rigid object is known, we want to minimize E by exploring only the possible regions R and R^c that result from projecting the surface S onto the image plane. For a calibrated camera, these regions are functions of the transformation g only. Solving for the transformation that minimize E can be undertaken via gradient descent over the parameters λ, as described below.

2.3 Gradient Flow

The partial differentials of E with respect to the pose parameters λ_i's can be computed using the chain-rule:

$$\begin{aligned} \frac{dE}{d\lambda_i} = \int_{\hat{c}} \Big(r_O\big(I(\mathbf{x})\big) - r_B\big(I(\mathbf{x})\big)\Big) \left\langle \frac{\partial \hat{c}}{\partial \lambda_i}, \hat{\mathbf{n}} \right\rangle d\hat{s} + \int_R \left\langle \frac{\partial r_O}{\partial \hat{c}}, \frac{\partial \hat{c}}{\partial \lambda_i} \right\rangle d\Omega \\ + \int_{R^c} \left\langle \frac{\partial r_B}{\partial \hat{c}}, \frac{\partial \hat{c}}{\partial \lambda_i} \right\rangle d\Omega, \end{aligned} \tag{2}$$

where $\hat{s}$ is the arc-length parametrization of the silhouette $\hat{c}$ and $\hat{\mathbf{n}}$ the (outward) normal to the curve at $\mathbf{x}$.

Using the arc-length s of C and the direct $\frac{\pi}{2}$-rotation matrix J, one has

$$\left\langle \frac{\partial \hat{c}}{\partial \lambda_i}, \hat{\mathbf{n}} \right\rangle d\hat{s} = \left\langle \frac{\partial \hat{c}}{\partial \lambda_i}, J\frac{\partial \hat{c}}{\partial \hat{s}} \right\rangle d\hat{s} = \left\langle \frac{\partial \pi(C)}{\partial \lambda_i}, J\frac{\partial \pi(C)}{\partial s} \right\rangle \frac{ds}{d\hat{s}} d\hat{s} = \left\langle \frac{\partial \pi(C)}{\partial \lambda_i}, J\frac{\partial \pi(C)}{\partial s} \right\rangle ds. \tag{3}$$

Let $\mathcal{J}$ denote the Jacobian of $\pi(\mathbf{X})$ with respect to the spatial coordinates, one has $\mathcal{J} = \frac{1}{Z^2}\begin{bmatrix} Z & 0 & -X \\ 0 & Z & -Y \end{bmatrix}$. From (3), one gets

$$\begin{aligned}\left\langle \frac{\partial \hat{c}}{\partial \lambda_i}, \hat{\mathbf{n}} \right\rangle d\hat{s} &= \left\langle \mathcal{J}\frac{\partial \mathbf{X}}{\partial \lambda_i}, J\mathcal{J}\frac{\partial \mathbf{X}}{\partial s} \right\rangle ds = \left\langle \frac{\partial \mathbf{X}}{\partial \lambda_i}, \mathcal{J}^T J\mathcal{J}\frac{\partial \mathbf{X}}{\partial s} \right\rangle ds \\ &= \frac{1}{Z^3}\left\langle \frac{\partial \mathbf{X}}{\partial \lambda_i}, \begin{bmatrix} 0 & Z & -Y \\ -Z & 0 & X \\ Y & -X & 0 \end{bmatrix} \frac{\partial \mathbf{X}}{\partial s} \right\rangle ds = \frac{1}{Z^3}\left\langle \frac{\partial \mathbf{X}}{\partial \lambda_i}, \frac{\partial \mathbf{X}}{\partial s} \times \mathbf{X} \right\rangle ds.\end{aligned} \tag{4}$$

In the equation above, the point $\mathbf{X}$ belongs to the occluding curve C. A necessary condition for a point $\mathbf{X}$ to belong to the occluding curve is that $\langle \mathbf{X}, \mathbf{N} \rangle = 0$ (since the associated vector $\mathbf{X}$, with origin at the center of the camera, corresponds to the projection/viewing direction and is tangent to the surface S at $\mathbf{X}$). The vector $\mathbf{t} = \frac{\partial \mathbf{X}}{\partial s}$ is the tangent to the curve C at the point $\mathbf{X}$. Since the vectors $\mathbf{t}$ and $\mathbf{X}$ belong to the tangent plane to S at $\mathbf{X}$, one has $\frac{\partial \mathbf{X}}{\partial s} \times \mathbf{X} = \|\mathbf{X}\| \mathbf{N} \sin(\theta)$, with $\theta = \widehat{(\mathbf{t}, \mathbf{X})}$ the angle between $\mathbf{t}$ and $\mathbf{X}$. For $\mathbf{X} \in C$, one has

$$\frac{\partial}{\partial s}\langle \mathbf{X}, \mathbf{N} \rangle = \underbrace{\left\langle \frac{\partial \mathbf{X}}{\partial s}, \mathbf{N} \right\rangle}_{=0} + \left\langle \frac{\partial \mathbf{N}}{\partial s}, \mathbf{X} \right\rangle = 0 = \langle d\mathbf{N}(\mathbf{t}), \mathbf{X} \rangle = \mathrm{II}(\mathbf{t}, \mathbf{X}).$$

Hence, since the second fundamental form $\mathrm{II}(\mathbf{t}, \mathbf{X}) = 0$, the vectors $\mathbf{t}$ and $\mathbf{X}$ are conjugate (see [28]). Hence, using the Euler formula, one can show that $K \sin^2\theta = \kappa_X \kappa_t$, where K is the Gaussian curvature, and κ_X and κ_t denote the normal curvatures in the directions $\mathbf{X}$ and $\mathbf{t}$ at $\mathbf{X} \in S$, respectively. Plugging into Equation (4), one gets

$$\left\langle \frac{\partial \hat{c}}{\partial \lambda_i}, \hat{\mathbf{n}} \right\rangle d\hat{s} = \frac{\|\mathbf{X}\|}{Z^3}\sqrt{\frac{\kappa_X \kappa_t}{K}} \left\langle \frac{\partial \mathbf{X}}{\partial \lambda_i}, \mathbf{N} \right\rangle ds. \tag{5}$$

The energy in (1) is valid for most functions r_O and r_B used in the literature. However, in this paper, we chose

$$r_O = -\log(\sigma_O) - \frac{(I(\mathbf{x}) - \mu_O)^2}{\Sigma_O} \quad \text{and} \quad r_B = -\log(\sigma_B) - \frac{(I(\mathbf{x}) - \mu_B)^2}{\Sigma_B} \tag{6}$$

as in the region-based active contour technique presented in [11]. Using these functions, the two last terms in Equation (2) collapse for $\mu_{\mathrm{O/B}} = \frac{\int_{R/R^c} I(\hat{\mathbf{x}}) d\Omega}{\int_{R/R^c} d\Omega}$ and $\Sigma_{\mathrm{O/B}} = \frac{\int_{R/R^c}\left(I(\hat{\mathbf{x}}) - \mu_{\mathrm{O/B}}\right)^2 d\Omega}{\int_{R/R^c} d\Omega}$.[3]

[3] For gray-scale images $\mu_{\mathrm{O/B}}$ and $\Sigma_{\mathrm{O/B}}$ are scalars. For color images, $\mu_{\mathrm{O/B}} \in \mathbb{R}^3$ and $\Sigma_{\mathrm{O/B}} \in \mathbb{R}^{3\times 3}$. Texture can also be used, see [15].

Thus, the flow becomes a simple line integral on C

$$\frac{dE}{d\lambda_i} = \int_C \Big(r_O\big(I(\pi(\mathbf{X}))\big) - r_B\big(I(\pi(\mathbf{X}))\big) \Big) \cdot \frac{\|\mathbf{X}\|}{Z^3} \sqrt{\frac{\kappa_X \kappa_t}{K}} \left\langle \frac{\partial \mathbf{X}}{\partial \lambda_i}, \mathbf{N} \right\rangle ds. \quad (7)$$

For $i = 1, 2, 3$ (i.e., λ_i is a translation parameter)

$$\left\langle \frac{\partial \mathbf{X}}{\partial \lambda_i}, \mathbf{N} \right\rangle = \left\langle \frac{\partial \mathbf{R}\mathbf{X_0} + \mathbf{T}}{\partial \lambda_i}, \mathbf{N} \right\rangle = \left\langle \frac{\partial \mathbf{T}}{\partial \lambda_i}, \mathbf{N} \right\rangle = N_i.$$

For $i = 4, 5, 6$ (i.e., λ_i is a rotation parameter), one can similarly show that $\left\langle \frac{\partial \mathbf{X}}{\partial \lambda_i}, \mathbf{N} \right\rangle = \langle \mathcal{M}_i, \mathbf{N} \rangle$, with $\mathcal{M}_i$ the ith column of $\mathcal{M} = \mathbf{R} \begin{bmatrix} 0 & Z_0 & -Y_0 \\ -Z_0 & 0 & X_0 \\ Y_0 & -X_0 & 0 \end{bmatrix}$.

3 Experiments

We now report a few experimental results obtained for both synthetic and real datasets. Four different 3D models of rigid objects (see Figure 1) were used to perform segmentation and tracking tasks that highlight the robustness of our technique to **initialization**, **noise** and **missing or imperfect information**. To save computational time, we used the approximation $\sqrt{\frac{\kappa_X \kappa_t}{K}} \simeq 1$ in our implementation of Equation (7), which still decreased the energy E. Also, we note that the shape of the objects, notably the Horse and the Van Gogh Bust, cannot readily be described in terms of geometric primitives (lines, ellipses, etc.) or even algebraically, and thus do not satisfy the working hypotheses of standard pose estimation techniques ([2,3,5,6]).

3.1 Robustness to Initialization

Figure 2 shows segmentation results (and 3D coordinates recoveries) obtained using our approach for a synthetic color image. Figure 3 shows results for diverse natural color images. Despite initializations that are quite far from the truth (e.g., large errors in translation or angular position) accurate segmentations are obtained.

(a) (b) (c)

Fig. 2. Robustness to initialization - Segmentation of a synthetic color image. (a): Initialization; (b): Intermediate step of the evolution; (c): Final result.

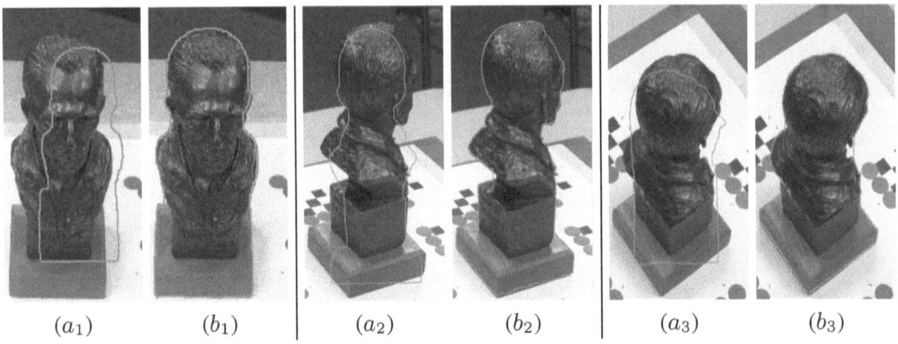

Fig. 3. Robustness to initialization - Segmentation of natural color images. (a_n)'s: Challenging initializations (e.g., large error in translation or angular positions - green curve). (b_n's): Final results with the proposed approach (green curve).

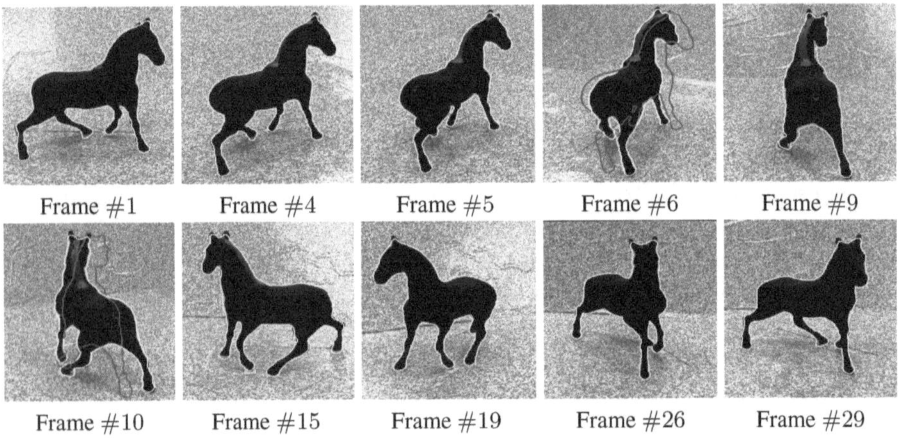

Fig. 4. Robustness to initialization - Tracking a natural sequence *Yellow contours*: final results after convergence; *Red contours*: Initializations from the result of the preceding image (see text for our tracking scheme). The aspect of the object changes drastically throughout the sequence. The position of the object undergoes large changes between successive images.

Figure 4 shows tracking results obtained for a real sequence. The sequence is composed of 32 images of a rigid toy Horse. The images were taken from discrete positions of a calibrated camera that underwent a complete rotation around the object. The camera "jumps" between successive images creating large changes in the pose of the object that need to be recovered (e.g., changes in the angular position of the camera can exceed $15°$ between frames). Tracking this sequence would be challenging for many 3D pose estimation techniques available in the literature: Many techniques using local features such as points or edges (e.g., [1,3]) are likely to be thrown off by the textured/noisy background (false features) and get trapped in local minima. The sequence was tracked with our technique, using a very simple scheme: For each image, initialization was performed using the pose parameters corresponding to the minimum of the energy

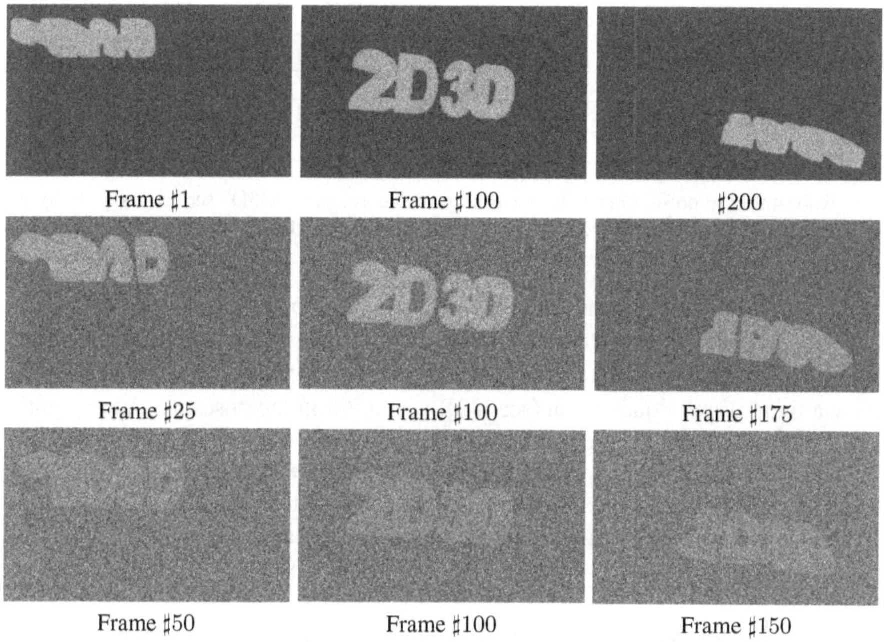

Fig. 5. Robustness to noise. Visual tracking results for the sequences involving the "2D3D" logo (Green curves). *First row:* Tracked sequence with Gaussian noise of standard deviation $\sigma = 10\%$. *Second row:* Tracked sequence for $\sigma = 30\%$. *Third row:* Tracked sequence for $\sigma = 100\%$.

obtained for the preceding image, and our approach was run until convergence. Despite the difficulties described above, very satisfying tracking performances were observed. This highlights the robustness of the technique to initialization since the large cameras jumps are accommodated and the method is not trapped in local minima. We note that region-based active contour techniques such as [10], would lead to satisfying segmentations on this particular sequence. However, these approaches would not also determine the pose of the object, which is valuable information for tracking applications.

3.2 Robustness to Noise

To test the robustness of our technique to noise, a sequence of 200 images was constructed by continuously transforming the 3D model of the "2D3D" logo and projecting it into the image plane using the parameters of a simulated calibrated camera (e.g., focal length f=200). The translation parameters, rotation axis and angle were continuously varied (e.g., the total angle variation over the sequence exceeded 160°) to ensure a large variation of the aspect and position of the object throughout the sequence. From the basic sequence obtained, diverse level of Gaussian noise were added of standard deviation ranging from $\sigma = 10\%$ to $\sigma = 100\%$ (See Figure 5).

Typical visual results obtained using our approach (and the tracking scheme above) are reproduced in Figure 5. For all noise levels, which can be rather large (e.g., in the case $\sigma = 100\%$ object and background are barely distinguishable), tracking was

Noise Level	mean error (in %)	std. dev. error (in %)	max error (in %)
$\sigma = 10\%$	T: 0.85 ; R: 0.96	T: 0.23 ; R: 0.45	T: 1.43 ; R: 2.60
$\sigma = 30\%$	T: 0.97 ; R: 1.09	T: 0.21 ; R: 0.47	T: 1.50 ; R: 2.94
$\sigma = 60\%$	T: 0.95 ; R: 1.30	T: 0.30 ; R: 0.52	T: 2.39 ; R: 2.60
$\sigma = 100\%$	T: 1.02 ; R: 2.12	T: 0.39 ; R: 0.87	T: 2.18 ; R: 4.36

Fig. 6. Robustness to noise. Quantitative tracking results for the "2D3D" sequences with diverse levels of noise. Table displaying %-**absolute** error statistics over the 200 images of the sequences.

maintained throughout the whole sequence. Figure 6 reproduces the results of the pose estimation procedure. For each image, percent *absolute* errors with respect to the ground-truth were computed for both the translation and rotation as: Error $= \frac{\|\mathbf{v}_{\text{measured}} - \mathbf{v}_{\text{truth}}\|}{\|\mathbf{v}_{\text{truth}}\|}$, with $\mathbf{v}$ a translation or quaternion (see [32]) vector. From the pose estimation point of view, the method appears to perform quite well: average error and standard deviation computed over the 200 frames of each sequence rarely exceed 2% and 1%, respectively, for *both* translation and rotation. This highlights the accuracy and reliability of the method, and suggests that it is quite resilient to large amounts of noise (very little deterioration of the results is observed with increasing noise levels).

3.3 Robustness to Missing/Imperfect Image Information

To test the robustness to missing information, we created two sequences by adding two different occlusions in the basic sequence featuring the "2D3D" model (see Figure 7). The first occlusion is a grey rectangle that can mask more than $2/3$ of the "2D3D" logo. The second occlusion is the word "SHAPE" written in black letters that can mask the object at several places. Gaussian noise of standard deviation 30% was also added to both resulting sequences. Figure 7 presents the results of tracking the sequences of 200

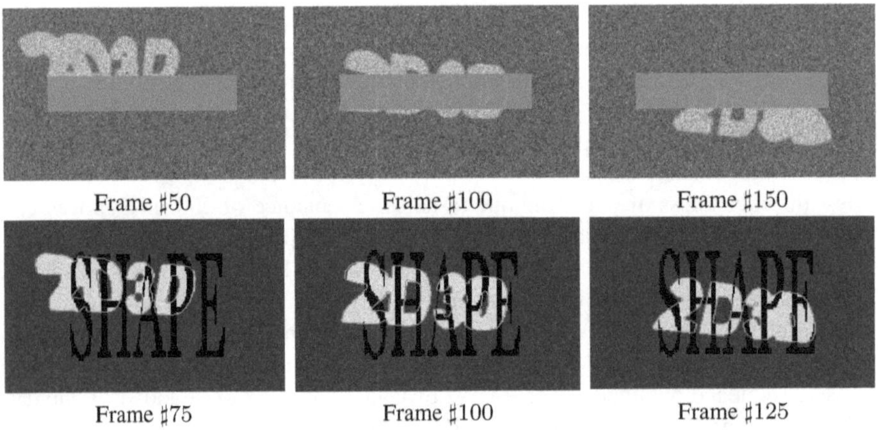

Fig. 7. Robustness to missing information. Tracking results (green curves) for the "2D3D" sequences with occlusions. *First row:* Rectangular occlusion. *Second row:* Word "SHAPE" as occlusion. Gaussian noise with $\sigma = 30\%$ was added.

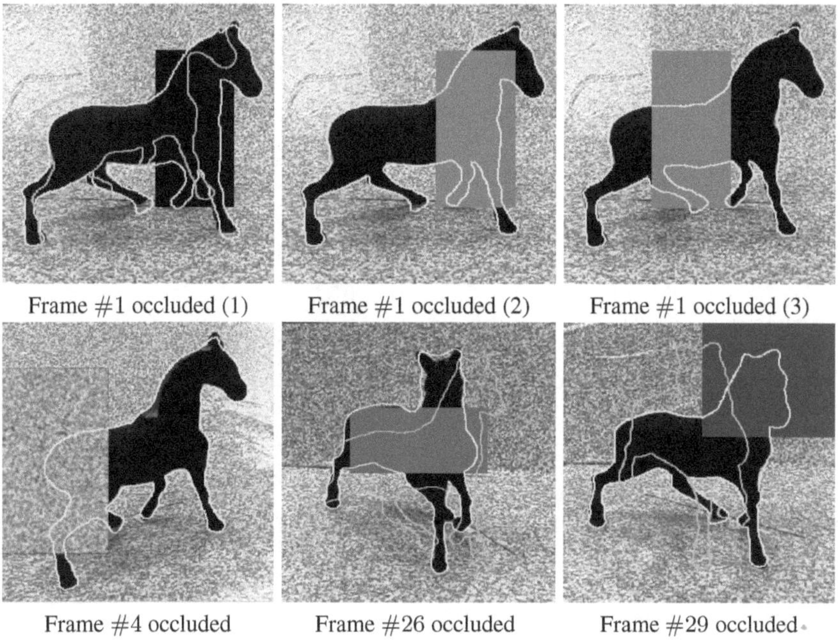

Fig. 8. Robustness to missing information. Segmentation results with occlusions. *Cyan contours*: Some of the initializations tested (note the large errors in angular position); *Yellow contours*: Final results (almost identical to results in Figure 4 with no occlusion).

frames with our approach. One notes that despite the occlusions (and noise), accurate segmentations are obtained: In particular, missing letters or parts are accurately localized and reconstructed. Track was maintained throughout both sequences. For the first sequence mean %-absolute-error (over the 200 frames) in the transformation parameters was 1.08% for translation ($\mathbf{T}$) and 1.57% for rotation ($\mathbf{R}$) - standard deviation 0.45% for $\mathbf{T}$ and 0.75% for $\mathbf{R}$. For the second sequence mean %-absolute-error was 0.87% for $\mathbf{T}$ and 1.19% for $\mathbf{R}$ (std. dev. 0.34% for $\mathbf{T}$ and 0.53% for $\mathbf{R}$).

In Figure 8, we used images extracted from the Horse sequence and occluded different parts of the Horse body (e.g., the legs that have valuable information about its angular position). Diverse pose parameters quite far from the truth were used as initializations (e.g., angular position could be off by more than $30°$). Despite the occlusions with various pixel intensities or texture (and poor initializations), very convincing segmentations were obtained. Also, the positions of the object in the camera referential were accurately recovered. As can be noticed by comparing with Figure 4, the results in the presence of occlusion are very comparable to the ones without occlusion.

In Figure 9, we present segmentation results where the background and object are difficult to distinguish based on pixel statistics only (due to specularities on the object and similar colors in object and background). The results obtained with the (infinite dimensional) active contour flow of [11], which is the region-based segmentation technique underlying our approach, are not satisfying since the contour leaks into the background. Robust results are obtained using our approach.

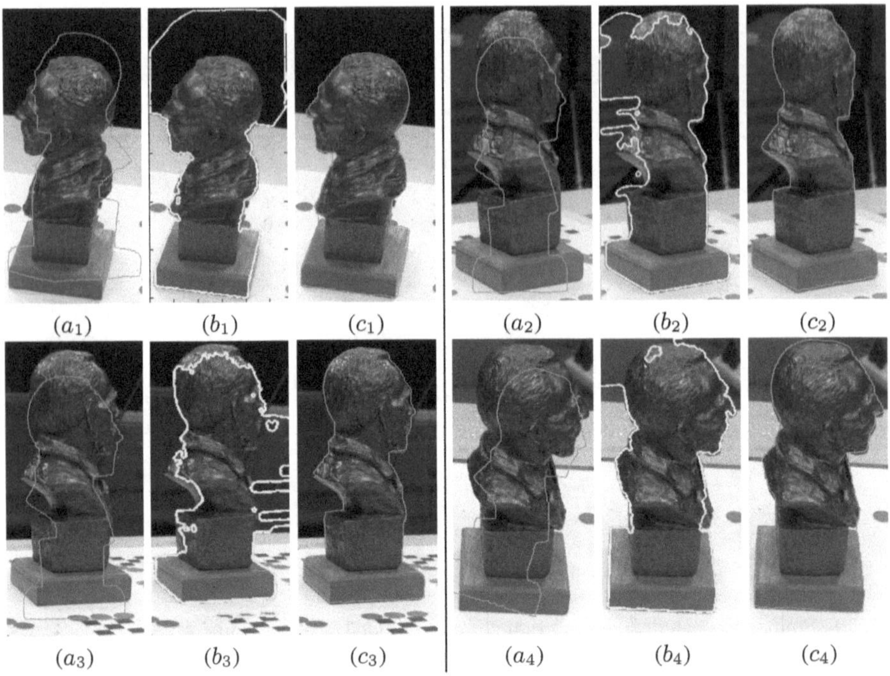

Fig. 9. Robustness to imperfect information. Comparative segmentation results. (a_n's): Initializations. (b_n's): Final results obtained with (infinite dimensional) active contour flow as in [11], which is the region-based segmentation technique underlying our approach. (c_n's): Final results with our approach. In these images, statistical distinction between object and background is difficult due to specularities on the object and similar colors in object and background.

The experiments of Figures 7, 8 and 9 would pose a major challenge to most region-based active contour techniques, even using shape priors [16,18,19]: Statistics only are not sufficient to segment the images, and the aspect of the object changes drastically from one image to the other. Hence, a large catalogue of 2D shapes would need to be learnt to achieve similar performances using the method in [16,18,19], for instance.

4 Conclusion and Future Work

In this work, we presented a region-based approach to the 3D pose estimation problem. This approach differs from other 3D pose estimation algorithms since it does not rely on local image features. Our method allows one to employ global image statistics to drive the pose estimation process. This confers a satisfying level of robustness to noise and initialization to our framework and bypasses the need to establish correspondences between image and object features, contrary to most 3D pose estimation approaches.

Furthermore, the approach possesses the typical qualities of a region-based active contour technique with shape prior, such as robustness to occlusion or missing information, without the need to evolve an infinite dimensional contour. Also, the prior knowledge of the shape of the object is compactly represented by a unique 3D model, instead of a dense catalogue of 2D shapes.

The main advantage of the proposed technique is that it enables to locate the object not only in 2D images (a typical task handled by GAC approaches) but also in the world (a typical task handled by 2D-3D pose estimation algorithms). This makes the method particularly suitable for tracking applications involving a unique calibrated camera.

A possible direction for future research is to extend the proposed approach to include the knowledge of multiple 3D shapes. The method in [17] (where evolution of parameters in the shape space is performed in addition to pose parameters) could be adapted to the problem at hand. It is expected that the resulting framework will allow to learn the possible deformations of the object and lead to robust performances for non-rigid registration and tracking tasks.

Acknowledgement. This work was supported in part by grants from NSF, AFOSR, ARO, MURI, as well as by a grant from NIH (NAC P41 RR- 13218) through Brigham and Womens Hospital. This work is part of the National Alliance for Medical Image Computing (NAMIC), funded by the National Institutes of Health through the NIH Roadmap for Medical Research, Grant U54 EB005149. Information on the National Centers for Biomedical Computing can be obtained from http://nihroadmap.nih.gov/bioinformatics. Allen Tannenbaum is also supported with a Marie Curie Grant through the Technion, Israel.

References

1. Quan, L., Lan, Z.D.: Linear n-point camera pose determination. IEEE Transactions on Pattern Analysis and Machine Intelligence 21, 774–780 (1999)
2. Dhome, M., Richetin, M., Lapreste, J.T.: Determination of the attitude of 3d objects from a single perspective view. IEEE Trans. Pattern Anal. Mach. Intell. 11, 1265–1278 (1989)
3. Marchand, E., Bouthemy, P., Chaumette, F.: A 2d-3d model-based approach to real-time visual tracking. Image and Vision Computing 19, 941–955 (2001)
4. Zerroug, M., Nevatia, R.: Pose estimation of multi-part curved objects. In: ISCV 1995: Proceedings of the International Symposium on Computer Vision, p. 431 (1995)
5. Rosenhahn, B., Perwass, C., Sommer, G.: Pose estimation of free-form contours. IJCV 62, 267–289 (2005)
6. Drummond, T., Cipolla, R.: Real-time tracking of multiple articulated structures in multiple views. In: Vernon, D. (ed.) ECCV 2000. LNCS, vol. 1843, pp. 20–36. Springer, Heidelberg (2000)
7. Caselles, V., Kimmel, R., Sapiro, G.: Geodesic active contours. IJCV, 22, 61–79 (1997)
8. Kichenassamy, S., Kumar, S., Olver, P., Tannenbaum, A., Yezzi, A.: Conformal curvature flow: From phase transitions to active vision. Archives for Rational Mechanics and Analysis 134, 275–301 (1996)
9. Zhu, S.C., Yuille, A.L.: Region competition: Unifying snakes, region growing, and Bayes/MDL for multiband image segmentation. IEEE Trans. PAMI 18, 884–900 (1996)
10. Chan, T., Vese, L.: Active contours without edges. IEEE TIP 10, 266–277 (2001)
11. Paragios, N., Deriche, R.: Geodesic active regions: A new paradigm to deal with frame partition problems in computer vision. Journal of Visual Communication and Image Representation 13, 249–268 (2002)

12. Dambreville, S., Yezzi, A., Niethammer, M., Tannenbaum, A.: A variational framework combining level-sets and thresholding. In: BMVC, pp. 266–280 (2007)
13. Michailovich, O., Rathi, Y., Tannenbaum, A.: Image segmentation using active contours driven by the bhattacharyya gradient flow. IEEE TIP, 2787–2801 (2007)
14. Kim, J., Fisher, J., Yezzi, A., Cetin, M., Willsky, A.: Nonparametric methods for image segmentation using information theory and curve evolution. In: Proc. ICIP, vol. 3, pp. 797–800 (2002)
15. Paragios, N., Deriche, R.: Geodesic active regions for supervised texture segmentation. In: ICCV (2), pp. 926–932 (1999)
16. Leventon, M., Grimson, E., Faugeras, O.: Statistical shape influence in geodesic active contours. In: Proc. IEEE CVPR, pp. 1316–1324 (2000)
17. Tsai, A., Yezzi, T., Wells, W., Tempany, C., Tucker, D., Fan, A., Grimson, E., Willsky, A.: A shape-based approach to the segmentation of medical imagery using level sets. IEEE Trans. on Medical Imaging 22, 137–153 (2003)
18. Cremers, D., Kohlberger, T., Schnoerr, C.: Shape statistics in kernel space for variational image segmentation. Pattern Recognition 36, 1292–1943 (2003)
19. Dambreville, S., Rathi, Y., Tannenbaum, A.: Shape-based approach to robust image segmentation using kernel pca. In: IEEE Conference on Computer Vision and Pattern Recognition, pp. 977–984 (2006)
20. Osher, S., Fedkiw, R.: Level Set Methods and Dynamic Implicit Surfaces. Springer, Heidelberg (2003)
21. Turk, G., Levoy, M.: Zippered polygon meshes from range images. In: Siggraph, pp. 311–318 (1994)
22. Yezzi, A., Soatto, S.: Structure from motion for scenes without features. In: Proc. IEEE CVPR, vol. 1, pp. 171–178 (2003)
23. Yezzi, A., Soatto, S.: Stereoscopic segmentation. International Journal of Computer Vision (IJCV) 53, 31–43 (2003)
24. Faugeras, O.D., Keriven, R.: Variational principles, surface evolution pdes, level set methods and the stereo problem. INRIA Tech. report 3021, 1–37 (1996)
25. Unal, G., Yezzi, A., Soatto, S., Slabaugh, G.: A variational approach to problems in calibration of multiple cameras. Trans. Pattern Analysis and Machine Intelligence 29, 1322–1338 (2007)
26. Rosenhahn, B., Brox, T., Weickert, J.: Three-dimensional shape knowledge for joint image segmentation and pose tracking. IJCV 73, 243–262 (2007)
27. Schmaltz, C., Rosenhahn, B., Brox, T., Cremers, D., Weickert, J., Wietzke, L., Sommer, G.: Region-based pose tracking. In: Pattern Recognition and Image Analysis, pp. 56–63 (2007)
28. DoCarmo, M.P.: Diferential Geometry of Curves and Surfaces. Prentice Hall, Englewood Cliffs (1976)
29. Lepetit, V., Fua, P.: Monocular model-based 3d tracking of rigid objects: A survey. Fundations and Trends in Computer Graphics and Vision 1, 1–89 (2005)
30. Forsyth, D., Ponce, J.: Computer Vision. Prentice Hall, Englewood Cliffs (2003)
31. Hartley, R., Zisserman, A.: Multiple view geometry in computer vision. Cambridge University Press, Cambridge (2000)
32. Ma, Y., Soatto, S., Kosecka, J., Sastry, S.: An invitation to 3D vision. Springer, Heidelberg

Linear Time Maximally Stable Extremal Regions

David Nistér and Henrik Stewénius

[1] Microsoft Live Labs
[2] Google Switzerland

Abstract. In this paper we present a new algorithm for computing Maximally Stable Extremal Regions (MSER), as invented by Matas et al. The standard algorithm makes use of a union-find data structure and takes quasi-linear time in the number of pixels. The new algorithm provides exactly identical results in true worst-case linear time. Moreover, the new algorithm uses significantly less memory and has better cache-locality, resulting in faster execution. Our CPU implementation performs twice as fast as a state-of-the-art FPGA implementation based on the standard algorithm.

The new algorithm is based on a different computational ordering of the pixels, which is suggested by another immersion analogy than the one corresponding to the standard connected-component algorithm. With the new computational ordering, the pixels considered or visited at any point during computation consist of a single connected component of pixels in the image, resembling a flood-fill that adapts to the grey-level landscape. The computation only needs a priority queue of candidate pixels (the boundary of the single connected component), a single bit image masking visited pixels, and information for as many components as there are grey-levels in the image. This is substantially more compact in practice than the standard algorithm, where a large number of connected components must be considered in parallel. The new algorithm can also generate the component tree of the image in true linear time. The result shows that MSER detection is not tied to the union-find data structure, which may open more possibilities for parallelization.

1 Introduction

Extraction of invariant regions has recently been the focus of intense study [1,2,3,4,5,6,7,8,9,10,11,12], supporting a wide variety of applications such as recognition, image retrieval, mosaicing, 3D reconstruction, tracking, robot navigation and more. Testing in a common framework was enabled by Mikolajczyk et al. [5] with a dataset that allows testing the repeatability of a region detector under disturbances such as blur, viewpoint change, zoom, rotation, lighting changes and JPEG compression.

Maximally Stable Extremal Regions (MSER) described by Matas el al [3] have become one of the commonly used region detector types, partly because of their high repeatability and partly because they are somewhat complementary to many other commonly used detectors [7,4,13]. They are viable with a relatively

D. Forsyth, P. Torr, and A. Zisserman (Eds.): ECCV 2008, Part II, LNCS 5303, pp. 183–196, 2008.

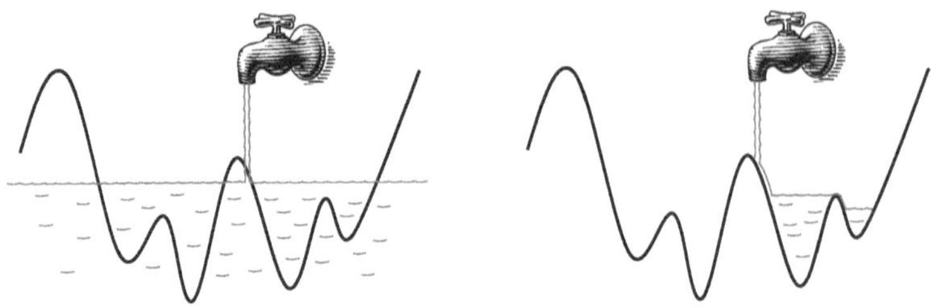

Fig. 1. The two immersion analogies. On the left, the standard immersion. On the right, the immersion corresponding to the new linear time algorithm.

small number of regions per image, and are therefore well suited for large scale image retrieval tasks [14,15]. They have also been used in recognition [16] as well as tracking [11] and have been extended to color [17] and volumetric images [18]. The detection has even been implemented on FPGA [19].

The standard algorithm for computing MSER follows the same lines as a very popular flooding simulation algorithm for computing a watershed segmentation, suggested in [20]. The watershed segmentation has a long history and has been intensely studied, see [21] for a review. In the immersion analogy, Fig 1, the grey-level profile of the image is imagined as a landscape height-map. The water level is raised gradually until the whole landscape is immersed. In the standard immersion analogy, the level is raised equally at all places in the landscape (image). That is, one either thinks of the landscape as porous so that the water level will be equal everywhere, or equivalently, imagines that a hole is pierced in each local minimum of the landscape, allowing water to enter. The algorithm keeps tracks of the connected components of water (connected components of pixels) using a union-find data structure with path-compression. The union-find data structure with path-compression supports quasi-linear time in the number of pixels [21,22,23]. More precisely, the time required can be shown to be bounded by $O(n\alpha(n))$, where n is the number of pixels and $\alpha(n)$ is the inverse of the Ackermann function, whose value is smaller than 5 if n is of the order 10^{80}. Hence the expression 'quasi-linear' is well motivated, although a true linear time algorithm for the union-find problem to our knowledge requires limiting assumptions such as for example the existence of a computer with $log(n)$ word length and incremental growing of the data structure [24].

In this paper, an algorithm is presented that can compute MSER and the so-called the component tree (of connected components as they evolve during the flooding), all in true linear time in the number of pixels. Perhaps even more importantly, the algorithm works with a single connected component of pixels, resulting in less memory usage and faster execution. The algorithm is natural and simple, sharing many properties with the previous algorithm, but with one critical difference. Instead of flooding the image with the same water level everywhere, the flooding occurs as if the landscape was opaque and water is being poured on at some arbitrarily selected point (pixel), see Fig. 1. The water first fills up the basin where water is poured on, and then spills over to other parts

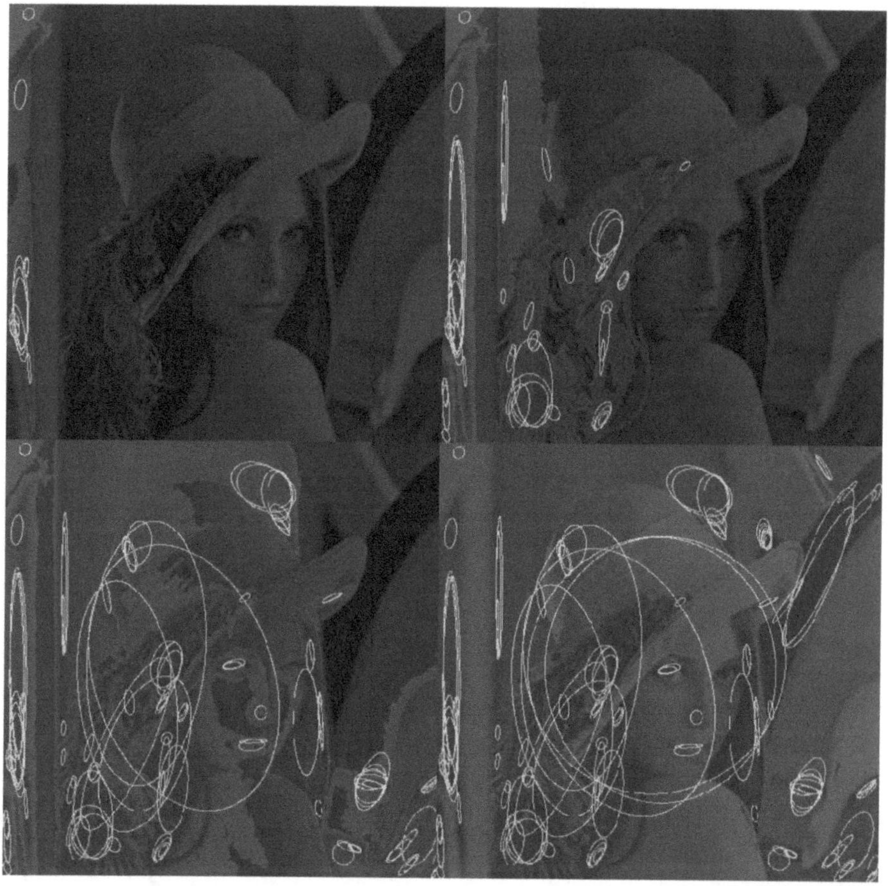

Fig. 2. Typical progress stages for the algorithm on the first pass. The top left pixel is used as the source pixel. The flooding occurs, with preference for dark regions, MSER are detected in the process, and eventually the whole image is flooded. The process will then be repeated with preference for bright regions.

of the landscape as they become accessible to the current body of water. That is, the flooding occurs in a very physically plausible manner. The water adapts to the actual landscape and remains one connected component connected to the point where it is being poured onto the landscape. The algorithm keeps track of the 'downhill stream' of water, which at any time amounts to information for at most as many pixel components as there are grey-levels in the image. When the downhill stream finds a minimum, the water starts filling it up, and when it fills back up, the corresponding pixels are processed for MSER.

Thus, the progress of the algorithm can be roughly described as a physical flood-fill adapting to the landscape, see Fig 2. In the following section, we will first give an algorithm overview, followed by algorithm details and a description of a recommended actual implementation.

2 Algorithm Overview

In this section we will give a high-level description of the algorithm. For simplicity, we will work on the dark to bright pass. Note however that in reality, the MSER detection algorithm is simply carried out twice, one from dark to bright and one from bright to dark (with the landscape flipped upside down).

2.1 Basic Definitions

Let the image I consist of n pixels indexed by the variable x. Let each pixel be assigned the grey-level value $f(x)$, taken from a set of m linearly ordered grey-level values. Let also the pixels be assigned some neighboring relation encoded by the function $N(x)$, where $N(x)$ denotes the set of pixels neighboring pixel x. An example is a two-dimensional image with four-connected pixels taking on grey-levels $[0, \ldots, 255]$. The generalization to arbitrary undirected finite connected graphs is straightforward.

A *path* from a pixel x to y is a sequence of pair-wise neighboring pixels starting with x and ending with y. A *connected set* X is a set of pixels that has a path entirely within X between any pair of pixels in X. A *minimum* is a connected set X of pixels with equal grey-level values, such that no pixel in X is connected to a pixel outside X with equal or smaller grey-level value. A *sub-level set*, parameterized by grey-level y, is the set of pixels $\{x \in I \mid f(x) \leq y\}$ with grey-level smaller than or equal to y. As we trace through the sub-level sets by varying the grey-level from dark to bright, we get a sequence of sets on which we can (in principle) perform connected component analysis on. The way the connected components evolve defines a tree, which we will refer to as the *component tree*, defined roughly as follows. When a new component appears from one grey-level

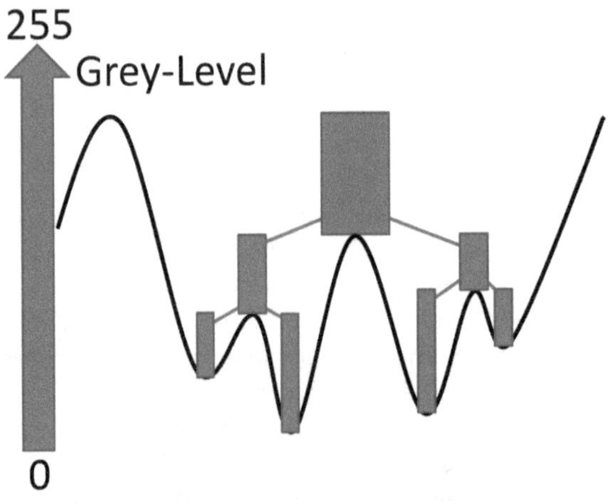

Fig. 3. The component tree with corresponding grey-levels. The nodes of the tree are drawn as vertical boxes where the vertical extent of the box shows the grey-level life span of each component.

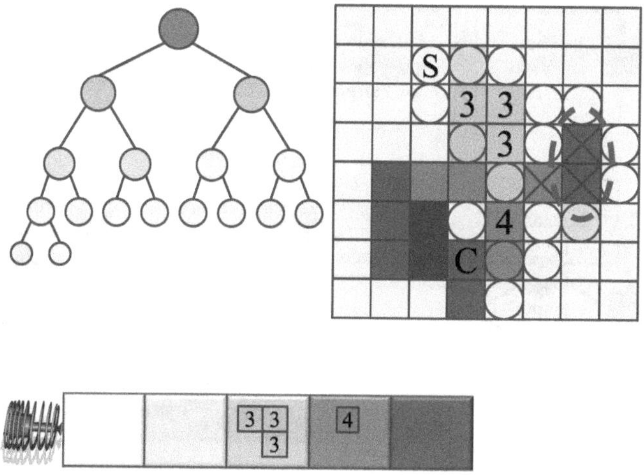

Fig. 4. The data-structures. Pixel status is shown on the top right. The source pixel is marked S. The current pixel is marked C. The pixels waiting in the heap of boundary pixels (left) are marked with circles. The pixels that belong to a component on the component stack (bottom) are marked with their component number. A detected MSER is shown by a dashed ellipse. Pixels that have been fully processed are marked with an X. Note how the stack holds components with decreasing grey-level.

to the next (in the sense that we get a component that consist entirely of pixels that were not in the previous sub-level set), this component of pixels is a minimum, which is made into a leaf node. When, as we increase the grey-level threshold, two or more components become joined into one, the joined component is assigned a new node and made into a parent of the original nodes. This process continues until the whole image is one component, which corresponds to the root node.

As the components evolve, we can also keep track of component information, which will differ dependent on what we wish to extract. For example, it could consist of a linked list of the pixels in the component. It could also just consist of the first and second order moments of the regions, which can in principle be encoded by six sums $\sum 1, \sum x, \sum y, \sum xx, \sum xy, \sum yy$ over the pixels in the component (although for numerical reasons, it is better to center these moments on something close to the centroid of the region). This is the output we need if we wish to turn the MSER into elliptical regions at the end of detection. An *MSER* is a component for which the relative growth-rate attains a minimum, see Matas [3] for further details.

2.2 Basic Algorithm

We will now describe the algorithm from an abstract point of view. See Section 2.3 on how to accomplish the steps efficiently, and Section 3 for analysis. The algorithm needs the following data-structures:

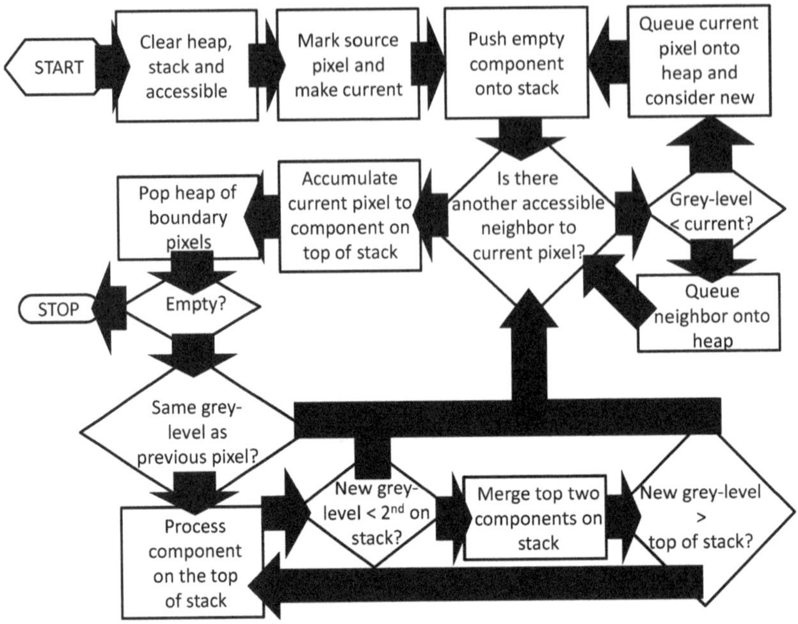

Fig. 5. State graph for the algorithm

- A binary mask of accessible pixels. These are the pixels to which the water already has access.
- A priority queue of boundary pixels, where priority is minus the grey-level. These pixels can be thought of as partially flooded pixels in the sense that water has access to the pixel in question, but has either not yet entered, or not yet explored all the edges out from the pixel. Along with the pixel id, an edge number indicating the next edge to be explored can be stored.
- A stack C of component information. Each entry holds the pixels in a component and/or the first and second order moments of the pixels in the component, as well as the size history of the component and the current grey-level at which the component is being processed. The maximum number of entries on the stack will be the number of grey-levels.

During execution of the algorithm, the components in the component info stack C may not correspond to components in the component tree. Rather, there will a number of components representing the 'down-stream' of water streaming downhill towards a minimum, where each component is the set of pixels at a single grey-level that is part of the down-stream. A single component represents the pixels covered by the water currently filling back out of a minimum. The algorithm in a sense has two states, one where the down-stream is flowing downhill in search of a minimum, and one where a minimum has been found, and the water level is currently rising out of it.

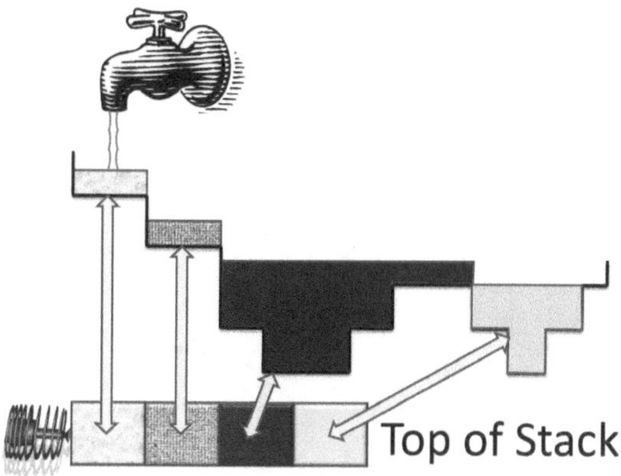

Fig. 6. The state of the component stack just before a merge of two components. The water has filled up one basin (dark blue component) and returned out of it to spill over into a second (light blue component) and the bright component is just about to merge with the dark component. Notice the textured components that are waiting as part of the 'downhill stream' of water.

To execute the algorithm, a pixel from which flooding will proceed is arbitrary chosen. This pixel can be thought of as the point at which water is being poured on, and the output result will be the same regardless of which pixel is selected, so we may simply pick the upper left corner of the image. We will refer to this as the source pixel. The algorithm is as follows, see also Fig 5:

1. Clear the accessible pixel mask, the heap of boundary pixels and the component stack. Push a dummy-component onto the stack, with grey-level higher than any allowed in the image.
2. Make the source pixel (with its first edge) the current pixel, mark it as accessible and store the grey-level of it in the variable *current_level*.
3. Push an empty component with *current_level* onto the component stack.
4. Explore the remaining edges to the neighbors of the current pixel, in order, as follows: For each neighbor, check if the neighbor is already accessible. If it is not, mark it as accessible and retrieve its grey-level. If the grey-level is not lower than the current one, push it onto the heap of boundary pixels. If on the other hand the grey-level is lower than the current one, enter the current pixel back into the queue of boundary pixels for later processing (with the next edge number), consider the new pixel and its grey-level and go to 3.
5. Accumulate the current pixel to the component at the top of the stack (water saturates the current pixel).
6. Pop the heap of boundary pixels. If the heap is empty, we are done. If the returned pixel is at the same grey-level as the previous, go to 4.
7. The returned pixel is at a higher grey-level, so we must now process all components on the component stack until we reach the higher grey-level. This is done with the *ProcessStack* sub-routine, see below. Then go to 4.

Fig. 7. For a ridge pixel, it is important for theoretical correctness of the algorithm that water is not allowed to 'spill' simultaneously into several directions. Therefore, if a lower pixel is found as any of the neighbors of a pixel, the current pixel has to be stacked for later and the newly found pixel tended to fully before proceeding to the other neighbors.

The *ProcessStack* sub-routine is as follows:
Sub-routine *ProcessStack(new_pixel_grey_level)*

1. Process component on the top of the stack. The next grey-level is the minimum of *new_pixel_grey_level* and the grey-level for the second component on the stack.
2. If *new_pixel_grey_level* is smaller than the grey-level on the second component on the stack, set the top of stack grey-level to *new_pixel_grey_level* and return from sub-routine (This occurs when the new pixel is at a grey-level for which there is not yet a component instantiated, so we let the top of stack be that level by just changing its grey-level.
3. Remove the top of stack and merge it into the second component on stack as follows: Add the first and second moment accumulators together and/or join the pixel lists. Either merge the histories of the components, or take the history from the winner. Note here that the top of stack should be considered one 'time-step' back, so its current size is part of the history. Therefore the top of stack would be the winner if its current size is larger than the previous size of second on stack.
4. If(*new_pixel_grey_level>top_of_stack_grey_level*) go to 1.

Here, the implementation of how to process a component follows the same lines as the standard algorithm, and depends on the information one wants about an MSER. It entails determining if a relative growth rate minimum has occurred, and if so, harvesting the required information from the component and declaring an MSER detected. The data-structure of a component holds information about the previous sizes of the component, and at which grey-levels they occurred.

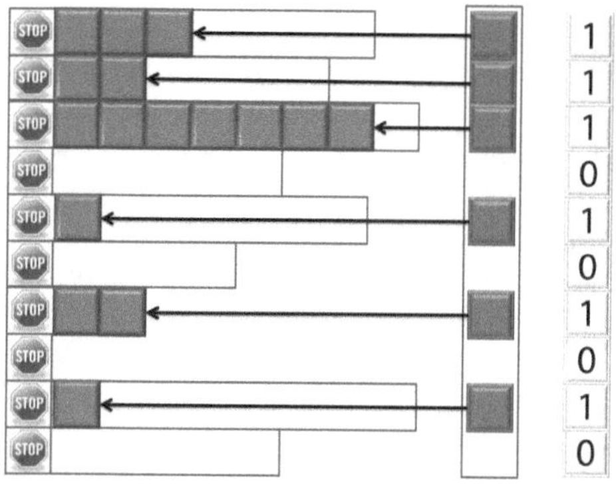

Fig. 8. The heap of boundary pixels can be efficiently implemented by a binary bitmask with number of grey-level bits, where bits are set if there are pixels at that grey-level, and a system of stacks, one for each grey-level.

It is immaterial to the result which order is used to explore the edges of a particular pixel, but it is important that if a lower grey-level pixel is found at the end of one of the edges, this pixel is tended to before any further edges are explored, see Fig 7. The reason is that otherwise water could enter a 'ridge' pixel that has several edges to lower grey-level pixels from distinct basins, and water would 'spill' into both basins simultaneously if we are not careful to process one of the edges first and fully saturate it before moving to the next edge. In practice, it is possible that this would not be a common problem, it would perhaps just make the algorithm 'myopic' in the sense that water could 'tunnel' through ridges, but by carefully saturating the edges in the correct order, our result will adhere exactly to the strict definition of MSER.

2.3 Implementation Details

In this section we give some implementation details. For efficiency, we implemented the heap of boundary pixel as a system of stacks of pixels at various grey-levels, see Fig 8. A bitmask is keeping track of which out of the 256 grey-levels have pixels waiting. This allows us to use a single instruction to find which is the smallest (or largest) occupied grey-level by using a machine instruction that retrieves the id number of the least significant or most significant bit set (for *x86* platforms *bsf* and *bsr*, or the more general 64-bit compliant *_BitScanForward,_BitScanReverse*). This is of course limited to as many grey-levels as the processor word-length, but for 256 grey-levels we simply chain multiple such instructions. This is essentially a way to get the hardware to do the work, and lower the time it takes to push and pop from the heap. This time is inherently constant with regards to the number of pixels, and logarithmic in the number of grey-levels. We also perform an explicit check for current grey-level before doing the general pop. This can be done since the grey-level is never smaller than the current. When it is equal, which is common, some work can be saved.

The heap could be dynamically allocated and implemented as linked lists, but we prefer to count the number pixels at each grey-level in a single pre-sweep of the image, just like the standard algorithm does. This allows us to both pre-allocate and use fixed arrays for the stacks without stacks ever running into one another. The total number of entries in the stacks is the number of pixels plus the number of grey-levels (due to one stop-element for each stack).

Finally, to avoid having to perform explicit checks for the image boundary, a border of one pixel around the image is used, and the accessible mask is always set when a sweep starts. This can actually be implemented without explicitly sweeping the mask except for in a detector initialization step, since the sweep sets the whole mask. We then simply flip the border bits and invert the meaning of the mask on the dark-to-bright sweep, finally leaving the mask in its original state.

3 Analysis

An informal analysis for the algorithm is straightforward. It is easy to see that only one copy of a pixel can ever be on the heap of boundary pixels, and that a pixel can only return onto the heap at most as many times as it has neighbors. Moreover, the components that are processed must have at least one new pixel in them, so the total amount of component processing is linear in the number of pixels. As it takes at most $\log(m)$ time to enter or pop a pixel from the heap, the worst case execution time is therefore bounded by

$$O((n+e)\log(m)), \tag{1}$$

where n is the number of pixels, m the number of grey-levels, and e is the number of edges in the image graph (such as $e \approx 2n$ for four-connected images). If we regard m and the connectivity as constants, this simply amounts to $O(n)$, linear time.

The memory usage is roughly a four-byte integer per pixel, which allows both the heap of boundary pixels, the accessible mask bit and the edge number to be stored while allowing a sufficient number of bits for the pixel coordinates. Storage for only m components is required (apart of course from the storage required by detected MSER, which would clearly be needed for any algorithm.

4 Experimental Validation

As the results of our detector are the same as the original MSER detector, which has been found in several studies to have excellent repeatability, our experimental focus is on execution time. As a reference, the recent paper [19] reports on an FPGA implementation with careful attention to detail, but based on the standard union-find algorithm, performing at 25 fps on images up to 350×350

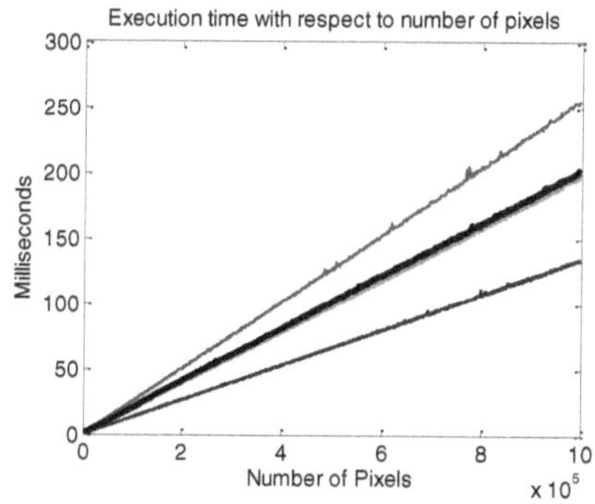

Fig. 9. Execution time as a function of image size on square images ranging from one pixel to one mega-pixel. Curves are shown, from fastest to slowest: for a blank image and an image filled with ramps (blue and black, almost identical), a pixel-size checkerboard pattern (thin green line), natural image (thick black line) and a random image (thin red line). Note the completely linear timings, which include both passes.

pixels or 54 fps detection on image streams with 320 × 240 pixels resolution. These execution times are for a single pass, so with both passes (bright-to-dark and dark-to-bright), this corresponds to 1.5 or 2 mega-pixel per second. Our implementation of the new algorithm, which is relatively carefully written but not fully optimized, is running at least twice as fast as this, at around 5 mega-pixel per second on a single core consumer grade CPU.

We ran experiments with single-threaded code on a laptop with 3GB of RAM and an Intel T7400 2.16GHz processor. To try to show the 'best' and 'worst' realistic performance of the algorithm, we give execution times with respect to the number of pixels in Fig 9 for various synthetic images and natural images. To avoid process scheduling noise, results were produced by running five independent times

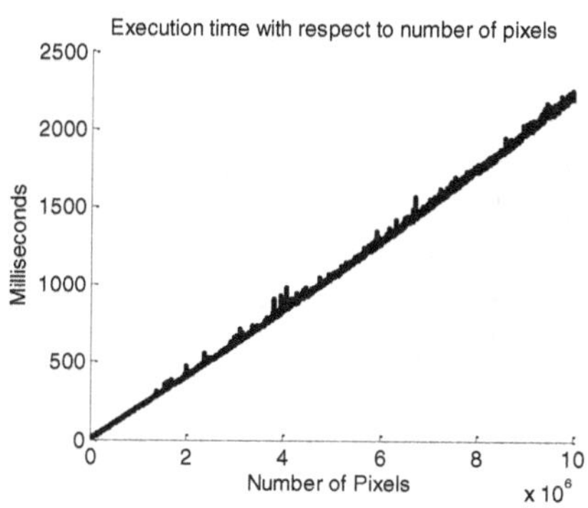

Fig. 10. Actual execution times up to ten mega-pixel. A tiny amount of curvature can be perceived, but the performance is very close to the theoretical linear.

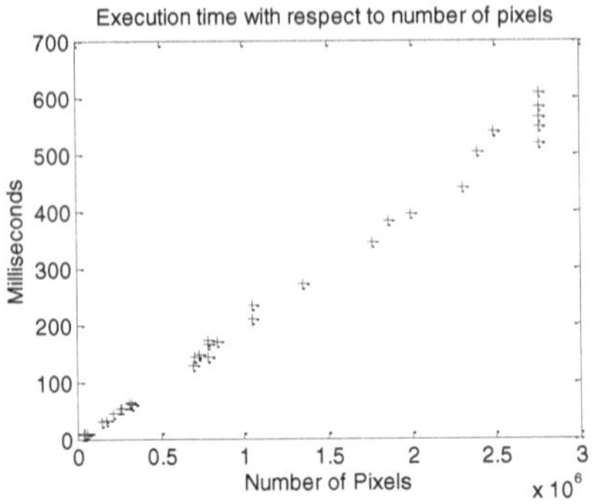

Fig. 11. A scatter plot of execution times with respect to number of pixels for a number of randomly selected photographs with random resolution. Note the clear linear trend of around five mega-pixel per second.

and taking the fastest run. Blank or smooth images execute the fastest, while completely random images execute the slowest. There is approximately a factor two difference between the two, which likely is connected to how many times each pixel is put back on the heap. It is worth noting that random images and checkerboard images are particularly challenging to the standard algorithm since it causes half as many concurrent components as there are pixels in the image. A natural image is typically somewhere in between. To provide a reproducible reference speed-curve, the natural image curve in this figure was produced by running on the 512×512 Lena image wrapped endlessly. The linearity persists to very large images, Fig 10 shows the timings up to ten mega-pixel.

A more general measurement of execution time for natural images is shown in Fig 11, where the execution times for randomly selected images of varying size are given. Note the clear linearity and the typical speed of roughly 5 mega-pixel per second on natural images.

5 Conclusions

We have described a new MSER detection algorithm, which runs in linear time and behaves like a true flood-fill, meaning a flood-fill where the next pixel filled must be the lowest point accessible so far. The performance tests validate the linear run-times and indicate the speed on a single core consumer grade CPU to be at around five mega-pixels per second for natural images, roughly the pixel rate of a PAL TV signal and more than twice as fast as a very recent FPGA implementation. The algorithm shows that MSER detection does not necessarily need the union-find implementation commonly used for watershed segmentation. This allows true linear time and opens up for further studies into parallelization of the algorithm.

References

1. Tuytelaars, T., Gool, L.V.: Matching widely separated views based on affine invariant regions. International Journal of Computer Vision 59(1), 61–85 (2004)
2. Schmid, C., Mohr, R., Bauckhage, C.: Evaluation of interest point detectors. International Journal of Computer Vision 37(2), 151–172 (2000)
3. Matas, J., Chum, O., Urban, M., Pajdla, T.: Robust wide baseline stereo from maximally stable extremal regions. In: British Machine Vision Conference, pp. 384–393 (2002)
4. Mikolajczyk, K., Schmid, C.: Scale and affine invariant interest point detectors. International Journal of Computer Vision 60(1), 63–86 (2004)
5. Mikolajczyk, K., Tuytelaars, T., Schmid, C., Zisserman, A., Matas, J., Schaffalitzky, F., Kadir, T., Gool, L.V.: A comparison of affine region detectors. International Journal of Computer Vision 65(1–2), 43–72 (2005)
6. Kadir, T., Zisserman, A., Brady, M.: An affine invariant salient region detector. In: European Conference on Computer Vision, pp. 404–416 (2004)
7. Lowe, D.: Distinctive image features from scale invariant keypoints. International Journal of Computer Vision 60(2), 91–110 (2004)
8. Lindeberg, T.: Feature detection with automatic scale selection. International Journal of Computer Vision 30(2), 77–116 (1998)
9. Triggs, B.: Detecting keypoints with stable position, orientation and scale under illumination changes. In: European Conference on Computer Vision, vol. 4, pp. 100–113 (2004)
10. Rothganger, F., Lazebnik, S., Schmid, C., Ponce, J.: 3d object modeling and recognition using local affine-invariant image descriptors and multi-view spatial constraints. International Journal of Computer Vision 66(3), 231–259 (2006)
11. Donoser, M., Bischof, H.: Efficient maximally stable extremal region (mser) tracking. In: IEEE Conference on Computer Vision and Pattern Recognition (CVPR), pp. 553–560 (2006)
12. Brown, M., Lowe, D.: Invariant features from interest point groups. In: British Machine Vision Conference, pp. 656–665 (2002)
13. Harris, C., Stephens, M.: A combined corner and edge detector. In: Proceedings of the 4th Alvey Vision Conference, pp. 147–151 (1988)
14. Sivic, J., Zisserman, A.: Video google: A text retrieval approach to object matching in videos. In: International Conference on Computer Vision, vol. 2, pp. 1470–1477 (2003)
15. Nistér, D., Stewénius, H.: Scalable recognition with a vocabulary tree. In: IEEE Conference on Computer Vision and Pattern Recognition (CVPR), vol. 2, pp. 2161–2168 (2006)
16. Obdrzalek, S., Matas, J.: Object recognition using local affine frames on distinguished regions. In: British Machine Vision Conference, vol. 1, pp. 113–122 (2002)
17. Forssén, P.E.: Maximally stable colour regions for recognition and matching. In: IEEE Conference on Computer Vision and Pattern Recognition (2007)
18. Donoser, M., Bischof, H.: 3d segmentation by maximally stable volumes (msvs). In: ICPR 2006: Proceedings of the 18th International Conference on Pattern Recognition, pp. 63–66 (2006)
19. Kristensen, F., MacLean, W.: Fpga real-time extraction of maximally-stable extremal regions. In: IEEE International Symposium on Circuits and Systems (2007)
20. Vincent, L., Soille, P.: Watersheds in digital spaces: an efficient algorithm based on immersion simulations. IEEE Transactions on Pattern Analysis and Machine Intelligence 13, 583–598 (1991)

21. Roerdink, J., Meijster, A.: The watershed transform: definitions, algorithms and parallelization strategies. Fundamenta Informaticae 41, 187–228 (2000)
22. Couprie, M., Najman, L., Bertrand, G.: Quasi-linear algorithms for the topological watershed. Journal of Mathematical Imaging and Vision 22(2), 231–249 (2005)
23. Tarjan, R.: Data Structures and Network Algorithms. SIAM, Philadelphia (1983)
24. Gabow, H., Tarjan, R.: A linear-time algorithm for a special case of disjoint set union. Journal of Computer and System Sciences 30(2), 209–220 (1985)

Efficient Edge-Based Methods for Estimating Manhattan Frames in Urban Imagery

Patrick Denis[1], James H. Elder[1], and Francisco J. Estrada[2]

[1] York University
{pdenis,jelder}@yorku.ca
[2] University of Toronto
strider@cs.utoronto.ca

Abstract. We address the problem of efficiently estimating the rotation of a camera relative to the canonical 3D Cartesian frame of an urban scene, under the so-called "Manhattan World" assumption [1,2]. While the problem has received considerable attention in recent years, it is unclear how current methods stack up in terms of accuracy and efficiency, and how they might best be improved. It is often argued that it is best to base estimation on all pixels in the image [2]. However, in this paper, we argue that in a sense, less can be more: that basing estimation on sparse, accurately localized edges, rather than dense gradient maps, permits the derivation of more accurate statistical models and leads to more efficient estimation. We also introduce and compare several different search techniques that have advantages over prior approaches. A cornerstone of the paper is the establishment of a new public groundtruth database which we use to derive required statistics and to evaluate and compare algorithms.

1 Introduction

The problem of single-view 3D reconstruction is of enormous theoretical and practical interest. Applications range from urban security to architectural and urban planning systems, to self-localization systems for mobile devices. Given the prevalence of security cameras in urban environments throughout the world, single-view methods could be of great value in the refining of 3D databases such as Google Earth and Microsoft Virtual Earth.

In general, single view reconstruction is an ill-posed problem. The problem becomes more tractable under the so-called "Manhattan-World" assumption [1,2]: that the surfaces of interest are rectangles aligned with a 3D Cartesian frame.

A key step in exploiting the Manhattan World assumption is the estimation of the Manhattan directions, i.e., the 3D rotation of the camera relative to the Manhattan frame [1,2]. This is the problem we address here.

The specific **contributions** we make here fall into four categories. First, we introduce a new public database, with appropriate groundtruthing, that can be used to design and evaluate algorithms. Second, we establish accurate statistics required to ground estimation algorithms. Third, we devise new algorithm

D. Forsyth, P. Torr, and A. Zisserman (Eds.): ECCV 2008, Part II, LNCS 5303, pp. 197–210, 2008.

variants that have potential advantages in accuracy and speed over previous approaches. Finally, we conduct a comprehensive evaluation of six approaches in terms of both accuracy and efficiency. The results suggest that basing estimates on sparse but accurate edge maps, rather than the dense gradient fields used in other methods [2,3,4], leads to a substantial improvement in both accuracy and computational efficiency.

2 Prior Work

Coughlan & Yuille [1,2] considered the problem of estimating the Manhattan frame using a calibrated camera. They based their estimation on the dense pattern of pixel gradients in the image, using a mixture model in which each pixel gradient has a probability distribution over five possible causes: (1) The pixel is not an edge. (2-4) The pixel is an edge that belongs to one of the three Manhattan-World directions. (5) The pixel is an edge that does not belong to any of the Manhattan-World directions.

Coughlan & Yuille searched over the space of camera rotations for the camera pose that maximized the likelihood of the observed gradient data under their mixture model. Although the gradients were represented in the continuous domain (not Houghed), their coarse-to-fine search technique was discrete.

A number of efforts have been made since to improve on this work, for example, to allow simultaneous calibration of the camera [3,5,4,6], and to extend the Manhattan assumption to more general scene models [3,6]. These methods also replace Coughlan & Yuille's coarse-to-fine discrete search technique, either with an EM method [5,4,6] or stochastic (particle-based) search strategy [3].

There are a number of potential limitations in these methods. Some use heuristics to initialize search, based on dominant orientations [5] or RANSAC methods [6], and it is unclear whether these heuristics will reliably direct the search toward correct solutions. The EM methods [5,4,6] assume that deviations of Manhattan edges and lines from the expected orientations are normally distributed, however this appears to not be the case ([2], and see Figure 3 of this paper). Computing probabilities for dense gradient fields over the entire image [2,4] entails substantial computational cost. This problem is multiplied by complicated iterative search techniques [4]. Subsampling and thresholding the gradient map [4] in order to contain the overall computation time may sacrifice accuracy. While Coughlan & Yuille made an effort to base their method on statistics of scenes, their models were simplified, and subsequent studies have generally been less concerned with accurate statistical models.

In this paper we will address these limitations and evaluate a number of design decisions, using an edge-based method for estimating the Manhattan frame from a calibrated camera. We show how using edges as sparse features allows an accurate statistical model to be derived and employed. We use a new public ground truth database to quantitatively evaluate and compare six algorithms, including the Manhattan World algorithm [2]. Prior quantitative evaluations of

Manhattan frame estimates have been limited, and to our knowledge no quantitative comparisons between algorithms have previously been made.

While there are some advantages to the simultaneous estimation of camera parameters, allowing camera parameters to vary freely also reduces the number of scenes providing sufficient constraints for reliable estimation of the Manhattan frame. In many practical situations camera parameters are known, can be independently estimated, or can be adequately approximated by nominal values. In this paper we assume a calibrated camera, and focus on the issue of how to maximize accuracy and efficiency of Manhattan frame estimation.

3 Approach

There are two main components to our approach: (1) Construction of the ground truth database and camera calibration, and (2) Design of the algorithms to be evaluated.

3.1 Ground Truth Database

We created two databases of 640×480-pixel images of urban Toronto scenes using a standard digital camera (Panasonic Lumix DMC-LC80). The first database, used to calibrate the camera, consisted of 10 images (3 indoor and 7 outdoor) taken with the camera held in a generic attitude with respect to the Manhattan world frame. The second database, used to train and evaluate algorithms, consisted of 102 images (45 indoor and 57 outdoor). The camera was held in a natural attitude. (Posthoc analysis revealed a mean and maximum absolute deviation of the image y-axis from estimated gravitational vertical of 4 deg and 19 deg, respectively.) These 102 images were further randomly divided into training and test sets of 51 images each.

To establish ground truth, we developed an interactive MATLAB program that allows a user to identify lines in the image with sub-pixel precision and to indicate the Manhattan directions with which they are associated (Figure 1).

Vanishing points were first inferred for the calibration database from these ground truth lines using the Gauss Sphere method of Collins & Weiss [7], and nominal values for the internal camera parameters. We then used a standard non-linear search to solve for the focal length and principal point that make the three Manhattan directions for each image mutually orthogonal in 3D space. We assumed a simplified camera model with no skew and unit aspect ratio. The focal length was estimated to be 6.053 ± 0.009mm, and the (x, y) coordinates of the principal point were estimated to be $(-12.9 \pm 4.4, 11.0 \pm 6.4)$ pixels, relative to the image centre.

Vanishing points for the training and test databases were subsequently calculated with the estimated camera parameters [7]. Since each Manhattan direction was estimated independently, the resulting ground truth Manhattan frames were not exactly orthogonal. We therefore fit an orthogonal frame to the three independently-estimated vectors for each image [8].

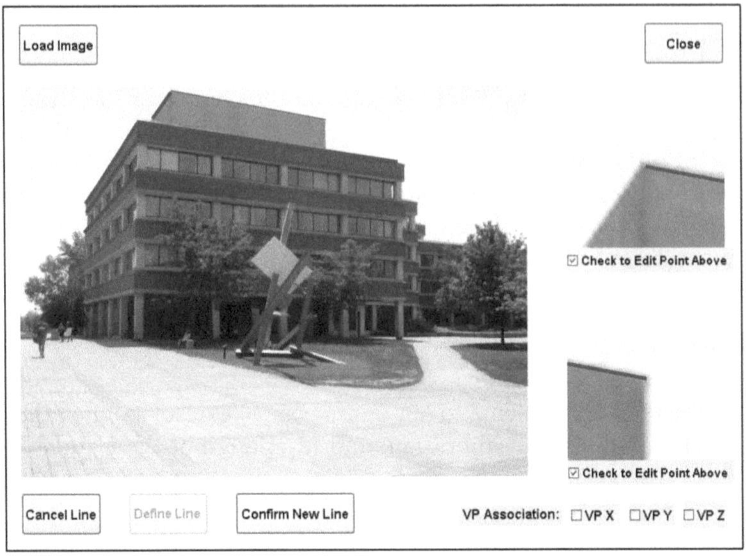

Fig. 1. Ground truth labeling tool

This ground truth database of images, including the estimated Manhattan lines and directions, is available for public use through our website at `http://www.elderlab.yorku.ca/YorkUrbanDB`, in the hope that it will encourage more rapid development as well as quantitative evaluation and comparison of algorithms.

3.2 Mixture Model

We employ a mixture model similar to that used by Coughlan & Yuille ([1], [2]) but applying only to the set of edges E detected and localized to subpixel precision by a standard multiscale edge detector [9]. This reduces the size of the input space by slightly more than a factor of 10.

Each edge $E_{\boldsymbol{u}}$ consists of a pixel location $\boldsymbol{u}$ and orientation $\phi_{\boldsymbol{u}}$, and is assumed to be generated by one of four causes:

- $m_{\boldsymbol{u}} = v$ Linear scene structure aligned with the vertical Manhattan direction
- $m_{\boldsymbol{u}} = h_1, h_2$ Linear scene structure aligned with one of the two horizontal Manhattan directions
- $m_{\boldsymbol{u}} = b$ Background scene structure not aligned with any Manhattan direction

Our goal is to use the edge data to estimate the rotation Ψ' of the coordinate frame of the urban scene (the "Manhattan frame") relative to the camera frame.

As in [2], we assume conditional independence between edges and use a maximum likelihood estimator:

$$\Psi^* = \text{argmax}_\Psi \sum_{\boldsymbol{u}} \log P(E_{\boldsymbol{u}}|\Psi) \tag{1}$$

where

$$P(E_{\boldsymbol{u}}|\Psi) = \sum_{m_{\boldsymbol{u}}} P(E_{\boldsymbol{u}}|m_{\boldsymbol{u}}, \Psi)P(m_{\boldsymbol{u}}). \tag{2}$$

Estimating the Manhattan frame thus requires a reasonable model for the conditional probability $P(E_{\boldsymbol{u}}|m_{\boldsymbol{u}}, \Psi)$ of observing a particular edge $E_{\boldsymbol{u}}$ given that it was generated by a specific cause $m_{\boldsymbol{u}}$, and for the expected proportion $P(m_{\boldsymbol{u}})$ of edges generated by each cause.

3.3 Error Model

In [1], the conditional probability $P(E_{\boldsymbol{u}}|m_{\boldsymbol{u}}, \Psi)$ of observing a particular edge $E_{\boldsymbol{u}}$ given that it was generated by one of the Manhattan causes $m_{\boldsymbol{u}} \in \{v, h_1, h_2\}$ was assumed to be a function only of the angular deviation $\Delta\phi(\boldsymbol{u}, m_{\boldsymbol{u}}, \boldsymbol{\Psi})$ of the observed edge orientation $\phi_{\boldsymbol{u}}$ from the expected direction (Figure 2a). The distribution of $\phi_{\boldsymbol{u}}$ given a background cause was assumed to be uniform. Coughlan & Yuille employed a simple box function to model all three Manhattan conditional distributions. Here we use our ground truth database to determine more accurate models of these distributions.

One of the difficulties in estimating these distributions lies in determining which edges are associated with the hand-labeled lines in the ground truth database. Hand-labelling all of these edges would be prohibitively time-consuming. Instead, consider the density of edges in the image as a function of their distance from the ground truth lines in the database (Figure 2b). The data show a uniform background density of 0.1 edges/pixel, on which is superimposed a peaked distribution of edges generated by the line. We infer from this distribution that the sub-pixel localized edges generated by ground truth lines lie within a distance $x_{\boldsymbol{u}}$ of roughly 1 pixel from the generating line, and thus

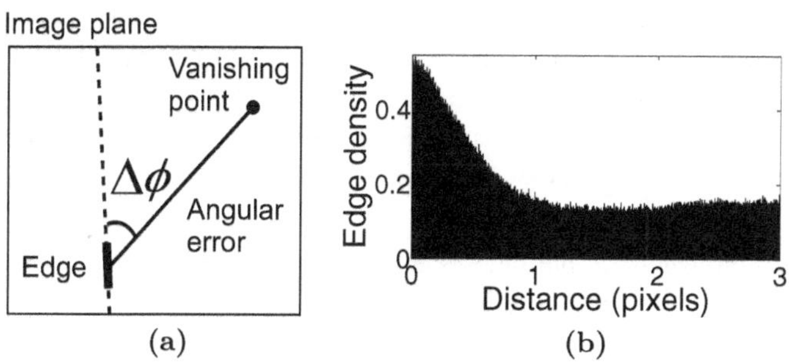

Fig. 2. (a) Angular deviation $\Delta\phi$ of a Manhattan edge from the expected direction, in the image plane. (b) Edge density histogram.

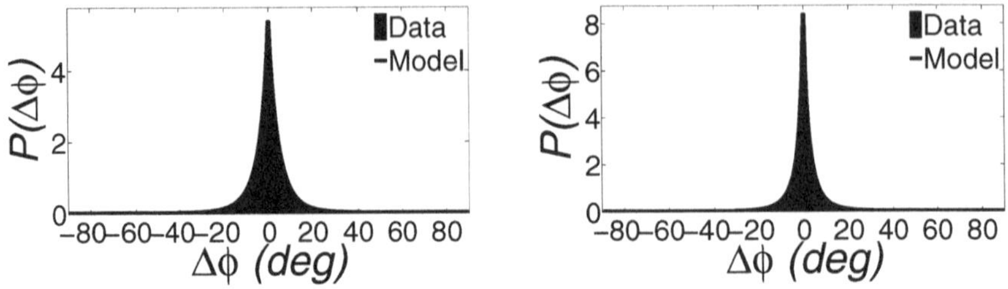

Fig. 3. Empirical densities of image-based angular deviation measure $\Delta\phi$ for edges lying within 1 pixel of horizontal (right) and vertical (left) Manhattan lines. Distributions are fit with a mixture model: see text for details.

confine our analysis to these edges. The conditional distributions of the angular deviation $\Delta\phi_{\boldsymbol{u}}$ for these edges, for both horizontal and vertical Manhattan directions, are shown in Figure 3.

We assume that these distributions arise from a mixture of edges generated by the nearby Manhattan line and edges generated by background structure. Letting $M \in \{v, h_1, h_2\}$ represent the Manhattan direction with which the line is associated, the density $P(\Delta\phi|x_{\boldsymbol{u}} < 1)$ can be represented as

$$P(\Delta\phi|x_{\boldsymbol{u}} < 1) = \lambda P(\Delta\phi|m_{\boldsymbol{u}} = M, \Psi) + (1 - \lambda)P(\Delta\phi|m_{\boldsymbol{u}} = b). \quad (3)$$

We model the former as a Generalized Laplace distribution and the latter by a uniform distribution:

$$P(\Delta\phi|m_{\boldsymbol{u}} = M, \Psi) = \frac{1}{Z}\exp\left(-|\Delta\phi_{\boldsymbol{u}}/b|^{\alpha}\right) \quad (4)$$

and

$$P(\Delta\phi|m_{\boldsymbol{u}} = b) = U(-90, 90) \quad (5)$$

Maximum likelihood parameters for the mixture model are shown in Table 1. Two interesting observations emerge: (1) The Manhattan densities $P(\Delta\phi|m_{\boldsymbol{u}} = M, \Psi)$ are highly leptokurtic ($\alpha < 1$), far from being Normal distributions $\alpha = 2$) as commonly assumed [4] and even further from the boxcar distributions employed by Coughlan & Yuille. (2) The dispersion of the vertical Manhattan density is one third that of the horizontal. This suggests that the orientations of edges generated by vertical structure are more accurate than for horizontal. One possible explanation is that vertical structure may have higher contrast, for example when silhouetted against the sky.

Table 1. Maximum likelihood parameters for mixture model representing deviation of nearby edges from Manhattan lines. See text for details.

	λ	b (deg)	α
Horizontal	0.91	4.0	0.84
Vertical	0.94	1.7	0.65

While Coughlan & Yuille measured error as angular deviation in the image domain, others ([5], [6]) have measured error $\Delta\phi$ as the angle between the vanishing direction and the interpretation plane of the edge in the Gauss Sphere. In the absence of error, this angle should be 0. Using the above method we have also derived estimates of the conditional distributions under this measure of deviation. For lack of space we do not show the distributions here, but we will evaluate and compare the two methods in Section 4.

3.4 Prior Model

To estimate the proportion of edges generated by each causal factor v, h_1, h_2, b, we first compute the density of the angular deviation measure $\Delta\phi$ (in the image) of *all* edges relative to both horizontal and vertical ground truth vanishing points (Figure 4). Note that these densities have peaks at 0 deg due to edges generated by the corresponding Manhattan structure, *and* at 90 deg, due to edges generated by orthogonal Manhattan structure.

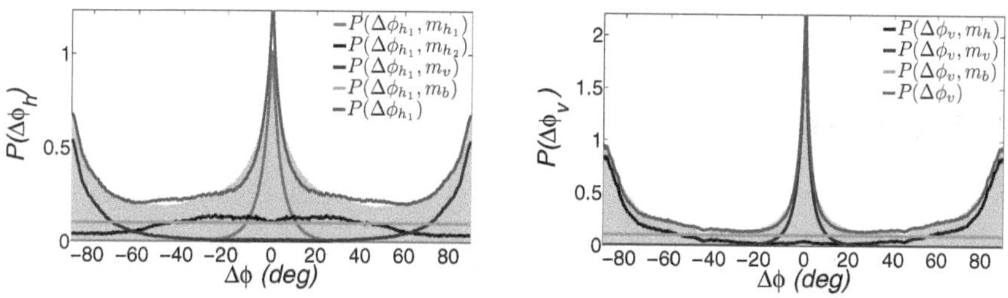

Fig. 4. Density of the angular deviation measure $\Delta\phi$ of all edges relative to both horizontal (right) and vertical (left) ground truth vanishing points. Density is fit with a mixture of densities representing the contributions of edges from all causal factors. See text for details.

To represent this latter contribution, we generalize our notation to let $\Delta\phi_M$, where $M \in \{v, h_1, h_2\}$, denote the deviation measure relative to each of the three Manhattan directions. Now $P(\Delta\phi_M|m_{\boldsymbol{u}})$ represents the conditional distribution of $\Delta\phi$ measured relative to the vanishing point for M, given an edge generated by $m_{\boldsymbol{u}} \in \{v, h_1, h_2, b\}$. We approximate the empirical distributions for $m_{\boldsymbol{u}} \in \{v, h_1, h_2\}, m_{\boldsymbol{u}} \neq M$ by measuring $\Delta\phi$ relative to a Manhattan direction, for edges within $x_{\boldsymbol{u}} < 1$ pixel of a *different* Manhattan direction. Now the total density $P(\Delta\phi_M)$ for all edges can be represented as a mixture from all 4 causes:

$$P(\Delta\phi_M) = \sum_{m_{\boldsymbol{u}} \in \{v, h_1, h_2, b\}} P(\Delta\phi_M|m_{\boldsymbol{u}})P(m_{\boldsymbol{u}}) \quad (6)$$

where the $P(m_{\boldsymbol{u}})$ represent the proportion of edges generated by each causal factor, serving as the mixing parameters in this model. Fitting the mixture to the empirical distribution for all edges thus provides an estimate of these proportions.

Table 2. Comparison between the edge priors estimated here and the gradient priors estimated by Coughlan & Yuille [1]

Model	MW Priors	Our Priors
$P(m_{\boldsymbol{u}} = h_1)$	0.2	0.23
$P(m_{\boldsymbol{u}} = h_2)$	0.2	0.23
$P(m_{\boldsymbol{u}} = v)$	0.2	0.23
$P(m_{\boldsymbol{u}} = b)$	0.4	0.32

Table 2 lists our estimates and compares them to estimates of *gradient* priors made by [1]. The estimates are in reasonable agreement.

3.5 Search Strategies

Our statistical models allow calculation of the likelihood $P(E|\Psi)$ of observing a set of edges E given a camera rotation Ψ relative to the Manhattan frame. Our task now is to estimate the Ψ^* that maximizes this probability.

Coughlan & Yuille [2] employed a coarse-to-fine search strategy over a discretized space representing the three Euler angles composing Ψ. The resolution of the discretization potentially limits the accuracy of the method. Here we explore two alternative search methods in the continuous domain.

Quasi-Newton Method. We first employ a method based upon the BFGS Quasi-Newton gradient descent technique [10]. First, the likelihood $P(E|\Psi)$ is evaluated for some number r of initial guesses $\Psi_i, i \in [1...r]$. Then, for the n most probable guesses, the BFGS Quasi-Newton method is deployed to compute n refined estimates, and the estimate maximizing $P(E|\Psi)$ is selected as the final estimate.

Initial guesses assumed the y-axis of the image plane to be aligned with gravitational vertical, and rotations about the vertical axis were uniformly sampled between 0 and 90 deg. Based on experiments using the training database, values for r and n were selected to achieve maximum speed while remaining within 10% of the achievable accuracy. Suitable values were found to be $r = 5, n = 2$.

EM Method. Schindler & Dellaert [4] proposed a different continuous alternative to Coughlan & Yuille's discrete search strategy, based on Expectation-Maximization (EM). The EM technique had been applied before, for the purpose of estimating individual (not necessarily orthogonal) vanishing directions [5].

As applied to the problem of estimating the Manhattan frame, the EM algorithm alternates between an E-step, in which an estimate is made of the probability $P(m_{\boldsymbol{u}}|E_{\boldsymbol{u}}, \Psi)$ over possible causes $m_{\boldsymbol{u}}$ of each edge $E_{\boldsymbol{u}}$, and an M-step, in which a new estimate of the camera rotation Ψ is made based upon the causal probabilities computed in the E-step. The E-step is straightforward, since, given an estimate Ψ_t of Ψ, $P(m_{\boldsymbol{u}}|E_{\boldsymbol{u}}, \Psi_t) \propto P(E_{\boldsymbol{u}}|m_{\boldsymbol{u}}, \Psi_t)P(m_{\boldsymbol{u}})$ can be calculated directly using the densities and distributions estimated in Section 3. The M-step, on the other hand, is non-trivial, since no closed-form estimator of Ψ is evident. By assuming the conditional error distributions $P(\Delta\phi|m_{\boldsymbol{u}} = M, \Psi), M \in \{v, h_1, h_2\}$ to be Gaussian, Schindler & Dellaert [4] reduce the M-step to a nonlinear least-squares problem, which they solve using iterative nonlinear optimization techniques. However, we know from Section 3 that these distributions are highly *non*-Gaussian.

Here we consider two methods for performing the M-step. In the first, we again use a BFGS Quasi-Newton gradient descent technique [10] to update the estimate of Ψ:

$$\Psi^{t+1} = \arg\max_{\Psi} \sum_{\boldsymbol{u}} \sum_{m} P(m_{\boldsymbol{u}}|E_{\boldsymbol{u}}, \Psi_i) \log(P(E_{\boldsymbol{u}}|m_{\boldsymbol{u}}, \Psi)P(m_{\boldsymbol{u}})) \tag{7}$$

$$= \arg\min_{\Psi} \sum_{\boldsymbol{u}} \sum_{m \in \{h_1, h_2, v\}} P(m_{\boldsymbol{u}}|E_{\boldsymbol{u}}, \Psi_t) \left| \frac{\Delta\phi_{m,\boldsymbol{u}}}{b_m} \right|^{\alpha_m} \tag{8}$$

where $\Delta\phi_{m,\boldsymbol{u}}$ is the angular deviation of the edge at site $\boldsymbol{u}$ from vanishing point m, and b_m and α_m are the Generalized Laplace parameters for that deviation.

As for the Quasi-Newton Method (Section 3.5), the likelihood $P(E|\Psi)$ was first evaluated for some number r of initial guesses. Then, for the n most probable guesses, the above EM method was deployed for t iterations. Based on experiments using the training database, suitable values were found to be $r = 15, n = 3, t = 15$.

As for the Schindler & Dellaert method, the above method nests two iterative procedures, and therefore is unlikely to be efficient. We therefore also considered a quasi-EM approach in which we simplify the M-step by decoupling the estimation of the three Manhattan directions. In particular, we have adapted the fast, closed-form method of Collins & Weiss [7] to estimate each vanishing point direction in the Gauss Sphere independently. To estimate a new vector for a particular Manhattan direction, this adaptation simply requires weighting the interpretation plane normal associated with each edge $E_{\boldsymbol{u}}$ in the image by the probability $P(m_{\boldsymbol{u}}|E_{\boldsymbol{u}}, \Psi)$, computed in the E-step, that the edge was generated by this cause.

Since the vanishing directions are estimated independently, the frame will no longer be orthogonal. At the end of every M-step, we therefore re-orthogonalize the frame by fitting an orthogonal Manhattan frame to the independently-estimated vectors.

As for the other search methods, we empirically estimated the parameters r, n and t that would yield the highest achievable accuracy. We found near-optimal values of $r = 2, n = 1$ and $t = 7$ iterations when measuring error in the image domain, and $r = 1, n = 1$ and $t = 8$ iterations when measuring error in the Gauss Sphere (Section 3.3). However, we also observed erratic convergence behaviour in both cases, suggesting that our approximation of the M-step had destroyed the crucial convergence properties of the EM algorithm.

4 Results

We evaluated six algorithms, summarized in Table 3, on our groundtruth test database. The MW algorithm is the original Manhattan World algorithm of Coughlan & Yuille [2], with code graciously provided by James Coughlan. We list the remaining algorithms in descending order of similarity to the original Manhattan World algorithm. The Edge-Based Newton MW Params algorithm is our edge-based variant of the MW method, using the same statistical distributions as the original algorithm, but using a Quasi-Newton search method instead of the coarse-to-fine search employed by Coughlan & Yuille. The Edge-Based Newton method replaces the Manhattan statistics with our own estimated distributions. The Edge-Based EM algorithm uses the EM method with a nested Quasi-Newton procedure in the M-step (Section 3.5). Finally, the Edge-Based Quasi-EM and Quasi-EM GS algorithms use the quasi-EM method employing the method of Collins & Weiss to update the estimate of the Euler angles in the M-step (Section 3.5).

Empirical accuracy for the six algorithms is shown in Figure 5a. We found that the Edge-Based Newton method performed about 35% better than the original Manhattan World (MW) method, despite using less than 10% of the input data, and even when using the original Manhattan World statistical models (Edge-Based Newton MW Params). We believe this improvement derives from two sources. The first is the higher information content and lower redundancy of the sub-pixel-localized edges [9] underlying the edge-based method, relative to the gradient features used in MW. Given perfect statistical models, noisier gradient features would be downweighted to avoid error. However, in a

Table 3. Properties of the six algorithms evaluated. See text for a description of each algorithm. For each algorithm we list: (1) the image feature used as a cue to estimate the Manhattan frame, (2) the source of the statistical distributions underlying the method, (3) the domain in which the deviation of each feature from the ideal is measured, (4) whether the search space is discretized, and (5) the search method.

Algorithm	**Feature**	**Statistics**	**Domain**	**Search Space**	**Search Method**
MW	Gradients (dense)	Coughlan & Yuille	Image	Discrete	Coarse-to-Fine
Edge-Based Newton MW Params	Edges (sparse)	Coughlan & Yuille	Image	Continuous	Quasi-Newton
Edge-Based Newton	Edges (sparse)	Current study	Image	Continuous	Quasi-Newton
Edge-Based EM	Edges (sparse)	Current study	Image	Continuous	EM
Edge-Based Quasi-EM	Edges (sparse)	Current study	Image	Continuous	Quasi-EM
Edge-Based Quasi-EM GS	Edges (sparse)	Current study	Gauss Sphere	Continuous	Quasi-EM

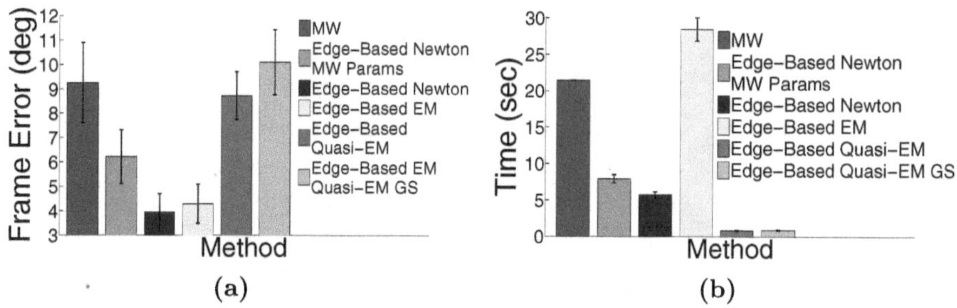

Fig. 5. (a) Average error and (b) average run time for the six algorithms tested. See Table 3 for a definition of each model. Error bars show standard error of the mean.

practical world, models are imperfect, and it is better to exclude noisy features that contain little independent information. The second advantage of the Edge-Based Newton method is the more accurate search algorithm, which is not limited by the discretized search space and discrete search strategy employed in MW.

Using the correct statistics led to a further 35% improvement in accuracy, with the result that the Edge-Based Newton method is more than twice as accurate as the original MW algorithm. Since our estimated priors are fairly close to those of MW, we believe these additional benefits derive mainly from the more accurate likelihood distributions we employ.

We found that replacing the Quasi-Newton search method with the Edge-Based EM method had relatively little influence on accuracy. However, a noticeable and significant loss in accuracy was observed when Manhattan directions were independently estimated in the M-step (Quasi-EM methods), presumably reflecting the loss of convergence properties due to this approximation.

Figure 5b shows the average time taken for each 640×480 pixel image in our test database, on a 1.83 GHz Core 2 Duo Intel CPU. All methods have been implemented in MATLAB except for the MW algorithm which was implemented in Python and C++. The Manhattan World method [2] and the edge-based EM method are relatively slow, requiring 20-30 seconds per image. Note that the EM

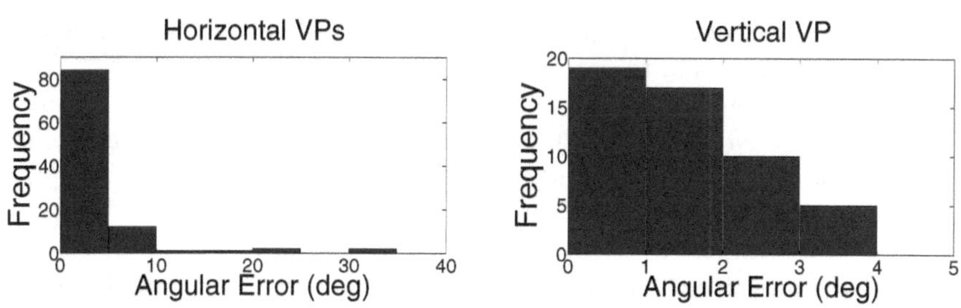

Fig. 6. Distribution of errors for the Edge-Based Newton algorithm over the 26 images in our test database

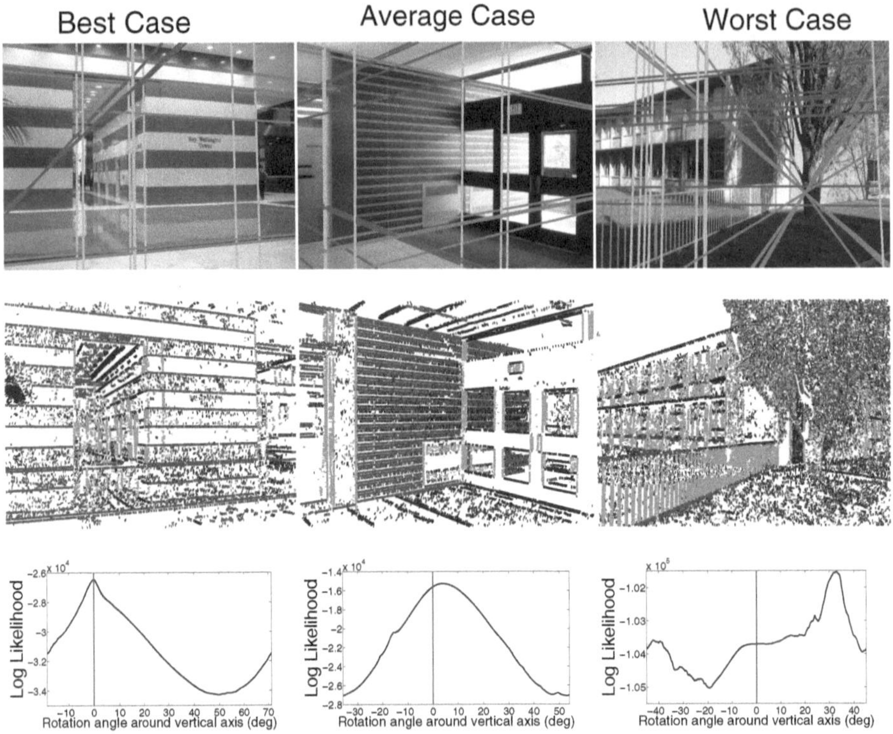

Fig. 7. Example results of our Edge-Based Newton method. Top Row: Lines are drawn through the 10 edges most likely to be generated by each Manhattan direction $m_u \in \{v, h_1, h_2\}$, i.e., maximizing $P(\Delta\phi|m_u, \Psi^*)$. Middle Row: Edge pixels $\boldsymbol{u}$ for which the most probable model $m_{\boldsymbol{u}}^* \in \{v, h_1, h_2\}$ is a Manhattan direction. Bottom Row: log likelihood distributions $\log P(E|\boldsymbol{\Psi})$ as a function of rotation around the ground truth vertical axis. 0 (red line) indicates ground truth frame.

method run on denser gradient maps, as proposed in the literature (e.g., [4]) would likely run even slower. We find the Edge-Based Newton and Quasi-EM methods run much faster, on the order of 5 seconds or less per image. However, since convergence of the Quasi-EM methods is unreliable, the Edge-Based Newton algorithm is the method of choice, providing the most accurate results of all methods evaluated, and running 4-6 times faster than previous methods from the literature.

Given these findings, we consider further the distribution of errors observed for the Edge-Based Newton algorithm. Figure 6 shows histograms of horizontal and vertical vanishing point errors over the 51 test images. While the great majority of estimates are very accurate

Figure 7 shows results of the top-performing Edge-Based Newton method on three example images (best, average and worst cases) from the test database. The grid-like scene structure in the first (best) image produces a sharp peak in the likelihood very near the ground truth solution. The average case image exhibits a greater variety of orientations, resulting in a broader likelihood with greater deviation from the ground truth. In our last (worst) image, several factors conspire to produce a horizontal error of 33 deg. Although there is sufficient visible vertical structure, only one of the horizontal directions is adequately represented. A tree not only obscures a large portion of the Manhattan structure in the scene, but also contributes edges forming a central nexus that masquerades as a vanishing point. Finally, an oblique fence contributes another false direction. In sum, the scene does not conform particularly well to the Manhattan assumption. The result is a likelihood function that is peaked very far from the ground truth direction, resulting in a large error.

This failure cannot be avoided by changing the search algorithm, since the likelihood distribution itself is misleading. Allowing inference of more than 3 vanishing points [5,4] could in theory prevent the oblique fence from biasing estimation of the Manhattan frame, however this would not solve the problem of capture by the tree. We find that the maximum log likelihood is negatively correlated with the frame error, and thus could be used as a threshold or a confidence measure for the applicability of the Manhattan model for a particular image [2].

5 Conclusions

We have introduced a new public database that may be used to evaluate and compare methods for estimating Manhattan frames in urban imagery. We have found that basing the estimation of the Manhattan frame on sparse, accurately-localized edges, rather than dense gradient maps, leads to substantially faster and more accurate performance. We believe the gain in accuracy is due mainly to the greater informativeness, accuracy and independence of the sub-pixel-localized edges [9] underlying the edge-based method, relative to the gradient features used in previous methods [2,3,4]. Using edges as input features also reduces the input size by a factor of 10, thereby allowing the deployment of better search methods, leading to further improvements in accuracy.

Several authors have proposed EM-based search methods. We find that a similar EM method based upon edges is reasonably accurate but slow. Methods using denser gradient maps [5,4,6] would likely be even slower. Here we have also explored faster quasi-EM methods in which approximate estimates of the Euler angles are made during the M-step, however convergence of these algorithms is found to be unreliable.

In contrast, we find an edge-based method that uses a Quasi-Newton procedure to directly estimate the Euler angles in the image domain yields the most accurate results, with double the accuracy of the original Manhattan world method, and running 4-6 times faster than methods previously proposed.

References

1. Coughlan, J.M., Yuille, A.L.: Manhattan world: Compass direction from a single image by bayesian inference. In: Seventh International Conference on Computer Vision, vol. 2, pp. 941–947. IEEE, Los Alamitos (1999)
2. Coughlan, J.M., Yuille, A.L.: Manhattan world: Orientation and outlier detection by bayesian inference. Neural Computation 15(5), 1063–1088 (2003)

3. Deutscher, J., Isard, M., MacCormick, J.: Automatic camera calibration from a single manhattan image. In: Heyden, A., Sparr, G., Nielsen, M., Johansen, P. (eds.) ECCV 2002. LNCS, vol. 2353, pp. 175–188. Springer, Heidelberg (2002)
4. Schindler, G., Dellaert, F.: Atlanta world: An expectation maximization framework for simultaneous low-level edge grouping and camera calibration in complex man-made environments. In: IEEE Conference on Computer Vision and Pattern Recognition, vol. 1, pp. I–203 – I–209. IEEE, Los Alamitos (2004)
5. Košecká, J., Zhang, W.: Video compass. In: Seventh European Conference on Computer Vision, pp. 476–490 (2002)
6. Wildenauer, H., Vincze, M.: Vanishing point detection in complex man-made worlds. In: 14th IEEE International Conference on Image Analysis and Processing, pp. 615–622. IEEE, Los Alamitos (2007)
7. Collins, R., Weiss, R.: Vanishing point calculation as a statistical inference on the unit sphere. In: Third International Conference on Computer Vision, pp. 400–403. IEEE, Los Alamitos (1990)
8. Kanatani, K.: Geometric Computation for Machine Vision. Oxford University Press, Inc., New York (1993)
9. Elder, J.H., Zucker, S.W.: Local scale control for edge detection and blur estimation. IEEE Transactions on Pattern Analysis and Machine Intelligence 20(7), 699–716 (1998)
10. Avriel, M.: Nonlinear Programming: Analysis and Methods. Prentice-Hall Inc., Englewood Cliffs (1976)

Multiple Component Learning for Object Detection

Piotr Dollár[1,2], Boris Babenko[2], Serge Belongie[1,2], Pietro Perona[1], and Zhuowen Tu[3]

[1] Electrical Engineering California Inst. of Tech.
{pdollar,perona}@caltech.edu
[2] Comp. Science & Eng. Univ. of CA, San Diego
{bbabenko,sjb}@cs.ucsd.edu
[3] Lab of Neuro Imaging Univ. of CA, Los Angeles
zhuowen.tu@loni.ucla.edu

Abstract. Object detection is one of the key problems in computer vision. In the last decade, discriminative learning approaches have proven effective in detecting rigid objects, achieving very low false positives rates. The field has also seen a resurgence of part-based recognition methods, with impressive results on highly articulated, diverse object categories. In this paper we propose a discriminative learning approach for detection that is inspired by part-based recognition approaches. Our method, Multiple Component Learning (MCL), automatically learns individual component classifiers and combines these into an overall classifier. Unlike previous methods, which rely on either fairly restricted part models or labeled part data, MCL learns powerful component classifiers in a weakly supervised manner, where object labels are provided but part labels are not. The basis of MCL lies in learning a set classifier; we achieve this by combining boosting with weakly supervised learning, specifically the Multiple Instance Learning framework (MIL). MCL is general, and we demonstrate results on a range of data from computer audition and computer vision. In particular, MCL outperforms all existing methods on the challenging INRIA pedestrian detection dataset, and unlike methods that are not part-based, MCL is quite robust to occlusions.

1 Introduction

Computer vision has recently witnessed rapid development in the areas of category detection (detection and localization) and recognition (categorization). In detection, approaches that learn classifiers from simple low-level features and large amounts of data can achieve low false positive rates for quasi-rigid objects [1,2,3]. In recognition, part-based methods, particularly patches-as-parts approaches [4,5,6], have proven effective at differentiating highly articulated categories. In this paper we propose a detection algorithm that is inspired by both lines of work. Through this combination, our system learns a robust, component-based classifier that achieves low false positive rates on articulated objects.

D. Forsyth, P. Torr, and A. Zisserman (Eds.): ECCV 2008, Part II, LNCS 5303, pp. 211–224, 2008.

Fig. 1. Response of first 5 learned components classifiers on randomly selected INRIA pedestrian test images (best viewed in color). At most one box is displayed per component after non-maximal suppression and thresholding. Three components correspond to semantically meaningful parts (head-magenta, left foot-red, right foot-yellow); 2 correspond to the region between between the legs. The components were learned with *no component labels* provided during training. See Sec. 4.3 for details.

Our approach is based on training a discriminative set classifier through a combination of boosting and weakly supervised learning, specifically Multiple Instance Learning (MIL). The resulting method, Multiple Component Learning (MCL), learns individual component classifiers and combines these into an overall object classifier. Object labels in the form of bounding boxes are provided, but component labels are not. The approach is general, and we demonstrate results on a range of data from computer audition and computer vision, including state of the art results on the challenging INRIA pedestrian detection dataset.

We begin with a review of related work below, followed by a thorough description of our method in Sec. 2. Next we present a theoretical justification of our approach in Sec. 3, where we also review related work from the machine learning literature. Finally, we present experimental results in Sec. 4.

1.1 Related Work

A number of discriminative detection systems that learn from simple low-level features and large amounts of data have been proposed; typically their focus is either on the learning aspect [1,3] or the design of appropriate features [2]. These methods require large amounts of labeled data to learn invariance to articulation, occlusion and intra-class variations. As mentioned, they have proven particularly successful for detecting rigid objects, achieving low false positive rates. MCL is similar to these methods in some regards, the key difference being that built into MCL is the domain knowledge that objects are composed of parts, that the mutual position of parts is somewhat variable and that parts may not all be visible. MCL is therefore much better suited for detecting articulated objects, *e.g.* pedestrians, and remains robust in the presence of occlusion.

Part-based approaches have a rich history; one of the earliest approaches dates back to 1973 [7]. A number of different ways of extracting parts from images have since been proposed. One approach involves designing part detectors by hand [8] or providing a system with labeled part examples [9,10]. Other than the obvious disadvantage of being labor intensive, these methods are restricted to using a limited number of possibly sub-optimal parts. An alternative approach involves searching for repeatedly occurring elements using different criteria such as frequency of appearance in the training data [11], lowering an empirical risk function [12], or increasing mutual information [13]. Unlike in MCL, the recurring elements in these methods are fairly simple, including edge fragments [11], Gaussian models [12], or image fragments [13].

A simple but effective method of extracting parts is to crop small image patches, either using an interest point operator [4,14] or by dense sampling [6]. The patches can be vector quantized to form 'codebooks' [4], and an image can then be represented using a 'bag of words' model [6]. Spatial information can be encoded using pairwise relationships [14] or with a spatial voting scheme [15]. Alternatively, a generative model can be used to model the object, *e.g.* the constellation model and its variants [4,16,17] have proven robust and capable of operating with little training data. Though effective in recognition tasks, these patch-based methods are limited by simple or fixed part appearance models, often relying on a patch distance measure in a predefined feature space, *e.g.* using normalized correlation or other patch descriptors. In MCL the component classifiers are learned from low level features, and although MCL requires more training data, the approach can lead to higher accuracy models.

Most closely related to our work is [18], which uses a formalism called latent SVMs to simultaneously learn part and object models. The resulting system is effective, though very different from our own. We emphasize that, as far as we are aware, aside from [18] MCL is the first part-based method that uses rich part appearance models without relying on part labels during training.

2 Multiple Component Learning

There are three primary challenges that we address in order to come up with our discriminative component-based object model. The first of these is how to learn a component classifier when only an object label is given. The second is how to learn *diverse* component classifiers given a method for learning a single component classifier. Finally, given multiple diverse component classifiers, we must combine these properly into an overall classifier for the object of interest. In this work we present a unified and effective solution to these challenges.

In order to learn component classifiers we turn to weakly supervised learning methods developed for object detection, where positive training images contain the object of interest, but, unlike the fully supervised case, the object location in each image is unknown [4,17,19]. Observe that learning a component classifier from images of objects where the components are in unspecified locations is an analogous problem. Thus, we can use weakly supervised learning to learn a single component classifier; specifically we use multiple instance learning (MIL) [20].

To learn a diverse collection of component classifiers we turn to boosting [21]. In boosting, multiple weak learners, each of which may have fairly high error, are combined into a single strong classifier with a low overall error. Weak classifiers are trained sequentially with the weights of the training samples adjusted so that incorrectly classified examples receive more weight. Boosting is ideally suited both for learning diverse classifiers and combining them into an overall classifier.

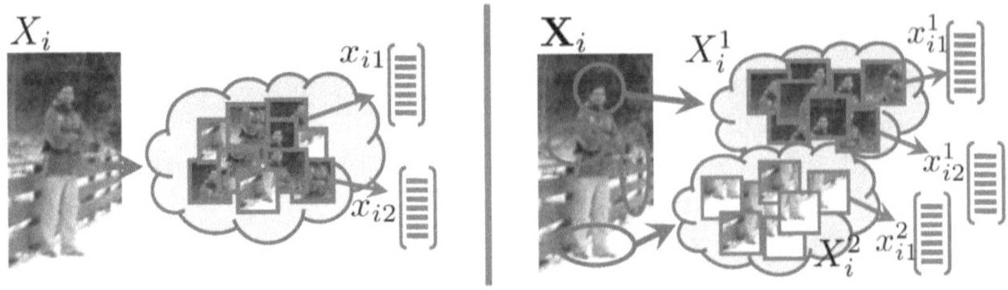

Fig. 2. We model an image as a *set* or *sequence of sets* of patches. **Left:** A pedestrian represented as a *set* of m densely sampled patches. A feature vector x_{ij} is computed for each patch, forming a set $X_i = \{x_{i1}, \dots, x_{im}\}$. **Right:** Pedestrians have limited articulation, so each component appears in a limited region of the image. Prior to training, p overlapping regions are randomly generated (details in Sec. 4). For each region we extract patches, generating an ordered *sequence of sets* $\mathbf{X}_i = [X_i^1, \dots, X_i^p]$.

The key to our approach is that we use component classifiers, trained using weakly supervised learning, as the weak classifiers in a boosting framework.

We refer to the individual constituent classifiers learned with weakly supervised learning as components rather than parts, as they need not always correspond to semantically meaningful object parts (see Fig. 1). We name our approach *Multiple Component Learning* (MCL) to emphasize that we learn a classifier composed of a number of constituent classifiers.

We begin by laying out the problem formulation and introducing some notation, followed by an overview of MIL in Sec. 2.2. The foundation of our work is in the development of a practical set classification algorithm in Sec. 2.3.

2.1 Problem Formulation and Notation

We assume that the image of an object is composed of multiple components, each of limited spatial extent, arranged in a particular pattern. For a given object, only some of its components will be visible, and their position, scale, rotation, *etc.* may vary. If a component is present and visible, there will be an image patch that contains it; the challenge is that, a priori, it isn't known which patches contain relevant components.

We model an object as a collection of patches, or more formally as a *set* of patches. To model spatial relationships, alongside the appearance of each patch we keep track of its location (details discussed in Sec. 4). For objects with limited articulation, each component can appear in a limited region of the image, and it becomes useful to model the image as an ordered *sequence of sets*, one set for each fixed image region. See Fig. 2 for an illustration.

We now review the framework for standard discriminative learning, then move on to discriminative learning for sets and sequences of sets. Formally, in *supervised learning* data samples are represented by feature vectors from some data space $\mathcal{X}$, *e.g.* $\mathcal{X} = \mathbb{R}^d$. The goal is to learn a function $f : \mathcal{X} \rightarrow \mathcal{Y}$, where $\mathcal{Y} = \{0, 1\}$ in binary classification. The learner is given a training data set

$\{(x_1, y_1) \ldots, (x_n, y_n)\}$, where $x_i \in \mathcal{X}$ and $y_i \in \mathcal{Y}$, and outputs a function f that predicts the labels of novel data points.

Supervised learning for *sets* can be defined in an analogous way. The training data has the form $\{(X_1, y_1), \ldots, (X_n, y_n)\}$, where each $X_i = \{x_{i1}, \ldots, x_{im}\}$, $x_{ij} \in \mathcal{X}$ and $y_i \in \{0, 1\}$. We emphasize that the X_i are sets, *i.e.* the elements of X_i are unordered, and we denote this by $X_i \in \mathcal{X}^m$. Although MCL allows for sets of varying cardinality, for notational simplicity we assume a fixed size m. The goal is to learn a function $\mathcal{F} : \mathcal{X}^m \rightarrow \{0, 1\}$ that generalizes well to unseen data. Technically speaking, set learning is a special case of supervised learning with the data space being $\mathcal{X}^m$. For the case of *sequences of sets*, each training instance $\mathbf{X}_i$ represents an ordered sequence of p (possibly overlapping) sets $\mathbf{X}_i = [X_i^1, \ldots, X_i^p]$, where $X_i^k \in \mathcal{X}^{m_k}$ represents a set like before. The goal is to learn a function $\mathcal{F} : [\mathcal{X}^{m_1}, \ldots, \mathcal{X}^{m_p}] \rightarrow \{0, 1\}$ that generalizes well.

2.2 Multiple Instance Learning

Here we give an overview of multiple instance learning (MIL), which will serve as the basis for learning a single component classifier. In MIL, training data is given in sets, just as in MCL (in MIL terminology 'bag/instance' are used in place of 'set/element'). Labels are defined for instances, and a set is labeled positive if any instance in the set is positive and negative otherwise. Using the notation from above, a MIL classifier has the following form:

$$F(X_i) = \begin{cases} 1 & \text{if } \exists j \text{ s.t. } f(x_{ij}) = 1 \\ 0 & \text{otherwise} \end{cases} \tag{1}$$

The goal is to minimize the error of F on the training data. Note that F depends entirely on the *instance classifier* $f : \mathcal{X} \rightarrow \{0, 1\}$. Thus, although the training data is given in sets X, the goal of MIL is to learn a classifier f that operates on instances. $F(X_i)$ is essentially the maximum of $f(x_{ij})$ over j. If f outputs a probability or confidence in $[0, 1]$, then we can define a 'soft' version of F:

$$\hat{F}(X_i) = softmax_j(f(x_{ij})), \tag{2}$$

where 'softmax' is some differentiable real-valued approximation of the max over $[0, 1]$. We use the model of MIL defined in Equation (2) throughout this paper.

Note that MIL can be considered a special case of set learning where the form of $\mathcal{F}$ is restricted to the form given in Equation (1). We summarize the difference between standard classification, MIL and MCL in Fig. 3.

The term MIL was originally coined by Dietterich *et al.* [20] in their study of drug activity prediction, and in earlier work, Keeler *et al.* [22] trained a digit recognition system using unaligned image data. More recently, Viola *et al.* [19] introduced a boosting variant of MIL called MIL-BOOST which is effective and robust. Throughout this work we use an extended version of MIL-BOOST described in [23]. Some of these extensions are necessary for MCL, *e.g.* weighted sets, others improve performance, *e.g.* the generalized mean softmax model.

Fig. 3. Three learning paradigms, and how each applies to object detection. **Left:** Ground truth in supervised learning includes a label for each training instance. **Middle:** For MIL, labels are given only for sets, where a set is positive if it contains at least one positive instance, *e.g.* an image is labeled positive if the target object is present. Like in the standard case, the goal is to learn a function f that classifies instances. **Right:** In MCL data is likewise given in sets, but the goal is to learn a *set* classifier $\mathcal{F}$. Note that in this case $\mathcal{X}^m$ represents a set of candidate component patches.

2.3 MCL Derivation

We are now ready to present the details of MCL. Again, the idea is to use component classifiers, trained using MIL, as the weak learners in a boosting framework.

We begin with a brief review of AdaBoost [21]. Given N labeled training examples (x_i, y_i) with $x_i \in \mathcal{X}$ and $y_i \in \mathcal{Y}$, and an initial distribution $D_1(i)$, AdaBoost combines a number of weak classifiers h_t to learn a strong classifier $H(x) = sign(\sum_{t=1}^{T} \alpha_t h_t(x))$. Here $\mathcal{Y} = \{-1, 1\}$. Boosting has excellent generalization abilities and a strong theoretical foundation.

A close inspection of AdaBoost reveals that it makes no assumption about the form of the input space $\mathcal{X}$. The only requirement is that there exists a routine for training a weak classifier $h_t : \mathcal{X} \to \{-1, 1\}$ that has error less than $\frac{1}{2}$ for any arbitrary weighing D_t of the training data. Therefore AdaBoost can be trained on sets as long as the weak classifiers are suited for set classification.

We use MIL to train each weak classifier, or component, inside the boosting framework. Recall that MIL, given input data organized in sets $X_i \in \mathcal{X}^m$, learns a function $\hat{F} : \mathcal{X}^m \to [0, 1]$ of the form $\hat{F}(X_i) = \text{softmax}_j(f(x_{ij}))$. It is possible, using the extended version of MIL-BOOST, to train $\hat{F}$ using an arbitrary distribution over the training data [23]. Inserting $\hat{F}$ for h in AdaBoost, and adjusting the form of the input data to be in sets, along with some other details, gives rise to MCL-AdaBoost; see Fig. 4 for details.

Note that MIL is asymmetric and only learns components from the positive class; to also learn components from the negative class we can train MIL with the positive/negative labels swapped, and keep the better of the two components.

Given: N labeled training examples (X_i, y_i) with $y_i \in \{-1, 1\}$ and $X_i = \{x_{i1}, \ldots, x_{im}\}$, and initial distr. of weights $D_1(i) = \frac{1}{N}$ over the examples.

For $t = 1, \ldots, T$:

- Train a MIL classifier $\hat{F}_t : \mathcal{X}^m \rightarrow [0, 1]$ using distribution D_t. Let $\hat{F}'_t(X_i) = (2 \times \mathbf{1}[\hat{F}_t(X_i) > th] - 1)$, where $th = .5$ or is chosen to minimize ϵ_t.
- Calculate error of $\hat{F}'_t : \epsilon_t = \sum_{i=1}^{N} D_t(i)\mathbf{1}(y_i \neq \hat{F}'_t(X_i))$.
- Set $\alpha_t = -\frac{1}{2} \log(\epsilon_t/(1 - \epsilon_t))$.
- Set $D_{t+1}(i) = D_t(i) \exp\left(-\alpha_t y_i \hat{F}'_t(X_i)\right)/Z_t$, where $Z_t = 2\sqrt{\epsilon_t(1 - \epsilon_t)}$ is a normalization factor.

Output the MCL classifier: $\mathcal{F}(X) = \text{sign}\left(\sum_{t=1}^{T} \alpha_t \hat{F}'_t(X)\right)$

Fig. 4. MCL-AdaBoost

Derivation of MCL with other boosting algorithms, such as RealBoost or GentleBoost [21], are similar to the derivation for MCL-AdaBoost. These require MIL to return real valued outputs, which is the case for MIL-BOOST. In our experiments MCL-RealBoost slightly outperforms MCL-AdaBoost; however, for simplicity we chose to use MCL-AdaBoost (which we abbreviate to MCL).

If each input is a *sequence of sets*, a simple variation of MCL proves effective. The training examples have the form $\mathbf{X}_i = [X_i^1, \ldots, X_i^p]$, where each X_i^k is a set. During each phase of boosting, for each value of $k < p$, we train MIL using X_i^k. Of the p candidates trained this way, we keep the one with lowest error and discard the rest. To make training more efficient, $T_1 < p$ classifiers can be selected instead. If $T_1 = T$, the diversity of the components comes from training each on different data as opposed to the reweighing step of boosting; however, boosting is still used to combine them. Details are omitted for space.

In the description of MCL given here, we assume that the input data has been organized into sets or sequences of sets and that MIL-BOOST is essentially a black box. Given this, MCL, like AdaBoost, only requires one parameter, the number of components T. In practice, we must also specify how to convert input data into sets, and the parameters to use when training MIL-BOOST, the most important being the feature space to use and the number of weak classifiers for MIL-BOOST. Moreover, the features used to capture spatial relations must be specified. These considerations are domain dependent, and we shall discuss them in more detail in the experiments sections.

3 Theoretical Foundations of MCL

Set learning can be approached as classifying distributions, where the elements of the set are observations sampled independently from an unknown density function. Examples include using statistical tests on histograms of quantized feature vectors to determine if two sets belong to the same distribution [6] and methods in computer audition that fit Gaussian Mixture Models (GMM) to the set distributions [24]. Alternatively, sets can be viewed as collections of elements where the order of the elements is unknown; however, an ordering is assumed to exist and part of the challenge is to recover it. A good example of this style of approach is kernel methods that implicitly or explicitly try to find correspondences between elements [25,26].

Like many existing set learning algorithms, MCL assumes that an ordering of the elements in a set exists but is not known. If the ordering was given during training of MCL, the components could be learned using supervised learning as opposed to weakly supervised learning (see Fig. 4). However, unlike existing set learning algorithms, MCL does not require a distance measure between elements, nor does it assume the elements are generated from a simple model. Instead, learning occurs *both* at the instance level and at the set level. In this sense MCL is fundamentally different from existing work.

Here we give a more formal derivation of MCL. We begin by defining the problem of discriminative set classification in a way that does not rely on any assumptions at the instance level. The resulting formulation is intractable; however, imposing additional constraints leads to a more tractable approximation, resulting in MCL. This derivation will make explicit the assumptions made by MCL. It will help put MCL into context, and finally, it will allow us to explore how at least some of these assumptions could be lifted in the future.

A note on notation: the uppercase script letters $\mathcal{F}, \mathcal{G}$ refer to classifiers that operate on sets $X \in \mathcal{X}^m$ and lowercase letters f, g refer to classifiers over instances $x \in \mathcal{X}$, where typically $\mathcal{X} = \mathbb{R}^d$. There exist multiple approaches to learning *instance classifiers*, *e.g.* boosting, support vector machines, neural networks, *etc.* Here we describe set classifiers in terms of instance classifiers.

A set $X_i = \{x_{i1}, \ldots, x_{im}\}$ is unordered, so a set classifier $\mathcal{F}(X_i)$ should not depend on the indexing of the elements x_{ij}. Let $[x_{ij_1}, \ldots, x_{ij_k}]$ denote the object obtained by stacking k elements in the set X_i in fixed order $j_1, \ldots, j_k$ (with possible repetitions and omissions). A general form for a set classifier is:

$$\mathcal{F}^k(X_i) = \begin{cases} 1 & \text{if } \exists j_1, \ldots, j_k \text{ s.t. } f([x_{ij_1}, \ldots, x_{ij_k}]) = 1 \\ 0 & \text{otherwise} \end{cases} \tag{3}$$

In other words $\mathcal{F}^k$ is defined in terms of some instance classifier f, such that $\mathcal{F}^k(X_i) = 1$ iff there exists some ordering of the elements in X_i that satisfies $f([x_{ij_1}, \ldots, x_{ij_k}]) = 1$. Although fairly general, this form of $\mathcal{F}^k$ is not tractable for large k as computing $\mathcal{F}^k(X_i)$ involves a search. In the worst case, evaluating $\mathcal{F}^k(X_i)$ requires computing f over all m^k orderings of X_i of length k.

A crucial observation is that training $\mathcal{F}^k$ reduces to training a MIL classifier. For the case of $k = 1$ this reduction is direct since $\mathcal{F}^1(X) \equiv F(X)$, where $F(X)$ is the standard MIL formulation given in Equation (1). Thus we can use existing MIL approaches to learn the instance classifier f. The reduction to MIL for $k > 1$ is slightly more complex. Given a set X, we can exhaustively generate a set X^*, where each element in X^* is created by concatenating k elements $x_{j_1}, \ldots, x_{j_k} \in X$ into the vector $[x_{j_1}, \ldots, x_{j_k}]$. Close inspection of Equations (1) and (3) shows that $\mathcal{F}^k(X) \equiv F(X^*)$. Unfortunately this reduction is exponential in k, since by construction X^* contains m^k elements.

The smaller the value of k the more tractable learning and computing $\mathcal{F}^k$ becomes, but also the less information $\mathcal{F}^k(X_i)$ can use about X_i. To compensate for the loss of information from using small k, T different classifiers $\mathcal{F}_i^k$ can be combined using an instance classifier g. Define $\mathcal{G}^k$ as follows:

$$\mathcal{G}^k(X) = g\left(\mathcal{F}_1^k(X), \ldots, \mathcal{F}_T^k(X)\right) \quad (4)$$

The formulation of set learning given in Equation (3) is strictly more general than the form given in Equation (4); however, a potentially intractable search has been replaced with T smaller ones. Computing $\mathcal{G}^k(X)$ is $O(Tm^k)$, which is tractable for small k, and $\mathcal{G}^k$ can still depend on all elements in X regardless of k. Equation (4) thus represents a reasonable model for a set classifier.

MCL learns $\mathcal{G}^1$, with g being an additive model. Specifically, MIL and AdaBoost are used to sequentially learn and combine $\mathcal{F}_1^1, \ldots, \mathcal{F}_T^1$ into an overall set classifier. We now consider the approximations used to operationalize $\mathcal{F}^k$ into MCL. First, $\mathcal{G}$ is used as a tractable alternative to $\mathcal{F}$. Additive models have been shown to be robust and general [21], so their use for g is reasonable. With appropriate choice of a MIL training procedure, the components $\mathcal{F}_i^1$ are chosen from a rich hypothesis class and can thus be quite general. Perhaps the most severe approximation is use of $k = 1$ since each component $\mathcal{F}_i^1$ is limited to depend on a single instance. Further experiments could reveal if this is limiting in practice, *e.g.* for object detection. Assuming that m is not too large, deriving MCL using $k = 2$ or 3 would be straightforward given the reduction of $\mathcal{F}^k$ to MIL described above; however, it is outside the scope of this work.

4 Experiments

We experiment with data coming from three domains: (1) handwriting identification, (2) speaker identification and (3) pedestrian detection. We compile datasets for the first two domains; for the third we use the INRIA pedestrian dataset [2]. Finally, we discuss the role of data alignment in Sec. 4.4.

The first two experiments are meant to show the generality and power of MCL. We compare to two representative approaches. The first is AdaBoost [21] applied to a standard, not set-based, representation of the data. AdaBoost has access to identical features as MCL. We also compare to a simple bag of features (BOF) method: first we use k-means to quantize the feature space (k chosen via cross-validation), next we compute a k-bin histogram for each set, finally these histograms are input into AdaBoost. Learning occurs at the set level but not at the instance level (since the distances used for clustering are fixed).

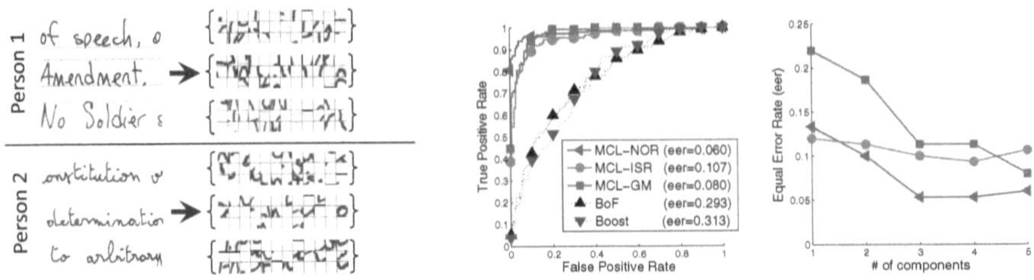

Fig. 5. Results for *writer identification*, see text for details

4.1 Handwriting Identification

In the first experiment we apply MCL to the offline, text-independent handwriting identification problem (recognizing the author by his/her handwriting) [27]. Intuitively, a person's handwriting has a number of distinct features which are independent of the actual text, hence a set representation is appropriate.

Our data consists of images of 4 pages of text handwritten by 2 people. Each training example is a randomly sampled 70x50 window of text. For AdaBoost and MCL we used Haar features [1]; for BOF we use Euclidean distance on pixels to perform clustering. For each training example, 25x25 patches were randomly sampled to generate the sets. We trained MCL with three versions of MIL-BOOST, altering the softmax model (see Equation 2 and [23] for details). MCL significantly outperforms both AdaBoost and BOF. Results are shown in Fig. 5; the last plot shows performance as a function of number of components T.

4.2 Speaker Identification

In this experiment our goal was to discriminate between the voices of John Kerry and George W. Bush (speaker identification is analogous to writer identification). We retrieved audio from a 2004 presidential debate[1] and cropped out roughly 6 minutes of each person speaking. The audio for each candidate was parsed into 360 second length clips, a quarter of which were used for training.

We used MFCC features which are commonly used in speaker recognition [24]. For AdaBoost one feature vector was generated per clip. For MCL and BOF, we generated a set representation of each clip by randomly sampling 78 short windows, and computed 30-dimensional MFCC features for each window. AdaBoost performed very poorly (ROC curve is outside range of plot), while MCL with two variations of MIL-BOOST significantly outperformed BOF when trained with enough components T. Results are shown in Fig. 6.

4.3 Pedestrian Detection

We now present results on the INRIA pedestrian dataset [2]. A number of recent methods have targeted this data [2,3,29], We compare our results to each, using

[1] Available publicly at
`http://www.archive.org/details/US_Pres_Debate_Sep.30.2004`

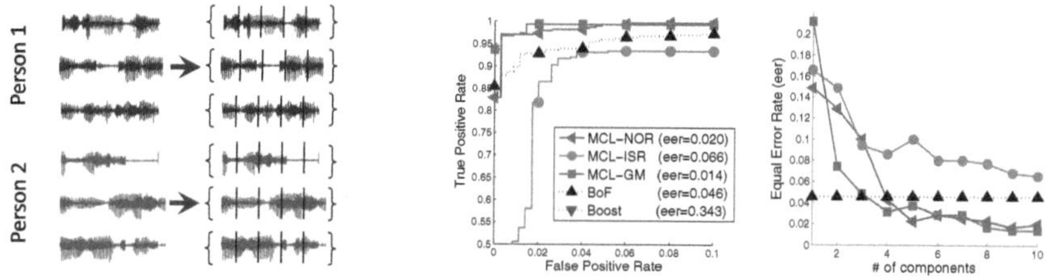

Fig. 6. Results for *speaker identification*, see text for details

the training and evaluation methodology presented in [2], except to [29] as it appears the results reported in that work are inaccurate[2].

Here we provide details for training MCL on this dataset. Training windows are 128x64, and for each of these we extract ~4K overlapping patches. We compute ~10K random Haar features [1] per patch, using the original grayscale image, as well as gradient magnitude and 6 channels of gradient quantized by orientation. We use MIL-BOOST with the generalized mean softmax model to select 256 stump classifiers. All 2416 positive sets and ~10K negative sets are used for training each MIL. We use the sequence of sets version of MCL (see Fig. 2). Initially 50 binary masks in the shape of ellipses are randomly generated, and one MIL classifier is trained per mask, using sets generated from patches only in the masked region. The original binary masks are random and likely suboptimal; after training each MIL, we compute a new mask based on the MIL probability response images on the training positives and then retrain. This mask refinement step improves results. From among the 50 MIL candidate classifiers, AdaBoost is used to select the 20 best components. Next, we bootstrap ~10K new negative windows from the training images from the false positives reported by MCL. The entire process is repeated 4 times to form a cascade. Our final MCL classifier is composed of 80 components (selected from 200 candidates).

We present a simple yet powerful method of incorporating spatial information into our MCL classifier. First, we compute the responses of the component classifier on the training images, giving one 128x64 probability map per training image for each component. Then, instead of using F (the max response over the set) as the weak classifier to AdaBoost, we use *Haar features computed over these probability maps.* Intuitively, single rectangle Haars capture absolute component location, multiple rectangle Haars capture relative component locations. Using these features, AdaBoost automatically learns the spatial relationships governing the components; no parameters need to be set in an ad-hoc manner. Using spatial information instead of the max was important for obtaining good results.

Training takes 2 weeks on a modern PC (majority of time is spent in MIL-BOOST training). Our classifier is composed of ~20K Haar features (256 per MIL), compared to ~6K in the Viola & Jones real-time face detector [1]; however, they are organized into multiple shallow cascades as opposed to a single deep cascade,

[2] See `http://www.cs.sfu.ca/ mori/research/papers/sabzmeydani_shapelet_cvpr07.html`

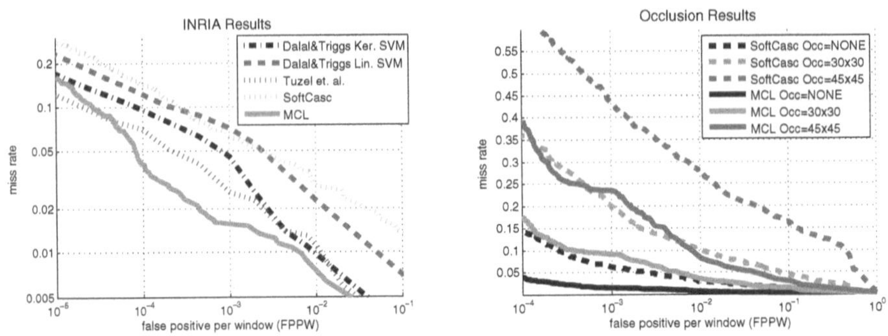

Fig. 7. Results on *pedestrian detection.* **Left**: MCL outperforms all reported results. At a false positive per window rate (FPPW) of 10^{-4}, a commonly used reference point, MCL has a miss rate of ~4%, compared to ~7% for [3] and ~10% for [2]. For comparison we also implemented the *SoftCascade* approach described in [28], using the same candidate Haar features we use in MCL. Consistent with previously reported results, a cascade of Haars performs poorly. **Right**: Results on artificially generated occlusion. We overlaid random 30x30 or 45x45 patches into random locations in the pedestrian test images. We then regenerated the ROC curves for MCL as well as SoftCascade (which, like [2,3], is not part-based). MCL on data with 30x30 performs similarly to SoftCascade on unoccluded data, and as the amount of occlusion increases, the gap between MCL and SoftCascade further increases.

so evaluation is slower. Simple feature sharing strategies should increase speed by an order of magnitude, reducing evaluation time to a few seconds per image.

The first five learned components are shown in Fig. 1. ROC curves comparing our method with Dalal and Triggs [2] and Tuzel *et al.* [3] are shown in Fig. 7, left. Part-based approaches have a natural advantage when occlusions are present. We show the robustness of MCL to occlusions in Fig. 7, right.

4.4 Role of Data Alignment

We perform a final experiment to show that using *aligned* data results in higher accuracy (part) classifiers. Note that simultaneous alignment of articulated objects is often impossible without relying on a part-based model.

We labeled the head and feet in the 2416 INRIA training pedestrians. For each part, we trained a classifier using patches sampled from each pedestrian in 1 of 3 ways: at the *labeled* part location, at the *mean* part location, and in a region around the mean part location (*mil*). We used AdaBoost for the first two cases and the extended version of MIL-BOOST for the third, using 256 Haars selected from an identical pool in each case. During testing, we applied each classifier to a region around the mean part location and recorded the maximum probability.

Results against bootstrapped negatives are shown in Fig. 8. Not surprisingly, *labeled* outperformed *mean*, showing alignment is beneficial. Additionally, *mean* performed better when we took the max probability over the region during testing rather than rely on the probability at the mean part location (*mean**). Also,

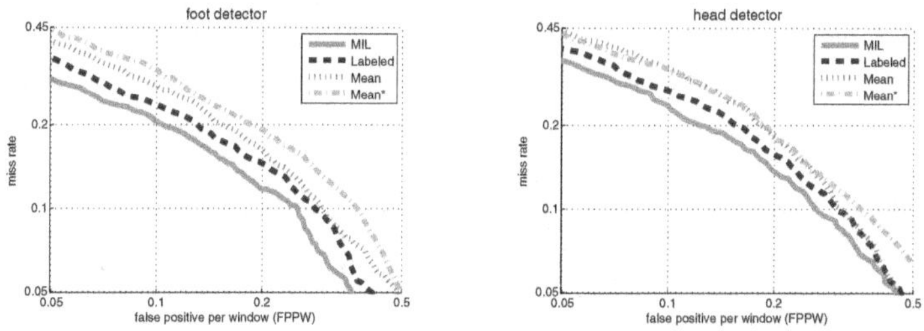

Fig. 8. Effects of alignment on training part classifiers, see text for details

consistent with the findings of [19], *mil* outperformed *labeled* even though it was trained in a weakly supervised manner (presumably our labeling is imperfect).

Together, these results make a strong case that data alignment is highly beneficial. As articulated objects typically cannot be fully aligned everywhere, MCL has an inherent advantage over methods that are not part-based.

5 Conclusion and Future Work

In this paper we presented a part-based object detection system called Multiple Component Learning (MCL). MCL combines elements of patch-based recognition and discriminative detection techniques to produce a high accuracy object detector that works well for articulated objects and is robust to occlusion. Unlike previous approaches, MCL learns powerful component classifiers without labeled part data. We showed results on data from computer vision and computer audition, including state of the art results in pedestrian detection.

In future work we plan on speeding up MCL, and seeing if tradeoffs in accuracy are necessary to achieve near real-time speeds. Additionally, results could be further improved by adapting MIL-BOOST to better suit the needs of MCL.

Acknowledgments

PD and BB were funded by NSF IGERT Grant DGE-0333451. SB was funded by NSF Career Grant #0448615 and the Alfred P. Sloan Research Fellowship. ZT was funded by NIH Grant U54RR021813 entitled Center for Computational Biology.

References

1. Viola, P., Jones, M.: Fast multi-view face detection. In: CVPR (2001)
2. Dalal, N., Triggs, B.: Hist. of oriented gradient for human det. In: CVPR (2005)
3. Tuzel, O., Porikli, F., Meer, P.: Human Detection via Classification on Riemannian Manifolds. In: CVPR (2007)
4. Weber, M., Welling, M., Perona, P.: Unsupervised learning of models for recognition. In: Vernon, D. (ed.) ECCV 2000. LNCS, vol. 1842, pp. 18–32. Springer, Heidelberg (2000)

5. Fergus, R., Perona, P., Zisserman, A.: Object class recognition by unsupervised scale-invariant learning. In: CVPR (2003)
6. Jurie, F., Triggs, B.: Creating efficient codebooks for vis. recog. In: ICCV (2005)
7. Fischler, M., Elschlager, R.: The Representation and Matching of Pictorial Structures. IEEE Transactions on Computers 100(22), 67–92 (1973)
8. Brunelli, R., Poggio, T.: Face recog.: features vs. templates. PAMI 15(10) (1993)
9. Mohan, A., Papageorgiou, C., Poggio, T.: Example-based object detection in images by components. PAMI 23(4), 349–361 (2001)
10. Mikolajczyk, K., Schmid, C., Zisserman, A.: Human Detection Based on a Probabilistic Assembly of Robust Part Detectors. In: Pajdla, T., Matas, J(G.) (eds.) ECCV 2004. LNCS, vol. 3021, pp. 69–82. Springer, Heidelberg (2004)
11. Amit, Y., Geman, D.: A computational model for visual selection. Neural Computation 11, 1691–1715 (1999)
12. Bar-Hillel, A., Hertz, T., Weinshall, D.: Object class recognition by boosting a part-based model. In: CVPR (2005)
13. Vidal-Naquet, M., Ullman, S.: Object recognition with informative features and linear classification. In: ICCV (2003)
14. Agarwal, S., Roth, D.: Learning a sparse repr. for object det. In: Heyden, A., Sparr, G., Nielsen, M., Johansen, P. (eds.) ECCV 2002. LNCS, vol. 2353, pp. 113–127. Springer, Heidelberg (2002)
15. Leibe, B., Leonardis, A., Schiele, B.: Robust Object Detection with Interleaved Categorization and Segmentation. IJCV, 1–31 (2005)
16. Leung, T., Burl, M., Perona, P.: Finding faces in cluttered scenes using random labeled graphmatching. In: ICCV, pp. 637–644 (1995)
17. Crandall, D.J., Huttenlocher, D.P.: Weakly supervised learning of part-based spatial models for visual object recognition. In: Leonardis, A., Bischof, H., Pinz, A. (eds.) ECCV 2006. LNCS, vol. 3951, pp. 16–29. Springer, Heidelberg (2006)
18. Felzenszwalb, P., McAllester, D., Ramanan, D.: A discriminatively trained, multiscale, deformable part model. In: CVPR (2008)
19. Viola, P., Platt, J.C., Zhang, C.: Multiple instance boosting for object detection. In: NIPS (2005)
20. Dietterich, T.G., Lathrop, R.H., Lozano-Perez, T.: Solving the multiple-instance problem with axis parallel rectangles. Artificial Intelligence (1997)
21. Friedman, J., Hastie, T., Tibshirani, R.: Additive logistic regression: a statistical view of boosting. Technical report, Stanford University (1998)
22. Keeler, J.D., Rumelhart, D.E., Leow, W.K.: Integrated segmentation and recognition of hand-printed numerals. In: NIPS (1990)
23. Babenko, B., Dollár, P., Tu, Z., Belongie, S.: Simultaneous learning and alignment: Multi-instance and multi-pose boosting. Technical Report CS2008, UCSD (2008)
24. Reynolds, D., Rose, R.: Robust text-indep. speaker ident. using gaussian mixture speaker models. In: IEEE Trans. on Speech and Audio Processing, pp. 72–83 (1995)
25. Grauman, K., Darrell, T.: Efficient Image Matching with Distributions of Local Invariant Features. In: CVPR (2005)
26. Huttenlocher, D., Klanderman, G., Rucklidge, W.: Comparing images using the Hausdorff distance. PAMI 15(9), 850–863 (1993)
27. Plamondon, R., Srihari, S.: Online and off-line handwriting recognition: a comprehensive survey. PAMI 22, 63–84 (2000)
28. Zhang, C., Viola, P.: Multiple-instance pruning for learning efficient cascade detectors. In: NIPS (2007)
29. Sabzmeydani, P., Mori, G.: Det. peds. by learning shapelet ftrs. In: CVPR (2007)

A Probabilistic Approach to Integrating Multiple Cues in Visual Tracking

Wei Du and Justus Piater

University of Liège, Department of Electrical Engineering and Computer Science
Montefiore Institute, B28, B-4000 Liege, Belgium
{wei.du,justus.piater}@ulg.ac.be

Abstract. This paper presents a novel probabilistic approach to integrating multiple cues in visual tracking. We perform tracking in different cues by interacting processes. Each process is represented by a Hidden Markov Model, and these parallel processes are arranged in a chain topology. The resulting Linked Hidden Markov Models naturally allow the use of particle filters and Belief Propagation in a unified framework. In particular, a target is tracked in each cue by a particle filter, and the particle filters in different cues interact via a message passing scheme. The general framework of our approach allows a customized combination of different cues in different situations, which is desirable from the implementation point of view. Our examples selectively integrate four visual cues including color, edges, motion and contours. We demonstrate empirically that the ordering of the cues is nearly inconsequential, and that our approach is superior to other approaches such as Independent Integration and Hierarchical Integration in terms of flexibility and robustness.

1 Introduction

From a Bayesian perspective, visual tracking is viewed as a problem of inferring target states over time based on image features or cues. Various types of cues have been used to characterize different object properties, including color [1], texture [2], points [3], edges [4], motion [5] and contours [6]. As no single cue remains reliable in all situations, the integration of multiple cues has proved successful at increasing the robustness of tracking algorithms. For instance, color and appearance are more sensitive to the lighting conditions than gradient features such as edges and points. Therefore, when the scene is subject to fast illumination changes, edges and points may provide complementary information that helps localize the targets being tracked.

Basically, multi-cue tracking is a fusion problem. Each cue tells a story about the targets of interest, and this information is processed and fused to estimate target states. One important issue is how to model the dependence between different cues, which in turn determines the manner in which the cues are combined. Many methods assume that the cues are conditionally independent [7,8,9,10], while more sophisticated methods model existing dependencies explicitly by e.g. graphical models [11].

D. Forsyth, P. Torr, and A. Zisserman (Eds.): ECCV 2008, Part II, LNCS 5303, pp. 225–238, 2008.

This paper presents a novel probabilistic approach to integrating multiple cues in visual tracking. In contrast to previous work, we perform tracking in different cues by individual but interacting processes, each of which is represented by a Hidden Markov Model (HMM). Chain models are used to link these parallel HMMs and to represent the dependence between these processes. The resulting Linked Hidden Markov Models (lHMMs)[1] naturally allow the use of two powerful inference algorithms, particle filtering and Belief Propagation (BP). By combining them in a unified framework, a Sequential Auxiliary Particle Belief Propagation algorithm is devised to perform inference in multi-cue tracking. In particular, a target is tracked in each cue by a particle filter, and the particle filters in different cues interact with each other via a message passing scheme.

Our approach has several advantages. First, it allows different target representations to be used in different cues so that each tracking process can be implemented separately and efficiently. Second, the approach is highly modular, facilitating the combination, addition and removal of vastly different cues. One only needs to define the potential function between each pair of neighboring cues according to their target representations. Third, the chain topology of the lHMMs reduces the complexity of the integration framework with respect to more elaborate graphical models. Due to the bidirectional propagation of information along the chain, the order of the cues in the chain is largely unimportant. We confirmed empirically that changing this order hardly affects the tracking results.

The rest of the paper is organized as follows. Section 2 discusses related work and highlights our contributions. Section 3 describes the lHMMs that model the multi-cue tracking problem. Section 4 formulates the problem and introduces the inference algorithm. Experimental results are presented in Section 5.

2 Related Work

Numerous approaches to multi-cue tracking have been reported in the literature. They differ in the way the cues are integrated and in the cues adopted. For example, Birchfield combined gradient and color cues for head tracking [7]. Triesch et al. used democratic integration to compute the consensus between the multiple cues [8]. Taylor et al. fused color, edge and texture cues in a Kalman-filter framework [2].

Particle filters, also known as Condensation [6] in the computer vision community, have achieved great success in solving tracking problems. Conventional particle filters maintain the target distributions over time based on a single observation model such as color [1] or contours [6]. It has been shown that multiple cues can easily be fused within a particle-filter framework. By assuming that the cues are conditionally independent, multi-cue observations were integrated by the product [9,13,14] or the sum [15] of the likelihoods in different cues. Based

[1] The term *Linked Hidden Markov Models* was introduced by Brand as a way of modeling two interacting processes [12]. Here, we borrow this term and extend it to multiple interacting processes.

on the same independence assumption, Leichter et al. combined Condensation-based or explicit PDF-yielding trackers by fusing only the trackers' output estimates [10]. Although not explicitly stated, the approach required each tracker to integrate information from all the other trackers, implying a fully-connected architecture. On the other hand, Pérez et al. propagated target distributions from cue to cue in a fixed order, hoping that downstream cues resolve the ambiguities introduced by the upstream cues [5]. Similar ideas were introduced under the names of hierarchical particle filters [4] and cascades of cues [16]. One downside of this strategy is that the performance depends on the order in which the cues are incorporated. Generally, heuristics are required to design this order. Wu et al. used factorial HMMs to model the dependency between color and contour cues and proposed a so-called co-inference algorithm [11].

Inspired by some of the cited work [10,5,11], this paper presents a general framework for multi-cue integration. Similarly to Leichter et al. [10], we combine particle filter-based trackers, each of which exploits a different cue. However, the dependencies between these trackers are explicitly modeled using lHMMs. The lHMMs link a set of parallel HMMs in a chain topology, which largely reduces both the architectural and algorithmic complexities. In doing so, we synchronously infer the target states in different cues. Unlike Pérez et al. [5] who propagated information from cue to cue in one direction, the undirected links in the chains in our lHMMs enable bidirectional message passing between the cues, relaxing the dependency on the ordering of the cues. The suggested approach is more general than e.g. that of Wu et al. [11] as it allows the integration of an arbitrary number of cues.

The combination of particle filters and BP was originally motivated for inference under non-linear and non-Gaussian models, resulting in Nonparametric Belief Propagation (NBP) [17,18]. Hua et al. first formulated the inference in temporally evolving graphical models and proposed a Sequential Belief Propagation (SBP) algorithm [19]. Briers et al. addressed the computational issue in the same sequential-inference context [20]. Auxiliary particle filters and the unscented approximation were used to sample particles, avoiding the need for Gibbs sampling. The resulting Sequential Auxiliary Particle Belief Propagation (SAPBP) reduced the computation from quadratic in the number of particles, as in NBP, to linear, which is desirable for online inference.

In this paper, we adopt a simplified SAPBP algorithm to solve the multi-cue tracking problem. The algorithm integrates the temporal evolution of each cue and the inter-cue correlations into a coherent framework. While independent temporal transition kernels are used for each cue, target states in different cues are related through messages passed along a chain. Contrary to the original SAPBP algorithm, we do not use the unscented approximation, as the inter-cue pairwise potentials are linear. Four visual cues are selected to demonstrate the effectiveness of our approach including color, edges, motion and contours. The general framework facilitates a customized combination of different cues in different situations, which is particularly desirable from the implementation point of view. Extensive experiments on tracking various objects in both indoor and

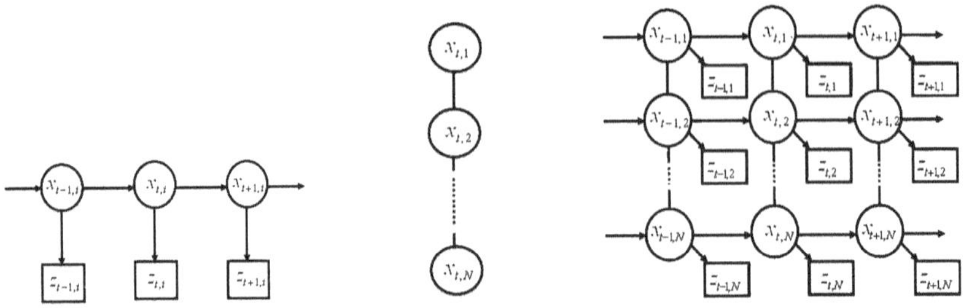

(a) *HMM* (b) *Chain* (c) *lHMMs*

Fig. 1. (a) HMM representing the tracking process in one cue. (b) Chain model representing the dependencies between different cues at a time instant. (c) Linked HMMs representing the interacting processes in different cues.

outdoor environments show that our approach is more flexible and robust than other approaches such as *independent integration* and *hierarchical integration.*

3 Linked Hidden Markov Models

Suppose M visual cues are used and each cue is associated with a different target state. Let $x_{t,i}$ be the target state in the ith cue at time t and $z_{t,i}$ the associated image observation, $i = 1, \ldots, M$. Given these definitions, tracking in the ith cue is formulated as

$$p(x_{t,i}|z_i^t) \propto p(z_{t,i}|x_{t,i}) \int p(x_{t,i}|x_{t-1,i}) p(x_{t-1,i}|z_i^{t-1}) \mathrm{d}x_{t-1,i}, \tag{1}$$

where $z_i^t = \{z_{1,i}, \ldots, z_{t,i}\}$, $p(z_{t,i}|x_{t,i})$ is the image likelihood and $p(x_{t,i}|x_{t-1,i})$ is the temporal transition kernel in the cue. This Bayesian formulation can be represented by a HMM, shown in Fig. 1(a).

In multi-cue tracking, the dependencies between target states at different cues must be taken into consideration. We use graphical models to capture these dependencies. Theoretically, fully-connected graphical models are required since all states in different cues may depend on each other. However, we choose to adopt a chain model (Fig. 1(b)) to reduce the model complexity and to simplify the inference algorithm. Linking the HMMs into chains results in so-called Linked Hidden Markov Models or lHMMs, shown in Fig. 1(c). The lHMMs represent the interacting processes in different cues.

Let $X_t = \{x_{t,1}, \ldots, x_{t,M}\}$ denote the multi-cue target state and $Z_t = \{z_{t,1}, \ldots, z_{t,M}\}$ the multi-cue image observation. Thus, tracking in the multiple cues amounts to the recursive inference of X_t, formulated as

$$p(X_t|Z^t) \propto p(Z_t|X_t) \int p(X_t|X_{t-1}) p(X_{t-1}|Z^{t-1}) \mathrm{d}X_{t-1}, \tag{2}$$

where $Z^t = \{Z_1, \ldots, Z_t\}$.

Direct inference using Eq. 2 is intractable due to the high dimensionality of the state space and the non-Gaussian nature of the target distributions. Therefore, we infer each $p(x_{t,i}|Z^t)$, $i = 1, \ldots, M$, collaboratively by a set of particle-filter-based processes, as detailed below.

4 Multi-cue Tracking by Interacting Processes

4.1 Problem Formulation

We first consider the inference in the chain model in Fig. 1(b). BP performs inference in graphical models by first computing messages and then computing beliefs. The chain topology admits a two-pass message-passing implementation. Specifically, the local messages passed from top to bottom are defined by

$$m_{i,i+1}(x_{t,i+1}) \propto \int p(z_{t,i}|x_{t,i}) m_{i-1,i}(x_{t,i}) \psi_{i,i+1}(x_{t,i}, x_{t,i+1}) \mathrm{d}x_{t,i}, \tag{3}$$

where $\psi_{i,j}$ is the potential function that describes the dependency between node i and j. The definition of this potential function depends on the target representations in the two neighboring cues and will be discussed later. Likewise, the messages passed from bottom to top have a symmetric form,

$$m_{i,i-1}(x_{t,i-1}) \propto \int p(z_{t,i}|x_{t,i}) m_{i+1,i}(x_{t,i}) \psi_{i-1,i}(x_{t,i-1}, x_{t,i}) \mathrm{d}x_{t,i}. \tag{4}$$

Then, the belief of $x_{t,i}$ is obtained by

$$p(x_{t,i}|Z_t) \propto p(z_{t,i}|x_{t,i}) m_{i-1,i}(x_{t,i}) m_{i+1,i}(x_{t,i}). \tag{5}$$

Note that Eqs. 3–5 are slightly different for the nodes at the ends of the chain.

In the sequential context, the message and belief equations for the lHMMs in Fig. 1(c) have similar forms as Eqs. 3, 4 and 5, adding only the terms of the temporal priors,

$$m_{i,i+1}(x_{t,i+1}) \propto \int p(z_{t,i}|x_{t,i}) p(x_{t,i}|Z^{t-1}) m_{i-1,i}(x_{t,i}) \psi_{i,i+1}(x_{t,i}, x_{t,i+1}) \mathrm{d}x_{t,i} \tag{6}$$

$$m_{i,i-1}(x_{t,i-1}) \propto \int p(z_{t,i}|x_{t,i}) p(x_{t,i}|Z^{t-1}) m_{i+1,i}(x_{t,i}) \psi_{i-1,i}(x_{t,i-1}, x_{t,i}) \mathrm{d}x_{t,i} \tag{7}$$

$$p(x_{t,i}|Z^t) \propto p(z_{t,i}|x_{t,i}) p(x_{t,i}|Z^{t-1}) m_{i,i-1}(x_{t,i}) m_{i,i+1}(x_{t,i}), \tag{8}$$

where the temporal prior is

$$p(x_{t,i}|Z^{t-1}) = \int p(x_{t,i}|x_{t-1,i}) p(x_{t-1,i}|Z^{t-1}) \mathrm{d}x_{t-1,i}.$$

Thus, the sequential inference in the lHMMs consists of passing messages using Eqs. 6 and 7 followed by belief updating using Eq. 8. Due to the lack of analytic representations of the above formulations, Monte Carlo approximations are required and can be obtained by using importance sampling techniques. A technical issue is the design of proper importance functions. We address this by adapting Sequential Auxiliary Particle Belief Propagation [20].

4.2 Sequential Auxiliary Particle Belief Propagation

The Monte Carlo approximation to Eqs. 6–8 requires the sampling of particles from products of individual terms, that is, from $p(x_{t,i}|Z^{t-1})m_{i-1,i}(x_{t,i})$ in Eq. 6, from $p(x_{t,i}|Z^{t-1})m_{i+1,i}(x_{t,i})$ in Eq. 7, and from $p(x_{t,i}|Z^{t-1})m_{i-1,i}(x_{t,i})m_{i+1,i}(x_{t,i})$ in Eq. 8. This sampling task requires an efficient way of combining, at each cue, the temporal prior and the incoming messages from the neighboring cue or cues.

Assume there are K terms in the product and each term consists of N particles. In order to compute the product of these terms, each term is represented as a mixture of Gaussians, and the computation amounts to multiplying K Gaussian mixtures of N components, which has a complexity of $O(N^K)$. Gibbs sampling is generally used for approximation and reduces the complexity to $O(\kappa K N^2)$, where κ is the number of iterations [17,18]. However, the chain topology of our lHMMs makes the Gibbs sampler ineffective since the number of terms K is only 2 for the messages and 3 for the beliefs. In the following, we will refer to the full multiplication of the terms as the full-product NBP in contrast to the traditional Gibbs-sampling NBP.

For real-time applications such as tracking, a more efficient method is required. By assuming that the terms in the product are independent, the complexity of this computation can be further reduced to $O(KN)$ by using SAPBP.

SAPBP adopts a two-step sampling procedure that first samples a component label from each term to construct an importance function and then samples particles from the constructed importance functions. An auxiliary variable $\theta_k \in \{1, \ldots, N\}$, $k = 1, \ldots, K$, is introduced to denote the component label in the kth term. As the component labels in different terms are independent, i.e., $p(\theta_1, \ldots, \theta_K) = \prod_{k=1}^{K} p(\theta_k)$, the importance function for sampling θ_k is chosen as

$$q(\theta_k = l_k) \propto \int \phi(x) f_k^{l_k}(x) \mathrm{d}x,$$

where $\phi(x)$ is the likelihood term and $f_k^{l_k}(x)$ is the l_kth component in the kth term. Sampling θ_k from this importance function is analogous to an auxiliary particle filter and is thus computationally efficient.

In our lHHMs, this label-sampling procedure is much simpler than in a general graphical model since the products contain few terms. For instance, to compute the messages in Eq. 6, we need to sample from $p(x_{t,i}|Z^{t-1})m_{i-1,i}(x_{t,i})$. The importance functions for sampling component labels from the above two terms are given by

$$q(\theta_1 = l_1) \propto \int p(z_{t,i}|x_{t,i}) p(x_{t,i}|x_{t-1,i}^{l_1}) \mathrm{d}x_{t,i},$$

$$q(\theta_2 = l_2) \propto \int p(z_{t,i}|x_{t,i}) \psi_{i,i-1}(x_{t,i}, x_{t,i-1}^{l_2}) \mathrm{d}x_{t,i},$$

where x_i^l denotes the lth particle for the state x_i. Similarly, to compute the beliefs in Eq. 8, the three importance functions for sampling component labels from $p(x_{t,i}|Z^{t-1})m_{i-1,i}(x_{t,i})m_{i+1,i}(x_{t,i})$ are given by

$$q(\theta_1 = l_1) \propto \int p(z_{t,i}|x_{t,i}) p(x_{t,i}|x_{t-1,i}^{l_1}) \mathrm{d}x_{t,i},$$

$$q(\theta_2 = l_2) \propto \int p(z_{t,i}|x_{t,i}) \psi_{i,i-1}(x_{t,i}, x_{t,i-1}^{l_2}) \mathrm{d}x_{t,i},$$

$$q(\theta_3 = l_3) \propto \int p(z_{t,i}|x_{t,i}) \psi_{i,i+1}(x_{t,i}, x_{t,i+1}^{l_3}) \mathrm{d}x_{t,i}.$$

After a label is sampled for each of the K terms, a Gaussian distribution is formed by the product of these individual (Gaussian) components and is used as an importance function $q(x_i|\theta_1, \dots, \theta_K)$, from which a particle is generated. In this way, N particles are sampled for each product. These sampled particles are used to approximate the messages and beliefs in Eqs. 6–8. The normalized weights for the messages (for Eq. 6 only) and beliefs (for Eq. 8) are given by

$$w_{i,i+1}^{t,n} = \frac{p(z_{t,i}^n|x_{t,i}^n) p(x_{t,i}^n|x_{t-1,i}^{\theta_1}) w_{i-1,i}^{t,\theta_2} \psi_{i,i-1}(x_{t,i}^n, x_{t,i-1}^{\theta_2})}{q(\theta_1) q(\theta_2) q(x_{t,i}^n|\theta_1, \theta_2)},$$

$$w_{t,i}^n = \frac{p(z_{t,i}^n|x_{t,i}^n) p(x_{t,i}^n|x_{t-1,i}^{\theta_1}) w_{i-1,i}^{t,\theta_2} \psi_{i,i-1}(x_{t,i}^n, x_{t,i-1}^{\theta_2}) w_{i+1,i}^{t,\theta_3} \psi_{i,i+1}(x_{t,i}^n, x_{t,i+1}^{\theta_3})}{q(\theta_1) q(\theta_2) q(\theta_3) q(x_{t,i}^n|\theta_1, \theta_2, \theta_3)}.$$

The intuition behind this algorithm is that, to combine the temporal priors and the incoming messages, we ignore the dependencies between them and sample independently from them by taking into account the anticipated "merit", i.e., the likelihoods evaluated after applying the temporal-transition kernel (for particles sampled from the temporal priors) or the inter-cue potentials (for particles sampled from the incoming messages). Ignoring the dependencies between the terms in a product may cause some loss of accuracy. However, in our case, this loss is unlikely to be dramatic since there are only two or three terms in each product. Moreover, the above sampling procedure takes into account the common dependence on the underlying targets as it is guided by the likelihood term. A comparison of our SAPBP algorithm with the full-product NBP will be shown in Section 5.

Our algorithm treats each cue equally with no explicit preference. However, as pointed out by Sun et al. [21], the asymmetric message passing in BP guarantees that the information is propagated mainly from high-confidence cues to low-confidence cues due to the smaller entropy of the messages in this direction. An extreme example is when the appearance of the target changes completely. Then, the color cue still "contributes" by propagating mostly uniformly distributed beliefs. Although this cue is not informative, it will not affect the tracking results at other cues.

4.3 Cues and Inter-cue Potentials

For this paper, we chose four simple and complementary cues including color, edges, motion and contours. The general framework of our approach allows an easy customization for various scenarios by combining different cues.

Color. Two different color histogramming methods are implemented. The first one is the traditional color histogramming proposed by Pérez et al. [1]. To model the spatial layout of the color distribution, a multi-part color model is obtained by splitting the tracked region into subregions, each with an individual color model. The second method is spatiogramming proposed by Birchfield et al. [22], which models the spatial layout of each color bin by a Gaussian distribution.

Edge. Edge orientation histogramming is used to disambiguate confusing colors in the background. Edges are detected using the horizontal and vertical Sobel operators, and the gradient magnitudes and orientations are computed. The detected edges are then histogrammed into orientation bins weighted by their strengths [4].

Motion. In the case of a static camera, the absolute image difference between each pair of successive frames provides motion information about the targets being tracked. A motion histogram is constructed by treating this absolute image difference as a grayscale image and histogramming the pixels in the regions of interest into intensity bins. To compute detection likelihoods, each candidate histogram is then compared to a uniform reference histogram [5].

Contour. The gradient information along the contour of a target is collected. A set of points are sampled along the contour and the distances to the closest edge points are computed and converted to a likelihood density.

In the color, edge and motion cues, we extract the histogram associated with each particle and compare it to a pre-learned reference histogram. This procedure was implemented using integral histograms [23]. In the contour cue, the Condensation algorithm [6] was implemented using a distance transformation. As the construction of the integral histograms and the distance maps is cheap with cost proportional to the size of the region of interest, we are able to evaluate a large number of particles in real time. In general, edge and motion cues alone are not discriminative, as edge orientation histograms do not capture the spatial distribution of the edges, and motion histograms favor equally all regions with moving objects. Nevertheless, integrating them does help disambiguate clutter in color and contour cues.

We model the targets of interest in the first three cues by rectangular regions, and in the contour cue by parametrized shape models. Ellipses and circles are adopted to model the simple targets in our experiments. Thus, a target is separately represented by

$$x_t^{\text{color}} = x_t^{\text{edge}} = x_t^{\text{motion}} = [u_t, v_t, s_t^u, s_t^v], \ x_t^{\text{contour}} = [u_t, v_t, s_t^u, s_t^v, \theta_t],$$

where $[u_t, v_t]$ is the translation of the target, $[s_t^u, s_t^v]$ is the scale change, and θ_t is the relative rotation when elliptic models are used in the contour cue. Representing the same target in different cues differently allows a separate and efficient implementation of each tracking process.

Given the target representations in different cues, we can define the potential functions ψ between each pair of neighboring cues. When the two neighboring cues share the same target representations, i.e., when color, edge and motion cues are neighboring to each other, then the potential between them is given by

$$\psi_{i,j}(x_{t,i}, x_{t,j}) = G(x_{t,i} - x_{t,j}; 0, \Sigma_{i,j}),$$

where $G(\cdot)$ is the Gaussian kernel function and $\Sigma_{i,j}$ is the covariance matrix. When different target representations are used in the two neighboring cues, i.e., when the contour cue is neighboring to the color, edge or motion cues, a similar potential function is defined by ignoring θ_t in x_t^{contour}. This inter-cue potential function models the mutual influence between the target states at different cues, and can be similarly defined for other target representations.

5 Results

We tested the performance of our approach on sequences of various objects taken in both indoor and outdoor environments. Here, the constant-velocity motion model was used for all cues in all experiments. We manually initialized the targets of interest in the first frame of each sequence and learned the reference models. The reference models were updated gradually with exponentially forgetting the past models, or were kept unchanged when dramatic changes to the models indicated occlusions [24].

The first experiment was tracking a dish under poor illumination, shown in Fig. 2. Four cues were integrated in the order of color (color histogram), edge, motion and contour. This sequence is difficult in that it contains both occlusions and fast appearance changes due to the mirror surface of the dish and the shadows. Thus, the color cue is only discriminative when the observed appearance is similar to the reference model. The motion cue helps the tracker concentrate on the areas with moving objects. As the edge and contour cues are based on gradient features, they are more reliable and play an important role during tracking. No single cue alone was able to track the dish for the entire sequence.

The second experiment was tracking a pedestrian in one of the PETS sequences and tracking a vehicle in a traffic sequence, shown in Fig. 3. The edge, color (color histogram) and motion cues were integrated in the given order for both sequences. The PETS sequence contains fast illumination changes and there is a camera rotation in the middle of the traffic sequence in order to keep the vehicle inside the view. Both problems violated the fundamental assumption of the motion cue that absolute image differences are introduced by moving objects only. However, thanks to the bidirectional information propagation between the cues, the poor information in the motion cue did not break the whole tracking system. In this experiment, the color cue alone is sufficient to track the targets most of the time. The motion cue helps overcome the partial occlusions and the edge cue helps localize the targets more precisely.

In the third experiment, we tested how dependent our approach was on the ordering of the cues. Two face sequences were selected to perform this test. We integrated all four cues for the first sequence and left out the motion cue for the second due to the motion of the camera. Spatiograms were used to implement

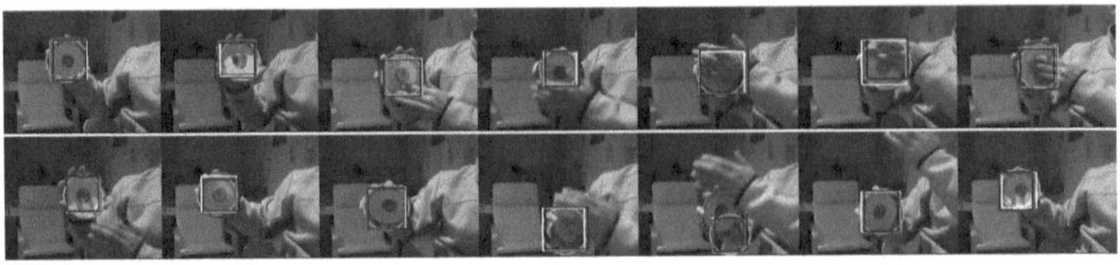

Fig. 2. Results of tracking a dish under poor illumination. The circles and the rectangles are the estimates of the contour (green), color (white), motion (blue) and edge (red) cues. The estimates by the four cues are slightly different from each other due to the Monte Carlo simulation. The same colors are used in the following experiments.

Fig. 3. Results of tracking a pedestrian in one of the PETS sequences (top) and a vehicle in a traffic sequence (bottom)

Fig. 4. Results of face tracking. The person's face rotates and undergoes dramatic appearance changes due to the lighting of a projector. Four cues are integrated in the order of color, edge, motion and contour.

the color cue. Figures 4 and 5 show the tracking results in the two sequences for one particular cue ordering. Similar results were obtained by varying the orders. Figure 6 shows the differences between the estimated target centers in the color cue under different orderings for the first sequence. Figure 7 shows the instantaneous particle approximations to the target distributions in different cues under different orderings for the second sequence. It can be seen that empirically, reordering the cues causes little difference in the tracking results. Nevertheless, setting the contour cue at one end of the chain makes the implementation of the algorithm clearer as x_t^{contour} contains the extra parameter θ_t and this information cannot be provided by other cues.

In the fourth experiment, we compared our approach to two typical particle-filter based multi-cue integration methods. The first selected method assumes that the cues are conditionally independent and fuses them by computing the product of the likelihoods in different cues [9,13,14]. We refer to it as *independent*

Fig. 5. Results of face tracking. The lady's face rotates, tilts and scales. Due to the motion of the camera, only three cues are integrated in the order of color, edge and contour.

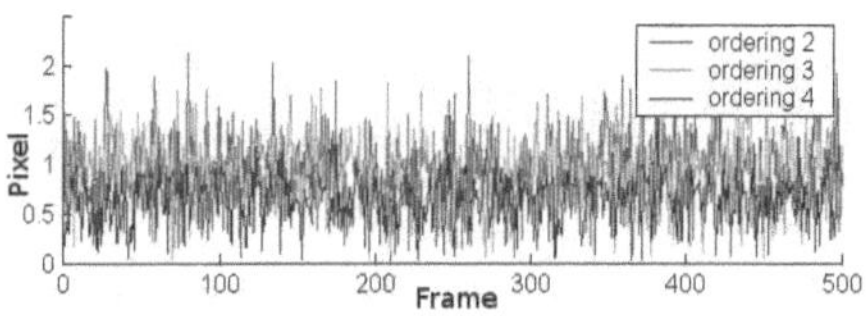

Fig. 6. The differences between the estimated target centers in the color cue under different orderings for the sequence shown in Fig. 4. Four different orderings were tested: (1) color, edge, motion and contour; (2) motion, color, contour and edge; (3) edge, contour, color and motion; (4) color, edge, contour and motion. The pixel distances between the target centers under the last three orderings and under the first ordering are computed respectively and are plotted in different colors.

Fig. 7. The instantaneous particle approximations to the target distributions in different cues for the sequence shown in Fig. 5. Results of different orderings are shown in different rows. From left to right, the ordering for the top row is edge, color and contour, for the middle row, color, contour and edge, and for the bottom row, color, edge and contour.

integration. This independence assumption is often violated, making the trackers unstable. The second selected method propagates target distributions from cue to cue in a fixed order [5]. We refer to it as *hierarchical integration.* Here, the problem is that background clutter occurs in different cues at different times, rendering the design of a fixed integration order difficult. An inappropriate order often causes the propagation of wrong information from cue to cue. In the sequence shown in Fig. 8, both of these classical methods failed. *Independent integration* was thrown off by conflicts between color and contour cues. For *hierarchical integration*, all shown cue orderings failed because misleading evidence was propagated downstream at some point. Moreover, no single cue was able to track the target by itself, except for the contour cue, which was inaccurate and brittle. In contrast, our approach exploited the dependencies between the cues and enabled bidirectional information propagation between the cues, leading to increased reliability. During system design, all cues were treated equally; no hierarchy needed to be imposed.

In the last experiment, we compared the precision and the efficiency of the SAPBP algorithm with the full-product NBP, shown in Fig. 9 and Fig. 10. It can be seen that both methods produced almost identical results, but our

Fig. 8. Comparison of our approach with alternative methods. **First row:** Results by our approach using cues in the order of color (spatiogram), edge, motion and contour. **Second row:** *hierarchical integration* under three different orderings: (1) color→edge→motion→contour (green); (2) contour→edge→motion→color (white); (3) color→contour→edge→motion (blue). Only the estimates of the last cues in the hierarchies under the different orderings are shown. **Third row:** *independent integration*; **Fourth row:** four single-cue trackers. Note that the contour tracker by itself was able to follow the target but with much less accuracy compared to our approach.

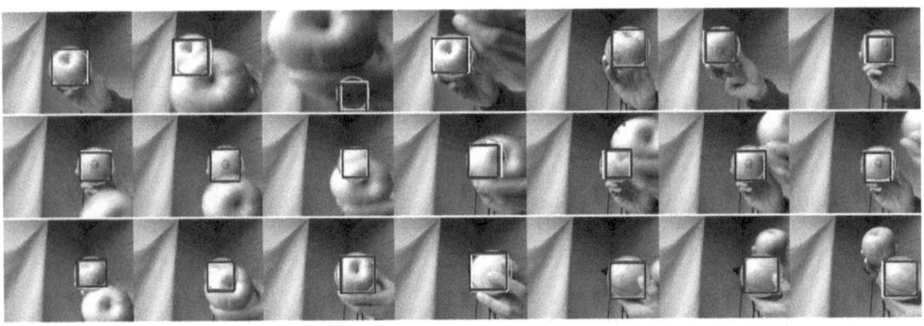

Fig. 9. Results on apple tracking. In the first row, the tracker was disturbed by a severe occlusion but recovered afterwards thanks to the color cue. In the second row, the tracker successfully handled a short occlusion by a distractor apple. When the occlusion lasted long and the scale of the distractor matched that of the target, the tracker accidentally jumped to the distractor apple, shown in the last row.

approach runs on average about 15 times faster than the full-product NBP. As the number of particles increases, more efficiency will be gained as the complexity of our approach increases linearly while that of the full-product NBP increases cubically. A comparison of different topologies of cue connections including the chain, the ring and the fully-connected models was also performed, shown in Figure 10. As the latter two models contain loops, a loopy version of our SAPBP algorithm was implemented that iterates 5 times in each step. On the ring and

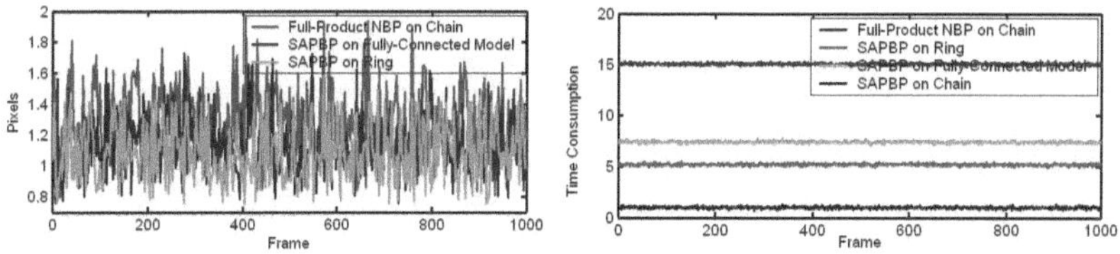

Fig. 10. The comparison of our approach, SAPBP on the chain, with the full-product NBP on the chain, SAPBP on the ring and on the fully-connected model. The left-hand figure plots the distances between the estimated target centers in the color cue by our approach and by the three other methods. As the maximum distance is no larger than 2 pixels, we conclude that they produced identical results. The right-hand figure plots the time consumptions of all the methods in each frame of the sequence.

fully-connected models, we obtained results essentially identical to on our chain model but on average 5.2 and 7.4 times slower, respectively.

6 Conclusions

This paper presents a systematic approach to integrating multiple cues in visual tracking. The strength and beauty of the approach lies in its unprejudiced treatment of each individual cue, which permits efficient inference based on linked HMMs and a Sequential Auxiliary Particle Belief Propagation algorithm. The simultaneous cues are arranged in a simple chain topology. Empirically, the ordering of the cues along the chain is inconsequential. In a sense, each cue is a black box and only exchanges messages with its neighbors. It doesn't have to know what other cues are being used and how they are implemented. Therefore, the general architecture allows the easy combination of an arbitrary number of cues in an arbitrary ordering. Our experiments confirmed a robustness superior to two competing approaches.

In this work, we deliberately chose simple cues and selectively combined them to highlight the flexibility of our approach. Other cues such as keypoints or even audio can easily be incorporated without any changes to the integration framework. The only issue is to define the inter-cue pairwise potentials, which depend solely on the target representations in the cue and its neighbors.

References

1. Pérez, P., Hue, C., Vermaak, J., Gangnet, M.: Color-based probabilistic tracking. In: Heyden, A., Sparr, G., Nielsen, M., Johansen, P. (eds.) ECCV 2002. LNCS, vol. 2350, pp. 661–675. Springer, Heidelberg (2002)
2. Taylor, G., Kleeman, L.: Fusion of multimodal visual cues for model-based object tracking. In: acra (2003)
3. Özuysal, M., Lepetit, V., Fleuret, F., Fua, P.: Feature harvesting for tracking-by-detection. In: Leonardis, A., Bischof, H., Pinz, A. (eds.) ECCV 2006. LNCS, vol. 3953, pp. 592–605. Springer, Heidelberg (2006)

4. Yang, C., Duraiswami, R., Davis, L.: Fast multiple object tracking via a hierarchical particle filter. In: ICCV, Beijing, China (2005)
5. Pérez, P., Vermaak, J., Blake, A.: Data fusion for visual tracking with particles. Proceedings of the IEEE 92(3), 495–513 (2004)
6. Isard, M., Blake, A.: Condensation – conditional density propagation for visual tracking. International Journal of Computer Vision 29(2), 5–28 (1998)
7. Birchfield, S.: Elliptical head tracking using intensity gradients and color histograms. In: CVPR, Santa Barbara, CA, pp. 232–237 (1998)
8. Triesch, J., Malsburg, C.v.: Self-organized integration of adaptive visual cues for face tracking. In: The Fourth IEEE International Conference on Automatic Face and Gesture Recognition (2000)
9. Giebell, J., Gavrila1, D., Schnörr, C.: A bayesian framework for multi-cue 3d object tracking. In: Pajdla, T., Matas, J(G.) (eds.) ECCV 2004. LNCS, vol. 3024, pp. 241–252. Springer, Heidelberg (2004)
10. Leichter, I., Lindenbaum, M., Rivlin, E.: A general framework for combining visual trackers: The black boxes approach. IJCV 67(3), 343–363 (2006)
11. Wu, Y., Huang, T.: Robust visual tracking by integrating multiple cues based on co-inference learning. International Journal of Computer Vision 58(1), 55–71 (2004)
12. Brand, M.: Coupled hidden Markov models for modeling interacting processes. Technical report, MIT Media Lab Perceptual Computing (1997)
13. Brasnett, P., Mihaylova, L., Canagarajah, N., Bull, D.: Particle filtering with multiple cues for object tracking in video sequences. In: Proceeding of SPIE-Image and Video Communications, vol. 5685, pp. 430–441 (2005)
14. Wang, H., Suter, D.: Efficient visual tracking by probabilistic fusion of multiple cues. In: ICPR, HongKong (2006)
15. Spengler, M., Schiele, B.: Towards robust multi-cue integration for visual tracking. In: MVA, vol. 14, pp. 50–58 (2003)
16. Gavrila, D.M., Munder, S.: Multi-cue pedestrian detection and tracking from a moving vehicle. International Journal of Computer Vision (2007)
17. Sudderth, E., Ihler, A., Freeman, W., Willsky, A.: Nonparametric belief propagation. In: IEEE Conference on Computer Vision and Pattern Recognition, Madison, WI, vol. 2, pp. 605–612 (2003)
18. Isard, M.: Pampas: Real-valued graphical models for computer vision. In: IEEE Conference on Computer Vision and Pattern Recognition, Madison, WI, vol. 1, pp. 613–620 (2003)
19. Hua, G., Wu, Y.: Multi-scale visual tracking by sequential belief propagation. In: IEEE Conference on Computer Vision and Pattern Recognition, Washington, DC, vol. 1, pp. 826–833 (2004)
20. Briers, M., Doucet, A., Singh, S.: Sequential auxiliary particle belief propagation. In: The Eighth International Conference on Information Fusion (2005)
21. Sun, J., Zheng, N., Harry, S.: Stereo matching using belief propagation. IEEE Transactions on Pattern Analysis and Machine Intelligence 25(7), 787–800 (2003)
22. Birchfield, S.T., Rangarajan, S.: Spatiograms versus histograms for region-based tracking. In: IEEE Conference on Computer Vision and Pattern Recognition, San Diego, CA (2005)
23. Porkili, F.: Integral histogram: A fast way to extract histograms in cartesian spaces. In: CVPR, San Diego, CA (2005)
24. Wang, H., Suter, D., Schindler, K.: Effective appearance model and similarity measure for particle filtering and visual tracking. In: Leonardis, A., Bischof, H., Pinz, A. (eds.) ECCV 2006. LNCS, vol. 3953, pp. 606–618. Springer, Heidelberg (2006)

Fast and Accurate Rotation Estimation on the 2-Sphere without Correspondences

Janis Fehr, Marco Reisert, and Hans Burkhardt

Chair of Pattern Recognition and Image Processing
Institute for Computer Science
Albert-Ludwigs-University, Freiburg, Germany
fehr@informatik.uni-freiburg.de

Abstract. We present a refined method for rotation estimation of signals on the 2-Sphere. Our approach utilizes a fast correlation in the harmonic domain to estimate rotation angles of arbitrary size and resolution. The method is able to achieve great accuracy even for very low spherical harmonic expansions of the input signals without using correspondences or any other kind of a priori information. The rotation parameters are computed analytically without additional iterative post-processing or "fine tuning".

The theoretical advances presented in this paper can be applied to a wide range of practical problems such as: shape description and shape retrieval, 3D rigid registration, robot positioning with omni-directional cameras or 3D invariant feature design.

1 Introduction

Methods involving functions on the 2-Sphere have recently received growing attention in a wide range of different computer vision and pattern recognition problems. For example in the area of shape description and retrieval, various invariant shape-features which have been proposed are based on spherical functions [1] [2] [3] [4]. For biomedical applications, many algorithms including invariant 3D features [5] or rigid 3D registration for 3D volume data analysis have their mathematical foundations in the rotation group $SO(3)$ which directly implies the use of functions on spheres. In all these cases, a fast and accurate rotation estimation would provide a helpful tool for further feature design and descriptors. Finally, in robotics, positioning algorithms which use images of omni-directional cameras [6] [7] [8] directly rely on a precise rotation estimation on the 2-Sphere.
Problem statement: given any two real valued signals f_1 and f_2 on a 2-sphere which are considered to be equal or at least similar under some rotational invariant measure ($\sim_R$):

$$f_1 \sim_R f_2, R \in SO(3) \tag{1}$$

the goal is to estimate the parameters of an arbitrary rotation R as accurate as possible without any additional information other than f_1 and f_2.

Obviously, the computational cost of a direct matching approach - testing all possible rotations R - is way too high. Especially when we are considering arbitrary resolutions of the rotation parameters.

D. Forsyth, P. Torr, and A. Zisserman (Eds.): ECCV 2008, Part II, LNCS 5303, pp. 239–251, 2008.

1.1 Related Work

Recently, there have been proposals for several different methods which try to overcome the direct matching problem. Here, we are only considering methods which provide full rotational estimates (there are many methods covering only rotations around the z-axis) without correspondences.

A direct nonlinear estimation (DNE) which is able to retrieve the parameters for small rotations via iterative minimization techniques was introduced in [7]. However, this method fails for larger rotations and was proposed only for "fine tuning" of pre-aligned rotations. Most other methods use representations in the spherical harmonic domain to solve the problem. The possibility to recover the rotation parameters utilizing the spherical harmonic shift theorem (SHIFT) [9] has been shown in [10]. This approach also uses an iterative minimization and was later refined by [8]. Again, the estimation accuracy is limited to small rotations.

The basic method which we extend in this paper is a fast full correlation (FCOR) in the spherical harmonic domain. This approach was first suggested by [11], stating a fast correlation in two angles followed by a correlation in the third Euler angle in an iterative way. This method was later extended to a full correlation in all three angles by [12]. This approach allows the computation of the correlation directly from the harmonic coefficients via FFT, but was actually not intended to be used to recover the rotation parameters. Its angular resolution directly depends on the range of the harmonic expansion - making high angular resolutions rather expensive. But FCOR has been used by [7] to initialize the DNE and SHIFT "fine tuning" algorithms. The same authors used a variation of FCOR (using inverse Spherical Fourier Transform [13] in stead of FFT) in combination with SHIFT [6] to recover robot positions from omni-directional images via rotation parameter estimation.

1.2 Contributions

The contribution of this work is twofold: first, we introduce novel theoretical extensions (3.2, 3.3) of the previous methods which allow accurate estimations for even very low spherical harmonic expansions without any additional iterative post-processing or "fine tuning". Secondly, we show how the rather complex theoretical methods can be implemented efficiently - previous publications gave very little insight into many practical implementation problems.

The remainder of this paper is structured as follows: first we revise the mathematical foundations of our approach in section (2). In section (3) we show the actual rotation estimation in all theoretical details, followed by methods for the practical implementation in section (4). Finally, we show detailed experiments in section (5).

2 Mathematical Foundations

Let us start with a very brief introduction of the basic mathematical tools and conventions used throughout the rest of the paper. For more detailed introductions please refer to [14] and [9].

Spherical Harmonics. Spherical Harmonics (SH) [14] form an orthonormal base on the 2-sphere. Analogical to the Fourier Transform, any given real valued signal f on a sphere with its parameterization over the angles Θ, Φ (latitude and longitude of the sphere) can be represented by an expansion in its harmonic coefficients:

$$f(\Theta, \Phi) = \sum_{l=0}^{\infty} \sum_{m=-l}^{m=l} \hat{f}_{lm} Y_m^l(\Theta, \Phi) \tag{2}$$

where l denotes the band of expansion, m the order for the l-th band and $\hat{f}_{lm}$ the harmonic coefficients. The harmonic base functions $Y_m^l(\Theta, \Phi)$ are calculated as follows:

$$Y_m^l(\Theta, \Phi) = \sqrt{\frac{2l+1}{4\pi} \frac{(l-m)!}{(l+m)!}} \cdot P_m^l(\cos\Theta) e^{im\Phi} \tag{3}$$

where P_m^l is the associated Legendre polynomial.

The harmonic expansion of a signal f will be denoted by $\hat{f}$ with corresponding coefficients $\hat{f}_{lm}$. In our case, where we are only considering signals on a discrete grid in $\mathbb{R}^3$, the $\hat{f}_{lm}$ can be computed via point-wise multiplication $(\cdot)$ of the 3D data grid with pre-computed discrete approximations of the harmonic base functions of fixed radii (more details in section (4)):

$$\hat{f}_{lm} = \sum_{\mathbb{R}^3} Y_m^l \cdot f \tag{4}$$

Rotations in SH. Throughout the rest of the paper we will use the Euler notation in zyz-convention denoted by the angles φ, θ, ψ with $\varphi, \psi \in [0, 2\pi[$ and $\theta \in [0, \pi[$ to parameterize the rotations $R \in SO(3)$ (short hand for $R(\varphi, \theta, \psi) \in SO(3)$). Rotations $R(\varphi, \theta, \psi) \cdot f$ in the Euclidean space find their equivalent representation in the harmonic domain in terms of the so called Wigner D-Matrices, which form an irreducible representation of the rotation group $SO(3)$. For each band l, $D^l(\varphi, \theta, \psi)$ (or short handed $D^l(R)$) defines a band-wise rotation in the SH coefficients. Hence, a rotation in the Euclidean space can be estimated in the harmonic domain (with a maximum expansion b), by

$$R \cdot f \approx \sum_{l=0}^{b} \sum_{m=-l}^{l} \sum_{n=-l}^{l} D_{mn}^l(R) \hat{f}_{lm} Y_m^l \tag{5}$$

In section (4.1), we further discuss practical issues concerning the computation of the D-matrices.

3 Estimating Rotation Parameters

We use several steps to recover the rotation parameters. First, we apply a fast correlation in the harmonic domain, gaining a similarity measure as stated in (1). Within the harmonic domain, we are then extending the angular resolution of the correlation function and minimize ambiguities of the Euler representation before we recover the full correlation matrix via FFT.

3.1 Fast Correlation in SH

We are following the fast correlation method which was introduced in [12] and later used in a similar way by [8]. The full correlation function $c : SO(3) \rightarrow \mathbb{R}$ of two signals f and g under the rotation $R \in SO(3)$ on a 2-sphere is given as:

$$c(R) := \int_{S^2} f \cdot (R \cdot g) \quad d\phi d\theta d\psi \tag{6}$$

Using the DFT Convolution Theorem and substituting f and g with their SH expansions (5, 4), leads to

$$c(R) = \sum_{lmn} \overline{D^l_{mn}(R)} \hat{f}_{lm} \overline{\hat{g}_{ln}} \tag{7}$$

The actual "trick" to obtain the fast correlation is now to factorize the original rotation $R(\varphi, \theta, \psi)$ into $R = R_1 \cdot R_2$, choosing $R_1(\xi, \pi/2, 0)$ and $R_2(\eta, \pi/2, \omega)$ with $\xi = \varphi - \pi/2, \eta = \pi - \theta, \omega = \psi - \pi/2$.

Using the fact that

$$D^l_{mn}(\varphi, \theta, \psi) = \mathrm{e}^{-im\varphi} d^l_{mn}(\theta) \mathrm{e}^{-in\psi} \tag{8}$$

where d^l is a real valued and so called "Wigner (small) d-matrix" (see section (4.1)), and

$$D^l_{mn}(R_1 \cdot R_2) = \sum_{h=-l}^{l} D^l_{nh}(R_1) D^l_{hm}(R_2) \tag{9}$$

we can rewrite

$$D^l_{mn}(R) = \sum_{h=-l}^{l} d^l_{nh}(\pi/2) d^l_{hm}(\pi/2) \mathrm{e}^{-i(n\xi + h\eta + m\omega)} \tag{10}$$

Substituting (10) into (7) provides the final formulation for the correlation function regarding the new angles ξ, η and ω:

$$c(\xi, \eta, \omega) = \sum_{lmhm'} d^l_{mh}(\pi/2) d^l_{hm'}(\pi/2) \hat{f}_{lm} \overline{\hat{g}_{lm'}} \mathrm{e}^{-i(m\xi + h\eta + m'\omega)} \tag{11}$$

The direct evaluation of this correlation function is of course not possible - but it is rather straight forward to obtain the Fourier transform of (11), hence eliminating the missing angle parameters:

$$\hat{c}(m, h, m') = \sum_{l} d^l_{mh}(\pi/2) d^l_{hm'}(\pi/2) \hat{f}_{lm} \overline{\hat{g}_{lm'}} \tag{12}$$

Finally, the correlation $c(\xi, \eta, \omega)$ can be retrieved via inverse FFT of $\hat{c}$

$$c(\xi, \eta, \omega) = FFT^{-1}(\hat{c}(m, h, m')), \tag{13}$$

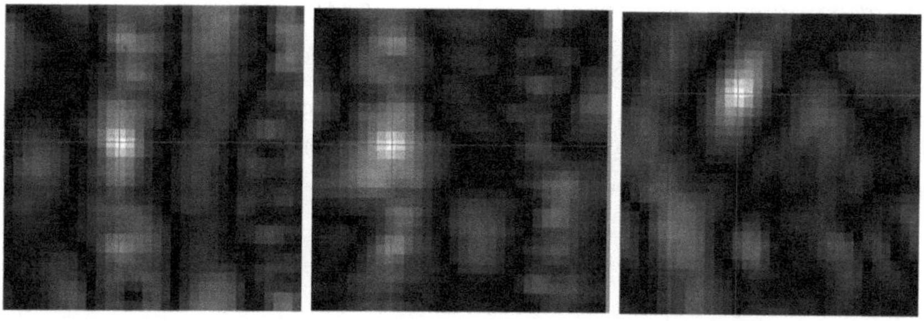

Fig. 1. Ortho view of a resulting 3D correlation grid in the (ξ, η, ω)-space with a maximum spherical harmonic expansion to the 16th band, $\phi = \pi/4, \theta = \pi/8, \psi = \pi/2$. From left to right: xy-plane, zy-plane, xz-plane.

revealing the correlation values on a sparse grid in a three dimensional (ξ, η, ω)-space. Figure (1) shows a resulting correlation grid for a sample rotation of random input data. Given our fast correlation, the basic idea is now to simply search the 3D grid for the position of the maximum correlation and compute the rotation parameters (φ, θ, ψ) from the (ξ, η, ω)-angles associated with this position.

3.2 Increasing the Angular Resolution

Let us take a closer look at Figure (1): first of all, it appears (and our experiments clearly support this assumption) that the fast correlation function has a clear and stable maximum in a point on the grid. This is a very nice property, and we could now simply recover the corresponding rotation parameters which are associated with this maximum position. But there are still some major problems: The image in Figure (1) appears to be quite coarse - and in fact, the parameter grids for expansions up to the 16th band have the size of $33 \times 33 \times 33$ since the parameters m, m', h in (12) are running from $-l, \ldots, l$. Given rotations up to $360°$, this leaves us in the worst case with a overall estimation accuracy of less than $15°$.

In general, even if our fast correlation function (13) would perfectly estimate the maximum position in all cases, we would have to expect a worst case accuracy of

$$Err_{corr} = 2 \cdot \frac{180°}{2l} + \frac{90°}{2l} \tag{14}$$

Hence, if we would like to achieve an accuracy of $1°$, we would have to take the harmonic expansion roughly beyond the 180th band. This would be computationally expensive. Even worse, since we are considering discrete data, the signals on the sphere are band-limited. So for smaller radii, higher bands of the expansion are actually not carrying any valuable information.

Due to this resolution problem, the fast correlation has so far only been used to initialize iterative algorithms [6][7].

Sinc Interpolation. Now, instead of increasing the sampling rate of our input signal by expanding the harmonic transform, we have found an alternative way to increase the correlation accuracy: interpolation in the frequency domain.

In general, considering the Sampling Theorem and given appropriate discrete samples a_n with step size Δ_x of some continuous 1D signal $a(x)$, we can reconstruct the original signal via sinc interpolation [15]:

$$a(x) = \sum_{n=-\infty}^{\infty} a_n \text{sinc}(\pi(x/\Delta_x - n)) \tag{15}$$

with

$$\text{sinc}(x) = \frac{\sin(x)}{x} \tag{16}$$

For a finite number of samples (15) changes to:

$$a(x) = \sum_{k=0}^{N} a_k \frac{\sin(\pi(x/\Delta_x - k))}{N\sin(\pi(x/\Delta_x - k)/N)} \tag{17}$$

This sinc interpolation features two nice properties[15]: it entirely avoids aliasing errors and it can easily be applied in the discrete Fourier space. Given the DFT coefficients α_r of the discrete signal $a_n, n = 0, 1, \ldots, N-1$, sinc interpolation is simply a zero padding of the spectrum between $a_{N/2}$ and $a_{(N/2)-1}$.

Returning to our original correlation problem, it is easy to see that the (m, h, m')-space in (12) actually is nothing else but a discrete 3D Fourier spectrum. So we can directly apply the 3D extension of (17) and add a zero padding into the (m, h, m')-space. This way, we are able to drastically increase the resolution of our correlation function at very low additional cost (see section (4) for implementation issues as well as suitable pad sizes). Figure (2) shows the effect of the interpolation on the correlation matrix for different pad sizes p.

It has to be noted, that even though the sinc interpolation implies some smoothing characteristics to the correlation matrix, the maxima remain fixed to singular positions in the grid.

Finally, we are now theoretically able to reduce the worst case accuracy to arbitrarily small angles for any given band.

$$Err_{corr}^{pad} = 2 \cdot \frac{180°}{2l+p} + \frac{90°}{2l+p} \tag{18}$$

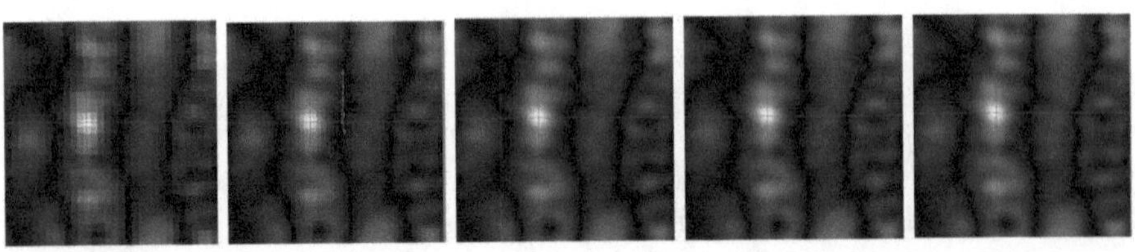

Fig. 2. Same experiment as in Fig. (1) but with increasing size of the sinc interpolation padding. From left to right: $p = 0, p = 16, p = 64, p = 128, p = 256$.

Of course, the padding approach has practical limitations - inverse FFTs are becoming computationally expensive at some point. But as our experiments show, resolutions below one degree are possible even for very low expansions.

3.3 Euler Ambiguities

After applying the sinc interpolation and the inverse FFT, (13) provides us with a fine parameter grid. The final obstacle towards the recovery of the rotation parameters inherits from the Euler parameterization used in the correlation function. Unfortunately, Euler angle formulations cause various ambiguities and cyclic shift problems.

One minor problem is caused by the fact that our parameter grid range is from $0, \ldots, 2\pi$ in all dimensions, while the angle θ is only defined $\theta \in [0, \pi[$. This causes two correlation peaks at $\theta = \beta$ and $\theta = 2\pi - \beta$ for an actual rotation of $\theta = \beta$. We avoid this problem by restricting the maximum search to $\theta \in [0, \pi[$, hence neglecting half of the correlation grid.

The formulation of the correlation function also causes further cyclic shifts in the grid representation of the Euler angles. This way, the zero rotation $R(\phi = 0, \theta = 0, \psi = 0)$ does not have its peak at the zero position $c(0,0,0)$ of the parameter grid as one would expect. For a more intuitive handling of the parameter extraction from the grid, we extend the original formulation of (12) and use a shift in the frequency space in order to normalize the mapping of $R(\pi, 0, \pi)$ to $c(0,0,0)$:

$$\hat{c}(m, h, m') = \sum_{l} d^l_{mh}(\pi/2) d^l_{hm'}(\pi/2) \hat{f}_{lm} \overline{\hat{g}_{lm'}} \cdot i^{m+2h+m'} \tag{19}$$

Rotation Parameters. Finally, we are able to retrieve the original rotation parameters. For a given correlation peak at the grid position $c(x, y, z)$, with maximum harmonic expansion b and padding p the rotation angles are:

$$\phi = \begin{cases} \pi + (2\pi - x\Delta) & \text{for } x\Delta > \pi \\ \pi - x\Delta & \text{otherwise} \end{cases} \tag{20}$$

$$\theta = \begin{cases} (2\pi - y\Delta) & \text{for } y\Delta > \pi \\ y\Delta & \text{otherwise} \end{cases} \tag{21}$$

$$\psi = \begin{cases} \pi + (2\pi - z\Delta) & \text{for } z\Delta > \pi \\ \pi - z\Delta & \text{otherwise} \end{cases} \tag{22}$$

$$\text{with} \quad \Delta = 2\pi/(b+p)$$

The resulting rotation estimates return very precise and unique parameter sets. Only one ambiguous setting has to be noted: for $\theta = 0, \pi$ all zyz-Euler formulations which hold $\phi + \psi = 2\pi$ encode the very same rotation (see Figure (3)). This is actually not a problem for our rotation estimation task, but might be quite confusing especially in the case of numerical evaluation of the estimation accuracy.

Fig. 3. $\phi\psi$-plane for of the correlation matrix with $\theta = 0$

3.4 Normalized Cross-Correlation

In many cases, especially when one tries to estimate the rotation parameters between non identical objects, it is favorable to normalize the (cross-)correlation results. We follow an approach which is widely known from the normalized cross-correlation of images: First, we subtract the mean from both functions prior to the correlation and then divide the results by the variances:

$$c_{norm}(R) := \int_{S^2} \frac{(f - \bar{f}) \cdot (R \cdot (g - \bar{g}))}{\sigma_f \sigma_g} \quad d\phi d\theta d\psi \tag{23}$$

Analogous to Fourier transform, we obtain the expected values $\bar{f}$ and $\bar{g}$ directly from the 0th SH coefficient. The variances σ_f and σ_g can be estimated directly from the band-wise energies:

$$\sigma_f \approx \sum_l |\hat{f}_l| \tag{24}$$

4 Implementation

An actual implementation of our algorithm can be quite challenging. In this section, we give some hints on how the methods can be implemented very efficiently and what pitfalls one should avoid. The main problem is, that even small implementation errors can easily produce correlation matrices which appear to be correct at first sight, but are in fact containing wrong estimates. Then, the search for errors in the $\hat{c}(m, h, m')$-space can be very tedious.

Harmonic Base Functions. The Spherical Harmonic base functions can be pre-computed off-line for a fixed radius and be stored in a file. The actual computation is straight forward (see [14]), but often not necessary because most common math tools like MATLAB or the GNU-scientific-library are already providing SH implementations.

Sinc Interpolation Padding. The implementation of the inverse FFT in (13) combined with the frequency space padding requires some care: we need an inverse complex to real FFT with an in-place mapping (the grid in the frequency space has the same size as the resulting grid in $\mathbb{R}^3$). Most FFT implementations are not providing such an operation. Due to the symmetries in the frequency space not all complex coefficients need to be stored, hence most implementations are using reduced grid sizes. We can avoid the tedious construction of such a reduced grid from $\hat{c}$ by using an inverse complex to complex FFT and taking only the real part of the result. In this case, we only have to shuffle the coefficients of $\hat{c}$, which can be done via simple modulo operations while simultaneously applying the padding. We rewrite (19) to:

$$\hat{c}(a, b, c) = \sum_l d^l_{mh}(\pi/2) d^l_{hm'}(\pi/2) \hat{f}_{lm} \overline{\hat{g}_{lm'}} \cdot i^{m+2h+m'} \tag{25}$$

where

$$s \coloneqq 2bp, \quad a \coloneqq (m+s+1) \bmod s, \quad b \coloneqq (h+s+1) \bmod s, \quad c \coloneqq (m'+s+1) \bmod s$$

Concerning the pad size: due to the nature of the FFT, most implementations achieve notable speedups for certain grid sizes. So it is very useful to choose the padding in such a way, that the overall grid size has, e.g., prime factor decompositions of mostly small primes [16].

4.1 Computation of the Wigner d-Matrices

The greatest implementation challenge is the actual computation of the Wigner d-Matrices. Even though, the d-Matrices for (25) can be pre-computed (we are always only considering $d^l_{mn}(\theta)$ with a fixed $\theta = \pi/2$), we still need an efficient implementation. In a direct approach, the d-Matrices can be computed by the sum

$$d^l_{mn}(\theta) = \sum_t (-1)^t \frac{\sqrt{(l+m)!(l-m)!(l+n)!(l-n)!}}{(l+m-t)!(l-n-t)!t!(t+n-m)!} \cdot \cos(\theta/2)^{2l+m-n-2t} \cdot \sin(\theta/2)^{2t+n-m} \tag{26}$$

over all t which lead to non-negative factorials [9]. It is easy to see that the constraints on t are causing the computational complexity to grow with the band of expansion. To overcome this problem, [17] introduced a recursive method for the d-Matrix computation. We are applying a closely related approach inspired by [18], where we retrieve d-matrices from recursively computed D-matrices.

Recursive Computation of the Wigner D-Matrices. Given D^l for the first two bands $l = 0$ and $l = 1$,

$$D^0(\phi,\theta,\psi) := 0$$

$$D^1(\phi,\theta,\psi) := \begin{pmatrix} e^{-i\psi}\frac{1+\cos(\theta)}{2}e^{-i\phi} & \frac{-\sin(\theta)}{\sqrt{2}}e^{-i\phi} & e^{i\psi}\frac{1-\cos(\theta)}{2}e^{-i\phi} \\ e^{-i\psi}\frac{\sin(\theta)}{\sqrt{2}} & \cos(\theta) & -e^{i\psi}\frac{\sin(\theta)}{\sqrt{2}} \\ e^{-i\psi}\frac{1-\cos(\theta)}{2}e^{i\phi} & \frac{\sin(\theta)}{\sqrt{2}}e^{i\phi} & e^{i\psi}\frac{1+\cos(\theta)}{2}e^{i\phi} \end{pmatrix}$$

we can compute D^l via band-wise recursion:

$$D^l_{mn} = \sum_{m,m',n,n'=-l}^{l} D^1_{m'n'} D^{l-1}_{(m-m')(n-n')} \cdot < l-1, m|1, m', l, m-m' > \cdot < l-1, n|1, n', l, n-n' > \tag{27}$$

where $< l, m|l', m', l'', m'' >$ denotes Clebsh-Gordan coefficients [9] known from angular momentum theory. Using (10), we finally obtain:

$$d^l_{mn}(\theta) = D^l_{mn}(0,\theta,0) \tag{28}$$

5 Experiments

Unlike previous publications [6][7][8], which only performed a small set of experiments with a fixed number of predefined example rotations, we evaluate our methods with a series of large scale experiments on real word data.

We use the "Princeton Shape Benchmark" (PSB) [1] dataset (which contains about 1800 3D objects) for our experiments. Figure (4) shows the volume rendering of a sample object. If not mentioned otherwise, all experiments have the same basic setup: for each parameter set, we evaluate the error statistics of 100 random rotations of random objects. We generate the rotations over all possible angles $\varphi, \psi \in [0, 2\pi[$ and $\theta \in [0, \pi[$ with a resolution of $0.001 \approx 0.1°$. Note that an error of $1° \approx 0.017$. All given error rates are the combined errors of all three angles.

Fig. 4. Volume rendering of an "ant"-object from the PSB

5.1 Rotating Objects in the Harmonic Domain

In this first series of experiments, we extract a harmonic expansion with a fixed radius around the object center and then rotate this expansion using (5).

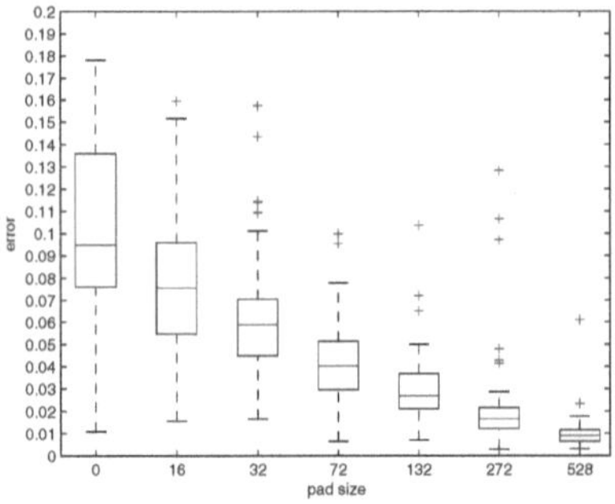

Fig. 5. Estimation errors with $b = 24$ and increasing pad size p

Pad Size. In a first experiment, we are able to show the effect of our padding method on the estimation accuracy. Figure (5) clearly shows the correlation of the pad size and the expected error. It is also evident, that we are able to achieve a precision of less than 1 degree while the experimental errors are found to be well within the theoretical bounds given in (18).

Maximum Band. The next two experiments investigate the practical influence of the maximum expansion band on the estimation errors. Figure (6) strongly supports our initial assumption, that the original formulation is not able to achieve accurate estimates for low expansions. Our method on the other hand, achieves very low error rates even for extremely low expansions with $b = 2$.

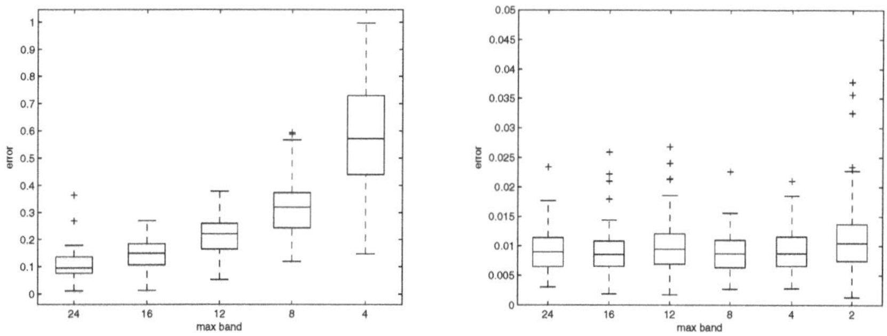

Fig. 6. Estimation errors with increasing maximum expansions. Left: $p = 0$. Right: $p > 512$ (p is not fix due to the padding to optimal FFT sizes). Note that the experiment with $p = 0, b = 2$ is left out because the result was so poor that it did not fit into the the chosen error scale.

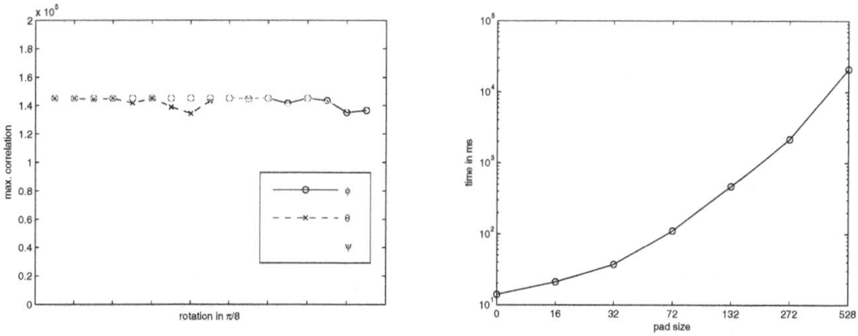

Fig. 7. Left: Maximum correlation for separate rotations in each angle. Right: Computational costs in ms on a standard PC.

Rotational Invariance and Computational Costs. are investigated in the last two experiments (figure (7)) of the first series. We rotate the object in $\pi/8$ steps in every angle to show that the correlation maximum is stable and indeed independent of the rotation. The computational complexity is largely dominated by the costs for the inverse FFT, hence growing with the pad size. So accuracy comes at some cost but reasonable accuracy can still be achieved well within 1 second.

5.2 Rotating Objects in $\mathbb{R}^3$

The results of figure (6) suggest, that the maximum expansion band has no influence on the quality of the rotation estimation - of course, this is only true if we are considering input signals which are limited to the very same maximum band. This is very unlikely for very low bands in the case of real data.

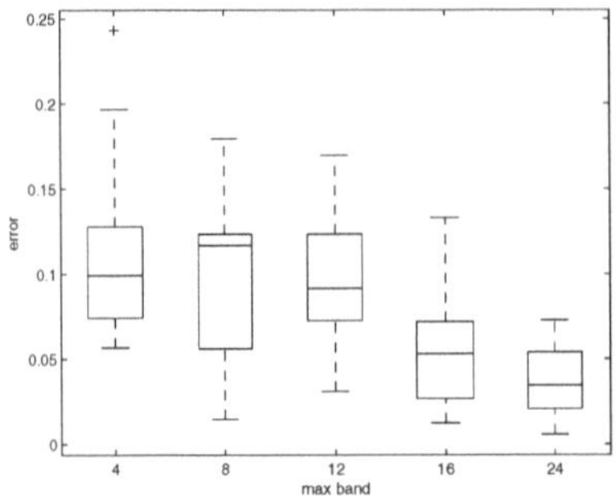

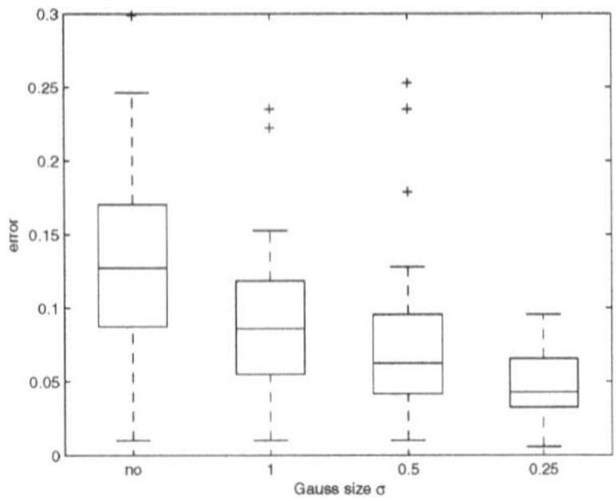

Fig. 8. Accuracy for rotations in $\mathbb{R}^3$. Left: Influence of the maximum band b, $p > 512, \sigma = 0.25$. Right: Correct sampling does matter! Without Gaussian smoothing and with different values for σ, $p > 512, b = 24$.

In order to evaluate the actual influence of the maximum expansion band, we need to rotate the objects in $\mathbb{R}^3$ and extract a second harmonic expansion after the rotation. As mentioned before, the usability of our sinc interpolation approach is limited to correctly sampled (concerning the Sampling Theorem) input signals. Hence, one must not expect to obtain precise rotation estimates for low band expansions, which act as a low pass filter, of high frequent input signals. Luckily, for most input data, we are not depending on the high frequent components in order to find the maximum correlation. Hence, we can apply a low pass filter (Gaussian) on the input data prior to the harmonic expansion. Figure (8) shows the impact of the maximum band and smoothing for rotations in $\mathbb{R}^3$. Overall, the estimation results are slightly worse than before, but are still quite reasonable.

6 Conclusions

We presented a fast and accurate rotation estimation on the 2-Sphere. We were able to show that our method is capable to estimate arbitrary rotations directly in the harmonic domain without any further iterative "fine tuning". We also introduced several implementation techniques, which allow a robust and efficient realization of the estimation algorithm.

Despite the theoretic nature of this paper, we emphasize the practical relevance of the rotation estimation for e.g. robotics or feature design.

References

1. Shilane, P., Min, P., Kazhdan, M., Funkhouser, T.: The princeton shape benchmark. In: Shape Modeling International, Genova, Italy (2004)
2. Kazhdan, M., Funkhouser, T., Rusinkiewicz, S.: Rotation invariant spherical harmonic representation of 3d shape descriptors. In: Symposium on Geometry Processing (2003)

3. Saupe, D., Vranic, D.V.: 3d model retrieval with spherical harmonics and moments. In: Radig, B., Florczyk, S. (eds.) DAGM 2001. LNCS, vol. 2191, pp. 392–397. Springer, Heidelberg (2001)
4. Vranic, D.V.: An improvement of rotation invariant 3d shape descriptor based on functions on concentric spheres. In: IEEE International Conference on Image Processing (ICIP 2003), vol. 3, pp. 757–760 (2003)
5. Ronneberger, O., Fehr, J.: Voxel-wise gray scale invariants for simultaneous segmentation and classification. In: Kropatsch, W.G., Sablatnig, R., Hanbury, A. (eds.) DAGM 2005. LNCS, vol. 3663, pp. 85–92. Springer, Heidelberg (2005)
6. Makadia, A., Daniilidis, K.: Rotation recovery from spherical images without correspondencesameesh. IEEE Transactions on Pattern Analysis and Machine Intelligence 28 (2006)
7. Makadia, A., Sorgi, L., Daniilidis, K.: Rotation estimation from spherical images. In: International Conference on Pattern Recognition, Cambridge (2004)
8. Makadia, A., Daniilidis, K.: Direct 3d-rotation estimation from spherical images via a generalized shift theorem. In: IEEE Conference on Computer Vision and Pattern Recognition, Madison (2003)
9. Brink, D., Satchler, G.: Angular Momentum, 2nd edn. Clarendon Press, Oxford (1968)
10. Burel, G., Henoco, H.: Determination of the orientation of 3d objects using spherical harmonics. Graph. Models Image Process. 57, 400–408 (1995)
11. Crowther, R.A.: The molecular replacement method. In: Rossmann, M. (ed.) Gordon and Breach, New York, pp. 173–178 (1972)
12. Kovacs, J.A., Wriggers, W.: Fast rotational matching. Acta Crystallogr (58), 1282–1286 (2002)
13. Healy Jr., D., Rockmore, D., Kostelec, P., Moore, S.: Ffts for the 2-sphere - improvements and variations. The Journal of Fourier Analysis and Applications 9, 341–385 (2003)
14. Groemer, H.: Geometric Applications of Fourier Series and Spherical Harmonics. Cambridge University Press, Cambridge (1996)
15. Yaroslavsky, L.: Boundary effect free and adaptive discrete signal sinc-interpolation algorithms for signal and image resampling. Appl. Opt. 42, 4166–4175 (2003)
16. Frigo, M., Johnson, S.G.: The design and implementation of FFTW3. Proceedings of the IEEE 93, 216–231 (2005); special issue on Program Generation, Optimization, and Platform Adaptation
17. Trapani, S., Navaza, J.: Calculation of spherical harmonics and wigner d functions by fft. applications to fast rotational matching in molecular replacement and implementation into amore. Acta Crystallogr, 262–269 (2006)
18. Reisert, M., Burkhardt, H.: Irreducible group representation for 3d shape description. In: Franke, K., Müller, K.-R., Nickolay, B., Schäfer, R. (eds.) DAGM 2006. LNCS, vol. 4174, pp. 132–142. Springer, Heidelberg (2006)

A Lattice-Preserving Multigrid Method for Solving the Inhomogeneous Poisson Equations Used in Image Analysis

Leo Grady

Siemens Corporate Research
Department of Imaging and Visualization
755 College Road East
Princeton, NJ 08540

Abstract. The inhomogeneous Poisson (Laplace) equation with internal Dirichlet boundary conditions has recently appeared in several applications ranging from image segmentation [1, 2, 3] to image colorization [4], digital photo matting [5, 6] and image filtering [7, 8]. In addition, the problem we address may also be considered as the generalized eigenvector problem associated with Normalized Cuts [9], the linearized anisotropic diffusion problem [10, 11, 8] solved with a backward Euler method, visual surface reconstruction with discontinuities [12, 13] or optical flow [14]. Although these approaches have demonstrated quality results, the computational burden of finding a solution requires an efficient solver. Design of an efficient multigrid solver is difficult for these problems due to unpredictable inhomogeneity in the equation coefficients and internal Dirichlet boundary conditions with unpredictable location and value. Previous approaches to multigrid solvers have typically employed either a data-driven operator (with fast convergence) or the maintenance of a lattice structure at coarse levels (with low memory overhead). In addition to memory efficiency, a lattice structure at coarse levels is also essential to taking advantage of the power of a GPU implementation [15,16,5,3]. In this work, we present a multigrid method that maintains the low memory overhead (and GPU suitability) associated with a regular lattice while benefiting from the fast convergence of a data-driven coarse operator.

1 Introduction

The inhomogeneous Poisson (Laplace) equation with internal Dirichlet boundary conditions has recently appeared in several applications ranging from image segmentation [1,2,3] to image colorization [4], digital photo matting [5,6] and image filtering [7]. In addition, the problem we address may also be considered as the generalized eigenvector problem associated with Normalized Cuts [9], the linearized anisotropic diffusion problem [10, 11, 8] solved with a backward Euler method, visual surface reconstruction with discontinuities [12, 13] or optical flow [14]. Although these approaches have demonstrated quality results, the computational burden of finding a solution on high-resolution 2D or 3D images demands an efficient solver that is not currently available in the literature.

D. Forsyth, P. Torr, and A. Zisserman (Eds.): ECCV 2008, Part II, LNCS 5303, pp. 252–264, 2008.

Although some of these algorithms are framed in a general graph setting, they have been almost exclusively employed in a rectilinear coordinate system that results in the use of a (widely) banded Laplacian matrix. Despite the banded structure of the matrix representing the discrete Laplacian operator on a grid, traditional fast Laplace/Poisson solvers are inappropriate due to the inhomogeneity of the diffusion constants (i.e., edge weights, in a graph context). These diffusion constants can vary by many orders of magnitude at neighboring pixels if, for example, a high-contrast image gradient is nearby. Unfortunately, the inhomogeneity of the diffusion constants is a very important feature of the behavior of anisotropic diffusion and related algorithms that cannot be eliminated for the sake of numerical efficiency. The solution of Poisson equations on domains with uniform diffusion constants may be performed quite efficiently (e.g., [17]).

Multigrid approaches are widely considered to be the best methods for solving a Poisson equation in a homogeneous domain, resulting in linear time algorithms [18, 19, 20]. Traditional multigrid approaches to solving the Poisson equation fall into two broad categories: Geometric and Algebraic. Classic geometric multigrid algorithms apply when the matrix represents a known grid, using bilinear interpolation to project the coarse-level solutions. Unfortunately, bilinear interpolation causes smoothing of coarse solutions across object boundaries (represented by small diffusion constants) in an image, causing a poor convergence rate. In contrast, algebraic multigrid [21, 22] uses the diffusion constants to generate problem-specific interpolation operators and coarsened matrices. Unfortunately, the coarsened matrices produced via algebraic multigrid are not guaranteed to represent a banded (lattice) sparsity structure. Note that some work has been addressed at finding matrix-dependent geometric multigrid methods (see [20] and references therein). For high-resolution images or 3D volumes in which each pixel corresponds to one equation, the banded structure of the Laplacian matrix offers significant storage and processing savings. Similarly, coarsening strategies have been explored in the past for graph-based algorithms (specifically, Markov Random Fields), but the explosion of non-local connectivity at high levels has made these approaches unworkable for high-resolution images/volumes (e.g., [23, 24]).

Both multiresolution methods (e.g., [25]) and multigrid numerical schemes have a long history in computer vision. Since at least as early as the work of Terzopoulos [26, 27], multigrid methods have been employed in computer vision in the context of visual surface reconstruction, shape-from-shading and optical flow. These early techniques employed the injection restriction operator and bilinear interpolation prolongation operator common to geometric multigrid methods. Although these operators are efficient and straightforward to implement, the convergence time for problems with strongly varying diffusion constants can be slower than necessary because bilinear interpolation smoothes the coarse-level solution over discontinuities (e.g., object boundaries in an image). Additionally, these restriction/prolongation operators are not adjoint to each other, which violates the Galerkin construction that is known theoretically to produce fast convergence [19]. Acton [28] applied similar multigrid operators to the anisotropic diffusion problem. Later work by Terzopoulos (and Szeliski) extended these ideas to handle discontinuities in surface reconstruction [12] and produce a parallel multigrid algorithm [13]. The technique for addressing discontinuities employed the concept of molecular inhibition across nodal discontinuities. Although similar in spirit to the multigrid technique presented here, these nodal molecules did not guarantee a lattice structure at coarse levels.

The anisotropy of the optical flow problem has prompted investigations into data (matrix) driven multigrid operators. Ghosal and Vaněk [14] proposed using a smoothed aggregate restriction/prolongation operator inspired by the algebraic multigrid concept introduced by Ruge of *strongly coupled nodes* [29]. Although similar to the technique that we present here in the sense that a data-driven aggregation operator is proposed, an important difference between our work and [14] is that their aggregation technique will not preserve a lattice structure at coarse levels. Additionally, our method is parameter-free. Algebraic multigrid techniques have also been applied in computer vision by Kimmel and Yavneh [30], using a more traditional algebraic multigrid approach for designing prolongation/restriction operators. However, as above, our method has the important distinction from this work that we preserve the lattice structure at each coarse level and employ no parameters for the construction of the prolongation/restriction operators.

In this work, we introduce the Maximally Connected Neighbor coarsening operator which both uses data-driven operators to produce fast multigrid convergence and preserves the memory-efficient lattice structure at higher levels. In Section 2, we describe the Maximally Connected Neighbor multigrid method and show that the coarse-level operators maintain the interpretation as the (inhomogeneous) Laplacian matrix of a grid. In Section 3 we compare the performance of our multigrid method with preconditioned conjugate gradients and the multigrid method of [31]. We finalize conclusions in Section 4.

2 Method

The Poisson equation that we consider will be written using graph notation due to the graph formulation of many of the algorithms in question. Despite this graph formulation, it is important to recognize that the "graph formulation" on a lattice is equivalent to a finite differences discretization of the continuous Poisson equation on a square domain with Neumann boundary conditions on the borders of the image.

We begin by fixing our notation for a graph. A **graph** consists of a pair $G = (V, E)$ with **vertices (nodes)** $v \in V$ and **edges** $e \in E \subseteq V \times V$. An edge, e, spanning two vertices, v_i and v_j, is denoted by e_{ij}. A **weighted graph** assigns a value to each edge called a **weight**. The weight of an edge, e_{ij}, is denoted by $w(e_{ij})$ or w_{ij}. The **degree** of a vertex is $d_i = \sum w(e_{ij})$ for all edges e_{ij} incident on v_i. The following will also assume that our graph is connected and undirected (i.e., $w_{ij} = w_{ji}$). In the language of continuous PDEs, the edge weights may be identified with diffusion constants, which are assumed here to have arbitrary positive values, i.e., $w_{ij} > 0$. The graphs considered in this paper will be taken to represent a Cartesian lattice in which every pixel (voxel) is uniquely identified with a node and each node is connected via an edge to each of its neighboring nodes with either a 4-connected or 8-connected structure in 2D or a 6-connected or 26-connected structure in 3D.

2.1 Problem Formulation

In the greatest generality, we will consider finding a solution to the equation

$$(T + L)x = f, \tag{1}$$

in which x is the desired solution (assigning one value to each node), f is an arbitrary function (assigning one value to each node), T is a diagonal matrix with an arbitrary nonnegative function on the diagonal, $t = \mathrm{diag}(T)$, and L represents the Laplacian matrix defined as

$$L_{ij} = \begin{cases} d_i & \text{if } i = j, \\ -w_{ij} & \text{if } v_i \text{ and } v_j \text{ are adjacent nodes,} \\ 0 & \text{otherwise,} \end{cases} \tag{2}$$

where L_{ij} is indexed by vertices v_i and v_j. Additionally, we would like to permit a set of internal Dirichlet boundary conditions at a set of nodes $V_B \subseteq V$ for which the solution of x_i, $v_i \in V_B$ is prescribed by function $x_i = h_i$ if $v_i \in V_B$. These internal boundary conditions occur frequently in the applications we consider such as segmentation [1, 2, 3], colorization [4], matting [5, 6] and filtering [7]. Note that (1) represents a discrete Poisson equation if $T = 0$ and a discrete Laplace equation if $f = 0$. In the following sections we will employ the notational convenience of $G = (T + L)$.

Although we motivate this work in terms of a linear system solver, note also that (1) may be viewed as applying to generalized eigenvector problems of the form

$$Lx = \lambda Tx, \tag{3}$$

in which λ represents an eigenvalue and $f = 0$. Problems of the form in (3) have been studied extensively with respect to the Normalized Cuts algorithm [9].

In the various imaging applications mentioned above the edge weights (diffusion constants) are used to encode the image structure. This is a common feature of graph based algorithms for image analysis and several weighting functions are commonly used in the literature [9, 32, 33]. Although this paper is concerned with solving linear systems of the form in (1) with arbitrary positive weights, an example of a very common weighting function is the Gaussian weighting given by

$$w_{ij} = \exp\left(-\beta(g_i - g_j)^2\right), \tag{4}$$

where g_i indicates the image intensity at pixel i and β represents a free parameter.

2.2 Incorporation of the Dirichlet Boundary Conditions

Since we are primarily concerned with maintaining a lattice structure, the internal Dirichlet boundary conditions can present a problem. Therefore, our first procedure will be to enforce the boundary conditions via a diagonal matrix. Specifically, we replace our problem in (1), including internal Dirichlet boundary conditions at nodes V_B with the problem

$$\left((T + L) + \alpha\mathrm{diag}(b)\right)x = f + \alpha h, \tag{5}$$

in which b is an indicator vector denoting membership in the boundary set (i.e., $b_i = 1$ if $v_i \in V_B$ and $b_i = 0$ otherwise) and h represents the boundary values ($h_i = 0$ if $v_i \notin V_B$). As $\alpha \to \infty$ the boundary values are enforced exactly with this construction. Since this method of enforcing the boundary conditions simply results in a modified diagonal matrix, T, and modified right hand side, f, we will persist in using the notation $Gx = f$ to represent the problem (1) with the understanding that any internal Dirichlet boundary conditions are enforced.

Algorithm 1. Basic steps of a multigrid method

`multigrid(L,x,f)`
while not converged to tolerance **do**
  Relax $G^0 x^0 = f^0$ with v iterations (relaxation e.g., via Gauss-Seidel)
  $r^0 = f - G^0 x^0$ (compute the residual)
  $r^1 = Rr^0$ (restrict the residual)
  $c = 0$ (initialize the correction to zero)
  `multigrid_recursion(1,c,`r^1`)`
  $x^k = x^k + Pc$ (update the solution with the prolongated correction)
end while

2.3 Multigrid Methods

Standard relaxation methods, such as Jacobi iteration or Gauss-Seidel relaxation, have been shown to perform well in correcting high-frequency error [18]. However, low-frequency error requires an excessive number of iterations to produce convergence with these methods. The multigrid method works by producing a correction of the current solution that is derived from a coarsened form of the system. The principle is that the coarse-grid updates correct low-frequency errors, while fine grid relaxations correct high-frequency errors. In order to produce the coarse grid correction, the multigrid method is applied recursively, forming what is known as a **V-cycle**, described by Algorithm 1. Denote the relaxation iterations parameter by v, the residual vector with r, the current (intermediate) solution at level k as x^k, the coarsened operator at level k as L^k, the restriction operator as R and the prolongation operator as P. Since our prolongation/restriction operators are linear, we view them as matrices. Starting with $k = 0$, the multigrid method is described in pseudocode in Algorithm 1 and the recursive function is given in Algorithm 2.

Given the above formalism for multigrid, the main issues that must be addressed in order to design a multigrid method are:

1. Determining the sampling structure of coarse nodes.
2. Specifying the restriction operator.
3. Specifying the prolongation operator.
4. Producing a coarsened operator.

The design of restriction, prolongation and coarse operator typically follows the Galerkin construction

$$G^{k+1} = RG^k P, \tag{6}$$

$$P = R^T, \tag{7}$$

which is known to have favorable theoretical properties [19]. Consequently, the specification of the restriction operator is sufficient to fully describe the prolongation and coarsened operators of a multigrid method.

The sampling structure of the coarse nodes used in the geometric multigrid has been to select every second node in each direction along the axes. Geometric

Algorithm 2. Recursive call of a multigrid method

`multigrid_recursion(`k, y, d`)`
if $k = k_{\text{MAX}}$ **then**
    Solve $G^k y = d$ exactly
else
    Relax $G^k y^k = d^k$ with v iterations (relaxation)
    $r^k = d^k - G^k y^k$ (compute the residual)
    $r^{k+1} = Rr^k$ (restrict the residual)
    $c = 0$ (initialize the correction to zero)
    `multigrid_recursion(`$k+1, c, r^{k+1}$`)` (continue recursion)
end if
$y^k = y^k + Pc$ (update the solution with the prolongated correction)

multigrid has a tremendous advantage for high-resolution images/volumes because the sampling structure and coarsened operator also represent a lattice and therefore the storage (and smoothing) at coarse levels has minimal indexing overhead. In addition to memory efficiency, a lattice structure at coarse levels is useful because it is possible to take advantage of the power of a GPU implementation [15, 16, 5, 3]. Unfortunately, matrix-independent geometric multigrid is inappropriate for inhomogeneous domains since the projection operators (bilinear interpolation) do not take the weights (diffusion constants) into account, permitting the smoothing of a coarse solution over areas of low diffusivity. The advantage of algebraic multigrid is that the weights are used to produce the prolongation/restriction and coarse operators, which does not smooth a coarse solution over areas of low diffusivity. Unfortunately, the algebraic multigrid procedure has a matrix-driven sampling structure that leads to coarse operators which have an arbitrary sparsity structure, forcing a high memory consumption and limiting possible application to high resolution images/volumes. Our approach in this paper is to suggest a new coarsening operator that adopts the lattice-preserving sampling structure of geometric multigrid while still using a matrix-driven (inhomogeneity-driven) restriction operator, chosen in such a way as to preserve a lattice structure at higher levels. Therefore, one may consider our method as a matrix-dependent geometric multigrid solver.

In the next section, we detail the proposed restriction operator and show that the resulting coarsened operators still maintain a lattice sparsity structure.

2.4 Maximally Connected Neighbor Coarsening

If R has only a single '1' entry on each row, then the coarse-level operator, RGR^T, represents the Laplacian (plus diagonal) of a coarse-level graph, where each node on the fine-level node is subsumed into a large node on coarse-level. Such a prolongation/restriction operator is termed an *agglomerative* operator [19] since it may be viewed as grouping together fine-level nodes into a single coarse-level node. The advantage of an agglomerative operator is that the coarse-level matrix is guaranteed to be the Laplacian of a graph, and therefore the approach is guaranteed to be stable. In contrast, the coarse-level operator RGR^T that uses weighted bilinear interpolation as the prolongation operator (as suggested by [31]) is not guaranteed to produce the Laplacian of some graph on the coarse level, and therefore exhibits instability.

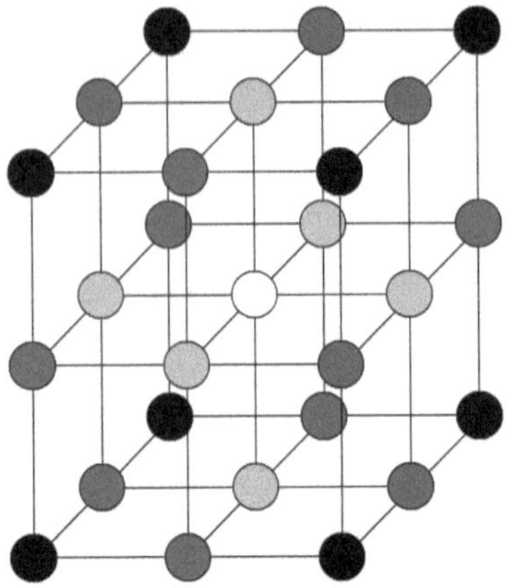

Fig. 1. Coarse-lattice neighbors in 3D. Black nodes correspond to coarse-lattice samples, dark-gray nodes to 1st order neighbors, light-gray nodes to 2nd order neighbors and white nodes to 3rd order neighbors.

In the context of the present applications, an agglomerative operator should exhibit two properties:

1. Coarse-level boundaries should correspond to fine-level boundaries (i.e., the agglomeration should represent a rough presegmentation).
2. The coarse-level graph topology should be regular (i.e., be a lattice), so that it may be stored/processed efficiently.

We now define the maximally connected neighbor (MCN) restriction operator to meet these objectives.

We will restrict our attention to 4- or 8- connected lattices in 2D, 6-, 10- or 26-connected lattices in 3D or, more generally, $\leq (3^p - 1)$-connected lattices in p-D. Without loss of generality, we will provide our definitions for a 3D lattice. We begin by first writing each node in terms of its 3D coordinates (starting with $(0, 0, 0)$ and moving in the positive directions). Define a node as belonging to the k-th order distance set, $\mathcal{N}^k \subset V$, if the number of odd digits in its coordinates are equal to k. Our coarse sampling set will consist of all nodes in $\mathcal{N}^0$. Note that, for p-D, $(\mathcal{N}^0 \cup \mathcal{N}^1 \cup \ldots \cup \mathcal{N}^p) = V$. The k-th order distance sets are represented pictorially in Figure 1. A node in the $\mathcal{N}^k$th distance set will be referred to as a **kth-order neighbor** of a coarse node.

We now proceed to define the maximally connected neighbor concept. Define the operator

$$\begin{aligned} M(v_i \in \mathcal{N}^k) &= v_j, \\ \text{s.t.} \quad v_j &\in \mathcal{N}^{k-1}, \\ \exists\, e_{ij} &\in E, \\ w_{ij} &\geq w_{ib}, \ \forall\, v_b. \end{aligned} \tag{8}$$

In the case of a tie (in terms of weight) for maximal neighbors of a node, we define $M(v_i) = v_j$ if $(i - j)$ is minimum for all eligible nodes having equal, maximal weight with v_i. We term v_j the **maximally connected neighbor** of v_i. This definition of $M(\cdot)$ may be applied recursively until a coarse node (a node in $\mathcal{N}^0$) is reached. Let $M^*(v_i)$ denote this recursive operator, i.e., $M^*(v_i) = M(M(\ldots M(v_i))) = v_j \in \mathcal{N}^0$.

Given the set of coarse nodes, $c_i \in \mathcal{N}^0$, we may therefore represent the $M(\cdot)$ operator in matrix form as:

$$R_{v_i c_j} = \begin{cases} 1 & \text{if } M^*(v_i) = c_j, \\ 0 & \text{else.} \end{cases} \tag{9}$$

2.5 Coarse Graphs

We first show that the coarse-level matrix, $RGR^T = R(L+T)R^T$, is in fact a matrix of the same form as G. Clearly, $RTR^T = T_c$ results in a T_c that is also diagonal with a nonnegative diagonal. Consequently, our concern is whether or not $L_c = RLR^T$ is also a coarse Laplacian matrix.

Any graph Laplacian matrix may be decomposed [34] into

$$L = A^T C A, \tag{10}$$

where C is an $m \times m$ diagonal matrix, indexed by edge, e_i, such that $C_{ii} = w_i$ and A is the $m \times n$ edge-node **incidence matrix** defined as

$$A_{e_{ij} v_k} = \begin{cases} +1 & \text{if } i = k, \\ -1 & \text{if } j = k, \\ 0 & \text{otherwise,} \end{cases} \tag{11}$$

for every vertex v_k and edge e_{ij}, where e_{ij} has been arbitrarily assigned an orientation. Note that neither P nor A contain information about the edge weights. Therefore, if $B = AP$ is an incidence matrix for some graph, then

$$L_c = RLR^T = RA^T CAR^T = B^T CB, \tag{12}$$

is also a graph Laplacian, which is necessarily isomorphic to a graph [34] (i.e., the matrix is a unique representation of a graph).

This result follows from the following lemma:

Lemma 1. *For $m \times n$ edge-node incidence matrix A and $n \times q$ matrix Q, the product $B = AQ$ is an edge-node incidence matrix if Q is a binary matrix (i.e., all values are zero or one) where each row has one and only one nonzero entry.*

Proof. A matrix is an incidence matrix if each row has no greater than one $\{-1, 1\}$ pair, with all other entries being zero.

Consider the ith row of A, a_i, which must contain either all zeros or a single $\{-1, 1\}$ pair. If all zeros, then the corresponding row of B, B_i is all zeros. If there is a $\{-1, 1\}$ pair, denote the nonzero entries by A_{ij} and A_{ik}, respectively. Since each row of Q contains a single '1', then B_i will either be all zeros (if $Q_{js} = Q_{ks} = 1$ for some column s) or will contain a $\{-1, 1\}$ pair (if $Q_{js} = Q_{kr} = 1$ for $s \neq r$). □

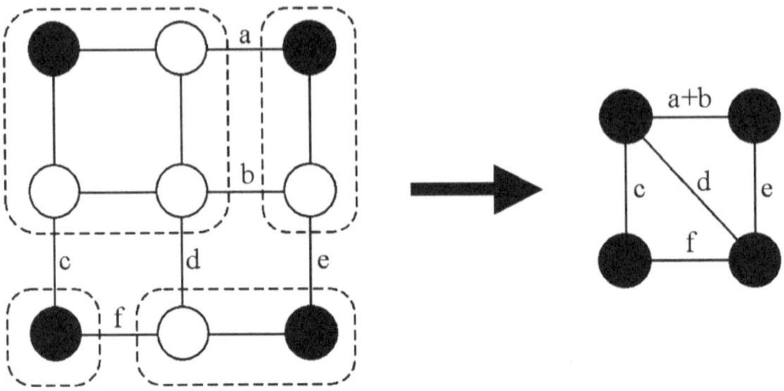

Fig. 2. Coarsening weights from a fine lattice to a coarse lattice. Black circles represent selected coarse nodes while open circles indicate fine-level nodes. Dashed lines show groups of fine-level nodes that have a common maximally connected neighbor. The graph on the right indicates the coarse-level weighting derived from the fine-level MCN groupings. Although a single diagonal edge has been introduced in the coarse-level graph, the coarse-level graphs will have connectivity no greater than an 8-connected lattice (or a 26-connected lattice in 3D). Weights on internal edges in the MCN grouping have no influence on the coarse-level weighting and are therefore unlabeled to avoid confusion.

We now proceed to show that the coarse level graph will maintain a lattice structure. Given the formulation of the coarse-level Laplacian matrix in (12), the weight between two coarse nodes on the upper-level graph is given by

$$w_{ij} = \sum_{M^*(v_s)=v_i, M^*(v_q)=v_j, \forall v_s, v_j} w_{sq}. \tag{13}$$

Therefore, the weight between neighboring coarse nodes is nonzero only if $\exists\ e_{ij} \in E$, s.t. $M^*(v_i) \neq M^*(v_j)$. A k-th order neighbor may be mapped to a set of 2^k coarse nodes through $M^*(v_i)$, which we may denote as the **reach** of v_i through M^*, $H(v_i)$. Given two neighboring nodes, v_i and v_j, of kth and $(k-1)$th order respectively, we note that $H(v_j) \subset H(v_i)$. Since $|H(\cdot)|$ is maximal for a node $v_i \in \mathcal{N}^k$, and contains all nodes in the hypercube surrounding v_i, only coarse-level nodes at the corners of this hypercube may become connected. Consequently, if a fine-level lattice is connected within a coarse-level hypercube, i.e., $(3^p - 1)$-connected in p-D, then the coarse-level lattice will remain $(3^p - 1)$-connected in p-D for every subsequent coarse level. Figure 2 illustrates an example of the MCN coarsening in which a fine-level four-connected 2D lattice is coarsened to an eight-connected 2D lattice.

2.6 Efficient Implementation

A description of an efficient method for effecting the three operators (restriction, coarse-level and prolongation) is given by:

1. Find M^* of each node in the fine-level graph by examining kth-level neighbors from $k = 0$ to $k = P$, where P represents the data dimension.

2. The restriction may be effected by summing together the solution at each fine-level node that maps through M^* to the same coarse-level node to produce the coarse-level solution.
3. Produce the weights for the coarse-level graph, L_c, from (13), i.e., for every two neighbors with different M^* nodes, add the weight between them to the weight between their M^* nodes.
4. Produce the diagonals for the coarse-level graph, T, by summing together the diagonal at each fine-level node that maps through M^* to the same coarse-level node.
5. The prolongation operator may be efficiently implemented by copying the solution at a coarse-level node to each of the fine-level nodes that were mapped by M^* to that coarse node.

Note that each of these steps require linear computational complexity. In the next section, we examine the speed effects of our method with respect to an optimized conjugate gradients and the pertinent multigrid method described in [31].

3 Results

The generalized inhomogeneous Poisson equation in (1) has found application for several problems, as reviewed above. The multigrid method will produce the same solution to the linear system, only faster. Consequently, it is sufficient to benchmark the method for speed on any model problem of the form in (1) if it is run to the same level of convergence as the competing numerical methods. The application chosen for the experiments was the segmentation method of [2]. Five high-resolution images were seeded and segmented using [2], followed by progressive downsampling of both the images and seeds to lower resolution. Our goal was to replicate the same segmentation problem at multiple resolutions.

We compared our multigrid approach for solving (1) to a conjugate gradients method and the multigrid method of [31] on 2D images of increasing size. The conjugate gradients method was heavily optimized and employed an incomplete Cholesky preconditioner that was modified to keep the dropped values along the diagonal (see [35] for more details). Both of the multigrid methods were run in a standard V-cycle with two Gauss-Seidel smoothing iterations for the multigrid method of [31] and one Gauss-Seidel smoothing iteration for the MCN multigrid. All three methods were run until the norm of the residual of the solution was within a tolerance value that was kept constant for all experiments. The computations used to provide a solution to (1) were all performed with double precision. Note that the conjugate gradients algorithm increased computation time with roughly $\mathcal{O}\left(n^{1.4}\right)$ complexity, despite the visualization which may make it appear to increase with a linear dependence on resolution. The MCN multigrid did in fact increase computation with a roughly linear dependence on resolution.

For each image at each resolution, all three algorithms were applied and the total time reported for execution on an Intel Xeon 2.4GHz machine with 3GB of RAM. The error bars represent one standard deviation of the runtime of the image set. While both multigrid approaches significantly outperform the conjugate gradients approach, the MCN multigrid is faster than the multigrid method of [31] by approximately five

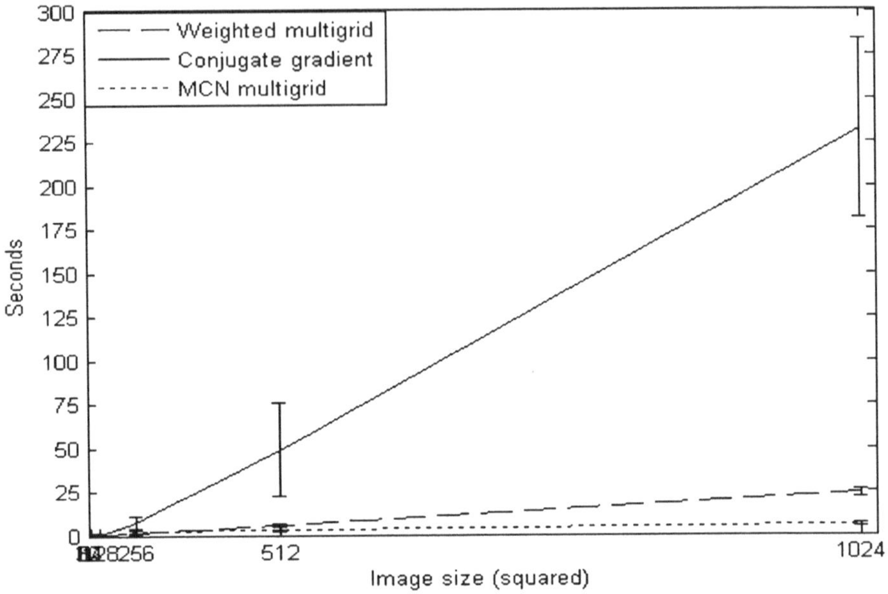

Fig. 3. Comparison of the conjugate gradient method (with incomplete Cholesky preconditioner) to the proposed multigrid method for solving (1) when segmenting images [2]. All images were square, with the length of one side given by the x-axis.

times. Due to the lattice structure of our MCN multigrid method, a GPU could easily be applied to produce additional speed gains [15].

4 Conclusion

We have presented a multigrid method for solving problems of the general form given in (1). Problems with this structure have become increasingly important in image processing applications such as segmentation [1, 2, 3], colorization [4], matting [5, 6] and filtering [7, 8]. Similar systems have also been studied in the context of visual reconstruction [12] and optical flow [14].

Previous approaches to multigrid solvers have typically employed either a matrix-dependent coarsening operator with fast convergence or the maintenance of a lattice structure at coarse levels with low memory overhead (see [20] for some exceptions). A coarse-level lattice structure is essential for memory demands (and fast processing) of high-resolution images or volumes. In addition to memory efficiency, a lattice structure at coarse levels is also essential to taking advantage of the power of a GPU implementation [15, 16, 5, 3]. However, matrix-dependent restriction/prolongation operators are essential for fast convergence of the multigrid method in the presence of discontinuities. The Maximally Connected Neighbor multigrid algorithm presented here both preserves a lattice structure at higher levels as well as employs the matrix-dependent restriction/prolongation operators that result in fast convergence. Our experiments compared our method with optimized preconditioned conjugate gradients code and another Laplacian-specific multigrid method. Our method strongly outperforms the speed of both methods when applied to real images. Future work includes a GPU implementation of our fast inhomogeneous Poisson solver which is expected to produce even greater efficiency gains and the technique of recombining iterants [36] to further accelerate the convergence speed of the algorithm.

Acknowledgments

The author would like to thank Oren Livne for several useful discussions.

References

1. Grady, L., Schwartz, E.L.: Isoperimetric graph partitioning for image segmentation. IEEE Trans. on Pat. Anal. and Mach. Intel. 28(3), 469–475 (2006)
2. Grady, L.: Random walks for image segmentation. IEEE Trans. on Pat. Anal. and Mach. Intel., 1768–1783 (November 2006)
3. Bhusnurmath, A.: Applying Convex Minimization Techniques to Energy Minimization Problems in Computer Vision. Ph.D thesis, U. Pennsylvania (2008)
4. Levin, A., Lischinski, D., Weiss, Y.: Colorization using optimization. In: Proc. of SIGGRAPH, vol. 23, pp. 689–694. ACM, New York (2004)
5. Grady, L., Schiwietz, T., Aharon, S., Westermann, R.: Random walks for interactive alpha-matting. In: Proc. of VIIP, pp. 423–429. ACTA Press (September 2005)
6. Levin, A., Lischinski, D., Weiss, Y.: A closed form solution to natural image matting. In: Proc. of CVPR 2006, New York (June 2006)
7. Grady, L., Schwartz, E.: Anisotropic interpolation on graphs: The combinatorial Dirichlet problem. Technical Report CNS-TR-03-014, Boston University (2003)
8. Zhang, F., Hancock, E.R.: Graph spectral image smoothing. In: Escolano, F., Vento, M. (eds.) GbRPR. LNCS, vol. 4538, pp. 191–203. Springer, Heidelberg (2007)
9. Shi, J., Malik, J.: Normalized cuts and image segmentation. IEEE Trans. on Pat. Anal. and Mach. Intel., 888–905 (August 2000)
10. Perona, P., Malik, J.: Scale-space and edge detection using anisotropic diffusion. IEEE Trans. on Pat. Anal. and Mach. Intel. 12(7), 629–639 (1990)
11. Desbrun, M., Meyer, M., Schröder, P., Barr, A.H.: Implicit fairing of irregular meshes using diffusion and curvature flow. In: Proc. of CGIT, pp. 317–324 (August 1999)
12. Terzopoulos, D.: The computation of visible-surface representations. IEEE Trans. on Pat. Anal. and Mach. Intel. 10(4), 417–438 (1988)
13. Szeliski, R., Terzopoulos, D.: Parallel multigrid algorithms and computer vision applications. In: Mandel, J., et al. (eds.) Proc. of the Fourth Copper Mountain Conf. on Multigrid Methods, pp. 383–398. SIAM, Philadelphia (1989)
14. Ghosal, S., Vaněk, P.: A fast scalable algorithm for discontinuous optical flow estimation. IEEE Trans. on Pat. Anal. and Mach. Intel. 18(2), 181–194 (1996)
15. Bolz, J., Farmer, I., Grinspun, E., Schröder, P.: Sparse matrix solvers on the GPU: Conjugate gradients and multigrid. ACM Trans. on Graphics, 917–924 (July 2003)
16. Krüger, J., Westermann, R.: Linear algebra operators for GPU implementation of numerical algorithms. ACM Trans. on Graphics, 908–916 (July 2003)
17. Sapiro, G.: Inpainting the colors. Technical Report IMA Preprint Series #1979, University of Minnesota, Institute for Mathematics and its Applications (2004)

18. Briggs, W.L., Hensen, V.E., McCormick, S.F.: A Multigrid Tutorial, 2nd edn. SIAM, Philadelphia (2000)
19. Trottenberg, U., Oosterlee, C.W., Schuller, A.: Multigrid. Academic Press, San Diego (2000)
20. Wesseling, P.: An Introduction to Multigrid Methods. R.T. Edwards (2004)
21. Dendy, J.E.: Black box multigrid. J. of Computational Physics 48, 366–386 (1982)
22. Brandt, A.: Algebraic multigrid theory: The symmetric case. Appl. Math. Comput. 19, 23–56 (1986)
23. Gidas, B.: A renormalization group approach to image processing problems. IEEE Trans. on Pat. Anal. and Mach. Intel. 11(2), 164–180 (1989)
24. Pérez, P., Heitz, F.: Restriction of a Markov random field on a graph and multiresolution statistical image modeling. IEEE Trans. on IT 42(1), 180–190 (1996)
25. Burt, P.J., Adelson, E.H.: The Laplacian pyramid as a compact image code. IEEE Trans. on Comm. COM-31(4), 532–540 (1983)
26. Terzopoulos, D.: Multilevel computational processes for visual surface reconstruction. Comput. Vision, Graphics, and Image Processing 24, 52–96 (1983)
27. Terzopoulos, D.: Image analysis using multigrid relaxation methods. IEEE Trans. on Pat. Anal. and Mach. Intel. 8(2), 129–139 (1986)
28. Acton, S.T.: Multigrid anisotropic diffusion. IEEE Trans. on Image Proc. 7(3), 280–291 (1998)
29. Ruge, J., Stüben, K.: Algebraic multigrid. In: McCormick, S. (ed.) Multigrid Methods. Frontiers in Applied Mathematics, vol. 3, pp. 73–130. SIAM, Philadelphia (1987)
30. Kimmel, R., Yavneh, I.: An algebraic multigrid approach for image analysis. SIAM J. of Sci. Comput. 24(4), 1218–1231 (2003)
31. Grady, L., Tasdizen, T.: A geometric multigrid approach to solving the 2D inhomogeneous Laplace equation with internal boundary conditions. In: Proc. of ICIP 2005, vol. 2, pp. 642–645. IEEE, Los Alamitos (2005)
32. Boykov, Y., Jolly, M.P.: Interactive graph cuts for optimal boundary & region segmentation of objects in N-D images. In: Proc. of ICCV 2001, pp. 105–112 (2001)
33. Black, M.J., Sapiro, G., Marimont, D.H., Heeger, D.: Robust anisotropic diffusion. IEEE Trans. on Image Proc. 7(3), 421–432 (1998)
34. Biggs, N.: Algebraic Graph Theory. Cambridge Tracts in Mathematics, vol. 67. Cambridge University Press, Cambridge (1974)
35. Barrett, R., Berry, M., Chan, T.F., Demmel, J., Donato, J.M., Dongarra, J., Eijkhout, V., Pozo, R., Romine, C., Van der Vorst, H.: Templates for the Solution of Linear Systems: Building Blocks for Iterative Methods. SIAM, Philadelphia (1993)
36. Brandt, A., Mikulinksy, V.: On recombining iterants in multigrid algorithms and problems with small islands. SIAM J. of Sci. Comput. (1995)

SMD: A Locally Stable Monotonic Change Invariant Feature Descriptor

Raj Gupta and Anurag Mittal

Indian Institute of Technology, Madras, India
gupta.raj@gmail.com, amittal@cse.iitm.ernet.in

Abstract. Extraction and matching of discriminative feature points in images is an important problem in computer vision with applications in image classification, object recognition, mosaicing, automatic 3D reconstruction and stereo. Features are represented and matched via descriptors that must be invariant to small errors in the localization and scale of the extracted feature point, viewpoint changes, and other kinds of changes such as illumination, image compression and blur. While currently used feature descriptors are able to deal with many of such changes, they are not invariant to a generic monotonic change in the intensities, which occurs in many cases. Furthermore, their performance degrades rapidly with many image degradations such as blur and compression where the intensity transformation is non-linear. In this paper, we present a new feature descriptor that obtains invariance to a monotonic change in the intensity of the patch by looking at orders between certain pixels in the patch. An order change between pixels indicates a difference between the patches which is penalized. Summation of such penalties over carefully chosen pixel pairs that are stable to small errors in their localization and are independent of each other leads to a robust measure of change between two features. Promising results were obtained using this approach that show significant improvement over existing methods, especially in the case of illumination change, blur and JPEG compression where the intensity of the points changes from one image to the next.

1 Introduction

Extraction and matching of distinctive feature points in images has been a major focus of research in the Computer Vision community for quite some time. Such an approach has been used in many applications such as mosaicing, classification, object recognition, automatic 3D reconstruction and stereo matching among others. The basic idea is to determine certain feature points in images that have certain properties that allow them to be distinguished from other points, either in the same image or in other images. Then, certain properties of the region around the point are used in order to transform this region to a normalized region that should remain the same under some (affine) transformation of the original patch. Finally, certain features are extracted from these normalized regions which form the feature "descriptor". These feature descriptors are matched between two feature points to determine the similarity between them.

Many methods have been proposed in the literature for feature point extraction and subsequent affine normalization. Popular methods include the Harris corner detector

D. Forsyth, P. Torr, and A. Zisserman (Eds.): ECCV 2008, Part II, LNCS 5303, pp. 265–277, 2008.

and its affine normalization [2, 3], the Hessian-affine detector [2], 'Maximally Stable Extremal Regions'(MSER) [4], edge and intensity-extrema based detectors [5, 6] and a 'salient regions' detector [7]. All of these feature point extractors have different strengths and weaknesses and yield different number of points depending on the image. An evaluation of the performance of these detectors was presented in [8]. MSER was consistently shown to outperform others in repeatability and matching scores using SIFT but produced lesser number of features, while hessian-affine and harris-affine produced more features than other methods while giving good repeatability and matching scores.

Once the features have been detected and normalized, a descriptor is determined for matching. Popular descriptors include SIFT (Scale-Invariant Feature Transform) [9], Shape Context [10], GLOH (Gradient Location Orientation Histogram) [11], SURF (Speeded up Robust Features) [12], PCA-SIFT[13], differential invariants and spin images among others. A performance evaluation of these descritors was presented in [11], where it was shown that the SIFT-based descriptors such as SIFT and GLOH perform the best while Shape Context, which is also a based on histogram of gradients/edges, comes quite close. More recently, Moreels and Perona [14] have reported results for matching in 3D objects where they show the best performance for hessian-affine detector combined with SIFT for viewpoint changes and harris-affine with SIFT and hessian-affine with Shape Context for lighting change and camera focal length change respectively.

The remarkable outperformance of SIFT can possibly be attributed to the fact that since it uses a statistical measure (*histogram*) of the gradients, it is relatively robust to small errors in feature localization and normalization, and small changes in the shape of the feature due to viewpoint or other changes. Furthermore, it normalizes the gradients which yields a method that is invariant to a *linear* change in intensities. However, while this descriptor has these interesting properties, it is not invariant to a non-linear change in intensities which often occurs in practice. This can happen, for instance, due to gamma correction, a non-linear camera response function especially near saturation and low light [1, 21, 6], small specular reflections, different illumination in different parts of an object, and image effects such as blur and image compression.

To deal with such effects, several papers have proposed the use of orders between pixels rather than the intensities themselves [15, 16]. These methods transform the intensity space to an "order" space that captures the order of a pixel with respect to its neighbors and develops a binary pattern from such orders. Statistical matching of histograms of these binary patterns has shown extremely good performance in some applications such as texture classification[16] and face recognition[17]. Although this may be a good monotonic illumination-invariant scheme for textures and faces, the relatively large space of the binary patterns makes it unsuitable for feature point description where the patch size is limited. Also, the intensity information is totally lost in the process and this can make the descriptor susceptible to Gaussian noise. Also, this gives equal weightage to high gradient and low gradient regions, which can be undesirable (SIFT gives weightage proportional to the gradient value).

Mittal and Ramesh[18] proposed an approach that utilizes a combination of intensity and order in order to develop a change measure that is more robust to Gaussian noise and weighs higher gradients more compared to lower gradients while still maintaining invariance to monotonic changes. However, while such a matching technique can be used in some applications such as stereo matching, such an approach cannot handle errors in point localization, scale and shape changes etc. very well that is needed in a feature matching application. In order to localize and normalize the points more accurately so that such an approach can be used, Gupta and Mittal[19] develop a feature point detector that detects feature points at the intersection of two lines. However, although they report quite high performance numbers, the number of features points detected by such an approach is rather low and this approach is not suitable for many applications where such linear structures are not present. Another work that is somewhat related to ours is that of [20] who use the idea of comparing pixel values for some random points around a keypoint in order to drop this keypoint down a randomized tree for recognition. They warp the keypoints in a given image in order to obtain numerous possible patch realizations under viewpoint changes (for wide-baseline point matching) and the problem is posed as a classification problem of a given keypoint in the second image to belong to one of the keypoints in this (first) image.

In this paper, we present a new feature descriptor that obtains invariance to a monotonic change in the intensities while at the same time works with any of the feature detectors used in the literature. We look at orders between certain pixels in the patch and the feature descriptor consists of point pairs. A penalty is awarded if there is an order change for a point pair between the two patches and such penalties for different pairs are summed in order to determine the "difference" between the two features. The point pairs have the property that the points in the pair are relatively stable in their intensity order with respect to both intensity noise and localization error. In order to obtain invariance to Gaussian noise, the points in a pair are chosen such that they have a certain minimum difference between their intensities. On the other hand, robustness to changes in the scale and localization of the feature point is obtained by picking point pairs such that moving the points a certain distance in their neighborhood does not change the order of the intensities of the pair. Furthermore, we allow a point to repeat only a certain number of times in the pairs in order to improve the independence between the different point pairs. Two features are matched by comparing the orders of the pixel pairs. The method was found to be extremely robust and on a standard dataset, it yielded results that are significantly superior to currently used methods. This makes the method highly suitable for many applications.

2 Basic Goals of Our Feature Matching Approach

We have several goals for feature comparison. First, the approach must be invariant to a monotonic change in the intensities. Second, it must be robust to noise in the pixel intensities as well as feature point localization and distortion. Third, the method must be reasonably efficient. Towards these goals, we extract certain point pairs for which the order will be tested across the feature points. Such points must have the following properties:

1. They must have a minimum intensity difference between them. This is needed so that the order between these pixels does not change with some amount of noise in the pixel intensities.

2. The order between these pixels should not change if there is some error in the localization of these points.
3. The different point pairs must not repeat the same points too many times so that the tests for the different point pairs are more or less independent of each other.

Computation of optimal points according to all of the above criteria appears difficult. However, we show that it is possible to do so quite efficiently using the concept of extremal regions and distance transforms. This is discussed next.

3 The Feature Descriptor

3.1 Computation of Extremal Regions

The first step in our algorithm is the computation of extremal regions. Extremal regions are regions that have intensities above or below a given threshold. Given that the points in the point pairs must have a given difference of intensity δ_I between them, we compute extremal regions with two thresholds T_1 and T_2 such that $T_1 - T_2 = \delta_I$:

$$\mathcal{R}^+ = Thresh^+(I, T_1)$$
$$\mathcal{R}^- = Thresh^-(I, T_2)$$

where $Thresh^+(I, T)$ is the set of all points in the Image I that are above a given threshold T and $Thresh^-(I, T)$ is the set of all points I below T.

Such regions are computed over a range of values of T_1 (T_2 is determined automatically as $T_1 - \delta_I$). As pointed out by Matas et.al. [4], the set of all extremal regions can be computed in time $O(n \log \log n)$ where n is the number of pixels in the image, using methods based on the union-find algorithm[21].

As should be obvious, all points in $\mathcal{R}^+$ are greater than all points in $\mathcal{R}^-$ by atleast an intensity difference of δ_I. The next step is to find points in $\mathcal{R}^+$ and $\mathcal{R}^-$ that are as far as possible from the boundaries. This will ensure that the points are stable with respect to localization errors.

3.2 Computation of Point Pairs

Given a pair of extremal regions $\mathcal{R}^+$ and $\mathcal{R}^-$, we wish to compute points that are as far as possible from the boundaries of these regions. This can be done quite efficiently using the distance transform. We use the one based on the Euclidean distance measure[1]. Points that have high distances from the boundaries are selected as possible candidates and points from $\mathcal{R}^+$ are matched with points from $\mathcal{R}^-$. Once a pair is selected for a given region pair, we do not want to select points close to these points. To obtain this behavior, we mark the selected points themselves as boundary points so that the points that are selected next are those that have the largest distance not only from the original boundaries but from the already selected points as well. This procedure is repeated till we cannot obtain points that have a certain minimum "distance" from the boundaries.

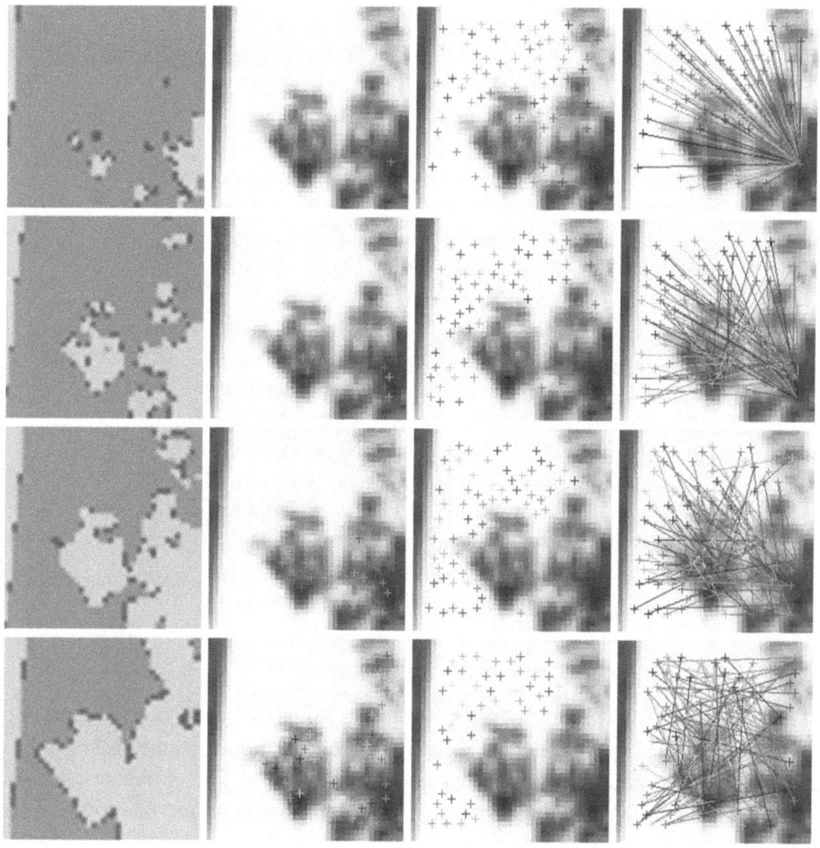

Fig. 1. An Example of pair extraction at different levels. First column shows the extracted extremal regions where the brighter gray is the "Min" extremal region, darker gray is the "Max' extremal region and black are the boundary points. The second and third columns show the extraction of some points in the min and max regions respectively, super-imposed on the original patch. Finally, the last column shows the pairs formed at each level, again super-imposed on the original patch. These are combined by throwing away pairs having common points in order to obtain the final pairs shown in Fig. 2.

The above mentioned procedure yields a set of points for a given extremal region pair ($\mathcal{R}^+$,$\mathcal{R}^-$) obtained at a certain threshold T_1. This procedure is repeated for different values of T_1 (For the experiments in this paper, we have used 25 such values ranging from 10 to 240 at a gap of 10 each.). Now, given the set of point pairs obtained from all of the above mentioned steps, we select the most stable ones based on the "distance" value associated with the points as per the distance transform. The minimum of the distance values of the two points in a pair is taken as the *stability factor* for that pair. Now, using a greedy approach, we simply select the most stable point-pairs one by one while taking care that any one point in the patch does not have too many close-by points in the already existing point-pairs (For the experiments in this paper, we allow

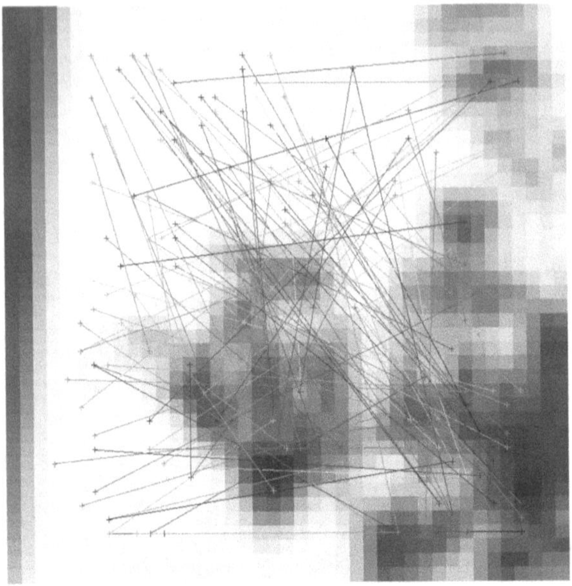

Fig. 2. An Example of pairs extracted in a patch

a point to be taken a maximum of 3 times for building the point pairs). This is done to ensure some independence between the different point pairs and to allow the pairs to spread out in the patch so as to obtain discriminability. The above procedure can be done quite efficiently by again using a distance transform image taking the selected points as the model points. The output of this procedure is a set of point pairs along with their stability factors. Features which do not produce a certain minimum number of stable pairs are thrown out as being unreliable for matching. This set of extracted point pairs, thus, forms our feature descriptor.

Fig. 1 shows the above process on an image patch, where the different rows show the computation of the extremal regions and point pairs at different thresholds. These are combined in the end in order to obtain the final point pairs shown in Fig. 2.

It is not hard to see that our algorithm yields a reasonably good set of pairs that is also optimal in a certain sense. Suppose there exists a point pair such that a spatial perturbation of the points in the pair *and* some point-wise noise in the intensity values of the pixels does not lead to an order change. Suppose the maximum value of change in the intensity of each pixel is ΔI and the spatial perturbation possible is δx. Then, if our thresholding levels were continuous and δ_I approximately two times ΔI, then this point pair will be observed by our method with a stability factor of atleast Δx when the thresholds are set such that the lower threshold value is slightly higher than the lower intensity point and the higher threshold value is just below the higher intensity point. Our algorithm simply tries to pick point pairs from such set that are the most stable and more or less independent, while capturing as much of the patch structure as possible.

4 Matching

Given the feature descriptor, we now describe how one may match any two features. For each of the features, we have a set of points pairs along with their stability factors $\{(p_i^1, p_i^2, s_i), i = 1 \ldots n\}$. For these point pairs, we test if the order of the pixels has changed in the other patch. Then, we simply calculate a weighted sum of the order flips, giving each point pair a weight that depends on its stability factor. Since higher stability points are very important for stable matching and should be given a higher weightage (recall that the stability factor is the minimum distance that any of the two points has to move in order to possibly have an order flip), we propose using the square of the stability factor (s^2) as the weight for a pair that has stability factor s. The pairs from both the feature points are combined in order to obtain the final weighted matching score:

$$M = \frac{\Sigma_{i=1}^n s_i^2 \operatorname{sgn}(I_o(p_i^1) - I_o(p_i^2))}{\Sigma_{i=1}^n s_i^2} \tag{1}$$

where $I_o(p)$ is the intensity of point p in the patch "other" than the one in which the point pair was computed (i.e. if the pair was computed in the first patch, then I_o is the intensity in the second patch and if the pair was computed in the second patch, then I_o is the intensity in the first patch). sgn is the *sign* function:

$$\operatorname{sgn}(x) = \begin{cases} 1 & \text{if } x > 0 \\ 0 & \text{if } x = 0 \\ -1 & \text{if } x < 0 \end{cases} \tag{2}$$

We assume in Eq. 1 that the pair (p_i^1, p_i^2, s_i) is stored such that the first point has higher intensity than the second in the original patch in which this pair was computed. We also note here that as opposed to many other methods where only the feature descriptors are matched directly, we also use the underlying image patches for comparison.

5 Results and Experiments

We tested our algorithm on the standard dataset from the evaluation papers of Mikolajczyk and Schmid [11] and Mikolajczyk, Tuytelaars et. al. [8]. In [11], different feature descriptors are evaluated for changes in scale, rotation, viewpoint (affine transformation), blur, jpeg compression and illumination. GLOH and SIFT performed better than others and almost similar to each other, while Shape Context came quite close. Most of the experiments were done using the Harris-affine and Hessian-affine feature detector since these detectors give the most number of points. It was also stated that the relative performance of different descriptors remains almost the same for different feature detectors.

We follow the approach of this paper in order to evaluate our descriptor by plotting the recall vs. 1 - precision. This is similar to an *ROC* curve. Recall is the number of the correct matches found divided by the total number of correct matches present while precision is the number of correct matches found divided by the total number of matches found by the descriptor. The curve is obtained by varying the threshold for each descriptor. As in [11], we show results using only the Harris-affine and Hessian-affine feature detectors. Since we throw out some features as being unreliable for matching, the set of

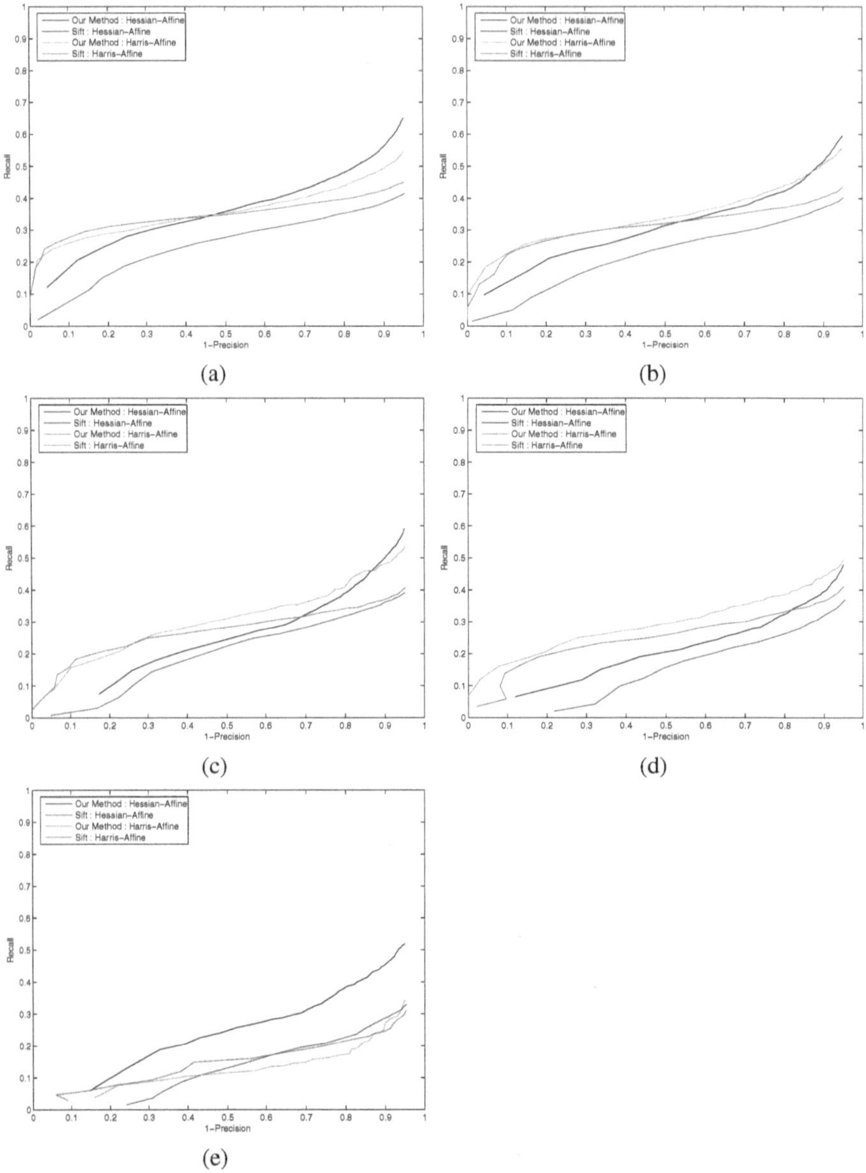

Fig. 3. Comparison of our matching approach with SIFT for different amounts of illumination change (leuven) using the Harris-Affine and Hessian-Affine detectors. The amount of lighting change increases from (a) to (e) with (e) having lighting change of approximately 60%. The number of correct correspondences existing in the images were: 3902 and 2205 for Hessian-Affine and Harris-Affine respectively for (a), 2952 and 1511 for (b), 2453 and 994 for (c), 1777 and 668 for (d) and 1197 and 299 for (e). The images in this standard dataset may be obtained from http://www.robots.ox.ac.uk/ vgg/research/affine.

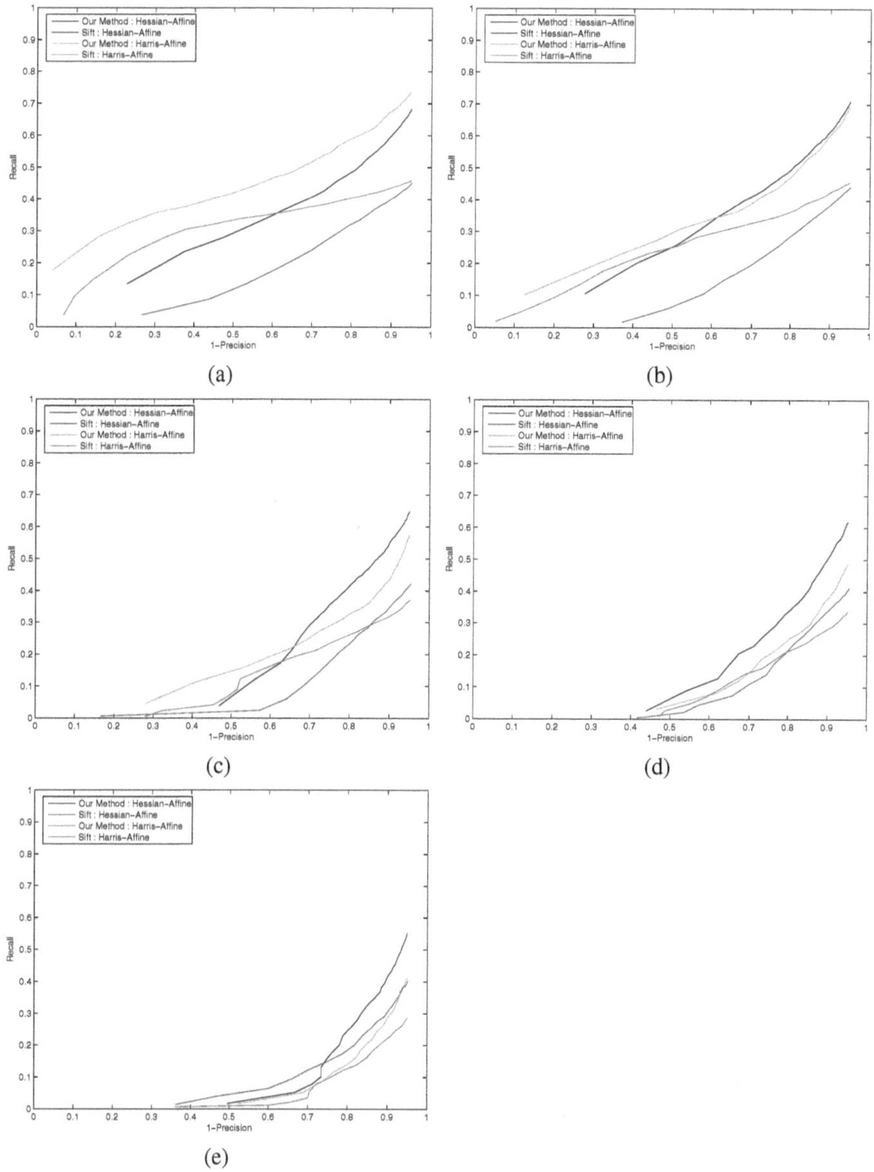

Fig. 4. Comparison of the performance of our method with SIFT for different amounts of image blur (Bikes) using the Harris-Affine and Hessian-Affine detectors. The amount of blurring increases from a values of 2 to 6 in plots (a) to (e). The number of correct correspondences existing in the images were: 4150 and 2372 for Hessian-Affine and Harris-Affine respectively for (a), 4127 and 2253 for (b), 3147 and 1827 for (c), 2548 and 1478 for (d) and 1640 and 888 for (e).

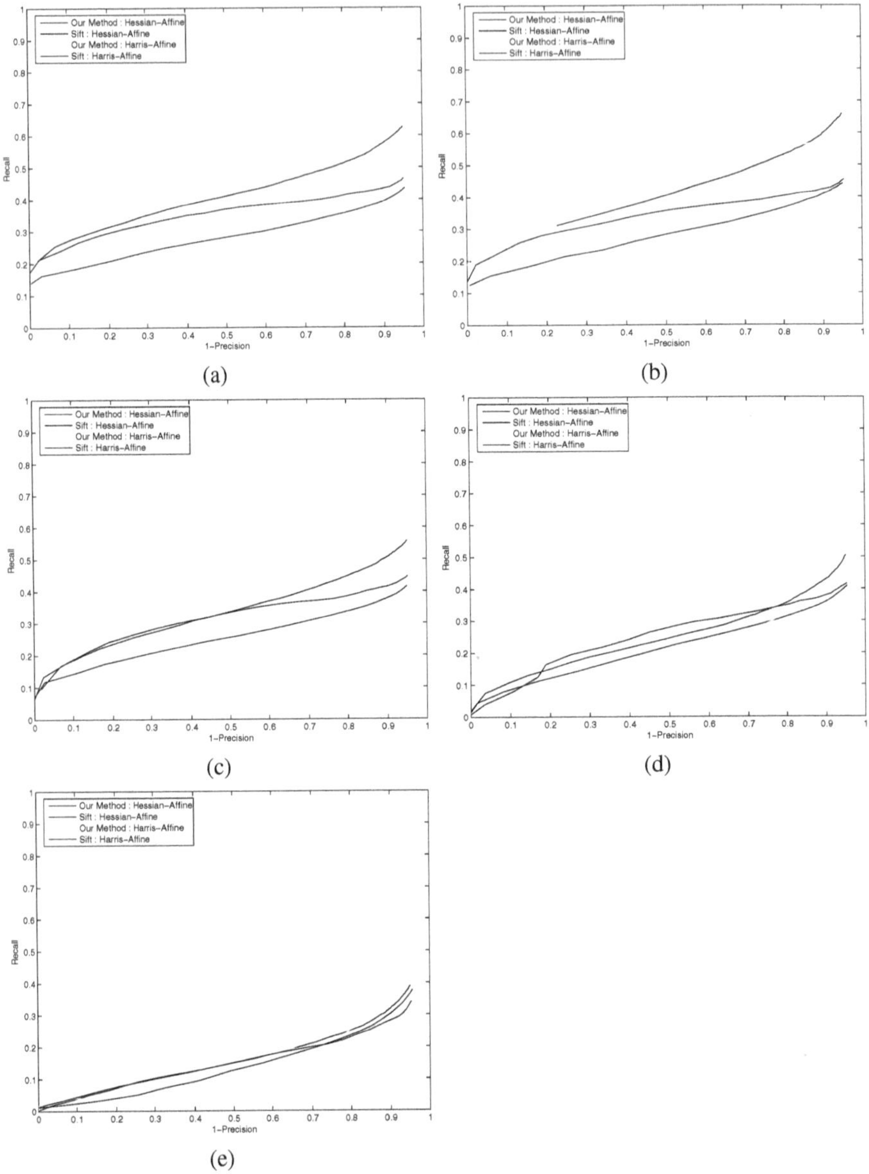

Fig. 5. Comparison of the performance of our method with SIFT for different amounts of JPEG compression (ubc) using the Harris-Affine and Hessian-Affine detectors. The compression ratios for the different plots are (a) 60 %, (b) 80 %, (c) 90 %, (d) 95, % and (e) 98 %. The number of correct correspondences existing in the images were: 10876 and 4053 for Hessian-Affine and Harris-Affine respectively for (a), 10751 and 3939 for (b), 10492 and 3548 for (c), 9243 and 2699 for (d) and 8043 and 2565 for (e).

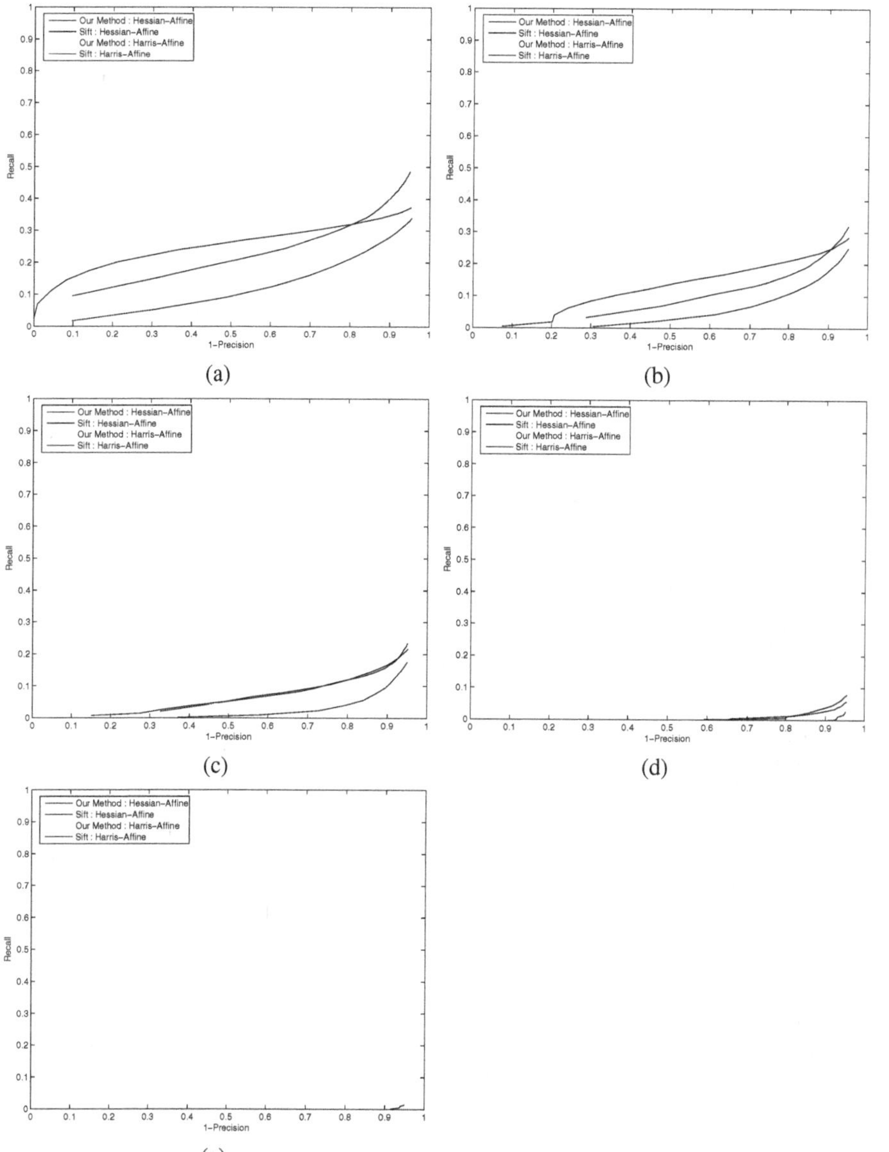

Fig. 6. Comparison of the performance of our method with SIFT for different amounts of affine/viewpoint changes (graffiti) using the Harris-Affine and Hessian-Affine detectors. The viewpoint change angle for the different plots are approximately (a) 20°, (b) 30°, (c) 40° , (d) 50°, and (e) 60°. The number of correct correspondences existing in the images were: 504 and 228 for Hessian-Affine and Harris-Affine respectively for (a), 497 and 214 for (b), 398 and 191 for (c), 219 and 107 for (d) and 135 and 73 for (e).

feature points that we use is a subset of the total feature points extracted. The reduction in feature points is approximately 50% for all the images combined, although it varies for different images. All comparisons are done using this reduced set of feature points[1]. Also, since the performance of different histogram-based approaches has been shown to be quite similar to each other by several authors, we only compare our method with SIFT in this paper.

Fig.s 3-6 show the results of our algorithm on the different images of the dataset provided and used by [11] and [8]. In Fig. 3, we show comparative results as a function of illumination changes, while Figs. 4 and 5 show the results as a function of image blur and compression. Finally, in Fig. 6, we show the performance as a function of affine transformation. In all of these plots, we have shown the results using both Harris-affine and Hessian-affine.

As can be seen from the plots, we improve substantially compared to SIFT on images which had a substantial transformation in the intensity values of the pixels. This happens not only under illumination changes but also when there is an image blur or compression. Since SIFT can only do a linear correction in the intensities, our out-performance on such images is not surprising. Even though blur and image compression do not exactly follow a monotonic change transformation model, they do not deviate far from such a model. Furthermore, since the points pixed for testing by our method are *stable* and hence lie away from points whose blurring or compression effects could cause an order flip, such image degradations do not affect us substantially. Our performance under viewpoint changes was slightly better than SIFT for Hessian-affine and almost the same as SIFT for Harris-affine points and we do not seem to have any particular advantage in this case.

It was also observed that the performance gap between our method and SIFT was larger for Hessian-affine points as compared to Harris-affine points. This could be due to the higher localization accuracy of Hessian-affine as noted by several previous authors. This helps our method more compared to SIFT as histogram-based approaches can handle more localization error as compared to our approach. On the other hand, if the feature point is localized relatively well, then we are able to get more discriminative structural information than just a statistical measure of the gradients in different regions. This, along with our invariance to a monotonic change in intensities, probably explains the superior performance of our approach compared to SIFT when the patches are better localized. A drawback of our approach, on the other hand, is that we are not able to obtain many stable points on patches that have very high frequency gradient changes. On such kind of patches, histogram-based methods would probably give better results.

6 Conclusion

We have presented an approach for feature description and matching that is invariant to a monotonic change in intensities while being robust to Gaussian noise and errors in feature localization and normalization. The method can be implemented quite efficiently since it depends on the fast algorithms that are available for extremal region extraction and distance transform computation. We obtained a significant improvement

[1] The degradation in performance of the method without using this feature reduction is around 10% for our method and around 5% for SIFT.

in performance compared to existing techniques such as SIFT on a standard dataset, especially in cases where there is a substantial transformation in the pixel intensities. Thus, the method holds great promise for many applications of feature matching such as image classification, object recognition, mosaicing and 3D reconstruction.

References

[1] Fabbri, R., da Fontoura Costa, L., Torelli, J.C., Bruno, O.M.: 2D Euclidean distance transform algorithms: A comparative survey. ACM Computer Survey (2008)

[2] Mikolajczyk, K., Schmid, C.: An Affine Invariant Interest Point Detector. In: Heyden, A., Sparr, G., Nielsen, M., Johansen, P. (eds.) ECCV 2002. LNCS, vol. 2350, pp. I-128. Springer, Heidelberg (2002)

[3] Schaffalitzky, F., Zisserman, A.: Viewpoint Invariant Texture Matching and Wide Baseline Stereo. In: ICCV, pp. II, 636–643 (2001)

[4] Matas, J., Chum, O., Urban, M., Pajdla, T.: Robust Wide Baseline Stereo from Maximally Stable Extremal Regions. In: IVC, September 10, 2004, pp. 761–767 (2004)

[5] Tuytelaars, T., Van Gool, L.J.: Content-Based Image Retrieval Based on Local Affinely Invariant Regions. In: Huijsmans, D.P., Smeulders, A.W.M. (eds.) VISUAL 1999. LNCS, vol. 1614, pp. 493–500. Springer, Heidelberg (1999)

[6] Tuytelaars, T., Van Gool, L.J.: Wide Baseline Stereo Matching based on Local, Affinely Invariant Regions. BMVC (2000)

[7] Kadir, T., Zisserman, A., Brady, M.: An Affine Invariant Salient Region Detector. In: Pajdla, T., Matas, J(G.) (eds.) ECCV 2004. LNCS, vol. 3021, pp. 228–241. Springer, Heidelberg (2004)

[8] Mikolajczyk, K., Tuytelaars, T., Schmid, C., Zisserman, A., Matas, J., Schaffalitzky, F., Kadir, T., Van Gool, L.J.: A Comparison of Affine Region Detectors. IJCV 1-2, 43–72 (2005)

[9] Lowe, D.G.: Distinctive Image Features from Scale-Invariant Keypoints. IJCV 2, 91–110 (2004)

[10] Mori, G., Belongie, S., Malik, J.: Efficient Shape Matching Using Shape Contexts. PAMI 11, 1832–1837 (2005)

[11] Mikolajczyk, K., Schmid, C.: A Performance Evaluation of Local Descriptors. PAMI 10, 1615–1630 (2005)

[12] Bay, H., Tuytelaars, T., Van Gool, L.J.: SURF: Speeded Up Robust Features. In: Leonardis, A., Bischof, H., Pinz, A. (eds.) ECCV 2006. LNCS, vol. 3951, pp. 404–417. Springer, Heidelberg (2006)

[13] Ke, Y., Sukthankar, R.: PCA-SIFT: a more distinctive representation for local image descriptors. In: CVPR, pp. II, 506–513 (2004)

[14] Moreels, P., Perona, P.: Evaluation of Features Detectors and Descriptors Based on 3D Objects. IJCV, 263–284 (July 3, 2007)

[15] Zabih, R., Woodfill, J.: Non-parametric local transforms fo computing visual correspondence. In: ECCV, pp. 151–158 (1994)

[16] Ojala, T., Pietikainen, M., Maenpaa, T.: Multiresolution Gray-Scale and Rotation Invariant Texture Classification with Local Binary Patterns. PAMI, 971–987 (July 7, 2002)

[17] Ahonen, T., Hadid, A., Pietikainen, M.: Face Description with Local Binary Patterns: Application to Face Recognition. PAMI, 2037–2041 (December 12, 2006)

[18] Mittal, A., Ramesh, V.: An Intensity-augmented Ordinal Measure for Visual Correspondence. In: CVPR, pp. I: 849–856 (2006)

[19] Gupta, R., Mittal, A.: Illumination and Affine- Invariant Point Matching using an Ordinal Approach. In: ICCV, pp. 1–8 (2007)

[20] Lepetit, V., Fua, P.: Keypoint Recognition Using Randomized Trees. PAMI, 1465–1479 (September 9, 2006)

[21] Cormen, T.H., Leiserson, C.E., Rivest, R.L., Stein, C.: Introduction to Algorithms (2005)

Finding Actions Using Shape Flows

Hao Jiang and David R. Martin

Computer Science Department, Boston College, Chestnut Hill, MA 02467, USA
{hjiang,dmartin}@cs.bc.edu

Abstract. We propose a novel method for action detection based on a new action descriptor called a *shape flow* that represents both the shape and movement of an object in a holistic and parsimonious manner. We find actions by finding shape flows in a target video that are similar to a template shape flow. Shape flows are largely independent of appearance, and the match cost function that we propose is invariant to scale changes and smooth nonlinear deformation in space and time. The problem of matching shape flows is difficult, however, yielding a large, non-convex, integer program. We propose a novel relaxation method based on *successive convexification* that converts this hard program into a vastly smaller linear program: By using only those variables that appear on the 4D lower convex hull of the matching cost volume, most of the variables in the linear program may be eliminated. Experiments confirm that the proposed shape flow method can successfully detect complex actions in cluttered video, even with self-occlusion, camera motion, and intra-class variation.

1 Introduction

An action can be characterized by the movement and deformation of a shape. A *flow line*, which is the space-time line formed by a tracked object point over time, provides a compact representation of motion. An object's *shape flow*, which we define simply as an assembly of flow lines, represents not only the object's motion but also its shape and deformation over time. Flow lines have previously been used for motion visualization [5]. We propose the shape flow as a representation for actions.

Consider the shape flows shown in Fig. 1. One can readily identify complex actions based on the shape flow alone, suggesting that it is a discriminative and descriptive representation. The shape flow itself is quite simple to compute. The challenge, which is the focus of this paper, is how to use the shape flow as a representation for actions. In particular how can we efficiently search for a template shape flow in a cluttered single-view video? And how can we do the search in a manner that is invariant to scale changes and nonlinear shape deformations, and also tolerant of occlusion and intra-class variation? Although shape flow matching is certainly an NP-hard problem because of the loopy relations between flow lines, we show that a novel linear relaxation based on *successive convexification* [6] yields both an efficient and accurate matching algorithm for finding actions in video.

There is much related work on detecting actions in video. The first dimension of related work consists of methods that use multi-view stereo to reliably access 3D spatial information. Parameswaran & Chellappa [12] use multiple cameras to capture joint

D. Forsyth, P. Torr, and A. Zisserman (Eds.): ECCV 2008, Part II, LNCS 5303, pp. 278–292, 2008.

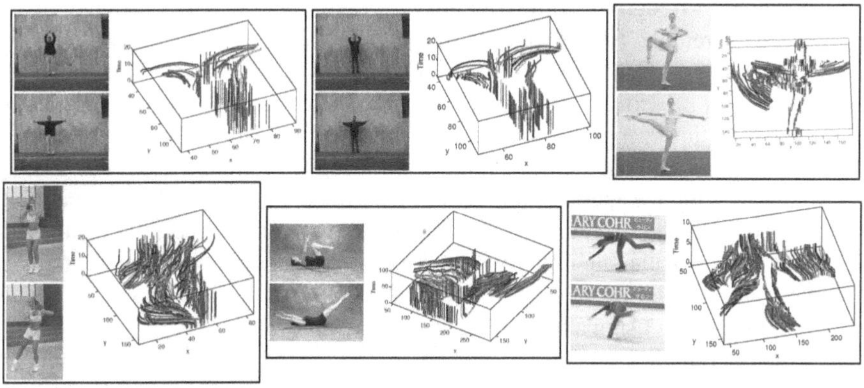

Fig. 1. Example shape flows and first/last frames for a variety of actions. Even though individual flow lines are noisy, the shape flow represents the holistic shape *and* movement of the object reliably. In this paper, we show how to efficiently search for a template shape flow in a target video. *Please note that the figures in this paper are best viewed in color.*

locations via motion capture, while Yilmaz & Shah [20] use manually labeled joint locations. Weinland et al. [18] use multiple cameras to achieve reliable background subtraction, using 3D silhouettes of foreground objects to describe actions. These methods are only tangentially related to this paper, however, as they rely on multi-view stereo.

The second dimension of related work includes methods that rely on background subtraction to formulate action features from silhouettes of foreground objects either in 2D or in 3D. Weinland et al. [18] use 3D silhouettes from multi-view stereo, as mentioned above. Both Yilmax & Shah [19] and Blank et al. [2] use background subtraction in 2D images to describe actions as 3D space-time volumes. Bobick & Davis [3] project space-time silhouettes into the image to get 2D action silhouettes. In this paper, we address the problem of finding actions in cluttered single-view video with a moving camera and moving background objects. Silhouette based features, which rely on high quality background subtraction, are difficult to use in this regime.

The third dimension of related work involves using labeled joint trajectories to detect human actions. In this class-specific approach, joint locations may be labeled in different ways. Parameswaran & Chellappa use motion capture and manual labeling for joint locations in 2D and 3D in their work [12]. Yilmaz & Shah [20] use manually labeled joint trajectories in 3D, and Sheikh & Shah [17] use manually labeled joint trajectories in 2D. Recognizing actions is challenging even with clean joint trajectories. Although automatic human pose tracking is recently much improved [11,13] and so could be used to extract (unclean) joint trajectories from unlabeled video, we pursue a non-parametric approach of analyzing the whole motion field rather than the motion of distinguished high-level feature points.

The final dimension of related work consists of non-parametric action models, and contrasts methods that rely on sparse descriptors located at interest points versus methods that use a dense motion or gradient field. These methods typically operate on uncalibrated single-view video, do not require background subtraction, and do not require manual labeling of foreground objects. Laptev & Lindeberg [9] have extended the notion of Harris extrema to find space-time interest points. Schuldt et al. [14] use motion and gradient histograms around these interest points to recognize actions using SVMs. Scovanner et al. [15] extend the popular SIFT descriptor to space-time volumes for the purpose of action recognition. Instead of constructing features that are tolerant to clutter, one may instead use dense feature fields along with clutter-tolerant matching. Efros et al. [4] recognize actions using optical flow fields from single frames, carefully smoothed and rectified. Shechtman & Irani [16] cleverly match space-time volumes directly, without any explicit motion estimate. Ke et al. [7] have extended that work using superpixels and part-based matching. Laptev & Prez [10] match histograms of gradients from space-time cubes using a boosted cascade.

These non-parametric methods represent a recent trend toward using volumetric space-time descriptors that combine shape *and* motion information, along with a matching framework that tolerates both foreground deformation and background clutter. We adopt this general approach to finding actions, although our proposed shape flow descriptor is neither an interest point method nor a dense field approach. As is clear in Fig. 1, the shape flow preserves much information about both the shape and movement of an object in a manner that is largely independent of the object's appearance. This enables us to match both shape and motion in a uniform framework.

The outline of the paper is as follows. In §2 we describe a simple method to compute shape flows. In §3, we formulate the shape flow matching problem and show how this NP-hard problem may be solved efficiently using a novel relaxation scheme. We present experimental results in §4 and conclude in §5.

2 The Shape Flow of an Action

We desire a compact yet expressive descriptor of both the shape and motion of an object. In addition, we seek a general method that may be applied to any object class, so we do not attempt to track distinguished points (such as joints) or to impose a shape model a priori. Instead, we seek a non-parametric shape and motion representation. We adapt the technique of flow fields from motion visualization [5]. If points on an object may be tracked in 2D video, then their positions through time form a flow field of lines in 3D. Flow lines will inevitably be individually unreliable, but the collection of flow lines—the *shape flow*, as shown in Fig. 1—is a compact and descriptive action representation.

We use a greedy scheme to compute flow lines based on iterative conditional modes (ICM) [1] to estimate the sparse point motion between adjacent frames. ICM is based on an MPEG-like local motion estimation search; it computes motion vectors that both minimize an image match cost and maximize the the motion consistency of neighboring points. We apply ICM to edge pixels that surpass a Canny detector threshold; neighborhood relations are defined by the Delaunay triangulation of these edge pixels. The resultant sparse motion field is then interpolated across Delaunay cells to produce a dense motion field. The frame-by-frame motion fields produced by this procedure are then simply concatenated to form 3D flow lines in the space-time video volume. There are no constraints to prevent flow lines from intersecting.

This method for computing flow lines is designed to produce flow lines that are good enough on average to generate a coherent flow field, since our robust matching

(a) Template shape flow (b) Template graph (c) Target video flow

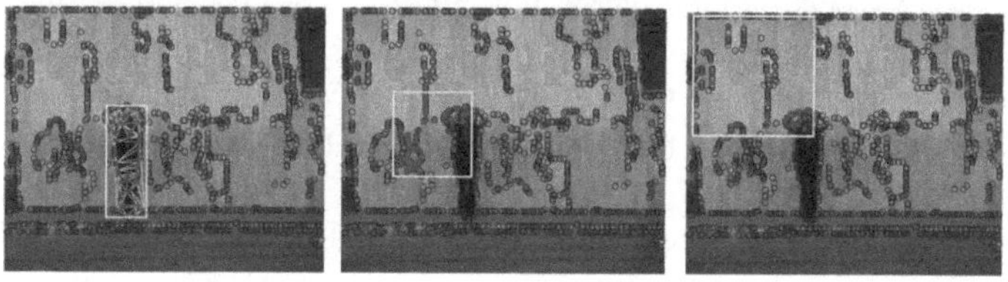

(d) This paper (e) DP (f) ICM

Fig. 2. Matching shape flows. A template shape flow (a) and its neighbor relation graph (b) are matched to a target video (c). Note that stationary points in (a) and (c) produce flow lines orthogonal to the page, which are depicted as dots. The matching result using the proposed method of this paper (d) is superior to the result achieved either by dynamic programming (e) or iterative conditional modes (f). In (d-f), the blue dots denote the start points of target candidate flow lines.

procedure tolerates flow line errors. Although the method is greedy, one can see from Fig. 1 that the flow lines are actually quite clean. More accurate but computationally more expensive methods such as [8] could be used to improve the flow line extraction.

Although the collection of flow lines that we compute is dense, for computational considerations we sample the field in both the template and the target. The spatial localization of flow lines in a shape flow is therefore somewhat imprecise. However, any localization error introduced by this sampling process is overwhelmed by variation caused by viewpoint changes and intra-class shape and motion variation. We now turn to the shape flow matching procedure.

3 Matching Shape Flows for Action Detection

In this section, we address the problem of matching a shape flow to a target video. We desire the matching to be scale invariant, operate in significant clutter, and also to be tolerant to spatial and temporal deformations caused by viewpoint variation and intra-class variation. The problem is made more difficult by the fact that the relationships between flow lines in a shape flow, which are defined by a Delaunay triangulation, form a loopy graph. We present a novel relaxation method that can accomplish effective and efficient matching using a compact linear program.

3.1 An Example

Before describing the matching procedure, we present the matching example shown in Fig. 2 in order to elucidate both the problem statement and the shape flow representation. Figs. 2(a,b) show the top-down view (projected along the time axis) of a template shape flow for a person waving both arms. Note that the shape flow consists of a collection of flow lines (a) related by a neighborhood graph (b) given by the Delaunay triangulation of the flow line start points. Fig. 2(c) shows the top-down view of the flow lines in the target video volume, which are sampled from edge pixels in the first frame. Note that the target is a different person at a different scale, and that, individually, the target flow lines differ significantly from those in the template.

The problem is to find an assignment between flow lines in the template and flow lines in the target so that (1) the paired flow lines are similar, and (2) the spatial relationships between flow lines as given by the neighbor graph are respected. Fig. 2(d) shows the matching result using the method proposed by this paper. Not only is the target detected and well localized, but the assignment of flow lines between the template and target is sensible.

Our proposed method's success relies on the combination of matching the loopy relation graph and doing a robust global search. We can see the result of removing one or the other of these elements. Fig. 2(e) shows the result of our method if the loopy graph is replaced by a chain. In this case, without cycles, the matching may be done efficiently using dynamic programming (DP). Fig. 2(f) shows the result of using the fully loopy graph, but using ICM instead of our proposed linear relaxation. The DP and ICM results are poor, despite the fact that both methods were given the advantage of having the template pre-scaled to the target scale.

3.2 The Matching Problem

The template shape flow consists of a set of flow lines. These flow lines originate from randomly selected edge points on the object. Pairwise neighbor relationships between flow lines are given by the Delaunay triangulation of the flow line start points. Thus, the neighbor graph has cycles. All of the flow lines in the template have the same temporal extent, but vary in length due to motion.

The target search domain is a space-time video volume having the same temporal extent as the template action. The search is performed over a randomly selected subset of flow lines anchored by edge points in the first frame of the volume. The goal is to find a consistent assignment of flow lines in the template to flow lines in the target. Matched flow lines should be similar, and the spatial arrangement in the target should match that of the template. We assume that there is little change in rotation between the template and target, but that there may be a large change in scale.

Let $\mathcal{S}$ denote the template point set and $\mathcal{N}$ the set of neighbor point pairs in the template. We formulate shape flow matching as the following optimization problem:

$$\min_{\mathbf{t},l} \left\{ \sum_{\mathbf{s}\in\mathcal{S}} c(\mathbf{s}, \mathbf{t}(\mathbf{s}), l) \; + \; \lambda \sum_{\{\mathbf{p},\mathbf{q}\}\in\mathcal{N}} ||(\mathbf{p}-\mathbf{q}) - (l\cdot\mathbf{t}(\mathbf{p}) - l\cdot\mathbf{t}(\mathbf{q}))|| \right\} \tag{1}$$

where $\mathbf{t}(\mathbf{s})$ maps template point $\mathbf{s}$ to the target point, and $c(\mathbf{s}, \mathbf{t}, l)$ is the cost of matching $\mathbf{s}$ and $\mathbf{t}$ with scale l. The first term in the energy function is the cost of matching individual flow lines. The second term penalizes assignments that violate the relative positions of flow lines given by the neighbor graph $\mathcal{N}$. λ weighs the relative importance

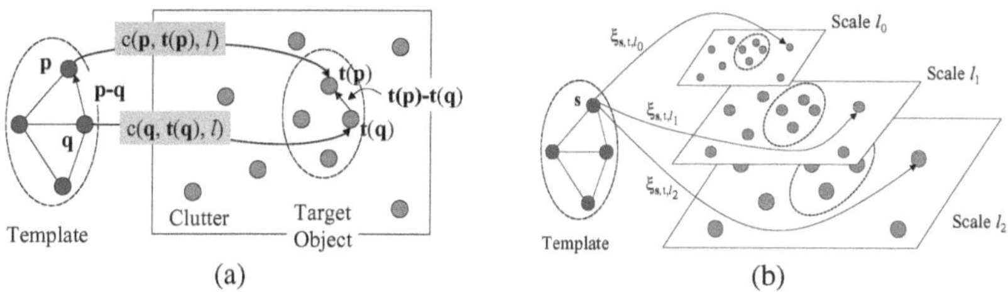

Fig. 3. (a) Matching a template shape flow to a target video frame. See Eq. 1. (b) Linearizing the problem with respect to scale. By quantizing the range of scales, the nonlinear scale factor is replaced by scale-specific indicator variables ξ. This quantization of scale is not a problem because the matching cost can tolerate some scale difference between the template and target.

of the two terms. We wish to find an assignment $\mathbf{t}(\cdot)$ and scale l so that the template shape flow matches a target shape flow with small cost in a spatially consistent manner.

The matching cost c between a pair of flow lines is computed as follows. Flow lines are normalized to a common spatial diameter (but regularized to account for stationary flow lines, which have zero diameter). Each flow line has the same number of sample points, given by the number of frames in the template. The matching cost is given simply by the Euclidean distance between the two vectors formed by the normalized flow lines' 2D spatial coordinates. Because of the spatial normalization, the flow line matching cost is scale invariant. One could use scale-specific matching costs, but we found this simpler method to be sufficient.

As stated, this optimization problem is difficult to solve. It is discrete, nonlinear, and highly non-convex. In addition, the loopy relation graph precludes efficient search. We are, however, able to solve this optimization problem effectively. In the following subsections, we show how to remove the nonlinearities, how to deal with the non-convexity, and how to dramatically reduce size of the search space, yielding a fast and small linear program solution.

3.3 Linearization

As stated, the optimization problem is a nonlinear integer program. The first step in transforming the problem into a linear program is to remove the nonlinearities. As shown in Fig. 3, we may remove the nonlinear scale factor from the energy function by searching over a range of m discrete scales $L = \{l_0, l_1, ..., l_{m-1}\}$. We use $m = 7$ scales, equally spaced between half and double the template's natural scale. Let $\xi_{\mathbf{s},\mathbf{t},l}$ be binary indicator variables that take the value 1 when template point $\mathbf{s}$ matches target point $\mathbf{t}$ at scale l. The nonlinear scale factor is replaced by the scale-specific indicator variables ξ. Quantizing scale in this manner works because the matching tolerates small differences in scale between the template and target shape flows. The first term of Eq. 1 may therefore be linearized as follows:

$$\sum_{\mathbf{s}\in\mathcal{S},\ \mathbf{t}\in\mathcal{T}(\mathbf{s}),\ l\in L} c(\mathbf{s},\mathbf{t},l)\cdot\xi_{\mathbf{s},\mathbf{t},l} \tag{2}$$

where $\mathcal{T}(\mathbf{s})$ is the target candidate point set for template point $\mathbf{s}$. In order to linearize the second term in the energy function, we use the L_1 vector norm and a standard linear program trick to linearize the absolute value function. By introducing some pairs of auxiliary variables, the second term may be linearized as follows:

$$\lambda \sum_{\{\mathbf{p},\mathbf{q}\}\in\mathcal{N}} (u^+_{\mathbf{pq}} + u^-_{\mathbf{pq}} + v^+_{\mathbf{pq}} + v^-_{\mathbf{pq}}) \tag{3}$$

where $u^+_{\mathbf{pq}}$, $u^-_{\mathbf{pq}}$, $v^+_{\mathbf{pq}}$ and $v^-_{\mathbf{pq}}$ are non-negative auxiliary variables. Let $x(.)$ and $y(.)$ be functions that extract the x and y coordinates of a point. The auxiliary variables follow the constraints $\forall\{\mathbf{p},\mathbf{q}\} \in\mathcal{N}$:

$$u^+_{\mathbf{pq}} - u^-_{\mathbf{pq}} = u_{\mathbf{p}} - u_{\mathbf{q}} - x(\mathbf{p}) + x(\mathbf{q}) \qquad v^+_{\mathbf{pq}} - v^-_{\mathbf{pq}} = v_{\mathbf{p}} - v_{\mathbf{q}} - y(\mathbf{p}) + y(\mathbf{q}) \tag{4}$$

and where $u_{\mathbf{p}}$ and $v_{\mathbf{p}}$ are x and y coordinates of the scaled target point $l \cdot \mathbf{t}(\mathbf{p})$:

$$u_{\mathbf{s}} = \sum_{\mathbf{t}\in\mathcal{T}(\mathbf{s}),\, l\in L} l \cdot x(\mathbf{t}) \cdot \xi_{\mathbf{s},\mathbf{t},l} \qquad v_{\mathbf{s}} = \sum_{\mathbf{t}\in\mathcal{T}(\mathbf{s}),\, l\in L} l \cdot y(\mathbf{t}) \cdot \xi_{\mathbf{s},\mathbf{t},l} \tag{5}$$

If the objective function is optimized given these constraints, at least one of $u^+_{\mathbf{pq}}$ or $u^-_{\mathbf{pq}}$ and at least one of $v^+_{\mathbf{pq}}$ or $v^-_{\mathbf{pq}}$ vanishes. The proof of this fact is by contradiction: If it were not the case, then we could subtract the smaller value in each pair from u^+ and v^-, producing a feasible solution at lower cost. Based on this observation, $u^+_{\mathbf{pq}} + u^-_{\mathbf{pq}} = |u_{\mathbf{p}} - u_{\mathbf{q}} - x(\mathbf{p}) + x(\mathbf{q})|$ and $v^+_{\mathbf{pq}} + v^-_{\mathbf{pq}} = |v_{\mathbf{p}} - v_{\mathbf{q}} - y(\mathbf{p}) + y(\mathbf{q})|$, and therefore $u^+_{\mathbf{pq}} + u^-_{\mathbf{pq}} + v^+_{\mathbf{pq}} + v^-_{\mathbf{pq}} = ||(\mathbf{p}-\mathbf{q}) - (l \cdot \mathbf{t}(\mathbf{p}) - l \cdot \mathbf{t}(\mathbf{q}))||$. The original energy minimization problem of Eq. 1 is now linear.

Additional constraints enforce that the matching is an assignment, where each template point matches a single target point:

$$\sum_{\mathbf{t}\in\mathcal{T}(\mathbf{s}),\, l\in L} \xi_{\mathbf{s},\mathbf{t},l} = 1,\ \xi_{\mathbf{s},\mathbf{t},l} = 0 \text{ or } 1,\ \forall \mathbf{s} \in \mathcal{S} \tag{6}$$

And we constrain each pair in the assignment to select the same scale w, so that the template scales uniformly:

$$\sum_{\mathbf{t}\in\mathcal{T}(\mathbf{s}),\, l\in L} l \cdot \xi_{\mathbf{s},\mathbf{t},l} = w,\ \forall \mathbf{s} \in \mathcal{S} \tag{7}$$

The matched target points are given by:

$$\hat{\mathbf{t}}_x = \sum_{\mathbf{t}\in\mathcal{T}(\mathbf{s}),\, l\in L} x(\mathbf{t}) \cdot \xi_{\mathbf{s},\mathbf{t},l} \qquad \hat{\mathbf{t}}_y = \sum_{\mathbf{t}\in\mathcal{T}(\mathbf{s}),\, l\in L} y(\mathbf{t}) \cdot \xi_{\mathbf{s},\mathbf{t},l} \tag{8}$$

The original nonlinear integer problem has been transformed into a linear integer program without any approximations: The linearized version solves the original problem exactly, apart from the effects of quantizing scale. The problem is still integer, however, and still non-convex. The following subsection addresses both of these issues.

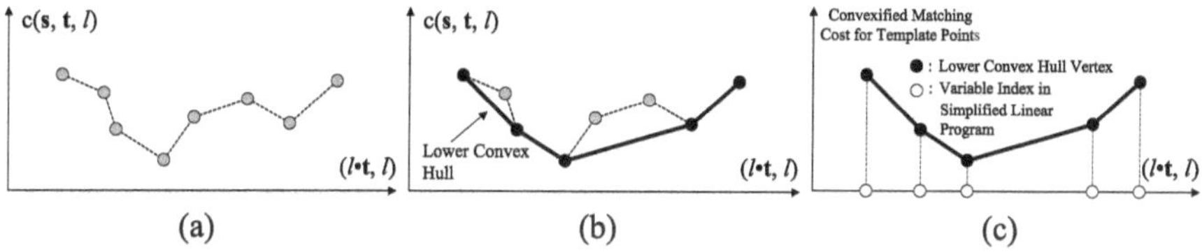

Fig. 4. Convexification using the 4D lower convex hull. The actual cost surface of the integer program is non-convex (a). Relaxing the ξ yields a linear program, the solution for which lies on the lower convex hull (b) of the original cost surface. Any variables corresponding to points above this surface are redundant in the linear program, so they may be pruned (c). In our formulation, the cost surface is 3D, and so the lower convex hull is 4D. As the dimensionality increases, pruning becomes proportionally more effective.

3.4 Relaxation and Convexification: The Lower Convex Hull Property

Due to the computational expense of integer programming, we relax the indicator variables $\xi_{\mathbf{s},\mathbf{t},l}$ so that instead of taking discrete binary values they take continuous values in the unit interval. In this now linear program, the ξ correspond to a kind of likelihood that a template flow line matches a target flow line at a particular scale. The program is very large, however: The number of variables $\xi_{\mathbf{s},\mathbf{t},l}$ is proportional to the number of template points, the number of candidate points, and the number of discrete scales. We can greatly reduce the number of ξ variables without changing the solution of the linear program by taking advantage of the following property.

Property. We need keep only those $\xi_{\mathbf{s},\mathbf{t},l}$ that correspond to the vertices of the 4D lower convex hull of the point clouds $(l \cdot \mathbf{t}, l, c(\mathbf{s}, \mathbf{t}, l))$ with respect to the last coordinate for each $\mathbf{s}$.

As depicted in Fig. 4, the solution to a linear program may be found by exploring the lower convex hull of the cost surface. Variables corresponding to interior points, depicted as gray dots in Fig. 4(b), are algebraically redundant because they may be represented as linear combinations of points on the hull. These "interior variables" may be safely removed from the linear program without affecting its solution.

This convexification procedure can reduce the size of the linear program and therefore greatly speed up its solution—by a factor of 100 or more in our experiments in this problem domain. Note that if $c(\mathbf{s}, \mathbf{t}, l)$ is convex over $(l \cdot \mathbf{t}, l)$ for each $\mathbf{s}$ in the original integer program, the relaxed linear program will find the global optimum.

3.5 Iterative Refinement

In general, the cost surface $c(\mathbf{s}, \mathbf{t}, l)$ is non-convex over the domain $(l \cdot \mathbf{t}, l)$. Because of the convexification, and because of the interactions between variables from $\mathcal{N}$, the linear relaxation gives an approximate solution. The magnitude of the approximation error depends on the size of the domain, however, because the convex approximation will be more accurate when the domain is smaller. This suggests an iterative approach known as *successive convexification* [6].

Initially, the search window, or *trust region*, covers the entire target domain: Every point in the target frame is a possible matching candidate for each template point at

(a) Sampled frames from the 1000-frame *fitness* video.

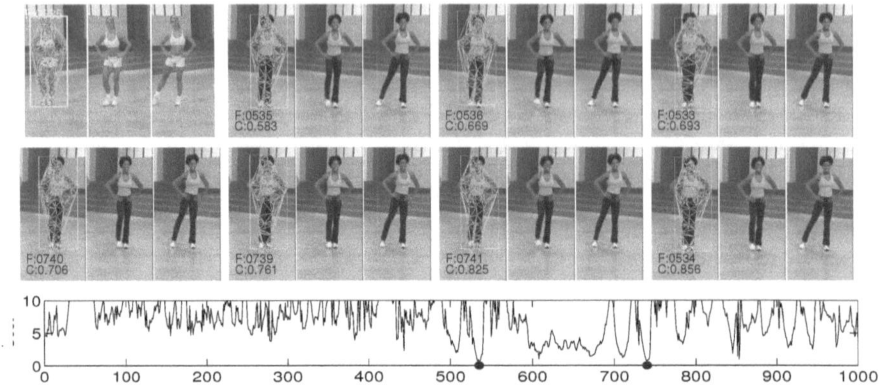

(b) Template #1, top 7 matches, and per-frame match cost.

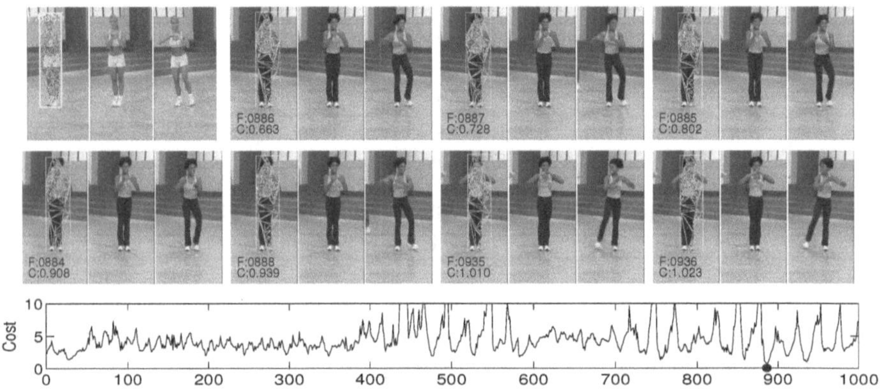

(c) Template #2, top 7 matches, and per-frame match cost.

Fig. 5. Detection of two 15-frame actions in the *fitness* video. The graphs show the per-frame match cost for the template; blue dots mark the locations of true positives. For each template and match, there are three images: (1) the first frame of the action with the template's neighbor graph, frame number F, and match cost C; (2) the second frame of the action; and (3) the last frame of the action. In (b), the top 6 matches correspond to the 2 true positives for a *right leg out* action. In (c), the top 5 matches correspond to the 1 true positive for a *right arm out, left knee out* action. In these examples, which include significant camera motion, there are no errors.

every scale. In this case, the matching will still likely find the correct target, but the convexification introduces localization errors. Given a solution, we may iteratively shrink the trust region over the domain (t, l), each time repeating the convexification and linear relaxation procedure. As the trust region shrinks, the lower convex hull fits the original cost surface more accurately, and the template is localized in the target more accurately. At most 5 iterations are required, and the program shrinks geometrically in size at each iteration.

4 Empirical Evaluation

We show experimental results for the proposed action detector on a wide range of videos and on the action dataset of Blank et al. [2]. The output of the optimization described above is the best match for a template in each frame of the target video, whether that match is good or bad. We require a method of ranking the matches by quality, across frames, to generate a top-list for action detection.

4.1 Scoring Action Matches and Ranking Results

The energy function that we minimize to find an action in a video frame is effective at locating the best match within a frame. However, the energy values cannot be meaningfully compared across frames. There are two reasons for this. First, the flow line match cost in the linear program is totally scale invariant, which is too lenient for a cross-frame match score. And second, the energy will be artificially low when the template is matched against a partially similar action; for example, a template of a person waving one arm will match well to a target waving two arms, but should not be scored high.

(a) Sampled frames from the 3501-frame *sign* video.

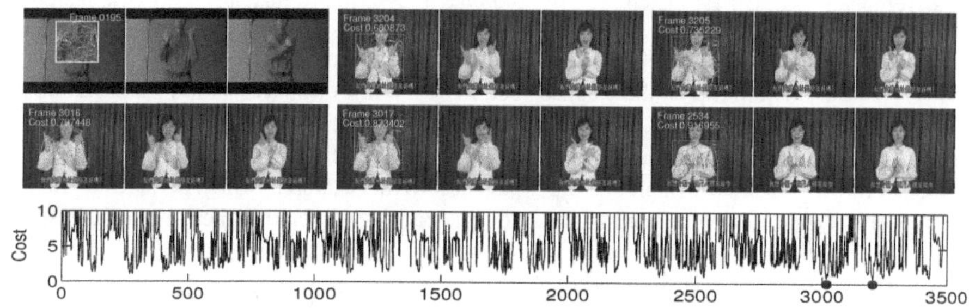

(b) Template #1, top 5 matches, and per-frame match cost.

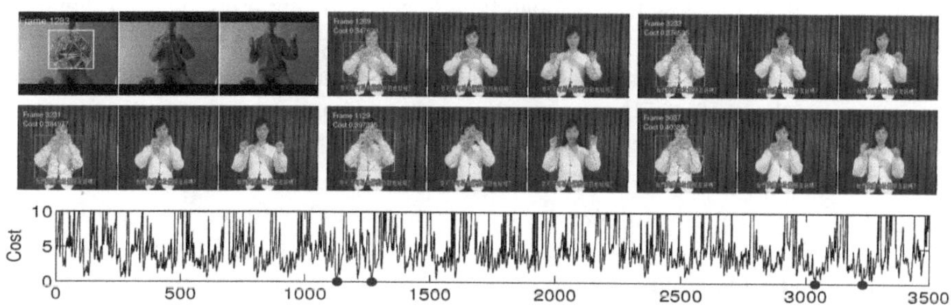

(c) Template #2, top 5 matches, and per-frame match cost.

Fig. 6. Compare with Fig. 5. In (b), the top 4 matches correspond to the 2 true positives for a 7-frame *exchange* action. In (c), the top 5 matches correspond to the 4 true positives for a 15-frame *message* action; the *message* template is partitioned into left/right halves. In these examples, which involve self-occlusion, there are no errors.

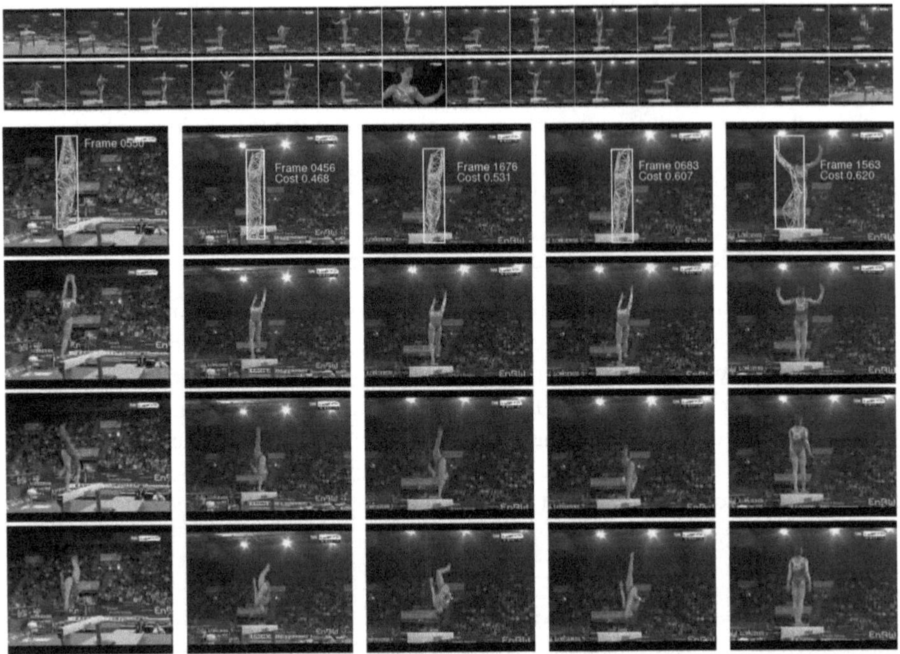

(a) Template (b) 1st match (c) 2nd match (d) 3rd match (e) 4th match

Fig. 7. Detection of a complex 15-frame action in the 2091-frame *gymnastics* video. The action is the first half or a forward flip on the balance beam. The top row of images shows frames sampled uniformly in time from the video. Column (a) shows the action template along with three frames (beginning, middle, end) from the action. Columns (b-e) show the top 4 matches. The sequence contains 3 true positives, which are the top 3 matches. In this example, which involves rapid and complex object motion as well as camera motion and background clutter, there are no errors.

To address these issues, we formulate a more robust similarity measure between the template shape flow and the matched target shape flow. First, we normalize the flow lines within each shape flow by the mean flow line length, and translate each flow line start point to the origin. The distance between these two bundles of normalized and shifted flow lines is defined as the average minimum distance between individual flow lines across bundles, which is a symmetric measure. The distance between individual flow lines is again defined as the Euclidean distance between the flow lines' spatial coordinate vectors.

As stated, this shape flow distance measure discards all information regarding the relative position of flow lines. For most actions, for ranking results, this is not a problem. However, consider the highly symmetric actions shown in Fig. 6(c). Because the flow lines on the left arm of one action will have low distance to the flow lines on the right arm of the other action, the total distance will be very low. Consequently, for such actions, we break the symmetry by partitioning the template into halves. Normalization is still performed on the entire template, but the distance is computed for each part separately. The total distance for a partitioned template is simply the average distance of the parts, weighted by the number of flow lines in each part. All of the results in this

Fig. 8. The top 30 detections for a 8-frame *swing* template action in an 8000-frame video *golf*. There are 10 hits (with duplicates) of 14 true positives and 11 false positives, yielding 71% recall and 63% precision. This is a difficult video with few true positives compared to negatives, and much camera motion.

paper use a whole single-part template; only the action in Fig. 6(c) required a (left-right) template partition.

4.2 Results

We first present action detection results for single actions in four extended videos. Fig. 5 shows results for the *fitness* video; Fig. 6 for the *sign* video; Fig. 7 for the *gymnastics* video; and Fig. 8 for the *golf* video. In all cases, the top matches are determined using the shape flow distance measure described in the previous subsection. These videos involve fast object motion, camera motion, and background clutter. In addition, the template and target are always of different people, which introduces scale variation, pose variation, and intra-class variation.

Fig. 5 shows match results for two 15-frame actions in the 1000-frame *fitness* video. All true positives from the video appear at the front of the list of top matches. Fig. 6 shows the same perfect result for the detection of two actions in the 3501-frame *sign* video. In both videos, there are multiple positives per true positive because the top list ranks matches from all frames in the video. The *fitness* video involves a moving camera and background clutter; the *sign* video involves background clutter (there are many

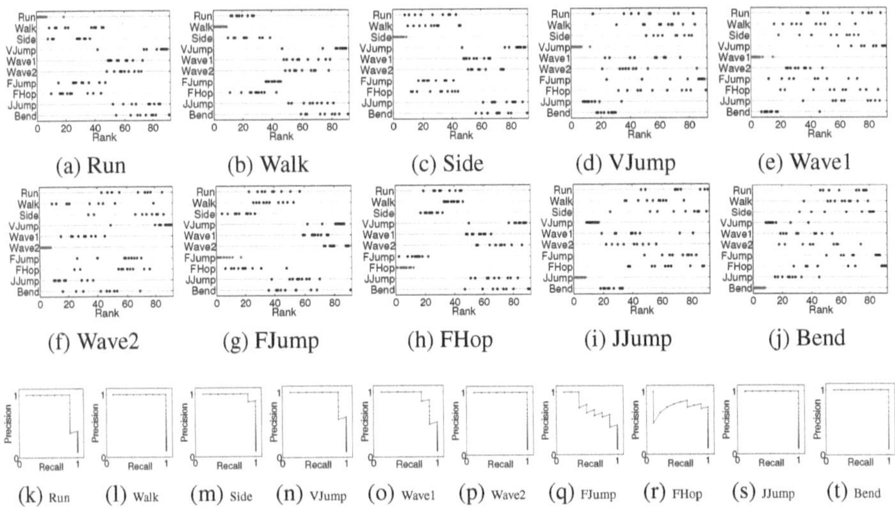

(a) Run (b) Walk (c) Side (d) VJump (e) Wave1

(f) Wave2 (g) FJump (h) FHop (i) JJump (j) Bend

(k) Run (l) Walk (m) Side (n) VJump (o) Wave1 (p) Wave2 (q) FJump (r) FHop (s) JJump (t) Bend

Fig. 9. Detecting actions in the video dataset of Blank et al. [2]. The dataset consists of 93 single action video clips for 10 actions. The 10 actions are: running (*Run*), walking (*Walk*), side stepping (*Side*), jumping in place (*VJump*), waving one arm (*Wave1*), waving two arms (*Wave2*), forward jumping (*FJump*), forward hopping (*FHop*), jumping jack (*JJump*), and bending (*Bend*). We use a single action template for each action class. Graphs (a-j) show, for each action, all 92 clips (template clip excluded) sorted by match score when matched against that action template; the y-axis places each clip into its ground truth category. Most of the same category target clips (red dots) are ranked first, which is the goal. Graphs (k-t) show the corresponding precision recall (PR) curves for the 10 actions. We achieve high precision and recall for 8 of 10 actions; the *FJump* and *FHop* actions are extremely similar visually, and panels (g,h) show that they get confused with each other.

stationary flow lines in the background) as well as complex self occlusion. Despite these challenges, the actions are detected successfully.

Fig. 7 shows match results for a complex 15-frame action in the 2091-frame *gymnastics* video. After applying non-minima suppression in the time dimension to the per-frame match scores, the three true positives appear as the top three matches. Non-minima suppression is not necessary, but prunes duplicate matches from the top list. Despite background clutter, constant camera motion, and significant intra-class variation, the proposed method is again successful.

Fig. 8 shows results for a 8-frame *swing* action in the 8000-frame *golf* video. This is the most challenging of the four videos, and demonstrates some match failures. The top 30 matches are shown over the entire sequence. There are 10 hits of 14 true positives with 11 false positives, yielding 71% recall and 63% precision (accounting for duplicate hits). This sequence involves much camera motion, a variety of individuals as targets, and highly variable background clutter. In addition, there are many distractor frames in which there is no relevant object present.

We also report results for the action recognition dataset of Blank et al. [2] in Fig. 9. The dataset consists of 93 single action video clips for 10 actions performed by various subjects. We extract a single 15-frame shape flow template for each action by randomly choosing a 15-frame sequence from a randomly chosen clip. Each template is then matched against each frame in the set of test clips (excluding the template clip). The match cost for a clip is taken as the minimum match cost across frames in the clip. Fig. 9 shows that the top matching clips have the correct class, yielding high performance precision recall curves for 8 of 10 classes; the *FJump* and *FHop* classes are not distinguished well by our detector, but that is because they are visually extremely similar. The equal precision-recall point across the dataset is 90%. Excluding the *FJump* and *FHop* categories, the equal PR point is at 95%.

5 Conclusion

We have proposed a novel representation for actions called the *shape flow*, which incorporates both the shape *and* motion of an object in an assembly of easily computed flow lines. We have formulated action detection and localization as an energy minimization problem, yielding a difficult nonlinear, non-convex integer program. We show how to linearize this energy function, and how to use successive convexification of the energy cost surface to relax the program and vastly reduce the size of the linear program without approximation. Experimental results on difficult videos and a standard action detection dataset confirm that the proposed method can accommodate complex object motion and self-occlusion, camera motion, background clutter, scale changes, pose variation, and intra-class variation. The method is non-parametric, and requires neither background subtraction nor accurate trajectories. We believe that this method is a useful tool for automatic action detection.

References

1. Besag, J.: On the Statistical Analysis of Dirty Pictures. Journal of the Royal Statistical Society, Series B 48(3), 259–302 (1986)
2. Blank, M., Gorelick, L., Shechtman, E., Irani, M., Basri, R.: Actions as Space-Time Shapes. In: ICCV (2005)
3. Bobick, A., Davis, J.: The Recognition of Human Movement Using Temporal Templates. IEEE Trans. PAMI 23(3), 257–267 (2001)
4. Efros, A., Berg, A., Mori, G., Malik, J.: Recognizing Action at a Distance. In: ICCV (2003)
5. Interrante, V., Grosch, C.: Visualizing 3D Flow. IEEE Computer Graphics and Applications 18(4), 49–53 (1998)
6. Jiang, H., Drew, M.S., Li, Z.N.: Matching by Linear Programming and Successive Convexification. IEEE Trans. on PAMI 29(6), 959–975 (2007)
7. Ke, Y., Sukthankar, R., Hebert, M.: Event Detection in Crowded Videos. In: ICCV (2007)
8. Kenwright, D., Mallinson, G.: A 3-D Streamline Tracking Algorithm Using Dual Stream Functions. Visualization, 62–68 (1992)
9. Laptev, I., Lindeberg, T.: Space-Time Interest Points. In: ICCV (2003)
10. Laptev, I., Prez, P.: Retrieving Actions in Movies. In: ICCV (2007)
11. Mori, G., Ren, X.F., Efros, A., Malik, J.: Recovering Human Body Configurations: Combining Segmentation and Recognition. CVPR, 326–333 (2004)
12. Parameswaran, V., Chellappa, R.: Human Action-Recognition Using Mutual Invariants. CVIU 98(2), 295–325 (2005)

13. Ramanan, D., Forsyth, D.A., Zisserman, A.: Tracking People by Learning Their Appearance. IEEE Trans. on PAMI 29(1), 65–81 (2007)
14. Schuldt, C., Laptev, I., Caputo, B.: Recognizing Human Actions: A Local SVM Approach. In: ICPR (2004)
15. Scovanner, P., Ali, S., Shah, M.: A 3-Dimensional SIFT Descriptor and its Application to Action Recognition. ACM Multimedia (September 2007)
16. Shechtman, E., Irani, M.: Space-Time Behavior Based Correlation. In: CVPR (2005)
17. Sheikh, Y., Shah, M.: Exploring the Space of an Action for Human Action Recognition. In: ICCV (2005)
18. Weinland, D., Ronfard, R., Boyer, E.: Free Viewpoint Action Recognition using Motion History Volumes. In: CVIU (November/December 2006)
19. Yilmaz, A., Shah, M.: Actions as Objects: A Novel Action Representation. In: CVPR (2005)
20. Yilmaz, A., Shah, M.: Recognizing Human Actions in Videos Acquired by Uncalibrated Moving Cameras. In: ICCV (2005)

Cross-View Action Recognition from Temporal Self-similarities

Imran N. Junejo, Emilie Dexter, Ivan Laptev, and Patrick Pérez

INRIA Rennes - Bretagne Atlantique
35042 Rennes Cedex - France

Abstract. This paper concerns recognition of human actions under view changes. We explore self-similarities of action sequences over time and observe the striking stability of such measures across views. Building upon this key observation we develop an action descriptor that captures the structure of temporal similarities and dissimilarities within an action sequence. Despite this descriptor not being strictly view-invariant, we provide intuition and experimental validation demonstrating the high stability of self-similarities under view changes. Self-similarity descriptors are also shown stable under action variations within a class as well as discriminative for action recognition. Interestingly, self-similarities computed from different image features possess similar properties and can be used in a complementary fashion. Our method is simple and requires neither structure recovery nor multi-view correspondence estimation. Instead, it relies on weak geometric properties and combines them with machine learning for efficient cross-view action recognition. The method is validated on three public datasets, it has similar or superior performance compared to related methods and it performs well even in extreme conditions such as when recognizing actions from top views while using side views for training only.

1 Introduction

Visual recognition and understanding of human actions has attracted much of attention over the past three decades [1,2] and remains an active research area of computer vision. A good solution to the problem holds a huge potential for many applications such as the search and the structuring of large video archives, video surveillance, human-computer interaction, gesture recognition and video editing. Previous work demonstrated the high difficulty of the problem associated with the large variation of human action data due to the individual variations of people in expression, posture, motion and clothing; perspective effects and camera motions; illumination variations; occlusions and disocclusions; and distracting effects of scenes surroundings. In addition, actions frequently involve and depend on manipulated objects adding another layer of variability. As a consequence, current methods often resort to restricted and simplified scenarios with simple backgrounds, a few simple kinematic action classes, static cameras and limited view variations.

D. Forsyth, P. Torr, and A. Zisserman (Eds.): ECCV 2008, Part II, LNCS 5303, pp. 293–306, 2008.

View variations originate from the changing and frequently unknown positions of the camera. Similar to the multi-view appearance of static objects, the appearance of actions may drastically vary from one viewpoint to another. Differently to the static case, however, the appearance of actions may also be affected by the dynamic view changes of the moving camera. Multi-view variations of actions pose substantial challenges for computer vision algorithms.

To address the multi-view problem, [3,4] employ the use of epipolar geometry. Point correspondences between actions are assumed to be known for imposing fundamental matrix constraints and performing view-invariant action recognition. [5] proposes a quasi view-invariant approach, requiring at least 5 body points lying on a 3D plane or that the limbs trace a planar area during the course of an action. Recently [6] showed that fundamental ratios can be used for view-invariant action recognition as well. However, obtaining automatic and reliable point correspondences for daily videos with natural human actions is a very challenging and currently unsolved problem prohibiting the application of above mentioned methods in practice. One alternative to the geometric approach is to represent actions by samples recorded for different views. [7,8] create a database of poses seen from multiple viewpoints. Extracted silhouettes from a test action are matched to this database to recognize the action being performed. The drawback of these methods is that each action needs to be represented by many training samples recorded for a large and representative set of views. Another method [9] performs a full 3D reconstruction from silhouettes seen from multiple deployed cameras. This approach requires a setup of multiple cameras or training on poses obtained from multiple views, which again restricts the applicability of methods in practice.

In this work we address view-independent action recognition from a different perspective and avoid many assumptions of previous methods. In contrast to the geometry-based methods above we require neither identification of body parts nor the estimation of corresponding points between video sequences. Differently to the previous view-based methods we do not assume multi-view action samples neither for training nor for testing.

Our approach builds upon self-similarities of action sequences over time. For a given action sequence we compute distances between action representations for all pairs of time-frames and store results in a Self-Similarity Matrix (SSM). We claim SSMs to be approximately invariant under view changes of an action. The intuition behind this claim is the following. If body poses of an action are similar at moments t_1, t_2, the value of $\mathrm{SSM}(t_1, t_2)$ will be low for any view of that action. On the contrary, if the body poses are different at t_1, t_2, the value of $\mathrm{SSM}(t_1, t_2)$ is likely to be large for most of the views. Fig. 1 illustrates this idea with an example of a golf swing action seen from two different views. For this example we compute SSMs using distances of points on the hand trajectory illustrated in Fig. 1(a,c). Close trajectory points **A**, **B** remain close in both views while the distanced trajectory points **A** and **C** have large distances in both projections. The visualizations of SSMs computed for both sequences in Fig. 1(b,d) have a striking similarity despite the difference in the projections of the action.

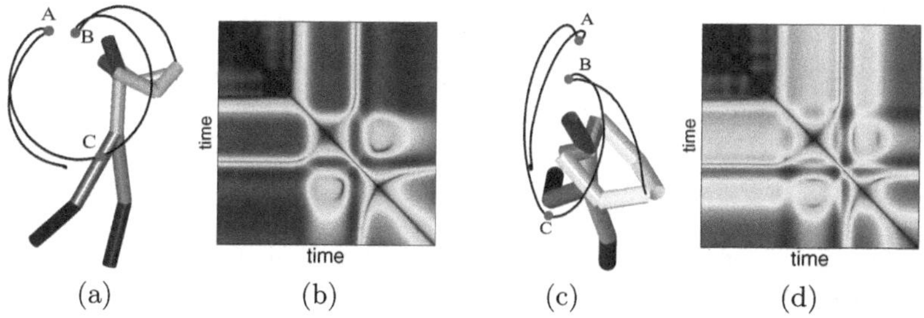

Fig. 1. (a) and (c) demonstrate a golf swing action seen from two different views, and (b) and (d) represent their computed self-similarity matrices (SSM), respectively. Even though the two views are different, the structure or the patterns of the computed SSMs is very similar.

In the rest of the paper we operationalize SSMs for human action sequences and deploy SSMs for view-independent action recognition. In particular, we observe similar properties of SSMs computed for different image features and use such SSMs in a complementary fashion. The rest of the paper is organized as follows. In the next section we review related work. Section 2 gives a formal definition of SSM using alternative image feature. Section 3 describes the representation and training of action sequences based on SSMs. In Section 4 we test the method on three public datasets and demonstrate the practicality and the potential of the proposed method. Section 5 concludes the paper.

1.1 Related Work

The methods most closely related to our approach are that of [10,11,12,13]. Recently for image and video matching [10] explored *local* self-similarity descriptors. The descriptors are constructed by correlating the image (or video) patch centered at a pixel to its surrounding area by the sum of squared differences. The correlation surface is transformed into a binned log-polar representation to form a local descriptor used for image and video matching. Differently to this method, we explore the structure of similarities between *all* pairs of time-frames in a sequence. The main focus of our work is on the use of self-similarities for view-invariant action recognition which was not addressed in [10].

Our approach has a closer relation to the notion of video self-similarity used by [11,12]. In the domain of periodic motion detection, Cutler and Davis [12] track moving objects and extract silhouettes (or their bounding boxes). This is followed by building a 2D matrix for the given video sequence, where each entry of the matrix contains the absolute correlation score between the two frames i and j. Their observation is that for a periodic motion, this similarity matrix will also be periodic. To detect and characterize the periodic motion, they use the Time-Frequency analysis. Following this, [11] use the same construct of the self-similarity matrix for gait recognition in videos of walking people. The periodicity of the gait creates diagonals in the matrix and the temporal symmetry of the gait cycles are

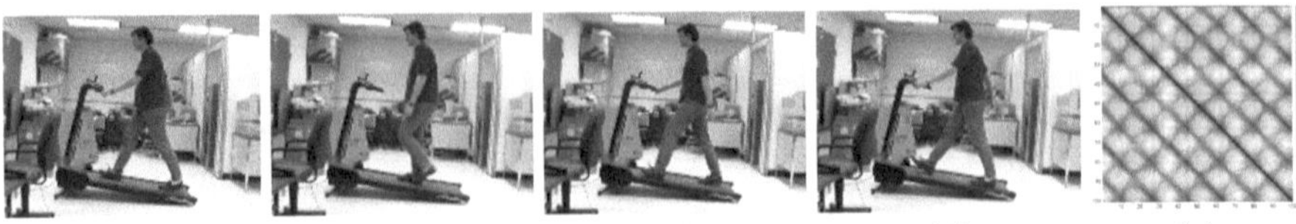

(a) (b) (c) (d) (e)

Fig. 2. (a)-(d) are four images from a sequence of a walking person. (e) represents the SSM obtained for this sequence by [12].

represented by the cross-diagonals. In order to compare sequences of different length, the self-similarity matrix is subdivided into small units. Both of these works focus primarily on videos of walking people for periodic motion detection and gait analysis. The method in [13] also concerns gait recognition using temporal similarities between frames of different image sequences. None of the methods above explores the notion of self-similarity for view-invariant action recognition.

2 Self-Similarity Matrix (SSM)

In this section we define self-similarity matrices for different image features and illustrate SSMs computed for several action classes and multiple views.

For a sequence of images $\mathcal{I} = \{\mathcal{I}_1, \mathcal{I}_2, \ldots \mathcal{I}_T\}$ in discreet (x, y, t)-space the square symmetric distance matrix $\mathcal{D}(\mathcal{I})$ in $\mathbb{R}^{T\times T}$ is defined as an exhaustive table of *distances* between image features taken by pair from the set $\mathcal{I}$:

$$\mathcal{D}(\mathcal{I}) = [d_{ij}]_{i,j=1,2,\ldots,T} = \begin{bmatrix} 0 & d_{12} & d_{13} & \ldots & d_{1T} \\ d_{21} & 0 & d_{23} & \ldots & d_{2T} \\ \vdots & \vdots & \vdots & & \vdots \\ d_{T1} & d_{T2} & d_{T3} & \ldots & 0 \end{bmatrix} \tag{1}$$

where d_{ij} represents a distance between the frames $\mathcal{I}_i$ and $\mathcal{I}_j$. The diagonal corresponds to comparing a frame to itself, hence, always zero. The exact structure or the patterns of $\mathcal{D}(\mathcal{I})$ depends on the features and the distance measure used for computing the entries d_{ij}. For example, after tracking walking people in a video sequence, [11,12] compute d_{ij} as the absolute correlation between two frames, an example of which is shown in Fig. 2. The computed matrix patterns (cf. Fig. 2(e)) have a significant meaning for their application - the diagonals in the matrix indicate periodicity of the motion.

In this work, to compute d_{ij}, we use the Euclidean distance to measure the distance between the different features that we extract from an action sequence. This form of $\mathcal{D}(\mathcal{I})$ is then known in the literature as the Euclidean Distance Matrix (EDM)[14].

Before describing the features that we use, some words about the importance of matrix $\mathcal{D}$ are in order. From morphometrics and isometric reconstruction

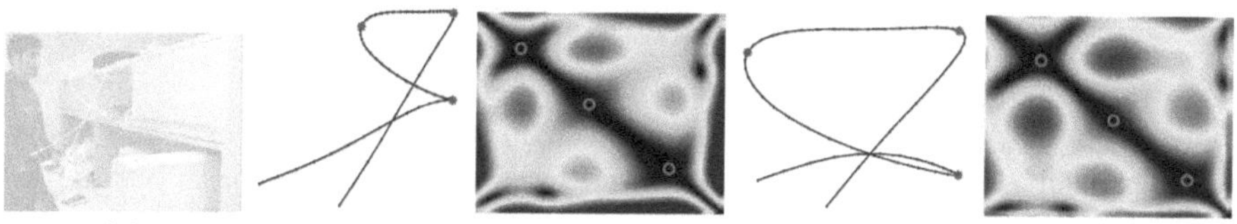

(a) (b) (c) (d) (e)

Fig. 3. Comparison of the proposed SSM with [16]: Two actors perform the action of opening a cabinet door from different viewpoints, where the hand trajectory is shown in (b) and (d). The computed SSM for these two actions are shown in (c) and (e), respectively. The dynamic instances (as proposed by [16]), marked in red '*' in (b) and (d), represent valleys in the corresponding SSM, depicted by magenta circle in (c) and (e), respectively. The spread of each valley depends on the peak-width of the corresponding dynamic instance.

to non-linear dimensionality reduction, this matrix has proven to be a useful tool for a variety of applications. For example, Isomap [15], a popular non-linear dimensionality reduction method, starts by computing distances between all pairs of images. These computed distances represent an adjacency matrix where each image represents a node in the graph.

To get an intuitive understanding of the proposed method, a comparison of SSM with the notion of "dynamic instances" proposed by Rao et al.[16] is illustrated in Fig. 3. The authors of [16] argue that continuities and discontinuities in position, velocity and acceleration of a 3D trajectory of an object are preserved under 2D projections. For an action of opening a cabinet door, performed by two different actors from considerably different viewpoints, these points are depicted in Fig. 3. Fig. 3(c)(e) shows the SSMs computed for these two actions, where red color indicates higher values and dark blue color indicates lower values. The dynamic instances, red '*' in Fig. 3(b)(d), correspond to valleys of different area/spread in our plot of SSM (cf. Fig. 3(c)(e)), marked by magenta circles along the diagonal of the matrix. The exact spread of these valleys depend on the width of the peaks in the spatio-temporal curvature of the actions, as shown in Fig. 3(b)(d). However, whereas [16] captures local discontinuities in the spatio-temporal curvature, the SSM captures more information about other dynamics of the actions represented in the off-diagonal parts of the matrix. We will argue that for actions recorded from different viewpoints the patterns of SSM are stable and discriminative.

SSMs are fairly robust, handle missing (or noisy) data robustly, and are fairly easy to compute [14]. The computation of SSM is flexible in the sense that we can choose to compute SSMs from a variety of different features depending on the available data. Below we describe some of the features we use to compute the SSM.

2.1 Trajectory-Based Self-similarities

If a subset of points distributed over the person body are tracked over some time, the mean Euclidean distance for k pairs of points at any two frames of the sequence can be computed as

$$d_{ij} = \frac{1}{k} \sum_{k} \|x_i^k - x_j^k\|_2 \qquad (2)$$

where x^k, x_j^k indicate positions of points on the track k at frames i and j. We denote the self-similarity matrix computed from (2) by SSM-pos. In our experiments with motion capture dataset, we track 13 joints on a person performing different actions [17], as shown in the Fig. 4(a). In order to remove the effect of translation, without loss of generality, the points are centered to their centroid so that their first moments are zero. The remaining scale normalization is achieved by $\mathbf{x}_i = \frac{\mathbf{x}_i'}{\|\mathbf{x}_i'\|}$, where $\mathbf{x}_i'$ represent the joints being tracked in frame i and $\mathbf{x}_i$ represent their normalized coordinates.

In addition to the SSM-pos, we also compute similarities based on the first and the second derivative of the 2D positions, i.e. the velocity and the acceleration features. Similarities computed by these features are denoted by SSM-vel and SSM-acc, respectively.

In this work we assume point tracks to be provided by an external module such as KLT point tracker. Our method is not restricted to any particular subset of points as far as the points are distributed over moving body parts. SSMs can be accumulated from any number of tracks with arbitrary length and starting time.

2.2 Image-Based Self-similarities

Next to the trajectories, alternative image features can be used to construct additional $\mathcal{D}(\mathcal{I})$ for the same image sequence. To describe spatial appearance of a person at each image frame we compute Histograms of Oriented Gradients (HoG) features [18]. This descriptor, originally used to perform human detection, characterizes the local shape by capturing the gradient structure. In our implementation, we use 4 bin histograms for each of 5×7 blocks defined on a bounding box around the person in each frame. d_{ij} is then computed as the Euclidean distance between two HoG vectors corresponding to the frames $\mathcal{I}_i$ and $\mathcal{I}_j$. We denote SSMs computed using HoG features by SSM-hog.

In addition to HoG features, we also test the proposed method by considering the estimated optical flow vector as an input feature. The optical flow is computed by Lucas and Kanade method [19] on person-centered bounding boxes using two consecutive frames. We consider v_x and v_y components of optical flow and define three features by concatenating responses of v_x, v_y and (v_x,v_y) into descriptor vectors. The SSMs computed for these three features are denoted as SSM-ofx, SSM-ofy and SSM-of respectively. We measure d_{ij} using Euclidean distance between the flow descriptor vectors corresponding to the two frames $\mathcal{I}_i$ and $\mathcal{I}_j$. In practice, we enlarge and resize bounding boxes in order to avoid border effects on the flow computation and ensure the same size of the flow vectors along an action sequence. We resize the height to a value equal to 150 pixels and the width is set to the greatest value for the considered sequence.

Examples of SSMs computed for different image features are shown in Fig. 4. Fig. 4(a) contains example actions from the CMU motion capture (mocap)

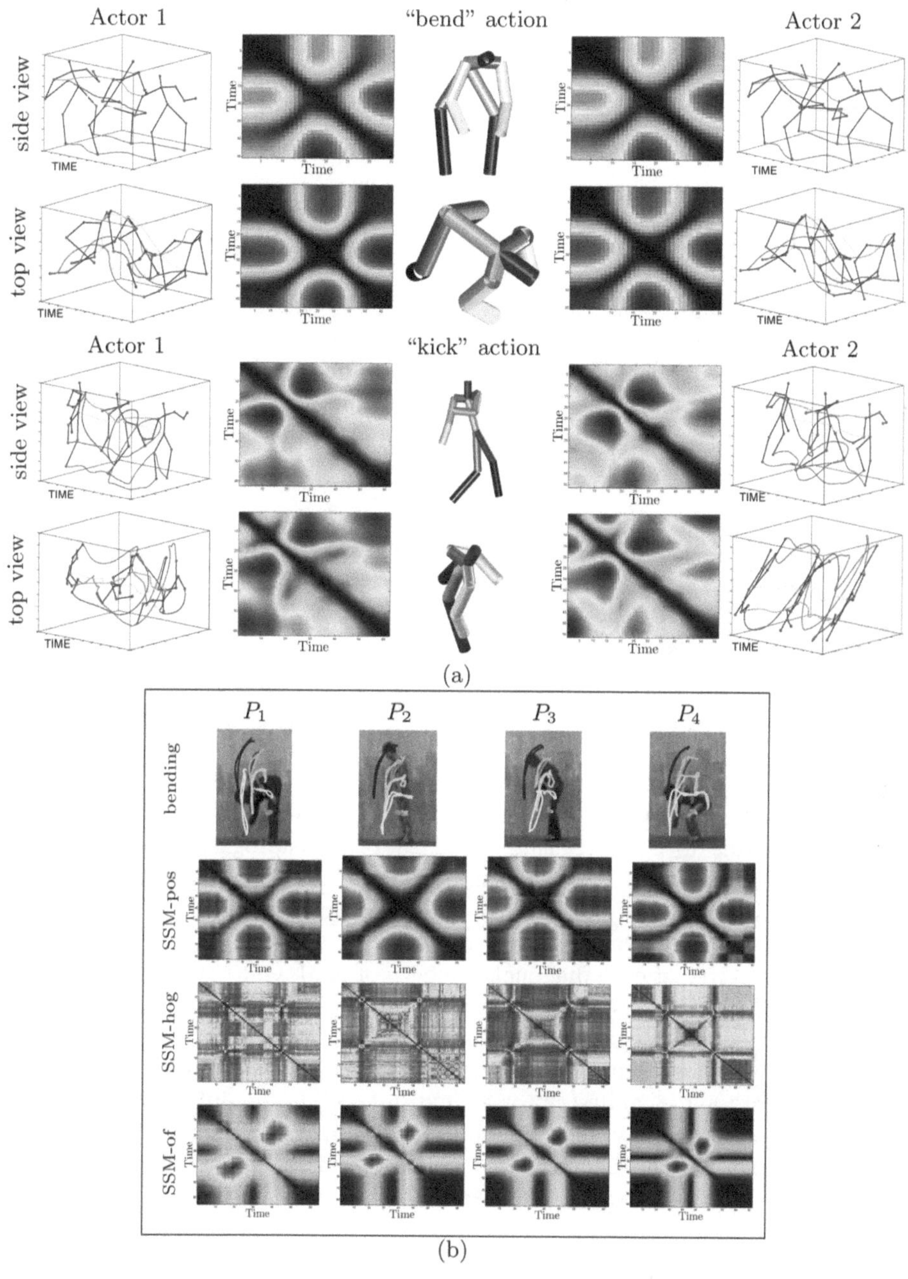

Fig. 4. (a) Examples from CMU mocap dataset. Columns 1 and 5 represent two actors while columns 2 and 4 represent corresponding SSM-pos, respectively. Different rows represent different actions and viewing angles. Note the stability of SSMs over different views and persons performing the same action. (b) Examples from Weizman video dataset [20]. Rows 2, 3, 4 represent SSM-pos, SSM-hog and SSM-of respectively for four bending actions in row 1. Note the similarity column-wise.

dataset projected onto different views. Column 1 and 5 of Fig. 4(a) represent two different actors while columns 2 and 4 represent their computed SSM-pos, respectively. The first two rows represent a bending action performed by two actors and projected onto two considerably different views. The last two rows, similarly, represent a football kick action for two actors and two different views. Note the similarity of SSMs computed for actions of the same class despite the changes of the actor and the considerable changes of views. Note also the visual difference of SSMs between two action classes. Fig. 4(b) illustrates SSMs obtained for the bending action from the video dataset [20]. Row 2 shows SSM-pos computed using point tracks illustrated in the row 1. Rows 3 and 4 show SSM-hog and SSM-of for the same sequences respectively. Note the similarity of SSMs for the same feature type and for the different action instances. SSMs for different feature types do not look similar since different features capture different properties of the action. This suggests the use of SSMs computed for different features in a complementary manner.

3 SSM-Based Action Description and Recognition

As illustrated in the previous section, SSMs have view-stable and action-specific structure. Here we aim to capture this structure and to construct SSM-based descriptors for subsequent action recognition. We pay attention to the following properties of SSM: (i) absolute values of SSM may depend on the variant properties of the data such as the projected size of a person in the case of SSM-pos; (ii) changes in temporal offsets and time warping may effect the global structure of SSM; (iii) the uncertainty of values in SSM increases with the distance from the diagonal due to the increasing difficulty of measuring self-similarity over long time intervals; (iv) SSM is a symmetric positive semidefinite matrix with zero-valued diagonal.

Due to (ii)-(iii) we choose a local representation and compute patch-based descriptors centered at each diagonal element i of SSM. Our patch descriptor has a log-polar block structure as illustrated in Fig. 5. For each of the 11 descriptor blocks j we compute 8-bin histogram of SSM gradient directions within a block and concatenate the normalized histograms into a descriptor vector h_i. When constructing a joint local descriptor for multiple SSMs computed for k different features, we concatenate k corresponding descriptors from each SSM into a single vector. The representation for a video sequence is finally defined by the sequence of local descriptors $H = (h_1, ..., h_n)$ computed for all diagonal elements of SSM.

3.1 Temporal Multi-view Sequence Alignment

Before addressing action recognition, we validate our representation on the problem of multi-view sequence alignment. We consider two videos recorded simultaneously for the side and the top views of a person in action as shown in Fig. 6(a). To further challenge the alignment estimation, we apply a nonlinear time transformation to one of the sequences. To solve alignment, we (i) compute SSM-of

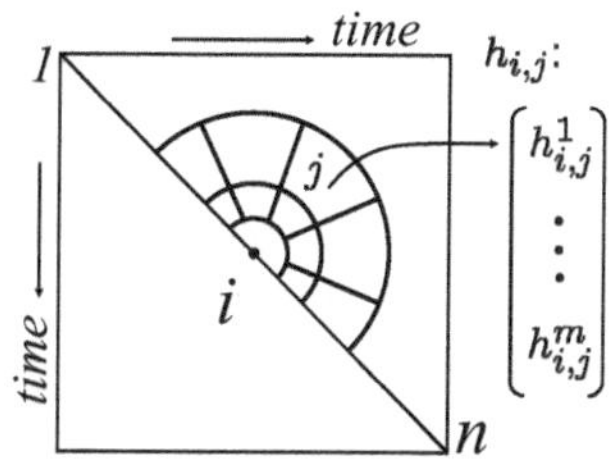

Fig. 5. Local descriptors for SSM are centered at every diagonal point $i = 1...n$ and have log-polar block structure. Histograms of m gradient directions are computed separately for each of 11 blocks j and are concatenated into a descriptor vector h_i.

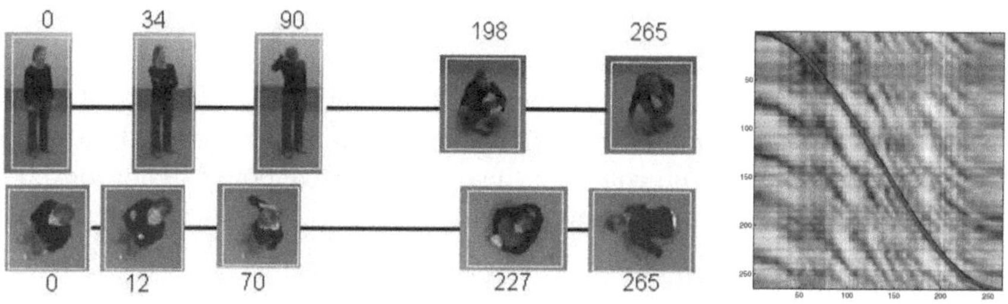

(a) (b)

Fig. 6. Temporal sequence alignment. (a): Two sequences with the side and the top views of the same action are represented by corresponding key-frames. The lower sequence has been time warped according to $t' = a\cos(bt)$ transformation. (b): Alignment of two sequences in (a) using SSM-based action descriptions and Dynamic Programming (red curve) recovers the original warping (blue curve) almost perfectly despite substantial view variations.

for both image sequences, (ii) represent videos by the sequences of local SSM descriptors H^1, H^2 as described above, (iii) and finally align sequences H^1 and H^2 by Dynamic Programming. The estimated time transformation is illustrated by the red curve in Fig. 6(b) and does almost perfectly recover the ground truth transformation (blue curve) despite the drastic view variation between image sequences.

3.2 Action Recognition

To recognize action sequences we follow recently successful bag-of-features approaches [21,22] and represent each video as a bag of local SSM descriptors H. We then apply either Nearest Neighbour Classifier (NNC) or Support Vector Machines (SVM) to train and classify instances of action classes. In the case of NNC, we assign a test sequence H_{tst} with the label of a training sequence H^i_{tr} with $i = \text{argmin}_j D_{NN}(H_{tst}, H^j_{tr})$ minimizing the distance over all training sequences. The distance D_{NN} is defined by the greedy matching of local descriptors as described in [21]. We apply NNC to datasets with a limited number of samples.

For SVMs we construct histograms of visual words and use them as input for SVM training and classification according to [23]. Visual vocabulary is obtained by k-means clustering of 10000 local SSM descriptors h from the training set into $k = 1000$ clusters. Each feature is then assigned to the closest (we use Euclidean distance) vocabulary word and the histogram of visual words is computed for each image sequence. We train non-linear SVMs using χ^2 kernel and adopt one-against-all approach for multi-class classification.

For all recognition experiments in the next section we report results for n-fold cross-validation and make sure the actions of the same person do not appear in the training and in the test sets simultaneously.

4 Experimental Results

In this section we evaluate SSM-based action descriptors for the task of multi-view action recognition. The first experiment in Section 4.1 aims to validate the approach in controlled multi-view settings using motion capture data. In Section 4.2 we demonstrate and compare the discriminative power of our method on the standard single-view action dataset [20]. We finally evaluate the performance of the method on the comprehensive multi-view action dataset [9] in Section 4.3.

4.1 Experiments with CMU MoCap Dataset

To simulate multiple and controlled view settings we have used 3D motion capture data from CMU dataset (*http://mocap.cs.cmu.edu*). Trajectories of 13 points on the human body were projected to six cameras with pre-defined orientation with respect to the human body (see Fig. 7(a)). We have used 164 sequences in total corresponding to 12 action classes. To simulate potential failures of the visual tracker we also randomly subdivided trajectories into parts with the average length of 2 seconds. Fig. 7(b) demonstrates results of NNC action recognition when training and testing on different views using SSM-pos, SSM-vel and SSM-acc. As observed from the diagonal, the recognition accuracy is the highest when training and testing on the same views while the best accuracy (95.7%) is achieved for cam5 (frontal view). Interestingly, the recognition accuracy changes slowly with substantial view changes and remains high across top and side views. When training and testing on all views, the average accuracy is 90.5%.

4.2 Experiments with Weizman Actions Dataset

To asses the discriminative power of our method on real video sequences we apply it to the standard single-view video dataset with nine classes of human actions performed by nine subjects [20](see Fig. 8(top)). On this dataset we compute NNC recognition accuracy when using either image-based self-similarities in terms of SSM-of-ofx-ofy-hog or trajectory-based SSM. Given the low resolution of image sequences in this dataset, the trajectories were acquired by [17] via semi-automatic tracking of body joints. Recognition accuracy achieved by our

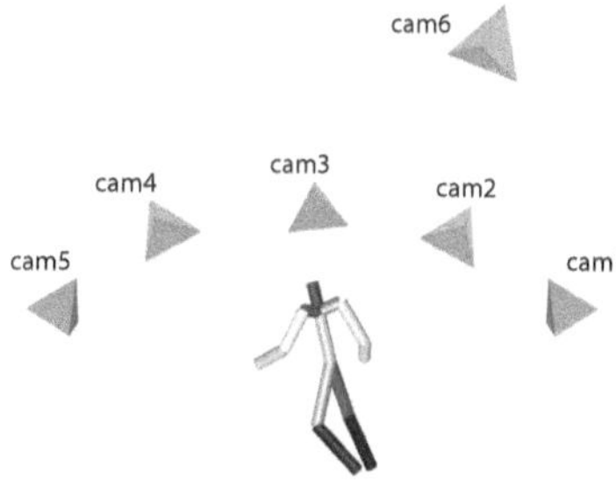

(a)

test views / training views

	cam1	cam2	cam3	cam4	cam5	cam6	All
cam1	**92.1**	89.0	76.2	71.3	73.2	84.8	81.1
cam2	87.2	**92.7**	83.5	72.6	64.6	78.7	79.9
cam3	78.7	83.5	**89.0**	90.9	67.7	61.0	78.5
cam4	78.0	75.6	88.4	**90.9**	72.6	63.4	78.2
cam5	81.1	73.8	76.8	83.5	**95.7**	80.5	81.9
cam6	86.0	88.4	73.8	76.2	78.0	**91.5**	82.3
All	90.9	90.2	87.8	90.9	92.7	90.9	**90.5**

(b)

Fig. 7. CMU dataset. (a): A person figure animated from the motion capture data and six virtual cameras used to simulate projections in our experiments. (b): Accuracy of the cross-view action recognition using SSM-pos-vel-acc.

	bend	jack	jump	pjump	run	side	skip	walk	wave
bend	100.0	0.0	0.0	0.0	0.0	0.0	0.0	0.0	0.0
jack	0.0	100.0	0.0	0.0	0.0	0.0	0.0	0.0	0.0
jump	0.0	0.0	77.8	0.0	0.0	0.0	11.1	11.1	0.0
pjump	0.0	0.0	0.0	100.0	0.0	0.0	0.0	0.0	0.0
run	0.0	0.0	0.0	0.0	100.0	0.0	0.0	0.0	0.0
side	0.0	0.0	0.0	0.0	0.0	100.0	0.0	0.0	0.0
skip	0.0	0.0	20.0	0.0	10.0	0.0	70.0	0.0	0.0
walk	0.0	0.0	0.0	0.0	0.0	0.0	0.0	100.0	0.0
wave	0.0	0.0	0.0	0.0	0.0	0.0	0.0	0.0	100.0

(a)

	bend	jack	jump	pjump	run	side	skip	walk	wave
bend	100.0	0.0	0.0	0.0	0.0	0.0	0.0	0.0	0.0
jack	0.0	100.0	0.0	0.0	0.0	0.0	0.0	0.0	0.0
jump	0.0	0.0	66.7	11.1	0.0	11.1	11.1	0.0	0.0
pjump	0.0	0.0	0.0	100.0	0.0	0.0	0.0	0.0	0.0
run	0.0	0.0	0.0	0.0	100.0	0.0	0.0	0.0	0.0
side	0.0	0.0	0.0	0.0	0.0	100.0	0.0	0.0	0.0
skip	0.0	0.0	12.5	0.0	0.0	0.0	87.5	0.0	0.0
walk	0.0	0.0	0.0	0.0	0.0	0.0	0.0	100.0	0.0
wave	0.0	0.0	0.0	0.0	0.0	0.0	0.0	0.0	100.0

(b)

Fig. 8. (Top): Example frames for Weizman action dataset [20] with image sequences for nine classes of actions. (a)-(b) confusion matrices corresponding to NNC action recognition using image-based self-similarities (a) and trajectory-based self-similarities (b).

method for image-based and trajectory-based self-similarities is 94.6% and 95.3% respectively and the corresponding confusion matrices are illustrated in Fig. 8(a)-(b). The recognition results are high for both types of self-similarity descriptors and outperform 92.6% achieved in by a recent trajectory-based method in [17]. Higher recognition rates on the same dataset have been reported e.g. in [24].

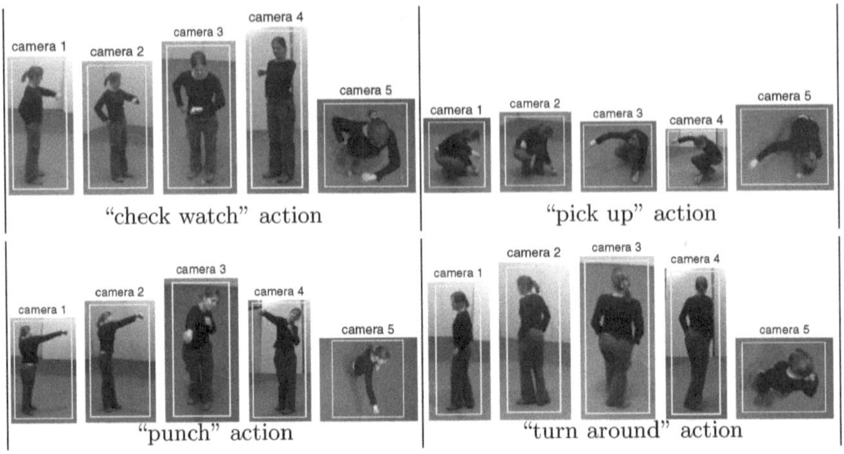

Fig. 9. Example frames for four action classes and five views of the IXMAS dataset

(a) test views (rows: training views)

training views	cam1	cam2	cam3	cam4	cam5	All
cam1	**76.4**	77.6	69.4	70.3	44.8	67.2
cam2	77.3	**77.6**	73.9	67.3	43.9	67.4
cam3	66.1	70.6	**73.6**	63.6	53.6	65.0
cam4	69.4	70.0	63.0	**68.8**	44.2	63.9
cam5	39.1	38.8	51.8	34.2	**66.1**	45.2
All	74.8	74.5	74.8	70.6	61.2	**72.7**

(a)

	check-watch	cross-arms	scratch-head	sit-down	get-up	turn-around	walk	wave	punch	kick	pick-up
check-watch	83.3	0.0	0.7	1.3	0.7	1.3	8.0	0.7	0.0	0.0	4.0
cross-arms	0.0	94.0	2.0	1.3	0.7	0.7	0.0	0.7	0.0	0.0	0.7
scratch-head	0.0	0.0	68.7	2.0	9.3	2.0	1.3	4.7	10.0	2.0	0.0
sit-down	0.7	4.7	3.3	55.3	1.3	20.0	3.3	0.7	10.7	0.0	0.0
get-up	2.0	3.3	7.3	0.7	59.3	0.7	0.0	23.3	2.7	0.7	0.0
turn-around	3.3	1.3	0.0	27.3	0.0	56.7	3.3	2.0	2.7	0.0	3.3
walk	10.0	0.7	0.0	2.7	0.7	2.7	68.7	1.3	1.3	0.0	12.0
wave	3.3	0.7	6.7	2.0	14.7	0.0	0.7	63.3	8.7	0.0	0.0
punch	0.7	0.0	6.0	6.0	0.7	2.7	0.0	1.3	74.0	8.7	0.0
kick	0.0	0.0	0.0	0.0	0.0	0.0	0.0	0.0	0.0	100.0	0.0
pick-up	2.0	0.0	0.0	2.7	0.7	4.7	13.3	0.7	0.0	0.0	76.0

(b)

	All-to-All
hog	57.8%
of	65.9%
of+ofx+ofy	66.5%
of+hog	71.9%
of+hog+ofx+ofy	**72.7%**

(c)

	cam1	cam2	cam3	cam4	cam5
This paper	**76.4%**	**77.6%**	**73.6%**	**68.8%**	**66.1%**
Weinland et al. [9] 3D	65.4%	70.0%	54.3%	66.0%	33.6%
Weinland et al. [9] 2D	55.2%	63.5%	—	60.0%	—

(d)

Fig. 10. Results for action recognition on IXMAS dataset. (a): Recognition accuracy for cross-view training and testing. (b): confusion matrix for action recognition in “all-training all-testing” setting. (c): relative performance of combined self-similarity descriptors. (d): Comparison with [9] for “camN-training camN-testing” setup.

4.3 Experiments with IXMAS Dataset

We finally present results for IXMAS video dataset [9] with 11 classes of actions performed three times by each of 10 actors and recorded simultaneously from 5 different views. Sample frames for all cameras and four action classes are illustrated in Fig. 9. Given the relatively large number of training samples, we apply SVM classification to image-based self-similarity descriptors in terms of

SSM-oh-ofx-ofy-hog. Fig. 10(a) illustrates recognition accuracy for cross-view training and testing. Similar to results on CMU dataset in Section 4.1, here we observe high stability of action recognition over view changes, now using visual data only. The method achieves reasonable accuracy even for top views when using side-views for training only. Fig. 10(c) illustrates recognition scores for different types of self-similarities and their combinations. We can observe the advantage of SSM-of over SSM-hog, however, the best results are achieved when combining self-similarities for several complementary features. In comparison to other methods, our method outperforms both 2D and 3D based recognition methods in [9] for all test scenarios as shown in Fig. 10(d). We may add that our method relies on the rough localization and tracking of people in the scene and, hence, relies on weaker assumptions compared to [9] that uses human silhouettes.

5 Conclusion

We propose a self-similarity based descriptor for view-independent action recognition. Experimental validation on several datasets using different types of self-similarities clearly confirms the stability of our approach to view variations. The proposed method does not rely on the structure recovery nor on the correspondence estimation, but makes only mild assumptions about the rough localization of a person in the frame. This lack of strong assumptions is likely to make our method applicable to action recognition beyond controlled datasets when combined with the modern techniques for person detection and tracking. We plan to investigate this direction in the future work.

References

1. Moeslund, T., Hilton, A., Krüger, V.: A survey of advances in vision-based human motion capture and analysis. CVIU 103, 90–126 (2006)
2. Wang, L., Hu, W., Tan, T.: Recent developments in human motion analysis. Pattern Recognition 36, 585–601 (2003)
3. Yilmaz, A., Shah, M.: Recognizing human actions in videos acquired by uncalibrated moving cameras. In: Proc. ICCV, pp. I:150–157 (2005)
4. Syeda-Mahmood, T., Vasilescu, M., Sethi, S.: Recognizing action events from multiple viewpoints. In: Proc. EventVideo, pp. 64–72 (2001)
5. Parameswaran, V., Chellappa, R.: View invariance for human action recognition. IJCV 66, 83–101 (2006)
6. Shen, Y., Foroosh, H.: View invariant action recognition using fundamental ratios. In: Proc. CVPR (2008)
7. Li, R., Tian, T., Sclaroff, S.: Simultaneous learning of nonlinear manifold and dynamical models for high-dimensional time series. In: Proc. ICCV (2007)
8. Ogale, A., Karapurkar, A., Aloimonos, Y.: View-invariant modeling and recognition of human actions using grammars. In: Proc. W. on Dyn. Vis., pp. 115–126 (2006)
9. Weinland, D., Boyer, E., Ronfard, R.: Action recognition from arbitrary views using 3D exemplars. In: Proc. ICCV (2007)
10. Shechtman, E., Irani, M.: Matching local self-similarities across images and videos. In: Proc. CVPR (2007)

11. Benabdelkader, C., Cutler, R., Davis, L.: Gait recognition using image self-similarity. EURASIP J. Appl. Signal Process 2004, 572–585 (2004)
12. Cutler, R., Davis, L.: Robust real-time periodic motion detection, analysis, and applications. PAMI 22, 781–796 (2000)
13. Carlsson, S.: Recognizing walking people. In: Vernon, D. (ed.) ECCV 2000. LNCS, vol. 1842, pp. 472–486. Springer, Heidelberg (2000)
14. Lele, S.: Euclidean distance matrix analysis (EDMA): Estimation of mean form and mean form difference. Mathematical Geology 25, 573–602 (1993)
15. Tenenbaum, J., de Silva, V., Langford, J.: A global geometric framework for non-linear dimensionality reduction. Science 290, 2319–2323 (2000)
16. Rao, C., Yilmaz, A., Shah, M.: View-invariant representation and recognition of actions. IJCV 50(2), 203–226 (2002)
17. Ali, S., Basharat, A., Shah, M.: Chaotic invariants for human action recognition. In: Proc. ICCV (2007)
18. Dalal, N., Triggs, B.: Histograms of oriented gradients for human detection. In: Proc. CVPR, pp. I: 886–893 (2005)
19. Lucas, B., Kanade, T.: An iterative image registration technique with an application to stereo vision. In: Image Understanding Workshop, pp. 121–130 (1981)
20. Gorelick, L., Blank, M., Shechtman, E., Irani, M., Basri, R.: Actions as space-time shapes. PAMI 29, 2247–2253 (2007)
21. Laptev, I., Caputo, B., Schüldt, C., Lindeberg, T.: Local velocity-adapted motion events for spatio-temporal recognition. CVIU 108, 207–229 (2007)
22. Niebles, J., Wang, H., Li, F.: Unsupervised learning of human action categories using spatial-temporal words. In: Proc. BMVC (2006)
23. Marszałek, M., Schmid, C., Harzallah, H., van de Weijer, J.: Learning object representations for visual object class recognition. In: The PASCAL VOC 2007 Challenge Workshop, in conjunction with ICCV (2007)
24. Ikizler, N., Duygulu, P.: Human action recognition using distribution of oriented rectangular patches. In: Workshop on Human Motion, pp. 271–284 (2007)

Window Annealing over Square Lattice Markov Random Field

Ho Yub Jung, Kyoung Mu Lee, and Sang Uk Lee

Department of EECS, ASRI, Seoul National University, 151-742, Seoul, Korea
hoyub@diehard.snu.ac.kr, kyoungmu@snu.ac.kr, sanguk@ipl.snu.ac.kr

Abstract. Monte Carlo methods and their subsequent simulated annealing are able to minimize general energy functions. However, the slow convergence of simulated annealing compared with more recent deterministic algorithms such as graph cuts and belief propagation hinders its popularity over the large dimensional Markov Random Field (MRF). In this paper, we propose a new efficient sampling-based optimization algorithm called WA (Window Annealing) over squared lattice MRF, in which cluster sampling and annealing concepts are combined together. Unlike the conventional annealing process in which only the temperature variable is scheduled, we design a series of artificial "guiding" (auxiliary) probability distributions based on the general sequential Monte Carlo framework. These auxiliary distributions lead to the maximum a posteriori (MAP) state by scheduling both the temperature and the proposed maximum size of the windows (rectangular cluster) variable. This new annealing scheme greatly enhances the mixing rate and consequently reduces convergence time. Moreover, by adopting the integral image technique for computation of the proposal probability of a sampled window, we can achieve a dramatic reduction in overall computations. The proposed WA is compared with several existing Monte Carlo based optimization techniques as well as state-of-the-art deterministic methods including Graph Cut (GC) and sequential tree re-weighted belief propagation (TRW-S) in the pairwise MRF stereo problem. The experimental results demonstrate that the proposed WA method is comparable with GC in both speed and obtained energy level.

1 Introduction

In vision problems, optimization or energy minimization techniques are widely used over various models. In areas such as segmentation and stereo vision, the minimum energy states are often posed as Bayesian solutions. And especially for large pairwise MRF problems, a variety of methods were introduced with notable impacts. Particularly, algorithms such as GC and loopy belief propagation (BP) allow a very accurate estimation of maximum a posteriori (MAP) states with reasonable computation time [1,2]. However, GC and BP have certain limitations over the size of cliques, computational speed, continuous state space, global constraints and non-submodular energy functions [3,4,5,6,7]. These issues are becoming more important in the vision community as we build more complex

D. Forsyth, P. Torr, and A. Zisserman (Eds.): ECCV 2008, Part II, LNCS 5303, pp. 307–320, 2008.

models. Researchers are now devising variants of GC and BP by modifying and combining several techniques for a more robust optimization scheme. However, building a BP and GC based algorithm that can overcome all of these drawbacks does not seem an easy task.

On the other hand, Monte Carlo methods, such as Gibbs sampler and SW cut, are free from most of these aforementioned issues [8]. Monte Carlo simulated annealing is one of the first methods introduced for MRF energy minimization and Gibbs sampler, data-driven MCMC (DDMCMC), window Gibbs (W-Gibbs) and Swendsen-Wang cut (SW-cut) were shown to be effective sampling techniques [9,10,11,8]. Combined with the annealing process, they can even optimize MRF with larger cliques as long as the state's probability density is known up to a normalizing constant. The only drawback is their unacceptable lengthy time to converge compared to GC and BP. Because of its slow convergence rate, the Monte Carlo methods have lost their popularity over problems with very high dimensional distributions. Thus, by today's standards, they are often disregarded as under-performing alternates to GC and BP.

A fast Monte Carlo based algorithm, then, can deal with all the drawbacks of GC and BP while achieving comparable energy minimization. In order to build Monte Carlo algorithms that can compete with GC and BP in speed of minimization, we deride away from the Metropolis-Hasting framework of DDMCMC sampler and SW-cut. Instead we examine the applicability of sequential Monte Carlo (SMC), which is a general Monte Carlo technique over any sequentially varying distributions (auxiliary distributions) [12,13]. In this paper, the annealing process will be seen as a special case of SMC over temperature varying distributions [13]. Being such, we explore varying distribution according not only to temperature, but also to cluster size, and introduce the window annealing (WA) scheme.

The proposed algorithm takes full advantage of the fact that it is not essential for intermediate auxiliary distributions to be built with reversible Markov chains that converge to a known distribution. A cluster scheme can be made simpler and faster for Markov chains over auxiliary distributions, unlike the SW-cut that has a more computationally extensive clustering scheme in order to satisfy the detailed balance constraint of the chains. However, by alleviating this constraint, we can greatly reduce computation time while still maintaining the advantages of cluster sampling. For a 4-neighbor pairwise smooth stereo energy function, the proposed WA scheme runs as fast as the state-of-the-arts energy minimization algorithms such as sequential tree re-weighted BP (TRW-S) and GC, producing a comparable low energy state. It also outperforms the existing MCMC-based methods including SW-cut and Gibbs sampling. Moreover, we will show that the proposed WA method can be applied effectively to the minimization of the continuous state MRF problem, and global constraint problems.

The next section will reformulate the reversible MCMC and traditional annealing framework that will be the basis of the proposed algorithm. Section 3 will introduce the fast Pseudo Window Metropolis Sampling and Pseudo Window Gibbs Sampling schemes for the auxiliary distributions and we will show

how they can be formulated into a Window Annealing scheme. Furthermore, a faster computation of the energy functions of candidate states will be discussed by proposing integral image techniques. In the experiment section, the performance of the proposed WA scheme will be compared to GC and BP for the pairwise MRF stereo problem. Few examples for continuous state problem and global constraint energy minimization will also be presented.

2 Related Works

A brief introduction to MCMC and simulated annealing methods will be addressed in this section which will mirror the proposed WA scheme. First, we will describe the square lattice structure of digital images for a pixelwise labelling problem. A deterministic scan reversible MCMC scheme will be described to sample from a target distribution. In the final subsection, simulated annealing will be presented as a special case of SMC and we will explain how the reversibility in an annealing process can be compromised.

2.1 Square Lattice MRF

A typical digital image is consist of pixels that are arranged in a squared grid structure. Thus, a pixel-wise labelling vision problem is usually formulated by a square lattice type undirected graphical model, denoted by $G_\circ =< V, E >$ with node V and edge E. And a label l in a discrete or continuous label set L is assigned to each vertex $v \in V$, producing a state (labeling) x.

If we have N nodes and Q labels, then the number of all possible states derived from this set up will be $P = Q^N$. The goal is to find the state that minimizes the associated energy φ among $\{x_1, x_2, ...x_P\}$. This problem is equivalent to finding the MAP state of the following Gibbs distribution.

$$\pi(x) = \frac{1}{Z} e^{-\varphi(x)}, \tag{1}$$

where Z is the partition function or the normalizing constant.

2.2 Reversible MCMC for Target Distributions

Sampling high dimension distribution (1) usually requires some form of MCMC such as Gibbs or SW-cut. In this section we will describe a simple MCMC scheme that will maintain the distribution (1) with enough samples. A basic reversible ergodic MCMC that converges to a desired distribution can be constructed through the Metropolis-Hasting acceptance rule [14,15]. In addition to satisfying the irreducibility and aperiodicity properties, if samples are filtered by the following acceptance probability, the Markov chain satisfies the detailed balance condition, enforcing the sample distribution to converge toward the target distribution $\pi(x)$.

$$\mathrm{A}(x', x) = \min \left\{ 1, \frac{\pi(x')q(x|x')}{\pi(x)q(x'|x)} \right\}. \tag{2}$$

This is the acceptance probability of going from state x to x'. $q(x|x')$ is the proposal probability of going to state x from state x' before the acceptance probability is applied.

There are numerous ergodic MCMC schemes that apply the above acceptance rule. Although more complex methods such as SW-cut or DDMCMC can be employed, simple deterministic scan Metropolis and Gibbs samplers suffice in our framework to provide the final distribution at the end of annealing.

The following describes the Deterministic Scan Metropolis Sampler (DSMS):

Deterministic Scan Metropolis Sampler: DSMS

1. Repeat for specified number of samples,
2. For $i = 1$ to $i = N$ repeat for current state x.
3. From an uniform distribution, pick a label l'.
4. Assign l' to v_i with following acceptance probability, $\mathrm{A} = \min\left(1, \frac{\pi(x')}{\pi(x)}\right)$.

In the above MCMC, the next state is determined by assigning a label from a uniform distribution to the currently chosen vertex. Here, the proposal probability can be written as $q(x|x') = \alpha\beta$, where α is a constant probability of choosing the vertex and β is also a constant probability of choosing a particular label. Therefore, the acceptance probability in (2) becomes $\mathrm{A} = \min\left(1, \frac{\pi(x')}{\pi(x)}\right)$.

Note that, however, one of the drawbacks of the DSMS is the inefficiency due to its high rejection sampling rate. In contrast, by taking samples from conditional probability, the Gibbs sampling method can eliminate the rejection step in DSMS [9] as follows.

Deterministic Scan Gibbs Sampler: DSGS

1. Repeat for specified number of samples,
2. For $i = 1$ to $i = N$.
3. Sample $v_i \sim \pi(v_i|v_1 ... v_{i-1}, v_{i+1} ... v_N)$.

However, in general, sampling from a continuous distribution can be problematic with Gibbs sampler, unless we have an easy to sample distribution such as mixture of Gaussian.

2.3 Simulated Annealing

The goal of simulated annealing is to estimate the MAP state of a target distribution (1) by introducing a series of auxiliary distributions that are controlled by temperature such that

$$\pi_0(x) = \lim_{T \to 0} \pi_T(x) = \lim_{T \to 0} \frac{1}{Z} e^{-\varphi(x)/T}. \tag{3}$$

The greater than zero temperature parameter T is adjusted to create a flatter version of the target distribution and slowly cool to zero or near zero temperature [16]. The MAP state is obtained by sampling π_T sequentially.

In many simulated annealing strategies, the Markov chain transition kernel $K_T(x'|x)$ at time T is assumed to satisfy the following detailed balance equation,

$$\pi_T(x)K_T(x'|x) = \pi_T(x')K_T(x|x'), \tag{4}$$

for all states x and x', and the Markov chain then converges to the desired distribution π_T at temperature T [14,15]. The Metropolis Hasting rule can be applied to enforce the detailed balance of the Markov chains as in the previous subsection.

However, note that the simulated annealing process as a whole does not satisfy the detailed balance condition because as the temperature changes, the distribution also changes. And the detailed balance between two different distributions cannot be satisfied. This means that upholding detailed balance during an annealing process has little advantage. Thus, by disregarding the detailed balance, we could design simpler and faster MCMC kernels that are better suited for estimating the MAP state (or the minimum energy state).

3 Window Annealing (WA)

Another way to interpret simulated annealing is to regard it as a single particle SMC sampling with temperature varying inhomogeneous MCMC transition kernels [13]. SMC's transition kernels need not to be reversible, Markov or even ergodic [17,18]. Consequently, it is not necessary for SMC to have MCMC kernels that converge to a known distribution or even to an invariant distribution. Thus, for the auxiliary MCMC kernels, the detailed balance condition (4) is a coincidental but unnecessary condition for a general SMC and annealing framework. We can hope to build better "guiding" distributions and sampling kernels toward the $\pi_0(x)$ by incorporating other auxiliary variables in addition to temperature.

In this section we introduce MCMC transition kernels that mimic the Metropolis sampling rule. Here, the key idea is to come up with an MCMC scheme that is fast mixing and assume a similar shape as the target distribution, but not necessarily converge to a known function because the detailed balance of Markov chain is not being enforced. The gained freedom allows more flexibility in designing the MCMC. Specifically, we can design a cluster sampling scheme in the same spirit as SW-cut but without the tedious formulation of reversibility and detailed balance.

3.1 Non-reversible Pseudo Metropolis Sampling

The modified versions of Metropolis and Gibbs sampling schemes are introduced in this subsection. First, by combining window-based cluster sampling strategy and DSMS, we can design the pseudo Window Metropolis sampler (PWMS) as follows.

Pseudo Window Metropolis Sampler: PWMS

1. Given $\pi(x)$ and the maximum window size n.
2. Repeat for specified number of samples,

3. From an uniform distribution, pick height h, width w and position shifts of window cluster (w_s, h_s) s.t.;
$(w, h) \in \{0, 1, ..., n\} \times \{0, 1, ..., n\}$,
$(w_s, h_s) \in \{0, 1, ..., w-1\} \times \{0, 1, ..., h-1\}$.
4. Using h_s, w_s, h, and w divide the graph (image) into subregions such that the vertices are divided into m rectangular window clusters, $V = \{V_1, V_2, ..., V_m\}$. See Figure 1.
5. Repeat for $j = 1$ to $j = m$, for current state x.
6. From an uniform distribution, pick a label l' from L.
7. Assign l' to V_j with following acceptance probability, $\mathrm{A} = \min\left(1, \frac{\tilde{\pi}(x')}{\tilde{\pi}(x)}\right)$, where $\tilde{\pi}(x) = Z\pi(x)$

The above PWMS algorithm clusters the graph (image) into rectangular grids under maximum integer size n, and it samples using the acceptance rule similar to the Metropolis scheme. However, violation of the reversibility occurs when the current state has inhomogeneous labels for V_j and the next state has homogenous labels for V_j. Therefore, the transition probability of going back from the current state to the previous state is zero since the sampling move is done only through homogenous labeling, i.e. $V_j = l'$. Without satisfying the reversibility and the detailed balanced condition (4), PWMS will not converge to the distribution $\pi(x)$.

Nevertheless, the PWMS scheme will converge to an invariant yet unknown distribution, since it satisfies the irreducibility and aperiodicity constraints [19]. The chain can visit every other state since it involves the 1×1 window sampling, which is a simple single node sampler. With always positive values of $\tilde{\pi}(\cdot)$, the acceptance probability, $\min(1, \tilde{\pi}(x')/\tilde{\pi}(x))$ is always positive. Thus, it can be asserted that there is a positive probability of visiting all states in finite steps and there is a positive probability of jumping out of any cycle.

Similarly to DSGS, the rejection sampling problem of PWMS can be eliminated by replacing the Metropolis sampling step with the Gibbs sampling. We call it as pseudo Window Gibbs sampler (PWGS). In PWGS, all the nodes in a selected window cluster can have either homogeneous labels or current labels, that is V_j can be assigned among $Q + 1$ possible states, where Q is the number of labels.

Pseudo Window Gibbs Sampler: PWGS

1. Given $\pi(x)$ and the maximum window size n.
2. Repeat for specified number of samples,
3. From an uniform distribution, pick height h, width w and position shifts of window cluster (w_s, h_s) s.t.;
$(w, h) \in \{0, 1, ..., n\} \times \{0, 1, ..., n\}$,
$(w_s, h_s) \in \{0, 1, ..., w-1\} \times \{0, 1, ..., h-1\}$.
4. Using h_s, w_s, h, and w divide the graph (image) into subregions such that the vertices are divided into m rectangular window clusters, $V = \{V_1, V_2, ..., V_m\}$. See Figure 1.

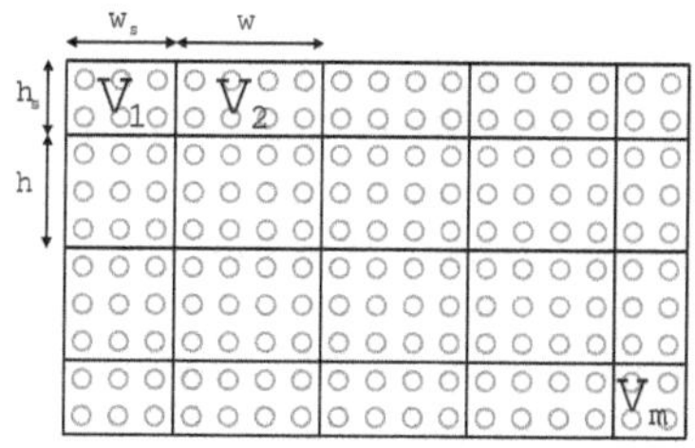

Fig. 1. An example of window clustering of a square lattice MRF model. Using the four random variables (w, h, w_s, h_s), the MRF is clustered into window shapes $V = \{V_1, V_2, ..., V_m\}$.

5. Repeat for $j = 1$ to $j = m$, for current state x.
6. Make the following possible next candidate states,
 $x = (V_1, ..., V_{j-1}, V_j, ..., V_m)$, the current state.
 $x_1 = (V_1, ..., V_{j-1}, V_j = l_1, ..., V_m)$.
 $x_2 = (V_1, ..., V_{j-1}, V_j = l_2, ..., V_m)$.
 ...
 $x_Q = (V_1, ..., V_{j-1}, V_j = l_Q, ..., V_m)$.
7. Sample next state from following discrete distribution.
 $\{\frac{1}{Z_c}\tilde{\pi}(x), \frac{1}{Z_c}\tilde{\pi}(x_1), ..., \frac{1}{Z_c}\tilde{\pi}(x_Q)\}$, where $\tilde{\pi}(x) = Z\pi(x)$ and
 $Z_c = \tilde{\pi}(x) + \sum_{i=1}^{Q} \tilde{\pi}(x_i)$.

3.2 Window Annealing Algorithm

In this subsection, we introduce the WA energy minimization algorithm. The key idea of WA is to design efficient auxiliary guiding distributions toward $\pi_0(x)$ in (3) by incorporating the temperature and maximum window size variables in the annealing process. The following pseudo codes represent the proposed WA algorithm in which samples are drawn from auxiliary distributions composed of series of PWMS and PWGS-based Markov chains.

Window Annealing: WA

1. Given $\pi(x)$, specify the schedules for the maximum window size n and temperature T such that
 $\mathcal{T} = \{T_k, T_{k-1}, ..., T_2, T_1, T_0\}$ where $T_i \geq 0$, $T_0 = 0$.
 $\mathcal{N} = \{n_k, n_{k-1}, ..., n_2, n_1, n_0\}$ where $n_i \geq 1$, $n_0 = 1$.
 Initialize, $i = k$, and state x.
2. Repeat while $i \geq 0$.
3. Sample next state by PWMS or PWGS with $\pi_{T_i}(x)$ and maximum window size n_i. Pass the final configuration x to the next iteration.
4. Reduce $i = i - 1$.

Here, we assume that if the maximum window sizes n_i are similar, the sample distributions will be similar as well. Therefore, the sequence of n_i is reduced by one for a gradual change toward the target distribution. In contrast to the previous annealing schemes where only the temperature variable was changed over the distribution, in the proposed WA, the window size scheduling controls the non-reversible Markov chain kernels for the intermediate distributions. Notice that if all n_i are set to one, then WA becomes same as the conventional simulated annealing with Metropolis rule.

3.3 Implementation of WA for Pairwise MRF

Most of the computational load of WA comes from the calculation of the probability density in step 7 of PWMS or PWGS. The efficient computation of the proportional distribution $\tilde{\pi}_{T_i}(\cdot)$ is an important issue for computational speed up. Thus, we propose to use the integral image technique (sum area tables) [20] to calculate the energy function that is associated with the proposal state probabilities $\tilde{\pi}_{T_i}(x')$ for a significant computation reduction. The rest of the subsection explains the efficient aggregation of energy functions in a pairwise MRF, but it can be easily extended to higher order clique MRFs.

If $\tilde{\pi}_{T_i}(x')$ of PWMS and $\tilde{\pi}_{T_i}(x_1), ..., \tilde{\pi}_{T_i}(x_Q)$ of PWGS were to be calculated naively, the energy function $\varphi(x')$ and $\varphi(x_1), ...\varphi(x_Q)$ have to be calculated with the computational complexity $O(N)$ where N is the number of nodes. However, since $\tilde{\pi}_{T_i}(x')$ and $\tilde{\pi}_{T_i}(x_1), ..., \tilde{\pi}_{T_i}(x_Q)$ are only required to be calculated up to a normalizing constant, actual computational load is much less than that. When states x' has the same label assignment as the current state x except for the window cluster V_j, then $\varphi(x')$ can be represented as $c + e' + \varphi(V_j = l')$, where c is the energy of node set $(V \setminus V_j)$. e' is the pairwise clique potentials between nodes in V_j and $(V \setminus V_j)$. Thus, $\tilde{\pi}_{T_i}(x')$ can be calculated in $O(N_v)$ where N_v is a number of nodes in V_j cluster. A more general description for different sized cliques are given in [11].

Furthermore, if the nodes are arranged in a square lattice structure, $\varphi(V_j = l')$ can be obtained efficiently from a precalculated integral images of uniformly labelled states $\varphi(V = l_1), ..., \varphi(V = l_Q)$. The unary and pairwise costs in V_j can be aggregated in $O(1)$ using integral images and the pairwise clique cost e' can be added on together. And by applying the equivalent approach, the calculation of PWGS's $\tilde{\pi}_{T_i}(x_1), ..., \tilde{\pi}_{T_i}(x_Q)$ can also be reduced by same order. However, for obtaining the current state energy $\varphi(x)$, the computational complexity will still be $O(N_v)$ since the integral images cannot be pre-made for each accepted state.

4 Experiments

The comparisons of energy minimization techniques have been frequently studied over the pairwise MRF stereo model [8,21,22]. Thus, it is natural that the same test should be a good indicator of comparative performance of the proposed algorithm. The proposed WA has been tested over the pairwise MRF stereo images

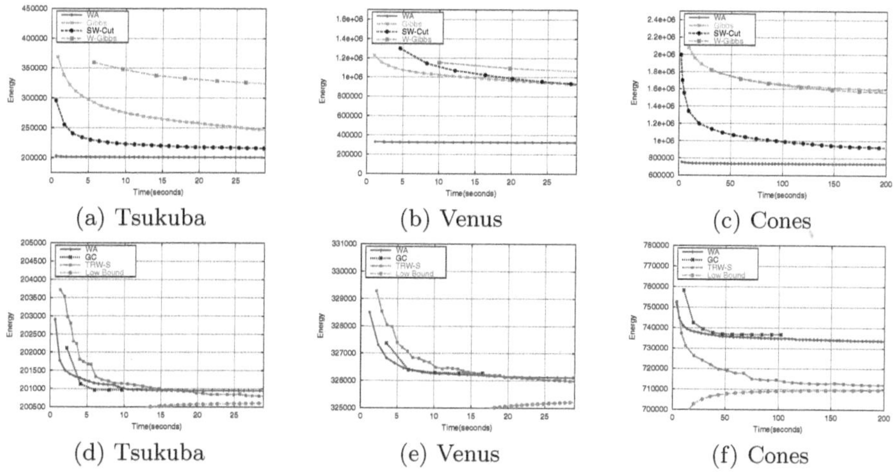

(a) Tsukuba (b) Venus (c) Cones

(d) Tsukuba (e) Venus (f) Cones

Fig. 2. Energy versus time graphs for Tsukuba, Venus, and Cones stereo pairs respectively using Potts, truncated linear and truncated quadratic discontinuity model. (a) (b) and (c) compare Gibbs sampler, W-Gibbs, and SW-cut with proposed window annealing (WA). (d), (e), and (f) compares α expansion GC and TRW-S with WA. The comparison between simulated annealing and deterministic methods are separated for better viewing.

provided by [23,24,25]. We compared the energy versus time with well-known GC (α expansion), TRW-S, Gibbs sampler, window-Gibbs (W-Gibbs) and SW-cut. GC and TRW-S were tested using the programs provided by [22,2,26,27,28,29]. The lower bound of TRW-S indicates the lowest energy reachable with the given target distribution. The three Monte Carlo based methods, Gibbs sampler, W-Gibbs and SW-cut were implemented based on their respective papers [9,11,8]. To the best of our knowledge, we are the first to implement W-Gibbs sampler over pairwise MRF stereo. Although the efficiency of W-Gibbs has not improved over the traditional Gibbs sampler, we included it on the comparative test because of its operative similarity to WA. Comparative graphs for Monte Carlo methods were presented separately from GC and TRW-S for better visual comparison.

We also implemented WA on a continuous state MRF stereo problem, where the disparity label for a pixel is a real number between the *max* and *min* disparities. Such continuous state MRF stereo has been proposed in the non-parametric BP article [30], but was not applied to real/complete stereo pairs. To the best of our knowledge, WA is the first effective algorithm applied to the continuous MAP-MRF stereo problem.

Finally, energy function with global constraints were implemented using WA. These types of energy functions cannot be solved with α expansion and message passing algorithms. A simple binary segmentation problem is used to demonstrated the flexibility of proposed method over general energy functions. All computations were on a 3.4GHz desktop.

4.1 Pairwise Discrete MRF Stereo

The matching cost for the stereo pair was calculated by the Birchfield and Tomasi cost over grey scale [31]. The Potts model, truncated linear and truncated quadratic models were used for the smoothness cost. The energy function for pairwise stereo MRF is shown below.

$$\varphi(x) = \sum_{p \in V} D(p) + \sum_{(p,q) \in N} V(p,q). \tag{5}$$

The matching cost $D(p)$ is a unary clique potential. The horizontal and vertical pairwise cliques consist of the smoothness constraint, denoted as $V(p,q)$. Energy is minimized with PWGS based WA and the energy versus time (seconds) graphs are constructed as shown in Figure 2.

The initial states of all Monte Carlo based methods were uniformly set to zero disparity. The temperature scheduling for Gibbs sampler, W-Gibbs and SW-cut were constantly remained equal to 1. We found that such scheduling minimized the energy fastest during our test period (30 to 200 seconds). On the other hand, the temperature scheduling for the WA algorithm was constantly set to zero with no instances of high temperature. The high temperature seemed to hinder WA from reaching a lower energy in a short period. Zero temperature scheduling provided the fastest convergence rate within the first few seconds. In our window size scheduling, the initial maximum size of the window n_k was set to be 50 and reduced gradually by 1 until it reached 1. The size was then increased to 50 again and decreased down by one as before. This was repeated in a series of cycles for a predetermined iteration.

Figure 2 shows the comparative evaluation (energy vs. time) results on Tsukuba, Venus and Cones stereo images using Potts, truncated linear, and truncated quadratic model, respectively. We can observe that the proposed WA method exhibits the best performance among the Monte Carlo based methods as depicted in Figures 2 (a)-(c). Although it is an accepted view in the previous studies that the sampling-based methods are generally many times slower than GC or BP-based methods [8,2], our results in Figures 2 (d)-(f) show that the performance of the proposed sampling-based WA method is comparable to those of GC (α expansion) and TRW-S. This is a noteworthy progress over previous annealing frameworks. Additionally, the qualitative results for the Tsukuba disparity at roughly the same computational time are shown in Figure 3. Not much difference can be found among the results of WA, GC and TRW-S, while SW-cut, Gibbs and W-Gibbs samplers showed visibly worse performances.

4.2 Pairwise Continuous MRF Stereo

For the continuous disparity MRF model, the unary costs were calculated from the Birchfield and Tomasi cost of discrete MRF model by piecewise constant interpolation. The energy was minimized with Potts, truncated linear and truncated quadratic regularity conditions over Venus stereo image pairs. Since PWGS is cumbersome over continuous state sampling, WA algorithm using PWMS was

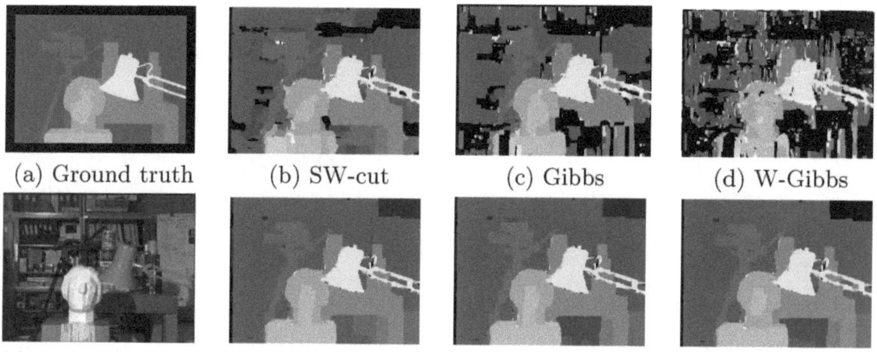

(a) Ground truth (b) SW-cut (c) Gibbs (d) W-Gibbs

(e) Reference image (f) GC (α expansion) (g) TRW-S (h) WA

Fig. 3. Resulting disparity maps on Tsukuba. (a) the ground truth disparity map; The resulting disparity maps of (b) SW-cut, (c) Gibbs, (d) W-Gibbs, (f) GC (α expansion), (g)TRW-S, and (h)WA (PWGS).

tested for this experiment. Unlike the discrete state MRF, the temperature variable played a more crucial role in the minimization process. It was no longer sufficient to start out with a temperature scheduling constantly equal to zero as with discrete MRF. Instead, the temperature T_i started out as 1 and gradually lowered toward zero within the fixed number of iterations. The results of our algorithm on Venus images with different smoothness constraints are shown in Figure 4. The computational time took 1 to 2 minutes. In Figure 4, it is interesting to see that truncated linear constraint maintained the discreteness of the disparity even in a continuous domain awhile the other models gave comparatively smoother result.

4.3 Unsupervised Binary Segmentation with Global Constraints

MRFs with global constraints are one of the energy functions that cannot be solved directly by alpha-expansion and belief propagation [32]. In this experiment, we have applied the proposed WA to the binary segmentation problem that is modeled by an energy function with global constraint as follows.

$$\varphi(x) = \sum_{(p,q)\in N} V(p,q) + c \cdot H(R_1, R_2) + d \cdot S(R_1, R_2). \quad (6)$$

Minimization of the above energy function divides an image into two disjoint regions R_1 and R_2 while making the difference of histograms of two regions be maximal and maintaining smoothness and size constraint. $V(p,q)$ is a simple pair-wise discontinuity cost. $H(R_1, R_2)$ is a histogram similarity measure between regions R_1 and R_2, calculated by sum of bin by bin absolute difference. $S(R_1, R_2)$ is a region size constraint that outputs the maximum size between R_1 and R_2. c and d are constants. $H(R_1, R_2)$ and $S(R_1, R_2)$ are termed global constraint. They are global because they are the function of whole state. Otherhand,

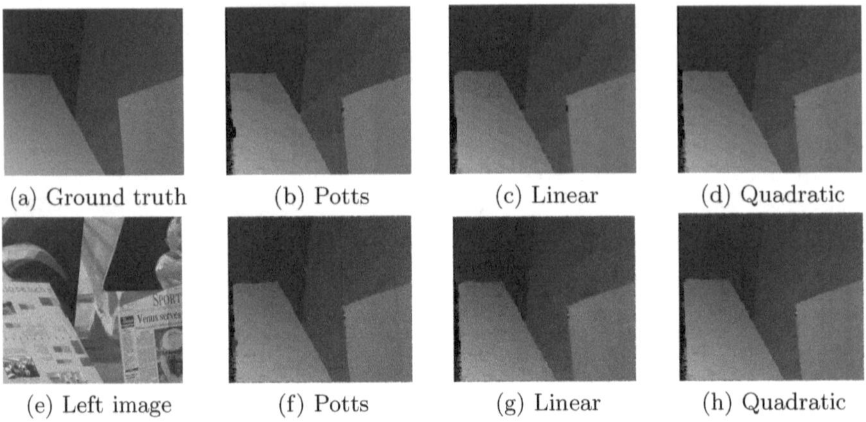

(a) Ground truth (b) Potts (c) Linear (d) Quadratic

(e) Left image (f) Potts (g) Linear (h) Quadratic

Fig. 4. (a) and (e) shows the stereo ground truth and reference image. (b), (c), and (d) are the result of TRW-S for the discrete MRF model respectively using Potts, truncated linear, and truncated quadratic discontinuity cost. (f), (g), and (h) are the results of WA (PWMS) for the continuous MRF stereo with Potts, truncated linear, and truncated quadratic discontinuity cost, respectively.

(a) (b) (c) (d)

Fig. 5. (a) to(d) show the unsupervised binary segmentation result of global constraint energy minimization

pair-wise potential is a function of two nodes, thus the size of $V(p,q)$ clique is 2. Similarly, $H(R_1,R_2)$ and $S(R_1,R_2)$ are cliques functions of size N, where N is the number of pixels/nodes. These types of energy functions are usually minimized with local methods such as snake and level set where the results change significantly with initial states, although level set and snake does not usually include pair-wise terms. However, global constraint energy function was able to be minimized with PWGS based WA. See Figure 5 (a) to (d). Computational time was less than 1 minute.

5 Conclusion and Future Work

In this paper, we proposed a new efficient cluster annealing scheme called WA (window annealing) for the estimation of MAP state over a square lattice MRF. By incorporating both temperature and maximum window size as the control variables for the annealing process, we can achieve faster energy minimization.

The proposed WA was tested over pairwise MRF stereo problem, and the experimental results demonstrated that the performance of WA was quite competitive with those of GC and TRW-S. Furthermore, the proposed window annealing has numerous advantages over the existing methods. It can minimize any energy function if the state's probability density is known up to a normalizing constant. We have illustrated the general applicability of proposed method over continuous state and global constraint energy minimization where GC and TRW-S cannot be applied. Such problems are generally solvable by Monte Carlo methods such as Gibbs sampler, except existing annealing techniques are impractically slow compared to more constrained methods such as GC. However, by proposing a faster Monte Carlo method, an annealing process with its wider applicability can now become a practical choice of approach.

Acknowledgment

This research was supported in part by the Defense Acquisition Program Administration and Agency for Defense Development, Korea, through the Image Information Research Center under the contract UD070007AD, and in part by the MKE (Ministry of Knowledge Economy), Korea under the ITRC (Information Technolgy Research Center) Support program supervised by the IITA (Institute of Information Technology Advancement) (IITA-2008-C1090-0801-0018).

References

1. Sun, J., Zheng, N.N., Shum, H.Y.: Stereo matching using belief propagation. PAMI 25 (2003)
2. Boykov, Y., Veksler, O., Zabih, R.: Fast approximate energy minimization via graph cuts. PAMI 23 (2001)
3. Potetz, B.: Efficient belief propagation for vision using linear constraint nodes. CVPR (2007)
4. Rother, C., Kolmogorov, V., Minka, T., Blake, A.: Cosegmenation of image pairs by histogram matching- incorporating a global constraint into mrfs. CVPR (2006)
5. Rother, C., Kolmogorov, V., Lepistsky, V., Szummer, M.: Optimizing binary mrfs via extended roof duality. CVPR (2007)
6. Kohli, P., Mudigonda, P., Torr, P.: p^3 and beyond: Solving energies with higher order cliques. CVPR (2007)
7. Lan, X., Roth, S., Huttenlocher, D., Black, M.J.: Efficient belief propagation with learned higher-order markov random fields. In: Leonardis, A., Bischof, H., Pinz, A. (eds.) ECCV 2006. LNCS, vol. 3952, pp. 269–282. Springer, Heidelberg (2006)
8. Barbu, A., Zhu, S.C.: Generalizing swendsen-wang cut to sampling arbitrary posterior probabilities. PAMI 27 (2005)
9. Geman, S., Geman, D.: Stochastic relaxation, gibbs distributions, and the bayesian restoration of images. PAMI 6 (1984)
10. Tu, Z., Zhu, S.C.: Image segmentation by data-driven markov chain monte carlo. PAMI 24 (May 2002)
11. Zhu, S.C., Liu, X.W., Wu, Y.N.: Exploring texture ensembles by efficent markov chain monte carlo: Toward a trichromacy theory of texture. PAMI 22(6) (2000)

12. Liu, J.S.: Monte carlo strategies in scientific computing. Springer, Heidelberg (2001)
13. Moral, P.D., Doucet, A., Jasra, A.: Sequential monte carlo for bayesian computation. Bayesian Statics (2006)
14. Metropolis, N., Rosenbluth, A.W., Rosenbluth, M.N.: Equations of state calculations by fast computing machines. Journal of Chemical Physics 21 (1953)
15. Hasting, W.K.: Monte carlo sampling methods using markov chains and their applications. Biometrika 57 (1970)
16. Kirkpatrick, S., Gelatt, C.D., Vechi, M.P.: Optimization by simulated annealing. Science 220 (1983)
17. Jasra, A., Stephens, D.A., Holmes, C.C.: On population-based simulation for static inference. Statistics and Computing 17 (2007)
18. Neal, R.M.: Annealed importance sampling. Technical Report 9805, University of Toronto (1998)
19. Andrieu, C., Freitas, N., Doucet, A., Jordan, M.: An introduction to mcmc for machine learning. Machine Learning 50 (2003)
20. Crow, F.: Summed-area tables for texture mapping. SIGGRAPH (1984)
21. Tappen, M., Freeman, W.: Comparison of graph cuts with belief propagation for stereo, using identical mrf parameters. In: ICCV (2003)
22. Szeliski, R., Zabih, R., Scharstein, D., Veksler, O., Kolmogorov, V., Agarwala, A., Tappen, M., Rother, C.: A comparative study of energy minimization methods for markov random fields. In: Leonardis, A., Bischof, H., Pinz, A. (eds.) ECCV 2006. LNCS, vol. 3952, pp. 16–29. Springer, Heidelberg (2006)
23. Scharstein, D., Szeliski, R.: A taxonomy and evaluation of dense two-frame stereo correspondence algorithms. IJCV (2002)
24. http://vision.middlebury.edu/stereo/
25. Scharstein, D., Szeliski, R.: High-accuracy stereo depth maps using structured light. CVPR (2003)
26. Kolmogorov, V., Zabih, R.: What energy functions can be minimized via graph cuts? PAMI 26 (2004)
27. Boykov, Y., Kolmogorov, V.: An experimental comparison of min-cut/max-flow algorithms for energy minimization in vision. PAMI 26 (2004)
28. Wainwright, M.J., Jaakkola, T.S., Willsky, A.S.: Map estimation via agreement on trees: Message-passing and linear-programming approaches. IEEE Trans. Information Theory 51(11) (2005)
29. Kolmogorov, V.: Convergent tree-reweighted message passing for energy minimization. PAMI 28 (2006)
30. Sudderth, E.B., Ihler, A.T., Freeman, W.T., Willsky, A.S.: Nonparametric belief propagation. MIT LIDS Technical Report P-2551 (2002)
31. Birchfield, S., Tomasi, C.: A pixel dissimilarity measure that is insensiitive to image samplin. PAMI 20 (1998)
32. Rother, C., Kolmogorov, V., Minka, T., Blake, A.: Cosegmentation of image pairs by histogram matchin- incorporating a global constraint into mrfs. CVPR (2006)

Unsupervised Classification and Part Localization by Consistency Amplification

Leonid Karlinsky, Michael Dinerstein, Dan Levi, and Shimon Ullman

Weizmann Institute of Science, Rehovot 76100, Israel
{leonid.karlinsky,michael.dinerstein,dan.levi,
shimon.ullman}@weizmann.ac.il

Abstract. We present a novel method for unsupervised classification, including the discovery of a new category and precise object and part localization. Given a set of unlabelled images, some of which contain an object of an unknown category, with unknown location and unknown size relative to the background, the method automatically identifies the images that contain the objects, localizes them and their parts, and reliably learns their appearance and geometry for subsequent classification. Current unsupervised methods construct classifiers based on a fixed set of initial features. Instead, we propose a new approach which iteratively extracts new features and re-learns the induced classifier, improving class vs. non-class separation at each iteration. We develop two main tools that allow this iterative combined search. The first is a novel star-like model capable of learning a geometric class representation in the unsupervised setting. The second is learning of "part specific features" that are optimized for parts detection, and which optimally combine different part appearances discovered in the training examples. These novel aspects lead to precise part localization and to improvement in overall classification performance compared with previous methods. We applied our method to multiple object classes from Caltech-101, UIUC and a sub-classification problem from PASCAL. The obtained results are comparable to state-of-the-art supervised classification techniques and superior to state-of-the-art unsupervised approaches previously applied to the same image sets.

1 Introduction

The goal of this paper is unsupervised classification, including discovery of a new category, learning a model of geometric arrangement of object parts and their appearance, and obtaining object and part localization, from a set of unlabeled images, which contains non-class images mixed with some unknown (usually small) percent of class images. The class instances may be uncropped, unaligned and of small size relative to the background.

The problem of unsupervised object classification has gained considerable recent interest [1,2,3,4,5,6,7,8,9,10,11,12], however, this task is still far from being completely solved. In this study we present a novel methodology to approach the problem. A common approach is to start from some limited, manageable set of

D. Forsyth, P. Torr, and A. Zisserman (Eds.): ECCV 2008, Part II, LNCS 5303, pp. 321–335, 2008.

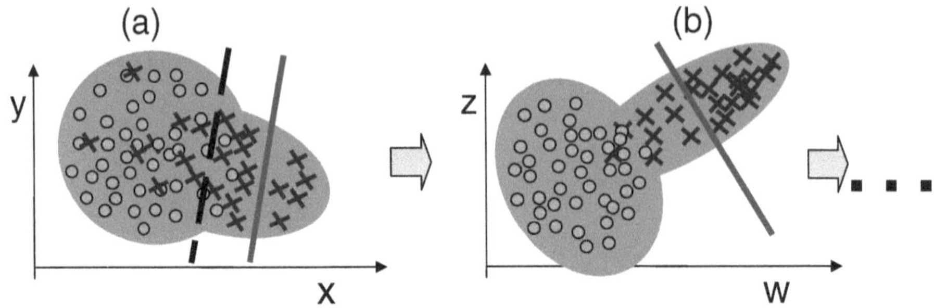

Fig. 1. Illustration of the feature re-extraction approach. (a) In the initial feature space (x-y) it is difficult to separate class (blue crosses) and non class (green circles) examples. In this feature space, the best separating hyperplane (which the unsupervised classification seeks to determine) is marked by the dashed line. Instead, our method identifies a subset of sure class examples separated from the rest (red solid line). (b) Using these examples, the method extracts a new feature set (z-w), in which a larger set of class examples can be identified. The process then continues iteratively. Each iteration uses new features, rather than previous features (or their combinations).

initial features $\mathcal{F}$, for example, a set of local descriptors extracted around image interest points or clusters extracted from such descriptors [1, 2, 11, 12, 13, 14, 15, 16, 5, 6, 7, 8, 9]. The set of features can be optimized by selecting a subset of the most useful features $\mathcal{F}_1 \subset \mathcal{F}$, or sometimes combinations of features in $\mathcal{F}_1$ are used as new features [2, 8, 14]. However, there is no guarantee that the choice of initial features will in general be sufficient for complete separation. In contrast, we approach the problem as a combined iterative search for features and a classifier. We do not use the initial feature set to obtain the final class separation, but only for identifying a subset of sure class examples which can be reliably separated from the rest (Fig. 1a). This goal is achieved by unsupervised training of a classifier that combines both appearance and part-geometry information. The extracted class examples are used to guide the subsequent extraction of new features, which were not a part of the initial feature set. It is not a-priori clear that this iterative approach will continue to improve classification: if the intermediate classification results are partly incorrect, their use could lead the process astray and cause deteriorating performance. In this work, we demonstrate that in the proposed algorithm, the constructed features become more class-specific as the computation evolves, and the class vs. non-class separation continuously improves (Fig. 1b), reaching a final high level of performance even compared with recent supervised methods.

We develop two main tools that allow the iterative combined search. One is the incremental discovery of part specific features, which combine different part appearances discovered in the training examples. The other is a novel star-like class-geometry model of object parts, which differs from the similar past models [3, 4, 5, 15, 16, 17, 13] and which can be learned efficiently without supervision in very noisy conditions. These two aspects are described briefly below, and explained in more detail in Sections 2.1 and 2.2.

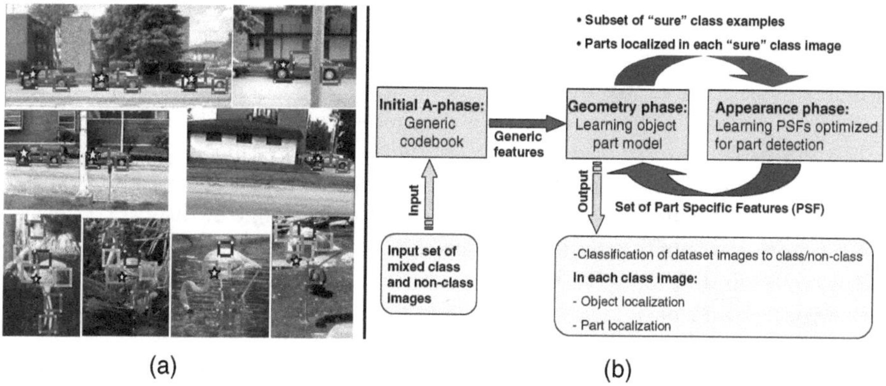

Fig. 2. (a) Example results of unsupervised object and part localization on two datasets (UIUC cars, flamingo). The yellow star is the detected model center location (see text), color coded rectangles are examples of detected object parts (for each object several out of about 150 modeled parts are shown). (b) Schematic diagram of the UCA algorithm.

Feature learning. most unsupervised approaches [1, 2, 5, 6, 7, 8, 9, 11, 12], including ours, start from some generic set of features $\mathcal{F}$. During learning, when a particular class is considered, the approaches select a subset $\mathcal{F}_1 \subset \mathcal{F}$ of so-called Class Specific Features (CSF), which coincide better with the class compared with the background or other classes. In contrast, our method extracts and learns a new set of features, termed Part Specific Features (PSF). The PSF are optimized to have higher detection scores at specific locations on the class objects, and at the same time to have lower scores at incorrect locations on the same objects and in non-object detections. Different part specific features have been used successfully in a number of supervised approaches, such as k-fan [17] and semantic hierarchy [18], and were shown to be useful for both object and part localization. Constructing such features in an unsupervised manner is challenging; our method is the first unsupervised method that learns and uses such features, resulting in improved object and part detection and localization.

Geometry learning. Past supervised and unsupervised classification methods can be categorized by their modeling of object geometry. In bag-of-feature methods [2, 9], geometry is ignored. Methods, such as [2, 8, 1, 14], extend the bag-of feature approach by using feature combinations. In [3, 4, 5, 11, 13, 15, 16, 17] object geometry is modeled by the spatial distribution of each feature in the object reference frame. A geometric part model is useful for classification, but it is challenging to construct such a model in an unsupervised setting. Most previous unsupervised methods therefore do not use a full geometric model [2,6,7,8,9,11,12]. Our method uses star-like geometry. It has several differences compared with similar past models. The method is not restricted by a small number of parts as in [3,4], unlike [13,15,16,17] it does not require any supervision, unlike [3,5,15,11] it models distribution of feature locations on the background, unlike [5,12] it does not rely on non-geometric pLSA [2] for internal supervision, and unlike [10,6,7], it is not based on prior image segmentation. These differences are explained in more detail in Sections 2 and 2.1.

In terms of class vs. background classification performance, our method outperforms the state-of-the-art unsupervised methods [2,5,7,10,11,12] on 18 classes from the Caltech 101, Weizmann horses and UIUC cars datasets. Surprisingly, the method is also comparable in performance to existing state-of-the-art supervised (and weakly supervised) methods applied to the same datasets. We further demonstrate how our method can be used to separate different object views on the cars class from the PASCAL challenge 2007 dataset. As the method achieves precise object and part localization, it provides a basis for top-down segmentation, as illustrated in supplementary material.

The rest of the paper is organized as follows. Section 2 presents an overview followed by a detailed description of each of the method stages. Section 3 presents results obtained on various datasets, together with an analysis and comparison with previously reported results. Conclusions are discussed in Section 4.

2 The Consistency Amplification Method

Our approach alternates between model learning and data partitioning. Given an image set S, an initial model (learned using initial features) is used to induce an initial partitioning by identifying highly likely class members. The initial partitioning is then used to improve both the appearance and geometrical aspects of the model, and the process is iterated. In this manner the process exploits intermediate classification results at a given stage to guide the next stage. Each stage leads to an improved consistency between the detected features and the model, which is why the process is termed Unsupervised Consistency Amplification (UCA). Each UCA iteration consists of two phases of learning: the feature learning Appearance-phase (A-phase) followed by the part model learning Geometry-phase (G-phase), explained in detail in sections 2.2 and 2.1 respectively. The approach and the order of the phases are summarized in Fig. 2b.

Initial Appearance-phase. In all our experiments, we use a generic codebook of SIFT descriptors of 40×40 patches for the initial (appearance) features. This codebook, denoted by $\mathcal{F}_0$, is computed by a standard technique [19] from all the images in given set S. The codebook descriptors are compared to the descriptors at all points of all the images in S and storing the points of maximal similarity (either one or several, see below) in each image.

Geometry-phase. The detection of parts using the generic features is usually noisy, due to detections in non-class images, and at some incorrect locations in the class images. The goal of the geometric part model learning is to distinguish between the correct and incorrect detections, based on consistent geometric relations between features. This is accomplished by the G-phase of the algorithm, which is also used for the selection of the most useful features and the automatic assignment of each of their detections in every image in S to either object or background model. In contrast with [3,4,5,15,11] that use uniform distribution of features on the background, we model the background by a distribution of the same family as the class object distribution, which allows to prevent the spurious geometric background consistency from being accounted for by the learned class model. In our experiments we found that modeling the background distribution is better than assuming uniformity with mean performance gain of $12 \pm 7\%$ EER in the first iteration of the UCA that uses initial generic appearance features. The learned background model is then discarded after the learning and is not used for classifying new images. Thus, the model used in the G-phase is a mixture of two stars, one for object and the other for background. It is learned without supervision from all the images in S using a novel graphical model formulation explained in detail in Section 2.1. After the geometric structure has been learned, a subset $H \subset S$ of images which contain class objects with high confidence is selected. In these images the object centers and parts are localized. Unlike [5,12] that learn the geometric constraints using only a set of objects identified by the non-geometric pLSA [2], our method identifies and localizes objects, and learns their part geometry, jointly and explicitly from the entire data.

Appearance-phase. Each part-specific feature constructed in the A-phase represents an object part by extracting several typical appearance patches of the part, from different images. Part-patches can be extracted, because the locations of the parts in the images of the subset H are already estimated from the previous G-phase. An optimal subset of these part patches is learned by a discriminative model described in section 2.2. The set of all part specific features extracted during the A-phase is denoted by $\mathcal{F}$.

Computing the output. After the G-phase at each iteration, the learned model is applied to produce classification, as well as object and part localization results for either the given dataset or an unseen test set. This is done without introducing any supervision to the system. The way we apply the learned model to test images is described in detail in section 2.3.

2.1 The Geometry Phase

We first describe the G-phase model, and then explain how it is learned from the data. The main goal of the G-phase is to identify the most likely locations of objects and their parts in all images of the given set S and to estimate a subset $H \subset S$ of images which contain class objects with high confidence. The G-phase models the data by a generative probabilistic graphical model depicted in Fig. 3a. Let the image set S have N unlabelled images: $S = \{I_1, I_2, \dots I_N\}$ and the current feature set $\mathcal{F}$ consist of M features: $\mathcal{F} = \{F_1, F_2, \dots F_M\}$. In the G-phase of the first UCA iteration these features are a codebook of generic SIFT descriptors $\mathcal{F}_0$, and in the following UCA iterations these are the learned PSFs. During the G-phase each feature is associated with an object part or the background. Denote the detected location of feature F_m in image I_n by X_m^n (the G-phase uses a single (maximal) detected location per feature in each image, see extension below.) The G-phase model independently generates observed samples:

$Data = \{ (F_m, I_n, X_m^n) \mid 1 \leq n \leq N, 1 \leq m \leq M \}$ The probability of observing a specific image $\Pr(I = I_n)$ is taken to be uniform. The overall observed data likelihood under the G-phase model can be written as:

$$\begin{aligned} \Pr(Data) \propto \prod_{n=1}^{N} \prod_{m=1}^{M} \sum_{C_m^n=1}^{2} \int_{L_m^n} \Pr(C_m^n \mid I_n) \Pr(F = F_m \mid C_m^n) \\ \Pr(L_m^n \mid I_n, C_m^n) \Pr(L_F = X_m^n \mid F_m, C_m^n, L_m^n) dL_m^n \end{aligned} \quad (1)$$

The meaning of the product inside the integral in eq. 1 is that each data sample (F_m, I_n, X_m^n) observed in image I_n for the feature F_m is independently generated as follows. First, the latent discrete binary "class" variable C_m^n is drawn with probability $\Pr(C_m^n = k \mid I_n) = \alpha_k^n$, independent of the feature F_m. $C_m^n = 1$ means that I_n contains a class object and F_m is generated from the class model. $C_m^n = 2$ means that F_m is generated from the background model, because either I_n does not contain an object or F_m was not detected consistently with the class model. After learning, the value α_k^n is the likelihood of class k (either object or background) in image I_n. Next, the latent location variable L_m^n is drawn from a Gaussian distribution $\Pr(L_m^n \mid I_n, C_m^n = k) = N(\mu_k^n, \Sigma_k^n)$. L_m^n represents the image position of the center of the star model (chosen by C_m^n), which generates the feature F_m in image I_n. Note that for every feature detected in image I_n that has chosen the class k, there is a separate variable L_m^n, but all of these variables are generated from the same distribution specific to I_n. Next, the observed feature variable F draws its value F_m from the distribution $\Pr(F = F_m \mid C_m^n = k) = \beta_k^m$ which depends on the chosen class k, but is independent of the image I_n. After learning, the value β_k^m is the likelihood of feature F_m to be consistent with the geometric model of class k. Finally, the observed feature location variable L_F draws its value X_m^n from a linear Gaussian distribution $\Pr(L_F = X_m^n \mid F_m, C_m^n = k, L_m^n) = N(L_m^n + \rho_k^m, \Lambda_k^m)$. This distribution models the uncertainty of the offset ρ_k^m of the feature F_m from the L_m^n - center of the star model chosen by C_m^n. It is specific to the feature F_m and the chosen class k and is independent of the specific image I_n.

To summarize, the parameters of the model are α, β, μ, Σ, ρ and Λ, all of them are learned by soft EM as described further below. A schematic drawing illustrating the data generation process and the meaning of the main model parameters is shown in Fig. 3b. The model uses a star-like geometry, but an important difference between the current model and past star model formulations is worth noting. In contrast with [17, 16, 15], that have a single reference point or k-fan per image, in our model there exists a separate reference point (center) random variable for each part, drawn, however, from the same distribution specific to the given image. This allows the features detected in the same image to be updated individually: features assigned to the class update the class star and features assigned to the background update the background star, both the assignments and the updates are soft. Although it may sound technical, it has fundamental importance, since, as we saw in our experiments, in different class images, different subsets of features are geometrically consistent with the object model. It is interesting to note that the transition from the standard star-model

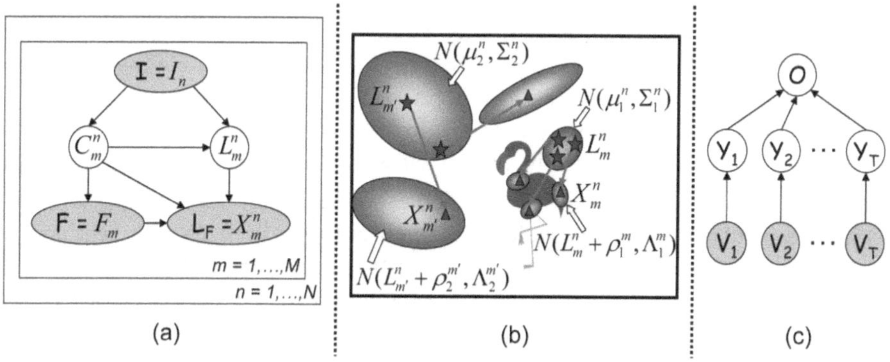

Fig. 3. The probabilistic models used by UCA. Shaded ellipses are observed variables, unfilled are hidden (latent). (a) Graphical representation of the G-phase generative model. (b) Generating the object and background. The object model illustrated in red, the background in green. Each generates centers, denoted by ★ (star) and features denoted by ▲ (triangle). The ellipses denote the uncertainty in position. X_m^n denotes the detected location of feature F_m. Every feature F_m detected on the object is generated using its own star center point L_m^n, but all these L_m^n are generated from the same distribution $N(\mu_1^n, \Sigma_1^n)$ specific to the image. As illustrated, the learned object distributions are tighter than the background distributions. (c) Graphical representation of the "Continuous Noisy OR" discriminative model.

to our version is entirely analogous to the transition from Naive Bayes (NB) to pLSA. In the NB there is only a single class node generating the entire feature vector of an image, while in pLSA each feature has a separate topic node generated from an image specific distribution. The pLSA is more flexible than NB and was found useful for unsupervised classification [2, 9]. Similarly we found that the modified star is useful in modeling feature geometry in the unsupervised setting.

Learning: The model is learned from the data using the soft EM algorithm. The EM update equations are provided in the supplementary material. As mentioned above, the data samples fitted by our model are of the form (F_m, I_n, X_m^n). In order to incorporate the features' detection scores into the learning process, we weight each sample by its score. Namely, the sample (F_m, I_n, X_m^n) is weighted by R_m^n - the similarity of F_m with I_n at location X_m^n. The parameters of the model: α, β, μ, ρ and Λ, are initialized at random and EM is run until convergence. In order to have the object model learned with respect to $C_m^n = 1$, in all our experiments, we initialize $|\Sigma_1^n| << |\Sigma_2^n|$ for every n. During EM iterations, this initialization causes the object feature detections to tend to update the $C_m^n = 1$ model, since they usually appear in more tight and repeatable configurations (i.e. fit a star with smaller center uncertainty Σ_1^n). At the same time, background feature detections, that are usually loosely scattered all over the image, will tend to coincide with the $C_m^n = 2$ model. This method of the initialization of all the parameters (including Σ) was identical throughout all our experiments.

Identifying the set of high-likelihood images: After the EM converges, the value of α_1^n (the image probabilities to belong to the class) shows a strong separation between a subset of class images and all the non-class images, see figure 6a. As a result, by the end of G-phase it becomes possible to identify a subset of high-likelihood class candidate images H. In all our experiments, we marked an image I_n as high likelihood class candidate if $\alpha_1^n > \eta \cdot \max_n(\alpha_1^n)$, where $\eta = 0.85$ was chosen empirically and used throughout all the experiments. Examples of objects automatically identified and localized by the G-phase of the first UCA iteration are shown in Fig. 6b. These are examples of the first G-phase output, obtained using the initial generic features. As can be seen, the localized object model centers appear at similar locations within the object in the different images. In the A-phase, these points are used to extract stacks of corresponding fragments, which are used to construct the part specific features - the CNOR part-detectors, as explained in the next section.

2.2 The Appearance Phase

In the G-phase we learned the position of each part relative to the object model center, and detected this center in the images belonging to H. We localize each part in these images by assuming it is located at the learned relative position ρ_1^m from the center located at μ_1^n. In the A-phase we learn for each part a detector trained to distinguish image patches in correct part locations from patches in incorrect ones. The detector is trained using the detected part locations as positive examples and all other locations on the images of H as negative examples. The constructed part detectors form the new feature set $\mathcal{F}$ for the G-phase of subsequent UCA iteration. We next describe the novel probabilistic discriminative model used by the part detector, the Continuous Noisy OR (CNOR), and how this model is trained.

For each part m, corresponding to F_m above, we extract a set of appearances in the following way. In each class candidate image $I_n \in H$, we take the 40x40 image patch at position $\mu_1^n + \rho_1^m$, where μ_1^n is the location of the learned object center in I_n and ρ_1^m is the learned offset of part m from the object center. The accumulated set of image patches is the candidate set of part appearances: $A_m = \{Z_1^m, \ldots, Z_T^m\}$. The next step is to select a subset of appearance representatives $R_m \subseteq A_m$, and learn to optimally combine their detection evidence in order to reliably detect the object part. Both tasks are achieved simultaneously by training the CNOR model, depicted in Fig.3c. Let P be an arbitrary image patch taken from an arbitrary location L in a new image. The binary variable O^P is set to $O^P = 1$ iff L and P are the location and appearance of part m respectively. The probability of O^P is discriminatively modeled as:

$$\Pr(O^P|V;\Theta) = \sum_Y \Pr(O|Y) \cdot \prod_{t=1}^{T} \Pr(Y_t|V_t;\theta_t, R_m) \tag{2}$$

The $\Theta = \{R_m, \theta_1, \ldots, \theta_T\}$ are the learned parameters of the model. $V = \{V_t\}$, where V_t is the output of a continuous SIFT similarity measure between P and

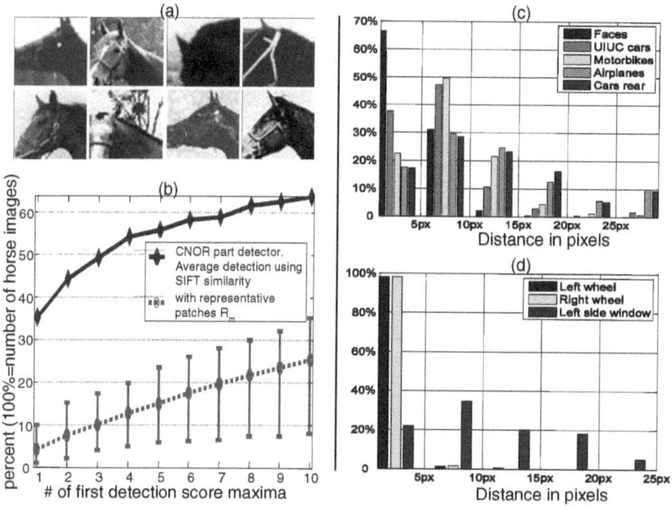

	Object localization error STD, % of object size
Faces	2.6
UIUC cars	4.6
Motor-bikes	5
Air-planes	8
Cars rear	8

Fig. 4. (I) Example of a CNOR part detector. (a) Selected representative appearances. (b) Cumulative histograms of detections within 15px from ground truth location relative to the number of detection score local maxima used. (II) Evaluation of object and part localizations obtained by the UCA method showing a distribution of localization error in pixels relative to manually marked ground truth. 10px is less then 7% of object size in all the sets in the figure. (c) Object localization, 100% = number of class images. (d) Part localization on UIUC cars, each part was detected in about 95% of objects. In the graph 100% = number of part detections.

$Z_t \in A_m$. Note that we do not explicitly model $\Pr(V)$, which can be a complex distribution. $Y = \{Y_t\}$, where Y_t is a latent binary variable representing the detection of appearance Z_t with:

$$\Pr(Y_t = 1|V_t; \theta_t, R_m) = \begin{cases} \frac{1}{1+e^{-\alpha_t(V_t-\tau_t)}} & Z_t \in R_m \\ 0 & Z_t \in A_m \backslash R_m \end{cases} \tag{3}$$

Here $\theta_t = \{\tau_t, \alpha_t\}$ are the parameters of the sigmoid in 3. $Y_t = 1$ becomes likely if patch P exceeds a similarity threshold τ_t with the representative patch $Z_t \in R_m$, with α_t representing the uncertainty of τ_t. If $Z_t \in A_m \backslash R_m$ (meaning Z_t is not a chosen representative) then V_t and Y_t have no effect on $\Pr(O^P|V;\Theta)$. Finally, $\Pr(O^P|Y)$ is a deterministic "or" of Y:

$$\Pr(O^P = 1|Y) = \begin{cases} 1 & \exists t.Y_t = 1 \\ 0 & otherwise \end{cases} \tag{4}$$

The entire model can intuitively be described as follows: the part is detected ($O_p = 1$) whenever P is "sufficiently" similar to at least one of the part m's representative patches in R_m.

To learn the model parameters, a training set of image patches E is constructed by taking all 40×40 image patches (on a fixed step grid) from all the images in the current class candidate set H. For each patch $P \in E$ the observed data vector is constructed as: $D^P = \langle V^P, O^P \rangle$ where $V^P = \{V_t^P\}$ is computed by measuring similarity between P and A_m patches and $O^P = 1$ iff $P \in A_m$ (and $O^P = 0$ otherwise). Finally the training data for the CNOR model of part m is: $D = \{D^P | P \in E\}$. By treating the correct part appearances (A_m) as positive examples and all other appearances (either other parts of the object or background patches) as negative examples, the object part detector is trained for correct localization of the part. To limit the number of representatives, the learning objective is to find the Minimum Description Length (MDL) parameters Θ, in other words, to find Θ that maximize a combined score of the model complexity (number of representatives) and model performance (data likelihood). We solve this learning problem using the Structural EM (SEM) algorithm optimizing the Bayesian Information Criterion (BIC) score [20]:

$$BIC = \sum_{P \in E} \log(\Pr(O^P | V^P; \Theta)) - \frac{\log T}{2} \cdot |R_m| \tag{5}$$

The SEM algorithm iterates between two stages. The first stage is, given a set of representatives R_m, to find the optimal values for the $\{\theta_t\}$ parameters. This stage is solved using the EM algorithm. It is computed efficiently, since in our model each EM iteration has linear time complexity. The second stage is, given the current assignment of $\{\theta_t\}$, to estimate an improved R_m. This is achieved by running several iterations of a greedy search over subsets of $R_m \subseteq A_m$, where at every step of the search a current subset R_m is modified by either adding or removing one element.

After learning, the set of part m detections is obtained by identifying first few local maxima of the probability $\Pr(O^P | V; \Theta)$ computed for all patches P in a given image. Selected representatives for an example part are shown in Fig. 4a. The resulting CNOR part detectors for all parts m, are a significantly more reliable set of object features then the initial generic set of features, as demonstrated in Fig. 4b and in section 3, and it provides a general method for reliable part detection for both supervised and unsupervised classification.

2.3 Applying the Learned Model to Classify and Localize Objects and Parts

To compute the classification score for an unseen test image, and to localize the class objects in it, we use the learned ρ_1^m (offsets from object star center) and Λ_1^m (STDs for these offsets) parameters in a voting scheme similar to [13] as follows. For each part detector, a number (five in our experiments) of highest-scoring locations at each image are marked. To identify the object star's center location, each detection X votes for a center location, by placing a Gaussian mask with STD Λ_1^m around the expected location $X - \rho_1^m$. After all the detectors voted, the point with the maximal accumulated vote determines the location of the object star's center in each image (in case there are multiple objects, several local maxima that exceed a global threshold are taken). The accumulated vote value at the detected center point serves as the object detection score in the image. These scores are then used to create the ROC that tests the separation between the class and the non-class images in the results section 3. The object localization results of our method are evaluated in Fig. 4c. The parts are localized by "back-projection" as in [13]. Each part detector that voted into one of the selected object center locations (with one of its five detections) is declared as 'detected' and is marked in the image. The accuracy of our part localization is demonstrated in figures 2a and 5 and evaluated in Fig. 4d. Marking all the detected parts in the image can be used for a top-down segmentation of the detected object (see examples in supplementary material). The details of the top-down segmentation are outside the scope of the current discussion.

3 Results

To test the performance of the UCA method, it was applied to the task of fully unsupervised classification and object and part localization on 18 different object classes. The list of the classes, the parameters of the datasets and ROC EERs obtained by the UCA are summarized in Table 1. The results show that our method obtains superior performance over the existing unsupervised methods in challenging conditions such as small objects relative to the background (e.g. UIUC cars, Caltech101 cars, flamingo), small percent of class images in the set (e.g. schooner, guitars), significant inter-class variability due to non-rigid deformations (e.g. bonsai, horses, crab, flamingo, starfish) and significant lack of alignment (e.g. UIUC cars, faces, PASCAL car views). Examples of the classes and object and part localizations obtained by UCA are shown in figures 2a and 5. Fig. 4c,d shows quantitative evaluation of automatic object and part localization by UCA compared to hand generated ground truth on several dataset. The background images for each dataset were chosen randomly out of Caltech backgrounds set containing 900 images. To challenge our method, we tested it on different class vs. non-class mixes, namely 10%, 20%, 30% and 50%. This is compatible with experimenting with Google data, since manual validation done by [5] showed that on average, above 25% of images returned by Google image search are good examples. For every dataset, increasing percent of class images above the percent reported in the table gives even better results. The UIUC cars dataset contained only the 170 non-cropped and non-aligned test images of the original set (and equal amount of random background images), the training images of the original set are cropped, so to make the task harder they were not used. The Caltech-5 datasets (from [3]) were tested in order to compare with past unsupervised approaches that were tested on the same data, namely [2,12,5,10,7]. To ensure that the chosen Caltech101 classes are sufficiently hard, 9 of the 11 tested Caltech101 classes are the ones with lowest reported performance by [7] (average of entries for these classes on [7]'s confusion matrix diagonal is 49%). Note that unlike [7], we do not use color information in our scheme.

An important characteristic of the UCA is its ability to deal with a low percentage of class images in the dataset. Methods such as [5, 12] that apply the

Table 1. Summary of fully unsupervised classification results obtained by the UCA method. For all datasets, the EER STD for UCA was $\leq 2\%$ (computed by cross-validation). For motorbikes, airplanes and cars-rear, the class images were randomly chosen from larger sets and remaining images were also used for testing the learned models obtaining 1.3%, 2.3% and 2.8% average EER respectively. The average EER of UCA on the Caltech-5 datasets was 2.65%. Results of other unsupervised methods reported for Caltech-5 were: average EER of 4.08% [12], 7.35% [5] and 11.38% [2] and average multiclass detection rate of 5.4% [7]. Results of leading supervised methods on Caltech-5 are comparable to our unsupervised result: average EER of 2.25% [15] and 1% [21] ([21] did not test on cars-rear class). The object size relative to the background for each dataset was approximated from several characteristic images.

Dataset	Origin	Total number of images	% class images in the set	Object size rel. to bgrnd.	UCA EER, %
Horses	Weizmann	646	50%	35%	2.7
Cars	UIUC	340	50%	3.5%	3.1
Car views	PASCAL	201	50%	40%	10.7
Motorbikes	Caltech-5	900	50%	30%	2.3
Airplanes	Caltech-5	900	50%	20%	2.7
Faces	Caltech-5	900	50%	20%	2.4
Cars-rear	Caltech-5	900	50%	20%	3.2
Bonsai	Caltech101	256	50%	35%	4.1
Ewer	Caltech101	283	30%	34%	2.3

Dataset	Origin	Total number of images	% class images in the set	Object size rel. to bgrnd.	UCA EER, %
Butterfly	Caltech101	303	30%	27%	6
Cars	Caltech101	600	20%	12%	1.1
Crab	Caltech101	365	20%	24%	10.9
Starfish	Caltech101	430	20%	11%	12.6
Laptop	Caltech101	405	20%	32%	4.3
Flamingo	Caltech101	335	20%	13%	7.3
Watch	Caltech101	1139	20%	40%	3.9
Guitars	Caltech101	650	10%	35%	8.2
Schooner	Caltech101	650	10%	27%	6.69

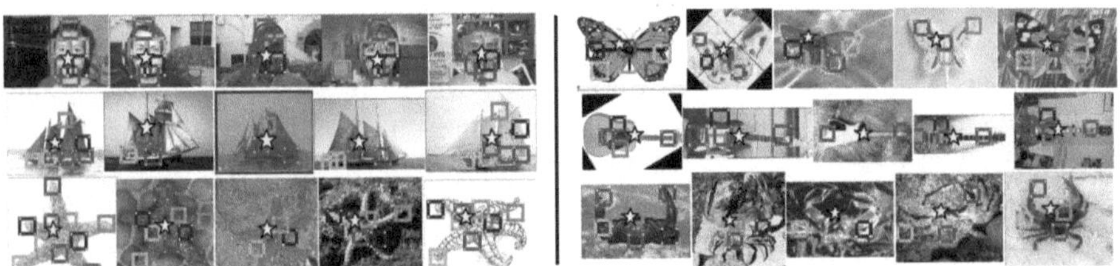

Fig. 5. More examples of unsupervised object and part localizations obtained by the UCA method. See explanation in fig. 2a.

pLSA method of [2] as a pre-processing step to identify and localize class examples may fail on such datasets. This was validated by testing the pLSA method with eight topics (optimal number proposed by [5]) on the schooner dataset that contains only 10% class images. From the N examples with maximal score in the class topic, less then 40% were class examples ($N = 10, 20, \ldots, 100$).

In the PASCAL car views experiment, we tested the ability of UCA to separate related sub-classes. In particular, out of the PASCAL 2007 training images, images depicting frontal and side views of cars were extracted. The scale of the images was normalized by vertical size and large background areas around each car was taken to make the set un-cropped and un-aligned. The UCA was then applied to this set in order to separate the views. Furthermore, when applied

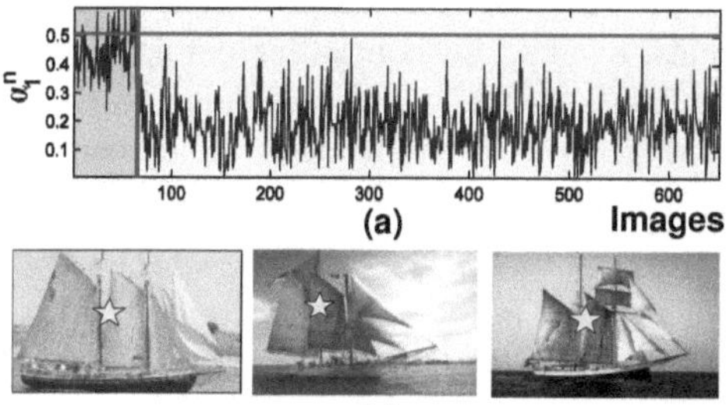

(b)

	Iter. 2 EER %	Iter. 3 EER %	Iter. 4 EER %	Iter. 5 EER %
Starfish	16.3	15.1	12.7	~
Crab	17.8	14.9	12.3	10.9
Flamingo	11.9	7.4	~	~
Guitars	17.4	8.2	~	~
Bonsai	8.6	3.9	~	~

(c)

Fig. 6. Improvements due to consistency amplification. (a) Example of class vs. background separation obtained by the first iteration for the schooners class. Yellow part are the class images and the rest are backgrounds (the ordering is only for illustration purposes). The horizontal line shows the adaptive threshold $\eta \cdot Max$ used to select the set H of high likelihood class examples for the next UCA iteration. (b) Examples of the objects identified and localized. (c) Table of EER improvement with UCA iterations. Only the classes that ran for more then two iterations are shown. The iterations continue until the set H stops growing. The average EER of the first iteration (that used the generic features and not PSFs) was 30% for these classes. This illustrates the low (relative to the PSFs) consistency between the generic features and the class objects.

on a set of about 700 images containing all the car views, UCA successfully learned the "frontal cars" subclass with similar EER to the two view experiment. Applying pLSA to the same set has yielded high error (32% EER). The ability of our method to separate similar sub-classes and specifically different views of the same class can also be useful in supervised learning applications. If a given training set of images of the same class can be automatically separated into a meaningful set of (inherently similar) subclasses, then it can greatly facilitate the learning task, by allowing the modeling of each subclass separately.

4 Conclusions

The UCA method has a number of basic advantages compared with previous unsupervised classification methods. First, the overall classification results are higher than obtained previously, and remain high even when class examples are sparsely distributed within the dataset. Surprisingly, on the tested classes, results of the unsupervised method are as good as leading supervised methods. Second, the method obtains precise object localization, indicated by a repeatable reference point on each detected object. Third, precise locations of the parts participating in the model are also made available. Fourth, the method is capable of separating similar classes and sub-classes, such as different views of the same class. The main novel aspects of the UCA method are the following. The model is iteratively improved by exploiting intermediate classification results, consistently improving the performance. A novel geometric model is used, which can be efficiently learned from the entire dataset, and therefore improve the methods ability to capture geometric consistencies, even when consistent configurations are sparse. The model uses a part detection scheme, which is trained to detect object parts with diverse appearances in their correct position. The resulting detections are therefore more reliable, providing precise part localization and improved overall performance.

Acknowledgments. This work was supported by EU IST Grant FP6-2005-015803 and ISF Grant 7-0369.

References

1. Fritz, M., Schiele, B.: Towards unsupervised discovery of visual categories. In: Franke, K., Müller, K.-R., Nickolay, B., Schäfer, R. (eds.) DAGM 2006. LNCS, vol. 4174, pp. 232–241. Springer, Heidelberg (2006)
2. Sivic, J., Russell, B.C., Efros, A.A., Zisserman, A., Freeman, W.T.: Discovering objects and their localization in images. In: ICCV, pp. 370–377 (2005)
3. Fergus, R., Perona, P., Zisserman, A.: Object class recognition by unsupervised scale-invariant learning. CVPR (2), 264–271 (2003)
4. Fergus, R., Perona, P., Zisserman, A.: A visual category filter for google images. In: Pajdla, T., Matas, J(G.) (eds.) ECCV 2004. LNCS, vol. 3021, pp. 242–256. Springer, Heidelberg (2004)
5. Fergus, R., Fei-Fei, L., Perona, P., Zisserman, A.: Learning object categories from google's image search. In: ICCV, pp. 1816–1823 (2005)
6. Russell, B.C., Freeman, W.T., Efros, A.A., Sivic, J., Zisserman, A.: Using multiple segmentations to discover objects and their extent in image collections. CVPR (2006)
7. Cao, L., Fei-Fei, L.: Spatially coherent latent topic model for concurrent object segmentation and classification. In: ICCV (2007)
8. Nowozin, S., Tsuda, K., Uno, T., Kudo, T., Bakir, G.H.: Weighted substructure mining for image analysis. CVPR (2007)
9. Li, L.J., Wang, G., Fei-Fei, L.: Optimol: automatic object picture collection via incremental model learning. CVPR (2007)
10. Ahuja, N., Todorovic, S.: Discovering hierarchical taxonomy of categories and shared subcategories in images. In: ICCV (2007)
11. Liu, D., Chen, T.: Semantic-shift for unsupervised object detection. In: CVPR Workshop (2006)
12. Liu, D., Chen, T.: Unsupervised image categorization and object localization using topic models and correspondences between images. In: ICCV (2007)
13. Leibe, B., Leonardis, A., Schiele, B.: Combined object categorization and segmentation with an implicit shape model. In: ECCV (2004)
14. Quack, T., Ferrari, V., Leibe, B., Gool, L.V.: Efficient mining of frequent and distinctive feature configurations. In: ICCV (2007)
15. Loeff, N., Arora, H., Sorokin, A., Forsyth, D.: Efficient unsupervised learning for localization and detection in object categories. In: NIPS (2005)
16. Sudderth, E.B., Torralba, A., Freeman, W.T., Willsky, A.S.: Learning hierarchical models of scenes, objects, and parts. In: ICCV (2005)

17. Crandall, D.J., Huttenlocher, D.P.: Weakly supervised learning of part-based spatial models for visual object recognition. In: Leonardis, A., Bischof, H., Pinz, A. (eds.) ECCV 2006. LNCS, vol. 3951, pp. 16–29. Springer, Heidelberg (2006)
18. Epshtein, B., Ullman, S.: Semantic hierarchies for recognizing objects and parts. CVPR (2007)
19. Jurie, F., Triggs, B.: Creating efficient codebooks for visual recognition. In: ICCV (2005)
20. Friedman, N.: The bayesian structural em algorithm. UAI, 129–138 (1998)
21. Dorko, G., Schmid, C.: Object class recognition using discriminative local features. INRIA (2005)

Simultaneous Visual Recognition of Manipulation Actions and Manipulated Objects

Hedvig Kjellström, Javier Romero, David Martínez, and Danica Kragić

Computational Vision and Active Perception Lab
School of Computer Science and Communication
KTH, SE-100 44 Stockholm, Sweden
{hedvig,jrgn,davidmm,dani}@kth.se

Abstract. The visual analysis of human manipulation actions is of interest for e.g. human-robot interaction applications where a robot learns how to perform a task by watching a human. In this paper, a method for *classifying manipulation actions in the context of the objects manipulated*, and *classifying objects in the context of the actions used to manipulate them* is presented. Hand and object features are extracted from the video sequence using a segmentation based approach. A shape based representation is used for both the hand and the object. Experiments show this representation suitable for representing generic shape classes. The action-object correlation over time is then modeled using conditional random fields. Experimental comparison show great improvement in classification rate when the action-object correlation is taken into account, compared to separate classification of manipulation actions and manipulated objects.

1 Introduction

Manipulation actions, i.e. hand actions for picking up objects, doing something with them and putting them down again, is an important class of hand activity not well studied in computer vision. The analysis of human manipulation is of interest for work-flow optimization, automated surveillance, and programming by demonstration (PbD) applications, in which a robot learns how to perform a task by watching a human do the same task.

An important cue to the class of a manipulation action is the object handled; for example, seeing a human bring a cup towards his/her face brings us to believe that he/she is drinking, without actually seeing the fluid. Similarly, a strong cue to the class of the object involved is the action; for example, a cup is to some extent defined as something you drink from. Therefore, it is beneficial to *simultaneously* recognize manipulation actions and manipulated objects.

A manipulation action is here defined as beginning with the picking-up of an object and ending with the putting-down of that object. Only one-hand actions are considered, although this is not a limitation to the method in a formal sense. In a video sequence of the action, the human head and hand are segmented and tracked using skin color, and objects are segmented as being in the neighborhood of the hand and moving with it.

D. Forsyth, P. Torr, and A. Zisserman (Eds.): ECCV 2008, Part II, LNCS 5303, pp. 336–349, 2008.

The action state space in each frame is the image position of the hand relative to the face, and the shape of the hand, represented with a gradient orientation histogram pyramid [1,2]. Section 5 shows the inclusion of hand shape in the state space to greatly improve action recognition compared to only hand position.

Objects in this application are "graspable" i.e. fairly rigid, so shape is a good object descriptor. We use pyramids of gradient orientation histograms for representation of object shape as well as hand shape. Experiments in Section 5 show this representation to lead to a state-of-the-art classification performance on the NORB dataset [3] which contains objects of this type. Specific for our application is that the classification method has access to several views of the object over the course of the action, something that improves the recognition. Section 3 describes the feature extraction.

There are implicit, complex interdependencies in the object and action data. The sequence of object viewpoints, as well as occlusion from the hand depend on the action; i.e. what the hand is doing with the object. Similarly, the hand shape depends on the size, shape and surface structure of the object in the hand. These dependencies are difficult to model, which leads us to use a discriminative sequential classifier, conditional random fields (CRF) [4], that does not model the data generation process.

On a semantic level, there are also action-object dependencies of the type "drink"–"cup", "drink"–"glass", "write"–"pen", "draw"–"pen" and so on, which can be explicitly modeled within the CRF framework. The action-object dependence can be modeled on a per-frame basis using a factorial CRF (FCRF) [5]. However, it might be the case that the dependencies between particular frames are weaker than the dependence beween the action and the object as whole. To model sequence-level dependence, we introduce a CRF stucture which we call *connected hierarchic CRF:s (CHCRF)*. This is detailed in Section 4.

Experiments in Section 5 show three things. Firstly, CRF structures with many degrees of freedom, such as structures with hidden nodes or large data connectivity, perform worse than simple structures when the amount of training data is limited. Secondly, the correlated action-object recognition outperform separate classification, and CHCRF:s perform better than FCRF:s on the action-object classification task. Last, the information on actions implicit in the object data is redundant to the information on objects in the action data.

The primary contribution of this paper is the idea of recognizing manipulation actions and manipulated objects in context of each other, while secondary contributions are the definition of the CHCRF and the representation of object shape using pyramids of gradient orientation histograms.

2 Related Work

Actions. In the last few years, considerable research effort has been spent on the analysis of human motion from video [6]. For the purpose of detecting atomic actions in video, Laptev and Pérez [7] use a boosted classifier of spatiotemporal volumes of optical flow. This approach is robust to significant changes in scale, appearance and viewpoint. Our method differs from [7] in that it is possible to model actions with a longer extension in time, possibly with a sub-structure. Furthermore, since our analysis involves the manipulated object, the hand needs to be located, putting different constraints on our feature extraction.

The analysis of hand motion is most often applied to gesture recognition for human-computer interfaces or sign language recognition [8]. These applications are often characterized by low or no occlusion of the hands from other objects, and a well defined and visually disparate set of hand poses; in the sign language case the gestures are designed to be easily separable to simplify fast communication. In contrast, the manipulation actions which we investigate suffer from large intra-class variability and sometimes occlusion of parts of the hand from the manipulated objects.

Feature extraction for hand action classification often means tracking the hand in 2D [9] or 3D [10,11]. However, to be able to handle low video frame-rates we prefer to use a segmentation-based method for human pose recovery not relying on time-incremental estimation [12,13,14].

Objects. Object recognition is a vast area of research and can be regarded as one of the core problems in computer vision. We do not make an attempt to review the whole field, but focus on contextual object recognition and the representation of shape in object recognition.

The caption of an image says something about what objects can be expected in it. When labeling images according to object content, any captions should therefore be taken into account. Caption-guided object detection can be used to segment the image into object regions and associate them with object labels [15], or to automatically label or cluster a large set of unlabeled images with captions given a smaller set of labeled images with captions [16].

In [17,18,19], the scene itself, the "gist" of the image, is used to guide object recognition. The scene itself is a strong prior cue as to which objects can be expected and where they are most likely to be found. CRF:s have also been used [19,20] to automatically learn sub-structure; the relations between different parts of the object or between different objects and the scene.

Earlier work on contextual object recognition [21,22] has focused on functional object recognition; objects are then classified in terms of their function. This is similar in spirit to our contextual recognition; object classes are here defined by how the objects are used (in which action context they appear), and classes of manipulation actions are defined by the class (or classes) of objects that are involved in the action.

Modeling shape is difficult; an important tradeoff is the sparseness of the representation (from silhouettes [13], via edge maps [14,23], to maps of gradient orientation) versus the robustness towards differences in lighting and fine object texture. We use pyramids of localized histograms of gradient orientation, a representation robust to small position and fine texture differences, while containing more texture information than e.g. edge maps.

Actions and objects. Moore et al. [24] provide a Bayesian framework for recognizing scenes, objects in it, and actions being performed on the objects. A system such as this could be the framework for our method: While their system keeps track the hands of a human, objects handled by the human, and which actions are performed on which objects, our method is concerned with classifying a single action and object.

Wu et al. [25] continue along this line, learning a dynamic Bayesian network model that represent *temporal sequences* of actions and objects involved in the actions. The features used for classification are RFID tags attached to the objects and the human hand. Again, our method could be incorporated in such a system, replacing the RFID-based classificiation with simultaneous classification of objects and actions from video.

Earlier on, Mann and Jepson [26] use a force-dynamic bottom-up approach to describe the interaction between hands and objects in scenes. In contrast, we use a statistical formulation, since the underlying process generating the video sequences in our case is far too complex to be modeled deterministically.

3 Features for Classification

Extraction of image features could be done in a variety of ways [6] depending on the purpose of the feature extraction. Features representing the human motion and appearance as well as object motion and appearance are of interest. As opposed to many other action recognition applications, like video annotation [7], it is here necessary to obtain the location of the human hand to find the manipulated object.

Considering the low framerate, large motion blur and low resolution of the hands of our video sequences, and that articulated hand tracking is a difficult problem [10,11], we use a segmentation based approach. The hand and face of the human is localized using skin color segmentation [27] and hand and face masks are extracted from the skin mask using connected components detection or with an $\alpha\beta$-filter when hand and face blobs merge.

Other cues than skin color, e.g. combinations of spatial or spatiotemporal filters, can of course also be exploited for the localization of hands and face.

The object involved in the manipulation action is in the human's right hand. To focus the attention of the object classification onto only that object, an object segmentation mask is also obtained, right of the hand in the image (based on the assumption that the object is in that area if grasped by the human). While the position of the area is automatically obtained from the hand position, the area shape is selected to fit the object in the first frame of the sequence, and is then held constant throughout the sequence.

Manipulation actions are here defined as beginning with a pick-up event and ending with a put-down event. With a fully automatic object segmentation, it is possible in each time-step to detect whether there is an object in the hand or not, so that the temporal segmentation can be done automatically.

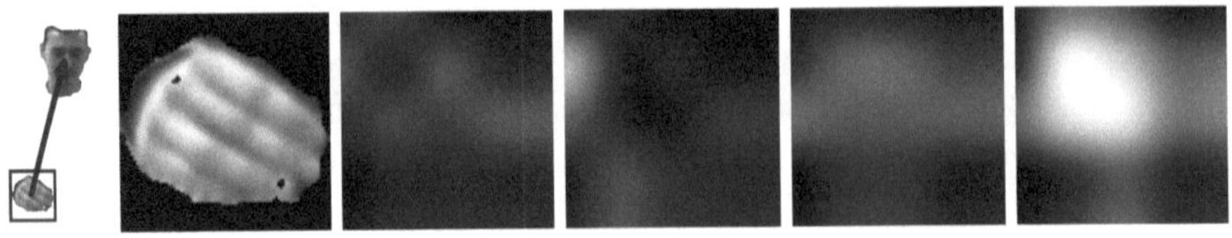

(a) Pos (b) Image (c) $\pi/8$ (d) $3\pi/8$ (e) $5\pi/8$ (f) $7\pi/8$

Fig. 1. Features used for action classification. a) 2D position p_t of the centroid of the right hand segment relative to centroid of the face segment. b) Hand image $\mathbf{I}_t^a$. c–f) Gradient orientation histograms from $\mathbf{I}_t^a$, with $B = 4$ bins, on level $l = 1$ in the pyramid of $L = 3$ levels. c) Bin 1, orientation $\pi/8$. d) Bin 2, orientation $3\pi/8$. e) Bin 3, orientation $5\pi/8$. f) Bin 4, orientation $7\pi/8$.

Action features. We seek a representation of hand motion that captures both the global hand position and the articulated hand pose over time. The global position of the hand at time t is represented with the 2D position p_t^a of the centroid of the hand mask relative to the centroid of the face mask (Figure 1a).

The local articulated hand pose is represented using gradient orientation histograms, frequently used for representation of human shape [1,2]. Gradient orientation $\boldsymbol{\Phi}_t \in [0, \pi)$ is computed from the segmented hand image $\mathbf{I}_t^a$ (Figure 1b) as $\boldsymbol{\Phi}_t = \arctan(\frac{\partial \mathbf{I}_t^a}{\partial y} / \frac{\partial \mathbf{I}_t^a}{\partial x})$ where x denotes downward (vertical) direction and y rightward (horizontal) direction in the image.

From $\boldsymbol{\Phi}_t$, a pyramid with L levels of histograms with different spatial resolutions are created; on each level l, the gradient orientation image is divided into $2^{L-l} \times 2^{L-l}$ equal partitions. A histogram with B bins is computed from each partition. Figures 1c–f show histograms at the lowest level of the pyramid.

The hand pose at time t is represented by the vector x_t^a which is the concatenation of the position p_t^a and all histograms at all levels in the pyramid. The length of x_t^a is thus $2 + B\sum_{l=1}^{L} 2^{2(L-l)}$. The performance of the classifier is quite insensitive to choices of $B \in [3, 8]$ and $L \in [2, 4]$; in our experiments in Section 5 we use $B = 4$ and $L = 3$. Before concatenation, p_t^a is normalized so that the standard deviations of the two dimensions of x_t^a originating from p_t^a have the same standard deviation in the training set (Section 5) as the remaining dimensions. The sequence of poses over the sequence is $\mathbf{x}^a = \{x_t^a\}, t = 1, \ldots, T$.

Object features. The objects considered in this application are all "graspable", i.e. more or less rigid. For example, a cup is graspable, but water is not. Shape can therefore be expected to be a good object class descriptor, while local descriptors like SIFT features [28] are unsuitable for our purposes.

Contrary to in many other object recognition applications, e.g. labeling of images according to object content, there is no search for object location involved. For manipulated objects, the position of the hand grasping the object gives an indication of the expected object location, and the recognition problem becomes one of classifying the given region. However, there might be deviations in position and orientation of the object within this region, as well as deviations in

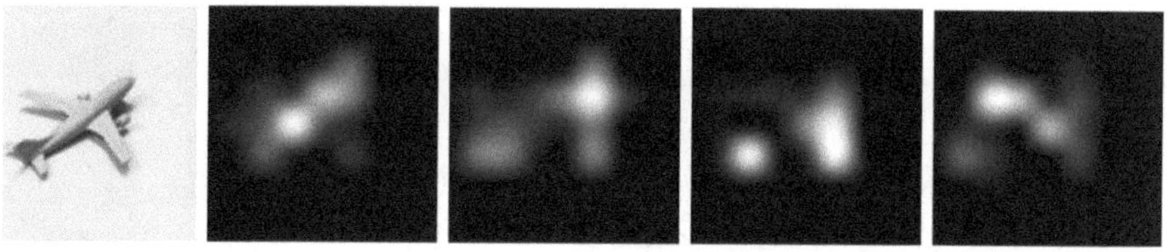

(a) Image (b) $\pi/8$ (c) $3\pi/8$ (d) $5\pi/8$ (e) $7\pi/8$

Fig. 2. Features used for object classification. a) Object image $\mathbf{I}^o_t$. b–e) Gradient orientation histograms from $\mathbf{I}^o_t$, with $B = 4$ bins, on level $l = 1$ in the pyramid of $L = 4$ levels. b) Bin 1, orientation $\pi/8$. c) Bin 2, orientation $3\pi/8$. d) Bin 3, orientation $5\pi/8$. e) Bin 4, orientation $7\pi/8$.

size, shape and color of different instances within each object class. The descriptor must therefore be insensitive to these intra-class variations, but still capture inter-class variations.

We select the same gradient orientation histogram representation as for the hand shape, a description that captures the shape of the object, with a certain insensitivity to absolute greylevels and small displacements of object parts (Figure 2). Note that this representation is not invariant to e.g. in-plane rotations; this is deliberate, since global orientation is indicative of object class in our application. The object at time t is represented by x^o_t, the concatenation of all histograms at all levels in the pyramid. In Section 5 this representation is evaluated on the problem of recognizing generic object categories, and it is found to be robust to intra-class variability in shape, orientation, position, rotation and lighting conditions, while maintaining a good inter-class discriminability.

Another factor specific to this object recognition application is that the data consist of not only one, but a sequence of object views. In most cases, parts of the object are also occluded by the human hand grasping it. The change in orientation of the object with respect to the camera during the sequence and the occlusion from the hand are descriptive of the object, since they reflect the way this object class is used by the human; they can be termed "typical view sequences" and "typical occlusions". Thus, the classifier should take the whole sequence of object views into account. Each measurement is therefore described by a sequence of descriptors $\mathbf{x}^o = \{x^o_t\}, t = 1, \ldots, T$.

Correlation between action and object features. As discussed in the introduction, the shape of the hand encoded in $\mathbf{x}^a$ gives cues about the object as well, since humans grasp different types of objects differently, due to object function, shape, weight and surface properties. Similarly, the view change in $\mathbf{x}^o$ over the course of the sequence is correlated with the type of action performed with the object. This representation of the correlation between manipulation actions and manipulated objects is *implicit* and difficult to model accurately, but should be taken into account when modeling the simultaneous action-object recognition.

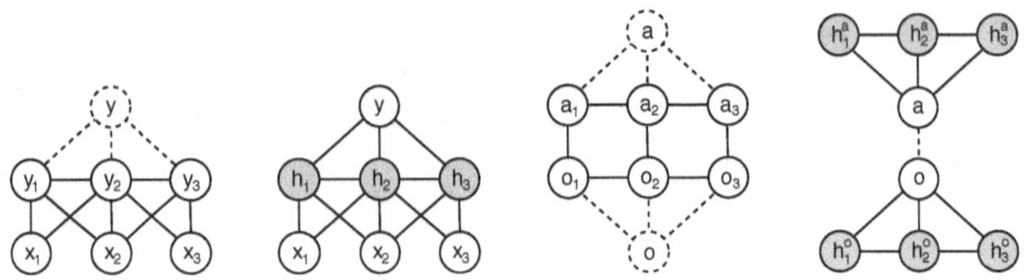

(a) CRF (b) HCRF (c) FCRF (d) CHCRF

Fig. 3. Different CRF structures used for action-object recognition with pre-segmented data. Dotted edges and nodes indicate that the weights θ associated with these are constrained during the training of the CRF. a) Linear-chain CRF [4]. b) Hidden CRF [20]. c) Factorial CRF [5], data layer not shown. d) Connected hierarchic CRF:s, data layer not shown.

4 Discriminative Classification Using CRF:s

Since we can expect complex dependencies within our action data $\mathbf{x}^a$ and object data $\mathbf{x}^o$ over time, a discriminative classifier which does not model the data generation process is preferable over a generative sequential classifier like a hidden Markov model (HMM) [29]. We thus employ conditional random fields (CRF:s) [4] which are undirected graphical models that represent a set of state variables $\mathbf{y}$, distributed according to a graph $\mathcal{G}$, and conditioned on a set of measurements $\mathbf{x}$. Let $C = \{\{\mathbf{y}_c, \mathbf{x}_c\}\}$ be the set of cliques in $\mathcal{G}$. Then,

$$P(\mathbf{y}|\mathbf{x};\theta) = \frac{1}{Z(\mathbf{x})} \prod_{c \in C} \Phi(\mathbf{y}_c, \mathbf{x}_c; \theta_c) \quad (1)$$

where Φ is a potential function parameterized by θ as

$$\Phi(\mathbf{y}_c, \mathbf{x}_c; \theta_c) = e^{\sum_k \theta_{c,k} f_k(\mathbf{y}_c, \mathbf{x}_c)} \quad (2)$$

and $Z(\mathbf{x}) = \sum_{\mathbf{y}} \prod_{c \in C} \Phi(\mathbf{y}_c, \mathbf{x}_c; \theta_c)$ is a normalizing factor. The feature functions $\{f_k\}$ are given, and training the CRF means setting the weights θ using belief propagation [4].

For linear-chain data (for example a sequence of object or action features and labels), $\mathbf{y} = \{y_t\}$ and $\mathbf{x} = \{x_t\}$, $t = 1, \ldots, T$ as shown in Figure 3a. This means that the cliques are the edges of the model, which gives

$$P(\mathbf{y}|\mathbf{x};\theta) = \frac{1}{Z(\mathbf{x})} \prod_{t=2}^{T} \Phi(y_{t-1}, y_t, \mathbf{x}; \theta_t) \quad (3)$$

with a potential function

$$\Phi(y_{t-1}, y_t, \mathbf{x}; \theta_t) = e^{\sum_k \theta_{t,k} f_k(y_{t-1}, y_t, \mathbf{x})} . \quad (4)$$

Each state y_t can depend on the whole observation sequence $\mathbf{x}$ – or any subpart of it, e.g. the sequence $\{x_{t-C}, \ldots, x_{t+C}\}$, C being the *connectivity* of the model.

A CRF returns an individual label for each time-step, which means that the time-sequential data $\mathbf{x}$ to be classified do not have to be segmented prior to classification. However, if a segmentation is readily available, as in our case, the robustness of the classification is increased by assuming that for all labels within the segment, $y_t = y \, \forall \, t \in [1, T]$. With pre-segmentation, the model thus becomes $P(y|\mathbf{x}; \theta) \equiv P(\mathbf{y} = [y, \ldots, y]|\mathbf{x}; \theta)$. This is illustrated by the dotted layer in Figure 3a. We will formalize this below.

The introduction of a hidden layer, where each hidden label represents the classification of a sub-part of the sequence (Figure 3b), has been shown [20] to improve the recognition rate in situations where there is such a sub-structure present in the data. This is indeed the case in our object and action recognition problems. With an observable label y and a set of hidden labels $\mathbf{h}$ that form a time-chain, the probabilistic model of a hidden CRF (HCRF) becomes

$$P(y|\mathbf{x}; \theta) = \frac{1}{Z(\mathbf{x})} \sum_{\mathbf{h}} \prod_{t=1}^{T} \Phi(y, h_t, \mathbf{x}; \theta_t^h) \prod_{t=2}^{T} \Phi(y, h_{t-1}, h_t, \mathbf{x}; \theta_t^{hh}) \tag{5}$$

with potential functions

$$\Phi(y, h_t, \mathbf{x}; \theta_t^h) = e^{\sum_k \theta_{t,k}^h f_k(y, h_t, \mathbf{x})}, \quad \Phi(y, h_{t-1}, h_t, \mathbf{x}; \theta_t^{hh}) = e^{\sum_k \theta_{t,k}^{hh} f_k(y, h_{t-1}, h_t, \mathbf{x})}. \tag{6}$$

Both inference and parameter estimation can be done using exact methods, provided that there are no loops in the hidden layer [20]. However, the introduction of hidden parameters leads to a non-convex optimization problem, which means that the parameter estimation procedure requires more data to converge, and might also reach local optima. Note also that an HCRF requires pre-segmented data; for continuous classification, other structures with hidden layers have been presented, like the latent-dynamic CRF (LDCRF) [30]. CRF and HCRF performance for pre-segmented data is compared in the experiments in Section 5.

The CRF in Figure 3a is in fact a special case of the HCRF in Figure 3b, where the weights θ are restricted so that *A*) $\mathbf{h}$ are not hidden, $h_t = y \, \forall \, t \in [1, T]$; and *B*) equal weight is given to each timestep, $\theta_{t_1}^h = \theta_{t_2}^h \, \forall \, t_1, t_2 \in [1, T]$.

"Early fusion": factorial CRF. In Section 3 we argue that there are correlations between action observations $\mathbf{x}^a$ and object observations $\mathbf{x}^o$ implicit in the data. We make use of this correlation on the data level by not imposing a simplified model on the data generation process and instead using a discriminative classifier, CRF. However, there is also an *explicit*, semantic correlation between actions and objects on the label level, as discussed in the introduction. This correlation can be modeled in two ways, which we denote "early" and "late fusion". Early fusion corresponds to modeling the correlation on a per-frame basis, i.e. the correlations between the labels a_t and o_t for each frame of the action, using a factorial CRF (FCRF) [5]. Figure 3c shows an FCRF with two states, action class a_t and object class o_t, in each time-step t. The conditional dependence on data is omitted in the figure for visibility. The cliques in this model are

the within-chain edges $\{a_{t-1}, a_t\}$ and $\{o_{t-1}, o_t\}$, and the between-chain edges $\{a_t, o_t\}$. The probability of $\mathbf{a}$ and $\mathbf{o}$ is thus defined as

$$P(\mathbf{a}, \mathbf{o}|\mathbf{x}; \theta) = \frac{1}{Z(\mathbf{x})} \prod_{t=1}^{T} \Phi(a_t, o_t, \mathbf{x}; \theta_t) \prod_{t=2}^{T} \Phi(a_{t-1}, a_t, \mathbf{x}; \theta_{a,t}) \Phi(o_{t-1}, o_t, \mathbf{x}; \theta_{o,t}) . \tag{7}$$

The weights θ are obtained during training using loopy belief propagation [5]. As for the linear-chain CRF, pre-segmentation means that $a_t = a, o_t = o \forall t \in [1, T]$ and that the distribution over the two dotted layers in the FCRF can be expressed as $P(a, o|\mathbf{x}; \theta) \equiv P(\mathbf{a} = [a, \ldots, a], \mathbf{o} = [o, \ldots, o]|\mathbf{x}; \theta)$.

"Late fusion": connected hierarchic CRF:s. Late fusion corresponds to modeling the correlation on a sequence level. The assumption is here that it is the action label a that is correlated with the object label o, not the labels a_t and o_t of a particular frame. The structure of a CRF for "late fusion", called *connected hierarchic CRF:s (CHCRF)* is shown in Figure 3d. In the most general case the two linear-chain layers are hidden, and the probability of an action a and an object o conditioned on the data $\mathbf{x}$ is

$$\begin{aligned} P(a, o|\mathbf{x}; \theta) &= \frac{1}{Z(\mathbf{x})} \Phi(a, o, \mathbf{x}; \theta^{ao}) \sum_{\mathbf{h}^a} \prod_{t=1}^{T} \Phi(a, h_t^a, \mathbf{x}; \theta_t^a) \prod_{t=2}^{T} \Phi(a, h_{t-1}^a, h_t^a, \mathbf{x}; \theta_t^{aa}) \\ &\quad \sum_{\mathbf{h}^o} \prod_{t=1}^{T} \Phi(o, h_t^o, \mathbf{x}; \theta_t^o) \prod_{t=2}^{T} \Phi(a, h_{t-1}^o, h_t^o, \mathbf{x}; \theta_t^{oo}) \\ &= \frac{\Phi(a, o, \mathbf{x}; \theta^{ao})}{\sum_{a,o} \Phi(a, o, \mathbf{x}; \theta^{ao})} P(a|\mathbf{x}; \theta^a) P(o|\mathbf{x}; \theta^o) \end{aligned} \tag{8}$$

in analog with Eq (5). To make the training more efficient, the data dependency in the first term is omitted, becoming $\mathcal{K}(a, o) = \frac{\Phi(a,o;\theta^{ao})}{\sum_{a,o} \Phi(a,o;\theta^{ao})}$, the *co-occurrence rate* of action label a and object label o in the training data. The individual probabilities $P(a|\mathbf{x}; \theta^a)$ and $P(o|\mathbf{x}; \theta^o)$ are estimated as in Eq (3), Eq (5), or using any classification method that returns probability estimates. The parameters of the action and object classifiers are learned separately.

5 Experiments

Evaluation of the object features. The object feature representation was first evaluated on its own, without the CRF framework. For this we used the NORB dataset [3], which contains 5 different classes of rigid objects; "animals", "humans", "airplanes", "trucks", and "cars" with 10 instances of each, 5 for test and 5 for training.The database contains stereo views of each object from 18 different azimuths and 9 elevations in 6 different lighting conditions. Only the normalized-uniform part of the dataset, designed to test classification performance, was used. (The other part of the dataset is designed to test object detection, a task we do not claim to address with this method.)

To evaluate the suitability of the feature representation for modeling shape *categories*, a support vector machine (SVM) [31] was trained with feature vectors x^o extracted from the NORB training images as described in Section 3. Table 1 left shows the results compared to others. Our object view representation together with a standard classifier reached the same classification accuracy as a state-of-the-art method for object categorization [3], which indicates that the gradient orientation histograms capture the specifics of a shape class, while allowing a significant variability among instances of that class. In comparison, training on the raw image downsampled to a size of 32×32 led to twice the classification error (a surprisingly good result, as noted in [3], given that the task is object categorization, not instance recognition). Furthermore, we note that the incorporation of stereo does not add much to the accuracy.

Certain robustness towards differences in color and lighting, as well as small position errors of the object segmentation mask, is also desireable. In [3], this was tested by adding "jitter", i.e. small transformations to both the training and test set. However, this arguably tested how the methods performed with a larger test set, rather than how they could handle noise that was not seen before (not present in the training data). Therefore, we did a variant of this experiment where we added jitter to *only the test set* (Table 1 right). First, the overall brightness of each test image was varied. Our feature representation was very robust to this noise, which is expected since it relies solely on the gradient orientations and not on their value. In comparison, the raw image classification error grew much quicker. Then, the test images were shifted vertically and horizontally in a random manner. The feature representation was more sensitive to this noise, but less so than the raw image representation.

Simultaneous action and object recognition. For the purpose of testing the simultaneous action-object recognition, the OAC (Object-Action-Complex) dataset was collected. The dataset consists of 50 instances each of three different action-object combinations; "look through binoculars", "drink from cup", and "pour from pitcher". The actions were performed by 7 men and 3 women, 5 times each. The classes were selected so that the object and action data are complementary: two of the actions, "look through" and "drink from" are similar, while "cup" and "pitcher" are similar.

Only one instance of each object was used, so the full object representation generated a perfect classification performance with all parameter settings. To simulate the performance in the more general object category recognition case,

Table 1. Results on the normalized-uniform NORB dataset, percent error. Left: Classification error percentage compared to methods presented in [3]. Right: Generalization test; robustness to different amount of jitter in test data (training data unaltered).

	Mono	**Stereo**
Hist + SVM	6.4	6.2
Raw + SVM	12.6 [**3**]	—
Conv Net 80	—	6.6 [**3**]

		Brightness			**Shift**		
	±0	**±10**	**±20**	**±30**	**±3**	**±6**	**±9**
Hist + SVM	6.4	6.4	7.1	8	10.3	18.1	29.2
Raw + SVM	12.6 [**3**]	15.8	18	21	20.8	35.1	48.6

(a) Instance of "look through binoculars"

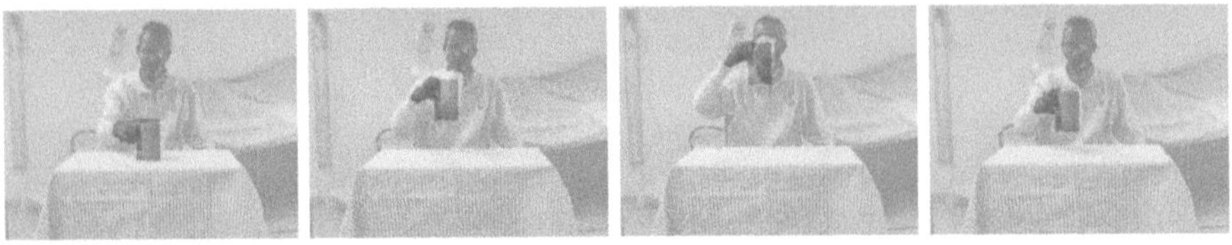

(b) Instance of "drink from cup"

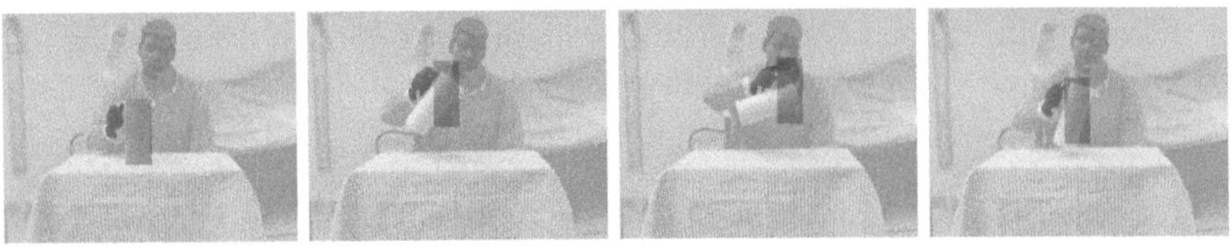

(c) Instance of "pour from pitcher"

Fig. 4. The three classes of the OAC dataset (for one person, instance each). Training and testing was performed in a jackknife manner, where the 15 sequences of one person at a time was used as a test data, the 135 sequences of the other 9 persons as training set.

all spatial information was removed, by using only $L = 1$ level in the gradient orientation histogram pyramid, and by normalizing the object segmentation window with respect to aspect ratio (i.e. scale all object segments to squares). The experiments below are not indicative of the object classification, but rather of the action classification and the benefits of combining object and action classification for manipulation action applications.

Table 2 shows separate action and object classification with different parameter settings. The baseline for actions is a representation without spatial information – only 2D pos $\mathbf{p}^a$ over time, and for objects a representation without temporal information – only the first frame x_1^o. Two things can be noted. Firstly, while the object classification does well without the temporal information, the

Table 2. Experiments with separate action and object (H)CRF classification with different connectivity $\mathcal{C}$, percent error on the format "median (max)" of 9 runs

	CRF conn 0	**CRF conn 1**	**CRF conn 2**	**HCRF conn 0**	**Baseline**
Actions	5.3 (9.3)	6 (12.7)	10.7 (18.7)	14.7 (21.3)	17.3 (20) **(2D)**
Objects	8 (8.7)	11.3 (13.3)	20.7 (28)	24.7 (28.7)	8.7 **(1:st fr)**

Table 3. Experiments comparing late (CHCRF) and early (FCRF) fusion, percent error on the format "median (max)" of 9 runs. Connectivity 0 everywhere.

	Connected Hiearchic CRF:s	Factorial CRF	Baseline (sep CRF:s)
Actions	3.3 (4.7)	5.3 (8)	5.3 (9.3)
Objects	3.3 (4.7)	6 (12)	8 (8.7)

Table 4. How important is the implicit, data-level correlation to the overall action-object classification? Classification where none, either, or both cues are stripped baseline versions, percent error on the format "median (max)" of 9 runs. Late (CHCRF) fusion, connectivity 0 everywhere.

	Actions	Actions Baseline (only 2D pos)
Objects	3.3 (4.7)	2.7 (8)
Objects Baseline (SVM 1:st fr)	4.7 (8.7)	14.7 (18)

action is to a large degree determined by spatial information, i.e. the shape of the hand. Secondly, when comparing to similar experiments in [20] where HCRF:s outperformed CRF:s, we draw the conclusion that it depends on the amount of training data. With relatively little data, a model with fewer parameters will perform better. In the applications we are considering, the less training data needed, the better, since data collection takes time and the system should be adaptable to different environments.

When comparing late and early fusion (Table 3) we see that while late fusion (CHCRF) greatly improves the classification, there is only a marginal improvement in the classification with early fusion (FCRF). There are two possible interpretations which might both be true: *A*) the per-frame correlation of actions and objects is simply a bad model of reality; *B*) the larger set of parameters defining the FCRF leads to a worse performance with our relatively small training set. As a side note, the CHCRF object and action classification rates are identical since the co-occurrence matrix $\mathcal{K}$ (Eq (8)) is diagonal in this example. A non-unique action-object mapping (e.g. "drink"–"glass" and "drink"–"cup") would lead to differences in action and object classification rate.

How important is the implicit, data-level correlation compared to the explicit, semantic correlation modeled in the FCRF and CHCRF? To test this, late fusion was performed with full data, $\mathbf{x} = [\mathbf{x}^a, \mathbf{x}^o]$, correlation in either cue removed, $\mathbf{x} = [\mathbf{x}^a, x_1^o]$ or $[\mathbf{p}^a, \mathbf{x}^o]$, and all correlation removed, $\mathbf{x} = [\mathbf{p}^a, x_1^o]$. From Table 4, it seems that the information on objects in the action data $\mathbf{x}^a$ and the information on actions in the object data $\mathbf{x}^o$ is largely redundant; the classification performance is not affected to any greater extent by using *either* $\mathbf{p}^a$ or x_1^o, but if *all* correlation data are removed, the classification is seriously affected.

6 Conclusions

A method for simultaneous sequential recognition of manipulation actions and manipulated objects was presented, employing CRF:s trained with object and hand shape over the course of the action. Two different CRF structures for fusion of action and object classification were compared; FCRF:s, which model the correlation on a per-frame level, and CHCRF:s, a structure introduced in this paper, which correlate the action and object classifcation as whole. CHCRF:s outperformed FCRF:s on the task of correlated classification of action and object sequences, probably because the action and object data are not correlated on a per-frame basis.

CRF structures with many degrees of freedom, such as HCRF:s, or CRF:s with high data connectivity, were found to perform worse than simpler structures with relatively small amounts of training data, although they have higher descriptive power. Thus, in applications where the training data are limited, the complexity of the model should be selected so that the training procedure will converge with the amount of training data at hand. Moreover, the implicit information on actions in the object data and on objects in the action data was found to be redundant. Thus, removing the correlated data in one cue or the other can be done without affecting the overall classification, while removing the correlated data in both cues will have serious effects on the classification rate.

The representation of object shape using a pyramid of gradient orientation histograms was shown to give state-of-the-art classification results on the NORB dataset, indicating that the representation is robust to intra-class differences in quite general shape categories, while capturing inter-class differences.

In the future, we plan to incorporate the method for correlated action and object classification into a PbD framework such as [24,25].

Acknowledgments. This research has been supported by the EU through the project PACO-PLUS, FP6-2004-IST-4-27657, and by the Swedish Foundation for Strategic Research.

References

1. Freeman, W.T., Roth, M.: Orientational histograms for hand gesture recognition. In: IEEE Int. Conf. Automatic Face and Gesture Recognition (1995)
2. Shakhnarovich, G., Viola, P., Darrell, T.: Fast pose estimation with parameter sensitive hashing. In: ICCV, vol. 2, pp. 750–757 (2003)
3. LeCun, Y., Huang, F.J., Bottou, L.: Learning methods for generic object recognition with invariance to pose and lighting. In: CVPR, vol. 2, pp. 97–104 (2004)
4. Lafferty, J., McCallum, A., Pereira, F.: Conditional random fields: Probabilistic models for segmenting and labeling sequence data. In: ICML (2001)
5. Sutton, C., Rohanimanesh, K., McCallum, A.: Dynamic conditional random fields: Factorized probabilistic models for labeling and segmenting sequence data. In: ICML (2004)
6. Moeslund, T.B., Hilton, A., Krüger, V.: A survey of advances in computer vision-based human motion capture and analysis. CVIU 104, 90–126 (2006)

7. Laptev, I., Pérez, P.: Retrieving actions in movies. In: ICCV (2007)
8. Pavlovic, V.I., Sharma, R., Huang, T.S.: Visual interpretation of hand gestures for human-computer interaction: A review. PAMI 19, 677–695 (1997)
9. Yang, M., Ahuja, N., Tabb, M.: Extraction of 2d motion trajectories and its application to hand gesture recognition. PAMI 24, 1061–1074 (2002)
10. Stenger, B., Thayananthan, A., Torr, P.H.S., Cipolla, R.: Model-based hand tracking using a hierarchical bayesian filter. PAMI 28, 1372–1384 (2006)
11. Sudderth, E., Mandel, M.I., Freeman, W.T., Willsky, A.S.: Visual hand tracking using non-parametric belief propagation. In: IEEE Workshop on Generative Model Based Vision (2004)
12. Sminchisescu, C., Kanujia, A., Metaxas, D.: Conditional models for contextual human motion recognition. CVIU 104, 210–220 (2006)
13. Belongie, S., Malik, J., Puzicha, J.: Shape matching and object recognition using shape contexts. PAMI 24, 509–522 (2002)
14. Sullivan, J., Carlsson, S.: Recognizing and tracking human action. In: Heyden, A., Sparr, G., Nielsen, M., Johansen, P. (eds.) ECCV 2002. LNCS, vol. 2350, pp. 629–644. Springer, Heidelberg (2002)
15. Carbonetto, P., de Freitas, N., Barnard, K.: A statistical model for general contextual object recognition. In: ECCV, vol. 1, pp. 350–362 (2004)
16. Quattoni, A., Collins, M., Darrell, T.: Learning visual representations using images with captions. CVPR (2007)
17. Torralba, A.: Contextual priming for object detection. IJCV 53, 169–191 (2003)
18. Murphy, K., Torralba, A., Freeman, W.T.: Using the forest to see the trees: A graphical model relating features, objects, and scenes. In: NIPS (2003)
19. Torralba, A., Murphy, K., Freeman, W.T.: Contextual models for object detection using boosted random fields. In: NIPS (2004)
20. Quattoni, A., Wang, S., Morency, L., Collins, M., Darrell, T.: Hidden conditional random fields. PAMI 29, 1848–1852 (2007)
21. Rivlin, E., Dickinson, S.J., Rosenfeld, A.: Recognition by functional parts. CVIU 62, 164–176 (1995)
22. Stark, L., Bowyer, K.: Generic Object Recognition using Form and Function. World Sci. Ser. Machine Perception and Artificial Intelligence 10 (1996)
23. Jurie, F., Schmid, C.: Scale-invariant shape features for recognition of object categories. CVPR, 2, 90–96 (2004)
24. Moore, D.J., Essa, I.A., Hayes, M.H.: Exploiting human actions and object context for recognition tasks. In: ICCV (1999)
25. Wu, J., Osuntogun, A., Choudhury, T., Philipose, M., Rehg, J.M.: A scalable approach to activity recognition based on object use. In: ICCV (2007)
26. Mann, R., Jepson, A.: Towards the computational perception of action. CVPR (1998)
27. Argyros, A.A., Lourakis, M.I.A.: Real time tracking of multiple skin-colored objects with a possibly moving camera. In: ECCV, vol. 3, pp. 368–379 (2004)
28. Lowe, D.G.: Distinctive image features from scale-invariant keypoints. IJCV 60, 91–110 (2004)
29. Rabiner, L.R.: A tutorial on hidden Markov models and selected applications in speech recognition. Proceedings of the IEEE 77(2), 257–286 (1989)
30. Morency, L.P., Quattoni, A., Darrell, T.: Latent-dynamic discriminative models for continuous gesture recognition. In: CVPR (2007)
31. Chang, C.C., Lin, C.J.: LIBSVM: a library for support vector machines (2001), software available at `http://www.csie.ntu.edu.tw/~cjlin/libsvm`

Active Contour Based Segmentation of 3D Surfaces

Matthias Krueger, Patrice Delmas, and Georgy Gimel'farb

Dept. of Computer Science, Tamaki Campus
The University of Auckland, Auckland, New Zealand
mkru007@aucklanduni.ac.nz

Abstract. Algorithms incorporating 3D information have proven to be superior to purely 2D approaches in many areas of computer vision including face biometrics and recognition. Still, the range of methods for feature extraction from 3D surfaces is limited. Very popular in 2D image analysis, active contours have been generalized to curved surfaces only recently. Current implementations require a global surface parametrisation. We show that a balloon force cannot be included properly in existing methods, making them unsuitable for applications with noisy data. To overcome this drawback we propose a new algorithm for evolving geodesic active contours on implicit surfaces. We also introduce a new narrowband scheme which results in linear computational complexity. The performance of our model is illustrated on various real and synthetic 3D surfaces.

1 Introduction

1.1 Motivation

Recently, there has been an increasing interest in generalising methods that are well-known and understood for planar images, e.g. isotropic and anisotropic diffusion [1,2], scale space analysis [2,3] or image inpainting [4], onto images painted on curved surfaces. The goal is to adequately process images from, for instance, 3D scanners or surface reconstruction algorithms.

However, due to the Gauss' famous *theorema egregium* (see e.g. [5]) surfaces with a non-zero Gaussian curvature cannot be mapped onto the plane without metric distortions. Moreover, looking at particular applications, recent research [6,7,8] shows that multi-modal approaches, combining 2D (texture) and 3D (shape) information, perform better than either 2D or 3D techniques alone in face biometrics. Thus, it is necessary to study object segmentation approaches, in particular active contour models, that incorporate both 2D and 3D information.

Object segmentation is one of the foremost problems in computer vision, and Spira and Kimmel [9] first attempted to apply active contours , namely geodesic active contours (GAC), to images on parametric surfaces. While working on parametric surfaces is an intuitive approach, it has a number of drawbacks. In Sect. 4 we will explain in more detail the most serious one: the problems

D. Forsyth, P. Torr, and A. Zisserman (Eds.): ECCV 2008, Part II, LNCS 5303, pp. 350–363, 2008.

arising when a balloon term pushing the contour with a constant force is added. Without a balloon term, however, the application of the method is limited to data involving no or very little noise.

Krueger et al. [10] proposed an equation for GACs on implicitly represented surfaces. Their approach avoids the problems of the parametric approach, but suffers from low computational efficiency. Applying Cheng et al.'s [11] framework for geometric curve evolution on implicit surfaces, Krueger et al. [10] obtain an algorithm with quadratic complexity. For large datasets the algorithm results in processing times which are infeasible in practice (see Sect. 5).

In this article we take up the implicit approach of Krueger et al. [10] and propose a new numerical scheme to reduce its computational complexity to linear. Our efficient implementation in conjunction with the versatility and robustness of the implicit surface representation makes our model applicable to a wide range of synthetic and natural surfaces.

1.2 Related Work

Active contours (also known as "snakes") were introduced in the pioneering work by Kass et al. [12]. Derived from the gradient flow of an energy functional, a snake is a curve that deforms its shape to minimise the total internal and external energy. The internal energy imposes regularity on the curve while the external energy attracts the curve to desired image features like edges. Later, Caselles et al. [13] introduced the GAC, based on an energy functional $\mathcal{E}$ that integrates a positive edge indicator function f along the contour C:

$$\mathcal{E}(C) = \int_C f(C)\, ds \,. \tag{1}$$

This formalisation yields the evolution equation:

$$\partial_t C = (\kappa f - \langle \nabla f, \nu \rangle)\, \nu \,, \tag{2}$$

where $\partial_t C := \frac{\partial C}{\partial t}$, κ, and ν are the time derivative, the curvature, and the unit normal field, respectively. The function f originally had small values for large image gradients and higher values elsewhere. Equation (2) is the gradient descent flow of the energy functional in (1), i.e. a curve evolving according to (2) decreases its energy as fast as possible. It converges towards a steady state if a sufficiently stable minimum of (1) is reached. One major drawback of snakes, their dependency on their parameterisation, was overcome by applying a geometric flow, namely a weighted curvature flow, to regularise the evolving curve. Caselles et al. [13] applied the level sets method [14] allowing the curve to change its topology during the evolution. They further proposed a general GAC model including a balloon force which shrinks or inflates the curve by a constant speed c:

$$\partial_t C = (f \cdot (\kappa - c) - \langle \nabla f, \nu \rangle)\, \nu \,. \tag{3}$$

Spira and Kimmel [9] applied the GAC model (2) to segment images painted on parametric surfaces. The evolution equation in this generalised setting,

$$\partial_t C = (\kappa_g f - \langle \nabla f, \hat{\nu} \rangle)\, \hat{\nu} \,, \tag{4}$$

is similar to (2). Here, κ_g denotes the geodesic curvature and $\hat{\nu}$ is the unit co-normal vector of C on the curved surface M. The co-normal $\hat{\nu}$ is defined as the normalised cross product of the normal ν and the tangent vector field C_s. The geodesic curvature is the tangential component of the curvature. However, it is beyond the scope of this paper to cover the geometric framework of curves on surfaces (see e.g. [9] and references therein). Spira and Kimmel [9] projected (4) onto the parameterisation plane of M and solved the resulting PDE applying the level sets framework [14].

Bogdanova et al. [2] pursued a similar approach. They considered GAC on three particular parametric surfaces, namely, the sphere, hyperboloid, and paraboloid, and applied their results to omnidirectional images. In contrast to Spira and Kimmel [9], Bogdanova et al. [2] derived the flow equations directly from an adapted version of the GAC energy functional (cf. (1)).

Cheng et al. [11] studied basic geometric curve evolution models on implicit surfaces, including the geodesic curvature, advection and normal flows.

Jin et al. [15] studied segmentation on surfaces in a slightly different context. To reconstruct a 3D scene from multiple 2D views in a variational framework, they simultaneously evolved both a level set surface and a radiance pattern of the surface. Discontinuities of the radiance were modeled by curves on the surface. Jin et al. [15] applied Cheng et al.'s [11] framework to evolve curves on the surface. However, their model is only 2.5D because the actual image segmentation according to a Mumford-Shah model takes place after projecting the surface to the original 2D views. Their approach is not applicable to general images on implicit surfaces since in general 2D projections are not available.

Krueger et al. applied Cheng et al.'s [11] work to propose a GAC model for images on implicit surfaces as succinctly outlined in [10]. We will review this approach in more detail in Sect. 2.

1.3 Contribution of the Paper

In this paper we present a new framework for feature extraction from 3D surface data using GACs. Working with implicitly defined surfaces we avoid numerical difficulties inherent to the parametric approach. Most importantly, in contrast to Spira's work [9] a balloon force can be integrated easily in our approach, making it applicable to real data which are usually noisy. To reduce the computational complexity of the implicit approach to linear we propose a new narrowband algorithm. Our experimental results show that in contrast to previous works our model can be applied to most of the complex surfaces in practice. To underline that, we present results of applying our 3D GAC model to human faces as well as other intricate, non-synthetic surfaces. To the best of our knowledge, this has not been done before.

The paper is organized as follows. Section 2 reviews Cheng et al.'s [11] framework for geometric curve evolution on implicit surfaces and the GAC model proposed by Krueger et al. [10]. Section 3 gives details on the numerical implementation and introduces the new narrowband scheme. Theoretical considerations comparing our approach with the known works on parametric surfaces are made in Sect. 4. Section 5 presents results of our experiments on synthetic and natural surfaces, and Sect. 6 concludes the paper.

2 Geodesic Active Contours on Implicit Surfaces

After embedding a curve C as an isocontour in a 2D scalar function ϕ, the evolution of the curve is controlled by solving a suitable PDE for ϕ (see e.g. [14,16]). The GAC in (3) corresponds to the following level set PDE:

$$\partial_t \phi = \underbrace{cf|\nabla\phi|}_{I} + \underbrace{\langle \nabla f, \nabla \phi \rangle}_{II} + \underbrace{f\kappa|\nabla\phi|}_{III} , \tag{5}$$

where the terms I, II, and III denote a motion in normal direction, an advection flow, and a weighted mean curvature flow, respectively. In order to generalise (5) to implicit surfaces, Krueger et al. [10] applied Cheng et al.'s framework [11] for geometric curve evolution on implicit surfaces. The latter forming the backbone of the model discussed in this paper is briefly reviewed below.

Given an implicit surface $M = \{x \in \mathbb{R}^3 | \psi(x) = 0\}$, an important question is how to describe a curve on M. Since there is no parameterisation at hand, the level set method has to be extended to this case. Cheng et al. [11] proposed to describe a curve on an implicit surface by intersecting the surface with another implicit surface given by the level set function ϕ. This is shown for two example surfaces in Fig. 1.

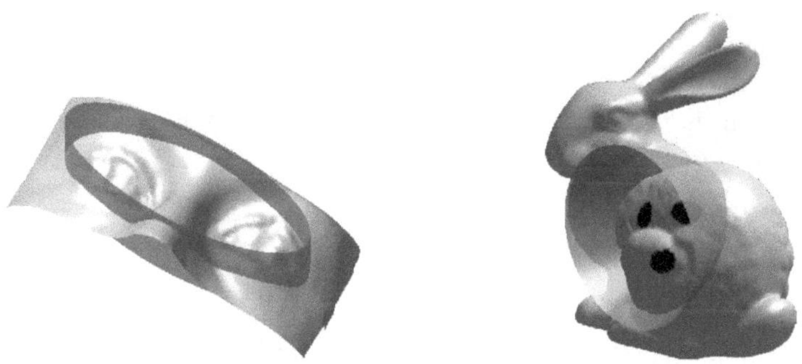

(a) (b)

Fig. 1. Initialisation of a curve on an implicitly given face (a) and bunny (b) by intersecting the surface with a cylinder and a paraboloid, respectively

It is well known [14,16] that the curvature flow equation in the level set setting is

$$\partial_t \phi = \left[\nabla \cdot \left(\frac{\nabla \phi}{|\nabla \phi|} \right) \right] |\nabla \phi| , \tag{6}$$

with the term in the square brackets being the curvature. In order to adapt this equation to the surface setting, Cheng et al. [11] projected all differential operators involved onto the surface M using the projection operator $P_w(v) = v - \frac{1}{|w|^2}\langle v, w\rangle w$, where v and w are vectors in $\mathbb{R}^3$. With $\nabla\psi(x)$ being orthogonal to M at point x, the operator $P_{\nabla\psi(x)}$ projects a vector onto the tangent plane T_xM of M at x. Cheng et al. [11] also showed that for a real valued function u defined on $\mathbb{R}^3$, the surface gradient ∇_M is given by $\nabla_M u = P_{\nabla\psi}\nabla u$. The surface divergence can be handled analogously. Projecting equation (6) onto the surface M yields the following formula for the geodesic curvature flow:

$$\partial_t\phi = \underbrace{\left[\nabla\cdot\left(\frac{P_{\nabla\psi}\nabla\phi}{|P_{\nabla\psi}\nabla\phi|}|\nabla\psi|\right)\right]\frac{1}{|\nabla\psi|}}_{\kappa_g}|P_{\nabla\psi}\nabla\phi| , \tag{7}$$

where κ_g is the geodesic curvature. Similar considerations lead Cheng et al. [11] to the equations for the geodesic advection flow under a static vector field V and the constant normal flow with velocity V_0. Those are given by $\partial_t\phi = \langle P_{\nabla\psi}V, \nabla\phi\rangle$ and $\partial_t\phi = V_0|P_{\nabla\psi}\nabla\phi|$, respectively.

Applying Cheng et al.'s [11] framework, Krueger et al. [10] generalized the GAC model (5) to implicit surfaces. With a surface M embedded in a cube $\Omega \subset \mathbb{R}^3$, a greyscale image I is a mapping $I : M \rightarrow \mathbb{Z}^+$. To apply Cheng et al.'s [11] method, the image I has to be smoothly extended to at least a certain band surrounding M (see e.g. [16]). As in the classical GAC model, the image I should be smoothed before calculating an edge indicator function. If a textured image is mapped onto the surface, this can be achieved by applying a low pass 2D filter to the image. Alternatively, images can be smoothed directly on an implicit surface [1] or by applying a 3D filter to the volumetric image I.

If I_σ is the smoothed image, the edge indicator function can be defined as $f = \frac{1}{1+|\nabla I_\sigma|^p}$, $p \in \{1, 2\}$. Note, that f can be defined in terms of geometric quantities such as the surface curvature as well (see our experiments in Sect. 3.1). Then the equation for GACs on implicit surfaces is

$$\partial_t\phi = \underbrace{cf|P_{\nabla\psi}\nabla\phi| + \langle P_{\nabla\psi}\nabla f, \nabla\phi\rangle}_{I} + \underbrace{\mu f\kappa_g|P_{\nabla\psi}\nabla\phi|}_{II} . \tag{8}$$

where the parameters c and μ determine the influence of the balloon force and the regularity terms, respectively. The image gradient term has a normalised weight. Then the evolving contour on M is given by the zero level set $C(t) = \{x \in \Omega|\phi(x,t) = \psi(x) = 0\}$. In Sect. 3 we study the numerical implementation of (8) in more detail.

3 Numerical Implementation

The general numerical framework for implementing evolution equations on implicitly given surfaces was introduced succinctly by Cheng et al. [11]. Krueger et al. [10] illustrated the numerical procedures for implementing (8) in more detail. State-of-the-art techniques for solving level set equations [16] were used to obtain a stable numerical algorithm.

Analysing (8), we note the term II is diffusive and can be approximated by central differences. However, term I is of Hamilton-Jacobi (HJ) type, thus the derivatives have to be computed using Essentially Non Oscillating (ENO) or Weighted Essentially Non Oscillating (WENO) schemes to avoid numerical instabilities [14,16]. Following Krueger et al. [10], we approximate the hyperbolic term I by a Lax-Friedrichs (LF) scheme with global flux and the time derivative by forward differencing.

Cheng et al. [11] proposed to reinitialise the level set function in two stages:

1. Transforming ϕ to a signed distance function on M (see (9)),
2. Enforcing ϕ to have its level sets perpendicular to M (see (10)).

Both stages were achieved by iterating a few steps of the adequate partial differential equation, namely

$$\partial_t \phi = \text{sign}\,(\phi_0)(1 - |P_{\nabla\psi} \nabla\phi|) \tag{9}$$

and

$$\partial_t \phi + \text{sign}(\psi) \left\langle \frac{\nabla\psi}{|\nabla\psi|}, \nabla\phi \right\rangle = 0\,. \tag{10}$$

In our experiments we observed that fifth-order WENO schemes are necessary to accurately reinitialize the level set function. For the GAC equation (8) second-order ENO proved to be sufficiently accurate.

3.1 The Optimised Framework

Cheng et al. [11] localised the numerical calculations to a neighbourhood of the surface, obtaining quadratic complexity. In this paper, we propose a new algorithm with linear complexity for evolving GACs on implicit surfaces. Essentially, it extends the approaches introduced in [10,11,17].

Let $\widetilde{\psi}$ be a signed distance function for M, which can be computed efficiently by the fast marching method [18]. We also assume ϕ is close to a signed distance function on M at a time t_0, e.g. $t_0 = 0$. We retain $\widetilde{\phi} := \phi(., t_0)$ (see below). The main idea is to define a function θ approximating the distance of a grid point in space from the evolving curve at time t_0. We define θ to be the Euclidean norm of $\widetilde{\psi}$ and $\widetilde{\phi}$, i.e. $\theta(x) := \sqrt{|\widetilde{\psi}(x)|^2 + |\widetilde{\phi}(x)|^2}$. Since both $\widetilde{\psi}$ and $\widetilde{\phi}$ are close to the signed distance functions in $\mathbb{R}^3$ and on the surface M, respectively, the level sets of θ resemble circular tubes along the curve C.

Having defined θ, we localise the level set equations as in [17]: Let $0 < \beta < \gamma$ be two constants which determine the radii of the inner and the outer narrowband. For our setting (with unit grid size $\Delta x = 1$ and fifth-order schemes), we obtained good results with $\beta = 2$ and $\gamma = 4$ (see Fig. 2). Only intricate surfaces with a comparably low grid resolution – such as the Stanford bunny in our experiment in Sect. 5 – required a larger narrowband with $\beta = 3$ and $\gamma = 6$ as suggested in [17] for the use of fifth order WENO schemes.

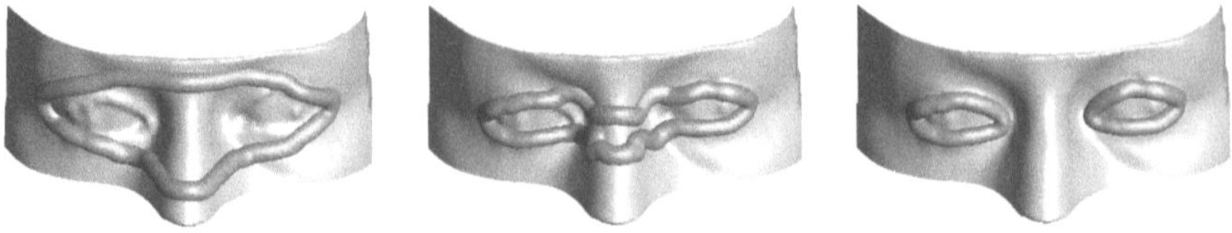

(a) (b) (c)

Fig. 2. Evolving narrowband on a face surface

Our algorithm for evolving GAC on implicit surfaces consists of the following steps:

Step 0. *Initialise distance functions* $\widetilde{\psi}$ *and* $\widetilde{\phi}$.
Since M is static, this has to be done only once for $\widetilde{\psi}$. However, the curve evolves and thus $\widetilde{\phi}$ has to be recomputed regularly (see Step 3).

Step 1. *Compute the narrowband* $T = \{x : \theta(x) \leq \gamma\}$ *and the extended narrowband* $N = \{x : \exists \mathbf{e} \in E : \theta(x + \mathbf{e}) \leq \gamma\}$ *with* E *the canonical basis of* $\mathbb{R}^3$.
θ only depends on $\widetilde{\psi}$ and $\widetilde{\phi}$. The narrowband has to be updated only if ϕ was reinitialised before and thus $\widetilde{\phi}$ has changed (see Step 3).

Step 2. *Update the level set function* ϕ *by solving a modified equation (11) in* T.
In order to avoid boundary stability issues, we localise the equation by multiplying with a cut-off function λ which equals one in the inner narrowband ($\theta \leq \beta$) and smoothly decreases to zero with $\theta \to \gamma$. A particular choice of λ is given in [17]. Then the modified equation is as follows:

$$\partial_t \phi = \lambda(\theta)\Big(cf|P_{\nabla\psi}\nabla\phi| + \langle P_{\nabla\psi}\nabla f, \nabla\phi\rangle + \mu f \kappa_g |P_{\nabla\psi}\nabla\phi|\Big)\,. \qquad (11)$$

Step 3. *Reinitialise if necessary, then return to Step 1.*
The level set function ϕ is reinitialised iff one of the three criteria is met:
1. The evolving curve reaches the tube $\{x : \theta(x) = 1\}$ of radius 1, (this ensures the updated narrowband T does not outreach the current extended narrowband N)
2. $|P_{\nabla\psi}\nabla\phi|$, differs from 1 significantly , or
3. The maximum number of steps is reached.

Reinitialisation is achieved by iterating a few steps of Eqs. (9) and (10) on the extended narrowband N. Since both these equations are hyperbolic, no boundary conditions are needed. Afterwards, we retain $\widetilde{\phi} = \phi(.,t)$ and return to Step 1.

4 Implicit vs. Parametric Surfaces

Below we illustrate, that GAC in the parametric framework introduced by Spira et al. [9] have a serious intrinsic drawback.

Previous work on GAC on non-Euclidean geometries using parametric surface representations [9,2] start from adapted versions of functional (1) to derive generalized equations for the GAC such as (4). Note, that no balloon force is included in this equation. In fact, an analysis of the respective equations shows that incorporating a balloon force in the parametric setting is problematic and results in a non-uniformly scaling equation. Without a balloon force however, GACs mostly fail when data are noisy, unless the contour is initialised very close to the desired object. In Sect. 5 we will support the theoretical considerations of this section with experimental results.

Starting from Spira et al.'s framework [9], for the sake of clarity we assume that the metric tensor g_{ij} is conformally equivalent to the Euclidean metric tensor, i.e. $g_{ij} = k \cdot \delta_{ij}$. This corresponds to a simple rescaling of an image. Then the equation for GAC on parametric surfaces reduces to

$$\partial_t \phi = f \frac{\kappa}{k} |\nabla \phi| + \frac{1}{k} \langle \nabla f, \nabla \phi \rangle \ , \tag{12}$$

where κ is the curvature of the evolving curve in the parameterisation plane. Comparing to the PDE (5) for a standard GAC, the curvature and the advection term scale equally with the factor k^{-1}. In fact, after adaption of the time step, the transformed equation (12) describes the same curve evolution as the original equation ((12) with $k = 1$). The situation is different with the balloon force term: in Spira's [9] framework the equation for the geodesic normal flow on parametric surfaces is given as $\partial_t \phi = -f |\nabla_M \phi|$.[1] After applying $g_{ij} = k \cdot \delta_{ij}$ again, this equation becomes $\partial_t \phi = -k^{-\frac{1}{2}} f |\nabla \phi|$. Note that relative to (5) with $c = 1$, this equation has the scaling factor $k^{-\frac{1}{2}}$. Therefore, adding a balloon term to the GAC equation (12) results in a PDE which does not scale uniformly.

While this is not a problem for standard 2D images with a constant metric, it has undesirable consequences if the metric is variable. In this case the PDE is ill-posed as the ratio between the weights of the balloon and other terms varies. Imagine a model surface with $k = 1$ in certain parts and a bigger $k > 1$ in other parts. Then, the balloon term and the further terms are weighted equally in the area with $k = 1$ (since $k^{-1} = k^{-\frac{1}{2}}$ for $k = 1$), while the balloon term is weighted higher in the area with $k > 1$ (as $k^{-\frac{1}{2}} > k^{-1}$ for $k > 1$). In the next section we will illustrate how this effect causes undesirable segmentation results.

The above problem is elegantly avoided in the implicit framework. Since implicit surfaces are embedded in the Euclidean space, they naturally inherit its flat metric. Thus, the ratio between the coefficients in the GAC equation remains constant and the above problems do not occur.

5 Experimental Results

We have tested our model on numerous synthetic and natural surfaces using an Intel Pentium M 1.6 GHz Notebook endowed with MATLAB®.

[1] Note, that the respective formula as stated in [9], $\partial_t \phi = -f g^{\frac{1}{2}} |\nabla_M \phi|$, is erroneous. This was acknowledged in personal communication by Prof. Ron Kimmel.

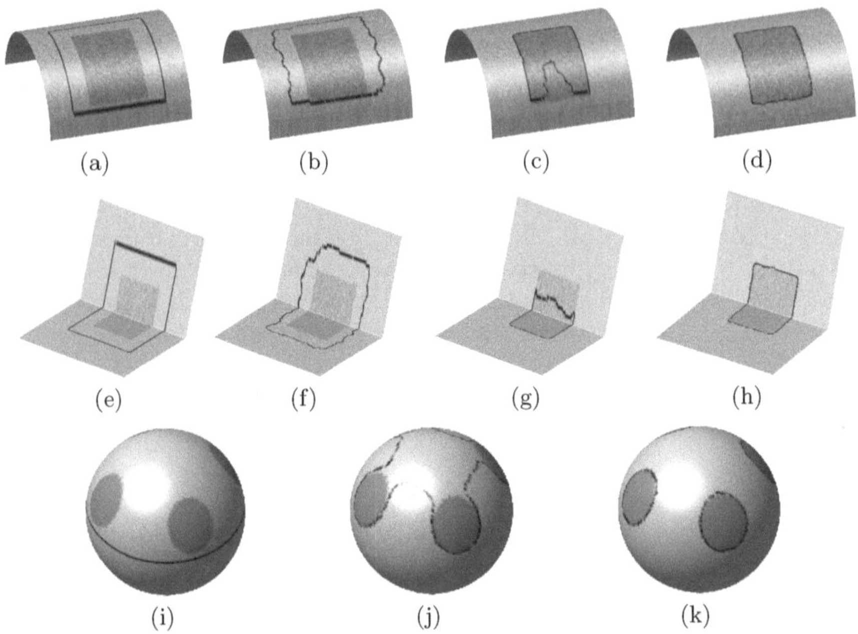

Fig. 3. (a) and (e): initial contours on a halfpipe and a notebook surface, resulting contours using Spira's parametric approach without (b,f) and with (c,g) an added balloon force. (d) and (h): results of our implicit method, (i)-(k): GAC evolution on an implicit sphere.

5.1 Comparison with Existing Methods

With our first experiments, we aimed to point out the main drawback of the GAC model [9], as analysed in Sect. 4. We could show that this drawback is overcome by our model. Both test surfaces, a halfpipe and a plane with a kink (we also call it 'notebook' surface), were parametrised as graphs $z = f(x)$. As the GAC model by Spira [9] includes no balloon force it failed in our tests with noisy texture data (see Figs. 3(b) and (f)). Then we added a balloon term to the equation (see the remarks in Sect. 4). Figures 3(c) and (g) show that the contour failed to converge to the desired object boundary, in fact it did not converge at all. In the surface areas with a larger slope – corresponding to large metric coefficients g_{ij} – the balloon term is weighted too high and thus the contour passes the object edges. We remark, that this is a systematic and reproducable effect.

Further note the following: The metric tensors of the surfaces in Fig. 3 are given by

$$(g_{ij}) = \begin{pmatrix} 1 + (\partial_x f)^2 & 0 \\ 0 & 1 \end{pmatrix} . \tag{13}$$

Table 1. Comparison of GACs using implicit and parametric surface representation: textures contain additive Gaussian noise with given variance values. The texture for the plane and the pyramid was similar to that of the notebook surface. The implicit algorithm with quadratic complexity was implemented as in [10]. BF means Balloon force, the parameterisation plane had the size of the first two dimensions of the implicit grid: e.g. 100 × 100 × 30 grid means 100 × 100 parameterisation plane. Uniform time step for the parametric model was $\tau = 0.4$, for the implicit models we chose $\tau = 0.4$ for the plane and the pyramid and $\tau = 1.0$ otherwise. GAC parameters: $c = 0.4$ and $\mu = 1$.

			quadr. impl.		lin. impl.			parametric	
Surface	Grid size	Noise	l^∞-err.	time	l^∞-err.	time	BF	l^∞-err.	time
Plane	100 × 100 × 30	0.003	1.18	752 s	1.2	191 s	yes	1.1	171 s
Pyramid	100 × 100 × 65	none	0.91	529 s	0.89	132 s	no	0.95	257 s
Halfpipe	130 × 140 × 75	0.003	1.73	3061 s	1.77	657 s	yes/no	fails	
Notebook	150^3	0.001	1.67	2445 s	1.65	468 s	yes/no	fails	
Sphere	100^3	0.001	0.92	1474 s	0.94	360 s	yes/no	not applicable	

Thus, the metrics of the surfaces vary only in one direction and the contours pass only the edges in the respective directions while converging to the others correctly (Figs. 3(c) and (g)). The desired results, as obtained by our method with implicit surface representations, are depicted in Figs. 3(d) and (h).

These experiments and the theoretical considerations in Sect. 4 show that using the parametric framework [9] a balloon force cannot be integrated properly into the GAC model. If the metric varies, the nonuniform scaling behaviour of the equation leads systematically to changing weights in the equation and undesired segmentation results.

Let us regard Table 1: the test in the plane shows that when both the parametric and the implicit framework can be applied, the quality of the results is comparable. Then we mapped a simple texture without noise to a pyramid surface and compared Spira's method without a balloon term to ours with a balloon term. While the accuracy results were similar, our method performed faster then Spira's due to much less needed iteration steps. The experiment on a sphere illustrates a further advantage of our framework: many surfaces, including the full sphere, cannot be parametrised without singularities, thus Spira's method is not applicable (cf. his experiments in [9], where the poles of the sphere are cut). In contrast, given an implicit sphere we can evolve GAC on it easily, as shown in Figs. 3(i)-(k).

Compared to the quadratic algorithm for GAC on implicit surfaces as proposed in [11] and applied in [10], the tests in Table 1 show a considerable speedup while maintaining the accuracy. Our tests confirm that the optimized framework presented in this paper has linear complexity and processing times basically depend on the length of the evolving curve.

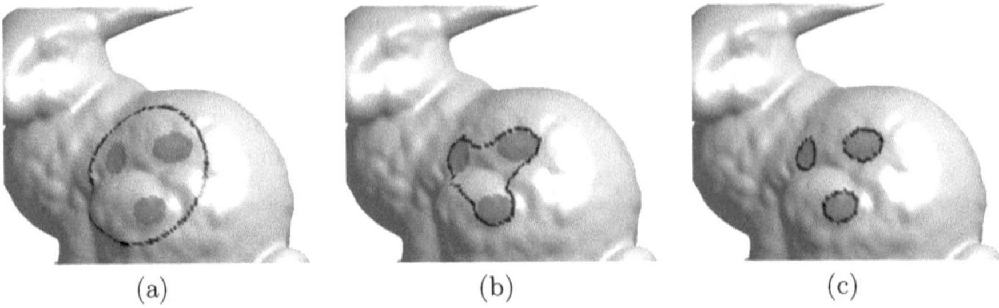

Fig. 4. Our GAC model applied to the Stanford bunny

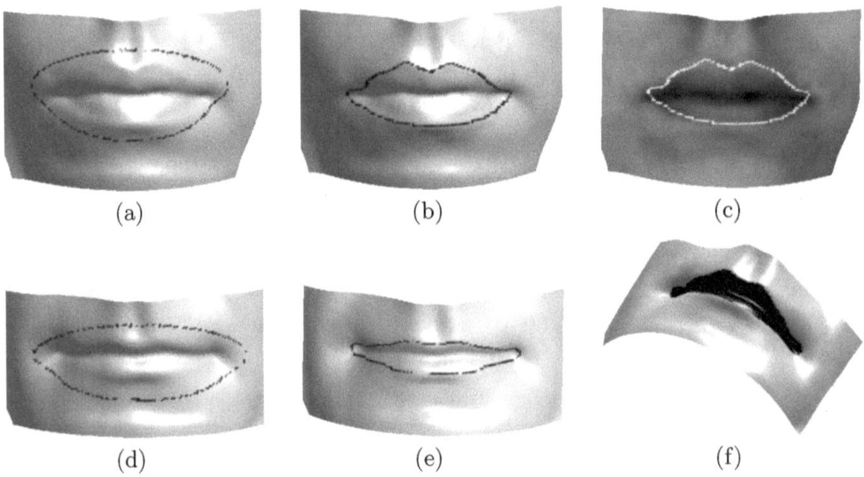

Fig. 5. 3D lip contour segmentation: initial (a,d) and final (b,e) contour, surface with added texture (c) and segmented lip area (f)

5.2 Application to 3D Face Feature Segmentation

One of the prospective application areas for our GAC model on surfaces is feature extraction from scanned 3D human faces, e.g. 3D lip contour segmentation. Feature extraction from 3D faces is important for face recognition and biometrics, yet remains a relatively young and unexplored field. However, recent research [6,7,8] emphasizes that multi-modal approaches, combining 2D and 3D information, perform better than either 2D or 3D methods alone.

As for 3D lip segmentation, we have so far tested our method with promising results on the Face Database of the MPI for Biological Cybernetics in Tuebingen, Germany [19] (see Fig. 5 for two examples and Table 2 for processing times). A detailed qualitative and quantitative evaluation is subject of our future work.

Note, that in this subsection on applications we did not compare our method to Spira et al.'s [9] as the data were noisy and a balloon force was needed. According to our analysis in Sect. 4 and our experiments in the last subsection this is not possible in the parametric framework proposed in [9].

Table 2. Performance of the proposed implicit GAC

Surface	Grid size	Parameters c	μ	Time step	Iterations	time
Bunny	$135 \times 160 \times 160$	0.4	1.0	0.4	600	480 s
Mouth1	$94 \times 143 \times 70$	0.03	0.15	1.0	900	296 s
Mouth2	$77 \times 147 \times 65$	0.03	0.15	1.0	1000	300 s
Both Eyes	$150 \times 66 \times 82$	0.035	0.25	1.0	1500	559 s
Left Eye	$150 \times 66 \times 82$	0.035	0.25	1.0	800	159 s
Right Eye	$150 \times 66 \times 82$	0.035	0.25	1.0	500	120 s
Ear	$123 \times 82 \times 114$	0.04	0.5	0.7	2000	724 s

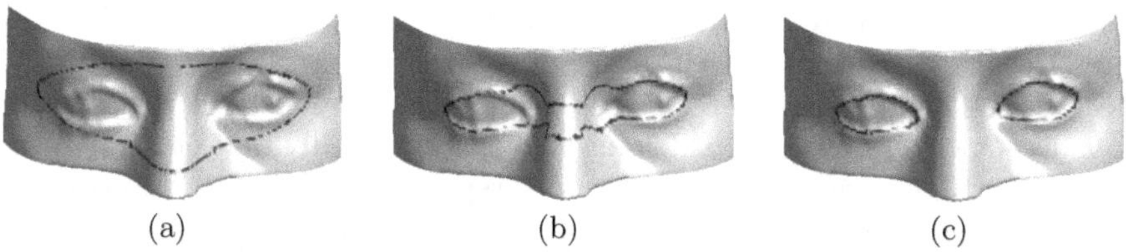

Fig. 6. 3D eye contour segmentation: Contour after 0, 800 and 1500 iterations

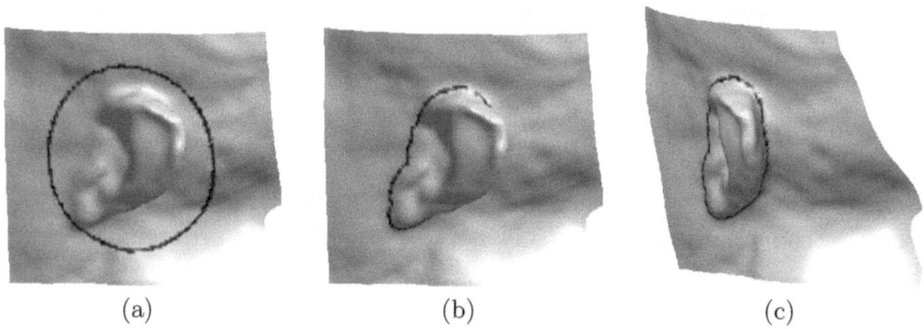

Fig. 7. 3D Ear segmentation: Initial (a) and final (b) contour, different view (c)

For the face experiments we defined the stopping term f in (8) by means of the mean curvature of the face surface instead of an image gradient, e.g. $f = \frac{1}{1+\tilde{H}}$ with $\tilde{H}$ a rescaled version of the mean curvature. For the lip contour experiments in particular, we designed a slightly different stopping term basing on the arccos-function. Details will be reported elsewhere. No texture data was needed to apply our method. It will be interesting to research, how our current results can be improved by incorporating the texture data.

Other face features like the eyes can be segmented by our GAC model as well. However, it is problematic that in 3D face data the eyes are often closed or of poor data quality. In Fig. 6 we show how we segmented the eyes. Note, that the results are good despite major artefacts in both eyes. The efficiency of the algorithm can be increased by segmenting both eyes separately rather than segmenting them in one step, provided the initial contours are sufficiently close to the eyes (see Table 2).

Given a full 3D head model, 3D ear segmentation could be an interesting future application of our method (cf. [8]). Due to lack of suitable 3D head data, we applied our model to one ear of the Max Planck bust [20] (see Fig. 7). Note, that this surface is very difficult to parametrise while easier to represent implicitly using algorithms such as [21,22].

6 Conclusion

This paper has proposed a new method for evolving geodesic active contours on implicitly represented surfaces. The proposed numerical scheme reduces the computational complexity of Cheng et al.'s [11] framework from quadratic to linear. This is a necessary prerequisite for applying the model to larger datasets. We have shown theoretically and confirmed experimentally that our approach is stable on virtually every surface that can be represented implicitly. Moreover, we have identified restrictions of the published works on parametric surfaces where adding a balloon term leads to a non uniformly scaling equation causing undesired segmentation results. The strictly implicit approach and the integration of a balloon term makes our model a natural generalization of the standard GAC to curved surfaces. Initial results on 3D human face data sets indicate that our 3D active contour model is a promising avenue for 3D face feature segmentation.

Acknowledgements

M. Krueger's research is funded by The New Zealand International Doctoral Research Scholarship and The University of Auckland International Doctoral Scholarship.

References

1. Bertalmio, M., Memoli, F., Cheng, L.T., Sapiro, G., Osher, S.: Variational problems and partial differential equations on implicit surfaces: Bye bye triangulated surfaces? In: Osher, S., Paragios, N. (eds.) Geometric Level Set Methods in Imaging, Vision, and Graphics, pp. 381–398. Springer, New York (2003)
2. Bogdanova, I., Bresson, X., Thiran, J., Vandergheynst, P.: Scale-space analysis and active contours for omnidirectional images. IEEE Trans. Image Processing 16, 1888–1901 (2007)
3. Kimmel, R.: Intrinsic scale space for images on surfaces: The geodesic curvature flow. Graphical Models And Image Processing 59, 365–372 (1997)
4. Wu, C., Deng, J., Zhu, W., Chen, F.: Inpainting images on implicit surfaces. In: Proc. Pacific Graphics Conf. 2005, University of Macau, pp. 142–144 (2005)
5. do Carmo, M.: Differential Geometry of Curves and Surfaces. Prentice-Hall, Englewood Cliffs (1976)

6. Chang, K.I., Bowyer, K.W., Flynn, P.J.: An evaluation of multimodal 2D+3D face biometrics. IEEE Trans. Pattern Anal. Machine Intell. 27, 619–624 (2005)
7. Tsalakanidou, F., Malassiotis, S., Strintzis, M.G.: Integration of 2D and 3D images for enhanced face authentication. In: Proc. IEEE 6th Int. Conf. Automatic Face and Gesture Recognition (FG 2004), Seoul, Korea, pp. 266–271 (2004)
8. Yan, P., Bowyer, K.W.: Biometric recognition using 3D ear shape. IEEE Trans. Pattern Anal. Machine Intell. 29, 1297–1308 (2007)
9. Spira, A., Kimmel, R.: Geometric curve flows on parametric manifolds. J. Comp. Physics 223, 235–249 (2007)
10. Krueger, M., Delmas, P., Gimel'farb, G.: Towards feature extraction on implicit surfaces using geodesic active contours. In: Proc. Image and Vision Computing New Zealand, Hamilton, New Zealand, pp. 294–299 (2007)
11. Cheng, L.T., Burchard, P., Merriman, B., Osher, S.: Motion of curves constrained on surfaces using a level-set approach. J. Comp. Physics 175, 604–644 (2002)
12. Kass, M., Witkin, A., Terzopoulos, D.: Snakes: Active contour models. Int. J. Computer Vision 1, 321–331 (1987)
13. Caselles, V., Kimmel, R., Sapiro, G.: Geodesic active contours. Int. J. Computer Vision 22, 61–79 (1997)
14. Osher, S., Sethian, J.: Fronts propagating with curvature-dependent speed: algorithms based on Hamilton-Jacobi formulations. J. Comp. Physics 79, 12–49 (1988)
15. Jin, H., Yezzi, A., Soatto, S.: Mumford-Shah on the move: Region-based segmentation on deforming manifolds with application to 3D reconstruction of shape and appearance from multi-view images. J. Math. Imaging Vision 29, 219–234 (2007)
16. Osher, S., Fedkiw, R.: Level sets methods and dynamic implicit surfaces. Springer, New York (2003)
17. Peng, D., Merriman, B., Zhao, H.K., Osher, S., Kang, M.: A PDE based fast local level set method. J. Comp. Physics 155, 410–438 (1999)
18. Sethian, J.: A fast marching level set method for monotonically advancing fronts. Proc. Nat. Acad. Sciences 93, 1591–1595 (1996)
19. Troje, N., Buelthoff, H.H.: Face recognition under varying poses: The role of texture and shape. Vision Research 36, 1761–1771 (1996)
20. Princeton suggestive contours gallery, `http://www.cs.princeton.edu/gfx/prog/sugcon/models/`
21. Carr, J.C., Beatson, R.K., Cherrie, J.B., et al.: Reconstruction and representation of 3D objects with radial basis functions. In: Proc. SIGGRAPH 2001, Los Angeles, USA, pp. 67–76 (2001)
22. Mauch, S.: A fast algorithm for computing the closest point and distance transform. Technical Report, California Institute of Technology (2000)

What Is a Good Nearest Neighbors Algorithm for Finding Similar Patches in Images?

Neeraj Kumar[1,*], Li Zhang[2], and Shree Nayar[1]

[1] Columbia University
[2] University of Wisconsin-Madison

Abstract. Many computer vision algorithms require searching a set of images for similar patches, which is a very expensive operation. In this work, we compare and evaluate a number of nearest neighbors algorithms for speeding up this task. Since image patches follow very different distributions from the uniform and Gaussian distributions that are typically used to evaluate nearest neighbors methods, we determine the method with the best performance via extensive experimentation on real images. Furthermore, we take advantage of the inherent structure and properties of images to achieve highly efficient implementations of these algorithms. Our results indicate that vantage point trees, which are not well known in the vision community, generally offer the best performance.

1 Introduction

Finding similar patches in images is a critical, yet expensive, step in a wide variety of computer vision tasks. For an image with N patches, an exhaustive search by scanning through the entire image for all N patches takes $O(N^2)$ time, which can take several minutes or even hours. For example, Figs. 1a and 1b each show results of a single search for patches (blue boxes) similar to the selected 21x21 patches (red boxes). Even though these searches were performed on relatively small images (roughly 800x600), they took 2.125 s and 1.375 s, respectively, to compute using brute-force scans. Finding the closest 10 neighbors for *all* patches in just these two images would take over 250 hours each!

However, by treating each image patch as a point in a high-dimensional space, we can use a Nearest Neighbors (NN) algorithm to compute the exact same results in a fraction of the time. Although a large number of general-purpose NN algorithms have been developed over the years [1,2,3,4,5], applying them directly to image search will not achieve the best results – images have distinct structure and unique properties that can be taken advantage of, to greatly improve search performance. In this work, we optimize many NN approaches specifically for finding similar patches in images and experimentally evaluate them to compare their performance. The following are the main contributions of our work:

* Supported by the National Defense Science & Engineering Graduate Fellowship.

D. Forsyth, P. Torr, and A. Zisserman (Eds.): ECCV 2008, Part II, LNCS 5303, pp. 364–378, 2008.

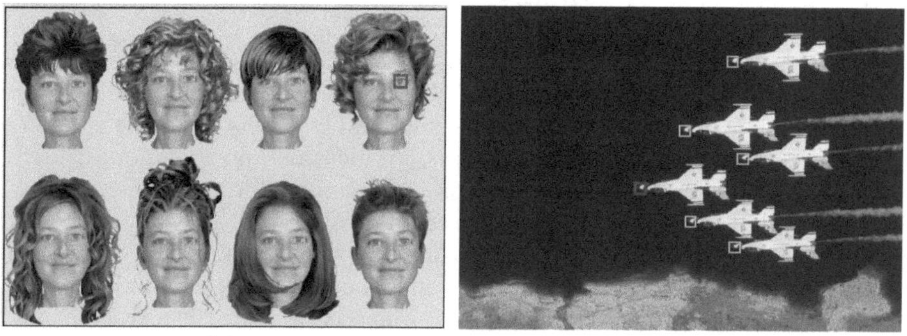

(a) (b)

Fig. 1. The closest neighbors of the 21x21 patches outlined in red are shown in blue. The matching eyes in (a) took 2125 ms to compute using an exhaustive search, but only 1.25 ms (1700X faster) using our image-optimized implementation of the *vp*-tree, one of the methods discussed in this paper. Similarly, finding the nosetips in (b) took 1375 ms using brute force, but only 3.43 ms (401X faster) using the *vp*-tree.

- We describe several NN methods, some of which have not been widely used in computer vision. We restrict ourselves to exact approaches which guarantee that all NNs are found – approximate methods are not discussed.
- We use several techniques to significantly improve the efficiencies of the different NN algorithms (in some cases exploiting the inherent properties of images). These include precomputing or using look-up tables for computing distances, as well as using priority queues for certain types of searches.
- We evaluate the off-line (construction) and on-line (search) performances of the different methods using a set of real images. We study the behaviors of the algorithms with respect to various parameters, including patch size, image size, number of NNs, search range, and number of input images. We find that the *vantage point tree*, which has not been widely used in vision, has the best overall performance. (Our image-optimized implementation of this method computes identical results for the searches in Figs. 1a and 1b in only 1.25 ms and 3.43 ms – significantly faster than exhaustive search.)

The results of our work can be used to significantly reduce the computation times of several vision algorithms. For recognition, several approaches use local patches to match objects across images or videos [6,7,8]. In [9], similar patches within a neighborhood are used to find objects with similar shapes but very different appearances. Similarity search also lies at the core of many other vision problems. In [10], a weighted average of similar patches is used to denoise an image, while in [11], similar patches and a continuity constraint are used to synthesize a large image texture from a small exemplar. Patch similarity has also been used to perform super-resolution [12].

2 The Nearest Neighbors Problem

The Nearest Neighbors (NN) problem has been studied in many different fields, and is the topic of many surveys (see [13,14] for recent examples). However, these surveys focus mostly on taxonomizing a large variety of methods and do not offer performance results. In contrast, we focus on approaches most useful for finding similar image patches, and present many performance results.

The NN problem refers to finding the set of points $\mathbf{P_c}$ within a dataset $\mathbf{P}$ that are "closest" to a query point q, as measured by some distance function $d(p, q)$. Although points usually belong to some subset of $\mathbb{R}^n$ and d is often one of L_1, L_2, or L_∞, the only requirement typically imposed is that d is a true distance function, i.e., it satisfies

Symmetry: $d(a, b) = d(b, a)$;
Non-Negativity: $d(a, a) = 0$ and $d(a, b) > 0, a \neq b$;
Triangle Inequality: $d(a, b) \leq d(a, c) + d(b, c)$.

These properties form the basis of all the methods we will describe. In particular, the triangle inequality allows us to compute bounds on various distances without having to evaluate the distance function itself. To make maximum use of this property, all the approaches that we will look at organize the dataset into a hierarchical tree data structure. In these trees, the data points are stored in the leaf nodes and the internal nodes serve only to prune the list of leaves returned by a query. Particular methods differ in what is stored at each internal node and how the tree is constructed, but there are several commonalities between them.

2.1 Tree Construction

Tree construction begins by assigning all points to the root node, and then recursively partitioning the points into one of several children of the node (the number of children, how they are chosen, and how the points are partitioned all vary depending on the particular method). This process continues until some termination criteria are met. Two common criteria are the maximum leaf size (the leaf contains fewer than a given number of points) and the maximum leaf radius (all points within the leaf are contained within a hyperball of given radius). Either criterion, or some combination of both, can be used for each tree type.

2.2 Tree Searches

There are two types of searches that we would like to perform:
ϵ-close neighbors (ϵ-NN): Find the set of points $\mathbf{P_c} \subset \mathbf{P}$ such that $p_i \in \mathbf{P_c} \Leftrightarrow d(p_i, q) \leq \epsilon$. This is also called *range search* in the existing literature.
k-nearest neighbors (k-NN): Find the set of k points $\mathbf{P_c} \subset \mathbf{P}$ such that $\forall p_i \in \mathbf{P_c}$ and $\forall p_j \notin \mathbf{P_c}, d(p_i, q) \leq d(p_j, q)$.

We now describe how NN searches are typically run on a given tree (see [13] for more detailed explanations). To perform an ϵ-NN search, the tree is traversed

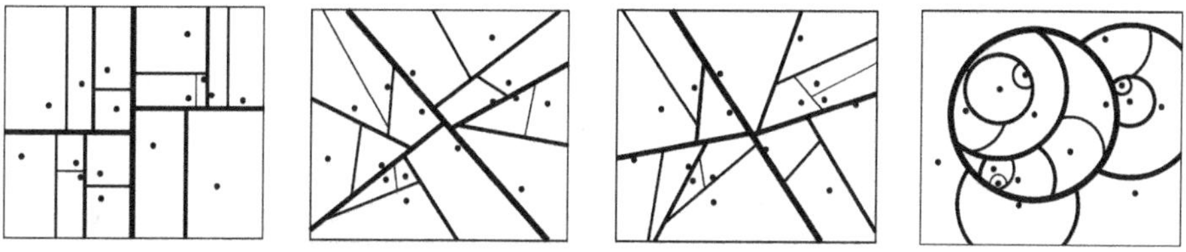

(a) *k*d-Tree (b) PCA Tree (c) Ball Tree (d) *vp*-Tree

Fig. 2. The partitioning of a 2D point set using different types of nearest neighbor trees, all with a maximum leaf size of 1 and a branching factor of 2. Line thickness denotes partition order (thicker lines were partitioned first). Note the very different structures created by the methods, which result in very different search speeds.

from the root, recursively exploring all children that intersect a hyperball of radius ϵ around the query point. Determining the list of intersecting children is done using the triangle inequality and some data stored in each node (dependent on the particular method). Once a list of leaf nodes are found, each contained data point must be verified as being within ϵ of the query. This is required because many leaves might be too close to be pruned by the triangle inequality, but still contain many points that are farther away than ϵ. This step is usually the most expensive one because it requires a large number of distance computations.

To perform k-NN searches, we could simply run a number of ϵ-NN searches with increasing ϵ, but this is a wasteful strategy. Instead, a better approach is to first locate the leaf node where the query point would be contained, and then use backtracking to explore nearby nodes. We maintain a sorted list of the k closest points found so far, as well as the maximum radius around the query point that has been fully explored. Skipping nodes that are farther away than the distance of the kth closest point found so far, we add each candidate node's points to our list, sorted by distance. This process has the effect of reducing the distance of the kth neighbor. When this distance becomes less than the maximum explored radius, the search is complete (because all remaining points must be at a greater distance than the current maximum).

Note that while ϵ-NN search performance does not depend on the order in which nodes are explored (because all matches must be found), search order greatly affects the performance of k-NN searches. Since the distance of the kth element of the list is used to prune candidate nodes, the faster we reduce this distance, the more nodes we prune. We will discuss how to take advantage of this observation in Sect. 4.3.

3 Methods for Finding Nearest Neighbors

We now describe various methods for finding nearest neighbors. To give a more intuitive idea of how the different methods operate, a simple 2D dataset partitioned using each of the methods is shown in Fig. 2.

3.1 *k*d-Trees

Despite having been invented over three decades ago, Bentley's *k*d-tree [1] remains one of the most commonly used algorithms for finding nearest neighbors today. Tree construction is simple: At each node, the points are recursively partitioned into two sets by splitting along one dimension of the data, until one of the termination criteria is met. The important choices to be made in tree construction are determining which dimension to split on and the split value. Choosing the dimension with maximum variance leads to smaller trees [15], while the split value is usually chosen to be the median value along the split dimension. This results in a balanced partitioning of points into axis-aligned hyper-rectangles, which get smaller in regions with many points (e.g., Fig. 2a).

The *k*d-tree exhibits many favorable properties and has proven to be quite efficient in practice for low-dimensional data. However, it has a few drawbacks:

1. The number of neighbors for each leaf node grows exponentially with dimension, causing search to quickly devolve into a linear scan.
2. The node divisions are always axis-aligned, regardless of the data distribution. This often results in poor search performance.

In real applications, the first problem is typically skirted by relaxing the requirement that all close neighbors be found. The Best Bin First (BBF) approach [16] is one such technique. It is based on the observation that the vast majority of the neighboring cells usually do not contain a nearest neighbor. It therefore searches the candidate cells in ascending order of their distance to the query, and terminates the search early to save computations. In [6], it is claimed that this method produces 95% of the correct neighbors at 1% the cost of an exhaustive search for one particular application. However, we note that this does not *guarantee* that all nearest neighbors are found, which can cause problems for various applications (e.g., as described in [17]).

3.2 PCA Trees

Sproull [2] attempts to remedy the axis-alignment limitation of *k*d-trees by applying Principal Components Analysis [18] at each node to obtain the eigenvector corresponding to the maximum variance and splits the points along that direction. This is equivalent to rotating the points about their mean such that their maximum variance now lies along the primary axis. We have implemented this variant as well, calling it the *PCA Tree*. For an example of what a tree partitioned using this method looks like, see Fig. 2b.

3.3 Ball Trees

The *k*d-tree and its variants can be termed "projective trees," meaning that they categorize points based on their projection into some lower-dimensional space. In contrast, all our remaining methods are "metric trees" – structures that organize points based on some metric defined on pairs of points. Thus, they don't require points to be finite-dimensional or even in a vector space.

The first type of metric tree that we will look at are *ball trees* [3]. In their original form, each node's points are assigned to the closest center of the node's two children. The children are chosen to have maximum distance between them, typically using the following construction at each level of the tree. First, the centroid of the points is located, and the point with the greatest distance from this centroid is chosen as the center of the first child. Then, the second child's center is chosen to be the point farthest from the first one.

The resulting division of points can be understood as finding the hyperplane that bisects the line connecting the two centers, and perpendicular to it (e.g., see Fig. 2c). Note that in this construction, there is no constraint on the number of points assigned to either node and the resulting trees can be highly unbalanced. While unbalanced trees are larger (and take longer to construct) than their balanced counterparts, this does not mean that they will be slower to search. On the contrary, such trees might be significantly faster if they capture the true distribution of points in their native space.

3.4 k-Means

While the previous description of ball trees is probably familiar to members of the machine learning community, vision researchers will no doubt have noticed its similarity to the k-means method [19]. This algorithm also assigns points to the closest of k centers, although it does so by iteratively alternating between selecting centers and assigning points to the centers until neither the centers nor the point partitions change. The resulting structure is equivalent to a Voronoi partition of the points [20], which simplifies to the hyperplane described in the previous section for the case of $k = 2$.

As originally described, the k-means method is a simple non-hierarchical clustering method that requires careful selection of both k and the initial centers to avoid local minima and bad partitions. Linde et al. extend this method to a hierarchical structure [4] where k now defines the branching factor between successive levels of the tree. This variant, which is the one we have implemented, has faster construction and search times because fewer distance evaluations need to be performed at each level. We also note that if the centers are initialized using the procedure described in the previous section, very few iterations have to be run for the centers to converge.

3.5 Vantage Point Trees

We turn now to a metric tree that uses a single "ball" at each level – the vantage point tree (*vp*-tree) [5]. Rather than partitioning points on the basis of relative distance from multiple centers (as was the case with ball trees and k-means), the *vp*-tree splits points using the absolute distance from a single center. This approach can be visualized as partitioning points into "hypershells" of increasing radius, e.g., Fig. 2d. The center of each node can be chosen randomly, as the centroid of the points, or as a point on the periphery (to maximize the distance between points). The number and thickness of the "hypershells" can also be chosen in various ways, and we explore them experimentally in Sect. 5. As we shall see, these methods perform the best for finding similar image patches.

3.6 Other Methods

Another common NN method [21] is based on using the L_∞ norm to quickly find all points within the hypercube of size 2ϵ around the query point. While this method works very well for applications which require strong correspondences between individual dimensions of points, it is less suited for cases where the overall distance (e.g., as measured using the L_2 metric) is more important. This is because in higher dimensions the volume of a hyperball of radius ϵ is much smaller than the volume of the enclosing hypercube. Thus, only a small fraction of the returned points will actually be within the search distance required.

Finally, we mention a class of methods that attempt to solve a relaxed version of the NN problem. Rather than finding the closest points, they allow the algorithm to return points that are within a factor of $(1+\epsilon)$ of the true closest distance, for some $\epsilon \geq 0$. This "approximate nearest neighbor" problem has proven slightly easier to attack, and there exist many approaches to this problem (e.g., the ANN [22] and Locality Sensitive Hashing [23] libraries). A recent work [24] provides a performance analysis of these methods, albeit applied to SIFT [6] descriptors, not image patches. However, they do not cover all of our optimizations, nor do they consider the *vp*-tree, which we found to be most effective. Finally, exact methods are preferred for many applications, where the potential of missing real results or getting too many false results can cause problems.

4 Efficient Implementation of Algorithms

Since our primary goal in this work is to use NN methods for finding similar patches in images, we can exploit various properties of images to develop very efficient implementations of these algorithms.

4.1 Computing L_p Norms Using Lookup Tables

The definition of what constitutes points as being "close" is critical. One commonly-used family of metrics are the L_p norms. For points $\mathbf{a}$ and $\mathbf{b}$ of dimensionality D, these are defined as $d(\mathbf{a},\mathbf{b}) = \left(\sum_i^D |a_i - b_i|^p\right)^{\frac{1}{p}}$. The L_p norms for $p=1$ (Manhattan distance) and $p=2$ (Euclidean distance) are especially common when comparing image patches. Here, we take advantage of the fact that images typically contain 8-bit quantized values. We use a 2-dimensional lookup table to store the result of $|a-b|^p$ for all possible values of a and b. This can be done during program initialization and saves D subtractions and $D \cdot (p-1)$ multiplications per distance function evaluation (converting them to array lookups, which are much faster on most hardware). Likewise, the expensive p-th root calculation can be stored in a lookup table if D and p are small enough.

4.2 Pre-calculating Distances within Each Leaf

During searches, a large portion of time is spent in evaluating the distance function for each point in returned leaf nodes. Once again, we take advantage of the triangle inequality. For each leaf, we precompute distances of each contained point to the leaf center, and store the points in increasing order of distance. At search time, we use the triangle inequality to find the upper and lower bounds of the list of points that are within our chosen distance and only calculate the distance for these points. While this might appear to be equivalent to adding one more level to a tree during construction, observe that this optimization is run at search time and is thus more flexible – it can be applied for different search distances ϵ, regardless of the maximum leaf radius chosen during construction. This simple procedure helps immensely. On average, we reduce the number of distance calculations in ϵ-NN searches by 38.5%.

4.3 Priority Queues for k-NN Searches

For k-NN searches, the order in which nodes are explored is extremely important. Therefore, we maintain a priority queue of nodes to visit, sorted by minimum distance from the center of each node to the query point. We first traverse the tree down to the leaf node corresponding to the query point, adding all possible branches at each node to our queue. Then, we repeatedly pick the closest node from our queue and expand it, again adding its children to our queue. This process continues until the closest node from the queue is farther away than the distance of the kth neighbor found so far. The benefit of this optimization is that we explore the closest nodes to our query first, which in turn fills our queue faster with a tighter distance bound. Although this is similar in spirit to the BBF criteria [16] mentioned earlier, note that we still guarantee exact results. Also, our optimization can be applied to all tree types, not just k-d trees.

5 Performance Evaluation

All of the methods we have described have similar asympototic search times. However, in practice, their actual performance is dominated by the structure of the constructed trees. This structure is determined by the distribution of the input data. It is well known that images exhibit complex distributions which are very different from simple uniform or Gaussian ones [25]. Since there are no analytical ways to compute the search performance of trees built from image patches, we empirically measure performance on a set of 15 real-world images.

Fig. 3. The image dataset used in the performance evaluation

5.1 Methodology

Thumbnails of the images from our dataset (mostly collected from `flickr.com`) are shown in Fig. 3. They were chosen to represent a wide variety of image types, e.g., indoor, outdoor, portraits, etc. Note that even with this small number of test images, there are still millions of patches to search through. To reduce computation times, we downsampled all images to smaller sizes (longest side to 640 pixels) and converted them to grayscale. We also removed all uninformative "flat" patches from the images, using the procedure described in Appendix A.

We use the number of distance comparisons as our cost unit for all tests, since this is typically the most expensive basic operation. To put cost values into perspective, we show the ratios of distance function evaluations performed in a naive brute force scan to those performed during search on a given tree type. When comparing search speeds, note that the brute force cost is independent of the query and the values of k or ϵ used for search. So, it is the relative search speeds of different methods that matters more than the exact numbers.

5.2 Comparison of Tree Types

The first task is to establish which tree type has the best construction and search performance. The two main parameters guiding tree construction are the maximum leaf size and the branching factor (or bin size), and so we construct many trees by sampling this 2D parameter space. At each parameter setting, we build a separate tree for each image and each tree type. Since our main goal at this point is just to compare the different NN methods, we use only non-flat 7x7 patches of each image. (We perform in-depth tests with different patch sizes in the next section.) For each tree, we then run ϵ- and k-NN searches for each patch of the same image ($\epsilon = 38.5$ and $k = 20$). Searching for every patch ensures that we do not introduce systematic biases into our results due to sampling issues. Also, the performance results are averaged over all trees built using the same parameters to get more accurate results.

Table 1 summarizes the construction and search performance results for all tree types. Note that we tried splitting vantage point trees using either constant branching factors or constant bin sizes, marked as *vp*-tree (k) and *vp*-tree (δ), because the two have different performance characteristics (other tree types did not show significant differences). This table allows us to pick the best parameter settings for each tree type.

Table 1. Comprehensive results. The splits are given as branching factors for ball trees, k-means, kd-trees, and vp-trees (k), and as bin sizes for PCA trees and vp-trees (δ). Construction costs are given as average number of distance calculations per patch. Search speeds are given as improvement over brute-force search. Optimal values are shown in **bold**.

Method	Leaf Size	Split	Cons. Cost	ε-NN Search Speed	k-NN Search Speed
Ball Tree	2	2	76.52	14.23	0.50
		3	65.81	14.74	0.62
		4	65.20	14.19	0.68
	16	2	69.80	**46.23**	**15.84**
		3	56.61	45.37	15.41
		4	56.26	44.17	15.42
	128	2	53.30	44.77	15.37
		3	46.44	44.31	14.87
		4	**46.33**	44.64	15.14
k-Means	2	2	303.92	4.47	0.50
		4	289.93	6.17	0.69
	16	2	271.60	19.20	7.62
		4	270.40	20.98	8.36
	128	2	**221.37**	35.09	12.69
		4	228.73	38.90	13.89
		8	298.72	**42.70**	**15.20**
PCA Tree	2	2	6.59	22.10	**10.26**
		4	8.25	25.50	9.81
		8	7.79	17.70	4.42
	16	2	5.50	26.60	10.13
		4	7.55	**27.72**	9.63
		8	7.61	17.41	4.24
	128	2	**4.32**	25.11	8.00
		4	6.24	25.80	8.37
		8	6.99	17.69	4.34

Method	Leaf Size	Split	Cons. Cost	ε-NN Search Speed	k-NN Search Speed
kd-Tree	2	1	**0.00**	4.45	1.88
	16	1	**0.00**	14.69	6.23
	128	1	**0.00**	**15.82**	**6.31**
vp-Tree (k)	2	2	16.28	38.54	1.69
		4	9.07	38.48	1.89
		8	7.49	34.83	1.57
		16	6.28	31.55	1.50
	16	2	13.97	44.66	15.84
		4	7.13	46.07	16.49
		8	5.00	**48.95**	17.61
		16	4.02	41.95	**19.12**
	128	2	11.05	33.80	11.85
		4	5.95	36.75	12.72
		8	4.00	37.50	12.53
		16	**3.00**	38.13	12.28
vp-Tree (δ)	2	2	5.71	45.29	10.27
		4	9.30	42.09	16.36
		8	9.60	36.96	16.94
	16	2	4.36	**50.45**	**18.27**
		4	7.94	49.36	18.21
		8	8.78	39.85	15.72
	128	2	**3.01**	44.25	13.05
		4	5.85	42.55	13.32
		8	6.64	36.23	12.31

To more easily compare the search performance of different tree types, we show the construction and search results for each tree in Figs. 4a and 4b, respectively, using their best parameter settings in Table 1. Note that kd-trees require no distance computations during construction, and so their construction cost shows up as 0. Also, PCA trees require many iterations of eigenvector decomposition, making their actual construction cost much higher than shown. We see that kd-trees can be built the fastest, followed by vp-trees. However, the search results in Figure 4b show that kd-trees are the slowest to search, while vp-trees and ball trees are the fastest. Although the slow search performance of kd-trees in high dimensions is well known [21], we would like to draw the reader's attention to the low construction cost and fast search performance of our image-optimized version of the vp-tree, a method which is not in widespread use in the vision community. We believe its superior performance for both construction and search is due to the following two properties:

- Only one distance function evaluation is required at each level of the vp-tree compared to multiple ones for ball trees and k-means.
- The "hypershell" partitions of vp-trees can capture clusters of similar image patches using fewer nodes than the "hyperplane" partitions of other methods. This results in smaller trees with leaf nodes at a lower average height.

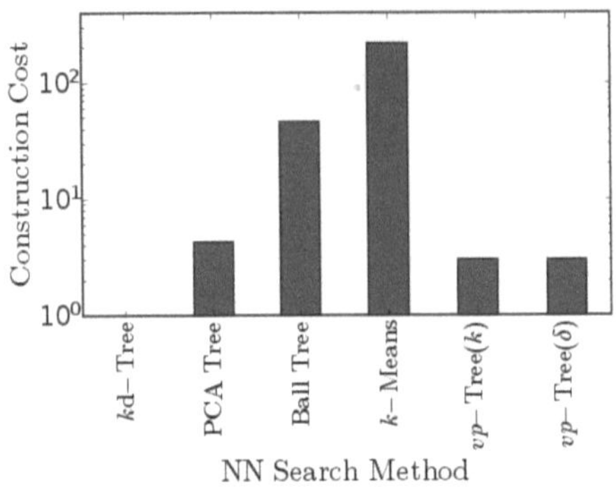

(a) Summary of Construction Costs

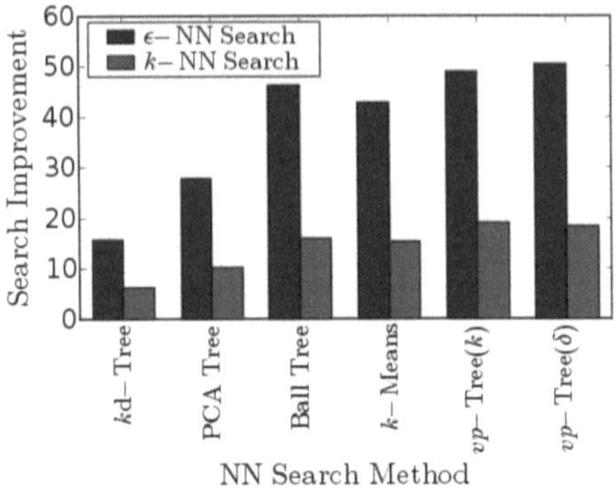

(b) Summary of Search Performance

Fig. 4. Summaries of the (a) construction costs and (b) search performance for different types of NN trees. The construction costs are given as average number of distance computations per input patch, plotted on a logarithmic scale. The search speeds are given as ratio of time taken for brute force scan versus time taken using the given tree type. All numbers are for best parameter settings (see Table 1 for full listings). Note that although ball trees and *vp*-trees have similarly good search performance, the latter has a much smaller construction cost.

5.3 Performance Along Various Dimensions

We now want to see how search performance varies as different parameters are changed, holding the rest constant. Since it would be infeasible to run these tests on all tree types at all parameter settings, we choose the best-performing trees from our comparison in the previous section. For k-NN tests, we use the *vp*-tree (k) with both the max leaf node size and branching factor set to 16. For ϵ-NN tests, we use the *vp*-tree (δ) with max leaf node size of 16 and bin size of 2.

Figure 5a shows search performance as a function of search distances ϵ and k, on the top and bottom, respectively. Note that both are plotted on a log-log scale, and so the linear portions of the graphs indicate an exponential drop-off in performance as either k or ϵ are increased. This results from the fact that the triangle inequality becomes progressively less useful as we search for increasingly dissimilar patches. It suggests that the search range should be kept as small as possible to obtain the best performance. Applications which require large search ranges will hence have to pay a steep penalty in terms of search speed.

Figure 5b plots the search performance as a function of patch size, on a semilog scale. To ensure a fair comparison of search distances at different patch sizes WxW, the actual search range ϵ used was computed as $\epsilon = W \cdot \epsilon'$, where the "average per-pixel distance" ϵ' was empirically chosen to be 5.5. We can see that as our patches get bigger, ϵ-NN search performance improves, but k-NN search performance detoriates. These are both effects of the increasing distances between patches in higher dimensions. Thus, a fixed "average per-pixel distance" will contain fewer points when the patch size increases, causing ϵ-NN search speeds to improve. However, the greater distance between patches also means

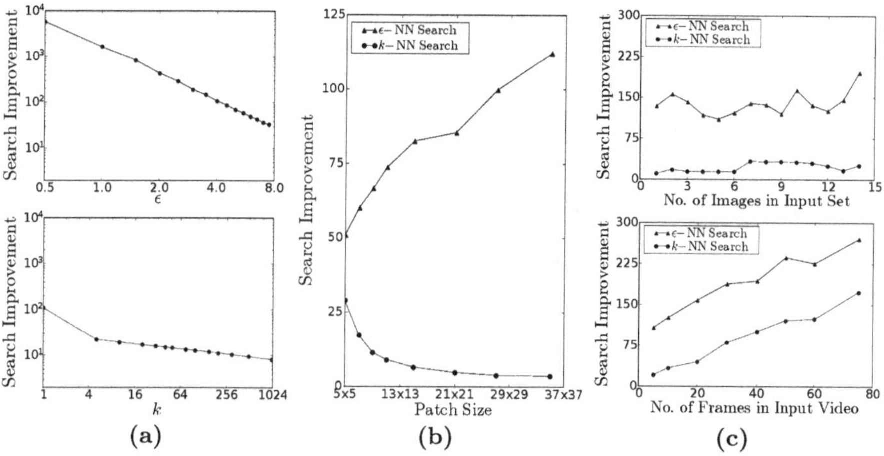

Fig. 5. Search performance of the *vp*-tree plotted as a function of (a) search distances in log-log, for (top) ϵ-NN and (bottom) k-NN searches; (b) patch size in semi-log; and (c) number of input images (top) and video frames (bottom). The exponential performance drop-off in (a) suggest that for high performance, small search ranges should be used (application allowing). As patch sizes go up, (b) shows k-NN searches slowing down and ϵ-NN searches speeding up – higher dimensionality causes the closest k neighbors to become harder to distinguish from other patches, but there are also less patches within the same "average per-pixel distance." In (c), we see that since the images in the test set did not have enough similarity for the *vp*-tree to take advantage of, the performance does not increase with the number of images (top graph). In contrast, performance improves dramatically with the number of video frames since frames taken from the same video clip tend to have strong similarity (bottom graph).

that the k most similar patches will be at greater distances from the query – making the triangle inequality less useful for pruning patches.

So far, we have used single images to construct and search each of our trees. This is useful if we need to search for patches within only one image. However, for many applications we would like to find patches across multiple images or video frames. Figures 5c top and bottom show the search performance for each of these cases, respectively. These results were computed by using a subset of the input images to build a *vp*-tree, and searching using all patches from one of the remaining images. In the case of multiple images, we arbitrarily chose the subsets from our set of images (as shown in Figure 3). We constructed several different subsets, and averaged their performance results. The search performance for this test varies drastically as we increase the number of images used (especially for ϵ-NN searches). This suggests that there was not much similarity between the images in our test set. Thus, the *vp*-tree's search performance scales roughly as that of brute force searches. Note, however, that *vp*-tree searches are still significantly faster than brute force.

The video datasets were constructed by sampling a 30fps, 5 second video at fixed intervals to use as inputs for building a *vp*-tree. As in the previous

Table 2. Summary of results. The *vp*-tree performs well in all respects.

Method	Construction Performance	ε-NN Search Performance	*k*-NN Search Performance
*k*d-Tree	Excellent	Poor	Poor
PCA Tree	Poor		
Ball Tree		Excellent	Excellent
k-Means	Poor		
vp-Tree	Excellent	Excellent	Excellent

case, this tree was then searched using one of the other frames from the video. The steady increase in the video search speeds confirms our hypothesis about similarity leading to better search performance. These frames share much more similarity than a set of arbitrary images, and the vantage point tree is able to take advantage of this fact. Thus, finding similar patches from a set of images becomes progressively faster as the dataset size grows, especially if the images have some similarity. In the future, it would be useful to see if large sets of arbitrary images contain enough similarity to show similar increases in performance.

6 Concluding Remarks

In this work, we have extensively evaluated a number of different approaches for solving the nearest neighbors problem. In particular, we have focused on using these approaches for finding similar image patches. Since this is a common step in a wide range of applications ranging from object recognition and texture synthesis to image denoising and compression, solving this problem efficiently will lead to faster solutions for all of these disparate tasks. Our main findings about using various NN methods for finding similar patches can be summarized as follows (also see Table 2):

- Vantage point trees have the best overall construction and search performance. Furthermore, their increasing speeds with larger datasets makes them very powerful for applications that involve large images or videos.
- Ball trees have similar search speeds, but much longer construction times.

References

1. Bentley, J.L.: Multidimensional binary search trees used for associative searching. Comm. ACM 18(9), 509–517 (1975)
2. Sproull, R.F.: Refinements to nearest-neighbor searching in k-dimensional trees. Algorithmica 6(4), 579–589 (1991)
3. Omohundro, S.M.: Five balltree construction algorithms. Technical Report 89-063 (1989)
4. Linde, Y., Buzo, A., Gray, R.M.: An algorithm for vector quantizer design. IEEE Transactions on Communications 28(1), 84–94 (1980)

5. Yianilos, P.N.: Data structures and algorithms for nearest neighbour search in general metric spaces. In: Symposium on Discrete algorithms, pp. 311–321 (1993)
6. Lowe, D.: Distinctive image features from scale-invariant keypoints. IJCV (2003)
7. Sivic, J., Zisserman, A.: Video google: A text retrieval approach to object matching in videos. In: ICCV, Washington, DC, p. 1470. IEEE Computer Society, Los Alamitos (2003)
8. Nister, D., Stewenius, H.: Scalable recognition with a vocabulary tree. CVPR (2006)
9. Shechtman, E., Irani, M.: Matching local self-similarities across images and videos. CVPR (2007)
10. Buades, A., Coll, B., Morel, J.M.: A non-local algorithm for image denoising. CVPR (2005)
11. Efros, A.A., Freeman, W.T.: Image quilting for texture synthesis and transfer. SIGGRAPH (2001)
12. Freeman, W.T., Jones, T.R., Pasztor, E.C.: Example-based super-resolution. IEEE Comput. Graph. Appl. 22(2), 56–65 (2002)
13. Chávez, E., Navarro, G., Baeza-Yates, R., Marroquín, J.L.: Searching in metric spaces. ACM Computing Surveys 33(3), 273–321 (2001)
14. Shakhnarovich, G., Darrell, T., Indyk, P.: Nearest-Neighbor Methods in Learning and Vision. MIT Press, Cambridge (2006)
15. Bentley, J.L.: K-d trees for semidynamic point sets. In: Symposium on Computational geometry, pp. 187–197 (1990)
16. Beis, J.S., Lowe, D.G.: Shape indexing using approximate nearest-neighbour search in high-dimensional spaces. CVPR (1997)
17. Hays, J., Efros, A.A.: Scene completion using millions of photographs. ACM Transactions on Graphics (SIGGRAPH) 26(3) (2007)
18. Hotelling, H.: Analysis of a complex of statistical variables with principal components. Journal of Educational Psychology 24, 417–441 (1933)
19. MacQueen, J.: Some methods for classification and analysis of multivariate observations. In: Symposium on Mathematical Statistics and Probability, pp. 281–297 (1967)
20. Edelsbrunner, H.: Algorithms in combinatorial geometry. Springer, New York (1987)
21. Nene, S.A., Nayar, S.K.: A Simple Algorithm for Nearest Neighbour Search in High Dimensions. IEEE Transactions on Pattern Analysis and Machine Intelligence 19(9), 989–1003 (1997)
22. Arya, S., Mount, D.M., Netanyahu, N.S., Silverman, R., Wu, A.Y.: An optimal algorithm for approximate nearest neighbor searching in fixed dimensions. J. ACM 45(6), 891–923 (1998)
23. Indyk, P., Motwani, R.: Approximate nearest neighbors: towards removing the curse of dimensionality. In: Symposium on Theory of Computing, pp. 604–613 (1998)
24. Mikolajczyk, K., Matas, J.: Improving descriptors for fast tree matching by optimal linear projection. In: ICCV (2007)
25. Ruderman, D.L.: The statistics of natural images. Network: Computation in Neural Systems 5, 517–548 (1996)
26. Viola, P., Jones, M.: Rapid object detection using a boosted cascade of simple features. CVPR (2001)

A Low-Variance Patch Removal

We remove "flat" patches, which are those with low variance: $\sigma^2 < \sigma^2_{min}$, and all pixels close to the mean: $L_\infty(patch, \mu) < dist_{min}$. Here, μ and σ^2 are the mean and variance of the patch of size WxW. We empirically define $\sigma^2_{min} = 3.7 \cdot W$ and $dist_{min} = 5$. This test can be efficiently evaluated using integral images [26].

Learning for Optical Flow Using Stochastic Optimization

Yunpeng Li and Daniel P. Huttenlocher

Department of Computer Science, Cornell University, Ithaca, NY 14853
{yuli,dph}@cs.cornell.edu

Abstract. We present a technique for learning the parameters of a continuous-state Markov random field (MRF) model of optical flow, by minimizing the *training loss* for a set of ground-truth images using simultaneous perturbation stochastic approximation (SPSA). The use of SPSA to directly minimize the training loss offers several advantages over most previous work on learning MRF models for low-level vision, which instead seek to maximize the likelihood of the data given the model parameters. In particular, our approach explicitly optimizes the error criterion used to evaluate the quality of the flow field, naturally handles missing data values in the ground truth, and does not require the kinds of approximations that current methods use to address the intractable nature of maximum-likelihood estimation for such problems. We show that our method achieves state-of-the-art results and requires only a very small number of training images. We also find that our method generalizes well to unseen data, including data with quite different characteristics than the training set.

1 Introduction

Optical flow is among the most widely studied problems in low-level vision. While the development of better matching and regularization criteria as well as more effective optimization techniques has significantly advanced the state of the art [1,2,3,4,5,6,7,8,9], the parameters of these methods are generally set by hand on the same data that is used for evaluation. Although there has recently been some work on learning for optical flow, such as [10], the relative lack of investigation of learning techniques stands in sharp contrast with higher-level vision such as object recognition, where learning techniques are ubiquitous. Hand tuning of models not only involves substantial human effort, more importantly it limits our understanding of how well optical flow methods generalize to unseen data and thus will perform in practice. This is illustrated by the recently developed Middlebury optical flow evaluation database [11], where methods that perform best on the classical "Yosemite" sequence tend not to perform as well on new imagery for which they were not hand tuned.

In this paper, we present a continuous-state Markov random field (MRF) [12,13,14] based model for optical flow. This is in contrast to most recent applications of MRF methods in computer vision, for problems such as stereo, where relatively small sets of discrete states are used. We learn the parameters of this model using simultaneous perturbation stochastic approximation (SPSA) [15] to minimize the *training loss* – that is the error on the training data under some error function. In particular, we measure

D. Forsyth, P. Torr, and A. Zisserman (Eds.): ECCV 2008, Part II, LNCS 5303, pp. 379–391, 2008.

training loss using the average end-point error (AEPE) [16] which is one of the error metrics employed in the Middlebury evaluation [11].

Directly optimizing for training loss as opposed to a maximum likelihood approach (e.g. as pursued in [10]) has a number of advantages. First, the likelihood of the data under models with loopy spatial dependency is intractable to compute. In order to obtain the maximum likelihood estimate, one thus has to resort to approximation techniques such as estimating the mode (i.e. the maximum *a posteriori* estimate) [17], sampling [10], or some type of local training [18]. These approximations, however, tend to be imprecise and may lead to noisy and unreliable estimates, as noted in [17]. Moreover, from a statistical learning point of view, the maximum likelihood estimate is not well justified for problems such as optical flow that have *structured* outputs, such as a label at each pixel, as opposed to a single overall right-or-wrong answer. Thus an attractive alternative approach is to minimize some specific loss function, or error metric, on the training data (e.g., [19]) as we discuss further in Section 3.

Learning model parameters that minimize the loss on the training data is a challenging optimization problem, since the relationship between the target function (i.e. the training loss) and the model parameters cannot be determined analytically except in some special cases (e.g. [20]). SPSA [15] is a convenient choice in this situation, since it only requires the target function to be smooth and to have non-vanishing gradient (with respect to the parameters being optimized), a rather generous condition that is usually satisfied, but does not require the analytical form or the true gradient to be known. Hence it can be used to optimize for a wide range of loss functions, including the commonly used error metrics on which optical flow quality is judged such as AEPE. Given the large number of problems in computer vision that have structured outputs and the breadth of applicability of SPSA [21], the approach that we develop here is likely to be of broader interest for other problems in computer vision.

We evaluate our method in the standard setup of supervised learning, namely by training the model on a set of sequences with ground truth and testing it on a different set that does not include any of the training sequences. This allows one to assess the generalization power of the learnt model. We compare our results to those of previous methods and show that our model both generalizes well to unseen data and achieves state-of-the-art performance for optical flow.

The rest of the paper is organized as follows. We provide some background and discuss related work in Section 1.1, and then define our model for optical flow in Section 2. The learning method is described in more detail in Section 3, and the experimental results are presented in Section 4. We conclude in Section 5.

1.1 Background and Related Work

Optical flow is a highly challenging low-level vision problem due to inherent ambiguity in local regions of the image, known as the aperture problem [22]. This is further complicated by phenomena such as motion discontinuity, untextured regions, and sensor noise. To address these issues, many early methods utilize local support windows over which some matching cost is aggregated (e.g. [23]). Window-based approaches approaches, however, suffer from the generalized aperture problem [24], namely that they are either too small to provide sufficient support or too big that they span over motion boundaries. Although this can be alleviated by using parametric and mixtures models (e.g. [24]), support windows have not proven to be good for accurately estimating the motion of non-rigid bodies undergoing deformation. Moreover, purely local methods are susceptible to erroneous matches in poorly-textured regions even with the aid of support windows. Global models for optical flow, first proposed in [25], compute the flow field by minimizing a global energy function. The energy function is usually composed of a data term that encourages agreement between frames and a spatial term (i.e. regularization) that enforces consistency of the flow field. Markov random fields (MRF) are closely related to energy-based models [19] in that the node and clique potentials of an MRF are often defined in terms of energy or cost functions in general exponential families. Thus equivalence can be drawn between the negative log-posterior of MRF models and global energy functions. MRF models, however, have explicitly-defined topology and are convenient to model higher-order and non-local (i.e. long-range) interactions (e.g. in [26,10]), which would otherwise be more difficult to express.

Learning for optical flow is a challenging subject, which has not been studied extensively. A major difficulty for learning optical flow models is the scarcity of ground-truth data. Despite the challenges, considerable progress has been made over the past decades that has improved our understanding of the problem. The robust estimation framework introduced in [5] makes a key observation that the brightness constancy and spatial smoothness assumptions are often violated near motion boundaries, and hence robust energy functions such as the Lorentzian should be used instead of quadratics to account for these violations. Although that work proposes a variety of robust function forms, it does not attempt to automatically estimate their parameters. Probability distributions of optical flow are studied in [27], where they are used to represent uncertainties and account for errors. Nevertheless, the parameters of the models used to compute optical flow are still set by hand. In [28], linear bases of parameterized models are learnt from examples using principal component analysis. Pioneering for its time, these models mainly target certain specific motion types and are not designed for general motion estimation. The work most closely related to ours is that of [10], where field-of-expert (FoE) models [29] are learnt from ground-truth flow fields inferred from range-scan data. To our knowledge, it is the first work to employ supervised learning technique for general optical flow estimation. However, their model differs from ours in several important respects, including the use of decoupled large-clique (5×5) filters and an approximation to maximum likelihood estimation rather than considering training loss.

Simultaneous perturbation stochastic approximation (SPSA) is a stochastic optimization method that iteratively minimizes a given target function. At each iteration all model parameters are simultaneously perturbed at random, and the loss function is evaluated at the perturbed positions in the parameter space to estimate its pseudo-gradient with respect to the parameter vector. This information is then used to determine the direction of descent and to subsequently update the model parameters. Since exact convergence is difficult to determine, the algorithm is usually run for either a fixed number iterations or until the reduction in loss becomes insignificant.

The SPSA algorithm was first proposed in [15] as a gradient-free stochastic optimization method, and it is closely related to the classical finite-difference stochastic approximation (FDSA) algorithm [30] since both methods estimate the gradient of the loss function by measuring its values only and hence avoid the necessity to know its closed-form derivatives. However, by simultaneously perturbing all model parameters, SPSA requires substantially fewer measurements of the loss function and hence achieves faster convergence rates [15,21]. The gradient-free SPSA is well-suited for problems where the input-output relationship of the system is difficult to determine. Since its introduction, SPSA has been applied to optimize for a variety of engineering systems ranging from traffic control, weapon targeting, to buried object localization [21]. Nevertheless, the method has received less attention in the vision community. To our knowledge, it has not previously been applied to learning parameters for low-level vision problems.

2 An MRF Model for Optical Flow

We model the optical flow computation as a labeling problem on a continuous-state Markov random field, where each node p, representing a pixel, receives a 2-dimensional vector label $\mathbf{w}_p \in \mathbb{R}^2$ indicating its flow (i.e. apparent motion). In our MRF model, each node is connected to nodes that are either adjacent or two pixels away in both the horizontal and vertical directions. Hence the resulting MRF consists of linear 3-cliques (i.e. 3-node complete subgraphs) that are either horizontally or vertically oriented. The 3-clique MRF topology allows us to model both the first derivative (i.e. gradient) and the second derivative (i.e. curvature) of the flow field, which are both important motion statistics. On the other hand, the clique size is small enough that it doesn't pose a severe computational burden.

Let $\mathcal{V}$ be the set of nodes and $\mathcal{C}$ be the set of cliques of the graph. As is well known, the posterior of a labeling $\mathbf{w}$ given data I decomposes into the product of maximal clique potentials and node potentials,

$$p(\mathbf{w}|I;\boldsymbol{\theta}) = \frac{1}{Z(\boldsymbol{\theta})} \prod_{c\in\mathcal{C}} \phi_c^{\boldsymbol{\theta}}(\mathbf{w}_c) \prod_{p\in\mathcal{V}} \phi_p^{\boldsymbol{\theta}}(\mathbf{w}_p, I), \tag{1}$$

where $\boldsymbol{\theta}$ represents the parameters of the model and $Z(\boldsymbol{\theta})$ is the partition function. The notations $\mathbf{w}_c$ and $\mathbf{w}_p$ denote the labeling over clique c and node p respectively. Recall that the MRF is continuous, and thus $\mathbf{w}$ is also in a continuous vector space. As is a common practice, we represent the distribution in the general exponential family so that $\phi_c^{\boldsymbol{\theta}}(\mathbf{w}_c) = \exp(-f_c^{\boldsymbol{\theta}}(\mathbf{w}_c))$ and $\phi_p^{\boldsymbol{\theta}}(\mathbf{w}_c, I) = \exp(-g_p^{\boldsymbol{\theta}}(\mathbf{w}_p, I))$. That is, $f_c^{\boldsymbol{\theta}}$ and $g_p^{\boldsymbol{\theta}}$ are the energy functions for the spatial term and the data term respectively, and the total energy of a labeling (i.e. flow field) $\mathbf{w}$ given input I can be written as

$$\begin{aligned} E(\mathbf{w};\boldsymbol{\theta}, I) &= -\log p(\mathbf{w}|I;\boldsymbol{\theta}) - \log Z(\boldsymbol{\theta}) \\ &= \sum_{c\in\mathcal{C}} f_c^{\boldsymbol{\theta}}(\mathbf{w}_c) + \sum_{p\in\mathcal{V}} g_p^{\boldsymbol{\theta}}(\mathbf{w}_p, I). \end{aligned} \tag{2}$$

Therefore minimizing the energy $E(\mathbf{w};\boldsymbol{\theta}, I)$ is equivalent to finding the maximum *a posteriori* (MAP) labeling over the MRF. Although exact minimization of the energy is generally intractable due to the loopy graph structure, methods based on gradient

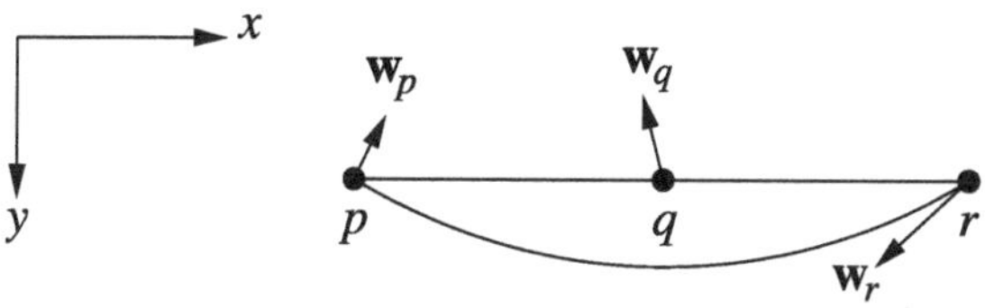

Fig. 1. A horizontal 3-clique over pixels (nodes) p, q, and r. Each pixel has a 2-D vector label indicating the flow value (i.e. the apparent motion).

descent can be used to obtain an approximate solution (which we shall return to in Section 2.2).

The input for our model is a pair of images, i.e. $I = (I_0, I_1)$. Without loss of generality, we assume that the flow is computed for I_0.

2.1 Energy Functions

We use robust energy functions as proposed in [5]; in particular, we choose the family of Lorentzian functions for their ability to maintain spatial consistency and brightness agreement while being tolerant to motion discontinuity in the spatial term and outliers in the data term. The energy function for the spatial term is defined as

$$f_c^{\boldsymbol{\theta}}(\mathbf{w}_c) = \lambda_S \cdot \rho(\sqrt{||\beta_1 \mathbf{d_1}||^2 + ||\beta_2 \mathbf{d_2}||^2}) \tag{3}$$

where λ_S, β_1, and β_2 are model parameters, the function

$$\rho(\cdot) = \log(1 + \frac{1}{2}|\cdot|^2) \tag{4}$$

is the standard Lorentzian function, and

$$\begin{aligned} \mathbf{d_1} &= \mathbf{w}_r - \mathbf{w}_p \\ \mathbf{d_2} &= \mathbf{w}_p - 2\mathbf{w}_q + \mathbf{w}_r, \end{aligned} \tag{5}$$

with $\mathbf{w}_p = (u_p, v_p)^T$ denoting the flow vector at pixel p. Here p, q, and r are the three pixels (i.e. nodes) belonging to the linear 3-clique c. If the clique c is horizontally oriented, they are the left, middle, and right pixel of the clique respectively and hence $\mathbf{d_1}$ and $\mathbf{d_2}$ are the discrete first and second partial derivatives (up to a constant factor) of the flow field along the horizontal direction; the case for clique c being vertically oriented is analogous. Figure 1 illustrates the layout of a horizontal clique and the vector labels of its nodes. Notice that the spatial energy function (and hence the clique potential of the MRF) is symmetric with respect to x and y directions, and therefore its value is unchanged if the coordinates are rotated by 90° or have the two axes switched. The spatial energy function is also isotropic (i.e. has perfect rotational symmetry) in the motion domain, since only the *magnitude* of relative motion is computed. Thus rotating the flow vectors by the same angle everywhere would result in no change of the spatial energy.

While the form of our energy function is similar to filter based models such as [29,10,31], it differs in a subtle but important way. Those models use a linear combination of functions over individual filter responses, which implicitly assumes that the filters are independent of each other. In our case, both the first and second derivative filters are inputs to the same non-linear robust function. Hence the influence of one derivative is reduced if the other is already large (due to the robust spatial term), thereby avoiding double-penalties at motion boundaries.

The energy function for the data term is defined in terms of the difference between a pixel in I_0 and its matching position in I_1 under the flow field $\mathbf{w}$,

$$g_p^{\boldsymbol{\theta}}(\mathbf{w}_p, I) = \lambda_D \cdot \rho(\beta_D || I_1(p + \mathbf{w}_p) - I_0(p)||), \tag{6}$$

where p is used synonymously with its 2-vector coordinates $(x_p, y_p)^T$ on the image grid, ρ is the same Lorentzian function as defined in Equation 4, and λ_D and β_D are model parameters. For color images, $I(p)$ is simply a 3-vector of the RGB values at position p in image I. Although using more psychophysically motivated color spaces such as Lab or XYZ may yield better matching models, it is beyond the scope of this work. Since in general $p + \mathbf{w}_p$ does not fall on integer grid positions, its value in I_1 is sampled using bilinear interpolation.

2.2 Optical Flow Estimation

To estimate optical flow, we perform approximate MAP inference on the MRF by minimizing the energy function in Equation 2 using gradient descent. Computing the gradient of the spatial term energy $E_S = \sum_{c \in C} f_c^{\boldsymbol{\theta}}(\mathbf{w}_c)$ is straightforward since it has an analytical form in $\mathbf{w}$, and the gradient at each pixel is given by

$$(\nabla_{\mathbf{w}} E_S)_p = \sum_{c \in C : p \in c} \nabla_{\mathbf{w}_p} f_c^{\boldsymbol{\theta}}(\mathbf{w}_c). \tag{7}$$

Since the data term (Equation 6) involves the image input, it is not a closed-form expression with respect to $\mathbf{w}$. Nevertheless, by using the chain rule, the gradient of the data term energy $E_D = \sum_{p \in \mathcal{V}} g_p^{\boldsymbol{\theta}}(\mathbf{w}_p, I)$ can be written as

$$(\nabla_{\mathbf{w}} E_D)_p = \nabla_{I_1(p+\mathbf{w}_p)} g_p^{\boldsymbol{\theta}}(\mathbf{w}_p, I) \nabla_{\mathbf{w}_p} I_1(p + \mathbf{w}_p). \tag{8}$$

The value of $\nabla_{I_1(p+\mathbf{w}_p)} g_p^{\boldsymbol{\theta}}(\mathbf{w}_p, I)$ is readily available, since it is analytical in $I_1(p + \mathbf{w}_p)$ (cf. Equation 6). Moreover, $\nabla_{\mathbf{w}_p} I_1(p + \mathbf{w}_p)$ is simply the image gradient of I_1 at position $p + \mathbf{w}_p$. (To see this, let $\mathbf{z} = (x, y)^T = p + \mathbf{w}_p$, i.e. $\mathbf{z}$ is the coordinates of the matching position of p. Since $\nabla_{\mathbf{w}_p} \mathbf{z} = \mathbf{1}$ according to the definition of $\mathbf{z}$, it follows that $\nabla_{\mathbf{w}_p} I_1(p + \mathbf{w}_p) = \nabla_{\mathbf{z}} I_1(p + \mathbf{w}_p) \nabla_{\mathbf{w}_p} \mathbf{z} = \nabla_{\mathbf{z}} I_1(p + \mathbf{w}_p)$.) We approximate the image gradient using the $\frac{1}{2}(-1, 0, 1)$ derivative filters and bilinear interpolation.

As is standard in the literature, the input images are preprocessed with a low-pass filter. In our case, we use a small Gaussian kernel with $\sigma = 0.25$. For performing gradient descent we use limited memory BFGS [32,33], which has faster converge speed than steepest descent. We also employ hierarchical coarse-to-fine strategy for optical flow computation [2,3], since it is well known to produce globally more consistent and hence more accurate flow estimations. As is commonly done, all flow values are initialized to zero at the beginning of the optimization.

3 Learning the Parameters

As noted above, we take the approach of learning models that yield low training loss (e.g., [19]) rather than those that maximize the likelihood of the training data. There are several advantages to this approach. First, as discussed in Section 1 maximum likelihood estimation is intractable for labeling problems on large loopy graphs leading to the use of a number of approximation techniques. Second, maximum likelihood estimation may not be consistent with the error measure that one would like to optimize, which we discuss further here. Finally, directly minimizing training loss is well suited to ground truth from real scenes, as opposed to synthetic data, which generally contain pixels at which the data is unknown.

We now turn to the second of these issues, training models that optimize for an appropriate error measure. If we regard the training data as a sample drawn from some unknown distribution characterizing the domain of the problem, then lower training loss implies lower expected generalization loss (i.e. error rate on unseen testing data) for a given class of models. The maximum likelihood estimate, other other hand, does not take the specific error metric into account. Thus even if the correct (zero error) output is assigned a high likelihood, it does not necessarily discriminate between bad outputs (i.e. those with high loss) and reasonably good ones (i.e. those that are not completely correct but nevertheless have low loss). For instance, suppose there are two different loss functions for optical flow, each designed to suit a different need. One of them heavily penalizes non-smooth flows in the uniform region but does not mind having blurred motion boundaries, whereas the other does just the opposite. Given this scenario, there is little reason to believe that the same model should be optimal for both loss functions. By minimizing the training loss one finds model parameters best suited to the particular loss function. Although in principle one could instead learn the parameters using maximum likelihood and then take the error metric into account during the inference stage, this would pose additional challenges of what optimization problem to solve that took both the model and the metric into account, and whether such a problem could be solved efficiently.

The third issue that is naturally handled by minimizing the training loss is that of incomplete ground-truth. Currently available ground-truth data of real, as opposed to synthetic, motion sequences contains a non-trivial number of pixels with unknown flow values due to phenomena such as occlusion. While this may be improved to some degree with the gathering of additional data, it is an inherent problem that some pixels will have unknown values. Handling pixels with missing ground-truth values is easy with pixel-based loss functions such as AEPE, as such pixels can simply be excluded from consideration. Computing the likelihood of data with missing values, on the other hand, is not so straightforward because of spatial interdependencies.

3.1 Training Loss Minimization Using SPSA

Thus we seek parameters $\boldsymbol{\theta} = (\beta_1, \beta_2, \beta_D, \lambda_S, \lambda_D)^T$ that minimize the average training loss $L(\boldsymbol{\theta})$. [1] As noted above we do this using SPSA [15]. SPSA is an iterative pseudo-gradient descent algorithm that updates its solution $\hat{\boldsymbol{\theta}} = (\hat{\theta}_1, \ldots, \hat{\theta}_i, \ldots, \hat{\theta}_m)^T$ at each step by

$$\hat{\boldsymbol{\theta}}_{k+1} = \hat{\boldsymbol{\theta}}_k - a_k \hat{\boldsymbol{g}}_k(\hat{\boldsymbol{\theta}}_k), \tag{9}$$

where a_k is the step size at iteration k and $\hat{\boldsymbol{g}}_k(\hat{\boldsymbol{\theta}}_k)$ is the pseudo-gradient of the loss function L. The pseudo-gradient is obtained using two-sided simultaneous perturbation,

$$\left(\hat{\boldsymbol{g}}_k(\hat{\boldsymbol{\theta}}_k)\right)_i = \frac{l(\hat{\boldsymbol{\theta}}_k + c_k \boldsymbol{\Delta}_k) - l(\hat{\boldsymbol{\theta}}_k - c_k \boldsymbol{\Delta}_k)}{2c_k \left(\boldsymbol{\Delta}_k\right)_i}, \tag{10}$$

where $l(\cdot)$ is some noisy measurement of the true loss $L(\cdot)$ ($l = L$ if the measurement is noise-free), $\boldsymbol{\Delta}_k$ is a user-defined m-dimensional random perturbation vector satisfying certain conditions [15], and c_k is a scalar factor. The gain sequences a_k and c_k both decrease over time, and are given by $a_k = a/(A+k)^\alpha$ and $c_k = c/k^\gamma$ [15]. In our case, we used the recommended values $(0.602, 0.101)$ for (α, γ) and set (A, a, c) to $(50, 5, 0.001)$ following the guidelines given in [34].

Since all the parameters in our model are some type of scale parameters and have a natural domain of $(0, \infty)$, we transform them into the logarithm space during learning so that $\hat{\boldsymbol{\theta}} = \log \boldsymbol{\theta} \in \mathbb{R}^m$. Although the most commonly used distribution for the random perturbation vector $\boldsymbol{\Delta}$ is the Bernoulli ± 1 for each component, we find that the Bernoulli distribution is somewhat overly restrictive on the possible directions of descent. Thus we instead sample each component of $\boldsymbol{\Delta}$ uniformly at random from the union of intervals $[-1-\delta, -1+\delta] \cup [1-\delta, 1+\delta]$ with $\delta = 0.99$. Note that the distribution has no probability mass at around zero, which is a condition that the perturbation vector is required to satisfy [15]. Since measuring training loss given parameters is deterministic and can be considered essentially noise-free, we require that the loss function (i.e. training error) decreases monotonically with time. Hence a solution $\hat{\boldsymbol{\theta}}_{k+1}$ is rejected, i.e. remains the same as $\hat{\boldsymbol{\theta}}_k$, if the loss $L(\hat{\boldsymbol{\theta}}_{k+1})$ is greater than $L(\hat{\boldsymbol{\theta}}_k)$. We also observe that most common types of parametric motions, such as affine transformation and divergence, result in large first derivative, but much smaller second derivative, of the optical flow field. Thus a large magnitude of the second derivative should reasonably produce more energy (hence lower probability) than that of the first derivative. To this end, we impose the constraint $\beta_2 \geq \beta_1$ to reflect this prior knowledge. If the constraint becomes violated during learning, we simply swap β_1 with β_2 and resume. This to some extent resembles a restart, a common mechanism used by stochastic methods to depart from undesired local optima. Finally, we run the SPSA algorithm multiple times and choose from the solutions the one with the lowest training loss. This helps to reduce the variance in the performance of learnt parameters.

---

[1] One could reduce the dimensionality of $\boldsymbol{\theta}$ by 1 by observing that only the ratio between λ_S and λ_D is relevant under MAP inference. This, however, has little effect on the learning process, since the dimensionality of the space of optimal solutions is also reduced. Thus we do not carry out this explicit reduction in our learning formulation.

4 Experimental Results

To evaluate our method, we trained our model on the "other" data set [2] from the Middlebury optical flow web site [11] and tested its performance on the "eval" data set from the same web site. For learning, we use the average end-point error (AEPE) [16,11] as the loss function. We initialized all parameters of the model to one and ran the SPSA algorithm for 300 iterations. The procedure was repeated 5 times (i.e. five models trained), and the model with the lowest training loss was chosen and used for testing on unseen data. [3] Among the multiple runs, the losses (i.e. training error) of the three best models are within 5% of each other while the other two have losses about 15% higher than the best model. This shows that the results obtained by SPSA is quite reliable, especially with multiple trials, given that the initial loss is many times higher. Figure 2 shows a plot of the training errors against the number of iterations for all five trials.

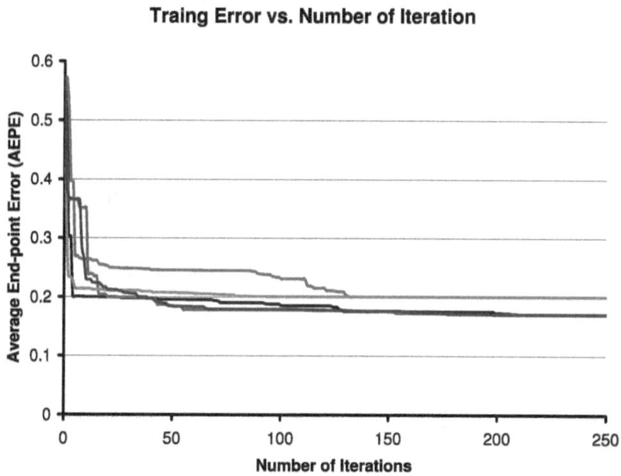

Fig. 2. Average training error in terms of AEPE plotted against the number of iterations over multiple trials of running the SPSA algorithm. The error rate generally decreases rapidly at the beginning and much more slowly afterward, indicating a reasonably good solution can be obtained with relatively few iterations.

For evaluation, we report error rates in both average angular error (AAE) [35] and average end-point error (AEPE). Table 1 shows the performance of our learnt model on the eight sequences for flow evaluation from the Middlebury optical flow web page [11]. The error rates of some of the well-known methods are also shown for comparison. These results demonstrate that our model achieves state-of-the-art performance, surpassing the previous methods on most of the benchmark sequences. We want to emphasize that the performance of our model is achieved using parameters trained on a

[2] Only sequences with ground-truth flow are used. We excluded "Venus" from the training set, since it is a stereo sequence.

[3] Note that the choice of the model is part of the learning procedure and is completely based on the training data, without any knowledge of the data on which the model is evaluated.

Table 1. Performance on the eight evaluation sequences from the Middlebury optical flow page [11], measured in terms of both average angular error (upper row) and average end-point error (lower row). The lowest error rates for each sequence are shown in bold fonts.

Method\Sequence	Army	Mequon	Schefflera	Wooden	Grove	Urban	Yosemite	Teddy	**Average**
Our model	**6.84**	**8.47**	**12.5**	**8.40**	3.88	**6.32**	2.56	7.29	**7.03**
	0.18	**0.57**	**0.84**	**0.52**	**1.12**	1.75	0.13	**1.32**	**0.804**
- max-likelihood	6.86	9.11	15.1	8.60	**3.84**	10.1	2.15	10.3	8.26
	0.18	0.65	0.99	0.62	1.28	1.98	0.11	1.81	0.953
Bruhn *et al.* [9]	10.1	9.84	16.9	14.1	3.93	6.77	**1.76**	**6.29**	8.71
	0.28	0.69	1.12	1.07	1.24	1.56	**0.10**	1.38	0.930
Black/Anandan [5]	7.83	9.70	13.7	10.9	4.67	8.00	2.61	8.58	8.25
	0.21	0.65	0.93	0.76	1.40	2.04	0.15	1.68	0.978
Horn/Schunk [25]	8.01	9.13	14.2	12.4	4.69	8.35	4.01	9.16	8.74
	0.22	0.61	1.01	0.78	1.27	**1.42**	0.16	1.51	0.873
Lucas/Kanade [23]	13.9	24.1	20.9	22.2	18.9	22.0	6.41	25.6	19.25
	0.39	1.67	1.50	1.57	2.95	3.30	0.30	3.80	1.94

Table 2. Error rates on the three training sequences, given in the form of AAE/AEPE, with "–" indicating result not available

Method\Sequence	Dimetrodon	RubberWhale	Hydrangea
Our model	2.92/0.152	5.22/0.149	2.43/0.198
Bruhn *et al.*[9]	10.99/0.43	–	–
Black/Anandon [5]	9.26/0.35	–	–
Lucas/Kanade [23]	10.27/0.37	–	–

different set of sequences, which includes none of those used for evaluation. In other words, the parameters are learnt completely without any knowledge of the testing data. Thus the results demonstrate that our model has good generalization power. In addition we also trained our model using the approximate maximum likelihood scheme of [17] (second method in Table 1), so as to compare it with SPSA (first method in Table 1). The results show that the model learnt with SPSA has better overall performance, demonstrating the effectiveness of SPSA learning for optical flow.

For completeness, we show in Table 2 the error rates on the training sequences. Results from other methods are quoted from [11] whenever available. One can see that our training data includes only three sequences, which is in contrast with the several hundred used in [10]. The available training sequences are also significantly less comprehensive in terms of the variation of appearance and motion than are the test sequences. For instance the training data does not have any synthetic sequences, which do appear in the evaluation set. Thus learning with this limited training data is especially challenging. Nonetheless, our model obtained under such adverse circumstances performs well on unseen data.

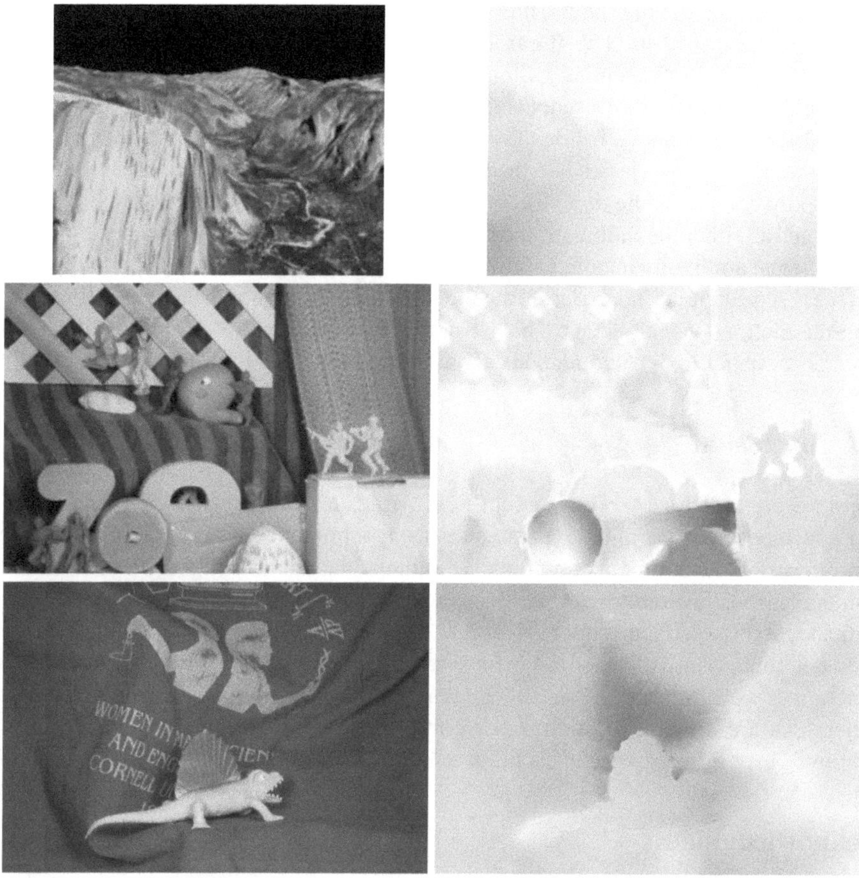

Fig. 3. Output of our model for the sequences "Yosemite" (top), "Army" (middle), and "Dimetrodon" (bottom). Left: A frame of the image sequence. Right: Estimated flow. Observe the predominantly smooth flow field of "Yosemite" and "Dimetrodon" in contrast to the large amount of motion discontinuity in "Army".

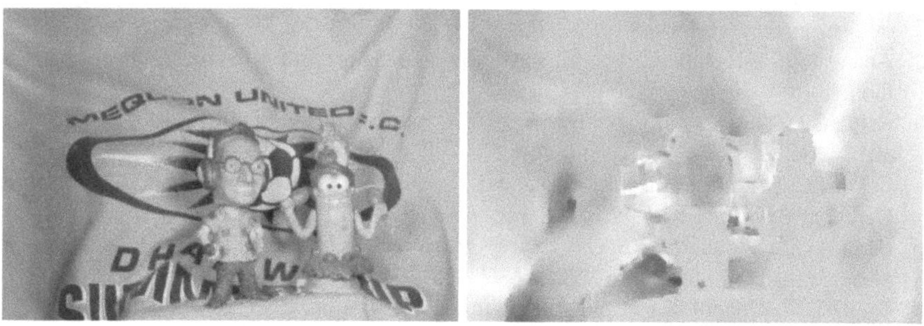

Fig. 4. Output of our model for the sequence "Mequon". Left: A frame of the image sequence. Right: Estimated flow. Most of the errors occur in the shadows around the cartoon models, due to their high relative motion with respect to the background.

Figure 3 displays some sample flow fields produced by our model, color coded using the scheme described in [11]. It can be seen that our model is capable of producing smooth flows (as in "Yosemite") while preserving motion boundaries (as in "Army"). Figure 4 shows the result of a sequence on which our model did not perform particularly well. Most of the errors lie inside the shadows (cast by the two cartoon models), which have rather high motion relative to the background on which they lie. Since optical flow is generally defined as the *apparent* motion, the flow inside a shadowed region can be interpreted as either the motion of the background or that of the shadow itself. Thus this is an inherent ambiguity in optical flow, which also occurs with transparent and specular surfaces. A principled approach to dealing with these phenomena is to estimate the multiple motions in such areas. This has attracted a fair amount of investigation from researchers (e.g. [36,5]), and remains an interesting topic for future work.

5 Conclusion

We have presented a Markov random field based model for estimating optical flow and a technique for learning its parameters using simultaneous perturbation stochastic approximation. Experiments on publicly available benchmark data sets show that our results compare favorably with previous methods and achieve the state-of-the-art performance. Moreover, our model is learnt from a separate set of training sequences that does not contain any of those used for evaluation. This demonstrates that our model generalizes well to unseen data. Since many low-level vision problems involve parameters that are difficult to optimize deterministically, the learning technique that we employed here may well prove useful in other research areas.

Acknowledgments

This work was supported in part by NSF grant IIS-0713185.

References

1. Nagel, H.H., Enkelmann, W.: An investigation of smoothness constraints for the estimation of displacement vector fields from image sequences. PAMI 8(5), 565–593 (1986)
2. Anandan, P.: A computational framework and an algorithm for the measurement of visual motion. IJCV 2(3), 283–310 (1989)
3. Bergen, J.R., Anandan, P., Hanna, K.J., Hingorani, R.: Hierarchical model-based motion estimation. In: ECCV, London, UK, pp. 237–252. Springer, Heidelberg (1992)
4. Cohen, I.: Nonlinear variational method for optical flow computation. In: Scandinavian Conference on Image Analysis (1993)
5. Black, M.J., Anandan, P.: The robust estimation of multiple motions: Parametric and piecewise-smooth flow fields. CVIU 63(1), 75–104 (1996)
6. Szeliski, R., Coughlan, J.: Spline-based image registration. IJCV 22(3) (1997)
7. Alvarez, L., Weickert, J., Sanchez, J.: Reliable estimation of dense optical flow fields with large displacements. International Journal of Computer Vision 39(1), 41–56 (2000)
8. Ye, M., Haralick, R.M., Shapiro, L.G.: Estimating piecewise-smooth optical flow with global matching and graduated optimization. PAMI 25(12), 1625–1630 (2003)

9. Bruhn, A., Weickert, J., Schnörr, C.: Lucas/kanade meets horn/schunck: combining local and global optic flow methods. IJCV 61(3), 211–231 (2005)
10. Roth, S., Black, M.J.: On the spatial statistics of optical flow. IJCV 74(1), 33–50 (2007)
11. Baker, S., Scharstein, D., Lewis, J.P., Roth, S., Black, M.J., Szeliski, R.: A database and evaluation methodology for optical flow. In: ICCV (2007)
12. Besag, J.E.: Spatial interaction and the statistical analysis of lattice systems. J. Royal Stat. Soc., B 36(2) (1974)
13. Geman, S., Geman, D.: Stochastic relaxation, Gibbs distributions, and the Bayesian restoration of images. PAMI 6(6), 721–741 (1984)
14. Szeliski, R.: Bayesian modeling of uncertainty in low-level vision. IJCV 5(3) (1990)
15. Spall, J.C.: Multivariate stochastic approximation using a simultaneous perturbation gradient approximation. IEEE Transactions on Automatic Control 37, 332–341 (1992)
16. Otte, M., Nagel, H.H.: Optical flow estimation: Advances and comparisons. In: ECCV, Secaucus, NJ, USA, pp. 51–60. Springer, New York (1994)
17. Scharstein, D., Pal, C.: Learning conditional random fields for stereo. CVPR (2007)
18. Sutton, C., McCallum, A.: Piecewise training of undirected models. In: UAI (2005)
19. LeCun, Y., Huang, F.J.: Loss functions for discriminative training of energy-based models. In: AIStats (2005)
20. Tappen, M.F.: Utilizing variational optimization to learn Markov random fields. CVPR (2007)
21. Spall, J.C.: Overview of the simultaneous perturbation method for efficient optimization. Hopkins APL Technical Digest 19, 482–492 (1998)
22. Bertero, M., Poggio, T., Torre, V.: Ill-posed problems in early vision. In: Proceedings of IEEE (1988)
23. Lucas, B.D., Kanade, T.: An iterative image registration technique with an application to stereo vision. In: IJCAI, pp. 674–679 (1981)
24. Jepson, A., Black, M.J.: Mixture models for optical flow computation. CVPR, 760–761 (1993)
25. Horn, B.K.P., Schunk, B.G.: Determining optical flow. Artificial Intelligence 17, 185–203 (1981)
26. Geman, S., Graffigne, C.: Markov random field image models and their applications to computer vision. In: Intl. Congress of Mathematicians (1986)
27. Simoncelli, E.P., Adelson, E.H., Heeger, D.J.: Probability distributions of optical flow. In: CVPR, Mauii, Hawaii, pp. 310–315. IEEE Computer Society, Los Alamitos (1991)
28. Fleet, D.J., Black, M.J., Yacoob, Y., Jepson, A.D.: Design and use of linear models for image motion analysis. IJCV 36(3), 171–193 (2000)
29. Roth, S., Black, M.J.: Fields of experts: A Framework for Learning Image Priors. CVPR (2005)
30. Kiefer, J., Wolfowitz, J.: Stochastic estimation of the maximum of a regression function. Annuals of Mathematical Statistics 23(3), 462–466 (1952)
31. Weiss, Y., Freeman, W.F.: What makes a good model of natural images? CVPR (2007)
32. Nocedal, J.: Updating quasi-newton matrices with limited storage. Mathematics of Computation 35, 773–782 (1980)
33. Liu, D.C., Nocedal, J.: On the limited memory BFGS method for large scale optimization. Mathematical Programming 45(3, (Ser. B)), 503–528 (1989)
34. Spall, J.C.: Implementation of the simultaneous perturbation algorithm for stochastic optimization. IEEE Trans. Aerosp. Electon. Syst. 34(3), 817–823 (1998)
35. Barron, J.L., Fleet, D.J., Beauchemin, S.S.: Performance of optical flow techniques. IJCV 12(1), 43–77 (1994)
36. Fleet, D.J., Jepson, A.D.: Computation of component image velocity from local phase information. IJCV 5(1), 77–104 (1990)

Region-Based 2D Deformable Generalized Cylinder for Narrow Structures Segmentation

Julien Mille[1], Romuald Boné[1], and Laurent D. Cohen[2]

[1] Laboratoire d'Informatique, Université François Rabelais de Tours
64 avenue Jean Portalis, 37200 Tours, France
[2] CEREMADE, CNRS UMR 7534, Université Paris Dauphine
Place du Maréchal de Lattre de Tassigny, 75775 Paris, France
julien.mille@univ-tours.fr

Abstract. In this paper, we present a region-based deformable cylinder model, extending the work on classical region-based active contours and gradient-based ribbon snakes. Defined by a central curve playing the role of the medial axis and a variable thickness, the model is endowed with a region-dependent term.This energy follows the narrow band principle, in order to handle local region properties while overcoming limitations of classical edge-based models. The energy is subsequently transformed and derived in order to allow implementation on a polygonal line deformed with gradient descent. The model is used to extract path-like objects in medical and aerial images.

1 Introduction

In order to segment objects with respect to region homogeneity criterion, a large family of region-based active contour models has been developed over the last decade. Extending the initial boundary-based snake by Kass *et al* [1], region-based active contours exhibit desirable property of holding global features about the region of interest [2]. Parametric implementations include the anticipating snake [3], the statistical snake [4] and the mixed model of [5] whereas implicit contours based on the level set framework include the active contours without edges [6], region descriptors [7] and the geodesic active regions [8].

The problem of recovering narrow structures using deformable models was addressed for instance in [9] in the context of cortex segmentation. The ribbon snake model [10] [11] [12], inspired by the work in [13], was developed for road extraction in aerial images. A different kind of model was proposed recently for vessels in [14]. By its geometrically constrained formulation, the ribbon snake model is well suited to extract path-like objects. Initial ribbon snakes are boundary-based, since they attach the model by means of a gradient-dependent term integrated over the contour. In presence of images suffering from noise and lack of salient boundaries between neighbouring structures, embedding a region term in the energy functional allows to describe homogeneity over the whole shape.

As a result, we extend the ribbon snake, which is actually a 2D generalized cylinder (GC), to region-based segmentation. Our GC is made up of a central

D. Forsyth, P. Torr, and A. Zisserman (Eds.): ECCV 2008, Part II, LNCS 5303, pp. 392–404, 2008.

planar curve, acting as a medial axis, and a variable thickness. This defines a band over which a region energy is formulated. Region-based active contours usually consider both inner and outer statistics [6]. Similarly, we use the narrow band principle [15] to define outer regions. The GC is subsequently deformed by coupled evolution of the curve and the thickness guided by energy minimization. When dealing with region integrals, the combined use of Green-Riemann theorem and calculus of variations enables to derive the Euler-Lagrange equations. Moreover, the minimal path approach of Cohen and Kimmel [16] provides a reliable initialization for the GC, preventing it from being trapped in erroneous local minima.

The paper is organized as follows. Section 2 presents the 2D deformable GC, describing its geometrical formulation and its associated energies. The model has a particular shape, which allows transformation of the region integral leading to a suitable form for polygonal implementation. Section 3 spends detailing the transformation of the integrals encountered in the region energy and its mathematical derivations leading to the Euler-Lagrange equations. Section 4 deals with the discretization and implementation on a polygonal curve. It also presents our utilization of the minimal path approach. Eventually, section 5 presents the experimental results obtained on medical and aerial images.

2 The Deformable Generalized Cylinder

We first recall the basics of active contours which we will rely on. The continuous active contour model is represented as a plane curve Γ with position vector $\mathbf{c}$:

$$\begin{aligned} \Gamma : \Omega &\longrightarrow \mathbb{R}^2 \\ u &\longmapsto \mathbf{c}(u) = (x(u), y(u))^T \end{aligned} \tag{1}$$

where x and y are continuously differentiable with respect to the arbitrary parameter u (different from arc length s) varying in the normalized domain $\Omega = [0, 1]$. Segmentation of an object of interest is performed by finding the curve minimizing an energy functional E, which has the general form:

$$E(\Gamma) = \int_\Omega \mathcal{L}(u, x, x', x'', y, y', y'')du \tag{2}$$

where $\mathcal{L}$ is usually made up of internal terms regularizing the curve and external terms attaching the curve to the image data. If the curve is a local minimizer of E, the associated Euler-Lagrange equation is verified. This implies the variational derivative of E with respect to $\mathbf{c}$, which may be split over coordinates x and y using calculus of variations:

$$\begin{cases} \dfrac{\delta E}{\delta x} = \dfrac{\partial \mathcal{L}}{\partial x} - \dfrac{d}{du}\dfrac{\partial \mathcal{L}}{\partial x'} + \dfrac{d^2}{du^2}\dfrac{\partial \mathcal{L}}{\partial x''} \\ \dfrac{\delta E}{\delta y} = \dfrac{\partial \mathcal{L}}{\partial y} - \dfrac{d}{du}\dfrac{\partial \mathcal{L}}{\partial y'} + \dfrac{d^2}{du^2}\dfrac{\partial \mathcal{L}}{\partial y''} \end{cases} \tag{3}$$

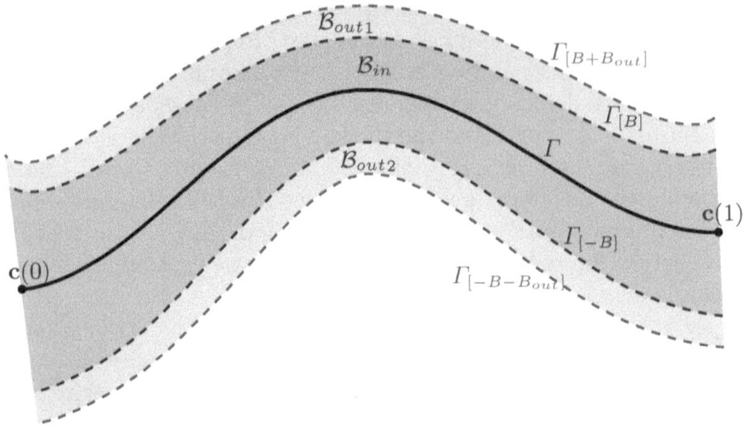

Fig. 1. Inner and outer bands for region energy

According to eq. 2 and 3, it is convenient to compute the variational derivative of E when it holds only boundary integrals. This is the case with traditional gradient-based snakes [1]. However, the increasing use of region terms inspired by the Mumford-Shah functional [17] has proven to overcome the limitations of uniquely gradient-based models, especially when dealing with data sets suffering from noise and lack of contrast. When dealing with region integrals, more advanced mathematical tools should be used, like Green-Riemann theorem to convert domain integrals into boundary integrals [2][18][19], or the shape gradient method [20][21]. When using implicit contours [22], a more practical approach consists in handling $E(\psi)$ instead of $E(\Gamma)$, where Γ is the zero level set of ψ. The Euler-Lagrange equation of E with respect to ψ is then easily determined, like in the Chan-Vese model [6].

The deformable GC presented in this paper is region-based. Whether they are implemented using parametric contours or level sets, usual region-based deformable models are represented as closed curves, splitting the image plane into inner and outer domains. Unlike these, the GC is based on an open curve Γ and a positive real-valued thickness function $B : \Omega \rightarrow \mathbb{R}$. The coupled curve and thickness define an inner band-shaped region $\mathcal{B}_{in}$, whose medial axis is Γ and width is $2B$, as depicted in fig. 1. Region $\mathcal{B}_{in}$ is bounded by curves $\Gamma_{[B]}$ and $\Gamma_{[-B]}$, whose respective position vectors are defined as follows:

$$\begin{aligned} \mathbf{c}_{[B]}(u) &= \mathbf{c}(u) + B(u)\mathbf{n}(u) \\ \mathbf{c}_{[-B]}(u) &= \mathbf{c}(u) - B(u)\mathbf{n}(u) \end{aligned} \tag{4}$$

where $\mathbf{n}$ is the oriented unit normal of Γ. One may notice that $\Gamma_{[B]}$ and $\Gamma_{[-B]}$ would be parallel curves of Γ (also known as "offset curves" [23]) if B was constant. The goal is to make both curve Γ and thickness B evolve in order to match a homogeneity criterion over $\mathcal{B}_{in}$. Obviously, $\Gamma_{[B]}$ and $\Gamma_{[-B]}$ implicitly follow the evolution of Γ and remain "symmetric" with respect to Γ, which implies a strong geometrical and topological constraint on the region. Thus, the deformable GC is dedicated to the extraction of thin shapes, like vascular structures in medical data or roads in aerial images.

The evolution of medial axis and thickness consists in determining for each boundary point $\mathbf{c}_{[B]}(u)$ (and similarly $\mathbf{c}_{[-B]}(u)$) whether it should move forward or backward. This decision is easier to make when an outer region is also considered, in the extent that image features at $\mathbf{c}_{[B]}(u)$ can be compared to the image statistics of both inner and outer regions. The deformable model divides the image into the objet of interest and the background, which contains usually many other objects and is therefore non-uniform. In many cases, we believe that the background is uniform only in the vicinity of the object, i.e. in narrow bands both sides apart of the ribbon. Following the principle in [15], we define two outer bands $\mathcal{B}_{out1}$ and $\mathcal{B}_{out2}$. We introduce a constant band thickness B_{out}, so that $\mathcal{B}_{out1}$ is bounded by curves $\Gamma_{[B]}$ and $\Gamma_{[B+B_{out}]}$, as shown in fig. 1. Similarly, $\mathcal{B}_{out2}$ is bounded by $\Gamma_{[-B]}$ and $\Gamma_{[-B-B_{out}]}$.

Rigorously, since Γ is an open curve, position vector $\mathbf{c}$ is not differentiable at end-points $\mathbf{c}(0)$ and $\mathbf{c}(1)$, making their tangent and normal vectors undefined. From a practical point of view, we may assume that Γ is a portion of a longer curve. As regards implementation, this requires careful discretization of derivatives using finite differences. An important property resulting from the definition in eq. 4 is that the velocity of boundary curves depends on the curvature of Γ. We denote ℓ the length element of curve Γ, i.e. the magnitude of its velocity vector, and κ the curvature:

$$\kappa = \frac{\mathbf{c}' \times \mathbf{c}''}{\|\mathbf{c}'\|^3} = \frac{x'y'' - x''y'}{\ell^3}$$

where $'$ means differentiation with respect to u. The velocity vector of curve $\Gamma_{[B]}$ is expressed as a function of the velocity vector of Γ, as well as its curvature and normal. Using the identity $\mathbf{n}' = -\kappa\mathbf{c}'$, we have:

$$\mathbf{c}_{[B]}{}' = \mathbf{c}' + B'\mathbf{n} + B\mathbf{n}' = (1 - B\kappa)\mathbf{c}' + B'\mathbf{n} \tag{5}$$

To the GC, we associate the energy functional E, weighted sum of the internal energy E_{smooth} and the external region energy E_{region}:

$$E(\Gamma, B) = \omega E_{\text{smooth}}(\Gamma, B) + (1 - \omega)E_{\text{region}}(\Gamma, B) \tag{6}$$

The user-provided coefficient ω weights the influence of E_{smooth} over E_{region}. It controls the elastic properties of the GC. Since E depends both on Γ and B, the GC minimizing E should verify two coupled Euler-Lagrange equations:

$$\frac{\delta E}{\delta \mathbf{c}} = \mathbf{0} \qquad \frac{\delta E}{\delta B} = 0 \tag{7}$$

The internal energy is the first-order regularizing term:

$$E_{\text{smooth}}(\Gamma, B) = \int_{\Omega} \left\| \frac{d\mathbf{c}}{du} \right\|^2 + \left(\frac{dB}{du} \right)^2 du \tag{8}$$

Minimizing this energy encourages a regular curve and a smoothly varying band thickness. We can reasonably assume that if Γ and B are smooth, then so will be all resulting parallel curves of Γ. We define a region energy over the bands, following the work in [15]:

$$E_{\text{region}}(\Gamma, B) = \iint\limits_{\mathcal{B}_{in}} g_{in}(\mathbf{x})d\mathbf{x} + \iint\limits_{\mathcal{B}_{out1}} g_{out1}(\mathbf{x})d\mathbf{x} + \iint\limits_{\mathcal{B}_{out2}} g_{out2}(\mathbf{x})d\mathbf{x} \tag{9}$$

where functions g_{in}, g_{out1} and g_{out2} hold some uniformity criterion with respect to image data I, depending on the application. Details are given in section 4. For the moment, they should be thought of as functions penalizing non-homogeneity, the most intuitive one for grayscale images being the intensity variance, as proposed by Chan and Vese [6].

An important issue is that, as expressed in eq. 9, E_{region} does not depend explicitly on position vector $\mathbf{c}$ nor band thickness B, making impossible the derivation leading to the Euler-Lagrange equation. In what follows, we express E_{region} in a suitable form implying possible derivation and implementation on a parameterized curve discretized as a polygon. From now on, we will use the general notation $J(f, R)$ to represent the integral of function f over any region R.

3 Derivation of Region Energy

In this section, we show that the domain integrals appearing in eq. 9 can be expressed as functions of $\mathbf{c}$ and B. The proof is based on Green-Riemann theorem, stating that for every region R, if $(P(x,y), Q(x,y))$ is a continuously differentiable $\mathbb{R}^2 \to \mathbb{R}^2$ vector field, then:

$$\iint\limits_{R} \frac{\partial Q}{\partial x} - \frac{\partial P}{\partial y}\, dxdy = \int_{\partial R} Pdx + Qdy$$

In order to apply the theorem on $J(f, R)$, we choose P and Q as follows:

$$Q(x,y) = \frac{1}{2}\int_{-\infty}^{x} f(t,y)dt \qquad\qquad P(x,y) = -\frac{1}{2}\int_{-\infty}^{y} f(x,t)dt$$

The theorem expects that ∂R should be at least piecewise smooth. Let us consider the more general case of a region R bounded by portions of curves Γ_1 and Γ_2 and line segments $[\mathbf{c}_1(u_0)\mathbf{c}_2(u_0)]$ and $[\mathbf{c}_1(u_1)\mathbf{c}_2(u_1)]$, as depicted in fig. 2. Starting from $\mathbf{c}_1(u_0)$ and travelling along ∂R in a counter-clockwise direction, the integral of f over R is expressed using Green's theorem:

$$\begin{aligned} J(f,R) = &\int_{u_0}^{u_1} {x_1}'P(\mathbf{c}_1) + {y_1}'Q(\mathbf{c}_1)du + \int_0^1 {x_3}'P(\mathbf{c}_3) + {y_3}'Q(\mathbf{c}_3)du \\ &+ \int_{u_1}^{u_0} {x_2}'P(\mathbf{c}_2) + {y_2}'Q(\mathbf{c}_2)du + \int_0^1 {x_4}'P(\mathbf{c}_4) + {y_4}'Q(\mathbf{c}_4)du \end{aligned} \tag{10}$$

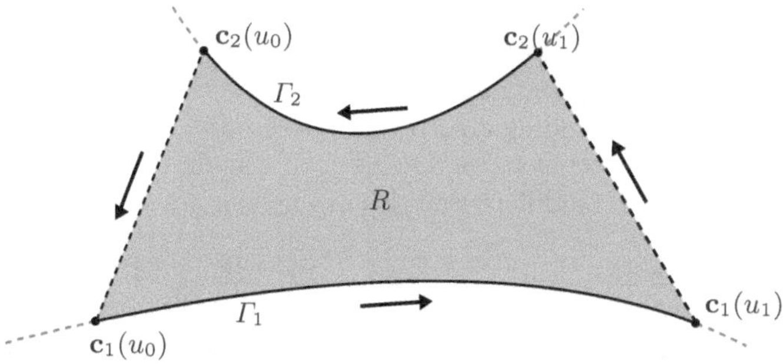

Fig. 2. Region R bounded by two portions of curves Γ_1 and Γ_2. Arrows indicate counter-clockwise direction for Green's theorem.

where position vectors $\mathbf{c}_3 = (x_3, y_3)^T$ and $\mathbf{c}_4 = (x_4, y_4)^T$ are parameterizations of line segments $[\mathbf{c}_1(u_1)\mathbf{c}_2(u_1)]$ and $[\mathbf{c}_2(u_0)\mathbf{c}_1(u_0)]$, respectively:

$$\mathbf{c}_3(u) = (1-u)\mathbf{c}_1(u_1) + u\mathbf{c}_2(u_1) \qquad \mathbf{c}_4(u) = (1-u)\mathbf{c}_2(u_0) + u\mathbf{c}_1(u_0)$$

Considering a family of curves $\{\Gamma_\alpha\}_{0\leq\alpha\leq 1}$ interpolating from Γ_1 to Γ_2, we have:

$$\mathbf{c}_1 - \mathbf{c}_2 = \int_0^1 \frac{d}{d\alpha}\{(1-\alpha)\mathbf{c}_2 + \alpha\mathbf{c}_1\}\, d\alpha$$

The joint use of the previous relation and integration by parts in eq. 10 enables to express $J(f, R)$ directly as a function of f, $\mathbf{c}_1$ and $\mathbf{c}_2$:

$$J(f,R) = \int_{u_0}^{u_1}\int_0^1 f((1-\alpha)\mathbf{c}_2 + \alpha\mathbf{c}_1)(\mathbf{c}_1 - \mathbf{c}_2) \times ((1-\alpha)\mathbf{c}_2{}' + \alpha\mathbf{c}_1{}')d\alpha du \quad (11)$$

This expression is intuitively understood since $(1-\alpha)\mathbf{c}_2 + \alpha\mathbf{c}_1$ sweeps all curves between Γ_1 and Γ_2 as α varies from 0 to 1. The cross product corresponds to the area of the infinitesimal quadrilaterals spanned by $\mathbf{c}_1 - \mathbf{c}_2$ and $(1-\alpha)\mathbf{c}_2{}' + \alpha\,\mathbf{c}_1{}'$. We apply this result on our GC to calculate $J(f, \mathcal{B}_{in})$, replacing Γ_1 and Γ_2 with $\Gamma_{[B]}$ and $\Gamma_{[-B]}$, respectively. Introducing a variable thickness $b = B(2\alpha - 1)$, we obtain:

$$J(f, \mathcal{B}_{in}) = \int_\Omega \int_{-B}^{B} f(\mathbf{c} + b\mathbf{n})\ell(1 - b\kappa)dbdu \quad (12)$$

A similar development is applied on bands $\mathcal{B}_{out1}$ and $\mathcal{B}_{out2}$, where b varies in the intervals $[B(u), B(u) + B_{out}]$ and $[-B(u) - B_{out}, -B(u)]$, respectively:

$$\begin{aligned} J(f, \mathcal{B}_{out1}) &= \int_\Omega \int_{B(u)}^{B(u)+B_{out}} f(\mathbf{c} + b\mathbf{n}) \left\|\frac{d\mathbf{c}}{du}\right\| (1 - b\kappa)dbdu \\ J(f, \mathcal{B}_{out2}) &= \int_\Omega \int_{-B(u)-B_{out}}^{-B(u)} f(\mathbf{c} + b\mathbf{n}) \left\|\frac{d\mathbf{c}}{du}\right\| (1 - b\kappa)dbdu \end{aligned} \quad (13)$$

The goal of these formulas is twofold. They allow to express explicitly E_{region} according to Γ and B, enabling the calculation of the variational derivatives. Moreover, they allow computation of $J(f, \mathcal{B}_{in})$ if Γ is discretized as a polygonal line, without using a flood-filling algorithm to determine inner pixels. In order to build a gradient descent scheme, we now perform the derivation of the energies, keeping eq. 3 in mind. The differentiations of E_{smooth} are straigthforward:

$$\frac{\delta E_{\text{smooth}}}{\delta \mathbf{c}} = -2\frac{d^2\mathbf{c}}{du^2} \qquad \frac{\delta E_{\text{smooth}}}{\delta B} = -2\frac{d^2 B}{du^2} \tag{14}$$

Regarding the region energy, the variational derivative with respect to $\mathbf{c}$ is:

$$\frac{\delta E_{\text{region}}}{\delta \mathbf{c}} = \frac{\delta J(g_{in}, \mathcal{B}_{in})}{\delta \mathbf{c}} + \frac{\delta J(g_{out1}, \mathcal{B}_{out1})}{\delta \mathbf{c}} + \frac{\delta J(g_{out2}, \mathcal{B}_{out2})}{\delta \mathbf{c}}$$

and similarly for B. We found more practical to compute these derivatives over the form given by Green's theorem in eq. 10. Since $\mathbf{c}(0)$ and $\mathbf{c}(1)$ are static, they do not intervene in the derivative expression. Hence, we write:

$$\frac{\delta J(f, \mathcal{B}_{in})}{\delta \mathbf{c}} = \frac{\delta J_1}{\delta \mathbf{c}} - \frac{\delta J_2}{\delta \mathbf{c}}$$
$$J_1 = \int_{\Omega} x_{[B]}{}' P(\mathbf{c}_{[B]}) + y_{[B]}{}' Q(\mathbf{c}_{[B]}) du$$
$$J_2 = \int_{\Omega} x_{[-B]}{}' P(\mathbf{c}_{[-B]}) + y_{[-B]}{}' Q(\mathbf{c}_{[-B]}) du$$

We find that $\delta J_1/\delta \mathbf{c}$ has both normal and tangential components:

$$\begin{aligned}\frac{\delta J_1}{\delta \mathbf{c}} &= f(\mathbf{c}_{[B]}) \left(\ell(B\kappa - 1) - \frac{BB'\ell'}{\ell^2} + \frac{B'^2 + BB''}{\ell} \right) \mathbf{n} \\ &+ BB' \left((1 - B\kappa)\nabla f(\mathbf{c}_{[B]}) \cdot \mathbf{t} + B'\nabla f(\mathbf{c}_{[B]}) \cdot \mathbf{n} \right) \mathbf{n} \\ &+ f(\mathbf{c}_{[B]}) B'(1 - B\kappa)\mathbf{t}\end{aligned} \tag{15}$$

The derivative $\delta J_2/\delta \mathbf{c}$ is trivially obtained from eq. 15, replacing B with $-B$. In the implementation part, the tangential component may be ignored, according to the principle of geodesic active contours [24], stating that the curve geometry is only affected by normal displacement whereas tangential motion only affects parameterization. We use a polygonal discretization which is resampled after each deformation step. Resampling keeps vertices almost evenly spaced along the curve and hence compensates the absence of tangential motion. The derivative with respect to B is obtained from eq. 12, which does not depends on B' nor B''. Hence:

$$\begin{aligned}\frac{\delta J(f, \mathcal{B}_{in})}{\delta B} &= \frac{\partial}{\partial B} \left\{ \int_{-B}^{B} f(\mathbf{c} + b\mathbf{n}) \left\| \frac{d\mathbf{c}}{du} \right\| (1 - b\kappa) db \right\} \\ &= \ell(f(\mathbf{c}_{[B]})(1 - B\kappa) + f(\mathbf{c}_{[-B]})(1 + B\kappa))\end{aligned}$$

In the same way, the derivatives of $J(g_{out1}, \mathcal{B}_{out1})$ and $J(g_{out2}, \mathcal{B}_{out2})$ are calculated by replacing band thicknesses appropriately, according to eq. 13.

4 Implementation

4.1 Polygonal Implementation

The curve is discretized as a polygonal line made up of n linked vertices $\{\mathbf{p}_1, ..., \mathbf{p}_n\}$. Functions g_{in}, g_{out_1} and g_{out_2} use global features such as area or mean intensity. Computed over $\mathcal{B}_{in}$, these global quantities are computed according to the following template formula, which is a discrete implementation of eq. 12:

$$J(f, \mathcal{B}_{in}) \approx \sum_{i=1}^{n} \sum_{b=-B_i}^{B_i} f(\mathbf{p}_i + b\mathbf{n}_i)\ell_i(1 - b\kappa_i) \tag{16}$$

where B_i, ℓ_i, $\mathbf{n}_i$ and κ_i are the discretized band thickness, length element, curvature and normal at vertex $\mathbf{p}_i$, using finite differences. The same computations are performed over regions $\mathcal{B}_{out1}$ and $\mathcal{B}_{out2}$ by discretizing eq. 13. Vertex coordinates and corresponding thicknesses are iteratively modified using gradient descent of eq 7:

$$\begin{aligned} \mathbf{p}_i^{(t+1)} &= \mathbf{p}_i^{(t)} - \Delta t \; \delta E / \delta \mathbf{c}|_{\mathbf{c}=\mathbf{p}_i} \\ B_i^{(t+1)} &= B_i^{(t)} - \Delta t \; \delta E / \delta B|_{B=B_i} \end{aligned} \tag{17}$$

The evolution of the discrete GC is sumarized by the following steps, repeated until a steady state is reached:

- Move all vertices according to eq. 17a
- Update normal vectors, areas and mean intensities
- Modify all thicknesses according to eq. 17b
- Resample the curve
- Update normal vectors, areas and mean intensities

4.2 Initialization by Minimal Path

Curve points $\mathbf{c}(0)$ and $\mathbf{c}(1)$, represented by vertices $\mathbf{p}_1$ and $\mathbf{p}_n$ respectively, are given by the user. They remain motionless in our approach, in the sense that the user clicks the desired end-points of the shape. Given these, the simplest way to initialize the medial curve would be a straight line. As is, the GC is sensitive to the initial location, as the energy is minimized through gradient descent. A straight line is problematic as soon as the region of interest is highly curved. The minimal path approach by Cohen and Kimmel [16] provides a reliable initialization for the central curve regardless of the shape geometry. Based on the fast marching algorithm [25][26], this approach finds the minimal path $\mathcal{C}^*$ between the two end-points, given a potential-dependent metric:

$$\mathcal{C}^* = \underset{\mathcal{C}}{\operatorname{argmin}} \left\{ \int_0^1 (w + P(\mathcal{C}(s)))ds \right\} \text{s.t.} \begin{cases} \mathcal{C}(0) = \mathbf{c}(0) \\ \mathcal{C}(1) = \mathbf{c}(1) \end{cases}$$

where s is the arc length parameter, w is a regularizing term weighting the length and P is the potential. A global search is performed, avoiding wrong local minima. We use the generated path as the initial guess for curve Γ, so that $\Gamma^{(0)} = \mathcal{C}^*$. The polygonal line $[\mathbf{p}_1, \mathbf{p}_2, ...\mathbf{p}_n]$ is generated by downsampling the path ($\approx$ a vertex every three pixels). The best path between the two end-points should be uniform and its mean intensity should be close to the one of the end-points. Hence, we chose the following potential:

$$P(\mathbf{x}) = \left(I(\mathbf{x}) - \frac{I(\mathbf{c}(0)) + I(\mathbf{c}(1))}{2} \right)^2 + \|\nabla I(\mathbf{x})\|^2$$

For RGB colour images, we use an extension of this potential, based on normalized RGB intensities. The end-points are intendedly kept static, so that a global minimum with respect to our criterion may be determined. Moreover, providing the end-points is an intuitive way of indicating the target path-like structure.

5 Experiments

We tested the proposed method on a set of images containing path-like objects of interest. First, we applied the deformable ribbon on vascular structures in grayscale medical images, mainly MRI data and angiograms. On these images in particular, we used the minimal variance principle by Chan and Vese [6]:

$$g_{in}(\mathbf{x}) = (I(\mathbf{x}) - \mu(\mathcal{B}_{in}))^2$$
$$g_{out1}(\mathbf{x}) = (I(\mathbf{x}) - \mu(\mathcal{B}_{out1}))^2 \qquad g_{out2}(\mathbf{x}) = (I(\mathbf{x}) - \mu(\mathcal{B}_{out2}))^2$$

where μ is the mean intensity of a given region, computed according to eq. 16. We considered a mean value for each outer region, although their respective average intensities were rather close. An important issue raised by the narrow band region energy is the similarity between inner and outer statistics that may arise when the GC, including outer bands, is initialized inside a uniform region. In such cases, we found necessary to add a bias, acting like a balloon force [27], in order to increase B in proportion to the similarity between inner and outer means. Secondly, we tested the ribbon method on aerial images, in order to extract portions of rivers. On color images, we used homogeneity measures based on chromaticity, i.e. normalized RGB components. Letting I_r, I_g and I_b be the red, green and blue intensities of a given pixel, the normalized red and green components are:

$$\overline{I_\mathrm{r}} = \frac{I_\mathrm{r}}{I_\mathrm{r} + I_\mathrm{g} + I_\mathrm{b}} \qquad \overline{I_\mathrm{g}} = \frac{I_\mathrm{g}}{I_\mathrm{r} + I_\mathrm{g} + I_\mathrm{b}}$$

The homogeneity measure is based on the variance of intensity features, independently considered:

$$g_\mathrm{in}(\mathbf{x}) = \left(\overline{I_\mathrm{r}}(\mathbf{x}) - \overline{\mu_\mathrm{r}}(\mathcal{B}_{in})\right)^2 + \left(\overline{I_\mathrm{g}}(\mathbf{x}) - \overline{\mu_\mathrm{g}}(\mathcal{B}_{in})\right)^2$$

where $\overline{\mu_\mathrm{r}}$ and $\overline{\mu_\mathrm{g}}$ are the average normalized red and green intensities of the corresponding region. g_{out1} and g_{out2} are expressed in a similar way. We found this measure to be more efficient than the L2 distance in RGB space. Indeed, normalized RGB is less sensitive to changes in illumination than can arise in a visually homogeneous region.

In order to justify the use of a region-based criterion, we provide a comparison with the same GC now endowed with an edge term, basing ourselves on the ribbon snake model of [10] and [11]. The edge-based GC is obtained by replacing the region energy with a combination of edge and balloon energies:

$$E_{\text{edge}}(\Gamma, B) = \int_{\Omega} k(\mathbf{c}_{[B]}) + k(\mathbf{c}_{[-B]})\, du$$
$$E_{\text{balloon}}(\Gamma, B) = \int_{\Omega}\int_{-B}^{B} \ell(1 - b\kappa)\, dbdu = 2\int_{\Omega} B\ell\, du$$

where k is an edge-detecting function (e.g. $k(\mathbf{x}) = -\|\nabla I(\mathbf{x})\|$ for grayscale images). The balloon energy is the area of the inner band, obtained from eq. 12 with $f = 1$. The variational derivatives, used as descent directions, were calculated as follows (note that we neglect the influence of Γ on the balloon energy):

$$\frac{\delta E_{\text{edge}}}{\delta \mathbf{c}} \approx \nabla k(\mathbf{c}_{[B]}) + \nabla k(\mathbf{c}_{[-B]})$$
$$\frac{\delta E_{\text{edge}}}{\delta B} = (\nabla k(\mathbf{c}_{[B]}) - \nabla k(\mathbf{c}_{[-B]})) \cdot \mathbf{n}$$
$$\frac{\delta E_{\text{balloon}}}{\delta B} = 2\ell$$

The minimization of the edge energy makes the curve and thickness evolve in order to stabilize on high-gradient magnitude pixels, whereas the balloon energy results in a constant speed added to the thickness. The balloon term is weighted with a negative coefficient, in order to increase band thickness and make the GC inflate.

Results are shown in fig. 3. For each image, column (a) holds the original image. Column (b) depicts the initial configuration of the GC, used both for region and edge-based evolution. Columns (c) and (d) represents the final states of the GC, with region and edge energies respectively. In all tests, the GC exhibited strong topological and geometrical prior, preventing it from flowing into neighbouring narrow structures having more or less the same intensity/colour than the desired shape. Consequently, the deformable GC is suitable for applications where strong prior knowledge about the shape and position of the target is available. We found our region-based GC to be less sensitive to initial location than the edge-based model, as the edge energy makes the GC collapse to one side of the object in some parts. Moreover, increasing the balloon influence can make the band pass through the actual contours in other parts.

The constant outer band thickness B_{out}, the initial thickness $B^{(0)}$ and the internal energy weight ω are important parameters. We found that values ranging from 4 to 50 were suitable for B_{out}. A too small thickness may make the outer region statistics irrelevant and cause the GC boundary curves to pass through the actual boundaries, whereas the advantage of the narrow band principle is

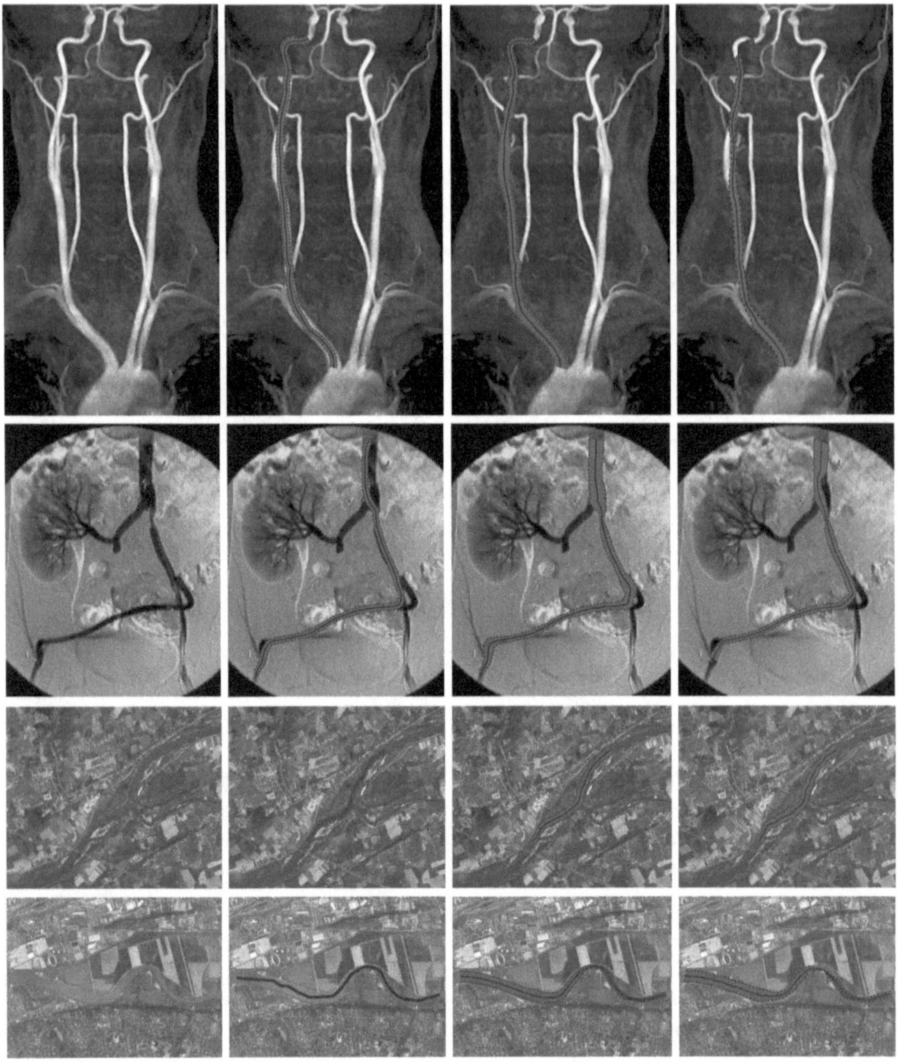

(a) (b) (c) (d)

Fig. 3. Segmentation of angiograms and aerial images (©GoogleEarth 2008): original image (a), initial GC (b), final region-based GC (c) and final edge-based GC (d)

lost if the thickness is too high. The initial inner thickness $B^{(0)}$ was typically set to 4 pixels as well. For ω, values ranging from 0.4 to 0.7 were sufficient to maintain boundary curves smooth. Segmentation process on a 1000×720 RGB image (shown on the last column of fig. 3) took approximately 1.5s with C++ implementation running on an Intel Core 2 Duo 2GHz PC (including the computational cost due to the construction of the minimal path).

6 Conclusion and Perspectives

In this paper, we presented a region-based deformable GC model, extending the work on classical region-based active contours and gradient-based ribbon snakes. Defined by a central curve playing the role of the medial axis and a variable thickness, the GC was endowed with a region-dependent energy following the narrow band principle. This energy was subsequently transformed and derived in order to allow implementation on a polygonal line deformed with gradient descent. The model proved to be efficient for recovering tubular objects, exhibiting geometrically constrained behaviour.

We may consider further investigations on this work. The deformable GC may be extended to three-dimensional space. The first 3D extension would consist in a space curve surrounded with a variable thickness, as in [28], which would be dependent on both curve parameter and angle, in order to model a deformable tube. The second 3D extension would consist in a surface homeomorphic to a plane endowed with a position-dependent thickness, which would be used to extract sheet-like structures. Substantial work has to be performed on the mathematical derivation and implementation of a suitable region energy on these 3D structures.

References

1. Kass, M., Witkin, A., Terzopoulos, D.: Snakes: active contour models. International Journal of Computer Vision 1(4), 321–331 (1988)
2. Jacob, M., Blu, T., Unser, M.: Efficient energies and algorithms for parametric snakes. IEEE Transactions on Image Processing 13(9), 1231–1244 (2004)
3. Ronfard, R.: Region based strategies for active contour models. International Journal of Computer Vision 13(2), 229–251 (1994)
4. Ivins, J., Porrill, J.: Active region models for segmenting textures and colours. Image and Vision Computing 13(5), 431–438 (1995)
5. Cohen, L., Bardinet, E., Ayache, N.: Surface reconstruction using active contour models. In: SPIE Conference on Geometric Methods in Computer Vision, San Diego, CA, USA (1993)
6. Chan, T., Vese, L.: Active contours without edges. IEEE Transactions on Image Processing 10(2), 266–277 (2001)
7. Jehan-Besson, S., Barlaud, M., Aubert, G.: DREAM2S: Deformable regions driven by an eulerian accurate minimization method for image and video segmentation. International Journal of Computer Vision 53(1), 45–70 (2003)
8. Paragios, N., Deriche, R.: Geodesic active regions and level set methods for supervised texture segmentation. International Journal of Computer Vision 46(3), 223–247 (2002)
9. Davatzikos, C., Prince, J.: An active contour model for mapping the cortex. IEEE Transactions on Medical Imaging 14(1), 65–80 (1995)
10. Mayer, H., Laptev, I., Baumgartner, A.: Multi-scale and snakes for automatic road extraction. In: Burkhardt, H.-J., Neumann, B. (eds.) ECCV 1998. LNCS, vol. 1406, pp. 720–733. Springer, Heidelberg (1998)

11. Laptev, I., Mayer, H., Lindeberg, T., Eckstein, W., Steger, C., Baumgartner, A.: Automatic extraction of roads from aerial images based on scale space and snakes. Machine Vision and Applications 12(1), 23–31 (2000)
12. Pérez, P., Blake, A., Gangnet, M.: Jetstream: probabilistic contour extraction with particles. In: IEEE International Conference on Computer Vision (ICCV), Vancouver, Canada (2001)
13. Fua, P., Leclerc, Y.: Model driven edge detection. Machine Vision and Applications 3, 45–56 (1990)
14. Li, H., Yezzi, A.: Vessels as 4-D curves: global minimal 4-D paths to extract 3-D tubular surfaces and centerlines. IEEE Transactions on Medical Imaging 26(9), 1213–1223 (2007)
15. Mille, J., Boné, R., Makris, P., Cardot, H.: 2D and 3D deformable models with narrow band region energy. In: IEEE International Conference on Image Processing (ICIP), San Antonio, USA, pp. 57–60 (2007)
16. Cohen, L., Kimmel, R.: Global minimum for active contour models: a minimal path approach. International Journal of Computer Vision 24(1), 57–78 (1997)
17. Mumford, D., Shah, J.: Optimal approximation by piecewise smooth functions and associated variational problems. Communications on Pure and Applied Mathematics 42(5), 577–685 (1989)
18. Zhu, S., Yuille, A.: Region competition: unifying snakes, region growing, Bayes/MDL for multiband image segmentation. IEEE Transactions on Pattern Analysis and Machine Intelligence 18(9), 884–900 (1996)
19. Chakraborty, A., Staib, L., Duncan, J.: Deformable boundary finding in medical images by integrating gradient and region information. IEEE Transactions on Medical Imaging 15(6), 859–870 (1996)
20. Aubert, G., Barlaud, M., Faugeras, O., Jehan-Besson, S.: Image segmentation using active contours: calculus of variations or shape gradients? SIAM Journal on Applied Mathematics 63(6), 2128–2154 (2003)
21. Debreuve, E., Gastaud, M., Barlaud, M., Aubert, G.: Using the shape gradient for active contour segmentation: from the continuous to the discrete formulation. Journal of Mathematical Imaging and Vision 28(1), 47–66 (2007)
22. Malladi, R., Sethian, J., Vemuri, B.: Shape modeling with front propagation: a level set approach. IEEE Transactions on Pattern Analysis and Machine Intelligence 17(2), 158–175 (1995)
23. Farouki, R., Neff, C.: Analytic properties of plane offset curves. Computer Aided Geometric Design 7(1-4), 83–99 (1990)
24. Caselles, V., Kimmel, R., Sapiro, G.: Geodesic active contours. International Journal of Computer Vision 22(1), 61–79 (1997)
25. Adalsteinsson, D., Sethian, J.: A fast level set method for propagating interfaces. Journal of Computational Physics 118(2), 269–277 (1995)
26. Sethian, J.: A fast marching level set method for monotonically advancing fronts. Proceedings of the National Academy of Science 93(4), 1591–1595 (1996)
27. Cohen, L., Cohen, I.: Finite element methods for active contour models and balloons for 2D et 3D images. IEEE Transactions on Pattern Analysis and Machine Intelligence 15(11), 1131–1147 (1993)
28. Terzopoulos, D., Witkin, A., Kass, M.: Symmetry-seeking models and 3d object reconstruction. International Journal of Computer Vision 1(3), 211–221 (1987)

Pose Priors for Simultaneously Solving Alignment and Correspondence*

Francesc Moreno-Noguer, Vincent Lepetit, and Pascal Fua

Computer Vision Laboratory
École Polytechnique Fédérale de Lausanne (EPFL)
CH-1015 Lausanne, Switzerland

Abstract. Estimating a camera pose given a set of 3D-object and 2D-image feature points is a well understood problem when correspondences are given. However, when such correspondences cannot be established *a priori*, one must simultaneously compute them along with the pose. Most current approaches to solving this problem are too computationally intensive to be practical. An interesting exception is the SoftPosit algorithm, that looks for the solution as the minimum of a suitable objective function. It is arguably one of the best algorithms but its iterative nature means it can fail in the presence of clutter, occlusions, or repetitive patterns. In this paper, we propose an approach that overcomes this limitation by taking advantage of the fact that, in practice, some prior on the camera pose is often available. We model it as a Gaussian Mixture Model that we progressively refine by hypothesizing new correspondences. This rapidly reduces the number of potential matches for each 3D point and lets us explore the pose space more thoroughly than SoftPosit at a similar computational cost. We will demonstrate the superior performance of our approach on both synthetic and real data.

1 Introduction

Estimating the pose of a calibrated camera from 3D-to-2D point correspondences between a 3D model and an image is a fundamental problem in Computer Vision. When matches are given, this is known as the Perspective-n-Point (PnP) problem and many effective methods have been proposed. However, as shown in Fig. 1, there are cases where it is difficult to establish the required correspondences, for example because the scene contains repetitive patterns or because the 3D points are simply salient features on a CAD model without associated texture.

In such cases, it becomes necessary to simultaneously compute the pose and establish the correspondences. As shown in Fig. 1, SoftPosit [1], arguably the most computationally effective algorithm able to do this based on geometry alone, tends to fail in presence of large amounts of clutter, occlusions, or repetitive patterns. Other methods can avoid these issues by using a RANSAC-style [2] approach [3,4] or searching for clusters in the pose space [5,6,7,8], but quickly become computationally intractable with realistic numbers of features.

* This work was supported by the Swiss National Science Foundation and by funds of the European Commission under the IST-project 034307 DYVINE.

D. Forsyth, P. Torr, and A. Zisserman (Eds.): ECCV 2008, Part II, LNCS 5303, pp. 405–418, 2008.

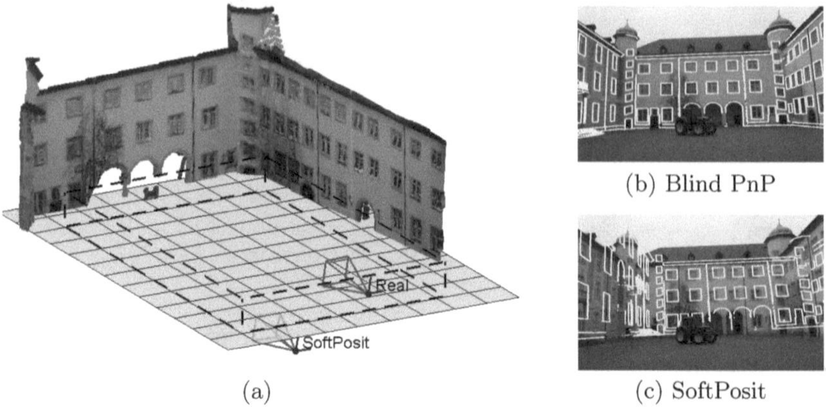

Fig. 1. Recovering the pose in a scene with repetitive patterns. (a) 3D model of the scene. The "Real" camera, and the pose recovered by SoftPosit [1] are indicated. Blind PnP is able to retrieve the "Real" pose. (b) Model reprojected after estimating the pose using Blind PnP. (c) Model reprojected after estimating the pose using SoftPosit.

In this paper, we propose an approach that is much more robust to occlusion, clutter, and repetitive patterns than SoftPosit without increase in computational complexity. We refer to our algorithm as *Blind PnP* because it solves the PnP problem without being given the correspondences. It relies on the fact that, in practice, prior information on the camera pose is often available. For example, in the case depicted by Fig. 1, the camera must be pointing towards the building and be above ground level. We model such a prior as a Gaussian Mixture Model (GMM) and use each component of the GMM to initialize a Kalman filter. We then explore the space of possible correspondences within a subset of potential matches and keep the hypothesis that yields the smallest reprojection error.

As illustrated by Fig. 2, the Kalman filter guides and speeds up the matching process, while preventing gross errors. As a result, the complexity of our method is of order $\mathcal{O}(Gn^3 M \log N)$, where G is the number of components in the GMM, M the number of 3D points, N the number of image features, and n the number of 2D correspondences that can be potentially matched to each 3D point when exploiting the prior. This turns out to be comparable to the complexity of SoftPosit, which is $\mathcal{O}(MN^2)$. However, since n is typically much smaller than N, our approach can perform a more exhaustive search for a similar amount of computation and therefore be less sensitive to distractors such as clutter.

Of course, further increases in speed could be achieved by considering image clues. For example, with the repetitive patterns of Fig. 1, it may not be possible assign to a 3D feature a single image feature on the sole basis of local image information but this information can nevertheless be used to restrict the number of potential matches. This is entirely compatible with the framework we propose.

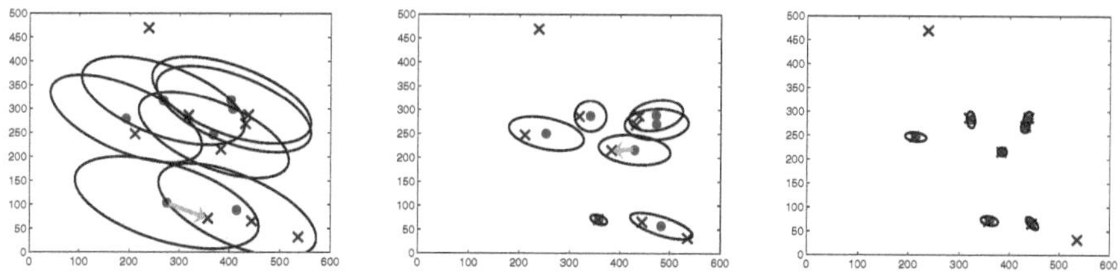

Fig. 2. Blind PnP. **Left:** The dots represent the projected 3D points and the ellipses the corresponding uncertainty regions given the pose prior. The crosses represent the 2D points. Even before hypothesizing any correspondence, the prior considerably reduces the set of potential matches. **Center:** Given a single matching hypothesis represented by the arrow, the number of possibilities for the second match is further reduced. **Right:** This is even truer of the third match and, once it has been hypothesized, the uncertainty ellipses become so small that finding additional matches becomes easy.

2 Related Work

Simultaneously recovering poses and correspondences has traditionally been achieved by hypothesizing small sets of k 3D-to-2D correspondences between the M 3D points and N 2D points, and validating the poses they generate using the remaining points [3]. Using the RANSAC algorithm [2] the computational complexity is $\mathcal{O}(MN^k \log N)$, which is prohibitive when the number features becomes large. This can be partially addressed by means of heuristic criteria to terminate the search as soon as a "good solution" is found [9,10,4]. Nevertheless, the algorithms remain exceedingly slow for realistic values of k, M and N.

Another approach to reducing the computational burden is to focus on specific regions of the search space. This can be done in a data-driven manner using indexing methods [11,12,13]. These techniques initially learn 2D feature groupings of 3D objects and store the associated vectors in data structures, such as Hash tables [11,12], or kd-trees [13]. At runtime, feature vectors are extracted from the test images and used to access the database to extract the 3D-to-2D hypotheses, which are then used for pose estimation. These techniques, however, can only handle relatively small pose spaces.

Specific subspaces of the search space can also be preselected in a model driven fashion, for example through a pose clustering approach [5,6,7,8,14]. These techniques consider all possible 3D-to-2D matches, and the poses they generate are represented in a 6D pose space. High-probability clusters are then extracted from this space, under the assumption they will contain only correctly hypothesized matches. Finally, the hypotheses within these clusters are further considered for verification and pose estimation. In fact, pose clustering techniques are similar to the hypothesize-and-test algorithms discussed above, in the sense that they initially take into account all possible matches. As a consequence they also lead to computational explosion when considering large sets of 3D-to-2D correspondences. For instance, one of the most efficient algorithms within this group [7], has a complexity of $\mathcal{O}(MN^3)$.

Other approaches iteratively solve for pose and correspondence [1,15,16], by optimizing a global cost function. Among these, SoftPosit [1] is the one that stands out because of its accuracy and efficiency. It combines an iterative pose estimation technique, with an algorithm to assign correspondences, resulting in an $\mathcal{O}(MN^2)$ complexity. One limitation of such an algorithm is that the global minimum can not be guaranteed. This is alleviated by randomly initializing it at different initial guesses. Yet, certain configurations of the data or situations with large amounts of occlusion and clutter, still cause SoftPosit to fail.

Our approach differs from the previous ones in that we approximate the 6D pose space as a GMM, meaning that we define several initial pose guesses with an associated uncertainty. Each Gaussian component is independently verified, with the advantage that only a very small subset of 3D-to-2D correspondences needs to be hypothesized. In contrast, the "hypothesis-and-test" and "pose clustering" approaches consider all possible hypotheses. On the other hand, the iterative algorithms, although started from many initial guesses, do not exploit the fact that given one such guess, the set of plausible 3D-to-2D correspondences is much reduced. These key differences allow Blind PnP to be as efficient as the SoftPosit algorithm while retaining the robustness of exhaustive techniques.

3 Pose Priors for Alignment and Correspondence

As shown in Fig. 2, we use the pose priors to restrict the number of possible 3D-to-2D matches. Hypothesizing one such match decreases the pose uncertainty, further reducing the possibilities for the second match, and even more so for the third. Once three correspondences have been hypothesized, the uncertainty becomes so small that we can estimate a pose and match other 3D points to 2D points that are close enough to their projection using that estimate. We can then evaluate its quality using a suitable objective function, iterate over all plausible triplets of correspondences, and retain the best one.

3.1 Formalization

We assume we are given a set $\{\mathbf{x}_1, \ldots, \mathbf{x}_M\}$ of M 3D model points and a set $\{\mathbf{u}_1, \ldots, \mathbf{u}_N\}$ of N 2D image points, coming for instance from a keypoint detector. Some of the 2D points correspond to 3D points and some do not. Similarly, the projections of some of the 3D points may not be among the 2D points. We suppose as well we are given a set of pose prior samples. Using Expectation Maximization we represent them as a Gaussian Mixture Model that includes G Gaussian components with 6-vectors $\{\mathbf{p}_1, \ldots, \mathbf{p}_G\}$ for the means, and 6×6 covariance matrices $\{\mathbf{\Sigma}_1^p, \ldots, \mathbf{\Sigma}_G^p\}$ [1]. Fig. 3 shows an example of such pose priors.

Our goal is to find both the correct pose $\mathbf{p}$ and as many 2D-to-3D correspondences as possible. Let $\mathsf{Matches}$ be a set of $(\mathbf{x}, \mathbf{u})$ pairs that represents those

[1] We use exponential maps to parametrize the pose rotation, yielding to 6 values to parameterize the complete pose.

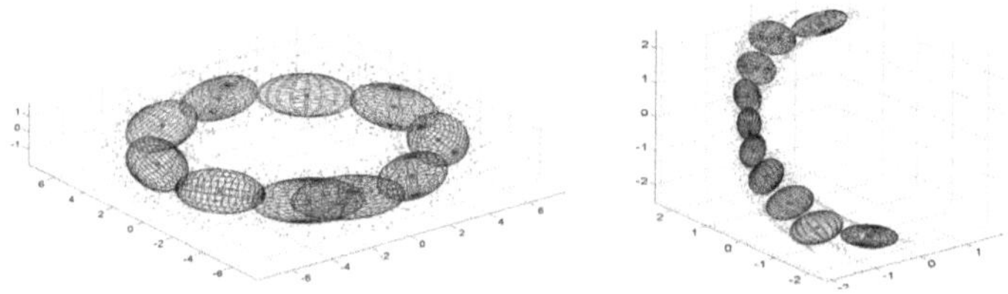

Fig. 3. Example of pose samples and the corresponding Gaussian components for a case where the pose is constrained to be inside a torus surrounding the object and pointing to it. For simplicity, we represent the translation (left) and rotation (right) spaces separately and do not represent the cross-covariances between the two spaces.

correspondences and NotDetected be the subset of 3D points for which no match can be established. We want to find the correct $\mathbf{p}$ and Matches by minimizing

$$\text{Error}(\mathbf{p}) = \sum_{(\mathbf{x},\mathbf{u})\in\mathsf{Matches}} \|\mathbf{u} - \text{Proj}(\mathbf{p};\mathbf{x})\| + \left|\mathsf{NotDetected}\right|\mathcal{T}\,, \tag{1}$$

where $\text{Proj}(\mathbf{p};\mathbf{x})$ is the 2D projection of the 3D point $\mathbf{x}$ given pose $\mathbf{p}$, and $\mathcal{T}$ is a penalty term that penalizes unmatched points and whose value will be set as discussed below. $|\cdot|$ accounts for the cardinality of a set.

We next turn to the steps of our approach. The pseudocode is given in Fig. 4.

3.2 Algorithmic Steps

Limiting the possibilities for a first match. We want to minimize (1) using the pose priors. Each Gaussian component gives an initial $\mathbf{p}_g$ pose estimate and an associated covariance $\mathbf{\Sigma}_g^p$. We use this covariance to restrict the possibility of matches between the 3D and 2D points, by computing the projections $\mathbf{v}_i$ and the covariances $\mathbf{\Sigma}_i^v$ of the $\mathbf{x}_i$ 3D points. Error propagation yields

$$\mathbf{\Sigma}_i^v = \mathbf{J}(\mathbf{x}_i)\,\mathbf{\Sigma}_g^p\,\left(\mathbf{J}(\mathbf{x}_i)\right)^\top\,, \tag{2}$$

where $\mathbf{J}(\mathbf{x}_i)$ is the Jacobian of $\text{Proj}(.;\mathbf{x}_i)$. This defines a search region for the point $\mathbf{x}_i$, and we only consider the 2D points $\mathbf{u}_j$ such that

$$\left(\mathbf{v}_i - \mathbf{u}_j\right)^\top\,\mathbf{\Sigma}_i^v\left(\mathbf{v}_i - \mathbf{u}_j\right) \leq \mathcal{M}^2 \tag{3}$$

as potential matches for $\mathbf{x}_i$. $\mathcal{M}$ is a threshold chosen to achieve a specified degree of confidence, based on the cumulative chi-squared distribution. In our experiments we use $\mathcal{M}$=2 and 3, which yield 86% and 99% of confidence, respectively. This reduces the search space from the entire image to a small elliptical region.

Updating the Pose and its Uncertainty from the First Correspondence. Among 3D points $\mathbf{x}_i$ that have at least one potential match, we begin by those that have the fewest and consider each one in turn. Hypothesizing a

```
BestSolution ← ⊥
For each Gaussian component g
| BuildMatchesTriplets(p_g, Σ^p_g, ∅, BestSolution)
Best pose and matches are in BestSolution

Function BuildMatchesTriplets(p, Σ^p, Hypotheses, BestSolution)
| If all x_i already appear in Hypotheses
| | Return
| For each x_i that do not appear in Hypotheses
| | v_i ← Proj(p; x_i) ;  Σ^v_i = J(x_i) Σ^p (J(x_i))^T
| | S_i ← {u_j | (v_i − u_j)^T Σ^v_i (v_i − u_j) ≤ M^2}
| m ← argmin_i |S_i|
| For each u_j ∈ S_m
| | p^+ ← K(u_j − v_m); Σ^{p+} ← (I − KJ(x_m)) Σ^p
| | If already 2 matches in Hypotheses
| | | Find optimal matches using p^+ and Σ^{p+} and put them in OptimalMatches
| | | Compute pose p' and Error from matches in OptimalMatches
| | | If Error < TerminationThreshold
| | | | Early Termination with pose p' and matches in OptimalMatches
| | | If Error < Error( BestSolution )
| | | | BestSolution ← (p', OptimalMatches )
| | Else
| | | BuildMatchesTriplets(p^+, Σ^{p+}, Hypotheses ⊕ [x_m ↔ u_j], BestSolution)
| If number of consecutive not detected points in Hypotheses < Threshold
| | BuildMatchesTriplets(p, Σ^p, Hypotheses ⊕ [x_m is not detected], BestSolution)
End function
```

Fig. 4. Pseudo-code of the Blind PnP

correspondence between $\mathbf{x}_i$ and $\mathbf{u}_j$ reduces the pose uncertainty. We use standard Kalman filter equations to update the $\mathbf{p}_g$ pose estimate and its covariance $\boldsymbol{\Sigma}_g$ and write

$$\mathbf{p}_g^+ = \mathbf{p}_g + \mathbf{K}\left(\mathbf{u}_j - \text{Proj}(\mathbf{p}_g; \mathbf{x}_i)\right) \ , \tag{4}$$

$$\boldsymbol{\Sigma}_g^{p+} = \left(\mathbf{I} - \mathbf{K}\mathbf{J}(\mathbf{x}_i)\right) \boldsymbol{\Sigma}_g^p \ , \tag{5}$$

where $\mathbf{K}$ is the Kalman Gain and $\mathbf{I}$ the Identity matrix.

Building Sets of Three Correspondences. After this update, the pose covariance is much smaller and the number of potential hypotheses for the second match will be highly reduced. The process is repeated one last time to select a third hypothesis. In general, the innovation term in (4) then becomes almost negligible. This is consistent with the fact that only three 3D-to-2D matches are needed to compute the camera pose.

Finding Optimal Correspondences. Therefore, after three match hypotheses, the pose does not evolve anymore and we can easily match the remaining 3D points by projecting them onto the image and assigning to them the closest of the N feature points. This assignment can be done in $\mathcal{O}(\log N)$ [17] for each point. The 3D points projected farther apart from any 2D feature than a fixed distance $\mathcal{T}$ are considered as not detected. Following a similar discussion as for $\mathcal{M}$ in (3), $\mathcal{T}$ is set to $9\sigma^2$, with σ^2 being the variance of the Normal image noise. This ensures with 99% probability that a 2D feature and a 3D projected point actually match.

Iterating and Terminating. We repeat the process above by iterating over the Gaussian components, and the 3D points $\mathbf{x}_i$ with their constrained set of potential 2D correspondents given by (3). We stop if the error in (1) drops below a threshold, taken sufficiently small to reduce the risks of false positives.

Dealing with Non-Detected 3D Points. We next explain how to deal with 3D points lacking of a 2D match. Once a pose has been computed from three correspondences, we handle this possibility by considering as not detected the 3D points projected farther apart from any 2D image feature than a fixed distance $\mathcal{T}$. However this is not sufficient because the points in the initial sets of three matches may have also not been detected. After iterating over the potential 2D correspondents $\mathbf{u}_j$ for $\mathbf{x}_i$, we consider the case that $\mathbf{x}_i$ may not have been detected, skip it, and consider the next best 3D point for the next correspondences.

But postponing the making of hypotheses may increase dramatically the algorithm complexity. Fortunately this can be avoided because the probability to consecutively pick r non-detected 3D points quickly decreases when r increases. This probability can be modeled as a sampling without replacement process:

$$P(\mathbf{x}_{i_1} \in \mathsf{NotDetected}, \ldots, \mathbf{x}_{i_r} \in \mathsf{NotDetected}) = \frac{N_{\text{nd}}!\ (M-r)!}{(N_{\text{nd}}-r)!\ M!}\,, \tag{6}$$

where $N_{\text{nd}} = \left|\mathsf{NotDetected}\right|$ is the number of non-detected 3D points. When this probability drops below a fixed threshold —we use 5% in all our experiments—, we stop considering the 3D points as non-detected. Note that the number of non-detected N_{nd} is not known *a priori*. However, conservatively setting this number to a sufficiently large and fixed value —we assume a ratio of 60% of outliers in all our experiments— prevents erroneously stopping the search when the actual number of non detected points is smaller.

4 Complexity Analysis

We now describe the complexity of our algorithm. Our goal is to estimate the number R of 3D-to-2D hypotheses that must be drawn in order to ensure with probability λ that at least one of them is outlier-free.

Estimating the pose requires three correct correspondences, which are both detected and correctly assigned to a 2D feature. The probability of this to happen for three randomly chosen model points is:

$$P_{\mathsf{3detected}} = \frac{M_d}{M} \cdot \frac{M_d - 1}{M-1} \cdot \frac{M_d-2}{M-2} \qquad P_{\mathsf{3correct\ assign}} = \frac{1}{n_1\, n_2\, n_3} \approx \frac{1}{n^3}$$

where $M_d = p_d M$ stands for the number of detected model points, p_d is the ratio of detected model points, n_1 is the number of 2D points in the first uncertainty region, n_2 is the number of 2D points in the uncertainty region after the first hypothesis, and n_3 the same but after the second hypothesis. Since the uncertainty regions progressively shrink, we have $n_1 > n_2 > n_3$, but for simplicity we write the right-most approximation, where n is usually much smaller than the total number N of 2D points. From the previous equations we get the probability that a set of three matches has only inlier correspondences:

$$P_{\text{3correct matches}} = P_{\text{3detected}} \times P_{\text{3correct assign}} \approx \frac{M_d^3}{M^3 n^3} = \left(\frac{p_d}{n}\right)^3 . \quad (7)$$

We can then write the probability of taking R consecutive incorrect samples as $(1 - P_{\text{3correct matches}})^R$. Hence, the number of samples R that need to be drawn to ensure with probability λ that at least one of them is correct is such that $(1 - P_{\text{3correct matches}})^R \leq \lambda$, which leads[2] to

$$R \approx \left(\frac{n}{p_d}\right)^3 \log\left(\frac{1}{1-\lambda}\right) . \quad (8)$$

Each time 3 model points are matched to 3 image features, the pose is recomputed in constant time using a Kalman Filter. All the model points are then projected onto the image in time $\mathcal{O}(M)$ and the correctness of the pose is verified by assigning image points to the projected model points. The assignment is done in time $\mathcal{O}(\log N)$ following [17]. This process has to be repeated in the worst case for all the G Gaussian components. Thus, the final cost of our algorithm is:

$$\left(\frac{n}{p_d}\right)^3 \log\left(\frac{1}{1-\lambda}\right) \times \mathcal{O}(G) \times \mathcal{O}(M) \times \mathcal{O}(\log N) = \mathcal{O}(Gn^3 M \log N) \quad (9)$$

where G and n are much smaller than N. Observe that with the probabilistic partition of the pose space, the complexity of our algorithm is significantly smaller to the $\mathcal{O}(MN^3 \log N)$ complexity of the general hypothesize-and-test approach [2,4], or the $\mathcal{O}(MN^3)$ of the traditional pose clustering algorithms [7] which still consider all the possible 3D-to-2D correspondences.

The computational cost of our approach is comparable to the $\mathcal{O}(MN^2)$ complexity of SoftPosit. Consider for instance a case of a 3D model with M=80 and N=100. Typical values for G and n are 20 and 5, respectively. This would mean that a hypothesize-and-test approach would require 3.6×10^8 operations to solve the problem, a pose clustering would require 8.0×10^7 operations, while SoftPosit and Blind PnP would require 8.0×10^5 and 9.2×10^5 operations, respectively.

From (9), it can be seen that the best efficiency of the Blind PnP is achieved when image points are more or less evenly spread over the image. In this situation, the number n of image points falling into each ellipse of uncertainty tends to be small, and so is the complexity of the algorithm. In contrast, a hypothetical image where all the points were concentrated in a small region, would be the worst situation for our algorithm —its performance would converge to that of the hypothesize-and-test algorithms. Nevertheless, this situation is both unusual, and undesirable for any pose estimation algorithms because it would lead to many ambiguities in the 3D-to-2D correspondences.

[2] Since $(p_d/n)^3 << 1$ we use the approximation $\log(1-x) \approx -x$ for small x.

5 Results

We evaluated our approach on both synthetic and real data by comparing it to SoftPosit. To ensure fairness, we initialized SoftPosit with the means of the GMM components we use to represent pose priors, instead of randomly. Since there is no obvious way to exploit the pose covariances in SoftPosit, we ran it several times using different values of a parameter β that controls the uncertainty of the initial pose [1]. To eliminate spurious local minima, we used the same stopping criterion for SoftPosit as for Blind PnP, which is to stop the search only if the residual in Eq. 1 drops below a very small threshold, which is much stricter than the one in [1]. Therefore, we compare our method to a Softposit version that was enhanced in order to take advantage of the pose priors.

To estimate convergence rates given the true camera rotation $\mathbf{R}_{true}$ and translation $\mathbf{t}_{true}$, we first computed the relative error of the estimated rotation $\mathbf{R}$ by $E_{rot}(\%) = \|\mathbf{q}_{true} - \mathbf{q}\|/\|\mathbf{q}\|$, where $\mathbf{q}$ and $\mathbf{q}_{true}$ are the normalized quaternions corresponding to the rotation matrices. Similarly, the relative error of the estimated translation $\mathbf{t}$ was taken to be $E_{trans}(\%) = \|\mathbf{t}_{true} - \mathbf{t}\|/\|\mathbf{t}\|$. Finally, we considered a pose to be correct if both $E_{rot}(\%)$ and $E_{trans}(\%)$ were below 10%.

5.1 Synthetic Experiments

We peformed numerous MonteCarlo simulations to account for clutter, occlusions and different 3D point configurations. Following [1], each simulation was characterized by the number M of 3D model points, the fraction $p_d \in \{0.4, 0.6, 0.8\}$ of occluded 3D points with no corresponding 2D point, and the percentage $p_c \in \{0.6, 0.8\}$ of clutter, that is, 2D points with no corresponding 3D point. For each set of parameter values, we performed $n_t = 50$ random and independent trials. For each trial, we ran SoftPosit three times with different values of $\beta \in \{10^{-4}, 10^{-3}, 10^{-2}\}$. The Blind PnP results of this section were obtained with the Mahalanobis boundary $\mathcal{M}$ on (3) set to 2. For comparison purposes, we also ran our algorithm with $\mathcal{M}=3$ and obtained virtually indistinguishable convergence rates but an increase in computation time by a factor of 2 to 3.

Fig. 5 depicts the three kinds of 3D points configurations—randomly distributed, repetitive structures, and lying on a CAD model of a castle—along with the regions of possible camera locations we used. The camera optical axes were

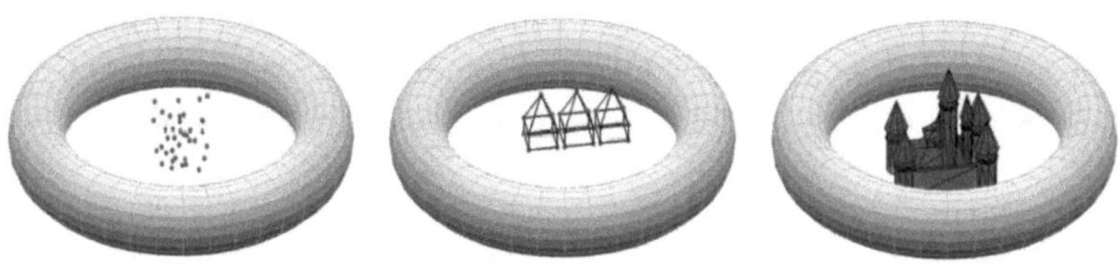

Fig. 5. 3D point configurations and possible locations for the camera centers. **Left:** From 20 to 80 randomly distributed points. **Middle:** 27 points forming repetitive structures. **Right:** 100 points laying on the CAD model of a castle.

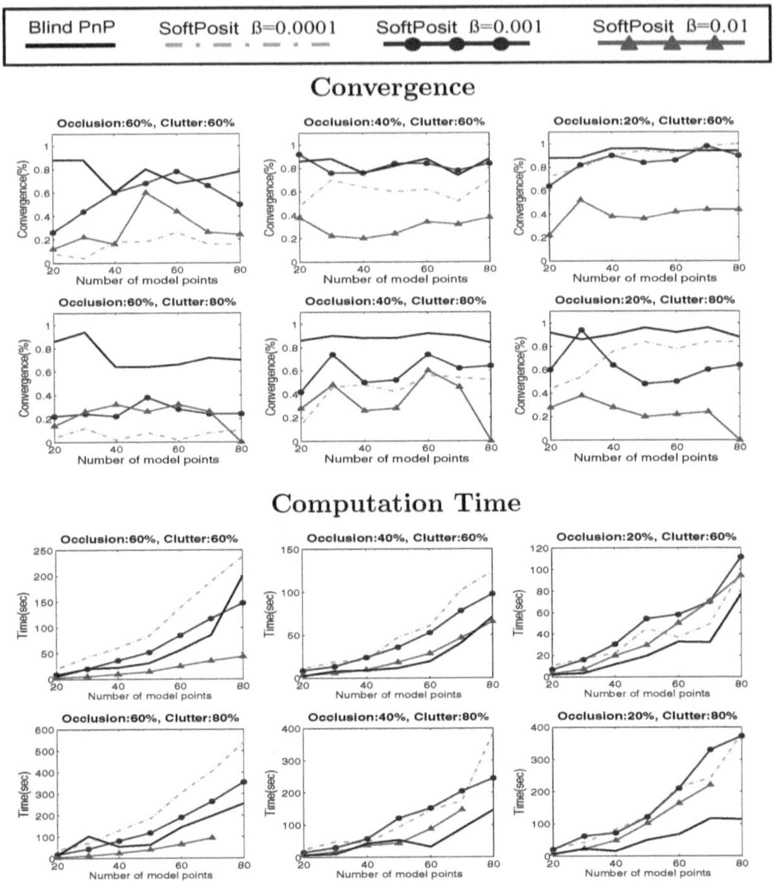

Fig. 6. Synthetic experiment using the randomly distributed points of Fig. 5. Top row: Convergence rate. Bottom row: Computation time per trial.

allowed to point anywhere on the 3D model. The complete 6D space of possible poses was approximated by a 20-component GMM. The 2D points were produced by generating random poses within the 6D space and projecting the model onto a 640×480 image using a virtual calibrated camera with an effective focal length of f=800, and a principal point at $(u_c, v_c)=(320, 240)$. Normal noise with standard deviation σ=2 was then added to the image coordinates of those projections, and clutter points with random coordinates were created.

Fig. 6 depicts the results on the random configuration of 3D points, with $M \in \{20, 30, \ldots, 80\}$. In terms of convergence, for 60% of clutter, Blind PnP and SoftPosit perform similarly. However, when the percentage of clutter increases to 80%, Blind PnP outperforms SoftPosit consistently. Computation times using `MATLAB` implementations of both algorithms are comparable, thus confirming the results of our complexity analysis. The advantage of Blind PnP becomes even more marked when dealing with 3D points that form repetitive structures, such

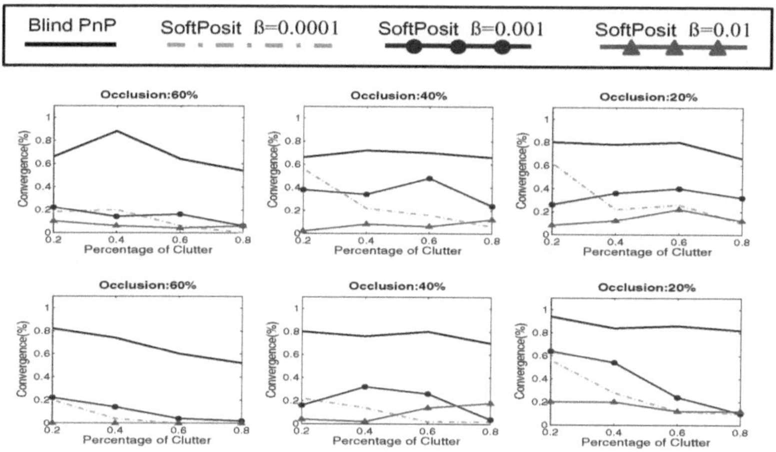

Fig. 7. Convergence rates when using the structured points of Fig. 5. Top row: Points forming repetitive structures. Bottom row: Points laying on the CAD model of a castle.

as the two rightmost sets of Fig. 5. As shown in Fig. 7, Blind PnP outperforms SoftPosit independently of the occlusion or clutter levels in both cases.

5.2 Real Data

We used several images of the scenes of Fig. 8 and 10 to validate our approach on real data. Before summarizing the results on these data sets, we first briefly present our approach to extracting 3D model points and 2D image features.

Extracting 3D and 2D Feature Points. A common approach to selecting 3D features is to pick corners and line intersections in the 3D model [1,7,?,14]. This implies that the object has crisp edges, which may not always be true. We instead manually registered the 3D model to one of the images from which we extracted feature points using the SIFT keypoint detector [?]. We then obtained our 3D points by simply backprojecting these 2D points to the model. A potential problem with this approach is that SIFT can return many thousands keypoints per image, which is far more than we can handle: As shown in Fig. 6 Blind PnP and SoftPosit can handle approximately about M=100 to 150 points in a reasonable amount of time, which is already more than earlier approaches.

We reduced the number of detected features using the simple clustering approach depicted by Fig. 8. The initially numerous keypoints were grouped into a few clusters using a k-means algorithm based on proximity. Then, within each cluster, a percentage of the keypoints with the largest gradient were selected. We also ran this simple procedure to extract 2D features from the images on which we tested our algorithm and found that it consistently reduced the number of keypoints from a few thousands to about a hundred in a stable way.

Fig. 8. Feature extraction. **Left:** A few thousand keypoints extracted from the image are clustered using a k-means algorithm. **Middle:** Only the most salient keypoints in each cluster are retained. **Right:** The same procedure is run on a different image and yields many of the same keypoints.

Occluded Indoor Scene. For the scene of Fig. 8, we acquired 17 different images by following a roughly circular path. We created the 3D model shown as a wireframe in Fig. 9 and registered it to one these images using Image Modeler (©2005 Realviz s.a.). The method described above produced $M=100$ 3D points in that image and N between 110 and 125 2D points in all others. The camera pose prior was modeled by a $G=20$ component GMM. The camera locations were assumed to be within a quarter-hemisphere, bounded by the two walls and the table while the optical axis was allowed to point anywhere on the 3D model.

To validate the algorithms under occlusion, we ran many trials for each image by synthetically removing all the keypoints falling within parts of the images, such as the black triangles of Fig. 9. As before, we ran SoftPosit using $\beta=\{10^{-4},10^{-3},10^{-2}\}$ and retained the best result. Similarly, to improve the convergence rate of Blind PnP on this noisy real-world data, we ran it first with Mahalanobis boundaries $\mathcal{M}=2$ and ran it again with $\mathcal{M}=3$ if it failed to converge. This improved our rates by more than 30% with only a limited computational overhead since the more expensive computations were only performed when absolutely necessary. In the images on the left side of Fig. 9, we reproject the model using the pose recovered by either Blind PnP or SoftPosit. To quantify these results, we also registered manually the 3D model in the test images and computed convergence rates. The graph on the right side of Fig. 9 summarizes these results for all trials on all images and confirms that Blind PnP is much more robust than SoftPosit to increasing occlusion levels.

Repetitive Outdoor Scene. Finally, we come back to the scene of Fig. 1 and its repetitive structure. The data set was composed of 8 images of an inner patio with windows, arches, and other elements evenly distributed over the scene. The 3D model was obtained using a laser range-finder and was manually registered to one of the images in order to obtain $M=120$ 3D points. In the other images, around $N=100$ 2D features were detected. Fig. 1(a) shows the boundaries of the $40\times 30\times 2$ m^3 volume defining the prior for camera locations. The camera was allowed to point any point of the 3D model. This pose space was approximated by $G=\{20,30,40,60,100\}$ Gaussian components. Fig. 10 shows 3 samples of model reprojection after recovering the pose using the Blind PnP and SoftPosit. By

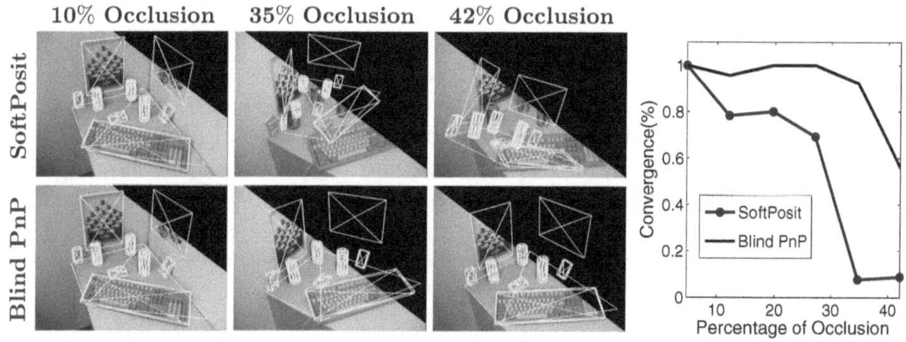

Fig. 9. Indoor results. Upper row images: 3D model reprojected using the SoftPosit pose. Lower row images: Using the Blind PnP pose. Graph: The convergence rate as a function of the percententage of occluded 3D points.

Fig. 10. Experiment 5. Upper row: SoftPosit. Lower row: Blind PnP. The images and 3D model for this experiment are courtesy of PhD Christoff Strecha

simply initializing the Blind PnP twice, with $\mathcal{M}=\{2,3\}$, we were able to retrieve the correct camera pose in all 7 test images using only $G=20$ components. By contrast, SoftPosit failed to retrieve correct poses for 5 of that 7 images even when it was initialized up to $G=100$ times, and with $\beta=\{10^{-4}, 10^{-3}, 10^{-2}\}$.

6 Conclusion

We have shown that introducing very weak priors on camera pose leads to efficient, robust, and simultaneous estimation of pose and of 3D-to-2D correspondences. The priors are represented by a GMM whose components we use to initialize a Kalman filter. The filter is used to restrict the set of 2D points that can be associated to the 3D features and its covariance is progressively reduced as successive matches are hypothesized. This gives us robustness to clutter, occlusions, and repetitive patterns. The approach presented here relies solely in geometry. In practice, even when one-to-one correspondences cannot be reliably established between model and image, image clues can be used to restrict the set of possible matches. Furthermore, when dealing with videos, the pose priors we use could be made much stronger based on motion clues. In future work, we will incorporate these additional sources into our algorithm, which should yield significant speed increases and make Blind PnP fit for real-time applications.

References

1. David, P., DeMenthon, D., Duraiswami, R., Samet, H.: Softposit: Simultaneous pose and correspondence determination. IJCV 59, 259–284 (2004)
2. Fischler, M.A., Bolles, R.C.: Random sample consensus: a paradigm for model fitting with applications to image analysis and automated cartography. Commun. ACM 24, 381–395 (1981)
3. Grimson, W.E.L., Meyer-Arendt, J.R.: Object recognition by computer: the role of geometric constraints, vol. 31. MIT Press, Cambridge (1992)
4. Grimson, W.E.L.: The combinatorics of heuristic search termination for object recognition in cluttered environments. IEEE PAMI 13, 920–935 (1991)
5. Breuel, T.: Fast recognition using adaptive subdivisions of transformation space. In: Proc. IEEE CVPR, pp. 445–451 (1992)
6. Cass, T.A.: Polynomial-time object recognition in the presence of clutter, occlusion, and uncertainty. In: Proc. ECCV, pp. 834–842 (1992)
7. Olson, C.F.: Efficient pose clustering using a randomized algorithm. IJCV 23, 131–147 (1997)
8. Stockman, G.: Object recognition and localization via pose clustering. Comput. Vision Graph. Image Process 40, 361–387 (1987)
9. Ayache, N., Faugeras, O.D.: Hyper: a new approach for the recognition and positioning to two-dimensional objects. IEEE PAMI 8, 44–54 (1986)
10. Grimson, W.E.L., Lozano-Pérez, T.: Localizing overlapping parts by searching the interpretation tree. IEEE PAMI 9, 469–482 (1987)
11. Lamdan, Y., Wolfson, H.J.: Geometric hashing: A general and efficient model-based recognition scheme. In: Proc. IEEE ICCV, pp. 238–249 (1988)
12. Burns, J.B., Weiss, R.S., Riseman, E.M.: View variation of point-set and line-segment features. IEEE PAMI 15, 51–68 (1993)
13. Beis, J.S., Lowe, D.G.: Indexing without invariants in 3D object recognition. IEEE PAMI 21, 1000–1015 (1999)
14. Jurie, F.: Solution of the simultaneous pose and correspondence problem using gaussian error model. CVIU 73, 357–373 (1999)
15. Wunsch, P., Hirzinger, G.: Registration of cad-models to images by iterative inverse perspective matching. In: Proc. ICPR., vol. 78 (1996)
16. Beveridge, J.R., Riseman, E.M.: Optimal geometric model matching under full 3D perspective. CVIU 61, 351–364 (1995)
17. Arya, S., Mount, D., Netanyahu, N., Silverman, R., Wu, A.: An optimal algorithm for approximate NN searching fixed dimensions. J. ACM 45, 891–923 (1998)
18. David, P., DeMenthon, D.: Object recognition in high clutter images using line features. In: Proc. IEEE ICCV, pp. 1581–1588 (2005)
19. Lowe, D.G.: Distinctive image features from scale-invariant keypoints. IJCV 15, 91–110 (2004)

Latent Pose Estimator for Continuous Action Recognition

Huazhong Ning[1], Wei Xu[2], Yihong Gong[2], and Thomas Huang[1]

[1] ECE, U. of Illinois at Urbana-Champaign, USA
{hning2,huang}@ifp.uiuc.edu
[2] NEC Laboratories America, Inc., USA
{xw,ygong}@sv.nec-labs.com

Abstract. Recently, models based on conditional random fields (CRF) have produced promising results on labeling sequential data in several scientific fields. However, in the vision task of continuous action recognition, the observations of visual features have dimensions as high as hundreds or even thousands. This might pose severe difficulties on parameter estimation and even degrade the performance. To bridge the gap between the high dimensional observations and the random fields, we propose a novel model that replace the observation layer of a traditional random fields model with a latent pose estimator. In training stage, the human pose is not observed in the action data, and the latent pose estimator is learned under the supervision of the labeled action data, instead of image-to-pose data. The advantage of this model is twofold. First, it learns to convert the high dimensional observations into more compact and informative representations. Second, it enables transfer learning to fully utilize the existing knowledge and data on image-to-pose relationship. The parameters of the latent pose estimator and the random fields are jointly optimized through a gradient ascent algorithm. Our approach is tested on HumanEva [1] – a publicly available dataset. The experiments show that our approach can improve recognition accuracy over standard CRF model and its variations. The performance can be further significantly improved by using additional image-to-pose data for training. Our experiments also show that the model trained on HumanEva can generalize to different environment and human subjects.

1 Introduction

Recognizing human actions in videos with natural environments provides an infrastructure for a wide range of applications spanning visual surveillance, systems for entertainment, human-computer interfaces, and so on. It is a specific example of a widely studied problem of sequence labeling that arises in several scientific fields. A well understood and widely used probabilistic model for this problem is Hidden Markov Model (HMM) [2]. But HMMs make strict assumption that observations are conditionally independent given class labels, and cannot represent multiple interacting features and long-range dependencies of the observations. Thus, it largely limits the applicability of the HMM models.

D. Forsyth, P. Torr, and A. Zisserman (Eds.): ECCV 2008, Part II, LNCS 5303, pp. 419–433, 2008.

Conditional random fields models [3] relax the assumption of observation independency and has exhibited success to discriminate actions in both segmented and unsegmented video sequences [4,5]. In [4], a latent-dynamic conditional random field (LDCRF) model is proposed to capture both extrinsic dynamics and intrinsic structure of head and gaze aversion gestures. The poses of 3D head or eye gaze are robustly estimated using a view based appearance model, and the model recognizes the gestures based on the pose representations. However, in action recognition where articulated human body is involved, the explicit pose of the body parts could not be reliably estimated by currently existing tracking algorithms. In other words, it is impractical to use LDCRF to recognize complex human actions in the same way as in [4]. An possible solution is to directly feed the visual features (*e.g.*, block SIFT [6] and silhouette[7]), instead of human poses, to the random fields as in [5] where a chain conditional random fields (CRF) model [3] is used. But the dimension of visual features is usually as high as hundreds or even thousands, compared to 10-50 dimensions of human poses, and their discriminative power to represent human actions is also worse than human poses. This might increase the complexity of parameter estimation in random fields while degrade the performance.

In this paper, we propose a novel model to bridge the gap between the high dimensional visual features and the random fields. As we know, human pose is a very compact and informative representation for human images. And recently, the multi-modal image-to-pose relationship has been successfully exploited using discriminative models, such as linear/nonlinear regression [7], Bayesian mixture of experts (BME) [6], and so on. Therefore, we propose a model that replaces the observation layer of the random fields with an image-to-pose discriminative model. Fig. 1 illustrates the graphical structure of our model. This image-to-pose layer learns to convert the high dimensional observations into more compact and informative representations (such as poses of articulated human body) under the supervision of labeled action data. We call it *latent pose estimator*, because it is not explicitly estimated from labeled image-to-pose data. Actually, the human poses are not observed in action training data. Of course, additional image-to-pose data, if available, can also be added to the learning process. In practice, the output of the latent pose estimator can be human pose (image-to-pose data are enough) or any other compressed representations (not enough or no pose data), but we still use the term "pose estimator". We call our model *latent pose conditional random fields (LPCRF)*. In this paper, our model is based on LDCRF, but it can be extended to any other CRF variations. Our experiments show that the use of latent pose estimator can improve recognition accuracy over traditional CRF or LDCRF even without using any image-to-pose data for training.

The structure of our LPCRF model not only maintains the strength of the LDCRF model in that it captures both extrinsic dynamics and internal substructure, but also enables the model to recognize actions from high dimensional visual features without increasing the complexity of the feature functions. Due

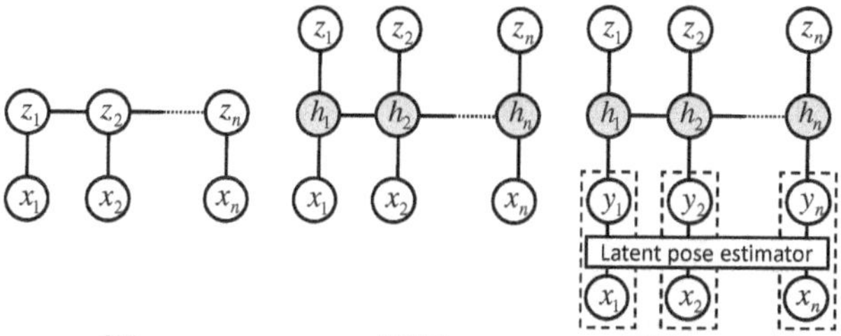

Fig. 1. Graphical structures of our LPCRF model and two existing models: CRF [3] and LDCRF [4]. In these models, $\mathbf{x}$ is a visual observation, z is the class label (*e.g.*, walking or hand waving) assigned to $\mathbf{x}$, and h represents a hidden state of human actions (*e.g.*, left-to-right/right-to-left walking). The subscripts index the frame number of the video sequence. In our LPCRF model, the observation layer of the random fields is replaced with a *latent pose estimator* that learns to compress the high dimensional visual features $\mathbf{x}$ into a compact representation (like human pose) $\mathbf{y}$. Our model also enables transfer learning to utilize the existing knowledge and data on image-to-pose relationship. The dashed rectangles means that $\mathbf{y}$'s are technically deterministic functions of $\mathbf{x}$ when the parameters of the latent pose estimator are fixed.

to this advantage, our model can recognize human actions with complex articulations in both segmented and unsegmented video sequences.

Another advantage of our LPCRF model is that it enables transfer learning [8] to fully utilize the existing knowledge and data on image-to-pose relationship. Firstly, our model can be incrementally built on an existing discriminative pose estimator that is well-trained on either synthetic or real image-to-pose datasets. More specifically, we use the parameters of an existing pose estimator (if available) to initialize the latent pose estimator. This allows a growing algorithm to acquire new knowledge while keeping old knowledge and to reduce the redevelopment and learning costs. Secondly, newly available image-to-pose data, if available, can be naturally added to the learning process to refine the model parameters without modifying the structure of our model.

The literature that is closely related to our research includes action recognition based on either 2D/3D silhouettes [9,10], 2D body configurations [11], motion trajectories [12], or body kinematics [13]. These works rely on representations of explicit pose or compact silhouette to recognize actions. These representations are generated by a fixed component which cannot be adapted to the given task of action recognition. While in our system, the latent pose estimator is treated as a trainable feature extractor that can be improved with more action data. The benefit of adaptation is clearly demonstrated in our experiments.

2 Latent-Dynamic Conditional Random Fields

First, we briefly introduce the LDCRF model[4] that was proposed to solve the problem of labeling unsegmented video sequences. It incorporates hidden state variables into the traditional CRF [3] to model the sub-structure of human actions, and combines the strengths of CRFs and HCRFs [14] to capture both extrinsic dynamics and intrinsic structure.

Given a sequence of observations $X = \{\mathbf{x}_1, \mathbf{x}_2, \cdots, \mathbf{x}_n\}$, the LDCRF model predicts a sequence of class labels $\mathbf{z} = \{z_1, z_2, \cdots, z_n\}$. Each $\mathbf{x}_i$ is a visual representation for i-th frame. Each $z_i \in \mathcal{Z}$ is the class label of frame $\mathbf{x}_i$, where $\mathcal{Z}$ is the set of all possible class labels of the concerned actions. To model the sub-structure of the actions (*e.g.*, left-to-right/right-to-left walking), a sequence of hidden variables $\mathbf{h} = \{h_1, h_2, \cdots, h_n\}$ is incorporated into the CRF model. Fig. 1 illustrates the graphical structure of the $\mathbf{h}$ layer in LDCRF. These hidden variables are not observed in the training examples, but are instantiated in the predicting stage. The transitions among them model both the substructure patterns of individual actions and the external dynamics between actions.

The LDCRF is restricted to have disjoint sets of hidden states associated with each class label. Denote $\mathcal{H}_z$ the set of hidden states for class z, and $\mathcal{H}_{\mathbf{z}} = \{\mathbf{h} | h_j \in \mathcal{H}_{z_j}, j = 1 \cdots n\}$ the set consisting of all the possible sequences of hidden states compatible with the label sequence $\mathbf{z}$. Let $\mathcal{H} = \sum_{\mathbf{z}} \mathcal{H}_{\mathbf{z}}$ all the possible hidden state sequences. With these definitions, the LDCRF model is defined as

$$P(\mathbf{z}|X, \Phi) = \sum_{\mathbf{h}} P(\mathbf{z}|\mathbf{h}, X, \Phi) P(\mathbf{h}|X, \Phi) = \sum_{\mathbf{h} \in \mathcal{H}_{\mathbf{z}}} P(\mathbf{h}|X, \Phi) \tag{1}$$

where Φ is set of parameters of the model. The second equation is due to the fact that, by definition, $P(\mathbf{z}|\mathbf{h}, X, \Phi) = 1$ for any $\mathbf{h} \in \mathcal{H}_{\mathbf{z}}$, otherwise 0.

Assume the graph for a video sequence is a simple chain. According to the fundamental theorem of random fields [15], the joint distribution over the hidden state sequence $\mathbf{h}$ given X has an exponential form

$$P(\mathbf{h}|X, \Phi) = \frac{1}{K_\Phi(X, \mathcal{H})} \exp\left(\sum_j V_\Phi(j, h_j, X) + \sum_j E_\Phi(j, h_{j-1}, h_j, X) \right), \tag{2}$$

where $K_\Phi(X, \mathcal{H})$ is the observation dependent normalization,

$$K_\Phi(X, \mathcal{H}) = \sum_{\mathbf{h} \in \mathcal{H}} \exp\left(\sum_j V_\Phi(j, h_j, X) + \sum_j E_\Phi(j, h_{j-1}, h_j, X) \right), \tag{3}$$

summarizing over all hidden state sequences in $\mathcal{H}$. And $\Phi = \{\lambda_1, \lambda_2, \cdots, \mu_1, \mu_2, \cdots\}$ is the set of model parameters. $V_\Phi(j, h_j, X)$ and $E_\Phi(j, h_{j-1}, h_j, X)$ are sum of feature functions on individual vertex j and edge $(j-1, j)$, respectively,

$$V_\Phi(j, h_j, X) = \sum_k \lambda_k s_k(j, h_j, X), \tag{4}$$

$$E_\Phi(j, h_{j-1}, h_j, X) = \sum_k \mu_k t_k(j, h_{j-1}, h_j, X). \tag{5}$$

where s_k and t_k are feature functions. s_k are state functions that depend on a single hidden variable of a sub-action in the model, and t_k are transition functions that depend on pairs of hidden variables. We also use V_{ja} and E_{jab} as short representations of $V_\Phi(j, a, X)$ and $E_\Phi(j, a, b, X)$.

3 Latent Pose Conditional Random Fields

Our latent pose conditional random fields (LPCRF) model is a generalization of CRF and LDCRF. Fig. 1 illustrates its graphical structure. The latent pose estimator learns to convert an observation vector $\mathbf{x}$ into a more compact and informative representation $\mathbf{y}$, and the model recognizes human actions based on the pose sequence $Y = \{\mathbf{y}_1, \mathbf{y}_2, \cdots, \mathbf{y}_n\}$. Later we denote the latent pose estimator as $P(\mathbf{y}|\mathbf{x}, \Theta)$ in probabilistic form or $\mathbf{y} = \Psi(\mathbf{x}, \Theta)$ in deterministic form, where Θ is the set of parameters of the latent pose estimator and is jointly optimized with the random fields using a gradient ascent algorithm.

3.1 Formulation of our LPCRF Model

Using the notations and definitions for the LDCRF model in Section 2, our model is defined as

$$P(\mathbf{z}|X, \Omega) = P(\mathbf{z}|Y, \Phi) = \sum_{\mathbf{h} \in \mathcal{H}_{\mathbf{z}}} P(\mathbf{h}|Y, \Phi) \tag{6}$$

where $Y = \Psi(X, \Theta)$ is the optimal estimation of the latent pose estimator given observations X and parameters Θ, and $\Omega = \{\Phi, \Theta\}$ represents all the model parameters. The joint distribution over the hidden state sequence $\mathbf{h}$ given Y still has an exponential form

$$P(\mathbf{h}|Y, \Phi) = \frac{1}{K_\Phi(Y, \mathcal{H})} \exp\left(\sum_j V_\Phi(j, h_j, Y) + \sum_j E_\Phi(j, h_{j-1}, h_j, Y) \right), \tag{7}$$

where $K_\Phi(Y, \mathcal{H})$ is the observation dependent normalization in Eqn. 3.

If the parameters Θ for the latent pose estimator are fixed, our LPCRF model collapses into an LDCRF model. If each class label $z \in \mathcal{Z}$ is constrained to have only one hidden sub-action, *i.e.*, $|\mathcal{H}_z| = 1$, the LDCRF model further collapses into a CRF model. Hence, our LPCRF model is a more general framework of CRF and LDCRF. However, our LPCRF model is essentially different from both CRF and LDCRF in some aspects. In our model, input features used by the random fields are trainable and are jointly optimized with the random fields, while in CRF and LDCRF, the input features are fixed and cannot be tuned for the given recognition task. The latent pose estimator encodes the knowledge of multimodal image-to-pose relationship and provides optimal feature representation for action recognition. This knowledge can be acquired from existing well-trained models (if available) and adapted for action recognition in the learning process. In all, the latent pose estimator is seamlessly integrated and globally optimized with the random fields.

The model parameters $\Omega = \{\Phi, \Theta\}$ are learned from training data consisting of labeled action sequences $(X^{(t)}, \mathbf{z}^{(t)})$. The labeled image-to-pose data $(\mathbf{x}^{(t)}, \mathbf{y}^{(t)})$, if available, can also be utilized as auxiliary data. The optimal parameters Ω^* is obtained by maximizing the objective function:

$$L(\Omega) = \underbrace{\sum_t \log P(\mathbf{z}^{(t)}|X^{(t)}, \Omega)}_{L_1(\Omega)} \underbrace{-\frac{1}{2\sigma^2}\|\Phi\|^2}_{L_2(\Omega)} + \eta \underbrace{\sum_t \log P(\mathbf{y}^{(t)}|\mathbf{x}^{(t)}, \Theta)}_{L_3(\Omega)} \tag{8}$$

where the first term, denoted as $L_1(\Omega)$, is the conditional log-likelihood of the action training data. The second term $L_2(\Omega)$ is the log of Gaussian prior $P(\Phi) \sim \exp\left(-\frac{1}{2\sigma^2}\|\Phi\|^2\right)$ with variance σ^2 and it prevents Φ from drifting too much. And the third term $L_3(\Omega)$ is the conditional log-likelihood of the image-to-pose training data. η is a constant learning rate. Note that our model enables the image-to-pose data to be naturally added to the learning process.

3.2 The Latent Pose Estimator

The image-to-pose relation is highly non-linear. Fortunately, close observation of human images shows that human appearance changes very fast as the human global orientation changes, while the appearance changes relatively slowly in a fixed orientation. Therefore, we may assume that the image-to-pose distribution in a fixed orientation can be well modelled by a single or a combination of linear regressor(s). This leads us to use the Bayesian mixtures of experts (BME) [16] to model the multi-modal image-to-pose distributions. Suppose $\mathbf{x}$ is the visual feature of the image and $\mathbf{y}$ is the human pose, the model with M experts is:

$$p(\mathbf{y}|\mathbf{x}, \Theta) = \sum_{i=1}^{M} g(\mathbf{x}, \nu_i) p(\mathbf{y}|\mathbf{x}, T_i, \Lambda_i) \tag{9}$$

where

$$g(\mathbf{x}, \nu_i) = \frac{e^{\nu_i^T \mathbf{x}}}{\sum_j e^{\nu_j^T \mathbf{x}}} \tag{10}$$

$$p(\mathbf{y}|\mathbf{x}, T_i, \Lambda_i) \sim \mathcal{N}(\mathbf{y}; T_i\mathbf{x}, \Lambda_i) \tag{11}$$

Here $\Theta = \{\nu_i, T_i, \Lambda_i | i = 1, 2, \cdots, M\}$ consists of the parameters of the BME model. The *expert* $p(\mathbf{y}|\mathbf{x}, T_i, \Lambda_i)$ is an Gaussian distribution with mean $T_i\mathbf{x}$ and covariance matrix Λ_i. The mixing proportions of the experts, $g(\mathbf{x}, \nu_i)$, are *input dependent* and work like gates that can competitively switch-on multiple experts for some input domains, allowing multi-modal conditionals. They can also pick a single expert for unambiguous inputs by switching-off other experts. Given the input $\mathbf{x}$ and parameters Θ, the optimal pose estimation of the BME model is

$$\mathbf{y}^* = \int \mathbf{y} P(\mathbf{y}|\mathbf{x}, \Theta) d\mathbf{y} = \sum_{i=1}^{M} g(\mathbf{x}, \nu_i) T_i \mathbf{x} \tag{12}$$

The BME model has exhibited success in representing image-to-pose distributions in the literature [6]. But our LPCRF model is not limited to the BME representation. We can replace the BME model with any other differentiable discriminative models without major modifications to our LPCRF structure.

3.3 Learning Model Parameters

We use an iterative gradient ascent algorithm to search for the optimal model parameters that maximize the objective function, *i.e.*, $\Omega^* = \arg\max_{\Omega} L(\Omega)$. At each iteration, Ω can be updated by $\Omega \leftarrow \Omega + \xi \frac{\partial L(\Omega)}{\partial \Omega}$ or any other Quasi-Newton methods, where ξ is the learning rate. First, we compute $\partial L(\Omega)/\partial \Phi$ for estimation of Φ.

$\partial L(\Omega)/\partial \Phi$ can be obtained by summarizing the gradients of $\log P(\mathbf{z}|X, \Omega)$ with respect to Φ over all training samples (X, z). According to the chain rule, the gradient $\partial \log P(\mathbf{z}|X, \Omega)/\partial \phi$ for a parameter $\phi \in \Phi$ is

$$\frac{\partial \log P(\mathbf{z}|X, \Omega)}{\partial \phi} = \sum_{j,a} \frac{\partial \log P(\mathbf{z}|X, \Omega)}{\partial V_{ja}} \frac{\partial V_{ja}}{\partial \phi} + \sum_{j,a,b} \frac{\partial \log P(\mathbf{z}|X, \Omega)}{\partial E_{jab}} \frac{\partial E_{jab}}{\partial \phi} \quad (13)$$

Given Eqn. 4 and 5, the gradients $\partial V_{ja}/\partial \phi$ and $\partial E_{jab}/\partial \phi$ are the corresponding feature function. And the gradients $\partial \log P(\mathbf{z}|X, \Omega)/\partial V_{ja}$ and $\partial \log P(\mathbf{z}|X, \Omega)/\partial E_{jab}$ can be expressed in terms of the marginal probabilities over individual vertex j or edge $(j-1, j)$,

$$\frac{\partial \log P(\mathbf{z}|X, \Omega)}{\partial V_{ja}} = P(h_j = a|\mathbf{z}, Y, \Phi) - P(h_j = a|Y, \Phi) \quad (14)$$

$$\frac{\partial \log P(\mathbf{z}|X, \Omega)}{\partial E_{jab}} = P(h_{j-1} = a, h_j = b|\mathbf{z}, Y, \Phi) - P(h_{j-1} = a, h_j = b|Y, \Phi) \quad (15)$$

These marginal probabilities can be efficiently calculated using belief propagation [3] due to the chain structure of the random fields.

Learning the parameters Θ involves computation of $\partial L_1(\Omega)/\partial \Theta$ and, if image-to-pose training data are available, $\partial L_3(\Omega)/\partial \Theta$. The gradient $\partial L_3/\partial \Theta$ is the sum of the gradients of the BME model that can be found in [16]. To obtain $\partial L_1(\Omega)/\partial \Theta$, we rely on the chain rule to compute

$$\frac{\partial \log P(\mathbf{z}|X, \Omega)}{\partial \Theta} = \sum_{j,a} \frac{\partial \log P(\mathbf{z}|X, \Omega)}{\partial V_{ja}} \frac{\partial V_{ja}}{\partial \mathrm{vec}(Y)} \frac{\partial \mathrm{vec}(Y)}{\partial \Theta} \quad (16)$$

for specific training sample (X, z). Here, $\mathrm{vec}(A)$ is the stacked columns of A, and Y is the optimal estimation of the latent pose estimator (by Eqn. 12) given X and current estimation Θ. In Eqn. 16, we assume the transition functions t_k are independent of Y. The gradient $\partial V_{ja}/\partial \mathrm{vec}(Y)$ depends on the definition of state functions. Note that the chain rule decouples the gradient $\partial \log P(\mathbf{z}|X, \Omega)/\partial \Theta$ into three simple factors. The value of the factor $\partial \log P(\mathbf{z}|X, \Omega)/\partial V_{ja}$ has already been computed by Eqn. 14 in estimation of Φ and is back propagated to the latent pose layer for estimation of Θ. The feature functions are entirely isolated into a single factor $\partial V_{ja}/\partial \text{vec}(Y)$. And the third factor consists of all the calculations on the latent pose estimator. Thus, the decoupling allows flexible deployment of feature functions and latent pose estimator without increasing the complexity of gradient computations.

Computation of the product of $\partial V_{ja}/\partial \text{vec}(Y)$ and $\partial \text{vec}(Y)/\partial \Theta$ is straightforward. Note that $L_1(\Omega)$ does not explicitly depend on the covariance matrixes of the BME model, *i.e.*, Λ_i cannot be updated from action data. We use image-to-pose data (if available) to update Λ_i, or directly use the Λ_i of an existing well-trained BME model, or, if both are not available, just assume constant covariance matrixes.

3.4 Feature Functions

In our model, the feature functions are differentiable with respect to the latent poses so that gradient ascent algorithm are applicable. Our feature functions resemble the well-known *logit model* and are widely used in the literature [4,3]. We have $|\mathcal{H}| \times |\mathcal{H}|$ transition functions and each corresponds to a hidden variable pair (h, h'), denoted by $t_{hh'}$:

$$t_{hh'}(j, h_{j-1}, h_j, Y) = \delta(h, h_{j-1})\delta(h', h_j). \tag{17}$$

The corresponding parameters $\mu_{hh'}$ form an $|\mathcal{H}| \times |\mathcal{H}|$ matrix that is essentially a transition matrix. It models both the external dynamics between actions and the internal substructures of individual actions [4].

To make the computation tractable, the state functions do not model the dependency of the entire observation sequence but, instead, depend only on a window around the current frame. In other words, $s_k(j, h_j, Y) = s_k(j, h_j, \tilde{Y}_j)$, where $\tilde{Y}_j = [\mathbf{y}_{j-w}, \cdots, \mathbf{y}_{j+w}]$ is a window around the j-th frame with window size $2w+1$. Assume $\mathbf{y}_j$ has dimension d. We have $|\mathcal{H}| \times d(2w+1)$ state functions, and each corresponds to a pair (h, l), where $1 \leq l \leq d(2w+1)$. The state function $s_{hl}(j, h_j, Y) = \delta(h, h_j)\tilde{Y}_j(l)$, where $\tilde{Y}_j(l)$ is the l-th entry in $\tilde{Y}_j$. With the window feature, $\partial V_{ja}/\partial Y$ is a sparse matrix $\left[\mathbf{0}^{d\times(j-w-1)}, \tilde{\lambda}_a, \mathbf{0}^{d\times(n-j-w)}\right]$, where $\tilde{\lambda}_a$ is a $d \times (2w+1)$ matrix consisting of the parameters λ_{al} corresponding to all s_{al} for a fixed $a \in \mathcal{H}$. Thus, Eqn. 16 can be further simplified.

3.5 Inference

Given the model parameters Ω^* learned from the training data, prediction of a new test sequence X is to estimate the most probable sequence labels $\mathbf{z}^*$ that maximizes our model

$$\mathbf{z}^* = \arg\max_{\mathbf{z}} P(\mathbf{z}|X, \Omega^*) = \arg\max_{\mathbf{z}} P(\mathbf{z}|Y^*, \Phi^*) \tag{18}$$

where Y^* is the optimal pose estimation given X and Θ^*.

The sequence labels $\mathbf{z}^*$ can be estimated using either Viterbi path [3] or marginal probabilities [4]. We choose the later for this paper. We compute for each frame i the marginal probabilities $P(h_i = a|Y^*, \Phi^*)$ for all $a \in \mathcal{H}$. The marginal probabilities are summed according to the sets of hidden states $\mathcal{H}_z$. The label z_i^* associated with the set having maximum summed marginal probabilities is assigned to frame i. As we mentioned in Section 3.3, $P(h_i = a|Y^*, \Phi^*)$ can be efficiently computed using belief propagation. Thus, the inference operation is also very efficient.

4 Experimental Results

HumanEva dataset. We test the effectiveness of our approach on a real human motion dataset–HumanEva–made publicly available by the Brown Group [1]. The dataset was captured simultaneously using a calibrated marker-based motion capture system and multiple high-speed video capture systems. It contains multiple subjects performing a set of predefined actions with repetitions, and was originally partitioned into *Train*, *Validate*, and *Test* sub-sets. Fig. 4(a) (top) gives a sample image. We choose 4 actions[1]: *Walking*, *Box*, *Jog*, and *Gestures*, performed by subjects S1, S2 and S3, captured by cameras C1, C2, and C3. There are 56,261 video frames in total. Note that our model will recognize actions under different view points.

The original motion data provided by HumanEva were (x, y, z) locations of the body parts in the world coordinate system. There is a total of 10 parts: torso, head, upper and lower arms, and upper and lower legs. In this work, we discard the internal parameters of the human body model (like limb length), and convert the (x, y, z) locations to global orientation of torso and relative orientation of adjacent body parts. Each orientation is represented by 3 Euler angles.

Visual features for image representation. We choose the bag-of-words model [17] to represent the human images, because it is resistant to a large misalignment of the human region in the detection window that may pose difficulties to many other representations. For each video frame, the human window is detected by human detector or background subtraction and is scaled to a fixed size. In the human window, we extract a set of local descriptors, called *Appearance and Position Context (APC) descriptor* [18]. The bag-of-words representation of this frame is a 300-bin histogram of the APC descriptors.

Comparison of different models. We run several experiments in order to compare CRF and LDCRF with our LPCRF model under different configurations. The training data include all videos in both *Train* and *Validate* subsets with 25,645 frames, and the testing data include the videos in the *Test* subset with 30,616 frames. The window size w for the feature function is set to 3, *i.e.*, the context of observations of 7 frames is explored at current positions. All models have 4 states, each corresponding to an action class. The number of hidden

[1] The *ThrowCatch* action is not selected, because we failed to extract video frames from S1_Validate and S3_Train/Validate, and have frequent frame drops in S1_Train and S2_Train/Validate, even using the tool provided by the HumanEva dataset itself.

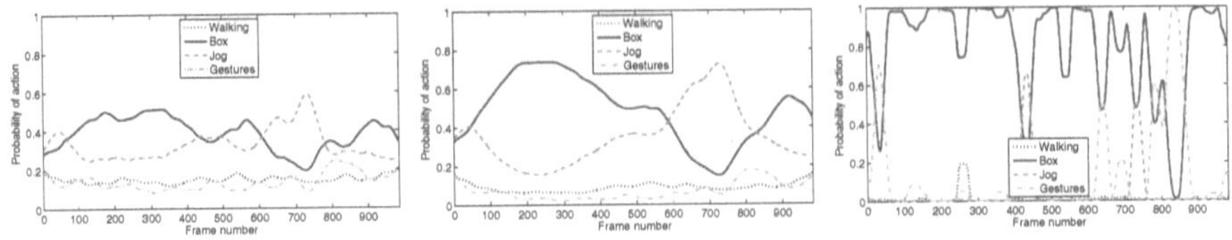

(a) CRF (b) LDCRF (c) $LPCRF_{pure}$

Fig. 2. Probability of actions on sequence S2_Box_Test_C3. Both CRF (a) and LDCRF (b) have much uncertainty in labeling this sequence, achieving recognition rate of 69.5% and 70.5%, respectively. Our $LPCRF_{pure}$ model (c) significantly improves the recognition rate to 85.6%, even without using pose knowledge. Best viewed in color.

states for each action class is chosen as 3 for both LPCRF and LDCRF. We use 5 experts for the BME model. In testing, each video frame is labelled as one of the 4 actions that has maximum marginal probabilities. Here, the recognition rate is computed based on frames, not on sequences, because each sequence is very long and this metric is also appropriate for unsegmented sequences.

Our LPCRF model is tested under 4 configurations. 1) $LPCRF_{init}$: the parameters Θ of the latent pose estimator are initialized using the parameters of an existing BME model that is well-trained on 5,000 frames randomly selected from the *Train* subset, and then are further updated iteratively on action videos. 2) $LPCRF_{fix}$: Θ are initialized using the same BME model but are fixed in the learning process. Essentially, it is an LDCRF trained on pose sequences that are obtained by an existing BME model. 3) $LPCRF_{xy}$: the model is randomly initialized, but uses training data including both action videos and 1,000 extra image-to-pose pairs selected from the *Train* subset. 4) $LPCRF_{pure}$: it is also randomly initialized but trained purely on action data, *i.e.*, it does not utilize image-to-pose knowledge and data, and uses a setup as CRF and LDCRF.

In Table 1, Figs. 2, **??**, and 3, we analyze the the performance of the six models. Table 1 gives the confusion matrixes and average recognition rates of these models. From them, we note that: 1) The $LPCRF_{pure}$ outperforms CRF and LDCRF without using any pose-to-image data. This demonstrates the effectiveness using a trainable feature extractor to compress the high dimensional observations into more informative and compact feature vectors. This trainable feature extractor decreases the complexity of the random fields, and, in turn, increases the performance. 2) With additional image-to-pose data, $LPCRF_{init}$ and $LPCRF_{xy}$ achieve the best performance, which demonstrates that our model effectively transfers the knowledge on image-to-pose relationship to the new vision task of action recognition. 3) $LPCRF_{fix}$ is essentially a LDCRF but it outperforms the latter. It concludes that the estimated human poses, compared with the original visual features, are more powerful in action recognition. 4) $LPCRF_{init}$ achieves a higher performance than $LPCRF_{fix}$, although both are initialized by the same BME model. This means that further updating of the

Table 1. Confusion matrixes of six models. The last row of each table gives model name and the average recognition rate. $LPCRF_{init}$ achieves the best performance 95.0%.

	wlk	*box*	*jog*	*gest*
wlk	**95.0**	0.1	4.9	0
box	0.4	**79.4**	10.9	9.3
jog	4.3	3.9	**89.7**	2.1
gest	1.6	25.0	0	**73.4**
	CRF: 85.0			

wlk	*box*	*jog*	*gest*
96.3	0	3.7	0
0	**75.0**	16.2	8.8
0	1.4	**97.3**	1.3
0.2	21.4	0	**77.4**
LDCRF: 87.2			

wlk	*box*	*jog*	*gest*	
97.8	0	0.2	2.0	*wlk*
0.9	**92.5**	5.8	0.8	*box*
0.4	0.2	**99.4**	0	*jog*
2.4	7.9	0.1	**89.6**	*gest*
$LPCRF_{init}$: **95.0**				

	wlk	*box*	*jog*	*gest*
wlk	**94.8**	0	0.5	4.7
box	1.0	**89.2**	8.6	1.2
jog	0.2	0.4	**99.4**	0
gest	2.9	11.8	0.3	**85.1**
	$LPCRF_{fix}$: 92.3			

wlk	*box*	*jog*	*gest*
95.6	0	0.7	3.7
0.8	**89.5**	6.5	3.2
0.5	0	**99.4**	0.1
0	10.9	0.6	**88.5**
$LPCRF_{xy}$: 93.4			

wlk	*box*	*jog*	*gest*	
95.5	0.4	2.2	1.9	*wlk*
2.2	**79.5**	7.6	10.7	*box*
4.5	0.9	**94.6**	0	*jog*
5.5	11.3	0.4	**82.8**	*gest*
$LPCRF_{pure}$: 88.5				

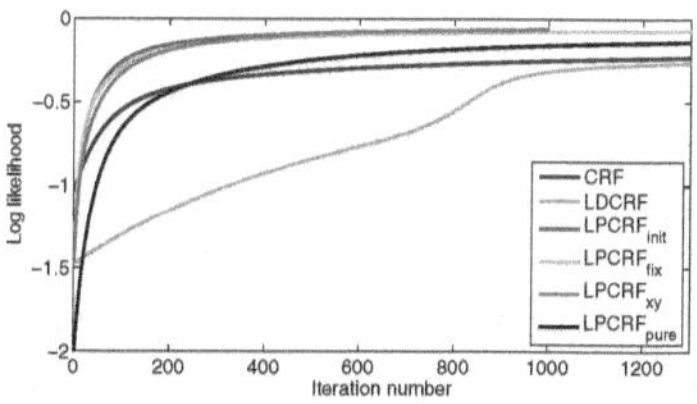

(a) Log likelihood

model	CRF	LDCRF	LPCRF *init*	LPCRF *fix*	LPCRF *xy*	LPCRF *pure*
Train Time(s)	2.5	3.2	22.1	2.8	22.3	22.1
Test Time(s)	1.62	2.14	2.08	2.08	2.08	2.08

(b) Run time

Fig. 3. (a) Log likelihood versus the iteration number for the six models. Note that $LPCRF_{init}$, $LPCRF_{fix}$, and $LPCRF_{xy}$ converge in about 400 iterations, and CRF, LDCRF, and $LPCRF_{pure}$ in about 1000 iterations. (b) Training time for one iteration and testing time for labeling the entire testing set. Best viewed in color.

latent pose estimator using action data does improve the performance, even after it is well-initialized. 5) Finally, LDCRF outperforms CRF on this dataset, which is also consistent with the conclusion in [4]. Fig. 2 shows the probability of actions on sequence S2_Box_Test_C3 assigned by models CRF, LDCRF, and $LPCRF_{pure}$. CRF and LDCRF have much uncertainty in labeling this sequence, and the uncertainty is largely reduced by our $LPCRF_{pure}$ model.

Fig. 3(a) gives the log-likelihood versus the iteration number for the six models. $LPCRF_{init}$, $LPCRF_{fix}$, and $LPCRF_{xy}$ converge quickly in about 400 iterations with auxiliary knowledge or data on human pose, compared to 1,000 iterations required by the other three that are randomly initialized. On the other hand, our LPCRF model (except $LPCRF_{fix}$) requires training time for each iteration as much as 8 times of that by CRF and LDCRF (see Fig. 3(b)). Most of the extra time is used for updating the parameters of latent pose estimator.

Table 2. Testing of upper bound performance. The performance of LDCRF (**Left**) trained on the ground truth human poses is supposed to be the upper bound of our LPCRF model (**Right**) trained on video data. Both models are trained on the corresponding *Train* subset and tested on the *Validate* subset. The last row gives the average recognition rate. *wlk*: Walking, *gest*: Gestures.

	wlk	*box*	*jog*	*gest*
wlk	**100**	0	0	0
box	0	**98.5**	0	1.5
jog	0	0	**100**	0
gest	19.5	1.4	0	**79.1**
LDCRF: 95.2%				

wlk	*box*	*jog*	*gest*	
99.6	0	0.2	0.2	*wlk*
0	**98.9**	0	1.1	*box*
0	1.0	**99.0**	0	*jog*
27.4	2.1	6.8	**63.7**	*gest*
$LPCRF_{init}$: 91.7%				

But our model achieves testing speed as fast as LDCRF and slightly slower than CRF. It takes about 2.1 seconds to label the entire testing set. Note that the time given in Fig. 3(b) does not include the time for feature extraction. And our program is pure MATLAB code and run on a laptop with 2GHz dual core CPU.

Upper bound performance. The performance of a conditional random fields model trained directly on the ground truth human poses is expected to be the upper bound of our model trained on video sequences. This is because the regression error of the latent pose estimator might degrade the performance, and our model reaches the upper bound if the latent pose estimator works perfectly. To demonstrate how much our model can achieve, we train an LDCRF on the ground truth human poses in the *Train* subset and test it on the *Validate* subset (note that the ground truth is unavailable in the *Test* subset), and train and test an $LPCRF_{init}$ model on the same subsets but using video data. Table 2 shows the performance in confusion matrix format. Our $LPCRF_{init}$ model trained on video achieves a recognition rate 91.7% that is close to the upper bound 95.2% given by LDCRF that is trained on ground truth human poses. Note that these rates are not comparable with those in Table 1 because here we use a different setup of training and testing sets.

Label unsegmented videos. It is natural to apply conditional random fields models to label unsegmented video sequences. Because there are no individual sequences in HumanEva consisting of all studied actions, we manually generate a new sequence by concatenating four action clips chosen from S1_Test_C1. Inference is done on the entire sequence. Our LPCRF models achieve a recognition rate higher than 97%, while the rates are 74.2% and 75.2% for CRF and LDCRF, respectively. See details in the first part of Table 3.

To further verify the generalization ability of our model, we apply the six models that are trained on HumanEva (setup is as above) to label some free-style action videos. These videos are unsegmented with resolution 640 × 480. And the scenes and subjects are completely different from that in HumanEva. Fig. 4(a) gives two sample images, selected from the HumanEva (top) and free style videos (bottom), respectively. In each free style video, a subject performs the four

Table 3. Accuracy of labeling unsegmented video sequences. The models are trained on the *Train* and *Validate* subsets of HumanEva. **First part**: a manually concatenated sequence with four action clips. **Second part**: two free-style videos with scenes and subjects completely different from HumanEva. See video illustrations in the supplemental materials.

seq	frm #	CRF	LDCRF	$LPCRF_{init}$	$LPCRF_{fix}$	$LPCRF_{xy}$	$LPCRF_{pure}$
S1_Test_C1	1207	74.2	75.2	**99.8**	99.0	99.5	97.2
Free style 1	910	71.4	64.8	80.1	80.9	**86.4**	81.7
Free style 2	1060	33.7	34.0	**76.4**	75.3	64.9	72.8

HumanEva

Free style

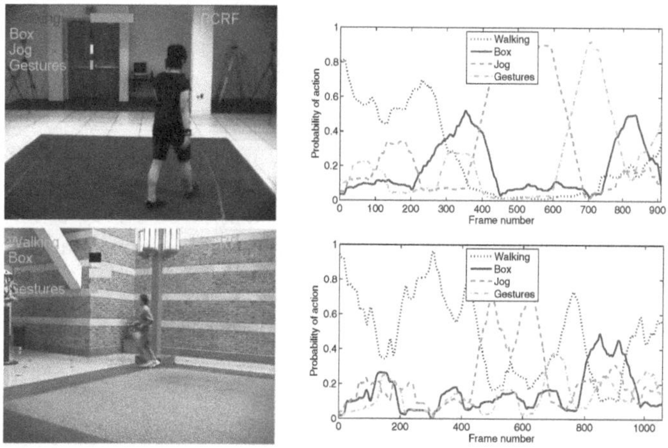

(a) Sample images

(b) Probability of actions

Fig. 4. (a) Two sample images selected from HumanEva (**top**) and free-style videos (**bottom**), respectively. Colored bars indicate action probabilities. (b) Probability of actions on free-style sequence 1 (**top**) assigned by $LPCRF_{xy}$ and on free-style sequence 2 (**bottom**) assigned by $LPCRF_{init}$. See better illustration in supplemental videos.

actions in arbitrary order, with arbitrary number of repetitions, and from arbitrary view angles. Challenges of these videos include: 1) varying view angle of actions; 2) varying scale of subjects; 3) a plant and pillar in the background having color very close to the clothes; and 4) rich textures on the wall.

The second part of Table 3 gives the labeling accuracy by the six models. The corresponding action probability of these two free-style sequences is shown in Fig. 4(b). Again, our LPCRF models achieve a significant improvement over CRF and LDCRF. Especially, $LPCRF_{init}$ produces very promising results. This shows that image-to-pose knowledge obtained from one dataset could be utilized to label video sequences outside of this dataset. We use the background subtraction code provided by HumanEva dataset to locate the human in these two sequences. It does not work reliably under this more challenging environment. Nonetheless, our models trained on HumanEva dataset still achieve reasonable performance on these videos. We expect that a more accurate human detector will largely improve the performance.

Discussions. Our model achieves best performance when the human is close to camera and image-to-pose knowledge is available. Thus, it is more appropriate to recognize subtle actions and be applied in HCI. We did not verify it on videos where the subjects are tiny and/or blurred (*e.g.*, surveillance videos). In this paper, our model is applied to human action recognition, but it can be easily generalized to other applications such as sign language translation, or in a more general form, used for application-specific feature compression.

5 Conclusion

We proposed a novel model to bridge the gap between the high dimensional observations and the random fields. This model replaces the observation layer of random fields with a latent pose estimator that learns to convert the high dimensional observations into more compact and informative representations under the supervision of labeled action data. The structure of our model also enables transfer learning to utilize the existing knowledge and data on image-to-pose relationship. Our model is tested on a real human motion dataset, and achieves significant improvement over the standard CRF and its variations.

References

1. Sigal, L., Black, M.J.: Humaneva: Synchronized video and motion capture dataset for evaluation of articulated human motion. Tech. Report, Brown Univ. (2006)
2. Rabiner, L.R.: A tutorial on hidden markov models and selected applications in speech recognition. Proceedings of the IEEE 77(2), 257–286 (1989)
3. Lafferty, J.D., Mccallum, A., Pereira, F.C.N.: Conditional random fields: Probabilistic models for segmenting and labeling sequence data. In: ICML, pp. 282–289 (2001)
4. Morency, L.P., Quattoni, A., Darrell, T.: Latent-dynamic discriminative models for continuous gesture recognition. CVPR (2007)
5. Sminchisescu, C., Kanaujia, A., Metaxas, D.: Conditional models for contextual human motion recognition. Comput. Vis. Image Underst. 104(2), 210–220 (2006)
6. Sminchisescu, C., Kanaujia, A., Metaxas, D.: Bm^{3}e: Discriminative density propagation for visual tracking. PAMI (2007)
7. Agarwal, A., Triggs, B.: Recovering 3d human pose from monocular images. IEEE Transactions on Pattern Analysis and Machine Intelligence (2006)
8. Caruana, R.: Multitask learning. Machine Learning 28(1), 41–75 (1997)
9. Lv, F., Nevatia, R.: Single view human action recognition using key pose matching and viterbi path searching. CVPR (2007)
10. Blank, M., Gorelick, L., Shechtman, E., Irani, M., Basri, R.: Actions as space-time shapes. In: ICCV, pp. 1395–1402 (2005)
11. Ramanan, D., Forsyth, D.A.: Automatic annotation of everyday movements. In: NIPS (2004)
12. Cuntoor, N.P., Chellappa, R.: Epitomic representation of human activities. CVPR (2007)

13. Yilmaz, A., Shah, M.: Recognizing human actions in videos acquired by uncalibrated moving cameras. In: ICCV (2005)
14. Quattoni, A., Collins, M., Darrell, T.: Conditional random fields for object recognition. Neural Information Processing Systems (2004)
15. Hammersley, J.M., Clifford, P.: Markov field on finite graphs and lattices (unpublished manuscript, 1971)
16. Jordan, M., Jacobs, R.: Hierarchical mixtures of experts and the em algorithm. Neural Computation 6, 181–214 (1994)
17. Fei-Fei, L., Perona, P.: A bayesian heirarchical model for learning natural scene categories. In: Proc. CVPR (2005)
18. Ning, H., Xu, W., Gong, Y., Huang, T.: Discriminative learning of visual words for 3d human pose estimation. CVPR (2008)

Relevant Feature Selection for Human Pose Estimation and Localization in Cluttered Images

Ryuzo Okada* and Stefano Soatto

Computer Science Department, University of California, Los Angeles
{okada,soatto}@cs.ucla.edu

Abstract. We address the problem of estimating human body pose from a single image with cluttered background. We train multiple local linear regressors for estimating the 3D pose from a feature vector of gradient orientation histograms. Each linear regressor is capable of selecting relevant components of the feature vector depending on pose by training it on a pose cluster which is a subset of the training samples with similar pose. For discriminating the pose clusters, we use kernel Support Vector Machines (SVM) with pose-dependent feature selection. We achieve feature selection for kernel SVMs by estimating scale parameters of RBF kernel through minimization of the radius/margin bound, which is an upper bound of the expected generalization error, with efficient gradient descent. Human detection is also possible with these SVMs. Quantitative experiments show the effectiveness of pose-dependent feature selection to both human detection and pose estimation.

1 Introduction

Human detection and pose estimation have numerous applications such as automated surveillance, driver assistance for automobiles, human-computer interfaces, and are an active area of research. These tasks are challenging because clothing, lighting conditions, and pose change the appearance of humans significantly. To cope with this problem, an image descriptor based on histograms of oriented gradients (HOG), which are computed on a uniform grid of overlapping local patches, has been used as a feature vector successfully for both human detection [1,2,3] and human pose estimation [4,5,6]. The descriptor also encodes unwanted background clutter into the feature vector, some of the feature vector components are irrelevant to the task. It is well known that such irrelevant components increase the complexity of classifiers and regressors and decrease generalization capacity, and that feature selection techniques [7] can be used for solving this problem.

One important aspect of relevant feature selection in detecting humans and estimating their pose is the fact that what features are relevant depends on pose (Fig. 1). Feature components that are relevant for some poses, are irrelevant to others, and this aspect has not received much attention in the literature. To address this issue, we propose a piecewise linear regression method where multiple local linear regressors approximate the nonlinear mapping function from HOG-based feature vectors to 3D poses. We train the

* Current affiliation: Corporate R&D Center, Toshiba Corporation.

D. Forsyth, P. Torr, and A. Zisserman (Eds.): ECCV 2008, Part II, LNCS 5303, pp. 434–445, 2008.

Fig. 1. Relevant feature depending on pose. The diagonal gradient orientation in the left figure (shown by the diagonal white line segment in the white rectangle) is a relevant component of the feature vector for describing the pose of the right arm while the same component is irrelevant to the pose shown in the right figure.

local linear regressor for each pose cluster which contains a subset of training samples with similar poses. Since the linear regressor implicitly performs feature selection [4], relevant features are selected automatically for each pose cluster. Note that piecewise regression approaches have been used for pose estimation in a different context [8,9], where multiple regressors are employed for describing one-to-many mapping from an image feature to 3D poses, e.g. multiple different poses have similar silhouettes, and such ambiguity is solved by taking temporal consistency into account. In our method, we discriminate the pose cluster from the others to select the linear regressor to be used for 3D pose estimation. For this purpose, we train a Support Vector Machine (SVM) with feature selection for each cluster, which enables pose-dependent feature selection.

This cluster discrimination process is applicable to human detection as well by adding non-human training samples and training SVMs to discriminate each pose cluster from the non-human ones.

We achieve feature selection for each SVM using a RBF kernel by estimating scale parameter of the RBF kernel for each component of the feature vector separately. This is a problem of hyperparameter estimation (or model selection) for SVMs, and is solved by minimizing the radius/margin (R/M) bound [10] based on a gradient descent method [11]. Although this feature selection method for SVMs has not been used for vision problems, it reduces generalization error effectively because the R/M bound is an approximation to an upper bound of the expected generalization error. Furthermore, we point to a new way to efficiently compute the gradient of the R/M bound.

Related works: Agarwal et al. [4] recover frontal poses of the upper human body by regression from the feature vector based on HOG encoded by non-negative matrix factorization (NMF) to suppress unwanted background. In their experiments, the error of pose prediction using NMF encoding is similar to the one obtained by linear regression without NMF encoding. This is because the linear regressor implicitly performs feature selection. Bissacco et al. [12] estimate human full body pose based on multi-dimensional boosting regression which enables relevant feature selection from a preselected set of Haar-like features. Shakhnarovich et al. [6] find k-NN samples by a fast parameter (pose) sensitive hashing algorithm and estimate pose by locally weighted regression using the k-NN samples. They assume that the background is simple and stationary and that the human body is segmented from the background. Many approaches have been proposed to deal with background clutter, from body parts detection [13,14,15,16],

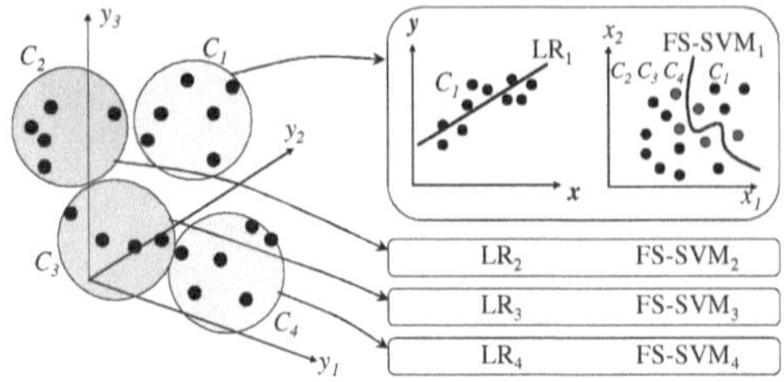

Fig. 2. Training piecewise linear regressors. The pose space is divided into several clusters by k-means. For each cluster, a linear regressor and an SVM are trained with feature selection.

which is particularly difficult in a single view because of self-occlusion, to silhouette extraction [8,17,18,19], which limits the applicability to fixed background, and edge-based template matching [9], which may be unstable under significant background clutter.

For Human detection, AdaBoost is used as a classifier to select relevant features from a set of, e.g., Haar-like wavelet features [20], gradient response in several directions [21], and SIFT-like gradient orientation features [2,3]. Support Vector Machines (SVMs) are also used for human detection with implicit feature selection (linear SVM) [1], and "filter-type" feature selection methods which preselect relevant features independently of the SVMs based on another machine learning method capable of feature selection, such as AdaBoost [22], or based on a heuristic that relevant features move decision boundaries significantly when they are removed [23]. Although the feature selection method is different and pose estimation is very coarse (front/left/right), [22] is similar to our approach in that it discriminates the pose clusters based on pose-dependent feature selection.

2 Piecewise Linear Regression

For recovering 3D human pose from a static image, a regression approach has been proposed in [8,4,17], where a mapping function from a feature vector extracted from the static image to a pose vector is learned using a set of labeled training samples. We take this type of regression based approach using the feature vector based on the histograms of oriented gradients (HOG) [4,5].

The feature vector consists of the histograms computed on a uniform grid of overlapping local patches to describe the contents of an image window. The histogram for each image patch encodes local shape and position information, while the coarse grid and orientation histogram is insensitive to variation in appearance and small misalignments. This representation also encodes unwanted background clutter into the feature vector and some of the components are irrelevant to human pose. More importantly, the relevant component is dependent on pose as shown in Fig. 1 because the relative position of the arms, legs, and torso in the image window vary depending on pose.

Considering this fact, we propose a piecewise linear regression method for recovering 3D pose from a feature vector $\boldsymbol{x} \in R^L$ (see Fig. 2). In this paper, we represent the

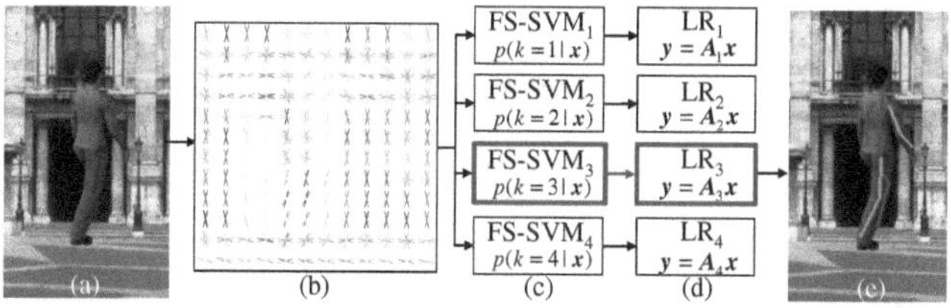

Fig. 3. Overview of our method for predicting 3D pose. (a) Given a test image, (b) the feature vector is computed in the window. (c) The pose cluster which the current pose belongs to is determined by SVM classifiers. (d) The 3D pose is recovered using the linear regressor of the selected cluster (k=3). (e) The recovered pose is shown by red and green line segments, where green represents the right leg and arm.

3D pose by a pose vector $\boldsymbol{y} \in R^P$ which is a collection of 3D locations of major joints. Given a set of training samples $\{(\boldsymbol{y}_i, \boldsymbol{x}_i)|i = 1 \cdots N\}$, we divide them into several clusters in the pose space by k-means so that each cluster contains similar poses, and train a linear regressor for each cluster C_k, which maps a feature vector $\boldsymbol{x}$ to a pose vector $\boldsymbol{y}$:

$$\boldsymbol{y} = \boldsymbol{A}_k \boldsymbol{x} + \boldsymbol{\epsilon}_k, \tag{1}$$

where $\boldsymbol{A}_k$ and $\boldsymbol{\epsilon}_k$ are a weight matrix and a residual error vector for a cluster C_k, respectively. It is important for our approach of pose-dependent feature selection that each pose cluster contains similar poses. We verified by preliminary experiments that each pose cluster generated by k-means (e.g. 6 clusters for walking sequences) was already local enough for approximating the nonlinear mapping function by multiple linear regressors, and that Expectation-Maximization type algorithms [9], which simultaneously optimize regressors and partition of clusters, did not improve generalization performance significantly. Although a kernel-based nonlinear regressor can be used with feature selection, we choose to use the linear regressor because of the following reasons: (1) The linear regressor is capable of implicit feature selection in the sense that it automatically reduces the weights with respect to irrelevant features through the estimation of the weight matrix $\boldsymbol{A}_k$. (2) The relation between the pose vectors and the feature vectors within a cluster is simple enough to be described by a linear function. (3) Training is much simpler than the kernel regressors with feature selection. Since similar poses in each cluster share the relevant features, each linear regressor achieves pose-dependent feature selection. Next, for each cluster, we train a Support Vector Machine (SVM) [10] with capabilities of feature selection [11] and probability output [24] in order to discriminate the cluster from the others. Feature selection is useful for discriminating the clusters as well because the feature vector contains cluttered background and the relevant features vary depending on the cluster to be discriminated.

For predicting a pose given a test image with a window circumscribing a subject, we first extract a HOG-based feature vector of the window (see Fig. 3). Secondly, we determine the cluster that the current pose of the subject belongs to by selecting the cluster with the highest probability output $p(k|\boldsymbol{x})$ of the SVM classifier. Thirdly, we predict the 3D pose using the linear regressor of the selected cluster.

2.1 Linear Regression Method

To train a linear regressor which recovers a pose vector $\boldsymbol{y}$ from a feature vector $\boldsymbol{x}$, we estimate the weight matrix $\boldsymbol{A}_k$ for each cluster C_k using training samples in C_k by minimizing prediction error with a regularization term $R(\cdot)$ to control overfitting:

$$\boldsymbol{A}_k = \arg\min_{\boldsymbol{A}_k} \left\{ \sum_{(\boldsymbol{y}_i,\boldsymbol{x}_i)\in C_k} \|\boldsymbol{A}_k\boldsymbol{x}_i - \boldsymbol{y}_i\|^2 + R(\boldsymbol{A}_k) \right\}. \tag{2}$$

This is equivalent to the MAP estimation in the probabilistic regression framework assuming a Gaussian distribution on the residual error vector. The first term, the data fidelity term, in the objective function corresponds to the likelihood, and the regularization term corresponds to the prior $p(\boldsymbol{A}_k)$.

For a Gaussian prior $p(\boldsymbol{A}_k) \sim \prod_l exp(-\nu\|\boldsymbol{a}_l\|^2)$, where $\boldsymbol{a}_l$ denotes the l-th column vector of $\boldsymbol{A}_k$, the regularizer takes the form $R(\boldsymbol{A}_k) \equiv \lambda\|\boldsymbol{A}_k\|_F^2$ and the solution gives a linear ridge regressor. The ridge regressor is capable of performing feature selection in the sense that the columns $\|\boldsymbol{a}_l\|$ weigh irrelevant components l, ideally belonging only to the background, automatically reducing their effect through the optimization (2).

Relevance Vector Machine (RVM) regression [25] results in taking a prior of the form $p(\boldsymbol{A}_k) \sim \prod_l \|\boldsymbol{a}_l\|^{-\nu}$, which gives a sparse solution because the prior is sharply peaked at $\|\boldsymbol{a}_l\| = 0$ and pushes the weights of the irrelevant components to zero. For achieving sparsity, a Laplace prior of the form $p(\boldsymbol{A}_k) \sim \prod_l exp(-\nu\|\boldsymbol{a}_l\|)$ is often used and an ϵ-insensitive loss function is used as the data fidelity term in SVM regression [26]. In this paper, we use the RVM regressor because it gives a more sparse solution while maintaining good generalization performance. Note that the SVM regressor can produce sparse solutions comparable to the RVM regressor by employing "reduced-set" post-processing [27].

Since similar poses contained in each cluster share relevant features, the linear regressor is capable of selecting such relevant features depending on the poses in the cluster.

2.2 Cluster Discrimination Using SVM

In this section, we describe the SVM classifiers for discriminating a cluster C_k from the other clusters. For simplicity of description, we drop the subscript k referring to the cluster C_k in the rest of this section.

Feature selection is useful for discriminating the clusters for the same reason as in the regression case. We introduce feature selection into the kernel SVM based on automatic relevance determination (ARD), which is achieved by tuning the scale parameters $\boldsymbol{\gamma} \in R^L \geq \mathbf{0}$ of the ARD Gaussian kernel

$$K(\boldsymbol{x}_i, \boldsymbol{x}_j) = \exp\left(-\frac{1}{2}\sum_l \gamma_l |x_{il} - x_{jl}|^2\right). \tag{3}$$

Although a linear SVM is capable of implicit feature selection, the kernel SVM performs better than the linear SVM in our experiments. Tuning the kernel scale parameters

γ, which are hyperparameters, is achieved by minimizing an estimate of the generalization error such as the leave-one-out (LOO) error. The LOO error has an upper bound referred to as the radius/margin (R/M) bound [10].

$$\gamma = \arg\min_{\gamma} \frac{1}{N} R^2 \|\boldsymbol{w}\|^2, \tag{4}$$

where $2/\|\boldsymbol{w}\|$ is a margin between the classes to be discriminated and R is the radius of the smallest sphere that contains all the vectors $\boldsymbol{z}_i = \phi(\boldsymbol{x}_i)$ in a projected (high dimensional) feature space. Here the inner product in $\boldsymbol{z}$-space is computed by the kernel function: $(\boldsymbol{z}_i, \boldsymbol{z}_j) = K(\boldsymbol{x}_i, \boldsymbol{x}_j)$. This means tuning the hyperparameters that maximize the margin while making the sphere as small as possible.

Let $t_i \in \{-1, +1\}$ be a target value. $t_i = +1$ denotes a training sample $(\boldsymbol{y}_i, \boldsymbol{x}_i)$ is in the cluster C_k and otherwise $t_i = -1$. $\boldsymbol{w}$ is the solution of the following SVM problem with hard margin[1]:

$$\min \frac{1}{2}\|\boldsymbol{w}\|^2 \quad s.t. \; t_i(\boldsymbol{w}^T \boldsymbol{z}_i + b) \leq 1 \;\; \forall i. \tag{5}$$

R is the solution of the following problem:

$$\min R^2 \quad s.t. \; \|\boldsymbol{z}_i - \boldsymbol{c}\| < R^2 \;\; \forall i. \tag{6}$$

To find the optimum value of γ by minimizing the R/M bound, gradient descent is shown to be an efficient method [11]. We eliminate irrelevant features that have a small value of γ_l during the optimization process of (4) every 5 iterations of the Broyden-Fletcher-Goldfarb-Shanno (BFGS) quasi-Newton algorithm [28].

To compute the derivative of the R/M bound, we need the derivatives of $\|\boldsymbol{w}\|^2$ and R^2 with respect to every scale parameter γ_l. This requires computing the entire kernel matrix for each γ_l (or storing it in memory to avoid re-computation), as done in [11,29].

Remark 1. The derivative of the margin $\|\boldsymbol{w}\|^2$ with respect to the scale parameter γ_l is computed using the support vectors only because the Lagrange multipliers α_i of the solution (5) are non-zero only for the support vectors:

$$\frac{\partial \|\boldsymbol{w}\|^2}{\partial \gamma_l} = \frac{1}{2} \sum_{\{i|\alpha_i \neq 0\}} \sum_{\{j|\alpha_j \neq 0\}} \alpha_i \alpha_j t_i t_j K(\boldsymbol{x}_i, \boldsymbol{x}_j) |x_{il} - x_{jl}|^2. \tag{7}$$

The derivatives of the radius R^2 can be computed similarly using the feature vectors corresponding to the non-zero Lagrange multipliers β_i of the solution (6).

Although this remark is not mentioned in the original papers [11,29], it reduces computational cost significantly, i.e., the computational complexity of (7) is $O(LN_{sv}^2)$ while that of the original formulation is $O(LN^2)$, where N, N_{sv}, and L are the numbers of training samples, support vectors, and feature dimensions, respectively. Hence the

[1] A non-separable case can be treated as an SVM problem with hard margins by converting the SVM problem using L_2 soft margin [11].

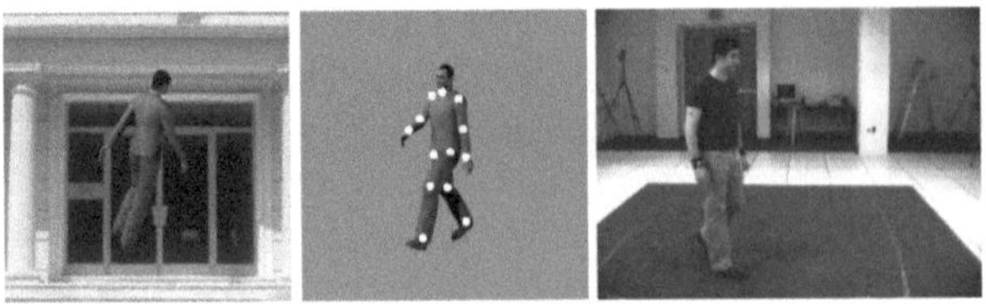

Fig. 4. Training images. The training images are synchronized with pose data. The synthesized dataset consists of cluttered background images (left) and clean background images (middle), where the white dots shows 13 joint locations used for encoding 3D pose. The real dataset (right) includes three subjects.

computational cost of (7) is smaller than that of the original formulation by a factor of $(N_{sv}/N)^2$.

To obtain the posterior probability $p(k|\boldsymbol{x})$ that the pose represented by a feature $\boldsymbol{x}$ belongs to the cluster with index k, the mapping from the SVM output to the posterior probability is approximated by a sigmoid function [24]. The parameters of the sigmoid function are tuned by cross validation. Given a feature vector $\boldsymbol{x}$ extracted from a test image, we compute the probability $p(k|\boldsymbol{x})$ for each cluster and determine that the pose represented by a feature $\boldsymbol{x}$ belongs to the cluster with the highest probability.

2.3 Human Detection

The framework presented thus far for pose discrimination can be applied to human detection as well, by adding the non-human training samples. They are negative samples for training each of the SVMs: The target value is $t_i = +1$ for the training sample $(\boldsymbol{y}_i, \boldsymbol{x}_i)$ in the cluster C_k, and $t_i = -1$ for the non-human samples.

3 Experiments

Pose prediction: We have conducted experiments on a synthetic database and a real database (see Fig. 4). In the synthetic dataset, human poses are randomly generated in a subspace constructed by PCA using the walking sequences extracted from the CMU Motion Capture Database[2]. Human images corresponding to each pose are rendered by a human model rendering software, Poser, with cluttered background of natural images and with uniform background. For real dataset, we used HumanEva-I dataset [30].

The human pose is represented by a 39-dimensional pose vector $\boldsymbol{y}$ which is a collection of 13 major joint locations as shown in Fig. 4 (middle). The feature vector is histograms of oriented gradients. We compute the orientation of gradients in $[0, \pi]$ and construct the histograms using 8 orientation bins in 3x3 spatial cells which comprise a spatial block. We used uniformly spaced 4x4 blocks overlapping with neighbor blocks by the length of a cell and obtain a 1152 dimensional feature vector for a image window.

Fig. 5 shows pose prediction errors with respect to different regression methods on the synthetic dataset with clean and cluttered background. The prediction errors

[2] http://mocap.cs.cmu.edu

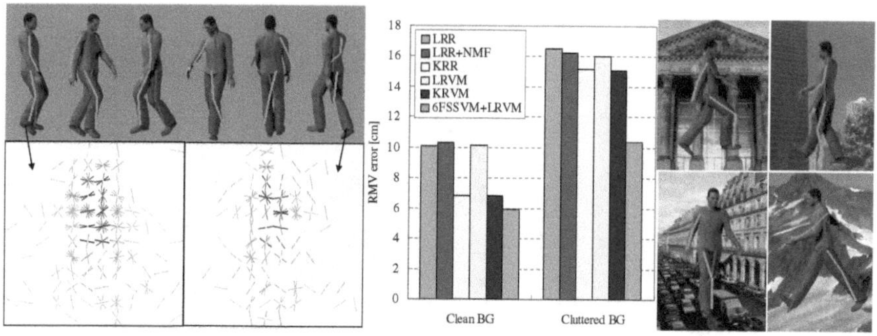

(a) (b) (c) (d)

Fig. 5. Results of pose estimation on the synthetic dataset. (a) Mean poses of 6 pose clusters drawn by red and green (right arm and leg) line segments on the closest training image, and relevant components of the feature vector for the left and right pose clusters (components with dark color have high weights $\|\boldsymbol{a}_l\|$). (b) Pose prediction errors with respect to different regression methods on clean and cluttered background. (c) Example of successful pose estimates. (d) Typical example of misestimation.

are measured in terms of the RMS deviation from the ground truth for the 3D locations of the 13 major joints. The number of training and test samples are 4000 and 1000, respectively. In the case of cluttered background, the prediction errors with respect to any single regressor (linear ridge regressor (LRR), LRR with NMF encoding (NMF+LRR) [4], RBF-kernel ridge regressor (KRR), linear RVM regressor (LRVM), and RBF-kernel RVM regressor (KRVM)) are equally large. Since our method is capable of reducing the disturbance of background clutter (see Fig. 5(a) bottom) by pose-dependent feature selection using 6 linear RVM regressors and 6 ARD kernel SVM classifiers[3] (6FSSVM+LRVM), the prediction error of our method is almost as good as that for the clean background with a linear regressor (LRR or LRVM). The problem of multiple posterior modes in mapping form the HOG-feature to the 3-D pose is not severe on clean backgrounds. However, such multi-modality is a typical failure mode in the presence of cluttered background as shown in Fig. 5 (bottom right), where the left arm and leg are misestimated as the right ones. Note that the prediction error of 6FSSVM+LRVM for clean background is smaller than those of the nonlinear regression methods. This shows that the number of pose clusters, which we have determined to be 6 experimentally, are sufficient for approximating the nonlinear mapping function from HOG feature to 3D pose.

Fig. 6 shows pose prediction errors and sample pose estimates for the real image dataset of HumanEva-I. In this dataset, image sequences captured by calibrated cameras are synchronized with motion capture data. We used walking sequences of three subjects, S1, S2 and S3, captured by a camera C1 only. The training subset of the dataset contains 590, 438, and 435 frames for S1, S2, and S3, respectively. We add feature vectors slightly shifted in position and scale to increase tolerance to misalignment of the test window, and the total number of training samples is 4389 to train a

[3] LIBSVM [31] is used with modification for solving the problems (5) and (6).

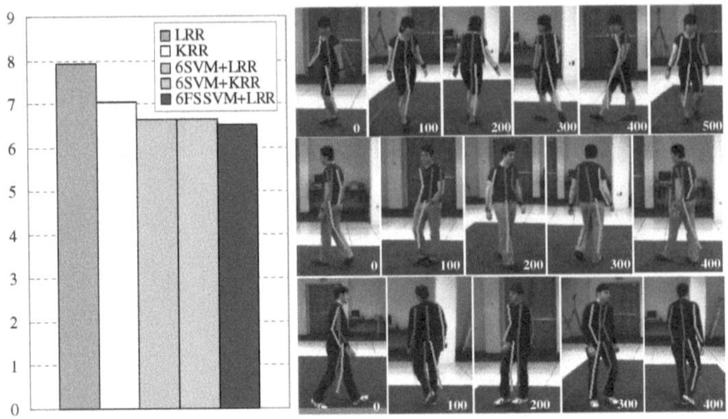

Fig. 6. Results of pose estimation on HumanEva-I dataset. The left bar graph shows pose prediction errors with respect to different regression methods. The right images show samples of predicted poses for 3 subjects drawn in the red and green (right arm and leg) line segments. (The head top joint (headDistal) is not used because of error in the dataset.)

Table 1. Mean errors of relative joint positions on the test subset of HumanEva-I dataset. The mean errors of our method without using background subtraction are smaller than those of Poppe's method [5] based on background subtraction for the subject S1, S2 and total average.

Method	S1	S2	S3	Average [mm]
Poppe [5]	41.24	39.56	55.27	42.85
Ours	41.19	35.03	37.69	37.98

linear ridge regressor (LRR), an RBF-kernel ridge regressor (KRR), 6 SVMs with LRRs (6SVM+LRR) or KRRs (6SVM+KRR), and 6 feature-selection SVMs with LRRs (6FSSVM+LRR). For testing, we used the validation subset (ground truth is known) and the test window circumscribing the subject is given based on the known 3D position of the subject. Pose-dependent feature selection by 6SVM+LRR decreases the prediction error. It is competitive with 6SVM+KRR where feature selection is not performed. The prediction error further decreases by introducing pose-dependent feature selection for both SVMs and regressors (6FSSVM+LRR). Although the background in the test window varies according to the 3D location of the subject, improvement by 6FSSVM+LRR is limited because the background clutter in HumanEva-I dataset is much simpler than that of the synthetic dataset. Training the kernel SVMs with feature selection requires solving the problems (5) and (6) many times, and it took about 3 hours for each SVM in this experiments using HumanEva-I dataset on a standard PC with 2.13 GHz Intel CPU and 2 GB memory. Training time for each linear RVM regressor was about 5 seconds. For estimating the pose given a feature vector, it takes 25 ms on average.

We compare our method with the recent method proposed by Poppe [5] using HOG-based feature vector on the test subset of HumanEva-I dataset. Training data used in this experiment are the same as [5]. His method uses background subtraction to locate

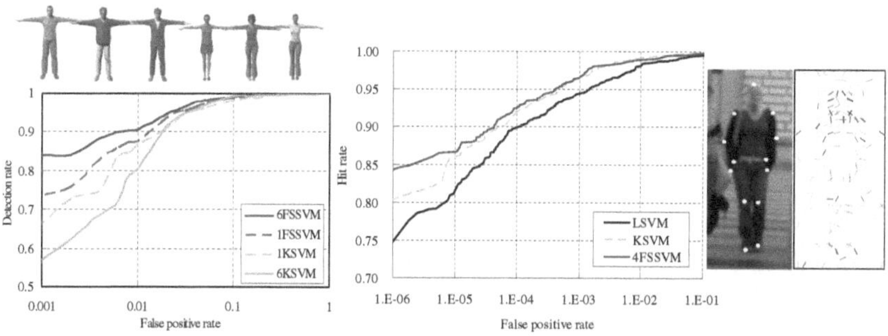

(a) Synthetic dataset (b) INRIA dataset

Fig. 7. Results of human detection. (a) Human models used for generating training and test images (top), and ROC curves of different classifiers (bottom). (b) ROC curves of our detector and the linear SVM [1] (left). The mean pose of a pose cluster is drawn on the closest training image (middle). Pose is represented by 13 points which are shown by white dots. Estimated kernel scale parameters γ_l for each feature component are shown by short line segments (right). Black color means a high value of γ_l.

the subject accurately in each image while our method uses the SVM-based human detector described in section 2.3, whose negative training samples are randomly sampled image patches without overlapping with human region more than 50%. Although we can search the entire image for obtaining global 3D position of the subject, we apply our human detector locally on a $5\times5\times7$ grid covering $40\times40\times150$ cm (wide search range in the camera direction) around the previous global position to reduce the computational cost. We obtain the current 3D global position as the mean of the weighted positions where the human detector is applied. The weights are the probability outputs of the human detector. In the initial frame, we give a rough estimate of 3D global position by watching the image since we do not know the ground truth in the test subset. Table 1 shows the mean errors of joint positions relative to the pelvis (torsoDistal) joint. Although our method does not perform background subtraction, the mean errors of our method are smaller than those of Poppe's method.

Human detection: Fig. 7(a) shows results of human detection using the synthetic dataset consisting of 6 subjects with different clothing and poses. The pose variation is the same as the case of pose estimation and each subject has 1000 samples. We use 5000 samples of the left 5 subjects in Fig. 7(a) and 1500 non-human samples for training. For testing, we use 1000 samples of the subject on the right and 1000 samples of non-human images which are not included in the training samples. Introducing the feature selection on a single SVM (1FSSVM) improves the performance and pose-dependent feature selection using 6 SVMs (6FSSVM) outperforms the other methods. The prediction error for the method using 6 kernel SVMs (6KSVM) is worse than 1KSVM due to the lack of samples for training each of the 6 kernel SVMs, which are much fewer than those of 1KSVM.

We test our human detector on the real images of the INRIA person dataset [1][4] to illustrate the flexibility of our approach that uses pose-dependent relevant features. Since

[4] `http://pascal.inrialpes.fr/data/human/`

our method requires pose information to build pose clusters, we added pose information to the positive (human) images by locating the 13 positions, such as head, elbows, knees an so on, by hand (see Fig. 7(b) middle). We trained our detector using 4 pose clusters, a linear SVM and a kernel SVM with the default setting of [1] on 1,246 positive and randomly sampled 12,180 negative samples plus hard negative samples for retraining, which are false positives in the first training. Fig. 7(b) shows the ROC curves and an example of estimated kernel scale parameters for a pose cluster. Our detector improves performance by 12% and 8% at 10^{-6} false positive rate compared with linear SVM and kernel SVM, respectively.

4 Conclusions

We presented a method for human detection and pose estimation from a static image with background clutter. We estimated the 3D pose from a feature vector consisting of local gradient orientation histograms by first discriminating the pose clusters using the kernel SVMs with feature selection and then estimating the pose using linear regressors. We performed feature selection for kernel SVMs by estimating scale parameters of ARD RBF kernel through minimization of the R/M bound which is an upper bound of the expected generalization error. For minimizing the R/M bound with gradient descent, we computed its gradient efficiently by using support vectors only. The process of discriminating the pose clusters by using the SVMs is applicable for human detection by training the SVMs with non-human samples. Both SVMs and linear regressors are capable of pose-dependent feature selection, which was shown to be effective by the quantitative experiments where our method was compared with other recent methods and outperformed them.

References

1. Dalal, N., Triggs, B.: Histograms of oriented gradients for human detection. In: Proc. of CVPR, vol. 2, pp. 886–893 (2006)
2. Mikolajczyk, K., Schmid, C., Zisserman, A.: Human detection based on a probabilistic assembly of robust part detectors. In: Pajdla, T., Matas, J(G.) (eds.) ECCV 2004. LNCS, vol. 3021, pp. 69–81. Springer, Heidelberg (2004)
3. Zhu, Q., Avidan, S., Yeh, M.C., Cheng, K.T.: Fast human detection using a cascade of histograms of oriented gradients. In: Proc. of CVPR, vol. 2, pp. 1491–1498 (2006)
4. Agarwal, A., Triggs, B.: A local basis representation for estimating human pose from cluttered images. In: Proc. of ACCV, vol. 1, pp. 50–59 (2006)
5. Poppe, R.: Evaluating example-based pose estimation: experiments on the HumanEva sets. In: Computer Vision and Pattern Recognition (CVPR 2007) workshop on Evaluation of Articulated Human Motion and Pose Estimation (EHuM2) (2007)
6. Shakhnarovich, G., Viola, P., Darrell, T.: Fast pose estimation with parameter-sensitive hashing. In: Proc. of ICCV, vol. 2, pp. 750–757 (2007)
7. Blum, A., Langley, P.: Selection of relevant features and examples in machine learning. Artificial Intelligence 97(1-2), 245–271 (1997)
8. Agarwal, A., Triggs, B.: Monocular human motion capture with a mixture of regressors. In: IEEE Workshop on Vision for Human-Computer Interaction, pp. 1–8 (2005)

9. Thayananthan, A., Navaratnam, R., Stenger, B., Torr, P.H.S., Cipolla, R.: Multivariate relevance vector machines for tracking. In: Leonardis, A., Bischof, H., Pinz, A. (eds.) ECCV 2006. LNCS, vol. 3953, pp. 124–138. Springer, Heidelberg (2006)
10. Vapnik, V.N.: Statistical Learning Theory. Wiley-Interscience, Chichester (1998)
11. Keerthi, S.S.: Efficient tuning of SVM hyperparameters using radius/margin bound and iterative algorithms. IEEE Trans. on Neural Networks 13(5), 1225–1229 (2002)
12. Bissacco, A., Yang, M.H., Soatto, S.: Fast human pose estimation using appearance and motion via multi-dimensional boosting regression. In: Proc. of CVPR, pp. 1–8 (2007)
13. Date, N., et al.: Real-time human motion sensing based on vision-based inverse kinematics for interactive applications. In: Proc. of ICPR, vol. 3, pp. 318–321 (2004)
14. Sigal, L., Bhatia, S., Roth, S., Black, M.J., Isard, M.: Tracking loose-limbed people. In: Proc. of CVPR, vol. I, pp. 421–428 (2004)
15. Sigal, L., Black, M.J.: Predicting 3D people from 2D pictures. In: Proc. of Conf. Articulated Motion and Deformable Objects, pp. 185–195 (2006)
16. Wren, C.R., Azarbayejani, A., Darrell, T., Pentland, A.P.: Pfinder: Real-time tracking of the human body. IEEE Trans. on PAMI 19(7), 780–785 (1997)
17. Agarwal, A., Triggs, B.: Recovering 3D human pose from monocular images. IEEE Trans. on PAMI 28(1), 44–58 (2006)
18. Okada, R., Stenger, B., Kondoh, N.: A video motion capture system for interactive games. In: Proc. of MVA, pp. 186–189 (2007)
19. Sminchisescu, C., Kanaujia, A., Li, Z., Metaxas, D.: Learning to reconstruct 3D human motion from bayesian mixtures of experts. a probabilistic discriminative approach. Technical Report CSRG-502, University of Toronto (2004)
20. Viola, P., Jones, M., Snow, D.: Detecting pedestrians using patterns of motion and appearance. In: Proc. of ICCV, pp. 734–741 (2003)
21. Sabzmeydani, P., Mori, G.: Detecting pedestrians by learning shapelet features. In: Proc. of CVPR, pp. 1–8 (2007)
22. Zehnder, P., Koller-Meier, E., Gool, L.V.: A hierarchical system for recognition, tracking and pose estimation. In: Bengio, S., Bourlard, H. (eds.) MLMI 2004. LNCS, vol. 3361, pp. 329–340. Springer, Heidelberg (2005)
23. Evgeniou, T., Pontil, M., Papageorgiou, C., Poggio, T.: Image representations for object detection using kernel classifiers. In: Proc. of ACCV, pp. 687–692 (2000)
24. Platt, J.: Probabilistic outputs for support vector machines and comparison to regularize likelihood methods. In: Smola, A., Bartlett, P., Schoelkopf, B., Schuurmans, D. (eds.) Advances in Large Margin Classifiers, pp. 61–74 (2000)
25. Tipping, M.E.: Sparse bayesian learning and the relevance vector machine. Journal of Machine Learning Research 1, 211–244 (2001)
26. Micchelli, C.A., Pontil, M.A.: On learning vector-valued functions. Neural Computation 17(1), 177–204 (2005)
27. Scholkopf, B., Smola, A.J.: Learning with Kernels: Support Vector Machines, Regularization, Optimization, and Beyond. MIT Press, Cambridge (2001)
28. Shanno, D.F., Phua, K.H.: Minimization of unconstrained multivarite functions. ACM Transactions on Mathematical Software 6, 618–622 (1980)
29. Chapelle, O., Vapnik, V., Bousquet, O., Mukherjee, S.: Choosing multiple parameters for support vector machines. Machinc Learning 46, 131–159 (2002)
30. Sigal, L., Black, M.J.: Humaneva: Synchronized video and motion capture dataset for evaluation of articulated human motion. Technical Report CS-06-08, Brown Univ. (2006), `http://vision.cs.brown.edu/humaneva/`
31. Chang, C.C., Lin, C.J.: LIBSVM: a library for support vector machines (2001), `http://www.csie.ntu.edu.tw/~cjlin/libsvm`

Determining Patch Saliency Using Low-Level Context

Devi Parikh[1], C. Lawrence Zitnick[2], and Tsuhan Chen[1]

[1] Carnegie Mellon University, Pittsburgh, PA, USA
[2] Microsoft Research, Redmond, WA, USA

Abstract. The increased use of context for high level reasoning has been popular in recent works to increase recognition accuracy. In this paper, we consider an orthogonal application of context. We explore the use of context to determine which low-level appearance cues in an image are salient or representative of an image's contents. Existing classes of low-level saliency measures for image patches include those based on interest points, as well as supervised discriminative measures. We propose a new class of unsupervised contextual saliency measures based on co-occurrence and spatial information between image patches. For recognition, image patches are sampled using a weighted random sampling based on saliency, or using a sequential approach based on maximizing the likelihoods of the image patches. We compare the different classes of saliency measures, along with a baseline uniform measure, for the task of scene and object recognition using the bag-of-features paradigm. In our results, the contextual saliency measures achieve improved accuracies over the previous methods. Moreover, our highest accuracy is achieved using a sparse sampling of the image, unlike previous approaches who's performance increases with the sampling density.

1 Introduction

Determining image patches of high saliency has recently received significant attention. The goal of saliency detection is to identify the image patches that are most informative of the image contents. A standard method for finding these patches is the use of interest point detectors based on local low-level image statistics [1–6]. Another class of saliency measures are discriminative in nature [7–12], where a patch is considered salient if it is informative from a classification perspective. The usefulness of a patch may be based on the mutual information between the presence of the patch and the scene categories [7] or the probability of misclassification of a patch [8]. Using these techniques, a relatively small number of patches can be sampled while still achieving high recognition accuracy.

In this paper, we explore the use of contextual information that is typically used for higher level reasoning for the low-level task of selecting informative or salient patches in an image. We consider a patch to be salient if it is predictive or representative of the other patches in the image. The relationships of image patches are modeled using co-occurrence and spatial information. Unlike

D. Forsyth, P. Torr, and A. Zisserman (Eds.): ECCV 2008, Part II, LNCS 5303, pp. 446–459, 2008.

previous saliency measures that rely only on local information, our approach incorporates contextual information using the patch statistics across the entire image. For recognition, the sampling of patches is performed using weighted random sampling based on patch saliency. In addition, we propose a sequential sampling approach based on increasing the maximum likelihood of patches given the set of previously selected patches.

Our saliency measure is evaluated within a bag-of-features framework [13–17]. This simple approach has shown good performance in a variety of recognition tasks and allows us to focus on the specific contributions of our paper. The bag-of-features approach consists of three components: a method for sampling patches from an image, a method for assigning the patches to a discrete patch vocabulary, and a method for classifying the resulting global descriptor. In this paper, we only address the first task of patch sampling, and use standard approaches for the other two components. A vocabulary of patch appearances, to which each patch is assigned, is constructed using K-means clustering [18, 19]. Classification is accomplished using an SVM classifier over the histogram of patch assignments [7, 15].

We compare our proposed contextual saliency measures to a variety of existing measures including interest points, discriminative approaches and random sampling. These measures are evaluated on both scene and object recognition tasks. In contrast to previous results that show recognition accuracy increases with the density of the sampling [7], the contextual measures achieve maximal accuracy using a sparse sampling. Moreover, in our experiments the accuracy of using contextual measures with sparse sampling is better than dense sampling using other methods.

The rest of the paper is organized as follows: Previous works are discussed in the following section. We describe our proposed contextual saliency measures in Section 3 and sampling methods in Section 4. In Section 5 we briefly describe the existing saliency measures used for comparison, followed by a description of the experimental setup in Section 6. Results and some discussion are provided in Section 7 and conclusions in Section 8.

2 Previous Work

Several works have explored the role of saliency measures for classification tasks. Nowak *et al.* [7] compare the interest operators to random dense sampling and find that random sampling performs comparable or superior to interest operators. Jurie *et al.* [20], apart from proposing a novel clustering approach to form codebooks, evaluate a discriminative saliency measure used for the feature selection problem. They find that when using smaller codebooks discriminative feature selection can be used to improve accuracies. However, using the full codebook for classification typically resulted in better performance. A related class of works [8, 21, 22] is visual search, where similar notions of saliency are important. The essence of visual search lies in the notion of active exploration, in which saliency maps are dynamically updated or areas are marked for further exploration.

Most existing approaches [1, 4, 15, 17, 18, 23, 24, 19] for selecting a sparse set of image patches are based on interest point detectors [1–6]. These include those based on edge cornerness [2], difference of Gaussian convolutions [1], stable extremal regions [5] or local entropy [3]. While these measures are useful for obtaining reliable correspondences or matching, they do not relate directly to image understanding via classification or recognition. The strategies of dense sampling [13, 16, 25] or random sampling [12] have shown to provide comparable or even better performance than interest points [7]. Biologically inspired saliency measures following the "feature integration theory" [26] extract regions of the image that stand out from their surrounding as being salient [27, 28]. [3, 29] are based on a similar notion and [30] consider features to be salient if they are rare. While this is a plausible explanation to predict task-independent attention, they do not take into account task-dependencies. Walther *et al.* [31] incorporate such task dependencies and combine the biologically plausible saliency map of [28] with interest point operators [1] to show improved performances.

Using high-level contextual information for better image understanding has received significant attention in recent works [32–40]. Most of these approaches use context as a post-processing step to prune out false positives [32, 33], aid in detection by eliminating unlikely locations of objects [32, 34–37], or ensure semantically consistent labels to regions of an image [32, 33, 38–40].

3 Proposed Contextual Saliency Measures

Our goal is to select a sparse set of image patches that are most informative for classification. We propose that the patches, which are representative or predictive of other patches in the image, are also the patches most useful for classification. We measure the predictiveness of a patch using a contextual saliency measure based on co-occurrence and spatial information. As stated earlier, we examine our measure of saliency within the bag-of-features framework. In this framework, classification is achieved by selecting a set of image patches and assigning them to codewords. A histogram of codewords is then constructed and used for classification. In this paper, we address the first task of selecting image patches. We describe the standard method of K-means for codebook creation and Support Vector Machines for classification in Section 7.

For each image patch x_i, we compute a patch descriptor y_i. This descriptor can vary based on the application and properties of particular datasets. In this paper we examine two descriptors. The first is a 4×4 vector of average color values over a patch. This descriptor is useful for scenarios in which color information is important, such as in scene recognition. For object recognition in which edge information is more useful than color, we use the standard SIFT descriptor [1].

The codebook W consists of m descriptor templates. Each patch x_i in an image is assigned to a codeword w_a in the codebook. These assignments may be soft or hard with α_{ia} being the probability of patch x_i being assigned to codeword w_a:

$$\alpha_{ia} = p(y_i|w_a) = \frac{1}{Z}\mathcal{N}(y_i; w_a, \sigma_w)$$

$\mathcal{N}$ is the standard normal distribution with mean w_a and variance σ_w. The value of Z is set so that the values of α_{ia} sum to one for all a, i.e. $\sum_{a=1}^{m} \alpha_{ia} = 1$. For hard assignments $\alpha_{ia} = 1$ for the codeword w_a that lies closest to y_i and $\alpha_{ia} = 0$ otherwise.

For each patch x_i in an image, we want to assign a saliency measure $\mathcal{S}_i$. In the following two sections we propose two saliency measures based on contextual information.

3.1 Occurrence-Based Contextual Saliency

Our measures of contextual saliency are based on how well each individual patch can predict the occurrence of other patches in the image. Our first saliency measure $\mathcal{S}^o$ uses co-occurrence information between codewords in images. Given a set of n patches in an image, we define the saliency of a patch x_i equal to the average likelihoods of the image patches conditioned upon y_i.

$$\mathcal{S}_i^o = \frac{1}{n}\sum_{j=1}^{n}\sum_{a=1}^{m} \alpha_{ia} p(x_j|w_a) \tag{1}$$

The value of $p(x_j|w_a)$ is computed by marginalizing over all possible codeword assignments for x_j

$$p(x_j|w_a) = \sum_{b=1}^{m} \alpha_{jb} p(w_b|w_a) \tag{2}$$

The value of $p(w_b|w_a)$ is the empirical conditional probability of observing codeword w_b given the codeword w_a has been observed somewhere in the image. These are learnt through MLE counts from the training images. Given hard assignments of patches to codewords, the two summations over m can be removed from equations (1) and (2).

Computing the above measure can be computational expensive, especially if the codebook size and number of patches is large. One method for reducing the computational complexity is to rearrange equations (1) and (2) as:

$$\mathcal{S}_i^o = \sum_{a=1}^{m} \alpha_{ia} \frac{1}{n}\sum_{j=1}^{n}\sum_{b=1}^{m} \alpha_{jb} p(w_b|w_a) \tag{3}$$

The value $\Phi_a = \frac{1}{n}\sum_{j=1}^{n}\sum_{b=1}^{m} \alpha_{jb} p(w_b|w_a)$ can then be pre-computed for each a, resulting in:

$$\mathcal{S}_i^o = \sum_{a=1}^{m} \alpha_{ia} \Phi_a \tag{4}$$

3.2 Location-Based Contextual Saliency

The previous contextual saliency measure was based solely on co-occurrence information without knowledge of the patch's location. In this section we propose a saliency measure that includes spatial information. The location of a patch in an image is modeled using a Gaussian Mixture Model with $c = 9$ components. For our experiments the Gaussian means are centered in a 3×3 grid evenly spaced across an image with standard deviations in each dimension equal to half the distance between the means. We define the value β_{iu} as the likelihood of x_i belonging to component l_u of the GMM, $u \in \{1, \ldots, c\}$, and $\sum_{u=1}^{c} \beta_{iu} = 1, \forall i$.

Similar to equation (1), we define our location-based contextual saliency measure $\mathcal{S}^l$ as

$$\mathcal{S}_i^l = \frac{1}{n} \sum_{j=1}^{n} \sum_{a=1}^{m} \sum_{u=1}^{c} \alpha_{ia} \beta_{iu} p(x_j | w_a, l_u) \tag{5}$$

The value of $p(x_j | w_a, b_u)$ is computed as

$$p(x_j | w_a, b_u) = \sum_{b=1}^{m} \sum_{v=1}^{c} \alpha_{jb} \beta_{jv} p(w_b, l_v | w_a, l_u) \tag{6}$$

The value of $p(w_b, l_v | w_a, l_u)$ is the empirical conditional probability of observing word w_b at location l_v given word w_a occurred at location l_u. These are learnt through MLE counts from the training images.

Similar to equation (4), we may pre-compute the values

$$\Psi_{au} = \frac{1}{n} \sum_{j=1}^{n} \sum_{b=1}^{m} \sum_{v=1}^{c} \alpha_{jb} \beta_{jv} p(w_b, l_v | w_a, l_u)$$

and find $\mathcal{S}_i^l$ as

$$\mathcal{S}_i^l = \sum_{a=1}^{m} \sum_{u=1}^{c} \alpha_{ia} \beta_{iu} \Psi_{au} \tag{7}$$

Since our proposed saliency measures are dependent on the codeword assignments of other image patches, a significant number of patches need to be sampled from the image for the measures to be reliable. However, we will only select a subset of these image patches for use in classification. While it may seem advantageous to use all the patches for classification, as we show later in our results, using a subset of the patches can actually lead to improved recognition rates.

As can be seen, there is no dependence of the saliency measures $\mathcal{S}^o$ or $\mathcal{S}^l$ on the class labels of the images, making the proposed contextual saliency measures unsupervised.

4 Sampling Strategies

Using the equations above we can compute a saliency measure for each patch in an image. In this section we discuss three methods for selecting a subset of these patches for use in classification. Let us assume s patches are desired for classification out of a possible n.

4.1 Sampling by Sorting

A naive approach to sampling is to pick the s patches with highest saliency. However, due to strong correlations in natural images, neighboring patches often have similar appearances, and would hence share similar saliency values. The result of using this technique is many neighboring patches being selected that convey similar information. As a consequence, classification rates may suffer.

4.2 Random Sampling

One method to reduce the odds of sampling neighboring or redundant patches is to use a weighted random sampling. The saliency map may be normalized to form a distribution over the patches from which samples can be drawn. This allows for patches with higher saliency to be sampled with a higher probability, without any one region dominating. This allows for a good balance between exploiting the highly salient regions, and exploring the rest of the image for other salient regions.

4.3 Sequential Sampling

The last strategy sequentially selects patches by considering the patches previously selected. Specifically, we pick the patch that is most predictive of the patches that were not highly likely given at least one of previously picked patches. Let us consider the saliency measures of equations (1) and (5), which compute the probability of $p(x_j|x_i)$ equal to $\sum_a \alpha_{ia} p(x_j|w_a)$ and $\sum_a \sum_u \alpha_{ia}\beta_{iu} p(x_j|w_a, l_u)$ respectively. Then given a set of previously picked patches $\{\acute{x}_1, \ldots, \acute{x}_t\}$ we compute our saliency measure as

$$\mathcal{S}_i(\acute{x}_1, \ldots, \acute{x}_t) = \frac{1}{n} \sum_{j=1}^{n} \max\left(p(x_j|x_i); p(x_j|\acute{x}_1), \ldots, p(x_j|\acute{x}_t)\right) \tag{8}$$

A each iteration, the patch with highest saliency is selected. This sequential approach selects patches that give the highest average increase in maximum predicted probability for the patches in the image. As a result, patches that convey similar information as those already chosen are unlikely to be selected.

5 Existing Saliency Measures

In our experiments, we compare our proposed contextual saliency measures with three classes of existing saliency measures. The first baseline measure is the uniform saliency measure across the entire image, where patches are sampled randomly from the image. Equivalently, this can be thought of as computing a distribution over the codewords using a normalized histogram. The distribution over codewords is then randomly sampled. The second measure is an interest-point based saliency measure. More specifically, we apply the Harris corner detector [2] to the image, and used its response at every location in the image as

Fig. 1. Example images from the (top) outdoor scene category dataset [41] and (bottom) Pascal-01 object recognition dataset [43]

the saliency map. We also provide experiments using the patches found from the SIFT detector [1]. Finally, we compare against a discriminative saliency measure. The discriminative measure considers a patch to be salient if the mutual information of the patch and the class labels is high. More specifically, if $\mathcal{M}(w_a)$ is the mutual information of the ath word with the class labels, the measure is defined as $\mathcal{S}_i^D = \sum_{a=1}^{m} \alpha_{ia}\mathcal{M}(w_a)$.

6 Experimental Setup

We evaluate our proposed contextual saliency measure for the tasks of scene and object recognition using the bag-of-features approach. In both scenarios we construct a codebook of feature descriptors using the standard K-means clustering technique with $K = 1000$. Classification is accomplished by first assigning each sampled patch to a codeword. A histogram of codewords is created as input into a Support Vector Machine (SVM) classifier. We use a Gaussian kernel SVM in all our experiments. We also experimented with adaptively thresholded histograms by picking thresholds that maximize mutual information with the class labels as suggested by Nowak *et al.* [7]. However, the results were comparable, so we only report experiments using the normalized histograms in our experiments. Experiments using the nearest-neighbor classifier were also tested. The results were consistently inferior, and hence are not included here.

6.1 Scene Recognition

We evaluate our approaches on the outdoor scene category dataset from Torralba *et al.* [41]. Example images from this dataset are shown in Fig. 1 (top). It contains images from 8 categories: coast, mountain, forest, open country, street, inside city, tall buildings and highways. There are a total of 2866 256×256 color-images. For scene recognition our 48 dimensional descriptor consists of the average color values sampled in a 4×4 grid. The patches were sampled evenly across the image on a 64×64 grid. The patch scale was set so that neighboring patches overlap by 75%. Each sampled patch is given a soft assignment to the codewords using equation (3) with $\sigma_w = 30$. Similar to Torralba *et al.* [42], we use 100 images per scene category for training, and the rest as testing.

6.2 Object Recognition

Our experiments on object recognition use the Pascal-01 [43] dataset which contains 4 object categories: cars, bicycles, motorbikes and people. Example images from the dataset are shown in Fig. 1 (bottom). A training set of 684 images and a test set of 689 images is defined. Since object recognition is more dependent on image gradients than color, we use SIFT [1] as our descriptor. The descriptor was sampled on a 64×64 uniformly spaced grid. The scale of the sampled patches was set so that horizontally neighboring patches overlap by 75%. In this scenario we used hard assignments of patches to codewords.

7 Results

7.1 Comparing Saliency Measures

Our first experiments test our contextual saliency measures and those described in Section 5 on the scene and object recognition datasets. Recognition accuracies are plotted relative to the density of samples used for classification in Fig. 2. The weighted random sampling strategy is used in all cases. For comparison, representative reported accuracies on these datasets are 84% on the outdoor scene recognition dataset by Torralba *et al.* [42] and $\sim$ 88% on the Pascal-01 object recognition dataset by Nowak *et al.* [7]. The highest accuracies achieved by the contextual saliency measures are 84% and 86% on the scene recognition task for $\mathcal{S}^l$ and $\mathcal{S}^o$ respectively, and 85% and 90% for the object recognition task. We also tested our algorithm using higher resolution grids for scene recognition. The highest accuracies for $\mathcal{S}^o$ were 55%, 68% and 81% for 8×8, 16×16 and 32×32 sampled grids respectively.

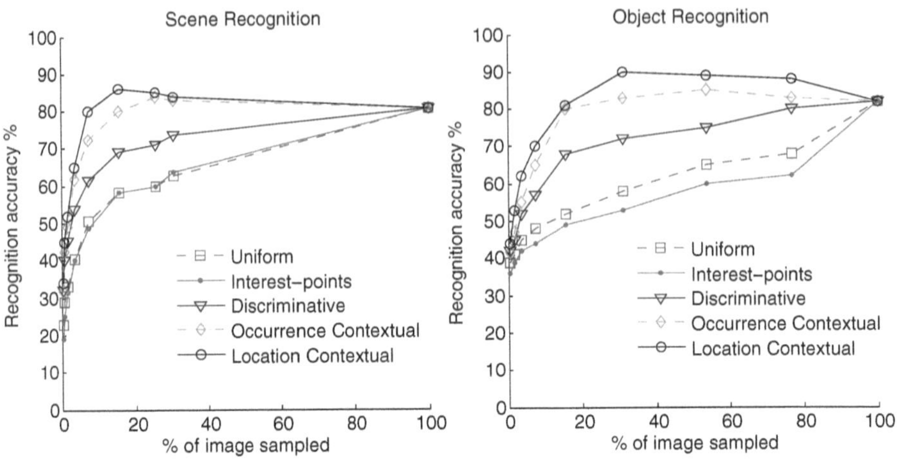

Fig. 2. Scene (left) and object (right) recognition accuracies for different saliency measures. The weighted random sampling strategy is used in all cases.

The contextual saliency measures have the best performance on both datasets. The discriminative measure using mutual information is next followed by similar results for both interest points and random saliency measures. We also ran experiments using the complete set of interest points found using the SIFT detector. On average the SIFT detector found 926 interest points and a recognition accuracy of 71% was achieved on the object recognition dataset. The performance of the saliency measures not using context increase monotonically with the density of the sampling. This is consistent with observations made by Nowak *et al.* [7]. However, with the contextual saliency measures a sparser sampling results in higher accuracy. This indicates that a sparser sampling is desirable not only for computational efficiency, but also higher recognition performance. With respect to the saliency measures, the usefulness of spatial information varies based on the dataset. The spatial information provides a larger performance boost for object recognition. We speculate that this is due to the increased spatial ambiguity of SIFT descriptors as compared to color descriptors. In scene recognition, color descriptors such as blue patches that correspond to sky or green patches that correspond to grass are strongly correlated with certain image locations. As a result, the spatial information may be redundant.

The various saliency maps for a set of sample images are shown in Fig. 3. In the scene recognition examples, objects that are unlikely given the scene category typically have lower saliency measures. As can be seen in Fig. 3(b), the saliency maps for the object recognition datasets have high saliency values even for the backgrounds. We believe this is due to several reasons: a strong correlation of objects and background, the higher entropy of the SIFT descriptor and the use of hard assignments.

The higher performance of contextual saliency measures over discriminative saliency measures may seem un-intuitive at first, since the discriminative saliency

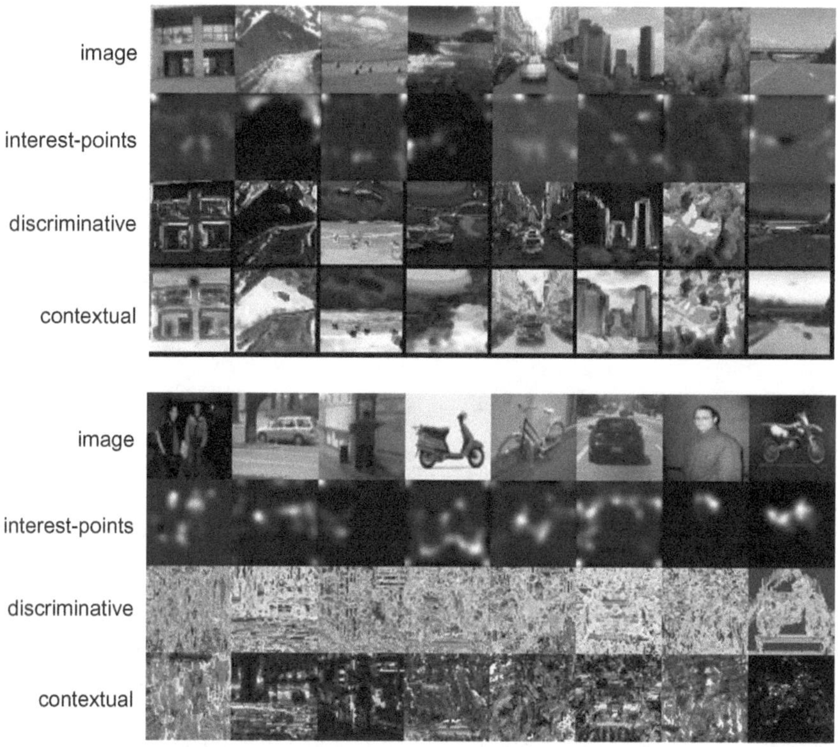

Fig. 3. Example saliency maps for images for the (top) scene recognition and (bottom) object recognition tasks using different classes of saliency measures. Maps are normalized to lie between 0 (least salient patch) and 1 (most salient patch).

Fig. 4. Red patches on the car in the highway image (left) and the white patches from the light behind the trees in the forest image (right) are considered to be salient by the discriminative measure because they occur pre-dominantly in sunset coast and snow-covered mountain images respectively. However, the contextual saliency measure incorporates the context of the rest of the scene and thus considers the road, sky and tress to be salient instead.

measure is supervised and is optimized specifically for recognition accuracies. However, it should be noted that the discriminative saliency measure ignores the rest of the scene, or the context in which the patch is present. This can lead

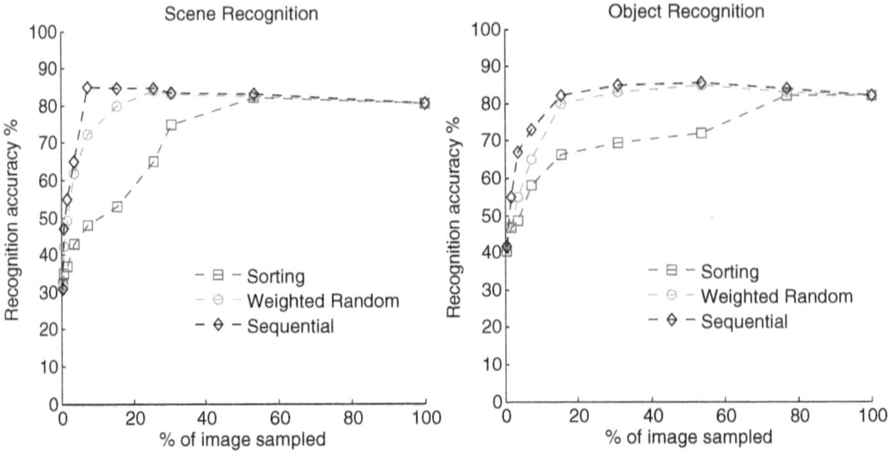

Fig. 5. Scene (left) and object (right) recognition accuracies for different sampling strategies. The occurrence-based contextual saliency measure $\mathcal{S}^o$ is used in all cases.

to undesirable artifacts. For instance, in a scene recognition task, red/orange patches may be considered to be salient by the discriminative measure since they occur pre-dominantly only in sunset (coast) images. Consider a highway test image that has a red car present in it, as seen in Fig. 4 (left). All the patches on the car will be considered highly salient by the discriminative saliency measure even though they are not representative of the scene. The saliency measure using context would identify that the red patches on the car are not representative of the image, and would not use them for classification. As a result, the contextual saliency measure is more likely to ignore clutter in the scene, resulting in higher accuracies. A similar result can be seen in Fig. 4 (right).

7.2 Comparing Sampling Strategies

To compare the different sampling strategies described in Section 4, we work with the occurrence-based contextual saliency measure. The scene and object recognition accuracies using the different sampling strategies are shown in Fig. 5. We can see that for scene recognition, the sorting strategy is much worse than the weighted random sampling. The features used for the scene recognition task are raw color patches, and hence neighboring patches in an image have very similar features and hence very similar saliency measures. While sequential sampling does not give higher accuracies, it reaches the peak accuracy using fewer patches than the weighted random sampling. Similar trends are seen for object recognition with the sequential and weighted random sampling results being more comparable. This may be due to the lower correlation of neighboring SIFT features as compared to the color descriptors used for scene recognition. Examples of how the saliency maps are updated after each iteration of the sequential sampling are shown in Fig. 6.

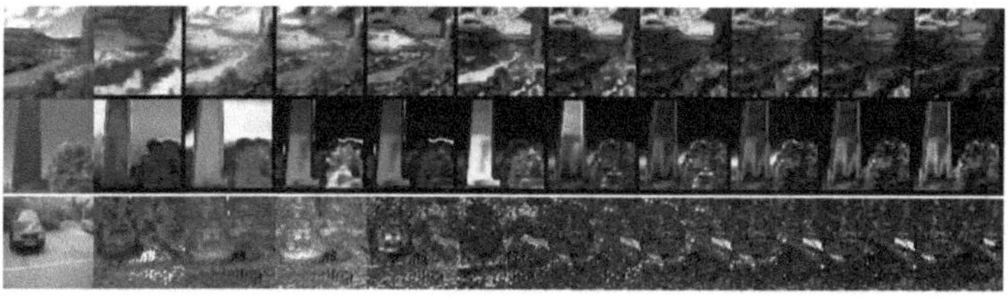

Fig. 6. Illustration of sequential sampling. Left: original image; Subsequent columns: saliency map being updated at each iteration; Top two rows: scene recognition; Bottom row: object recognition.

7.3 Discussion

Typically, contextual information is used for high-level reasoning about interactions between objects. In this paper, we demonstrate how contextual information may also be useful for low-level applications such as measuring patch saliency. While low-level contextual reasoning lacks semantic object information, even color or texture patches can supply useful contextual information as also shown in [44].

Discriminative saliency measures capture classification specific statistics of the patches. Our proposed contextual saliency measures capture contextual information of the entire image to determine saliency of patches. Both these aspects are complementary, and are both important to select representative patches that give good recognition accuracies. For the data sets we experimented with so far, the contextual information was more critical than the discriminative information, such as the example shown in Figure 4. However, one can imagine scenarios where the reverse is true. The balance between the two is task and domain dependent. A natural future direction, hence, is to combine discriminative and contextual information to design the optimum saliency measure for a given task. This is related to subjectiveness in the notion of saliency itself. Salient regions may be considered to be those that are representative of the image (as we do in this work), or those that are rare or unusual and hence draw attention. If we consider a more generic definition of saliency as being informative, it leads us back to the notion of task and domain dependency.

While our contextual saliency measure is unsupervised it is still dataset specific. That is, training images are needed to learn the co-occurrence statistics of the codewords. Other methods such as the use of interest points or random sampling may be better suited for applications in which the statistics of the images may not be known beforehand.

8 Conclusions

In this paper we propose two measures of saliency using contextual information. The first measure relies on co-occurrence information between codewords, while the second measure includes spatial information. We test our saliency measures against several others using the bag-of-features paradigm. Our experiments show improved results over other saliency measures on both scene and object recognition datasets. In contrast to previous works that produce results with accuracies that monotonically increase with sampling density, the contextual saliency measures produce optimal results with a sparse sampling.

References

1. Lowe, D.: Distinctive image features from scale-invariant keypoints. In: IJCV (2004)
2. Harris, C., Stephens, M.: A combined corner and edge detector. In: AVC (1988)
3. Kadir, T., Brady, M.: Saliency, scale and image description. In: IJCV (2001)
4. Mikolajczyk, K., Schmid, C.: Scale and affine invariant interest point detectors. In: IJCV (2004)
5. Matas, J., Chum, O., Urban, M., Pajdla, T.: Robust wide baseline stereo from maximally stable extremal regions. In: BMVC (2002)
6. Schmid, C., Mohr, R., Bauckhage, C.: Evaluation of interest point detectors. In: IJCV (2000)
7. Nowak, E., Jurie, F., Triggs, B.: Sampling strategies for bag-of-features image classification. In: ECCV (2006)
8. Moosmann, F., Larlus, D., Jurie, F.: Learning saliency maps for object categorization. In: ECCV International Workshop on The Representation and Use of Prior Knowledge in Vision (2006)
9. Walker, K., Cootes, T., Taylor, C.: Locating salient object features. In: BMVC (1998)
10. Fritz, G., Seifert, C., Paletta, L., Bischof, H.: Entropy based saliency maps for object recognition. In: ECOVISION (2004)
11. Serre, T., Riesenhuber, M., Louie, J., Poggio, T.: On the role of object-specific features for real world object recognition in biological vision. In: BMVC (2002)
12. Vidal-Naquet, M., Ullman, S.: Object recognition with informative features and linear classification. In: ICCV (2003)
13. Leung, T., Malik, J.: Representing and recognizing the visual appearance of materials using three-dimensional textons. In: IJCV (2001)
14. Lazebnik, S., Schmid, C., Ponce, J.: Affine-invariant local descriptors and neighborhood statistics for texture recognition. In: ICCV (2003)
15. Csurka, G., Dance, C., Fan, L., Willamowski, J., Bray, C.: Visual categorization with bags of keypoints. In: ECCV workshop on Statistical Learning in Computer Vision (2004)
16. Winn, J., Criminisi, A., Minka, T.: Object categorization by learned universal visual dictionary. In: ICCV (2005)
17. Fergus, R., Fei-Fei, L., Perona, P., Zisserman, A.: Learning object categories from googles image search. In: ICCV (2005)
18. Sivic, J., Zisserman, A.: Video Google: A text retrieval approach to object matching in videos. In: ICCV (2003)
19. Sivic, J., Russell, B., Efros, A.A., Zisserman, A., Freeman, B.: Discovering objects and their location in images. In: ICCV (2005)
20. Jurie, F., Triggs, B.: Creating efficient codebooks for visual recognition. In: ICCV (2005)

21. Ye, Y., Tsotsos, J.K.: Where to look next in 3d object search. In: ISCV (1995)
22. Viola, P., Jones, M.: Robust real-time object detection. In: IJCV (2001)
23. Grauman, K., Darrell, T.: Efficient image matching with distributions of local invariant features. In: CVPR (2005)
24. Fei-Fei, L., Perona, P.: A bayesian hierarchical model for learning natural scene categories. In: CVPR (2005)
25. Agarwal, A., Triggs, B.: Hyperfeatures – multilevel local coding for visual recognition. In: ECCV (2006)
26. Treisman, A.M., Gelade, G.: A feature-integration theory of attention. Cognitive Psychology (1980)
27. Itti, L., Koch, C., Niebur, E.: A model of saliency-based visual attention for rapid scene analysis. In: PAMI (1998)
28. Koch, C., Ullman, S.: Shifts in selective visual attention: towards the underlying neural circuitry. Human Neurobiology (1985)
29. Sebe, N., Lew, M.: Comparing salient point detectors. Pattern Recognition Letters (2003)
30. Hall, D., Leibe, B., Schiele, B.: Saliency of interest points under scale changes. In: BNVC (2002)
31. Walther, D., Rutishauser, U., Koch, C., Perona, P.: On the usefulness of attention for object recognition. In: ECCV (2004)
32. Hoiem, D., Efros, A., Hebert, M.: Putting objects in perspective. In: CVPR (2006)
33. Torralba, A., Murphy, K., Freeman, W.: Contextual models for object detection using boosted random fields. In: NIPS (2005)
34. Torralba, A., Sinha, P.: Statistical context priming for object detection. In: ICCV (2001)
35. Murphy, K., Torralba, A., Freeman, W.: Using the forest to see the trees: a graphical model relating features, objects, and scenes. In: NIPS (2003)
36. Bose, B., Grimson, E.: Improving object classification in far-field video. In: ECCV (2004)
37. Torralba, A., Murphy, K., Freeman, W., Rubin, M.: Context-based vision system for place and object recognition. AI Memo (2003)
38. Rabinovich, A., Vedaldi, A., Galleguillos, C., Wiewiora, E., Belongie, S.: Objects in context. In: ICCV (2007)
39. Parikh, D., Zitnick, C.L., Chen, T.: From appearance to context-based recognition: Dense labeling in small images. In: CVPR (2008)
40. Singhal, A., Luo, J., Zhu, W.: Probabilistic spatial context models for scene content understanding. In: CVPR (2003)
41. Oliva, A., Torralba, A.: Modeling the shape of the scene: a holistic representation of the spatial envelope. In: IJCV (2001)
42. Torralba, A.: Outdoor scene category dataset, `http://people.csail.mit.edu/torralba/code/spatialenvelope/`
43. Pascal01, `http://pascallin.ecs.soton.ac.uk/challenges/voc/`
44. Shotton, J., Winn, J., Rother, C., Criminisi, A.: Textonboost: joint appearance, shape and context modeling for multi-class object recognition and segmentation. In: Leonardis, A., Bischof, H., Pinz, A. (eds.) ECCV 2006. LNCS, vol. 3951. Springer, Heidelberg (2006)

Edge-Preserving Smoothing and Mean-Shift Segmentation of Video Streams

Sylvain Paris

Adobe Systems, Inc.

Abstract. Video streams are ubiquitous in applications such as surveillance, games, and live broadcast. Processing and analyzing these data is challenging because algorithms have to be efficient in order to process the data on the fly. From a theoretical standpoint, video streams have their own specificities – they mix spatial and temporal dimensions, and compared to standard video sequences, half of the information is missing, *i.e.* the future is unknown. The theoretical part of our work is motivated by the ubiquitous use of the Gaussian kernel in tools such as bilateral filtering and mean-shift segmentation. We formally derive its equivalent for video streams as well as a dedicated expression of isotropic diffusion. Building upon this theoretical ground, we adapt a number of classical algorithms to video streams: bilateral filtering, mean-shift segmentation, and anisotropic diffusion.

1 Introduction

This paper proposes a coherent approach to analyzing and filtering video streams. We develop tools that process video frames as soon as they are available and broadcast the result immediately. This scenario encompasses a wealth of practical applications such as surveillance, live preview, interactive simulations, and games. A major constraint imposed by these tasks is the requirement for instant results. For instance, latency cannot be tolerated in games – the displayed video must instantaneously follow the player inputs. This motivates our strict assumption that no future data are known when we process a frame; only the current and past data are available. In this context, we focus on edge-preserving smoothing and image segmentation because they are two techniques at the core of widely used applications such as vignetting correction [1], noise estimation and removal [2, 3, 4], object selection [5, 6], and stylization [7, 8, 9, 10].

Many video-processing tools are inherently off-line because they require the entire video sequence to be provided as input to the algorithm [2, 3, 11, 12, 13, 14, 15, 16, 17, 18, 19, 20]. Nonetheless, a few on-line methods exist [9, 10] but they often resort to a frame-by-frame approach. As a consequence, these techniques can be used only on high-quality data with stable filters such as the bilateral filter. Otherwise, temporal incoherence may appear. Intuitively, the frame-by-frame methods under-exploit the available information because they process each frame separately although all the previous frames are known. We address this issue by using past data when processing the current frame.

Our paper is based on a theoretical study that explicitly focuses on data smoothing. Although it is of limited interest when applied to pixel values because it blurs the image, smoothing is the basis of powerful tools. Bilateral filtering has been shown to

D. Forsyth, P. Torr, and A. Zisserman (Eds.): ECCV 2008, Part II, LNCS 5303, pp. 460–473, 2008.

be a Gaussian convolution in a higher-dimensional space [21]. Mean-shift segmentation is known to be driven by a Gaussian kernel [22, 14] while anisotropic diffusion based on partial differential equations (PDE) can be seen as a refinement over isotropic diffusion [23]. These results motivate and structure our work: we first derive smoothing operators for video streams and then build upon them to obtain a set of new tools adapted to video streaming.

Overview and Contributions. First, we review related work and describe how isotropic diffusion and Gaussian convolution can be derived from gradient minimization in image space (Section 2). Then, we formally study video streams and characterize the specificity of the time axis as mixed boundary conditions. We build upon this study to extend two fundamental operators to video streams: isotropic diffusion and Gaussian convolution. We show that a first-order PDE in the temporal domain is the equivalent of the second-order PDE that defines isotropic diffusion in image space. We also demonstrate that an exponential decay over time corresponds to the Gaussian kernel classically used in image space. The theoretical contribution of our paper is the relationship between the temporal PDE and kernel, and their spatial counterpart (Section 3). With this result, we naturally extend a number of algorithms to video streams. We demonstrate anisotropic diffusion, bilateral filtering, and mean-shift segmentation (Section 4). These algorithms run in real time with low memory consumption and achieve temporal coherence on par with off-line methods (Section 5).

2 Background

2.1 Theoretical Scale-Space Studies

A few articles analyze spatio-temporal data from a scale-space standpoint [24, 25, 26, 27]. These papers define multiscale representation of images from various axioms such as non-enhancement of local extrema. Our theoretical results are related and we will discuss the links as they appear. But from an application perspective, these studies are concerned by the intrinsic structure of the data whereas we focus on typical computer vision objectives such as edge-preserving smoothing and image segmentation.

2.2 Video Applications

In this section, we review practical applications and categorize them according to how they deal with the temporal axis t.

Frame-by-Frame Techniques. These methods filter each frame as an isolated image. Formally, they use a Dirac peak centered on the current frame as temporal kernel. If the processing time is shorter than the delay between frames, then real-time video processing can be achieved as demonstrated by Winnemöller *et al.* [9] and Chen *et al.* [10] with bilateral filtering implemented on graphics hardware. The downside is that temporal coherence is not ensured because the filter has no knowledge of the adjacent frames. Only data with limited noise can be processed this way.

Spatio-Temporal Techniques. To ensure temporal coherence, spatio-temporal methods filter along the time axis in addition to the x and y dimensions. *Omniscient* approaches process a given frame assuming past and future data to be known whereas *causal* techniques rely only on past data.

Omniscient methods process a frame using both past and future data. One approach is to handle the t axis as another spatial dimension and process the data as a *video volume* [17] (also known as *video cube* [28]). These techniques involve a 3D Gaussian kernel [12,14] or a PDE [17] dealing similarly with space (xy) and time (t). Another option is to differentiate space and time. This strategy is often used for denoising [2,3,20,16] to favor temporal filtering over spatial smoothing in order to preserve image sharpness. All omniscient techniques are inherently limited to off-line processing since the last frame must be known before the algorithm starts. Moreover, with long sequences, the amount of data to handle can be arbitrarily large, requiring specific memory management techniques such as tiling and out-of-core processing.

Causal methods are the most related to ours since they filter a given frame using only past data. Kim and Woods [29], and Patti *et al.* [30] use Kalman filtering to aggregate data over time. But, because of the ordering imposed by the Kalman filter, pixels have an asymmetric spatial neighborhood which can incur visible defects. Bennett and McMillan [31] use an exponential decay over time to display motion trails in time-lapse videos. For bilateral filtering, Chen *et al.* [10] also apply an exponential decay to remove artifacts due to their sampling strategy.

A contribution of our paper is to motivate the temporal exponential decay in a general context and to apply it to several applications. In comparison, previous work [31, 10] introduces it as a heuristic for specific cases and limit its use to a single task.

2.3 Background on Image Processing

Smoothing an image I aims for reducing its variations. This can be expressed as reducing the norm of the gradients of I, that is, as minimizing the integral $\int (I_x^2 + I_y^2)$ where $I_x = \partial I / \partial x$. Isotropic diffusion and Gaussian convolution are two classical and equivalent ways to achieve this minimization.

Isotropic Diffusion. One option to minimize $\int (I_x^2 + I_y^2)$ is to apply an iterative descent scheme. That is, I evolves such that the gradients are progressively reduced. The "evolution time" is represented by a variable e. Intuitively, each iteration transforms $I(e)$ into $I(e + \delta e)$ which gradients are smaller. This process can be formalized by considering infinitesimal steps δe. The image evolution is then defined by the derivative $I_e = \partial I / \partial e$. Assuming this quantity known, a possible implementation of the filter is iterating $I(e + \delta e) = I(e) + \delta e \times I_e$. The rest of this section explains how to use the Euler-Lagrange formula to define I_e.

For an energy of the form $\int \Phi(\alpha, f, f_\alpha)\, \mathrm{d}\alpha$ where α is a scalar, f a function of α, and Φ a function of α, f, and $f_\alpha = \partial f / \partial \alpha$, the following partial derivative equation (PDE) is satisfied at stationary points: $\frac{\partial \Phi}{\partial f} - \frac{\partial}{\partial \alpha}\left(\frac{\partial \Phi}{\partial f_\alpha}\right) = 0$. Using this property, an evolution scheme is defined to progressively transforms f to minimize $\int \Phi$. The variable e denotes the "evolution time". At $e = 0$, f is not transformed, and f evolves as e increases. The evolution is defined by the derivative of f with respect to e:

$$f_e = \frac{\partial}{\partial \alpha}\left(\frac{\partial \Phi}{\partial f_\alpha}\right) - \frac{\partial \Phi}{\partial f} \tag{1}$$

For multiple independent variables α_1, α_2..., the equation is extended by adding corresponding second-order terms.

The above Euler-Lagrange formula is applied with $\Phi(..., f_{\alpha_1}, f_{\alpha_2}) = f_{\alpha_1}^2 + f_{\alpha_2}^2$ and $f(\alpha_1, \alpha_2) \equiv I(x, y)$. Only the second-order terms are non-zero, leading to isotropic diffusion (*a.k.a.* the heat equation): $I_e = I_{xx} + I_{yy}$.

Gaussian Convolution. Using $G(\sigma) = (2\pi\sigma^2)^{-1} \exp\left(-(x^2 + y^2)/2\sigma^2\right)$ for the Gaussian kernel and $\otimes$ for the convolution operator, $G(\sqrt{2e}) \otimes I$ is known to be the only solution of the heat equation [23, 32]. As a consequence, Gaussian convolution is equivalent to isotropic diffusion and also minimizes the gradient norms.

Summary. The following equation summarizes the relationship between gradient minimization, Gaussian convolution, and the heat equation.

$$\text{minimize } \textstyle\int \left(I_x^2 + I_y^2\right) \tag{2a}$$

$$G(\sigma) \otimes I \qquad \text{with: } \sigma = \sqrt{2e} \tag{2b}$$

$$I_e = I_{xx} + I_{yy} \tag{2c}$$

Equations 2b and 2c are equivalent and are derived from 2a. In practice, a finite value of σ and e is selected in order not to produce a constant (*i.e.*, infinitely smooth) image, thereby not fully minimizing (2a). One of our contributions is to derive a similar relationship for video streams.

3 Isotropic Diffusion and Gaussian Convolution for Video Streams

Gaussian convolution and isotropic diffusion are classical image-space operators upon which a number of powerful tools are built. For instance, anisotropic diffusion is an extension of isotropic diffusion and the bilateral filter has been shown to be a Gaussian convolution in the intensity-space domain [21]. In this section, we derive equivalent operators for video streams. First, we discuss the specificity of the time axis and then derive new operators that minimize the video-stream gradients.

3.1 Time Axis in Video Streams

We consider a frame of a video stream V at time t_0. All the previous frames have already been processed, that is, we know $V(t)$ for all $t < t_0$. In the video-streaming context, data are displayed or broadcast immediately after being processed and thus cannot be modified. We use the notation $\overline{V}(t)$ for these processed, fixed data at $t < t_0$. Conversely, future frames at $t > t_0$ are unknown and cannot be accessed. These properties make the time dimension fundamentally different from the x and y axes. For instance, when processing a pixel $V(x)$, both $V(x - \epsilon)$ and $V(x + \epsilon)$ can be accessed and modified.

Mixed Boundary Conditions. When processing the current data $V(t)$, we treat the past data $\overline{V}(t_0 - \epsilon)$ with $\epsilon > 0$ as hard constraints, that is, we have Dirichlet boundary conditions on the negative side. On the positive side, we cannot compute any derivative since $V(t_0 + \epsilon)$ is unknown. It means that right derivatives will never appear, or equivalently, be set to 0. Thus, we have Neumann boundary conditions on the positive side.

Discrete Modeling. A continuous model in which t is continuous variable and V a smooth variable is not a suitable approach to cope with the streaming setting. By continuity of V, we could write $V(t) = \lim_{\epsilon \to 0} \overline{V}(t - \epsilon)$ that would enforce the current frame to be always equal to the previous frame which is already known. In a continuous setup, we would always process data on the domain boundary where constraints are expressed. Therefore, although continuous modeling can be useful for image-space analysis [23], we prefer a discrete approach that lets us work at a distance from the domain boundaries. Such a choice has also been done in scale-space studies [26, 27].

3.2 Minimizing the Gradients

Gaussian convolution (Eq. 2b) and isotropic diffusion (Eq. 2c) cannot be applied directly to video streams because they do not take the temporal structure of video streams into account. For instance, we cannot use a Gaussian kernel along the time axis because it requires the use of future frames which are not available. Nevertheless, we can still formulate our goal as minimizing variations, *i.e.* gradient norms. We seek a scheme that minimizes $\int \left(V_x^2 + V_y^2 + cV_t^2\right)$ where c is a constant that weights the temporal metric versus the spatial metric. The x and y components are not affected by the video-stream structure, *i.e.* Equation 2 still holds. In the following discussion, we omit these terms for clarity and focus on the temporal term.

As discussed in the previous section, we use a discrete formulation for the time dimension, *i.e.* we minimize $\sum c\,[V(t) - \overline{V}(t-1)]^2$ where $t \in \mathbb{Z}$ is an integer indexing the frames and $V(t) - \overline{V}(t-1)$ is a backward difference, the discrete equivalent to a left derivative with the major difference that it involves the data V and the boundary constraint $\overline{V}$. The forward difference cannot be computed since the frame $t+1$ is unknown at time t. Recall also that at a given time t, the past data $\overline{V}(t-1)$ are fixed and that only the current frame $V(t)$ is processed.

Discrete Euler-Lagrange. With $\alpha \in \mathbb{Z}$ an integer variable, f a function of the variable α, $\Delta_\alpha f(\alpha) = f(\alpha+1) - f(\alpha)$ the discrete derivative of f, and Φ a function of α, f, and $\Delta_\alpha f$, the Euler-Lagrange formula becomes:

$$f_e = \Delta_\alpha\left(\frac{\partial \Phi}{\partial(\Delta_\alpha f)}\right) - \frac{\partial \Phi}{\partial f} \tag{3}$$

This discrete formula is known to be equivalent to the continuous one (Eq. 1). We refer to the work of Guo *et al.* [33] for a detailed formal proof.

Minimizing Temporal Variations. The temporal term $V(t) - \overline{V}(t-1)$ is not a standard discrete derivative because it involves $\overline{V}$ which is a boundary condition, only $V(t)$ is unknown and can vary. We use the notation $\overline{\Delta}_t V(t-1) = V(t) - \overline{V}(t-1)$. As a consequence, $\sum c\,[\overline{\Delta}_t V(t-1)]^2$ is a zeroth-order term, and the Euler-Lagrange formula (Eq. 3) results in a first-order scheme $V_e = -c\,\overline{\Delta}_t V$.

Diffusion in Video Streams. With the spatial terms and using Δ^2 for the discrete second derivative, we obtain the equivalent of the heat equation (Eq. 2c) for video streams:

$$V_e = \Delta_x^2 V + \Delta_y^2 V - c\overline{\Delta}_t V \tag{4}$$

This formula defines isotropic diffusion for video streams. The structure of the t axis induces a major change in the diffusion process with a first-order term instead of the second-order term classically found along spatial dimensions.

Intuition The temporal term acts as an attachment term. It adds to current values a portion of the difference with the previous data. The more we apply the diffusion equation, the more it moves the current data closer the values of the previous frame.

3.3 Integrating the Diffusion Equation

To obtain the equivalent of the Gaussian kernel, we integrate the diffusion equation (4) against the evolution time e. For clarity, we use the notation $V(t,e)$ from now on. For instance, $V(t,0)$ is the input frame at time t.

Spatial Dimensions. The spatial part of Equation 4 is similar to the image-space heat equation (Eq. 2c). As a consequence, the spatial component of the video-stream kernel is a Gaussian $G(\sqrt{2e})$ similarly to the image-space case.

Temporal Dimension. We focus on the temporal part: $V_e = -c\,[V(t,e) - \overline{V}(t-1,e)]$ (Eq. 4). It is a first-order differential equation with V as unknown since $\overline{V}(t-1,e)$ is constant. By imposing the value at $e=0$, the solution is:

$$V(t,e) = \overline{V}(t-1,e) + \exp(-c\,e)[V(t,0) - \overline{V}(t-1,e)] \tag{5}$$

With $q = 1 - \exp(-c\,e)$, we rewrite Equation 5 to express the output at t as a function of the input at t and the output at $t-1$:

$$V(t,e) = (1-q)\,V(t,0) + q\,\overline{V}(t-1,e) \tag{6}$$

We recursively apply Equation 6 at time $t-s$ for $s>0$ to remove the dependency on the past output frames. Since $q<1$ when $e>0$, we have $\lim_{s\to+\infty} q^s = 0$ and:

$$V(t,e) = (1-q)\sum_{s\geq 0} q^s\,V(t-s,0) \tag{7}$$

Equation 7 corresponds to a convolution between the input frames $V(\cdot,0)$ and a truncated exponential decay D:

$$V(t,e) = \big[D \otimes V(\cdot,0)\big](t) \tag{8a}$$

$$\text{with} \quad D : t \mapsto k\,\exp(-t/\lambda)\,H(t) \tag{8b}$$

where $k = 1 - q = \exp(-c\,e)$ is a normalization constant, H the Heaviside function ($H(t) = 1$ if $t \geq 0$ and 0 otherwise), and λ obtained by comparing two successive frames: $\lambda = -1/\log\big(1 - \exp(-c\,e)\big)$.

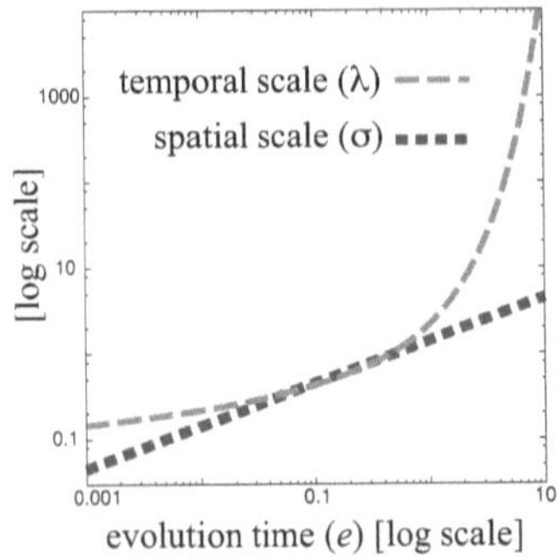

Fig. 1. Spatial and temporal scales as functions of the evolution time e for $c = 1$.

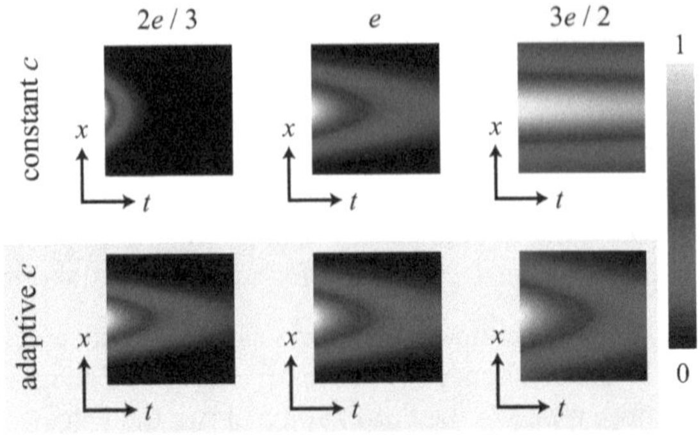

Fig. 2. Variation of the kernel shape with the evolution time e. Without adjustment, the temporal scaling is more important than the spatial one (top row). We can enforce a uniform scaling by adapting the metric parameter c (bottom row).

Scale. The σ and λ parameters respectively control the spatial and temporal scales of the kernel. Their behavior with the respect to the evolution time e is different as illustrated in Figure 1. When using the diffusion operator (Eq. 4), varying e does not uniformly scale the spatio-temporal kernel: the t dimension is more altered than the xy axes (Fig. 2-top). Nonetheless, the metric constant c can be adjusted to obtain a uniform scaling. For instance, one can define the scale σ and the aspect ratio $\gamma = \sigma/\lambda$. From Equation 2b, we get $e = \sigma^2/2$, and from the definition of λ: $c = -2\log(1 - \exp(\sigma/\gamma))/\sigma^2$. Using these settings, one can uniformly scale the kernel (Fig. 2-bottom).

Practical Recursive Algorithm. The spatio-temporal convolution can be made efficient since it is separable. Several techniques such as Deriche's recursive scheme [34] exist for the spatial Gaussian part. For the temporal part, we use the recursive formula (Eq. 6). It performs an exact computation based only on the current and last frames. This aspect makes our approach practical since only the current frame $V(t)$ and the previous one $\overline{V}(t-1)$ are needed at a given time. As a consequence, our algorithm has low memory requirements and the data always fit in memory independently of the video length. Figure 3 provides the pseudo-code for the convolution. In comparison, omni-

FASTSPATIOTEMPORALCONVOLUTION
· input: $V(t,0)$, $\overline{V}(t-1,e)$, e, c
· internal variables: q, $\tilde{V}$
· output: $V(t,e)$

1. Decay computation: $q = 1 - \exp(-ce)$
2. Space convolution: $\tilde{V} = G(\sqrt{2e}) \otimes V(t,0)$
3. Time recursion: $V(t,e) = (1-q)\tilde{V} + q\overline{V}(t-1,e)$

Fig. 3. Pseudo-code of the spatio-temporal convolution

scient methods are either limited to short sequences on the order of a few seconds or resort to using complex memory management strategies such as tiling. The differential formulation (Eq. 4) enjoys the same low-memory benefits.

3.4 Summary and Discussion

We have shown that gradients of video streams are minimized by a PDE that extends the classical second-order heat equation with a first-order temporal term. After integration, this PDE is equivalent to a spatio-temporal kernel made of a spatial Gaussian and a temporal exponential decay. The difference between the spatial dimensions and the time axis stem from the previously processed frames and the missing future data which impose boundary constraints onto the smoothing process. The following relationship and Figure 4 summarize this result.

$$\text{minimize} \sum \left[(\Delta_x V)^2 + (\Delta_y V)^2 + (\overline{\Delta}_t V)^2 \right] \tag{9a}$$

$$G(\sigma) D(\lambda) \otimes V \tag{9b}$$

$$V_e = \Delta_x^2 V + \Delta_y^2 V - c\overline{\Delta}_t V \tag{9c}$$

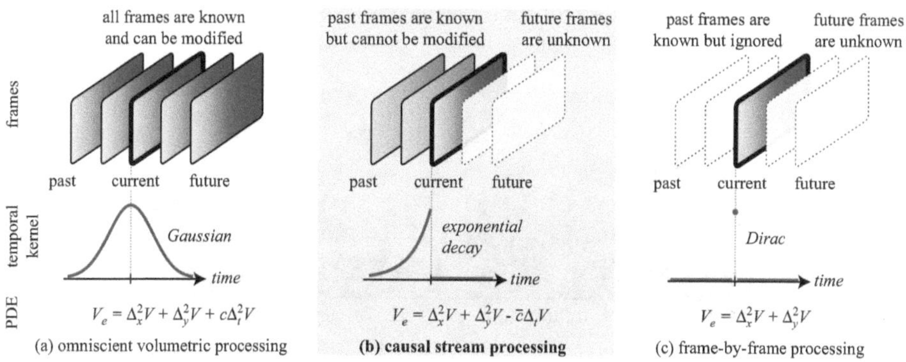

Fig. 4. Our approach (b) uses past data to ensure temporal coherence unlike frame-by-frame techniques (c) that rely only the quality of the input and the stability of the filter to achieve coherence. Compared to omniscient techniques (a), we do not assume that the future frames are known, thereby enabling on-line processing.

Link to Existing Work. The spatio-temporal PDE and kernel has been described by Lindeberg using scale-space axioms [26, 27]. Our contribution is to derive them from a simple smoothing problem, *i.e.* gradient minimization. The other major aspect of our work is to extend this result to edge-preserving smoothing and image segmentation as described in the following section.

4 Applications

Using the previously defined PDE and kernel directly on the pixel values is of limited interested because it blurs the video content. For instance, a naive approach would be to apply temporal smoothing as a post-process to an edge-preserving filter to enforce temporal coherence. This produces blurry results as shown in the companion video. Nonetheless, smoothing can be used to build several useful tools which demonstrate the value of our theoretical study. Videos are provided in the supplemental material.

4.1 Bilateral Filtering

The bilateral filter [35, 36, 37] is an edge-preserving filter that has proven to be an effective tool in computational photography [38]. For an image I, the output at a pixel $\mathbf{p}$ is the normalized average of the adjacent pixels $\mathbf{q}$ weighted by two Gaussian functions, $G(\sigma_s, \cdot)$ and $G(\sigma_i, \cdot)$, accounting for the spatial and intensity distances respectively:

$$\frac{1}{W} \sum_{\mathbf{q}} G(\sigma_s, \|\mathbf{p} - \mathbf{q}\|)\, G(\sigma_i, |I(\mathbf{p}) - I(\mathbf{q})|)\, I(\mathbf{q})$$

Our adaption to video streams is based on the result by Paris and Durand who expressed this filter as a Gaussian convolution in the space-intensity domain [21]. We build upon the result of the previous section and adapt the bilateral filter to video streams by convolving with our spatio-temporal kernel instead of the original Gaussian. Figure 5 shows a sample result produced by our algorithm.

(a) sample noisy input frame (b) bilateral filter output

Fig. 5. Bilateral filter result. The companion video demonstrates the stability of our approach in presence of noise.

4.2 Mean-Shift Segmentation

Mean shift is a method to segment images made popular by Comaniciu and Meer [22]. Each pixel is associated to a feature point and the produced segmentation can be seen as the modes of the density of feature points estimated with a Gaussian kernel [22, 14]. We extend this approach to video streams by estimating the feature-point density with our causal spatio-temporal kernel. With this approach, modes span space and time and temporal coherence is naturally achieved (Fig. 6).

4.3 PDE-Based Anisotropic Diffusion

A large body of work exists on smoothing images with anisotropic diffusion using PDEs [23]. From a high-level standpoint, one can interpret these filters as variants of

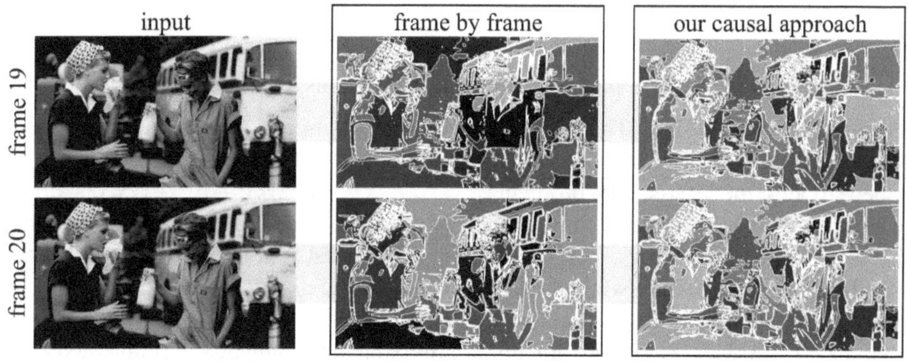

Fig. 6. Mean-shift segmentation. The clusters are color-coded. The frame-by-frame approach produces new clusters at each frame (indicated by new colors). A heuristic to "link" clusters across frames would not be satisfying since boundaries are not coherent. In comparison, our algorithm achieves a coherent segmentation. This is better seen in the companion video.

the heat equation (Eq. 2c) that steer the diffusion to avoid blurring the main image contours. We demonstrate our approach on the Perona-Malik filter [39]. In image space, this filter is similar to the heat equation with the gradients weighted by a stopping function $g(\cdot)$: $I_e = \mathrm{div}(g(\|\nabla I\|)\nabla I)$ where *div* is the divergence $\Delta_x + \Delta_y$. Based on our result (Eq. 9c), we propose the following equivalent for video streams:

$$V_e = \mathrm{div}(g(\|\nabla V\|)\nabla V) - c\,g(\|\nabla V\|)\overline{\Delta}_t V$$

The divergence term is equivalent to the image-space case. The last term is specific to video streaming. A sample output is shown in Figure 7.

(a) sample input frame (b) output

Fig. 7. Close-up on anisotropic diffusion results. We applied a strong smoothing effect to make it more visible. The full sequence is provided in supplemental material.

4.4 Possible Extensions

In the following sections, we discuss uses of our approach that requires studies beyond the scope of this paper. We plan to pursue these directions as future work.

Poisson Reconstruction. This technique is commonly used to reconstruct in the least-square sense an image from an approximated gradient field (u, v, w) [40, 41, 42]. For video streams, the least-square problem is: $\left[(\Delta_x V - u)^2 + (\Delta_y V - v)^2 + (\overline{\Delta}_t V - w)^2\right]$, which can be solved similarly to Equation 9a with the discrete Euler-Lagrange formula: $V_e = \Delta_x^2 V - \Delta_x u + \Delta_y^2 V - \Delta_y v - \overline{\Delta}_t V + w$. The first four terms correspond to the standard Poisson equation while the last two are specific to video streaming.

We experimented with this equation on panorama stitching [42], pasting [41], and Retinex [40]. At best, we found only minor improvements compared to frame by frame, while the lack of motion compensation induces defects at moving boundaries with Retinex. Unfortunately, optical-flow techniques have limitations that make them unsuitable for video streaming as discussed in the following paragraph.

Optical Flow. In general, we could benefit from optical flow to account for the scene motion. However, high-accuracy algorithms are computationally expensive and fast methods such as the one used in MediaPlayer have limited precision [43]. Thus, the current state of the art in motion estimation does not allow for stream processing yet. However, we believe that it is a promising research direction for the future. Real-time and accurate flow computation would have many applications. In our case, it would allow for steering the temporal component $\overline{\Delta}_t V$ according to the optical flow, thereby yielding results at boundaries as accurate as the ones in uniform areas.

5 Results

Setup. We have implemented the described applications in C++, compiled them with GCC 4.0.1 with optimization turned on, and run the tests on an Intel Xeon 3GHz. The Gaussian part of the kernel is implemented using a downsampled 5-tap approximation. The bilateral filter and the mean-shift segmentation are set with $\sigma_\mathrm{s} = 32$ and $\sigma_\mathrm{i} = 2.5$ considering that the intensity range spans $[0..100]$. Anisotropic diffusion uses axis-projected gradients as suggested by Perona and Malik [39]. Frame rates reported in Table 1 are averaged over 100 frames, not including disk operations. Memory consumption excludes system libraries. Videos are provided in the supplemental material.

Performance. Our approach yields memory efficient algorithms that achieves real-time or near real-time performance on gray-level videos at DVD resolution while maintaining temporal coherence (Table 1). There is no limit on the video length, the reported

Table 1. Frame rate and memory consumption. Input stream is 640×360 in gray levels.

	frame-by-frame	our approach
bilateral filter	29.1Hz (3.8MB)	28.4Hz (3.9MB)
mean shift	59.5Hz (1.1MB)	52.9Hz (1.3MB)
Perona-Malik	14.6Hz (8.8MB)	11.7Hz (9.7MB)

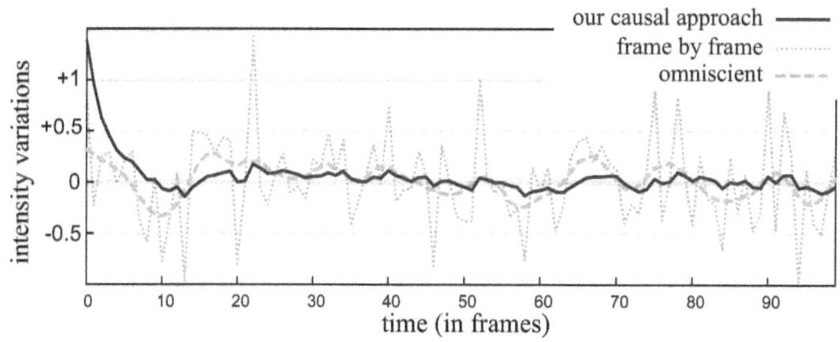

Fig. 8. Temporal stability of the Perona-Malik filter. Our approach exhibits a transient start but rapidly achieves more stable results than frame-by-frame and even omniscient methods because it relies on the past frames which have already been smoothed. We added noise up to $\pm 5\%$ of the intensity range. All filters have been applied with $e = 2$.

performance and memory usage can be sustained for an arbitrary duration. As future work, we plan to leverage graphics hardware to process color data.

Stability. Temporal stability is a key issue, especially in uniform areas where incoherence yields undesirable flickering. We evaluated the stability of our approach compared to frame-by-frame and omniscient methods using the Perona-Malik filter. We processed 100 uniform frames with uncorrelated noise added and kept the evolution time e fixed to ensure a fair comparison. Figure 8 shows the intensity plot of the center pixel. As expected, frame-by-frame results lack coherence. Our causal approach exhibits a transient start because there is no past data available. Then, it becomes slightly more stable than the omniscient output. This property stems from our use as hard constraints of the past data which are already smoothed whereas the omniscient process exploits more data but smoothes them at the same time, *i.e.* in the first iterations, past and future frames are still noisy and not as useful as the smoothed past frame in our case.

Discussion. Compared to omniscient algorithms which are inherently limited to off-line processing, our approach enables on-line tasks. Nonetheless, if on-line processing is not required, omniscient approaches offer more flexibility such as multi-pass analysis [12] since they work in a less constrained scenario. Compared to frame-by-frame results, the computation and memory overhead is limited (Table 1). For bilateral filtering and anisotropic diffusion, our tests showed that our approach yields more stable results on noisy images. Chen *et al.* [10] used this property to further speed up their hardware implementation of the bilateral filter. Our contribution is to formally motivate the technique that Chen described as a heuristic. For mean-shift segmentation, unlike other approaches, our method can coherently process arbitrary long videos.

6 Conclusions

We described a formal approach to processing video streams. We did not assume any knowledge about the future data and relied on the past frames to ensure temporal coherence. This translates into boundary constraints on the time axis which we used to derive smoothing operators dedicated to video streaming. Based on this result, we revisited bilateral filtering, anisotropic diffusion, and mean-shift segmentation, and demonstrated that they can be applied on video streams in real time. Our tests show that our causal approach is suitable for applications such as coherent denoising for surveillance and real-time segmentation for on-line analysis. In particular, it should be preferred to frame-by-frame methods for noisy data and segmentation tasks.

Acknowledgement. The author thanks his Adobe and MIT colleagues for their feedback.

References

1. Zheng, Y., Lin, S., Kang, S.B.: Single-image vignetting correction. In: Proc. of the conf. on Computer Vision and Pattern Recognition, vol. 1, pp. 461–468. IEEE, Los Alamitos (2006)
2. Bennett, E.P., McMillan, L.: Video enhancement using per-pixel virtual exposures. ACM Transactions on Graphics 24, 845–852 (2005); Proc. of the ACM SIGGRAPH conf.
3. Bennett, E.P., Mason, J.L., McMillan, L.: Multispectral bilateral video fusion. IEEE Transactions on Image Processing 16, 1185–1194 (2007)
4. Liu, C., Freeman, W.T., Szeliski, R., Kang, S.: Noise estimation from a single image. In: Proc. of the Computer Vision and Pattern Recognition Conf. IEEE, Los Alamitos (2006)
5. Li, Y., Sun, J., Shum, H.Y.: Video object cut and paste. ACM Transactions on Graphics 24, 595–600 (2005); Proc. of the ACM SIGGRAPH conf.
6. Wang, J., Bhat, P., Colburn, R.A., Agrawala, M., Cohen, M.F.: Video cutout. ACM Transactions on Graphics 24 (2005); Proc. of the ACM SIGGRAPH conf.
7. Bae, S., Paris, S., Durand, F.: Two-scale tone management for photographic look. ACM Transactions on Graphics 25, 637–645 (2006); Proc. of the ACM SIGGRAPH conf.
8. DeCarlo, D., Santella, A.: Stylization and abstraction of photographs. In: Proc. of the ACM SIGGRAPH conf. (2002)
9. Winnemöller, H., Olsen, S.C., Gooch, B.: Real-time video abstraction. ACM Transactions on Graphics 25, 1221–1226 (2006); Proc. of the ACM SIGGRAPH conf.
10. Chen, J., Paris, S., Durand, F.: Real-time edge-aware image processing with the bilateral grid. ACM Transactions on Graphics 26 (2007); Proc. of the ACM SIGGRAPH conf.
11. Wang, J., Xu, Y., Shum, H.Y., Cohen, M.F.: Video tooning. ACM Transactions on Graphics 23, 294–302 (2004); Proc. of the ACM SIGGRAPH conf.
12. Wang, J., Thiesson, B., Xu, Y., Cohen, M.F.: Image and video segmentation by anisotropic mean shift. In: Proc. of the European Conf. on Computer Vision (2004)
13. Bousseau, A., Neyret, F., Thollot, J., Salesin, D.: Video watercolorization using bidirectional texture advection. ACM Transactions on Graphics 26 (2007); Proc. of the ACM SIGGRAPH conf.
14. Paris, S., Durand, F.: A topological approach to hierarchical segmentation using mean shift. In: Proc. of the IEEE conf. on Computer Vision and Pattern Recognition (2007)
15. Buades, A., Coll, B., Morel, J.M.: A non local algorithm for image denoising. In: Proc. of the conf. on Computer Vision and Pattern Recognition (2005)
16. Chen, J., Tang, C.K.: Spatio-temporal Markov random field for video denoising. In: Proc. of the IEEE conf. on Computer Vision and Pattern Recognition (2007)
17. Drori, I., Leyvand, T., Fleishman, S., Cohen-Or, D., Yeshurun, H.: Video operations in the gradient domain. Technical report, Tel-Aviv University (2004)
18. Chuang, Y.Y., Agarwala, A., Curless, B., Salesin, D., Szeliski, R.: Video matting of complex scenes. ACM Transactions on Graphics 21 (2002); Proc. of the ACM SIGGRAPH conf.

19. DeMenthon, D.: Spatio-temporal segmentation of video by hierarchical mean shift analysis. In: Proc. of the Statistical Methods in Video Processing Workshop (2002)
20. Zitnick, C.L., Jojic, N., Kang, S.B.: Consistent segmentation for optical flow estimation. In: Proc. of the International Conf. on Computer Vision (2005)
21. Paris, S., Durand, F.: A fast approximation of the bilateral filter using a signal processing approach. In: Proc. of the European Conf. on Computer Vision (2006)
22. Comaniciu, D., Meer, P.: A robust approach toward feature space analysis. IEEE Transactions on Pattern Analysis Machine Intelligence 24, 603–619 (2002)
23. Aubert, G., Kornprobst, P.: Mathematical problems in image processing: Partial Differential Equations and the Calculus of Variations. Applied Mathematical Sciences, vol. 147. Springer, Heidelberg (2002)
24. Koenderink, J.J.: Scale-time. Biological Cybernetics 58 (1988)
25. ter Haar Romeny, B.M., Florack, L.M.J., Nielsen, M.: Scale-time kernels and models. In: Proc. of the conf. on Scale-Space and Morphology in Computer Vision (2001)
26. Lindeberg, T., Fagerström, D.: Scale-space with causal time direction. In: Proc. of European Conf. on Computer Vision (1996)
27. Lindeberg, T.: Linear spatio-temproal scale-space. In: Proc. of the International Conf. on Scale-Space Teory in Computer Vision (1997)
28. Klein, A., Sloan, P.P., Finkelstein, A., Cohen, M.F.: Stylized video cubes. In: Proc. of the ACM SIGGRAPH Symposium on Computer Animation (2002)
29. Kim, J., Woods, J.W.: Spatiotemporal adaptive 3-D Kalman filter for video. IEEE Transactions on Image Processing 6 (1997)
30. Patti, A.J., Tekalp, A.M., Sezan, M.I.: A new motion-compensated reduced-order model Kalman filter for space-varying restoration of progressive and interlaced video. IEEE Transactions on Image Processing 7 (1998)
31. Bennett, E.P., McMillan, L.: Computational time-lapse video. ACM Transactions on Graphics 26 (2007); Proc. of the ACM SIGGRAPH conf.
32. Koenderink, J.J.: The structure of images. Biological Cybernetics 50 (1984)
33. Guo, H.Y., Li, Y.Q., Wu, K.: Difference discrete variational principle, Euler-Lagrange cohomology and symplectic, multisymplectic structures. ArXiv Math. Physics e-prints (2001)
34. Deriche, R.: Recursively implementating the Gaussian and its derivatives. Technical Report RR-1893, INRIA (1993)
35. Aurich, V., Weule, J.: Non-linear gaussian filters performing edge preserving diffusion. In: Proc. of the DAGM Symposium (1995)
36. Smith, S.M., Brady, J.M.: SUSAN – a new approach to low level image processing. International Journal of Computer Vision 23, 45–78 (1997)
37. Tomasi, C., Manduchi, R.: Bilateral filtering for gray and color images. In: Proc. of the International Conf. on Computer Vision, pp. 839–846. IEEE, Los Alamitos (1998)
38. Paris, S., Kornprobst, P., Tumblin, J., Durand, F.: A gentle introduction to bilateral filtering and its applications. In: Course at the ACM SIGGRAPH conf. (2007)
39. Perona, P., Malik, J.: Scale-space and edge detection using anisotropic diffusion. IEEE Transactions Pattern Analysis Machine Intelligence 12, 629–639 (1990)
40. Horn, B.K.P.: Determining lightness from an image. Computer Graphics and Image Processing 3 (1974)
41. Pérez, P., Gangnet, M., Blake, A.: Poisson image editing. ACM Transactions on Graphics 22 (2003); Proc. of the ACM SIGGRAPH conf.
42. Levin, A., Zomet, A., Peleg, S., Weiss, Y.: Seamless image stitching in the gradient domain. In: Proc. of the European Conf. on Computer Vision (2006)
43. Baker, S., Scharstein, D., Lewis, J., Roth, S., Black, M., Szeliski, R.: A database and evaluation methodology for optical flow. In: Proc. of the International Conf. on Computer Vision (2007)

Deformed Lattice Discovery Via Efficient Mean-Shift Belief Propagation

Minwoo Park[1], Robert T. Collins[1], and Yanxi Liu[1,2]

[1] Department of Computer Science and Engineering
[2] Department of Electrical Engineering
The Pennsylvania State University, University Park, PA 16802
{mipark,rcollins,yanxi}@cse.psu.edu

Abstract. We introduce a novel framework for automatic detection of repeated patterns in real images. The novelty of our work is to formulate the extraction of an underlying deformed lattice as a spatial, multi-target tracking problem using a new and efficient Mean-Shift Belief Propagation (MSBP) method. Compared to existing work, our approach has multiple advantages, including: 1) incorporating higher order constraints early-on to propose highly plausible lattice points; 2) growing a lattice in multiple directions simultaneously instead of one at a time sequentially; and 3) achieving more efficient and more accurate performance than state-of-the-art algorithms. These advantages are demonstrated by quantitative experimental results on a diverse set of real world photos.

1 Introduction

Repeated patterns, presented as imagery or otherwise, provide fundamental cues for both human and machine perception [1,2]. At one extreme, mathematicians have generalized these concepts into perfectly periodic patterns generated by a single fundamental region (lattice unit) under a finite number of crystallographic groups [3,4]. In the computer vision and computer graphics communities, on the other hand, these patterns are usually regarded as *textures* with a random nature, composed of deformed versions of one or more basic *texture elements* [5,6,7,8].

Ample evidence can be found that repeated textures are not merely random collections of texture elements (Figures 4 and 5). Instead, these textures often exhibit geometric, topological and statistical regularities [7], especially those texture patterns that originate from human designs or from nature and biology. Using wallpaper texture patterns [9] as an example, all texture elements are related to each other by a pair of linearly independent vectors t_1, t_2, or composition of them, that are the shortest basis vectors [3] forming the boundary of the respective *lattice units*. We fully acknowledge the importance of a robust lattice detection tool for automatic discovery and quantification of texture regularity in real images.

Instead of treating texture elements as isolated individuals, the underlying topological lattice structure of a near-regular texture (NRT) was first introduced by Liu et al for texture analysis and manipulation [10,11,7]. Subsequently, Hays et al [12] developed the first deformed lattice detection algorithm for real images without pre-segmentation; and Lin and Liu [13,14] developed the first deformed lattice tracking algorithm for dynamic NRTs.

D. Forsyth, P. Torr, and A. Zisserman (Eds.): ECCV 2008, Part II, LNCS 5303, pp. 474–485, 2008.

The idea behind [12] is simply to look for the t_1, t_2 neighbors of a randomly selected interest point in the image. If a sufficient number of such points look like their respective t_1, t_2 neighbors (lower order similarity) and also share their t_1, t_2 neighbors' directions/orientations (higher order correspondences) with other interest points in the image, those points and their neighboring relations are confirmed to be part of the final lattice. With the found correspondence, the slightly deformed lattice can be straightened out and a new round of lattice discovery begins, so the extracted lattice grows bigger and bigger. Formulating the lattice detection problem as a higher order correspondence problem adds computational robustness against geometric distortions and photometric artifacts in real images. However, although Hays et al [12] produces impressive results, there are several serious drawbacks preventing its wider applicability. First, local correlation-based peak finding is used as a last resort for finding points of interest, which is both time consuming and sensitive to noise, occlusion and transform discontinuity in the image (Figures 4 and 5). Second, the method is based on finding the eigenvalues of a $n^2 \times n^2$ sparse matrix (n is the number of potential texture elements), which is cumbersome computationally. Third, the algorithm only examines one of the t_1 and t_2 vectors at a time, and is thus less robust against misleading repetitions and prone to wasting time on interest points that do not lead to legitimate t_1, t_2 neighbors. Our proposed method overcomes these weaknesses.

Our work is partially inspired by Lin and Liu [13,14], who treat lattice detection as a spatial tracking problem. Different from [13,14], we do not require an initial texton to be given (by the user) nor do we use affine template matching for the spiral, outward growth of the deformed lattice. Instead, we propose to formulate the detection of the underlying deformed lattice in an unsegmented image as a spatial, multi-target tracking problem, using a recently published, fast, Mean-shift Belief Propagation (MSBP) method [15]. The MSBP algorithm only examines the values of the belief surface within its local kernel window, thus there is no need to generate the entire belief surface, yielding great computational savings over non-parametric belief propagation and discrete belief propagation [15]. Inferencing on a Markov Random Field (MRF) for this type of texture (NRTs) is a natural and effective choice given the topological degree-4 graph structure of the underlying lattices of deformed wallpaper patterns [12,14]. In addition, there is a perfect conceptual match between the iterative nature of seeking the underlying lattice of a repeated pattern in an image and performing mean-shift on the implicit belief surface (marginal density) generated by the belief propagation algorithm on a graphical model.

Compared with [12], our proposed approach offers significant improvements in accuracy, robustness and efficiency for automatic lattice detection by: 1. incorporating higher order constraints early-on to propose highly plausible lattice points; 2. growing the lattice along multiple directions simultaneously instead of one at a time; 3. achieving a deterministic algorithm linearly dependent on the size of the lattice structure, instead of quadratic or higher order on the number of texels found, as in [12]. Quantified experimental results on an extensive set of diverse real images demonstrate these advantages (Figures 4 and 5).

2 Lattice Fitting Via Mean-Shift Belief Propagation

The main idea underlying lattice finding using belief propagation is that identifying the location of texture elements in a repeated pattern is made easier and more robust if multiple elements are searched for jointly, rather than one at a time. This is so because the location of each element is constrained by its neighboring elements, such that finding some of them provides the knowledge constraining where the others may be. In an extreme case, if you know where each of the four neighbors of a texture element is, you have a good idea about where the central texture element should be, even if it is occluded or otherwise hard to find. The key to leveraging this insight is to encode the topological structure of texture elements explicitly so that it can be used effectively to perform inference over the joint space of spatial constraints.

A Markov Random Field (MRF) specifies a factorization of the joint distribution of a set $\mathbf{X}$ of random variables. An MRF can be represented as an undirected graph $\mathbf{G} = (\mathbf{N}, \mathbf{E})$, where each node in $\mathbf{N}$ represents a random variable in set $\mathbf{X}$ and each edge in $\mathbf{E}$ represents a statistical dependency between random variables in $\mathbf{X}$. In the present context, the value of a random variable is the location of the center of a texture element, and the spatial dependency between variables represents how one element's position constrains the positions of other elements in the pattern.

Another piece of information to be represented for each texture element is the appearance similarity of the texture elements at given image locations. Since the lattice indicates a repeated pattern, the appearance of each of the texture elements can be described by a single reference appearance model. Determining the location of an element thus requires combining these two sources of information: the compatibility of the texture elements at various image locations with what we think a texture element should look like, and the compatibility of an element location with the spatial constraints provided by other elements.

Determining the location of one particular element given estimated positions of all other elements is infeasible if it requires brute force evaluation of the marginal distribution of that single random variable's value, since that will lead to $O(n \times n^{k-1})$ computation time where k is the number of nodes in the graph and n is the size of the hidden variable space (for simplicity we can think of it as $2D$ location in the current example). Thus determining where each of the elements is in the pattern requires $O(kn^k)$ computation time.

Fortunately, the joint probability over the pattern state x and image measurement z in an MRF can be factored as

$$p(x_1, ..., x_N, z_1, ..., z_N) = \prod_{(i,j)} \psi(x_i, x_j) \prod_s \phi(x_s, z_s) \tag{1}$$

where ψ and ϕ are functions specifying pairwise compatibility (spatial constraints between elements) and joint compatibility (appearance similarity of elements at given image locations), respectively. The belief propagation algorithm takes advantage of this factorization to perform inference on the graph efficiently. Cost of computation for estimating the state of all texture elements is reduced from

$O(kn^k)$ to $O(kn^2)$. However, if the hidden variable state space is large, BP can still be very expensive. BP also is not feasible for hidden variable spaces with continuous values.

We have developed an efficient and effective inference method called Mean Shift Belief Propagation (MSBP) [15] that works iteratively with local samples and weights. Since mean-shift is equivalent to finding a local mode within a Parzen window estimate of a density function, we can use mean-shift as a non-parametric mode-seeking mechanism operating on weighted samples generated within the belief propagation framework. Geometrically, the MSBP process can be visualized as performing mean-shift on the implicit belief surface (marginal density) generated by the belief propagation algorithm. Since MSBP only examines the values of the belief surface within its local kernel window, there is no need to generate the entire belief surface, yielding computational savings over non-parametric belief propagation and discrete belief propagation.

3 Our Approach

Our approach for deformed lattice detection is composed of three phases (Figure 1). Phase I has two steps: interest point generation and lattice-unit proposal. Phase II is lattice inference via spatial tracking using MSBP, followed by verification of the found lattice. Phase III is regularized thin-plate spline warping, where the deformed lattice is rectified into a regular lattice. Phase III is a transition phase, from which a new round of lattice inference starts again (Figure 1).

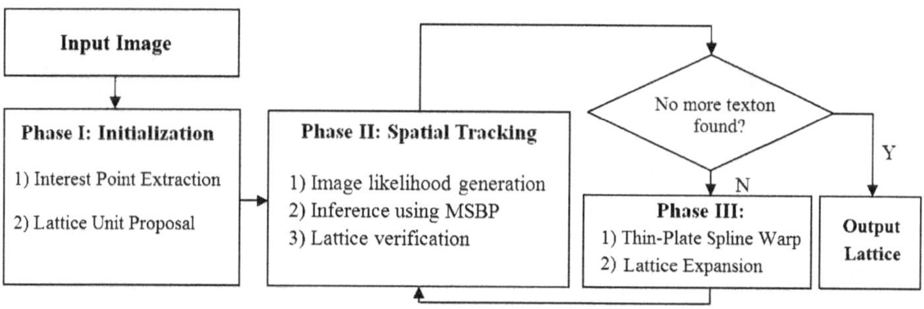

Fig. 1. Flowchart of our algorithm. There are three phases: initialization, spatial tracking and incremental warping (transition phase).

3.1 Phase I: Initialization

Interest point extraction. Our approach starts by extracting a set of interest points in the image. Any interest point detector can be used in this stage. The goal is to extract enough feature points to expose the repeated image substructures reliably without an overwhelming number of false positives for the subsequent lattice finder. For this initial implementation, we use KLT corner

Fig. 2. Our proposed algorithm proceeds from (a) to (e): (a) is input, (b) is the output of phase I, (c) and (d) are results from phase II and III, and (e) is the final result (refer also to Figure 1).

features [16]. Since KLT generates a sorted list based on corner strength, the top N features are passed to the next phase as candidate interest points. We set $N = 300$ for all images in our experiments.

Lattice-unit proposal. Our lattice-unit proposal step differs significantly from [12] in two ways. First, each lattice-unit found is composed of the current point under consideration and its *two* nearest matched neighbors, forming an L-shaped (t_1, t_2)-vector pair (Figure 3), as opposed to the previous practice of considering t_1 and t_2 sequentially [12]. Second, the final lattice-unit proposal is generated by a consensus vote of all potential t_1, t_2-vector pairs. This is equivalent to bringing higher-order constraints into the proposal stage so that unnecessary computation of infeasible lattice points is avoided early on, rather than waiting to prune hopeless lattice points at a later stage [12].

A "candidate neighbor" list L_c^i is kept for each interest point p_i, denoting good potential matches within its neighborhood. The goodness is measured by normalized cross correlation (NCC) between the candidate match and the reference intensity template centered at interest point p_i. Let the list of candidate neighbors of point p_i be L_c^i=$\{p_k | k = 1, \cdots, n, k \neq i\}$ and suppose points p_a and p_b are chosen from this list. Then we have $t_1 = p_a - p_i$ and $t_2 = p_b - p_i$ as a potential (t_1, t_2) candidate pair for the ith point p_i. We collect all the possible (t_1, t_2)-vector pairs from the lists, $L_c^i, i = 1, \cdots, N$ to form L_{tv}, a list of all candidate (t_1, t_2)-vector pairs. If there is regularity in the image, a set of spatially and appearance-wise consistent (t_1, t_2)-vector pairs should exist.

Fig. 3. The results of t_1, t_2-vector proposal : The green L-shape is the t_1, t_2-vector proposal, and the red ones are its supporting members (inlier votes). The images are cropped to emphasize the area of interest.

We measure the spatial consistency of two (t_1, t_2)-vector pairs, (t_1^i, t_2^i) and (t_1^j, t_2^j), using the normalized error term defined below:

$$E(t_1^i, t_2^i, t_1^j, t_2^j) = max(\left\|t_1^i - t_1^j\right\|_2 / \|t_1^i\|_2, \left\|t_2^i - t_2^j\right\|_2 / \|t_2^i\|_2) \qquad (2)$$

where $\|\|_2$ is L_2 vector norm.

We regard a (t_1, t_2)-vector pair as an inlier if E is smaller than ε. This ε is a tunable knob that can be used to gauge the algorithm's tolerance to geometric irregularity in the found lattice (larger values allow more irregularity). We set $\varepsilon = 0.2$ throughout our experiments. The candidate (t_1, t_2)-vector pair with the largest number of inliers is selected as the best t_1, t_2 proposal.

3.2 Phase II: Lattice Expansion - Spatial Tracking with MSBP

We treat discovery of the underlying lattice of unknown texture elements in an image as a multitarget tracking problem. Without knowing the target to start with, Phase I of our algorithm proposes a lattice unit as a potential target. Now we can state our multi-target tracking problem as: given a single lattice unit **u** and a proposed estimate of the (t_1, t_2)-vectors, a lattice of elements can be extracted by translating **u** spatially via $t1, t2$ in multiple directions to predict its potential locations and then "tracking" those new locations by combining the predictions with image data to produce refined location estimates. Since the lattice may be geometrically distorted in the image, it is dangerous to try to predict the whole lattice of elements at once. Instead, an initial seed lattice is predicted and refined, and then gradually grown outwards into a larger and larger lattice, while the image is progressively unwarped to "straighten out" geometric deformations discovered in the gradually growing lattice structure.

Given its efficiency, we have used the MSBP algorithm [15] as our inference engine for refining predicted texture element locations (Section 2). An initial lattice of size 3 by 3 is built from the (t_1, t_2) proposal generated in Phase I, and an initial texture element template T_0 centered at the origin of the proposed (t_1, t_2)-vectors with size $min(\|t_1\|_2, \|t_2\|_2)$ by $min(\|t_1\|_2, \|t_2\|_2)$ is used to generate an image likelihood map via NCC. This image likelihood map is then taken as a prior density function on the image location of texture elements, and the joint compatibility function (observation model) in the lattice MRF is given by

$$\phi(\mathbf{x}_{[i,j]}, z_{[i,j]}) = \exp(-\alpha(1 - z_{[i,j]})), z_{[i,j]} = NCC(T_0, I(\mathbf{x}_{[i,j]})) \quad (3)$$

where $\mathbf{x}_{[i,j]}$ is the 2D location of node $[i, j]$ at the ith row and j^{th} column in the lattice, $I(\mathbf{x}_{[i,j]})$ is an image patch centered at the location of node $[i, j]$ and T_0 is the initial texture element's appearance template. Equation (3) is of a form typical for data compatibility functions that measure likelihood by appearance similarity. Parameter α is a fixed constant that can be set empirically.

The second kind of function for the MRF is the pairwise compatibility function that specifies the spatial constraints between neighboring pairs of texture elements. In the context of lattice tracking, the pairwise compatibility function governs the geometric characteristics of (t_1, t_2)-vector pairs in the lattice. We define our pairwise compatibility function as

$$\begin{aligned}&\psi(\mathbf{x}_{[i,j]}, \mathbf{x}_{[i,j\pm1]})\\&= \exp(-\beta \times E(\overrightarrow{\mathbf{x}^{(it)}_{[i,j]}\mathbf{x}^{(it)}_{[i,j\pm1]}}, \overrightarrow{\mathbf{x}^{(it)}_{[i,j]}\mathbf{x}^{(0)}_{[i+1,j]}}, \overrightarrow{\mathbf{x}^{(0)}_{[i,j]}\mathbf{x}^{(0)}_{[i,j\pm1]}}, \overrightarrow{\mathbf{x}^{(0)}_{[i,j]}\mathbf{x}^{(0)}_{[i+1,j]}})^2)\end{aligned} \quad (4)$$

$$\begin{aligned}&\psi(\mathbf{x}_{[i,j]}, \mathbf{x}_{[i\pm1,j]})\\&= \exp(-\beta \times E(\overrightarrow{\mathbf{x}^{(it)}_{[i,j]}\mathbf{x}^{(it)}_{[i\pm1,j]}}, \overrightarrow{\mathbf{x}^{(it)}_{[i,j]}\mathbf{x}^{(0)}_{[i,j+1]}}, \overrightarrow{\mathbf{x}^{(0)}_{[i,j]}\mathbf{x}^{(0)}_{[i\pm1,j]}}, \overrightarrow{\mathbf{x}^{(0)}_{[i,j]}\mathbf{x}^{(0)}_{[i,j+1]}})^2)\end{aligned} \quad (5)$$

where E is given by equation (2) and measures the consistency of a hypothetical pair of lattice element vectors $(t_1^{(it)}, t_2^{(it)})$ at iteration it with the original proposed vectors $(t_1^{(0)}, t_2^{(0)})$. Equation (4) with subscripts $[-]$ and $[+]$ is used for left-right and right-left message passing respectively. Equation (5) with subscripts $[-]$ and $[+]$ is used for up-down and down-up message passing respectively. The β parameter is a fixed parameter that can be set empirically. Because the error term is normalized, a fixed β parameter can be used for all images regardless of spatial scale of the lattice elements. We used $\alpha = \beta = 5$ in all our experiments. Using the compatibility equations defined above, MSBP [15] is performed. The use of MSBP is critical for speeding up the inference process in real applications, as otherwise the inference process is very slow.

Once the optimization via MSBP converges for this intermediate stage of lattice growth, verification of the converged texture element positions is performed.

This is necessary because propagation of incorrect information to other nodes in the graph may corrupt the optimization process. Verification gives us a safety measure, telling us if we can trust the correspondences between the deformed image lattice and a hypothetical regular lattice, which will be used in Phase III. The general idea behind verification is that we would like to require a converged location to correspond to a significant local maximum or "peak" in the likelihood image. Rather than a hard-coded threshold, we use the region of dominance idea introduced by Liu. et al [10] to determine if an estimated texture element position can be trusted. If the current estimated location is a dominant peak within its neighboring region (that is, if it is a local maximum with a significantly high likelihood score, but is not located close to another local maximum with an even higher score) we select it as a peak location.

Possible mis-alignment at a certain iteration of our algorithm is acceptable because it can be fixed as the iterative fitting procedure repeats. After one iteration of Phase II and Phase III, the lattice structure is expanded from 3 by 3 to 5 by 5, and the growing process continues until each side of the lattice hits a boundary of the image or no more textons are found. A flow diagram of the overall algorithm and the results of subsequent iterations are shown in Figures 1 and 2 respectively.

3.3 Phase III: Regularized Thin-Plate Spline Warping

Once a partial lattice is found in an image, it is natural and useful to relate this possibly deformed lattice to its regular origin: the wallpaper structures. It is natural since the detected degree-4 lattice has the same topological structure as the regular wallpaper patterns [3]. It is useful since straightening out (rectifying) the deformed lattice and its neighborhood helps the iterative algorithm to expand its search for larger and larger lattice structures in the image reliably (Figure 2). We achieve this un-warping step using regularized thin-plate spline (TPS) warping, similar to what is done in both [12,17]. The practical benefits of this phase include: 1) it allows us to deal with deformation discontinuity in the scene; 2) the initial texton proposal can be re-used throughout the procedure, hence preventing the template drift problem; and 3) it enables us to keep the same initial regularized lattice model at all times.

Warping of the found lattice is performed sequentially with template matching on the rectified image, and the results fed back into Phase II for spatial tracking. This process repeats until the growing lattice reaches the edge of the image or there is no more lattice to track (Fig. 2). Although good results can be achieved without warping on patterns where the local distortion is changing gradually, the unwarping process produces improved results when there is an abrupt discontinuity in the distorted pattern, for example, at a fold or crease in the surface.

The main advantage comes from the coupling of BP with MRF and regularized TPS warping. As BP converges, the inference engine provides the deformation correspondences explicitly to the regularized TPS procedure. As the regularized TPS warping rectifies the image, it provides better observation and compatibility constraints in the MRF, resulting in enhanced correspondences on deformed patterns.

4 Experimental Results

We have tested our proposed algorithm on 32 real world images (Figures 4 and 5). Our approach is successful in finding lattice structures in real images, even when

Fig. 4. Sample lattice detection Results: The input images (leftmost), the results of [13,14] (second left), the results of [12] (second right), and our results (rightmost). Note that the comparison with [13,14] is not a fair comparison since they are designed to start interactively. For a complete set of images tested, see our website `http://vision.cse.psu.edu/MSBPLattice.htm`

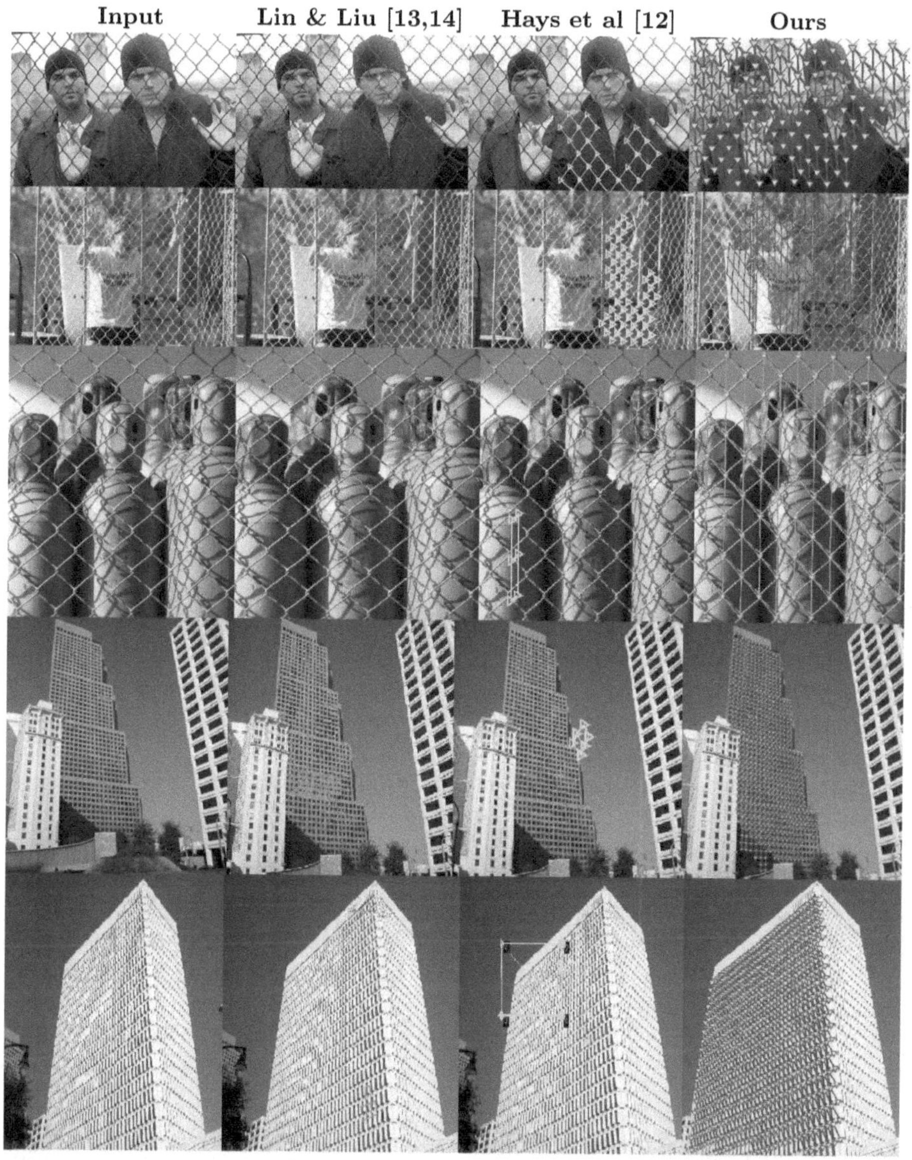

Fig. 5. Sample lattice detection Results: The input images (leftmost), the results of [13,14] (second left), the results of [12] (second right), and our results (rightmost). Note that the comparison with [13,14] is not a fair comparison since they are designed to start interactively. For a complete set of images tested, see our website `http://vision.cse.psu.edu/MSBPLattice.htm`

the scene contains textons with irregular appearances, such as chain link fences where each texture element is dominated by the scattered background. Quantitative evaluations of lattice detection rate and running time on this dataset are as follows: detection rate of Lin and Liu [13,14] is $20 \pm 21\%$, of Hays et al[1] is $47 \pm 38\%$, and of our algorithm is $81 \pm 19\%$. The detection rate is computed by the ratio of the number of detected textons over the number of ground truth textons. The ground truth is manually obtained by two human coders. The average ratio of run times of the Hays' algorithm versus ours is 10.66 ± 9.6. Since [13,14] is a semi-automatic detector, we report the average running time ratio of [12] and ours only. In addition, [12] failed earlier on 5 images which are excluded when computing average time ratio. Running time ratio is defined by the ratio of the time used by [12] over the time used by our algorithm to detect the lattice on each of the 32 test images.

5 Conclusions

We formulate the detection of an underlying deformed lattice from real images as a spatial, multi-target tracking problem using a recently published, efficient MSBP method [15]. The main contributions of our work are: 1) incorporating higher order constraints early-on to propose highly plausible potential interest points as lattice vertices; 2) coupling an MRF optimization procedure with regularized thin-plate spline warping, where MRF with MSBP and regularized TPS warping correct and support each other to yield reliable results in challenging situations; and 3) providing a new lattice detection algorithm with a deterministic time complexity linear in the number of textons in the scene. The quantitative results show that our algorithm is more accurate and significantly faster than other state-of-the-art lattice detection algorithms on diverse real world images.

Acknowledgments

This work was partially funded under NSF grant IIS-0535324 and NSF grant IIS-0729363. We thank J. Hays and W.C. Lin for sharing their code, and K. Brocklehurst for helping with the ground truth labeling.

References

1. Gibson, J.J.: The Perception of the Visual World. Houghton Mifflin, Boston (1950)
2. Julesz, B.: Visual pattern discrimination. IRE Transactions on Information Theory 8(2), 84–92 (1962)
3. Coxeter, H.: Introduction to Geometry, 2nd edn. Wiley, New York (1980)
4. Weyl, H.: Symmetry. Princeton University Press, Princeton (1952)

[1] The algorithm of [12] includes an element of randomness; it runs for several iterations, and takes the best result according to a modified A-score. A-score, originally introduced in [7], is the average per-pixel standard deviation among the final, aligned texels. The modification is the inclusion of $\sqrt{n}$ in the divisor in order to bias the A-score toward more complete lattices [12].

5. Malik, J., Belongie, S., Shi, J., Leung, T.: Textons, contours and regions: cue integration in image segmentation. In: The Proceedings of the Seventh IEEE International Conference on Computer Vision, vol. 2, pp. 918–925 (1999)
6. Forsyth, D.: Shape from texture without boundaries. In: 7th European Conference on Computer Vision, pp. 43–66 (2002)
7. Liu, Y., Lin, W.C., Hays, J.: Near-regular texture analysis and manipulation. ACM Transactions on Graphics 23(3), 368–376 (2004)
8. Ghanem, B., Ahuja, N.: Phase based modelling of dynamic textures. In: IEEE 11th International Conference on Computer Vision, pp. 1–8 (2007)
9. Schattschneider, D.: The plane symmetry groups: their recognition and notation. American Mathematical Monthly 85, 439–450 (1978)
10. Liu, Y., Collins, R.T., Tsin, Y.: A computational model for periodic pattern perception based on frieze and wallpaper groups. IEEE Transactions on Pattern Analysis and Machine Intelligence 26(3), 354–371 (2004)
11. Liu, Y., Tsin, Y., Lin, W.C.: The promise and perils of near-regular texture. International Journal of Computer Vision 62(1-2), 145–159 (2005)
12. Hays, J., Leordeanu, M., Efros, A., Liu, Y.: Discovering texture regularity as a higher-order correspondence problem. In: 9th European Conference on Computer Vision (2006)
13. Lin, W.C., Liu, Y.: Tracking Dynamic Near-regular Textures under Occlusions and Rapid Movements. In: 9th European Conference on Computer Vision (2006)
14. Lin, W.-C., Liu, Y.: A lattice-based MRF model for dynamic near-regular texture tracking. IEEE Transactions on Pattern Analysis and Machine Intelligence 29(5), 777–792 (2007)
15. Park, M., Liu, Y., Collins, R.T.: Efficient Mean Shift Belief Propagation for Vision Tracking. In: Computer Vision and Pattern Recognition, Anchorage, Alaska (2008)
16. Shi, J., Tomasi, C.: Good features to track. In: Computer Vision and Pattern Recognition, 1994, pp. 593–600 (1994)
17. Belongie, S., Malik, J., Puzicha, J.: Shape matching and object recognition using shape contexts. IEEE Transactions on Pattern Analysis and Machine Intelligence 24(4), 509–522 (2002)

Local Statistic Based Region Segmentation with Automatic Scale Selection

Jérome Piovano and Théodore Papadopoulo

Odyssée Project Team, INRIA Sophia Antipolis - Méditerranée, France
{Jerome.Piovano,Theodore.Papadopoulo}@sophia.inria.fr

Abstract. Recently, new segmentation models based on local information have emerged. They combine local statistics of the regions along the contour (inside and outside) to drive the segmentation procedure. Since they are based on local decisions, these models are more robust to local variations of the regions of interest (contrast, noise, blur, ...). They nonetheless also introduce some new difficulties which are inherent to the fact of basing a global property (the segmentation) on pure local decisions. This papers explores some of those difficulties and proposes some possible corrections. Results on both 2D and 3D data are compared to those obtained without these corrections.

1 Introduction

Image segmentation has been widely studied in the last decades, and still remains a challenging task. Many approaches exist in the literature to segment regions [16], [2], [10]. This paper focuses on those based on active contours [12], [9], [18], [21], [7]. This choice is driven by two main reasons: first, a framework for segmentation from local statistics has already been proposed [20], [3], [17], and conveniently serves as basis for the work presented in this paper. Second, the phenomenons that are presented in this paper are inherent to any segmentation procedure that will be based on local decision. Consequently, other optimization strategies such as graph-cuts or MRF will suffer from the same problems exposed here, provided that they can be applied (which is not necessary true as those methods tend to impose restrictions to the criterion so as to be applicable).

Active contours based strategies can be classified in two distinct categories: edge-based models and region-based ones. Edge-based models consist in evolving a contour in homogeneous areas, and locally stopping it when it reaches high image gradients [4] [13], [26]. These models have several advantages like local coherence and robustness to region inhomogeneities, but they also have important drawbacks that make them inefficient on noisy images [7], or when the contour initialization is not completely inside or outside the region to segment. Moreover, they are only able to segment regions with sharp edges, and this can result in "leaks" when the region edges are smoother. In [26], the authors managed to

D. Forsyth, P. Torr, and A. Zisserman (Eds.): ECCV 2008, Part II, LNCS 5303, pp. 486–499, 2008.

improve the initialization problem, by creating a vector flow driving the active contour to high image gradients, but the sensitivity to noise still remains.

Region-based models consist in evolving a contour in order to separate significantly the statistics of the regions it delimits [5], [24], [23], [19], [7]. For example, separating bright areas from dark ones, or extracting areas matching a certain histogram [14]. These models take account of the whole region and are thus more robust to noise and to initialization than edge-based models. However, they are inefficient in handling image with regions whose statistics are spatially varying across the domain.

The authors of [19] introduced a new model that is the addition of both the edge-based model and the region-based model. This model combines the advantages of the two previous ones by adding the two different evolution terms. As a consequence, it requires some parameters that may be difficult to adjust.

One difficulty comes from the fact that segmentation models often relies on *global* criteria [1], whereas images are most of the time subject to *local* variations (in intensity, contrast, noise ...). In opposition, it is admitted that the human visual system is more sensitive to local variations of the information [25], and the Adelson illusion is one of the many example showing this property. This explains why it is often difficult to create segmentation models that are able to extract regions matching our perception. Moreover, medical images suffer from some alterations such as locally diffuse contours, biases and trends across the image, that makes usual segmentation models most of the time inefficients.

To overcome these problems, new segmentation models have recently emerged to segment regions by computing their statistics locally, in the neighborhood of the active contour [20], [3], [17]. These methods introduce a neighborhood parameter. Its effect on the segmentation is very interesting: when a large neighborhood is chosen, the model is equivalent to the standard region-based model, whereas a small neighborhood results in the model that behaves like a standard edge-based one. Thus, these models can be seen as an elegant unification of edge-based and region-based models. However, fixing the neighborhood parameter introduces some new phenomenons, which are explored in this paper.

However, basing a segmentation on local decisions brings a whole new set of difficulties: locally the information might be insufficient to make the contour evolve, and it might need to be looked at different scales through a scale space approach; local decisions in neighboring regions might be contradictory, and the notions of *inside* and *outside* might be different depending on the initialization and position of the active contour. The second part of this paper explores these problems and describes two rules that help in defining a meaningful segmentation procedure based on local contrast information.

This paper is organized as follows: first we recall the local statistics region model, and explain in detail how to compute efficiently these statistics. Second, we show some important drawbacks of this method and our solutions to overcome them. Finally, we compare our model in 2D and 3D with standard models and run experiments on synthetic and real data.

[1] Either a contour strength or global distributions of pixel values inside regions.

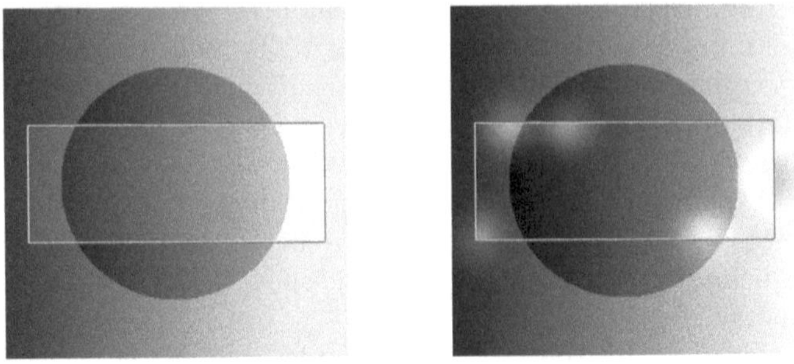

Fig. 1. Inside (yellow) and outside (blue) statistics : left computed in the whole region; right computed at the neighborhood of the active contour

2 Region Segmentation Using Local Statistics

2.1 The Model

Let us recall the segmentation model introduced in [20]. Assume that the image is covered by a set of regions Ω_i, $i = 1 \dots n$. The model is based on modeling these regions through local Gaussian statistics computed on a Gaussian windows. Denoting respectively by $\mu_i(x)$ and $\sigma_i(x)^2$, the local mean and local variance of the region Ω_i at point x, these quantities are defined by (Fig. 1):

$$\begin{cases} \mu_i(x) = \dfrac{\int_{\Omega_i} g_\rho(x-y)I(y)dy}{\int_{\Omega_i} g_\rho(x-y)dy} \\ \sigma_i(x)^2 = \dfrac{\int_{\Omega_i} g_\rho(x-y)\left(I(y)-\mu_i(x)\right)^2 dy}{\int_{\Omega_i} g_\rho(x-y)dy} . \end{cases} \tag{1}$$

where I is a given image and g_ρ a Gaussian kernel of standard deviation ρ.

The model is actually a spatially localized version of the maximum a posteriori segmentation for the Gaussian distribution [7], and results in the minimization of the following energy:

$$E(\Omega_1, \cdots, \Omega_N) = \sum_i \int_{\Omega_i} -\log p_i(I(x)|x,\rho)dx + \nu|C| , \tag{2}$$

with

$$-\log p_i(I(x)|x,\rho) = \frac{(I(x)-\mu_i(x))^2}{2\sigma_i(x)^2} + \frac{1}{2}\log(2\pi\sigma_i(x)^2) , \tag{3}$$

where p_i is the probability for a point x with intensity $I(x)$ to belong to the local region characterized by the local mean $\mu_i(x)$ and local variance $\sigma_i(x)^2$; and $|C|$ the total length of the interface between regions, which stand for a regularization term.

Following [20], the evolution speed is obtained by differentiating Eq. (2) in the level-set notation [21], using the shape gradient method [1]. For the bi-partitioning case, this lead to the following gradient descent:

$$\frac{\partial \Phi}{\partial t}(x) = \frac{(I(x) - \mu_{in}(x))^2}{2\sigma_{in}(x)^2} - \frac{(I(x) - \mu_{out}(x))^2}{2\sigma_{out}(x)^2} + \log \frac{\sigma_{in}(x)^2}{\sigma_{out}(x)^2} + \nu \nabla \left(\frac{\nabla \Phi}{|\nabla \Phi|} \right) ,$$

where μ_{in}, σ_{in}^2 and μ_{out}, σ_{out}^2 are respectively the local mean and local variance in the inside and outside regions.

2.2 An Efficient Implementation

All the local statistics used in the model are Gaussian convolutions, and can be computed very efficiently using the recursive filter implementation proposed in [8]. The complexity of all these convolutions remains in $O(n)$.

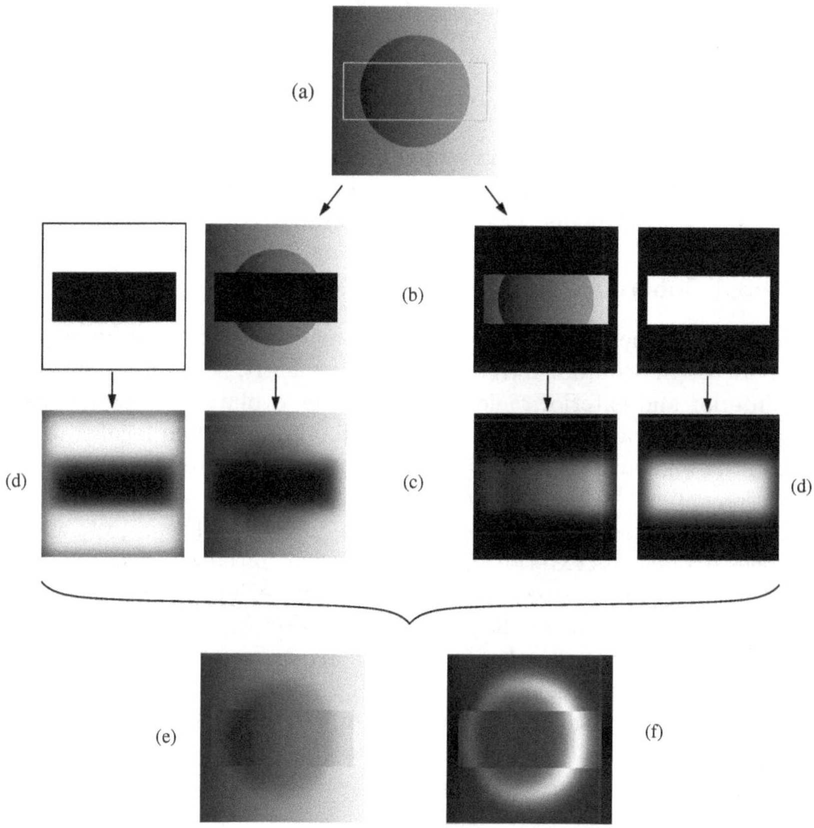

Fig. 2. Local statistics computation: The local statistics are computed thanks to a succession of convolutions (see text)

Figure 2 depicts a framework for computing these local statistics:

1. The regions delineated by the yellow contour (Fig. 2(a)) are first separated (Fig. 2(b)).
2. The local means (Fig. 2(e)) are then computed by blurring each region separately (Fig. 2(c)), and then normalized by a blurred characteristic function of the regions (Fig. 2(d)).
3. The local variances (Fig. 2(f)) are then computed similarly, using the local means thanks to the relation $V(X) = E(X^2) - E(X)^2$.

2.3 Advantages of Local Models

By estimating region statistics locally, the robustness of the standard global model is improved with respect to region inhomogeneities (Fig. 6). Indeed, region inhomogeneities are alterations of low frequencies that can be overcame by restricting the locality of the estimation of the region statistics. Furthermore, these models will favor significant local separation of regions: the neighborhood parameter is a term that will influence the sharpness of region edges.

3 Problems with Local Models

Unfortunately, the local models presented in the previous section introduce new sets of problems summarized in Fig. 3. some new drawbacks that were not present in global region-based models are also introduced.

3.1 Global Coherence Problem

Since only a small neighborhood in the regions defines the contour evolution, various parts of the contour may evolve independently, selecting their own notions of interior and exterior regions. Later in the evolution, these incoherence lead to the creation of singularities corresponding to unwanted local stable minima. When a singularity is formed, no topological change is possible, and the segmentation cannot be turned into something coherent anymore. This is illustrated in Fig. 3(c): the red circle segments the interior of the object, while the green circles segment its exterior according to their initializations. No topological change are possible between these two growing regions, and a singularity is indeed created. Other circles collapse under the effect of regularization, because of the absence of contribution due to the data attachment term in homogeneous areas. Note that this problem might also occur with simpler initializations, as soon as the contour crosses the edges of the object to segment.

3.2 Homogeneous Areas Problem

It is easy to understand that there is no data term contribution in homogeneous areas, as the local statistics in the region inside and outside are the same.

At convergence, the small circles in homogeneous areas collapse because of the regularization term, which leads to some kind of phantom contour, resulting

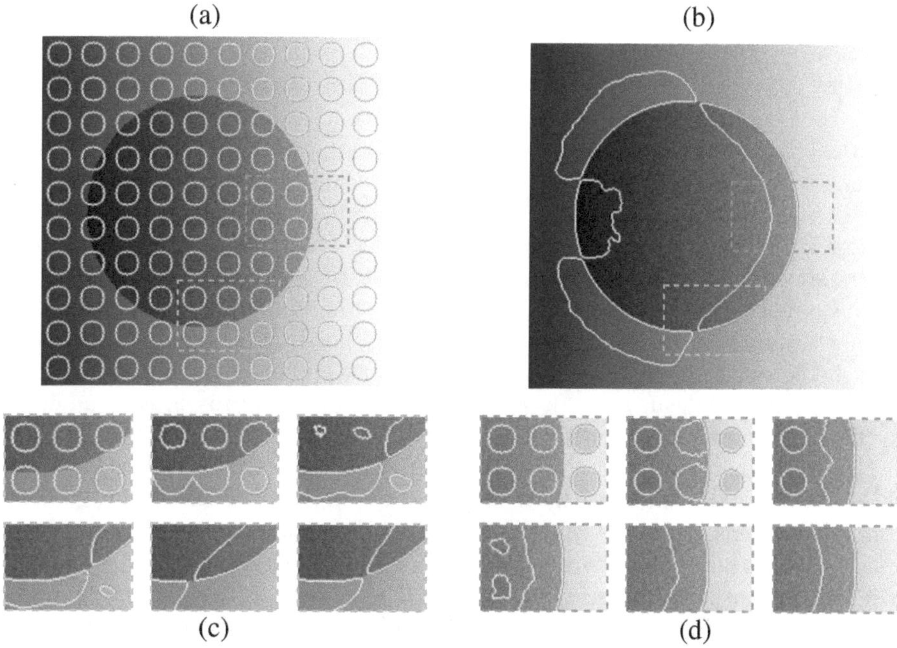

Fig. 3. (a) Initialization of the segmentation; (b) Convergence of the segmentation; (c) Evolution which leads to a problem of coherence in the segmentation; (d) Absence of evolution in homogeneous areas, leading to the creation of "layers"

in "segmentation layers" that are of course not desired (Fig. 3(d)). These layers come from the fact that the only contribution of the data attachment term is close to image edges, where the difference of local region statistics make the contour evolve. However, it can be observed that the "thickness" of these segmentation layers depends on the neighborhood parameter of the model. With a small neighborhood, the layers will be thin, whereas a large neighborhood will result in a wider layer. Unfortunately, increasing the neighborhood makes the model behave like global models, and thus loosing its good locality properties.

In Fig. 3(d), the left blue circles collapse because they are too far from an image edge. This edge proximity is linked to the neighborhood parameter of the model, and a larger neighborhood would have indeed made these circles evolve. Note that the right circles are collapsing under the effect of the data attachment term, as the "middle layer" influence their local interior mean.

4 Global Coherence Solution

Several works have been done to avoid the kind of local minima in Fig. 3(c).

In [11], the authors added some noise in the evolution to perturb the active contour, and thus prevent it from being stuck in local minima. Unfortunately, this stochastic evolution is not well adapted to our problem, as it would create as much incoherence in the evolution.

In [6], [22], the authors smoothed the evolution term of the active contour along the contour, in order to add some correlation in the evolution of neighbor points. This allows the contour to evolve in a more coherent way, avoiding local minima. Unfortunately, the local minima we tackle cannot be avoided in this way, as the evolution can be seen as coherent until a singularity is created.

However, if we make the assumption that objects to segment are characterized by boundaries where image intensity changes "in the same way" from the inside to the outside, we can use this constraint to maintain the coherence in the segmentation (Fig. 4). This constraint can be defined by having, for each point x of the contour, the local interior mean $\mu_{in}(x)$ whether inferior or superior to the local exterior mean $\mu_{out}(x)$. Thus, if the contour is locally in contradiction to this constraint, the evolution direction needs to be changed.

Normally, the gradient descent will ensure that the contour evolves in a way that maximizes the difference between $\mu_{in}(x)$ and $\mu_{out}(x)$ for each point x of the contour. In areas where the contour is locally in contradiction with the previous constraint, taking the *opposite* gradient direction will ensure that the evolution will lead to a situation where $\mu_{in}(x) = \mu_{out}(x)$. In essence, this process is somewhat similar to a form of *local simulated annealing* allowing a local increase of the energy to avoids unwanted local minima. However this increase is not stochastic but driven the constraint. This leads to the following evolution equation:

$$\begin{cases} \dfrac{\partial \Phi}{\partial t}(x) = \mathcal{S}(x) \left(\dfrac{(I(x) - \mu_{in}(x))^2}{2\sigma_{in}(x)^2} - \dfrac{(I(x) - \mu_{out}(x))^2}{2\sigma_{out}(x)^2} + log \dfrac{\sigma_{in}(x)^2}{\sigma_{out}(x)^2} \right) + \nu\kappa \\ \mathcal{S}(x) = \pm \operatorname{sign}\left(\mu_{in}(x) - \mu_{out}(x)\right) \end{cases}$$

where '$\pm$' depends on the type of the global constraint. The energy is therefore increased or decreased, depending on the sign of $\mathcal{S}(x)$.

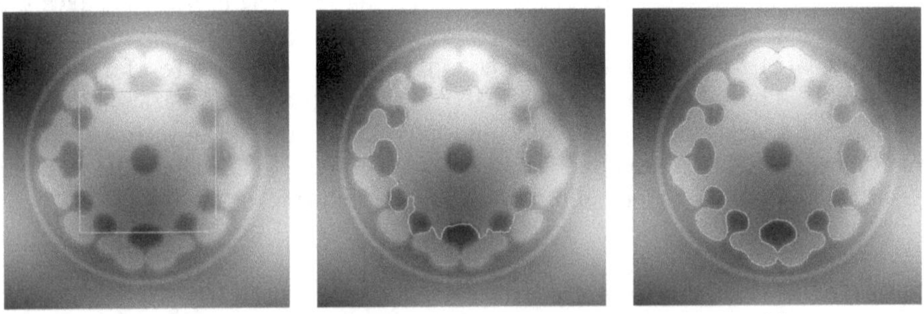

(a) (b) (c)

Fig. 4. (a) Initialization; Convergence without (b) and with (c) the constraint

In Fig. 4, the convergence of the active contour can be seen with and without this coherence constraint, on a synthetic image created as a simple cortex model with high intensity drifts. The initialization in figure 4.a is quite problematic, as it crosses the edges of the object to segment.

Without the constraint (Fig. 4(b)), the segmentation obtained is "locally correct", as it locally separates the interior and exterior of the contour. However, It has no sense if we look at the global scope, and this kind of segmentation is of course not acceptable.

With the constraint (Fig. 4(c)), the contour evolves in a more coherent way, and finally manage to extract the totality of the white matter, even under very strong inhomogeneities. Although this constraint restrict the nature of objects to segment, it is quite effective on objects that satisfy it. Actually this constraint is not so restrictive, as the majority of structures of interest in medical images can be characterized by it. Note that this simulated annealing can be used with multiple regions, by increasing the energy if the local features of the contour separating two different regions are not the ones desired.

5 Varying the Local Neighborhood

As explained in section 3.2, there is no evolution in areas where the local statistics inside and outside the contour are the same. This results in the creation of "layers" in the segmentation, in the neighborhood of image edges (Fig. 3.d). As mentioned earlier, the thickness of these layers is correlated with the neighborhood parameter of the model, but using large neighborhoods give results equivalent to standard global models.

Thus, we would like to use small neighborhoods close to an image edge, and large neighborhoods in homogeneous areas, depending on the proximity of the closest image edge (Fig. 6). The question is how to detect this edge proximity.

In [15] the authors tackled similar problems with the purpose of extracting edges and ridges from images. Their method is based on detecting features at various scales of the same image, and keeping the k most salient features depending on a saliency measure. It allows them to extract sharp edges as well as smooth edges by increasing the locality of the estimation of these features.

Our problem is somehow similar. If we assume that the neighborhood parameter is actually a scale used to compute statistics, the problem becomes one of finding *the most salient scale* for each point of the contour to make the contour locally evolve. As we want to favor sharp edges over smooth ones, and make the contour evolve even in locally homogeneous areas, the scale is chosen as *the smallest one inducing an evolution speed superior to a given threshold*:

$$\rho(x) = \inf_{\rho} \left(\left.\frac{\partial \Phi}{\partial t}\right|_{\rho} (x) > \epsilon \right) \qquad x \in \Gamma \quad , \tag{4}$$

where $\left.\frac{\partial \Phi}{\partial t}\right|_{\rho}$ is the evolution equation computed with a Gaussian windows of variance ρ, x a point of the contour Γ, and ϵ a constant corresponding to a threshold for the scale selection.

Fig. 5. Top: initialization and evolution of the contour; Bottom: Scales chosen at each point of the contour, with $\epsilon = 1$ and using 30 scales

By choosing this scale selection scheme, the contour will evolve even in homogeneous regions. Indeed, at each iteration and for each point, the optimal scale is found by slowly increasing the neighborhood from the minimum scale to the maximum, until it reaches some edge of an object, thus giving information about how the contour should locally evolve. Actually, the model depicted in section 2 is not restricted by choosing a unique size of Gaussian windows for all points, and we can indeed use different windows size depending on local features.

Figure 5 shows the evolution of the active contour (top row), as well as the different scales chosen at each point (bottom row). This image is interesting as it shows three fundamental problems in medical images: large intensity drifts (top corner of the triangle), noisy regions (right corner of the triangle) and smooth contours (left corner of the triangle). Where standard edge-based or region-based model would fail to provide good results, our model behaves quite well, choosing accurately the scales depending of the proximity and the "smoothness" of the triangle edges. The robustness of our model to noisy regions comes from the fact that the speed is normalized by the local variance, so the noisier the region is, the smaller the speed is, and the higher the scale chosen will be. This allows to choose larger scales if the contour is in a noisy regions, thus improving the accuracy of computation of local statistics.

6 Implementation

With the multiscale approach, various scales are incorporated in our model, so each region has to be blurred several times using different Gaussian kernels to obtain a scale space of local statistics. This allows to retrieve, for each point of the contour, the optimal scale using the criteria of Eq. 4.

For efficiency purpose, these local statistics are limited to only a few scales. In the experiments, the small scales are more important, so the variances take the values: 2, 4, 8, 16. The model is described by the following algorithm:

Algorithm 1. Segmentation using adaptive local statistics

repeat
  for all scales $\rho_j \in \{\rho_1, \ldots, \rho_n\}$ **do**
    for all regions $\Omega_i \in \{\Omega_1, \ldots, \Omega_n\}$ **do**
      Compute local mean $\mu_{i,j}$ of Ω_i at scale ρ_j
      Compute local variance $\sigma_{i,j}^2$ of Ω_i at scale ρ_j
    end for
  end for
  for all points in the active contour **do**
    Find optimal scale ρ_{opt} using Eq. (4) with $\mu_{i,j}$ and $\sigma_{i,j}^2$
    Compute speed with $\mu_{i,opt}$ and $\sigma_{i,opt}^2$
  end for
  Update contour
until convergence

Finding the correct scale from the set of local statistics is very fast, and the bottleneck remains the creation of the scale space of statistics at each iteration.

Table 1. Time execution of 500 iterations. '4 scales GPU' corresponds to time execution of the multi-threaded algorithm on a GPU (NVidia's Quadro FX570M).

Images	single scale	4 scales	4 scales GPU
2D (300^2)	13.17 sec	56.36 sec	7.61 sec
3D (256^3)	51mn 20sec	3h 4mn	25mn

In Table 1, the execution time of 500 iteration of the algorithm is shown on a 2D and a 3D image. It is compared with the single scale model [20], and with its implementation on a GPU (graphics processing unit), using the NVidia's Cuda Library[2]. Indeed, our implementation of the recursive filter is an optimized one, using multi-threaded routine to fully use the potentials of highly parallel computing device. This leads to reasonable times for both 2D and 3D image. Notice that our model remains faster than the ones that need some PDE computation at each iteration [24].

7 Results

Our model was tested on several synthetic and real images in 2D and 3D.

In Fig. 6, our model is compared to the global Gaussian one (Chan & Vese model), on a synthetic image created as a simple cortex model with large inhomogeneities.

[2] www.nvidia.com/cuda

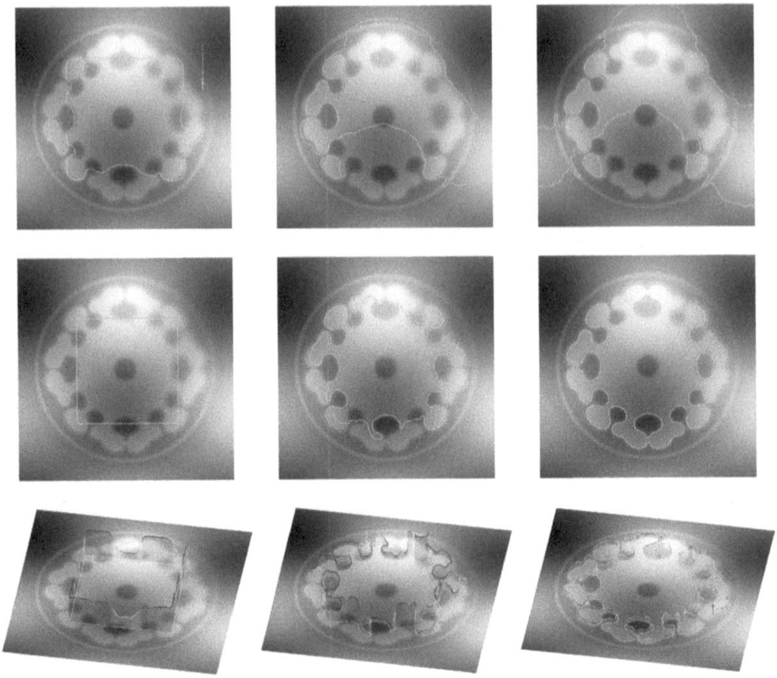

Fig. 6. Top row: Initialization and evolution of the contour with the global Gaussian model; Middle row: Initialization and evolution of the contour with our model; Bottom-row: Scales chosen a each point of the contour

The kind of initialization used here does not facilitate the global model as the contour crosses the edges of the structure to segment. Moreover, the distribution of intensity of the internal and external regions are the same. Due to its sensitivity to inhomogeneities, the global model (Fig. 6 top row) results in separating globally the bright areas from the dark ones, with no preference to sharp edges. Global non parametric models such as Parzen windows [14] give similar results. By contrast, our model (Fig. 6 middle row) behaves quite well. It extracts accurately the details of the region's shape, is more robust to inhomogeneities, and improves notably the behavior of the model to initialization. Note that the scales (Fig. 6 bottom row) are chosen accurately, depending on the proximity to the closest image edge. Due to the proximity of the initialization to the solution, only a few scales are needed.

Figure 7 shows the same kind of comparison with a real 3D anatomical MRI of a head. We initialize the contour in both model with a sphere centered in the middle of the brain (red contour in the Fig. 7 top left image). The background of the image being significantly darker than the brain, the global model (green contour in the Fig. 7 top left image) only manages to extract the skin of the head from the background, and does not retrieve important details of the brain such as sulcii or gyrii. It does not converge segmenting the white matter, and "leaks" in the cerebrospinal fluid (Fig. 7 middle row)).

As can be seen in the bottom row of Fig. 7, our model manages to extract a lot of details of the white matter of the brain, and does not leak in the cerebrospinal fluid thanks to its preference of sharper image edges (blue contour in the Fig. 7 top left image). Note that in both results, the same amount of regularization has been used.

The complexity of the Deriche recursive filter allows us to compute 3D convolutions in a reasonable time (3 pass of the image in 3D), and due to the number of blurs our model requires, a non recursive convolution scheme would have made it unusable in 3D.

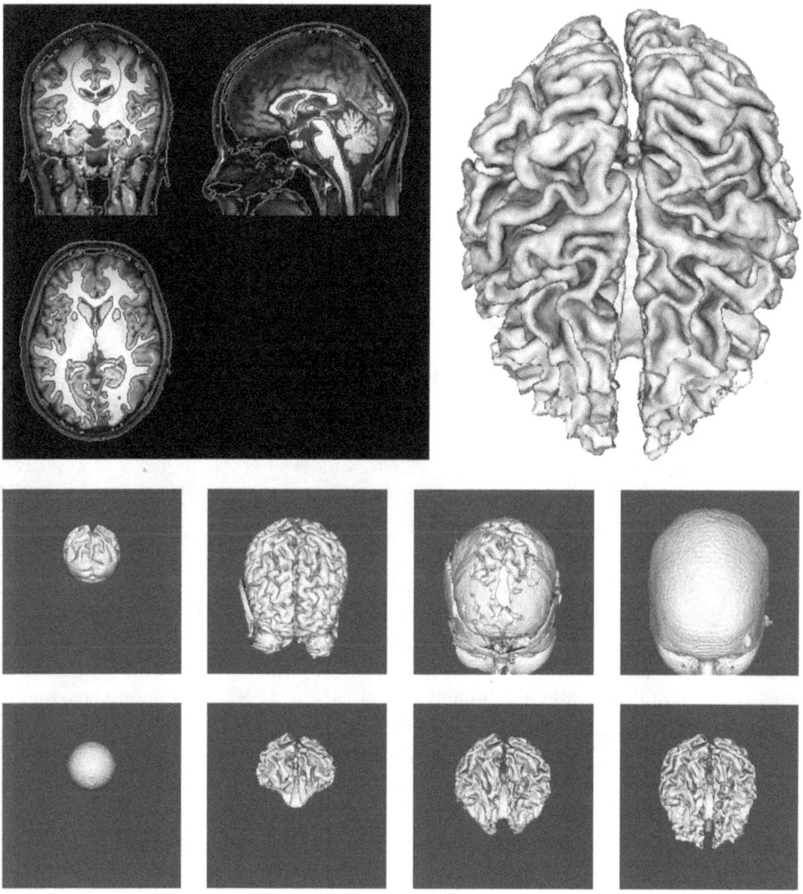

Fig. 7. Top Row: left, MRI with contours (see text), right, convergence of our model ; Middle Row: evolution of the global model ; Bottom Row: evolution of our model

8 Conclusion and Future Work

In this paper, the robustness of the region segmentation model has been improved using local statistics by making the model spatially coherent, and by locally selecting the correct scale at each point of the contour. Promising results were obtained in both 2D and 3D. This model has been used successfully to segment particular structures in medical images and we are currently looking at extending this model to non scalar images (color images or registered multimodal medical images).

References

[1] Aubert, G., Barlaud, M., Faugeras, O., Jehan-Besson, S.: Image segmentation using active contours: Calculus of variations or shape gradients? SIAM Journal of Applied Mathematics 63(6), 2128–2154 (2003)
[2] Boykov, Y., Jolly, M.-P.: Interactive graph cuts for optimal boundary and region segmentation of objects in n-d images. In: ICCV, pp. 105–112 (2001)
[3] Brox, T., Cremers, D.: On the statistical interpretation of the piecewise smooth Mumford-Shah functional. In: Sgallari, F., Murli, A., Paragios, N. (eds.) SSVM 2007. LNCS, vol. 4485, pp. 203–213. Springer, Heidelberg (2007)
[4] Caselles, V., Kimmel, R., Sapiro, G.: Geodesic active contours. Technical report, HP Labs (September 1994); A shorter version appeared at 5th ICCV 1995, Boston (1995)
[5] Chan, T., Vese, L.: Active contours without edges. IEEE Transactions on Image Processing 10(2), 266–277 (2001)
[6] Charpiat, G., Maurel, P., Pons, J.-P., Keriven, R., Faugeras, O.: Generalized gradients: Priors on minimization flows. The International Journal of Computer Vision 73(3), 325–344 (2007)
[7] Cremers, D., Rousson, M., Deriche, R.: A review of statistical approaches to level set segmentation: Integrating color, texture, motion and shape. International Journal of Computer Vision 72(2), 195–215 (2007)
[8] Deriche, R.: Recursively implementing the gaussian and its derivatives. Technical Report 1893, INRIA, Unité de Recherche Sophia-Antipolis (1993)
[9] Dervieux, A., Thomasset, F.: A finite element method for the simulation of Rayleigh-Taylor instability. Lecture Notes in Mathematics, vol. 771, pp. 145–159 (1979)
[10] Grady, L.: Random walks for image segmentation. IEEE Trans. on Pattern Analysis and Machine Intelligence 28(11), 1768–1783 (2006)
[11] Juan, O., Keriven, R., Postelnicu, G.: Stochastic motion and the level set method in computer vision: Stochastics active contours. International Journal of Computer Vision 69(1), 7–25 (2006)
[12] Kass, M., Witkin, A., Terzopoulos, D.: Snakes: Active contour models. The International Journal of Computer Vision 1(4), 321–331 (1987)
[13] Kichenassamy, S., Kumar, A., Olver, P., Tannenbaum, A., Yezzi, A.: Gradient flows and geometric active contour models. In: Proceedings of the 5th International Conference on Computer Vision, June 1995, pp. 810–815 (1995)
[14] Kim, J., Fisher, J., Yezzi, A., Cetin, M., Willsky, A.: Nonparametric methods for image segmentation using information theory and curve evolution. In: IEEE International Conference on Image Processing, September 2002, pp. 797–800 (2002)

[15] Lindeberg, T.: Feature detection with automatic scale selection. The International Journal of Computer Vision 30(2), 77–116 (1998)
[16] Morel, J.M., Solimini, S.: Variational Methods in Image Segmentation. In: Progress in Nonlinear Differential Equations and Their Applications, Birkhäuser, Basel (1995)
[17] Mory, B., Ardon, R., Thiran, J.-P.: Variational Segmentation using Fuzzy Region Competition and Local Non-Parametric Probability Density Functions. In: IEEE International Conference on Computer Vision (ICCV), Rio, Brazil (2007)
[18] Osher, S., Sethian, J.A.: Fronts propagating with curvature-dependent speed: Algorithms based on Hamilton–Jacobi formulations. Journal of Computational Physics 79(1), 12–49 (1988)
[19] Paragios, N., Deriche, R.: Geodesic active regions: a new paradigm to deal with frame partition problems in computer vision. Journal of Visual Communication and Image Representation, Special Issue on Partial Differential Equations in Image Processing, Computer Vision and Computer Graphics 13(1/2), 249–268 (2002)
[20] Piovano, J., Rousson, M., Papadopoulo, T.: Efficient segmentation of piecewise smooth images. In: Sgallari, F., Murli, A., Paragios, N. (eds.) SSVM 2007. LNCS, vol. 4485, pp. 709–720. Springer, Heidelberg (2007)
[21] Sethian, J.A.: Level Set Methods and Fast Marching Methods: Evolving Interfaces in Computational Geometry, Fluid Mechanics, Computer Vision, and Materials Sciences. Cambridge Monograph on Applied and Computational Mathematics. Cambridge University Press, Cambridge (1999)
[22] Sundaramoorthi, G., Yezzi Jr., A.J., Mennucci, A.C.: Sobolev active contours. The International Journal of Computer Vision 73(3), 345–366 (2007)
[23] Tsai, A., Yezzi, A.J., Willsky, A.S.: Curve evolution implementation of the Mumford-Shah functional for image segmentation, denoising, interpolation, and magnification. IEEE Transactions on Image Processing 10(8), 1169–1186 (2001)
[24] Vese, L.A., Chan, T.: A multiphase level set framework for image segmentation using the Mumford and Shah model. The International Journal of Computer Vision 50(3), 271–293 (2002)
[25] Wohrer, A.: Model and large-scale simulator of a biological retina with contrast gain control. Ph.D thesis, University of Nice Sophia-Antipolis (2008)
[26] Xu, C., Prince, J.L.: Snakes, shapes, and gradient vector flow. IEEE Transactions on Image Processing 7(3), 359–369 (1998)

A Comparative Analysis of RANSAC Techniques Leading to Adaptive Real-Time Random Sample Consensus

Rahul Raguram[1], Jan-Michael Frahm[1], and Marc Pollefeys[1,2]

[1] Department of Computer Science, The University of North Carolina at Chapel Hill
{rraguram,jmf,marc}@cs.unc.edu
[2] Department of Computer Science, ETH Zürich
marc.pollefeys@inf.ethz.ch

Abstract. The Random Sample Consensus (RANSAC) algorithm is a popular tool for robust estimation problems in computer vision, primarily due to its ability to tolerate a tremendous fraction of outliers. There have been a number of recent efforts that aim to increase the efficiency of the standard RANSAC algorithm. Relatively fewer efforts, however, have been directed towards formulating RANSAC in a manner that is suitable for real-time implementation. The contributions of this work are two-fold: First, we provide a comparative analysis of the state-of-the-art RANSAC algorithms and categorize the various approaches. Second, we develop a powerful new framework for real-time robust estimation. The technique we develop is capable of efficiently adapting to the constraints presented by a fixed time budget, while at the same time providing accurate estimation over a wide range of inlier ratios. The method shows significant improvements in accuracy and speed over existing techniques.

1 Introduction

The RANSAC (Random Sample Consensus) algorithm [1] is a simple, yet powerful, technique that is commonly applied to the task of estimating the parameters of a model, using data that may be contaminated by outliers. RANSAC estimates a global relation that fits the data, while simultaneously classifying the data into inliers (points consistent with the relation) and outliers (points not consistent with the relation). Due to its ability to tolerate a large fraction of outliers, the algorithm is a popular choice for a variety of robust estimation problems.

RANSAC operates in a hypothesize-and-verify framework: a minimal subset of the input data points is randomly selected and model parameters are estimated from this subset. The model is then evaluated on the entire dataset and its *support* (the number of data points consistent with the model) is determined. This hypothesize-and-verify loop is repeated until the probability of finding a model with better support than the current best model falls below a predefined threshold (typically 1%-5%). RANSAC can often find the correct solution even for high levels of contamination; however, the number of samples required to do so increases exponentially, and the associated computational cost is substantial.

D. Forsyth, P. Torr, and A. Zisserman (Eds.): ECCV 2008, Part II, LNCS 5303, pp. 500–513, 2008.

There have been a number of recent efforts aimed at increasing the efficiency of the basic RANSAC algorithm. Some of these strategies [2,3,4] aim to optimize the process of model verification, while others [5,6,7] seek to modify the sampling process in order to preferentially generate more useful hypotheses. While these efforts have shown considerable promise, none of them are directly applicable in situations where real-time performance is essential. Relatively fewer efforts have been directed towards the goal of formulating RANSAC in a manner that is suitable for real-time implementations. In particular, Nister [8] describes the preemptive RANSAC framework, where a *fixed* number of hypotheses are evaluated in a *parallel, multi-stage* setting. In this case the goal is to find, within a fixed time budget, the best solution from a restricted set of hypotheses. While the preemptive RANSAC framework facilitates real-time implementation, there exist a few limitations in the scheme. One of the primary limitations of preemptive RANSAC is its inherent non-adaptiveness to the data. The selection of a fixed number of hypotheses implicitly implies that a good prior estimate of the inlier ratio is available; in practice, this is often not the case. For low contamination problems, preemptive RANSAC is often slower than standard RANSAC, since it evaluates many more hypotheses than necessary. On the other hand, when the inlier ratio is too low, preemptive RANSAC is unlikely to find a good solution, since it does not test enough hypotheses.

There has been recent work on dealing with issues arising from degenerate data configurations [9,10]. While we do not address degeneracy specifically in the paper, we note that both techniques are extensions of RANSAC and accordingly can be improved by the proposed scheme in a similar way.

The goal of the current paper is two-fold. First, we provide a comparative analysis of the important RANSAC algorithms that have been proposed over the past few years [2,3,4,5,7,8,11]. We classify the various approaches based on the aspect of RANSAC that they seek to optimize, and evaluate their performance on synthetic and real data. Following the evaluation, we propose a powerful new framework for real-time robust estimation. One of the main challenges in real-time robust estimation lies in being able to adapt to the strict limitations presented by the fixed time budget. The framework we develop is able to elegantly cope with variations in the contamination level of the data, while at the same time providing accurate estimation at real-time speeds.

2 Survey of RANSAC Techniques

2.1 RANSAC

As mentioned earlier, RANSAC operates in a hypothesize-and-verify framework. Given a set $\mathcal{U}$ containing N tentative correspondences, RANSAC randomly samples subsets of size m from the data, where m is the minimal number of samples required to compute a solution, which is equivalent to the complexity of the geometric model. Once a model has been hypothesized from this minimal subset, it is verified against all data points in $\mathcal{U}$ and its support is determined.

This process is repeated until a termination criterion is met. The standard termination criterion for RANSAC is based on the minimum number of samples required to ensure, with some level of confidence η_0, that at least one of the selected minimal subsets is outlier-free. Given the true inlier ratio ε, the probability of selecting an uncontaminated sample of m inliers is ε^m. The probability of selecting k samples, each of which is contaminated with at least one outlier, is given by $(1-\varepsilon^m)^k$. Thus, the minimum number of samples that must be drawn in order to ensure that this probability falls below a $1-\eta_0$ threshold is

$$k \geq \frac{log(1-\eta_0)}{log(1-\varepsilon^m)} \tag{1}$$

Since the true inlier ratio ε is *a priori* unknown, a lower bound on this ratio can be found by using the sample which currently has the largest support. This estimate is then updated as the algorithm progresses. A number of recently proposed methods seek to optimize different aspects of the basic RANSAC algorithm. We discuss three strategies to optimize the model verification stage of RANSAC in Section 2.2. We also review in Section 2.3, methods that have been proposed to incorporate non-uniform sampling into the hypothesis generation stage. Finally, in Section 2.4, we discuss a very different RANSAC framework that operates in a real-time setting.

2.2 Optimizing Model Verification

While the number of samples drawn by RANSAC depends on the inlier ratio and the model complexity, an additional computational factor is the total number of correspondences N, since a hypothesis is verified against all data points. However, since most of the generated models are likely to be contaminated, these will be consistent only with a small fraction of the data. Thus, it is often possible to discard bad hypotheses early on in the verification process. Some of the early efforts to speed up RANSAC made use of precisely this observation.

2.2.1 The $T_{d,d}$ Test

Matas and Chum [2] design a pre-evaluation stage which attempts to quickly filter out bad hypotheses. Model verification is first performed using a subset of d randomly selected points (where $d \ll N$). The remaining $N-d$ points are evaluated only if the first d points are all inliers to the model. A setting of $d=1$ is recommended as the optimal value. Due to the randomized hypothesis evaluation procedure, a valid hypothesis may be mistakenly rejected by the $T_{d,d}$ test. Thus, one of the consequences of this approach is that it requires many more hypotheses than the original RANSAC. However, in general, the overall running time is reduced due to the preverification procedure.

2.2.2 Bail-Out Test

The idea of early termination of bad hypotheses was further extended by Capel in [3]. Given a randomly selected subset of n points, the number of inliers κ_n in this subset follows a hyper-geometric distribution: $\kappa_n \sim \mathrm{HypG}(\kappa, n, \overline{\kappa}, N)$, where $\overline{\kappa}$ is the total number of inliers for the current hypothesis. Given the current best hypothesis with inlier count $\overline{\kappa}_{best}$, consider a new hypothesis that has been partially evaluated against a subset of n data points. The goal is to determine the probability that having observed κ_n inliers so far, the total number of inliers for the new hypothesis, $\overline{\kappa}$, is greater than $\overline{\kappa}_{best}$. Since this is computationally expensive to calculate, a lower bound κ^n_{min} is derived on the number of inliers observed after evaluating n data points. If, at this point, the number of inliers is below the threshold, bail-out occurs and the hypothesis is discarded. For a given confidence bound P_{conf}, also note that the minimum bound κ^n_{min} may be approximated as a binomial distribution (for small values of n) or as a normal distribution (for large values of n).

2.2.3 WaldSAC

Most recently, Chum and Matas described an optimal randomized model verification strategy [4,12] based on Wald's theory of sequential decision making. The evaluation step is cast as an optimization problem which aims to decide whether a model is good (H_g) or bad (H_b), while simultaneously minimizing the number of verifications performed per model. Wald's Sequential Probability Ratio Test (SPRT) is based on the likelihood ratio

$$\lambda_j = \prod_{r=1}^{j} \frac{p(x_r|H_b)}{p(x_r|H_g)} \tag{2}$$

where x_r is equal to 1 if the rth data point is consistent with a given model, and 0 otherwise. $p(1|H_g)$ denotes the probability that a randomly chosen data point is consistent with a good model, and this can be approximated by the inlier ratio ε. Similarly, $p(1|H_b)$ is the probability that a randomly chosen data point is consistent with a bad model, and this can be modeled using a Bernoulli distribution with parameter δ. If, after evaluating j data points, the likelihood ratio becomes greater than some threshold A, the model is rejected. The decision threshold A is the only parameter of the SPRT test, and it can be set to achieve optimal running time (for more details we refer to [5]). When the level of contamination is known *a priori*, the WaldSAC strategy is provably optimal. In practice, however, inlier ratios have to be estimated during the evaluation process and the threshold A is adjusted to current estimates. An initial estimate of the parameter δ is obtained through geometric considerations, and this estimate is revised during the sampling process. The performance of the SPRT test is not significantly affected by the imperfect estimation of these parameters. A detailed comparative evaluation of the three methods described above is provided in Section 4. While we defer a discussion of the experimental results until later, we briefly note that in general, the performance of the bail-out and SPRT tests are comparable, both producing speed-ups ranging between 2-7 times compared to standard RANSAC.

2.3 Improving Hypothesis Generation

The RANSAC algorithm generates hypotheses by uniformly sampling the input data set. This implicitly implies that there is no *a priori* information available about the accuracy of the data points. In practice, this may be an overly pessimistic approach, since *a priori* information is often available, and can be used to generate better hypotheses.

2.3.1 PROSAC

Typically, correspondences between two or more images are obtained by the use of a local matching algorithm. A similarity function is evaluated over a number of points, and subsequently thresholded to obtain a set of tentative correspondences. Based on the assumption that points with high similarity are more likely to be inliers than points with low similarity, it may be possible to generate better hypotheses by sampling from a reduced set of points with high similarity scores. This fact is exploited in [7], where the correspondences are ordered based on their similarity scores, and progressively larger subsets of tentative correspondences are used to generate hypotheses. The Progressive Sample Consensus (PROSAC) algorithm is designed to draw the same samples as RANSAC, only in a different order. Consider a sequence of T_N samples of size m drawn by RANSAC from the set of all N correspondences. This sequence is denoted by $\{\mathcal{M}_i\}_{i=1}^{T_N}$. Let $\mathcal{U}_n$ denote a subset of $\mathcal{U}_N$ containing n points with the highest quality. The sequence $\{\mathcal{M}_i\}_{i=1}^{T_N}$ contains, on average, T_n samples containing points only from $\mathcal{U}_n$, where

$$T_n = T_N \frac{\binom{n}{m}}{\binom{N}{m}} \tag{3}$$

A recurrence relation for T_{n+1} can be obtained as: $T_{n+1} = \frac{n+1}{n+1-m} T_n$. Since this may not necessarily be an integer, T'_{n+1} is defined as $T'_{n+1} = T'_n + \lceil T_{n+1} - T_n \rceil$. The t-th sample in PROSAC is denoted by $\mathcal{M}_t$ and is generated as follows:

$$\mathcal{M}_t = \{\mathbf{u}_{g(t)}\} \bigcup \mathcal{M}'_t \tag{4}$$

where $g(t)$ is a growth function defined as $g(t) = \min\{n : T'_n \geq t\}$ and $\mathcal{M}'_t$ is a set containing $m-1$ points drawn at random from the set $\mathcal{U}_{g(t)-1}$. In practice, the PROSAC approach often achieves significant computational savings, since good hypotheses are generated early on in the sampling process. Two important points must, however, be noted. First, though PROSAC often succeeds in dramatically reducing the number of hypotheses required, this is data-dependent, and also hinges on the availability of a reasonable similarity function to rank correspondences. Secondly, we observe that in many cases, correspondences with high similarity scores often lie on the same spatial structure and are potentially in a degenerate configuration. Thus, when utilizing the PROSAC approach, it is advisable to ensure robustness to degenerate configurations [9,10].

2.3.2 Guided Sampling for MLESAC

A similar approach to PROSAC was proposed earlier by Tordoff and Murray in [6], wheer the Maximum Likelihood Estimation Sample Consensus (MLESAC) algorithm [11] was combined with non-uniform sampling of correspondences. The

Algorithm 1. Preemptive RANSAC

```
Generate all hypotheses (indexed by 1, ..., f(1))
for i = 1 to N do
  Score the hypotheses 1, ..., f(i) using data point i
  Reorder the set to retain the best f(i + 1) hypotheses
  if f(i + 1) = 1 then
    Break with the remaining hypothesis as the top one
  end if
end for
```

MLESAC algorithm is a generalization of RANSAC, which adopts the same sampling strategy but attempts to maximize the likelihood of the solution, as opposed to the number of inliers. While MLESAC assumes a uniform prior for the validity of a match, the *guided-sampling* approach of [6] uses the quality function of the feature matching algorithm to derive probabilities of match validity. These are then incorporated as priors in MLESAC. The technique was experimentally shown to reduce the number of iterations required by MLESAC by an order of magnitude.

2.3.3 Lo-RANSAC

One of the assumptions inherent in the standard termination criterion of RANSAC (equation (1)) is that a model computed from an uncontaminated sample is consistent with all inliers. In practice, this is often not the case, particularly when the data points are noisy. Chum et al. [5] define a locally optimized RANSAC variant to deal with this issue. By observing that a good model tends to find a significant fraction of the inliers, an *inner RANSAC* strategy is devised where a constant number of hypotheses (~ 20) are generated using only the set of inliers to the current best model. Since inner RANSAC operates on a set of inliers, it is not essential that hypotheses are generated from minimal subsets. In addition to providing a more robust fit, the inner RANSAC technique has the effect of improving the consensus score more rapidly than standard RANSAC, which causes the termination criterion (1) to be met earlier.

2.4 Preemptive RANSAC

As discussed earlier, though the recent variants of RANSAC show an appreciable decrease in runtime, they are still not directly applicable in real-time implementations. This is primarily due to the fact that in order to achieve an η_0 confidence in the solution, all these methods need to draw a significantly large number of samples, particularly for low inlier ratios. In the case of real-time systems the goal is to achieve the threshold when possible; however, when this is not the case, it is desirable to obtain the best solution given the time constraints. Preemptive RANSAC attempts to do precisely this: a fixed number of hypotheses are generated beforehand, and then compared against each other by scoring them in parallel. The preemption schemes described in Section 2.2 are *depth-first* in nature, meaning that a particular hypothesis is completely evaluated before moving on to the next hypothesis. Preemptive RANSAC uses a *breadth-first* approach, where all the hypotheses generated are evaluated on a subset of the data points. Subsequently, the hypotheses are reordered based on the results of this scoring procedure, and only a fraction of the hypotheses ar e evaluated on the next subset of data. A non-increasing preemption function $f(i)$, $i = 1, ..., N$ defines the number of hypotheses retained before evaluating the ith data point. The preemption function used in [8] is

$$f(i) = \lfloor M 2^{(-\lfloor \frac{i}{B} \rfloor)} \rfloor \tag{5}$$

where M is the number of hypotheses in the initial set and B is the number of data points that a hypothesis is evaluated against, before the preemption and reordering step takes place. This process continues until only one hypothesis remains, or all data points have been used. Recommended values for these parameters are $M = 500$ hypotheses and $B = 100$ data points per subset.

3 Adaptive Real-Time Random Sample Consensus

In this section, we describe a novel framework for Adaptive Real-Time Random Sample Consensus (ARRSAC), which is capable of providing accurate real-time estimation over a wide range of inlier ratios. From the discussion on preemptive RANSAC (Section 2.4), it can be seen that the number of hypotheses M in the initial candidate set is fixed *a priori*. In a real-time setting, this ensures that the runtime of the algorithm is bounded. However, fixing this number beforehand effectively places a restriction on the inlier ratio. When the true inlier ratio is high, preemptive RANSAC evaluates far too many samples. On the other hand, when the true inlier ratio is low, the likelihood that preemptive RANSAC finds a correct solution decreases drastically. The breadth-first scan advocated by preemptive RANSAC is indeed a natural formulation for real-time applications, since the primary constraint is the fixed time-budget. On the other hand, adapting the number of hypotheses to the contamination level of the data requires an estimate of the inlier ratio, which necessitates the adoption of a depth-first scan. The ARRSAC framework retains the benefits of both approaches (i.e, bounded run-time as well as adaptivity) by operating in a *partially depth-first* manner. Hypotheses for the initial candidate set are generated one-by-one, and are evaluated on the first data block. Bad hypotheses are discarded based on a depth-first preemptive strategy. Hypotheses that pass this verification procedure provide an estimate of the inlier ratio, which is used to determine the required number of hypotheses, keeping the upper limit fixed at M. Note that since the evaluation is performed on only a subset of the data (thus, partially depth-first), the estimated inlier ratio may be either above or below the true inlier ratio. In view of this fact, the ARRSAC algorithm allows the generation of additional hypotheses at a later stage in the evaluation process. While this discussion describes a method for adaptively determining the number of hypotheses, note that there is still an upper limit on the size of the hypothesis set. In such a scenario, it is then

Algorithm 2. The ARRSAC algorithm

Initialization
Set values for M (max. number of candidate hypotheses) and B (block size)
Set parameters for the SPRT test, calculate the initial value of A (see [4])
1. Generate initial hypothesis set (Algorithm 3)
k =total number of hypotheses generated in the initial stage
2. Preemptive evaluation
for $i = B + 1$ to N **do**
  Set p =number of hypotheses remaining, n =min($f(i), p/2$)
  Reorder and select hypotheses $h(1), ..., h(n)$
  if $n = 1$ **then**
    Break with the remaining hypothesis as the top one
  end if
  Score the hypotheses using data point i
  if (i mod B) = 0 **then**
    Calculate best inlier ratio $\hat{\varepsilon}$ and number of hypotheses M' (equation (1))
    M' =max(M, M')
    if $M' > k$ **then**
      Generate and evaluate $M' - k$ new hypotheses on i data points
      $k = M'$
    end if
  end if
end for

important to choose a strong set of candidate hypotheses, since this facilitates good performance at low inlier ratios. ARRSAC accomplishes this by adopting a non-uniform sampling approach, generating hypotheses from the highest quality matches. Note, however, that even in the absence of information to guide non-uniform sampling (or when this does not help), ARRSAC still has a bounded runtime. This is in contrast with strategies such as PROSAC, which in the worse case operates like RANSAC (and hence has a potentially unbounded runtime). The ability to sample non-uniformly from the data is crucial only when the inlier ratios are very low. As an additional strategy, once a hypothesis passes the verification test, it is also possible to generate new hypotheses by sampling from data points that are consistent with this good hypothesis. This idea is similar to the *inner RANSAC* technique in [5]. Since the sampling now is selecting points from inlier data, it is not essential that hypotheses are generated from minimal subsets. An important advantage of the competitive evaluation framework is that ARRSAC does not spend excessive time in local optimization, since the hypotheses generated in the inner RANSAC are forced to compete among themselves. This prevents the algorithm from spending excessive time in the refinement, particularly when the inlier ratio is high.

Finally, while the partially depth-first evaluation strategy described above serves to adaptively determine the number of hypotheses, it also provides an additional computational advantage. In the original preemptive RANSAC algorithm, hypotheses are evaluated on all B data points in a block before the reordering and preemption step takes place. As we have noted earlier, this

Algorithm 3. ARRSAC: Generating the initial hypothesis set

$k = 1$, $M' = M$, $count_{in} = 0$, Inner RANSAC flag: $flag_{in} = 0$
Set max number of inner RANSAC iterations: M_{in}
repeat
  if $flag_{in} = 0$ **then**
    Generate hypothesis $h(k)$ by selecting the k-th PROSAC sample
  else
    Generate hypothesis $h(k)$ by sampling (perhaps non-minimal) subset from $\mathcal{U}_{in}$
    $count_{in} = count_{in} + 1$
    if $count_{in} = M_{in}$, reset inner RANSAC parameters $(count_{in}, flag_{in})$
  end if
  Evaluate hypothesis $h(k)$ using Wald's SPRT
  if hypothesis $h(k)$ is rejected **then**
    Reestimate parameters of the SPRT (if required)
  else if hypothesis $h(k)$ is accepted and has the largest support so far **then**
    $flag_{in} = 1$, $count_{in} = 0$, set $\mathcal{U}_{in}$ = support of hypothesis $h(k)$
    Reestimate parameters of the SPRT
    Estimate inlier ratio $\hat{\varepsilon}$ and M' (equation (1)). Cap M' at a max of M
  end if
  $k = k + 1$
until $(k > M')$
Return k, set containing accepted hypotheses

constitutes an unnecessary overhead since most of the generated hypotheses are likely to be contaminated. Thus, in ARRSAC, bad hypotheses are discarded after partial evaluation and only the hypotheses that survive are selected for propagation to subsequent stages of evaluation. The preemption function (7) thus effectively defines an upper bound on the number of hypotheses required for the next stage of evaluation. The number of good hypotheses that need to be reordered and propagated in ARRSAC is typically much less than this bound.

To summarize the main contributions of the ARRSAC algorithm: the adoption of a partially depth-first strategy for hypothesis generation provides a great degree of control over the initial set. Adaptively determining the number of hypotheses ensures computational efficiency for higher inlier ratios. Furthermore, this strategy also allows more control over the quality of hypotheses that are included in the initial set, thereby increasing the likelihood of obtaining a good solution even for low inlier ratios. Finally, the use of optimized verification techniques is geared towards improving the overall efficiency of the algorithm. We note that unlike preemptive RANSAC, the proposed method makes no assumptions about the underlying data. The ability to efficiently adapt to the data and provide accurate real-time robust estimation makes ARRSAC easily deployable in real-time vision systems.

3.1 Algorithm Description

Generating the Initial Hypothesis Set. The ARRSAC algorithm adopts a PROSAC-based non-uniform sampling strategy to generate hypotheses. For

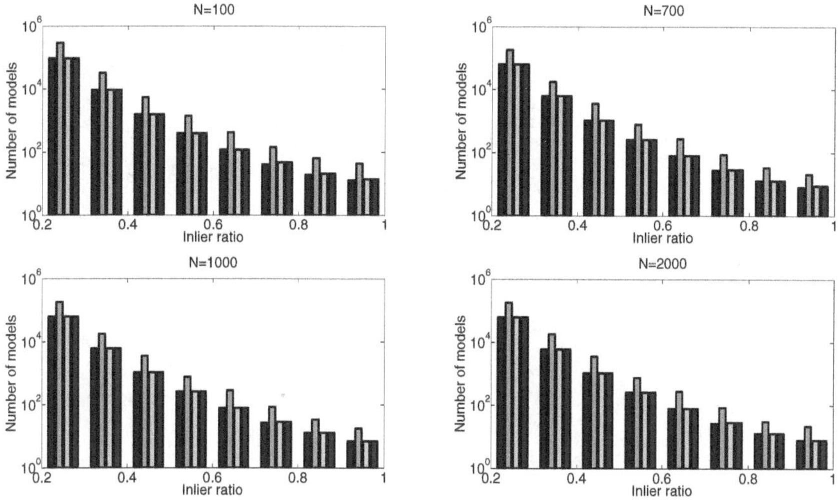

Fig. 1. Results of synthetic data evaluation. Number of models evaluated vs. inlier ratio for four different values of N. From left-right in each group: RANSAC, $T_{d,d}$, bail-out, WaldSAC. Note that the y-axis uses a log scale.

clarity, we also restrict our attention to a preemptive approach based on SPRT, though in principle any depth-first preemption scheme may be used. For the SPRT, the threshold parameter A depends on the inlier ratio ε and the Bernoulli parameter δ. These are initialized conservatively ($\varepsilon = 0.1$, $\delta = 0.05$) and updated as the processing continues. During the SPRT, a hypothesis is either rejected or accepted. In the first case, we continue to the next hypothesis, updating δ if required. In the second case, when a hypothesis is accepted and has the best support so far, a fixed number of new hypotheses, M_{in}, are generated by sampling non-minimal subsets from the inliers to this hypothesis. Furthermore, an accepted hypothesis provides an estimate $\hat{\varepsilon}$ of the inlier ratio; this is used to determine the new number of hypotheses required, M'. Note that we cap the maximum number of hypotheses in the initial set at M, so $M' \leq M$. It is also important to note that the estimate $\hat{\varepsilon}$ may be either greater than or less than the actual inlier ratio. However, the flexibility of the ARRSAC algorithm allows us to always generate additional hypotheses at a later stage, if required to do so.

Preemptive evaluation. The evaluation procedure in ARRSAC is based on the partially depth-first approach outlined earlier. While preemptive RANSAC uses the fixed preemption function (7), ARRSAC is more flexible in this regard. Thus, at any given point, there are typically far less than $f(i)$ hypotheses in the set, due to the fact that contaminated hypotheses are discarded. In practice, this turns out to be a very significant computational saving (see Section 4). After every evaluation of B data points, the inlier ratio $\hat{\varepsilon}$ is re-estimated, and the value of M' is updated. If, at any point, the inlier ratio is found to be an overestimate, additional hypotheses are evaluated as shown in Algorithm 2.

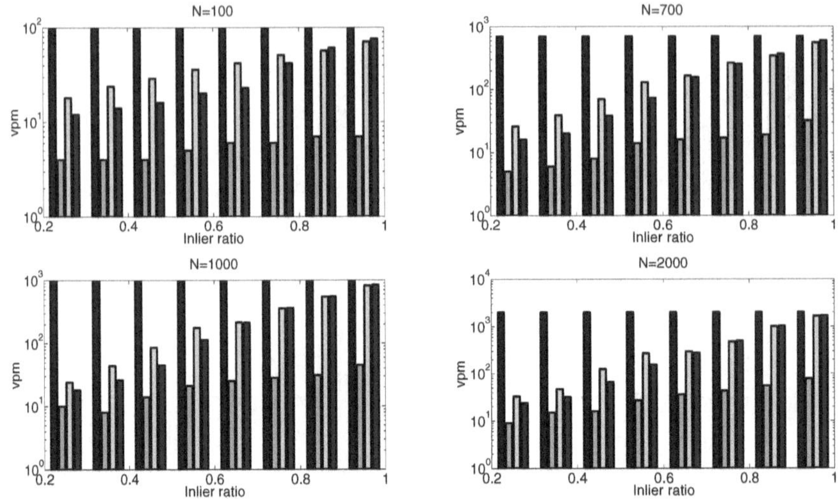

Fig. 2. Results of synthetic data evaluation. Number of verifications per model (vpm) vs. inlier ratio for four different values of N. From left-right in each group: RANSAC, $T_{d,d}$, bail-out, WaldSAC. Note that the y-axis uses a log scale.

4 Experimental Evaluation

In this section, we provide a detailed comparative evaluation of the approaches described in this paper. First, we compare the performance of the various depth-first preemptive approaches discussed in Section 2.2. For this experiment, we estimate the epipolar geometry for synthetic data over a wide range of parameter variations. Gaussian noise ($\mu = 0, \sigma = 0.2$) was added to the pixel locations. The results of the synthetic evaluation are shown in Figures 1 and 2. Figure 1 shows the variation in the number of models evaluated as a function of the inlier ratio ε, for four different values of N (total number of correspondences). Figure 2 shows the number of verifications per model, again as a function of ε and N. The results are averaged over 200 runs for each tested case. Figure 2 shows that the $T_{d,d}$ test succeeds in greatly reducing the number of verifications per model. On the other hand, it requires the generation of a much larger number of hypotheses. The performance of the bail-out and SPRT tests are comparable over the range of parameter variations, with the SPRT producing slightly greater speed-ups on average. However, for high inlier ratios, when the total number of models generated is small, the bail-out test often performs marginally better than SPRT, since the parameters in the SPRT are initialized conservatively.

We next evaluate the performance of preemptive RANSAC on synthetic data, over a range of parameter variations. The goal in this case is not to examine the number of verifications or the execution time (which are constant). Rather, we seek to examine the ability of preemptive RANSAC to cope with varying inlier ratios. Figure 3(a) shows the fraction of inliers found, as a function of ε and N. The main trend to be observed from the graph is that for lower inlier ratios

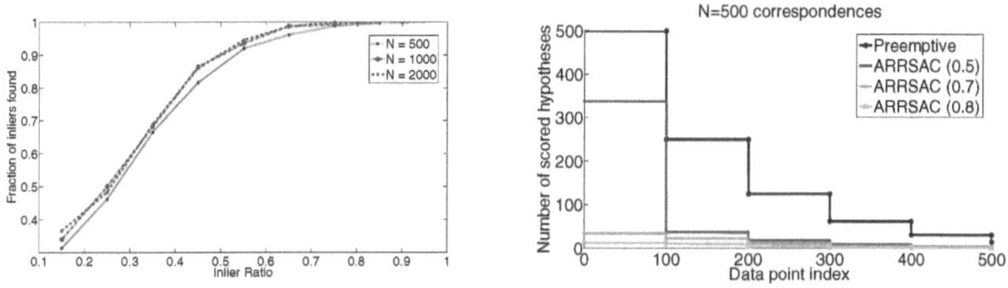

Fig. 3. (a) Variation in fraction of inliers found by preemptive RANSAC as a function of inlier ratio. (b) Number of hypotheses scored at each stage of preemption in preemptive RANSAC and ARRSAC, for four values of ε.

(< 0.5), preemptive RANSAC does not evaluate a sufficient number of hypotheses, and thus misses a significant number of inliers. This presents a drawback for the robust operation of a real-time system. Figure 3(b) compares the number of hypotheses retained at each stage of preemption for preemptive RANSAC and ARRSAC, for synthetic data at four values of ε. Note that for this experiment, it was assumed that a PROSAC-based ordering of the hypotheses was unavailable. It can be seen that even without PROSAC, the partially depth-first evaluation strategy used in ARRSAC causes an appreciable decrease in the number of hypotheses evaluated at each stage. Thus, the effective preemption function in ARRSAC always remains below that of equation (7), providing considerable computational savings. This demonstrates one of the advantages associated with being able to adapt to the contamination level of the data.

Finally, we provide a comprehensive comparative evaluation of the various RANSAC methods on real image data, and demonstrate the efficiency of the ARRSAC approach, in terms of accuracy and computational efficiency. A sample of the test images chosen for this experiment are shown in Table 1. These comprise of image pairs from well-known sequences (A-C), as well as images taken from various other sources (D-F, G, H-I, J). The images were selected to cover a wide range of inlier ratios and number of correspondences. Corresponding points in the image pairs were detected using Harris corner matching. The results for the evaluation are shown in Table 1 which lists, for each method and test sequence, the number of inliers found (I), the number of models evaluated (k), the number of verifications per model (vpm) and the relative speed-up obtained with respect to standard RANSAC (spd-up). A few important trends are noticeable from the table. Firstly, for images where the total number of models evaluated is low (A, B, C), it can be observed that preemptive RANSAC evaluates more hypotheses than standard RANSAC and is thus slower. On the other hand, ARRSAC is able to efficiently estimate the inlier ratio and accordingly limit the number of hypotheses, producing a considerable speed-up. Secondly, for images with low inlier ratios (D, E, F), preemptive RANSAC is often unable to find the correct solution since it evaluates too few hypotheses. In contrast, the adaptive hypothesis generation in ARRSAC enables accurate estimation of

Table 1. Evaluation results for ten selected real image pairs. The images show variation over a range of inlier ratios and number of correspondences. It can be observed from the above results that the ARRSAC approach produces significant computational speed-ups, while simultaneously providing accurate robust estimation in real-time. In practice, the ARRSAC technique stays well within the time budget, with estimation speeds ranging between 55-350 Hz. It can be seen from the table that the number of hypotheses evaluated by ARRSAC is always less than preemptive RANSAC. In addition, the correct epipolar geometry is always recovered.

		RANSAC	$T_{d,d}$	Bail-out	Wald	PROSAC	Pre-emptive	ARRSAC
A : $\varepsilon = 0.83$, $N = 1322$	I	884	884	885	889	885	933	1099
	k	112	156	118	148	5	500	6
	vpm	1322	531	361	594	1322	396	434
	spd-up	1.0	1.9	2.6	2.2	12.9	0.4	37.3
B : ($\varepsilon = 0.7$, $N = 795$)	I	514	512	514	517	514	527	559
	k	242	329	242	283	43	500	47
	vpm	795	149	65	132	795	374	218
	spd-up	1.0	2.4	4.1	3.9	7.7	0.3	11.1
C : ($\varepsilon = 0.65$, $N = 2162$)	I	1344	1342	1344	1345	1344	1372	1413
	k	208	947	208	211	17	500	17
	vpm	2162	59	67	52	2162	400	430
	spd-up	1.0	1.9	5.5	5.8	15.2	0.9	29.1
D : ($\varepsilon = 0.33$, $N = 1500$)	I	483	483	484	484	484	350	491
	k	15506	49090	15509	15514	201	500	241
	vpm	1500	22	37	28	1500	398	59
	spd-up	1.0	1.9	5.7	6.2	104.2	90.6	305.7
E : ($\varepsilon = 0.44$, $N = 420$)	I	174	174	173	177	177	134	183
	k	2714	4889	2719	2716	439	500	499
	vpm	420	10	24	13	420	305	70
	spd-up	1.0	2.0	4.4	5.7	5.3	11.5	20.2
F : ($\varepsilon = 0.43$, $N = 1325$)	I	557	557	557	557	557	488	582
	k	4353	11753	4366	4359	288	500	329
	vpm	1325	5	31	24	1325	396	182
	spd-up	1.0	2.1	4.7	5.3	6.9	12.3	15.1
G : ($\varepsilon = 0.56$, $N = 2875$)	I	1537	1524	1539	1537	1537	1539	1616
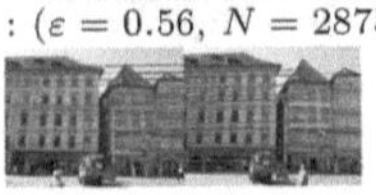	k	432	628	437	432	30	500	33
	vpm	2875	20	24	31	2875	400	456
	spd-up	1.0	2.2	3.1	4.2	16.6	1.3	47.2
H : ($\varepsilon = 0.67$, $N = 1986$)	I	1249	1247	1251	1252	1250	1237	1330
	k	335	1153	351	347	17	500	18
	vpm	1986	29	41	32	1986	400	467
	spd-up	1.0	2.4	4.2	4.9	22.2	1.1	60.9
I : ($\varepsilon = 0.76$, $N = 1508$)	I	945	934	948	951	949	960	1149
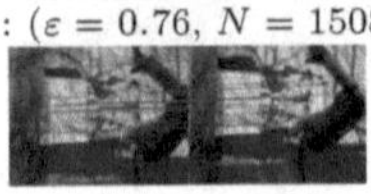	k	498	1644	505	502	58	500	65
	vpm	1508	43	31	33	1508	398	277
	spd-up	1.0	1.9	2.7	2.5	12.3	1.8	45.1
J : ($\varepsilon = 0.61$, $N = 1521$)	I	872	872	873	873	872	870	927
	k	362	1167	381	372	40	500	44
	vpm	1521	39	28	25	1521	398	338
	spd-up	1.0	2.2	6.1	6.3	9.8	1.9	26.1

the epipolar geometry even for very low inlier ratios. Thirdly, even in the case where a PROSAC-based sampling strategy does not provide any significant advantage (E), ARRSAC is still able to achieve a performance speed-up, due to the efficient preemption strategy employed (as illustrated in Fig 3(b)). As the results in Table 1 indicate, the ARRSAC algorithm is able to elegantly adapt to the data and ensure accurate robust estimation over a wide range of inlier ratios, while at the same time providing considerable computational savings.

5 Conclusion

In this paper, we presented a discussion and comparative analysis of a number of important RANSAC techniques. In addition, a powerful new formulation for adaptive real-time random sample consensus (ARRSAC) was proposed. In contrast to existing techniques for real-time robust estimation, the ARRSAC approach makes no limiting assumptions about the underlying data, and is capable of efficiently adapting to the contamination level of the data. The framework we develop is suitable for use in real-time applications with a limited time budget, while at the same time providing accurate robust estimation over a wide range of inlier ratios. Experimental results on synthetic and real data demonstrate very significant improvements in accuracy and speed over existing methods.

References

1. Fisher, M., Bolles, R.: Random sample consensus: A paradigm for model fitting with applications to image analysis and automated cartography. Comm. of the ACM 24(6), 381–395 (1981)
2. Matas, J., Chum, O.: Randomized RANSAC with $T_{d,d}$ test. Image and Vision Computing 22(10), 837–842 (2004)
3. Capel, D.: An effective bail-out test for RANSAC consensus scoring. In: Proc. BMVC, pp. 629–638 (2005)
4. Matas, J., Chum, O.: Randomized ransac with sequential probability ratio test. In: Proc. ICCV, pp. 1727–1732 (2005)
5. Chum, O., Matas, J., Kittler, J.: Locally optimized RANSAC. In: Michaelis, B., Krell, G. (eds.) DAGM 2003. LNCS, vol. 2781, pp. 236–243. Springer, Heidelberg (2003)
6. Tordoff, B., Murray, D.W.: Guided sampling and consensus for motion estimation. In: Heyden, A., Sparr, G., Nielsen, M., Johansen, P. (eds.) ECCV 2002. LNCS, vol. 2350, pp. 82–98. Springer, Heidelberg (2002)
7. Chum, O., Matas, J.: Matching with PROSAC - progressive sample consensus. In: Proc. CVPR, vol. 1, pp. 220–226 (2005)
8. Nister, D.: Preemptive RANSAC for live structure and motion estimation. In: Proc. ICCV, vol. 1, pp. 199–206 (2003)
9. Chum, O., Werner, T., Matas, J.: Two-view geometry estimation unaffected by a dominant plane. In: Proc. CVPR, pp. I: 772–779 (2005)
10. Frahm, J.M., Pollefeys, M.: Ransac for (quasi-)degenerate data (QDEGSAC). In: Proc. CVPR, pp. 453–460 (2006)
11. Torr, P., Zisserman, A.: MLESAC: A new robust estimator with application to estimating image geometry. CVIU, 138–156 (2000)
12. Chum, O., Matas, J.: Optimal randomized RANSAC. Pattern Analysis and Machine Intelligence (to appear)

Video Registration Using Dynamic Textures

Avinash Ravichandran and René Vidal

Center for Imaging Science, Johns Hopkins University, Baltimore, MD 21218, USA

Abstract. We propose a dynamic texture feature-based algorithm for registering two video sequences of a rigid or nonrigid scene taken from two synchronous or asynchronous cameras. We model each video sequence as the output of a linear dynamical system, and transform the task of registering frames of the two sequences to that of registering the parameters of the corresponding models. This allows us to perform registration using the more classical image-based features as opposed to space-time features, such as space-time volumes or feature trajectories. As the model parameters are not uniquely defined, we propose a generic method to resolve these ambiguities by jointly identifying the parameters from multiple video sequences. We finally test our algorithm on a wide variety of challenging video sequences and show that it matches the performance of significantly more computationally expensive existing methods.

1 Introduction

A classical problem in computer vision is aligning two images of the same scene, taken from different viewpoints. This problem is known as image registration, and its objective is to recover the spatial alignment between the two images.

There exists a vast amount of literature on the problem of image registration for both rigid and nonrigid scenes (see [1] and the references therein). For example, feature-based approaches to image registration proceed by first extracting features such as Harris Corners [2], SIFT [3], MOPS [4], etc., from the images. The features from the first image are then matched to the features from the second image. The alignment between the two images is recovered from these matches using methods such as RANSAC [5]. A homography or an affine transformation is then computed by fitting such a model to the inliers given by RANSAC.

The last few years have seen an increasing interest in the more challenging problem of registering video sequences. In addition to recovering the spatial transformation, video registration also involves recovering the temporal alignment between the two videos. Although this poses additional challenges with respect to image registration, one can still argue that image registration algorithms can be extended in a straightforward manner to solve the video registration problem. When there is no temporal lag/shift between the two video sequences, the problem reduces to aligning corresponding pairs of images from the two video sequences. That is, one can extract features from every frame of the two video sequences, match the features from a frame in the first sequence to the features from the corresponding frame in the second sequence, and then

D. Forsyth, P. Torr, and A. Zisserman (Eds.): ECCV 2008, Part II, LNCS 5303, pp. 514–526, 2008.

recover a spatial transformation by applying RANSAC to all features from all pairs of frames. When there is a temporal lag between the two videos, one could follow the same procedure. However, the correspondence between frames of the two video sequences is unknown. This correspondence can be found using all possible choices of frame pairs from the two video sequences. This exhaustive search approach will indeed lead to the optimal result. However, this comes at the price of the computational cost involved in solving this brute force optimization.

To circumvent this problem, Caspi et al. [6,7] use feature trajectories rather than features from every frame. Feature points are extracted from the first frame and tracked throughout the video. The registration problem is posed as a trajectory matching problem between the two video sequences. An optimization problem is then solved to jointly recover the spatial and temporal transformation between the video sequences. In some sense the tracking algorithm performs the matching between two consecutive frames, which is an easier problem compared to matching features across non corresponding frames. Ukrainitz et al. [8] on the other hand, work with dense space-time information instead of feature trajectories. The idea here is to maximize a global similarity measure between small space-time volumes extracted from the two videos using a costly optimization scheme.

Paper Contributions. In this paper, we propose a much simpler video registration framework, which requires neither feature tracking nor a costly optimization procedure. Instead, we reduce the spatial-temporal video registration problem to the spatial registration of a collection of image bases extracted from the video.

Similar to [9], we model the two video sequences as the output of two linear dynamical systems (LDSs). Since the two cameras observe the same dynamical scene, but from a different viewpoint, we assume that the dynamics of the image intensities in the two views are the same. Our first contribution is to propose a simple scheme for the joint identification of the two LDSs. The proposed method retains the sub-optimality of the original identification algorithm of Doretto et al. [10], but at the same time enforces the criterion that the dynamics of the two video sequences must be the same.

Since the parameters of the identified LDSs do not depend on the temporal alignment, the spatial registration of the two videos can be obtained by applying classical image registration techniques to the parameters of the LDSs. However, since the parameters of the LDSs are computed only up to change of basis, they need to be transformed to a canonical form before they can be compared. Our second contribution is to propose the use of the Jordan canonical form (JCF), together with a method for converting the identified parameters to the JCF. This converts all the parameters into the same basis and makes comparing the parameters more straightforward. The method is numerically stable, independent of a reference sequence and scales very well for an arbitrary number of sequences.

We finally propose a feature-based method for registering multiple video sequences using the dynamic texture model. Our approach does not work with space-time volumes [7] or feature trajectories [8], and does not rely on heuristics such as the appearance image [9]. Instead, we extract standard image based features from the parameters of the LDSs to perform the registration.

2 Video Registration Using Dynamic Textures

2.1 Dynamic Textures Framework

We first briefly review the dynamic texture model. Please see [10] for more details. Given a video sequence $\{I(t) \in \mathbb{R}^p\}_{t=1}^F$, we model it as the output of a LDS

$$z(t+1) = Az(t) + Bv(t) \tag{1}$$

$$I(t) = C^0 + Cz(t) + w(t). \tag{2}$$

The vector $z(t) \in \mathbb{R}^n$ represents the *hidden state* of the system at time t. Its evolution is controlled by the *dynamics matrix* $A \in \mathbb{R}^{n\times n}$ and the *input-to-state matrix* $B \in \mathbb{R}^{n\times q}$. The zero-mean Gaussian processes $v(t) \sim \mathcal{N}(0,Q)$ and $w(t) \sim \mathcal{N}(0,R)$ model the process noise and the measurement noise, respectively. The *appearance matrix* $C \in \mathbb{R}^{p\times n}$ maps the hidden state to the image and the vector $C^0 \in \mathbb{R}^p$ is the temporal mean of the video sequence. The order of the system is given by n and p is the number of pixels in the image.

The advantage of using this model is that it enables us to decouple the appearance of the video, represented by C, from the temporal evolution of the video represented by A. This property will allow us to recover the spatial registration independently from the temporal lag between the video sequences.

Given $\{I(t)\}_{t=1}^F$, the first step is to identify the parameters of the LDS. There are several choices for the identification of such systems from the classical system identification literature, e.g., subspace identification methods such as N4SID [11]. The problem with such methods is that, as the size of the output increases, these methods become computationally very expensive. Hence, traditionally the method of identification for dynamic textures has been a suboptimal solution proposed in [10]. This method is essentially a Principal Component Analysis (PCA) decomposition of the video sequence. Given the video sequence $\{I(t)\}_{t=1}^F$, the mean C^0 is first calculated. The parameters of the system are then identified using SVD to factorize the mean subtracted matrix as

$$\left[I(1) - C^0, \ldots, I(F) - C^0\right] = U(SV^\top) = CZ. \tag{3}$$

Given $Z = [z(1)\ldots z(F)]$, the parameter A is obtained as the least squares solution to the system of linear equations $A[z(1)\ldots z(F-1)] = [z(2)\ldots z(F)]$.

It is well known that if one performs the SVD, the factorization is unique up to an invertible transformation, i.e. the recovered factors are CP^{-1} and PZ, where $P \in \mathbb{R}^{n\times n}$ is an arbitrary invertible matrix. Hence the LDSs with parameters (A,B,C) and (PAP^{-1}, PB, CP^{-1}) both generate the same output process. This fact does not pose a problem when dealing with a single video sequence. However, if one wants to compare the parameters identified from multiple sequences, each set of identified parameters could potentially be computed with respect to a different basis. This poses a problem when one wants to compare the parameters say column by column. In order to address this issue, in the next section we outline a method to account for the change of basis. We propose to do this by converting the parameters into a canonical form. Once the parameters are in the canonical form, the task of comparing these parameters becomes straightforward.

2.2 Jordan Canonical Form for Parameter Comparison

There have been a few attempts to address the basis issue in prior work. In [12], Chan et al. used one sequence as the reference and converted the other sequence into the basis of the reference sequence. Vidal et al. in [13] used the diagonal form of the A matrix. The problem is that the resulting parameters in canonical form are complex, because the eigenvalues of A can be complex. To deal with this issue, the Reachability Canonical Form (RCF) was used in [9]. However, the RCF form uses the pair (A, B) to convert the system into canonical form. For applications of dynamic textures, such as registration and recognition, it is preferable to have a canonical form based on the parameters (A, C), because they model the appearance and the dynamics of the system. The matrix B, which models the input noise, is not that critical to describe the appearance of the scene. An obvious alternative is to use the Observability Canonical Form (OCF) [14]

$$A_c = \begin{bmatrix} -a_{n-1} & 1 & 0 & \cdots & 0 \\ -a_{n-2} & 0 & 1 & \cdots & 0 \\ \vdots & \vdots & \vdots & \ddots & \vdots \\ -a_1 & 0 & 0 & \vdots & 1 \\ -a_0 & 0 & \dots & 0 & 0 \end{bmatrix} \in \mathbb{R}^{n \times n} \quad \text{and} \quad C_c = \begin{bmatrix} 1 \, 0 \, 0 \cdots 0 \end{bmatrix} \in \mathbb{R}^{1 \times n}, \tag{4}$$

where $A^n + a_{n-1}A^{n-1} + \ldots + a_0 I = 0$ is the characteristic polynomial of A. However, it is well known that using the OCF can lead to numerical instabilities [14].

In order to overcome this issue, we propose to use a canonical form based on the Jordan real form. When A has $2p$ complex eigenvalues and $n - 2p$ real eigenvalues, the Jordan Canonical Form (JCF) is given by

$$A_c = \begin{bmatrix} \sigma_1 & \omega_1 & 0 & \cdots & 0 \\ -\omega_1 & \sigma_1 & 0 & \cdots & 0 \\ \vdots & \vdots & \ddots & 0 & 0 \\ 0 & 0 & 0 & \lambda_{2p-n-1} & 0 \\ 0 & 0 & \dots & 0 & \lambda_{2p-n} \end{bmatrix} \quad \text{and} \quad C_c = \begin{bmatrix} 1 \, 0 \, 1 \, 0 \dots 1 \end{bmatrix}, \tag{5}$$

where the eigenvalues of A are given by $\{\sigma_1 \pm i\omega_1, \ldots, \sigma_p \pm i\omega_p, \lambda_1, \ldots, \lambda_{n-2p}\}$. Note that the JCF is equivalent to the RCF or the OCF, but in a different basis.

Given any canonical form, we now outline the steps that we need to take in order to convert the identified parameters into the canonical form. Assume that we have the identified parameters (A, C). We need to find an invertible matrix P such that $(PAP^{-1}, \gamma^\top CP^{-1}) = (A_c, C_c)$, where the subscript c represents any canonical form. The vector $\gamma \in \mathbb{R}^p$ is an arbitrary vector chosen to convert the

LDS (A, C) with p outputs to a canonical form, which is defined for only one output. In our experiments we set it to be $\begin{bmatrix}1\ 1\ldots 1\end{bmatrix}^\top$, so that all rows of C are weighted equally. Now, the relationship between the A matrix and its canonical form A_c is a special form of the Sylvester equation

$$A_c P - PA = 0. \tag{6}$$

Vectorizing this equation, we obtain $(I \otimes A_c - A^\top \otimes I)\text{vec}(P) = 0$, where $\otimes$ represents the Kronecker product. If we also vectorize the equation relating the C matrices, $C_c P = \gamma^\top C$, we can solve for P from both equations as follows

$$\begin{bmatrix} I \otimes A_c - A^\top \otimes I \\ I \otimes C_c \end{bmatrix} \text{vec}(P) = \begin{bmatrix} 0 \\ C^\top \gamma \end{bmatrix}. \tag{7}$$

Once we have solved this equation we can convert the parameters into the canonical form using P. It should be noted that the JCF is unique only up to a permutation of the eigenvalues. However, if the eigenvalues are different, we can choose a predefined way to sort the eigenvalues, and obtain a unique JCF.

2.3 Joint Identification of Dynamic Textures

In this section, we will show that using a canonical form helps us to jointly identify the parameters of the LDSs associated with multiple video sequences of the same rigid or nonrigid scene, but taken by multiple (asynchronous) cameras. We will model such a collection of m video sequences $\{I_i(t) \in \mathbb{R}^{p_i}\}_{t=1,\ldots,F}^{i=1,\ldots m}$ as the output of m LDSs whose dynamics are the same. Specifically, the LDSs share the same A matrix and the same state $z(t)$, modulo a temporal shift $\tau_i \in \mathbb{Z}$, i.e.

$$z(t+1) = Az(t) + Bv(t) \quad \text{and} \quad I_i(t) = C_i^0 + C_i z(t+\tau_i) + w_i(t). \tag{8}$$

One possible way of identifying the parameters these m LDSs is to apply the identification algorithm described in §2.1 to each video sequence separately. More specifically, we let $\tilde{I}_i(t) = I_i(t) - C_i^0$, form the matrix $W_i = [\tilde{I}_i(1) \ldots \tilde{I}_i(F)]$, and calculate its SVD $W_i = U_i S_i V_i^\top$. The parameters of the LDSs are identified as $C_i = U(:, 1:n)$ and $Z_i = S(1:n, 1:n)V(:, 1:n)^\top$. The matrix A_i is then linearly obtained from Z_i. An immediate problem with this scheme is that, due to the presence of noise, viewpoint changes and the suboptimal identification, there is no guarantee that the A_i matrices for the m videos will be the same.

In order to enforce the same dynamics, we propose a simple method where all the videos are stacked into a single matrix W, which is factorized using the SVD as

$$W = \begin{bmatrix} \tilde{I}_1(1) \ldots \tilde{I}_1(F) \\ \vdots \\ \tilde{I}_m(1) \ldots \tilde{I}_m(F) \end{bmatrix} = USV^\top. \tag{9}$$

Although this seems to be the intuitively obvious thing to do, we will now show that this is indeed the correct thing to do. If for the sake of analysis we ignore the noise terms, we obtain the state evolution as $z(t) = A^t z_0$, where z_0 is the initial state of the system. Now if we consider a temporal lag $\tau_i \in \mathbb{Z}$ for the ith video sequence, then the evolution of the hidden state of the ith sequence is given by $z_i(t) = A^{\tau_i} z(t)$. Therefore, we can decompose W as follows

$$W = \begin{bmatrix} C_1 A^{\tau_1} z(1) & \dots & C_1 A^{\tau_1} z(F) \\ C_2 A^{\tau_2} z(1) & \dots & C_2 A^{\tau_2} z(F) \\ \vdots & & \\ C_m A^{\tau_m} z(1) & \dots & C_m A^{\tau_m} z(F) \end{bmatrix} = \begin{bmatrix} C_1 A^{\tau_1} \\ C_2 A^{\tau_2} \\ \vdots \\ C_m A^{\tau_m} \end{bmatrix} \begin{bmatrix} z(1) \, z(2) \dots z(F) \end{bmatrix}. \quad (10)$$

Note that the C_i matrix we recover from stacking all the videos is the true C_i matrix multiplied by A^{τ_i}. The problem is that A is unknown, and so we cannot directly compute C_i. To resolve this, consider the equation for the ith video

$$[\tilde{I}_i(1) \dots \tilde{I}_i(F)] = C_i A^{\tau_i} [z(1) \dots z(F)] = C_i A^{\tau_i} (A^{\tau_i})^{-1} [z(\tau_i + 1) \dots z(F + \tau_i)]. \quad (11)$$

We can see that the parameters we estimate are the original parameters of the system, but in a different basis. Therefore, by converting the parameters to the canonical form, we can remove the trailing A^{τ_i} and recover the original parameters in their canonical form. The details can be found in Algorithm 1. Having identified a dynamic texture model for all video sequences with a common A, and all the C matrices with respect to the same basis, in the next section we describe a method to register multiple video sequences using the joint identification framework.

2.4 Registering Video Sequences Using Dynamic Textures

Consider two video sequences $I_1(x, t)$ and $I_2(x, t)$, where x denotes the pixel coordinates, and $t = 1, \dots, F$. Assume that the video sequences are related by a spatial transformation H and a temporal lag τ, i.e. $I_1(x, t) = I_2(H(x), t + \tau)$. It follows from (8) that $\tau = \tau_1 - \tau_2$, $C_1^0(x) = C_2^0(H(x))$, and $C_1(x) = C_2(H(x))$. Therefore, the joint identification algorithm described in §2.3 allows us to decouple the recovery of the spatial alignment from the temporal lag. Based on this fact, we now propose a straightforward method to spatially register two video sequences using both the *mean image* C^0 and the n columns of the C matrix, which we will refer to as the *dynamic appearance images*. Once the spatial alignment has been recovered, we temporally align the two sequences using a simple line search in the temporal direction, i.e. $\tau = \arg\min_\tau \sum_t \|I_1(\mathbf{x}, t) - I_2(H(x), t + \tau)\|^2$.

Our algorithm to spatially register the two video sequences $I_1(t)$ and $I_2(t)$ proceeds as follows. We calculate the mean images C_1^0 and C_2^0, identify the system parameters (A, C_1) and (A, C_2) in the JCF, and convert every column of C_i into its image form. We use the notation C_j^i to denote the ith column of the jth sequence represented as an image. We use a feature-based approach to spatially register the two set of images $\{C_1^0, C_1^1, \dots, C_1^n\}$ and $\{C_2^0, C_2^1 \dots, C_2^n\}$. We extract SIFT features and a feature descriptor around every feature point

Algorithm 1. (Joint identification of multiple video sequences)

1 Given $I_i(t)$, first calculate C_i^0, the temporal image mean. Set $\tilde{I}_i(t) = I_i(t) - C_i^0$.
2 Compute C, Z using the rank n singular value decomposition of the matrix

$$W = \begin{bmatrix} \tilde{I}_1(1), \ \dots \ , \tilde{I}_1(F) \\ \vdots \\ \tilde{I}_m(1), \ \dots \ , \tilde{I}_m(F) \end{bmatrix} = USV^\top \quad Z = SV^\top, \quad C = U.$$

3 Compute $A = [z(2), \dots, z(F)][z(1), \dots, z(F-1)]^\dagger$.
4 Let $C_i \in \mathbb{R}^{p_i \times n}$ be the matrix formed by rows $\sum_{j=1}^{i-1} p_j + 1$ to $\sum_{j=1}^{i} p_j$ of C, and convert the pair (A, C_i) to canonical form.

Algorithm 2. (Registration of multiple video sequences)

1 Given $I_1(t)$ and $I_2(t)$, calculate the parameters A and (C_i^0, C_i).
2 Extract features and feature descriptors from $(C_j^i), j = \{1, 2\}, i = 0, \dots, n$.
3 Match features from C_1^i to C_2^i and also in the reverse direction. Retain the matches that are consistent across both directions and concatenate the feature point location from C_1^i into X_1 and it's corresponding match into X_2
4 Recover the Homography H using RANSAC such that $X_2 \sim HX_1$.
5 Calculate temporal alignment τ as $\tau = \arg\min_\tau \sum_t \|I_1(x,t) - I_2(H(x), t+\tau)\|^2$.

in the two sets of $n+1$ images. We match the features extracted from image C_1^i with those extracted from image C_2^i, where $i \in \{0, \dots, n\}$. We then concatenate the correspondences into the matrices $X_1 \in \mathbb{R}^{3\times M}$ and $X_2 \in \mathbb{R}^{3\times M}$. The corresponding columns of X_1 and X_2 are the location of the matched features in homogenous co-ordinates and M is the total matches from the $n+1$ image pairs. We then recover a homography H such that $X_2 \sim HX_1$. In order to recover the homography, we first run RANSAC and obtain the inliers from the matches. We then fit a homography using the non linear method outlined in [15]. Notice that, by following this approach, we weight the contribution of the correspondences from every image pair equally. This is because the best matches given by RANSAC could arise from the mean image or the dynamic appearance images or both. Hence we do not explicitly restrict the algorithm to use only the mean image or only the dynamic appearance images, as done in [9]. This choice now becomes automatic.

3 Experimental Results

In this section, we evaluate the different steps of our algorithm on a wide variety of sequences and compare its performance to existing algorithms.

Evaluation of the Jordan Canonical Form. To evaluate our canonical form, we first take a video sequence and identify the parameters of the system

Table 1. Parameter errors before and after converting to the canonical form

Transformation	Errors initially		Errors in JCF		Errors in RCF	
	$\|A-\tilde{A}\|_F$	$\|C-\tilde{C}\|_F$	$\|A-\tilde{A}\|_F$	$\|C-\tilde{C}\|_F$	$\|A-\tilde{A}\|_F$	$\|C-\tilde{C}\|_F$
Sign flip	2.568	7.063	0	0	0	0
Invertible	1.16e+02	1.51e+02	6.79e-10	1.31e-04	2.00e+06	1.08e+08
Orthogonal	2.61e+00	7.06e+00	8.04e-14	1.31e-08	1.48e+07	6.46e+08

(a)

(b)

C^{14} using JCF C^{15} using JCF C^{14} using RCF C^{15} using RCF

Fig. 1. Sample columns of the C matrix displayed as images (a) Sequence 1 from the parking set (b) Sequence 2 from the parking set

using the suboptimal approach. We apply different kinds of transformations to the parameters of the system, i.e. given (C, A) we form new parameters $(\tilde{C}, \tilde{A}) = (CP^{-1}, PAP^{-1})$. This simulates the ambiguities we encounter using the suboptimal approach. We then convert (C, A) and $(\tilde{C}, \tilde{A})$ to their canonical form (C_c, A_c) and $(\tilde{C}_c, \tilde{A}_c)$ respectively. Errors are then calculated between the parameters before and after converting it to the canonical form. We perform the experiment for 200 trials of each transformation and then calculate the mean error. The transformations are randomly generated from three different classes of transformations: an invertible matrix, an orthogonal matrix or a sign flip. By a sign flip transformation, we refer to a diagonal matrix with entries in $\{-1, 1\}$. The results of this experiment are summarized in Table 1. We see that for simple transformations such as the sign flip, both the canonical forms perform very well. However, when the transformation gets more involved, the errors from the RCF are higher than the initial errors, while the JCF is still able to perform well.

In order to qualitatively show the difference between the canonical forms, we show two sample columns from the C matrix of two video sequences from the parking set, which we will later show registration results on. Fig. 1 shows columns of C identified using both algorithms. One can see that the corresponding column images are spatially transformed versions of each other when we use the JCF. However, no such trend is observed when we use the RCF.

Evaluation of the Registration. For our experiments on registration, we tested our algorithm on all the sequences available at [16], one sequence from [17] and one from [18]. These test video sequences consist of 3 pairs of sequences of rigid scenes, 3 pairs of sequences of nonrigid scenes, and 2 other pairs of sequences. Our last category had one pair of sequences with a relatively large zoom factor and the other was from two non stationary cameras capturing different modalities.

We use the *composite video* in order to qualitatively asses the performance of the registration. The composite video is formed by taking the red and blue color channels from the first video and the green color channel from the second video. We show the composite video before and after registration so as to display both the initial and final alignments. Notice that green and purple regions in the composite video correspond to errors in the registration, or regions that exist in one image, but not in the other.

We first compare the methods of [9] and [6] to ours on the common sequences from [9] and [6], i.e. the flag, parking, and fountain sequences. We see from Fig. 2 that for all the 3 sequences the alignment we obtain is as good if not better than

Initial Alignment | Ravichandran [9] | Caspi [6] | Our Method

Fig. 2. Comparison of results from three methods: First Row: Flag sequence, Second Row: Parking sequence and Last Row: Fountain sequence

Fig. 3. Comparison with [6] (a) Light (b) Palm (c) Firework Sequences. Comparison with [7] (d) Wide Zoom Sequence. Comparison with [8] (e) Library Sequence.

Table 2. Percentage inliers from dynamic appearance images for different sequences

Type	Rigid			Non-Rigid			Other	
Sequence	Palm	Parking	Light	Fountain	Fireworks	Flag	Wide	Multimodal
Inlier Percentage	68.75	78.74	84.66	93.80	93.90	100.00	71.02	89.90

that obtained by the other methods. Moreover, when compared to [9], we are able to perform the registration with lower model orders for the LDSs. The model order for [9] was in the range 50-75, but in our method they were in the range of 20-30. In comparision with [6], our algorithm avoids expensive optimizations and is able to recover the same quality of alignment from much fewer images.

We now show additional results in Fig. 3. The sequences here exhibit different kinds of variations such as, variation in intensity, shape (non-rigid objects) and modality. We see that in all the cases we perform as well as state-of-the-art methods. We are able to register images with a large baseline transformation as shown in Fig. 3(d). Although we have not taken any explicit measure to account for multi-modality, we found that by using SIFT features we are able to register multimodal video sequences as shown in Fig. 3(e). Here, one sequence is captured using a normal camera and the other is captured using an infrared camera. For this case, one can notice that our result and the result from [8] do not look the same. This is because we do not perform any kind of fusion.

In order to further analyze our registration algorithm, we obtained the inliers from RANSAC and then calculated the percentage of inliers from the mean image and from the dynamic appearance images. This result can be seen in Table 2. The interesting fact about these numbers is that our algorithm automatically adapts the percentage of features obtained from the mean and dynamic appearance images. Notice that the percentage of inliers from the dynamic appearance images is 100% for the flag sequence. This is in agreement from the conclusions of [9], where the authors reported that by using the appearance image alone they were able to register the sequences the best. Thus we see that the algorithm performs very well on a large variety of sequences and our results are comparable to existing methods.

4 Discussion and Conclusion

We have proposed a method for registering video sequences based on the dynamic texture model. As compared to [6], we see that we are able to recover the spatial transformation independent of the temporal transformation. Our results show that our method performs equivalently to theirs. However our method reduces the number of frames we need to process. In the case of [6], they need to perform feature extraction, tracking, and trajectory matching for two sets of F frames. In our case, we only need feature extraction over two sets of $n+1 \ll F$ images. In addition, we have the computations for calculating the system parameters. Typically F/n for the sequences we have presented is in the range of 8-10, hence using our method gives an advantage with respect to number of frames we need to process. As compared to [9], we do not need to make the choice of whether to register using the mean or the dynamic appearance or both. Given the information extracted from the video sequences, the algorithm automatically makes this choice. This gives us a generic algorithm that can be applied to both rigid and non-rigid sequences. One other advantage is that we are able to register the video sequences with lower system orders. In short, we have presented an efficient method that works equally well compared to the state of the art.

The other two important contributions of this paper are the use of a joint system identification framework together with a canonical form representation. The joint identification and the canonical form are not only applicable to the case of registering video sequences, but also to the entire genre of algorithms based on the dynamic texture model. Our future work involves investigating the performance gain obtained by this framework on other problems such as dynamic texture recognition.

Acknowledgments

This work was partially supported by startup funds from JHU, by grants ONR N00014-05-10836, NSF CAREER 0447739, NSF EHS-0509101, and by contract JHU APL-934652.

References

1. Szeliski, R.: Image alignment and stitching: A tutorial. Fundamental Trends in Computer Graphics and Vision 2(1), 1–104 (2006)
2. Harris, C., Stephens, M.: A combined corner and edge detection. In: Proceedings of The Fourth Alvey Vision Conference (1988)
3. Lowe, D.: Distinctive image features from scale-invariant keypoints. International Journal of Computer Vision 20, 91–110 (2003)
4. Brown, M., Szeliski, R., Winder, S.: Multi-image matching using multi-scale oriented patches. In: CVPR, pp. 510–517 (June 2005)
5. Fischler, M.A., Bolles, R.C.: RANSAC random sample consensus: A paradigm for model fitting with applications to image analysis and automated cartography. Communications of the ACM 26, 381–395 (1981)
6. Caspi, Y., Irani, M.: Spatio-temporal alignment of sequences. IEEE Transactions on Pattern Analysis and Machine Intelligence 24(11), 1409–1424 (2002)
7. Caspi, Y., Simakov, D., Irani, M.: Feature-based sequence-to-sequence matching. International Journal of Computer Vision 68(1), 53–64 (2006)
8. Ukrainitz, Y., Irani, M.: Aligning sequences and actions by maximizing space-time correlations. In: European Conference on Computer Vision, pp. 538–550 (2006)
9. Ravichandran, A., Vidal, R.: Mosaicing nonrigid dynamical scenes. In: Workshop on Dynamic Vision (2007)
10. Doretto, G., Chiuso, A., Wu, Y., Soatto, S.: Dynamic textures. International Journal of Computer Vision 51(2), 91–109 (2003)
11. Overschee, P.V., Moor, B.D.: N4SID: Subspace algorithms for the identification of combined deterministic-stochastic systems. Automatica, Special Issue in Statistical Signal Processing and Control, 75–93 (1994)

12. Chan, A., Vasconcelos, N.: Probabilistic kernels for the classification of autoregressive visual processes. In: Conference on Computer Vision and Pattern Recognition, vol. 1, pp. 846–851 (2005)
13. Vidal, R., Ravichandran, A.: Optical flow estimation and segmentation of multiple moving dynamic textures. In: Conference on Computer Vision and Pattern Recognition, vol. II, pp. 516–521 (2005)
14. Rugh, W.J.: Linear System Theory, 2nd edn. Prentice Hall, Englewood Cliffs (1996)
15. Hartley, R., Zisserman, A.: Multiple View Geometry in Computer Vision, Cambridge (2000)
16. `http://www.wisdom.weizmann.ac.il/~vision/VideoAnalysis/Demos/Seq2Seq/`
17. `http://www.wisdom.weizmann.ac.il/~vision/VideoAnalysis/Demos/Traj2Traj`
18. `http://www.wisdom.weizmann.ac.il/~vision/SpaceTimeCorrelations.html`

Hierarchical Support Vector Random Fields: Joint Training to Combine Local and Global Features

Paul Schnitzspan, Mario Fritz, and Bernt Schiele

Computer Science Department, TU Darmstadt, Germany
{schnitzspan,fritz,schiele}@cs.tu-darmstadt.de

Abstract. Recently, impressive results have been reported for the detection of objects in challenging real-world scenes. Interestingly however, the underlying models vary greatly even between the most successful approaches. Methods using a global feature descriptor (e.g. [1]) paired with discriminative classifiers such as SVMs enable high levels of performance, but require large amounts of training data and typically degrade in the presence of partial occlusions. Local feature-based approaches (e.g. [2–4]) are more robust in the presence of partial occlusions but often produce a significant number of false positives. This paper proposes a novel approach called hierarchical support vector random field that allows 1) to combine the power of global feature-based approaches with the flexibility of local feature-based methods in one consistent multi-layer framework and 2) to automatically learn the tradeoff and the optimal interplay between local, semi-local and global feature contributions. Experiments show that both the combination of local and global features as well as the joint training result in improved detection performance on challenging datasets.

1 Introduction

The first goal of this paper is to propose a novel hierarchical framework that effectively combines the power of global feature-based models with the flexibility of local feature-based representations. The second goal is to derive an efficient and effective procedure to jointly train all model parameters in order to automatically learn the tradeoff and the interdependence of the different layers of the novel hierarchical model as well as between the local and global feature contributions. To achieve this, this paper leverages the ability of CRFs [5] to model neighborhood dependencies not only between local image features, but also between object sub-parts and parts using a multi-layer CRF. On the top-layer we incorporate a global object detector while on the layers below we employ smaller apertures in terms of object-parts and local features or sub-parts. The layers are connected via intra-layer potentials to benefit from simultaneous bottom-up and top-down propagation schemes. This allows to set up a joint and hierarchical model of local and global discriminative methods that augments CRFs to a multi-layer model with powerful unary classifiers.

D. Forsyth, P. Torr, and A. Zisserman (Eds.): ECCV 2008, Part II, LNCS 5303, pp. 527–540, 2008.

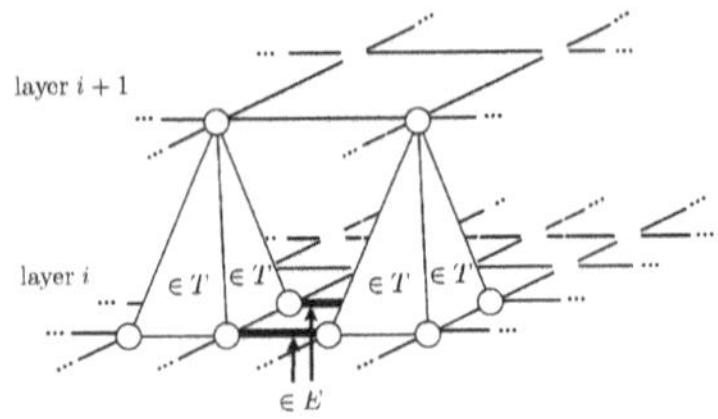

Fig. 1. Illustration of the model architecture. Two layers are connected via the ternary cliques T. The alternation between binary cliques E and ternary cliques T is key to the computationally feasability while a high degree of of interconnectedness is introduced.

The contributions of this paper are the following. First, we extend classical one-layer CRFs to multi-layer CRFs while maintaining computational tractability. Second, this work shows how to integrate local, semi-local and global information in a powerful model. Third, we extend CRFs to a consistent framework, which allows to jointly train the parameters of nonlinear classifiers and the CRF parameters. Fourth, we experimentally show the contributions of the various components of the model on challenging datasets. The paper is structured as follows. First, we refer to related work. In section 2 we introduce our multi-layer model, the respective potential functions and the parameters to be optimized. In section 3 we explain how we apply the model to object detection and verification. Finally, in section 4 we evaluate various aspects of our work on two different datasets.

Related Work. Related work [4, 6–9] addressing the combination of global and local features showed promising results and specifically improved performance compared to making use of only one type of features - either local or global. Especially the idea of [8] of integrating localized features of growing apertures in one model is relevant to our goals since meaningful object parts can be determined while ensuring global consistency. In our work we exploit a similar approach in terms of evidence aggregation, but additionally we are able to learn local neighborhood dependencies and address bi-directional interactions between entire objects and object parts. In contrast to [4, 7, 9] where global and local features are matched independently from each other, we combine bottom-up and top-down cues simultaneously showing improved performance.

Concerning CRF literature, [6, 10] proposed multi-layer CRFs to account for global consistency and due to that showed improved performance. In [6] a global scene potential is introduced to assert consistency of local regions. Thereby, the authors were able to benefit from integrating the context of a given scene. However, their model works with global priors set in advance and only uses learned local classifiers. Rather than to rely on priors alone, in our work, all parameters of the layers are trained jointly. In [10] a tree-structured CRF is proposed based on previously inferred segmentations of images. Thereby, the authors neglect direct local neighborhood dependencies, which our model learns jointly with long range dependencies.

[11–13] introduced two-layer CRFs, where the top layer deploys one node, which superimposes objects. That idea is especially interesting for object detection since the superimposed node manages local deformations of objects and therefore asserts consistency of object instances. Our work goes beyond those approaches by introducing multiple layers of evidence aggregation, which not only guarantees consistency of object instances, but also deploys different levels of information (from local to global) to our model.

CRF-based approaches like [11, 13–15] showed that incorporating powerful unary classifiers in CRFs is key to the overall performance, but those unary classifiers were trained independently from local neighborhood potentials. In contrast, we investigate the joint optimization of all aspects of our model and show that this joint optimization leads to improved performance compared to independent training. Taskar et al. [16] transformed the original problem introduced by Lafferty et al. [5] into its equivalent exponentially sized dual formulation but the latter lacks the intuitive nature of the primal problem and they only evaluated it on an OCR task. Due to the modularity of our approach we reduce the problem size while keeping the intuitive nature of the original formulation.

2 Hierarchical Support Vector Random Fields (hSVRF)

While global detectors have been shown to achieve impressive results in object detection for unoccluded object instances, part-based approaches tend to be more successful in dealing with partial occlusion. Since adjacent regions in images are not independent from each other, CRFs model these dependencies directly by introducing binary clique potentials. However, standard CRFs work on a very local level and long range dependencies are not addressed explicitly in simple one-layer models. Therefore, our approach incorporates SVMs and multiple layers of CRFs in one consistent framework in order to combine local neighborhood and long range dependencies. In the following we will describe how we set up the multi-layer model step by step starting from the simple one-layer case.

2.1 One-Layer CRF Model

We overlay the image X with a grid of nodes where each node is linked to the evidence in the image via unary ψ and binary ϕ potential functions. We denote the set of grid nodes by $y_i \in Y$ in which each y_i is associated to a certain region x_i in the image X. $e_{ij} = (y_i, y_j) \in E$ refers to the binary cliques connecting two adjacent nodes y_i and y_j. Each node $y_i \in Y$ will be assigned a label from $\{0, \ldots, p\}$ which indicates the parts of an object $\{1, \ldots, p\}$ or background $\{0\}$. We denote the set of all labels by $\mathcal{Y}$. Therefore, the factorization of the conditional probability distribution of the nodes Y given the image can be written as

$$p(Y|X) = \frac{1}{Z} \prod_{y_i \in Y} \psi(y_i, x_i) \prod_{e_{ij} \in E} \phi(y_i, y_j, x_i, x_j) \ . \tag{1}$$

Here, Z refers to a normalization factor called the partition function.

2.2 Multilayer CRF Model

As motivated before, one-layer CRFs act at a very local level and represent a single view on the data typically represented with unary and binary potentials. In order to overcome those local restrictions, we introduce multiple layers $l \in \{1, \ldots, N_L\}$ with associated unary potentials ψ^l and binary potentials ϕ^l, to enhance the model by evidence aggregation on a local ($l = 1$) to a global level ($l = N_L$). Different numbers of parts are deployed to different layers $\{0, \ldots, p^l\}$. We propose a connectivity between the layers as displayed in Figure 1, which provides a high degree of interconnectedness and yet results in a computationally tractable model, which is highly desirable for both inference and training. The key to this is the alternation between binary cliques $e_{ij} \in E^l$ and ternary cliques $t_{ijk} \in T^l$ that omit the introduction of higher (higher than third) order cliques. The conditional distribution for this multi-layer model resolves into:

$$\begin{aligned} p(Y|X) = \tfrac{1}{Z} \textstyle\prod_{l=1}^{L} \left[\prod_{y_i \in Y^l} \psi^l(y_i, x_i) \prod_{e_{ij} \in E^l} \phi^l(y_i, y_j, x_i, x_j) \right] & \quad \text{(intra-layer)} \\ \textstyle\prod_{l=1}^{L-1} \prod_{t_{ijk} \in T^l} \theta^l(y_i, y_j, y_k, x_i, x_j, x_k) & \quad \text{(inter-layer)} \end{aligned} \qquad (2)$$

where additional to the one-layer notation $\theta^l(\cdot, \cdot, \cdot, \cdot, \cdot, \cdot)$ denotes the ternary clique potentials that connect layer l to layer $l + 1$ using third-order cliques. T^l describes the set of all ternary cliques between layer l and layer $l + 1$ (see Figure 1 for illustration).

This model combines different views on the data by layer-specific potentials and the hierarchical structure accounts for longer range dependencies.

2.3 Potentials

As described in Eq. 1 and 2 the conditional probabilities factor into unary potentials ψ^l, binary potentials ϕ^l and additional ternary potentials θ^l required for the multi-layer model. Due to the flexibility of CRFs, the layer-specific feature functions $f^l(X)$, $g^l(X)$ and $h^l(X)$ for the unary, binary and ternary potentials respectively can be chosen freely. Those deployed in the experiments are detailed in section 3.1.

Unary Potentials. The discriminative power in the unary potentials is key to the overall performance of the CRF. In some cases, a CRF using less powerful classifiers such as the commonly used logistic regression can even be outperformed by an SVM employing no connectivity at all [14].

Therefore, we build our unary potentials on SVMs to leverage previous results on robust large margin classification. We adapt the one-against-all strategy which results in training one SVM for each class. $f^l(\cdot)$ refers to the feature function for the node features and ρ^l_c denotes the offset. Then, the potential of node y_i being of class c is defined as

$$\psi^l(y_i = c, x_i; \beta^l_c, \zeta^l_c, \rho^l_c) = \exp\left(\sum_{j \in \zeta^l_c} \left(\beta^l_c\right)_j K(f^l_j(X), f^l_i(X)) + \rho^l_c \right) . \qquad (3)$$

where ζ_c^l indexes the set of support vectors for class c and layer l and (β_c^l) refers to model parameters to be optimized. We used RBF kernels to define the kernel function $K(f_j^l(X), f_i^l(X)) = \exp\left(-\gamma \left\|f_j^l(X) - f_i^l(X)\right\|^2\right)$ with bandwidth parameter γ. Note, that this approach employs multiclass one-against-all SVMs.

Binary and Ternary Potentials. We define the binary and ternary potentials using a linear classification model, which is a popular choice in the CRF literature. For the binary potentials we set

$$\phi^l(y_i = c_1, y_j = c_2, x_i, x_j; u^l) = \exp\left(\left(u_{c_1c_2}^l\right)^T g_{ij}^l(X)\right) . \tag{4}$$

where y_i and y_j are two adjacent nodes and c_1 and c_2 refer to any label from $\{0, \ldots, p^l\}$. $g^l(\cdot)$ denotes the feature function for the binary potentials of layer l. $u_{c_1c_2}^l$ refers to the parameters to be trained. The ternary potentials are defined as

$$\theta^l(y_i = c_1, y_j = c_2, y_k = c_3, x_i, x_j, x_k; v^l) = \exp\left(\left(v_{c_1c_2c_3}^l\right)^T h_{ijk}^l(X)\right) . \tag{5}$$

where y_i, y_j, y_k belong to one three-wise connected clique t_{ijk} and $c_1, c_2 \in \{0, \ldots, p^l\}$ and $c_3 \in \{0, \ldots, p^{l+1}\}$. $h^l(\cdot)$ denotes the feature function for the ternary potentials at layer l. $v_{c_1c_2c_3}^l$ refers to the parameters to be optimized.

2.4 Parameter Learning and Inference

In contrast to CRF literature like [11, 13–15], we jointly optimize all model parameters. Given M training images $X^m, m = \{1, \ldots, M\}$ we optimize the conditional log-likelihood $\mathcal{L}(\beta, u, v) = \sum_{m=1}^{M} \log P(Y^m | X^m; \beta, u, v)$ via gradient descent for binary and ternary clique potentials. The unary potentials are trained with Newton optimization.

This joint training is facilitated by the primal SVM training proposed by Chapelle [17] that showed competitive results compared to common quadratic programming in the dual formalism. We make use of that idea and incorporate primal SVM training in the CRF framework.

Primal SVM Training. As described in [17] the constraints of the original primal optimization problem can be integrated with a loss function in the objective function, yielding an unconstrained optimization problem. As long as this loss function is differentiable with respect to the model parameters, the optimization can be solved by Newton optimization. Originally, the non-differentiable hinge loss is used for SVM training in the dual, but [17] showed competitive results using the differentiable quadratic loss or the Huber loss (a differentiable approximation of the hinge loss). The primal optimization problem for kernel SVMs is denoted by:

$$\min_{\beta_c^l} F = \min_{\beta_c^l} \left(\sum_{i,j \in \zeta_c^l} \left(\beta_c^l\right)_i \left(\beta_c^l\right)_j K(f_i(X), f_j(X)) + C \sum_{i=1}^{n} L\left(y_i, S_c^l\left(f_i^l(X)\right)\right) \right) . \tag{6}$$

where L denotes a suitable loss function and C the penalty term. The target function $S_c^l(\cdot)$ is of the form (representer theorem [18]):

$$S_c^l\left(f_i^l(X)\right) = \sum_{j \in \zeta_c^l} \left(\beta_c^l\right)_j K(f_j^l(X), f_i^l(X)) + \rho_c^l \ . \tag{7}$$

where $f_i^l(x)$ denotes a feature vector to be classified. $K(\cdot,\cdot)$ denotes the kernel function and $\left(\beta_c^l\right)$ refers to the parameters to be optimized (note that these are not Lagrange multipliers). We consider the differentiable quadratic loss:

$$L(y_i, S_c^l(f_i(X))) = \left(\max\left\{0; 1 - (\delta(y_i, c))\left(S_c^l\left(f_i^l(X)\right)\right)\right\}\right)^2 \ . \tag{8}$$

where $\delta(y_i, c) \in \{-1, 1\}$ refers to whether y_i belongs to class c (=1) or not (=-1).
[17] proposed to optimize the parameters $\left(\beta_c^l\right)$ with Newton optimization:

$$\beta_c^l \leftarrow \beta_c^l - \eta \left(H^l\right)^{-1} \frac{\partial F}{\partial \beta_c^l} \ . \tag{9}$$

where η denotes the learning rate and the Hessian H^l equals $2\left(\frac{1}{C}K + KI^0K\right)$ with kernel matrix K. I^0 is a diagonal matrix, where the entries are 1 for $\beta_c^l > 0$ and 0 otherwise. The number of non-zero entries equals the number of support vectors. In order to update the offset ρ_c^l the Hessian can be augmented by an additional row and column and the offset term can be concatenated with the parameters β_c^l (see [17] for details).

Joint Training of hSVRF. In order to account for joint training of the hSVRF parameters, we adapt the loss function $L(\cdot,\cdot)$ to consider unary SVM classifications as well as joint CRF classifications , which respects the entire multi-layer model. In that sense, object evidence, local neighborhood dependencies as well as longer range dependencies are taken into account to optimize the unary parameters. We achieve this by adapting the loss function to consider the belief of node y_i belonging to class c inferred with Loopy Belief Propagation [19].

$$L(y_i, b_c(y_i), S_c^l(f_i(X))) = \left[\max\left\{0; (1 - \delta(y_i, c) b_c(y_i))\left(1 - \delta(y_i, c) S_c^l\left(f_i^l(X)\right)\right)\right\}\right]^2 \tag{10}$$

where the belief $b_c(y_i)$ of node y_i belonging to class c ranges between -1 and 1. Whenever the CRF votes for the wrong class $((1 - \delta(y_i, c) b_c(y_i)) > 1)$ the original primal SVM loss function is amplified for calculating the Newton step. Otherwise $((1 - \delta(y_i, c) b_c(y_i)) < 1)$ the impact of the original primal loss function on the Newton step is reduced. Note, the Hessian is not affected by our changes in the loss function and $\frac{\partial F}{\partial \beta_c^l}$ can be computed similar to [17].

The parameters of the binary and ternary clique potentials can be optimized via gradient descent. Similar to [12], the gradient with respect to the binary parameters $\{u_{c_1c_2}^l\}$ of layer l can be expressed as

$$\frac{\partial \mathcal{L}}{\partial u_{c_1c_2}^l} = \sum_{e_{ij} \in E} \left(\delta(y_i, c_1)\delta(y_j, c_2) - b_{c_1c_2}(y_i, y_j)\right) g_{ij}^l(x) \ . \tag{11}$$

(a) (b)

Fig. 2. (a) Three-layer instantiation of our model. The evidence aggregation is sketched: Starting from local information like fragments of a wheel over whole wheels to entire objects at the top layer. (b) Example of the part assignment of training data (left: training image; middle: part assignment of middle layer; right: part assignment of bottom layer). Colors encode assignments of parts; dark blue indicates background.

where $\delta(\cdot,\cdot)$ refers to the Kronecker-delta and $b_{c_1c_2}(\cdot,\cdot)$ denotes the pairwise belief of two adjacent nodes belonging to class c_1 and c_2.

Analogously, the gradient with respect to the ternary clique parameters $v^l_{c_1c_2c_3}$ of layer l can be written as

$$\frac{\partial \mathcal{L}}{\partial v^l_{c_1c_2c_3}} = \sum_{t_{ijk}\in T} \left(\delta(y_i,c_1)\delta(y_j,c_2)\delta(y_k,c_3) - b_{c_1c_2c_3}(y_i,y_j,y_k)\right) h^l_{ijk}(x) \ . \quad (12)$$

where $b_{c_1c_2c_3}(\cdot,\cdot,\cdot)$ denotes the ternary beliefs of three connected nodes.

This concept for updating the parameters of our model alternates between the max margin notation of SVM training and the max likelihood formalism of CRFs. Although we show performance improvements with our scheme, this might be a restriction, since one unique optimization scheme is desirable. In future work we will investigate algorithms to overcome this restriction.

We use quadratic programming (the common SVM training) decoupled from the CRF to initialize the parameters β^l. The dual support vector coefficients α^l_c and parameters β^l_c are connected via $\left(\beta^l_c\right)_j = y_j \left(\alpha^l_c\right)_j$ as described in [17]. Given the starting solution for β^l we start the joint optimization by Newton optimization for unary classifiers and gradient descent for binary and ternary clique parameters.

Inference. Given the parameters (β^l), (u^l) and (v^l) we seek to infer probabilities of the nodes belonging to the different classes. Loopy Belief Propagation (LBP) [19] infers beliefs, that one node y belongs to class c while respecting the pairwise and three-wise dependencies of adjacent nodes.

3 Application to Computer Vision Tasks

To support our claims about the benefits of the local to global CRF model and the presented joint optimization, we evaluate the approach on two challenging computer vision tasks: object detection and hypothesis verification. But first we describe in detail how the method is adapted to the specific settings and show

how to obtain part annotations for the training phase. We consider a 3-layer instantiation of the presented model as visualized in Figure 2(a) and detailed below.

3.1 Feature Functions

Until now, we have not defined the feature functions $f^l(\cdot)$, $g^l(\cdot)$, $h^l(\cdot)$, that are specific to each layer in the CRF we propose. They link the potentials to the actual image evidence and account for local neighborhood and long range dependencies. We build on the concept of computing histograms of oriented gradients, that has been shown to be very successful on a local level, describing interest points [20], as well as on a global level [1], describing full objects in a holistic manner. However, due to the generality of our work, any suitable feature function can be deployed to our model.

Unary Potential Feature Functions. We calculate histograms of oriented gradients for a grid of non-overlapping 8×8 pixel regions and concatenate 4 neighboring histograms of gradients to one block descriptor as described in [1]. This results in a 36 dimensional feature for each node, that we define to be the unary feature function on the first level $f^1(\cdot)$. For the higher levels $f^2(\cdot), \ldots, f^L(\cdot)$ we successively double the number of considered blocks in horizontal and vertical directions until on the highest level, we encode the full object as in [1]. As illustrated in Figure 2(a), the motivation behind this scheme is to aggregate evidence for an object class from different spatial localities ranging from fragments (e.g. fragment of a wheel), parts (e.g. whole wheel) to a holistic view on the object (e.g. whole motorbike).

Binary Potential Feature Functions. Intuitively, binary potentials are responsible for modeling local dependencies by supporting or inhibiting label propagation to the neighboring nodes. In computer vision, simple pixel-based gradient-based measures are often used to inhibit propagation across potential object borders [15]. Our approach goes beyond that by taking into account the change in the gradient orientation histograms between the neighboring nodes.

$$g^l_{ij}(X) = \left(\left| f^l_i(X) - f^l_j(X) \right| \ , \ 1 \right)^T \ . \tag{13}$$

Here, we extended each difference by an offset for being capable eliminating small isolated regions.

Ternary Potential Feature Functions. Similar to the binary potentials, ternary potentials encode local dependencies, too. But furthermore, they act as a link between layers, facilitating propagation of information across locality and position in our model. Due to the computational tractability of the hierarchy we can propagate object evidence across layers and thereby manage efficient bottom-up and top-down reasoning during inference.

To allow the ternary potential to assess the compatibility of a particular labeling of a three-wise connected clique, we define the ternary potential feature function to be the stacked pairwise difference of the feature vector associated to the 3 relevant nodes. Since higher level nodes involve more HOG blocks and are higher dimensional than lower level ones, we calculate the average over connected blocks (denoted by operation $avg(\cdot)$) in order to fit the dimension of lower level nodes.

$$h^l_{ijk}(x) = \left(\mid f^l_i(x) - f^l_j(x) \mid , \mid f^l_i(x) - avg(f^{l+1}_k(x)) \mid , \mid f^l_j(x) - avg(f^{l+1}_k(x)) \mid , 1\right)^T. \tag{14}$$

where nodes i and j are on layer l and node k is on layer $l+1$.

3.2 Part Assignment

For optimizing the conditional log-likelihood during training, ground truth part labels are required for each training instance in order to be able to train the multiclass potentials of our model. While the labeling for the top (object) layer is given by a bounding box annotation or segmentation of the objects, the part annotation on the lower layers is not obvious. Inspired by [21] we obtain part labels in a data driven way by applying k-means clustering across images to infer part annotations. Instead of mere spatial clustering, we append to the image coordinates the features described in 3.1. In this fashion the importance is on the cluster appearance and the 2 coordinate dimensions act as regularization for the clustering to maintain a rough spatial layout. Despite the simple data-driven approach, we obtain a sensible partitioning of our training instances, that exposes appearance-based though well localized assignment of parts as exemplified in Fig. 2(b).

3.3 Object Detection and Verification

In this paragraph we show how we infer object locations of one object class. As described in section 2.4, LBP yields a label assignment across layers taking into account beliefs that nodes are associated with parts (bottom and middle layer), object (top layer) or background (all layers). Given a test image we could initialize our model at every pixel location for being able to infer all possible object hypotheses. However, to reduce computational effort we first deploy the bottom and middle layer of our model. This step produces a part map of the whole image while respecting the dependencies of the bottom and middle layer. From the training set we know possible part constellations and we search for those constellations in the part map of test images to infer hypotheses of object locations. This approach resembles the ISM voting of [4] despite that the evidence of parts of our model are inferred simultaneously and therefore the parts interact with each other. Thus, the evidence of one sub-part conditions the constellation of direct neighbors via links in the bottom layer and via the middle layer it also affects the evidence of further image regions. This step generates initial hypotheses that still need to be validated by the complete model. Since the part-based approach of the lower layers showed to yield a high recall, this approach makes sense as we first search for possible locations and then infer the complete

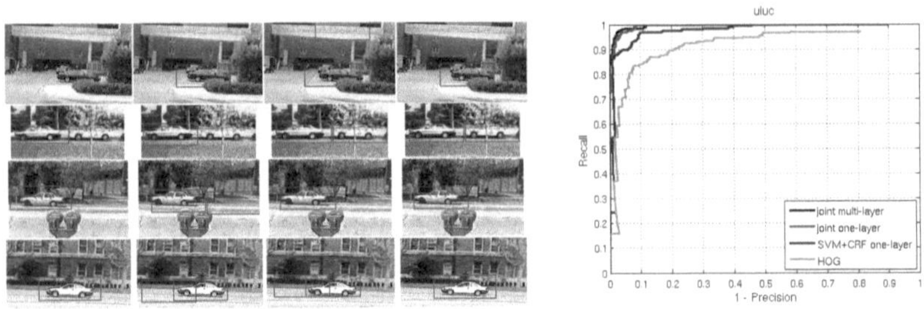

(a) (b)

Fig. 3. (a) Examples on the UIUC dataset. The columns show results at EER of HOG detector, one-layer separate training, one-layer joint training and multi-layer model. (b) UIUC detection performance of the different aspects.

model for the hypothized bounding boxes. The approach of coupling generative models with a discriminative verification stage has been shown to be fruitful [22]. In this spirit, we address a hypothesis verification task by only infering our model at hypothesized bounding boxes. LBP simultaneously infers beliefs of all nodes of our model. Since at the top layer we only deploy one node, we can directly use the belief of that node belonging to the object class as a score. For the layers underneath we compute object probabilities similar to the ISM [4] part voting scheme as described above and multiply them to the global belief. Thereby we distinguish between left and right facing objects and consider the maximum of the deduced scores.

4 Experiments

In all experiments we used SVM^{light} [23] for initial SVM training. Training the model took approximately 12 hours while we were able to infer 15 hypotheses per second.

Object Detection. For the detection task we evaluated our model on the UIUC single scale car dataset. We trained the whole model on 250 bounding boxes containing cars and 200 negative crops. This experiment contains performance measurements of i) only the global object detector, ii) the one-layer model of our approach while training the SVM and CRF parameters separately, iii) the one-layer model while training the parameters jointly and iv) the complete multi-layer model. For the part labeling during training we deployed k-means clustering with 8 means for the bottom layer and 4 means for the middle layer. For the one-layer model we used 8 means in the clustering step. The detection performance was evaluated on the 170 UIUC test images. Figure 3(b) compares the different aspects described in the previous sections. Both the joint training and the multi-layer approach consistently improved the performance. Especially note the large performance gap between the complete model (97,5% in equal error rate) and the HOG detector (87,0% in EER). Figure 3(a) shows some example

Table 1. Results of the detection task on the UIUC car dataset

Method	EER
Multi-layer	97,5
One-layer part-based joint training	96,0
One-layer part-based without joint training	93,0
Global object detector [1]	87,0
Mutch & Lowe [24]	99,9
Leibe et al. [4]	97,5
Hoiem & Winn [11]	approx. 93,5
Winn & Shotton [13]	approx. 92,9

images where the HOG detector can not detect all cars due to partial occlusion and the one-layer models infer false positives, while the multi-layer model detects all cars correctly. These results expose the benefits of joint training and integration of local to global information. Our model successfully learns the tradeoff between global vs. local object detection and improves the performance of both ideas by combining powerful global descriptors and flexible local feature approaches. Further note the performance improvement between training the SVM independently from the other CRF parameters (93,0%) and training them jointly (96,0%) for the one-layer model. This evaluation highlights the advantage of training all model parameters jointly as proposed in section 2. In Tab. 1 we compare our model to the state-of-the-art in object detection on this dataset. As it can be seen, we achieve competitive results compared to other well performing models. Only Mutch and Lowe [24] outperform our model while we obtained the same performance as [4]. Further, we outperform the CRF-based approaches of Hoiem and Winn [11] and Winn and Shotton [13].

Verification of HOG Detector Hypothesis. For the hypothesis verification task we evaluated our model on the PASCAL 2006 motorbikes dataset [25] containing challenging multiscale, partially occluded and multiview instances. Since we want to explore the combination of an initial detector with our model acting as a verification stage, we trained a HOG detector on the provided training set and generated initial hypotheses on the test set. We set the parameters to allow for high recall at the drawback of more false positives. We also trained our joint multi-layer model on the training set and calculated the score of our approach on the hypotheses of the HOG detector (see Figure 4(a)). Thereby, our multi-layer model achieved 43,7% in average precision (the common performance measure of [25]) improving the state-of-the-art by 4,7%. Note in particular that we outperformed the global HOG detector, that reported an average precision of 39%, which emphasizes the benefit of combining global and local features. The next best performance for the motorbikes is 37,1% achieved by the approach of [26] which we outperform by 6,6%. Furthermore, our model shows a high performance improvement (more than 10% in average precision) compared to the remaining approaches.Particularly, the high precision for high scores of bounding boxes is promising; with no false positives 16% of all motorbikes are extracted while none of the other state-of-the-art approaches obtained such high recall at perfect precision.

Fig. 5(a) shows precision-recall-curves from which the contributions of different aspects of our model to the overall performance gain can be deduced. Consistent

Method	AP	AP trained on sidev.
Multi-layer	43,7	36,0
Multi-layer not jointly trained	42,0	33,5
Decoupled multiple layers	-	32,2
One-layer model	-	27,7
One-layer not jointly trained	-	24,2
One-layer unary classifier	-	19,0
Dalal & Triggs [1]	39,0	30,0
Chum & Zisserman [26]	37,1	-
Laptev [27]	31,8	-
Viitaniemi & Laaksonen [28]	26,5	-
Shotton et al. [15]	17,8	-
Leibe et al. [4]	15,3	-

(a)

(b)

Fig. 4. (a) Results for the motorbike PASCAL06 challenge (AP = average precision). (b) Example images for detecting sideviews of motorbikes: (Left) one-layer part-based model; (middle) HOG sideviews; (right) joint multi-layer model.

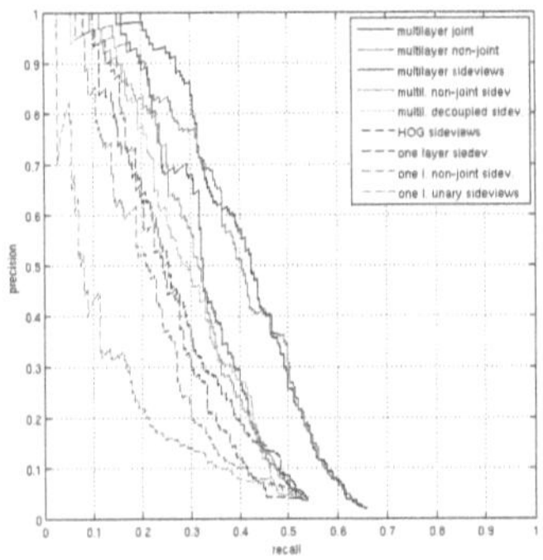

(a)

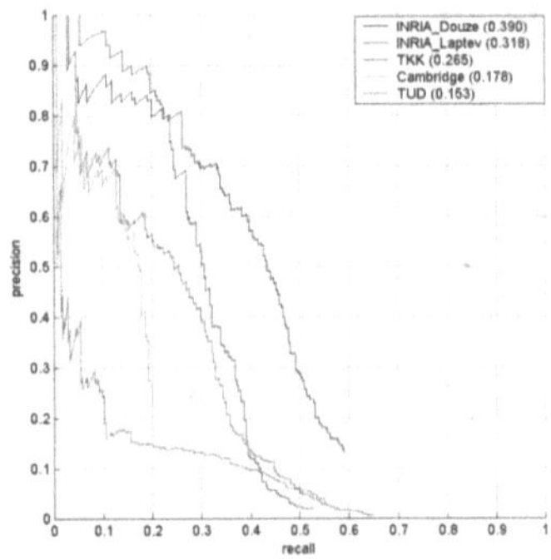

(b)

Fig. 5. (a) PASCAL06 detection performance of our model. (b) State of the art approaches on the PASCAL06 motorbikes.

with the results obtained on the UIUC database, the jointly trained multi-layer model improves the performance to 43,7% while the non-jointly trained model with fixed SVM coefficients obtained 42% in average precision.

Verification of Generative ISM Object Detector Hypothesis. For further testing the different aspects of our model, we decided to test our discriminative model on hypotheses obtained by the ISM-model. Since the latter model was shown to yield promising results for the subset of left or right facing instances, we trained our model on sideviews of motorbikes, but evaluated the aspects on the complete multiview dataset. Overall the ISM model extracted 4238 hypotheses and achieved an average precision of 15,3%. We trained our model on 100 rightfacing and respective mirrored left views and 200 randomly cropped background images. As it can be seen in Fig. 4(a) our multi-layer model could improve the performance compared to other settings of our approach.

Concerning the average precision performance measure the jointly trained multi-layer model (36,0%) significantly improved the results of the non-jointly trained model with fixed SVM parameters (33,5%), completely decoupled layers (32,3%), the HOG detector trained on sideviews (30%) and the one-layer settings of our model: jointly trained SVM and CRF parameters (27,7%), fixed SVM parameters (24,2%) and the unary classifier (SVM) (19,0%).

Fig. 4(b) shows some example detections for training on sideviews. Partially occluded objects can not be detected by the global detector, while the part based approach and our multi-layer model infer them correctly. Furthermore, false detections of the part based approach can be removed by the global detector for correct detections of our multi-layer model. However, the rear view of the motorbike (third row) can not be detected correctly due to the focus on sideviews. The measured improvements for joint training and the multi-layer approach are consistent with respect to both tested databases.

5 Conclusion

We present a novel multi-layer CRF which combines the power of global object detectors and flexible local feature approaches. Our model successfully learns the tradeoff between local and global feature contributions for improved performance. Furthermore, we show how SVM classifiers can be incorporated into this multi-layer CRF framework and how training can be performed jointly. Experiments show that performance improves consistently. Finally, we outperform the state-of-the-art on the challenging PASCAL06 motorbike detection task. In future, we will investigate to use more layers in our model and deploy different tractable hierarchies to it, which can be easily done due to the generality of our work. We will also explore different features for evidence aggregation.

Acknowledgments. We thank Stefan Roth for helpful discussions and Joris Mooij for making libDAI available online. This work has been funded, in part, by the GRK 1362 of the DFG and the EU project CoSy (IST-2002-004250).

References

1. Dalal, N., Triggs, B.: Histograms of Oriented Gradients for Human Detection. In: CVPR (2005)
2. Felzenszwalb, P.F., Huttenlocher, D.P.: Pictorial Structures for Object Recognition. IJCV (2005)
3. Fergus, R., Zisserman, A., Perona, P.: Object class recognition by unsupervised scale invariant learning. In: CVPR 2003 (2003)
4. Leibe, B., Seemann, E., Schiele, B.: Pedestrian Detection in Crowded Scenes. In: CVPR (2005)
5. Lafferty, J., McCallum, A., Pereira, F.: Conditional Random Fields: Probabilistic Models for Segmenting and Labeling Sequence Data. In: ICML (2001)
6. He, X., Zemel, R., Carreira-Perpinan, M.: Multiscale conditional random fields for image labeling. In: CVPR 2004 (2004)

7. Heisele, B., Ho, P., Wu, J., Poggio, T.: Face recognition: Component-based versus global approaches. In: CVIU 2003 (2003)
8. Lazebnik, S., Schmid, C., Ponce, J.: Beyond Bag of Features: Spatial Pyramid Matching for Recognizing Natural Scene Categories. In: CVPR (2006)
9. Zhang, D., Li, S.Z., Gatica-Perez, D.: Real-Time Face Detection Using Boosting in Hierarchical Feature Spaces. In: ICPR 2004 (2004)
10. Reynolds, J., Murphy, K.: Figure-ground segmentation using a hierarchical conditional random field. In: CRV 2007 (2007)
11. Hoiem, D., Rother, C., Winn, J.: 3D Layout CRF for Multi-View Object Class Recognition and Segmentation. In: CVPR 2007 (2007)
12. Kumar, S., August, J., Hebert, M.: Discriminative Random Fields. In: IJCV 2006 (2006)
13. Winn, J., Shotton, J.: The Layout Consistent Random Field for Recognizing and Segmenting Partially Occluded Objects. In: CVPR (2006)
14. Lee, C.H., Greiner, R., Schmidt, M.: Support Vector Random Fields for Spatial Classification. In: Jorge, A.M., Torgo, L., Brazdil, P.B., Camacho, R., Gama, J. (eds.) PKDD 2005. LNCS (LNAI), vol. 3721, pp. 121–132. Springer, Heidelberg (2005)
15. Shotton, J., Winn, J., Rother, C., Criminisi, A.: TextonBoost: Joint Appearance, Shape and Context Modeling for Multi-Class Object Recognition and Segmentation. In: Leonardis, A., Bischof, H., Pinz, A. (eds.) ECCV 2006. LNCS, vol. 3951, Springer, Heidelberg (2006)
16. Taskar, B., Guestrin, C., Koller, D.: Max margin markov networks. In: NIPS 2003 (2003)
17. Chapelle, O.: Training a Support Vector Machine in the Primal. In: Neural Computation (2007)
18. Kimeldorf, G.S., Wahba, G.: A correspondence between bayesian estimation on stochastic processes and smoothing by splines. Annals of Math. Stat. (1970)
19. Pearl, J.: Probabilistic Reasoning in Intelligent Systems: Networks of Plausible Inference. Morgan Kaufmann, San Francisco (1988)
20. Lowe, D.: Distinctive Image Features from Scale-Invariant Keypoints. In: IJCV 2004 (2004)
21. Bouchard, G., Triggs, B.: Hierarchical Part-Based Visual Object Categorization. In: CVPR (2005)
22. Fritz, M., Leibe, B., Caputo, B., Schiele, B.: Integrating Representative and Discriminant Models for Object Category Detection. In: ICCV (2005)
23. Joachims, T.: Making large-Scale SVM Learning Practical. In: Advances in Kernel Methods - Support Vector Learning (1999)
24. Mutch, J., Lowe, D.: Multiclass Object Recognition with Sparse, Localized Features. In: CVPR (2006)
25. Everingham, M., Zisserman, A., Williams, C.K.I., Van Gool, L.: The PASCAL Visual Object Classes Challenge 2006 (VOC 2006) Results (2006)
26. Chum, O., Zisserman, A.: An Exemplar Model for Learning Object Classes. In: CVPR (2007)
27. Laptev, I.: Improvements of object detection using boosted histograms. In: BMVC 2006 (2006)
28. Viitaniemi, V., Laaksonen, J.: Techniques for Still Image Scene Classification and Object Detection. In: ICANN (2006)

Scene Segmentation Using the Wisdom of Crowds

Ian Simon and Steven M. Seitz

University of Washington
{iansimon,seitz}@cs.washington.edu

Abstract. Given a collection of images of a static scene taken by many different people, we identify and segment interesting objects. To solve this problem, we use the distribution of images in the collection along with a new *field-of-view* cue, which leverages the observation that people tend to take photos that frame an object of interest within the field of view. Hence, image features that appear together in many images are likely to be part of the same object. We evaluate the effectiveness of this cue by comparing the segmentations computed by our method against hand-labeled ones for several different models. We also show how the results of our segmentations can be used to highlight important objects in the scene and label them using noisy user-specified textual tag data. These methods are demonstrated on photos of several popular tourist sites downloaded from the Internet.

1 Introduction

With billions of photos now online, community photo collections found on Flickr [1] and other photo sharing sites offer tremendous opportunities for computer vision research. Recently, much work in the field has been devoted to making use of such photo collections for visualization [2], hole filling [3], learning object category models [4], dense 3D scene reconstruction [5], and geolocation [6]. While these recent efforts have capitalized on the *quantity* and *variety* of online images to enable various applications, a second important source of information is the *distribution* of photos — i.e., which photos do people choose to take?

The distribution of photos in a large collection holds valuable semantic information about the content of the scene. In this paper, we seek to leverage this information to automatically identify and segment interesting objects in a scene. While extremely challenging to solve purely by analyzing pixels in a single image, this problem is much more tractable with a large image collection. For example, a robust interest operator is obtained by simply finding features (e.g. SIFT [7]) that appear in numerous photos. By identifying oft-photographed features, this operator tends to highlight the parts of the scene that people find most interesting. The fact that this works robustly is a powerful demonstration of *the wisdom of crowds* [8], where a community of people combine to provide information that is otherwise difficult to obtain.

D. Forsyth, P. Torr, and A. Zisserman (Eds.): ECCV 2008, Part II, LNCS 5303, pp. 541–553, 2008.

(a) (b) (c) (d)

Fig. 1. (a) A "bad" image of the Hagia Sophia. (b,c) Two "good" images of objects in the Hagia Sophia. Good images provide useful segmentation cues. (d) An image of the Trevi Fountain showing why photo collections of some scenes may not contain many good images. In this case, there are interesting statues in the façade, but it is difficult for photographers to get close enough to photograph them individually.

While detecting interesting features is straightforward via simple counting methods, identifying interesting *objects* is more challenging, as it necessitates segmentation — another difficult, ill-posed computer vision problem. To address this segmentation problem, we propose a new *field-of-view* cue for inferring perceptual grouping information from large photo collections. The key idea is very simple—it's based on the observation that people tend to take photos of "objects" as opposed to arbitrary regions of the scene, and these objects are usually framed within the field of view of the photo. Consequently, if two points are on the same object, those two points will likely appear together in many photos. We may therefore use the co-occurence of features in many images as a cue for grouping them together.

The main contribution of this paper is the introduction of this field-of-view cue and techniques for leveraging it for identifying and segmenting objects in images and point clouds. We also demonstrate the use of this analysis to display the relative importances of objects, automatically compute textual labels for these objects from noisy user-contributed Flickr tags (which are attached to images, not image regions), and browse a scene using an interactive map.

1.1 Related Work

The problem of decomposing a set of images into recurring objects in an unsupervised manner has been the subject of much recent work in computer vision, such as Fergus et al. [9], Sivic et al. [10] and Sudderth et al. [11]. However, to our knowledge, ours is the first paper to address this problem for static scenes, where objects always have the same 3D context and variation among images arises from differing camera positions and viewing directions. A closely related problem is addressed by Campbell et al. [12], who use a camera fixation cue that is similar to our field-of-view cue to segment a single 3D object from a set of images.

In a similar vein to our paper, Russell et al. [13] use grouping cues in an unsupervised approach to object category segmentation. They compute multiple segmentations for each image and then apply a latent topic model where the *segments* serve as documents, relying on the fact that each object will appear against several different backgrounds, so "bad" segments will contain multiple latent topics. In our case, however, as the scene is static, any image region will almost always appear in the same context. We instead rely on the field-of-view of the photographs to provide our grouping cues.

Some of our applications draw on other work in related areas. Simon et al. [14] use the distribution of photos in an online collection to find canonical images of a scene. In their work, these images are intended to serve as a summary of the scene, and they do not attempt to extract individual objects, nor do they segment the scene or any of the images. A similar problem is addressed by Epshtein et al. [15], who use the distribution of viewing frusta in a photo collection to organize the photos into a hierarchy. Barnard et al. [16] learn the joint distribution of image regions and text labels and use this distribution to predict labels for regions in new images. Instead of attempting to predict noisy tags, we find strong associations between clusters and tags and use them to select good labels from the set of tags submitted by Flickr users in Section 4.2.

1.2 Our Approach

As discussed in the previous section, much existing work deals with identifying and segmenting object categories from visual scenes, where each object may appear against different backgrounds in different images. We address a different question: how can one identify and segment interesting objects from a static scene? The fact that the background changes was important for previous work, and enabled a solution using latent topic models. While handling static scenes may seem like a simpler problem, the fact that the background is not changing is a challenge — different cues are needed.

We instead use information provided by the distribution of photos taken of the scenes. To segment the scenes, we use what could be called a *field-of-view* or *incidence* constraint. For scenes that contain individual objects of interest, we hypothesize that multiple photographers are going to take pictures "of" each

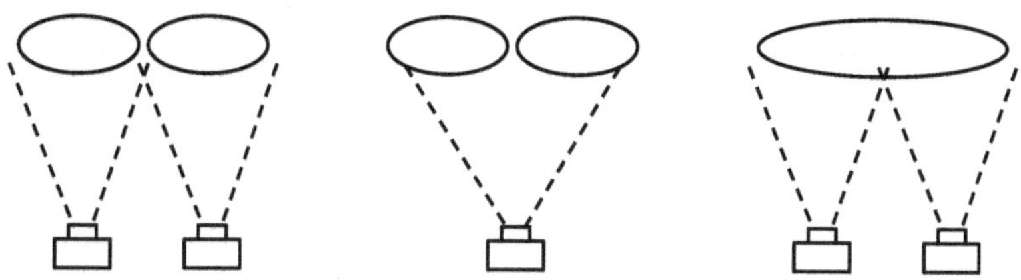

(a) (b) (c)

Fig. 2. (a) A case in which field-of-view cues and spatial cues agree, indicating a pair of objects. (b) A case in which field-of-view cues indicate a single object but spatial cues indicate a pair of objects. (c) A case in which field-of-view cues indicate a pair of objects but spatial cues indicate a single object.

object: pictures in which the object is prominent and takes up most of the frame. We call such images "good", and other images "bad". Good images, for our purposes, are ones which provide useful segmentation cues. Figure 1 shows a few example images of both types. When enough images are good, we find that field-of-view constraints can be used to accurately segment the scene into interesting objects. In the remainder of the paper, we give a simple scene model which takes advantage of field-of-view constraints and evaluate the effectiveness of this model on several different scenes. We also demonstrate the three applications of scene segmentation mentioned above: an "interestingness" viewer, an automatic object labeler that uses user-submitted Flickr tags, and an overhead map browser.

2 Algorithm

Given a set of images of a scene, we first follow the strategy of Snavely et al. [2]:

1. Use the SIFT keypoint detector [7] to extract feature regions from all images. These regions are represented using the SIFT descriptor.
2. For each pair of images, perform feature matching on the descriptors. Prune this set of matches by using RANSAC [17] to estimate a fundamental matrix, removing all inconsistent matches.
3. Organize the matches into tracks, or connected components of features, removing tracks that contain multiple features in the same image.
4. Perform structure from motion on the set of feature tracks, which returns a set of 3D point locations for all valid tracks, as well as camera parameters for each image.

We now operate on two structures, the (sparse) point-image incidence matrix, indicating which points appear in which images, and the set of 3D point locations. Let V be the set of images and X be the set of points, with $M = |V|$ and $N = |X|$. We use $x_j \in V_i$ if point j is visible in image i. Our goal is to compute a clustering C over the points X, where two points belong to the same cluster if they are part of the same object.

Our approach is to use both field-of-view cues and spatial cues to segment the scene. Figure 2 illustrates how these cues work at a basic level. Field-of-view cues encourage points seen in the same view to be part of the same object, while spatial cues encourage objects to be spatially localizable. We use image incidence to enforce field-of-view cues and a single 3D Gaussian distribution per object to enforce spatial cues. To construct a probabilistic model that takes advantage of both types of cues, we combine a mixture of 3D Gaussians (Figure 3a), which uses spatial cues only, and pLSA [18] (Figure 3b), which uses incidence cues only. Our combined model (Figure 3c) uses both types of cues.

We briefly review the probability distributions specified by a mixture of Gaussians and pLSA, and combine them into a single model. A mixture of Gaussians corresponds to the following distribution:

$$P(C, X|\pi, \mu, \Sigma) = \prod_j P(c_j|\pi)P(x_j|c_j, \mu, \Sigma) \tag{1}$$

$$P(c_j|\pi) \sim \text{Mult}(\pi)$$
$$P(x_j|c_j, \mu, \Sigma) \sim \mathcal{N}(\mu_{c_j}, \Sigma_{c_j})$$

In this model, there is a class variable c_j associated with each point x_j. The class variable is drawn from a multinomial distribution with parameter π, and the point locations are drawn from 3D Gaussians with parameters μ_{c_j} and Σ_{c_j}, where the point class c_j specifies which Gaussian to use. The pLSA model corresponds to the following distribution:

$$P(C, X|\theta, \Phi) = \prod_i \prod_{j|x_j \in V_i} P(c_{ij}|\theta_i)P(x_{ij}|c_{ij}, \Phi) \tag{2}$$
$$P(c_{ij}|\theta_i) \sim \text{Mult}(\theta_i)$$
$$P(x_{ij}|c_{ij}, \Phi) \sim \text{Mult}(\Phi_{c_{ij}})$$

In ordinary pLSA, x_{ij} would be a tally of the number of times word j appears in document i. However, in our case the words are 3D points, none of which can appear more than once in a single image. In the pLSA model, there is a class variable c_{ij} for each point-image *incidence*. In other words, a point can belong to different objects in different images. This is not really desirable in our case, so we restrict our model to use a single class variable per point. In addition, we introduce a spatial term. Our combined model corresponds to the following distribution:

$$P(C, X|\theta, \pi, \mu, \Sigma) = \left(\prod_i \prod_{j|x_j \in V_i} P(c_{ij}|\theta_i)\right) \times \left(\prod_j P(c_j|\pi)P(x_j|c_j, \mu, \Sigma)\right) \tag{3}$$
$$P(c_{ij}|\theta_i) \sim \text{Mult}(\theta_i)$$
$$P(c_j|\pi) \sim \text{Mult}(\pi)$$
$$P(x_j|c_j, \mu, \Sigma) \sim \mathcal{N}(\mu_{c_j}, \Sigma_{c_j})$$
$$c_{ij} = c_j$$

Instead of directly replacing the multinomial topic distributions from pLSA with 3D Gaussian distributions, we add a single class variable c_j for each 3D point x_j and tie the values of the incidence class variables c_{ij} with c_j. The resulting model is not strictly a valid Bayesian network, but could easily be converted to one in which the c_{ij} variables are eliminated and c_j is directly conditioned on all of the images in which the point appears. We find it is easier to think about tied variables, and the above joint density can easily be maximized (locally) using the EM algorithm [19].

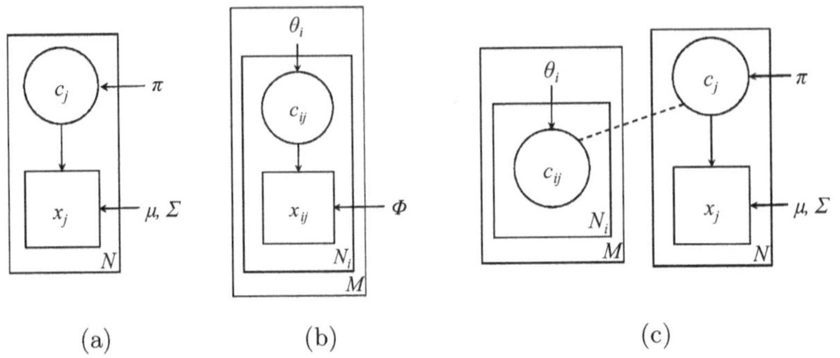

Fig. 3. Graphical models for (a) a mixture of Gaussians on 3D point locations, (b) the pLSA model, and (c) the combined model that uses both point location and image incidence information

3 Evaluation

We tested our model on six scenes, each of which contains between 300 and 3000 images downloaded from Flickr: Trafalgar Square, the Pantheon in Rome, Hagia Sophia, Trevi Fountain, Old Town Square in Prague, and Piazza Navona. These scenes all contain component objects which have names and can be identified visually. In order to test our segmentations, we created ground-truth clusterings for each scene manually, as follows. For each scene, we reconstructed a set of 3D points using the system of Snavely et al. [2]. We then assigned 3D points to clusters by manually selecting regions in images and grouping all points in the selected region with a particular cluster. Though any hand-labeling of this nature is somewhat arbitrary, we attempted to label objects as uncontroversially as possible. We also used Wikipedia [20] text and images to decide which objects should be included when there was some uncertainty. In other cases, there is a natural segmentation implied by the scene. For example, in the Trafalgar Square scene we labeled each building and statue as a separate object. Since the reconstructed scenes contain hundreds of thousands of 3D points, many of which don't belong to an easily nameable object, we only hand-label a small fraction of the points, and evaluate our algorithm on these points only. Note that the aforementioned manual steps were used only to create the ground truth used for evaluation purposes. The segmentation algorithms themselves are fully automatic.

For all scenes, we tested three different models: a mixture of 3D Gaussians, the pLSA model, and our combined model that uses both spatial and incidence cues. Each model was tested using multiple different values of k, the number of clusters. For each value of k, we used 5 different random initializations and ran 100 iterations of the EM algorithm for each, then kept the single result with highest joint probability for each value of k. We created hard cluster assignments by assigning each point to its most likely cluster (or, for pLSA, the cluster under which the point has highest probability). We evaluate clusterings using

Table 1. Median values of the VI clustering metric $VI(C, C^*)$ for each algorithm and scene, over multiple runs of EM. Lower values indicate the computed clustering is closer to the hand-labeled clustering.

	Trafalgar	Pantheon	Hagia Sophia	Trevi	Prague	Navona
mixture of Gaussians	1.15	1.36	0.63	**0.81**	0.35	0.68
pLSA	2.07	1.70	0.64	3.12	1.13	1.46
combined model	**0.69**	**0.38**	**0.53**	2.07	**0.20**	**0.45**

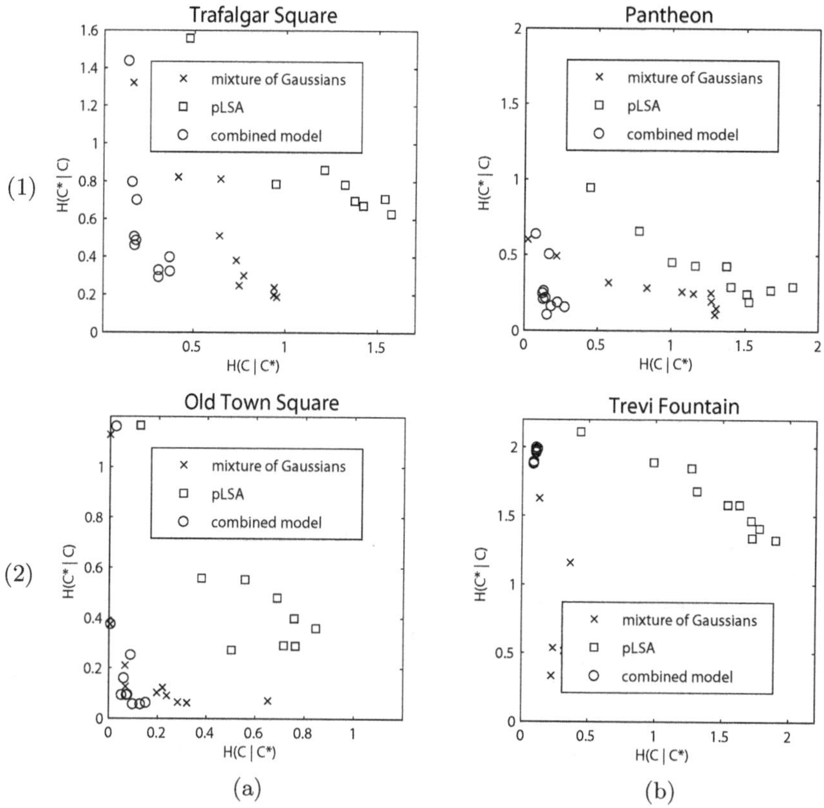

Fig. 4. Evaluation of clustering C against ground truth clustering C^* for the (1a) Trafalgar Square, (1b) Pantheon, (2a) Old Town Square, and (2b) Trevi Fountain datasets. The horizontal axis $H(C|C^*)$ is a measure of over-segmentation, and the vertical axis $H(C^*|C)$ is a measure of under-segmentation. Note that for the Trevi Fountain, both pLSA and the combined model are prone to undersegmentation, as most images of the Trevi Fountain are of the entire façade. This causes field-of-view cues to prefer larger objects, even when there are interesting details within the façade.

Meila's Variation of Information metric [21]. Given a ground truth clustering C^* and computed clustering C, the VI metric $VI(C, C^*) = H(C|C^*) + H(C^*|C)$ measures the amount of information lost and gained between the two clusterings.

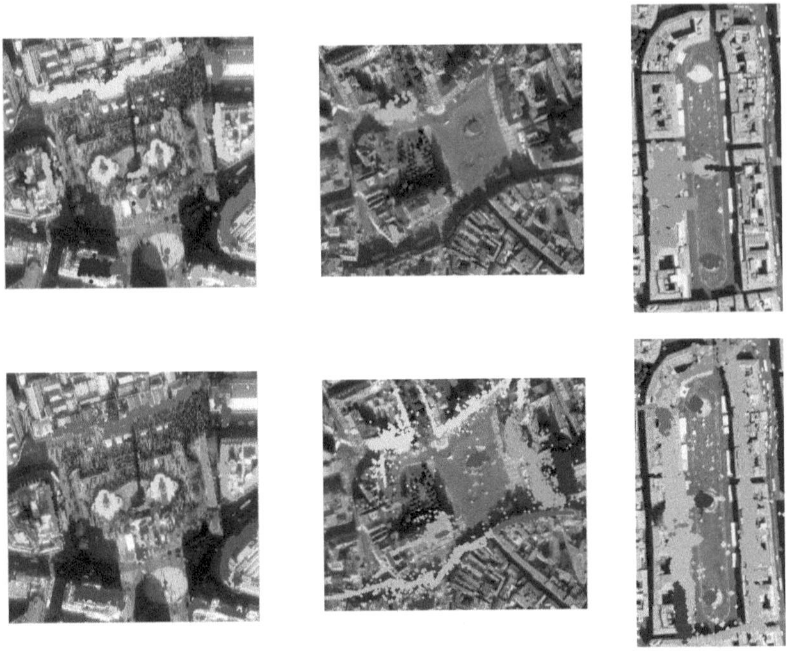

(a) (b) (c)

Fig. 5. Satellite view of ground truth (top row) and computed segmentations (bottom row) for (a) Trafalgar Square, (b) Old Town Square, and (c) Piazza Navona. Each color corresponds to a different cluster. For the Trafalgar Square scene, not all reconstructed points are shown in the computed segmentation to avoid clutter. In all figures in this paper, we use the same color for each cluster in each scene. As there is no explicit correspondence between ground-truth clusters and computed clusters, the ground-truth clusterings do not use the same color scheme.

For two identical clusterings, this value is zero. Also, by looking at the two conditional entropy terms separately, we can get a sense of how over- and under-segmented C is.

Table 1 contains median VI distances for each of the three algorithms and six scenes. The combined model performs best on all scenes except the Trevi Fountain, which suffers from undersegmentation as photographers cannot get close enough to the façade to take closeup images of the interesting objects. Figure 4 shows our over- and under-segmentation results on four of the scenes. In general, when the collection of photographs contains many images of interesting objects, our combined model does well. Since what we are testing is whether or not field-of-view cues provide additional information that is useful for segmentation, our results demonstrate that this is true for many scenes. Visualizations of the clusterings themselves are shown in Figures 5 and 6.

(a) (b) (c) (d)

Fig. 6. Image views of computed segmentations for (a,b) Trafalgar Square, (c) Old Town Square, and (d) Piazza Navona

4 Applications

4.1 Importance Viewer

Once we have computed a 3D point segmentation for a scene, we can use this segmentation to compute and display additional information about the scene. In this section, we describe two such applications: highlighting interesting objects in images and labeling image regions using noisy user-submitted Flickr tags.

Given an image in the collection, we want to identify regions which belong to objects that are important or interesting. We define an object as interesting if there are many photos of it in the collection. As this tends to overly reward large background objects, we also penalize objects for size. Our importance score is:

$$\text{imp}(c) = \alpha \frac{1}{|\Sigma_c|} \sum_i \theta_i(c) \tag{4}$$

Here, α is a scene normalization coefficient that enforces a fixed total importance, $\frac{1}{|\Sigma_c|}$ penalizes clusters proportional to the determinant of their covariance matrices, and $\sum_i \theta_i(c)$ rewards clusters for appearing in many images. To visualize importance in an image, we assign each pixel to its nearest feature point in the image and highlight the pixel with intensity proportional to the importance of the cluster to which this point is assigned, falling off as distance to this point increases. Some resulting importance images can be seen in Figure 7.

(a) (b) (c)

Fig. 7. Importance images computed for (a) Piazza Navona, (b) Prague, and (c) the Pantheon (top) and Trafalgar Square (bottom). Importance is indicated by color saturation, with different hues for different clusters.

4.2 Region Labeling

Flickr and most other photo sharing sites give users the ability to attach textual tags to entire images. (Flickr also provides functionality for leaving rectangular notes on image regions, but this feature is much less utilized.) In general, these tags are noisy and the majority of them do not correspond to actual objects in the scene. For many objects, however, we are able to compute accurate tags by examining tag-cluster co-occurrence statistics. To find good object tags, we first apply pLSA to the tags with fixed image-topic distributions θ computed from the point clustering. This gives us a distribution over tags for each cluster $P(t|c)$, which also gives us the joint distribution $P(c,t)$. For a particular cluster c and tag t, we compute the following score:

$$\text{score}(c,t) = P(c,t)\,(\log P(c,t) - \log P(c)P(t)) \tag{5}$$

This gives high scores to cluster-tag pairs that appear much more frequently than would be expected given just the marginals, which indicates that the cluster and tag are probably related. We then assign the tag with highest score to each cluster if the score exceeds a specified threshold, otherwise we assign no tag to the cluster. Region tagging results can be seen in Figure 8.

4.3 Interactive Map Viewer

Many systems have been created to support interactive browsing of the visual content of a scene [2,14,15,22]. Our segmentations allow for the possibility of

(a) (b) (c)

Fig. 8. Region tags automatically computed from our segmentation, and used to label two images of (a) Trafalgar Square, (b) Old Town Square, and (c) Piazza Navona. We compute these labels using noisy user-submitted tags downloaded from Flickr, automatically associating a single tag (or no tags) with each cluster. In these images, we manually moved the labels to make them more readable.

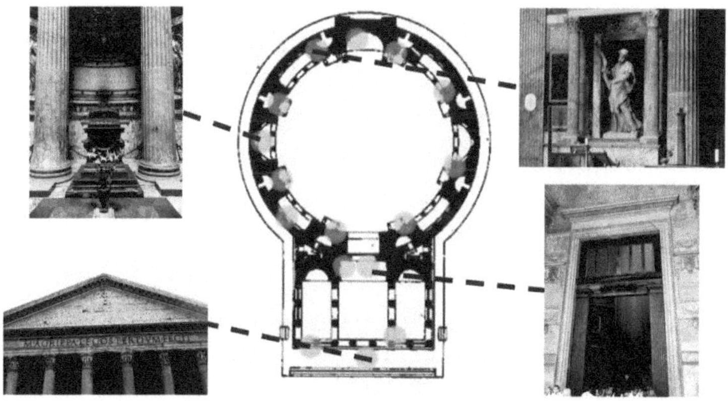

Fig. 9. An interactive floor plan viewer showing the Pantheon (center). By moving the mouse over one of the highlighted circles, the user can see an image of the object at that location (sample images shown on left and right).

object-centric interactive scene viewers. We have created a simple interactive viewer based on an overhead map or floor plan, shown in Figure 9.

To create the interactive floor plan, we first segment the scene using the algorithm from Section 2. We then manually align the scene points with an overhead view. Since we only want to include segments which can be localized at a reasonably-sized spot on the map, we remove all segments larger than a size threshold. To choose the representative image for each segment, we compute the Kullback-Leibler divergence between the distribution of scene points in each image and each cluster.

5 Conclusion

In this paper we have proposed a new field-of-view cue that can be used to extract objects from static 3D scenes, along with a probabilistic model that takes advantage of this cue and several applications of our method. We stress that the probabilistic model is intended to evaluate the usefulness of field-of-view cues, and not to provide a complete solution to the scene segmentation problem. Note in particular that we are not incorporating any of the more standard image segmentation cues such as intensity, color, contour, or region information, except through feature matching for estimating point correspondences. Including such additional terms and using more sophisticated models, like Latent Dirichlet Allocation [23] or Dirichlet process mixtures would likely further improve upon our results. Still, our model that takes into account only field-of-view and spatial proximity is able to achieve segmentations that are good enough to enable a variety of applications.

Acknowledgements

This work was supported in part by National Science Foundation grant IIS-0743635, the Office of Naval Research, the University of Washington Animation Research Labs, Adobe, Microsoft, Google, and an endowment by Rob Short and Emer Dooley.

References

1. Flickr, `http://www.flickr.com`
2. Snavely, N., Seitz, S.M., Szeliski, R.: Photo tourism: Exploring photo collections in 3D. In: SIGGRAPH Conference Proceedings, pp. 835–846 (2006)
3. Hays, J., Efros, A.A.: Scene completion using millions of photographs. In: SIGGRAPH Conference Proceedings, vol. 26(3) (2007)
4. Fergus, R., Fei-Fei, L., Perona, P., Zisserman, A.: Learning object categories from google's image search. In: Proceedings of the 10th IEEE International Conference on Computer Vision, vol. 2, pp. 1816–1823 (2005)
5. Goesele, M., Snavely, N., Curless, B., Hoppe, H., Seitz, S.M.: Multi-view stereo for community photo collections. In: Proceedings of the 11th IEEE International Conference on Computer Vision, pp. 1–8 (2007)
6. Hays, J., Efros, A.A.: IM2GPS: estimating geographic information from a single image. In: Proceedings of the IEEE Conference on Computer Vision and Pattern Recognition (2008)

7. Lowe, D.G.: Distinctive image features from scale-invariant keypoints. International Journal of Computer Vision 60(2), 91–110 (2004)
8. Surowiecki, J.: The Wisdom of Crowds. Random House, New York (2004)
9. Fergus, R., Perona, P., Zisserman, A.: Object class recognition by unsupervised scale-invariant learning. In: Proceedings of the IEEE Conference on Computer Vision and Pattern Recognition, vol. 2, pp. 264–271 (2003)
10. Sivic, J., Russell, B., Efros, A., Zisserman, A., Freeman, W.: Discovering objects and their localization in images. In: Proceedings of the 10th IEEE International Conference on Computer Vision, vol. 1, pp. 370–377 (2005)
11. Sudderth, E., Torralba, A., Freeman, W., Willsky, A.: Describing visual scenes using transformed objects and parts. International Journal of Computer Vision 77(1-3), 291–330 (2007)
12. Campbell, N., Vogiatzis, G., Hernandez, C., Cipolla, R.: Automatic 3D object segmentation in multiple views using volumetric graph-cuts. In: Proceedings of the British Machine Vision Conference, vol. 1, pp. 530–539 (2007)
13. Russell, B.C., Efros, A.A., Sivic, J., Freeman, W.T., Zisserman, A.: Using multiple segmentations to discover objects and their extent in image collections. In: Proceedings of the IEEE Conference on Computer Vision and Pattern Recognition, vol. 2, pp. 1605–1614 (2006)
14. Simon, I., Snavely, N., Seitz, S.M.: Scene summarization for online image collections. In: Proceedings of the 11th IEEE International Conference on Computer Vision, pp. 1–8 (2007)
15. Epshtein, B., Ofek, E., Wexler, Y., Zhang, P.: Hierarchical photo organization using geo-relevance. In: Proceedings of the 15th Annual ACM International Symposium on Advances in Geographic Information Systems, pp. 1–7. ACM, New York (2007)
16. Barnard, K., Duygulu, P., Forsyth, D., De Freitas, N., Blei, D.M., Jordan, M.I.: Matching words and pictures. Journal of Machine Learning Research 3(6), 1107–1135 (2003)
17. Fischler, M.A., Bolles, R.C.: Random Sample Consensus: a paradigm for model fitting with applications to image analysis and automated cartography. Communications of the ACM 24(6), 381–395 (1981)
18. Hofmann, T.: Probabilistic latent semantic analysis. Proceedings of Uncertainty in Artificial Intelligence, 289–296 (1999)
19. Dempster, A.P., Laird, N.M., Rubin, D.B.: Maximum likelihood from incomplete data via the EM algorithm. Journal of the Royal Statistical Society 39(1), 1–38
20. Wikipedia, `http://www.wikipedia.org`
21. Meilă, M.: Comparing clusterings: an information based distance. Journal of Multivariate Analysis 98(5), 873–895 (2007)
22. Ahern, S., Naaman, M., Nair, R., Yang, J.H.: World Explorer: visualizing aggregate data from unstructured text in geo-referenced collections. In: Proceedings of the 2007 Conference on Digital Libraries, pp. 1–10 (2007)
23. Blei, D.M., Ng, A.Y., Jordan, M.I.: Latent Dirichlet Allocation. Journal of Machine Learning Research 3(4-5), 993–1022 (2003)

Optimization of Symmetric Transfer Error for Sub-frame Video Synchronization

Meghna Singh[1], Irene Cheng[2], Mrinal Mandal[1], and Anup Basu[2]

[1] Department of Electrical and Computer Engineering
[2] Department of Computing Science
University of Alberta, Edmonton, Alberta, Canada

Abstract. In this work we present a method to synchronize video sequences of events that are acquired via uncalibrated cameras at unknown and dynamically varying temporal offsets. Unlike existing methods that synchronize videos of similar events (i.e., videos related to each other through the motion in the scene) up to an integer alignment, we establish sub-frame video synchronization. While contemporary synchronization algorithms implement a unidirectional alignment which biases the results towards a single reference sequence, we adopt a bi-directional or symmetrical alignment approach that results in a more optimal synchronization. To this end, we propose a novel symmetric transfer error which is dynamically minimized, and reduces the propagation of error from feature extraction and spatial mapping into temporal synchronization. The advantages of our approach are validated by tests conducted on (publicly available) real and synthetic sequences. We present qualitative and quantitative comparisons with another state-of-the-art algorithm. A unique application of this work in generating high-resolution 4D MRI data from multiple low-resolution MRI scans is described.

1 Introduction

Synchronization of video sequences plays a crucial role in applications such as super-resolution imaging [1][2], 3D visualization [3], robust multi-view surveillance [4] and mosaicking [5]. Most video synchronization algorithms deal with video sequences of the same scene and hence assume that the temporal offset between the video sequences does not change over time, i.e., a simple temporal translation is assumed. Video synchronization, however, is not limited to aligning sequences of the same scene, and can be extended to find the spatio-temporal alignment between related scenes, for applications such as video search, video comparison and enhanced video generation. In such scenarios, the temporal offset between the video sequences is dynamically changing, and cannot be estimated by a translational offset. The closest related works in synchronization of video sequences that are related by dynamically varying temporal offsets are limited to integer frame alignment of video sequences [6][7].

Another drawback identified with contemporary works is the asymmetric or unidirectional nature of the alignment, due to which error or uncertainty in

D. Forsyth, P. Torr, and A. Zisserman (Eds.): ECCV 2008, Part II, LNCS 5303, pp. 554–567, 2008.

feature extraction in the reference sequence propagates into the synchronization result. Our contribution lies in formulating the synchronization problem as the iterative minimization of a *symmetric transfer error* (STE), which allows us to compute *sub-frame accurate* synchronization of video sequences that have a *dynamically varying temporal offset* between them. In addition, the method of minimizing STE allows us to reduce the occurrence of singularities[1] in the synchronization.

The rest of this paper is organized as follows. Section 2 outlines related work in video synchronization, and justifies the necessity for this work. In Section 3, we present the symmetric transfer error and an optimization strategy to minimize it. In Section 4, we discuss our experimental setup and show comparative results with a rank-constraint based approach. Application of our work in 4D MRI visualization is presented in Section 5. Finally, summary and conclusions are given in Section 6.

2 Review of Prior Research

Past literature in video synchronization and temporal registration can be broadly classified into two categories — video sequences of the same scene or video sequences of similar scenes; differing primarily on the assumptions made with respect to the temporal offset between sequences. It can be seen from Table 1, that our work addresses all three scenarios (view-invariance, dynamic time shifts and sub-frame accuracy), while previous works have only addressed a subset of these scenarios.

2.1 Video Synchronization of Same Scene

In synchronizing videos of the same scene, the temporal offset is considered to be an affine transform [1], such as $t' = s.t + \Delta t$, where s is the ratio of frame rates and Δt is a fixed translational offset. Dai et al. [9] use 3D phase correlation between video sequences, whereas Tuytelaars et al. [10] compute synchronization by checking the rigidity of a set of five (or more) points. Tresadern et al. [11] also follow a similar approach of computing a rank-constraint based rigidity measure between four non-rigidly moving feature points. Lee et al. [4] derive the spatial relationship between multiple surveillance cameras viewing the same (or partially overlapping) scene by matching motion trajectories captured in the video sequences to planar models. Caspi et al. [1] recover the spatial and temporal relation between two sequences by minimizing the SSD error over extracted trajectories that are visible in both of the sequences. Carceroni et al. [12] extend [1] to align sequences based on scene points that need to be visible only in two

[1] Singularities [8] occur when multiple frames in the target sequence map to the same frame in the reference sequence. Such a situation can occur when the temporal speed of an event in one sequence is significantly slower than another sequence. However, in most cases this slowing down of the event should lead to a sub-frame mapping and not singularities.

Table 1. Review of related work. Legend: D.S.-Dynamic time shift, V.I.-View Invariant, S.A.-Sub-frame Accuracy.

Author	Scene	D.S.	V.I	S.A.	Author	Scene	D.S.	V.I	S.A.
Caspi et al. [2] [1]	Same	X	$\surd$	$\surd$	Tresadern et al. [11]	Same	X	$\surd$	$\surd$
Lei et al. [13]	Same	X	$\surd$	X	Carceroni et al. [12]	Same	X	$\surd$	$\surd$
Wolf et al. [14]	Same	X	$\surd$	X	Pooley et al. [15]	Same	X	$\surd$	$\surd$
Dai et al. [9]	Same	X	$\surd$	$\surd$	Tuytelaars et al. [10]	Same	X	$\surd$	X
Singh et al. [3]	Same	X	X	$\surd$	Lee et al. [4]	Same	X	$\surd$	X
Perperidis et al. [16]	Diff	$\surd$	X	X	Rao et al. [6]	Diff	$\surd$	$\surd$	X
Proposed	Diff	$\surd$	$\surd$	$\surd$	Giese et al. [17]	Diff	$\surd$	X	X

consecutive frames. Singh et al. [3] also build on [1] to develop parameterizable event models of discrete trajectories and align the event models based on an SSD measure for sub-frame synchronization. However, they do not address the problem of view-invariance and dynamic temporal offsets.

2.2 Video Synchronization of Different Scenes

When aligning video sequences of different scenes, albeit sequences correlated via motion, one has to factor in the dynamic temporal scale of activities in the video sequences. Giese and Poggio [17] approach alignment of activities of different people by computing a dynamic time warp between the feature trajectories. They do not address the problem when the activity sequences are from varying viewpoints and their approach is a one-to-one frame correspondence. Perperidis et al. [16] attempt to locally warp cardiac MRI sequences, by extending Caspi's work [1] to incorporate spline based local alignment. Though their approach does lead to good alignment of time-varying sequences, it has two main drawbacks: (i) the computation space for spline based registration is quite large and the authors need to compute points of inflexion in the cardiac volume change; and (ii) the alignment is still a one-to-one frame correspondence and not sub-frame accurate. Others, such as Rao et al. [6] use rank constraint as the distance measure in a dynamic time warping algorithm to align multiple sequences; this is the first work that can deal with video sequences of correlated activities, hence we will present comparisons with their rank constraint based (RCB) algorithm.

3 Proposed Approach

In this work we assume that the frame rates of the cameras are identical and fixed throughout the acquisition, the scene is planar and the backgrounds in both the scenes have sufficient static points that can be extracted using view-invariant feature detectors [18] to estimate the spatial relationship between the two scenes. These assumptions are not restrictive since the synchronization algorithm can be adapted to account for epipolar geometry of non-planar scenes and

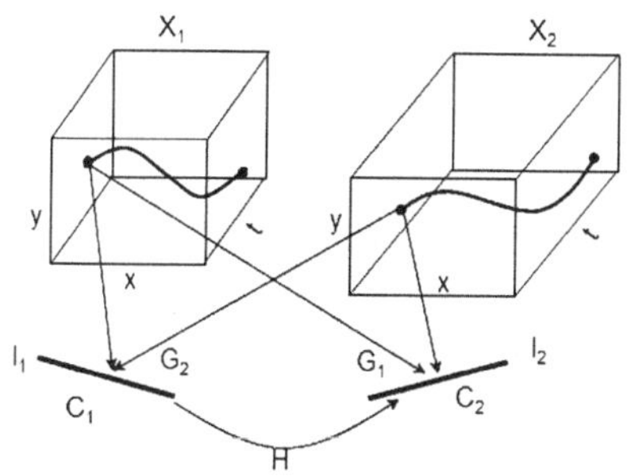

Fig. 1. Illustration of two distinct scenes acquired using two distinct cameras. The projections (ghosts) of scenes onto the reciprocal cameras are also shown.

robust correspondence algorithms can be used for wide-baseline cameras [19]. Additionally, we assume that a single feature trajectory of interest is available to us[2] such that the beginning and end points of the activity are marked in the trajectory, similar to the assumption made in [6].

Given that we have two cameras C_1 and C_2, as shown in Fig.1, that view two independent scenes of similar activities; C_1 views scene X_1 and acquires video I_1, and C_2 views scene X_2 and acquires a video sequence I_2. Features, F_1 and F_2, are extracted and tracked in both the acquired video sequences. A single feature trajectory is used to illustrate this in the figure. On their own these feature trajectories are discrete representations of the event in the scene, and we need to interpolate between these discrete representations. Most algorithms use linear interpolation, however, we use the method proposed by Singh et al. [3] to generate continuous models from discrete points. The continuous feature trajectories are represented as $\mathcal{F}_1$ and $\mathcal{F}_2$. Note that we store these continuous models as discrete matrices in computer memory, albeit the temporal resolution of the continuous models is much higher than the discrete models. Hence, we index into $\mathcal{F}$ using integer numbers.

The spatial relationship (homography H) between the two scenes is computed by using the Direct Linear Transform (DLT) algorithm to initialize the non-linear least squares (Levenberg-Marquardt) computation of the homography [20]. As the feature trajectories are dynamically offset from each other (as illustrated in Fig.1), we cannot directly optimize for both the homography and temporal offset as proposed by Caspi et al. [1]. Instead, we project the trajectory from Scene 1 to Scene 2, as if Camera C_2 did view Scene 1 (and vice versa). Note that the accuracy of this projection is subject to noise in the extracted trajectories as well as error in the homography computation. We call these projections the *ghosts* of the trajectories, represented as $\mathcal{G}$.

---

[2] In general videos, multiple object trajectories will be generated and an additional task of the synchronization algorithm will be to find corresponding feature trajectories in the multiple video sequences. This is an open problem in vision research and one that we will not address in this paper.

$$\mathcal{G}_1 = H_{1\to 2} \cdot \mathcal{F}_1$$
$$\mathcal{G}_2 = H_{2\to 1} \cdot \mathcal{F}_2 \qquad (1)$$

The aim of the synchronization algorithm is to temporally align a trajectory in Scene 1 with the ghost of the trajectory from Scene 2, and vice versa. Current dynamic offset synchronization algorithms, *e.g.* [6] and [17], synchronize discrete feature trajectories to a frame-by-frame correspondence by only computing a unidirectional alignment. For example, they assume Trajectory 1 to be the reference, and warp Trajectory 2 towards it. This unidirectional alignment biases the synchronization towards the reference sequence. Feature extraction and tracking errors in the reference sequence now propagate unchecked into the synchronization. We demonstrate that a more symmetric approach will not only mitigate such an error propagation, but will also result in better sequence synchronization. Next, we introduce our symmetric optimization approach.

3.1 Alignment as the Minimization of Symmetric Transfer Error

Feature trajectories and associated event models are computed independently for each sequence; introducing a variability in the sequence alignment, such that frame alignment from Sequence 1 to Sequence 2 is not identical to frame alignment from Sequence 2 to Sequence 1; i.e., if $\mathcal{W}_{1,2}$ is a mapping from $\mathcal{F}_1$ in Sequence 1 to the ghost $\mathcal{G}_2$ of Sequence 2 and $\mathcal{W}_{2,1}$ is a mapping from $\mathcal{F}_2$ in Sequence 2 to the ghost $\mathcal{G}_1$ of Sequence 1, then $\mathcal{W}_{1,2} \neq \mathcal{W}_{2,1}$. This asymmetry is illustrated with a hypothetical example in Fig.2, where $\mathcal{W}_{1,2}$ and $\mathcal{W}_{2,1}$ are two reciprocal alignments computed between the sequences. Asymmetry in mappings for real video sequences is shown in Fig.9. If indexing function $\mathcal{M}(\mathcal{W}_{1,2}, i)$ indexes the mapping $\mathcal{W}_{1,2}$ and returns the frame in Sequence 2 corresponding to the ith frame in Sequence 1 (similarly $\mathcal{M}(\mathcal{W}_{2,1}, i)$ indexes mapping $\mathcal{W}_{2,1}$ and returns the corresponding frame in this mapping), then the symmetric transfer error (STE) for the ith frame (where $i \in \mathbb{R}$) is defined as follows:

$$\mathcal{E}(i) = |\mathcal{M}(\mathcal{W}_{1,2}, i) - \mathcal{M}(\mathcal{W}_{2,1}, i)| \qquad (2)$$

Let us illustrate (2) with an example. Suppose in Fig.2, for the ith frame in Sequence 1, mapping $\mathcal{W}_{1,2}$ reports the corresponding frame in Sequence 2

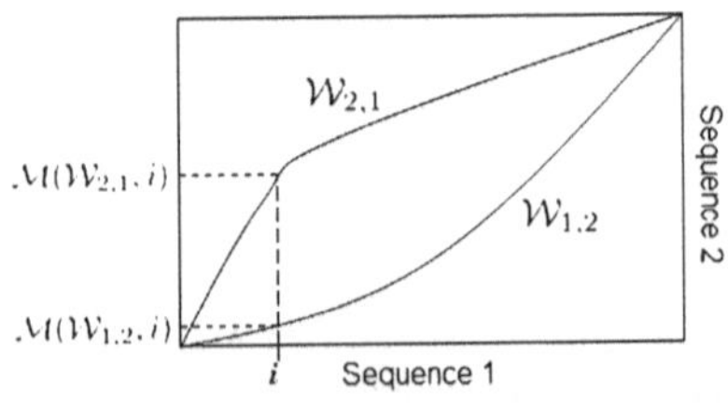

Fig. 2. Computing symmetric transfer error for a single frame 'i' in Sequence 1

to be the 7th frame, and mapping $\mathcal{W}_{2,1}$ reports the corresponding frame to be the 10th frame. Then, STE for the ith frame is $\mathcal{E}(i) = 3$. STE for the entire sequence is then defined as the summation in (3), where N is the length of $\mathcal{F}_1$ and $\mathcal{G}_1$ of Sequence 1 and M is the length of $\mathcal{F}_2$ and $\mathcal{G}_2$ of Sequence 2. Intuitively, minimizing STE (3) for synchronization is akin to computing an optimal compromise between the reciprocal mappings $\mathcal{W}_{1,2}$ and $\mathcal{W}_{2,1}$.

$$\mathcal{E} = \sum_{i=1..max(N,M)} |\mathcal{M}(\mathcal{W}_{1,2}, i) - \mathcal{M}(\mathcal{W}_{2,1}, i)| \ . \tag{3}$$

The mapping functions $\mathcal{W}_{1,2}$ and $\mathcal{W}_{2,1}$ are computed using a regularized dynamic time warping (DTW) approach. The implementation of DTW via dynamic programming, factors in boundary conditions, continuity and monotonicity of the mapping function. An important distinction to be made here is that the STE is not the same as a symmetric implementation of DTW (details on this distinction are provided online at [21]). While (4)-(7) detail how $\mathcal{W}_{1,2}$ is computed; the same apply to $\mathcal{W}_{2,1}$ with suitable substitutions made for $\mathcal{F}_2$ and $\mathcal{G}_1$. We build a cost matrix $\mathcal{D}(n,m)$ $\forall n \in [1..N]$ and $\forall m \in [1..M]$, where N is the length of event model $\mathcal{F}_1$ for Sequence 1 and M is the length of event model $\mathcal{F}_2$ for Sequence 2, as follows:

$$\mathcal{D}(n,m) = \|(\mathcal{F}_1(n) - \mathcal{G}_2(m))\|^2 + w(\|\partial\mathcal{F}_1(n) - \partial\mathcal{G}_2(m)\|^2) \ , \tag{4}$$

$$\partial\mathcal{F}(k) = \frac{\mathcal{F}(k) - \mathcal{F}(k-1) + [\mathcal{F}(k+1) - \mathcal{F}(k-1)]/2}{2} \ . \tag{5}$$

In (4), w is the weight assigned to the regularization function. The motivation behind regularization of the cost function is two fold: (i) it allows us to factor in a smoothness constraint on the warping and (ii) it also reduces the occurrence of singularities in the mapping. The mapping function $\mathcal{W}$, is computed by traversing the path of minimum cost in the cost matrix $\mathcal{D}$ (illustrated in Fig.3(a)), with n and m initialized to N and M respectively, as follows:

$$\mathcal{W}_{1,2}(n,m) = \mathcal{D}(n,m) + \min(\phi) \ , \tag{6}$$

$$\phi = [\mathcal{W}_{1,2}(n-1,m), \mathcal{W}_{1,2}(n-1,m-1), \mathcal{W}_{1,2}(n,m-1)] \ . \tag{7}$$

In (7) we consider a neighborhood of three frames (similar to [6]), however, this neighborhood can be extended. It can be seen from (3)-(6) that STE is dependent on the regularization weight w, and minimization of the STE with respect to w (8) optimizes the sequence synchronization. An example of STE values computed for varying values of w for two synthetic trajectories are shown in Fig.3(b).

$$\mathcal{W}_{\text{opt}} = arg \min_{w} \sum_{i=1..max(N,M)} |\mathcal{M}(\mathcal{W}_{1,2}, i) - \mathcal{M}(\mathcal{W}_{2,1}, i)| \tag{8}$$

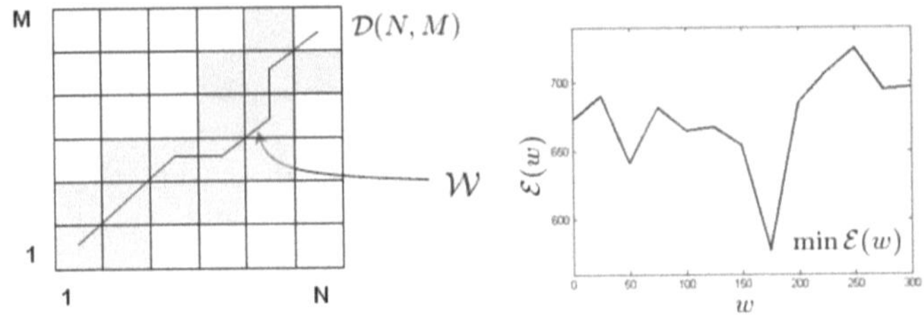

(a) (b)

Fig. 3. (a) Illustration of computing the mapping function $\mathcal{W}$ from the cost matrix $\mathcal{D}$. (b) Plot of STE $\mathcal{E}(w)$ versus regularization weight w for two synthetic sequences of length 100 and 140 frames. $min(\mathcal{E})$ is also indicated.

Equation (8) enforces the minimization of the difference between the two mappings $\mathcal{W}_{1,2}$ and $\mathcal{W}_{2,1}$ in Fig.2. The advantage of optimizing a symmetric measure, as opposed to an asymmetric measure, is validated experimentally on real video sequences. The pseudocode of the proposed algorithm is given in Table 2. For our experiments the variables in the pseudocode min_w, max_w, and $iter\_step$, were empirically determined and set to 0, 300 and 25 respectively.

4 Experiments and Comparative Analysis

We tested our method on both synthetic and real image sequences. We also implemented the rank-constraint based (RCB) algorithm as described in [6] that deals with aligning videos of similar events. While the RCB method cannot compute sequence alignment to sub-frame accuracy, we still compare our method with it for integer alignment. Our test cases and results are presented next. All experiments were run on a 3.2GHz Intel Pentium IV processor with 1GB RAM using MATLAB 7.04. Excluding the preprocessing time, the STE algorithm took 1.79 seconds to run through all iterations, to compute the optimal alignment between two synthetic sequences of length 100 and 140. The processing time for real sequences of length 84 and 174 was 1.93 seconds.

4.1 Synthetic Sequences

In synthetic tests, we generate planar trajectories, 100 frames long, using a pseudo-random number generator (modified version of Marsaglia's subtract with borrow algorithm). These trajectories are then projected onto two image planes using user defined camera projection matrices. The camera matrices are designed so that the acquisition emulates a homography, and are used only for generation purpose and not thereafter. A time warp is then applied to a section of one of the trajectory projections, such that its length now becomes 140 frames. Both the RCB and the STE methods are then applied to the synthetic trajectories to

Table 2. Pseudocode of the proposed algorithm

Pre-preprocessing
Extract feature trajectories F_1 and F_2
Compute Event models $\mathcal{F}_1$ and $\mathcal{F}_2$
Compute Homography H
Project Event models to derive ghosts $\mathcal{G}_1$ and $\mathcal{G}_2$
Set max_w, min_w, iter_step

Iterate to minimize STE
For w=min_w : iter_step : max_w
Compute regularized warp $\mathcal{W}_{1,2}$
1. Compute cost metric $\mathcal{D}(n,m) = \|(\mathcal{F}_1(n) - \mathcal{G}_2(m))\|^2 + w(\|\partial\mathcal{F}_1(n) - \partial\mathcal{G}_2(m)\|^2)$
2. Compute mapping as the path of minimum cost $\mathcal{W}_{1,2}(n,m) = \mathcal{D}(n,m) + \min(\phi)$
$\phi = [\mathcal{W}_{1,2}(n-1,m), \mathcal{W}_{1,2}(n-1,m-1), \mathcal{W}_{1,2}(n,m-1)]$
Compute regularized warp $\mathcal{W}_{2,1}$
Follow step 1-2 above for $\mathcal{F}_2$, $\mathcal{G}_1$
Compute Symmetric Transfer Error $\mathcal{E} = \sum_{i=1..max(N,M)} |\mathcal{M}(\mathcal{W}_{1,2}, i) - \mathcal{M}(\mathcal{W}_{2,1}, i)|$
$\mathcal{M}$ is a simple indexing function.
End

Find $\min \mathcal{E}(w)$
Report $\mathcal{W}_{\text{opt}}$ corresponding to $\min \mathcal{E}(w)$

compute the alignment between them. This process is repeated on 100 synthetic trajectories. Fig. 4 shows some synchronization results with synthetic trajectories. Additional synchronization results are available online at [21], where it can be seen that for simple trajectories, the STE and RCB methods result in comparable synchronization, close to the correct alignment. However, as the complexity of the synthetic trajectory begins to increase, the RCB method starts producing erroneous alignments while the STE method continues to compute synchronization that closely matches the actual synchronization. On average, the STE algorithm made 34% less errors in computing the synchronization when compared to the RCB method. To test for sub-frame alignment, we divided the projected trajectories into two parts, such that F_1 contains every even trajectory point and F_2 contains every odd trajectory point. The computed synchronization between these two trajectories by the STE method accurately placed frames in Sequence F_2 at 0.5 frame intervals from F_1.

Table 3. Synchronization Errors for RCB and STE methods for noisy trajectories

Noise σ^2	STE	RCB
0.0001	281	384
0.001	326	470
0.01	400	755
0.1	623	1199

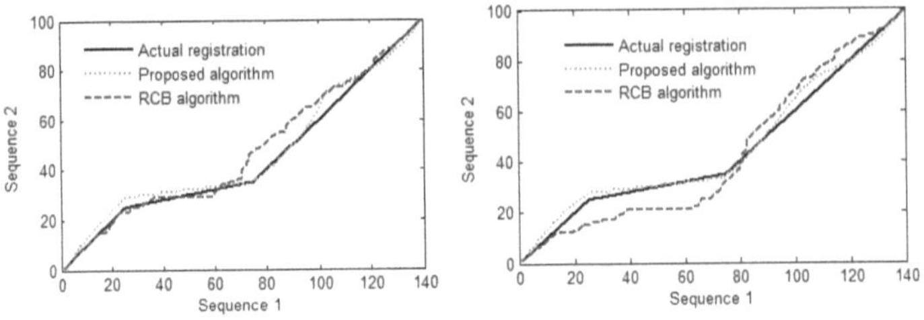

(a) (b)

Fig. 4. (a)-(b) Results of synchronization of synthetic trajectories using proposed approach and RCB approach

We also tested the effect of noisy trajectories on our synchronization approach. Normally distributed and zero mean noise with various values of variance (σ^2) was added to the synthetic feature trajectories. The results of synchronization of noisy trajectories with both the RCB and STE approach are shown in Table 3, where the sum of absolute differences between the actual and computed frame correspondence is reported as the synchronization error. The STE approach is affected slightly by the addition of noise, however, the performance of the RCB method degrades dramatically. Some representative alignment results are shown in Fig.5.

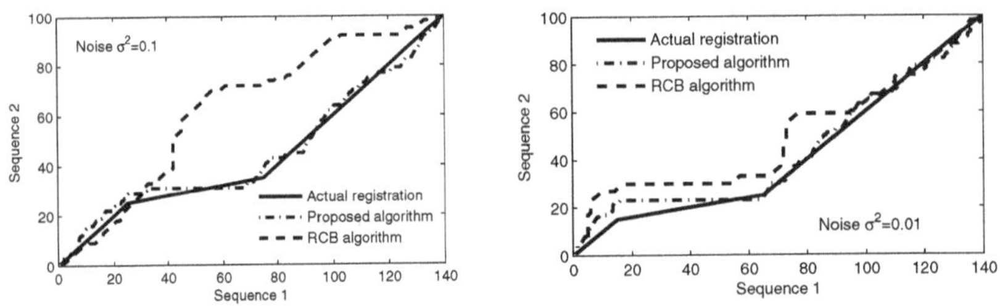

(a) (b)

Fig. 5. Performance of the STE and RCB methods for noisy trajectories with noise variance (a) 0.1 and (b) 0.01

4.2 Real Sequences

We used video sequences provided by Rao et al. at `http://server.cs.ucf.edu/vision/` and also acquired our own video sequences of activities similar to their data. Feature trajectories were available for the UCF video files. For our test video sequences, we provided an input template image of a coffee cup that was tracked in the video sequences to generate feature trajectories. Both the proposed and RCB synchronization methods were then applied to the real video data. In

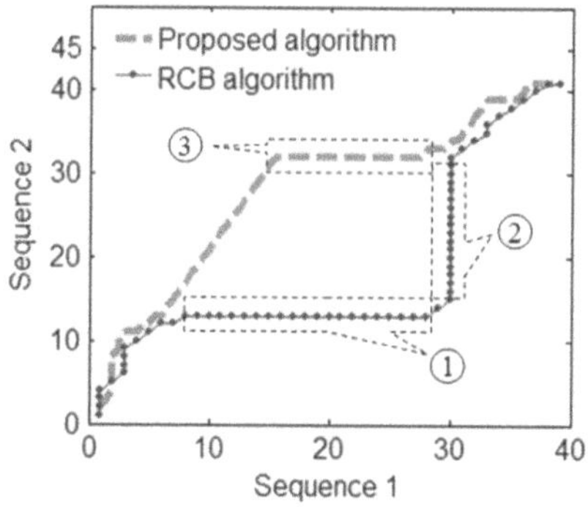

Fig. 6. Warping computed between Sequence 1 and Sequence 2 of realOUR.avi test files. Point(1)-(3) are singularities marked on the warp.

the real data tests, ground truth information is not available. However, a tentative ground truth alignment is computed by visual judgement. The test videos and results are available online at [21] as files **realUCF.avi, *realOUR.avi.*

Figure 7 shows the synchronization computed using the RCB algorithm – frames (a)-(d) of Sequence 1 are matched to frames (e)-(h) of Sequence 2. It can be seen from the position of the hand holding the cup in Fig.7 that the computed synchronization is clearly off by a few frames. Fig. 8 shows the synchronization computed by the STE algorithm, and the alignment is a close temporal match. The warping path computed between the two sequences by both the methods is shown in Fig.6. Points (1)-(3) marked on Fig.6 highlight the regions in the warp where multiple frames in Sequence 1 were warped to a single frame in Sequence 2 and vice versa —*i.e.*, singularities. It can be seen from these highlighted regions that the proposed method reduces the number and length of singularities.

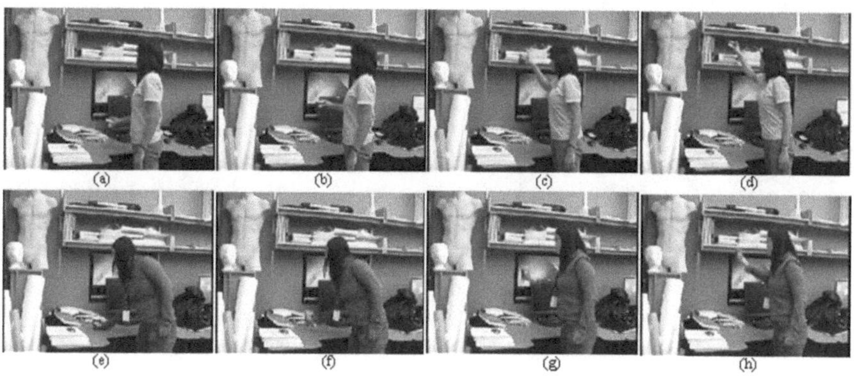

Fig. 7. Results of synchronization using a rank-constraint based RCB method. (a)-(d) Frames from Sequence 1, (e)-(h) Corresponding synchronized frames from Sequence 2.

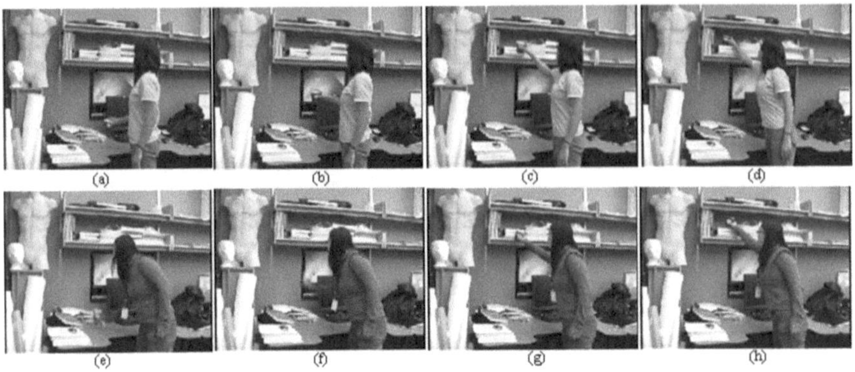

Fig. 8. Results of synchronization using proposed method. (a)-(d) Frames from Sequence 1, (e)-(h) Corresponding synchronized frames from Sequence 2.

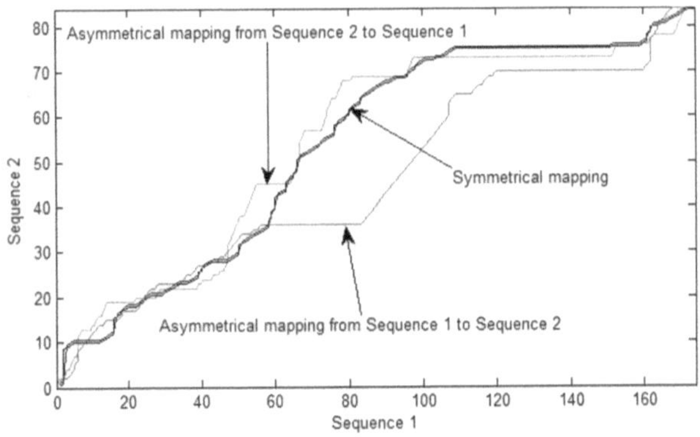

Fig. 9. Symmetric vs. Asymmetric Synchronization of realUCF video files

4.3 Symmetric vs. Asymmetric Synchronization

In the previous section we compared our STE based synchronization against the asymmetric synchronization proposed in [6]. The advantages of our approach in terms of sub-frame synchronization capability and reduction in singularities has been highlighted in the previously mentioned experiments. However, our approach and the RCB approach differ not only in the symmetry aspect, but also in the dynamic aspect of the cost function (4). In order to highlight the advantage of the dynamic optimization of the STE, we validate it against unidirectional asymmetric synchronization using a fixed warping cost function. The mappings computed using both the symmetric and asymmetric approach for realUCF.avi test video are shown in Fig.9. The synchronized output video files using both these approaches are available online at [21] as SymVsAsym_realUCF.avi. From the video files, the higher accuracy of a symmetric optimization over an asymmetric measure can be clearly seen.

5 Application – 4D MRI Registration

One of the motivations behind this work is to build a 4D (volume+time) representation of functional events in the body using 2D planar acquisitions, specifically swallowing disorders. In our experiments, a subject lies prone inside an MRI scanner and is fed small measured quantities of water (bolus) via a system of tubing (water is displayed as white in the MRI images because of the high Hydrogen content in it). For each swallow, a time series of 2D images is acquired on a fixed plane. The acquisition plane is then changed and another series of 2D images is acquired. We acquire three such video sequences corresponding to left, right and center MRI slice planes. The MRI video sequences are subjected to dynamic temporal offsets in the motion of the bolus. Also, since the acquisitions are at very low frame rates, limited by the technology to 4-7 fps, it is crucial to align the sequences to sub-frame accuracy. An important note, with regards to the repeatability of the swallowing motion, is that there are three stages in swallowing; the second and third stages are involuntary and anatomical movements are replicable, the first stage, when the tongue begins to push the bolus back into the pharynx is the only voluntary stage where variation in deformation is likely to take place. However, variability in the first stage can be satisfactorily controlled by limiting the volume of the bolus delivered to the subject (we limit it to 10cc) and by restricting the position of the subject (in our case - supine position with no head rotation). Our approach shows promising results in aligning the MRI sequences to generate a 4D representation. Implementation details of this application are discussed next.

The trailing and leading edges of the bolus are extracted from the MRI sequences using standard background separation techniques. The center of the trailing bolus is extracted using horizontal and vertical profiles, and is used to generate feature trajectories in the three sequences. After suitable event models have been computed for the trajectories, both the RCB and the STE algorithms are applied to the video sequences to compute synchronization. The results of synchronization with the STE and RCB algorithms are shown in Fig.10(a)-(b) and Fig.10(c) respectively. Fig.10(b) shows a few frames from the synchronization computed between the center and right MRI slices that demonstrate sub-frame alignment. Frame 7 of the right MRI sequence is mapped to frame 6.5 of the center MRI sequence. Visually it can be seen that this sub-frame alignment is quite accurate and the results are much better than those produced with the RCB algorithm shown in Fig.10(c). Video sequences comparing the results of both the algorithms are available online at [21] as files *MRI_sync*.avi*. Once the synchronization has been computed, 4D visualization of the MRI data is carried out with a 4D model which can be viewed online at [21]. We have also implemented an approach for registering multiple MRI video contours with 3D facial structure for an integrated 4D structure-multiple video visualization. A video of the implementation is also available online. Details on this part of the work will not be discussed here because of space limitations.

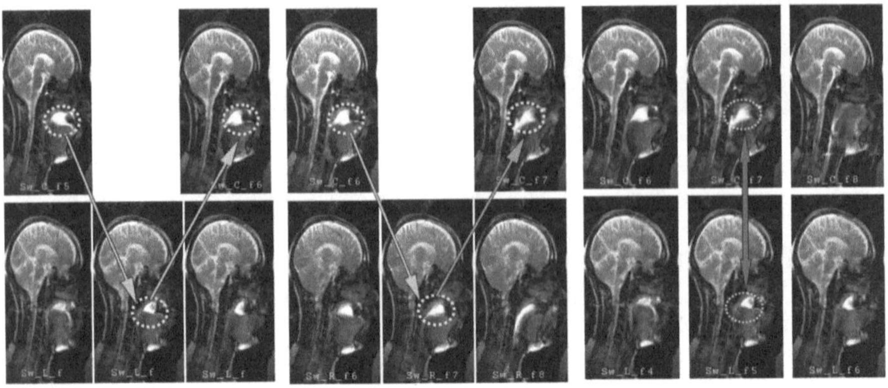

(a) (b) (c)

Fig. 10. (a)-(b) Synchronization computed between Center-Left and Center-Right MRI sequences respectively by proposed algorithm, (c) Synchronization computed between Center-Left MRI sequences by RCB algorithm

6 Summary and Conclusions

We proposed and successfully tested a novel method to synchronize video sequences that are related by varying temporal offsets. Our formulation of synchronization as the minimization of a symmetric transfer error (STE) resulted in synchronization that was not biased by the choice of the reference sequence. The regularized nature of the STE significantly reduced the occurrence of singularities and resulted in sub-frame synchronization. Comparative analysis with a rank-constraint based method demonstrated a marked improvement in video synchronization with our method. An application of the proposed method in 4D MRI visualization was also presented. In the future, we would like to incorporate multiple feature trajectories and extend this work to more general scenes that are not restricted by a homography relationship.

References

1. Caspi, Y., Irani, M.: Spatio-temporal alignment of sequences. PAMI 24, 1409–1424 (2002)
2. Caspi, Y., Simakov, D., Irani, M.: Feature-based sequence-to-sequence matching. IJCV 68, 53–64 (2006)
3. Singh, M., Basu, A., Mandal, M.: Event dynamics based temporal registration. IEEE Transactions on Multimedia 9, 1004–1015 (2007)
4. Lee, L., Romano, R., Stein, G.: Monitoring activities from multiple video streams: establishing a common coordinate frame. IEEE Transactions on Pattern Analysis and Machine Intelligence 22, 758–767 (2000)
5. Hess, R., Fern, A.: Improved video registration using non-distinctive local image features. In: CVPR, pp. 1–8 (2007)

6. Rao, C., Gritai, A., Shah, M., Mahmood, T.F.S.: View-invariant alignment and matching of video sequences. In: Proc. ICCV, pp. 939–945 (2003)
7. Rao, C., Shah, M., Syeda-Mahmood, T.: Invariance in motion analysis of videos. In: Proceedings of the ACM Int. Conf. on Multimedia, pp. 518–527 (2003)
8. Keogh, E., Pazzani, M.: Derivative dynamic time warping. In: SIAM International Conference on Data Mining (2001)
9. Dai, C., Zheng, Y., Li, X.: Subframe video synchronization via 3d phase correlation. In: ICIP, pp. 501–504 (2006)
10. Tuytelaars, T., VanGool, L.J.: Synchronizing video sequences. In: CVPR, pp. I: 762–768 (2004)
11. Tresadern, P., Reid, I.: Synchronizing image sequences of non-rigid objects. In: Proc. BMVC, vol. 2, pp. 629–638 (2003)
12. Carceroni, R.L., Padua, F.L.C., Santos, G.A.M.R., Kutulakos, K.N.: Linear sequence-to-sequence alignment. CVPR, I: 746–753 (2004)
13. Lei, C., Yang, Y.H.: Tri-focal tensor-based multiple video synchronization with subframe optimization. IEEE Trans. on IP 15, 2473–2480 (2006)
14. Wolf, L., Zomet, A.: Wide baseline matching between unsynchronized video sequences. IJCV 68, 43–52 (2006)
15. Pooley, D.W., Brooks, M.J., van den Hengel, A.J., Chojnacki, W.: A voting scheme for estimating the synchrony of moving-camera videos. In: ICIP, vol. 1, pp. I–413–16 (September 14-17, 2003)
16. Perperidis, D., Mohiaddin, R., Rueckert, D.: Spatio-temporal free-form registration of cardiac mr image sequences. Medical Image Analysis 9, 441–456 (2005)
17. Giese, M.A., Poggio, T.: Morphable models for the analysis and synthesis of complex motion patterns. Int. J. Comput. Vision 38, 59–73 (2000)
18. Mikolajczyk, K., Tuytelaars, T., Schmid, C., Zisserman, A., Matas, J., Schaffalitzky, F., Kadir, T., Gool, L.V.: A comparison of affine region detectors. Int. J. Comput. Vision 65, 43–72 (2005)
19. Matas, J., Chum, O., Urban, M., Pajdla, T.: Robust wide baseline stereo from maximally stable extremal regions. In: Proceedings of British Machine Vision Conference, September 2002, vol. I, pp. 384–393 (2002)
20. Hartley, R.I., Zisserman, A.: Multiple View Geometry in Computer Vision, 2nd edn. Cambridge University Press, Cambridge (2004)
21. Authors, www.ece.ualberta.ca/~meghna/ECCV08.html

Shape-Based Retrieval of Heart Sounds for Disease Similarity Detection

Tanveer Syeda-Mahmood and Fei Wang

IBM Almaden Research Center, 650 Harry Road, San Jose, CA 95120
{stf,wangfe}@almaden.ibm.com

Abstract. Retrieval of similar heart sounds from a sound database has applications in physician training, diagnostic screening, and decision support. In this paper, we exploit a visual rendering of heart sounds and model the morphological variations of audio envelopes through a constrained non-rigid translation transform. Similar heart sounds are then retrieved by recovering the corresponding alignment transform using a variant of shape-based dynamic time warping. Results of similar heart sound retrieval are demonstrated for various diseases on a large database of heart sounds.

Keywords: Sound pattern analysis, audio retrieval, curve analysis, healthcare application.

1 Introduction

The field of pattern recognition is becoming increasingly applicable to new modalities. In this paper we explore the application of computer vision techniques to an important class of audio signals, namely, heart sounds. Heart auscultation, i.e., listening to the sounds produced by the heart, is a common practice in the screening of heart disease. Although different diseases produce characteristic sounds, forming a diagnosis based on sounds heard through a stethoscope is a skill that takes years to perfect. Numerous studies have shown that as much as 87% of patients referred to cardiologists for evaluation are as a result of false alarms [2]. Thus software diagnostic tools that aid physicians in their diagnosis of heart sounds are needed.

Auditory discrimination of heart sounds is inherently difficult as these sounds are faint and lie at the lower end of the audible frequency range [1]. Although there are tools to visually render the sounds [3], we recently observed that the visual representation of heart sounds actually brings out the differentiating characteristics of various diseases more readily than the audio signal. Figure 1 illustrates this by recording the visual appearance of the audio signal within a single heart beat duration, from patients with various diseases. As can be seen, different diseases show characteristically different shape patterns. Further, it is also easier to spot the similarity in the sounds across patients with similar diseases through their visual representations. Figure 2 illustrates this for different patients diagnosed with the same disease. From these examples, it appears plausible that the disease similarity can be inferred by developing a measure for capturing the visual similarity of audio signals.

D. Forsyth, P. Torr, and A. Zisserman (Eds.): ECCV 2008, Part II, LNCS 5303, pp. 568–581, 2008.

In this paper we address the similarity retrieval of heart sounds using spatial representations of heart sounds. Specifically, we extract perceptual envelopes of audio shapes that capture the essential low frequency disease-specific information. We then model the morphological variations in the spatial envelopes from the same disease class, as a constrained non-rigid translation transform. Matching involves recovering the corresponding alignment transform using a new shape-based dynamic time warping algorithm. Results of heart sound retrieval are reported for various diseases on a large database of heart sounds. Our experiments demonstrate that our method outperforms state-of-the-art heart sound signal analysis methods when they are adapted for use in similar heart sound retrieval.

The paper makes several important contributions. It addresses, for the first time, the problem of similarity retrieval of heart sounds to aid physician decision support (a new use of content-based retrieval methods). Secondly, it demonstrates the advantages of visual analysis of heart sounds over conventional audio processing methods to enhance disease similarity detection. Further, it proposes the use of envelope curves of heart sounds as an adequate representation to discriminate between diseases. Finally, the audio envelop representation used here is invariant to a number of factors including heart rate differences in young and adults, number of heart beat samples, noise in recordings, etc.

The rest of the paper describes this approach in detail. In Section 2 we introduce the domain of heart sounds and the need for modeling non-rigid time deformations for detecting visual similarity of audio signals. In Section 3, we discuss related work. In Sections 4-5, we develop a general method of modeling the shape changes in audio signals corresponding to similar sounds and present a method of audio shape matching. In Section 6, we describe the various pre-processing steps for enabling the shape matching. Finally, in Section 7 we present results demonstrating the effectives of shape similarity measures for discovering audio similarity among heart sounds.

2 The Heart Sound Matching Problem

The Heart sounds are produced during the heart cycle, where events such as the motion of the valves and the movement of the blood generate vibrations [1]. In healthy adults, the first heart sound (S1) ('lub') and second heart sound (S2) ('dub') are produced by the closure of the AV valves and semi-lunar valves as shown in Figure 3. In addition to these normal sounds, a variety of other sounds may be present including S3 which may be heard at the beginning of the diastole, during the rapid filling of the ventricles and S4, which may be heard in late diastole during atrial contraction, regurgitation sounds, i.e. murmurs which are noise-like sounds that are heard between the two major heart sounds during systole or diastole, and adventitious sounds, or clicks.

Matching heart sounds is difficult as they are highly non-stationary signals of short duration. Further, the energy content of low-frequency vibrations is much higher than that of high-frequency vibrations. Thus much of the discriminating heart sound information is captured in their low frequency envelopes which must be extracted. Next the variations in heart beat must be taken into account. This requires extraction of single heart beats from signals which are not always periodic (particularly for cardiac arrhythmias). Further, the amplitude variations due to recording levels and varying digital stethoscope qualities affect the matching. Finally, due to variability in the systolic and diastolic phases across patients, both advancement as well as preponement of cardiac audio phases (sounds S1 and S3) is possible.

Figure 4 illustrates matching problem for heart sounds using two patients both diagnosed with Atrial Septal Defect (ASD). Figure 4a and b show the original sounds. Due to the sampling rate differences, a single heart beat corresponds to 6172 samples in Figure 4a while there are 34542 samples within a single heart beat for the sound in Figure 4b. The two signals, nevertheless sound very similar. Direct comparison of the two signals hardly reveals any similarity as can be seen through direct superposition in Figure 4c. Once the signals are normalized in both time and amplitude by scaling using the sampling rate and the range of 0 to 1, the similarity can be seen under a shift as shown in Figure 4d. Using cross-correlation, we can recover the shift to roughly align them as shown in Figure 4e. From this figure, it is clear that although a single fixed shift can bring them into rough alignment, matching peaks require different amounts of shift in different regions. Furthermore, the direction of the shift could be in opposite directions in different phases to correspond to the cases where the systolic and diastolic phases are affected differently (due to the disease). Figure 4f illustrates this non-uniformity in the time shift for the signals from Figure 4d which are already normalized in time (marked by arrows in the figure). Thus any measure of similarity in sounds for the same disease class needs to model non-uniformity in the scaling of time.

3 Related Work

While we are not aware of work on similarity retrieval of heart sounds, there is considerable work in audio signal analysis and retrieval in general [10,16,12,15,11], and heart sound signal analysis[6,7,4], in particular. Further, there is also work on computer-aided diagnosis of heart sounds [19,20] although no commercial systems exist in practice. Classical methods of audio analysis divide the audio into short-time segments and extract spatio-temporal features such as the zero-crossing rate (ZCR) [10], average energy in short-time segments, hidden Markov model features[14], etc. Methods based on frequency-domain information of the audio signal often form the speech spectrogram and extract features such as Mel-frequency cepstral coefficients (MFCC) [12], subband spectral flux, harmonicity prominence, spectral roll-off [15], beat frequency estimation [11], multi-resolution wavelets etc. Often, the spectral representations are generated from a moving overlapping window analysis of the time domain signal.

The short-time signal analysis methods are not suitable for modeling heart sounds due to (a) their highly non-stationary nature, (b) variability in heart rate, (c) disease-specific variations in the relative placement of S2,S2,S3,and S4 phases, and (d) due to the selective importance that needs to be given to the lower frequencies of the spectrum. Further, the choice of window size is not clear for heart sounds. Often, the segments picked are end points of S1 or S3 phases. However, for diseased cases, there is smearing of the S1, and S3 phases (eg. Mitral regurgitation). Further, additional sounds such as S2 and S4 may be present. In such cases, it is not clear how to appropriately extract a coherent segment for spatial-frequency analysis. The method of MFCC suffers from the same limitation due to overlapping window analysis. Further, MFCC is well-known to be sensitive to even small amounts of noise.

Automatic analysis of heart sounds has been attempted for abnormality detection within a single sound. Early work on phono-cardiograms found correlation between sounds and various heart defects [19]. However, the sounds were studied by isolating single heart beat duration. The predominant approach in heart sound similarity detection has been based on feature extraction and classification as is conventional for audio analysis. In [19], a neural network was trained on heart sounds to classify several valve-related disorders. However, the identification of disorders was based on classifying sound into intuitive groups such as normal, split sound, open snap, etc based on detection of S1 and S2 sounds and noting their separation. For many diseases, such as pericardial rub, the periods S1 and S2 are hardly distinct for detection and hence are not reliable for audio similarity detection. In [20] again, a neural net classifier was used in conjunction with wavelet processing for heart sound classification. However, the system was trained on a single sample for each disease using different heart beat cycles. Heart sound analysis using time-frequency representations has also been common including recent uses of MFCC to heart sounds [5].

In all these approaches, no explicit variational modeling across different disease instances was performed other than using bounds on the sizes of the S1 and S3 phases. Finally, most of these methods assume that a single period of the heart sounds can be obtained. However, in a long recording, particularly in the case of arrhythmias, finding periodic repetitions by simple correlation is insufficient.

The visually modeling of heart sounds is analogous to visual modeling of music spectrograms reported in [25] although the spectral features were different. Finally, several methods of curve matching have been developed in literature[23,24]. Methods exist to match two curves using curvature-scale space [24], and to dynamically warp the curves into one another using dynamic time warping [23]. The curve matching approaches, in general, are sensitive to spurious and missing data. This condition is quite likely in modeling disease instances as the S1 or S3 phases are prolonged. The dynamic time warping algorithms for curve matching have mostly used time order and time proximity constraints. Shape-based dynamic time warping using combined shape cues including angle of corners, and orientation of their bisectors in combination with proximity have not been proposed previously.

4 Representing Heart Sounds Using Perceptual Envelopes

In our approach, we model the heart sound (within a single heart beat) through a perceptual envelope. Various algorithms are available in literature for envelop extraction

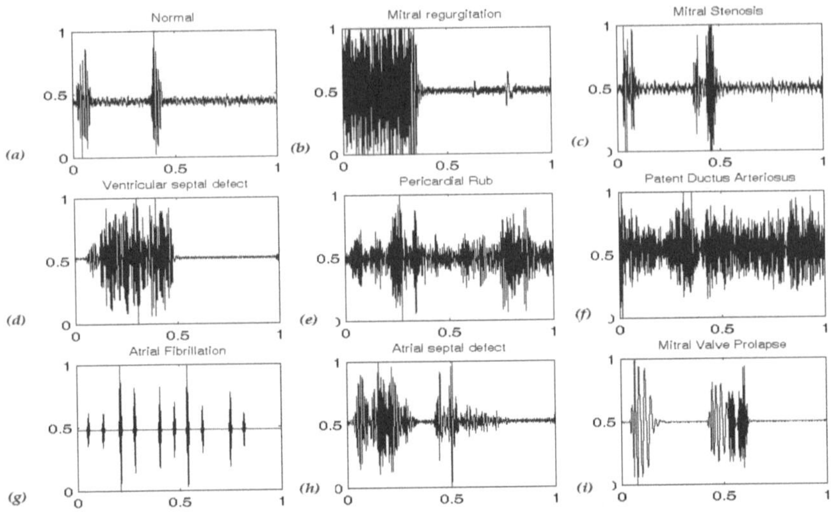

Fig. 1. Illustration of different visual appearances for heart sounds for different diseases indicating visual discrimination of heart sounds is possible. (a)-(c) (d)-(f) , (g)-(i) left to right, top to bottom.

from signals including homomorphic filtering [8]. While these algorithms are less sensitive to noise-related fluctuations, they frequently extract the low frequency component of the signal rather than render a faithful approximation of the perceptual envelope as can be seen from Figure 5.

To extract audio envelope curve, we perform noise filtering using wavelet filters to remove the hissing noise that comes from digital stethoscopes. We then form a line segment approximation to the audio signal. This is a standard split-and-merge algorithm for line segment approximation that uses two thresholds, namely, a distance threshold δ and a line length threshold l to recursively partition the audio signal into a set of line segments. Each consecutive pair of line segments then defines a corner feature C_i. With δ=0.01 and l=5, a faithful rendering of the audio signal is made possible through a line segment approximation while retaining only 10% of the samples. The thresholds for curve parameterization are not as critical here as the shape matching algorithm we present next, is robust to missing and spurious features. We define the *audio envelope (AE)* as the set of points $AE = \{P_i\} \quad where\ P_i = C_j \quad for\ some\ j \quad s.t.$

$$P_i(y) \geq P_{i-1}(y) \,\&\, P_i(y) \geq P_{i+1}(y) \text{ and } P_i(y) \geq baseline.$$

Only the values of the signal above the baseline are retained in the maxima envelope curve. Similarly, all peaks below the baseline for the minima envelope curve.

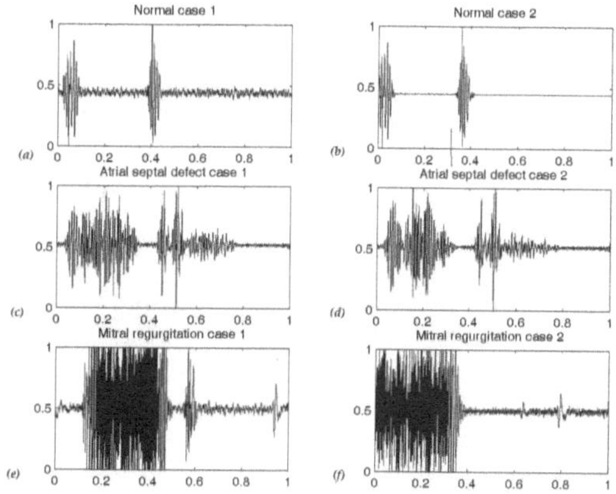

Fig. 2. Illustration of similarity retrieval for heart sounds. The top matches in each case are matching patients with the same disease. Caption runs from (a)-(b), (c)-(d), (e)-(f) , Left to right, top to bottom in the figure.

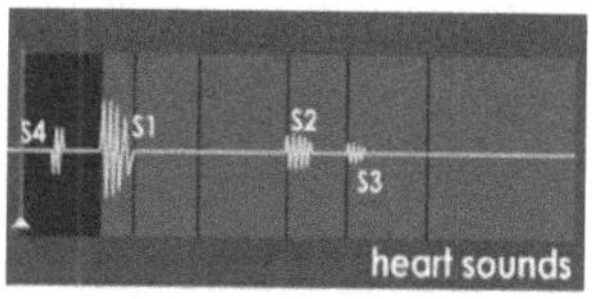

Fig. 3. Illustration of four major heart sounds

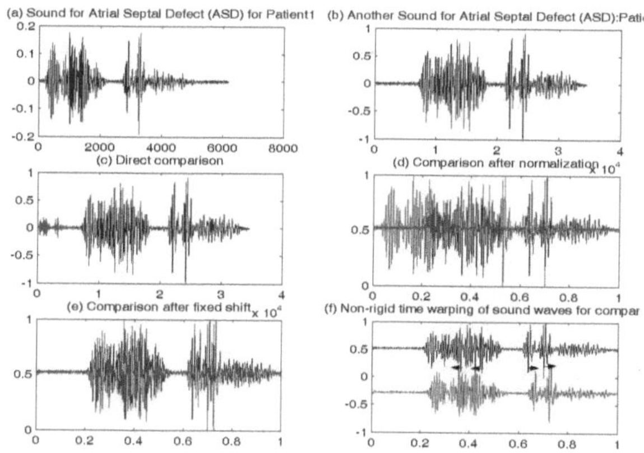

Fig. 4. Illustration of the alignment problem for heart sounds

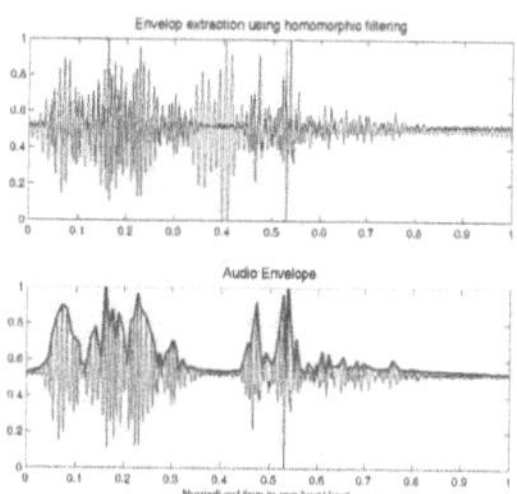

Fig. 5. Illustration of envelope curve extraction of sounds. (a) Homomorphic filtering-based low frequency approximation (shown in red), (b) perceptual envelope.

Figure 5b illustrates the envelope curve extracted from a heart sound. As can be seen, the boundary shape of the signal is well-captured in the envelope curve. Once the envelope curve f(t) is extracted, its shape can be represented by the curvature change points or corners on the envelope curve. The shape information at each corner is captured using the following parameters

$$S(\vec{f}(t_i))=<t_i,\vec{f}(t_i),\theta(t_i),\varphi(t_i)> \tag{1}$$

These features were chosen to facilitate matching of audio envelopes. Using the angle of the corner ensures that wider complexes are not matched to narrow complex as these can change the disease interpretation. The angular bisector, on the other hand, ensures that polarity reversals such as inverted waves can be captured. Here $\theta(t_i)$ is the included angle in the corner at t_i, and $\varphi(t_i)$ is the orientation of the bisector at corner t_i.

5 Modeling Morphological Variations

We now develop a model of morphological variations in the shape of heart sounds envelopes for different instances of a disease.

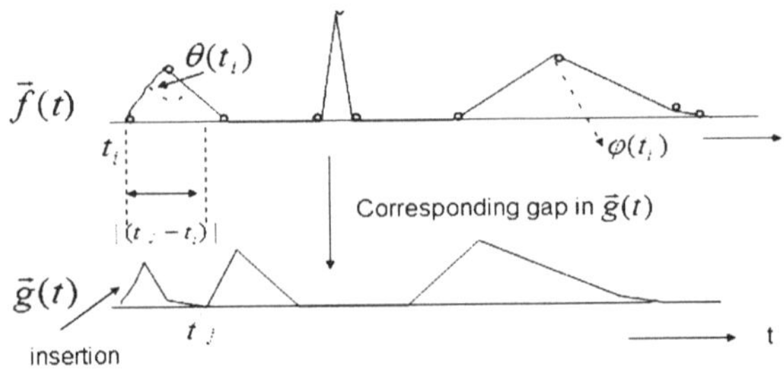

Fig. 6. Illustration of the shape matching problem for envelop curves

Referring to Figure 6, consider an envelop curve g(t) corresponding to a heart sound. Consider another curve f(t) that is a potential match to g(t), i.e. comes from a different patient diagnosed with the same disease. The curve f(t) is considered perceptually similar to g(t) if a non-rigid transform characterized by [a,b, Γ] can be found such that

$$|f'(t)-g(t)| \leq \delta \tag{2}$$

where | | represents the distance metric that measures the difference between f'(t) and g(t), the simplest being the Euclidean norm and

$$f'(t)=af(\Phi(t)) \quad \text{with} \quad \Phi(t)=bt+\Gamma(t) \tag{3}$$

where the (bt) is the linear component of the transform and Γ is the non-linear translation component. The parameters a and b can be recovered by normalizing in amplitude and time. We can normalize in amplitude by transforming f(t) and g(t) such that

$$\hat{f}(t) = \frac{f(t) - f_{\min}(t)}{f_{\max}(t) - f_{\min}(t)} \quad \text{and} \quad \hat{g}(t) = \frac{g(t) - g_{\min}(t)}{g_{\max}(t) - g_{\min}(t)} \tag{4}$$

so that a=1. To eliminate solving for b, we can normalize the time axis by dividing the heart rate. Suppose the sampling rate of points on the curve f(t) is F_S. Let a periodicity detection algorithm signal the heart rate period to be T_1 (Periodicity detection is discussed in Section 5). Then dividing by the heart period samples, we can ensure that that all time instants lie in the range [0,1]. Thus the time normalization can be easily achieved as:

$$\vec{f}(t) = \hat{f}(t/T_1) \quad \vec{g}(t) = \hat{g}(t/T_2) \tag{5}$$

where T1 and T2 are the heart beat durations of f(t) and g(t) respectively. With this time normalization, b=1. Such amplitude and time normalization automatically makes the shape modeling invariant to amplitude variations in audio recordings, as well as variations in heart rate across patients. Since the non-uniform translation Γ is a function of t, we can avoid computational overhead by recovering it at important fiducial points such as the corners, and recover the overall shape approximation by interpolation. Let there be K features extracted from $\vec{f}(t)$ as $F_K = \{(t_1, \vec{f}_1(t_1)), (t_2, \vec{f}_2(t_2))...(t_K, \vec{f}_K(t_K))\}$ at time $\{t_1, t_2, ..t_K\}$ respectively. Let there be M fiducial points extracted from $\vec{g}(t)$ as $G_M = \{(t'_1, \vec{g}_1(t'_1)), (t'_2, \vec{g}_2(t'_2))...(t'_M, \vec{g}_M(t'_M))\}$ at time $\{t'_1, t'_2, ..t'_M\}$ respectively. If we can find a set of N matching fiducial points $C_\Gamma = \{(t_i, t'_j)\}$, then the non-uniform translation transform Γ can be defined as:

$$\Gamma(t') = \begin{cases} t_i & if\, t = t'_j\, and\, (t_i, t'_j) \in C_\Gamma \\ t_r + \left(\dfrac{t_s - t_r}{t'_l - t'_k}\right)(t' - t'_k) & where\, (t_r, t'_k), (t_s, t'_l) \in C_\Gamma \end{cases} \tag{6}$$

and t'_k is the highest of {tj'} $\leq t'$ and t'_l is the lowest of {tj'} $\geq t'$ that have a valid mapping in C_Γ. Other interpolation methods besides linear (eg. spline) are also possible.

Using Equations 5 and 6, the shape approximation error between the two curves is then given by:

$$| f'(t) - g(t) | = | \vec{f}(t' = \Gamma(t')) - g(t') | \tag{7}$$

For each g(t), we would like to select Γ such that it minimizes the approximation error in (7) while maximizing the size of match C_Γ. Finding the best matching audio based on shape can then be formulated as finding the g(t) such that

$$g_{best} = \arg\min_{g} \mid \vec{f}(\Gamma(t')) - \vec{g}(t') \mid \tag{8}$$

while choosing the best Γ for each respective candidate match g(t).

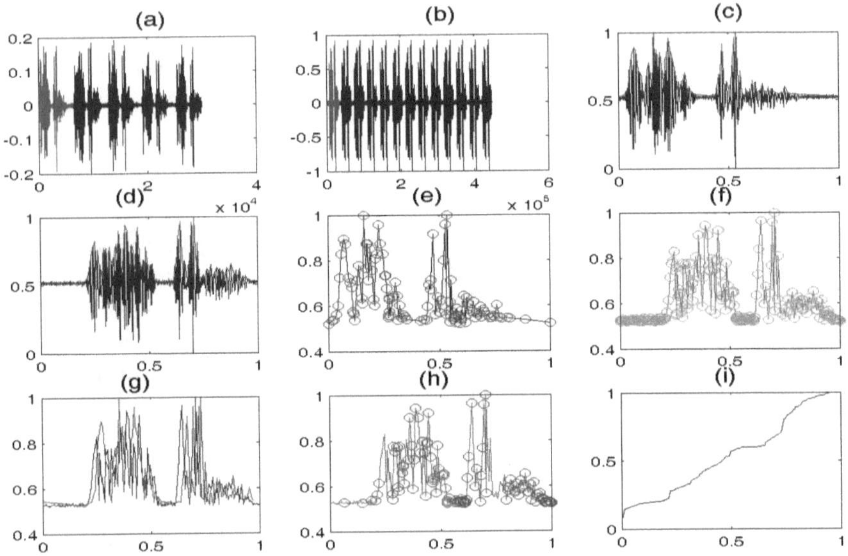

Fig. 7. Illustration of shape-based dynamic time warping

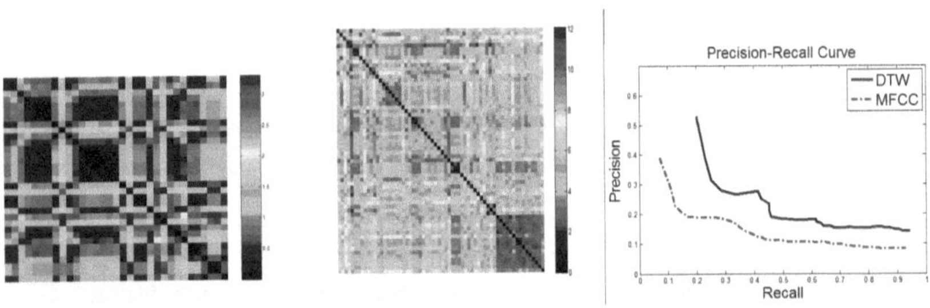

Fig. 8. Illustration of the comparison between shape-based time warping and MFCC algorithm for pair-wise discrimination. (a) Confusion matrix using our algorithm (b) Confusion matrix using MFCC.

5.1 Shape-Based Dynamic Time Warping

If we consider the feature set F_K, G_M extracted from the respective curves as sequences, the problem of computing the best Γ reduces to finding the best global subsequence alignment using the dynamic programming principle. The best global alignment maximizes the match of the curve fragments while allowing for possible gaps and insertions. Gaps and insertions correspond to signal fragments from feature set F_K that don't find a match in set G_M and vice versa. In fact, the alignment can be computed using a dynamic programming matrix H where the element H(i,j) is the cost of matching up to the ith and jth element in the respective sequences. As more features find a match, we want the cost to increase as little as possible. The dynamic programming step in our case becomes:

$$H_{i,j} = \min\begin{cases} H_{i-1,j-1} + d(\vec{f}(t_i), \vec{g}(t'_j)) \\ H_{i-1,j} + d(\vec{f}(t_i), 0) \\ H_{i,j-1} + d(0, \vec{g}(t'_j)) \end{cases} \tag{9}$$

with initialization as $H_{0,0} = 0$ and $H_{0,j} = \infty$ and $H_{i,0} = \infty$ for all 0< i <= K, and 0<j<=M. Here d(.) is the cost of matching the individual features described next. Also, the first term represents the cost of matching the feature point $\vec{f}(t_i)$ to feature point $\vec{g}(t'_j)$ which is low if the features are similar. The second term represents the choice where no match is assigned to feature $\vec{f}(t_i)$.

5.2 Shape Similarity of Envelop Curves

After the transformation Γ is recovered, the similarity between two envelop curves is given by the cost function $d(\vec{f}(t_i), \vec{g}(t'_j))$

$$d(\vec{f}(t_i), \vec{g}(t_j)) = \begin{cases} \sqrt{\begin{array}{l}(t_i - t'_j)^2 + (\vec{f}(t_i) - \vec{g}(t'_j))^2 + \\ (\theta(t_i) - \theta(t'_j))^2 + (\varphi(t_i) - \varphi(t'_j))^2\end{array}} & \text{if } \begin{array}{l} |(t_i - t'_j)| \le \lambda_1 \\ |(\vec{f}(t_i) - \vec{g}(t'_j)| \le \lambda_2 \\ |(\theta(t_i) - \theta(t'_j))| \le \lambda_3 \\ |(\varphi(t_i) - \varphi(t'_j))| \le \lambda_4 \end{array} \\ \infty \quad otherwise \end{cases} \tag{10}$$

The thresholds $(\lambda_1, \lambda_2, \lambda_3, \lambda_4)$ are determined through a training phase in which the expected variations per disease class are noted. The cost function $d(\vec{f}(t_i), 0)$ can be computed by substituting $t'_j = 0,\ \vec{g}(t'_j) = 0,$ and $\theta(t'_j) = 0, \varphi(t'_j) = 0$ in Equation 10. The cost function $d(0, \vec{g}(t'_j))$ can be similarly computed.

Thus using this approach two heart sounds are considered similar if enough number of fiducial points between query and target envelop curves can be matched using shape-based dynamic time warping. In general, due to the period estimation offset errors, the signals may have to be circularly shifted by a fixed translation for a rough alignment before the fine non-rigid alignment described above.

6 Feature Pre-processing

The above formulation assumed that a single heart beat duration signal could be isolated prior to initiating shape matching. We now describe the pre-processing steps that lead to such an isolation.

6.1 Periodicity Detection

While normal heart sounds show good repetition, the periodicity is surprisingly difficult to spot for abnormal heart sounds where the repetitions can be nested or irregular (arrhythmias). Simple auto-correlation is often insufficient for this purpose. A robust periodicity detector was, therefore, developed that treats periodicity detection as the problem of recovering a shift/translation that best aligns two self-similar curves. Consider a periodic curve g(t) with period T. Then by definition g(t) = g(t+kT) for all multiples k=0,1,2,.... Consider a candidate period τ. Form a curve f(t) by shifting g(t) by τ, i.e.

$$f(t) = \begin{cases} g(t-\tau) & if\ t \geq \tau \\ g(t) & otherwise \end{cases} \tag{11}$$

Then define a function $R(\tau)$ that records the number of curve features that can be verified to satisfy the periodicity condition based on the current estimate of the period as

$$R(\tau) = \frac{|\{g(t_i)\}|}{\max\{N-\tau, \tau\}} \quad s.t. |\, g(t_i) - f(t)_i \,| \leq \varepsilon \tag{12}$$

where N is the total number of points on the curve. The above function can be computed in linear time in comparison to the quadratic time for the autocorrelation function. The function $R(\tau)$ shows peaks at precisely those shifts which correspond to periodic repetitions of the curve. The most likely period is then taken as the smallest τ with the most integer multiples in the allowed range of heart beats (40- 180 beats/minute). In general, there may be more than one candidate for period, particularly when the periodic repetitions are nested. Our algorithm finds all such overlapping repetitions and tests the dynamic time warping algorithm with each such choice.

7 Results

The audio shape matching algorithm was tested on a large database of heart sounds. The data came from various teaching hospitals and reference CDs provided by digital stethoscope makers (Littman). Thus the ground truth labels for diseases were known during the evaluation. Currently, the collection has over several hundred heart sound examples for various kinds of murmurs, Mitral regurgitation, Mitral Stenosis, septal defects, Cardiomyopathy, etc. Each disease was represented by at least 6-10 patient samples in the database.

Figure 7 illustrates the steps involved in audio shape matching through a sample retrieval. Figure 7a shows the query audio signal and Figure 7b shows a matching audio sound from the database. The result of periodicity detection is shown overlaid in the respective figures in red. Figure 7c and d show the audio envelopes extracted from the respective signals normalized in amplitude and time over a single heart beat duration. Figure 7e and 7f show fiducial features extracted for matching from the query and database signals. Direct alignment of the two signals by cross-correlation is shown in Figure 7g. As can be seen, alignment errors are still present. The shape-based dynamic time warping alignment is indicated in Figurc 7i. The projection of fiducial features from the query (green circle) onto the matching database curve (in magenta) using the non-rigid alignment is indicated in Figure 7h. In this example, 110 of the 114 of query fiducial features from Figure 7e have found a match with fiducial features of the database curve. From this figure, it is clear that although a single fixed shift can bring them into rough alignment, matching the signals requires different amounts of shift in different regions. Furthermore, the direction of the shift could be opposite in different sections when the systolic and diastolic phases are affected differently due to the disease. Figure 2a-f illustrates three other examples of heart sound similarity retrieval. In each case, the query is shown on the left and the best matching heart sound is shown on the right.

7.1 Comparison with MFCC

We compared the performance of our algorithm with a traditional signal processing approach using Mel-frequency Cepstral Coefficients (MFCC) [2]. The MFCC method was chosen as it has been the most successful of the audio analysis approaches and has recently been used to classify heart diseases [5]. We implemented a version of MFCC in which we divided the raw audio signals (no periodicity detection) into short-time segments of 400 samples and MFCC coefficients were then extracted from each segment using the short-time Fourier transform (STFT). The coefficients of all segments were used to form the feature vector. Euclidean distance was used to find nearest matches.

We tested the two techniques for grouping sounds from similar diseases and discriminating sounds from different diseases using the pair-wise similarity matrix. Thus a technique that discriminates well would form bands along the diagonal in the similarity matrix. Figure 8a and 8b show the similarity matrices using our algorithm and using MFCC. As can be seen, the banding structure is clearer using our method in comparison to MFCC where the banding is evident mostly for normal heart sounds. Further, the outlier (the orange colored signal) that has no match in the database clearly stands out using the DTW alignment.

7.2 Precision Recall Performance

Finally, we evaluated the precision and recall of the two methods by using all available samples per disease as queries and retrieving matches from the database for various choices of K and retained top K matches. Precision and recall were defined as

$$Recall = \frac{\#\, correct\ matches\ selected}{\#\, of\ correct\ matches\ present} \qquad Precision = 1 - \frac{\#\, incorrect\ matches\ selected}{\#\ of\ matches\ returned}$$

The precision and recall values were averaged over the queries tested for the respective disease classes. Figure 8c shows the performance of the two methods on the entire audio database. Again, as can be seen, both precision and recall are higher using the more precise audio shape matching algorithm over MFCC. Due to the averaging over a large number of queries, the precision values are somewhat lower than expected here. Sample precision and recall values for some diseases averaged over all queries of the same disease class are shown in Table 1 below indicating actually good precision and recall for specific diseases. In general, we observed that the top K matches returned by our method often contained signals whose visual appearance were similar and the corresponding auscultation sounds heard were similar.

Table 1. Illustration of precision and recall

Disease	• DTW Precision	• DTW Recall	• MFCC Precision	• MFCC Recall
Mitral regurgitation	• 56.67%	• 70.83%	• 43.33%	• 50.0%
Patent Ductus Arteriosus	• 100%	• 75%	• 50.0%	• 37.5%
Pericardial Rub	• 67.2%	• 85.6%	• 54.4%	• 45.7%
Mitral stenosis	• 76.2%	• 78.3%	• 45.2	• 39.8%
Ventricular septal defect	• 65.4%	• 82.3%	• 54.2%	• 48.9%
Atrial Fibrillation (AF)	• 78.2%	• 84.5%	• 64.7%	• 54.3%

8 Conclusions

In this paper, we have presented a novel algorithm for shape based retrieval of heart sounds. Unlike existing work based on feature extraction and classification, we take the approach of nonrigid shape alignment for retrieval. The algorithm is independent of disease specifics and is potentially applicable to other audio signals where their visual appearance is sufficiently discriminatory. Our experiments demonstrate that this approach significantly outperforms the current state-of-the-art approached based on heart sound signal analysis.

References

[1] Erikson, B.: Heart sounds and murmurs: A practical guide. Mosby-Year, pp. 9–12 (1997)
[2] Pease, A.F.: If the heart could speak, `http://w4.siemens.de/FuI/en/archiv/pof/heft2_01/artikel19/index.html`
[3] `http://www.zargis.com/WhitePaperIntro.asp`

[4] See survey of papers on auscultation analysis at, http://www.bsignetics.com/bioengineeringpapers.htm
[5] Kamarulafizam, I., et al.: Heart Sound Analysis Using MFCC and Time Frequency Distribution. In: Proc. 3rd Intl. Conf. on Biomedical Eng., Kaula Lampur (2006)
[6] Alim, O.A., Hamdy, N., El-Hanjouri, M.A.: Heart diseases diagnosis using heart sounds. In: Proc. National Radio Science Conference (NRSC), pp. 634–640 (2002)
[7] Hebden, J.E., Torry, J.N.: Identification of aortic stenosis and mitral regurgitation by heart sound analysis. Computers in Cardiology 1997, 109–112 (September 7-10, 1997)
[8] Rezek, I., Roberts, S.J.: Envelope Extraction via Complex Homomorphic Filtering. Research Report TR-98-9 (June 1998)
[9] Univ. of Washington Medical School, http://depts.washington.edu/physdx/index.htm
[10] Rabiner, L., Shafer, R.: Digital Processing of Speech Signals. Prentice Hall, NJ
[11] Foote, J., Cooper, M.: Audio Retrieval by Rhythmic Similarity. In: Proceedings of ISMIR 2002, Paris, France (October 2002)
[12] Foote, J.T.: Content-based retrieval of music and audio. In: Proc of SPIE, vol. 3229, pp. 138–147 (1997)
[13] Lu, L., Zhang, H.-J., Jiang, H.: Content Analysis for Audio Classification and Segmentation. IEEE Trans. on Speech and Audio Processing 10(7), 504–516 (2002)
[14] Lu, L., Cai, R., Hanjalic, A.: Towards A Unified Framework for Content-based Audio Analysis. In: IEEE International Conference on Acoustics, Speech and Signal Processing (ICASSP) 2005, vol. II, pp. 1069–1072 (2005)
[15] CAi, R., Lu, L., Zhang, H.-J.: Using structure patterns of temporal and spectral feature in audio similarity measure. In: Proc. ACM Multimedia 2003, pp. 219–222 (2003)
[16] Cai, R., Lu, L., Hanjalic, A., Zhang, H.-J., Cai, L.-H.: A flexible framework for key audio effects detection and auditory context inference. IEEE Trans. Speech Audio Processing (May 2006)
[17] Guo, G., Li, S.Z.: Content-Based Audio Classification and Retrieval by Support Vector Machines. IEEE Transactions on Neural Networks 14(1) (January 2003)
[18] Jandre, F.C., Souza, M.N.: Wavelet analysis of phonocardiograms: differences between normal and abnormal heart sounds. In: Proceedings of the 19th Annual International Conference of the IEEE, October 30- November 2, 1997, vol. 4, pp. 1642–1644 (1997)
[19] Alim, O.A., et al.: Heart diseases diagnosis using heart sounds. In: Proc. 19th National Radio Science Conference, Alexandria (March 2002)
[20] Say, O., Dokur, Z., Olmez, T.: Classification of heart sounds by using wavelet transform. In: Say, O., Dokur, Z., Olmez, T. (eds.) 24th Annual Conference and the Annual Fall Meeting of the Biomedical Engineering Society EMBS/BMES Conference, vol. 1, pp. 128–129 (2002)
[21] Rezek, I., Roberts, S.J.: Envelope Extraction via Complex Homomorphic Filtering. Research Report TR-98-9 (June 1998)
[22] Elfeky, M.G., Aref, W.G., Elmagarmid, A.K.: WARP: time warping for periodicity detection. In: Fifth Intl. Conf. on Data Mining, p. 8 (November 2005)
[23] Wolfson, H.J.: On curve matching. IEEE Trans. PAMI 12, 483–489 (1990)
[24] Avants, B., Gee, J.: Continuous curve matching with scale-space curvature and extrema-based scale selection. In: Proc. Scale-space Methods in Computer Vision, p. 1079 (2003)
[25] Ke, Y., Hoiem, D., Sukthankar, R.: Computer vision for music identification. In: Proc. CVPR 2000 (2000)

Learning CRFs Using Graph Cuts

Martin Szummer[1], Pushmeet Kohli[1], and Derek Hoiem[2]

[1] Microsoft Research, Cambridge CB3 0FB, United Kingdom
[2] Beckman Institute, University of Illinois at Urbana-Champaign, USA

Abstract. Many computer vision problems are naturally formulated as random fields, specifically MRFs or CRFs. The introduction of graph cuts has enabled efficient and optimal inference in associative random fields, greatly advancing applications such as segmentation, stereo reconstruction and many others. However, while fast inference is now widespread, parameter learning in random fields has remained an intractable problem. This paper shows how to apply fast inference algorithms, in particular graph cuts, to learn parameters of random fields with similar efficiency. We find optimal parameter values under standard regularized objective functions that ensure good generalization. Our algorithm enables learning of many parameters in reasonable time, and we explore further speedup techniques. We also discuss extensions to non-associative and multi-class problems. We evaluate the method on image segmentation and geometry recognition.

1 Introduction

The availability of efficient and provably optimal inference algorithms, such as graph cuts [1] and its approximate extensions [2], has inspired progress in many areas of computer vision. For example, the efficiency of graph cut based algorithms enables interactive image [3,4] and real-time video segmentation tasks. The optimality guarantees of these algorithms allow computing the maximum a posteriori (MAP) solution of the model distribution.

The ability to compute the minimum energy solution has revealed that simple energy models (e.g., with one unary term and an isotropic smoothing penalty in grid labeling problems) are insufficient to model the complex structures inherent in computer vision problems [5,6]. Despite this knowledge, overly simplistic hand-tuned random field models continue to be common practice. We believe that the continued use of such impoverished models reflects, not a belief in the supremacy of such representations, but the absence of tractable machine learning techniques for large MRF and CRF problems. Currently, the most widely used learning algorithms include cross-validation and simple partition function approximations [7]. However, many works do not perform learning at all and rely on hand-tuned parameters.

In this paper, we describe an efficient and easy-to-implement technique that is capable of learning dozens of parameters from millions of pixels in minutes. Our algorithm is based on the structured support vector machine (SVMStruct)

D. Forsyth, P. Torr, and A. Zisserman (Eds.): ECCV 2008, Part II, LNCS 5303, pp. 582–595, 2008.

framework of Tsochantaridis *et al.* [8] and the maximum-margin network learning of Taskar *et al.* [9,10]. These works were not focused on computer vision problems. They did not explore graph cuts for learning, and instead chose approximate inference algorithms (such as sum-product loopy BP) which do not perform well on grid-graph problems encountered while dealing with images [6].

For a given parameterized energy function, our goal is to learn the parameters so that the ground truth has the lowest energy by the largest possible margin, or, if that is not possible, that the energy of the ground truth is as close as possible to that of the minimum energy solution. We begin with arbitrary initial parameters and an empty "contrastive" solution set. Then, iteratively, we find the minimum energy solution for the current parameters, add it to our alternative set, and find the parameters that maximize our objective function, considering only the ground truth and the current alternative set of solutions.

When the energy is a linear or quadratic function of the parameters (as is typically the case) and can be optimally minimized, we can estimate the parameters accurately in a highly efficient manner, typically requiring a few dozen iterations for problems involving dozens of unary and pairwise random-field parameters. We also show that, when our objective function is maximized, the sufficient statistics (those statistics necessary to compute the energy) of our solutions will closely match the sufficient statistics of the ground truth. Additionally, we describe several extensions of our technique, evaluating the suitability of alternative objective functions, and dealing with cases for which optimal energy minimization is not possible. We validate our algorithm and its extensions on the problems of image segmentation and multi-class geometry labeling.

Contributions. We present an efficient and practical algorithm to train random field models for images. The contributions of our paper include:

1. Use of graph-cuts to *efficiently* do maximum margin learning of parameters *exactly* for submodular MRFs and CRFs. Generalization to new images is ensured via a large margin regularizer. Approximations such as pseudolikelihood are not required.
2. Learning parameters of non-submodular problems using alpha-expansions.
3. Investigation of loss functions for segmentation and geometry labeling.
4. Formulation of the polygon-bounded image segmentation task, for the purpose of obtaining refined segmentations from rough user interactions.

2 Basic Learning Algorithm

Many computer vision tasks such as segmentation and stereo estimation can be modeled with random fields, which describe a joint probability distribution of all random variables, either jointly or conditionally on the input pixels. The probability is expressed as an exponentiated energy. The energy typically consists of a sum of terms, including node energies for each random variable and interaction energies that model dependencies between adjacent variables. Usually, the energies are parameterized as a linear function of these unary and pairwise features.

Applying this model requires 1) learning the parameters of the model from training data, and 2) inferring the most likely labels for the test data, using the learned parameters. Inference in this model amounts to an energy minimization problem, which can frequently be solved via graph cuts.

Parameter learning aims to find parameters that fit the training data and that also generalize to unseen test data. The learning is formulated as minimizing a loss function on the training set. A common loss function is the negative likelihood of the training labels, but for random fields this leads to an intractable optimization problem. Instead, we employ a tractable approach that uses other loss functions. In this section, we describe a simple algorithm that maximizes the difference between the energy of ground truth labeling and other labelings. In section 3, we refine the loss function to incorporate degrees of labeling error. This makes the method robust to outliers and ensures good generalization to test data. We begin by detailing the model and motivating the loss function.

2.1 The Random Field Model

Let $X = \{\boldsymbol{x}^{(n)}\}_n$ be the input collection of instances $\boldsymbol{x}^{(n)}$, with corresponding labelings $Y = \{\boldsymbol{y}^{(n)}\}_n$, where n indexes the instances. For example, an instance may be an image with $\boldsymbol{x}$ denoting the pixel values and $\boldsymbol{y}$ their segmentation labels (foreground or background). Our model is a conditional random field of the form

$$P(Y \mid X, \boldsymbol{w}) = \frac{1}{Z} \mathrm{e}^{-\sum_n E(\boldsymbol{y}^{(n)}, \boldsymbol{x}^{(n)}; \boldsymbol{w})}, \tag{1}$$

where $\boldsymbol{w}$ are parameters (weights), and Z is a normalizer (the partition function). A typical energy decomposes over nodes $\mathcal{V}$ (e.g. individual pixels) and edges $\mathcal{E}$ (e.g. pairs of adjacent pixels). We consider energies E that are linear in the parameters $\boldsymbol{w}$ and the set of node and edge features $\boldsymbol{\phi}$ such as:

$$E(\boldsymbol{y}, \boldsymbol{x}; \boldsymbol{w}) = \sum_{i \in \mathcal{V}} \boldsymbol{w}_1 \boldsymbol{\phi}_i^{(1)}(y_i, \boldsymbol{x}) + \sum_{(i,j) \in \mathcal{E}} \boldsymbol{w}_2 \boldsymbol{\phi}_{ij}^{(2)}(y_i, y_j, \boldsymbol{x}), \tag{2}$$

which we can abbreviate as $E(\boldsymbol{y}, \boldsymbol{x}; \boldsymbol{w}) = \boldsymbol{w}^T \boldsymbol{\phi}(\boldsymbol{x}, \boldsymbol{y})$. Thus, we have a log-linear model in the exponential family.

To infer labels of a test input $\boldsymbol{x}$, given parameters $\boldsymbol{w}$, we will find the maximum a posteriori (MAP) labelings $\boldsymbol{y}^* = \operatorname{argmax}_{\boldsymbol{y}} P(\boldsymbol{y} \mid \boldsymbol{x}, \boldsymbol{w})$.

During training we look for weights $\boldsymbol{w}$ that assign the training labels $\boldsymbol{y}^{(n)}$ a greater than or equal probability of any other labeling $\boldsymbol{y}$ of instance n, i.e.,

$$P(\boldsymbol{y}^{(n)} \mid \boldsymbol{x}^{(n)}, \boldsymbol{w}) \geq P(\boldsymbol{y} \mid \boldsymbol{x}^{(n)}, \boldsymbol{w}) \quad \forall \boldsymbol{y} \neq \boldsymbol{y}^{(n)} \; \forall n. \tag{3}$$

If this were successful, the inferred MAP labelings for the training data would equal the training labels (except for possible ties). We can cancel the normalizer Z from both sides of the constraints (Eq. 3) and express the constraints in terms of energies

$$E(\boldsymbol{y}^{(n)}, \boldsymbol{x}^{(n)}; \boldsymbol{w}) \leq E(\boldsymbol{y}, \boldsymbol{x}^{(n)}; \boldsymbol{w}) \quad \forall \boldsymbol{y} \neq \boldsymbol{y}^{(n)} \; \forall n. \tag{4}$$

Thus, we desire weights that give the training labels at least as low an energy as any other label configurations on the training data. However, the inequalities in (4) may have multiple or no solutions. We resolve this by finding the parameters which satisfy the inequalities with the largest possible energy margin γ, so that the ground truth labeling has the lowest energy relative to other labelings. This large margin concept serves to regularize the problem and provide generalization to unseen test data. The margin may be negative if the original inequalities had no solutions. Thus we have

$$\max_{\boldsymbol{w}:\|\boldsymbol{w}\|=1} \gamma \quad \text{such that} \tag{5}$$
$$E(\boldsymbol{y}, \boldsymbol{x}^{(n)}; \boldsymbol{w}) - E(\boldsymbol{y}^{(n)}, \boldsymbol{x}^{(n)}; \boldsymbol{w}) \geq \gamma \quad \forall \boldsymbol{y} \neq \boldsymbol{y}^{(n)} \; \forall n.$$

We have constrained the weight norm $\|\boldsymbol{w}\|$ to unity to prevent weights from growing without bounds.

2.2 An Efficient Learning Algorithm

Here we describe a simple but efficient learning algorithm based on the optimization (5). It is not feasible to solve that program, as it has an exponential number of constraints, one for each possible labeling $\boldsymbol{y}$ of each instance. Instead, we shall perform the optimization over a much smaller set of labelings $\{\boldsymbol{S}^{(n)}\}_n$. We explicitly enforce that the ground truth has lower energy than labelings in this set. We check whether this also holds true for the remaining labelings, by finding a labeling that has the lowest energy via graph cuts, or any other efficient method. If this labeling has lower energy than the ground truth, or doesn't achieve the margin, we add it to the constraint set. Finally, we update the parameters to satisfy the new constraint, and iterate. If the parameters do not change (for example, if we did not find any example better than the ground truth), then we stop. This is detailed in Figure 1. Each step of this algorithm is a standard task that can be solved efficiently. A key observation is that the MAP labeling in Step 1 can be translated to a graph cut problem for many vision tasks. Graph cut algorithms can quickly find global optima for a large class of so-called *submodular* energies [1,11].

Step 3 is written as a general optimization for clarity, but is in fact just a quadratic program (given in Section 3). Thus it is convex and is free from local minima. The overall procedure converges as there are only a finite number of labelings that can be added. Crucially, it converges in a low number of steps in practice, and even in the worst case, is proven to require only a polynomial number of steps if a global optimum can be found in step 1 (see [8]), which is true for the case of submodular energy functions. At convergence, all label configurations outside of $\boldsymbol{S}^{(n)}$ will be at least a margin distance away from the ground truth which is why it is important to study the choice of margin measure (loss function).

The constraint $\boldsymbol{w}_2 \geq 0$ in Eq. 6 is needed in the context of graph cuts to ensure that the problem remains submodular for energies $E(\boldsymbol{y}, \boldsymbol{x}; \boldsymbol{w})$ that would

BASIC ALGORITHM

Input:
- input-labeling pairs $\{(\boldsymbol{x}^{(n)}, \boldsymbol{y}^{(n)})\}$ training set
- empty set of competing low energy labelings: $\boldsymbol{S} = \emptyset$
- initial parameters: $\boldsymbol{w} = \boldsymbol{w}_0$

Repeat until $\boldsymbol{w}$ is unchanged (within a tolerance)

Loop over training instances n

1. Find the MAP labeling of instance n, using e.g. graph cuts
   $\boldsymbol{y}^* \leftarrow \text{argmin}_{\boldsymbol{y}} E(\boldsymbol{y}, \boldsymbol{x}^{(n)}; \boldsymbol{w})$
2. Add $\boldsymbol{y}^*$ to the constraint set
   $\boldsymbol{S}^{(n)} \leftarrow \boldsymbol{S}^{(n)} \cup \{\boldsymbol{y}^*\}$
3. Update the parameters $\boldsymbol{w}$ to ensure ground truth has the lowest energy

$$\max_{\boldsymbol{w}:\|\boldsymbol{w}\|=1} \gamma \quad \text{such that} \tag{6}$$
$$E(\boldsymbol{y}, \boldsymbol{x}^{(n)}; \boldsymbol{w}) - E(\boldsymbol{y}^{(n)}, \boldsymbol{x}^{(n)}; \boldsymbol{w}) \geq \gamma \quad \forall \boldsymbol{y} \in \boldsymbol{S}^{(n)} \ \forall n$$
$$\boldsymbol{w}_2 \geq 0$$

Fig. 1. Pseudocode for the basic random field learning algorithm

become non-submodular for $\boldsymbol{w}_2 < 0$ (we have assumed that all pairwise potentials $\boldsymbol{\phi}_{ij}^{(2)}(y_i, y_j, \boldsymbol{x})$ are non-negative and convex); for other exact optimization techniques (e.g. max-product belief propagation on junction trees, or branch and bound), this constraint can be lifted. Multiple parameter optima may exist, but will all result in the same energy for the training data.

Non-submodular Energy Functions. Some problems in computer vision involve non-submodular energy functions that are NP-hard and cannot generally be exactly minimized in polynomial time. Examples include multi-label problems such as multi-class geometric labelling [12] and stereo [2]. In these cases, efficient approximate inference can be used in Step 3 of the cutting plane algorithm in Figure 1. The choice of the optimization technique affects the convergence and correctness guarantees of this method [13].

The approximate algorithms commonly used in computer vision include (1) message passing algorithms such as tree-reweighed message passing (TRW) and loopy belief propagation (LBP), and (2) move-making algorithms such as alpha-expansion [2] and Fast-PD[14]. Algorithms such as TRW and Fast-PD work by formulating the energy minimization problem as an integer programming problem and then solving its linear relaxation. These algorithms in addition to giving an approximate solution, also produce a per-instance lower-bound of the energy which can be used to check how close the approximate solution is to the global minima. We could use this information to isolate good solutions and add them to the solution set. If for a particular set of weights, the lower bound is far from the energy of the resulting solution, we know that this set of parameter is not appropriate as it results in a difficult inference problem and would not lead to good solutions.

For our experiments with the nonsubmodular geometric labeling problem (Section 4.2), we chose to use alpha-expansion. We were motivated by the result [14] showing that alpha-expansion yields close to optimal results for the nonsubmodular Potts model (although our model is more complex).

Relation to Previous Work. This work presents a method for max-margin learning in loopy graphs such as grids (images). In contrast, [15] uses Viterbi on a linear chain, an easier problem not applicable to images. Tsochantaridis et al. [8] and Taskar et al. [10] rely on MAP inference but do not use graphcut for learning. Anguelov et al. [9] use a generic CPLEX solver, which can only work on small vision problems (equivalent to 30000 pixels, unlike 10^7 pixels in our case).

The work of Kumar et al. [7] is most related to us, as it does use graphcut. However, it attempts to use a max likelihood framework, which is not compatible with MAP inference, requiring a drastic approximation of the partition function by a single MAP configuration. Our choice, the large margin framework, allows us to solve the problem exactly with high probability (see Theorem 18 in [8] and [13]).

2.3 Parameter Learning for Image Segmentation

In Figure 2 we illustrate our method with one such problem: the binary image segmentation into foreground (cow) and background based on color models estimated from user-given patches and smoothness constraints. In this example, the energy has a single node term (negative log likelihood of the color given the label) and two interaction terms: a constant penalty and a contrast-based penalty C for neighboring pixels having different labels. Our energy function can be written as

$$E(\boldsymbol{x}^{(n)}, \boldsymbol{y}^{(n)}; \boldsymbol{w}) = w_1 \sum_{i \in V} - \log P(x_i^{(n)} \mid y_i^{(n)}) + \tag{7}$$

$$w_2 \sum_{(i,j) \in \mathcal{E}} 1_{y_i^{(n)} \neq y_j^{(n)}} + w_3 \sum_{(i,j) \in \mathcal{E}} 1_{y_i^{(n)} \neq y_j^{(n)}} C(x_i^{(n)}, x_j^{(n)}),$$

where 1 denotes an indicator function equaling one for differing labels. The indicator functions enforce a Potts model penalty on the labelling. The likelihood $P(x_i^{(n)} \mid y_i^{(n)})$ is computed from color models of the foreground and background. These models are built using user-specified cues or seeds. The working of the simple parameter learning algorithm for the image segmentation problem is illustrated in figure (2).

3 Objectives

In this section we consider alternative objectives that are more robust to noisy training data and have refined loss functions. Using the transformation $\|\boldsymbol{w}\| \leftarrow 1/\gamma$, we can write program (5) as a standard quadratic program (recall that $E(\boldsymbol{y}^{(n)}, \boldsymbol{x}^{(n)}; \boldsymbol{w})$ is linear in $\boldsymbol{w}$ (Eq. 2)).

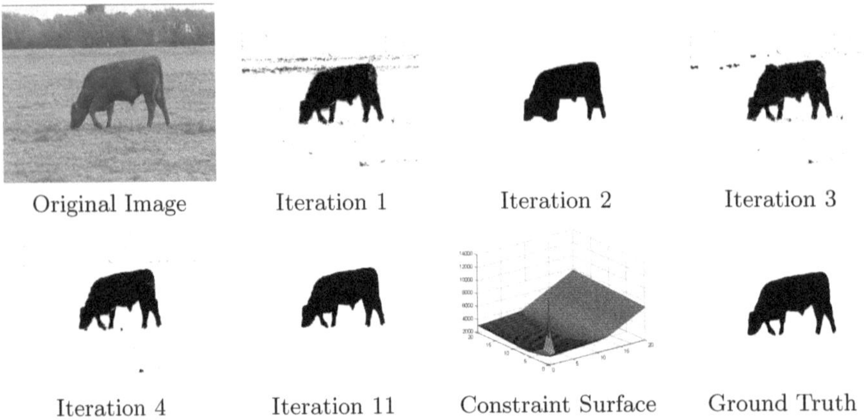

Fig. 2. Parameter Learning for Image Segmentation. The figure illustrates the working of the basic parameter learning algorithm on the image segmentation problem modelled with three parameters (7). The segmentations shown in the figure correspond to the results of using the parameters estimates in iterations 1, 2, 3, 4 and 11 (in order) of the learning algorithm. These solutions are added to the minimal solution set $\boldsymbol{S}^{(n)}$. The algorithm converges in 11 iterations. For the final iteration (11), we also show the plot of the 0-1 loss cost function surface on the smoothness w_2 and contrast w_3 parameters of the segmentation energy (7) keeping w_1 constant. The optimal values of the weights were $w_2 = 1.96$ and $w_3 = 6.94$.

$$\min_{\boldsymbol{w}} \frac{1}{2}\|\boldsymbol{w}\|^2 \quad \text{s.t.} \tag{8}$$
$$E(\boldsymbol{y}, \boldsymbol{x}^{(n)}; \boldsymbol{w}) - E(\boldsymbol{y}^{(n)}, \boldsymbol{x}^{(n)}; \boldsymbol{w}) \geq 1 \quad \forall \boldsymbol{y} \neq \boldsymbol{y}^{(n)} \ \forall n$$

To add robustness for noisy training data, we relax the margin constraints by adding slack variables ξ_n. Thus, the individual margins may be smaller, but that is discouraged through a slack penalty in the objective, regulated by the slack penalty parameter C.

$$\min_{\boldsymbol{w}} \frac{1}{2}\|\boldsymbol{w}\|^2 + \frac{C}{N}\sum_n \xi_n \quad \text{s.t.} \tag{9}$$
$$E(\boldsymbol{y}, \boldsymbol{x}^{(n)}; \boldsymbol{w}) - E(\boldsymbol{y}^{(n)}, \boldsymbol{x}^{(n)}; \boldsymbol{w}) \geq 1 - \xi_n \ \forall \boldsymbol{y} \neq \boldsymbol{y}^{(n)} \ \forall n$$
$$\xi_n \geq 0 \qquad \forall n$$

These are all standard large margin formulations. However, in the context of optimization over joint labelings $\boldsymbol{y}$, we now consider refinements to the standard margin.

3.1 Rescaled Margin

The energy constraints above enforce the same unit margin for all labelings $\boldsymbol{y}$ competing with the training data labeling $\boldsymbol{y}^{(n)}$, regardless of how similar $\boldsymbol{y}$ and

MARGIN RESCALED ALGORITHM
Input: as in Fig. 1 Repeat until $\boldsymbol{w}$ is unchanged (within a tolerance) Loop over training instances n 1. Run graph cuts to find the MAP labeling of instance n $\boldsymbol{y}^* \leftarrow \text{argmin}_{\boldsymbol{y}} E(\boldsymbol{y}, \boldsymbol{x}^{(n)}; \boldsymbol{w}) - \Delta(\boldsymbol{y}^{(n)}, \boldsymbol{y})$ 2. If $\boldsymbol{y}^* \neq \boldsymbol{y}^{(n)}$, add $\boldsymbol{y}^*$ to the constraint set $\boldsymbol{S}^{(n)} \leftarrow \boldsymbol{S}^{(n)} \cup \{\boldsymbol{y}^*\}$ 3. Update the parameters $\boldsymbol{w}$ to ensure ground truth has the lowest energy $$\min_{\boldsymbol{w}} \frac{1}{2}\|\boldsymbol{w}\|^2 + \frac{C}{N}\sum_n \xi_n \quad \text{s.t} \quad \forall \boldsymbol{y} \in \boldsymbol{S}^{(n)}\ \forall n \qquad (11)$$ $$E(\boldsymbol{y}, \boldsymbol{x}^{(n)}; \boldsymbol{w}) - E(\boldsymbol{y}^{(n)}, \boldsymbol{x}^{(n)}; \boldsymbol{w}) \geq \Delta(\boldsymbol{y}^{(n)}, \boldsymbol{y}) - \xi_n$$ $$\xi_n \geq 0$$ $$\boldsymbol{w}_2 \geq 0$$

Fig. 3. Pseudocode for the margin rescaled learning algorithm

$\boldsymbol{y}^{(n)}$ are (as long as they are not exactly identical). However, for vision problems it is sensible to adapt the margin according to how much competing labelings differ from the ground truth labeling. The idea is to enforce a relatively larger margin for labelings that are far from the ground truth. The desired difference between the ground truth $\boldsymbol{y}^{(n)}$ and a labeling $\boldsymbol{y}$ will be quantified via a loss function $\Delta(\boldsymbol{y}^{(n)}, \boldsymbol{y})$. For example, an appropriate loss may be the Hamming loss $\Delta(\boldsymbol{y}^{(n)}, \boldsymbol{y}) = \sum_i \delta(y_i^{(n)}, y_i)$. For a 0-1 loss function $\Delta(\boldsymbol{y}^{(n)}, \boldsymbol{y}) = \delta(\boldsymbol{y}^{(n)}, \boldsymbol{y})$ we reduce to the previous formulation.

Taskar et al. [16] proposed to rescale the margin to enforce the constraint

$$E(\boldsymbol{y}, \boldsymbol{x}^{(n)}; \boldsymbol{w}) - E(\boldsymbol{y}^{(n)}, \boldsymbol{x}^{(n)}; \boldsymbol{w}) \geq \Delta(\boldsymbol{y}^{(n)}, \boldsymbol{y}) - \xi_n \qquad (10)$$

for all $\boldsymbol{y} \neq \boldsymbol{y}^{(n)}$ and all n.

To construct a learning algorithm that takes loss into account, we again iteratively collect a set of competing labelings. Now however, we must find examples that violate these new constraints, which require us to look beyond minimum energy labelings with respect to E.

Importantly, the margin rescaled formulation still allows graph cuts to find the minimum energy solutions under Hamming loss, or any loss that decomposes the same way as the energy. The Hamming loss decomposes across nodes, which allows us to absorb the loss into the node energies, and define a refined energy function $E'(\boldsymbol{y}, \boldsymbol{x}; \boldsymbol{w}) = E(\boldsymbol{y}, \boldsymbol{x}; \boldsymbol{w}) - \Delta(\boldsymbol{y}^{(n)}, \boldsymbol{y})$. We then arrive at the algorithm in Figure 3. Note that the loss Δ is only incorporated during parameter learning. When inferring labels of test data, energy minima of E are found as before.

Generating multiple solutions using graph cuts. The objectives (programs (8)–(10)) require us to enforce constraints for all $\boldsymbol{y} \neq \boldsymbol{y}^{(n)}$. In step 1 (Figure 3) it is possible that graph cuts find a minimum energy labeling $\boldsymbol{y}^*$ equal to the ground truth $\boldsymbol{y}^{(n)}$. This would not provide any constraint on which to enforce the margin. If this occurred for all training instances, the parameters would not change and the algorithm would terminate without enforcing the margin. This problem can be overcome by adding multiple low-energy labelings at every iteration, instead of just one. The global minima of a submodular function can be found using graph cuts. Kohli and Torr [17] showed how exact min-marginal energies can be efficiently computed for submodular functions. We use these min-marginals to compute the N-best solutions for the energy minimization problem. The time taken for computing these multiple solutions is 3–4 times more than that taken for computing the MAP solution.

3.2 Rescaled Slack

Tsochantaridis *et al.* [8] proposed an alternative way to incorporate a loss into the margin constraints. They rescale the slack variables by the loss, requiring that $\forall \boldsymbol{y} \neq \boldsymbol{y}^{(n)}$:

$$E(\boldsymbol{y}, \boldsymbol{x}^{(n)}; \boldsymbol{w}) - E(\boldsymbol{y}^{(n)}, \boldsymbol{x}^{(n)}; \boldsymbol{w}) \geq 1 - \frac{\xi_n}{\Delta(\boldsymbol{y}^{(n)}, \boldsymbol{y})} \tag{12}$$

This formulation has the advantage that the loss Δ and slack penalty C share the same scale. In particular, it enforces the same default margin of 1 for all examples. In contrast, the margin-rescaled formulation requires large margins of labelings differing significantly from the ground truth, which could cause the algorithm to focus on assigning high energies to poor labelings, rather than assigning low energies to labelings close to ground truth. Unfortunately, graph cut optimization cannot directly find labelings that violate the constraints (12). Instead, we use graph cut to find a minimum-energy solution and check whether it violates the constraints.

3.3 Minimum Energy Loss

While the rescaled margin and rescaled slack methods are a large improvement over the 0-1 loss, they may still minimize training error poorly because they focus on the constraints that require the largest slack, rather than the minimum-energy solutions for a particular set of parameters. For instance, in the case of Taskar *et al.*, even if the ground truth is the minimum energy solution, if any high loss solution does not have a sufficiently high margin, the algorithm will change the parameters. The modification of Tsochantaridis *et al.* is better, since the same margin is required of all examples, but the cost function still will often concern itself with solutions that are not minimum-energy solutions (and, thus, would not be returned during inference). We, therefore, provide a new cost function that minimizes the training error more directly.

$$\min_{\boldsymbol{w}} \frac{1}{2}\|\boldsymbol{w}\|^2 + \frac{C}{N}\sum_n \xi_n + \frac{C_2}{N}\sum_n \Delta(\boldsymbol{y}^{(n)}, \hat{\boldsymbol{y}}^{(n)}) \tag{13}$$

where $\hat{\boldsymbol{y}}^{(n)} = \operatorname{argmin}_{\boldsymbol{y} \in \boldsymbol{S}^{(n)} \cup \{\boldsymbol{y}^{(n)}\}} E(\boldsymbol{y}, \boldsymbol{x}^{(n)}; \boldsymbol{w})$.

Table 1. Number of training iterations and test accuracy for each cost function using our graph cuts learning method to learn parameters for segmentation. The training data consisted of rough 6 and 12-point polygon segmentations. The method is highly efficient under all cost functions (an iteration of processing 50 images of size 320×240 takes about 45 seconds). Our new loss-augmented functions provide the best test accuracy.

Polygon Type	12 Point Polygon	6 Point Polygon
Cost Function	(Iterations, Accuracy)	(Iterations, Accuracy)
Maximum likelihood, unary only	(N/A, 79.7%)	(N/A, 64.3%)
Pseudo-likelihood	(26, 81.9%)	(21, 74.0%)
0-1 Loss	(13, 80.7%)	(-,-)
Rescaled Margin	(43, 83.0%)	(11, 74.5%)
Rescaled Slack	(13, 82.3%)	(13, 70.7%)
Rescaled Margin + Loss	(45, 85.7%)	(184, 76.1%)
Rescaled Slack + Loss	(90, 84.8%)	(118, 76.2%)

To optimize this objective, we iteratively find the minimum-energy solution for the given parameters and find the parameters that minimize this cost. Loosely speaking, this is a greedy search for minimum training loss, locally approximating the loss gradient with the gradient of the slack. For large C_2, this is guaranteed to converge to a local minimum in training loss.

4 Experiments

The goal of these experiments is to demonstrate the efficacy of the max-margin learning method for computer vision problems, and to study the behavior of different loss functions in this framework. To do this, we perform experiments on two tasks: segmentation and geometry labeling. For these tasks, there are too many parameters to tune by hand. Our results demonstrate that our method is highly efficient and that training with our new loss-augmented cost functions improves test accuracy considerably.

4.1 Refining Segmentations

While completely automatic segmentation is extremely difficult, refining a coarse user-provided segmentation with a nearly pixel-perfect segmentation is achievable and quite useful. For instance, in a photo-editing program, a person could draw a polygon around the object, and the computer would attempt to produce a pixel-accurate segmentation. We evaluate our algorithm on this task, with run-time and accuracy comparisons to other methods. See Figure 4 for some qualitative examples.

The input to our algorithm is an image and a 4, 6 or 12-point polygon that approximates the foreground region. We train on fifty segments from fifty separate images in the Berkeley Segmentation Dataset and test on fifty segments from a different fifty images. We define three unary potentials: a foreground prior,

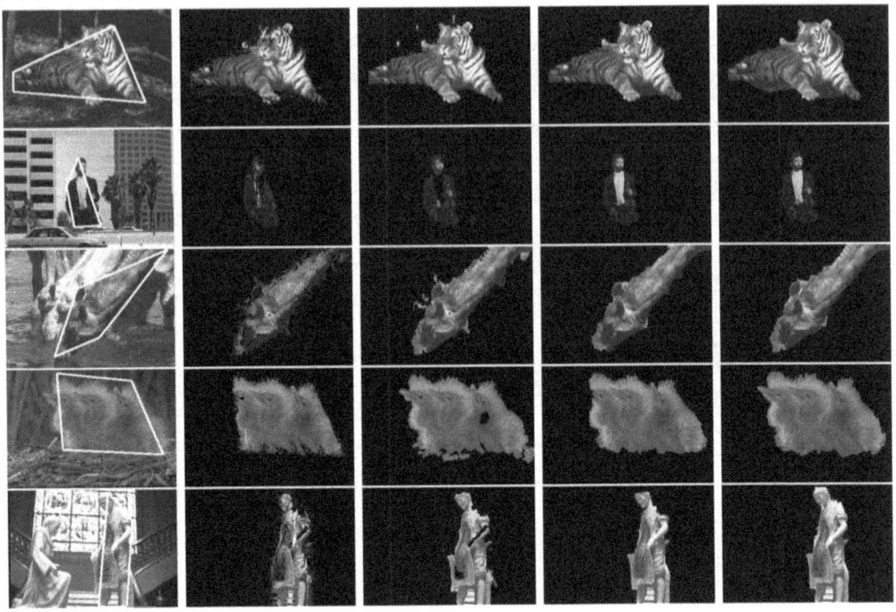

Fig. 4. Parameter learning for image segmentation. Given the input image and a 4-corner polygon denoting the approximate foreground region (column 1), the goal is to produce a refined pixel-perfect segmentation. Shown are the segmentation results from exact maximum likelihood learning using unary features (column 2), and after learning CRF parameters with our algorithm using rescaled margin (column 3), rescaled margin with augmented loss (column 4), and rescaled slack with augmented loss (column 5).

signed distance from the nearest polygon boundary, and a color histogram term. On an eight-connected graph, we define three associative pairwise terms that penalize connected pixels having different labels: constant, intensity contrast-based, and color contrast-based (Euclidean distance of A and B channels in Lab space). We learn separate parameters for horizontal, vertical, and diagonal edges to give a total of three unary and nine pairwise parameters. To evaluate, we define segmentation accuracy as the intersection to union ratio of the ground truth and resulting foreground regions, and report the average test accuracy.

Comparison of Cost Functions. In Table 1, we report accuracy when training based on the cost functions described in Section 3, using 12-point input polygons. The baseline case is to return the input polygon as the final foreground region (80.9% accuracy; note that this is based on a rough human segmentation and does not correspond to a learned model). As expected, the cost function based on 0-1 loss performs poorly (80.7%), while the margin-rescaling and slack-rescaling method offer some improvement (83.0% and 82.3%, respectively).

During training, we define loss as pixel error for the margin-rescaling method to ensure that we can reach the global minimum cost with our graph cut method. We cannot make the same guarantee when using the slack-rescaling method, regardless of the label loss function used, so we define loss as the intersection-union ratio, as is used to report test error. Note that more general (but much slower) minimization methods may be able to apply the slack-rescaling method more effectively. Training with our new loss-augmented cost functions provides significant further improvement to test accuracies of 85.7% and 84.8%, when using margin-rescaling and slack-rescaling, respectively. For comparison, we also trained a model using maximum likelihood including the three unary potentials only; this gave 79.7% accuracy (exact ML training including the pairwise terms is intractable). Note that these improvements of 4-5% over the baseline do not fully convey the large qualitative improvement that can be seen in Figure 4.

We also trained on coarser 6-point polygon inputs. This task is more challenging and accuracy decreases, as seen in table 1. For figure 4, even coarser 4-point polygons were used. We also performed experiments with the parameters learned by pseudolikelihood maximization. We found the training times to be similar to our method, but the resulting accuracies (74.0% for 6-point polygons, 81.9% for 12-point polygons) were below those achieved by our method with our proposed minimum energy loss functions.

4.2 Geometry Estimation

With this experiment, we seek to answer whether our algorithm remains tractable on a large-scale vision problem with several dozen parameters. Our task is to label each pixel in the image into "ground", "vertical", or "sky", using the geometric context dataset provided by Hoiem *et al.* [12].

To do this, we define 8 unary parameters to model the priors, general appearance, and image-specific appearance. The appearance terms are modeled as the logistic outputs of boosted decision tree classifiers trained separately on four sets of features described by Hoiem *et al.*: color, texture, position and shape, and perspective. For each image, we also estimate the color of each class within that image, based on an initial estimate of the geometry likelihoods. Our pairwise interaction terms model texture and color gradients between neighboring pixels and the relative positions. In total there are 72 interaction terms (4 directions, including diagonal, 9 pairwise labelings, and two types of gradients). We could learn, for example, that sky is unlikely to appear directly beneath vertical regions (although it can happen with overhanging objects) or that certain types of gradients are indicative of a boundary for particular geometric classes.

What we are chiefly interested in here, however, is whether our learning algorithm can do anything useful on this large-scale, multi-class learning problem.

Table 2. Accuracy on the geometry estimation problem

Objective	Unary only	Rescaled Margin	Rescaled Slack (Loss)	Rescaled Margin (Loss)
Accuracy	84.6%	87.7%	86.8%	87.0%

Fig. 5. Results of geometry ("ground", "vertical", or "sky") estimation based on our CRF learning method. Input images (top row), results based on unary energies alone (middle) and the learned CRF (bottom).

To handle the case of multiple classes, we replace our regular graph cut inference step with alpha-expansion graph cuts. Our results, shown in Figure 5 and Table 2 show that we are able to learn parameters that significantly outperform the unary classifier. Despite the large number of parameters, the learning required only about fifty iterations for each objective, indicating that our method is suitable for large-parameter problems.

5 Discussion and Conclusion

We have presented an efficient algorithm to train random field models (MRFs and CRFs) for images. These models are not tractable to train exactly using maximum likelihood. Instead, we start from a standard large-margin framework, and then leverage graphcut to perform inference and thereby arrive at a truly practical algorithm for the computer vision field.

Intuitively, for a learning problem involving many parameters a large set of minimum energy solutions will be needed to constrain the feasible parameter space of our convex program. We note that some experiments required up to 184 iterations, as our parameter learning algorithm only adds one solution to the solution set in each iteration. The rate of convergence of the learning algorithm can be improved by adding multiple samples from the posterior labeling distribution to the solution set in each iteration of the algorithm. We plan to explore this in future work.

References

1. Kolmogorov, V., Zabih, R.: What energy functions can be minimized via graph cuts? IEEE Trans. Pattern Anal. Mach. Intell. 26(2), 147–159 (2004)
2. Boykov, Y., Veksler, O., Zabih, R.: Fast approximate energy minimization via graph cuts. IEEE Trans. Pattern Anal. Mach. Intell. 23(11), 1222–1239 (2001)
3. Boykov, Y., Jolly, M.: Interactive graph cuts for optimal boundary and region segmentation of objects in n-d images. In: ICCV, pp. I: 105–112 (2001)
4. Rother, C., Kolmogorov, V., Blake, A.: "GrabCut": interactive foreground extraction using iterated graph cuts. ACM Transactions on Graphics 23(3), 309–314 (2004)
5. Barbu, A., Zhu, S.C.: Generalizing Swendsen-Wang to sampling arbitrary posterior probabilities. IEEE Trans. Pattern Anal. Mach. Intell. 27(8), 1239–1253 (2005)
6. Szeliski, R., Zabih, R., Scharstein, D., Veksler, O., Kolmogorov, V., Agarwala, A., Tappen, M.F., Rother, C.: A comparative study of energy minimization methods for Markov random fields. In: Leonardis, A., Bischof, H., Pinz, A. (eds.) ECCV 2006. LNCS, vol. 3952, pp. 16–29. Springer, Heidelberg (2006)
7. Kumar, S., August, J., Hebert, M.: Exploiting inference for approximate parameter learning in discriminative fields: An empirical study. In: EMMCVPR, pp. 153–168 (2005)
8. Tsochantaridis, I., Joachims, T., Hofmann, T., Altun, Y.: Large margin methods for structured and interdependent output variables. Jrnl. Machine Learning Research 6, 1453–1484 (2005)
9. Anguelov, D., Taskar, B., Chatalbashev, V., Koller, D., Gupta, D., Heitz, G., Ng, A.Y.: Discriminative learning of Markov random fields for segmentation of 3D scan data. In: CVPR, pp. 169–176 (2005)
10. Taskar, B., Chatalbashev, V., Koller, D., Guestrin, C.: Learning structured prediction models: a large margin approach. In: ICML, pp. 896–903 (2005)
11. Boros, E., Hammer, P.L.: Pseudo-boolean optimization. Discrete Applied Mathematics 123(1–3), 155–225 (2002)
12. Hoiem, D., Efros, A.A., Hebert, M.: Geometric context from a single image. In: ICCV (2005)
13. Finley, T., Joachims, T.: Training structural SVMs when exact inference is intractable. In: Proc. Intl. Conf. on Machine Learning (ICML), pp. 304–311 (2008)
14. Komodakis, N., Tziritas, G., Paragios, N.: Fast, approximately optimal solutions for single and dynamic MRFs. In: CVPR (2007)
15. Collins, M.: Discriminative training methods for hidden Markov models: theory and experiments with perceptron algorithms. In: ACL-02 conf. Empirical methods in natural lang. processing (EMNLP), pp. 1–8 (2002)
16. Taskar, B., Chatalbashev, V., Koller, D.: Learning associative Markov networks. In: ICML (2004)
17. Kohli, P., Torr, P.H.S.: Measuring uncertainty in graph cut solutions - efficiently computing min-marginal energies using dynamic graph cuts. In: Leonardis, A., Bischof, H., Pinz, A. (eds.) ECCV 2006. LNCS, vol. 3952, pp. 30–43. Springer, Heidelberg (2006)

Feature Correspondence Via Graph Matching: Models and Global Optimization

Lorenzo Torresani[1], Vladimir Kolmogorov[2], and Carsten Rother[1]

[1] Microsoft Research Ltd., Cambridge, UK
{ltorre,carrot}@microsoft.com
[2] University College London, UK
vnk@adastral.ucl.ac.uk

Abstract. In this paper we present a new approach for establishing correspondences between sparse image features related by an unknown non-rigid mapping and corrupted by clutter and occlusion, such as points extracted from a pair of images containing a human figure in distinct poses. We formulate this matching task as an energy minimization problem by defining a complex objective function of the appearance and the spatial arrangement of the features. Optimization of this energy is an instance of graph matching, which is in general a NP-hard problem. We describe a novel graph matching optimization technique, which we refer to as dual decomposition (DD), and demonstrate on a variety of examples that this method outperforms existing graph matching algorithms. In the majority of our examples DD is able to find the global minimum within a minute. The ability to globally optimize the objective allows us to accurately learn the parameters of our matching model from training examples. We show on several matching tasks that our learned model yields results superior to those of state-of-the-art methods.

1 Introduction

Feature correspondence is one of the fundamental problems of computer vision and is a key ingredient in a wide range of applications including object recognition, 3D reconstruction, mosaicing, motion segmentation, and image morphing. Several robust algorithms (see e.g. [1,2]) exist for registration of images of static scenes and for visual correspondence under rigid motion. These methods typically exploit powerful constraints (e.g. epipolar constraints) to reduce the search space and disambiguate the correspondence problem. However, such constraints do not apply in the case of non-rigid motion or when matching different object instances. A popular approach in these cases is to discard the information about the spatial layout of features, and to find correspondences using appearance only. For example, many object recognition methods (see e.g. [3,4]) represent images as orderless sets of local appearance descriptors, known as bags of features. Recent work [5] has suggested that for many correspondence problems, learned appearance-based models perform similarly or better than state-of-the-art structural models exploiting information about spatial arrangement of features. This is primarily due to the challenges posed by the optimization and training of structural models, which often require approximate solution of NP-hard problems. In this paper we contrast this theory, and demonstrate that a complex structural model for image matching

D. Forsyth, P. Torr, and A. Zisserman (Eds.): ECCV 2008, Part II, LNCS 5303, pp. 596–609, 2008.

can be learned and optimized successfully. We cast the visual correspondence problem as an energy minimization task by defining a complex image matching objective depending on (i) feature appearance, (ii) geometric compatibility of correspondences, and (iii) spatial coherence of matched features. Additionally, we impose a uniqueness constraint allowing at most one match per feature. We introduce a novel algorithm to minimize this function based on the dual decomposition approach (DD) from combinatorial optimization, see e.g. [6,7,8,9,10,11]. The DD method works by maximizing a lower bound on the energy function. The value of the lower bound can be used to gauge the distance from the global minimum and to decide when to stop the optimization, in the event the global minimum cannot be found. For the majority of our examples DD finds the global minimum in reasonable time, and otherwise provides a solution whose cost is very close to the optimum. In contrast, previously proposed optimization methods such as [12,13] often fail to compute good solutions for our energy function. Our experimental evaluation shows that the model and the algorithm presented in this paper can be applied to a wide range of image matching problems with results matching or exceeding those of existing algorithms [5,14].

1.1 Relation to Previous Work

Models for feature matching. Our technique is loosely related to algorithms that find correspondences by matching appearance descriptors under smooth spatial transformations (see e.g. [15,16]). However, unlike such approaches, our method does not make a parametric assumption about the transformation relating the input images, and thus can be used in a wider range of applications. Belongie et al. [14] inject spatial smoothness in the match by means of an iterative technique that alternates between finding correspondences using shape features, and computing a regularized transformation aligning the matching features. The shape descriptors are recomputed in each iteration after the warping. Since the objective is changed at each iteration, the convergence properties of this algorithm are not clear. Our approach is most closely related to the work of Berg et al. [17], and Leordeanu and Hebert [18], who formulate visual correspondence as a graph matching problem by defining an objective including terms based on appearance similarity as well as geometric compatibility between pairs of correspondences. Our model differs from those in [17,18] in several ways. The methods proposed in [17] and [18] handle outliers by removing low-confidence correspondences from the obtained solutions. Instead, we include in our energy an explicit occlusion cost, as for example previously done in [19]. Thus our algorithm solves for the outliers as part of the optimization. We add to the objective a spatial coherence term, favoring spatial aggregation of matched features, which reduces the correspondence error on our examples. We also show that geometric penalty functions defined in local neighborhoods provide more accurate correspondences than global geometric costs, such as those used in [17] and [18]. Finally, we use the method of Liu et al. [20] to learn the parameter values for the model from examples, thus avoiding the need of manual parameter tuning.

Graph matching optimization. Graph matching is a challenging optimization problem which received considerable attention in the literature (see [21] for a comprehensive survey of methods). Proposed techniques include the graduated assignment algorithm of

Gold and Rangarajan [22], spectral relaxation methods [18,12], COMPOSE method of Duchi et al. [13]. Maciel and Costeira [23] reduce the problem to concave minimization and apply the exact method in [24]. Torr [19] and Schellewald and Schnörr [25] use semi-definite programming (SDP) relaxation for graph matching. Among these papers, only [23] and [25] report obtaining optimal (or near optimal) solutions. The method in [23] was tested only on a single example with quadratic costs. We conjecture that on practical challenging instances this method will suffer from an exponential explosion[1]. As shown in [12], the SDP relaxation approach in [25] scales quite poorly and is too expensive for problems of reasonable size.

2 Energy Function

We now describe the energy function of our matching model. Let P' and P'' be the sets of features extracted from the two input images. We denote with $A \subseteq P' \times P''$ the set of potential assignments between features in the two sets. We will use the terms assignment and correspondence interchangeably to indicate elements of A. We represent a *matching configuration* between the two point sets as a binary valued vector $\boldsymbol{x} \in \{0,1\}^A$. Each correspondence $a \in A$ indexes an entry x_a in the vector $\boldsymbol{x}$. A correspondence a is active if $x_a = 1$, and it is inactive otherwise. We define an energy function $E(\boldsymbol{x})$ modeling our matching problem assumptions. This will allow us to formulate the matching task as minimization of $E(\boldsymbol{x})$. In this paper we consider matching problems where at most one active correspondence per feature is allowed. This requirement is known as the uniqueness constraint and it is commonly used in correspondence problems. In order to enforce this condition we define the constraint set M:

$$M = \{\boldsymbol{x} \in \{0,1\}^A \mid \sum_{a \in A(p)} x_a \leq 1 \quad \forall p \in P\} \tag{1}$$

where $P = P' \cup P''$ is the set of features from both images, and $A(p)$ is the set of correspondences involving feature p. The goal is to find the configuration $\boldsymbol{x} \in M$ minimizing $E(\boldsymbol{x})$. We define our energy as a weighted sum of four energy terms:

$$E(\boldsymbol{x}) = \lambda^{\text{app}} E^{\text{app}}(\boldsymbol{x}) + \lambda^{\text{occl}} E^{\text{occl}}(\boldsymbol{x}) + \lambda^{\text{geom}} E^{\text{geom}}(\boldsymbol{x}) + \lambda^{\text{coh}} E^{\text{coh}}(\boldsymbol{x}) \tag{2}$$

where $\lambda^{\text{app}}, \lambda^{\text{occl}}, \lambda^{\text{geom}}, \lambda^{\text{coh}}$ are scalar weights. We describe the energy terms below.

Function $E^{\text{app}}(\boldsymbol{x})$ favors correspondences between features having similar appearance. We define this function as a sum of unary terms:

$$E^{\text{app}}(\boldsymbol{x}) = \sum_{a \in A} \theta_a^{\text{app}} x_a \,. \tag{3}$$

For an assignment $a = (p', p'') \in A$, θ_a^{app} is the distance between appearance descriptors (such as Shape Context [14]) computed at points p' and p'' in the respective images. We have used different features depending on the task at hand (see sec. 4).

[1] The method in [23] first selects a *linear* function E^- which is an underestimator on the original objective function E, i.e. $E^-(\boldsymbol{x}) \leq E(\boldsymbol{x})$ for all feasible solutions $\boldsymbol{x}$. It then visits *all* feasible solutions $\boldsymbol{x}$ with $E^-(\boldsymbol{x}) \leq E(\boldsymbol{x}^*)$ where $E(\boldsymbol{x}^*)$ is the cost of the optimal solution. For each solution a linear program is solved.

The term $E^{\text{occl}}(\boldsymbol{x})$ imposes a penalty for unmatched features. We define $E^{\text{occl}}(\boldsymbol{x})$ to be the fraction of unmatched features in the smallest of the two feature sets. We can write this function as

$$E^{\text{occl}}(\boldsymbol{x}) = 1 - \frac{1}{\min\{|P'|, |P''|\}} \sum_{a \in A} x_a \tag{4}$$

by noting that $\sum_{a \in A} x_a$ is equal to the number of distinct matched features in P' and P'', $\forall \boldsymbol{x} \in M$. This result derives trivially from the uniqueness constraint.

The term $E^{\text{geom}}(\boldsymbol{x})$ is a measure of geometric compatibility between active correspondences. This term is similar to the distortion costs proposed in [17,18]. Note, however, that the energy terms used in these previous approaches include distortion costs for all pairs of matched features, which results in energy functions penalizing any deviation from a global rigid transformation. Instead, our function $E^{\text{geom}}(\boldsymbol{x})$ measures geometric compatibility of correspondences only for *neighboring* features. We demonstrate that this model permits more flexible mappings between the two sets of features and yields more accurate correspondences. We use a "neighborhood system" N to specify the pairs of correspondences involved in our measure of geometric compatibility. N consists of all correspondence pairs defined over neighboring features:

$$N = \{\langle (p', p''), (q', q'') \rangle \in A \times A \mid p' \in N_{q'} \vee q' \in N_{p'} \vee p'' \in N_{q''} \vee q'' \in N_{p''}\} \tag{5}$$

where N_p indicates the set of K nearest neighbors of p (computed in the set of feature p), and K is a positive integer value controlling the size of the neighborhood, which we call *geometric neighborhood size*. $E^{\text{geom}}(\boldsymbol{x})$ is computed over pairs of active correspondences in the set N:

$$E^{\text{geom}}(\boldsymbol{x}) = \sum_{(a,b) \in N} \theta_{ab}^{\text{geom}} x_a x_b \tag{6}$$

where:

$$\theta_{ab}^{\text{geom}} = \eta(e^{\delta_{a,b}^2/\sigma_l^2} - 1) + (1 - \eta)(e^{\alpha_{a,b}^2/\sigma_\alpha^2} - 1) \tag{7}$$

$$\delta_{(p',p''),(q',q'')} = \frac{\big|||p' - q'|| - ||p'' - q''||\big|}{||p' - q'|| + ||p'' - q''||} \tag{8}$$

$$\alpha_{(p',p''),(q',q'')} = \arccos\left(\frac{p' - q'}{||p' - q'||} \cdot \frac{p'' - q''}{||p'' - q''||}\right) \tag{9}$$

Intuitively, $\theta_{(p',p''),(q',q'')}^{\text{geom}}$ computes the geometric agreement between neighboring correspondences (p', p''),(q', q'') by evaluating how well the segment $\overline{p'q'}$ matches the segment $\overline{p''q''}$ in terms of both length and direction. The parameter η is a scalar value trading off the importance of preserving distances versus preserving directions.

The term $E^{\text{coh}}(\boldsymbol{x})$ favors spatial proximity of matched features. It incorporates our prior knowledge that matched features should form spatially coherent regions within each image, corresponding to common objects or parts in the image pair, in analogy to coherence on a pixel grid, used for example in image segmentation. We define the cost $E^{\mathrm{coh}}(\boldsymbol{x})$ as the fraction of neighboring feature pairs with different occlusion status (this can be viewed as an MRF Potts model over feature occlusion). We now show how to write this function directly in terms of solution $\boldsymbol{x}$. Let N_P be the set of pairs of neighboring features in the two images:

$$N_P = \{(p,q) \in (P' \times P') \cup (P'' \times P'') \mid p \in N_q \vee q \in N_p\}. \tag{10}$$

Then we can express $E^{\mathrm{coh}}(\boldsymbol{x})$ as a sum of unary and pairwise terms:

$$E^{\mathrm{coh}}(\boldsymbol{x}) = \frac{1}{|N_P|} \sum_{(p,q)\in N_P} V_{p,q}(\boldsymbol{x}) \tag{11}$$

where:

$$V_{p,q}(\boldsymbol{x}) = \sum_{a\in A(p)} x_a + \sum_{b\in A(q)} x_b - 2 \sum_{a\in A(p), b\in A(q)} x_a x_b\,. \tag{12}$$

$V_{p,q}(\boldsymbol{x})$ is equal to 0 if p, q are either both matched or both unmatched; $V_{p,q}(\boldsymbol{x})$ is equal to 1 otherwise.

Feature correspondence as graph matching. The problem defined above can be written as

$$\min_{\boldsymbol{x}\in M} E(\boldsymbol{x} \mid \bar{\theta}) = \sum_{a\in A} \bar{\theta}_a x_a + \sum_{(a,b)\in N} \bar{\theta}_{ab} x_a x_b \tag{13}$$

where the constraint set M is given by (1). This problem is often referred to as *graph matching* in the literature [22,5]. Features P' and P'' are viewed as vertices of the two graphs. Pairwise term $\bar{\theta}_{ab} x_a x_b$ with $a = (p', p'')$, $b = (q', q'')$ encodes compatibility between edges (p', q'), (p'', q'') of the first and second graph, respectively, while unary term $\bar{\theta}_a x_a$ measures similarity between vertices p', p''.

We now address the question of how to optimize problem (13). Unfortunately, this problem is NP-hard [22]. We propose to use the *problem decomposition* approach (or *dual decomposition* - DD) for graph matching. Details are given in the next section.

3 Problem Decomposition Approach

On the high level, the idea is to decompose the original problem into several "easier" subproblems, for which we can compute efficiently a global minimum (or obtain a good lower bound). Combining the lower bounds for individual subproblems will then provide a lower bound for the original problem. The decomposition and the corresponding lower bound will depend on a parameter vector $\boldsymbol{\theta}$; we will then try to find a vector $\boldsymbol{\theta}$ that maximizes the bound. This approach is well-known in combinatorial optimization; sometimes it is referred to as "dual decomposition" [6]. It was applied to quadratic pseudo-boolean functions (i.e. functions of binary variables with unary and pairwise terms) by Chardaire and Sutter [7]. Their work is perhaps the closest to the method in this paper. As in [7], we use "small" subproblems for which the global minimum can be computed exactly in reasonable time. Our choice of subproblems for graph matching, however, is different from [7]. In vision the decomposition approach is probably best known in the context of the MAP-MRF inference task. It was introduced by Wainwright et al. [8] who decomposed the problem into a convex combination of trees and proposed message passing techniques for optimizing vector $\boldsymbol{\theta}$. These techniques do not necessarily find the best lower bound. Schlesinger and Giginyak [9,10] and Komodakis et al. [11] proposed to use subgradient techniques [26,6] for MRF optimization, which guarantee to converge to a vector $\boldsymbol{\theta}$ giving the best possible lower bound.

3.1 Graph Matching Via Problem Decomposition

We now apply this approach to the graph matching problem given by eq. (13). We decompose (13) into subproblems characterized by vectors θ^σ, $\sigma \in I$ with positive weights ρ_σ. (These weights are chosen a priori, and may affect the speed of convergence of the subgradient method in section 3.3.) Here I is a finite set of subproblem indexes. We will require the vector $\boldsymbol{\theta} = (\theta^\sigma \mid \sigma \in I)$ to be a *ρ-reparameterization* of the original parameter vector $\bar{\theta}$ [8], i.e.

$$\sum_{\sigma \in I} \rho_\sigma \theta^\sigma = \bar{\theta} \tag{14}$$

For each subproblem σ we will define a lower bound $\Phi_\sigma(\theta^\sigma)$ which satisfies

$$\Phi_\sigma(\theta^\sigma) \leq \min_{\boldsymbol{x} \in M} E(\boldsymbol{x} \mid \theta^\sigma) \tag{15}$$

It is easy to see that the function

$$\Phi(\boldsymbol{\theta}) = \sum_{\sigma \in I} \rho_\sigma \Phi_\sigma(\theta^\sigma) \tag{16}$$

is a lower bound on the original function. Indeed, if $\boldsymbol{x}^*$ is an optimal solution of (13) then from (14)-(16) we get

$$\Phi(\boldsymbol{\theta}) \leq \sum_{\sigma \in I} \rho_\sigma \min_{\boldsymbol{x} \in M} E(\boldsymbol{x} \mid \theta^\sigma) \leq \sum_{\sigma \in I} \rho_\sigma E(\boldsymbol{x}^* \mid \theta^\sigma) = E(\boldsymbol{x}^* \mid \bar{\theta})$$

In section 3.2 we will describe the subproblems that we use. For each subproblem σ we will do the following: (1) define constraints on vector θ^σ; (2) define the function $\Phi_\sigma(\theta^\sigma)$; (3) specify an algorithm for computing $\Phi_\sigma(\theta^\sigma)$. In section 3.3 we will discuss how to maximize the lower bound $\Phi(\boldsymbol{\theta})$ using the subproblem solutions and, finally, how to obtain solution $\boldsymbol{x} \in M$ for our original problem.

3.2 Graph Matching Subproblems

Linear subproblem. In our first subproblem, which we denote by the index "L", we require all pairwise terms to be zero: $\theta^L_{ab} = 0$ for $(a,b) \in N$. In such case problem (13) can be solved exactly in polynomial time, for example using the Hungarian algorithm [27]. (This is often known as the *linear assignment problem.*) We define $\Phi_L(\theta^L) = \min_{\boldsymbol{x} \in M} E(\boldsymbol{x} \mid \theta^L)$. To compute this minimum, we converted the problem to an instance of a minimum cost circulation with unit capacities and ran the successive shortest path algorithm [27]. This solves the problem using $O(|P| + |A|)$ Dijkstra shortest path computations in graphs with $|P| + 1$ nodes and $O(|P| + |A|)$ edges.

Maxflow subproblem. In the second subproblem, which we denote by the index "M", we do not put any restrictions on the vector θ^M. To get a lower bound, we ignore the uniqueness constraint $\sum_{a \in A(p)} x_a \leq 1$ and leave only the discreteness constraint: $x_a \in \{0, 1\}$. If the function $E(\boldsymbol{x} \mid \theta^M)$ is submodular (i.e. coefficients θ^M_{ab} are non-positive for all pairwise terms $(a, b) \in N$), then we can compute a global minimum using a maxflow algorithm. With arbitrary θ^M_{ab} the problem becomes NP-hard [28]. We use the *roof duality* relaxation [29] to get a lower bound $\Phi_M(\theta^M)$ on the problem. It can be defined as the optimal value of the following linear program:

$$\Phi_M(\theta^M) = \min \sum_{a \in A} \theta^M_a x_a + \sum_{(a,b) \in N} \theta^M_{ab} x_{ab} \tag{17}$$

$$\text{subject to} \begin{cases} 0 \leq x_a \leq 1 & \forall\, a \in A \\ x_{ab} \leq x_a, \quad x_{ab} \leq x_b, \quad x_{ab} \geq x_a + x_b - 1, \quad x_{ab} \geq 0 & \forall\, (a, b) \in N \end{cases}$$

This relaxation can be solved in polynomial time by computing a maximum flow in a graph with $2(|A| + 1)$ nodes and $O(|A| + |N|)$ edges [28].

Local subproblems. For our last set of subproblems we use an exhaustive search to compute the global minimum (see [30] for details). Thus, we need to make sure that subproblems are sufficiently small. We use the following technique. For each point $p \in P$ we choose $N^d_p \subseteq P$ to be the set of K^d nearest points in the same image where K^d is a small constant, e.g. 2 or 3. (The superscript d stands for "decomposition".) We then consider the subproblem which involves only assignments in the set $A(N^d_p) = \{(p', p'') \in A \mid p' \in N^d_p \vee p'' \in N^d_p\}$ and the edges between those assignments. More precisely, we require vector θ^p corresponding to this subproblem to satisfy the following constraints: (i) $\theta^p_a = 0$ if $a \notin A(N^d_p)$, and (ii) $\theta^p_{ab} = 0$ if $a \notin A(N^d_p)$ or $b \notin A(N^d_p)$. These constraints imply that we can fix assignments $a \in A - A(N^d_p)$ to 0 when computing the minimum $\min_{\boldsymbol{x} \in M} E(\boldsymbol{x} \mid \theta^p)$. Then we get a graph matching problem where the set of points in one of the images is N^d_p.

3.3 Algorithm Summary and Properties of Decomposition

Lower bound optimization. In the previous section we described constraints on vector $\boldsymbol{\theta}$ and a lower bound $\Phi(\boldsymbol{\theta})$ consisting of $|P| + 2$ subproblems. It can be seen Φ is a concave function of $\boldsymbol{\theta}$. Furthermore, the constraints on $\boldsymbol{\theta}$ yield a convex set Ω. This set is defined by the reparameterization equation (14) and constraints on individual subproblems $\theta^\sigma \in \Omega_\sigma$ given by equalities $\theta^\sigma_i = 0$, where index i here may denote either an assignment ($i = a$) or an edge ($i = (a, b)$). Let $I_i \subseteq I$ be the subsets of subproblem indexes for which θ^σ_i is **not** constrained to be 0. Thus, assignment $a \in A$ is involved in subproblems $\sigma \in I_a$, and edge $(a, b) \in N$ is involved in subproblems $\sigma \in I_{ab}$. Similar to [7,9,10,11], we used a projected subgradient method [26,6] for maximizing $\Phi(\boldsymbol{\theta})$ over Ω. One iteration is given by $\theta := \mathcal{P}_\Omega(\boldsymbol{\theta} + \lambda \mathrm{g})$ where $\mathcal{P}_\Omega$ is the operator that projects a vector to Ω, g is a subgradient of $\Phi(\boldsymbol{\theta})$ and $\lambda > 0$ is a step size.

Projection. To project vector $\boldsymbol{\theta}$ to Ω, we first compute vector $\hat{\theta} = \sum_\sigma \rho_\sigma \theta^\sigma$ and then update $\boldsymbol{\theta}$ as follows: $\theta_i^\sigma := 0$ for $\sigma \in I - I_i$, and

$$\theta_i^\sigma := \theta_i^\sigma + \rho_\sigma \frac{\bar{\theta}_i - \hat{\theta}_i}{\sum_{\sigma' \in I_i} \rho_{\sigma'}^2} \qquad \forall \sigma \in I_i$$

Subgradient computation. A subgradient of function $\Phi(\boldsymbol{\theta})$ is given by $\mathbf{g} = \sum_{\sigma \in I} \rho_\sigma \mathbf{g}^\sigma$ where $\mathbf{g}^\sigma$ is a subgradient of function $\Phi_\sigma(\theta^\sigma)$. If the latter function is the global minimum of $E(\boldsymbol{x} \mid \theta^\sigma)$ (which is the case for $\sigma \in I - \{M\}$) then we can take $g_a^\sigma = x_a^\sigma$, $g_{ab}^\sigma = x_a^\sigma x_b^\sigma$ where $\boldsymbol{x}^\sigma$ is a global minimizer of $E(\boldsymbol{x} \mid \theta^\sigma)$. For the maxflow subproblem a subgradient can be computed as $\mathbf{g}^M = \boldsymbol{x}^M$ where $\boldsymbol{x}^M$ is an optimal solution of linear program (17). The method in [31] produces a half-integer optimal solution where $x_a^M \in \{0, 0.5, 1\}$ for all assignments a and x_{ab}^M is determined as follows: if $(x_a^M, x_b^M) \neq (0.5, 0.5)$ then $x_{ab}^M = x_a^M x_b^M$, otherwise $x_{ab}^M = 0$ if $\theta_{ab}^M \leq 0$ (i.e. the corresponding term is submodular) and $x_{ab}^M = 0.5$ if $\theta_{ab}^M > 0$.

Solution computation. To conclude the description of the method, we need to specify how to obtain solution $\boldsymbol{x} \in M$. We compute the solution in each iteration as follows: starting with labeling $\boldsymbol{x} = 0$, we go through local subproblems $\sigma \in I - \{L, M\}$ and assignments a involved in σ (in a fixed order), set $x_a = 1$ if $x_a^\sigma = 1$ and this operation preserves the uniqueness constraint on $\boldsymbol{x}$. (Here $\boldsymbol{x}^\sigma$ denotes a global minimum of subproblem σ.) We maintain the solution with the smallest energy computed so far, and output it as a result of the method.

Further implementation details, e.g. the choice of the step size λ, are given in [30].

Properties of decomposition. It is not necessary to use all subproblems described in section 3.2. The only requirement is that each assignment $a \in A$ and edge $(a, b) \in N$ must be covered by at least one subproblem. In [30] we discuss how the choice of subproblems affects the optimal bound achievable with the decomposition method. Due to lack of space, here we give only a summary: (i) the linear subproblem is not essential with our choice of local subproblems; (ii) the maxflow subproblem is not essential if $K \leq K^d$; (iii) our lower bound is the same as or tighter than the bound of the roof duality approach [28] applied to an equivalent *quadratic pseudo-boolean optimization* formulation of problem (13).

4 Experimental Results

In most of our experiments we learned problem-specific parameters of our energy model from ground truth correspondences. We applied Nonlinear Inverse Optimization [20] (NIO) to learn non-negative parameters $\{\lambda^{\mathrm{app}}, \lambda^{\mathrm{occl}}, \lambda^{\mathrm{geom}}, \lambda^{\mathrm{coh}}, \eta, \sigma_l^2, \sigma_\alpha^2\}$. We used DD within NIO to optimize the learning objective (see [30] for details). The learning procedure was initialized using default parameters corresponding to uniform values for the weights $\{\lambda^{\mathrm{app}}, \lambda^{\mathrm{occl}}, \lambda^{\mathrm{geom}}, \lambda^{\mathrm{coh}}\}$, $\eta = 0.5$, and variance values $\sigma_l^2 = 0.5$, $\sigma_\alpha^2 = 0.9$.

In our experiments we compare the following algorithms:
DD. We used $K^d = \min\{K, 4\}$, where K is the geometric neighborhood size. Motivated by results in sec. 3.3, we did not use the linear subproblem. We set $\rho_\sigma = 1$ for all other subproblems σ. We used a maximum of 10000 iterations, and stopped earlier if the gap between the lower bound and the cost became smaller than 10^{-6}.

FUSION. This technique was introduced in [32] for MRF optimization with multiple labels. We propose to use it for graph matching as follows. First, we generate 256 solutions by applying one pass of coordinate descent (ICM) to zero labeling using random orders. (Different orders of visiting assignments usually yield different solutions.) We then "fuse" together pairs of solutions using the binary tree structure until a single solution remains. Fusion of solutions $\boldsymbol{x}'$, $\boldsymbol{x}''$ is defined as follows. First, we fix all assignments $a \in A$ for which $\boldsymbol{x}'$ and $\boldsymbol{x}''$ agree, i.e. $x'_a = x''_a$. Then we convert the obtained graph matching problem to a quadratic pseudo-boolean optimization problem (see [30] for conversion details). Finally, we run the QPBO-PI method [33] starting either with labeling $\boldsymbol{x}'$ if $E(\boldsymbol{x}' \mid \bar{\theta}) < E(\boldsymbol{x}'' \mid \bar{\theta})$ or with $\boldsymbol{x}''$ otherwise. The produced solution $\boldsymbol{x}$ is guaranteed to have the same or smaller cost than the costs of $\boldsymbol{x}'$ and $\boldsymbol{x}''$.

BP. We converted graph matching to a quadratic pseudoboolean optimization problem and ran max-product belief propagation algorithm[2]. We also tested applying the roof duality approach instead of BP, but results were quite discouraging (see details in [30]).

SMAC. We ran the spectral relaxation method of Cour et al. [12], using the graduated assignment algorithm [22] for discretization. Since SMAC imposes affine constraints on the solution, we applied this algorithm only to datasets without outliers, where the one-to-one affine constraint is satisfied. In principle, SMAC could handle outliers by the introduction of dummy nodes. However, this would increase the number of variables and potentially make the problem harder to solve.

COMPOSE. We reimplemented the algorithm in [13]. The problem was cast as assigning a label from the set $A(p') \cup \{``occlusion''\}$ to each point $p' \in P'$. Min-marginals for the linear subnetwork were computed via $O(|A| + |P'|)$ calls to the Dijkstra algorithm. As in [13], we used Residual Belief Propagation (RPB) [34] with damping=0.3 for computing pseudo min-marginals for the "smoothness" subnetwork containing pairwise terms $\theta_{ab}x_a x_b$. However, in our experiments messages did not converge, so we set an additional termination criterion for RBP: we stop it after passing $20|N|$ messages. As in [13], we computed the configuration by looking at individual messages at each node. We did not use damping for the outer loop since otherwise the produced configurations usually did not satisfy the uniqueness constraint.

HUNG. As in [5], we also tested the Hungarian algorithm using an energy consisting only of linear terms. On problems with occlusions, we used our occlusion cost in addition to the appearance energy term, i.e. $E^{\text{HUNG}}(\boldsymbol{x}) = \lambda^{\text{app}} E^{\text{app}}(\boldsymbol{x}) + \lambda^{\text{occl}} E^{\text{occl}}(\boldsymbol{x})$.

4.1 Comparative Results

Hotel sequence: wide baseline matching. We first demonstrate our approach on the CMU 'hotel' sequence[3]. We used as features the same manual labeling of 30 landmark

[2] We used the code from http://www.adastral.ucl.ac.uk/~ vladkolm/papers/TRW-S.html
[3] Available at: http://vasc.ri.cmu.edu/idb/html/motion/hotel/index.html.

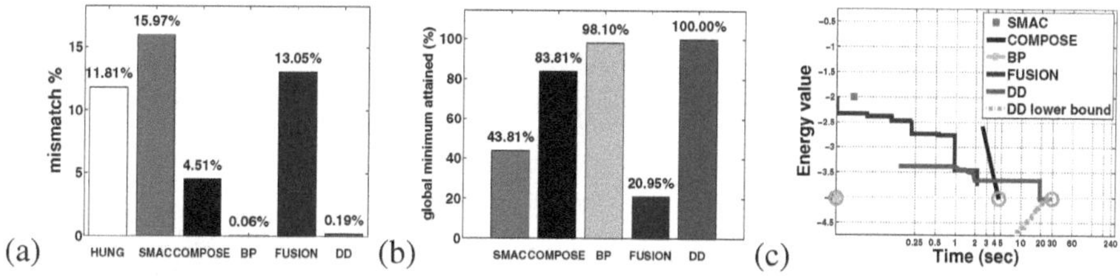

Fig. 1. Results on the Hotel sequence ($|P'| = |P''| = 30, |A| = 900$). (a) Mismatch percentages of HUNG and different optimizations applied to our energy model. (b) Frequency of convergence to global minimum. (c) Energy minimization versus time.

points employed in [5]. Since these points are visible in all frames and the motion is rigid, this matching problem is relatively simple. We include this dataset in our experiments as it was used in [5] to compare the performance of graph matching methods. We reproduce the experimental setup described in [5] using a subset of 105 frame pairs and adopting distances between Shape Context descriptors as unary terms. However, we replace the pairwise terms proposed in [5], with our geometric energy function $E^{\text{geom}}(\boldsymbol{x})$, using $K = 2$. Due to the absence of outliers, we remove $E^{\text{coh}}(\boldsymbol{x})$ from our energy and use a large constant value for λ^{occl}. We set the remaining parameters to default values, as defined above. We set $A = P' \times P''$. Figure 1(a) shows the matching error obtained by optimizing this model with different methods. We include in the plot also the performance of HUNG. Here BP and DD perform dramatically better than the other methods, with errors approaching 0%. Note that the error of our system is over 50 times smaller than the errors reported in [5]. On this dataset DD found always the global minimum within a minute (see Figure 1(b)). Figure 1(c) illustrates performance versus runtime on one image pair (frame 1 and 64). In this plot we indicate convergence to a global minimum with a green circle. BP does well on this sequence, nearly matching the minimization performance of DD, at a reduced cost. We also implemented the energy function described in [5]. This model uses Delaunay triangulation to define the graph topology in each image, and employs binary edge compatibility values in $\{0,1\}$. DD provided the best performance with this model, with a 5.7% error. This suggests that both our model and our optimization contribute to the improvement over the results reported in [5] where the best system had a matching error above 10%.

Matching MNIST digits. Here we describe experiments on images of handwritten digits from the MNIST dataset [35]. For training, we randomly sampled from this dataset one image pair for each of the 10 digits. We repeated the same procedure to generate a test set of 10 pairs of same digits. From each pair we extracted point sets P' and P'' by uniformly sampling 100 points along the Canny edges of each image, using the procedure described in [14]. We defined the unary potentials $\theta^{\text{app}}_{(p',p'')}$ to be the Euclidean distances between Shape Context descriptors computed at points p', p''. We formed the set of candidate assignments $A \in P' \times P''$ by selecting the 5 most similar features, in terms of Shape Context distance, for each point $p \in P$. We collected ground truth correspondences in the set $(P' \times P'')$ for each of the 20 image pairs. The parameters of our model were learned from the 10 training image pairs with NIO. Figure 2(a) shows that

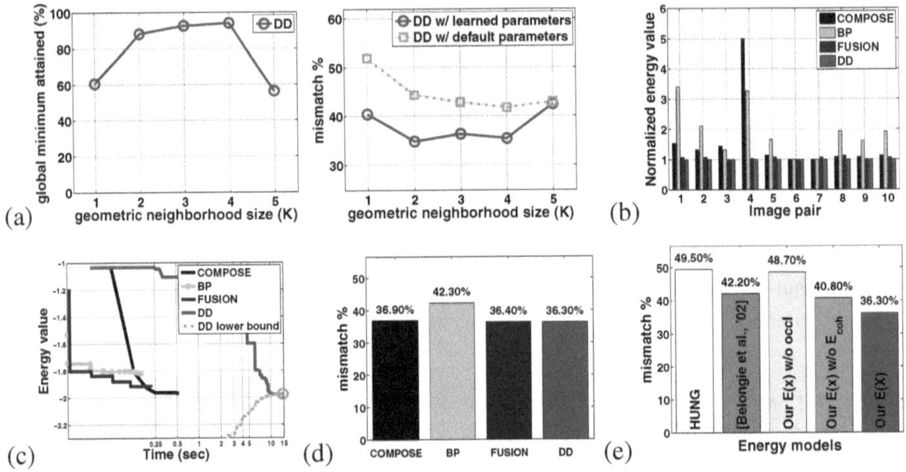

Fig. 2. Experimental results on MNIST digits ($|P'| = |P''| = 100, |A| = 695$, on average). (a) Correlation between learning accuracy and matching performance: the left plot shows the frequency of global minimum convergence during learning versus K; the right plot shows mismatch error on test set. (b) Normalized energy values. (c) Optimization performance versus runtime. (d) Mismatch error comparison between different optimization methods using our energy model. (e) Mismatch error using different energy models.

the matching accuracy on the test set critically depends on the ability to globally optimize the energies during the model learning stage. The left plot reports the frequency of convergence to a global minimum during learning, plotted as a function of K, the geometric neighborhood size. The second plot shows the test set matching error of DD with learned versus default parameters. Matching error here is measured as percentage of incorrect correspondences (as defined in [30]). We can see that the matching is much more accurate when using the parameters for which DD reached more frequently global optimality during learning. Interestingly, although the frequency of global minimum convergence increases slightly when varying K from 2 to 4, the matching error remains roughly the same. This suggests that geometric penalty terms defined over small neighborhoods are sufficient to spatially regularize the correspondences. Thus, models involving geometric costs defined over all pairs of matched features, such as those used in [17,18], may be unnecessarily restrictive for many applications, in addition to being more difficult to optimize.

Given these results, we have used the model learned with $K = 3$ for the MNIST experiments described below. Figure 2(b) shows the normalized energy values obtained by different optimization methods on the test set. For each family of results we performed an *additive* normalization so that for each image pair the energy of the best method becomes a fixed number. On 9 out of the 10 test image pairs, DD reaches global optimality, and provides the minimum energy value on all examples. FUSION, BP, and COMPOSE find the global minimum only on 2 images. FUSION finds solutions with energy values very close to those obtained by DD. COMPOSE and BP provide considerably higher

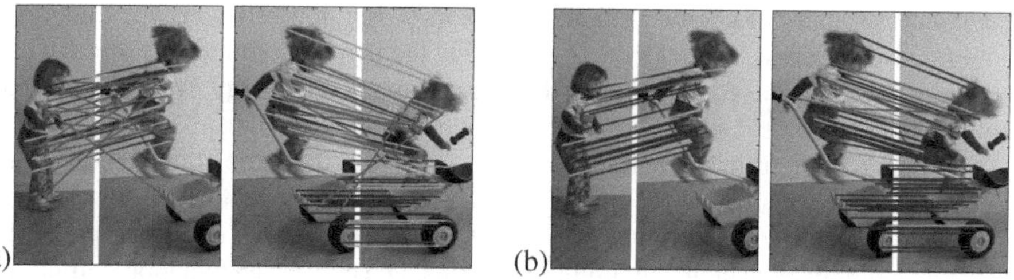

Fig. 3. Estimating human motion ($|P'| = 118$, $|P''| = 172$, $|A| = 1128$ on average). Correspondences computed with (a) the Hungarian method and (b) DD. Correct correspondences are shown in blue, missed assignments in green, and mismatches in red.

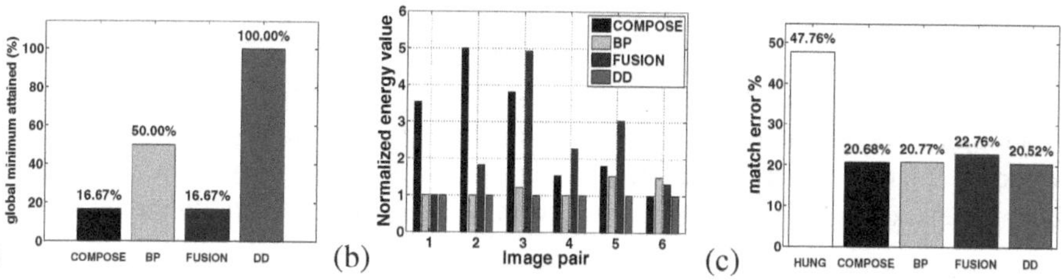

Fig. 4. Experiments on human motion frames. (a) Frequency of convergence to global minimum. (b) Normalized energy values. (c) Correspondence error.

energy values on some of the examples. Figure 2(c) shows minimization performance as a function of time, evaluated on a sample image pair. Figure 2(d) shows the correspondence accuracy obtained by optimizing our energy with the different methods. Again, we find that DD and FUSION yield the best accuracy. We also evaluated variations of the energy model defined in Equation (2) obtained by using only the linear tearms (HUNG), by dropping the spatial coherence term, and by forcing all points to be matched (implemented by fixing λ^{occl} to a large value). The parameters of these modified models were learned again with NIO, using DD for both training and testing. We see from Figure 2(e) that both the spatial coherence prior, as well as the occlusion cost, improve the matching accuracy. On these instances the simple appearance-based model used by HUNG gives poor accuracy. We also report the matching error given by the model and optimization method of Belongie et al., which was applied to MNIST digit examples in [14]. Our approach performs better than this state-of-the-art method.

Estimating long range non-rigid motion. In this subsection we describe results on the task of estimating large-disparity motion. For this experiment we used four (time-separated) video frames of a child jumping. We matched each image to every other image, for a total of six matches. The motion between any pair of these pictures is very large and highly non-rigid. There is self-occlusion created by the motion of arms and torso, and occlusion due to a tricycle positioned between the child and the camera. Feature points were extracted by running the Harris corner detector on each image. We used

Euclidean distances of geometric blur descriptors [17] computed at each feature point, both for selecting assignments in A (by choosing the five most similar features for each point $p \in P$) as well as for calculating the unary terms of our energy. We learned the parameters in our model by applying the NIO algorithm to ground truth correspondences of two image pairs from a separate sequence containing the same child walking. Here we report results using $K = 6$. Figure 3 shows two matching examples from this experiment and correspondences found with HUNG and DD. Note the ability of our system to cope well with occlusion and multiple motions. DD converged to a global minimum on all the image pairs in this experiment (see Figure 4(a,b)). Figure 4(c) reports the correspondence errors (including mismatches as well as missed assignments).

Additional results and experiments on another dataset are given in [30].

5 Conclusions

We have introduced novel models and optimization algorithms for feature correspondence. We believe to be the first to demonstrate graph matching techniques capable of reaching global optimality on various real-world image matching problems. As a future work, we plan to replace exhaustive search for local subproblems with a branch-and-bound method, as in [7]. We hope that this may speed up substantially the DD method.

Acknowledgements. We are grateful to Timothee Cour, Praveen Srinivasan, and Jianbo Shi for providing the software of their SMAC algorithm. We thank John Duchi, Gal Elidan and Danny Tarlow for sharing code and answering questions about the COMPOSE method. Thanks to Tiberio Caetano for providing the Hotel feature data.

References

1. Belhumeur, P.N.: A binocular stereo algorithm for reconstructing sloping, creased, and broken surfaces in the presence of half-occlusion. In: ICCV (May 1993)
2. Kolmogorov, V., Zabih, R.: Computing visual correspondence with occlusions using graph cuts. In: ICCV (2001)
3. Dorko, G., Schmid, C.: Selection of scale-invariant parts for object class recognition. In: ICCV (2003)
4. Sivic, J., Russell, B., Efros, A., Zisserman, A.: Discovering object categories in image collections. In: ICCV (2005)
5. Caetano, T.S., Cheng, L., Le, Q.V., Smola, A.J.: Learning graph matching. In: ICCV (2007)
6. Bertsekas, D.: Nonlinear Programming. Athena Scientific (1999)
7. Chardaire, P., Sutter, A.: A decomposition method for quadratic zero-one programming. Management Science 41(4), 704–712 (1995)
8. Wainwright, M.J., Jaakkola, T.S., Willsky, A.S.: MAP estimation via agreement on trees: Message-passing and linear-programming approaches. IEEE Trans. Information Theory 51(11) (2005)
9. Schlesinger, M.I., Giginyak, V.V.: Solution to structural recognition (MAX,+)-problems by their equivalent transformations. Part 1. Control Systems and Computers (1), 3–15 (2007)
10. Schlesinger, M.I., Giginyak, V.V.: Solution to structural recognition (MAX,+)-problems by their equivalent transformations. Part 2. Control Systems and Computers (2), 3–18 (2007)

11. Komodakis, N., Paragios, N., Tziritas, G.: MRF optimization via dual decomposition: Message-passing revisited. In: ICCV (2007)
12. Cour, T., Srinivasan, P., Shi, J.: Balanced graph matching. In: NIPS (2007)
13. Duchi, J., Tarlow, D., Elidan, G., Koller, D.: Using combinatorial optimization within max-product belief propagation. In: NIPS (2007)
14. Belongie, S.J., Malik, J., Puzicha, J.: Shape matching and object recognition using shape contexts. PAMI 24(4), 509–522 (2002)
15. Torr, P.H.S.: Geometric motion segmentation and model selection. Philosophical Transactions of the Royal Society, 1321–1340 (1998)
16. Sclaroff, S., Pentland, A.: Modal matching for correspondence and recognition. PAMI 17(6), 545–561 (1997)
17. Berg, A., Berg, T., Malik, J.: Shape matching and object recognition using low distortion correspondence. In: CVPR (2005)
18. Leordeanu, M., Hebert, M.: A spectral technique for correspondence problems using pairwise constraints. In: ICCV (2005)
19. Torr, P.H.S.: Solving Markov random fields using semi definite programming. In: AISTATS (2003)
20. Liu, C.K., Hertzmann, A., Popović, Z.: Learning physics-based motion style with nonlinear inverse optimization. ACM Trans. on Gr. 24(3), 1071–1081 (2005)
21. Conte, D., Foggia, P., Sansone, C., Vento, M.: Thirty years of graph matching in pattern recognition. Int. J. of Pattern Recognition and Artificial Intelligence 18(3), 265–298
22. Gold, S., Rangarajan, A.: A graduated assignment algorithm for graph matching. PAMI 18(4), 377–388 (1996)
23. Maciel, J., Costeira, J.: A global solution to sparse correspondence problems. PAMI 25(2), 187–199 (2002)
24. Cabot, A., Francis, R.: Solving certain nonconvex quadratic minimization problems by ranking the extreme points. Operations Research 18(1), 82–86 (1970)
25. Schellewald, C., Schnörr, C.: Probabilistic subgraph matching based on convex relaxation. In: EMMCVPR (2005)
26. Shor, N.Z.: Minimization methods for nondifferentiable functions. Springer, Heidelberg (1985)
27. Ahuja, R., Magnanti, T., Orlin, J.: Network Flows: Theory, Algorithms, and Applications. Prentice Hall, Englewood Cliffs (1993)
28. Boros, E., Hammer, P.: Pseudo-boolean optimization. Discr. Appl. Math. 123(1-3) (2002)
29. Hammer, P.L., Hansen, P., Simeone, B.: Roof duality, complementation and persistency in quadratic 0-1 optimization. Mathematicl Programming 28, 121–155 (1984)
30. Torresani, L., Kolmogorov, V., Rother, C.: Feature correspondence via graph matching: Models and global optimization. Technical Report MSR-TR-2008-101 (2008)
31. Boros, E., Hammer, P.L., Sun, X.: Network flows and minimization of quadratic pseudo-Boolean functions. Technical Report RRR 17-1991, RUTCOR (May 1991)
32. Lempitsky, V., Rother, C., Blake, A.: LogCut - efficient graph cut optimization for Markov random fields. In: ICCV (2007)
33. Rother, C., Kolmogorov, V., Lempitsky, V., Szummer, M.: Optimizing binary MRFs via extended roof duality. In: CVPR (2007)
34. Elidan, G., McGraw, I., Koller, D.: Residual belief propagation: Informed scheduling for asynchronous message passing. In: UAI (2006)
35. LeCun, Y., Bottou, L., Bengio, Y., Haffner, P.: Gradient-based learning applied to document recognition. Proceedings of the IEEE 86(11), 2278–2324 (1998)

Event Modeling and Recognition Using Markov Logic Networks*

Son D. Tran and Larry S. Davis

Department of Computer Science
University of Maryland, College Park, MD 20742 USA
{sontran,lsd}@cs.umd.edu

Abstract. We address the problem of visual event recognition in surveillance where noise and missing observations are serious problems. Common sense domain knowledge is exploited to overcome them. The knowledge is represented as first-order logic production rules with associated weights to indicate their confidence. These rules are used in combination with a relaxed deduction algorithm to construct a network of grounded atoms, the Markov Logic Network. The network is used to perform probabilistic inference for input queries about events of interest. The system's performance is demonstrated on a number of videos from a parking lot domain that contains complex interactions of people and vehicles.

1 Introduction

We consider the problem of event modelling and recognition in visual surveillance and introduce an approach based on Markov Logic Networks ([1]) that naturally integrates common sense reasoning with uncertain analyses produced by computer vision algorithms for object detection, tracking and movement recognition. We motivate and illustrate our approach in the context of monitoring a parking lot, with the goal of matching people to the vehicles they arrive and depart in.

There are numerous frameworks for event recognition. In declarative approaches (e.g. [2]), events are represented with declarative templates. Events are typically organized in a hierarchy, starting with primitive events at the bottom and composite events on top. The recognition of a composite event proceeds in a bottom-up manner. These approaches have several drawbacks. First, a miss or false detection of a primitive event, which occurs frequently in computer vision, especially in crowded or poorly illuminated conditions, often leads to irrecoverable failures in composite event recognition. Second, uncertainty is often not modelled and so these methods are generally not robust to typical errors in image analysis. In probabilistic frameworks, such as HMMs (e.g. [3]), or DBNs ([2]), events are represented with probabilistic models. Event recognition is usually performed using maximum likelihood estimation given observation sequences. While these approaches provide robustness to uncertainty in image analysis, their representations often lack flexibility (e.g. number of states or actors is fixed) and hence it is difficult to use them in dynamic situations.

* This research was funded in part by the U.S. Government VACE program.

D. Forsyth, P. Torr, and A. Zisserman (Eds.): ECCV 2008, Part II, LNCS 5303, pp. 610–623, 2008.

In general, the problems of noise or missing observations always exist in real world applications. Our contention is that common sense knowledge, specific to the domain under consideration, can provide useful constraints to reduce uncertainties and ambiguities. Having a good knowledge base (KB) and an effective reasoning scheme helps to improve event recognition performance.

Technically, we address uncertainty in observations and representational richness of event specification by a combination of logical and probabilistic models.

1. Domain common sense knowledge is represented using first order logic statements. Both negation and disjunction are allowed.
2. Uncertainty of primitive event detection is represented using detection probabilities. Uncertainty of logical relations (including event models or logical constraints) is represented with a real-valued weight set based on, for example, domain knowledge.
3. Logical statements and probabilities are combined into a single framework using Markov Logic Networks (MLN, [1]).

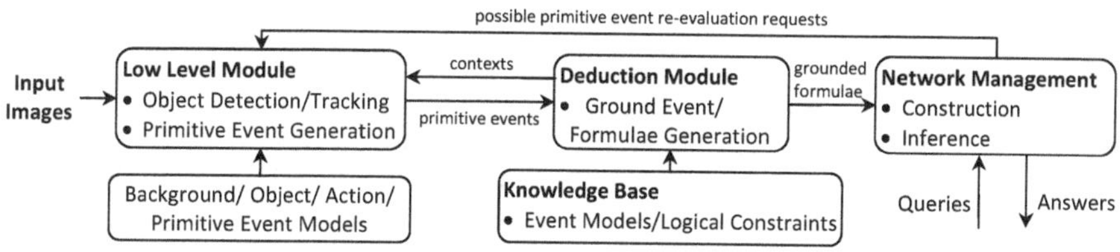

Fig. 1. Overview of our system

Our system maintains an undirected network of grounded atoms which correspond to events that have occurred in the video. At any moment, primitive events are detected with associated detection probabilities. They are then used to ground logical rules in the KB, which generally leads to generating more grounded events. Next, these grounded logical rules are added to the Markov network. The network parameters or structures are revised with these updates. The marginal probability of any (composite) event can be determined using probabilistic inference on this network. Fig. 1 shows an overview of our system.

2 Related Work

Visual event detection from video has a long history in computer vision. We review here only approaches that are most relevant to ours. Logic has been used for visual event recognition in a number of works. In [4], Rota et al presented an elegant treatment for representing activities using declarative models. Recognition was performed effectively using a constraint-satisfaction algorithm. In [5], Shet et al used a multi-valued default logic for the problem of identity maintenance. Default reasoning was conducted on a bi-lattice of truth values with prioritized default rules. Identity maintenance rules are prioritized mainly based on domain knowledge. A continuous bi-lattice was used in [6] for human detection. Here, instead of using a multi-valued logic, we use a combination of logic and probability to handle inexact inference including identity maintenance. Each level of prioritization can be mapped to our framework using a rule weight.

The combination of probability and (first order) logic has been pursued extensively in AI and led to the emergence of Statistical Relation Learning (SRL, [7]). SRL representations often involve unrolled or grounded graphical models (directed (Bayesian) as well as undirected (Markov) ones), which are constructed using a frame-based or a logic-based approach. They have been used for human activity recognition (although not in vision-based systems). In [8], Liao et al recognized human activities based on the information about locations they visited provided by GPS sensors worn by users (location-based). Probabilistic inference is performed on an unrolled Markov network formed from a Relational Markov Network, which essentially encodes high-level domain knowledge. Relation weights can be learned using a MAP estimation technique. In [9], Pentney et al recognized human activities based on objects used. The objects were RFID tagged and identified using RFID readers worn by the users (object-use based). Logical rules are grounded and linked to form a probabilistic network within a single time slice. In general, these approaches are intrusive. They require users to wear additional sensors. Therefore their application to general surveillance tasks is limited. Here, we work with visual input and use common sense knowledge to complement limitations in visual perception. Markov Logic Networks were used to construct a DBN for activity recognition by Biswas et al in [10]. That work only addressed inaccuracy in logic statements, while ours addresses a wider range of issues, including detection uncertainty, missing observations and identity maintenance.

Approaches that are based on probabilistic grammars for event recognition such as [11] typically use simpler rules than ours. For example, they do not allow existential quantifiers, which are needed for dealing with missing observations (Section 5.2). It is also difficult to express domain constraints such as "a car can only be driven by one person" using generative grammars. Furthermore, methods to perform probabilistic propagation are better understood for graphical models than for probabilistic grammars.

3 Sample Problem

We motivate our approach with the surveillance problem of monitoring a parking lot and determining which people enter or leave in which cars (Fig. 2). In a parking lot, cars of various shapes and sizes can park close together. Occlusion is not only unavoidable but sometime severe. This leads to many difficulties in tracking people, since their corresponding foreground blobs may change from complete to fragmented or become totally missing as they move between parked cars. It is often difficult to determine the exact moment a person enters a car, or even which car a person enters. Pure declarative, bottom-up approaches (e.g. [2], [4]) that rely on accurate capture of primitive events will not work well here. Probabilistic action recognition (e.g. HMMs([3])) might fail as well since, locally, an observation may be missing altogether. For more robust event recognition, we

Fig. 2. A frame from a parking lot sequence and its corresponding foreground regions detected using background subtraction. Here, parked cars introduce significant occlusion and door openings lead to many false alarms.

propose to use common sense knowledge about the domain under consideration. The knowledge base will contain rules that range from definite ones such as "if a car leaves, there must exist some person driving it" or "a person can drive only one car at any time" to weaker rules such as "people walking together usually enter the same car" or "if a person puts a bag into the trunk of a car, he or she is likely to enter that car". We represent these rules using first order logic augmented with probabilities to represent the degree of confidence for each rule. Additionally, the recognition of actions or primitive events such as "walking together" or "put a bag into a car" is uncertain. Probability theory provides a convenient method to handle these also. We then need an approach that combines logic and probabilistic elements in a coherent framework.

Briefly, we achieve this by using 1) first order logic formulae to represent domain knowledge, 2) a real-valued weight to represent the confidence in each logic rule, 3) probability to model uncertainty for primitive event and action recognition, 4) a probabilistic logic network, namely the Markov Logic Network, to connect (detected) ground atoms and to perform probabilistic inference (e.g. determine the probability that a person enters some car, given the input sequences). The following sections discuss in detail these aspects for our particular surveillance problem as well as for general surveillance contexts.

4 Background on Markov Logic Networks

Markov Logic Networks (MLN, [1]) are one type of the unrolled graphical models developed in SRL([7]) to combine logical and probabilistic reasoning. In MLN, every logic formula F_i is associated with a nonnegative real-valued weight w_i. Every instantiation of F_i is given the same weight. An undirected network, called a Markov Network, is constructed such that,

- Each of its nodes correspond to a ground atom x_k.
- If a subset of ground atoms $x_{\{i\}} = \{x_k\}$ are related to each other by a formula F_i, then a clique C_i over these variables is added to the network. C_i is associated with a weight w_i and a feature f_i defined as follows

$$\begin{aligned} f_i(x_{\{i\}}) &= 1, \text{ if } F_i(x_{\{i\}}) \text{ is true}, \\ &= 0, \text{ otherwise} . \end{aligned} \tag{1}$$

Thus first-order logic formulae in our knowledge base serve as templates to construct the Markov Network. This network models the joint distribution of the set of all ground atoms, X, each of which is a binary variable. It provides a means for performing probabilistic inference.

$$P(X = x) = \frac{1}{Z}\exp(\sum_i w_i f_i(x_{\{i\}})). \tag{2}$$

where Z is the normalizing factor, $Z = \sum_{X \in \mathbf{X}} \exp(\sum_i w_i f_i(x_{\{i\}}))$. If $\phi_i(x_{\{i\}})$ is the potential function defined over a clique C_i, then $log(\phi_i(x_{\{i\}})) = w_i f_i(x_{\{i\}})$.

Inference. Based on the constructed Markov network, the marginal distribution of any event given some evidence (observations) can be computed using probabilistic inference. Since the structure of the network may be very complex (e.g. containing undirected cycles), exact inference is often intractable. MCMC sampling is a good choice for approximate reasoning ([1]). In MLN, the probability that a ground atom X_i is equal to x_i given its Markov blanket (neighbors) B_i is

$$P(X_i = x_i | B_i = b_i) = \frac{\exp(\sum_{f_j \in F_i} w_j f_j(X_i = x_i, B_i = b_i))}{\exp(\sum_{f_j \in F_i} w_j f_j(X_i = 0, B_i = b_i)) + \exp(\sum_{f_j \in F_i} w_j f_j(X_i = 1, B_i = b_i))} \tag{3}$$

where F_i is the set of all cliques that contain X_i and f_j is computed as in Eq. 1.

Basic MCMC (Gibb sampling) is known to have difficulty dealing with deterministic relations, which are unavoidable in our case. It has been observed that using simulated tempering ([12]) gives better performance than the basic Gibb sampling ([12]). Simulated tempering is a MC method that is closely related to simulated annealing. However, instead of using some fixed cooling schedule, a random walk is also performed in the temperature space whose structure is predetermined and discrete ([12]). These moves aim at making the sampling better at jumping out of local minima.

5 Knowledge Representation

In this section, we will describe our approach to represent knowledge and its associated uncertainty. In our framework, object states and their interactions (including the so-called events, actions or activities as they are interchangeably referred to in other work, e.g. [2], [4]) are all represented with first order logic predicates. A predicate is intensional if its truth value (for a certain grounding of its arguments) can only be inferred (i.e. cannot be directly observed) ([13]). A predicate is extensional if its truth value can be directly evaluated by a low-level vision module. It is strictly extensional if this is the only means to evaluate it (i.e. it can only be observed and not inferred).

5.1 Logical Representation

In [1], the Markov network is constructed using an exhaustive grounding scheme, which can lead to an explosion in the number of ground atoms and network connections. Most of them are irrelevant and create significant difficulties for inference. A more efficient scheme was proposed in [14], which essentially grounded only clauses that can become unsatisfied using a greedy search. It is not clear if this approach could handle dynamic domains that involve, for example, time and location. Here, we represent our knowledge in the form of production rules, $production \rightarrow conclusion$, and use deduction to ground (and add to the Markov network) only literals (including both positive and negative atoms) that are possibly true.

In traditional deductive systems (e.g. [13]), production rules in the form of Horn clauses are used extensively. However, Horn clauses cannot represent negations and disjunctions, which are often required to capture useful commonsense knowledge. To increase our system's representational ability, we allow the following rule forms,

$(\bigwedge_i a_i) \rightarrow b$ Definite (i.e. Horn) clauses are used to define a composite event from sub-events (similar, for example, to multi-thread event definition in [2]), causal and explanatory relationships between observations and underlying actions (e. g. $use(Bowl) \rightarrow make(Cereal)$ or $at(Resstautrant) \rightarrow have(Dinner)$ in object-use based [9] and location-based frameworks [8])

$(\bigwedge_i a_i)(\bigwedge_j \neg b_j) \rightarrow c$ Many events can only be described with a rule that has negative preconditions, for example, $at(C, S, t) \wedge \neg stopped(C, t) \rightarrow violate(C, S, t)$ where C is a car and S is a stop sign. Identity maintenance ([5], [15]) also often leads to formulae with negative preconditions, for example, $own(H_1, Bag) \wedge take(H_2, Bag) \wedge \neg eq(H_1, H_2) \rightarrow theft(H_2, Bag)$.

$(\bigwedge_i a_i) \rightarrow \neg b$ This form is often used to describe an exclusion relation. For example, the rule "a person P belongs to only one group G" can be written as $belongto(G_1, P) \wedge \neg eq(G_1, G_2) \rightarrow \neg belongto(G_2, P)$.

$(\bigwedge_i a_i) \rightarrow (\bigvee_j b_j)$ Disjunctions are used when a single conclusion cannot be made. For example, $use(Cup) \rightarrow (drink(Coffee) \vee drink(Tea))$. When it fires, all atoms in the conclusion are added to the ground atom database. Disjunctions also arise from existential quantifiers (next section).

These forms, of course, are not the most general ones in First Order Logic. However, practically, they are sufficiently rich to represent a wide range of common sense knowledge and to capture complex events in surveillance domains.

5.2 Uncertainty Representation

Uncertainty is unavoidable in practical visual surveillance applications. We consider two classes of uncertainty: logical ambiguity and detection uncertainty. Their sources and ways to represent them are described below.

Incomplete or Missing Observations. Occlusion and bad imaging conditions (e.g. dark, shadowed areas of the scene) are two common conditions that prevent us from observing the occurrence of some actions. In some cases, even if a unique conclusion cannot be made, some weaker (disjunctional) assertion might still be possible. Rules with disjunctive effects are often needed then. For example, the statement "if a bag b is missing at some time interval t and location L, then someone must have picked it up" could be formalized as $missing(b, l, t) \rightarrow (\exists p\ passBy(p, l, t) \wedge pickUp(p, b, t))$. Here the action $pickUp(p, b, t)$ can be inferred when its direct detection is missed. This type of formulae involves an existential quantifier and will be expanded to a disjunction of conjunctive clauses when grounded. For example, suppose that $passBy(P_1, L, T)$ and $passBy(P_2, L, T)$ are true for two persons P_1 and P_2 (i.e. two persons P_1 and P_2 passed by when the bag went missing), then the grounding of this rule would be $missing(B, L, T) \rightarrow (passBy(P_1, L, T) \wedge pickUp(P_2, B, T)) \vee (passBy(P_2, L, T) \wedge pickUp(P_2, B, T))$. This expansion obviously is not suitable for infinite domains. However, in practice, most object domains are finite (e.g. number of people or cars is finite) therefore the expansion is feasible for surveillance. As evidence arrives, previously expanded domains may need to be updated (section 6.1).

Non-perfect Logical Statements. Common sense statements in the KB are not always true. We use a real-value weight to represent the confidence of each rule in the KB. Rules with absolute certainty, such as "a person can drive only one car at a time", are given an infinite weight. In practice, such a hard clause is "softened" with a maximum weight, $MAXW$, to facilitate the inference process. Rules that are almost always true, such as "a person interacts with only one car", are given strong weights. Weak weights are assigned to rules that describe exceptions (i.e. situations that are possibly true but not common such as "a driver might enter a car from the passenger side").

Extensional Evaluation Uncertainty. The evaluation of an extensional predicate, E, by the low-level vision module might return answers with absolute certainty or with some associated (detection) probability, $p_D(E = true)$. For the first case, whether the result is $true$ or $false$, we make E an evidence variable and add it to the Markov network. For the second case, a method to integrate E and its detection probability for high-level logical reasoning are needed.

One approach would be to add this grounded, single-atom clause, $(E, w \propto p_D)$ and its complement, $(E, w \propto 1 - p_D)$ to the Markov network. (Note that using only one of these clauses is not sufficient). This way, the marginal probability, $p(E = true)$, is fixed to p_D. However, evidence from other sources may change the probability $p(E = true)$, especially when E is not strictly extensional. Therefore, it would be better to add an observation variable O and use these two formulae: $(observe(O) \rightarrow E, w \propto p_D)$ and $(observe(O) \rightarrow E, w \propto 1 - p_D)$. The variable O has a fixed value that represents the corresponding measurement. It is specific to this grounding. The predicate $observe(O)$ will not take part in any logical deduction and is always assumed true. This formulation allows evidence from related sources (beside O) to have their effects on $p(E = true)$.

Extensional predicates can be of various kinds depending on the domain under consideration. Two classes and their associated uncertainty that we consider are object recognition and action detection (see section 7.1).

Identity Maintenance. Identity maintenance is necessary when there exist multiple identities that actually refer to the same object([5], [15]). In surveillance, it is caused by lack of visual information (appearance, shape...) to make unique identity connections across observation gaps. Our approach to solve this problem is similar to the one proposed in [15] for entity resolution in relational databases ([7]), with a slightly more concise formulation.

Identification of two objects A and B is represented by a predicate $eq(A, B)$. It comes with the following set of axioms (with infinite weights): 1) *Reflexive*, $eq(A, A)$; 2) *Symmetry*, $eq(A, B) \leftrightarrow eq(B, A)$; 3) *Transitivity*, $eq(A, B) \wedge eq(B, C) \rightarrow eq(A, C)$; 4) *Predicate Equivalence*, $P(X_1, Y) \wedge eq(X_1, X_2) \rightarrow P(X_2, Y)$, (for two-ary predicates but can be similarly stated for n-ary predicates).

The equivalence predicate can be extensionally evaluated or intensionally inferred. Extensional evaluation of $eq(A, B)$ is done using appearance matching. The probability $p(eq(A, B) = true)$ is calculated based on a matching score. Intensional deduction of $eq(A, B)$ can be done using the above axioms and commonsense rules in the KB. Several prioritized rules in [5], such as "possession of some special objects (e.g. car keys) determines owners' identity", can be used here, where each prioritizing level is mapped to a corresponding weight.

6 Network Construction

This section describes our deduction algorithm that uses the production rules in the KB (section 5.1) to deduce grounded atoms for the Markov network. Due to noise or incompleteness in observations, some events that have not actually occurred might get grounded and added to the ground atom database (ADB). Our procedure is thus a relaxed version of logical deduction and may not be logically consistent.

6.1 Deduction Algorithm

Typically, with definite clauses, deduction is performed via forward chaining. In our system, logic rules take richer forms that require us to additionally deal with negative preconditions and disjunctive conclusions. Following are several preliminaries for our algorithm.

Close World Assumption(CWA). Since it is usually not convenient and sometimes impossible to detect (consistently) events that are not happening, such as the $notstopped(C_i, t)$ event (for all cars at all time points), the CWA is used to check for negative preconditions: what is not currently known to be true is assumed false. Then, forward chaining is still used, but is divided into two phases: the first for rules that do not have negative preconditions and the second for the remaining rules. This is to delay, for example, the conclusion that $\neg a$ is true using the CWA until all possible ways of deducing a have been tried.

Context-dependent Preconditions. Consider the predicate $nearBy(P, loc, t)$ in the formula $happenAt(E, loc, t) \wedge nearBy(P, loc, t) \rightarrow witness(E, P)$. It would be cumbersome to evaluate $nearBy(P, loc, t)$ and add it to the ADB for all people

P, all locations loc and all times t. Instead, it should only be evaluated after $happenAt(E, loc, t)$ is true with specific bindings of loc and t. In this case, the satisfaction of the first precondition serves as the context that enables the lazy evaluation of the second one. Generally, we use lazy evaluation for an extensional predicate when it would be expensive to evaluate otherwise due to the large size of the domain (e.g. ones that involve time or location).

Disjunction Domains. Generally, in our system, disjunctions need no special treatment. However, when they are in the scope of an existential quantifier, domain expansion and several bookkeeping steps are required. In the missing bag example in section 5.2, the predicate $passBy(P, L, T)$ limits P to the set $\{P_1, P_2\}$ and the existential quantifier is expanded over the entirety of this domain. In general, we eliminate the existential quantifier by considering that the conclusion has two parts, one for defining the object domain ($passBy(P, L, T)$) and the other for describing the actual conclusion ($pickUp(P, B, T)$) In other words, our general production rule would be $precondition \rightarrow (\exists x\ domaindef_x \wedge conlusion_x)$. An empty clause $domaindef_x$ implies that the domain consists of all instantiations of x. During deduction, we may need to expand domains as new objects that satisfy domain predicates are discovered. In such cases, the previously grounded formula is replaced with the new one and the network is modified with the new clique.

Ground Atom and Formulae Deduction

- *Input* ADB - ground atom database; KB_{pos} - set of definite rules; KB_{neg} - set of rules that have negative preconditions
- *Output* ADB - with new ground atoms added; GS - the set of grounded clauses.

Repeat until no new ground atom is generated

1. Repeat until no new atom
   For $\forall R \in KB_{pos}$, instantiate R w. r. t. ADB and for each instantiation r,
   (a) If all context-independent preconditions are satisfied, then evaluate all context-dependent preconditions and add the newly evaluated atoms to ADB.
   (b) If all succeeded, get effects and add to ADB.
   (c) $GS \leftarrow GS \cup r$.
2. Similar to step 1 for $R \in KB_{neg}$ with CWA added during instantiation of rules.

Fig. 3. The algorithm for deducing new ground atoms

The deduction procedure is shown in Fig. 3. In step 1(a), when grounding a clause, if context-independent preconditions are satisfied then context-dependent predicates will be extensionally evaluated. Instances that are evaluated to true will be added to the ADB. In step 1(b), all atoms in the conclusion as well as their complemented literals (i.e. E and $\neg E$) are added to the ADB. If an existential quantifier is involved, we need to check and update, if necessary, its previously expanded domain. Step 2 essentially repeats step 1 with the addition of the CWA. For a precondition, $\neg E$, if we are unable to observe or deduce E, $\neg E$ is assumed true. If the related clause ends up being grounded (i.e. all other preconditions are evaluated to true) then the literal $\neg E$ is added to the ADB.

All ground clauses are then added to the Markov network. This construction procedure is performed whenever there is a new event generated. It can be done incrementally by deriving only deductions that originate from new events.

7 Implementation and Experiments

7.1 Implementation

We describe here some basic elements needed to address the parking lot application: object set, predicate set, their evaluation and the KB. Three types of objects are considered: cars (denoted as C_i), humans (H_i) and locations (L_i). Time is represented using atomic intervals with granularity of n_I frames (e.g. $n_I = 30$, approximately 2 seconds). Each primitive event or action is assumed to be true within one time interval. Below, time labels are omitted for clarity. Our vocabulary consists of the following predicates: extensional, context-independent, $inTrunkZone(C, H)$, $inLeftZone(C, H)$, $inRightZone(C, H)$, $disappear(H, L)$, $equal(H_1, H_2)$, $shakeHand(H_1, H_2)$ and $carLeave(C)$; extensional, context-depen-dent, $openTrunk(C, H)$; intensional: $enter(C, H)$ and $drive(C, H)$. Additionally, we have measurement objects and their corresponding predicates (Sec. 5.2).

Background subtraction, human detection and tracking (see e.g. [16]) techniques were first applied to identify and track object locations. The orientation and direction of each car were estimated simply using its corresponding foreground blob and parking lot layout. Fig. 4.1 shows the estimated layouts of the three detected cars during one experiment.

A spatial predicate, for example, $inTrunkZone(C, H)$, is generated when the foot location of person, H, intersects significantly with the trunk zone of the car, C, for a sufficiently long period of time; $disappear(H, L)$ is generated when we lose track of H. Identity maintenance predicates are evaluated using the distance between color histograms of the two participating objects. $shakeHand(H_1, H_2)$ is modeled by analyzing the connecting area between two standing separate persons. $openTrunk(C, H)$ is evaluated base on the motion pattern in the trunk area of car C. The rules that constitute our knowledgebase are listed in the appendix. The maximum weight, $MAXW$, is set to be proportional to the network's size (number of ground atoms [12]). The range $0 - MAXW$ is uniformly discretized to five levels corresponding to very strong, strong, medium, weak and very weak certainties. These values are assigned to rules according to our confidence in them, based on domain knowledge.

7.2 Experiments

We analyzed a set of parking lot videos that involve a number of people entering different cars as listed in Table 1. A typical scenario is as follows. Initially, three cars, C_1, C_2 and C_3, park next to each other. A person H_1 appears, walks up to C_2, opens its trunk (Fig. 4.2), puts something in, closes the trunk and then disappears between C_1 and C_2 (Fig. 4.3). Then two persons, H_2 and H_3, walk close to each other near the parked cars. They shake hands (Fig. 4.4) and

Table 1. Four sequences used in our experiments

	♯ of people	♯ of cars	Durations
seq 1	6	3	2 min 10 sec
seq 2	5	3	3 min
seq 3	4	2	1min 30 sec
seq 4	6	3	4 min

disappear between C_2, C_3 and around the left of C_3 respectively (Fig. 4.5). Person H_4 walks to C_1 and disappears from the passenger side of C_1 (Fig. 4.5). A person H_5 follows a similar path (Fig. 4.6). Person H_6 walks to the cars and disappears between C_2 and C_3 (Fig. 4.7). Then C_1 pulls out and leaves. Finally, C_2 and C_3 follow in order (Fig. 4.8). This scenario consists of a number of interesting interactions and we would like to query the system about who entered or drove which car. The challenges arise mostly due to noise and occlusion, which lead to loss of track well before a person enters a car.

Fig. 4. 1) Estimated layouts of the parked cars. 2-8) Key frames from the scenario. Foreground and human detection results are also shown. Irrelevant people and cars are removed.

As the scenario unfolds, new events are generated and our ground network evolves accordingly. We can query our system at any instant of time. Here, we ran queries after all cars had departed. Detection probabilities for $openTrunk(C_2, H_1)$ and $shakeHand(H_2, H_3)$ were respectively 0.9 and 0.5. Identity confusion is not significant so no related ground atom is generated. Fig. 5 shows the results for our queries, the probabilities $p(enter(c, h) = true)$ and $p(drive(c, h) = true)$ for all cars C_i's and human H_i's. These results can be explained intuitively as follows. Person H_1 disappeared between car C_1 and C_2, so he could have equally entered either of them (with $p = 0.5$). But since he was observed to open C_2's trunk, the probability that he entered C_2 increased to 0.86. Person H_2 could have entered either C_2 or C_3 (each with $p = 0.5$). But he

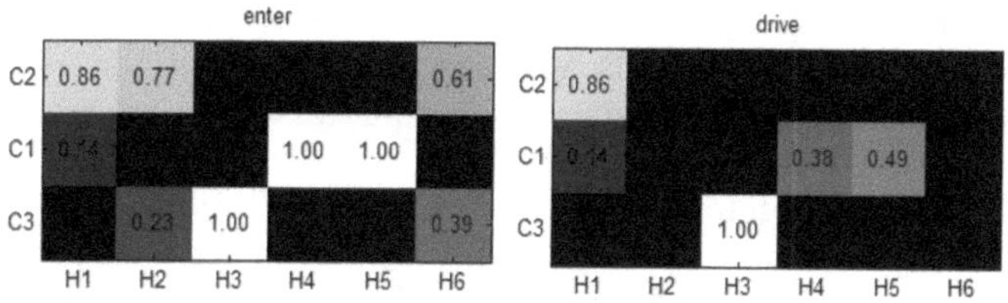

Fig. 5. Right, probabilities of entering cars for individual persons, $p(enter(c,h) = true)$. Left, probabilities of driving cars, $p(drive(c,h) = true)$. Darker means being lower in value.

had been detected shaking hands with H_3 (and so probably saying "goodbye"), who entered C_3 with high certainty. Hence, the probability of H_2 entering C_3 was reduced and entering C_2 was increased. However, since the detection probability was not very high ($p = 0.5$), the increase was not as much as for the first person. Persons H_3, H_4 and H_5 entered cars C_3, C_1 and C_1 respectively with high certainties since there was only a single car close to them when they disappeared. Person H_6 disappeared between cars C_1 and C_3. Since there is no further supporting evidence, the probabilities for entering C_1 and C_3 should be the same. The observed discrepancy is due mainly to sampling approximation.

H_1 and H_3 drove C_2 and C_3, respectively, with high certainty since they had been observed to enter their cars from the driver side and there was no competing alternative. For car C_1, H_4 and H_5 were observed to enter it from the passenger (right) side. The only person who could possibly enter it from the driver side was H_1. But it was strongly believed H_1 entered and drove C_2. Hence, no one entered C_1 from the driver side. This led to the conclusion that either H_4 or H_5 had moved from the passenger side to drive C_1. This reasoning was made possible because of the low-weight rule 5, which states that, although unlikely, drivers can enter a car from the passenger side (the driver side may be blocked). In general, this type of rule is added to increase the system's ability to handle exceptional cases. They bear some of the flavor of default reasoning in the sense that in the absence of supporting evidence for "normal" rules, rules with low weights will be activated to explain the situation under consideration. Note that rule 5 was also applied to other persons, such as H_2 and H_6, but produced negligible effects since competing hypotheses were much stronger. Therefore, the probabilities that H_2 and H_6 drove any car were still close to zero.

In the initial querying, our system was able to conclude that either H_4 or H_5 drove car C_1 but was unable to determine which of them did. Consider adding to the KB a very weak rule stating that among the persons entering a car from the passenger side, whoever enters it first is its driver (no new extensional predicate evaluation is needed). When we do this and re-evaluate the queries, $p(drive(C_1, H_4) = true)$ increased to 0.60 while $p(drive(C_1, H_5) = true)$ dropped to 0.23, which matched the observation that H_4 entered C_1 before H_5 and drove it away. This example is intended to illustrate that domain knowledge can be easily added to our system to improve its performance. This would not be so easy in a purely logical system, because a significant amount of consistency checking is required before admitting a new piece of knowledge. Also, it would not be straightforward in a purely probabilistic system, since adding complicated structures involves cumbersome changes to the internal structure of the system.

Our system is implemented using C++ and runs in Windows and Cygwin on an Intel Pentium Dual-Core 1.6 GHz machine. The average running time for background subtraction, human detection and object tracking, on 640x480 RGB frames, are 5 frames per second. Average running time when performing inference for all queries presented here is approximately fifteen seconds (maximum number of MCMC step is set to 5000).

8 Discussion

We described how a combination of a probabilistic graphical model, the Markov Logic network, and first-order logic statements can be used for event recognition in surveillance domains, where unobservable events and uncertainties in detection are common. Logic provides a convenient mechanism to utilize domain knowledge to reason about the unobservable. Probabilistic models give us a coherent framework to deal with uncertainty. The combination brings us the ability to capture interesting interactions in complex domains.

Obtaining a sufficient and efficient knowledge base (KB) is important to our recognition performance. This issue is also receiving increasing attention in the area of human activity understanding([9], [8]). Approaches that use efforts from open communities to build common sense KB have been proposed such as Open Common Sense and Open Indoor Initiatives ([17], [18]). Large databases of rules are now publicly available([17]). Automated rule extraction and weight learning from them have been attempted (e.g. [9]). At its current stage, for visual surveillance, rules and facts collected from such open databases are more often than not irrelevant and not useful. For example, a query for "parking lot" would output relations such as it is "a place for parked cars", "would be hot in the summer", "may or may not be empty"... ([17]). However, as these knowledge bases grow and, possibly, specialize ([18]), their application to our framework seems promising. Exploiting them is part of our future investigation.

References

1. Richardson, M., Domingos, P.: Markov Logic Networks. Machine Learning 62, 107–136 (2006)
2. Nevatia, R., Zhao, T., Hongeng, S.: Hierarchical language-based representation of events in video stream. In: Proc. of CVPRW on Event Mining, pp. 39–47. IEEE, Los Alamitos (2003)
3. Yamato, J., Ohya, J., Ishi, K.: Recognizing human action in time-sequential images using Hidden Markov models. In: Proc. CVPR 1992, pp. 379–385. IEEE, Los Alamitos (1992)
4. Rota, N., Thonnat, M.: Activity recognition from video sequences using declarative models. In: Proc. ECAI 2002, pp. 673–680 (2002)
5. Shet, V., Harwood, D., Davis, L.: Multivalued default logic for identity maintenance in visual surveillance. In: Leonardis, A., Bischof, H., Pinz, A. (eds.) ECCV 2006. LNCS, vol. 3954, pp. 119–132. Springer, Heidelberg (2006)

6. Shet, V., Neumann, J., Ramesh, V., Davis, L.: Bilattice-based logical reasoning for human detection. In: Proc. CVPR 2007, pp. 1–8. IEEE, Los Alamitos (2007)
7. Getoor, L., Taskar, B.: Intro. to Statistical Relational Learning. MIT Press, Cambridge (2007)
8. Liao, L., Fox, D., Kautz, H.: Location-based activity recognition using Relational Markov Networks. In: Proc. IJCAI 2005, pp. 773–778. Morgan Kaufmann, Inc., San Francisco (2005)
9. mins .1em Pentney, W., Popescu, A., Wang, S., Kautz, H., Philipose, M.: Sensor-based understanding of daily life via large-scale use of common sense. In: Proc. AAAI 2006 (2006)
10. Biswas, R., Thrun, S., Fujimura, K.: Recognizing activities with multiple cues. In: IEEE Works. on Human Motion, pp. 255–270 (2007)
11. Bobick, A., Ivanov, Y.: Action recognition using probabilistic parsing. In: CVPR 1998, pp. 196–202 (1998)
12. Kok, S., Singla, P., Richardson, M., Domingos, P.: The Alchemy system for statistical relational AI. Tech Report - CS Dept., Univ. of Washington (2005)
13. Minker, J., Seipel, D.: Disjunctive logic programming: A survey and assessment. In: Computational Logic, London, UK, pp. 472–511. Springer, Heidelberg (2002)
14. Singla, P., Domingos, P.: Memory-efficient inference in relational domains. In: Proc. AAAI 2006, AAAI Press, Menlo Park (2006)
15. Singla, P., Domingos, P.: Entity resolution with Markov Logic. In: Proc. ICDM, pp. 572–582. IEEE, Los Alamitos (2006)
16. Dalal, N., Triggs, B.: Histograms of oriented gradients for human detection. In: Proc. CVPR 2005, vol. 2, pp. 886–893. IEEE, Los Alamitos (2005)
17. Liu, H.: The ConceptNet Project, `http://web.media.mit.edu/~hugo/conceptnet/`
18. Stork, D.: The OpenMind Initiative, `http://www.openmind.org/`

A Logic Rules

List of rules and their corresponding weights used for the parking lot problem (see problem description in Sec. 3).

1. If a person disappears, he/she enters a nearby car, $w = \frac{4}{5}MAXW$
   $disappear(h) \rightarrow (\exists c\ (inLeftZone(c,h) \vee inRightZone(c,h)) \wedge enter(c,h))$
2. If a person opens the trunk of a car, he/she will (likely) enter that car
   $disappear(h) \wedge openTrunk(c,h) \rightarrow enter(c,h)$, $w = \frac{4}{5}MAXW$
3. A person enters only one car: $enter(c_1,h) \wedge \neg equal(c_1,c_2) \rightarrow \neg enter(c_2,h), w = MAXW$ (if h gets into c temporarily and gets out of it, h is not considered to have entered c, just being temporarily occluded)
4. A person entering a car c from the left (driver) side will (likely) drive c
   $inLeftZone(c,h) \wedge enter(c,h) \rightarrow drive(c,h)$, $w = \frac{4}{5}MAXW$
5. A person entering a car c from the right (passenger) side will (less likely) drive c
   $inRightZone(c,h) \wedge enter(c,h) \rightarrow drive(c,h)$, $w = \frac{1}{5}MAXW$
6. Two persons shaking hand with each other will (likely) not enter the same cars.
   $shakeHand(h_1,h_2) \wedge enter(c_1,h_1) \rightarrow \neg enter(c_1,h_2)$, $w = \frac{4}{5}MAXW$
7. For a car to drive away, it needs a driver: $carLeave(c) \rightarrow (\exists\ h\ enter(c,h) \wedge drive(c,h))$, $w = MAXW$
8. A car has only one driver: $drive(c,h_1) \wedge \neg equal(h_1,h_2) \rightarrow \neg drive(c,h_2), w = MAXW$
9. A person can drive only one car: $drive(c_1,h) \wedge \neg equal(c_1,c_2) \rightarrow \neg drive(c_2,h), w = MAXW$
10. A person has to enter a car to drive it: $drive(c,h) \rightarrow enter(c,h), w = MAXW$

Illumination and Person-Insensitive Head Pose Estimation Using Distance Metric Learning

Xianwang Wang, Xinyu Huang, Jizhou Gao, and Ruigang Yang

Center for Visualization & Virtual Environments,
University of Kentucky,
Lexington KY 40507, USA
{xwang,xhuan4,jgao5,ryang}@cs.uky.edu

Abstract. Head pose estimation is an important task for many face analysis applications, such as face recognition systems and human computer interactions. In this paper we aim to address the pose estimation problem under some challenging conditions, e.g., from a single image, large pose variation, and un-even illumination conditions. The approach we developed combines non-linear dimension reduction techniques with a learned distance metric transformation. The learned distance metric provides better intra-class clustering, therefore preserving a smooth low-dimensional manifold in the presence of large variation in the input images due to illumination changes. Experiments show that our method improves the performance, achieving accuracy within 2-3 degrees for face images with varying poses and within 3-4 degrees error for face images with varying pose and illumination changes.

1 Introduction

Human face analysis, due to its vast application from biometric authentication to human-computer interactions, is a very active topic in computer vision research. Head pose estimation is a central component for many of these applications. For example, face recognition systems require the capability of handling the significant pose variations. Zhao *et al.* [1] shows that the pose and illumination are the major factors affecting the performance of face recognition algorithms. The difference between individuals' face images taken under the same lighting conditions is smaller than the difference between two face images of same individual taken under varying lighting conditions. That is, image variation due to lighting changes is more significant than variation due to different personal identities [2]. While person-independent head pose estimation has been studied reasonably well in recent years [3,4,5], robust illumination- and person-independent head pose estimation remains a challenging problem. The main contribution of this paper is to introduce a new approach to address this problem.

There exists many methods for pose estimation from a single image (we exclude pose estimation from a video sequence in the context of this paper). These methods can be classified into five categories [4,5]: shape-based geometric analysis methods [6], appearance-based methods [7,8], model-based methods [9],

D. Forsyth, P. Torr, and A. Zisserman (Eds.): ECCV 2008, Part II, LNCS 5303, pp. 624–637, 2008.

template-based methods [10,11], and dimensionality reduction based methods. It is beyond the scope of this paper to discuss all the related work in pose estimation, and we will focus on the fifth category: dimensionality reduction (DR) techniques, which is also what our proposed method belongs to. Every DR approach is essentially to learn a distance metric that removes the data redundancy and leads to more compact feature representation. However, the problem of illumination variation is usually conceded or treated lightly as a pre-processing step. As we shown in Figure 1, we aim to deal with some very harsh illumination conditions. The lack of DR-based pose estimation under these difficult conditions is probably due to the fact that illumination changes are usually larger than that from different persons or small pose variations.

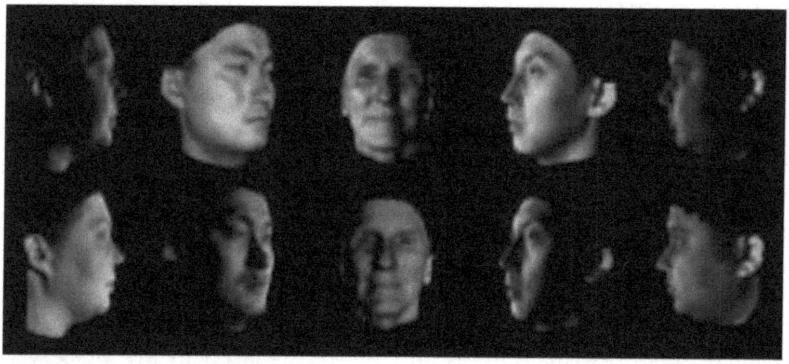

Fig. 1. Sample input images for pose estimation, notice the large variation in rotation and the harsh un-even illumination

In this paper we present a novel approach to estimate the pose from a single input image by *combining both the unsupervised metric learning technique and the supervised metric learning techniques.* Unlike previous DR-based pose estimation method that treats illumination as a part of pre-processing step, we develop a *unified* framework that does not require any pre-processing for illumination normalization or correction. This is possible by applying a *learned* distance transformation after the use of nonlinear DR techniques. This is different from prior approaches in which the original images are modified or filtered before applying nonlinear DR techniques. To the best of our knowledge this is the first time a DR-based method is capable of producing fairly accurate pose estimation (within a few degrees) under harsh illumination conditions as shown in Figure 1.

The rest of the paper is organized as follows. Section 2 reviews the related work on head pose estimation. Section 3 describes our proposed approach of pose estimation. Section 4 presents the experimental results, and the conclusion is made in section 5.

2 Related Works

In this section we review the research work of distance metric learning and head pose estimation approaches based on dimensionality reduction. Many methods have been developed. We discuss each one of them in the following sections.

2.1 Distance Metric Learning

The main idea for methods in this category is to learn metrics that separate data points from different classes, while keeping close data points within the same class. Xing *et al* [12] posed metric learning as a convex programming problem to satisfy the constraints of the better representation. Relevant Components Analysis (RCA) [13] learns a full ranked Mahalanobis distance metric using equivalence constraints. Neighborhood Components Analysis (NCA) [14] maximizes the leave-one-out cross validation to learn a distance metric for KNN classifier. Large Margin Nearest Neighbor(LMNN) [15] extends NCA through a maximum frame work. Local Fisher Discriminant Analysis (LFDA) [16] can be viewed as localized variant of FDA.

2.2 Manifold Learning

The goal for approaches in this category is to learn a low-dimensional manifold in which most "intrinsic information" (e.g., distance) are preserved. Popular approaches include ISOMAP [17], Locally Linear Embedding (LLE) [18], and Laplacian Eigenmaps (LE) [19]. ISOMAP preserves the geodesic inter-point distances, LLE preserves the distance based on locally linear combination of neighborhood, and LE preserves the distance described by a weighted connected graph constructed from neighborhood.

Raytchev *et al* [20] apply ISOMAP-based manifold learning technique for user-independent pose estimation and evaluate their method in comparison with the Linear Subspace and Locality Preserving Projections(LPP) [21].

Chen *et al.* [22] uses the face images of two specific head poses and estimates the head poses between them through classification-based nonlinear interpolation. This approach is based on the assumption that the face images of multiple view lie on a manifold in the original image feature space.

Fu and Huang [5] presented an appearance-based strategy for head pose estimation using supervised Graph Embedding(GE) analysis. The neighborhood weighted graph is first constructed in the sense of supervised LLE. The out-of-sample data points may be treated using the projection transformation solved in closed-form based on GE linearization. The K-nearest neighbor classification is then employed to estimate the head pose. Their method is successful with low pose estimation error. They consider face images with only pose variation, but not illumination change in their experiment.

Balasubramanian *et al.* [3] proposed the Biased Manifold Embedding (BME) framework for head pose estimation. The pose information of the given face

Table 1. performance comparison of different methods for head pose estimation

Method	Interval	Increment	Best result: error	Illumination (Yes/No)
ISOMAP [20]	$[-90° + 90°]$	15°	11°	No
Fisher manifold learning [22]	$[-10° + 10°]$		3°	No
BME with ISOMAP, LLE [3]	$[-90° + 90°]$	2°	3°	Yes$^{\diamond}$
BME with LE [3]	$[-90° + 90°]$	2°	2°	Yes$^{\diamond}$
LEA [5]	$[-90° + 90°]$	1°	2°	No

$^{\diamond}$LoG filter is used.

image data is used to compute a biased neighborhood of each point in the feature space, before determining the low-dimensional embedding. BME uses a Generalized Regression Regression Neural Network (GRNN) to learn the non-linear mapping for dealing with out-of-sample data points, and applies linear multivariate regression to estimate the pose. Essentially, BME proposes a more general model to modify the distance matrix, which unifies the other supervised manifold learning approaches [23,24]. The interested reader can refer to the discussion in [4] for details. Their experiment shows this method works well with person-independent face images with pose variation. BME uses LoG to remove the illumination effects, but LoG representation are not sufficient for pose estimation under a wide variety of lighting conditions, in particular hash lightings with shadows, as in our experiments.

The performance of representative DR-based approaches for head pose estimation are summarized in Table 1. It shows that LEA and BME are the current state-of-art techniques in pose estimation. But none of them treats illumination variations in a principled way, most of them do not discuss the effect of illumination.

3 Our Approach

In this section, we present our approach for pose estimation. We assume that there is a training face database, and each face image in the database is associated with a pose label. Our goal is to estimate the unknown pose label from an input face image that is not in the training database.

A good low-dimensional embedding for pose estimation should have the following properties: (1) Separation. The embedding from different poses are kept apart, and there is no overlap among them. Furthermore, the embeddings of different individuals with different illuminations but same pose should be close to each other, i.e., within a cluster. (2) Smoothness. The low-dimensional manifold should change smoothly according to the pose. Figure 2 shows an ideal 3-dimensional manifold embedding of 24 subjects with the same illumination while the pose angles vary from $-90°$ to $+90°$ with a granularity of 4 from our training database. In this figure, there are 46 clutters in total, each with unique color corresponding to a specific pose angle. Within each of clutters, there are 24 data points, which are embeddings of 24 faces from the same pose.

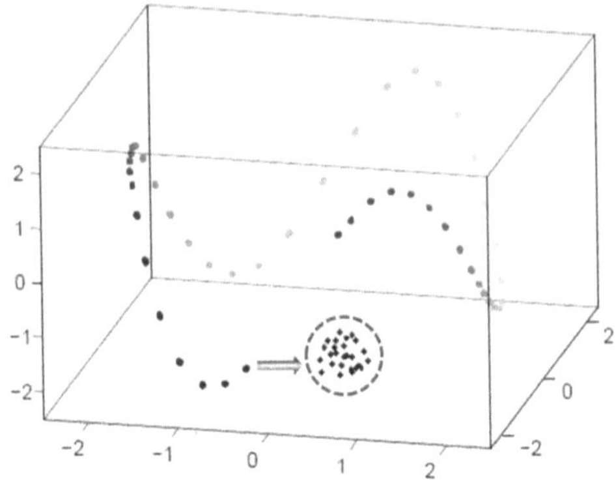

Fig. 2. The ideal 3-dimensional embedding of 24 subjects' face images with only pose variation between $[-90° + 90°]$ at 4° increments. The pose changes are represented by different colors.

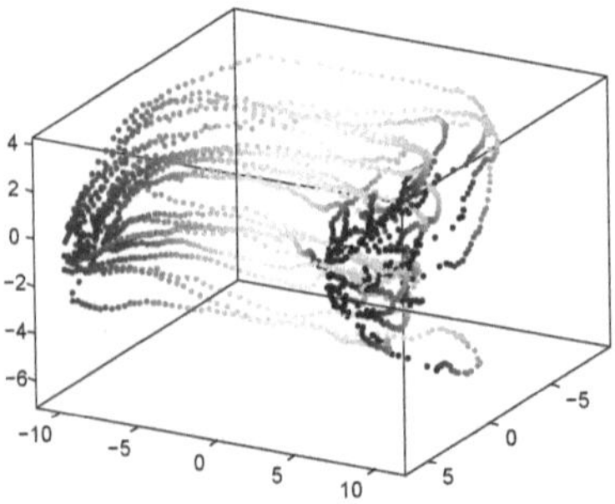

(a) ISOMAP embedding (NN = 200)

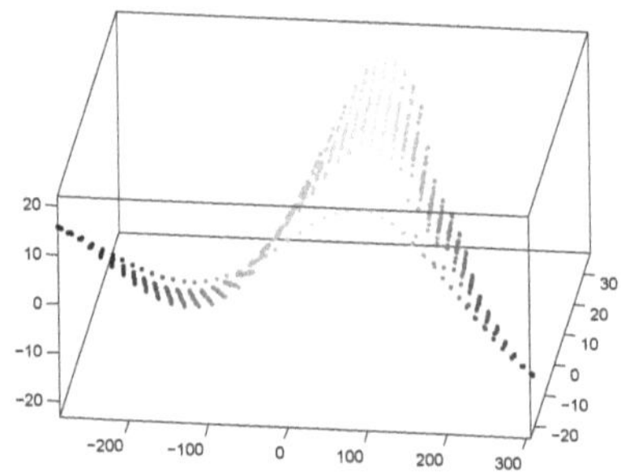

(b) Biased ISOMAP embedding (NN = 200)

Fig. 3. The 3-dimensional embedding of 24 subjects' face images with only pose variation between $[-90° + 90°]$ at 4° increments, using ISOMAP and BME. The pose changes are represented by different colors.

The 3D embedding of ISOMAP and BME with ISOMAP are shown in Figure 3. Figure 3(a) shows the ISOMAP embedding, in which 200 nearest neighbors (NN) are used. It maps face images of 24 subjects into 24 different pose manifolds. This is because ISOMAP can not find the nearest neighbors of each point accurately when there are multiple individuals in the training set. To deal with identity variations, BME finds the right nearest neighbors for each data point by the given pose labels. However, when there are illumination variations, especially large illumination variations, BME can not generate a good pose manifold either. As shown in Figure 4, we generate two manifolds with ISOMAP and biased ISOMAP from 10 subjects with pose angles varying from −90° to +90° with a granularity of 4 and illumination changes from 0° to +45° at 5° increments. It

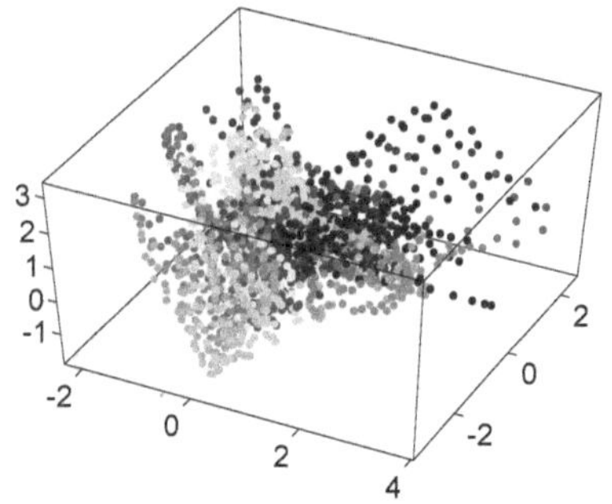

(a) ISOMAP embedding (NN = 200)

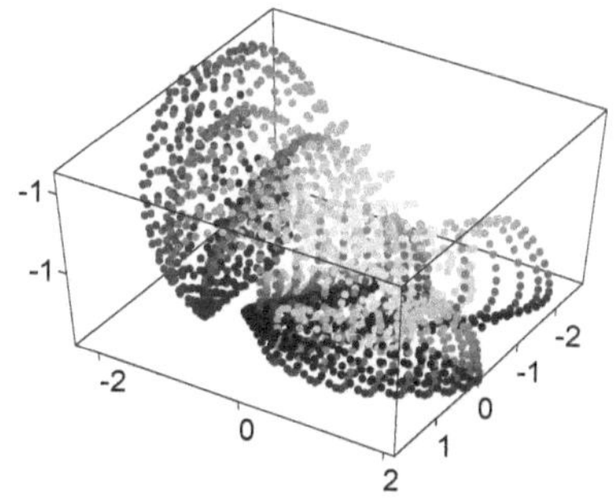

(b) Biased ISOMAP embedding (NN = 200)

Fig. 4. The 3-dimensional embedding of 10 subjects' face images with pose variation between $[-90° +90°]$ at $4°$ increments and illumination changes from $0°$ to $+45°$ at $5°$ increments, using ISOMAP and BME. The pose changes are represented by different colors.

is clear that there are many overlaps between pose angles in ISOMAP embedding (Figure 4(a)), as well as BME embedding (Figure 4(b)). Hence, it is quite difficult to estimate poses from the manifolds generated by ISOMAP and BME. This is because that 1) the computation of a nonlinear manifold relies on the distance between data points. For example, in order to compute a smooth pose manifold, the distance between face images under large illumination variation and the same pose should be small. 2) BME only uses pose labels to find the right nearest neighbors. The distances between data points remain unchanged. 3) the distortion caused by illumination variation is much larger than the distortion caused by the identity of individuals by comparing Figure 4(b) and 3(b). Therefore, the distance between data points should be modified in order to obtain a smooth manifold. This brings forth the need to develop pose estimation techniques that can work well with the face images that have large illumination changes and pose changes from many different individuals.

To obtain a good low-dimensional embedding, we propose an approach based on manifold learning techniques and supervised distance metric learning techniques for head pose estimation. We first construct the low-dimensional embedding using ISOMAP (Figure 4(a)). The low-dimensional embedding is then linearly mapped to the transformed feature space with modifying the distance between data points, using Local Fisher Discriminant Analysis (LFDA) [16] by pose labels. The combination of ISOMAP and Fisher Discriminant Analysis (FDA) was ever proposed in the work [25]. In their work, each data point is represented by a feature vector, which is its geodesic distance from other points. Then, FDA is applied to find an optimal projection direction for classification. The main difference between our approach and this extended ISOMAP is that LFDA was employed to refine the low-dimensional manifold and maintain pose class separation.

3.1 ISOMAP

The classical linear algorithms such as Principle Component Analysis (PCA) and Multidimensional Scaling (MDS) cannot always reveal the intrinsic distribution of a given complex data set. We therefore adopt ISOMAP for nonlinear dimensionality reduction [17]. The input is data matrix $X = (x_1, \cdots, x_N) \in \Re^{D\times N}$ containing N face images from the training data, where $x_i \in \Re^D (i = 1, 2, \ldots, N)$ is the D-dimensional samples of face images. In our case, D is equal to 1024, since we vectorized face images with the resolution 32×32. The output is the nonlinear embedding matrix $Y = (y_1, \cdots, y_N) \in \Re^{d\times N}$ of X, where $y_i \in \Re^d (i = 1, 2, \ldots, N)$ is the d-dimensional data points in the low-dimensional embedding. It first determines the neighbor relationship on the manifold M based on the pairwise Euclidean distance $d_X(i,j)$ between pairs of face images x_i,x_j. These neighbor relations are represented as a weighted graph G over the data points, with the edges of weights $d_X(i,j)$ between neighboring points. The pairwise geodesic distance $d_M(i,j)$ on the manifold are then estimated with the distance of the shortest path $d_G(i,j)$ in the graph G using Floyds or Djkstraas algorithm. The classical MDS is finally applied to the matrix of graph distances $D_G = d_G(i,j)$ to construct d-dimensional embedding Y.

3.2 Local Fisher Discriminant Analysis

We use LFDA to learn the matrix $P_{LFDA} \in \Re^{d\times d}$ that transforms $y_i (i = 1, 2, \ldots, N)$ to $z_i (i = 1, 2, \ldots, N)$. $z_i \in \Re^d (i = 1, 2, \ldots, N)$ is the d-dimensional data points in the transformed feature space. $z_i = P^T_{LFDA} y_i$ in the same pose angle are kept close together, while z_i from different pose angles are well separated.

LFDA evaluates within-class scatter and between-class scatter in a local manner by combining the idea of FDA and LPP [21]. The local within-class scatter matrix $\tilde{S}_B$ and the local between-class scatter matrix $\tilde{S}_W$ are defined as follows,

$$\tilde{S}_W = \frac{1}{2} \sum_{i,j=1}^{N} \tilde{W}^{(w)}_{i,j} (y_i - y_j)(y_i - y_j)^T \tag{1}$$

$$\tilde{S}_B = \frac{1}{2} \sum_{i,j=1}^{N} \tilde{W}^{(b)}_{i,j} (y_i - y_j)(y_i - y_j)^T \tag{2}$$

where

$$\tilde{W}^{(w)}_{i,j} = \begin{cases} W_{i,j}/N_c, & y_i \in c, y_j \in c \\ 0, & otherwise \end{cases} \tag{3}$$

$$\tilde{W}^{(b)}_{i,j} = \begin{cases} W_{i,j}(1/N - 1/N_c), & y_i \in c, y_j \in c \\ 1/N, & otherwise \end{cases}, \tag{4}$$

and $W_{i,j}$ is the affinity between in y_i and y_j that are ranged in $[0, 1]$. The discuss about the definition of $W_{i,j}$ can be found in [21,26]. LFDA considers

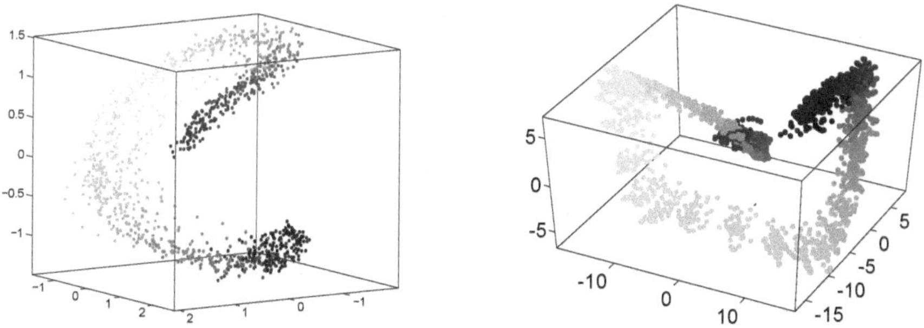

(a) The 3-dimensional embedding of 24 subjects' face images with only pose variation from $-90°$ to $+90°$ at $4°$ increments using our method (NN =200). The pose changes are represented by different colors.

(b) The 3-dimensional embedding of 10 subjects' face images with pose variation from $-90°$ to $+90°$ at $4°$ and illumination changes from $0°$ to $+45°$ at $5°$ increments using our method (NN =200). The pose changes are represented by different colors.

Fig. 5. The results of our method for face images with only pose variation and pose plus illumination variation

the maximization of the following objective to find the transformation matrix P_{LFDA},

$$J(P) = \frac{\|P^T \tilde{S_B} P\|}{\|P^T \tilde{S_W} P\|} \tag{5}$$

Noticing that J is invariant with respect to scale, we can formulate the objective into the constrained optimization problem as follows:

$$\begin{aligned} &\min_P -\tfrac{1}{2} P^T \tilde{S_B} P \\ &s.t.\ P^T \tilde{S_W} P = I \end{aligned} \tag{6}$$

The lagrangian corresponds to this optimization problem is,

$$\ell = -\frac{1}{2} P^T \tilde{S_B} P + \frac{1}{2}(P^T \tilde{S_W} P - I) \tag{7}$$

Using the KKT conditions, the problem is transformed into the following generalized eigenvalue problem,

$$\tilde{S_B} \varphi = \lambda \tilde{S_W} \varphi \tag{8}$$

Then, the LFDA transformation matrix is defined by the solution as follows,

$$P_{LFDA} = [\varphi_1, \varphi_2, \cdots, \varphi_d] \tag{9}$$

where $\{\varphi_i\}_{i=1}^d$ are generalized eigenvectors corresponding to the generalized eigenvalues $\lambda_1 \geq \lambda_2 \geq \cdots \geq \lambda_d$.

Algorithm 1. Pose Estimation Pipeline

1: Learning phase

Input: the training face images $x_i (i = 1, 2, \cdots, N)$ and their corresponding labels of pose angles ℓ_i.

Output: the nonlinear mapping between x_i and y_i, P_{LFDA}, the regression model for pose estimation.

(a): Get the low-dimensional embedding y_i using ISOMAMP.

(b): Learn the nonlinear mapping between x_i and y_i using GRNN.

(c): Find the transformation matrix P_{LFDA} from low-dimensional embedding y_i.

(d): Learn the regression model between y_i and its corresponding pose labels using RVM.

2: Testing phase

Input: the test face images $x_i (i = N + 1, N + 2, \cdots, NN)$.

Output: the pose labels for x_i.

(a): Compute its low-dimensional embedding y_i using learned nonlinear mapping.

(b): Map y_i to the transformed feature space using $z_i = P_{LFDA}^T y_i$.

(c): Estimate the pose angle by applying the learned regression model on z_i

We apply our method to the same data that we use in ISOMAP and BME methods to get the low-dimensional embedding, shown in Figure 5. Compared to the results of ISOMAP and BME in Figure 3 and Figure 4, our method is much better in clustering the face images in the same pose angle and separating the face images from different pose angles better than ISOMAP and BME methods in both pose-only variation case and pose+illumination variation case. For a new input face image, we can compute its low-dimensional embedding using the nonlinear mapping learned by Generalized Radial Neural Network (GRNN), and then project it to the transformed feature space by applying the linear transformation. Its pose angle will be finally estimated by Relevance Vector Machine (RVM) [27]. The detailed algorithm procedure using ISOMAP and LFDA for pose estimation is shown in algorithm 1.

4 Experiment and Results

4.1 Data Sets

To evaluate the performance of our approach, we employed the 3D face dataset from [28]. The pose changes horizontally from $-90°$ to $+90°$ at 2 degree increment. The illumination varies from $0°$ to $+45°$ at $1°$ increment, as shown in Figure 6. Other public face databases such as FERET, the CMU-PIE database,Yale Face database, and MIT database, are not used in our experiment, because none of them provide a precise measure for pose and illumination angles and also they do not contain face images with a wide variety of illumination

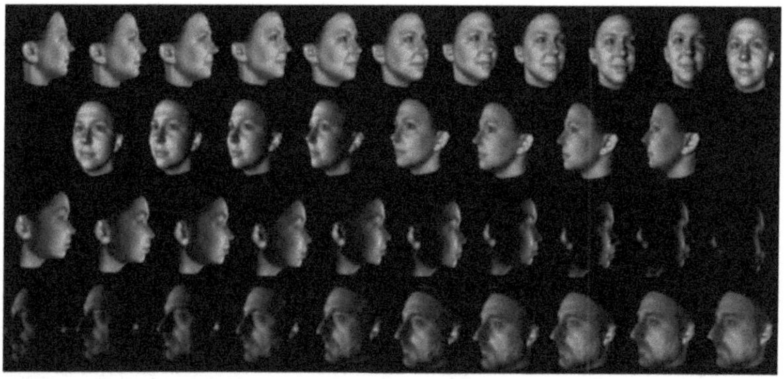

Fig. 6. Samples of face images with varying pose and illumination from 3D face scans

and pose changes [3]. To assess the robustness of our approaches, we perform the experiment in two cases: pose estimation for face images without and with illumination variation. We compared our method with the state-of-art pose estimation techniques, BME. The performance is analyzed with varying choices of the embedding dimensions (marked by "o" in the figures) and 200 neighbors.

4.2 Pose Estimation for Face Images without Illumination Variation

The experiment was performed over 24 subjects with pose angles varying from $-90°$ to $+90°$ at $2°$ increments and illumination of $22°$ using 8-fold cross-

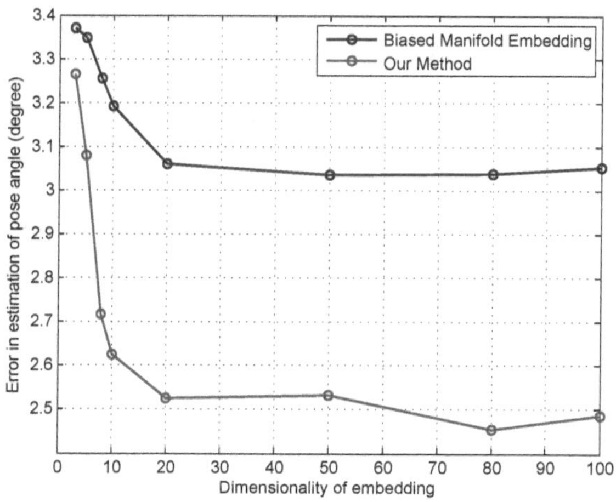

Fig. 7. Pose estimation results comparison of our method against BME with ISOMAP for the face images without illumination variation (NN=200) in different dimensionality. The red line indicates the results of our method.

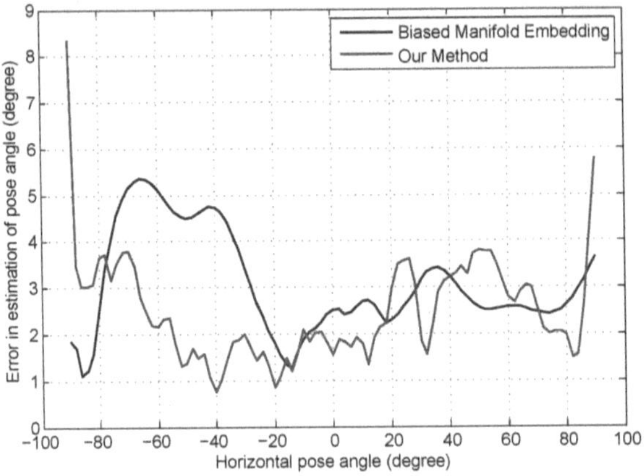

Fig. 8. Distribution of the average error of Our method against the BME framework in pose estimation for the face images without illumination variation. Each of the views is between $[-90°, +90°]$ at $2°$ increments. The red line indicates the results of our method.

validation. We use 1911 face images from 21 subjects (91 images per subject) as the training data in each fold, and then use the 273 images from the other 3 subjects as testing data. The images were down-sampled to 32×32 resolution. The results of the experiment are shown in Figure 7. The red line indicates the performance of our method, while the blue line shows the performance of the BME framework. The result shows the accuracy of our method is slightly better than that of the BME with ISOMAP. Figure 8 shows the head pose estimation error of our method against BME in each of the views in this pose angle interval.

4.3 Pose Estimation for Face Images with Illumination Variation

Due to memory limitation, this experiment was performed over 10 subjects, with pose angles varying from $-90°$ to $+90°$ at $4°$ increments and illumination variation from $0°$ to $-45°$ with a granularity of $5°$. We use leave-one-out cross-validation (LOOCV), i.e., we sequentially take out the face images of one individual and use all the remaining images of the other individuals as training data. The images are also down-sampled to 32×32 resolution. The experiment results are shown in Figure 9. The red line indicates the performance of our method, while the blue line shows the performance of the BME framework. It shows that our method significantly improves the head pose estimation performance compared to the BME framework. Figure 10 shows the head pose estimation error of our method against BME in each of the views in this pose angle interval (100-dimensional embedding) .

In both cases, our method performs well for the frontal and intermediate poses, but not for the profile (almost $-90°$ or $+90°$) views. This may be caused

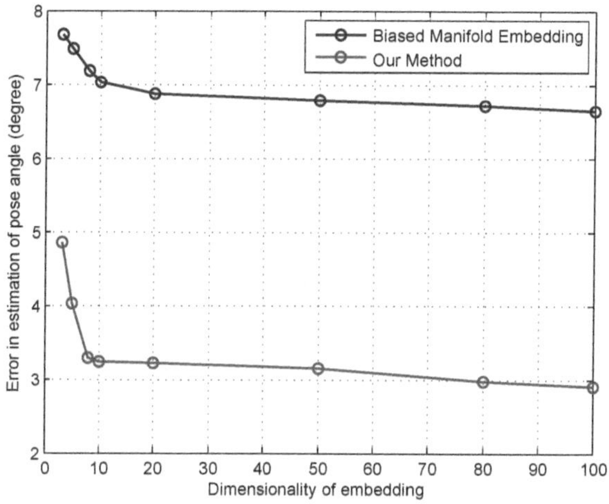

Fig. 9. Pose estimation results comparison of our method against BME with ISOMAP for the face images with illumination variation. The red line indicates the results of our method.

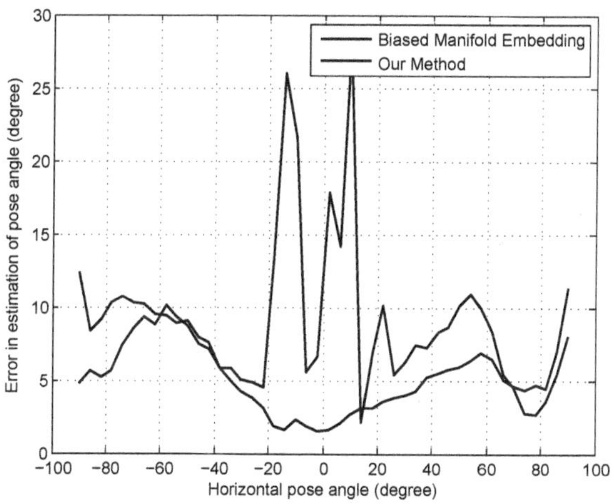

Fig. 10. Distribution of the average error of our method against the BME framework in pose estimation for the face images illumination variation. each of the views is between $[-90^\circ, +90^\circ]$ at 4° increments. The red line indicates the results of our method.

by the noisy features in the face images of these profile views, which results in some overlaps between the data points in the transformed feature space.

Comparing our results with those listed in Table 1, we can see that our method is comparable with the the state-of-art methods for face images with only pose variation. When the illumination also varies, our proposed method maintains the high accuracy for most cases and performs better that BME.

5 Conclusions

In this paper, we proposed an approach to illumination- and person-insensitive head pose estimation. We studied the limitation of related approaches in pose estimation for face images with large illumination variation, and addressed the problem by combining ISOMAP and LFDA. We conducted several experiments to evaluate our approach. The experiment results demonstrate that our method is robust to variation in dimensionality of embedding, illumination and identity of individuals. Our method can be easily extended to other manifold learning techniques, such as LLE and LE, and supervised distance metric learning techniques like RCA, NCA, and LMNN. Furthermore, our method can be used to estimate the illumination direction by using illumination labels in LFDA. Looking into the future, we will apply our approach in the application of face synthesis.

References

1. Zhao, W., Chellappa, R., Rosenfeld, A., Phillips, P.: Face recognition: A literature survey (2003)
2. Adini, Y., Moses, Y., Ullman, S.: Face recognition: The problem of compensating for changes in illumination direction. IEEE Transactions on Pattern Analysis and Machine Intelligence 19(7), 721–732 (1997)
3. Balasubramanian, V., Ye, J., Panchanathan, S.: Biased Manifold Embedding: A Framework for Person-Independent Head Pose Estimation. In: IEEE Conference on Computer Vision and Pattern Recognition, pp. 1–7 (2007)
4. Balasubramanian, V.N., Krishna, S., Panchanathan, S.: Person-Independent Head Pose Estimation using Biased Manifold Embedding. EURASIP Journal on Advances in Signal Processing (accepted, 2007)
5. Fu, Y., Huang, T.S.: Graph Embedded Analysis for Head Pose Estimation. In: International Conference on Automatic Face and Gesture Recognition, pp. 3–8 (2006)
6. Hu, Y., Chen, L., Zhou, Y., Zhang, H.: Estimating Face Pose by Facial Asymmetry and Geometry. In: IEEE International Conference on Automatic Face and Gesture Recognition, pp. 651–656 (2004)
7. Ba, S.O., Odobez, J.M.: A Probabilistic Framework for Joint Head Tracking and Pose Estimation. In: International Conference on Pattern Recognition, pp. 264–267 (2004)
8. Li, S.Z., Lu, X., Hou, X., Peng, X., Cheng, Q.: Learning Multiview Face Subspaces and Facial Pose Estimation using Independent Component Analysis. IEEE Transactions on Pattern Analysis and Machine Intelligence 14(6), 705–712 (2005)
9. Cootes, T.F., Edwards, G., Taylor, C.J.: Active Appearance Models. IEEE Transactions on Pattern Analysis and Machine Intelligence 23(6), 681–685 (2001)
10. McKenna, S., Gong, S.: Real-Time Face Pose Estimation. Real-Time Imaging, Special Issue on Visual Monitoring and Inspection 4(5), 333–347 (1998)
11. Bichsel, M., Pentland, A.: Automatic Interpretation of Human Head Movements. In: International Joint Conference on Artificial Intelligence, Workshop on Looking At People (1993)
12. Xing, E., Ng, A., Jordan, M., Russell, S.: Distance metric learning, with application to clustering with side-information (2003)

13. Bar-Hillel, A., Hertz, T., Shental, N., Weinshall, D.: Learning a Mahalanobis Metric from Equivalence Constraints. Journal of Machine Learning Research 6, 937–965 (2005)
14. Goldberger, J., Roweis, S.T., Hinton, G.E., Salakhutdinov, R.: Neighbourhood components analysis. In: NIPS (2004)
15. Weinberger, K., Blitzer, J., Saul, L.: Distance metric learning for large margin nearest neighbor classification. In: Advances in Neural Information Processing Systems, vol. 18, pp. 1473–1480 (2006)
16. Sugiyama, M.: Local fisher discriminant analysis for supervised dimensionality reduction. In: International Conference on Machine Learning (2006)
17. Tenenbaum, J.B., de Silva, V., Langford, J.C.: A Global Geometric Framework for Nonlinear Dimensionality Reduction. Science 290(5500), 2319–2323 (2000)
18. Roweis, S.T., Saul, L.K.: Nonlinear Dimensionality Reduction by Locally Linear Embedding. Science 290(5500), 2323–2326 (2000)
19. Belkin, M., Niyogi, P.: Laplacian eigenmaps for dimensionality reduction and data representation (2003)
20. Raytchev, B., Yoda, I., Sakaue, K.: Head Pose Estimation by Nonlinear Manifold Learning. In: International Conference on Pattern Recognition, pp. 462–466 (2004)
21. He, X., Niyogi, P.: Locality preserving projections. In: Advances in Neural Information Processing Systems, vol. 16 (2003)
22. Chen, L., Zhang, L., Hu, Y., Li, M., Zhang, H.: Head pose estimation using fisher manifold learning. In: Proceedings of the IEEE International Workshop on Analysis and Modeling of Faces and Gestures (AMFG 2003), pp. 203–207 (2003)
23. Ridder, D.D., Kouropteva, O., Okun, O., Pietikainen, M., Duin, R.: Supervised locally linear embedding (June 2003)
24. Li, C.G., Guo, J.: Supervised isomap with explicit mapping (2006)
25. Yang, M.H.: Extended isomap for classification. In: Proceedings of the 16th International Conference on Pattern Recognition, vol. 3, p. 30615 (2002)
26. Zelnik-Manor, L., Perona, P.: Self-tuning spectral clustering. In: NIPS (2004)
27. Tipping, M.: The relevance vector machine. In: Advances in Neural Information Processing Systems, San Mateo, CA. Morgan Kaufmann, San Francisco (2000)
28. Blanz, V., Vetter, T.: A Morphable Model for the Synthesis of 3D Faces. In: SIGGRAPH, pp. 187–194 (1999)

2D Image Analysis by Generalized Hilbert Transforms in Conformal Space*

Lennart Wietzke, Oliver Fleischmann, and Gerald Sommer

Kiel University, Department of Computer Science
Christian-Albrechts-Platz 4, 24118 Kiel, Germany
lw@ks.informatik.uni-kiel.de

Abstract. This work presents a novel rotational invariant quadrature filter approach - called *the conformal monogenic signal* - for analyzing i(ntrinsic)1D and i2D local features of any curved 2D signal such as lines, edges, corners and junctions without the use of steering. The *conformal monogenic signal* contains the *monogenic signal* as a special case for i1D signals and combines monogenic scale space, phase, direction/orientation, energy and curvature in one unified algebraic framework. The *conformal monogenic signal* will be theoretically illustrated and motivated in detail by the relation of the 3D Radon transform and the generalized Hilbert transform on the sphere. The main idea is to lift up 2D signals to the higher dimensional conformal space where the signal features can be analyzed with more degrees of freedom. Results of this work are the low computational time complexity, the easy implementation into existing Computer Vision applications and the numerical robustness of determining curvature without the need of any derivatives.

1 Introduction

Low level 2D image analysis is often the first step of many Computer Vision tasks. Therefore *local* signal features such as color information, gradient, curvature, orientation and phase determine the quality of subsequent higher level processing steps. It is important not to lose or to merge any of the original information within the local neighborhood, i.e. to span a orthogonal feature space. In this paper 2D images $f \in L_2(\Omega;\mathbb{R})$ with $\Omega \subset \mathbb{R}^2$ will be locally analyzed on a low level. 2D signals are classified into local regions $N \subseteq \Omega$ of different intrinsic dimension (see figure 1)

$$\begin{aligned}
\text{i0D} &:= \{f \in L_2(\Omega;\mathbb{R}) : f(\mathbf{x}_i) = f(\mathbf{x}_j)\ \forall \mathbf{x}_i, \mathbf{x}_j \in N\} \\
\text{i1D} &:= \{f \in L_2(\Omega;\mathbb{R}) : f(x,y) = g(x\cos\theta + y\sin\theta)\ \forall (x,y) \in N\} - \text{i0D} \\
\text{i2D} &:= L_2(\Omega;\mathbb{R}) - (\text{i0D} \cup \text{i1D}) .
\end{aligned}$$

The assumed local signal model which will be analyzed in this work is defined as a curve which can be locally approximated by a circle with arbitrary orientation and curvature

* We acknowledge funding by the German Research Foundation (DFG) under the project *SO 320/4-2*

D. Forsyth, P. Torr, and A. Zisserman (Eds.): ECCV 2008, Part II, LNCS 5303, pp. 638–649, 2008.

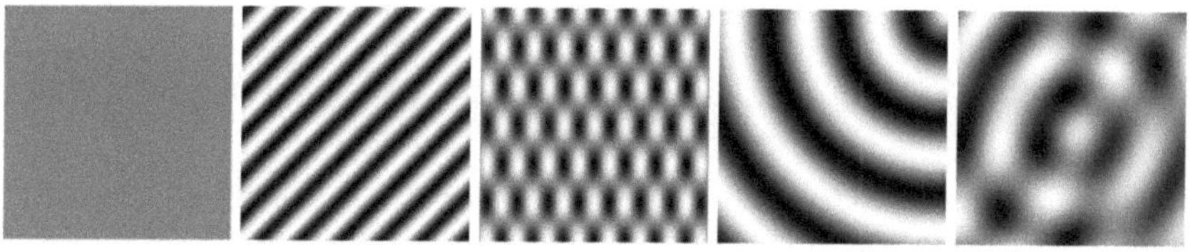

Fig. 1. From left to right: a constant signal (i0D), an arbitrary rotated 1D signal (i1D) and an i2D checkerboard signal consisting of two simple superimposed i1D signals. A curved i2D signal and two superimposed curved i2D signals. Note that all signals displayed here preserve their intrinsic dimension globally.

$$f(x,y) := a\ \cos\left(\left\|\begin{bmatrix}x\\y\end{bmatrix} - \frac{1}{\kappa}\begin{bmatrix}\cos\theta\\ \sin\theta\end{bmatrix}\right\| + \phi\right) \in \text{i1D} \cup \text{i2D} \tag{1}$$

with $a \in \mathbb{R}$ as the local amplitude, $\phi \in [0, 2\pi)$ as the local phase, $\kappa \in \mathbb{R}$ as the local curvature and $\theta \in [0, 2\pi)$ as the local direction/orientation of the signal for $\kappa \neq 0$. For the special case of $\kappa = 0$ the curved 2D signal degrades to an i1D function. Therefore the task is to solve an *inverse problem*, i.e. to determine local features such as amplitude a, phase ϕ, orientation θ and curvature κ of any curved signal such as lines, edges, corners and junctions. One important local structural feature is the phase ϕ which can be calculated by means of the Hilbert transform [1]. Furthermore all signals will be analyzed in *monogenic scale space* [2] since the Hilbert transform can only be interpreted for narrow banded signals consisting of only one frequency $f(x, y; s_s) := \mathcal{P}(x, y; s_s) * f(x, y)$ with $*$ as the convolution operator and s_s as the scale space parameter. The Poisson kernel of the applied low pass filter reads

$$\mathcal{P}(\mathbf{x}) := \mathcal{P}(\mathbf{x}; s_s) := \frac{s_s}{2\pi\left(s_s^2 + \|\mathbf{x}\|^2\right)^{\frac{n+1}{2}}} \quad n \in \mathbb{N},\ \mathbf{x} \in \mathbb{R}^n\ . \tag{2}$$

1.1 Related Work

Phase and energy of 1D signals can be analyzed by the *analytic signal* [1]. The generalization of the analytic signal to multidimensional signal domains has been done by the *monogenic signal* [3]. In case of 2D signals the *monogenic signal* delivers local phase, orientation and energy information restricted to the set of i1D signals. This work presents the generalization of the *monogenic signal* for 2D signals to analyze both i1D and i2D signals in one unified framework. The *conformal monogenic signal* delivers local phase, orientation, energy and curvature for i1D and i2D signals with the *monogenic signal* as a special case. The *monogenic signal* replaces the classical 1D Hilbert transform of the analytic signal by the generalized Hilbert transform [4]

$$R\{f\}(\mathbf{x}) := (\mathcal{Q} * f)(\mathbf{x}) := (h_n * \mathcal{P} * f)(\mathbf{x}) \quad \mathbf{x} \in \mathbb{R}^n,\ n \in \mathbb{N} - \{1\} \tag{3}$$

with $\mathcal{Q}$ as the conjugated Poisson kernel and h_n as the generalized Hilbert transform kernel

$$h_n(\mathbf{x}) := \frac{2}{A_{n+1}} \frac{\mathbf{x}}{\|\mathbf{x}\|^{n+1}} \quad \mathbf{x} \in \mathbb{R}^n,\ n \in \mathbb{N} - \{1\} \tag{4}$$

with A_{n+1} as the surface area of the unit sphere $\mathbb{S}^n$ in Euclidian space $\mathbb{R}^{n+1}$. To enable interpretation of the generalized Hilbert transform, its relation to the Radon transform is the key [5]. The generalized Hilbert transform can be expressed by a concatenation of the Radon transform, the inverse Radon transform and the well known classical 1D Hilbert transform. Note that the relation to the Radon transform is required solely for interpretation and theoretical results. Neither the Radon transform nor its inverse are ever applied to the signal in practice. Instead the generalized Hilbert transformed signal will be determined by convolution in spatial domain and the signal features can be extracted in a rotational invariant way.

2 Generalized Hilbert Transforms in Conformal Space

The feature space of the 2D *monogenic signal* is spanned by phase, orientation and energy information. This restriction correlates to the dimension of the associated 2D Radon space [5]. Therefore, the idea of this work is that the feature space of the 2D signal can only be extended by lifting up the original signal to higher dimensions. This is one of the main ideas of the *conformal monogenic signal.* In the following the 2D monogenic signal will be generalized to analyze also i2D signals by embedding the 2D signal into the 3D conformal space [6]. The 2D generalized Hilbert transform can be expressed by the 2D Radon transform which integrates all function values on lines [7]. This restriction to lines is one of the reasons why the 2D monogenic signal is limited to i1D signals (such as lines and edges) and can not be applied to corners and general curves. To analyze also i2D signals and to measure curvature $\kappa = \frac{1}{\rho}$, a 2D Radon transform which integrates on curved lines (i.e. local circles with radius ρ) is preferable. In 3D domain the Radon transform integrates on planes, although at first sight 3D planes are not related to 2D signals. But the idea is that circles form the intersection of a sphere (with center at $\left[0, 0, \frac{1}{2}\right]$ and radius $\rho = \frac{1}{2}$) and planes passing through the origin $(0, 0, 0)$ of 3D space. Since the generalized Hilbert transform can be extended to any dimension [8] and the 3D generalized Hilbert transform can be expressed by the 3D Radon transform, the 2D signal coordinates must be mapped appropriately to the sphere. This mapping must be conformal (i.e. angle preserving), so that angular feature interpretation of the 3D generalized Hilbert transform in conformal space is still reasonable. Analogous to the line parametrization by $(t, \theta) \in \mathbb{R} \times [0, \pi)$ of the 2D Radon transform [5], the planes of the 3D Radon transform are uniquely defined by the parameters $(t, \theta, \varphi) \in \mathbb{R} \times [0, 2\pi) \times [0, \pi)$. This new parametrization truly extends the interpretation space of the monogenic signal by one dimension. In contrast to the well known Monge patch embedding known from differential geometry [9], the original 2D signal will now be embedded into the conformal space.

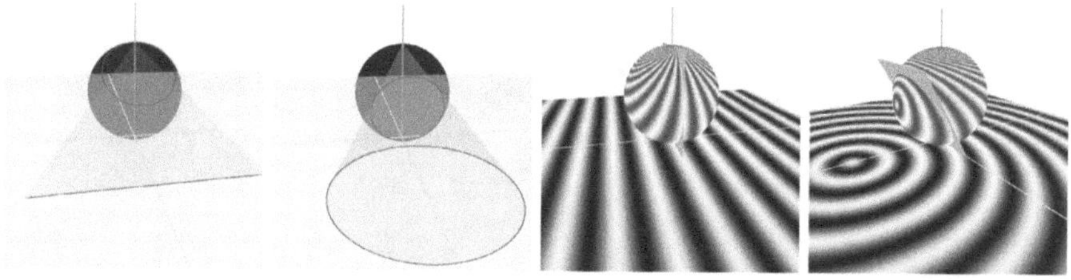

Fig. 2. Lines and circles of the 2D image domain are both mapped to circles on the sphere. Each circle on the sphere is uniquely defined by its parameterized intersection plane in conformal space. The third figure illustrates the monogenic signal as a special case.

2.1 The Conformal Space

The main idea is that the concept of lines in 2D Radon space becomes the concept of planes in 3D Radon space and the more abstract concept of hyperplanes in multidimensional space. These planes determine circles on the sphere in conformal space. Since lines and circles of the 2D signal domain are mapped to circles [6] on the sphere (see figure 2), the integration on these circles determines points in the 3D Radon space. The projection $\mathcal{C}$ known from complex analysis [6] maps the original 2D signal domain to the sphere and can be inverted by $\mathcal{C}^{-1}$

$$\mathcal{C}(x,y) := \frac{1}{x^2+y^2+1}\begin{bmatrix} x \\ y \\ x^2+y^2 \end{bmatrix}, \quad \mathcal{C}^{-1}(x,y,z) := \frac{1}{1-z}\begin{bmatrix} x \\ y \end{bmatrix}. \tag{5}$$

This mapping has the property that the 2D origin $(0,0)$ of a local coordinate system will be mapped to the south pole $(0,0,0)$ of the sphere in conformal space and both $-\infty, +\infty$ will be mapped to the north pole $(0,0,1)$ of the sphere. Lines and circles of the 2D signal domain will be mapped to circles on the sphere and can be determined uniquely by planes in 3D conformal space. The integration on these planes corresponds to points (t,θ,φ) in the 3D Radon space.

2.2 3D Radon Transform in Conformal Space

To interpret the *conformal monogenic signal*, the relation to the 3D Radon transform in conformal space must be taken into account. The 3D Radon transform is defined as the integral of all function values on the plane (see figure 2) defined by

$$\mathcal{R}\{c\}(t,\theta,\varphi) = \int_{\mathbf{x}\in\mathbb{R}^3} c(\mathbf{x})\delta_0(\mathbf{x}\begin{bmatrix} \sin\varphi\cos\theta \\ \sin\varphi\sin\theta \\ \cos\varphi \end{bmatrix} - t)d\mathbf{x}\,. \tag{6}$$

Since the signal is mapped on the sphere and all other points of the conformal space are set to zero, the 3D Radon transform actually sums up all points lying on the intersection of the plane and the sphere. For all planes this intersection

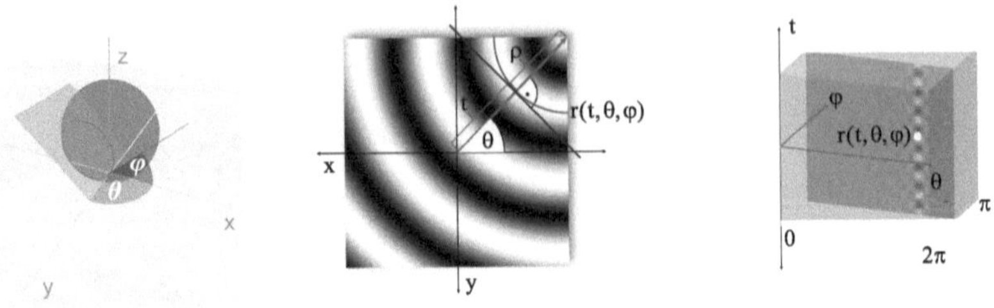

Fig. 3. Left figure: Each $(t = 0, \theta, \varphi)$ parameterized plane can be determined exactly by the generalized Hilbert transforms on the sphere. The interpretation of this parameter set delivers the features such as direction, phase and curvature of the original signal without any steering. Middle figure: Curved i2D signal with orientation θ and curvature $\kappa = \frac{1}{\rho}$. Right figure: Corresponding 3D Radon space representation of the i2D signal spanned by the parameters t, θ and φ. Since the Radon transform on circles directly on the plane of the original 2D signal is not possible, the Radon transform has to be done in higher dimensional 3D conformal space where circles correspond to planes.

can either be empty or a circle. The concept of circles in the conformal 3D Radon transform can be compared with the concept of lines known from the 2D Radon transform. Since lines in the 2D signal domain are also mapped to circles, the *conformal monogenic signal* can analyze i1D as well as curved i2D signals in one single framework. Recall the very important fact that every corner or curve can be locally approximated by a circle. The inverse 3D Radon transform exists and differs from the 2D case such that it is a *local* transformation [10].

$$\mathcal{R}^{-1}\{r\}(0,0,0) = -\frac{1}{8\pi^2} \int_{\theta=0}^{2\pi} \int_{\varphi=0}^{\pi} \frac{\partial^2}{\partial t^2} r(t,\theta,\varphi)|_{t=0}\, d\varphi\, d\theta\ . \tag{7}$$

That means the generalized Hilbert transform at $(0,0,0)$ is completely determined by all planes passing the origin (i.e. $t = 0$). In contrast, the 2D monogenic signal requires all integrals on all lines (t,θ) to reconstruct the original signal at a certain point and is therefore called a *global* transform. This interesting fact turns out from the definition of the inverse 3D Radon transform $\mathcal{R}^{-1}\{\cdot\}$. Therefore, the local features of i1D and i2D signals can be determined by the *conformal monogenic signal* at each test point of the original 2D signal without knowledge of the whole 3D Radon space.

2.3 The Conformal Monogenic Signal

To give the generalized Hilbert transform more degrees of freedom for signal analysis, the original 2D signal will be embedded in a applicable subspace of the 3D conformal space by the mapping

$$c(x,y,z) := \begin{cases} f(\mathcal{C}^{-1}(x,y,z)^T; s_s) & , x^2 + y^2 + \left(z - \frac{1}{2}\right)^2 = \frac{1}{4} \\ 0 & , \text{else} \end{cases} . \tag{8}$$

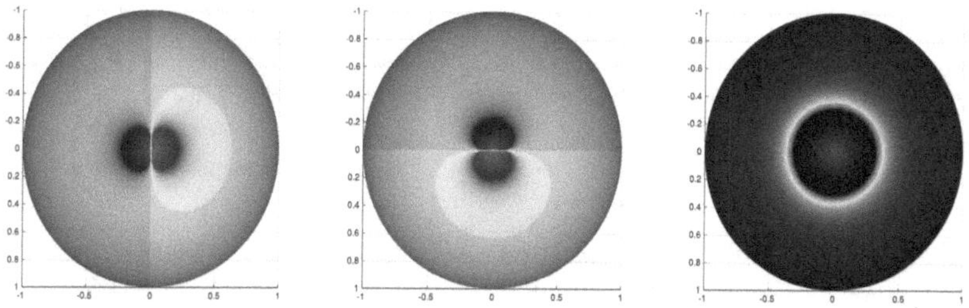

Fig. 4. From left to right: 2D convolution kernels in spatial domain of the *conformal monogenic signal* in x, y and z direction

Thus, the 3D generalized Hilbert transform can be applied to all points on the sphere. The center of convolution in spatial domain is the south pole $(0, 0, 0)$ where the test point of the 2D signal domain meets the sphere. At this point the 3D generalized Hilbert transform will be performed at the origin $(\mathbf{0})$ of the applied local coordinate system for each test point separately. The *conformal monogenic signal* [11] is defined as

$$f_{\mathrm{CMS}}(\mathbf{0}) := \left[c(\mathbf{0}), R_x\{c\}(\mathbf{0}), R_y\{c\}(\mathbf{0}), R_z\{c\}(\mathbf{0})\right]^T \tag{9}$$

and can be expressed by the classical 1D Hilbert transform kernel $h_1(\tau) := -\frac{1}{\pi\tau}$ [1], the 3D Radon transform and its inverse analogous to the monogenic signal in 2D [5]

$$\begin{bmatrix} R_x\{c\}(\mathbf{0}) \\ R_y\{c\}(\mathbf{0}) \\ R_z\{c\}(\mathbf{0}) \end{bmatrix} = \left[\mathcal{R}^{-1}\left\{\begin{bmatrix} \sin\varphi\cos\theta \\ \sin\varphi\sin\theta \\ \cos\varphi \end{bmatrix} h_1(t) * \mathcal{R}\{c\}(t, \theta, \varphi)\right\}(0, 0, 0)\right] \tag{10}$$

with $*$ as the 1D convolution operator. Compared to the 2D monogenic signal the *conformal monogenic signal* performs a 3D generalized Hilbert transformation in conformal space.

3 Interpretation

Analogous to the interpretation of the *monogenic signal* in [5], the parameters of the plane within the 3D Radon space determine the local features of the curved i2D signal (see figure 2). The *conformal monogenic signal* can be called the generalized monogenic signal for i1D and i2D signals, because the special case of lines and edges can be considered as circles with zero curvature. These lines are mapped to circles passing through the north pole in conformal space. The parameter θ will be interpreted as the orientation in i1D case and naturally deploys to direction $\theta \in [0, 2\pi)$ for the i2D case

$$\theta = \operatorname{atan2}\left(R_y\{c\}(\mathbf{0}), R_x\{c\}(\mathbf{0})\right) . \tag{11}$$

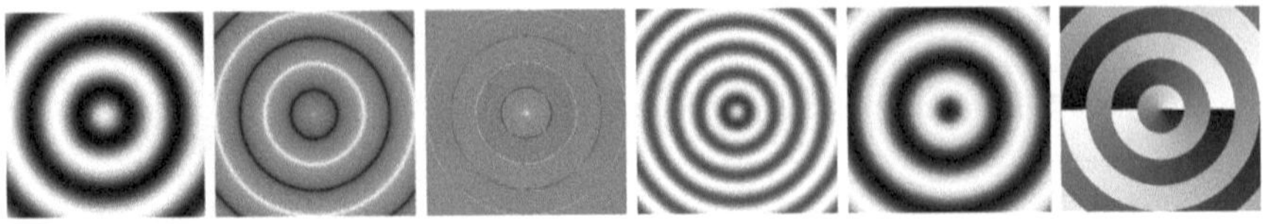

Fig. 5. From left to right: Original synthetic i2D signal $f(x,y) = a\cos(\sqrt{x^2+y^2}+\phi)$, *conformal monogenic signal* curvature and classical isophote curvature [12], *conformal monogenic signal* energy, phase and direction. Convolution mask size: 11×11 pixels. Rotational invariance and isotropic properties can be clearly seen

The energy of the signal is defined by

$$E = a^2 = c^2(\mathbf{0}) + R_x^2\{c\}(\mathbf{0}) + R_y^2\{c\}(\mathbf{0}) + R_z^2\{c\}(\mathbf{0}) \,. \tag{12}$$

The i1D and i2D curvature phase is defined by

$$\phi = \operatorname{atan2}\left(\sqrt{R_x^2\{c\}(\mathbf{0}) + R_y^2\{c\}(\mathbf{0}) + R_z^2\{c\}(\mathbf{0})}, c(\mathbf{0})\right) \,. \tag{13}$$

Note that all proofs are analogous to those for the 2D monogenic signal shown in [5].

3.1 Local Isophote Curvature

The parameter φ of the 3D Radon space corresponds to the isophote curvature κ [12] known from differential geometry

$$\varphi = \arctan \frac{\sqrt{R_x^2\{c\}(\mathbf{0}) + R_y^2\{c\}(\mathbf{0})}}{R_z\{c\}(\mathbf{0})}, \quad \kappa = \frac{-f_{xx}f_y^2 + 2f_xf_yf_{xy} - f_{yy}f_x^2}{\left(f_x^2 + f_y^2\right)^{\frac{3}{2}}} \tag{14}$$

Proof:
Let be $\gamma(t) := [\rho(\cos\theta + \cos t), \rho(\sin\theta + \sin t)]^T$ with $t \in [0, 2\pi)$ a parametrization of a circle in the 2D plane touching the origin $(0,0)$ with radius ρ and tangential orientation θ. This circle will be the model for the osculating circle touching the isophote curve of the 2D signal f at the origin $(0,0)$ of the local coordinate system for each test point. Therefore, $f(\gamma(t_1)) = f(\gamma(t_2))\ \forall t_1, t_2 \in [0, 2\pi)$. Define $\gamma_S(t) := \mathcal{C}(\gamma(t))$ as the projection of γ to the sphere $\mathbb{S}^2(m_S, \rho_S) := \left\{v \in \mathbb{R}^3 : \|v - m_S\| = \rho_S\right\}$ with the center $m_S = \left[0, 0, \frac{1}{2}\right]^T$ and the radius $\rho_S = \frac{1}{2}$. Furthermore define $f_S(\gamma_S(t)) := f(\mathcal{C}^{-1}(\gamma_S(t)))$. The conjugate Poisson kernel in $\mathbb{R}_+^{n+1}$ reads $\mathcal{Q}(x) := [\mathcal{Q}_x(x), \mathcal{Q}_y(x), \mathcal{Q}_z(x)]^T = (h_3 * \mathcal{P})(x)$ with $R\{f_S\}(\mathbf{x}) = (\mathcal{Q} * f_S)(\mathbf{x})$. The radius ρ of the osculating circle described by the parameterized curve γ reads

$$\rho = \frac{2R_z\{f_S\}(\mathbf{0})}{\sqrt{R_x^2\{f_S\}(\mathbf{0}) + R_y^2\{f_S\}(\mathbf{0})}} \,. \tag{15}$$

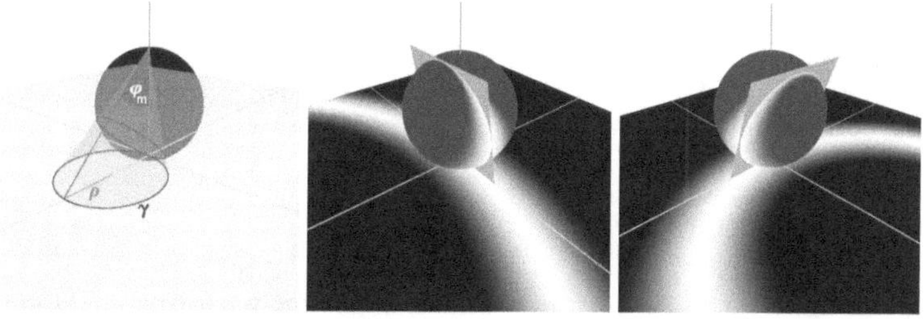

Fig. 6. Left figure: Visualization of the circle described by γ projected to $\mathbb{S}^2(m_S, \rho_S)$. This figure illustrates that $\frac{1}{\kappa} = \rho = \tan \varphi_m$.

Since the values of $f_S(x)$ will only be nonzero for $x \in \mathbb{S}^2(m_S, \rho_S)$, the integration can be restricted to the ball $\mathbb{B}^2(m_S, \rho_S) := \left\{v \in \mathbb{R}^3 : \|v - m_S\| \leq \rho_S\right\}$. Furthermore $f_S(x)$ is only nonzero for the circle projected on the sphere $M := \{\gamma_S(t) : t \in [0, 2\pi)\}$. Now let $\mathbb{S}^2(m, \rho)$ be the sphere whose intersection with $\mathbb{S}^2(m_S, \rho_S)$ results in M. Then the set M is a circle on the surface of $\mathbb{S}^2(m_S, \rho_S)$ and $\mathbb{S}^2(m, \rho)$. The integration over the volumes of $\mathbb{B}^2(m, \rho)$ and $\mathbb{B}^2(m_S, \rho_S)$ will be the same.

$$\int_{x \in \mathbb{R}^3_+} \mathcal{Q}(x) f_S(x)\, dx = \int_{x \in \mathbb{B}^2(m_S, \rho_S)} \mathcal{Q}(x) f_S(x)\, dx = \int_{x \in \mathbb{B}^2(m, \rho)} \mathcal{Q}(x) f_S(x)\, dx$$

According to the results from harmonic analysis [13] the convolution of a function in $\mathbb{R}^n$ with the Poisson kernel $\mathcal{P}$ in upper the half space $\mathbb{R}^{n+1}_+$ results in a harmonic function in $\mathbb{R}^{n+1}_+$. Therefore, $\mathcal{Q}$ is harmonic in $\mathbb{R}^{3+1}_+$. Using the *mean value theorem* for harmonic functions it follows that

$$\int_{x \in \mathbb{B}^2(m, \rho)} \mathcal{Q}(x)\, dx = k\, \mathcal{Q}(m) \tag{16}$$

with the components of $\mathcal{Q}$ written in spherical coordinates

$$\mathcal{Q}(m) = \begin{bmatrix} \mathcal{Q}_x(m) \\ \mathcal{Q}_y(m) \\ \mathcal{Q}_z(m) \end{bmatrix} = \frac{1}{\left[\|m\|^2 + s_s^2\right]^2} \begin{bmatrix} \sin \varphi_m \cos \theta_m \\ \sin \varphi_m \sin \theta_m \\ \cos \varphi_m \end{bmatrix} \tag{17}$$

with s_s as the scale space parameter. Since f_S is the signal model for the isophote curve of a signal in the plane, it is a curve consisting of constant values. Therefore, $f_S(x)$ will be constant for all $x \in M$ which results in

$$\int_{x \in \mathbb{B}^2(m, \rho)} \mathcal{Q}(x) f_S(x)\, dx = f_c \int_{x \in \mathbb{B}^2(m, \rho)} \mathcal{Q}(x)\, dx = f_c\, k\, \mathcal{Q}(m)\,. \tag{18}$$

With equation (15) it is now possible to determine $\frac{\sin\varphi_m}{\cos\varphi_m}$. Figure 6 illustrates that this is exactly $\frac{\rho}{2\,\rho_S}$. Since $\rho_S = \frac{1}{2}$ it follows that the radius of the local curvature can be determined by $\rho = \frac{\sin\varphi_m}{2\cos\varphi_m} = \frac{\sqrt{Q_x^2(m)+Q_y^2(m)}}{2Q_z(m)}$.

4 Implementation

Although in theory the *conformal monogenic signal* performs a 3D generalized Hilbert transform in conformal space this can be accelerated by simplifying to a faster 2D convolution directly on the sphere. Therefore the following implementation can be done in $O(n^2)$ with n as the convolution mask size in one dimension. Any 2D image can now be analyzed locally at every test point by the following algorithm

```
//Input:  double Image(double x,double y)
//Input:  double x,y (Local pixel test point for analysis)
//Input:  double Coarse > Fine > 0 (Bandpass filter parameters)
//Input:  double Size > 0 (Convolution mask size)
//Output: Direction, Phase, Curvature, Energy

double Coarse=0.2,Fine=0.1,Size=5;//e.g.
double rp=0,rx=0,ry=0,rz=0;
for(double cx = -Size;cx <= Size;cx++)
for(double cy = -Size;cy <= Size;cy++) {
    //Map points (cx,cy) to conformal space (u,v,w)
    double d = pow(cx,2) + pow(cy,2) + 1;
    double u = cx / d, v = cy / d, w = (d - 1) / d;
    //Generalized Hilbert transform in conformal space
    double uvw = pow(u,2) + pow(v,2) + pow(w,2);
    double pf  = pow(pow(Fine   ,2) + uvw,-2);
    double pc  = pow(pow(Coarse,2) + uvw,-2);
    double f   = Image(x + cx,y + cy);
    double c   = f * (pf - pc);
    rp += f * (Fine * pf - Coarse * pc);
    rx += u * c, ry += v * c, rz += w * c;
}
Curvature = sqrt(pow(rx,2) + pow(ry,2)) / rz;
Direction = atan2(ry,rx);
Phase     = atan2(sqrt(pow(rx,2) + pow(ry,2) + pow(rz,2)),rp);
//For energy a DC free convolution kernel must be used instead
Energy    = pow(rp,2) + pow(rx,2) + pow(ry,2) + pow(rz,2);
```

5 Experimental Results

On synthetic signals with known ground truth the average error of the feature extraction converges to zero with increasing refinement of the convolution mask

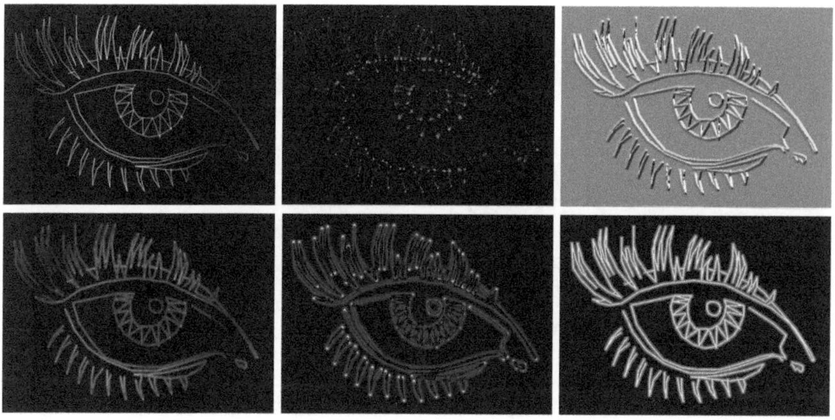

Fig. 7. Top row from left to right: Original signal, classical isophote curvature [12] and local direction information. Bottom row from left to right: *conformal monogenic signal* energy, curvature and phase. Note the observable illumination invariance of the conformal curvature and phase. Convolution mask size: 7×7 pixels.

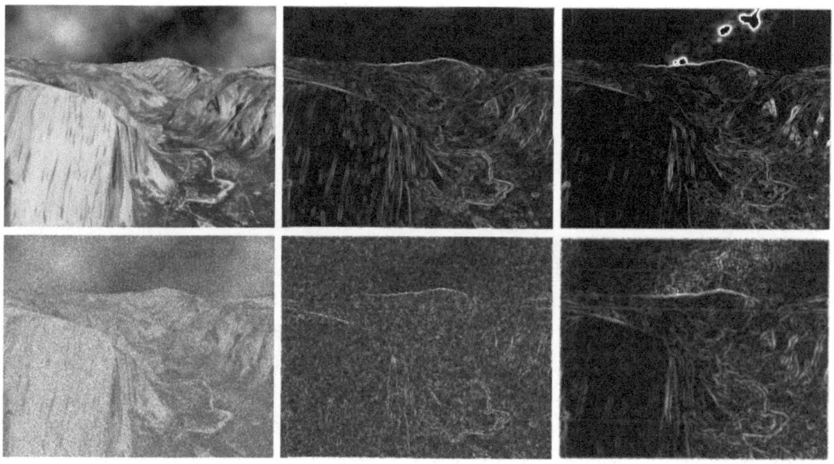

Fig. 8. Top row from left to right: Original Yosemite image, Sobel detector output and *conformal monogenic signal* curvature which delivers much more structural information (see cloudy sky). Bottom row from left to right: Noise degraded image (SNR=10dB), blurred Sobel output and *conformal monogenic signal* curvature. Convolution mask size: 7×7 pixels.

size. The advanced rotational invariance and isotropic behavior of the *conformal monogenic signal* curvature compared to the curvature delivered by the classical differential geometry approach can be seen clearly in figure 5. Under the presence of noise the *conformal monogenic signal* curvature performs more robust

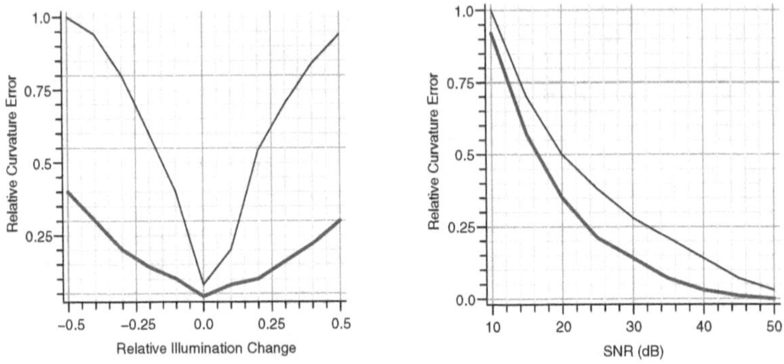

Fig. 9. Comparison of classical isophote curvature [12] (thin blue curve) and conformal curvature (thick red curve) errors. Left figure: Illumination change relatively to original signal. Right figure: Additive Gaussian noise. Convolution mask size: 5×5 pixels.

than e.g. the gradient based Sobel detector (see figure 8). Substituting the gradient and even the in [14] proposed phase constraints, the *conformal monogenic signal* curvature feature performs better in optical flow applications with an average angular error (AAE) of 1.99° compared to the best result $AAE = 2.67°$ achieved in [14] on the Yosemite sequence (see figure 8). Since the *conformal monogenic signal* combines all intrinsic dimensions in one framework it could be an interesting alternative for the gradient or the Laplace operator.

6 Conclusion

In this paper a new fundamental idea for locally analyzing 2D curved signals such as lines, edges, corners and junctions in one unified framework has been presented. It has been shown that the feature space can be extended by embedding 2D signals in higher dimensional conformal spaces in which the original 2D signal can be analyzed by generalized Hilbert transforms with more degrees of freedom. Without steering and in an rotational invariant way, local signal features such as phase, direction, energy and curvature can be determined in spatial domain by 2D convolution. The *conformal monogenic signal* can be computed efficiently and easily implemented into existing low level image processing steps of Computer Vision applications. Furthermore exact curvature can be calculated with all the advantages of rotational invariant local phase based approaches (robustness against noise and illumination changes) and without the need of any derivatives. Hence, lots of numerical problems of partial derivatives on discrete grids can be avoided. All results can be proofed mathematically and by experiments. More applications of the *conformal monogenic signal* such as object tracking [12] by the conformal isophote curvature and the extension to image sequences will be part of future work.

References

1. Hahn, S.L.: Hilbert Transforms in Signal Processing. Artech House Inc., Boston (1996)
2. Felsberg, M., Sommer, G.: The monogenic scale-space: A unifying approach to phase-based image processing in scale-space. Journal of Mathematical Imaging and Vision 21, 5–26 (2004)
3. Felsberg, M.: Low-Level Image Processing with the Structure Multivector, Technical Report No. 2016. Ph.D thesis, Kiel University, Department of Computer Science (2002)
4. Brackx, F., Knock, B.D., Schepper, H.D.: Generalized multidimensional Hilbert transforms in Clifford analysis. International Journal of Mathematics and Mathematical Sciences (2006)
5. Wietzke, L., Sommer, G., Schmaltz, C., Weickert, J.: Differential geometry of monogenic signal representations. In: Sommer, G. (ed.) RobVis 2008. LNCS, vol. 4931, pp. 454–465. Springer, Heidelberg (2008)
6. Needham, T.: Visual Complex Analysis. Oxford University Press, Oxford (1997)
7. Toft, P.: The Radon Transform - Theory and Implementation. Ph.D thesis, Technical University of Denmark (1996)
8. Delanghe, R.: Clifford analysis: History and perspective. Computational Methods and Function Theory 1(1), 107–153 (2001)
9. do Carmo, M.P.: Differential Geometry of Curves and Surfaces. Prentice-Hall, Englewood Cliffs (1976)
10. Bernstein, S.: Inverse Probleme. Technical report, TU Bergakdemie Freiberg (2007)
11. Wietzke, L., Sommer, G.: The Conformal Monogenic Signal. In: Rigoll, G. (ed.) DAGM 2008. LNCS, vol. 5096, pp. 527–536. Springer, Heidelberg (2008)
12. Lichtenauer, J., Hendriks, E.A., Reinders, M.J.T.: Isophote properties as features for object detection. CVPR (2), 649–654 (2005)
13. Axler, S., Bourdon, P., Ramey, W.: Harmonic Function Theory Graduate Texts in Mathematics, vol. 137. Springer, Heidelberg (2002)
14. Zang, D., Wietzke, L., Schmaltz, C., Sommer, G.: Dense optical flow estimation from the monogenic curvature tensor. In: Sgallari, F., Murli, A., Paragios, N. (eds.) SSVM 2007. LNCS, vol. 4485, pp. 239–250. Springer, Heidelberg (2007)

An Efficient Dense and Scale-Invariant Spatio-Temporal Interest Point Detector

Geert Willems[1], Tinne Tuytelaars[1], and Luc Van Gool[1,2]

[1] ESAT-PSI, K.U. Leuven, Belgium
{gwillems,tuytelaa,vangool}@esat.kuleuven.be
[2] ETH, Zürich, Switzerland

Abstract. Over the years, several spatio-temporal interest point detectors have been proposed. While some detectors can only extract a sparse set of scale-invariant features, others allow for the detection of a larger amount of features at user-defined scales. This paper presents for the first time spatio-temporal interest points that are at the same time scale-invariant (both spatially and temporally) and densely cover the video content. Moreover, as opposed to earlier work, the features can be computed efficiently. Applying scale-space theory, we show that this can be achieved by using the determinant of the Hessian as the saliency measure. Computations are speeded-up further through the use of approximative box-filter operations on an integral video structure. A quantitative evaluation and experimental results on action recognition show the strengths of the proposed detector in terms of repeatability, accuracy and speed, in comparison with previously proposed detectors.

1 Introduction

As video becomes a ubiquitous source of information, video analysis (*e.g.* [1]) and action recognition (*e.g.* [2,3]) have received a lot of attention lately. In this context, local viewpoint invariant features, so successful in the field of object recognition and image matching, have been extended to the spatio-temporal domain [4,5,6,7,8]. These extensions take the 3D nature of video data into account and localize features not only spatially but also over time.

Laptev and Lindeberg [5] were the first to propose such a spatio-temporal extension, building on the Harris-Laplace detector proposed by Mikolajczyk and Schmid [9]. They typically detect only a sparse set of features as a time-consuming iterative procedure has to be repeated for each feature candidate separately. Furthermore, the iterative procedure often diverges. As a result, detecting a low number of features is a necessity to keep the computation time under control.

Dollár *et al.* [6], on the other hand, claim that direct 3D counterparts to 2D interest point detectors are inadequate for the detection of spatio-temporal feature points, since true spatio-temporal corners are quite rare. They propose to select local maxima over space and time of a response function based on a spatial Gaussian convolved with a quadrature pair of 1D Gabor-filters along the time axis. However, their features are not scale-invariant. The size of these *cuboids* is determined by the user.

D. Forsyth, P. Torr, and A. Zisserman (Eds.): ECCV 2008, Part II, LNCS 5303, pp. 650–663, 2008.

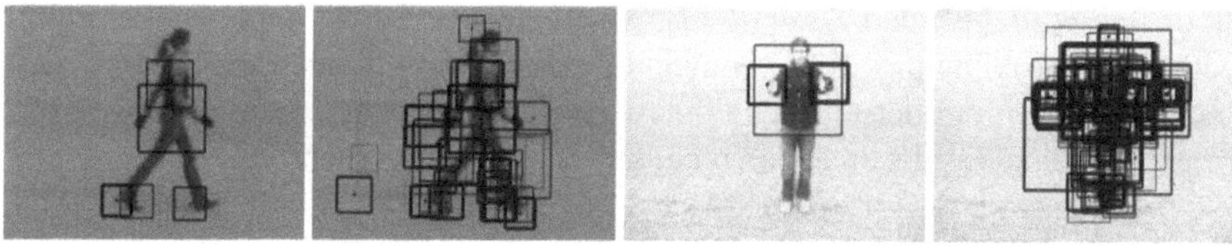

Fig. 1. The proposed scale-invariant spatio-temporal interest points (Hes-STIP). The density of features can be varied from very sparse (first and third image) to very dense (second and fourth image), simply by changing the threshold and with minimal effect on the computation time.

Oikonomopoulos *et al.* [8] have proposed a spatio-temporal extension of the salient region detector proposed by Kadir and Brady [10]. The features are scale-invariant yet sparse, as was also the case for the original spatial detector.

Recently, Wong and Cipolla [11] have developed a novel method for extracting spatio-temporal features using global information. Using their method based on the extraction of dynamic textures, only a sparse set of features is needed for action recognition. However, all input videos need to be preprocessed into samples containing one iteration of the action each.

Also related is the work of Ke *et al.* [7] on visual event detection. They build on the concept of integral video to achieve realtime processing of video data. However, rather than relying on interest points, they use dense spatio-temporal Haar-wavelets computed on the optical flow. Discriminative features are then selected during a training stage. This results in application dependent features which are, again, not scale-invariant.

Table 1 summarizes the most important properties of the previously mentioned detectors. The currently available spatio-temporal interest point (STIP) detectors [5,6,8] are computationally expensive and are therefore restricted to the processing of short or low resolution videos. The existing scale-invariant feature detectors [5,8] only yield a sparse set of features.

In this paper, we present a novel spatio-temporal feature detector which is the first to obtain a dense set of scale-invariant features (fig 1) in an efficient way. Our main contributions can be summarized as follows. First, we show that features can be localized both in the spatio-temporal domain and over both scales simultaneously when using the determinant of the Hessian as saliency measure. We thus remove the need for the iterative scheme in the work by Laptev and Lindeberg [5]. Second, building on the

Table 1. Comparison between spatio-temporal interest point detectors

detector	scale selection	feature set	efficient	app. independent
Laptev [5]	yes, iterative	sparse (rare)	no	yes
Dollár [6]	no	dense	no	yes
Ke [7]	no	dense	yes (box filters)	no
Oikonomopoulos [8]	yes	sparse (rare)	no	yes
proposed method	yes	dense	yes (box filters)	yes

work of Bay *et al.* [12] and Ke *et al.* [7], we create an efficient implementation of the detector by approximating all $3D$ convolutions using box-filters. Finally, we compare the repeatability of our detector with the two best known state-of-the-art detectors and show experimental results on action recognition and video synchronization.

2 Spatio-Temporal Interest Point Detection

In this section, we first briefly recapitulate some basic scale space terminology as well as the Harris-Laplace-based space-time interest points (HL-STIP) of Laptev and Lindeberg [5]. Next, we propose our Hessian-based spatio-temporal interest point (Hes-STIP) detector and discuss its advantage w.r.t. localisation and scale selection.

2.1 Some Scale Space Principles

Starting from a spatio-temporal signal $f(\cdot)$, a spatio-temporal scale space representation is obtained by convolving $f(\cdot)$ with a Gaussian kernel [13]

$$L(\cdot;\sigma^2,\tau^2) = g(\cdot;\sigma^2,\tau^2) * f(\cdot) \tag{1}$$

with σ and τ the spatial and temporal scales respectively. Building blocks of virtually any method working in scale space are the Gaussian derivatives

$$L_{x^k y^l t^m}(\cdot;\sigma^2,\tau^2) = \partial_{x^k y^l t^m} g(\cdot;\sigma^2,\tau^2) * f(\cdot) \tag{2}$$

The amplitude of these spatio-temporal derivatives decreases with scale. To obtain scale invariance, *scale-normalized derivatives* should be used, defined as

$$L^{norm}_{x^k y^l t^m}(\cdot;\sigma^2,\tau^2) = \sigma^{k+l}\tau^m L_{x^k y^l t^m}(\cdot;\sigma^2,\tau^2) \tag{3}$$

Working with scale-normalized derivatives ensures that the same values are obtained irrespective of the scale.

Scale selection refers to the process of selecting a characteristic scale [13]. This can be achieved by searching local extrema of some saliency measure. In this context, normalization factors $\sigma^{\gamma(k+l)}$ or $\tau^{\lambda m}$ are often used. This is known as γ*-normalization*. By playing with different values for γ and λ one can ensure that, at least for prototypical patterns such as a perfect Gaussian blob, the saliency measure reaches a local maximum and that the spatio-temporal extent of the detected features corresponds to some meaningful entity (*e.g.* the correct scale of the Gaussian blob).

2.2 Introduction to Space-Time Interest Points

Since our approach has some similarity with the space-time interest points proposed by Laptev and Lindeberg [5], we shortly describe their method that extends the 2D scale-invariant Harris-Laplace corner detector [14] into the spatio-temporal domain. To this end, a 3×3 spatio-temporal second-moment matrix

$$\mu = g(\cdot;\sigma_i^2,\tau_i^2) \star \begin{pmatrix} L_x^2 & L_xL_y & L_xL_t \\ L_xL_y & L_y^2 & L_yL_t \\ L_xL_t & L_yL_t & L_t^2 \end{pmatrix} \tag{4}$$

is defined, with σ_i and τ_i the spatial and temporal integration scales. The strength of each interest point at a certain scale is then computed by the (extended) Harris corner function

$$S = det(\mu) - k' trace^3(\mu) \tag{5}$$

with a typical value for $k' = 0.001$. To recover the spatio-temporal extent of $f(\cdot)$, scale selection is applied based on the γ-normalized Laplacian, defined as

$$(\nabla^2 L)^{\gamma norm} = \sigma^{2a}\tau^{2b}L_{xx} + \sigma^{2a}\tau^{2b}L_{yy} + \sigma^{2c}\tau^{2d}L_{tt} \tag{6}$$

In order to achieve an extremum at the correct scales for a prototype Gaussian blob, the normalizing parameters are set to $a = 1, b = 1/4, c = 1/2, d = 3/4$ [5].

To find points in scale-space that are both maxima of the Harris corner function (5) in space/time and extrema of the normalized Laplacian (6) over both scales, an iterative scheme must be used. First, interest points are detected for a sparsely distributed set of scales. Then each point is iteratively updated until convergence by alternating between scale optimization and re-detection of the position given the novel scales.

2.3 Hessian-Based Localization and Scale Selection

In this paper, we propose the use of the Hessian matrix for spatio-temporal feature detection:

$$H(\cdot, \sigma^2, \tau^2) = \begin{pmatrix} L_{xx} & L_{xy} & L_{xt} \\ L_{yx} & L_{yy} & L_{yt} \\ L_{tx} & L_{ty} & L_{tt} \end{pmatrix} \tag{7}$$

The strength of each interest point at a certain scale is then computed by

$$S = |det(H)| \tag{8}$$

This can be considered as a spatio-temporal extension of the saliency measure proposed by Beaudet for blob detection [15]. However, unlike the 2D case, a positive value of S does not guarantee all eigenvalues of $H(\cdot, \sigma^2, \tau^2)$ having the same sign. As a result, apart from blobs, also saddle points can give rise to local extrema. For most applications, the nature of the interest points does not really matter – as long as they can be detected reliably, i.e. with high repeatability. If for whatever reason only blobs should be found, one can easily check the sign of the eigenvalues and reject the saddle points in a postprocessing step.

Using the Hessian matrix, scale selection can be realized in various ways. First, we show how to use γ-normalization, similar to the work of Laptev and Lindeberg. Next, a different strategy is proposed, leading to a more efficient solution.

Scale Selection through γ-Normalization. Using γ-normalization, we alter the saliency measure so as to ensure that the 'correct' scales σ_0 and τ_0 are found for a perfect Gaussian blob $g(x, y, t; \sigma_0^2, \tau_0^2)$. At the center of this blob, the determinant is solely determined by the first term $L_{xx}L_{yy}L_{tt}$ as all other terms vanish. The γ-normalized determinant at the center can thus be written as

$$L_{xx}^{\gamma norm}L_{yy}^{\gamma norm}L_{tt}^{\gamma norm} = \sigma^{2p}\tau^{2q}L_{xx}L_{yy}L_{tt} \tag{9}$$

To find the extrema, we differentiate $det(H)^{\gamma norm}$ with respect to the spatial and temporal scale parameters σ^2 and τ^2, and set these derivatives equal to zero. Again, all terms but the first vanish at the center of the Gaussian blob. From this analysis, it follows that the local extrema $\tilde{\sigma}$ and $\tilde{\tau}$ coincide with the correct scales σ_0 and τ_0 if we set $p = 5/2$ and $q = 5/4$.

Note that these values are related to the values a, b, c, and d for the normalization of the Laplacian, namely $p = 2a + c$ and $q = 2b + d$. This reflects the fact that the determinant at the center of the Gaussian blob reduces to the product of two spatial second-order derivates and one temporal second-order derivative.

This way, a γ-normalized operator is obtained with $\gamma \neq 1$, as was also the case with the γ-normalized Laplacian (6). This implies, however, that the measure used for scale selection is *not* truely scale invariant, and as a result cannot be used to find local maxima over scales. With two different criteria to optimize, we are again bound to use an iterative method.

Simultaneous Localization and Scale Selection. In contrast with the normalized Laplacian, scale invariance *and* good scale selection can be achieved simultaneously with the scale-normalized determinant of the Hessian. Indeed, using $p = 2$ and $q = 1$ (*i.e.* $\gamma = 1$) in equation 9, we obtain the following relationship between the local extrema $(\tilde{\sigma}, \tilde{\tau})$ and the correct scales (σ_0, τ_0) for a Gaussian blob:

$$\tilde{\sigma}^2 = \frac{2}{3}\sigma_0^2 \quad \tilde{\tau}^2 = \frac{2}{3}\tau_0^2 \tag{10}$$

In general, it can be shown that, in D dimensions, the determinant of the scale-normalized Hessian, with $\gamma = 1$, reaches an extremum at the center of a Gaussian blob $g(\mathbf{x}; \boldsymbol{\sigma_0})$ with $\boldsymbol{\sigma_0} = [\sigma_{0,1}, \ldots, \sigma_{0,D}]$, for scales

$$\tilde{\boldsymbol{\sigma}} = \sqrt{\frac{2}{D}}\boldsymbol{\sigma_0} \tag{11}$$

Even though the detected scales $\tilde{\sigma}, \tilde{\tau}$ do not coincide with the correct scales σ_0, τ_0, they are related by a fixed scale factor. This makes it trivial to obtain the latter [1].

Since we now have a single, scale-invariant measure that can be used both for the localization as well as for the selection of the spatial and temporal scale, a non-iterative method can be used. To this end, we select local extrema over the $5D$ space defined by (x, y, t, σ, τ). We then multiply the scales of each interest point found with the factor $\sqrt{3/2}$ to obtain the real scales of the underlying signal. This brings a clear speed advantage over the iterative procedure of [5], avoids problems with convergence and allows for the extraction of any number of features simply by changing the threshold of the saliency measure.

[1] Note that the use of the determinant of the Hessian is crucial for the above method to work. Using a scale-normalized version of the Laplacian, no extrema are found unless for $\tilde{\boldsymbol{\sigma}} = 0$.

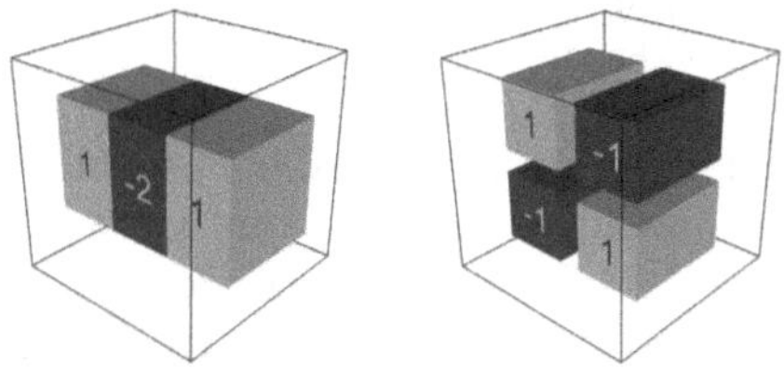

Fig. 2. The two types of box filter approximations for the $2 + 1D$ Gaussian second order partial derivatives in one direction (left) and in two directions (right)

2.4 Implementation Details

Integral Video. In the previous section, we showed that the use of the determinant of the Hessian as a saliency measure allows for the direct localisation of spatio-temporal interest points in a $5D$ space. Nevertheless, computing the determinant of the Hessian at many positions and many scales can become computationally prohibitive.

In [7], Ke *et al.* combine box-filters in order to obtain volumetric features. By using an *integral video* structure - the spatio-temporal generalization of integral images - these box-filters can be computed very efficiently and even allow for realtime action detection.

We also build on integral videos in order to make this part of the problem tractable. In a first step, a video containing F frames of dimension $W \times H$ is converted into an integral video structure where an entry at location $\mathbf{x} = (x, y, t)$ holds the sum of all pixels in the rectangular region spanned by $(0, 0) - (x, y)$, summed over all frames $[0, t]$. Using integral videos, the sum of values within any rectangular volume can be approximated with 8 additions, independent of the volume's size.

We approximate all Gaussian second-order derivatives very roughly with box-filter equivalents as was done in 2D by Bay *et al.* [12]. In total there are 6 different second order derivatives in the spatio-temporal domain: D_{xx}, D_{yy}, D_{tt}, D_{xy}, D_{tx} and D_{ty}, which can be computed using rotated versions of the two box-filters shown in figure 2.

Spatio-Temporal Search Space. Thanks to the use of the integral video structure and the box-filters, the scale spaces do not have to be computed hierarchically but can be efficiently implemented by upscaling the box-filters. Each octave is divided into 5 scales, with a ratio between subsequent scales in the range $1.2 - 1.5$ for the inner 3 scales.

The determinant of the Hessian is computed over several octaves of both the spatial and temporal scale. For a combination of octaves o_σ and o_τ, each pair of scales (σ_i, τ_i) results in a cube structure filled with Hessian-based strengths. Once all cubes have been filled, we use a non-maximum suppression algorithm to obtain all extrema within the obtained 5 dimensional search-space (x, y, t, σ, τ). Upon the detection of an extremum, a gradient descent is started to obtain *sub-pixel* accuracy in all 5 values.

Note that, although it is possible to search for local extrema in the 5 dimensional search-space, this is not always required by the application. Indeed, depending on the application, the temporal or spatial scale can be fixed, reducing the search space to 3 or 4 dimensions. Another strategy, similar to what is done in [16], is to consider only a small band around a chosen temporal scale.

Descriptor. To describe the interest points, we implemented an extended version of the SURF descriptor [12]. Around each interest point with spatial scale σ and temporal scale τ, we define a rectangular volume with dimensions $s\sigma \times s\sigma \times s\tau$ with s a user-defined magnification factor (typically 3). The volume is subsequently divided into $M \times M \times N$ bins, where M and N are the number of bins in the spatial and temporal direction respectively. The bins are filled by a weighted sum of uniformly sampled responses of the 3 axis-aligned Haar-wavelets d_x, d_y, d_t. For each bin, we store the vector $\boldsymbol{v} = (\sum d_x, \sum d_y, \sum d_t)$. We do not include the sums over the absolute values, as done in [12], as they proved to be of no significant benefit in our experiments while doubling the descriptor size.

If invariance to (spatial) rotation is required, we compute the dominant orientation as proposed by [12] except that, for the spatio-temporal case, all Haar-wavelets used in this step stretch out over the full length of the temporal scale of the interest point. Due to space limitations, we refer the reader to the technical report [17] for more in-depth information regarding the implementation issues of both the detector and the descriptor.

3 Quantitative Evaluation

3.1 Methodology

To quantitatively evaluate our detector, we measure the *repeatability*, i.e. how often does it extract the same spatio-temporal features in spite of various geometric and photometric transformations? To this end, we define an *overlap criterion* between two corresponding features defined by ellipsoids E_a and E_b, similar to the 2D overlap criterion for elliptical regions used in [9]. We define the *overlap error* ϵ_O as

$$\epsilon_O = 1 - \frac{V_{E_a} \cap V_{T.E_b}}{V_{E_a} \cup V_{T.E_b}} \tag{12}$$

where V_E represents the volume of an ellipsoid, and T stands for the geometric transformation between the two videos. Two spatio-temporal features are said to *correspond* if the overlap error ϵ_O is below a predefined threshold. The *repeatability* of a detector for a transformation T is then computed as the ratio between the number of corresponding features found and the minimum number of features detected in the common part of the two videos. As in [9], we rescale all features to a fixed size before computing the overlap, so as to prevent a bias towards larger regions. For our experiments, we set the overlap error threshold to 55%. This is equivalent to the threshold of 40% used in [9] for 2D detectors. Figure 3.1 shows what this means in practice. To put this threshold in perspective, we randomly create several sets of features inside a $3D$ volume where the number of features per set, the dimension of the volume and the scale ranges are chosen to be similar to the obtained data used in section 3.2. Using the specified threshold, repeatability between random sets lies around 6%.

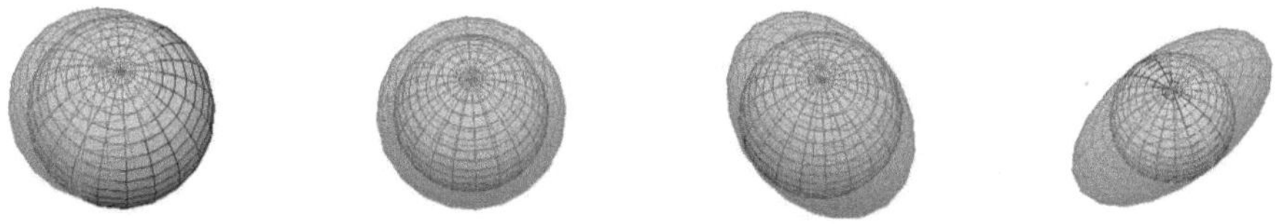

Fig. 3. Some examples of pairs of spatio-temporal ellipsoids with an overlap error equal to our threshold value of 55%. This corresponds to (from left to right) a position error with respect to the diameter of 12%, a uniform (spatial + temporal) scale change of 22%, a 2D spatial scaling of 35%, a 1D temporal scaling of 88%.

3.2 Discussion

We compare our Hes-STIP detector with the HL-STIP detector of [5] at multiple scales and cuboids [6] extracted both at a single scale and at multiple scales. For all of these, we have used executables made available by the respective authors with default parameters. The multi-scale versions were run at scales of 2, 4, 8 and 16, resulting in 16 scale combinations for space and time. Figure 4 shows the results. These were obtained based on artificial transformations of several randomly selected clips from the TRECVID 2006 dataset [2].

Scale changes The scale invariance of our detector stands out when we test the effect of spatial and temporal scale changes (fig. 4(a,b)). The repeatability degrades only slightly in spite of significant scale changes. The single-scale cuboids can only cope with minor scale changes - in fact, only as long as it stays within the error bounds of our overlap criterion. Extracting features at multiple scales only partially overcomes this problem: good repeatability scores are then obtained when the scale factor equals a power of 2, i.e. when the rescaling corresponds exactly to the multiscale resolution. However, for other scale factors, the results again drop significantly. The HL-STIPs only give moderate results.

In-plane rotation From figure 4(c), one can conclude that the Hes-STIPs are relatively sensitive to in-plane rotations (50% repeatability for rotations of 45 degrees), although not as much as the HL-STIPs (30% repeatability). The non-scale-invariant cuboids are more robust in this respect. This is in line with findings on 2D scale-invariant interest points: the higher complexity needed to increase the level of invariance comes at a price and reduces the robustness.

Compression The same also holds for the results under increased levels of compression (fig. 4(d)). However, we detect many more features than the other detectors. If we increase the threshold for our detector such that the number of extracted features is similar, the repeatability improves and comes closer to that of the non-scale-invariant cuboids (red line in figure 4(b)). HL-STIPs again perform quite poorly.

[2] Due to lack of space, we cannot show the results for all the clips. However, the overall trends were the same. For each transformation, the behaviour of one clip is shown.

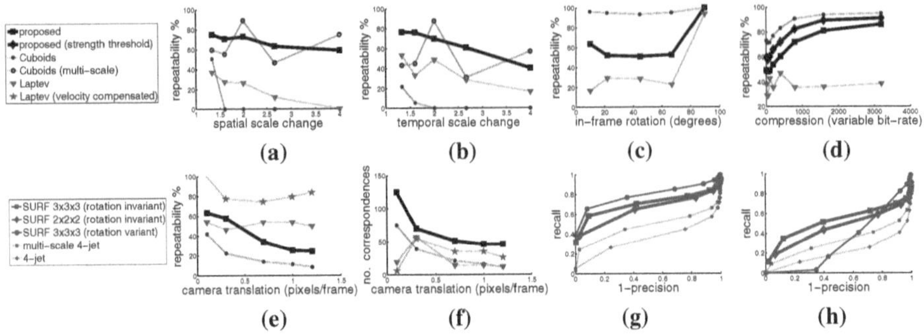

Fig. 4. Quantitative evaluation: repeatability scores for spatial scale changes (a), temporal scale changes (b), in-plane rotations (c), compression (d), and horizontal camera translation (e); total number of corresponding features found for camera translation (f); precision vs. recall of different descriptors for a spatial scale change of 1.5 (g) and a rotation of 22 degrees (h).

Camera motion Finally, we show results for a moving camera (figure 4(e)). This is simulated by gradually translating the image over time. Laptev developed a velocity compensated version of his detector to deal with this case [16], and this seems to give high repeatability scores indeed. Our detector performs only moderately, as could be expected since this type of transformation was not taken into account during the design of the detector. Still, it performs significantly better than cuboids. This is the only type of transformation where HL-STIPs outperform the other detectors - even without motion compensation. Nevertheless, since our method extracts a dense set of features the total number of correspondences found by our method is still higher than what is obtained by the other methods. This is illustrated in figure 4(f), where we plot the absolute number of corresponding features found.

3.3 Evaluation of the Descriptor

We also evaluate the quality of our descriptor and compare it with 4-jet [18] (both single- and multiple-scale). To this end, we compute *precision-recall* curves, i.e. for a varying threshold, how many of the matched features are actually correct ($\epsilon_O < 55\%$) (*precision*), and how many of the corresponding features have actually been matched (*recall*). Matching is performed based on the nearest neighbour ratio threshold measure. Here we show results for a spatial scale change of 1.5 (fig. 4(g)) and for a rotation of 22 degrees (fig. 4(h)).

Our descriptor clearly outperforms the jet-descriptors. $3 \times 3 \times 3$ subdivisions are more discriminative than $2 \times 2 \times 2$, albeit at the cost of a higher dimensional descriptor. Even larger descriptors do not result in significant further improvements anymore. Rotation invariance is valuable when rotations are to be expected. However, when no rotations are present, it has a negative impact on the results due to the increased complexity and reduced discriminativity.

	drink	eat	explore	groom	sleep
drink	83.3	0.0	0.0	16.7	0.0
eat	1.9	88.9	9.3	0.0	0.0
explore	0.0	1.4	83.6	12.3	2.7
groom	0.0	6.7	6.7	86.7	0.0
sleep	0.0	0.0	0.0	0.0	100.0

(a)

	walking	running	jogging	waving	clapping	boxing
walking	91.7	8.3	0.0	0.0	0.0	0.0
running	5.6	72.2	22.2	0.0	0.0	0.0
jogging	5.6	16.7	77.8	0.0	0.0	0.0
waving	0.0	0.0	0.0	100.0	0.0	0.0
clapping	0.0	0.0	0.0	11.1	86.1	2.8
boxing	0.0	0.0	0.0	8.3	13.9	77.8

(b)

Fig. 5. Action recognition results. (a) Confusion matrix for the mouse behaviour dataset [6]. The overall recognition rate is 87.12%. (b) Confusion matrix for the KTH human action dataset [2] using all 4 scenarios for training and testing. The accuracy is 84.26%.

4 Applications

4.1 Action Classification

Next, we test our detector and descriptor in the context of action classification. To this end, we extract spatio-temporal features over 5 octaves. A visual vocabulary is built based on the feature descriptors contained in the training set videos using k-means clustering. Then, a bag-of-words is computed for each video using a *weighted* approach, similar to the one proposed in [19], where each visual word t_k of the visual vocabulary $T = [t_1, \ldots, t_k, \ldots, t_K]$ is given the weight

$$t_k = \sum_{i=1}^{N} \sum_{j=1}^{M_i} \frac{1}{2^{i-1}} \tag{13}$$

with M_i the number of interest points that have visual word k as their i^{th} nearest neighbour. In practice, we use the $N = 4$ nearest neighbours. Finally, the histogram is normalized using the $L1$-norm. An SVM classifier [20] is trained with a χ^2-RBF-kernel using 5-fold cross-validation, followed by a single run over the joined training and validation sets. The results below show the classification performance on the test sets.

The Mouse Behaviour Dataset. [6] contains clips taken from seven 15-minute videos of the same mouse filmed in a vivarium while eating, drinking, exploring, grooming and sleeping. As the number of videos differs significantly between the 5 behaviours, weights inversely proportional to their occurrence are included while clustering the features of the training set. Roughly 37% of the clips of each action of the dataset have been designated as the test set, while the other clips are divided between training and validation set. The resulting confusion matrix, shown in figure 5(a), has an overall accuracy of 87.1% on the test set, compared to the 72% reported by Dollár *et al.* [6].

The KTH human action dataset. [2] contains six types of human actions performed by 25 people in 4 different scenarios. Some detected interest points for two actions are shown in figure 1. We divide the dataset in training, validation and test set as given

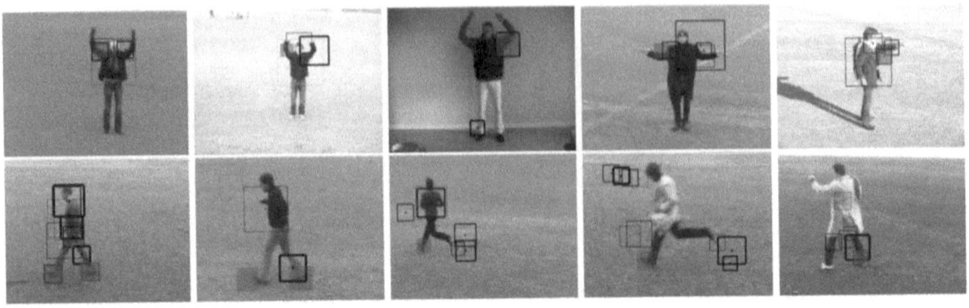

Fig. 6. In each row, features corresponding to a selected visual word are highlighted (in red). The same basic motions are recognized despite temporal and spatial variations. On the top row, the temporal invariance of our feature allows to cope with variable speeds of hand waving, yet also introduces some confusion with some instances of clapping and boxing. On the bottom row, the average temporal scale of the features varies from 8.9 for walking to 5.2 and 3.6 for jogging and running respectively. The same feature was also detected on the foot work by the boxing figure. (For visibility, only a very small number of detected features is shown.)

by [2]. As can be seen in the confusion matrix in figure 5(b), we obtain an accuracy of 84.26%, which is in line with state-of-the-art results reported by Schüldt *et al.* [2] (71.72%), Dollár *et al.* [6] (81.17%), Ke *et al.* [7] (62.96%), Niebles *et al.* [3] (81.50%) and Nowozin *et al.* [21] (87.04%). We outperform all but [21] who employed an extensive grid search in order to find the best model parameters. Further finetuning over all parameters could probably also improve our results further.

As expected, running and jogging turn out to be the most difficult to distinguish, followed by boxing and clapping. This may to some extent be explained by the temporal scale invariance of our features. On one hand, this temporal invariance brings robustness to actions performed at varying speed (as observed for instance in the waving action, where we get a 100% recognition accuracy). On the other hand, this increases the confusion between different actions sharing the same basic motions but performed at different speeds, such as the touch-down of the foot in walking, jogging and running (see also figure 6). By including the selected scales as part of the descriptor, this confusion can probably be reduced.

4.2 Synchronization between an Original Video and a Mashup

Next, we demonstrate the use of our features in the context of synchronizing a video mashup with the original source material. *Video mashups* are video clips that have been created by combining material from different sources. Here, we focus on a 3 minute video clip "Robocop vs. Neo" [3] which combines scenes from the movies "Robocop" and "The Matrix". As original source material, we downloaded the corresponding scene of the movie "The Matrix" [4]. Both video clips have different resolutions, are heavily compressed and contain aliasing artefacts due to ripping and recoding. Moreover,

[3] `http://www.youtube.com/watch?v=UFou895WluU` by AMDS Films

[4] `http://www.youtube.com/v/T8fDJpid7gg`

Fig. 7. Synchronization of a scene from the movie "The Matrix" and a scene from a mashup between "The Matrix" and "Robocop"

the mashup does not merely consist of concatenations of shots from both movies, but includes color changes, gamma corrections, scale changes, time-stretching, flipped & reversed shots as well as novel shots where elements from both movies are combined into new scenes (see fig. 7). As a result, global methods such as the correlation-based approach of [22] are bound to fail.

First, the interest point detector is run on the full videos. Then, shot cut detection is applied to split up the videos, after which approximate nearest neighbour matches between the features of each shot are computed. Finally, we check geometric consistency based on the random sampling scheme RANSAC, using a simple, linear model with 5 parameters (spatial and temporal translations and scale factors, i.e. d_x, d_y, d_t, $s_x = s_y$ and s_t).

An example of synchronized shots is shown in figure 7. Note the significant change in color and the different spatial resolution. Furthermore, additional elements have been added into the scene, while other details have been removed.

4.3 Computation Time

Table 2 gives an overview of some of the video sequences we processed and the needed computation time. For low-resolution videos, such as the KTH human action dataset,

Table 2. General information on all video sequences used in this paper together with their processing times. A quad CPU Opteron 275 with 6GB of memory was used for processing. The buildup of the integral videos is included in the timings. In the second column, the numbers $'N/M'$ denote that the search space extended over N temporal and M spatial octaves. The minimum strength threshold for detection was set to 0.001 (with 1.0 the maximum response at a perfect spatio-temporal Gaussian blob).

datasets	#octaves temporal/spatial	spatial resolution	#frames	#interest points	detection & description time	fps
mouse behaviour dataset [6] (70min)	5/5	(avg.) 244×180	105650	784303	2h43min	10.8
human action dataset[2] (3h14min)	5/5	160×120	291756	715279	3h15min	24.9
"Robocop vs Neo" (2,7min)	5/5	125×313	4084	39237	6min	11.3
scenes from "The Matrix" (1,7min)	5/5	240×320	2536	93970	25min	1.6
scenes from "The Matrix" (1,7min)	1/5	240×320	2536	54448	5min	7.9

detection and description can be done in realtime. Moreover, the needed computation time for feature detection is independent of the number of features. This is in contrast to HL-STIPs where the most time consuming step is the iterative procedure, which takes a time linear in the number of features.

Depending on the application, the number of temporal octaves to process can be adapted, and this further reduces the needed computation time. This is illustrated in the bottom two rows of table 2, where we give the difference in processing time and in the number of detected interest points between processing just one or five temporal octaves. Note that even with one octave we are still scale-invariant, albeit within a smaller range of scale changes.

5 Conclusion

In this paper, we have proposed a novel spatio-temporal interest point detector. First, we have shown that by using the determinant of the 3D Hessian matrix, it is possible to combine point localization and scale-selection in a direct way, therefore removing the need for an iterative scheme. Further, we have developed an implementation scheme using integral video, that allows for an efficient computation of scale-invariant spatio-temporal features. Our detector scores well in terms of repeatability and is on par with currently used spatio-temporal interest points. Finally, we have demonstrated its potential in the domain of action classification and video synchronization. Future work will aim at extending the current batch-mode approach towards a sliding-window framework, similar to Ke *et al.* [7]. Executables are available [5] to the community.

Acknowledgements. This work is supported by the European IST Programme DIRAC Project FP6-0027787, the European IST Programme CLASS Project FP6-0027978, the Research Fund K.U.Leuven and the Flemish Fund for Scientific Research (FWO).

References

1. Sivic, J., Zisserman, A.: Video Google: A text retrieval approach to object matching in videos. In: ICCV, vol. 2, pp. 1470–1477 (October 2003)
2. Schüldt, C., Laptev, I., Caputo, B.: Recognizing human actions: a local SVM approach. In: ICPR (2004)
3. Niebles, J.C., Wang, H., Fei-Fei, L.: Unsupervised learning of human action categories using spatial-temporal words. In: BMVC, Edinburgh, U.K (2006)
4. Laptev, I., Lindeberg, T.: Space-time interest points. In: ICCV, Nice, France (October 2003)
5. Laptev, I.: On space-time interest points. IJCV 64(2), 107–123 (2005)
6. Dollár, P., Rabaud, V., Cottrell, G., Belongie, S.: Behavior recognition via sparse spatio-temporal features. Visual Surveillance and Performance Evaluation of Tracking and Surveillance, 65–72 (2005)
7. Ke, Y., Sukthankar, R., Hebert, M.: Efficient visual event detection using volumetric features. In: ICCV, vol. I, pp. 166–173 (2005)

[5] `http://homes.esat.kuleuven.be/~gwillems/research/Hes-STIP`

8. Oikonomopoulos, A., Patras, I., Pantic, M.: Spatiotemporal salient points for visual recognition of human actions. IEEE Transactions on Systems, Man, and Cybernetics – Part B: Cybernetics 36(3), 710–719 (2006)
9. Mikolajczyk, K., Tuytelaars, T., Schmid, C., Zisserman, A., Matas, J., Schaffalitzky, F., Kadir, T., Gool, L.V.: A comparison of affine region detectors. IJCV 65(1-2), 43–72 (2005)
10. Kadir, T., Brady, M.: Scale, saliency and image description. IJCV 45(2), 83–105 (2001)
11. Wong, S.F., Cipolla, R.: Extracting spatiotemporal interest points using global information. In: ICCV, Rio de Janeiro, Brazil, pp. 1–8 (2007)
12. Bay, H., Tuytelaars, T., Van Gool, L.: Surf: Speeded-up robust features. In: ECCV, Graz, Austria (2006)
13. Lindeberg, T.: Feature detection with automatic scale selection. IJCV 30(2), 77–116 (1998)
14. Mikolajczyk, K., Schmid, C.: Scale and affine invariant interest point detectors. IJCV 60(1), 63–86 (2004)
15. Beaudet, P.: Rotationally invariant image operators. In: International Joint Conference on Pattern Recognition, pp. 579–583 (1978)
16. Laptev, I., Lindeberg, T.: Velocity adaptation of space-time interest points. In: ICPR, Cambridge, U.K (2004)
17. Willems, G., Tuytelaars, T., Van Gool, L.: An efficient dense and scale-invariant spatio-temporal interest point detector. Technical Report KUL/ESAT/PSI/0802, K.U. Leuven (2008)
18. Laptev, I., Lindeberg, T.: Local descriptors for spatio-temporal recognition. In: Int. Workshop on Spatial Coherence for Visual Motion Analysis
19. Jiang, Y.G., Ngo, C.W., Yang, J.: Towards optimal bag-of-features for object categorization and semantic video retrieval. In: CIVR, pp. 494–501 (2007)
20. Chang, C.C., Lin, C.J.: LIBSVM: a library for support vector machines (2001), `http://www.csie.ntu.edu.tw/~cjlin/libsvm`
21. Nowozin, S., Bakir, G., Tsuda, K.: Discriminative subsequence mining for action classification, pp. 1919–1923 (2007)
22. Yan, J., Pollefeys, M.: Video synchronization via space-time interest point distribution. In: Advanced Concepts for Intelligent Vision Systems, ACIVS 2004 (2004)

A Graph Based Subspace Semi-supervised Learning Framework for Dimensionality Reduction

Wuyi Yang, Shuwu Zhang, and Wei Liang

Digital Content Technology Research Center, Institute of Automation, Chinese Academy of Sciences, Beijing, China, 100190

Abstract. The key to the graph based semi-supervised learning algorithms for classification problems is how to construct the weight matrix of the p-nearest neighbor graph. A new method to construct the weight matrix is proposed and a graph based Subspace Semi-supervised Learning Framework (SSLF) is developed. The Framework aims to find an embedding transformation which respects the discriminant structure inferred from the labeled data, as well as the intrinsic geometrical structure inferred from both the labeled and unlabeled data. By utilizing this framework as a tool, we drive three semi-supervised dimensionality reduction algorithms: Subspace Semi-supervised Linear Discriminant Analysis (SSLDA), Subspace Semi-supervised Locality Preserving Projection (SSLPP), and Subspace Semi-supervised Marginal Fisher Analysis (SSMFA). The experimental results on face recognition demonstrate our subspace semi-supervised algorithms are able to use unlabeled samples effectively.

1 Introduction

Techniques for dimensionality reduction that allow one to represent the data in a lower dimensional space have attracted much attention in computer vision and pattern recognition. The supervised dimensionality reduction algorithms include Linear Discriminant Analysis (LDA) [1], Locality Preserving Projection (LPP) [2] for supervised case and Marginal Fisher Analysis (MFA) [3].

Labeled data are often, however, very time consuming and expensive to obtain, as they require the efforts of human annotators [4]. However, in many cases, it is far easier to obtain large numbers of unlabeled data. The problem of effectively combining unlabeled data with labeled data is therefore of central importance in machine learning [4]. The semi-supervised learning has attracted an increasing amount of attention recently, and several novel approaches have been proposed. Graph based semi-supervised learning algorithms [4-8] have attracted considerable attention in recent years. These algorithms consider the graph over all the samples as a prior to guide the decision making. Techniques of M. Belkin et al. [5] and Deng Cai et al. [8] are base on regularization.

The weight matrix is the key to the graph based semi-supervised learning algorithms for classication problems. In order to obtain the weight matrix, the

D. Forsyth, P. Torr, and A. Zisserman (Eds.): ECCV 2008, Part II, LNCS 5303, pp. 664–677, 2008.

previous methods [4-8] are based on the assumption that points that are near are likely to have the same label. However, we found that the assumption may not be tenable in some cases. For example, if the face images were captured under varying pose, illumination and expression, the face images that are near in the original image space are very likely to belong to different classes. So, in this case, the problem is how to obtain the weight matrix which can precisely represent the relationship of the points.

A new method to construct the weight matrix of the p-nearest neighbor graph is proposed and a graph based Subspace Semi-supervised Learning Framework (SSLF) is developed. By using the labeled data, we first find a projection subspace by applying the supervised algorithm and embed the labeled and unlabeled data into this subspace. In the subspace, the points of different classes are apart from each other, while the points of the same class are close to each other. Then, the weight matrix is obtained according to not only the relationship between the mapped points in the subspace but also the labels of the points. Using this method, the obtained weight matrix can more precisely represent the relationship of the points than the weight matrix which is obtained based on the relationship of the points in the original point space. SSLF, which exploits the geometric structure of the labeled and unlabeled data and incorporates it as an additional regularization term in the objective function, aims to find an embedding transformation which respects the discriminant structure inferred from the labeled data, as well as the intrinsic geometrical structure inferred from both the labeled and unlabeled data. By utilizing this framework as a tool, we drive three semi-supervised dimensionality reduction algorithms: Subspace Semi-supervised Linear Discriminant Analysis (SSLDA), Subspace Semi-supervised Locality Preserving Projection (SSLPP), and Subspace Semi-supervised Marginal Fisher Analysis (SSMFA).

The rest of this paper is organized as follows. In Section 2, we develop the basic framework for subspace semi-supervised learning where we formulate an objective function that utilizes both labeled and unlabeled data, and we extend the framework to non-linear dimensionality reduction scenarios by applying the kernel trick. In Section 3, we derive different algorithms for subspace semi-supervised learning that arise out of our framework. The experimental results are presented in Section 4. Finally, we conclude the paper in Section 5.

2 Subspace Semi-supervised Learning Framework

In this section, we introduce our subspace semi-supervised framework which respects both discriminant and geometrical structure in the data for dimensionality reduction.

2.1 Background

Given a set of labeled samples $\{\mathbf{x}_i\}_{i=1}^{l} \subset \mathbb{R}^m$ and the class label of the sample $\mathbf{x}_i$ is assumed to be $t_i \in \{1, \cdots, c\}$, dimensionality reduction aims to find a mapping function $f : \mathbf{x} \rightarrow \mathbf{y}$ that transforms $\mathbf{x} \in \mathbb{R}^m$ into $\mathbf{y} \in \mathbb{R}^d (d \ll m)$, where l is the number of samples and c is the number of classes.

Let $G_g = \{X_l, B\}$ be the intrinsic graph with vertex set $X_l = [\mathbf{x}_1, \mathbf{x}_2, \cdots, \mathbf{x}_l]$ and affinity matrix B which is a symmetric $l \times l$ matrix. Let $G_c = \{X_l, C\}$ be the constraint graph with the same vertex set X_l as that of G_g and constraint matrix C which is a symmetric $l \times l$ matrix. These two graphs are defined to characterize certain statistical or geometric properties of the sample set. The edge weights of the matrix B and C measure the similarities between vertex pairs. The purpose of graph embedding is to represent vertices as low dimensional vectors that preserve the similarities which are measured by the edge weights of the matrix B, and suppress the similarities which are measured by the edge weights of the matrix C.

Let $\{w_i\}_{i=1}^l$ be the one dimensional representation of the vertices and $\mathbf{w} = [w_1, w_2, \cdots, w_l]^T$. The general framework of graph embedding for dimensionality reduction finds the optimal $\mathbf{w}$ by minimizing

$$\mathbf{w}_{opt} = \arg\max_{\mathbf{w}} \frac{\mathbf{w}^T B \mathbf{w}}{\mathbf{w}^T C \mathbf{w}}. \tag{1}$$

With different choices of affinity matrix B and constraint matrix C, this framework will lead to many recently proposed manifold learning algorithms, e.g., ISOAMP [9], Laplacian Eigenmap [10], Locally Linear Embedding [11].

If we choose a linear function, i.e. $w_i = f(\mathbf{x}_i) = \mathbf{a}^T \mathbf{x}_i$, we have $\mathbf{w} = X_l^T \mathbf{a}$. The objective function (1) becomes

$$\mathbf{a}_{opt} = \arg\max_{\mathbf{a}} \frac{\mathbf{a}^T X_l B X_l^T \mathbf{a}}{\mathbf{a}^T X_l C X_l^T \mathbf{a}}. \tag{2}$$

The optimal $\mathbf{a}$ is the eigenvector corresponding to the maximum eigenvalue of eigen-problem:

$$X_l B X_l^T \mathbf{a} = \lambda X_l C X_l^T \mathbf{a}. \tag{3}$$

This approach is the linear extension of graph embedding. Many recently proposed linear dimensionality reduction algorithms, like Linear Discriminant Analysis [1], Locality Preserving Projection [2], and Marginal Fisher Analysis [3], can be interpreted in this framework with different choices of affinity matrix B and constraint matrix C.

2.2 The Objective Function

Our goal is to extend this framework and get our Subspace Semi-supervised Learning Framework (SSLF) which respects both discriminant and geometrical structure in the data for dimensionality reduction.

Given a data set $\{\mathbf{x}_1, \cdots, \mathbf{x}_l, \mathbf{x}_{l+1}, \cdots, \mathbf{x}_n\} \subset \mathbb{R}^m$ and a label set $L = \{1, \cdots, c\}$, the first l points $\mathbf{x}_i (i \leqslant l)$ are labeled as $t_i \in L$ and the remaining points $\mathbf{x}_u (l + 1 \leqslant u \leqslant n)$ are unlabeled, where c is the number of classes.

We first select a subspace R and the algorithm will be performed in this subspace. Suppose the number of dimensions of the subspace R is r, i.e. $\dim(R) = r$. Assume that the subspace R is spanned by the orthonormal vectors $\mathbf{u}_1, \cdots, \mathbf{u}_r$, i.e. $R = \text{span}[\mathbf{u}_1, \cdots, \mathbf{u}_r]$. We project both labeled and unlabeled data into the subspace R as follows:

$$\mathbf{x}_i \rightarrow \mathbf{z}_i = P^T \mathbf{x}_i, i = 1, \cdots, l, l+1, \cdots, n,$$

where $P = [\mathbf{u}_1, \cdots, \mathbf{u}_r]$. Let $Z = [\mathbf{z}_1, \mathbf{z}_2, \cdots \mathbf{z}_n]$, $X = [\mathbf{x}_1, \mathbf{x}_2, \cdots, \mathbf{x}_n]$ and $X_l = [\mathbf{x}_1, \mathbf{x}_2, \cdots, \mathbf{x}_l]$. Then $Z_l = PX_l$ and $Z = PX$. In the subspace R, the objective function (2) becomes:

$$\mathbf{a}_{opt} = \arg\max_{\mathbf{a}} \frac{\mathbf{a}^T Z_l B Z_l^T \mathbf{a}}{\mathbf{a}^T Z_l C Z_l^T \mathbf{a}} = \arg\max_{\mathbf{a}} \frac{\mathbf{a}^T P X_l B X_l^T P^T \mathbf{a}}{\mathbf{a}^T P^T X_l C X_l^T P^T \mathbf{a}}. \tag{4}$$

The optimal $\mathbf{a}$ is the eigenvector corresponding to the maximum eigenvalue of eigen-problem:

$$Z_l B Z_l^T \mathbf{a} = \lambda Z_l C Z_l^T \mathbf{a}. \tag{5}$$

The optimization problem of the regularized version of the objective function (4) can be written as follows:

$$\mathbf{a}_{opt} = \arg\max_{\mathbf{a}} \frac{\mathbf{a}^T Z_l B Z_l^T \mathbf{a}}{\alpha \cdot \mathbf{a}^T Z_l C Z_l^T \mathbf{a} + (1-\alpha) R(\mathbf{a})}. \tag{6}$$

The regularization term $R(\mathbf{a})$ provides us the flexibility to incorporate the geometrical structure of the data manifold.

We construct a p-nearest neighbors graph G to model the relationship between data points. Let W be the weight matrix of G. Ideally, the weight of the edge between points of different classes is zero, and the weight of the edge between near points of the same classes is allotted a big value. However, in some cases, we found that points that are near in the original data space are also very likely to belong to different classes. So, based on the assumption that points that are near in the original data space are likely to belong to the same class, we cannot obtain the weight matrix which can precisely represent the relationship of the points. So, we try to find a subspace. In the subspace, the mapped points of different classes are apart from each other, while the points of the same class are close to each other. Then, the weight matrix can be obtained based on the relationship of the mapped points in the subspace.

Let the column vectors $\mathbf{h}_1, \cdots, \mathbf{h}_{c-1}$ be the solutions of (5), ordered according to their eigenvalues, $\lambda_1 > \cdots > \lambda_{c-1}$. Then, we find a transformation matrix $H = [\mathbf{h}_1, \cdots, \mathbf{h}_{c-1}]$. In the subspace which is spanned by the columns of H, the mapped points of different classes are apart from each other, while the points of the same class are close to each other. The transformation matrix H maps all the n samples to a set of points $\{\mathbf{q}_i | \mathbf{q}_i = H^T \mathbf{z}_i, i = 1, \cdots, n\}$ in $\mathbb{R}^{c-1}$. Let $l(\mathbf{x}_i)$ be the class label of $\mathbf{x}_i$, and $N_p(\mathbf{q}_i) = \{\mathbf{q}_i^1, \mathbf{q}_i^2, \cdots, \mathbf{q}_i^p\}$ be the set of $\mathbf{q}_i$'s p-nearest neighbors. The weight matrix W is defined as follows:

$$W_{ij} = \begin{cases} \gamma, \text{ if } \mathbf{x}_i \text{ and } \mathbf{x}_j \text{ are labeled and they share the same label;} \\ \gamma, \text{ if } \mathbf{x}_i \text{ is labeled, } \mathbf{x}_j \text{ is unlabeled, } \mathbf{q}_i \in N_p(\mathbf{q}_j) \\ \quad \text{and for each } \mathbf{q}_s \in N_p(\mathbf{q}_j), \text{ if } \mathbf{x}_s \text{ is labeled, } l(\mathbf{x}_s) = l(\mathbf{x}_i); \\ \gamma, \text{ if } \mathbf{x}_j \text{ is labeled, } \mathbf{x}_i \text{ is unlabeled, } \mathbf{q}_j \in N_p(\mathbf{q}_i) \\ \quad \text{and for each } \mathbf{q}_s \in N_p(\mathbf{q}_i), \text{ if } \mathbf{x}_s \text{ is labeled, } l(\mathbf{x}_s) = l(\mathbf{x}_j); \\ 1, \text{ if } \mathbf{x}_i \text{ and } \mathbf{x}_j \text{ are unlabeled, and } \mathbf{q}_i \in N_p(\mathbf{q}_j) \text{ or } \mathbf{q}_j \in N_p(\mathbf{q}_i); \\ 0, \text{ otherwise.} \end{cases} \tag{7}$$

In order to discover both geometrical and discriminant structure of the data manifold, we incorporate the manifold structure of both labeled and unlabeled samples as the regularization term in the objective function (6). Thus the regularization term can be defined as follows:

$$\begin{aligned} R(\mathbf{a}) &= \frac{1}{2}\sum_{i,j}(\mathbf{a}^T\mathbf{z}_i - \mathbf{a}^T\mathbf{z}_j)^2 W_{ij} \\ &= \mathbf{a}^T Z(D-W)Z^T\mathbf{a} \\ &= \mathbf{a}^T PXLX^TP^T\mathbf{a} \end{aligned} \tag{8}$$

where D is a diagonal matrix; its entries are column sum of W, $D_{ii} = \sum_j W_{ij}$. $L = D - W$ is the Laplacian matrix. We get the objective function of our Subspace Semi-supervised Learning Framework (SSLF):

$$\mathbf{a}_{opt} = \arg\max_{\mathbf{a}} \frac{\mathbf{a}^T PX_lBX_l^TP^T\mathbf{a}}{\mathbf{a}^T(\alpha PX_lCX_l^TP^T + (1-\alpha)PXLX^TP^T)\mathbf{a}}. \tag{9}$$

The projective vector $\mathbf{a}$ that maximizes (9) is given by the maximum eigenvalue solution to the gerneralized eigenvalue problem

$$PX_lBX_l^TP^T\mathbf{a} = \lambda P(\alpha X_lCX_l^T + (1-\alpha)XLX^T)P^T\mathbf{a}. \tag{10}$$

Let the column vector $\mathbf{a}_1, \cdots, \mathbf{a}_d$ be the solutions of equation (10), ordered according to their eigenvalues, $\lambda_1 > \cdots > \lambda_d$. Thus, the embedding is as follows:

$$\mathbf{x}_i \to \mathbf{y}_i = A^T\mathbf{z}_i = A^TP^T\mathbf{x}_i = (PA)^T\mathbf{x}_i, A = [\mathbf{a}_1, \cdots, \mathbf{a}_d],$$

where $\mathbf{y}_i$ is a d-dimensional representation of the high dimensional data point $\mathbf{x}_i$. PA is the transformation matrix.

2.3 Subspace Selection

The selection of subspace R will influence the performance of the proposed algorithm. We can select the original m-dimensional space $\mathbb{R}^m$ as the subspace R to perform our algorithm, i.e. $R = \mathbb{R}^m$. The vector set $\{\mathbf{e}_i\}_{i=1}^m$ is one orthonormal basis of $\mathbb{R}^m$, where $\mathbf{e}_i$ is a m-dimensional vector with $\mathbf{e}_i(j) = 1$ if $i = j$; 0 otherwise. Then the matrix $P = [\mathbf{e}_1, \cdots, \mathbf{e}_n] = I$ is a $m \times m$ identity matrix.

The subspace R can also be the subspace spanned by the columns of X_l, i.e. $R = R(X_l) = \text{Span}[\mathbf{x}_1, \cdots, \mathbf{x}_l]$. Suppose the rank of X_l is r, i.e., $\text{rank}(X_l) = \dim(R(X_l)) = r$. Perform the singular value decomposition of X_l as $X_l = U\sum V^T$, where $U = [\mathbf{u}_1, \cdots, \mathbf{u}_r, \mathbf{u}_{r+1}, \cdots, \mathbf{u}_m]$ is the left orthonormal matrix, $\sum$ is the singular value matrix, and V is the right orthonormal matrix. Then the matrix $P = [\mathbf{u}_1, \cdots, \mathbf{u}_r]$.

2.4 Complete Algorithm

In summary of the discussion so far, the complete algorithm is given below:

1. **Select a subspace R.**
2. **Find a orthonormal basis of R :** Find a vector set $\{\mathbf{u}_i\}_{i=1}^r$ which forms a orthonormal basis of the subspace R, where r is the number of dimensions of the subspace R. The subspace R is spanned by the orthonormal vectors $\mathbf{u}_1, \cdots, \mathbf{u}_r$, i.e. $R = \text{span}[\mathbf{u}_1, \cdots, \mathbf{u}_r]$. The matrix $P = [\mathbf{u}_1, \cdots, \mathbf{u}_r]$ gives a vector space projection from $\mathbb{R}^m$ to the subspace R.
3. **Map data points into the subspace R:**
$$\mathbf{x}_i \rightarrow \mathbf{z}_i = P^T\mathbf{x}_i, i = 1, \cdots, l, l+1, \cdots, n.$$
4. **Construct the affinity matrix B and constraint matrix C:** According to the similarities between labeled sample pairs, construct the affinity matrix B and the constraint matrix C.
5. **Eigen-problem:** Compute the eigenvectors with respect to the eigenvalues for the gerneralized eigenvector problem:
$$Z_lBZ_l^T\mathbf{h} = \lambda Z_lCZ_l^T\mathbf{h}.$$
The column vectors $\mathbf{h}_1, \cdots, \mathbf{h}_{c-1}$ are the solutions ordered according to their eigenvalues, $\lambda_1 > \cdots > \lambda_{c-1}$. We obtain the transformation matrix $H = [\mathbf{h}_1, \cdots, \mathbf{h}_{c-1}]$.
6. **Construct the adjacency graph:** Map all the samples to the subspace which is spanned by the columns of H, construct the p-nearest neighbors graph matrix W as in (7) to model the relationship between nearby data points and calculate the graph Laplacian matrix $L = D - W$.
7. **Eigen-problem:** Compute the eigenvectors with respect to the eigenvalues for the gerneralized eigenvector problem:
$$PX_lBX_l^TP^T\mathbf{a} = \lambda P(\alpha X_lCX_l^T + (1-\alpha)XLX^T)P^T\mathbf{a}.$$
The column vectors $\mathbf{a}_1, \cdots, \mathbf{a}_d$ are the solutions ordered according to their eigenvalues, $\lambda_1 > \cdots > \lambda_d$. We obtain the transformation matrix $A = (\mathbf{a}_1, \cdots, \mathbf{a}_d)$.
8. **Linear graph embedding:** The data point can be embedded into d-dimensional subspace by
$$\mathbf{x} \rightarrow \mathbf{y} = A^T\mathbf{z} = A^TP^T\mathbf{x} = (PA)^T\mathbf{x}.$$

2.5 Kernel SSLF for Non-linear Dimensionality Reduction

When the data manifold is highly nonlinear, the linear method described above may fail to discover the intrinsic geometry. Here we show how SSLF can be extended to non-linear dimensionality reduction scenarios.

For a given nonlinear mapping ϕ , the input data in the space $\mathbb{R}^m$ can be mapped into the feature space H:

$$\phi : \mathbb{R}^m \to H, \mathbf{x} \mapsto \phi(\mathbf{x}). \tag{11}$$

The algorithm will be performed in the subspace $R(\Phi)$, where $\Phi = [\phi(\mathbf{x}_1), \cdots, \phi(\mathbf{x}_k)]$ and $k = l$ or n. Let $\{\beta_1, \cdots, \beta_r\}$ be one orthonormal basis of the subspace $R(\Phi)$, where r is the rank of Φ. It is easy to show that every vector $\beta_i, i = 1, \cdots, r$ can be linear expanded by

$$\beta_i = \sum_{j=1}^{k} \gamma_{ij} \phi(\mathbf{x}_j), i = 1, \cdots, r. \tag{12}$$

Denote $A = \Phi^T \Phi$, whose elements can be determined by the following kernel function:

$$A_{i,j} = \phi(\mathbf{x}_i)^T \phi(\mathbf{x}_j) = (\phi(\mathbf{x}_i) \cdot \phi(\mathbf{x}_j)) = K(\mathbf{x}_i, \mathbf{x}_j). \tag{13}$$

Calculate the orthonormal eigenvectors $\nu_1, \cdots, \nu_r$ of A corresponding to the r largest positive eigenvalues, $\lambda_1 > \cdots > \lambda_r$. Then the orthonormal vectors $\beta_1, \cdots, \beta_r$ can be obtain by

$$\beta_i = \lambda_i^{-\frac{1}{2}} \Phi \nu_i, i = 1, \cdots, r. \tag{14}$$

After the projection of the mapped samples $\phi(\mathbf{x}_i)$, $i = 1, \cdots, n$ into the subspace $R(\Phi)$, the transformed feature vectors $\mathbf{z}_i$, $i = 1, \cdots, n$ can be obtained by

$$\mathbf{z}_i = P^T \phi(\mathbf{x}_i), i = 1, \cdots, l, l+1, \cdots, n, \tag{15}$$

where $P = [\beta_1, \cdots, \beta_r]$. Let $\Lambda = diag(\lambda_i^{-\frac{1}{2}}, \cdots, \lambda_r^{-\frac{1}{2}})$ and $V = [\nu_1, \cdots, \nu_r]$. The equation (15) can be rewritten as:

$$\begin{aligned} \mathbf{z}_i &= [\beta_1, \cdots, \beta_r]^T \phi(\mathbf{x}_i) \\ &= \Lambda V^T \Phi^T \phi(\mathbf{x}_i) \\ &= \Lambda V^T [K(\mathbf{x}_1, \mathbf{x}_i), \cdots, K(\mathbf{x}_l, \mathbf{x}_i)]^T \end{aligned} \quad . \tag{16}$$

According to the similarities between labeled sample pairs, construct the affinity matrix B and the constraint matrix C. Compute the eigenvectors with respect to the $c-1$ largest eigenvalues for the eigenvector problem:

$$Z_l B Z_l^T \mathbf{h}_i = \lambda Z_l C Z_l^T \mathbf{h}_i, i = 1, 2, \cdots, c-1. \tag{17}$$

We obtain the transformation matrix $F = [\mathbf{h}_1, \cdots, \mathbf{h}_{c-1}]$. The transformation matrix F maps all the n samples to a set of points $\{\mathbf{q}_i | \mathbf{q}_i = F^T \mathbf{z}_i, i = 1, \cdots, n\}$

in $\mathbb{R}^{c-1}$. Construct the p-nearest neighbor graph matrix W as in (7) to model the relationship between nearby data points and calculate the graph Laplacian matrix $L = D - W$.

Compute the eigenvectors with respect to the d largest eigenvalues for the gerneralized eigenvector problem:

$$Z_l B Z_l^T \mathbf{a}_i = \lambda(\alpha Z_l C Z_l^T + (1-\alpha) Z L Z^T)\mathbf{a}_i, i = 1, \cdots, d. \tag{18}$$

Let $A = (\mathbf{a}_1, \cdots, \mathbf{a}_d)$. For a data point $\mathbf{x}$, the embedded data point $\mathbf{y}$ can be obtained by

$$\begin{aligned} \mathbf{y} &= A^T \Lambda V^T \Phi_l^T \phi(\mathbf{x}) \\ &= (V\Lambda A)^T [K(\mathbf{x}_1, \mathbf{x}), \cdots, K(\mathbf{x}_k, \mathbf{x})]^T. \end{aligned} \tag{19}$$

3 Instances of SSLF

SSLF provides an efficient framework for the graph embedding problem. With different choices of affinity matrix B and constraint matrix C, this framework will lead to three linear dimensionality reduction algorithms: Subspace Semi-supervised Linear Discriminant Analysis (SSLDA), Subspace Semi-supervised Locality Preserving Projection (SSLPP), and Subspace Semi-supervised Marginal Fisher Analysis (SSMFA).

3.1 SSLDA

Linear Discriminant Analysis (LDA) [1] seeks directions that are efficient for discrimination. Suppose l_k is the number of labeled data in the k-th class, then $l = \sum_{k=1}^{c} l_k$ .The objective function of LDA is as follows:

$$\mathbf{a}_{opt} = \arg\max_{\mathbf{a}} \frac{\mathbf{a}^T S_b \mathbf{a}}{\mathbf{a}^T S_t \mathbf{a}}, \tag{20}$$

where S_b is the between-class scatter matrix and S_t is the total scatter matrix. Define

$$W_{i,j}^{LDA} = \begin{cases} 1/l_k, & \text{if } \mathbf{x}_i \text{ and } \mathbf{x}_j \text{ both belong to the } k\text{-th class;} \\ 0, & \text{otherwise.} \end{cases} \tag{21}$$

With this W^{LDA}, it is easy to check that $S_b = X_l(W^{LDA} - \frac{1}{l}\mathbf{1}\mathbf{1}^T)X_l^T$ and $S_t = X_l(I - \frac{1}{l}\mathbf{1}\mathbf{1}^T)X_l^T$, where $\mathbf{1} = [1, \cdots, 1]^T$ is a l-dimensional vector and I is a $l \times l$ identity matrix. Please see [2] for detail derivation. With the choice of affinity matrix $B = W^{LDA} - \frac{1}{l}\mathbf{1}\mathbf{1}^T$ and constraint matrix $C = I - \frac{1}{l}\mathbf{1}\mathbf{1}^T$, the SSLF leads to the Subspace Semi-supervised Linear Discriminant Analysis (SSLDA). The between-class scatter matrix S_b has at most rank $c-1$. This implies that the multiplicity of $\lambda = 0$ is at least $m - c + 1$. Therefore, SSFDA can find at most $c - 1$ meaningful directions.

3.2 SSLPP

Let $N_k^+(\mathbf{x}_i)$ denote the set of k nearest neighbors of $\mathbf{x}_i$ among the data sharing the same label with $\mathbf{x}_i$. Define

$$W_{i,j}^{LPP} = \begin{cases} 1, & \text{if } \mathbf{x}_i \in N_k^+(\mathbf{x}_j) \text{ or } \mathbf{x}_j \in N_k^+(\mathbf{x}_i); \\ 0, & \text{otherwise.} \end{cases} \tag{22}$$

The objective function of LPP [2] is as follows:

$$\mathbf{a}_{opt} = \arg\min_{\mathbf{a}} \frac{\mathbf{a}^T X_l L^{LPP} X_l^T \mathbf{a}}{\mathbf{a}^T X_l D^{LPP} X_l^T \mathbf{a}} = \arg\max_{\mathbf{a}} \frac{\mathbf{a}^T X_l W^{LPP} X_l^T \mathbf{a}}{\mathbf{a}^T X_l D^{LPP} X_l^T \mathbf{a}}, \tag{23}$$

where D^{LPP} is a diagonal matrix whose entries are column sums of W^{LPP} and $L^{LPP} = D^{LPP} - W^{LPP}$ is the graph Laplacian. Please see [2] for details. With the choice of the affinity matrix $B = W^{LPP}$ and the constraint matrix $C = D^{LPP}$, the SSLF leads to the Subspace Semi-supervised Locality Preserving Projection (SSLPP).

3.3 SSMFA

In Marginal Fisher Analysis (MFA [3]), the intrinsic graph characterizes the intraclass compactness and connects each data point with its neighboring points of the same class, while the constraint graph connects the marginal points and characterizes the interclass separability. Let $N_k^-(\mathbf{x}_i)$ denote the set of k nearest neighbors of $\mathbf{x}_i$ among the data sharing the different label with $\mathbf{x}_i$.Define

$$W_{i,j}^{+} = \begin{cases} 1, & \text{if } \mathbf{x}_i \in N_{k_1}^+(\mathbf{x}_j) \text{ or } \mathbf{x}_j \in N_{k_1}^+(\mathbf{x}_i); \\ 0, & \text{otherwise.} \end{cases} \tag{24}$$

$$W_{i,j}^{-} = \begin{cases} 1, & \text{if } \mathbf{x}_i \in N_{k_2}^-(\mathbf{x}_j) \text{ or } \mathbf{x}_j \in N_{k_2}^-(\mathbf{x}_i); \\ 0, & \text{otherwise.} \end{cases} \tag{25}$$

Let D^+ be the diagonal matrix whose entries are column sums of W^+ and $L^+ = D^+ - W^+$. Let D^- be the diagonal matrix whose entries are column sums of W^- and $L^- = D^- - W^-$. The objective function of MFA is

$$\mathbf{a}_{opt} = \arg\max_{\mathbf{a}} \frac{\mathbf{a}^T X_l (D^- - W^-) X_l^T \mathbf{a}}{\mathbf{a}^T X_l (D^+ - W^+) X_l^T \mathbf{a}} = \arg\max_{\mathbf{a}} \frac{\mathbf{a}^T X_l L^- X_l^T \mathbf{a}}{\mathbf{a}^T X_l L^+ X_l^T \mathbf{a}} \tag{26}$$

Please see [3] for details. With the choice of the affinity matrix $B = L^-$ and the constraint matrix $C = L^+$, the SSLF leads to the Subspace Semi-supervised Marginal Fisher Analysis (SSMFA).

4 Experimental Results

In this section, the performance of SSLDA, SSLPP and SSMFA for face recognition is investigated. A face recognition task is handled as a multi-class classification problem. Each test image is mapped to a low-dimensional subspace via the embedding learned from training data, and then it is classified by the nearest neighbor criterion. We evaluate the performance of SSLDA in comparison with LDA and Semi-supervised Discriminant Analysis (SDA) [8], the performance of SSLPP with comparison to LPP and the performance of SSMFA in comparison with MFA

4.1 Datasets

In our experiments, we use the CMU PIE (Pose, Illumination and Expression) databases [12] for face recognition to evaluate the performance of the proposed algorithms. In all the experiments, preprocessing to locate the faces was applied. Original images were normalized (in scale and orientation) such that the two eyes were aligned at the same position. Then, the facial areas were cropped into the final images for matching. The size of each cropped image in all the experiments

Table 1. Recognition error rates (%) on CMU PIE database

Method	Subspace	G40/L3/P40		G40/L4/P40	
		Unlabeled error	Test error	Unlabeled error	Test error
LDA	–	40.6±8.8 (19)	41.6±10.3 (19)	31.4±10.2 (19)	36.1±8.6 (19)
SDA	–	45.5±7.0 (19)	43.7±7.5 (19)	38.6±10.5 (19)	40.5±8.5 (19)
SSLDA	$\mathbb{R}^m$	28.8±7.2 (19)	29.5±9.3 (19)	22.7±9.4 (19)	26.0±7.5 (19)
SSLDA	$R(X_l)$	**26.2±6.3** (19)	**26.3±7.5** (19)	**20.2±8.0** (19)	**24.2±6.0** (19)
LPP	–	38.0±8.3 (19)	38.4±10.3 (39)	30.4±10.7 (41)	35.2±9.2 (39)
SSLPP	$\mathbb{R}^m$	29.2±7.4 (28)	29.4±10.0 (29)	23.5±9.6 (19)	26.1±7.9 (19)
SSLPP	$R(X_l)$	**26.0±5.3** (38)	**26.1±7.5** (38)	**20.6±7.9** (35)	**24.5±5.3** (35)
FMA	–	37.7±7.9 (21)	38.3±10.7 (23)	31.5±11.4 (33)	35.9±9.1 (30)
SSFMA	$\mathbb{R}^m$	27.4±6.2 (21)	27.8±9.1 (20)	22.4±9.0 (22)	**24.9±6.7** (25)
SSFMA	$R(X_l)$	**26.7±5.5** (25)	**26.9±7.4** (23)	**21.3±7.2** (22)	25.3±5.6 (30)
Method	Subspace	G40/L5/P40		G40/L6/P40	
		Unlabeled error	Test error	Unlabeled error	Test error
LDA	–	28.5±10.6 (19)	28.8±11.2 (19)	23.1±7.6 (19)	22.2±6.2 (19)
SDA	–	37.3±10.0 (19)	36.6±10.7 (19)	33.9±7.8 (19)	30.4±7.4 (19)
SSLDA	$\mathbb{R}^m$	20.0±7.9 (19)	21.0±9.2 (19)	17.1±6.5 (19)	16.6±4.6 (19)
SSLDA	$R(X_l)$	**19.0±7.7** (19)	**19.7±8.7** (19)	**15.5±6.0** (19)	**15.4±4.3** (19)
LPP	–	27.1±10.3 (40)	27.5±10.7 (39)	22.1±6.7 (43)	21.1±5.8 (43)
SSLPP	$\mathbb{R}^m$	20.8±8.1 (25)	21.2±9.0 (28)	18.6±7.6 (28)	16.9±4.8 (30)
SSLPP	$R(X_l)$	**18.1±7.3** (39)	**19.1±8.1** (39)	**15.5±5.6** (39)	**15.2±4.1** (38)
FMA	–	28.3±9.7 (33)	28.4±9.7 (35)	23.2±8.5 (36)	22.0±5.2 (27)
SSFMA	$\mathbb{R}^m$	19.8±7.6 (19)	20.4±7.9 (19)	16.9±6.6 (19)	**15.9±4.2** (19)
SSFMA	$R(X_l)$	**19.1±7.3** (28)	**19.6±7.4** (27)	**16.5±6.6** (25)	16.2±4.3 (30)

is 32×32 pixels, with 256 gray levels per pixel. Thus, each image is represented by a 1024-dimensional vector in image space.

The CMU PIE face database contains more than 40,000 facial images of 68 people. The face images were captured under varying pose, illumination and expression. From the dataset which contains five near frontal poses (C05, C07, C09, C27, C29), we randomly choose 20 persons and 80 face images for each person in the experiments .

The image set is then partitioned into the gallery and probe set with different numbers. For ease of representation, Gm/Ll/Pn means m images per person are randomly selected for training and the remaining n images are for testing. Among these m images, l images are randomly selected and labeled which leaves other $(m-l)$ images unlabeled.

4.2 Face Recognition Results

In the experiments, we empirically set the parameters as $p = 5$, $\alpha = 0.8$ and $\gamma = 0.9$. The recognition error rates of different algorithms on CMU PIE databases

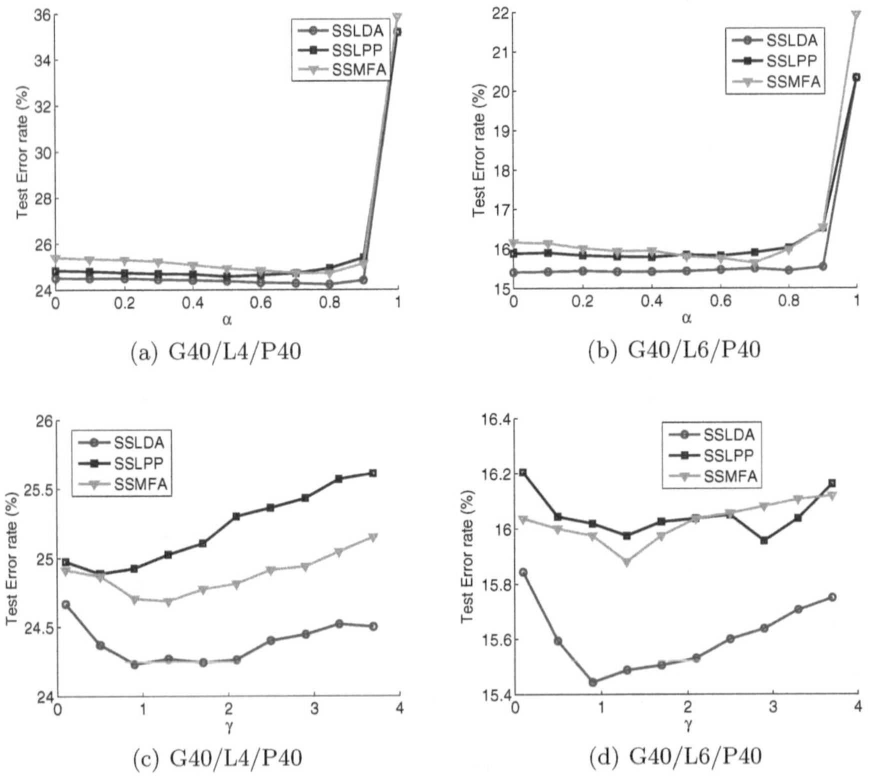

(a) G40/L4/P40 (b) G40/L6/P40

(c) G40/L4/P40 (d) G40/L6/P40

Fig. 1. Model selection for algorithms: error rate vs. different values of the parameters α and γ

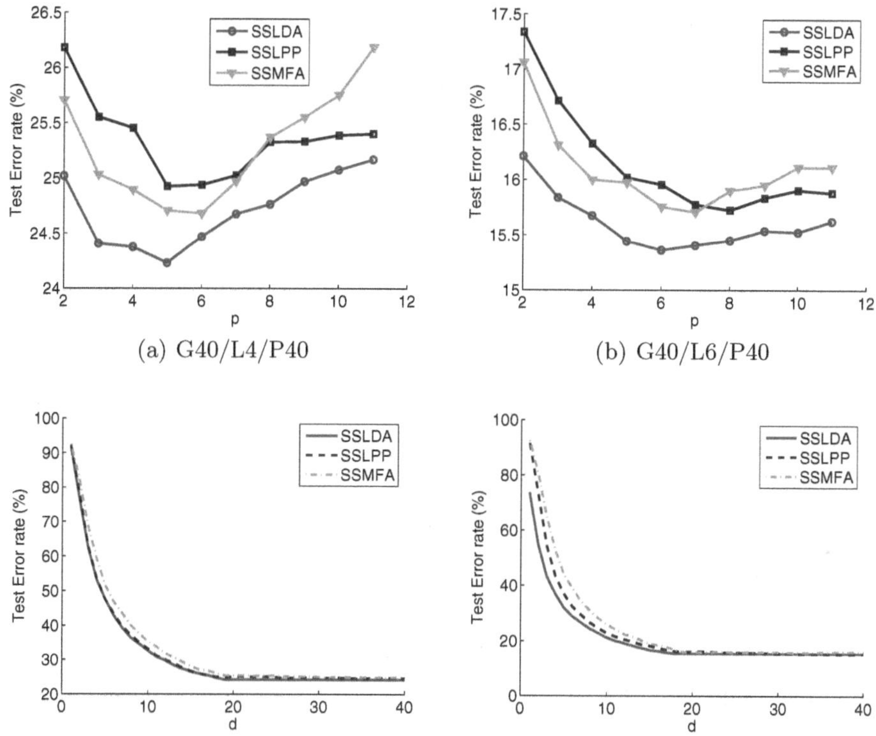

(c) G40/L4/P40 (d) G40/L6/P40

Fig. 2. Model selection for algorithms: error rate vs. different values of the parameters p and d

are shown in Table 1. In each case, the minimum error rate is boldfaced. For each Gm/Ll/Pn, we average the results over 20 random splits and report the mean as well as the standard deviation.

Dimensionality estimation is a crucial problem for most of the subspace learning based face recognition methods. The performance usually varies with the number of dimensions. We show the best results obtained by these subspace learning algorithms. The numbers in parentheses in Table 1 are the corresponding feature dimensions with the best results after dimensionality reduction.

From the results in Table 1, we can make the following comments: (1) In comparison with the performance of LDA, SDA, LPP and MFA, the proposed algorithms greatly improved the performance of face recognition by using the unlabeled data. (2) In the subspace $R(X_l)$ spanned by the labeled data, the proposed algorithms almost always achieved the best performance.

4.3 Model Selection

Model selection is a crucial problem in many learning problems. The learning performance may vary drastically with different choices of the parameters. In our algorithms, there are four parameters: α, γ , p, and d. Figure 1 and 2 show the performance of our algorithms with respect to different values of these parameters. As can be seen, our algorithm is not sensitive to α and γ. For the value of p, in order to discover the local geometrical and discriminant structure of the data space, it is usually set to a small number. If the number of dimensions of the learned subspace $d \geqslant c-1$, with the increase of d, there are little varieties in the performances of the different algorithms, where c is the number of classes.

5 Conclusions

In this paper, we propose a general Subspace Semi-supervised Learning Framework for dimensionality reduction. The framework aims to find an embedding transformation which respects the discriminant structure inferred from the labeled data, as well as the intrinsic geometrical structure inferred from both the labeled and unlabeled data. By utilizing this framework as a tool, we derive three linear dimensionality reduction algorithms: SSLDA, SSLPP, and SSMFA. Experimental results on face recognition demonstrate the subspace semi-supervised algorithms are able to use unlabeled samples effectively.

Acknowledgement

This work has been supported by the National Sciences & Technology Supporting Program of China under Grant No.2006BAH02A13 and No.2006BAH02A36.

References

1. Fukunaga, K.: Introduction to Statistical Pattern Recognition, 2nd edn. Academic Press, New York (1990)
2. He, X., Yan, S., Zhang, Y.H.P.N., Face, H.J.: recognition using laplacianfaces. IEEE Transactions on Pattern Analysis and Machine Intelligence 27(3), 328–340 (2005)
3. Yan, S., Xu, D., Zhang, B., Zhang, H.-J., Yang, Q., Lin, S.: Graph embedding and extension: A general framework for dimensionality reduction. IEEE Transactions on Pattern Analysis and Machine Intelligence 29(1), 40–51 (2007)
4. Zhu, X., Ghahramani, Z., Lafferty, J.: Semi-supervised learning using gaussian fields and harmonic functions. In: 20th International Conference on Machine Learning (2003)
5. Belkin, M., Niyogi, P., Sindhwani, V.: Manifold regularization: A geometric framework for learning from examples. Journal of Machine Learning Research 7(11), 2399–2434 (2006)
6. Sindhwani, V., Niyogi, P., Belkin, M.: Beyond the point cloud: from transductive to semi-supervised learning. In: ICML 2005, 22nd International Conference on Machine Learning (2005)
7. Zhou, D., Bousquet, O., Lal, T., Weston, J., Scholkopf, B.: Learning with local and global consistency. In: Advances in Neural Information Processing Systems (2003)

8. Deng Cai, X.H., Han, J.: Semi-supervised discriminant analysis. In: IEEE International Conference on Computer Vision (2007)
9. Tenenbaum, J., de Silva, V., Langford, J.: A global geometric framework for nonlinear dimensionality reduction. Science 290(5500), 2319–2323 (2000)
10. Belkin, M., Niyogi, P.: Laplacian eigenmaps and spectral techniques for embedding and clustering. In: Dietterich, T.G., Becker, S., Ghahramani, Z. (eds.) Advances in Neural Information Processing Systems, vol. 14. MIT Press, Cambridge (2002)
11. Roweis, S., Saul, L.: Nonlinear dimensionality reduction by locally linear embedding. Science 290(5500), 2323–2326 (2000)
12. Sim, T., Baker, S., Bsat, M.: The cmu pose, illuminlation, and expression database. IEEE Transactions on Pattern Analysis and Machine Intelligence 25(12), 1615–1618 (2003)

Online Tracking and Reacquisition Using Co-trained Generative and Discriminative Trackers

Qian Yu, Thang Ba Dinh, and Gérard Medioni

University of Southern California, Los Angeles, CA 90089-0273, USA
{qianyu,thangdin,medioni}@usc.edu

Abstract. Visual tracking is a challenging problem, as an object may change its appearance due to viewpoint variations, illumination changes, and occlusion. Also, an object may leave the field of view and then reappear. In order to track and reacquire an unknown object with limited labeling data, we propose to learn these changes online and build a model that describes all seen appearance while tracking. To address this semi-supervised learning problem, we propose a co-training based approach to continuously label incoming data and online update a hybrid discriminative generative model. The generative model uses a number of low dimension linear subspaces to describe the appearance of the object. In order to reacquire an object, the generative model encodes all the appearance variations that have been seen. A discriminative classifier is implemented as an online support vector machine, which is trained to focus on recent appearance variations. The online co-training of this hybrid approach accounts for appearance changes and allows reacquisition of an object after total occlusion. We demonstrate that under challenging situations, this method has strong reacquisition ability and robustness to distracters in background.

1 Introduction

Object tracking is challenging [1] due to appearance changes, which can be caused by varying viewpoints and illumination conditions. Appearance can also change relative to background due to the emergence of clutter and distracters. Also, an object may leave the field of view (or be occluded) and reappear. To address these difficulties, we aim to track an arbitrary object with limited initialization (labeled data) and learn an appearance model on-the-fly, which can then be used to reacquire the object when it reappears.

This tracking problem can be formulated in two different ways: generative and discriminative. Generative tracking methods learn a model to represent the appearance of an object. Tracking is then expressed as finding the most similar object appearance to the model. Several examples of generative tracking algorithms are Eigentracking [2], WSL tracking [3] and IVT [5]. To adapt to appearance changes, the object model is often updated online, as in [5]. Due to the fact that the appearance variations are highly non-linear, multiple subspaces [6] and non-linear manifold learning methods [7] have been proposed. Note that traditional generative tracking methods are trained based on object appearance without considering background information.

D. Forsyth, P. Torr, and A. Zisserman (Eds.): ECCV 2008, Part II, LNCS 5303, pp. 678–691, 2008.

Instead of building a model to describe the appearance of an object, discriminative tracking methods aim to find a decision boundary that can best separate the object from the background. Recently, many discriminative trackers are proposed [9,10,11] and demonstrate strong robustness to avoid distracters in the background. Support Vector Tracking (SVT)[8] integrates an offline trained Support Vector Machine (SVM) classifier into an optic-flow-based tracker. In order to update the decision boundary according to new samples and background, discriminative tracking methods with online learning are proposed in [10,9]. In [10], a confidence map is built by finding the most discriminative RGB color combination in each frame. However, a limited color feature pool restricts the discriminative power of this method. In [9], Avidan proposes to use an ensemble of online learned weak classifiers to label a pixel as belonging to either the object or the background. To accommodate object appearance changes, at every frame, new weak classifiers replace part of old ones that do not perform well or have existed longer than a fixed number of frames. Both methods [9,10] use features at the pixel level and rely on a mode seeking process (mean shift) to find the best estimate on a confidence map, which restricts the reacquisition ability of these methods. Oza and Russell [12] proposed an online boosting algorithm, which is applied to the visual tracking problem [13,14]. Due to the large number of features, either an offline feature selection procedure or an offline trained seed classifier is usually required in practice. Thus, for tracking methods based on online boosting, it is difficult to generalize to arbitrary object types

It has been shown that discriminative classifiers often outperform generative models [15] if enough training data is available. However, generative methods often have better generalization performance when the size of training data is small. Specifically, a simple generative classifier (naive Bayes) outperforms its discriminative counterpart (logistic regression) when the amount of labeled training data is small [16]. Recently, hybrid discriminative generative methods have opened a promising direction to benefit from both types of methods. Several hybrid methods [17,18,19,15] have been proposed in many application domains [17,18,19]. Most of them imbue generative methods with the discriminative power via "discriminative training" of a generative model. These methods train a model by optimizing a convex combination of the generative and discriminative log likelihood functions. Due to the asymmetry in training data, "discriminative training" of a generative model requires a parameter to govern the trade-off between generative and discriminative. Theoretical discussions in [15] show that an improper hybrid of discriminative generative model generates even worse performance than pure generative or discriminative methods.

We propose to use co-training to combine generative and discriminative models. Here, the online learning an appearance model of an arbitrary object with limited labeled data is treated as a semi-supervised problem. The co-training approach proposed by Blum and Mitchell [20] is a principled semi-supervise training method. The basic idea is to train two classifiers on two conditionally independent views of the same data (with a small number of exemplars) and then use the prediction from each classifier to enlarge the training set of the other. It is proved that co-training can find an accurate decision boundary, starting from a small quantity of labeled data as long as the two feature sets are independent [20]. Empirical results [21] show that co-training works well

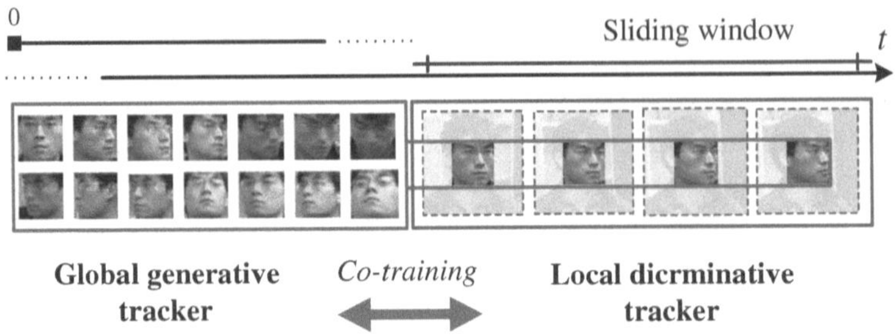

Fig. 1. Online co-training a generative tracker and a discriminative tracker with different life span (the area bounded by dashed red boxes indicates the background)

even in the case where the independence is not perfectly satisfied. In [22], the features used for classification are derived from PCA bases, which are obtained offline from training samples, and co-training is used to improve an offline learned object detector. More recently, Tang *et al.* [23] proposes to use co-training to online train two SVM trackers with color histogram features and HOG features. This method uses an incremental/decremental SVM solver[25] to focus on recent appearance variations without representing the global object appearance.

In order to represent the global object appearance, we propose to use a generative model, which contains a number of low dimension subspaces. This generative model encodes all the appearance variations that have been seen in a compact way. An online subspace updating algorithm is proposed to modify the subspaces adaptively. The descriptive power of the generative model increases as new samples are added. For the discriminative classifier, we use an incrementally learned SVM classifier [24] with histogram of gradient (HOG) [26] features. In practice, we find the number of support vectors grows quite fast when the appearance of object and background changes. Moreover, the adaption of the discriminative model to new appearance changes becomes more and more slow as samples are accumulated. To address these problems, we decrementally train the SVM to focus on recent appearance variations within a sliding window. The training data flow is shown in Figure 1. The image patches bounded with green boxes are samples used in the generative model. They contain all the object appearance variations since tracking starts. The red bounding boxes indicate the positive and negative training samples within the sliding window used in the discriminative classifier. The main advantage of this method is that it collaboratively combines the generative and discriminative models with complementary views (features) of the training data and it encodes the global object appearance variations, thus can handle reacquisition. Experiments show that our method has strong reacquisition ability and is robust to distracters in background clutter.

The rest of this paper is organized as follows. The overview of the co-training framework is presented in Section 2. The details of the generative tracker and the discriminative tracker are presented in Section 3 and Section 4. The experiments are shown in Section 5, followed by conclusions and future work.

2 Bayesian Inference with Co-training

We formulate the visual tracking problem as a state estimate problem in a similar way as [5,27]. Given a sequence of observed image regions $O_t = (o_1, \ldots, o_t)$ over time t, the goal of visual tracking is to estimate the hidden state s_t. In our case, the hidden state refers to an object's 2D position, scale and rotation. Assuming a Markovian state transition, the posterior can be formulated as a recursive equation

$$p(s_t|O_t) \propto p(o_t|s_t) \int p(s_t|s_{t-1})p(s_{t-1}|O_{t-1})ds_{t-1} \tag{1}$$

where $p(o_t|s_t)$ and $p(s_t|s_{t-1})$ are the observation model and state transition model respectively. $p(s_{t-1}|O_{t-1})$, which is represented as a set of particles and weights, is the posterior distribution given all the observations up to time $t-1$. The recursive inference in Eq.1 is implemented with resampling and importance sampling processes[27]. In our approach, the transition of the hidden state is assumed to be a Gaussian distribution as, $p(\mathbf{s}_t|\mathbf{s}_{t-1}) = \mathcal{N}(\mathbf{s}_t; \mathbf{s}_{t-1}, \Psi_t)$, where Ψ_t is a time variant diagonal covariance matrix. In this recursive inference formulation, $p(o_t|s_t)$ is the crucial part for finding the ideal posterior distribution. $p(o_t|s_t)$ measures the likelihood of observing o_t given one state of the object. Besides the 2D position, our state variables encode an object's rotation and scale. This reduces the appearance variations caused by such motion at the price of that more particles are needed to represent the distribution.

Our measurement of one observation comes from two independent models. One is the generative model, which is based on online constructed multi-subspaces. The other is the discriminative model, which is online trained with HOG features. The features used by these two models, namely intensity pattern and local gradient features, are complementary. After limited initialization, these two models are co-trained with sequential unlabeled data. It is worth noting that co-training is not a classification framework [21], but an automatic way to train with unlabeled data. In our approach, each model makes the decision based on its own knowledge and this information is used to train the other model. The final decision is made by the combined hybrid model. Due to the independence between the two observers, our observation model $p(o_t|s_t)$ can be expressed as a product of two likelihood functions from the generative $\mathcal{M}$ model and the discriminative model $\mathcal{C}$, $p(o_t|s_t) \propto p_{\mathcal{M}}(o_t|s_t)p_{\mathcal{C}}(o_t|s_t)$.

We adaptively adjust Ψ_t in the state transition model according to the tracking result at time t. If neither of the models accepts the new unlabeled data, we increase the covariance matrix Ψ_t and the number of samples. The extreme condition is that a very flat transition distribution is close to scanning the whole state space uniformly. This Bayesian formulation is very proper for the object tracking and reacquisition problem. By adaptively resampling, we can cover a large search region efficiently. Also, as partial appearance variations are compensated by the motion state, our method can deal with object rotation and scale changes. The key step in the co-training algorithm is to incrementally update both the generative and discriminative trackers, which are discussed in Section 3 and Section 4 respectively.

3 Generative Tracker with Multiple Linear Subspaces

The global appearance of one object under different viewpoints and illumination conditions is known to lie on a low dimension manifold. However, such a global appearance manifold is highly non-linear. In [7], a non-linear mapping from the embedding space to the input space is offline learned for tracking a specific object. Although the appearance manifold is globally non-linear, the local appearance variations can still be approximated as a linear subspace. Thus, we propose to incrementally learn a set of low dimension linear subspaces to represent the global appearance manifold. A multi-subspace representation is used in [6], where a fixed number of subspaces are offline built and are online updated with new samples.

Let $\mathcal{M} = \{\Omega_1, ..., \Omega_L\}$ represent the appearance manifold of one object and $\Omega_l, l \in [1, ..., L]$ denote the local sub-manifold. An appearance instance $\mathbf{x}$ is a d-dimension image vector. Let $\Omega_l = (\hat{\mathbf{x}}_l, U_l, \Lambda_l, n_l)$ denote one sub-manifold, where $\hat{\mathbf{x}}_l, U_l, \Lambda_l$ and n_l represent the mean vector, eigenvectors, eigenvalues and the size (number of samples) of the subspace respectively. For simplicity, we omit the subscript when this causes no confusion. Here, $\Lambda = \text{diag}(\lambda_1, \ldots, \lambda_n)$ with sorted eigenvalues of the subspace, $\lambda_1 \geq \lambda_2 \cdots \geq \lambda_n$. A η-truncation is usually used to truncate the subspaces, namely $m = \arg\min_i (\sum_i \lambda_i / \text{tr}(\Lambda) \geq \eta)$. From a statistical point of view, a subspace with m eigenbases can be regarded as a m-dimensional Gaussian distribution. Suppose Ω is a subspace with the first m eigenvectors, the projection of $\mathbf{x}$ on Ω is $\mathbf{y} = (y_1, \ldots, y_m)^T = U^T(\mathbf{x} - \hat{\mathbf{x}})$. Then, the likelihood of $\mathbf{x}$ can be expressed [28] as

$$p(\mathbf{x}|\Omega) = \left[\frac{\exp\left(-\frac{1}{2}\sum_{i=1}^{m}\frac{y_i^2}{\lambda_i}\right)}{(2\pi)^{m/2}\prod_{i=1}^{m}\lambda_i^{1/2}}\right] \cdot \left[\frac{\exp\left(-\frac{\varepsilon^2(\mathbf{x})}{2\rho}\right)}{(2\pi\rho)^{(d-m)/2}}\right] \qquad (2)$$

where $\varepsilon(\mathbf{x}) = |\mathbf{x} - UU^T\mathbf{x}|$ is the projection error, namely L_2 distance between the sample $\mathbf{x}$ and its projection on the subspace. The parameter $\rho = \frac{1}{d-m}\sum_{i=m+1}^{d}\lambda_i$ [28] or uses the $\frac{1}{2}\lambda_{m+1}$ as a rough approximation. By using Eq.2, we can evaluate the confidence of a sample from one subspace. As our generative model contains multiple subspaces (each subspace can be regarded as a hyper-ellipsoid), we maintain the neighborhood according to L_2 distance between the mean vectors of subspaces. To evaluate the confidence of one sample from such a generative model, we use the maximum confidence of the K-nearest (we use $K = 4$ in experiments) neighboring subspaces.

3.1 Online Subspace Learning

Given that samples are given in a sequential way, we aim to learn the low dimension linear subspaces incrementally. A new subspace is created with d_0 dimension, namely $d_0 + 1$ sequential samples form a new subspace. Local smoothness is guaranteed by a small d_0. A new subspace is created and added into the subspace pool. In order to represent a large number of sequential samples, we use a fixed number subspaces: if the number of subspaces exceeds a predetermined maximum, the most similar two subspaces are merged. The outline of the online subspace learning algorithm is shown in

Algorithm 1. Online Subspace learning algorithm

Input: $(\mathcal{I}, \mathcal{M}, d_0, L)$
$\mathcal{I} = \{I_1, \cdots, I_n, \cdots\}$: a sequence of samples $\mathcal{M} = \emptyset$: the appearance manifold
d_0: the initial dimension for each subspace L: the maximum number of subspaces
Output: $\mathcal{M} = (\Omega_1, \cdots \Omega_L)$: multi-local linear subspaces
while $\mathcal{I} \neq \emptyset$ **do**
  fetch $d_0 + 1$ samples and form a new subspace
  $\Omega_n \leftarrow (I_i, \cdots, I_{i+d_0})$
  if there exists an empty subspace **then**
    Add Ω_n to $\mathcal{M}$
  else
    $(p, q)^* = \arg\max Sim(\Omega_p, \Omega_q), p, q \in [1, \cdots, L], p \neq q$
    $\Omega_m = \Omega_p \cup \Omega_q$ and replace Ω_p and Ω_q with Ω_m
  end if
end while

Algorithm 1. In order to maintain the local property of the subspaces, merging only happens between neighboring subspaces. Merging of two subspaces and measuring the similarity between two subspaces are two critical steps in this algorithm.

Several methods have been proposed to incrementally update the eigenspaces. Only the method proposed by Hall *et al.* [29] takes into account the change of the mean of a subspace. This approach provides an exact solution to update an eigenspace and does not require storing original samples. Similar method was also used in [5,6] to update a subspace given new samples. We summarize Hall's method in [29] by using scatter matrixes to simplify the representation. Suppose there are two subspaces $\Omega_1 = (\mathbf{x}_1, U_1, \Lambda_1, N)$ and $\Omega_2 = (\mathbf{x}_2, U_2, \Lambda_2, M)$, which we are trying to merge to a new subspace $\Omega = (\bar{\mathbf{x}}, U, \Lambda, M + N)$. If the dimension of Ω_1 and Ω_2 are p and q, the dimension r of the merged subspace Ω satisfies: $\max(p, q) \leq r \leq p + q + 1$. The vector connecting the centers of the two subspaces does not necessarily belong to either subspace. This vector causes the additional one in the upper bound of r.

It is easy to verify that the scatter matrix S of the merged subspace Ω satisfies, $S = S_1 + S_2 + \frac{MN}{M+N}(\mathbf{x}_1 - \mathbf{x}_2)(\mathbf{x}_1 - \mathbf{x}_2)^T$. We aim to find a sufficient orthogonal spanning of S. Let $h_1(\mathbf{x})$ denote the residual vector of a vector $\mathbf{x}$ on Ω_1, $h_1(\mathbf{x}) = \mathbf{x} - U_1 U_1^T \mathbf{x}$. Note that $h_1(\mathbf{x})$ is orthogonal to U_1, *i.e.* $h(\mathbf{x})'U = 0$. Now, $U' = [U_1, \mathbf{v}]$ is a set of orthogonal bases to span the merged space, where $\mathbf{v} = GS\left(h_1\left(U_2, (\mathbf{x}_2 - \mathbf{x}_1)\right)\right)$ and $GS(\cdot)$ denote the Gram-Schmidt process.

Given the sufficient orthogonal bases, we can obtain the SVD decomposition of S.

$$U'^T S U' = \begin{bmatrix} \Lambda_1 & 0 \\ 0 & 0 \end{bmatrix} + \begin{bmatrix} G\Lambda_2 G^T & G\Lambda_2 \Gamma^T \\ \Gamma\Lambda_2 G^T & \Gamma\Lambda_2\Gamma^T \end{bmatrix} + \frac{MN}{M+N} \begin{bmatrix} gg^T & g\gamma^T \\ \gamma g^T & \gamma\gamma^T \end{bmatrix} = R\Lambda R^T \quad (3)$$

where $G = U_1^T(\mathbf{x}_2 - \mathbf{x}_1)$, $\Gamma = \mathbf{v}^T U_2$, $g = U_1^T(\mathbf{x}_2 - \mathbf{x}_1)$ and $\gamma = U'(\mathbf{x}_2 - \mathbf{x}_1)$. Now, the eigenvalue of the merged subspace is Λ in Eq.3 and the eigenvector U is simply $U'R$. Note that incrementally updating a subspace with one observation as in [6] is one special case of merging two subspaces using Eq.3.

3.2 Subspace Distance

The other critical step in Algorithm 1 is to determine the similarity between two subspaces. We use two factors to measure the similarity between two neighboring subspaces Ω_1, Ω_2, the canonical angles (principal angles) and the data-compactness.

Suppose the dimensions of two subspaces are $p, q, p \geq q$, then there are q canonical angles between the two subspaces. A numerical stable algorithm [30] computes the angles between all pairs of orthonormal vectors of the two subspaces as, $\cos\theta_k = \sigma_k(U_1^T U_2), k = 1, \cdots, q$, where $\sigma_k(\cdot)$ is the kth sorted eigenvalue computed by SVD. The consistency of two neighboring subspaces can be represented as follows.

$$Sim_1(\Omega_1, \Omega_2) = \prod_{k=q-d_0+1}^{q} \sigma_k(U_1^T U_2) \quad k = 1, \cdots, q \tag{4}$$

As the dimensionality of subspaces is larger than d_0, the initial dimension, we select the d_0 largest principal angles, which approximately measure the angle between two local subspaces. In a 3D space, the largest canonical angle between two 2D subspaces is equivalent to the angle between the two planes. In this case, we prefer to merge 2D patches with a small plane-to-plane angle. Note that the merge only happens between neighbor subspaces. The neighborhood is defined according to the mean vector L_2 distance. Merging subspaces with a small principal angle can avoid destroying the local structure of the appearance manifold.

The other factor to consider is data-compactness, which measures how much extra dimensionality is incurred by a merge operation. Suppose the dimension of two subspaces Ω_1, Ω_2 is $p, q, p \geq q$, the sorted eigenavalues of original merged subspace are $\Lambda_r = (\lambda_1, \ldots, \lambda_r), r = p+q+1$. The similarity based on data-compactness is defined as

$$Sim_2(\Omega_1, \Omega_2) = \sum\nolimits_{i=1}^{p} \lambda_i / \sum\nolimits_{i=1}^{r} \lambda_i \tag{5}$$

If Sim_2 is close to one, this indicates the merge operation does not incur any new dimension; on the contrary, if Sim_2 is small, this indicates the variations in Ω_1 and Ω_2 cannot use common eigenvectors to represent it. Combining the two factors in Eq.4 and Eq.5, the final similarity between two subspaces is defined in Eq.6.

$$Sim(\Omega_1, \Omega_2) = Sim_1(\Omega_1, \Omega_2) + w_d Sim_2(\Omega_1, \Omega_2) \tag{6}$$

where w_d is the weight to balance these two factors. We use $w_d = 0.2$ in experiments.

One example of online trained local subspaces is shown in Figure 2. A saddle-like 3D surface (shown in Figure 2(a) is generated with Gaussian noise. The 3D points are input to Algorithm 1 sequentially. The final subspaces are shown in Figure 2(b) with $L = 15, \eta = 0.995$. The initial dimension d_0 is one. Although the online built subspaces depend on the order of the samples, a compact representation of the samples can always be created as long as the data are input with local smoothness. The subspace updating operation dominates the complexity of the generative tracker. Merging two subspaces with dimension p, q requires Golub-Reinsch SVD $O(r^3)$ of $r = p + q + 1$ dimension square matrix. Since the dimension of each subspace is low, the total complexity is quite low. The low dimensionality of the local subspaces can guarantee both the local property and efficient computation.

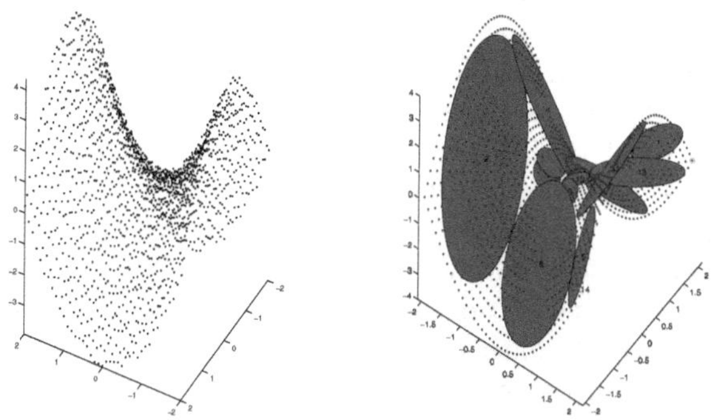

(a) 3D saddle surface (b) final subspaces built

Fig. 2. 3D example of incremental updating subspaces

4 Discriminative Tracker Using Online SVM

It has often been argued that SVM has better generalization performance than other discriminative methods on a small training set. For the discriminative model, we adopt an incremental SVM algorithm, *LASVM* [24], to train a classifier between object and background. SVM [31] is able to form the optimal separating function, which reduces to a linear combination of kernels on the training data, $f(\mathbf{x}) = \sum_j \alpha_j y_j K(\mathbf{x}_j, \mathbf{x}) + b$, with training samples $\mathbf{x_i}$ and corresponding label $y_i = \pm 1$.

In practice, this is achieved by maximizing the dual objective function $\max_\alpha W(\alpha)$ with $W(\alpha) = \sum_i \alpha_i y_i - \frac{1}{2}\sum_{i,j} \alpha_i \alpha_j K(x_i, x_j)$, subject to

$$\sum_i \alpha_i = 0, A_i \leq \alpha_i \leq B_i, \tag{7}$$

where $A_i = \min(0, Cy_i), B_i \max(0, Cy_i)$. Here, α is a vector of weights on y_i. A SVM solver can be regarded as updating α along some direction to maximize $W(\alpha)$. Let $g = (g_1, \ldots, g_n)$ denote the gradient of $W(\alpha)$

$$g_k = \frac{\partial W(\alpha)}{\partial \alpha_k} = y_k - \sum_i \alpha_i K(x_i, x_k) = y_k - \hat{y}(x_k) + b \tag{8}$$

LASVM suggests that optimization is faster when the search direction mostly contains zero coefficients. *LASVM* uses the search directions whose coefficients are all zero except for a single +1 and a single -1. The two non-zero coefficients, are called *τ-violating pair* (i, j) if $\alpha_i < B_i, \alpha_j > A_j$, and $g_i - g_j > \tau$, where τ is a small positive value. and *LASVM* selects the τ-violating pair (i, j) that maximizes the directional gradient $g_i - g_j$.

The LASVM algorithm contains two procedures named PROCESS and REPROCESS [24]. When a new sample x_k arrives, PROCESS forms a *τ-violating pair* (i, j), which contains x_k and another existing support vector, and updates the weights of this pair. Following PROCESS, REPROCESS selects a *τ-violating pair* from the set of support vectors and updates their weights. The new sample x_k may become a new support vector through PROCESS, while another support vector may need to switch out by REPROCESS. Both PROCESS and REPROCESS select *τ-violating pair* with the largest gradient. The complexity of such a selection grows linearly with the number of vectors. A finishing step, which runs REPROCESS multiple times to further remove as many *τ-violating pairs* as possible, is performed after online process. For tracking, the intermediate classifier is useful, hence we run this finishing step every 10 frames. Note that, since we do not need to look at the ignored vectors for incremental learning, *τ-violating pair* is only selected from the set of support vectors. For online tracking problem, many appearance variations and limited training samples may degrade the generalization ability of SVM. Also, in experiments, we find that the number of support vectors grows fast when the appearance of object and background changes. Thus, we propose to decrementally train the SVM and focus on recent appearance variations within a sliding window. REPROCESS in LASVM can be used to achieve the "unlearning" of old samples. For decremental learning, removing ignored vectors (when ignored vectors move out of the sliding window) will not change the decision boundary. However, the removal of a support vector will affect the decision boundary and some ignored vectors may become support vectors. In order to remove one support vector, we first zero its coefficient and put its coefficient into the closest vector to keep the constraint in Eq.7. We then apply REPROCESS multiple times to select τ-violating pairs in set of both ignored and support vectors and update the weights. The cost of decremental learning is that we need to store all samples within a sliding window.

5 Experiments

For both generative and discriminative models, we use image vectors of size 32×32 (for face) or 32×64 (for human and vehicle). For the generative model, η is set to 0.95-0.99 and the maximum number of subspaces is set to 5-10. The initial subspace dimension is 4, which is very low compared to the input space. Thus, every 5 frames, we form a new subspace, which is then inserted into the subspace pool. For the discriminative model, we use *LASVM* with R-HOG feature vectors, which are created from 16×16 blocks containing 8×8 cells. The strike size is 4 to allow overlapping HOG descriptors. Each cell has 9 bins oriented histogram; hence, we have 36-bin oriented histogram for a block. For a 32×64 window, the vector size is 2340. We use the linear kernel function in SVM. The number of support vectors varies between 50-150 for different sequences. We use a sliding window of 30 frames. We manually label the first 10 frames as the initialization for the two trackers. The Bayesian inference framework generates 600 particles. The co-trained hybrid model is implemented in C++. The combined tracker runs at around 2 fps on a P4 2.8GHz dual core PC. All testing sequences are 320×240 graylevel images.

During co-training, each learner labels the unlabeled data on which it makes a confident prediction based on its own knowledge. For this purpose, a threshold is needed for each learner. For the generative model, we set a threshold based on the log likelihood in Eq.2. To be more conservative, we use a second criteria: we find several local optima in the posterior distribution and if ratio ρ between the second optimum and the global optimum is small enough ($\rho \leq 0.7$), we accept the global optimum as a positive sample and all other samples that far enough from the global optimum are negative samples. For the discriminative model, due to the very limited training data, the positive and negative training data are usually well separated. Thus, we cannot adopt the way used in [21] to select the threshold. Instead, we select the confidence threshold so that at most 80% positive samples' confidence is above that threshold. This threshold is updated every 30 frames. The positive and negative samples labeled by the generative model will not be added to the discriminative model unless they are close to the decision boundary. To express the SVM confidence as a probability, we use the method in [32] to fit a sigmoid function that is updated every 30 frames.

5.1 Comparative Analysis

We compare our co-trained tracker with two generative methods, including (G1) IVT [5] and our multiple linear subspaces (G2) algorithm and three discriminative methods, including online selection of discriminative color (D1) [10], our online SVM method (D2) and ensemble tracking (E.T) [9]. G1 uses a single 15D linear subspace and updates it incrementally. Note that D1 does not consider tracking with large scale change and rotation. G1, G2, D2 and the co-trained tracker use the same parameters in CONDENSATION algorithm, but G1, G2 and D2 use self-learning to update their models.

We compare these methods with challenging data sets, which contain image sequences of various types of object, including face (seq1-seq2), human (seq3-seq5) and vehicle (seq6). The challenging conditions include significant illumination changes (seq1), abrupt camera motion and significant motion blur (seq2-seq5), viewpoint changes and/or pose variations (seq3-seq6), and also occlusions (seq4-seq6). To compare the robustness under the challenging conditions, we show how many frames these methods can track the objects before tracking failure, *i.e.* after this frame a tracker cannot recover without re-initialization. Table 1 shows the comparison between different methods. The number of frames and the number of frames where occlusion happens in each sequence are also shown in Table 1. The comparison demonstrates that the co-trained tracker performs more robustly than other methods. Note that D1 requires color information, thus it cannot process some sequences, which are indicated with "n/a". The visual results are shown in Figure 4, where the tracked objects and part of negative

Table 1. Comparison of different methods G1:IVT [5], G2: incremental learning multiple subspaces, D1: online selection of discriminative color features [10], D2: online SVM, E.T: ensemble tracking [9]. D1 uses color information, which is not available for Seq1 and Seq6.

	Frames	Occlusion	G1	G2	D1*	D2	E.T	Ours
Seq1	761	0	17	261	n/a	491	94	**759**
Seq2	313	0	75	282	313	214	44	**313**
Seq3	140	0	11	15	6	89	22	**140**
Seq4	338	93	33	70	8	72	118	**240**
Seq5	184	30	50	50	50	50	53	**154**
Seq6	945	143	163	506	n/a	54	10	**802**

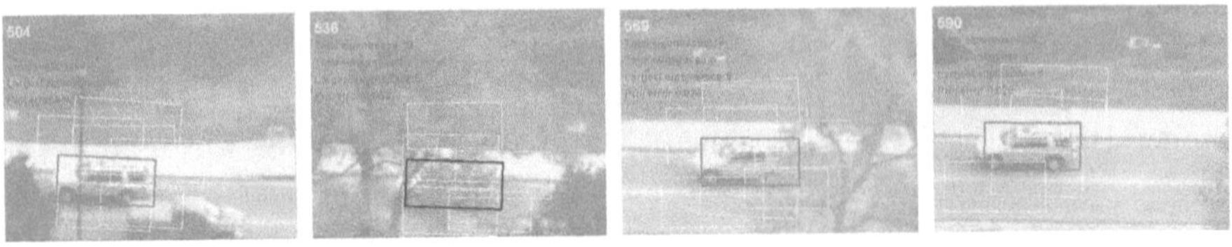

(a) the proposed method

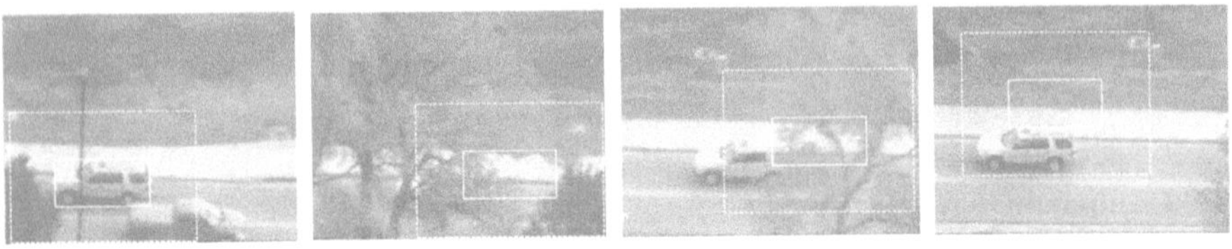

(b) ensemble tracking

Fig. 3. Tracking and reacquisition with long occlusion and clutter background

samples are bounded with green boxes and white boxes respectively. The red box indicates none of the models is updated in this frame. In experiments, we frequently find that the co-trained tracker has better self-awareness of current tracking performance and can safely enlarge the search range (by changing the diffusion dynamics) without being confused by distracters in the background. Also, the co-trained tracker successfully avoids drifting caused by varying viewpoints and illumination changes.

We compare with two other methods to demonstrate our method's reacquisition ability. One is ensemble tracking [9]. The other one is the online SVM tracker without decremental learning. In the experiments, we find that ensemble tracking can only reacquire the object after a short occlusion. Also, ensemble tracking cannot deal with rotation and scale change, which restricts its reacquisition ability. Without decremental learning, the number of support vectors increases to several hundreds quickly when the background becomes cluttered and the object appearance changes. This makes the online SVM tracker very slow. Meanwhile, the performance of the tracker goes down, exhibiting as drifting and being trapped by distracters.

We also compare our generative tracker G2 with G1 and another generative method G3, in indoor environments with few distracters. G3 uses 5 key 10D subspaces (corresponding to front, left profile, right profile, up and down), which are trained offline by manually labelling 150 samples into 5 categories; our generative method G2 uses at most 10 subspaces and each new subspace starts from 4-dimensions. The indoor sequence exhibits significant head pose variations, shown in Figure 5(b). We calculate projection errors of each method in Figure 5(a) (average of multiple runs) within 1000 frames before the other two trackers start to drift. Offline 5-key subspaces method shows large projection errors at some frames where the poses are not covered in offline samples. G1 is not able to adapt rapidly to new appearance variations after running a long sequence. Our generative method G2 can promptly adapt to appearance variations and show smaller projection errors consistently, though each subspace in our generative track has much smaller dimensionality than the other two methods. As we can see, the

(a) Tracking and reacquisition with abrupt motion and blur

(b) Tracking human with clutter background and distracters

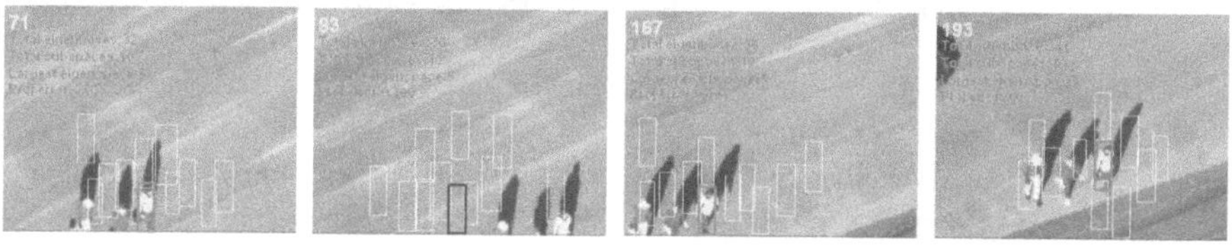

(c) Tracking and reacquisition with long leaving out of field of view

Fig. 4. Tracking various type of objects in outdoor environments

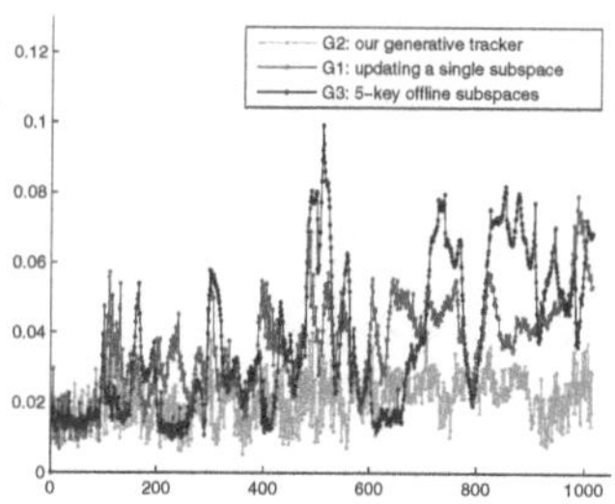

(a) Comparison of projection errors

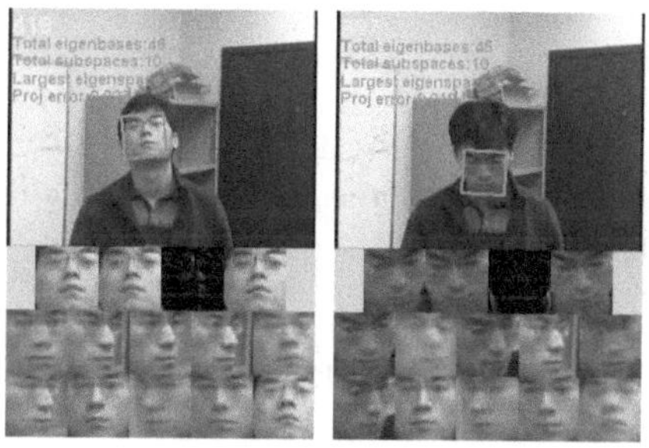

(b) Tracking with pose variations

Fig. 5. Comparison of generative methods in indoor environments

online trained subspaces approximately represent the poses that have been observed, though we do not train offline in different poses.

6 Summary and Future Work

We have proposed a co-training framework to combine one global generative tracker and one local discriminate tracker. Our generative tracker builds a compact representation of the complete appearance of an object by online learning a number of local linear subspaces. The discriminative tracker adopts the online SVM algorithm to focus on local appearance. By co-training, the two trackers can train each other on-the-fly with limited initialization. Though no offline training data is used, our method has strong reacquisition ability and robustness against the distracters in background. Extensive experiments demonstrate that our method can handle different challenging situations.

Our method relies on smoothness in appearance changes and cannot deal with abrupt appearance changes. Also, our method cannot explicitly handle the partial occlusion problem. Partial occlusions are often regarded as non-object by our method. This is a safe strategy to avoid updating the model with wrong appearance instances. It is worth noting that our method addresses online building appearance model for unknown types of objects. In the future, we expect to combine this method with offline learning for tracking one particular type of object. For tracking a particular type of objects, the robustness of tracking can be significantly improved by combining online training with offline training, as shown in [33] and [34]. We also expect to introduce collaborative part-based trackers under this co-training framework to deal with partial occlusions.

Acknowledgements. This research was funded, in part, by MURI-ARO W911NF-06-1-0094. The second author is also supported by Vietnam Education Foundation. We thank Dr. David Ross, Dr. Shai Avidan and Dr. Antoine Bordes for sharing the data and implementations of methods in [5], [9] and [24] respectively. We also thank Yuan Li for providing part of the data.

References

1. Yilmaz, A., Javed, O., Shah, M.: Object tracking: A survey. ACM Comput. Surv. (2006)
2. Black, M.J., Jepson, A.D.: Eigentracking: Robust matching and tracking of articulated objects using a view-based representation. IJCV 26, 63–84 (1998)
3. Jepson, A.D., Fleet, D.J., El-Maraghi, T.F.: Robust online appearance models for visual tracking. In: CVPR (2001)
4. Isard, M., Blake, A.: Contour tracking by stochastic propagation of conditional density. In: ECCV, pp. 343–356 (1996)
5. Lim, J., Ross, D., Lin, R., Yang, M.: Incremental learning for visual tracking. In: NIPS, pp. 793–800 (2004)
6. Lee, K.C., Kriegman, D.: Online learning of probabilistic appearance manifolds for video-based recognition and tracking. In: CVPR, pp. 852–859 (2005)
7. Elgammal, A.: Learning to track: Conceptual manifold map for closed-form tracking. In: CVPR, pp. 724–730 (2005)
8. Avidan, S.: Support vector tracking. PAMI (2004)
9. Avidan, S.: Ensemble tracking. In: CVPR, vol. 2, pp. 494–501 (2005)
10. Collins, R.T., Liu, Y., Leordeanu, M.: Online selection of discriminative tracking features. In: PAMI, vol. 27, pp. 1631–1643 (2005)
11. Nguyen, H.T., Smeulders, A.W.: Robust tracking using foreground-background texture discrimination. IJCV 2006 (2006)
12. Oza, N.C., Russell, S.: Online bagging and boosting. In: International Workshop on Artificial Intelligence and Statistics (2001)
13. Grabner, M., Grabner, H., Bischof, H.: Learning features for tracking. In: CVPR (2007)
14. Liu, X., Yu, T.: Gradient feature selection for online boosting. In: ICCV (2007)

15. Lasserre, J.A., Bishop, C.M., Minka, T.P.: Principled hybrids of generative and discriminative models. In: CVPR (2006)
16. Ng, A.Y., Jordan, M.I.: On discriminative vs. generative classifiers: a comparison of logistic regression and naive bayes. In: NIPS (2001)
17. Raina, R., Shen, Y., Ng, A.Y., McCallum, A.: Classification with hybrid generative/discriminative models. In: NIPS (2003)
18. Lin, R.S., Ross, D., Lim, J., Yang, M.H.: Adaptive discriminative generative model and its applications. In: NIPS (2004)
19. Yang, M., Wu, Y.: Tracking non-stationary appearances and dynamic feature selection. In: CVPR (2005)
20. Blum, A., Mitchell, T.: Combining labeled and unlabeled data with co-training. In: COLT (1998)
21. Levin, A., Viola, P., Freund, Y.: Unsupervised improvement of visual detectors using co-training. In: ICCV (2003)
22. Javed, O., Ali, S., Shah, M.: Online detection and classification of moving objects using progressively improving detectors. In: CVPR (2005)
23. Tang, F., Brennan, S., Zhao, Q., Tao, H.: Co-tracking using semi-supervised support vector machines. In: ICCV (2007)
24. Bordes, A., Ertekin, S., Weston, J., Bottou, L.: Fast kernel classifiers with online and active learning. In: JMLR 2005, pp. 1579–1619 (2005)
25. Cauwenberghs, G., Poggio, T.: Incremental and decremental support vector machine learning. In: NIPS (2001)
26. Dalal, N., Triggs, B.: Histograms of oriented gradients for human detection. In: CVPR, pp. 886–893 (2005)
27. Isard, M., Blake, A.: Condensation - conditional density propagation for visual tracking. IJCV, 5–28 (1998)
28. Moghaddam, B., Pentland, A.: Probabilistic visual learning for object representation. PAMI 19, 696–710 (1997)
29. Hall, P., Marshall, D., Martin, R.: Merging and splitting eigenspace models. In: IEEE PAMI, pp. 1042–1049 (2000)
30. Bjoerck, A., Golub, G.H.: Numerical methods for computing angles between linear subspaces. Mathematics of computation, 579–594 (1973)
31. Vapnik, V.N.: Statistical Learning Theory. John Wiley Sons, Chichester (1998)
32. Platt, J.C.: Probabilistic outputs for support vector machines and comparisons to regularized likelihood methods. In: Advances in Large Margin Classifiers (1999)
33. Li, Y., Ai, H., Yamashita, T., Lao, S., Kawade, M.: Tracking in Low Frame Rate Video: A Cascade Particle Filter with Discriminative Observers of Different Lifespans. In: IEEE CVPR (2007)
34. Kim, M., Kumar, S., Pavlovic, V., Rowley, H.: Face Tracking and Recognition with Visual Constraints in Real-World Videos. In: IEEE CVPR (2008)

Statistical Analysis of Global Motion Chains

Jenny Yuen[1,*] and Yasuyuki Matsushita[2]

[1] CSAIL, Massachusetts Institute of Technology, Cambridge MA 02139
[2] Visual Computing Group, Microsoft Research Asia, Beijing 100080, China
jenny@csail.mit.edu, yasumat@microsoft.com

Abstract. Multiple elements such as lighting, colors, dialogue, and camera motion contribute to the style of a movie. Among them, camera motion is commonly overlooked yet a crucial point. For instance, documentaries tend to use long smooth pans whereas action movies usually have short and dynamic movements. This information, also referred to as global motion, could be leveraged by various applications in video clustering, stabilization, and editing. We perform analyses to study the in-class characteristics of these motions as well as their relationship with motions of other movie types. In particular, we model global motion as a multi-scale distribution of transformation matrices from frame to frame. Secondly, we quantify the difference between pairs of videos using the KL-divergence of these distributions. Finally, we demonstrate an application modeling and clustering commercial and amateur videos. Experiments performed show advantage compared to the usage of some local motion-based approaches.

1 Introduction

Global motion plays an important role in defining and characterizing the style of a movie. Think about the style of Alfred Hitchcock's movies, where his characteristic talent manipulating the audience's fears and desires was known greatly for his technique in moving and placing his camera [28, 2]. Another example is the highly dynamic motion in *The Blair Witch Project*, where critics have described how the camera motion can be used to reveal a story through its startling images and lucid characterizations [19]. Moreover, global motion not only varies depending on the director style, but also its content. For example, the characteristic smooth and continuous camera motions in *National Geographic* documentaries are different from the fast and diverse mix of camera movements in an action movie like *Spiderman*.

Videos of different genres clearly contain different *global motion chains*, which is a sequence of global motions. In this work, we study the statistics of global motion *chains* in the spirit of analogous works analyzing image and local flow statistics [23, 11]. Unlike previous works that use global motions for some particular applications such as video classification, video stabilization and etc., this work focuses on the development of a framework to analyze global motion chains in two aspects (Figure 1):

* Part of this work was done while the author was visiting Microsoft Research Asia.

D. Forsyth, P. Torr, and A. Zisserman (Eds.): ECCV 2008, Part II, LNCS 5303, pp. 692–705, 2008.

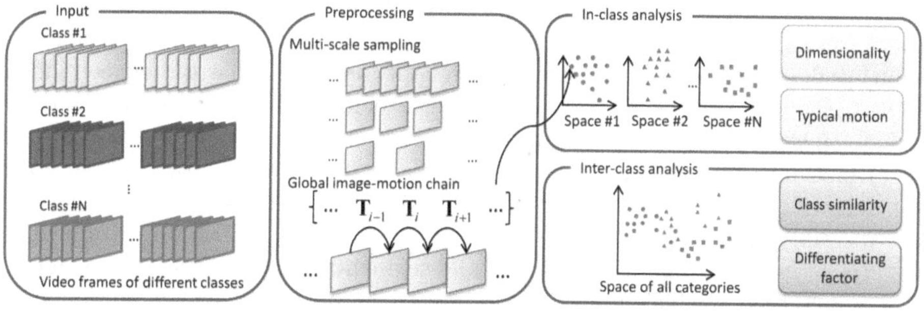

Fig. 1. Overview of the analysis of motion chains

1. *In-class Analysis.* We want to learn what the motion in a class of videos is like by finding its characteristic camera movements and learning about their intrinsic dimensionality.
2. *Inter-class Analysis.* We want to learn what makes camera motions of a class different from motions in another class and, if particular motions are better at distinguishing a class from another one.

Furthermore, we apply the learnt statistics to compare and cluster videos using KL-divergence as an application of the analyses. We show the results for three types of professional (action movies, basketball games, and documentaries) as well as unprofessional videos taken by numerous people. Finally, we compare the effectiveness of our feature with other motion-based ones when used for clustering and will demonstrate the differences and advantages of our feature in the area of video clustering.

1.1 Related Work

There has been a lot of work using global image motion but to the extent to our knowledge, none studying the statistics of global motion chains. The notion of global motion chains, or sequence of transformations between consecutive video frames in a video, has been widely used in the context of video stabilization [13, 20, 16]. These tend to be parameter-based with the objective of smoothing sequences for motion compensation. Motion compensated frames are obtained by warping the original video frames by a cascade of the original and smoothed transformation chains.

While not using the global motion chains, there are several previous works analyzing actions in video clips. These are classified in model-based approaches [21, 18,1,14,24] and nonparametric approaches [5,6,30]. The model-based approaches are designed for detecting predefined activities occurring in videos. These are shown to be useful for restricted conditions where the assumed models are valid. More recent trends in action analysis lies in non-parametric approaches.

Chomat and Crowley [5] use a set of spatio-temporal filters and evaluate the joint statistics of filter responses for the purpose of probabilistic action recognition. Efros *et al.* [6] use dense optical flow fields as a feature to perform a nonparametric action recognition. Zelnik-Manor and Irani [30] propose a multi-scale spatio-temporal feature which captures the similarity between video clips based on the behavior contained in the videos. Their feature is based on a local intensity gradient in space-time domain, and it successfully captures local motion occurring in videos. Wang *et al.* [29] developed various methods for detecting and classifying events in surveillance videos using hierarchical linear discriminant analysis (LDA) to cluster local motion attributes. Stauffer *et al.* [27] [9] use tracking for recognition and segmentation of objects in the video . These methods are efficient and robust in many applications such as surveillance or activity recognition and have been extensively tested in still camera settings. We further extend the problem domain and work with videos in a large number of settings and containing significant camera motion.

For the purpose of video classification and other applications, the global motion has been used by several researchers. Roach *et al.* [22] use a background image motion represented by (x, y) translation for the purpose of video genre classification. Affine camera motion is used as one of video features by Smith and Kanade [26] for video skimming. Kobla *et al.* [12] use the camera motion feature for identifying sports videos. Bouthemy and Ganansia [4] propose a method for finding shot changes in a video using global motion. Closest to our work, Fablet *et al.* [7] use global motion for video classification and retrieval. The global motion is represented by a temporal Gibbs random field to perform the temporal analysis. All these works show interesting applications for camera dynamics. We present this paper with a different focus: we want to learn and understand the statistics of motion chains of camera motions across different types of video.

Our work differs from these previous works because we focus on a multi-scale analysis of global motion chains. In particular, we are interested in observing the degree of information contained in the global motion chains across videos of different categories and, as an example, to evaluate to what extent classification can be done in a setting of varied short clips and a moving camera.

While, to the best of our knowledge, there has been no work studying the statistics of global motion chains, Roth and Black [23] propose a model to learn the spatial statistics of optical flow fields using 3D range data. They model the flow field as a *Field-of-Experts* and show numerous statistics of the velocity and orientation of this information. Furthermore, they apply these statistics as a prior for a more robust estimation of optical flow. This paper is analogous to Roth and Black's work, with the objective of learning the form, distribution, and characteristics of global motion chains, which has not yet been done.

1.2 Background

Let $V = \{I_1, ..., I_n\}$ be an image sequence composed of n frames where I_i is its i-th frame. Assume for now that only the camera moves while the environment captured in the video is static and planar. We refer to the term *global motion* between two consecutive frames as a 3×3 projective transformation matrix $\mathbf{T}$. By representing 2D pixel positions (x, y) in homogeneous coordinates $\mathbf{x} = (x, y, 1)^T$, the geometric transformation $I(x,y) \xrightarrow{\mathbf{T}} I'(x',y')$ can be written as $\mathbf{x}' \sim \mathbf{Tx}$, where $\sim$ denotes equality up to scale. The condition for the input video is relaxed for the case of an image sequence with moving objects in the scene by extracting the transformation matrix assuming that the majority of the objects in the captured scene remain static in the real world.

For our global motion estimation, we will assume the affine transformation model. Our reasoning for this assumption is justified by the fact that often the majority of image pixels will be background pixels and they will be far enough from the camera that could be approximated to being on a plane. The affine transformation can be written as:

$$\mathbf{x}' = \mathbf{Ax} = \begin{pmatrix} a_{11} & a_{12} & a_{13} \\ a_{21} & a_{22} & a_{23} \\ 0 & 0 & 1 \end{pmatrix} \mathbf{x}, \tag{1}$$

where $\mathbf{x}$ is the original position, $\mathbf{x}'$ is the new position, $\mathbf{A}$ is an affine transformation matrix. This will reduce the number of variables to six per matrix. We also define a vector representation of $\mathbf{T}(= \mathbf{A})$ in scan-line order as $\mathbf{t}$:

$$\mathbf{t} = [a_{11}, a_{12}, a_{13}, a_{21}, a_{22}, a_{23}]^T. \tag{2}$$

Global Motion Chains. In our analysis, we use ordered *sequences* of global motions which we will refer to as *global motion chains.* We represent the global motion chain of an image sequence centered at the i-th frame as an ordered list of transformation matrices of consecutive frames as:

$$\mathbf{C}_j = \left[[\mathbf{t}_{i-r}]^T, [\mathbf{t}_{i-(r+1)}]^T, \ldots, [\mathbf{t}_{i+r}]^T\right]^T, \tag{3}$$

where r is a positive constant that denotes the radius of the temporal window of the transformation chain $\mathbf{C}_i$, $\mathbf{t}_i$ is the vector representation of the transformation matrix $\mathbf{T}_i$ from frame i to frame $i+1$.

We further extend this notation to a multi-scale scheme where the sampling of frames is controlled by a frame skip rate s. The global motion chain with the frame skip rate s can be written as

$$\mathbf{C}_j^s = \left[[\mathbf{t}_{i-rs}^s]^T, [\mathbf{t}_{i-(r+1)s}^s]^T, \ldots, [\mathbf{t}_{i+rs}^s]^T\right]^T, \tag{4}$$

where, analogously, $\mathbf{C}_i^s$ is a global motion chain centered at the i-th frame and $\mathbf{t}_i^s$ is the vector representation of the transformation matrix $\mathbf{T}_i^s$ from frame i to frame $i+s$. The frame skip rate allows us to perform a multi-scale representation of the motion information used later in our study.

2 Analysis of Global Motion Chains

In this section we will describe a framework for comparing global motions. We are interested in learning how global motions vary per class and what their characterizing elements are. We model a class of videos as a distribution of

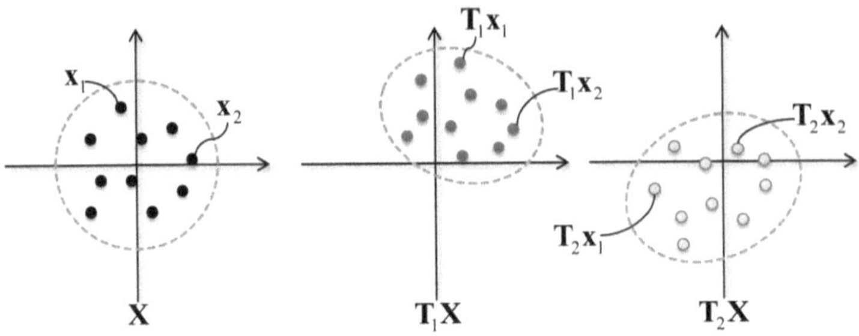

Fig. 2. Illustration of the similarity of global motions. Left: points are distributed in a two dimensional coordinate system. These points are transformed by $\mathbf{T}_1$ and $\mathbf{T}_2$ in the middle and right figures. The similarity of global motions $\mathbf{T}_1$ and $\mathbf{T}_2$ is defined by the sum of distances between corresponding points $\mathbf{X}$.

motion chains. In general, there will be a motion chain over a temporal window r for each frame in each video of a class (excluding some frames at the temporal edges of each video). We will further extend this to a multi-scale approach where we will subsample frames at different rates but the general idea of our model is a *distribution of motion chains.*

2.1 Similarity of Global Motion Chains

We will first define a similarity measure of global motion for general linear transformations in 2-D, e.g., rigid body, similarity, affine and projective. For this case, the similarity of two transformations has the nice and simple property that it can be expressed as the L_2 norm distance between the two transformations when defining the similarity as the amount of point displacement in a 2D plan. The derivation is as follows.

Given some disk in a 2D plane and distributed points on the disk (Figure 2 left). We define the similarity of two linear transformations by the sum of distances of the warped points. Given two linear transformations $\mathbf{T}_1$ and $\mathbf{T}_2$, the similarity $S(\mathbf{T}_1, \mathbf{T}_2)$ of these two transformations can be described as

$$\begin{aligned} S(\mathbf{T}_1, \mathbf{T}_2) &\stackrel{\text{def}}{=} \frac{1}{|D|} \int_D p(x,y) ||\mathbf{T}_1 (x \quad y)^T - \mathbf{T}_2 (x \quad y)^T ||_2^2 dx dy \\ &= \frac{1}{|D|} \int_D p(x,y) ||(\mathbf{T}_1 - \mathbf{T}_2)(x \quad y)^T ||_2^2 dx dy, \end{aligned} \tag{5}$$

where D represents a set of points distributed on the disk, $|D|$ is the number of points, $||\cdot||_2$ denotes the L_2 norm, and $p(x,y)$ is the probability density function of the point at (x,y). Assuming that $p(x,y)$ has a uniform distribution, as it is the case in a regular 2D images if we consider pixels as points, the above equation can be simplified to

$$S(\mathbf{T}_1, \mathbf{T}_2) = Z ||\mathbf{t}_1 - \mathbf{t}_2||_2^2, \tag{6}$$

where the constant Z is written as

$$Z = \frac{1}{|D|} \int_D p(x,y) ||(x \quad y)^T||_2^2 dx dy. \tag{7}$$

From this result, the similarity of two linear transformations can be computed by the L^2 norm. Consequently, the similarity $S_{\mathcal{C}}$ of a pair of global motion chains $\mathcal{C}$ and $\mathcal{C}'$ can be written as

$$S_{\mathcal{C}}(\mathcal{C}, \mathcal{C}') = \sum_i S(\mathbf{T}_i^{(\mathcal{C})}, \mathbf{T}_i^{(\mathcal{C}')}), \tag{8}$$

where $\mathbf{T}_i^{(\mathcal{C})}$ represents the i-th transformation matrix in $\mathcal{C}$.

The similarity metric can be augmented by modifying the probability density function $p(x, y)$. For example, if we assume non-uniform importance of pixels, i.e., regions of interest, $p(x, y)$ becomes non-uniform. While it is possible to use this generalized form of the similarity metric, we assume the uniform density of $p(x, y)$ for simplicity.

2.2 In-Class Analysis

In-class typical motions. The purpose of this section is to study what typical motion chains are like. Intuitively, we can think about how shaky camera motion could characterize handheld-captured videos, slow pans might be common in documentaries about nature, or highly dynamic movements in action movies.

To identify the typical transformation chains, we fit a mixture of Gaussians for each class distribution of transformation chains. The means of the Gaussian centers describe *typical* transformation chains in the class. We choose the number of Gaussians m according to the Bayesian information criterion (BIC) [25]:

$$m = \underset{m}{\operatorname{argmin}} \{ -\log P(data|m, \hat{\theta}_m) + \frac{k_m}{2} \log n \}, \tag{9}$$

where k_m is the number of parameters given m Gaussians, $\hat{\theta}_m$ is the maximum likelihood parameters for the distribution with m Gaussians, and n is the number of points in the distribution.

Space of global motion chains. We will also study the compactness in the distribution of each class of videos. For example, consider the extreme case for surveillance videos where the camera is static. In this case, the camera motion of this video can be compactly characterized by this static state (or viewed in our terms as an identity matrix since there is no global motion from frame to frame). On the other hand, if we have completely random motion, the information required to characterize the space will be larger than in the previous example.

To analyze the compactness of the in-class distribution of transformation chains, we project each distribution of points to their principal components and analyze the cumulative variance of the distribution as a function of the number of principal components selected.

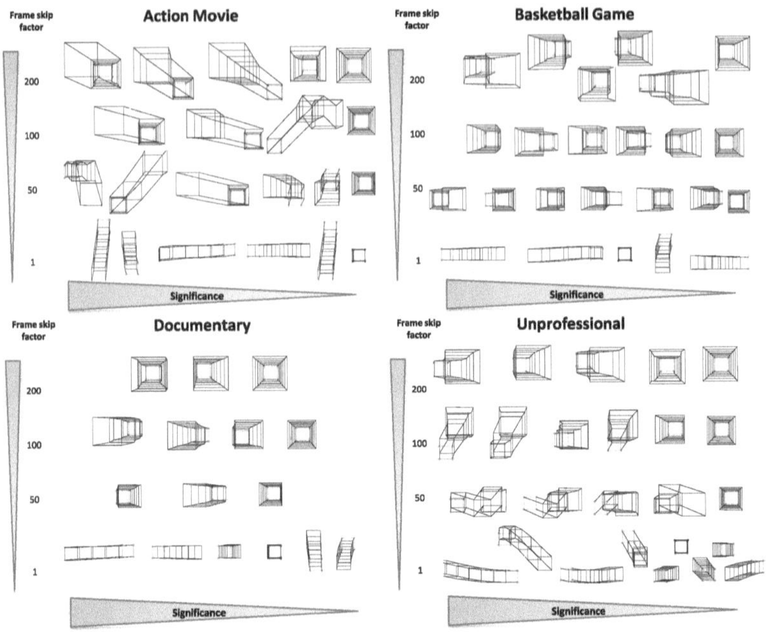

Fig. 3. Characteristic motions in each category of videos across different frame skip rates. The motions are ordered by descending significance (see Section 2.3) from left to right. Each frame in the motion chain is described as a rectangle undergoing an affine transformation from frame to frame. The first frame is denoted by the rectangle with the * in each corner.

2.3 Inter-Class Analysis

This section focuses on the development of a comparison metric between camera motion distributions. Based on the modeling of a group of movies of a particular class as a distribution of motion chains, we develop a metric to compare between two videos (or groups of videos). Moreover, exploiting the form of a Gaussian Mixture and the representation of a typical motion as a Gaussian component, we propose a metric for ranking characteristic motions in order of *significance*, which intuitively means we order typical motions with respect to how useful they are distinguishing the class they belong to from others.

Video Comparison. The divergence of two distributions is useful for comparing pairs of video classes, which in turn can be used for clustering of individual videos. In section 4 we show how we apply this metric for video clustering.

We compare pairs of distributions (each representing a class) and analyze how much one diverges from the other using an approximation of the KL-divergence of two mixtures of Gaussians proposed by Goldgerber *et al.* [8]:

$$KL(f||g) \approx \sum_{i=1}^{n} \alpha_i \min_j \left(KL(f_i||g_j) + \log \frac{\alpha_i}{\beta_j} \right), \tag{10}$$

where α_i and β_j are the mixing coefficients for f_i and g_j respectively and $KL(f_i||g_j)$ is the KL divergence approximation for two Gaussian mixture components f_i and g_j:

$$KL(f_i||g_j) \approx \frac{1}{2}\left(\log \frac{|\Sigma_{g_j}|}{|\Sigma_{f_i}|} + \mathrm{Tr}(\Sigma_{g_j}^{-1}\Sigma_{f_i}) + (\mu_{f_i} - \mu_{g_j})^T \Sigma_{f_i}^{-1}(\mu_{f_i} - \mu_{g_j}) \right). \tag{11}$$

Significance of a Typical Motion. Identifying the canonical motions that differentiate a class from others can be helpful in creating stronger classifiers. In this section we present a ranking method for typical motions to identify what are the motions that characterize each class and at the same time differentiate it from the rest.

As a consequence of equation (10), Goldberger *et al.* [8] propose a method for matching some Gaussian component from a mixture of Gaussians (MoG) f to another component $\pi(i)$ of another MoG g via the following matching function:

$$\pi(i, g) = \operatorname*{argmin}_j \Bigg(\frac{1}{2}(\log \frac{|\Sigma_{2,j}|}{|\Sigma_{1,i}|} + \mathrm{Tr}(\Sigma_{2,j}^{-1}\Sigma_{1,i}) + \tag{12}$$
$$(\mu_{1,i} - \mu_{2,j})^T \Sigma_{2,j}^{-1}(\mu_{1,i} - \mu_{2,j})) - \log \beta_j \Bigg).$$

Based on this matching-based approximation, we propose a way of ranking canonical motions of each distribution in a set of MoG G as

$$p(i) = \min_{g \in \{G-f\}} \Bigg(\frac{1}{2}(\log \frac{|\Sigma_{2,j_g}|}{|\Sigma_{1,i}|} + \mathrm{Tr}(\Sigma_{2,j_g}^{-1}\Sigma_{1,i}) + \tag{13}$$
$$(\mu_{1,i} - \mu_{2,j})^T \Sigma_{2,j_g}^{-1}(\mu_{1,i} - \mu_{2,j_g})) - \log \beta_{j_g} \Bigg).$$

This ranking is based on the matching-based approximation from equation 13 where, the larger a value is, the more divergent the typical motion (Gaussian component) is from other typical motions in the set of distributions (other video classes).

3 Experiments and Results

3.1 Dataset and Data Processing

We perform our analyses over two video datasets:

1. Dataset 1. Consists of commercially produced video and unprofessional video (home videos captured by various amateurs) divided into the following categories: action movie (180 minutes) , documentary (120 minutes), basketball game (120 minutes), and unprofessional (360 minutes).

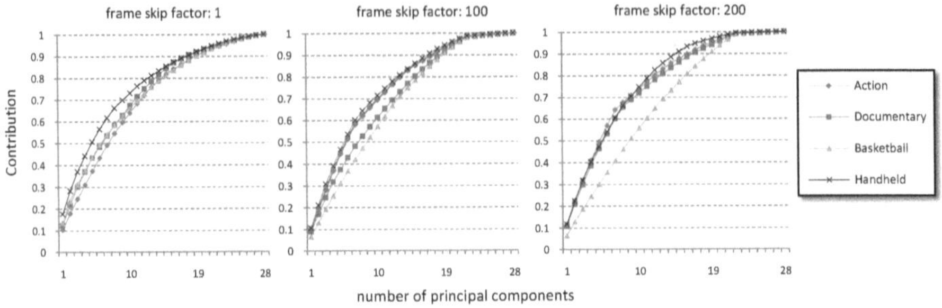

Fig. 4. Plots of cumulative variance as a function of principal components of motion chains. These plots show the size of space accounted by some number of principal components.

2. Dataset 2. Consists of 176 video clips downloaded from YouTube: 130 professional and 46 unprofessional

We extracted the similarity transformation matrices between each pair for consecutive video frames in the sequences sampled at the particular skip rates to construct transformation chains of a temporal window of radius $r = 7$ using the hierarchical Lucas-Kanade algorithm [15,3]. Using the similarity transformation we have 6 variables, therefore each transformation chain is a feature point in 42 $(= 6 \times 7)$ dimensions in our case.

3.2 In-Class Analysis

Figure 4 shows the cumulative variances of the unprofessional videos and the three professional classes of videos as a function of number of principal components used (ordered by eigenvalue). Each graph shows this relation across different levels (skip rates).

In these plots, it is observed that the cumulative variance of the unprofessional dataset (handheld) is consistently high among the four datasets. This suggests that the space of unprofessional motion-chain is more compact compared with others in these levels. This result is counter-intuitive because the unwanted high-frequency motions in unprofessionally-created movies are expected to create a larger variability. One interpretation for this result is that professionally-created movies have wider variety of global motions that are well-designed, while the unprofessional ones do not. This differentiates the variety of global motions, and it appears as a difference in size of spaces.

Figure 3 shows the motion means from each of the unprofessional videos as well as the three professional subclasses. Each row shows characteristic motions at a particular skip rate and in each row, the motions are ordered in order of significance.

3.3 Inter-Class Analysis

Figure 3 shows typical motions in each class ordered by significance (see equation 14). As expected, in all classes, the static camera motion chain was ranked the lowest in terms of significance. In other words, a non-moving camera shot exists in videos of all types and contributes little in characterizing the video or its class. In contrast, more complex motions serve to characterize and discriminate within classes:

- **Action Movies.** Characterized by a variety of horizontal and vertical motions. While some translations are carefully-generated continuous paths, there are some drastic changes in location and zooms. At a low level there are simple vertical and horizontal motions while at a higher skip rate, there are more zooms with highly dynamic motions. For instance, the camera can be focusing on a character and moving quickly when the focus changes to another character.
- **Basketball Games.** Contain very carefully guided curved camera paths with continuous zoom. Low level motions are general motions in horizontal and vertical directions as well as the lack of movement (static camera). High-level motions are mostly careful zooms with little translations. There are more horizontal motions at higher skip rates compared to other classes. This is probably due to the rapid and varied motions of players in a sport of this nature. Moreover, these effects are possible thanks to equipment not available in other classes like the unprofessional one.
- **Documentaries.** Motions are very simple and there are fewer characteristic translations with respect to the other classes studied. At a low level there are continuous motions vertically and horizontally while more complex motions happen at higher skip rates including zooms and carefully controlled paths. At very high skip rates, there are mostly zooms with non or very little translation. This is characteristic of documentaries when filming packs of animals in aerial views.
- **Unprofessional.** This class has several characteristic motions in all directions. Due to the limited translation capability of the cameraman in at a higher level, there is little translation at higher skip rates. In general there are less continuous zooms and unstable translations compared to the professionally-generated videos.

4 Application to Video Classification

We are interested in seeing to what extent clustering can be achieved only using global motion as a feature. We represent each video clip as a distribution of global motion chains and use the symmetric version of the KL-divergence approximation previously described in section 2 to compare pairs of global motion chain distributions to create an affinity matrix used to perform a normalized-cuts spectral clustering [17] . The similarity between two video clips with distributions of motion chains f_s and g_s at scale s respectively is therefore defined as follows:

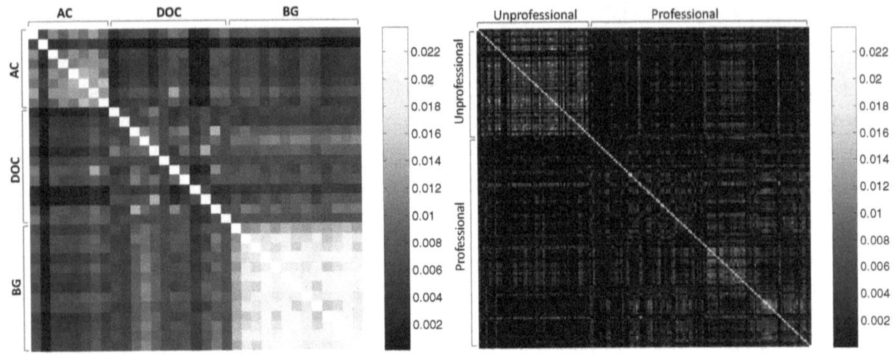

(a) Within professional videos (b) Professional and unprofessional videos

Fig. 5. Similarity matrices: (a) Similarity matrix for professional videos: action movies (AC), documentaries (DOC), and basketball games (BG), and (b) Similarity matrix for professional and unprofessional YouTube videos

$$S(v_f, v_g) \stackrel{\text{def}}{=} \frac{1}{\sum_{s=1}^{M} \{KL(f_s||g_s) + KL(g_s||f_s)\}}. \tag{14}$$

Figures 5 (a) and (b) show the affinity matrix for the professional videos in dataset 1 and for all of dataset 2 respectively. The brightness of the region is directly proportional to the affinity of the videos in that section. In dataset 1, basketball game clips are more similar to each (and easier to cluster) other than action movie and documentary clips. In the case of dataset 2, unprofessional clips are more similar to each other than professional videos.

Comparison to other Motion Features. We evaluate the effectiveness of our global motion chain feature representation in the dataset described in section 3. The 13 hours of video are cut into 3 minute long non-overlapping clips totalling 253 video clips.

We compare our motion chains feature with three other motion-based features: motion images described by [10], motion magnitude histograms, and motion direction histograms. Motion Images are 2D grids accumulating motion information at the corresponding parts of the canvas where the motion happened. The motion magnitude and direction histograms summarize the magnitude and direction respectively of optical flow vectors throughout the video clip.

We implement a version of motion images and the magnitude and direction histograms using optical flow over pairs of consecutive frames using the block matching implementation in OpenCV. The flow vector direction and magnitude information are summarized in normalized histograms and the values of each bin are used as feature values.

To compute the accuracy of the clustering results within each class, we calculate the ratio between the number of instances that were grouped together (if there is more than one grouping, we choose the largest grouping) and the total number of instances in the ground-truth of that class. Each of the 4 types of

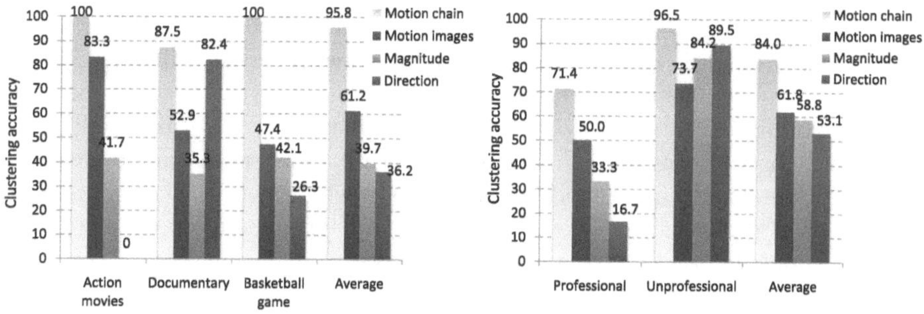

(a) Dataset 1

(b) Dataset 2

(a) Dataset 1

(b) Dataset 2

Fig. 6. Clustering accuracy over datasets 1 and 2

features were used to construct similarity matrices and spectral clustering was applied to them. Figure 6 (a) shows the classification rates amongst professional videos using the different features for dataset 1. Global motion performed far better than our competition in all the three classes with an average of advantage of at least 34.57%. The most challenging class was the documentaries one (DOC) where we achieved an 87% accuracy, still significant in comparison to the other methods tested.

Figure 6 (b) shows the performance of our feature when used to discriminate between professional and unprofessional videos in our dataset of YouTube videos. We are able to correctly group unprofessional videos with 96.48% accuracy and professional ones with 71.43% accuracy.

5 Discussion and Conclusion

We perform an analysis of the global motion distribution within various video classes and show how the space of unprofessional motions is more compact with respect to professionally-captured ones at a high level. At a low level, action movies are the least compact followed by basketball games and documentary films in that order. In addition, we learned that unprofessional movies have similar wide range of motion chains with respect to professional films while at a higher level, the camera motion of professional films is more controlled and less varied in comparison with unprofessional movies.

We also develop a way to compare camera motions which, as a result, we apply in a video clustering application. Our method is more efficient at clustering videos when compared to other motion-based features in the literature. Global motion features perform better in our dataset where there is a large variety of camera motions characterizing the genre of the video while other features such as global motion and other low-level ones based on local motion perform well in certain scenarios where the clip is short, the camera is static, and contains simple actions.

In conclusion, have performed a camera motion study and show that camera motion is indeed different across video classes and describe characteristics across classes and what differentiates them. We also show experimentally that global motion carries significant information about the movie type by comparing our feature to other features based on local motion, where the clustering results were consistently better using global motion as a feature. Other potential applications of this work include camera motion transfer between sequences to change styles, recognition of directors in videos, or simple characteristic motion extractors for learning about different genres or direction styles.

References

1. Black, M.J., Yacoob, Y.: Recognizing facial expressions in image sequences using local parameterized models of image motion. International Journal of Computer Vision 25(1), 23–48 (1997)
2. Blanford, S.: The Film Studies Dictionary. Oxford University Press, Oxford (2001)
3. Bouguet, J.Y.: Pyramidal implementation of the lucas kanade feature tracker: description of the algorithm. In: OpenCV Document, Intel, Microprocessor Research Labs (2000)
4. Bouthemy, P., Ganansia, F.: Video partitioning and camera motion characterization forcontent-based video indexing. In: Proceedings of International Conference on Image Processing, vol. 1, pp. 905–908 (1996)
5. Chomat, O., Crowley, J.L.: Probabilistic recognition of activity using local appearance. In: Proceedings of Computer Vision and Pattern Recognition, pp. 104–109 (1999)
6. Efros, A., Berg, A., Mori, G., Malik, J.: Recognizing action at a distance. In: Proceedings of International Conference on Computer Vision, vol. 2, pp. 726–733 (2003)
7. Fablet, R., Bouthemy, P., Pèrez, P.: Statistical motion-based video indexing and retrieval. In: Proceedings of RIAO Conference on Content-Based Multimedia Information Access (2000)
8. Goldberger, J., Gordon, S., Greenspan, H.: An efficient image similarity measure based on approximations of kl-divergence between two gaussian mixtures. In: Proceedings of International Conference on Computer Vision (2007)
9. Grimson, E., Stauffer, C., Romano, R., Lee, L.: Using adaptive tracking to classify and monitor activities in a site. In: Proceedings of Computer Vision and Pattern Recognition, vol. 2, pp. 22–29 (1998)
10. Haubold, A., Naphade, M.: Classification of video events using 4-dimensional time-compressed motion features. In: Computer Vision and Image Understanding, pp. 178–185 (2007)
11. Huang, J.: Statistics of natural images and models. Ph.D. Thesis (2000)
12. Kobla, V., DeMenthon, D., Doermann, D.: Identifying sports videos using replay, text, and camera motion features. In: Proceedings of the SPIE conference on Storage and Retrieval for Media Databases, pp. 332–343 (2000)
13. Litvin, A., Konrad, J., Karl, W.C.: Probabilistic video stabilization using kalman fitering and mosaicking. In: Proceedings of IS&T/SPIE Symposium on Electronic Imaging, Image and Video Communications, pp. 663–674 (2003)
14. Liu, F., Picard, R.W.: Finding periodicity in space and time. In: Proceedings of International Conference on Computer Vision, pp. 376–383 (1998)

15. Lucas, B.D., Kanade, T.: An iterative image registration technique with an application to stereo vision. In: Proceedings of International Joint Conference on Artificial Intelligence, pp. 674–679 (1981)
16. Matsushita, Y., Ofek, E., Tang, X., Shum, H.-Y.: Full-frame video stabilization. In: Proceedings of Computer Vision and Pattern Recognition, vol. 1, pp. 50–57 (2005)
17. Meila, M., Shi, J.: A random walks view of spectral segmentation. In: Artificial Intelligence and Statistics (2001)
18. Niyogi, S.A., Adelson, E.H.: Analyzing and recognizing walking figures in xyt. In: CVPR, pp. 469–474 (1994)
19. Ohanian, T.A., Phillips, M.E.: Digital Filmmaking: The Changing Art and Craft of Making Motion Pictures. Focal Press (2000)
20. Pilu, M.: Video stabilization as a variational problem and numerical solution with the viterbi method. In: Proceedings of Computer Vision and Pattern Recognition, vol. 1, pp. 625–630 (2004)
21. Polana, R., Nelson, R.C.: Detecting activities. In: Proceedings of Computer Vision and Pattern Recognition, pp. 2–7 (1993)
22. Roach, M.J., Mason, J.D., Pawlewski, M.: Video genre classification using dynamics. In: ICASSP, pp. 1557–1560 (2001)
23. Roth, S., Black, M.J.: On the spatial statistics of optical flow. International Journal of Computer Vision 74(1), 33–50 (2007)
24. Saisan, P., Doretto, G., Soatto, S., Wu, Y.N.: Dynamic texture recognition. In: Proceedings of Computer Vision and Pattern Recognition, vol. 2, pp. 58–63 (2001)
25. Schwartz, G.: Estimating the dimension of a model. The Annals of Statistics 5(2), 461–464 (1978)
26. Smith, M., Kanade, T.: Video skimming and characterization through the combination of image and language understanding techniques. In: Proceedings of Computer Vision and Pattern Recognition, pp. 775–781 (1997)
27. Stauffer, C., Grimson, E.: Learning patterns of activity using real-time tracking. IEEE Transactions on Pattern Analysis and Machine Intelligence 22(8), 747–757 (2000)
28. Walker, M.: Hitchcocks's Motifs. Amsterdam University Press (2005)
29. Wang, X., Ma, X., Grimson, E.: Unsupervised activity perception by hierarchical bayesian models. In: Proceedings of Computer Vision and Pattern Recognition, vol. 2, pp. 1–8 (2007)
30. Zelnik-Manor, L., Irani, M.: Statistical analysis of dynamic actions. IEEE Transactions on Pattern Analysis and Machine Intelligence 28(9), 1530–1535 (2006)

Active Image Labeling and Its Application to Facial Action Labeling

Lei Zhang[1], Yan Tong[2], and Qiang Ji[1]

[1] Electrical, Computer, and Systems Engineering Department, Rensselaer Polytechnic Institute
[2] Visualization and Computer Vision Lab, GE Global Research Center
zhangl2@rpi.edu, tongyan@research.ge.com, qji@ecse.rpi.edu

Abstract. For many tasks in computer vision, it is very important to produce the groundtruth data. At present, this is mostly done manually. Manual data labeling is labor-intensive and prone to the human errors. The training data it produces often lacks in both quantity and quality. Fully automatic data labeling, on the other hand, is not feasible and reliable. In this paper, we propose an interactive image labeling technique for efficient and accurate data labeling.

The proposed technique includes two parts: an automatic labeling part and a human intervention part. Constructed on a Bayesian Network, the automatic image labeler produces an initial labeling of the image. A person then examines the initial labeling and makes some minor corrections. The selected human corrections and the image measurements are then integrated by the Bayesian Network framework to produce a refined labeling. To minimize the human involvement, an active user feedback strategy is developed, through which the optimal user feedback is determined, so that the labeling errors in the subsequent re-labeling process can be maximally reduced. The proposed framework combines the advantages of the human input with those of the machine so that the reliable, accurate, and efficient data labeling can be achieved. We demonstrate the validity of the proposed framework for interactive labeling of facial action units. The proposed methodology, however, is not limited to labeling of facial action units. It can be easily extended to other areas such as interactive image segmentation.

1 Introduction

For a variety of classification tasks in computer vision including image segmentation, feature detection, and object recognition, image labeling is needed in order to create the training data for the classifiers. Producing image groundtruth is typically carried out manually. However, manual image labeling is labor-intensive, and with limited throughput. In addition, image labeling is an error-prone process due to various reasons, such as the labeler's errors or the imperfect description of classes. Thus, a second (or more) pass(es) of labeling often by a different human labeler is usually required in order to fix the errors and inconsistencies in the earlier pass(es). To alleviate these problems, various interactive or semi-automatic learning tools have been used. For example, Levin et al. [1] used co-training in the context of visual car detection to improve the accuracy of a classifier using the unlabeled examples. The active learning is another machine learning technique to address these problems. Active learning techniques are sequential learning methods that are designed to reduce the manual training costs in achieving

D. Forsyth, P. Torr, and A. Zisserman (Eds.): ECCV 2008, Part II, LNCS 5303, pp. 706–719, 2008.

adequate learning performance. It is being used increasingly to help reduce the amount of labeled training data by incorporating limited user feedback selectively and intelligently during training. In this paper, we will introduce a framework based on Bayesian Network for active labeling and demonstrate its use for facial action unit labeling.

Facial expression recognition represents an active research area in computer vision. Current research in facial expression recognition can be classified into two categories: facial expression recognition and facial action recognition. While the former focuses on recognizing the global facial patterns, the latter concentrates on recognizing the local facial motions. Since there are many different global facial expressions in real life, the current research in global facial expression recognition studies only a few prototype expressions (happiness, disgust, surprise, etc.). Local facial motion recognition focuses on recognizing the local facial actions as defined in Facial Action Coding System (FACS) [2]. FACS defines 44 facial Action Units (AUs) and 8 head pose action units at 5 asymmetric intensity levels, co-occurring in various combinations. Each AU represents a kind of facial muscular activity that produces facial appearance changes and is anatomically related to the contraction of a specific set of facial muscles. The FACS has been demonstrated to be a powerful means for detecting and measuring a large number of facial displays by observing a small set of visually discernable facial muscular movements [3]. Table 1 shows some commonly occurring AUs and their interpretations.

Table 1. A list of commonly occurring action units and their interpretations [4]

AU1 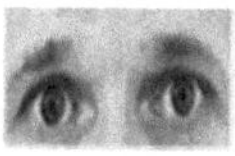 Inner brow raiser	AU2 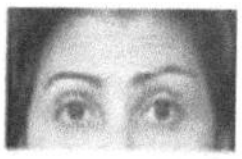 Outer brow raiser	AU4  Brow Lowerer	AU5 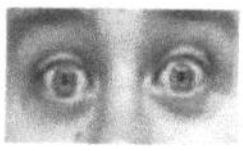 Upper lid raiser	AU6 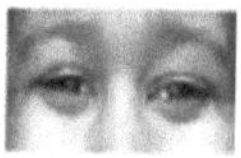 Cheek raiser
AU7  Lid tighten	AU9  Nose wrinkle	AU12  Lip corner puller	AU15  Lip corner depressor	AU17 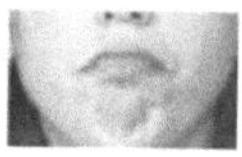 Chin raiser
AU23 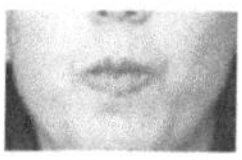 Lip tighten	AU24 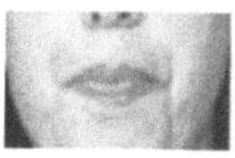 Lip presser	AU25 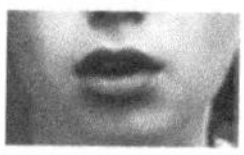 Lips part	AU27 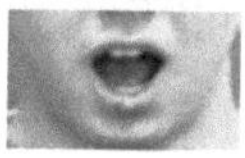 Mouth stretch	

Compared with recognizing the global facial expressions, recognizing the local facial actions is expression-independent, can capture subtle facial changes, and is free of human emotional judgment/interpretation. In addition, from the recognized local facial actions, it is possible to perform various high level understanding such as facial expression recognition, human behavior analysis, cognitive state recognition, etc.

For facial action recognition, AU labeling is required to create the training data. Manual AU labeling of facial expression, in particular spontaneous expressions, is very time consuming, error prone, and expensive. It usually requires years of training before a person can become a certified AU coder. In addition, the inter-variation among AU coders is often large. AU labeling becomes even more difficult when multiple AUs co-occur. Given these difficulties, recent researches have focused on developing automatic AU coding systems [3]. Despite much progress in automatic AU recognition, the current automatic AU recognition systems are limited mainly to recognizing AUs from the posed expressions and with mostly the frontal faces. They remain ineffective for recognizing AUs from the spontaneous facial expressions due to in part the subtlety of facial actions in the spontaneous expressions and in part the face occlusion/distortion from head movements.

To further push researches in spontaneous facial expression recognition, it is critical to have both sufficient and reliable AU-labeled facial expressions since supervised training of automatic facial action recognition systems require the groundtruth of AU labels. To achieve this goal, we propose an interactive AU labeling system that combines automatic AU measurements with limited yet selective human input to perform both accurate and fast AU labeling. We construct a unified probabilistic framework based on a Bayesian Network to incorporate both the automatic AU measurements and the human input. Furthermore, we also propose an active human intervention strategy based on the mutual information calculated from BN inference. This active human intervention strategy can provide the suggestion for the next human input so that the person can focus on certain errors with priority. It is potential to reduce the whole human involvement when these errors are first corrected. In addition, this framework can handle the uncertainty of the AU measurement by the automatic labeling technique and that of the human labeling. Compared with other interactive image labeling system, ours enjoys the following advantages: 1) incrementally allow the human input (from one or multiple human experts) to be incorporated at any stage; 2) systematically combine the human input with the automatic AU measurements, and 3) allow to determine the optimal human interaction so that human involvement can be kept to the minimum level.

2 Related Work

From the machine learning perspective, semi-automatic image labeling is a problem of semi-supervised active learning [5], which combines semi-supervised learning [6, 7, 8, 9, 10, 11, 12] with active learning [13, 14, 15, 16, 17, 18, 19]. The semi-supervised learning aims to design image classifier by making use of the unlabeled data, and is useful especially when the training data is limited; while the active learning mechanism aims to enlarge the useful information conveyed by the human feedback, and provides the annotators the most informative samples according to the current image classifier. One commonly used semi-supervised learning technique is the co-training technique [20, 21, 1], which aims to improve classifiers' learned information of the labeled data by maximizing their agreement on the unlabeled data.

The active learning proceeds sequentially, with the learning algorithm actively asking for the labels (categories) of some instances from a teacher (also referred to as membership queries). The objective is to ask the teacher to label the most informative instances in order to reduce the total labeling costs and accelerate the learning. For active learning, the key is the sample selection strategy, which, based on a predefined criterion, selects the next samples for the user to label. A better criterion means the corresponding sample will provide a larger amount of performance enhancement after being manually labeled. For classification, the most commonly used criterion is the close-to-boundary criterion. Other criteria have also been proposed. The expected information gain is a natural measure to select the next sample whose label is going to be asked for. However, it is not sufficient for guaranteeing a large reduction in the expected prediction error. Freund et al. [22] enhance the Query by Committee method to get high expected information gain. Shen et al. [23] incorporate multi-criteria (informativeness, representativeness and diversity) for active learning. Besides using the distance from the classification boundary, Kapoor [24] propose the variance and uncertainty of a Gaussian Process for object categorization as the active learning criteria. Raghavaan et al. [25] improves active learning by combining the instance level feedback with the feature level feedback.

In speech recognition, the concept of active labeling is introduced to minimize the number of mislabelings by combining the output of machine labeling results with the human feedback [26]. Specifically, active labeling aims to minimize the number of utterances to be checked again by first automatically selecting the ones that are likely to be erroneous or inconsistent with the previously labeled examples. The user feedback is then requested to correct the selected erroneous labeling. The corrected labels are then feedback to the automatic labeler to reclassify the utterances. Despite its use in speech recognition, active labeling has not yet received due attention in computer vision.

Although some human interventions are usually required for most of the AU recognition systems to achieve an accurate and reliable face detection and/or facial feature motion extraction for AU recognition [27], AU labeling is still performed either automatically or manually. In contrast, our interactive AU labeling system combines the advantage of automatic AU measurements (i.e., efficiency and objectivity) with the advantage of human labeling (i.e., reliability and accuracy) to perform both reliable and efficient AU labeling.

3 The Interactive AU Annotation System

3.1 Overview of the Image Labeling Method

Considering various shortcomings with the manual AU labeling, we propose to build an interactive AU annotation system by which the human annotators can interact with an automatic AU recognition system in labeling AUs, especially for spontaneous facial expressions. Figure 1 outlines the flowchart of the proposed system. Given an image, the automatic AU recognition system first performs an initial labeling of AUs in the image. Based on the result of the automatic labeling, the human needs to determine which AU (or which sets of AUs) should be changed next in order to reduce the involvement in future. To achieve this, we employ an active human involvement strategy, by which we actively decide the optimal human intervention in order to maximize its effectiveness. Given the identified AU, a human can then manually set its state. This completes the human input stage. The automatic AU recognition system then systematically and incrementally combines the current human input with the existing knowledge of AUs

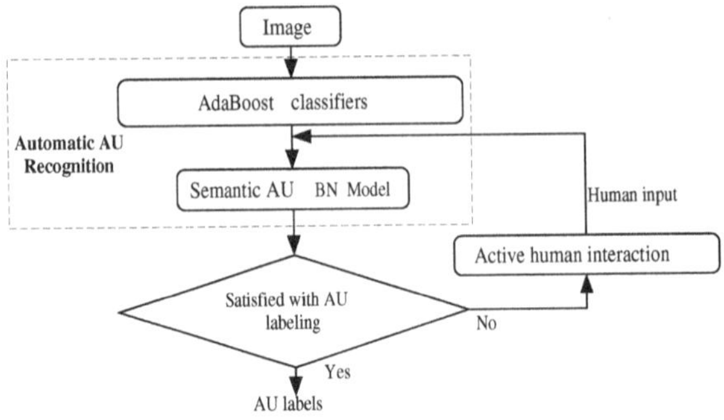

Fig. 1. The flowchart of the proposed interactive AU labeling system

(e.g. AU measurements and the previous human inputs) to correct the mislabeled AUs through a belief propagation. This process repeats until the human is satisfied with the AU labeling. The system can incrementally combine the image data and the human input to produce progressively improved AU labeling. In addition, using this active human labeling scheme, a human can focus only on the most important AUs and, therefore, can significantly reduce the human involvement. In this way, both the labeling speed and accuracy are improved.

3.2 Automatic AU Recognition

To perform an effective interactive labeling, we have developed an automatic AU recognition system. The system consists of two major components as shown in Figure 1. The first component is an image-based AU classification system based on Adaboost. The second component is a probabilistic AU model that captures the spatial and semantic dependencies among AUs. The final AU recognition results are based on combining the AdaBoost classification results with the AU model. In the paragraphs to follow, we summarize our automatic AU recognition system.

AU Classification Using AdaBoost. We first perform the face and eye detection from images. Given the knowledge of eye centers, the face region is normalized and convolved pixel by pixel by a set of multiscale and multiorientation Gabor filters. Then, the measurement for each AU is obtained through a computer vision technique based on Gabor wavelet features and AdaBoost classifiers similar to [28]. However, this frame-by-frame AdaBoost classification method is susceptible to image noise and inaccurate image alignment. In addition, it cannot perform effectively when multiple AUs co-occur or when facial deformations are subtle, which happens frequently for spontaneous facial expressions. Therefore, besides measuring each AU individually, it is more important to incorporate other related prior knowledge to help improve AU recognition under these difficult conditions.

AU Modeling with A Bayesian Network. Based on the previous study by Tong et al. [29], there are semantic relationships (the co-occurrence relationships and the

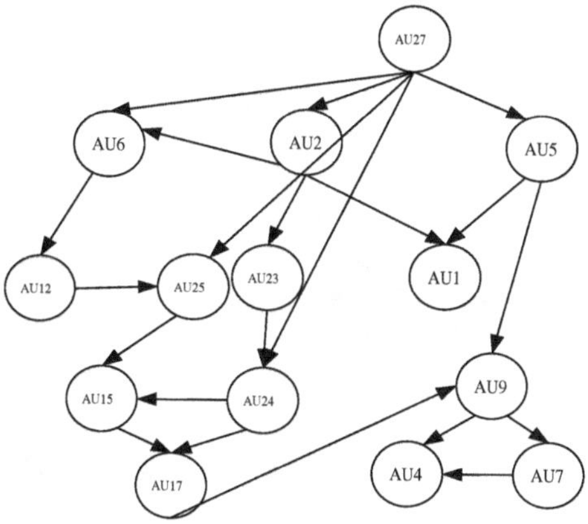

Fig. 2. A BN model for modeling the semantic relationships among AUs. Each white circle represents an AU, and the link between two white circles represents the semantic relationship between the two AUs.

mutually exclusive relationships) among AUs. These relationships exist due to either certain facial expressions or the underlying facial anatomy and physiology. Specifically, some AU combinations are physiologically impossible to occur together, while other AUs co-occur mainly because of certain facial expressions. For example, AU6 (cheek raiser) tends to happen with AU12 (lip corner puller)when smiling. On the other hand, AU25 (lips part) can hardly happen with AU24 (lip presser) simultaneously. We will exploit these relationships for automatic AU labeling. In particular, following the work by Tong et al. [29], we use a Bayesian Network (BN), as shown in Figure 2, to capture the co-occurrence and the mutually exclusive relationships among AUs. A BN is a directed acyclic graph (DAG) that represents a joint probability distribution among a set of variables, where the nodes represent the variables and the links among those nodes represent the conditional dependency among the variables.

Specifically, the nodes represent AUs and the links between the nodes capture the relationships (co-occurrence or mutual exclusiveness) among AUs. The model can be constructed and parameterized using the training data with a standard learning technique [30]. We employ this BN model as our base learning algorithm for their effectiveness in many learning tasks. Compared with other learning methods such as Support Vector Machine and Neural Network, BN has the capabilities of incorporating different types of human inputs through its hierarchical structure, incorporating the human input at any stage of the labeling, systematically combining the human inputs with the existing data, and exerting the impacts of human inputs on other entities through belief propagation.

To integrate the AU measurements and the user feedback, we extend the BN model in Figure 2 for interative labeling as shown in Figure 3. Specifically, we associate each AU node AU_i with two measurement nodes, which are represented by a grey circle and a black circle, respectively. The grey circle O_i represents the AU measurement obtained through the computer vision technique (i.e., AdaBoost classifier in this work),

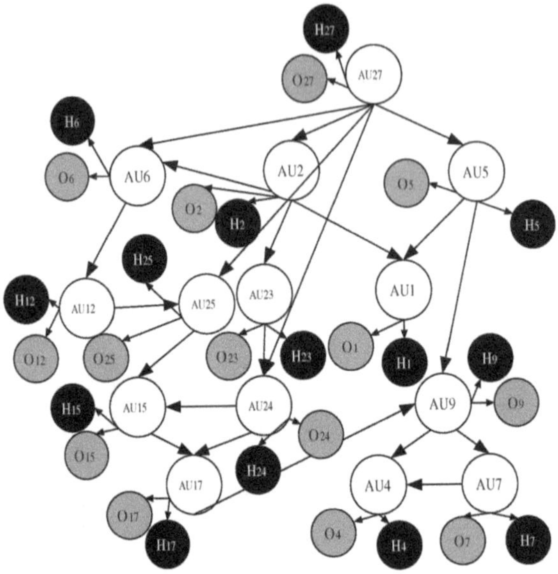

Fig. 3. An extended BN model for interactive labeling. The white circle represents each AU node that should be labeled. The grey circle represents the measurement for each AU node, which is obtained using the AdaBoost classifier. The black circle represents the human labeling for each AU node, which is obtained through interactive labeling.

and the link between AU_i and O_i represents the measurement uncertainty resulting from the computer vision technique. This uncertainty is obtained from analyzing the performance (accuracy) of the AdaBoost classifier for each AU. The black circle H_i represents the human input for this AU, and the link between AU_i and H_i represents the reliability and confidence of the human labeling. The confidence is determined by the person who is giving the human input. For a domain expert who is very sure about his/her judgment, the confidence is set as 1. For other persons, this confidence is set depending on the level of his/her expertise. In this way, the BN model systematically combines the objective image measurements of AUs, the subjective human labeling, and their uncertainties.

Given the model structure shown in Figure 2, we need to define the states for each node and learn the model parameters associated with each node. Each AU node has two states for representing its "presence" and "absence" states. Its parameter is characterized by the Conditional Probabilistic Tables $p(AU_i|pa(AU_i))$, where $pa(AU_i)$ denotes the parent configuration of AU_i. Given the training data (AU labels), these parameters can be learned [31]. For each measurement node, the parameters $p(O_i|AU_i)$ or $p(H_i|AU_i)$ are assigned manually based on the recognition accuracy of the computer vision technique or human confidence, respectively. Based on the BN model, the states of the AU nodes are inferred using the junction tree inference approach (see Section 3.3 for more details).

Active Human Input Determination. Given the AU label assignment by the automatic AU recognition, we need decide 1) which AU is mislabeled, and 2) among the mislabeled AUs, which AU's state shall be corrected next. An important observation for AU labeling is that it is not straightforward for a human to quickly identify which AU is mislabeled. Identifying the mislabeled AUs requires both time and expertise. It is even more difficult to decide which AU shall be corrected next in order to minimize the subsequent user feedback. To achieve this goal, it is important to identify the AU not based on its obvious mistake, but based on its potential to maximally correct other mislabeled AUs. We employ the active user feedback strategy for this purpose. The idea of the active user feedback is to select the mislabeled AUs to solicit the user's feedback so as to achieve the maximal reduction in the overall estimation uncertainties and minimize the misclassification by the BN model after incorporating the user's feedback. The most informative AU is the one that maximally reduces the overall estimation uncertainty.

For this purpose, we need develop a measure to quantify the utility of each AU. This measure reflects the estimated contribution of a certain AU to the reduction of uncertainty in the subsequent labeling process when its label is manually corrected. The mutual information between the overall model uncertainty and the user feedback is one of such measures. It measures the potential of the user feedback in reducing the overall uncertainty of the model. Mutual information is found to be quite effective in the relevancy feedback for image retrieval and is easy and quick to compute. Specifically, let H_k be the human labeling of the kth AU, and let $AU_1, \cdots, AU_N$ be all the AU nodes in the BN model, the mutual information between H_k and $AU_1, \cdots, AU_N$ can be computed as follows:

$$I(AU_1, \cdots, AU_N; H_k) = \sum_{AU_1, \cdots, AU_N} p(AU_1, \cdots, AU_N, H_k) log \frac{p(AU_1, \cdots, AU_N, H_k)}{p(AU_1, \cdots, AU_N)} \tag{1}$$

We use this mutual information measure to return a ranked list of AUs sorted in a decreasing order of importance. A person can then use this list to guide the selection of the next AU for correction.

3.3 Interactive AU Labeling through Probabilistic Inference

Given the BN model (Figure 3) for interactive AU labeling , the AU labeling is performed through a probabilistic inference by maximizing the joint probability of all AUs given their image measurements and all available human inputs, i.e.

$$AU_1^*, \cdots, AU_N^* = \underset{AU_1, \cdots, AU_N}{\text{argmax}} \; p(AU_1, \cdots, AU_N | O_1, \cdots, O_N, H_1, \cdots, H_K) \tag{2}$$

where $AU_1, \cdots, AU_N$ represent all the AUs that should be labeled, $O_1, \cdots, O_N$ represent all the image measurements that are obtained through the computer vision technique, and $H_1, \cdots, H_K$ represent all the available human inputs for the AUs. Based on the conditional independence in BN, the joint probability can be factorized as follows:

$$p(AU_1, \cdots, AU_N | O_1, \cdots, O_N, H_1, \cdots, H_K) \propto \prod_{j=1}^{N} p(AU_j | pa(AU_j)) \prod_{j=1}^{N} p(O_j | AU_j) \prod_{k=1}^{K} p(H_k | pa(H_k)) \tag{3}$$

where $pa(AU_j)$ represents the parent configuration of AU_j in the BN model and $pa(H_k)$ represents the kth AU with the human input.

In this way, the AUs can be labeled semi-automatically and incrementally by combining the subjective knowledge from the human coder and the objective image measurements. For efficiency, we only allow one human correction on one AU in each iteration. This single correction on one AU may affect the AU labeling results for all AUs through belief propagation in the BN.

4 Experimental Results

To demonstrate our proposed interactive AU labeling system, we perform the AU labeling experiments on two databases. The first database is the Cohn-Kanade DFAT-504 database [4]. This database is collected under controlled illumination and background, and has been widely used for evaluating facial AU recognition system. The Cohn-Kanade database only contains the frontal and posed facial expressions. On the other hand, it is also very important to study the labeling for spontaneous facial expressions. We therefore also evaluate our system on the second database, i.e., a spontaneous facial expression database. The second database consists of images collected from: (1) Multiple Aspects of Discourse (MAD) research lab at the University of Memphis [32], (2) Belfast natural facial expression database [33], and (3) videos obtained from the website (http://www.youtube.com/). In these images, the subjects are displaying various natural facial expressions (such as frown and confusion) with often subtle facial appearance changes. Furthermore, the head poses are not limited to the frontal view in these images. Due to these reasons, the automatic AU recognition from spontaneous facial expression is more challenging.

4.1 Interactive AU Labeling on Posed Facial Expressions

We first perform the interactive AU labeling on the posed facial expression images from the Cohn-Kanade database. We divided the database into 8 sections, each of which contains about 1000 images from different subjects. Each time, we randomly choose 7 sections for training the AdaBoost classifier for each AU and the BN model as shown in Figure 3. Then, we randomly select 50 samples from the remaining section for testing. Hence, there are totally 400 samples used for testing the automatic AU labeling, the AU labeling with arbitrary human input, and the AU labeling with active human input.

For the posed facial expression database, besides comparing with the fully automatic AU recognition, we also compare two methods of interactive AU labeling with two different ways of human intervention. In the first method, the human coder arbitrarily chooses a mislabeled AU for correction. The human coders in our experiments are domain experts with certain familiarity with AU recognition. They are allowed to use their experiences to select the mislabeled AU for correction. The second method uses the proposed active human input strategy to identify the mislabeled AU for correction. In each iteration, only one human intervention is allowed to correct one AU label for both methods. Table 2 and Figure 4 summarize the results for the two user feedback methods.

Table 2. Comparison of AU labeling accuracy on Cohn-Kanade database using the automatic AU recognition with only the image measurements (denoted by "BN"), the interactive labeling with the image measurements and the arbitrary human input, and the interactive labeling with the image measurements and the active human input, respectively

AU label		AU1		AU2		AU4		AU5		AU6		AU7		AU9	
		Pos #	Neg #	Pos #	Neg #	Pos #	Neg #	Pos #	Neg #	Pos #	Neg #	Pos #	Neg #	Pos #	Neg #
Total test samples		57	343	38	362	60	340	20	380	40	360	36	364	15	385
Error rate		FN	FP	FN	FP	FN	FP	FN	FP	FN	FP	FN	FP	FN	FP
BN		0.210526	0.0831	0.1205	0.0292	0.1826	0.0793	0.1852	0.0643	0.1081	0.0483	0.144737	0.058036	0.125	0.0188
Arbitrary input	iter1	0.017544	0.0219	0	0.0058	0.1167	0.0329	0.15	0.033	0.075	0.0152	0.055556	0.01497	0.066667	0.008
	iter2	0.052632	0.0063	0	0	0.05	0.0066	0.05	0.011	0.05	0.0061	0.027778	0.005988	0.066667	0.0027
	iter3	0	0	0	0	0.05	0.0033	0	0.0027	0	0	0	0.002994	0	0
	iter4	0	0	0	0	0	0	0	0	0	0	0	0.002994	0	0
	iter5	0	0	0	0	0	0	0	0	0	0	0	0	0	0
Active input	iter1	0.087719	0.0094	0.1053	0.0116	0.1167	0.0263	0.15	0.0192	0	0.0212	0.055556	0.011976	0.066667	0.0053
	iter2	0.017544	0.0031	0.0526	0	0.05	0	0.1	0.0082	0	0.003	0	0	0	0
	iter3	0.017544	0	0	0	0.0167	0	0.05	0	0	0.003	0	0	0	0
	iter4	0	0	0	0	0.0167	0	0	0	0	0	0	0	0	0
	iter5	0	0	0	0	0	0	0	0	0	0	0	0	0	0

AU label		AU12		AU15		AU17		AU23		AU24		AU25		AU27	
		Pos #	Neg #	Pos #	Neg #	Pos #	Neg #	Pos #	Neg #	Pos #	Neg #	Pos #	Neg #	Pos #	Neg #
Total test samples		64	336	26	374	53	347	23	377	24	376	147	253	30	370
Error rate		FN	FP	FN	FP	FN	FP	FN	FP	FN	FP	FN	FP	FN	FP
BN		0.1	0.0367	0.2609	0.0741	0.1619	0.0738	0.4091	0.0591	0.2857	0.0696	0.172881	0.089912	0.042857	0.0126
Arbitrary input	iter1	0.046875	0.0193	0.0385	0.0363	0	0.0312	0.1304	0.0162	0.125	0.0384	0.040816	0.022026	0.033333	0
	iter2	0	0.0032	0	0.0056	0	0.0156	0.0435	0.0108	0.0833	0.0219	0.013605	0.017621	0	0
	iter3	0	0.0096	0	0	0	0.0093	0.0435	0.0081	0	0.011	0	0.004405	0	0
	iter4	0	0.0064	0	0	0	0.0093	0	0	0	0.0055	0	0	0	0
	iter5	0	0	0	0	0	0	0	0	0	0	0	0	0	0
Active input	iter1	0.015625	0.0193	0.0385	0.0251	0.0189	0.0405	0.1739	0.0135	0	0.0329	0.07483	0	0	0
	iter2	0	0.0032	0	0.0168	0	0.0187	0.0435	0.0054	0	0.0082	0.020408	0	0	0
	iter3	0	0	0	0.0084	0	0.0125	0	0	0	0	0	0	0	0
	iter4	0	0	0	0	0	0.0062	0	0	0	0	0	0	0	0
	iter5	0	0	0	0	0	0	0	0	0	0	0	0	0	0

Table 2 reports the AU labeling accuracy for each AU individually by using different methods. In Table 2, "FN" and "FP" represent the false-negative rate (i.e. the error rate of positive samples) and false-positive rate (i.e. the error rate of negative samples) for each AU, respectively. From Table 2, we can compare how often the active input can drive the individual AU recognition error to zero before the arbitrary input can. The active input performs better in 14 cases, while the arbitrary input performs better in 7 cases. They perform the same in 7 cases. These results demonstrate the superiority of the active input to the arbitrary input. Compared with the arbitrary human input, less human intervention is involved using the proposed active human input strategy. Usually, only two human corrections will be sufficient to obtain the accurate AU labeling for all 14 AUs. It makes the AU labeling much more efficient than the manual labeling and much more accurate than the fully automatic labeling.

Figure 4 reports the average AU labeling performance for all AUs. In Figure 4, we can find that both false-negative rate and false-positive rate decrease significantly with the help of a few human corrections. For example, the average false-negative rate decreases from 17.9% (automatic AU labeling) to 6.4% and the average false-positive rate also decreases from 5.7% (automatic AU labeling) to 1.7% with only one human correction using the proposed interactive AU labeling method with active human input. The two curves are not separated significantly mainly due to two reasons. First, the automatic AU recognition already achieve very good accuracy. There are not many errors to correct. As a result, the difference of the corrected AU labels in these two methods

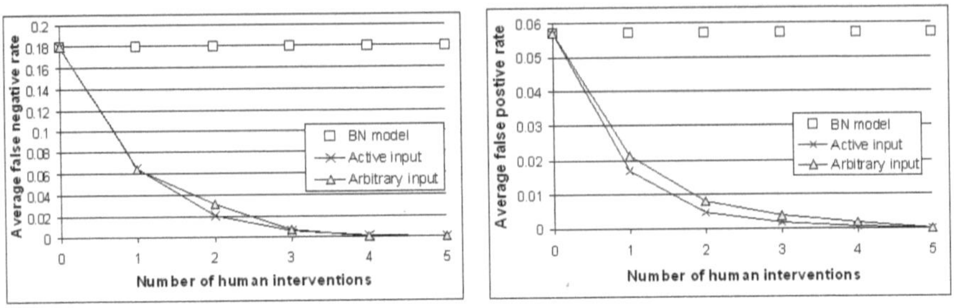

(a) (b)

Fig. 4. Average AU labeling performance on Cohn-Kanade database using the automatic AU recognition with only image measurements, the interactive labeling with image measurements and arbitrary human input, and the interactive labeling with image measurements and active user input, respectively. x axis represents the number of human corrections (iterations), while y axis represents the (a) average false-negative rate and (b) average false-positive rate, respectively

can only be a small fraction of the total AU samples. Second, the human coders also use their experiences to select the AU to be corrected. They somewhat try to select certain AUs with high priority for correction, too. This implicit use of domain knowledge also reduces the performance difference between two methods.

For the AUs that are hard to label using the automatic AU labeling, the improvement is impressive. For example, with one human correction, the false-negative rate of AU15 (lip corner depressor) decreases from 26.09% (automatic AU labeling) to 3.85%, and its false-positive rate also decreases from 7.41% (automatic AU labeling) to 2.51%. The false-negative rate of AU24 (lip presser) decreases from 28.57% (automatic AU labeling) to 0%, and its false-positive rate also decreases from 6.96% (automatic AU labeling) to 3.29% using the proposed interactive AU labeling method with active human input.

4.2 Interactive AU Labeling on Spontaneous Facial Expressions

Currently the public available facial expression databases provide a large number of training data for the posed facial expressions, whereas the resource of spontaneous databases is limited. Due to the increasing interest in the spontaneous facial expression recognition, it is more important and desirable to label the AUs for the spontaneous facial expression. In the second set of experiments, we evaluate the proposed interactive labeling method on a spontaneous facial expression database described previously. Specifically, we use 1300 images for training the automatic AU recognition network, and use different 300 images for testing AU labeling.

Table 3 and Figure 5 compare the AU labeling accuracies on the spontaneous facial expressions using the automatic labeling using BN inference with only the image measurements, and the interactive labeling using BN inference with both the image measurements and the proposed active human input. Table 3 reports the AU labeling accuracies for 12 AUs that commonly occur in the spontaneous facial expression database. Figure 5 reports the average AU labeling performance for all AUs.

Table 3. Comparison of AU labeling accuracies on a spontaneous facial expression database using BN inference with only the image measurements, and the interactive labeling with both the image measurements and the active human input

AU label		AU1		AU2		AU4		AU5		AU6		AU12	
		Pos #	Neg #	Pos #	Neg #	Pos #	Neg #	Pos #	Neg #	Pos #	Neg #	Pos #	Neg #
Total test samples		74	226	74	226	24	276	51	249	88	212	97	203
Error rate		FN	FP	FN	FP	FN	FP	FN	FP	FN	FP	FN	FP
BN		0.1493	0.0798	0.2239	0.0423	0.5714	0.0474	0.1111	0.0172	0.2361	0.0406	0.297619	0.067708
Active input	Iter1	0	0.028	0.0484	0.014	0.75	0.0039	0	0.0126	0.0882	0	0.060976	0.010695
	Iter2	0	0	0	0	0.5	0	0	0	0.0147	0	0.012195	0.005348
	Iter3	0	0	0	0	0	0	0	0	0	0	0	0
	Iter4	0	0	0	0	0	0	0	0	0	0	0	0

AU label		AU15		AU17		AU23		AU24		AU25		AU27	
		Pos #	Neg #	Pos #	Neg #	Pos #	Neg #	Pos #	Neg #	Pos #	Neg #	Pos #	Neg #
Total test samples		19	281	28	272	35	265	15	285	271	29	36	264
Error rate		FN	FP	FN	FP	FN	FP	FN	FP	FN	FP	FN	FP
BN		0.2353	0.0436	0.5263	0.0389	0.6429	0.0158	0.5556	0.0144	0.1635	0.1818	0.285714	0.045977
Active input	Iter1	0	0.054	0	0.0425	0.0909	0	0.625	0.0072	0.0455	0.0909	0.205882	0
	Iter2	0.0667	0.0252	0	0.0232	0	0	0	0	0.0038	0.0455	0	0
	Iter3	0	0	0	0.0039	0	0	0	0	0	0	0	0
	Iter4	0	0	0	0	0	0	0	0	0	0	0	0

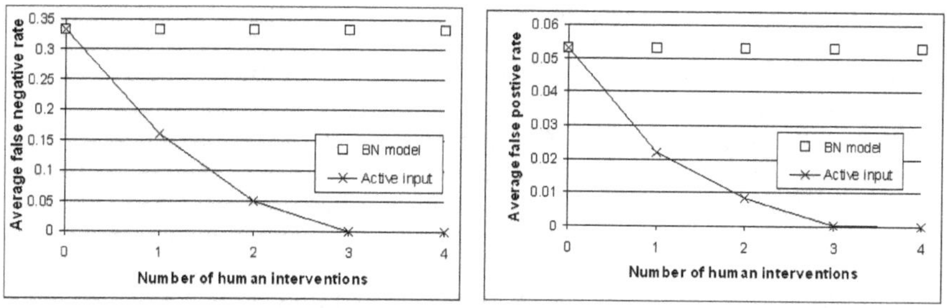

(a) (b)

Fig. 5. Average AU labeling performance on a spontaneous facial expression database using BN inference given only the image measurements, and the interactive labeling by active human input, respectively. x axis represent the number of human corrections (iterations), while y axis represents the (a) average false-negative rate and (b) average false-positive rate, respectively.

Due to the low intensity, non-additive effect, and individual difference of spontaneous facial action, the purely automatic AU labeling cannot achieve an accurate and reliable AU labeling. However, from Figure 5, we can see that an accurate AU labeling can be achieved with a few human interactions. For example, we can achieve about 5% false-negative rate and 1% false-positive rate with only two corrections. The improvement is especially significant for the AUs that are difficult for the automatic AU labeling. Compared with the automatic AU labeling, the false-negative rate of AU23 (lip tightener) decreases from 64.29% (automatic AU labeling) to 9.09%, and its false-positive rate also decreases from 1.58% (automatic AU labeling) to 0% with one human correction using the proposed interactive AU labeling method with active human input. These results demonstrate the effectiveness and efficiency of the proposed interactive labeling approach.

5 Conclusion

In this research, we introduce an interactive image labeling system for effective combination of the automatic image labeling with the limited human labeling. By using a BN as the engine for the automatic image labeling and adopting the active user feedback strategy, the system is effective for image labeling in several aspects: 1) it allows human inputs to be incrementally incorporated at any stage; 2) it allows to combine different human inputs systematically; and 3) it employs an active human input scheme to minimize the amount of human input, thereby improving both labeling efficiency and accuracy. We have applied the framework to the interactive facial action units labeling. The experiments demonstrate the effectiveness of the system in improving both the AU labeling accuracy and the efficiency. Besides, the proposed framework is generic enough to be applied to different tasks in computer vision, including image segmentation, image retrieval, and object recognition, etc..

References

1. Levin, A., Viola, P., Freund, Y.: Unsupervised improvement of visual detectors using co-training. In: Int'l. Conf. on Computer Vision, pp. 13–16 (2003)
2. Ekman, P., Friesen, W.V., Hager, J.C.: Facial Action Coding System: the Manual. Research Nexus, Div., Network Information Research Corp., Salt Lake City, UT (2002)
3. Pantic, M., Bartlett, M.: Machine analysis of facial expressions. In: Delac, K., Grgic, M. (eds.) Face Recognition, pp. 377–416. I-Tech Education and Publishing, Vienna (2007)
4. Kanade, T., Cohn, J.F., Tian, Y.: Comprehensive database for facial expression analysis. In: Proc. 4th IEEE Int'l Conf. Automatic Face and Gesture Recognition, pp. 46–53 (2000)
5. Zhou, Z.H., Chen, K.J., Dai, H.B.: Enhancing relevance feedback in image retrieval using unlabeled data. ACM Trans. on Information Systems 24(2), 219–244 (2006)
6. In Chapelle, O., Weston, J., Schölkopf, B. (eds.): Semi-supervised learning. MIT Press, Cambridge (2006)
7. Cozman, F., Cohen, I., Cirelo, M.: Semi-supervised learning of mixture models. In: ICML (2003)
8. Wang, F., Zhang, C.: Label propagation through linear neighborhoods. In: ICML (2006)
9. Weston, J., Leslie, C., Zhou, D., Elisseeff, A., Noble, W.S.: Semisupervised protein classification using cluster kernels. In: Advances in neural information processing systems, vol. 16. MIT Press, Cambridge (2004)
10. Xu, L., Schuurmans, D.: Unsupervised and semi-supervised multi-class support vector machines. In: Proc. of the 20th National Conf. on Artificial Intelligence (2005)
11. Zhou, Z.H., Li, M.: Semi-supervised regression with co-training. IJCAI (2005)
12. Zhu, X.: Semi-supervised learning with graphs. Doctoral dissertation, Carnegie Mellon University (2005)
13. Ackley, D.H., Littman, M.L.: Generalization and scaling in reinforcement learning. In: Touretzky, D.S. (ed.) Advances in Neural Information Processing Systems, vol. 2, pp. 550–557. Morgan Kaufmann, San Mateo (1990)
14. Boyan, J.A., Moore, A.W.: Generalization in reinforcement learning: Safely approximating the value function. In: Tesauro, G., Touretzky, D.S., Leen, T.K. (eds.) Advances in Neural Information Processing Systems, vol. 7. The MIT Press, Cambridge (1995)
15. Gullapalli, V.: Reinforcement learning and its application to control. Ph.D thesis, University of Massachusetts (1992)

16. Lin, L.J.: Reinforcement learning for robots using neural networks. Ph.D thesis, Carnegie Mellon University (1993)
17. Maes, P., Brooks, R.A.: Learning to coordinate behaviors. In: Proc. 8th National Conf. on Artificial Intelligence, pp. 796–802 (1990)
18. Zhang, W., Dietterich, T.G.: A reinforcement learning approach to job-shop scheduling. IJ-CAI (1995)
19. Chang, E.Y., Lai, W.C.: Active learning and its scalability for image retrieval. In: ICME, pp. 73–76 (2004)
20. Christoudias, C., Saenko, K., Morency, L., Darrell, T.: Co-adaptation of audio-visual speech and gesture classifiers. In: Int'l. Conf. on Multimodal Interfaces, pp. 84–91 (2006)
21. Blum, A., Mitchell, T.: Combining labeled and unlabeled data with co-training. In: Proc. of the Workshop on Computational Learning Theory, pp. 92–100 (1998)
22. Freund, Y., Seung, H., Shamir, E., Tishby, N.: Selective sampling using the query by committee algorithm. Machine Learning 28, 133–168 (1997)
23. Shen, D., Zhang, J., Su, J., Zhou, G., Tan, C.L.: Multi-criteria-based active learning for named entity recognition. In: ACL 2004: Proceedings of the 42nd Annual Meeting on Association for Computational Linguistics (2004)
24. Kapoor, A., Grauman, K., Urtasun, R., Darrell, T.: Active learning with gaussian processes for object categorization. In: ICCV (2007)
25. Raghavan, H., Madani, O., Jones, R.: Active learning with feedback on both features and instances. Journal of Machine Learning Research 7, 1655–1686 (2006)
26. Tur, G., Rahim, M., Hakkani-Tuk, D.: Active labeling for spoken language understanding. Eurospeech, 2782–2789 (2003)
27. Pantic, M., Rothkrantz, L.J.M.: Automatic analysis of facial expressions: The state of the art. IEEE Trans. PAMI 22(12), 1424–1445 (2000)
28. Bartlett, M.S., Littlewort, G.C., Frank, M.G., Lainscsek, C., Fasel, I.R., Movellan, J.R.: Automatic recognition of facial actions in spontaneous expressions. J. Multimedia 1(6), 22–35 (2006)
29. Tong, Y., Liao, W., Ji, Q.: Facial action unit recognition by exploiting their dynamic and semantic relationships. IEEE Trans. PAMI 29(10), 1683–1699 (2007)
30. Heckerman, D., Geiger, D., Chickering, D.M.: Learning bayesian networks: The combination of knowledge and statistical data. Machine Learning 20(3), 197–243 (1995)
31. Heckerman, D.: A tutorial on learning with bayesian networks. Technical Report MSR-TR-95-06, Microsoft Research, pp. 1–40 (1995)
32. Multiple Aspects of Discourse research lab, `http://madresearchlab.org/`
33. Douglas-Cowie, E., Cowie, R., Schroeder, M.: The description of naturally occurring emotional speech. In: Fifteenth Int'l Congress of Phonetic Sciences (2003)

Real Time Feature Based 3-D Deformable Face Tracking

Wei Zhang[1,*], Qiang Wang[2], and Xiaoou Tang[1]

[1] Dept. of Information Engineering, The Chinese University of Hong Kong, Hong Kong, China
{dylan,xtang}@ie.cuhk.edu.hk
[2] Microsoft Research Asia, Beijing, China
qiangwa@microsoft.com

Abstract. In this paper, we develop a novel framework for 3D tracking of the non-rigid face deformation from a single camera. The difficulty of the problem lies in the fact that 3D deformation parameter estimation becomes unstable when there are few reliable facial features correspondences. Unfortunately, this often occurs in real tracking scenario when there is significant illumination change, motion blur or large pose variation. In order to extract more information of feature correspondences, the proposed framework integrates three types of features which discriminate face deformation across different views: 1) the semantic features which provide constant correspondences between 3D model points and major facial features; 2) the silhouette features which provide dynamic correspondences between 3D model points and facial silhouette under varying views; 3) the online tracking features that provide redundant correspondences between 3D model points and salient image features. The integration of these complementary features is important for robust estimation of the 3D parameters. In order to estimate the high dimensional 3D deformation parameters, we develop a hierarchical parameter estimation algorithm to robustly estimate both rigid and non-rigid 3D parameters. We show the importance of both features fusion and hierarchical parameter estimation for reliable tracking 3D face deformation. Experiments demonstrate the robustness and accuracy of the proposed algorithm especially in the cases of agile head motion, drastic illumination change, and large pose change up to profile view.

1 Introduction

Reliable tracking of 3D deformable faces is still a challenging task in computer vision. The basic difficulty is that we need to estimate dozens of rigid and non-rigid 3D parameters from noisy image observation. The situation becomes even worse when the face is in profile view and almost half of the face region is occluded.

There are mainly two categories of 3D deformable face tracking algorithms. The first one is appearance based approach [1] which uses generative linear

* This work was done when Wei Zhang was visiting at Microsoft Research Asia.

D. Forsyth, P. Torr, and A. Zisserman (Eds.): ECCV 2008, Part II, LNCS 5303, pp. 720–732, 2008.

models of face appearance such as 2D Active Appearance Models (AAM) [2] and 3D Morphable Models [3] to capture the texture and shape variation of face respectively. The deformation parameters can then be efficiently estimated using gradient descent optimization. However, it is known that AAM has quite weak generalization capability because the general face texture space is too large to be captured by reasonable size training data. Therefore such approaches are limited to be used in conditions similar to the training data. The second one is feature based approach which uses aggregation of sparse facial features to represent face. The face variation is captured either by a multiple view 2D Active Shape Model(ASM) [4] or a complete 3D shape model [5] or a combination of both [6]. Compared with the appearance based algorithms, the feature based algorithms have better generalization capability. However, the local features used in ASM are semantic features whose correspondences in 3D model are manually defined. They are mainly located in regions around eyes, nose, and mouth which could be severely occluded when the face is in poses with pan angle larger than 45 degrees. The shortage of reliable semantic correspondences could make the tracking unstable.

In this paper, we propose a feature based framework for 3D tracking of the deformable face. In order to extract more discriminative information on face deformation across different views, we combine three types of features: the semantic features which are the same as those defined in ASM, the silhouette features which are dynamically specified and matched according to different poses, and the online tracking features which are obtained by the robust matching of image interest points. In addition to the feature augmentation, we show that hierarchical parameter estimation is quite important to progressively reject outlier feature correspondences and estimate both rigid and non-rigid 3D parameters. In our framework, the parameter estimation is formulated as an energy minimization problem under the constraint of a deformable 3D face model which usually converges in a few iterations, thus fulfills the real-time tracking task.

2 Algorithm Overview

Fig. 1 shows the flow diagram of the tracking algorithm at frame t: given the input image, the algorithm builds up the feature matches and estimates the 3D pose and deformation parameters in a hierarchical way: Firstly the initial 3D pose parameters are estimated from online interest point information, then the initial feature matches are refined by rejecting many outliers and the shape and pose parameters are estimated in an iterative way.

Since the shape and 3D pose estimation are based on feature matching between the neighboring frames, it is well known that such a strategy always suffers from accumulated drift. Our solution is the usage of key-frames and to anchor the current frame to them. The key-frames are selected online from the past estimated frames with high accuracy in which pose and shape parameters are deterministic. For the details of key-frame selection and matching, please refer to [7].

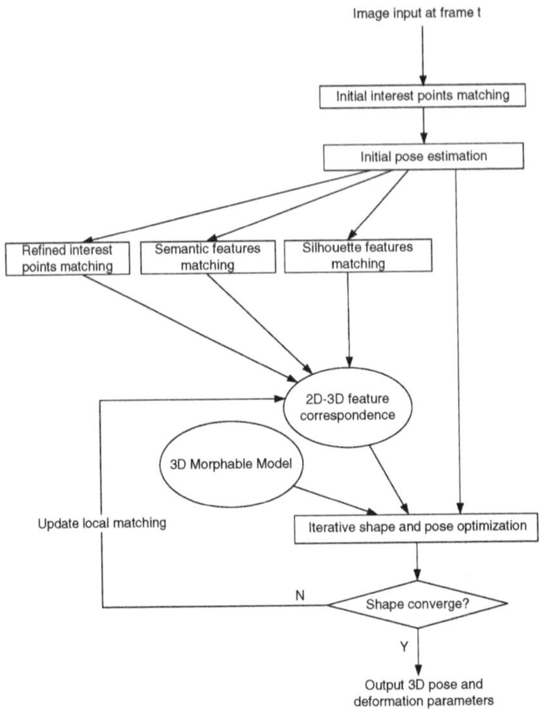

Fig. 1. Flow diagram of the proposed algorithm

Initial Pose Estimation. The initial 3D pose estimation is obtained from interest points matching among the current frame, previous frame and some selected key-frames, the 2D interest points in previous frame and key-frames can be back-projected into the 3D face model to get their 3D position. The initial pose in current frame can be estimated by the 2D-3D point correspondences in a Bayesian inference way. Details are given in Sec. 4.3.

Iterative Shape and Pose Optimization. Given the initial pose estimation, the matches for semantic features, silhouette features, and interest points are obtained and refined, their 3D correspondences in the model are also determined. Then the 3D pose and deformation parameters are estimated from the 2D-3D point correspondences. The above procedure is carried out iteratively until convergence is achieved. Details are given in Sec. 5.

Tracking Initialization. In order to make the whole tracking algorithm fully automatic, we need to estimate the 3D pose and deformation parameters in the first frame: the face is first detected by the face detector [8], then the facial features are extracted by the face alignment algorithm [9]. Such facial features are semantic features whose 3D correspondences in the face model are fixed, so we can apply the POSIT algorithm [10] to estimate the 3D pose using rigid semantic features and then estimate the 3D deformation parameters.

3 3D Deformable Shape

The shape of a face is defined by a 3D triangulated mesh. The shape vector is denoted as $S = (v_1^t, \ldots, v_n^t)$, where $v_i = (x_i, y_i, z_i)^t, (i = 1, \ldots, n)$ are the 3D coordinates of the i-th vertex. The deformation of a 3-D face is described by a linear model [3][1][11], such as PCA:

$$S = \mu + \sum_{i=1}^{k} \alpha_i \Phi_i, \qquad (1)$$

where μ represents the average shape from the training samples, Φ_i are orthogonal shape vectors, and α_i are scalar values indicating the contributions of the shape deformation from the i-th shape vector.

The deformable shape model is obtained from a collection of 3D face shape samples. Our collection includes 100 3D face samples with neutral expression from the USF 3D face database [3] and 2907 synthetic 3D face samples with expressions. The images used for synthesis come from the AR database [12] and XM2VTSbd database [13]. The synthetic face samples are generated as follows: firstly the semantic facial features on 2907 frontal face images are labeled which have various expressions, then a face modeling algorithm [14] is applied to approximate the most likely 3D face shape in neutral expression, finally the Face Animation Parameters (FAP) [15] are estimated to fit the deformation between 3D neutral face shape and 2D labeled facial features as the following form:

$$||P_w\left(\bar{S} + A\gamma\right) - \mathbf{s}||^2, \qquad (2)$$

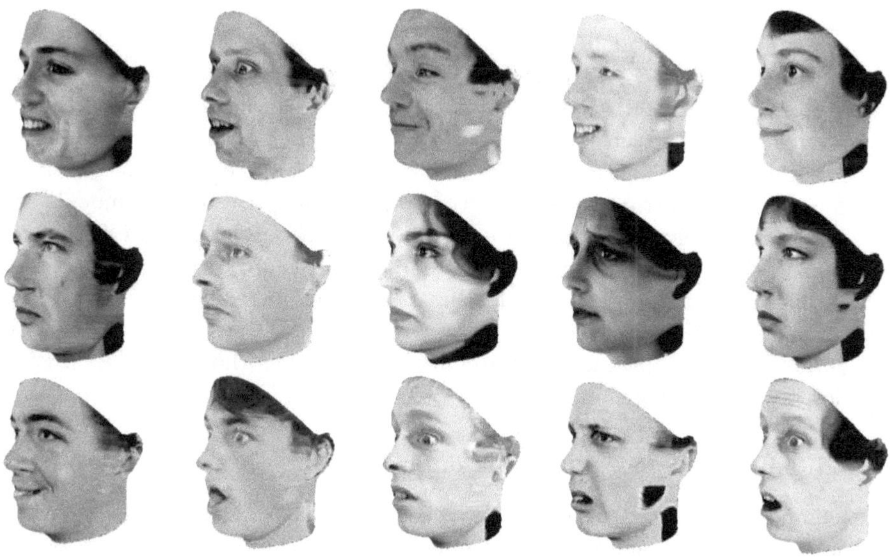

Fig. 2. Synthesized 3D face model with expression

where $\bar{S}$ is the approximated 3D neutral face shape, A is a block diagonal matrix where each block represents the linear facial deformation function for each FAP, γ is the vector of FAP, P_w is a 2D projection function with weak-perspective model, and $\mathbf{s}$ is the coordinates of labeled 2D facial features. Some of the 3D synthetic face samples are illustrated in Fig. 2.

The original 3D face models have 8955 vertices and 17535 triangular facets which are too many for the 3D tracking task in a relatively low resolution video sequence, so the original models are simplified to the models which contain only 180 vertices and 332 triangular facets. The 3D deformable face model is trained on the simplified models via PCA, where 42 shape variations are kept with 99% energy preserved.

4 Local Features Training and Matching

There are three types of local features in the proposed algorithm: the semantic features are defined as the ASM features [9] in inner face region, such features have constant correspondences in 3D model which does not change with 3D face pose; the silhouette features locate at the boundary of the face region and their 3D correspondences dynamically change according to the 3D face pose, similar features are used in [16] for 3D face reconstruction; the interest points are salient image features on the face, so they vary across images. The distribution of the appearance of the first two type of features can be learned from training samples in offline training. In the following subsections, we will describe the training and matching of the local features.

4.1 Semantic Feature Training and Matching

The view based local models for semantic features are constructed as follows: we render the 3D textured face models in 11 different views, where the roll angle ranges from $-60°$ to $60°$ and the pitch angle ranges from $-30°$ to $30°$, with a step angle of $30°$. Given the location of each feature in each view, one image patch centered at the location is extracted as positive sample and 24 neighboring image patches are extracted as negative samples. The local feature models are then discriminatively trained using a modified RankBoost algorithm [9]. In our experiment, the local feature models are built on three levels of image pyramid.

The matching of the semantic feature is carried out as follows: firstly the view of the feature model is selected by the current 3D pose estimation, then the feature match is obtained by searching in the neighborhood for the position with lowest feature model energy.

4.2 Silhouette Feature Training and Matching

The silhouette features are similar to the semantic features except that their 2D positions and 3D correspondences dynamically change according to face pose. So the training part of the silhouette feature is the same as that of the semantic feature and here we only focus on the determination of their 2D-3D correspondences.

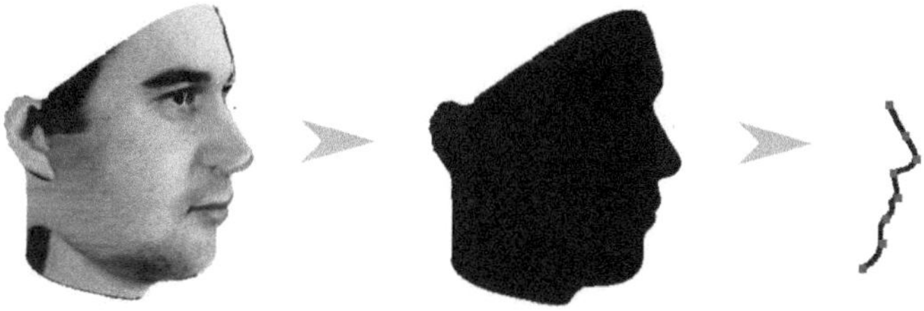

Fig. 3. Illustration of silhouette representations by sparse points

As shown in Fig. 3, we first construct a silhouette map by rendering the 3D face model with the current estimated pose and deformation parameters, then the boundary of the silhouette map is extracted and silhouette features are sampled along the extracted curve. For each sampled silhouette feature, we calculate its 3D correspondence as follows: first the triangular mesh whose projection contains the feature is retrieved and its vertices are obtained, then the coordinate of the 3D correspondence is calculated using the bary-centeric coordinate of the silhouette feature in the projected triangle.

Although some previous work uses silhouette information [17] for 3D motion estimation, the silhouette is often used in curve form and evolved using level-set techniques [18]. The cost of curve optimization is usually high and not suitable for real-time applications. Our point based silhouette representation is more flexible and can be efficiently integrated into a 3D pose and deformation estimation algorithm as described in Sec. 5. Furthermore, the appearance model of the silhouette can be trained to achieve more stable performance during curve searching.

4.3 Interest Point Selection and Matching

The interest point is selected according to its image saliency [19]. For the tracking at time t, the interest points are selected from the previous frame and some key-frames and then they are matched to current frame. Since the 3D pose and deformation parameters of previous frame and key-frames are assumed to be known, the 3D correspondences of the interest points can be calculated.

It should be noted that the online matching of interest points could have many outliers. With the estimate of current 3D pose and deformation parameters, we can accordingly reject the matches which violate the geometric constraints among the previous frame, current frame and key-frames. Such rejection technique is applied in the pose and deformation estimation process described in Sec. 5.

5 Hierarchical Shape and Pose Inference

The 3D face parameters to be estimated are denoted as $\{\alpha, Q\}$,where $Q = \{R, T\}$ represents the rigid motion of six parameters, three for object rotation and three for its translation; α is the coefficient of the linear model in Eq. (1). Given the

2D-3D correspondence between current input image and 3D morphable model, the simultaneous pose and deformation recovery is an ill-posed problem with the following objective function:

$$\{\hat{\alpha}, \hat{Q}\} = \underset{\alpha, Q}{arg\min}\Big(\sum \rho\left(||\Psi_c(x, Q, \alpha)||^2\right) + \sum \rho\left(||\Psi_s(x, Q, \alpha)||^2\right) + \sum \rho\left(||\Psi_i(x_j, Q, \alpha)||^2\right) + \beta \sum_{i=1}^{k} \alpha_i^2/\lambda_i^2\Big). \quad (3)$$

In the above equation, Ψ is an essential optimization function which denotes the 2D projection error between a 3D model point and its 2D correspondence x. The 3D points are represented by the deformation parameters α. The footnotes of Ψ_c, Ψ_s and Ψ_i represent the errors of the three types of features respectively. All types of correspondence errors are treated equally. ρ is a robust function [20] with a threshold T:

$$\rho(r) = \begin{cases} r & r < T \\ 2T & r \geq T \end{cases} \quad (4)$$

The last regularized item in Eq. (3) models the constraint of the deformation from 3D morphable model in Eq. (1). In our experiment's configuration, T indicates the sum of square projection error from 3D model.

As shown in [21], α and Q in Eq. (3) are coupled in the non-linear function Ψ. A two-stage optimization scheme is taken where each item in $\{\alpha, Q\}$ is fixed in turn for optimal inference. In general case, such a sequential optimization scheme often suffers from the non-trivial local optima. Hence, we make use of a hierarchial optimization strategy described in Sec. 5.2 to estimate the deformation and pose parameters efficiently.

5.1 Robust Initial Pose Estimation

Initial pose is estimated based on the rigid assumption for the inter-frame motion. Given the deformation parameters estimated in previous frame, the initial pose is obtained via minimization of the first item in Eq. (3). The optimization details is also in Sec. 5.2.

5.2 Two-Stage Parameters Estimation

Estimate Q Given α. Here, the cost function we need to minimize is:

$$E(Q) = \rho(||\mathbf{x} - P(Q(X))||^2). \quad (5)$$

where $Q(X)$ is the rigid transform with rotation matrix R and translation vector T, P is the perspective projection of the 3D vector $Q(X)$ in the image plane.

We employ a stochastic optimization approach extended from RANSAC [22], [7] to optimize Eq. (5) as follows: firstly, a number of samples are generated from the 2D-3D feature pairs set $\{\mathbf{x}, \mathbf{X}\}$, in which each sample consists of a minimum

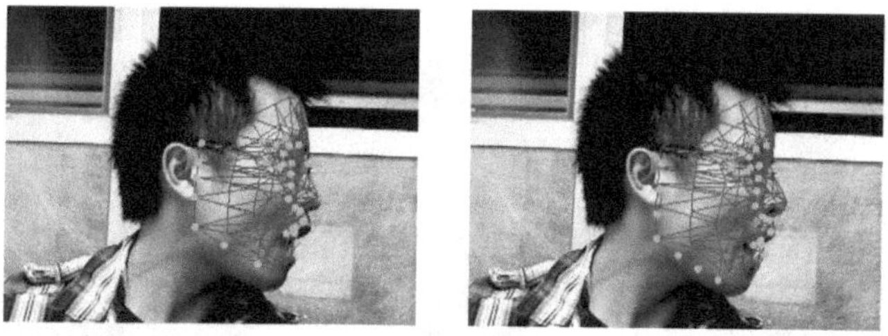

Fig. 4. Model fitting result given an initial pose

set of feature pairs that can recover the relative pose Q. The robust function ρ in Eq. (5) is evaluated. The Q associated with the sample of minimum cost is the optimization result. Here, we use the POSIT algorithm [10] to recover the pose Q with a minimum input of four 2D-3D feature pairs.

Estimate α Given Q. When the pose parameter Q is fixed, the cost function in Eq. (3) can be re-arranged with the following form:

$$E(\alpha) = \rho(||\mathbf{x} - P(H(\alpha))||^2) + \beta \sum_{i=1}^{k} \alpha_i^2/\lambda_i^2. \tag{6}$$

where H denotes the linear transform of all the 3-D model points from the shape parameter α. The first item in Eq. (6) is still a non-linear form with a high dimensional parameter α.

Here we deduce the minimization of the first item in Eq. (6) to the following objective function:

$$E_s(\alpha) = ||M\alpha - b||^2, \tag{7}$$

where matrix M is a linear transform matrix stemming from $\{Q, \Phi, x\}$. We also prove that the solution $\hat{\alpha}$ for Eq. (7) is also the optimal solution for Eq. (6) when the linear equation $\{M\alpha = b\}$ has the unique solution; otherwise, it is the solution with the assumption that the projection degenerates to the weak-perspective model, which is the assumption used in [21],[14]. The detailed inference is list in Appendix A.

In order to obtain the optimal shape parameters, we set the partial derivative of Eq. (6) to zero. The optimal parameters are obtained by solving the following linear equation:

$$\left(M^T M + \beta \Lambda\right) \alpha = M^T b. \tag{8}$$

6 Experiments

To verify the robustness and generalization ability of the proposed tracking method, the algorithm has been tested on the image sequences captured from a USB camera for four different persons. The difficulties for tracking in the image sequence stem from the out-of-plane motion, large expression change and

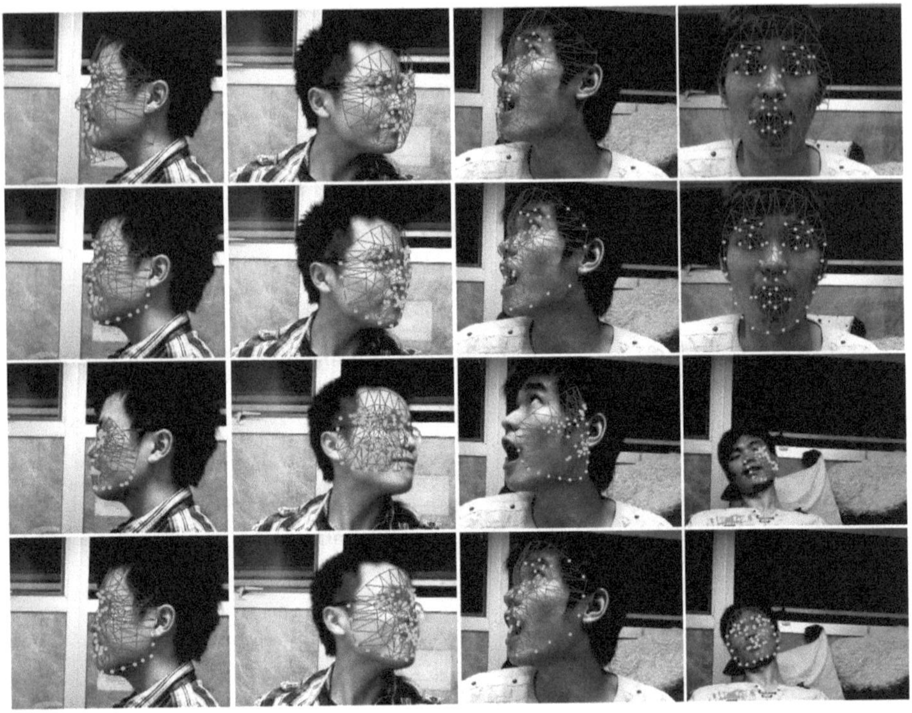

Fig. 5. The first and third rows are the tracking results without silhouette features and interest points respectively, the corresponding results of the proposed method are listed in the second and fourth rows

numerous occlusions. Aside from providing the tracking output, a set of comparison experiments are performed to demonstrate the rationality of the integration of complementary features. More tracking results for the captured image sequences can be obtained from the supplementary materials or downloaded from `http://mmlab.ie.cuhk.edu.hk/project.htm`.

In our experiments, the typical USB camera's resolution is set to $320 * 240$, and 100 interest points, 68 semantic features, and 15 silhouette features are used to achieve the results in this paper. As for the key-frame setting, we use two key-frames for robust estimation: one key-frame is simply the first input frame and another key-frame is selected from a pool of history frames whose parameters are reliably estimated and are moderately similar to the current frame. The threshold value of T in Eq. (4) takes the fixed value 6 for all the test samples.

The non-rigid face motion are shown by projecting the 3D morphable model into the image plane under the estimated 3D pose. Meanwhile, the semantic features and silhouettes features are also imposed on the image for visualization purpose. For better tracking result illustration, we use only half of those features for demonstration. In Fig. 4, the initial estimated pose from interest points matching is shown in the first image, the iterative shape and pose estimation is

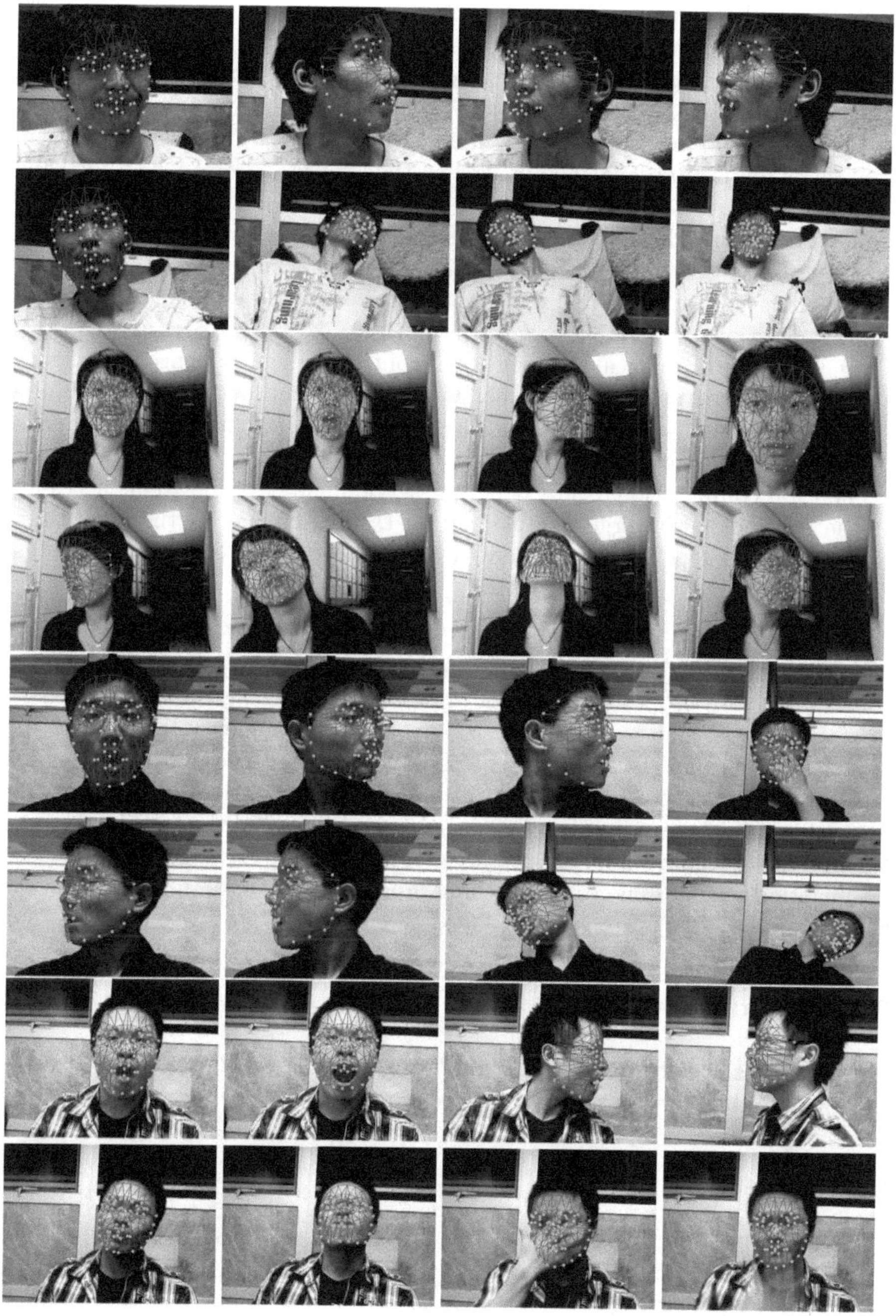

Fig. 6. Tracking results with live captured image sequences. The second image at the fifth row shows a bad fitting result on the excessive expression. This is due to the lack of extreme expression variation in our training database. The last two images illustrate the robustness of our algorithm with partial occlusion.

performed and the final tracking result is illustrated in the second image. The fitting process converges to the right solution within 10 iterations.

The tracking results in the first two rows of Fig. 5 demonstrate the importance of the silhouette features for face boundary localization especially in the near profile pose, where the available semantic features reduce by half with comparing to those of the frontal face. The other two types of features are kept the same for a fair comparison.

The tracking results in the last two rows of Fig. 5 illustrate the key role of the interest points on rigid motion estimation. The tracking without the interest points suffers from agile motion where the search of semantic points cannot reach accurate positions.

Fig. 6 shows some typical tracking results of three image sequences. Our tracking algorithm accurately localizes the facial components such as eyes, eyebrows, noses and mouthes, under the conditions with agile motion, large expression variation, and large occlusion.

The total computational time depends on the number of iterations necessary for the method to converge. On a Pentium-D 3.6G computer, the algorithm's speed is about 15 fps for the video with $320 * 240$ resolution.

7 Conclusion

In this paper, we propose an efficient approach for tracking the deformable 3D face from a single camera. By taking the advantage of complementary features integration, we obtain a large number of 2D-3D correspondences for the estimation of pose and deformation parameters. Aside from the low cost of computation, the usage of local features enhances the generalization ability of the tracking system. The non-linear shape and pose optimization is efficiently solved by a hierarchical optimization scheme. The whole tracking process is efficient, automatic, and the result achieved is accurate even for low quality video.

Acknowledgement

The authors sincerely thank Ming Liu, Yingze Wang and Ting Huang for the experimental implementation and proof-reading.

References

1. Xiao, J., Baker, S., Matthews, I., Kanade, T.: Real-time combined 2D+3D active appearance models. In: Proceedings of the 2004 IEEE Conference on Computer Vision and Pattern Recognition, vol. 2, pp. 535–542 (2004)
2. Matthews, I., Baker, S.: Active Appearance Models Revisited. International Journal of Computer Vision 60(2), 135–164 (2004)
3. Blanz, V., Vetter, T.: A morphable model for the synthesis of 3D faces. In: Proceedings of the 26th annual conference on Computer graphics and interactive techniques, pp. 187–194 (1999)

4. Zhou, Y., Zhang, W., Tang, X., Shum, H.: A Bayesian Mixture Model for Multiview Face Alignment. In: Proceedings of the 2005 IEEE Computer Society Conference on Computer Vision and Pattern Recognition, pp. 741–746 (2005)
5. Gu, L., Kanade, T.: 3D Alignment of Face in a Single Image. In: Proceedings of the 2006 IEEE Computer Society Conference on Computer Vision and Pattern Recognition, pp. 1305–1312 (2006)
6. Vogler, C., Li, Z., Kanaujia, A., Goldenstein, S., Metaxas, D.: The Best of Both Worlds: Combining 3D Deformable Models with Active Shape Models. In: IEEE 11th International Conference on Computer Vision, ICCV 2007, pp. 1–7 (2007)
7. Wang, Q., Zhang, W., Tang, X., Shum, H.: Real-Time Bayesian 3-D Pose Tracking. IEEE Transactions on Circuits and Systems for Video Technology 16(12), 1533–1541 (2006)
8. Xiao, R., Zhu, L., Zhang, H.: Boosting chain learning for object detection. In: Proceedings. Ninth IEEE International Conference on Computer Vision, pp. 709–715 (2003)
9. Yan, S., Li, M., Zhang, H., Cheng, Q.: Ranking prior likelihood distributions for Bayesian shape localization framework. In: Ninth IEEE International Conference on Computer Vision, pp. 51–58 (2003)
10. Dementhon, D., Davis, L.: Model-based object pose in 25 lines of code. International Journal of Computer Vision 15(1), 123–141 (1995)
11. Xin, L., Wang, Q., Tao, J., Tang, X., Tan, T., Shum, H.: Automatic 3D Face Modeling from Video. In: Proceedings of the Tenth IEEE International Conference on Computer Vision (ICCV 2005) (2005)
12. Martinez, A., Benavente, R.: The AR face database. CVC Technical Reportno 24 (1998)
13. Messer, K., Matas, J., Kittler, J., Luettin, J., Maitre, G.: XM2VTSDB: The Extended M2VTS Database. In: Second International Conference on Audio and Video-based Biometric Person Authentication, vol. 626 (1999)
14. Hu, Y., Jiang, D., Yan, S., Zhang, L., et al.: Automatic 3D reconstruction for face recognition. In: Sixth IEEE International Conference on Automatic Face and Gesture Recognition, pp. 843–848 (2004)
15. Ostermann, J.: Face animation in mpeg-4 (2002)
16. Romdhani, S., Vetter, T.: Estimating 3D shape and texture using pixel intensity, edges, specular highlights, texture constraints and a prior. In: IEEE Conference on Computer Vision and Pattern Recognition (2005)
17. Kass, M., Witkin, A., Terzopoulos, D.: Snakes: Active contour models. International Journal of Computer Vision 1(4), 321–331 (1988)
18. Brox, T., Rosenhahn, B., Cremers, D., Seidel, H.: High accuracy optical flow serves 3-D pose tracking: exploiting contour and flow based constraints. In: Leonardis, A., Bischof, H., Pinz, A. (eds.) ECCV 2006. LNCS, vol. 3952, pp. 98–111. Springer, Heidelberg (2006)
19. Shi, J., Tomasi, C.: Good features to track. In: Proceedings of the 1994 IEEE Computer Society Conference on Computer Vision and Pattern Recognition, pp. 593–600 (1994)
20. Torr, P., Davidson, C.: IMPSAC: Synthesis of importance sampling and random sample consensus. IEEE Transactions on Pattern Analysis and Machine Intelligence 25(3), 354–364 (2003)
21. Hill, A., Cootes, T., Taylor, C.: Active Shape Models and the shape approximation problem. Image and Vision Computing 14(8), 601–607 (1996)

22. Lu, C., Hager, G., Mjolsness, E.: Fast and globally convergent pose estimation from video images. IEEE Transactions on Pattern Analysis and Machine Intelligence 22(6), 610–622 (2000)

Appendix

Given $Q = \{R, T\}$ fixed, the projection error E_{proj} in Eq. (6) can be expanded to the following form

$$\sum_{i=1}^{n} \left\| f\frac{\sum_1^k \alpha_i \phi'_{3j,i} + t_0}{\sum_1^k \alpha_i \Phi'_{3j+2,i} + t_2} - v_{2i} \right\|^2 + \left\| f\frac{\sum_1^k \alpha_i \Phi'_{3j+1,i} + t_1}{\sum_1^k \alpha_i \Phi'_{3j+2,i} + t_2} - v_{2i+1} \right\|^2 \tag{9}$$

where $T = \{t_0, t_1, t_2\}$ and $\Phi' = \{\phi_{j,i}\}$ is obtained by applying rotation R to the eigen-vectors Φ of shape PCA. Consider the following the equation:

$$f\frac{\sum_1^k \alpha_i \phi'_{3j+l,i} + t_l}{\sum_1^k \alpha_i \Phi'_{3j+2,i} + t_2} = v_{2i+l}, l \in \{0, 1\}. \tag{10}$$

Eq. (10) can be re-arranged in a linear form as:

$$M_{2n,k}\alpha = \mathbf{b}_{2n} \tag{11}$$

where

$$\begin{aligned} m_{2i+l,j} &= f\phi'_{j,3i+l} - v_{2i+l}\phi'_{j,3i+2}, \\ b_{2i+l} &= v_{2i+l}t_2 - ft_l \end{aligned} \tag{12}$$

If Eq. (11) has the unique solution, for example, $Rank(M) = Rank(M|b)$, the original function Eq. (9) reaches its optimal value zero. Otherwise, the solution $\hat{\Lambda}$ for' Eq. (11) is the optimal solution in the meaning of MSE(minimum square error), we can denote the residual error as

$$\epsilon = \left\| M\hat{\Lambda} - b \right\|^2. \tag{13}$$

Therefore, the optimal value in Eq. (9) is constrained in the following range:

$$E_{proj} \in \left(\frac{\epsilon}{t_{max}}, \frac{\epsilon}{t_{min}} \right), \tag{14}$$

where t_{max} and t_{min} are the corresponding maximum and minimum values of the denominator item in Eq. (10), with the constraint that the value of Λ is only valid in the range determined by the corresponding eigen-value σ in the PCA. A small ϵ means a narrow range for the optimal value. Under the assumption of weak-perspective projection model, in which t_{max} equals t_{min} for all the points, the optimal values is equivalent in Eq. (9) and in Eq. (10). That means the solution in Eq. (10) is the solution with the weak perspective model assumption in the worst case.

Rank Classification of Linear Line Structure in Determining Trifocal Tensor

Ming Zhao and Ronald Chung*

Department of Mechanical & Automation Engineering
The Chinese University of Hong Kong
{mzhao,rchung}@mae.cuhk.edu.hk

Abstract. The problem we address is: given line correspondences over three views, what is the condition of the line correspondences for the spatial relation of the three associated camera positions to be uniquely recoverable? We tackle the problem from the perspective of trifocal tensor, a quantity that captures the relative positions of the cameras in relation to the three views. We show that the rank of the matrix that leads to the estimation of the tensor reduces to 7, 11, 15 respectively for line pencil, point star, and ruled plane, which are structures that belong to linear line space; and 12, 19, 23 for general ruled surface, general linear congruence, and general linear line complex. These critical structures are quite typical in reality, and thus the findings are important to the validity and stability of practically all algorithms related to structure from motion and projective reconstruction using line correspondences.

1 Introduction

We address the question of, given line correspondences over three views, what is the condition of the observed lines for the relative geometry of the three associated camera positions to be uniquely recoverable in space? There are at least two issues: what particular structure the lines in space belong to, and whether there are enough of the lines. In this work we deal with both the structure issue and the quantity issue, and we deal with them from the perspective of trifocal tensor – the $3 \times 3 \times 3$ quantity that captures the relative positions of the cameras in relation to the views. We investigate whether the observed lines are of a too specific structure in 3-D space to allow the trifocal tensor to be uniquely recoverable, and if they indeed are, by how much they are shy of a general enough structure. We also investigate, for each of the investigated line structures, how many lines are minimally needed to allow the full information of the line structure to be reflected in the image projections.

A set of observed lines in space are called *critical* if there are multiple projectively non-equivalent configurations of the camera positions that can picture the same image triplet of the lines. We shall examine the different classes of structure of the line set that are critical. In all cases, we assume that the camera positions are general with respect to the set of lines, meaning that the images

* Corresponding author.

D. Forsyth, P. Torr, and A. Zisserman (Eds.): ECCV 2008, Part II, LNCS 5303, pp. 733–744, 2008.

are not degenerate images like those having multiple lines projected to the same image line, a result of accidental alignment.

Trifocal tensor, an important quantity in multiple view geometry, is the counterpart of fundamental matrix. Trifocal tensor has drawn much attention in the last decade or so, for it can be computed directly from image correspondences, and its determination would allow camera matrices and in turn projective structure of the imaged scene to be reconstructed. Existing algorithms of determining trifocal tensor are mainly iterative and require initialization typically from a linear method. The degeneracy in using linear method to determine camera motion has been studied in a number of works. However, the works are mainly about the cases where point correspondences are used. There have been a few earlier works on the cases where line correspondences are used, but the treatment has been centered around the use of direct geometric primitives like the relative displacements and orientations of the cameras [1]. Since trifocal tensor has the essence of expressing the relative positions of all cameras in relation to the image observables in a single collective term, we believe the study would be simpler and the degeneracy be more easily quantified if the issue is pursued from the angle of trifocal tensor, and this work aims at doing that.

Line correspondences are in many ways as indispensable as point correspondences. There is evidence [2,3] that 3-D reconstruction from lines could be more accurate than from points. Moreover, linear line structures as the structures to be addressed in this paper do appear quite often in real scenes, as exemplified in Fig. 1. Putting it simply, linear line space includes three subspaces: line pencil as a family of lines that contain the same point and that intersect at a non-incident line (in other words, the lines are both co-planar and co-intersecting at a point on the plane); point-star (also named bundle of lines) as a family of lines that are concurrent at a point; ruled plane (also named field of lines) as a family of lines incident on a particular plane. More general than linear line space, linear line structures include linear line complex which are lines satisfying one linear constraint, linear line congruence satisfying two linear constraints, and linear ruled surface-lines satisfying three linear constraints. In Fig. 1(a), a typical linear line complex structure is shown. The Eiffel tower in Fig. 1(b) can also be approximated as several sets of line bundle and line fields that belong to linear line congruence. The building in Fig. 1(c) consists of a field of lines that are coplanar and they also belong to linear line congruence.

2 Previous Work

In this work we are concerned primarily with recovering camera displacement or camera motion from visual observations. Point and line features can both be used in motion estimation from multiple views. Certain algebraic properties of correspondences over multiple views are summarized in [4] and [5]. Critical configuration of point features, in particular, has already drawn much attention [6,7,8] in the literature, and the studies have been about the condition of the point features under which the camera motion cannot be determined uniquely. A

Fig. 1. Scenes consisting of (left) linear line complex; and linear line congruence ((middle) the sub-class: point-star, and (right) the sub-class: ruled plane) respectively

classification of multiple-view critical configurations is provided in [9], and [10] summarizes the latest results on the subject.

As for line features, several structure and motion estimation algorithms that use line correspondences have been proposed in the literature ([11], [12,13,14]). However, the ambiguity issue possibly involved in the use of line correspondences has not been given as much attention as that to the use of point correspondences. A pioneering work was done by Buchanan [15,16] who pointed out that $(3, 6, 5)$-congruence is a critical line set, and suggested a few line sets that defeat the Liu-Huang algorithm [17] of recovering motion and structure from line correspondences. Maybank [18] continued Buchanan's work by choosing to use Semple's representation to describe lines so as to let line congruence be parameterized by surfaces of lower degree. However, Semple's representation does not have a direct way of expressing the projection of 3-D lines to image lines. In [19] Navab and Faugeras addressed the analysis of the critical sets of lines in the Euclidean domain using two equations from Liu and Huang [17], and related the degeneracy analysis to the Liu-Huang algorithm. The treatment was however in terms of geometric primitives of the Euclidean space directly related to individual camera positions and orientations, not the collective quantity trifocal tensor. As discussed earlier, trifocal tensor is a concise quantity that captures how the positions and orientations of all cameras are related to image observations. There is value in treating the degeneracy issue in terms of trifocal tensor, and it is especially so when observables are not direct 3-D but only image data. In particular, as we shall see in this paper, the rank of the tensor estimation matrix involved in the trifocal tensor's determination could be a measure of the degeneracy of the observed line set, and the parallel measure in direct geometric primitives is not as accessible.

The work of Stein and Shashua [20] is among the very few that look at the degeneracy of line structures from the viewpoint of trifocal tensor. The work points out that the tensor estimation matrix has the rank reduced to 23 if the observed line set belong to the structure of linear line complex (LLC).

There are however line structures other than LLC in reality. This work aims at providing a comprehensive treatment of the various critical configurations of line structures that can be present in real scenes, and how they impact the determination of trifocal tensor from image data. The problem is challenging because line correspondences are in many ways less easy to handle than point correspondences, for the reason that their representation, say the Plücker's coordinates, contain nonlinear constraint (Plčker equality) over the coordinates.

3 Preliminaries and Notations

Plücker line coordinates. Given two 3-D points $\mathbf{M^T} \sim (\bar{\mathbf{M}}^\mathbf{T}|m)$ and $\mathbf{N^T} \sim (\bar{\mathbf{N}}^\mathbf{T}|n)$ of a 3-D line to be represented, the Plücker line coordinates of the line are the homogeneous 6-vector $\mathbf{L} \sim (\mathbf{a^T}|\mathbf{b^T})^T$, where $\mathbf{a} = \bar{\mathbf{M}} \times \bar{\mathbf{N}}$, $\mathbf{b} = m\bar{\mathbf{N}} - n\bar{\mathbf{M}}$. The Plücker coordinates also satisfy the Plücker equality $\mathbf{\Omega} = \mathbf{a}^T\mathbf{b} = 0$. As a general 3-D line has four degrees of freedom, with the Plücker representation, a line in 3-D can be represented as a point in $\mathcal{P}^4$ [18], or a point that lies on a non-singular quadric $\mathbf{\Omega}$ in $\mathcal{P}^5$, called the *Klein model* [21].

Camera line projection matrix. Suppose the camera matrix that projects 3-D point to image point is the 3×4 matrix $\mathbf{P} \sim (\bar{\mathbf{P}}|\mathbf{p})$, then the 3×6 line projection matrix [13,14] that projects 3-D line (in the Plücker coordinates as a 6-vector) to image line (in the projective coordinates of $\mathcal{P}^2$ as a 3-vector) is

$$\tilde{\mathbf{P}} \sim [det(\bar{\mathbf{P}})\bar{\mathbf{P}}^{-\mathbf{T}}|[\mathbf{p}]_\mathbf{x}\bar{\mathbf{P}}]. \tag{1}$$

4 Linear Structure

General 3-D lines form a quadruply infinite system $\mathcal{P}^4$ [22]. Overally speaking, specific family of lines in the 4-dimensional line space that possess 1 degree of freedom (DoF), 2 DoFs, and 3 DoFs are respectively called ruled surface, line congruence, and line complex in the terminologies of classical line geometry [21]. In particular, if the constraints that reduce the degrees of freedom of the line set are all linear, the above structures are called *linear ruled surface*, *linear line congruence*, and *linear line complex* respectively, and as a collection they are called **linear line structures**. In this work only line structures with linear constraint(s) are investigated; in other words we only consider linear line structures. On top of the above, if the Plücker equality that governs the 6 line coordinates happens to be linear, then the linear line structure is reduced to the **linear line space**, which has linear ruled surface reduced to the sub-structure *line pencil*, and linear line congruence reduced to *point-star* or *ruled plane*. Due to limitation of space, below we only describe linear line congruence and linear line complex in more details for illustration purpose.

4.1 Linear Line Congruence

Line congruence is defined by two independent equations, as the intersection of two complexes. Linear line congruence is line congruence with the two constraints being linear. It is the intersection of two distinct linear line complexes $\mathbf{L}_1$, $\mathbf{L}_2$. In Plücker coordinates with the norm scaled to be 1, a line $\mathbf{L} \sim (\mathbf{a}^T|\mathbf{b}^T)^T$ in a linear line congruence is bounded by the following: $\mathbf{a}_1^T\mathbf{L} = \mathbf{a}_2^T\mathbf{L} = 0$, $\mathbf{a}^T\mathbf{b} = 0$, $|\mathbf{L}| = 1$, where $\mathbf{a}_1$, $\mathbf{a}_2$ are constant vectors.

4.2 Linear Line Complex

Linear line complex is constrained by a single equation, leaving only 3 DoFs for any line in the line system. Linear line complex is line complex with the governing constraint being linear. It is the 3-dimensional linear manifold of lines defined by one linear equation. In Plücker coordinates with the norm scaled to be 1, a line $\mathbf{L} \sim (\mathbf{a}^T|\mathbf{b}^T)^T$ in a linear line complex is: $\mathbf{a}_1^T\mathbf{L} = 0,\ \mathbf{a}^T\mathbf{b} = 0,\ |\mathbf{L}| = 1$. If the 6-vector $\mathbf{a}_1$ satisfies the Plücker equality ($\boldsymbol{\Omega}_{\mathbf{a}_1,\mathbf{a}_1} = 0$), meaning that $\mathbf{a}_1$ is the Plücker coordinates of a 3-D line, then all lines of the linear line complex (LLC) meet the line $\mathbf{a}_1$ as its axis, and the LLC is called special or singular. Otherwise, the LLC is called general or regular. The degeneracy study of Stein and Shashua in [20] is about special linear line complex.

5 Critical Configurations of Line Structure in 3-View Analysis

In this work we examine how the estimation of the trifocal tensor from an image triplet would behave should the observed line correspondences over the three views be from particular line systems in 3-D. In particular, we examine how the rank of the tensor estimation matrix $\mathbf{A}$ would be affected, and how distant each of the particular line systems is from the necessary system that allows the trifocal tensor to be uniquely determined.

Line Projection. For any 3-D line $\mathbf{L}$ visible in three given views, its image projections $\mathbf{l}, \mathbf{l}', \mathbf{l}''$ in the three views are related to $\mathbf{L}$ in Plücker coordinates by:

$$\bar{\mathbf{K}} = \begin{bmatrix} \mathbf{l} \\ \mathbf{l}' \\ \mathbf{l}'' \end{bmatrix} = \begin{bmatrix} \tilde{\mathbf{P}} \\ \tilde{\mathbf{P}}' \\ \tilde{\mathbf{P}}'' \end{bmatrix} \mathbf{L} \tag{2}$$

where $\tilde{\mathbf{P}}, \tilde{\mathbf{P}}', \tilde{\mathbf{P}}''$ are the three cameras' line projection matrices.

In the above, $\mathbf{L}$ can be taken to include the coordinates of all N 3-D lines visible in the image data. In that case $\mathbf{L}$ will be a $6 \times N$ matrix, $\mathbf{l}, \mathbf{l}', \mathbf{l}''$ will be each a $3 \times N$ matrix, and the entire left side of the above equation, collectively referred to as the $\bar{\mathbf{K}}$ matrix, is about a $9 \times N$ matrix that captures all image line observations in all views of all N lines. As we shall advocate below, this $9 \times N$ observation matrix $\bar{\mathbf{K}}$ is all it takes to define the tensor estimation matrix $\mathbf{A}$ – the key to the determination of the trifocal tensor.

In the rest of paper we also need to access the individual rows of the image line observations $\mathbf{l}$ or $\mathbf{l}'$ or $\mathbf{l}''$ that are visible in View 1, 2, and 3 respectively. We shall define the N-vector $\mathbf{l}_i$ as the ith row of $\mathbf{l}$, and scalar l_{ij} as the jth entry of that row $\mathbf{l}_i$ of $\mathbf{l}$. In other words, $\mathbf{l} = [{\mathbf{l}_1}^T, {\mathbf{l}_2}^T, {\mathbf{l}_3}^T]^T$, and $\mathbf{l}_i = [l_{ij}]$ for all $i = 1, 2, 3$. Similarly, we have N-vector $\mathbf{l}'_i$ and scalar l'_{ij} for image line observations $\mathbf{l}'$ in the second view, and N-vector $\mathbf{l}''_i$ and scalar l''_{ij} for image line observations $\mathbf{l}''$ in the third view.

Tensor Estimation. Each line correspondence over three views gives two independent linear equations for the $3 \times 3 \times 3$ trifocal tensor [5] whose entries can be listed out in a column fashion as a 27×1 vector $\mathbf{t}$. Given N line correspondences, we have $2N$ linear equations for $\mathbf{t}$, which can be collectively captured by a $2N \times 27$ matrix $\mathbf{A}$. The tensor-equivalent vector $\mathbf{t}$ and the matrix $\mathbf{A}$ which we refer to as the *tensor estimation matrix* or simply *estimation matrix* are related by:

$$\mathbf{At} = \mathbf{0}$$

where

$$\mathbf{A} = \begin{bmatrix} & \vdots & \vdots \\ \mathbf{0}_{N\times 9} & l_{3i}\mathbf{B}_i & -l_{2i}\mathbf{B}_i \\ & \vdots & \vdots \\ \vdots & & \vdots \\ -l_{3i}\mathbf{B}_i & \mathbf{0}_{N\times 9} & l_{1i}\mathbf{B}_i \\ \vdots & & \vdots \end{bmatrix} \tag{3}$$

and $\mathbf{B}_i$ is the ith row ($i = 1, 2, \ldots, N$) of the $N \times 9$ matrix $\mathbf{B}$ defined by the image line observations in the following manner:

$$\mathbf{B}_i = [l'_{1i}l''_{1i},\ l'_{1i}l''_{2i},\ l'_{1i}l''_{3i},\ l'_{2i}l''_{1i},\ l'_{2i}l''_{2i},\ l'_{2i}l''_{3i},\ l'_{3i}l''_{1i},\ l'_{3i}l''_{2i},\ l'_{3i}l''_{3i}] \tag{4}$$

Notice that once the line correspondences are established, $\mathbf{B}$ is known, and so is the estimation matrix $\mathbf{A}$.

The rest of the work will at various places need to employ a rather simple property:

Rank Invariance property. Given any matrix $\mathbf{M} = [\mathbf{m}_i]$, where $\mathbf{m}_i$ is the i-th row of $\mathbf{M}$; then $\text{rank}([a_i\mathbf{m}_i]) \leq \text{rank}(\mathbf{M})$ for all set of scalars $a_i \in \mathcal{R}$.

5.1 Linear Line Space

Linear line space includes only line pencil (with 1 DoF in the line set), point-star (2 DoFs in the line set), and ruled-plane (2 DoFs in the line set).

As captured by Eqn. (2), with N 3-D lines included into the 3-D line set matrix $\mathbf{L}$, the image observations will collectively become a $9 \times N$ matrix $\bar{\mathbf{K}}$. Below we work out the rank property of the estimation matrix $\mathbf{A}$ through the rank property of this observation matrix $\bar{\mathbf{K}}$.

Our presentation will be conducted in this manner. We first show that if the $\bar{\mathbf{K}}$ matrix defined by the line correspondences owns either one of a few rank properties, the rank of the estimation matrix $\mathbf{A}$ will be capped accordingly at a certain value. This we present in three separate lemmas, one for line-pencil, another for point-star, and yet another for ruled plane. We then in a theorem show that the required rank properties of the $\bar{\mathbf{K}}$ matrix are indeed there should the observed line structure be a line-pencil, point-star, and ruled plane respectively.

Due to limitation of space we have to omit some of the proofs in this article.

Lemma 1. *For any* $9 \times N$ *matrix* $\mathbf{K} = [\mathbf{k}_1{}^T, \mathbf{k}_2{}^T, \mathbf{k}_3{}^T]^T$, *where* $\mathbf{k}_i$ *is each a* $3 \times N$ *matrix, suppose*

– *rank*$(\mathbf{k}_i) = 2$, *for* $1 \leq i \leq 3$,
– *rank*$([\mathbf{k}_i^T, \mathbf{k}_j^T]^T) = 2$, *for all* i, j *such that* $1 \leq i, j \leq 3$ *and* $i \neq j$,
– *rank*$(\mathbf{K}) = 2$.

Then if $\mathbf{K}$ *is taken as the image line observation matrix* $\bar{\mathbf{K}}$ *(and thus all* $\mathbf{k}_i$, $i = 1, 2, 3$, *as* $\mathbf{l}, \mathbf{l}', \mathbf{l}''$ *respectively) to calculate the estimation matrix* $\mathbf{A}$, *the rank of* $\mathbf{A}$ *is at most* 7.

Proof. The proof consists of 3 parts: one for the first 9 columns of $\mathbf{A}$, the other for its second 9 columns, and the third for its last 9 columns. In each part, we examine how many of the 9 columns are linearly independent and contribute to the overall rank of $\mathbf{A}$. Let $\mathbf{A}_{*,j}$ be the jth column of the estimation matrix $\mathbf{A}$, for all $j = 1, 2, \ldots, 27$.

I **<u>First 9 columns of A:</u>**
Consider the 1st to 9th columns of $\mathbf{A}$, i.e., the columns $\mathbf{A}_{*,1}$ to $\mathbf{A}_{*,9}$.

I(a) As pointed out by Eqn. (3), these columns are all about the $N \times 9$ sub-matrix $[-l_{3i}\mathbf{B}_i]$. Both the three sets of columns $\{\mathbf{A}_{*,1}, \mathbf{A}_{*,2}, \mathbf{A}_{*,3}\}$, $\{\mathbf{A}_{*,4}, \mathbf{A}_{*,5}, \mathbf{A}_{*,6}\}$, and $\{\mathbf{A}_{*,7}, \mathbf{A}_{*,8}, \mathbf{A}_{*,9}\}$ are about $[a_i[\mathbf{l}''_{1,i}, \mathbf{l}''_{2,i}, \mathbf{l}''_{3,i}]]$ for some a_i. By the Rank Invariance property stated above, the rank of each of such 3-column sets is no more than rank$([\mathbf{l}''_{1,i}, \mathbf{l}''_{2,i}, \mathbf{l}''_{3,i}])$ = rank$([\mathbf{l}_1''^T, \mathbf{l}_2''^T, \mathbf{l}_3''^T]^T)$ = rank$(\mathbf{k}_3)$ which is 2 by the first stated rank property of $\mathbf{K}$. In other words, in counting the linearly independent columns, generally we can rule out $\mathbf{A}_{*,3}$, $\mathbf{A}_{*,6}$, and $\mathbf{A}_{*,9}$, and keep only $\mathbf{A}_{*,1}$, $\mathbf{A}_{*,2}$, $\mathbf{A}_{*,4}$, $\mathbf{A}_{*,5}$, $\mathbf{A}_{*,7}$, and $\mathbf{A}_{*,8}$.

I(b) Similar to the above, by the first stated rank property of $\mathbf{K}$ and the rank invariance property, again it can be deduced that rank$([\mathbf{A}_{*,j}, \mathbf{A}_{*,j+3}, \mathbf{A}_{*,j+6}]) \leq$ rank$([\mathbf{l}_1'^T, \mathbf{l}_2'^T, \mathbf{l}_3'^T]^T)$ = rank$(\mathbf{k}_2) = 2$ for $j = 1, 2, 3$. In other words, $\mathbf{A}_{*,7}$ is dependent of $\mathbf{A}_{*,1}$ and $\mathbf{A}_{*,4}$, and $\mathbf{A}_{*,8}$ is dependent of $\mathbf{A}_{*,2}$ and $\mathbf{A}_{*,5}$.
With the above, within the first 9 columns of $\mathbf{A}$, we have at most 4 columns: $\mathbf{A}_{*,1}, \mathbf{A}_{*,2}, \mathbf{A}_{*,4}, \mathbf{A}_{*,5}$, that are linearly independent.

I(c) Next we shall show that $\mathbf{A}_{*,4}$ is linearly independent of $\{\mathbf{A}_{*,1}, \mathbf{A}_{*,2}\}$. Observing the second rank property of $\mathbf{K}$, we can pick $\{\mathbf{l}_1, \mathbf{l}_2\}$ as the basis of $\{\mathbf{l}_1, \mathbf{l}_2, \mathbf{l}_3, \mathbf{l}'_1, \mathbf{l}'_2, \mathbf{l}'_3, \mathbf{l}''_1, \mathbf{l}''_2, \mathbf{l}'_3\}$. In other words, we have

$$\begin{aligned}\mathbf{l}'_1 &= c_1\mathbf{l}_1 + c_2\mathbf{l}_2\\ \mathbf{l}'_2 &= c_3\mathbf{l}_1 + c_4\mathbf{l}_2\\ \mathbf{l}''_1 &= c_5\mathbf{l}_1 + c_6\mathbf{l}_2\\ \mathbf{l}''_2 &= c_7\mathbf{l}_1 + c_8\mathbf{l}_2\end{aligned} \tag{5}$$

for some $c_i \in \mathbf{R}$.
Suppose that $\mathbf{A}_{*,4}$ is linearly dependent of $\{\mathbf{A}_{*,1}, \mathbf{A}_{*,2}\}$. Then there should exist $X, Y \in \mathbf{R}$ such that $\mathbf{A}_{*,4} = X\mathbf{A}_{*,1} + Y\mathbf{A}_{*,2}$ or simply

$$[l'_{2i}l''_{1i}] = X[l'_{1i}l''_{1i}] + Y[l'_{1i}l''_{2i}]$$

where $[*]$ is a vector with $*$ as its column entry. Using Eqn. (5) to expand the above in terms of $\mathbf{l}_{1i}, \mathbf{l}_{2i}$ and equating the coefficients of both sides, the above equation can be reduced to

$$\begin{bmatrix} c_1c_5 & c_1c_7 \\ c_1c_6 + c_2c_5 & c_1c_8 + c_2c_7 \\ c_2c_6 & c_2c_8 \end{bmatrix} \begin{bmatrix} X \\ Y \end{bmatrix} = \begin{bmatrix} c_3c_5 \\ c_3c_6 + c_4c_5 \\ c_4c_6 \end{bmatrix}$$

which is an over-constraining set of equations for X, Y, showing that X, Y do not exist in general – a contradiction to the supposition. Thus $\mathbf{A}_{*,4}$ is independent of $\mathbf{A}_{*,1}, \mathbf{A}_{*,2}$.

I(d) By a similar "Proof by Contradiction" step, we can show that $\mathbf{A}_{*,5}$ is linearly dependent of $\{\mathbf{A}_{*,1}, \mathbf{A}_{*,2}, \mathbf{A}_{*,4}\}$.

Putting together all the above linear dependencies, $\mathbf{A}_{*,3}, \mathbf{A}_{*,5}, \mathbf{A}_{*,6}, \mathbf{A}_{*,7}, \mathbf{A}_{*,8}, \mathbf{A}_{*,9}$ are linearly dependent of $\{\mathbf{A}_{*,1}, \mathbf{A}_{*,2}, \mathbf{A}_{*,4}\}$, and $\text{rank}([\mathbf{A}_{*,1}, \dots, \mathbf{A}_{*,9}]) \leq 3$.

II <u>**Second 9 columns of A:**</u>
$\{\mathbf{A}_{*,1}, \dots, \mathbf{A}_{*,9}\}$ and $\{\mathbf{A}_{*,10}, \dots, \mathbf{A}_{*,18}\}$ are all about $l_{3i}\mathbf{B}_i$. Thus by the same reason as the first 9 columns of **A**, we have $\text{rank}([\mathbf{A}_{*,10}, \dots, \mathbf{A}_{*,18}]) = \text{rank}([\mathbf{A}_{*,1}, \dots, \mathbf{A}_{*,9}]) \leq 3$.
$\{\mathbf{A}_{*,1}, \mathbf{A}_{*,2}, \mathbf{A}_{*,4}, \mathbf{A}_{*,10}, \mathbf{A}_{*,11}, \mathbf{A}_{*,13}\}$ can be picked as the basis of the first 18 columns of **A**.

III <u>**Last 9 columns of A:**</u>
Again by the rank invariance property, the rank of the $2N \times 9$ matrix $\begin{bmatrix} [-l_{2i}\mathbf{B}_i] \\ [l_{1i}\mathbf{B}_i] \end{bmatrix}$ is no more than the rank of $[\mathbf{B}_i]$. In other words, the rank property of $\{\mathbf{A}_{*,19}, ..., \mathbf{A}_{*,27}\}$ is the same as that of $\{\mathbf{A}_{*,1}, ..., \mathbf{A}_{*,9}\}$ and $\{\mathbf{A}_{*,10}, ..., \mathbf{A}_{*,18}\}$. Thus, from the proof in Step **I**, $\mathbf{A}_{*,21}, \mathbf{A}_{*,23}, ..., \mathbf{A}_{*,27}$ are linearly dependent of $[\mathbf{A}_{*,19}, \mathbf{A}_{*,20}, \mathbf{A}_{*,22}\}$, and $\text{rank}([\mathbf{A}_{*,19}, \dots, \mathbf{A}_{*,27}]) \leq 3$.

Finally, we need to determine the linear dependency between the 3 column vectors $\mathbf{A}_{*,19}, \mathbf{A}_{*,20}, \mathbf{A}_{*,22}$ and the basis of $\{\mathbf{A}_{*,1}, \dots, \mathbf{A}_{*,18}\}$. Again, "Proof by contradiction" schemes similar to those above can be used to show that $\mathbf{A}_{*,19}$ is independent of $\{\mathbf{A}_{*,1}, \mathbf{A}_{*,2}, \mathbf{A}_{*,4}, \mathbf{A}_{*,10}, \mathbf{A}_{*,11}, \mathbf{A}_{*,13}\}$, but $\{\mathbf{A}_{*,20}, \mathbf{A}_{*,22}\}$ is dependent of $\{\mathbf{A}_{*,1}, \mathbf{A}_{*,2}, \mathbf{A}_4, \mathbf{A}_{*,10}, \mathbf{A}_{*,11}, \mathbf{A}_{*,13}, \mathbf{A}_{*,19}\}$. Due to page limit, we omit the elaboration of such steps here. To conclude, the rank of **A** is at most 7.

Lemma 2. *For any* $9 \times N$ *matrix* $\mathbf{K} = [\mathbf{k}_1{}^T\ \mathbf{k}_2{}^T\ \mathbf{k}_3{}^T]^T$, *suppose*

- $rank(\mathbf{k}_i) = 2$, *for* $1 \leq i \leq 3$,
- $rank([\mathbf{k}_i,\ \mathbf{k}_j]^T) = 3$, *for* $1 \leq i, j \leq 3$ *and* $i \neq j$,
- $rank(\mathbf{K}) = 3$.

Then if **K** *is taken as the image line observation matrix* $\bar{\mathbf{K}}$ *(and thus all* $\mathbf{k}_i$, $i = 1, 2, 3$, *as* $\mathbf{l}, \mathbf{l}', \mathbf{l}''$ *respectively) to calculate the estimation matrix* **A**, *the rank of* **A** *is at most* 11.

Lemma 3. *For any* $9 \times N$ *matrix* $\mathbf{K} = [\mathbf{k}_1{}^T\ \mathbf{k}_2{}^T\ \mathbf{k}_3{}^T]^T$, *suppose*

- *rank*$(\mathbf{k}_i) = 3$, *for* $1 \leq i \leq 3$,
- *rank*$([\mathbf{k}_i\ \mathbf{k}_j]^T) = 3$, *for* $1 \leq i \leq 3$, $1 \leq j \leq 3$ *and* $i \neq j$,
- *rank*$(\mathbf{K}) = 3$.

Then if $\mathbf{K}$ *is taken as the image line observation matrix* $\bar{\mathbf{K}}$ *(and thus all* $\mathbf{k}_i$, $i = 1, 2, 3$, *as* $\mathbf{l}, \mathbf{l}', \mathbf{l}''$ *respectively) to calculate the estimation matrix* $\mathbf{A}$, *the rank of* $\mathbf{A}$ *is at most* 15.

Theorem 1. *The rank of tensor estimation matrix is at most* 7, 11, 15 *for entities in the various structure classes of linear line space: at least* 4, 6, 8 *lines for line pencil, point-star, and ruled plane respectively.*

Proof. Eqn. (2) relates the image observations $[\mathbf{l}^T, \mathbf{l}'^T, \mathbf{l}''^T]^T$ to the 3-D line set $\mathbf{L}$ through the cameras' line projection matrices $\tilde{\mathbf{P}}, \tilde{\mathbf{P}}', \tilde{\mathbf{P}}''$. The relationship leads to the fact that

$$\text{rank}(\begin{bmatrix} \mathbf{l} \\ \mathbf{l}' \\ \mathbf{l}'' \end{bmatrix}) \leq \min(\text{rank}(\begin{bmatrix} \tilde{\mathbf{P}} \\ \tilde{\mathbf{P}}' \\ \tilde{\mathbf{P}}'' \end{bmatrix}), \text{rank}(\mathbf{L})) \tag{6}$$

Since $[\tilde{\mathbf{P}}^T, \tilde{\mathbf{P}}'^T, \tilde{\mathbf{P}}''^T]^T$ is of full column rank (=6) in general, we have rank $([\mathbf{l}^T, \mathbf{l}'^T, \mathbf{l}''^T]^T) = \text{rank}(\mathbf{L})$.

<u>**Line Pencil:**</u> $\text{rank}(\mathbf{L}) = 6 - 3 - 1 = 2$, with the "3" coming from three linear constraints on $\mathbf{L}$ and the "1" from the Plücker equality (which is degraded to be linear, or else the structure does not belong to linear line space). By the assumption that the projection is general, plus Eqn. (6), it can be deduced that $\text{rank}([\mathbf{l}^T, \mathbf{l}'^T, \mathbf{l}''^T]^T]) = 2$. Since $\text{rank}(\mathbf{l}) = \text{rank}(\mathbf{l}') = \text{rank}(\mathbf{l}'') = 2$, we have $\text{rank}([\mathbf{l}^T, \mathbf{l}'^T]^T) = \text{rank}([\mathbf{l}^T, \mathbf{l}''^T]^T) = \text{rank}([\mathbf{l}'^T, \mathbf{l}''^T]^T) = 2$. All constraints on image lines have thus been considered, and they give rise to the rank properties of $\bar{\mathbf{K}}$ required by Lemma 1. By Lemma 1, the rank of $\mathbf{A}$ is at most 7 for line pencil. As each line correspondence provides 2 linear constraints on $\mathbf{A}$, the minimum number of line correspondences needed is $\lceil 7/2 \rceil = 4$. Similar proofs can be derived for point star and ruled plane.

In the following subsections we tackle the cases of linear ruled surface, linear line congruence, and linear line complex. We adopt a presentation similar to that for linear line space.

5.2 Linear Ruled Surface

Theorem 2. *The rank of tensor estimation matrix is at most* 12 *if the available 3-view observations are about a set of at least* 6 *lines that belong to a general linear ruled surface.*

5.3 Linear Congruence

Lemma 4. *For any* $9 \times N$ *matrix* $\mathbf{K} = [\mathbf{k}_1{}^T, \mathbf{k}_2{}^T, \mathbf{k}_3{}^T]^T$*, suppose*

- $rank(\mathbf{k}_i) = 3$*, for* $1 \leq i \leq 3$*,*
- $rank([\mathbf{k}_i, \mathbf{k}_j]^T) = 4$ *for* $1 \leq i \leq 3$*,* $1 \leq j \leq 3$ *and* $i \neq j$*,*
- $rank(\mathbf{K}) = 4$.

Then if $\mathbf{K}$ *is taken as the image line observation matrix* $\bar{\mathbf{K}}$ *(and thus all* $\mathbf{k}_i$*,* $i = 1, 2, 3$*, as* $\mathbf{l}, \mathbf{l}', \mathbf{l}''$ *respectively) to calculate the estimation matrix* $\mathbf{A}$*, the rank of* $\mathbf{A}$ *is at most* 22.

Theorem 3. *The rank of tensor estimation matrix is at most* 19 *if the image observables are about a set of at least* 10 *lines that belong to a general linear congruence.*

Theorem 2 and 3 can be proved with aid from Lemma 3 and 4 respectively.

5.4 Linear Line Complex

Lemma 5. *For any* $9 \times N$ *matrix* $\mathbf{K} = [\mathbf{k}_1{}^T, \mathbf{k}_2{}^T, \mathbf{k}_3{}^T]^T$*, suppose*

- $rank(\mathbf{k}_i) = 3$*, for* $1 \leq i \leq 3$*,*
- $rank([\mathbf{k}_i, \mathbf{k}_j]^T) = 5$ *for* $1 \leq i \leq 3$*,* $1 \leq j \leq 3$ *and* $i \neq j$*,*
- $rank(\mathbf{K}) = 5$.

Then if $\mathbf{K}$ *is taken as the image line observation matrix* $\bar{\mathbf{K}}$ *(and thus all* $\mathbf{k}_i$*,* $i = 1, 2, 3$*, as* $\mathbf{l}, \mathbf{l}', \mathbf{l}''$ *respectively) to calculate the estimation matrix* $\mathbf{A}$*, the rank of* $\mathbf{A}$ *is at most* 26.

Theorem 4. *The rank of tensor estimation matrix is at most* 23 *if the image observables are about a set of at least* 12 *lines that belong to a general linear line complex.*

Proof. For linear line complex, $rank(\mathbf{L}) = 6 - 1 = 5$, with 1 coming from one linear constraint. From Eqn. (6), since projection is assumed to be general, $rank([\mathbf{l}^T, \mathbf{l}'^T, \mathbf{l}''^T]^T) = 5$. From $rank(\mathbf{l}) = rank(\mathbf{l}') = rank(\mathbf{l}'') = 3$, by Lemma 5, it can be concluded that the rank of $\mathbf{A}$ is at most 26 for linear line complex.

However, the above has not yet used the Plücker equality. Even though the Plücker equality is generally nonlinear here, it can be shown that by projecting $\mathbf{L}$ to images 1 and 2, 1 and 3, 2 and 3 respectively, the Plücker equality provides three additional independent linear equations on $\{l_{in}l'_{jn}l''_{kn}\}, i, j, k = 1, 2, 3, i \neq j \neq k, 1 \leq n \leq N$. Thus, three DoFs have to be deducted and the rank of $\mathbf{A}$ is at most $26 - 3 = 23$. As each line correspondence provides 2 linear constraints on T, the minimum number of line correspondences needed is $\lceil 23/2 \rceil = 12$. □

An alternative proof has been presented by Stein and Shashua in [20]. The proof here is of a more general nature because it applies not only to special linear line complex but general linear line complex.

6 Conclusion and Future Work

We have supplied a classification of the geometry of a set of lines in space in regard to the determination of trifocal tensor from three views of the lines. In particular, we have presented what rank the estimation matrix would have if the line set is a sub-class (linear ruled surface, linear line congruence, or linear line complex) of linear line structure, or more specifically a sub-structure (point-pencil, point-star, or ruled plane) of linear line space. Trifocal tensor determination is crucial to motion estimation and projective reconstruction, and ambiguity would be present if a less than general set of lines are unfortunately all that can be observed. Notice that in this work not only is the presence of ambiguity pointed out, the extent of the ambiguity is also specified. Such a study over a list of major line structures is unprecedented.

The whole concept can also be looked at from the opposite angle. By examining the rank property of the tensor estimation matrix, one can have a glimpse of what possible line structure the cameras are indeed observing even if it is an under-ranked case. And this can be done without even attempting motion estimation, 3-D structure recovery, or trifocal tensor determination.

We have conducted thorough synthetic and real image data experiments, and they all confirmed the validity of the findings.

Possible future work includes extension of the rank property study for non-linear line structures, and even different camera motions.

Acknowledgment

The work described in this paper was partially supported by a grant from the Research Grants Council of the Hong Kong Special Administrative Region, China (Project No. CUHK4195/04E). This work is affiliated with the Microsoft-CUHK Joint Laboratory for Human-centric Computing and Interface Technologies.

References

1. Navab, N., Faugeras, O.: Monocular pose determination from lines: Critical sets and maximum number of solutions. In: IEEE Conference on Computer Vision and Pattern Recognition, pp. 254–260 (1993)
2. Leung, M., Liu, Y., Huang, T.: Estimating 3d vehicle motion in an outdoor scene from monocular and stereo image sequences. In: IEEE Workshop on Visual Motion, October 7-9, 1991, vol. 91, pp. 62–68 (1991)
3. Liu, Y.: Rigid object motion estimation from intensity images using straight line correspondences. Ph.D. Thesis (1990)
4. Faugeras, O., Mourrain, B.: On the geometry and algebra of the point and line correspondences between n images. In: International Conference on Computer Vision, Washington, DC, USA, p. 951. IEEE Computer Society, Los Alamitos (1995)
5. Hartley, R.I., Zisserman, A.: Multiple View Geometry in Computer Vision, 2nd edn. Cambridge University Press, Cambridge (2004)

6. Shashua, A., Maybank, S.: Degenerate n point configurations of three views: Do critical surfaces exist. Technical Report 96-19, Hebrew Univ. of Jerusalem (November 1996)
7. Hartley, R.I.: Ambiguous configurations for 3-view projective reconstruction. In: Proceedings of the 6th European Conference on Computer Vision-Part I, London, UK, pp. 922–935. Springer, Heidelberg (2000)
8. Kahl, F., Hartley, R., Astrom, K.: Critical configurations for n-view projective reconstruction. In: Proceedings of the Computer Vision and Pattern Recognition, vol. 2, p. 158. IEEE Computer Society, Los Alamitos (2001)
9. Hartley, R., Kahl, F.: A critical configuration for reconstruction from rectilinear motion. In: Proceedings of the IEEE Conference on Computer Vision and Pattern Recognition, vol. 1, p. 511. IEEE Computer Society, Los Alamitos (2003)
10. Hartley, R., Kahl, F.: Critical configurations for projective reconstruction from multiple views. Int. J. Comput. Vision 71(1), 5–47 (2007)
11. Weng, J., Huang, T., Ahuja, N.: Motion and structure from line correspondences: Closed-form solution, uniqueness, and optimization. IEEE Transactions on Pattern Analysis and Machine Intelligence 14(3), 318–336 (1992)
12. Bartoli, A., Sturm, P.: The 3d line motion matrix and alignment of line reconstructions. Int. J. Comput. Vision 57(3), 159–178 (2004)
13. Bartoli, A., Hartley, R.I., Kahl, F.: Motion from 3d line correspondences: Linear and non-linear solutions. In: Proceedings of the IEEE Conference on Computer Vision and Pattern Recognition, vol. 1, p. 477. IEEE Computer Society, Los Alamitos (2003)
14. Bartoli, A., Sturm, P.: Structure-from-motion using lines: representation, triangulation, and bundle adjustment. Comput. Vis. Image Underst. 100(3), 416–441 (2005)
15. Buchanan, T.: Critical sets for 3d reconstruction using lines. In: Sandini, G. (ed.) ECCV 1992. LNCS, vol. 588, pp. 730–738. Springer, Heidelberg (1992)
16. Buchanan, T.: On the critical set for photogrammetric reconstruction using line tokens in p3(c). Geometriae Dedicata 44, 223–232 (1992)
17. Liu, Y., Huang, T.: A linear algorithm for motion estimation using straight line correspondences. In: Proceedings of the International Conference on Pattern Recognition, pp. I: 213–219 (1988)
18. Maybank, S.J.: The critical line congruence for reconstruction from three images. Applicable Algebra in Engineering, Communication and Computing 6, 89–113 (1993)
19. Navab, N., Faugeras, O.: The critical sets of lines for camera displacement estimation: A mixed euclidean-projective and constructive approach. Int. J. Comput. Vision 23, 17–44 (1997)
20. Stein, G.P., Shashua, A.: On degeneracy of linear reconstruction from three views: Linear line complex and applications. IEEE Trans. Pattern Anal. Mach. Intell. 1407, 862–878 (1998)
21. Helmut Pottmann, J.W.: Computational line geometry. Mathematics and visualization. Springer, Berlin (2001)
22. Semple, J., Kneebone, G.: Algebraic projective geometry. Clarendon Press, Oxford University Press (1952)

Learning Visual Shape Lexicon for Document Image Content Recognition

Guangyu Zhu, Xiaodong Yu, Yi Li, and David Doermann

University of Maryland, College Park, MD 20742, USA

Abstract. Developing effective content recognition methods for diverse imagery continues to challenge computer vision researchers. We present a new approach for document image content categorization using a lexicon of shape features. Each lexical word corresponds to a scale and rotation invariant shape feature that is generic enough to be detected repeatably and segmentation free. We learn a concise, structurally indexed shape lexicon from training by clustering and partitioning feature types through graph cuts. We demonstrate our approach on two challenging document image content recognition problems: 1) The classification of 4,500 Web images crawled from Google Image Search into three content categories — pure image, image with text, and document image, and 2) Language identification of 8 languages (Arabic, Chinese, English, Hindi, Japanese, Korean, Russian, and Thai) on a 1,512 complex document image database composed of mixed machine printed text and handwriting. Our approach is capable to handle high intra-class variability and shows results that exceed other state-of-the-art approaches, allowing it to be used as a content recognizer in image indexing and retrieval systems.

1 Introduction

Image content categorization has become a pressing problem in computer vision as we are facing phenomenal increase in the diversity of visual content. Content category recognition aims to reduce the semantic gap for ensuing tasks by providing usage-oriented content description that can be exploitable in individual applications. For vision systems involving high-volume, complex, and heterogeneous multimedia data, effective high-level content interpretation is essential prior to object detection or object category recognition at a finer level.

One pervasive form of information content is text. Once text content and the language are recognized, images containing text can be processed by an optical character recognition (OCR) system and indexed. Text-oriented content recognition provides a reliable alternative to object detection and recognition in a wide range of applications. Towards this end, however, there are a number of challenges to image content category recognition that are largely unsolved. In this paper, we focus on two important problems that are central to heterogeneous document image collections.

First, we consider the recognition of primary content of a general image as one of three content categories — pure image (*e.g.*natural image and human photos),

D. Forsyth, P. Torr, and A. Zisserman (Eds.): ECCV 2008, Part II, LNCS 5303, pp. 745–758, 2008.

Fig. 1. Examples of images returned by Google Image using the keyword "CD cover"

image with text (see examples in Fig. 1), or document image. Automated content categorization like this has a big impact in image search, and content-based image indexing and retrieval.

Second, we address the problem of recognizing the primary language of a document image in an unconstrained setting. This is a fundamental research challenge currently facing systems that need to automatically process diverse multilingual document images, such as Google Book Search [30] or an automated global expense reimbursement application [32], because almost all existing work on OCR requires that the script and/or language of the processed document be known [23]. The performance of language identification is crucial for the success of a broad range of tasks — from determining the correct OCR engine for text extraction to document indexing, translation, and search [33]. Progress in the field of language identification has focused almost exclusively on machine printed text. Document collections, however, often contain a diverse and complex mixture of machine printed and unconstrained handwritten content, and vary tremendously in font and style. Language identification on document images involving diverse content types, including unconstrained handwriting, is still an open research area [22] and to our best knowledge, no reasonable solutions have been presented in the literature.

We propose a novel approach for document image content recognition using image descriptors built from a lexicon of generic low-level shape features that are translation, scale, and rotation invariant. To construct a structural index among large number of features extracted from diverse content, we dynamically partition the space of shape primitives by clustering similar feature types. We formulate feature partitioning as a graph cuts problem with the objective to obtain a concise and globally balanced lexicon index by sampling from training data. Each cluster in the lexicon is represented by an exemplary lexical word, making association of feature type efficient. We obtain very competitive document image content categorization performance using a multi-class SVM classifier.

The structure of this paper is as follows. Section 2 reviews related work. In Section 3, we describe the algorithm for learning the shape lexicon and present a document image content recognition approach using the shape lexicon. We discuss experimental results in Section 4 and conclude in Section 5.

2 Related Work

In this following, we first review contour learning approaches, and then point out work related to us on whole-image categorization. We present a comprehensive overview of existing on script and language identification techniques and discuss their limitations on document images with unconstrained content.

2.1 Contour-Based Learning

Topological relationships among adjacent contours is an important aspect in many vision problems and have been exploited in work on object detection [1,9,24], and on perceptual grouping [12,17]. Ferrari *et al.* [8] proposed scale-invariant adjacent segments (kAS) features extracted from the contour segment network of image tiles, and use them in a sliding window scheme for object detection. By explicitly encoding both geometric and spatial arrangement among the segments, kAS descriptor demonstrates state-of-the-art performance in shape-based object detection, and outperforms descriptors based on interest points and histograms of gradient orientations [6]. However, kAS descriptor is not rotation invariant, because segments are rigidly ordered from left to right. This limits the repeatability of high-order kAS, and the best performance in [8] is reported when using 2AS.

2.2 Image Content Categorization

There has been little computer vision and image processing literature directly addressing the problem of content recognition for heterogeneous image repositories. However, several approaches based on different motivations have demonstrated good performance in tasks that involve diverse objects. Oliva and Torralba [21] developed a holistic image representation called Spatial Envelope for scene recognition using a set of discriminative energy spectrum templates that characterizes the dominant spatial structure of a scene. Another fairly intuitive approach is to treat blocks of text as texture [29]. One widely used rotation invariant feature for texture analysis is the Local Binary Patterns (LBP) proposed by Ojala *et al.* [20], which effectively captures spatial structure of local image texture in circular neighborhoods across angular space and resolution.

2.3 Language Identification

Prior research on script and language identification has largely focused on the domain of machine printed document images. These works can be broadly classified into three categories — statistical analysis of text lines [7,14,18,27,28], texture analysis [4,29], and template matching [11].

Statistical analysis using discriminating features extracted from text lines, including distribution of upward concavities [14,27], horizontal projection profile [7,28], and vertical cuts of connected components [18], has shown to be effective on homogeneous collection of printed documents. These approaches, however, do have a few major limitations. First, they are based on the assumption of uniformity among printed text, and require precise baseline alignment and word segmentation. Freestyle handwritten text lines are curvilinear, and in general, there are no well-defined baselines, even by linear or piecewise-linear approximation [15]. Second, it is difficult to extend these methods to a new language, because they employ a combination of hand-picked and trainable features and a variety of decision rules. In fact, most of these approaches require effective script identification to discriminate between selected subset of languages, and use different feature sets for script and language identification, respectively.

Script identification using rotation invariant multi-channel Gabor filters [29] and wavelet log co-occurrence features [4] were demonstrated on small blocks of printed text with similar characteristics. However, no results were reported on full-page documents that involve variations in layouts and fonts.

The current state-of-the-art script/language identification system was developed by Hochberg *et al.* [11] at Los Alamos National Laboratory, and is able to process 13 machine printed scripts without explicit assumptions of distinguishing characteristics for a selected subset of languages. The system determines the most likely script on a document page by probabilistic voting on matched templates. Each template pattern is of fixed size and is rescaled from a cluster of connected components. Template matching is intuitive and delivers impressive performance when the content is constrained (*i.e.*printed text in similar fonts). However, templates are not flexible enough to generalize across large variations in fonts or handwriting styles that are typical in diverse datasets [11]. From a practical point of view, the system also has to learn the discriminability of each template through labor-intensive training to achieve the best performance. This requires tremendous amount of supervision and further limits applicability.

3 Recognizing Image Content Using Shape Lexicon

Recognition of diverse visual content needs to account for large variations, because content appears in many forms and in many different contexts today. The scope of the problem is intuitively in favor of low-level shape primitives that can be detected repeatably. Rather than focusing on selection of class-specific features, our approach aims to distinguish intricate differences between content types collectively using the statistics of a large variety of generic, geometrically invariant feature types (lexical words) that are structurally indexed. Our emphasis on the generic nature of lexical words provide a different perspective to recognition that has traditionally focused on finding sophisticated features or visual selection models, which may limit generalization performance.

We explore the kAS contour feature recently introduced by Ferrari *et al.* [8], which consists of a chain of k roughly straight, connected contour segments. Specifically, we focus on the case of triple contour segments, which strike a balance between lower-order contour features that are not discriminative enough and higher-order ones that are less likely to be detected robustly.

3.1 Extraction of Contour Feature

Feature detection in our approach is very efficient since we perform computation locally. First, we compute edges using the Canny edge detector [5], which

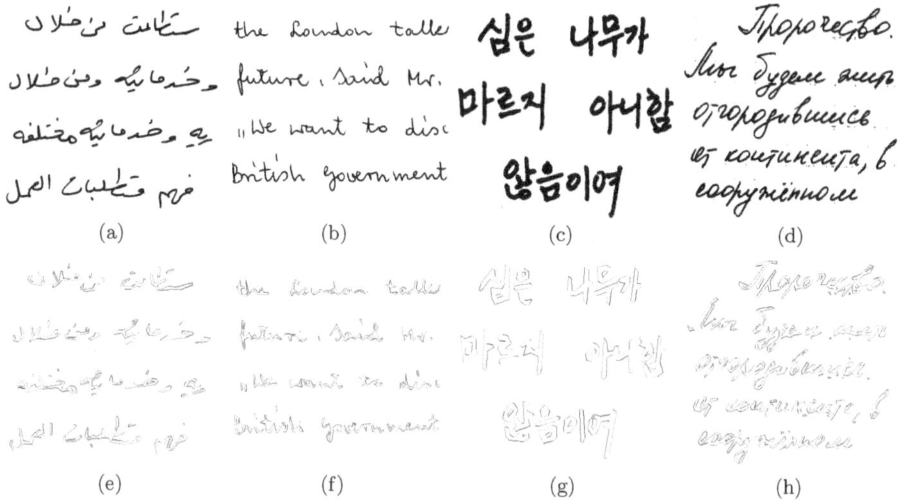

Fig. 2. Visual shape differences are captured locally by a large variety of neighboring contour features. (a)-(d) Examples of handwriting from four different languages. (e)-(h) Detected contour features by our approach, each shown in a random color.

consistently demonstrates good performance on text content and gives precise localization and unique response. Second, we group contour segments by connected components and fit them locally into line segments. Then, within each connected component, every triplet of connected line segments that starts from the current segment is extracted. Fig. 2 provides visualization of the quality of detected contour features by our approach using random colors.

Our feature detection scheme requires only linear time and space in the number of contour fragments n, and is highly parallelizable. It is much more efficient and stable than [8], which requires construction of contour segment network and depth first search from each segment, leading to $O(n \log(n))$ time on average and $O(n^2)$ in the worst case.

We encode object contours in a translation, scale, and rotation invariant fashion by computing orientations and lengths with reference to the first detected line segment. A contour feature C can be compactly represented by an ordered set of lengths and orientations of c_i for $i \in \{1, 2, 3\}$, where c_i denotes line segment i in C. This is distinct from the motivation of kAS descriptor that attempts to enumerate spatial arrangements of contours within local regions. Furthermore, kAS descriptor does not take into account of rotation invariance.

3.2 Measure of Dissimilarity

The overall dissimilarity between two contour features can be quantified by the weighted sum of the distances in lengths and orientations. We use the following generalized measure of dissimilarity between two contour features C_a and C_b

$$d^{(C_a, C_b, \boldsymbol{\lambda})} = \boldsymbol{\lambda}_{\text{length}}^T \mathbf{d}_{\text{length}} + \boldsymbol{\lambda}_{\text{orient}}^T \mathbf{d}_{\text{orient}}, \tag{1}$$

where vectors $\mathbf{d}_{\text{length}}$ and $\mathbf{d}_{\text{orient}}$ are composed of the distances between contour lengths and orientations, respectively. $\boldsymbol{\lambda}_{\text{length}}$ and $\boldsymbol{\lambda}_{\text{orient}}$ are their corresponding weight vectors, providing sensitivity control over the tolerance of line fitting. One natural measure of dissimilarity in lengths between two contour segments is their log ratio. We compute orientation difference between two segments by normalizing their absolute value of angle difference to π. In our experiments, we use a larger weighting factor for orientation to de-emphasize the difference in the lengths because they may be less accurate due to line fitting.

3.3 Learning the Shape Lexicon

We extract a large number of lexical words by sampling from training images, and construct an indexed shape lexicon by clustering and partitioning the lexical words. A lexicon provides a concise structural organization for associating large varieties of low-level features, and is efficient because it enables comparison to much fewer feature types.

Clustering Lexical Words. Prior to clustering, we compute the distance between each pair of lexical words and construct a weighted undirected graph $\mathcal{G} = (V, E)$, in which each node on the graph represents a word. The weight on an edge connecting two nodes C_a and C_b is defined as a function of their distance

$$w(C_a, C_b) = \exp(-\frac{d(C_a, C_b)^2}{\sigma_d^2}), \tag{2}$$

where we set parameter σ_d to 20 percent of the maximum distance among all pairs of nodes.

We formulate feature clustering as a spectral graph partitioning problem, for which we seek to group the set of vertices V into disjoint sets $\{V_1, V_2, \ldots, V_K\}$, such that by the measure defined in (1) the similarity among the vertices in a set is high and that across different sets is low.

More concretely, let the $N \times N$ symmetric weight matrix for all the vertices be W, where $N = |V|$. We define the degree matrix D as an $N \times N$ diagonal matrix, whose i-th element $d(i)$ along the diagonal satisfies $d(i) = \sum_j w(i,j)$. We use an $N \times K$ matrix X to represent a graph partition, *i.e.*$X = [X_1, X_2, \ldots, X_K]$, where each element of matrix X is either 0 or 1. We can show that the feature clustering formulation that seeks globally balanced graph partitions is equivalent to the normalized cuts criterion [26], and can be written as

$$\text{maximize } \epsilon(X) = \frac{1}{K} \sum_{l=1}^{K} \frac{X_l^T W X_l}{X_l^T D X_l}, \tag{3}$$

$$\text{subject to } X \in \{0,1\}^{N \times K}, \text{ and } \sum_j X(i,j) = 1. \tag{4}$$

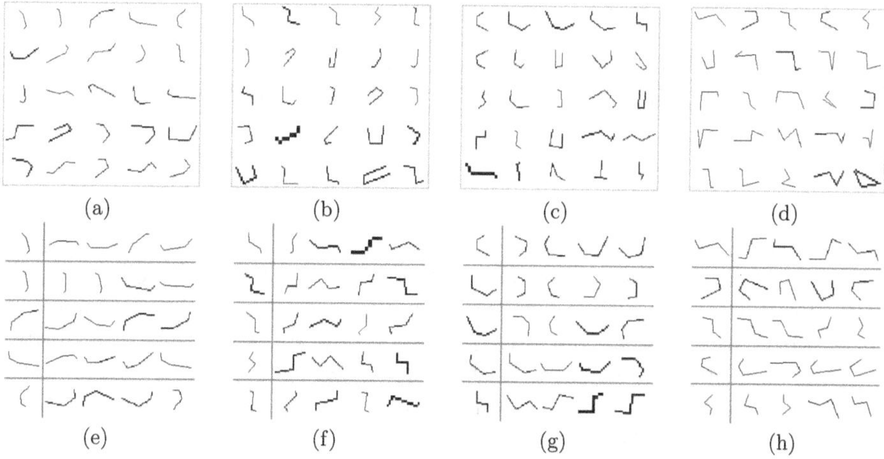

Fig. 3. The 25 most frequent exemplary lexical words in (a) Arabic, (b) Chinese, (c) English, and (d) Hindi document images, which very well capture the distinct features between languages. (e)-(h) show lexical words in the same cluster as the top 5 exemplary lexical words for each language, ordered by ascending distances to the center of their clusters. Scaled and rotated versions of feature types are clustered together.

Minimizing normalized cuts exactly is NP-complete. We use a fast algorithm [31] for finding its discrete near-global optima, which is robust to random initialization and converges faster than other clustering methods.

Organizing Features in the Lexicon. For each cluster, we select the feature instance closest to the center of the cluster as the exemplary lexical word. This ensures that an exemplary word has the smallest sum of squares distance to the other features within the cluster. In addition, each exemplary word is associated with a cluster radius, which is defined as the maximum distance from the cluster center to all the other features within the cluster. The constructed shape lexicon $\mathcal{L}$ is composed of all exemplary lexical words.

Fig. 3 shows the 25 most frequent exemplary lexical words for Arabic, Chinese, English, and Hindi, which are learned from 10 documents of each language. Distinguishing features between languages, including cursive style in Arabic, 45 and 90-degree transitions in Chinese, and various configurations due to long horizontal lines in Hindi, are learned automatically. Each row in Fig. 3(e)-(h) lists examples of lexical words in the same cluster, ordered by ascending distances to the center of their associated clusters. Through clustering, translated, scaled and rotated versions of feature types are grouped together.

Since each lexical word represents a generic local shape feature, intuitively a majority of lexical words should appear in images across content categories, even though their frequencies of occurrence deviate significantly. In our experiments, we find that 95.1% and 92.6% of lexical words in natural images also appear in images with text and document images, respectively. In addition, 86.3% of lexical word instances appear in document images across all 8 languages.

3.4 Constructing the Image Descriptor

We construct a shape descriptor for each image, which provides statistics of the frequency at which each feature type occurs. For each detected lexical word W from the test image, we compute the nearest exemplary word C_k in the shape lexicon. We increment the descriptor entry corresponding to C_k only if

$$d(W, C_k) < r_k, \tag{5}$$

where r_k is the cluster radius associated with the exemplary lexicon word C_k. This quantization step ensures that unseen features that deviate considerably from training features are not used for image description. In our experiments, we found that only less than 2% of the contour features cannot be found in the shape lexicon learned from the training data.

4 Experimental Results

Before we demonstrate our approach in image content categorization and language identification, we first quantitatively evaluate the discriminative power and geometrical invariance of shape lexicon in two object category recognition experiments using public shape databases.

4.1 Shape-Based Object Recognition

Kimia Database. One widely used shape database is the Kimia dataset [25], which contains 25 images from 6 categories. Several categories require rotation invariant matching for effective recognition. It has been tested in [25,10,2,16]. In this experiment, we use the χ^2 statistic as the measure of dissimilarity between two image descriptors, and the distance $D_{\chi^2}(h_a, h_b)$ between two normalized K-bin image descriptors h_a and h_b is defined as

$$D_{\chi^2}(h_a, h_b) = \frac{1}{2} \sum_{k=1}^{K} \frac{[h_a(k) - h_b(k)]^2}{h_a(k) - h_b(k)}. \tag{6}$$

Historically, results on Kimia database are reported as the number of 1st, 2nd, and 3rd nearest-neighbors that fall into the correct category. Our result is 25/25, 25/25, 24/25, which outperforms four other approaches shown in Table 1. This experiment demonstrates the discriminative power of shape lexicon descriptor, in addition to its rotation invariance.

MPEG-7 Shape Database. Our next experiment involves the MPEG-7 CE-Shape-1 database [13], which consists of 1,400 silhouette images: 70 shape categories, 20 images per category. The retrieval performance is measured using the "bullseye test", in which each image is used as a query and one counts the number of correct images in the top 40 matches. The retrieval score is the ratio of the number of correct hits to the best possible number of hits.

Table 1. Summary of results on Kimia database

Method	Top 1	Top 2	Top 3
Sharvit *et al.* [25]	23/25	21/25	20/25
Gdalyahu and Weinshall [10]	25/25	21/25	19/25
Belongie *et al.* [2]	25/25	24/25	22/25
Ling and Jacobs [16]	25/25	24/25	25/25
Our approach	**25/25**	**25/25**	**24/25**

It is important to note that most well-known shape descriptors [2,16] and geometric hashing schemes [3] tested on MPEG-7 database have much higher complexity than the shape lexicon descriptor. For instance, shape contexts (SC) [2] and the articulation-invariant inner-distance shape contexts (IDSC) [16] both require attaching a spatial histogram for every point on the shape, which describes the distribution of the relative positions of all remaining points. The correspondences between two point sets are solved among $O(n^2)$ descriptor pairs, where n is the number of points sampled on object contours. The $O(n^3)$ complexity for each matching can be prohibitively expensive in most practical problems unless the object has been precisely localized and effectively segmented.

We quantitatively evaluate the discriminability of different descriptors, as this understanding is fundamental for image indexing and retrieval problems when it comes to the choice of either a local or global representation. In our experiment, the shape lexicon descriptor obtains a retrieval score of 62.10%. The bullseye score with SC is 64.59%, and the IDSC has the highest score of 68.83% as inner-distance can more robustly handle articulation than Euclidean distance.

To help understand the performance, we further verify the results and discover that a large number of errors by the shape lexicon descriptor come from object categories with low complexity, where the discriminative shape parts have been smoothed away after line fitting. This is less a problem for SC and IDSC as they directly sample along the object contour. The performance of the shape lexicon descriptor is comparable with SC and IDSC for object categories that involve complex local part structures. This demonstrates that a descriptor using nonparametric distribution of a set of indexed local shape primitives can have discriminability fairly close to a competitive global descriptor, such as shape contexts. In addition, our approach runs efficiently in linear time and space, which is critical in high-volume image indexing and retrieval.

4.2 Image Content Category Recognition

To evaluate our approach for image content category recognition, we construct a 4,500-image dataset by crawling Web images from the Google Image search engine using a wide variety of keywords. Fig. 1 shows some examples of images with text returned by using the text keyword "CD cover". All the images are automatically downloaded by a script. Duplicate and junk images are manually inspected and removed to reduce the proportion of unrelated images.

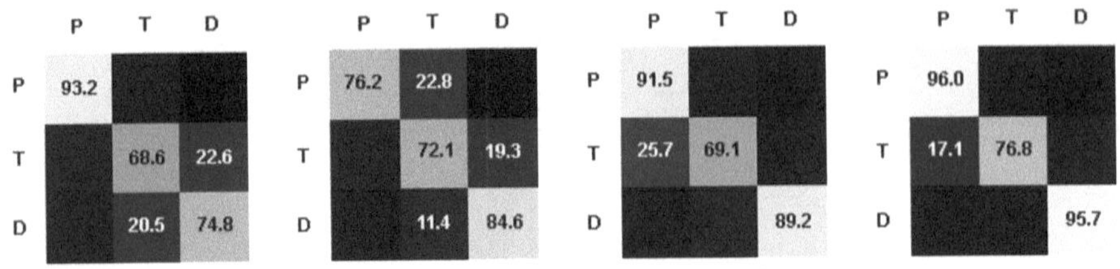

(a) Mean diagonal = 78.9% (b) Mean diagonal = 77.6% (c) Mean diagonal = 83.3% (d) Mean diagonal = 89.6%

Fig. 4. Confusion tables for image content category recognition using (a) Spatial envelope [21], (b) LBP [20], (c) kAS [8], (d) Our approach. (P: Pure image, T: Image with text, D: Document image).

We compare our approach with spatial envelope [21], local binary patterns (LBP) [20], and the state-of-the-art kAS descriptor [8], which are well-known approaches based on different views of whole-image characterization. Spatial envelope uses a holistic image representation without attempting to exploit localized information such as shape, whereas LBP is based on rotation-invariant texture analysis. Since 2AS gives the best performance among different kAS [8], we use it as the benchmark for kAS.

We use a multi-class SVM classifier trained with LIBSVM. The SVM classifier is trained using only 100 randomly selected images from each category, and used to test the rest images in the collection. For easy comparison, we set the dimensions of the image descriptor to 90 for both kAS and our approach in the following experiments.

The confusion tables for spatial envelope, LBP, kAS, and our approach are shown in Fig. 4. Spatial envelope demonstrates good performance in recognizing pure images, but it is not very effective for text content. Texture-based LBP gives balanced results for all the three content types. Our approach obtains the best performances for recognizing each content class, with a respectable mean diagonal of 89.6%. The only notable confusion occurs when distinguishing between image with text and pure image.

4.3 Language Identification

We use 1,512 document images of 8 languages (Arabic, Chinese, English, Hindi, Japanese, Korean, Russian, and Thai) from the University of Maryland multilingual database [15] and the IAM handwriting database DB3.0 [19] (see Fig. 7) for evaluation on language identification. Both databases are large public real-world collections, containing the source identity of each image in the ground truth. This enables us to construct a diverse dataset that closely mirrors the true complexities of heterogeneous document image repositories in practice.

We compare our approach with the state-of-the-art language identification system [11], which is based on template matching. We also include LBP and kAS in this experiment since they have demonstrated reasonable performance on

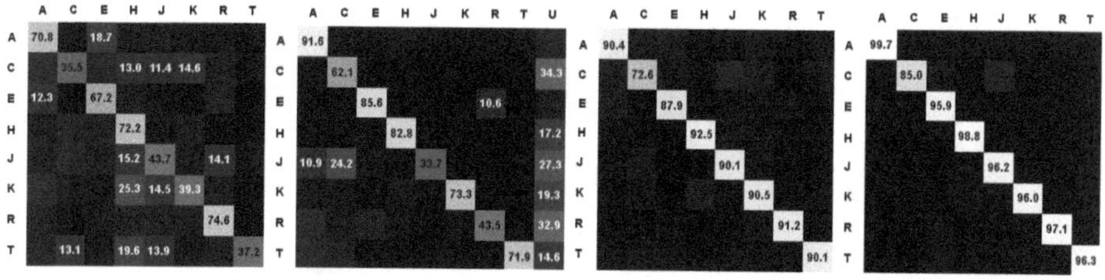

(a) Mean diagonal = 55.1% (b) Mean diagonal = 68.1% (c) Mean diagonal = 88.2% (d) Mean diagonal = 95.6%

Fig. 5. Confusion tables for language identification using (a) LBP [20], (b) Template matching [11], (c) *k*AS [8], (d) Our approach. (A: Arabic, C: Chinese, E: English, H: Hindi, J: Japanese, K: Korean, R: Russian, T: Thai, U: Unknown).

Table 2. Confusion table of our approach for the 8 languages

	A	C	E	H	J	K	R	T
A	99.7	0.3	0	0	0	0	0	0
C	1.4	85.0	4.0	1.0	6.7	1.0	0.7	0.2
E	1.6	0	95.9	0.2	0	1.1	0.6	0.6
H	0.2	0.2	0	98.8	0.8	0	0	0
J	0	1.3	1.0	0.2	96.2	1.3	0	0
K	0	0.8	0.1	1.9	0.5	96.0	0.5	0.1
R	0.5	0	2.0	0	0	0	97.1	0.4
T	0	0.3	1.6	0.9	0.6	0.3	0	96.3

diverse text contents. In this experiment, the SVM classifier is trained using the same pool of 50 randomly selected images from each category.

The confusion tables for LBP, template matching, *k*AS, and our approach are shown in Fig. 5. The performance of template matching varies significantly across languages. One significant confusion of template matching is between Japanese and Chinese since a document in Japanese may contain varying amount of Kanji (Chinese characters). Rigid templates are not flexible for identifying discriminative partial features, and the bias in voting decision towards the dominant candidate causes less frequently matched templates to be ignored. Another source of error that lowers the performance of template matching (mean diagonal = 68.1%) is undetermined cases (see the *unknown* column in Fig. 5(b)), where probabilistic voting cannot decide between languages with roughly equal votes. Texture-based LBP could not effectively recognize differences between languages on a diverse dataset because distinctive layouts and unconstrained handwriting exhibit irregularities that are difficult to capture using texture, and its mean diagonal is only 55.1%.

Our approach gives excellent results on all 8 languages, with an impressive mean diagonal of 95.6% (see Table 2 for all entries in the confusion table of our approach). *k*AS, with a mean diagonal of 88.2%, is also shown to be very effective. Neither method has difficulty generalizing across large variations such

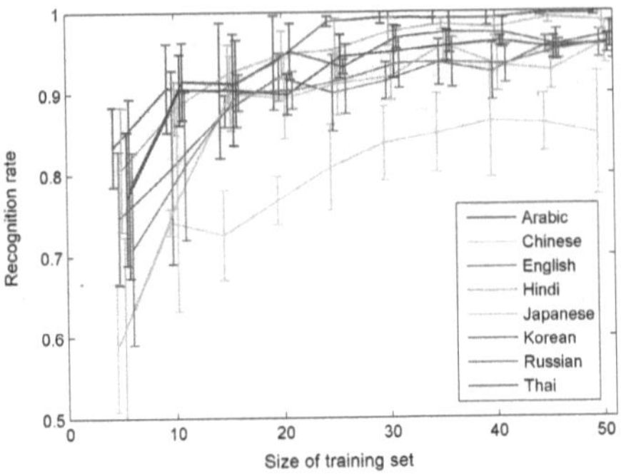

Fig. 6. Recognition rates of our approach for different languages as the size of training data varies. Our approach achieves excellent performance even using a small number of document images per language for training.

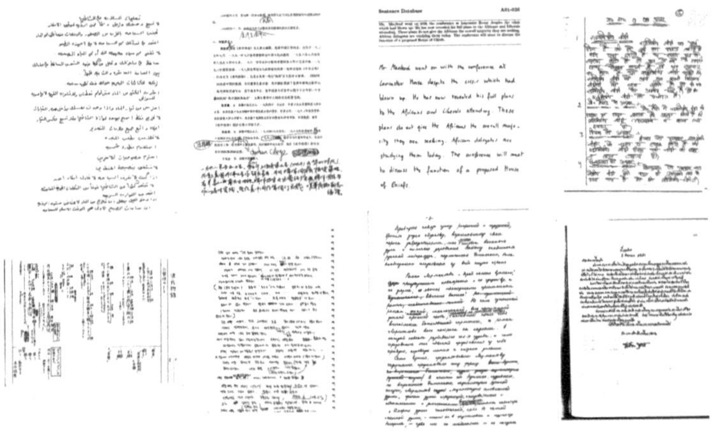

Fig. 7. Examples from the Maryland multilingual database [15] and the IAM handwriting DB3.0 database [19]. Languages in the top row are Arabic, Chinese, English, and Hindi, and those in the second row are Japanese, Korean, Russian, and Thai.

as font types or handwriting styles, which greatly impact the performances of LBP and template matching. This demonstrates the effectiveness of using generic low-level shape features when mid or high-level vision representations may not generalized or flexible enough for the task.

Fig. 6 shows the recognition rates of our approach as the size of training set varies. We observe very competitive language identification performance on this challenging dataset even when a small amount of training data per language class is used. In addition, our results on language identification are very encouraging from a practical point of view, as the training in our approach requires considerably less supervision than template matching. Our approach only needs the class label of each training image, and does not require skew correction, scale normalization, or segmentation.

5 Conclusion

In this paper, we proposed a novel approach for document image content categorization using a lexicon composed of a wide variety of local shape features. Each lexical word represents a characteristic structure that is generic enough to be detected repeatably and segmentation free. The lexicon provides a principled approach to structurally indexing and associating a vast number of feature types, and is learned from training data with little supervision. Our approach is fully extensible and does not require constructing explicit content models. In two challenging real world document image content recognition problems involving large-scale, highly variable image collections, our approach demonstrated excellent results and outperformed other state-of-the-art approaches. Our future work will be directed towards refining and evaluating the approach by further incorporating spatial co-occurrences of lexical words using a secondary lexicon.

References

1. Amit, Y., Geman, D.: A computational model for visual selection. Neural Computation 11, 1691–1715 (1999)
2. Belongie, S., Malik, J., Puzicha, J.: Shape matching and object recognition using shape contexts. IEEE Trans. Pattern Anal. Mach. Intell. 24(4), 509–522 (2002)
3. Biswas, S., Aggarwal, G., Chellappa, R.: Efficient indexing for articulation invariant shape matching and retrieval. In: Proc. CVPR, pp. 1–8 (2007)
4. Busch, A., Boles, W., Sridharan, S.: Texture for script identification. IEEE Trans. Pattern Anal. Mach. Intell. 27(11), 1720–1732 (2005)
5. Canny, J.: A computational approach to edge detection. IEEE Trans. Pattern Anal. Mach. Intell. 8(6), 679–697 (1986)
6. Dalal, N., Triggs, B.: Histograms of oriented gradients for human detection. In: Proc. CVPR, pp. 886–893 (2005)
7. Ding, J., Lam, L., Suen, C.: Classification of oriental and European scripts by using characteristic features. In: Proc. ICDAR, pp. 1023–1027 (1997)
8. Ferrari, V., Fevrier, L., Jurie, F., Schmid, C.: Groups of adjacent contour segments for object detection. IEEE Trans. Pattern Anal. Mach. Intell. 30(1), 1–16 (2008)
9. Fidler, S., Leonardis, A.: Towards scalable representations of object categories: Learning a hierarchy of parts. In: Proc. CVPR, pp. 1–8 (2007)
10. Gdalyahu, Y., Weinshall, D.: Flexible syntactic matching of curves and its application to automatic hierarchical classification of silhouettes. IEEE Trans. Pattern Anal. Mach. Intell. 21(12), 1312–1328 (1999)
11. Hochberg, J., Kelly, P., Thomas, T., Kerns, L.: Automatic script identification from document images using cluster-based templates. IEEE Trans. Pattern Anal. Mach. Intell. 19(2), 176–181 (1997)
12. Jacobs, D.: Robust and efficient detection of salient convex groups. IEEE Trans. Pattern Anal. Mach. Intell. 18(1), 23–37 (1996)

13. Latecki, L., Lakamper, R., Eckhardt, U.: Shape descriptors for non-rigid shapes with a single closed contour. In: Proc. CVPR, pp. 424–429 (2000)
14. Lee, D., Nohl, C., Baird, H.: Language Identification in Complex, Unoriented, and Degraded Document Images. Document Analysis Systems II (1998)
15. Li, Y., Zheng, Y., Doermann, D., Jaeger, S.: Script-independent text line segmentation in freestyle handwritten documents. IEEE Trans. Pattern Anal. Mach. Intell. 30(8), 1313–1329 (2008)
16. Ling, H., Jacobs, D.: Shape classification using the inner-distance. IEEE Trans. Pattern Anal. Mach. Intell. 29(2), 286–299 (2007)
17. Lowe, D.: Three-dimensional object recognition from single two-dimensional images. Artificial Intelligence 31(3), 355–395 (1987)
18. Lu, S., Tan, C.: Script and language identification in noisy and degraded document images. IEEE Trans. Pattern Anal. Mach. Intell. 30(2), 14–24 (2008)
19. Marti, U., Bunke, H.: The IAM-database: An English sentence database for offline handwriting recognition. Int. J. Document Analysis and Recognition 5, 39–46 (2006), `http://www.iam.unibe.ch/~fki/iamDB/`
20. Ojala, T., Pietikainen, M., Maenpaa, T.: Multiresolution gray-scale and rotation invariant texture classification with local binary patterns. IEEE Trans. Pattern Anal. Mach. Intell. 24(7), 971–987 (2002)
21. Oliva, A., Torralba, A.: Modeling the shape of the scene: A holistic representation of the spatial envelope. Int. J. Computer Vision 42(3), 145–175 (2001)
22. Plamondon, R., Srihari, S.: On-line and off-line handwriting recognition: A comprehensive survey. IEEE Trans. Pattern Anal. Mach. Intell. 22(1), 63–84 (2000)
23. Rice, S., Nagy, G., Nartker, T.: Optical Character Recognition: An Illustrated Guide to the Frontier. Kluwer Academic Publishers, Dordrecht (1999)
24. Rothwell, C., Zisserman, A., Forsyth, D., Mundy, J.: Planar object recognition using projective shape representation. Int. J. Computer Vision 16(5), 57–99 (1995)
25. Sharvit, D., Chan, J., Tek, H., Kimia, B.: Symmetry-based indexing of image database. J. Visual Commun. and Image Representation 9(4), 366–380 (1998)
26. Shi, J., Malik, J.: Normalized cuts and image segmentation. IEEE Trans. Pattern Anal. Mach. Intell. 22(8), 888–905 (2000)
27. Spitz, A.: Determination of script and language content of document images. IEEE Trans. Pattern Anal. Mach. Intell. 19(3), 235–245 (1997)
28. Suen, C., Bergler, S., Nobile, N., Waked, B., Nadal, C., Bloch, A.: Categorizing document images into script and language classes. In: Proc. ICDAR, pp. 297–306 (1998)
29. Tan, T.: Rotation invariant texture features and their use in automatic script identification. IEEE Trans. Pattern Anal. Mach. Intell. 20(7), 751–756 (1998)
30. Vincent, L.: Google Book Search: Document understanding on a massive scale. In: Proc. ICDAR, pp. 819–823 (2007)
31. Yu, S., Shi, J.: Multiclass spectral clustering. In: Proc. ICCV, pp. 11–17 (2003)
32. Zhu, G., Bethea, T.J., Krishna, V.: Extracting relevant named entities for automated expense reimbursement. In: Proc. ACM SIGKDD Int. Conf. Knowledge Discovery and Data Mining, pp. 1004–1012 (2007)
33. Zhu, G., Yu, X., Li, Y., Doermann, D.: Unconstrained language identification using a shape codebook. In: Proc. ICFHR, pp. 13–18 (2008)

Unsupervised Structure Learning: Hierarchical Recursive Composition, Suspicious Coincidence and Competitive Exclusion

Long (Leo) Zhu[1], Chenxi Lin[2], Haoda Huang[2], Yuanhao Chen[3], and Alan Yuille[1,4]

[1] Department of Statistics, University of California, Los Angeles
{lzhu,yuille}@stat.ucla.edu
[2] Microsoft Research Asia
{chenxi.lin,hahuang}@microsoft.com
[3] University of Science and Technology of China
yhchen4@ustc.edu
[4] Department of Psychology and Computer Science, UCLA

Abstract. We describe a new method for unsupervised structure learning of a *hierarchical compositional* model (HCM) for deformable objects. The learning is unsupervised in the sense that we are given a training dataset of images containing the object in cluttered backgrounds but we do not know the position or boundary of the object. The structure learning is performed by a bottom-up and top-down process. The bottom-up process is a novel form of hierarchical clustering which recursively composes proposals for simple structures to generate proposals for more complex structures. We combine standard clustering with the *suspicious coincidence principle* and the *competitive exclusion principle* to prune the number of proposals to a practical number and avoid an exponential explosion of possible structures. The hierarchical clustering stops automatically, when it fails to generate new proposals, and outputs a proposal for the object model. The top-down process validates the proposals and fills in missing elements. We tested our approach by using it to learn a hierarchical compositional model for parsing and segmenting horses on Weizmann dataset. We show that the resulting model is comparable with (or better than) alternative methods. The versatility of our approach is demonstrated by learning models for other objects (e.g., faces, pianos, butterflies, monitors, etc.). It is worth noting that the low-levels of the object hierarchies automatically learn generic image features while the higher levels learn object specific features.

1 Introduction

The goal of this paper is to learn a hierarchical compositional model (HCM) for deformable objects. The learning is unsupervised in the sense that we are given a training dataset of images containing the object in cluttered backgrounds but we do not know the position or boundary of the object. Unsupervised learning is

D. Forsyth, P. Torr, and A. Zisserman (Eds.): ECCV 2008, Part II, LNCS 5303, pp. 759–773, 2008.

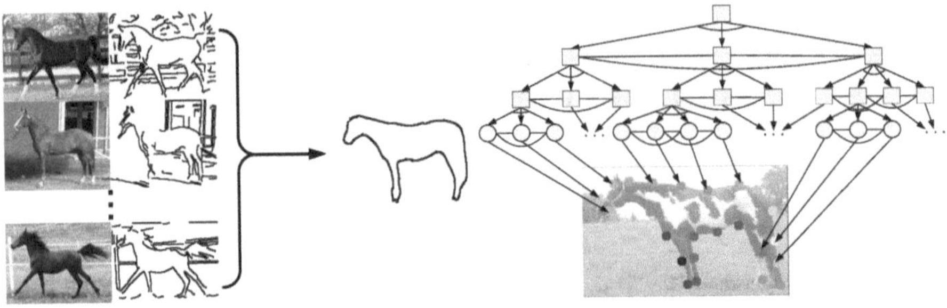

Fig. 1. Left:The learning is unsupervised in the sense that we are given a training dataset of images containing the object in cluttered backgrounds but we do not know the position or boundary of the object. Right: The hierarchical representation of the object. The boxes represent non-leaf nodes. The circles denote leaf nodes that directly relate to properties of the input image. The topology of the structure is not given, but learnt in an unsupervised way.

desirable since it avoids the need for time consuming hand labeling and prevents implicit biases. We apply the model to tasks such as object segmentation and parsing (matching of object parts).

Learning a hierarchical compositional model is very challenging since it requires us to learn the structure of the model (e.g. the relationships between the variables, and the existence of hidden variables) in addition to the parameters of the model. The difficulties are visually illustrated in figure 1: (i) the objects are deformable so there is ambiguity in the structure. (ii) there is cluttered background noise. (iii) parts of the object may be missing, (iv) the input is simple oriented edge features, which is a simple and highly ambiguous representation.

We now discuss some critical computational issues for unsupervised learning which motivate our hierarchical approach and contrast it to other unsupervised approaches for learning object models (e.g. [1], [2]). An abstraction of the structure learning problem is that we have a dataset of images each of which contain approximately M object features and N total features. In this paper, we are considering the case of $M = 100$ and $N = 5000$. Learning an object model requires solving a complicated correspondence problem to determine which features are object, which are background, and the spatial relationships between the object features. There are several strategies for addressing this correspondence problem. The first naive strategy is brute force enumeration which involves testing all N^M possibilities. This is only practical if M and N are small and the appearances of the features are sufficiently distinctive to enable many possible correspondences to be rejected. The second strategy is to learn the model in a greedy (non-hierarchical) way by sequentially growing subparts one by one [1,2]. This is practical in cases where brute force fails, but it still makes two assumptions: 1) sparseness assumption which requires M and N to be fairly small, e.g., $M = 6$ and $N = 100$ in [1] and 2) small ambiguity assumption which requires the appearances of the features to be somewhat distinctive. Neither of these two strategies are applicable if the features are edgelets, because M and N are both large and all edgelets have the same appearance which leads to big ambiguity. (One strategy is to use more powerful features, but we argue that this merely postpones the complexity problem). The purpose of this paper is to develop a more general unsupervised learning method without making strong assumptions of sparseness and low ambiguity. In other words, we try to provide a unified learning scheme for both generic features and object structures. Hence, we are driven to a third strategy, named as Recursive Composition, that creates a model by combining elementary structures to build a hierarchy.

Our strategy is based on a novel form of bottom-up hierarchical clustering. We assume that the object can be expressed in terms of recursive compositions of elementary structures (see the right panel of figure 1). We learn the structure by repeatedly combining proposals for substructures to make proposals for more complex structures. This process stops when we cannot generate any more proposals. A huge number of proposals are examined and stored at every level to avoid ambiguity since small substructure are not necessarily distinct enough between object and background. However, combining proposals in this way risks a combinatorial explosion (imagine that the number of combinations will grow exponentially as we go up to the upper levels). We avoid this by the use of two principles: (i) suspicious coincidences, and (ii) competitive exclusion. Suspicious coincidences eliminates proposals which occur infrequently in the image dataset (in less than 90 percent of the images). The competitive exclusion principle is adapted from ecology community where it prevents two animals from sharing the same environmental niche. In this paper, competitive exclusion eliminates proposals which seek to explain overlapping parts of the image.

The bottom-up clustering is followed by a top-down stage which refines and fills in gaps in the hierarchy. These gaps can occur for two reasons. Firstly, the competitive exclusion principle sometimes eliminates subparts of the hierarchy because of small overlaps. Secondly, gaps may occur at low-levels of the hierarchy, for example at the neck of a horse, because there are considerable local shape variations and so suspicious coincidences are not found. The top-down process can remove these gaps automatically by relaxing the two principles. For example, at higher levels the system discovers more regular relationships between the head and torso of the horse which provides context enabling the algorithm to fill in the gaps.

In summary, hierarchical bottom-up and top-down procedure allows us to grow the structure exponentially (the height of a hierarchy is $\log(M)$) and end up with a nearly complete structure (the resulting representation is dense and the size M could be big). We tested our approach by using it to learn a hierarchical compositional model for parsing and segmenting horses on Weizmann dataset [3]. We show that the resulting model is comparable with (or better than) alternative methods. The versatility of our approach is demonstrated by learning models for other objects (e.g., faces, pianos, butterflies, monitors, etc.).

2 Background: Hierarchical Structure Learning

Hierarchical design dates back to Fukushima's Neocognitron [4]. Recently, there have been many new developments including new representations and learning algorithms. For example, Geman et al. [5] propose a hierarchical object model designed by a compositional principle using an AND/OR graph. Inference algorithms have been invented for this type of model [6], which we will adapt and use to perform inference in this paper (Note [6] does not address any learning algorithm). Deep networks [7], another type of multi-layer system, has recently been proposed by Hinton et al. [7]. Ullman et al. [8] learns a hierarchical feature representation in a supervised manner. Poggio and his colleagues'[9] build a hierarchical structure for rapid object recognition motivated by mimicing the architecture of the visual cortex of the human brain. Ahuja and Todorovic's hierarchical representation [10] is based on segmented image regions. Fleuret and Geman [11] provide a coarse-to-fine strategy which starts from edgelets. Fidler and Leonardis[12]'s unsupervised learning approach is most related to our work (the main difference is that they treat rigid objects, have a less dense representation, and do not have a top-down stage). In summary, in several cases these models[5,6,9] are not learnt but are specified by the user. Some unsupervised methods [8] assume that background is clean and the object is roughly aligned. Unsupervised learning (in the sense of this paper) has been performed [1,2] (without a hierarchy) but for object models with limited number of object features (as discussed earlier). Most models [1,9,8,5,2] focus on recognition or categorization, but not parsing.

3 The Hierarchical Compositional Model (HCM) and the Inference Algorithm

3.1 The Hierarchical Compositional Model (HCM)

The graph structure for the hierarchical model is depicted in the right panel of figure 1 and defined as follows. We let V_r be the set of nodes of the graph with root node r. Each node $\nu \in V_r$ has a set of child nodes T_ν (a node is constrained to have a single parent), V_r^{LEAF} are the leaf nodes (the only ones which interact with the image). For any node $\nu \in V_r$, we can define a graph V_ν with root node ν containing the descendants of ν. The edges for the graph are of three types: (i) vertical *data edges* which relate the leaf nodes to the image, (ii) horizontal edges relating the children of each node to each other (described below), and (iii) vertical edges relating parents to children (specified by $\{T_\nu\}$). In this paper, all vertical edges are directed. The horizontal edges only connect the child nodes with the same parent. Moreover, we restrict the number of child nodes to be less than six, i.e., $|T_\nu| \leq 6$. This ensures that the size of the biggest clique of the hierarchy is small. We use the notation Ω_r to represent the graph (i.e. the set of nodes V_r and the set of edges).

A configuration of the graph is an assignment of state variables $z = \{z_\nu\}$ to all $\nu \in V_r$. In this paper, we set $z_\nu = (x_\nu, y_\nu, \theta_\nu, s_\nu)$, where (x, y), θ and s

denote image position, orientation, and scale respectively (the scale is the area of the image that the node represents). We use the notation $Z_\nu = \{z_\mu : \mu \in V_\nu\}$ to denote the state of node ν and all its descendent nodes.

We define a Gibbs distribution $P(z, d; \omega, \Omega)$ for the probability of the graph as follows:

$$P(z, d; \omega, \Omega) = \frac{1}{Z(\omega, \Omega)} \exp\{-E(z, d; \omega, \Omega)\}. \tag{1}$$

where $Z(\omega, \Omega)$ is the partition function, d is the image, ω denotes the parameters of the distribution and Ω denotes the graph structure (i.e. the nodes and the edges). The energy function $E(z, d; \omega, \Omega)$ is the sum of three terms corresponding to the three types of edges in the graph:

$$E_d(d, Z_r; \omega, \Omega) + E_h(Z_r; \omega, \Omega) + E_v(Z_r; \omega, \Omega), \tag{2}$$

where E_d, E_h, E_v are energy terms defined at the data, horizontal, and vertical edges. We now describe the specific choices used in this paper.

The *data term* E_d is given by:

$$E_d(d, Z_r; \omega, \Omega) = \sum_{\nu \in V_r^{LEAF}} f(d_\nu, z_\nu), \tag{3}$$

where V_r^{LEAF} is the set of the leaf nodes and $f(.,.)$ is the (negative) logarithm of a Gaussian defined over the grey-scale intensity gradient (i.e. $d_\nu = \nabla I_\nu$). It biases the leaf nodes to be located at image locations where the image gradient is large, and for their orientations to be perpendicular to the image gradient. This term is fixed and not learnt.

The *horizontal term* E_h is given by:

$$E_h(Z_r; \omega, \Omega) = \sum_{\nu \in V_r / V_r^{LEAF}} \sum_{(\mu,\rho,\tau):\mu,\rho,\tau \in T_\nu} g(z_\mu, z_\rho, z_\tau; \omega), \tag{4}$$

where $g(z_\mu, z_\rho, z_\tau; \omega)$ is the (negative) logarithm of Gaussian distribution defined on the *invariant triplet vector* (ITV) $l(z_\mu, z_\rho, z_\tau)$ constructed from (z_μ, z_ρ, z_τ) [2]. (The ITV is invariant to the translation, rotation, and scaling of the triple, which ensures that the full probability distribution is also invariant to these transformations). The summation is over all triples formed by the child nodes of each parent, see the right panel of figure (1). The parameters of the Gaussian are indicated by $\omega = (\mu, \sigma)$.

The *vertical term* E_v is used to hold the structure together by relating the state of the parent nodes to the state of their children. It is defined by:

$$E_v(Z_r; \omega, \Omega) = \sum_{\nu \in V_r / V_r^{LEAF}} h(z_\nu; \{z_\mu \text{ s.t.} \mu \in T_\nu\}), \tag{5}$$

where $h(.,.) = 0$ provided the average orientations and positions of the child nodes are equal to the orientation and position of the parent node – formally $(x_\nu, y_\nu, \theta_\nu) = (1/|T_\nu|) \sum_{\mu \in T_\nu} (x_\mu, y_\mu, \theta_\mu)$. If this equality does not hold, then

$h(.,.) = \kappa$, where κ is a large positive number. The scale variable $s_\nu = \sum_{\mu \in T_\nu} s_\mu$, hence the parent node represents the sum of the regions in the image represented by its children.

Observe that the nodes of the graph have the same variables at all levels (i.e. (x, y, θ, s)). This makes use of hierarchical independence assumption–the higher level interactions depend only on the summarization of all child nodes at the lower levels. More precisely, the child nodes only propagate a limited summary (center position, total size and orientation) of their state information to their parents. The choice of the vertical term E_v means that this information is simply the average of their node variables.

We stress that we can use equations (2,3,4,5) to calculate the energy for any subgraph by specifying the root node. This gives an effective way for computing the full energy, by combining the energy of the subgraphs, and is exploited during inference and learning. We can also exploit this hierarchy in order to compute other important properties, such as the partition function $Z(\omega, \Omega)$. Note, in figure 1, the vertical edges are directed and horizontal edges only connect the child nodes (less than six) with the same parent. Therefore, the partition function can be factorized into several components (corresponding to the cliques) whose sizes are small.

3.2 The Inference Algorithm for Parsing When HCM is Known

We now specify an efficient inference algorithm for the HCM. This algorithm will be used both when we are learning the model, see section (4), and when we are applying the model to new images to perform detection, segmentation, and parsing. We adapt a compositional algorithm that was designed for inference on AND/OR graphs [6] (Note their work does not present any learning algorithm which is the major topic of this paper). The algorithm has a bottom-up and a top-down stage. We only take the bottom-up stage which makes proposals for the states of the node variables Z_r. For brevity we omit the details. See [6] for more details.

4 Unsupervised Structure Learning

We now address the critical issue of how to learn the structure of the hierarchical model from a set of images which contain an example of the object with varying background. Formally, this requires us to estimate the structure parameters Ω and the distribution parameters ω. The task of the unsupervised learning is defined as :

$$
\begin{aligned}
(\omega^*, \Omega^*) &= \arg\max P(\omega, \Omega | d) && (6)\\
&= \arg\max P(d|\omega, \Omega) P(\omega, \Omega) && (7)\\
&= \arg\max \sum_z P(d|z, \omega) P(z|\omega, \Omega) P(\omega, \Omega) && (8)
\end{aligned}
$$

where $P(d|z,\omega)P(z|\omega,\Omega) = P(d,z|\omega,\Omega)$ is defined in equation (1). $P(\omega,\Omega)$ accounts for the prior distribution of the structure which plays a similar role of structure regularization. $P(\omega,\Omega)$ is factorized into $P(\omega)$ and $P(\Omega)$. The parameters indicated by ω are μ and σ in the gaussian functions. The prior on σ is of the form $\mathcal{N}(0,\beta I)$. The prior on μ^l at certain level l puts hard constraints on the range of the distance allowed between any two substructures. The prior of Ω is uniform distribution. The summation $\sum_z P(d|z,\omega)P(z|\omega,\Omega)P(\omega,\Omega)$ is used as a *score* function to measure the goodness of the fit of the structure. Intuitively, the score function tells us how frequently a structure encoded by (Ω,ω) appears in the training set. Our approach is based on a bottom-up process to propose a set of structures (Ω,ω) followed by a top-down process which refines the result by adjusting (Ω,ω).

Our algorithm makes several approximations to simplify the learning problem. Firstly, we will exploit the hierarchical nature of the model to estimate the parameters ω locally. In principle, we could use these local estimates to initialize an EM algorithm to determine the parameters globally. Secondly, while our algorithm proposes a structure Ω we cannot guarantee that it is the globally optimal structure. But our experimental results, see section (5), provide empirical evidence that these approximations are reasonable.

4.1 The Bottom-Up Process

The bottom-up process constructs hierarchical object models by composing them from more elementary components. We use two principles to prevent a combinatorial explosion of compositions and to ensure that our compositions result in desirable object models. The two principles are: (i) *suspicious coincidences* where we keep compositions which occur frequently in the images and reject compositions which do not, and (ii) *competitive exclusion* where we remove compositions which match to regions of the image which overlap with other compositions. The relative importance of these different principles is shown empirically in the results section, see table 2 and figure 6. For example, observe how competitive exclusion play a small role when the compositions are small but is of major importance as the compositions get large.

Our strategy proceeds by creating vocabularies of concepts at different levels in the hierarchy, where each concept is a hierarchical compositional model. See figure 2 for visual (symbolic) illustration. The concepts are generated by composing concepts at lower levels. The basic ideas are to detect all instances of the concepts at level l in the images (using the inference algorithm for each concept). We *form compositions* of these concepts by identifying sets of these concepts that appear in sub-regions of the images. We *cluster* these compositions based on the spatial relationships between their elements (the instances of the concepts at level $l-1$) to get a set of concepts at level l whose distribution parameters ω are estimated from the clusters. We run the inference algorithm on the images to find the instances of all the concepts. Then we use the *suspicious coincidence principle* to remove concepts which occur infrequently (i.e. have few instances). Next we use the *competitive exclusion principle* to remove concepts

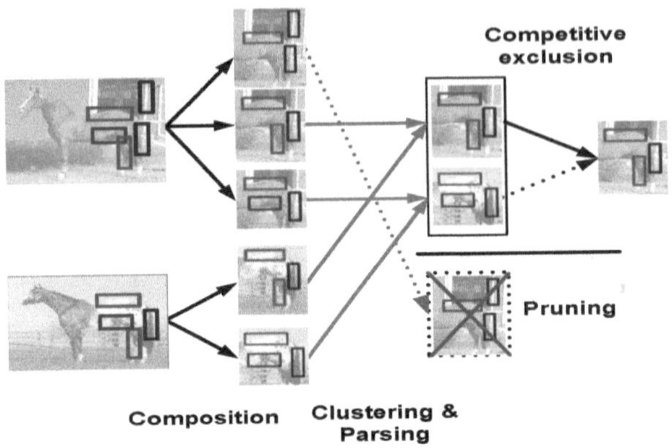

Fig. 2. This figure illustrates the procedure of the unsupervised learning. The first column shows the responses of the features from the "vocabulary" at level 1. the second column shows the compositions at step 1). the third column shows the clustering at step 2. The noise(non-regular) pattern bounded by dotted line is pruned out by step 4. The last column plots the results after step 5. At step 5, the compositions, which are constructed by different components, but parse the same areas in the image, are grouped together (the maximum is kept.

whose instances overlap with those of other concepts (with better scores). The remaining concepts form the vocabulary of concepts at level $l+1$. The process repeats until no new concepts are composed. For our applications (e.g. images containing a large object with variable background), it will terminate automatically in a small number of proposals when the hierarchy has reached a maximum size (i.e. there is no larger structure to be found).

The full procedure is described more formally in the next few paragraphs and is summarized by the pseudo-code in figure (3) and illustrated visually in figure 2. First we introduce some notation. A *concept* at level l is notated by $P^l_{\omega,\nu}$. It is a HCM with root node ν and ω represents the parameters of this model. A *vocabulary of concepts* at level l is the set $\{P^l_{\omega,\nu}\}$. We can parse the training dataset using the vocabulary of concepts at level l to get a set of *instances* of each concept $\{MP^l_{\nu,a}\}$, where $a = 1, ..., N^l_\nu$ indexes the set of instances. Each instance is represented by its state variables Z_ν.

We define the bottom-up process as follows. At each level $l-1$ we have a vocabulary of concepts $\{P^{l-1}_{\omega,\nu}\}$ and a set of instances of the concepts in the images $\{MP^{l-1}_{\nu,a}\}$. We then repeat the following steps.

1. Composition. We search through the images to find instances of triplets of concepts $MP^{l-1}_{\mu,\omega}, MP^{l-1}_{\rho,\omega}, MP^{l-1}_{\tau,\omega}$ within subregions of size S_{l-1} (for all μ, ρ, τ in the vocabulary). From each triplet instance we construct an instance P^l_ν where node ν has children $\mu, \rho, \tau \in T_\nu$.

2. Clustering. We cluster the instances P^l_ν by the second order moments of the shape descriptor (triplet vector) of the positions, scales and orientation $(x_\rho, y_\rho, \theta_\rho, s_\rho)$ of the children of ν (this clustering is done separately for all triples

Input: $\{MP^1\}$. Output: $\{MP_{\omega}{}^L\}$. $\oplus$ denotes the operation of combining two proposals.

- Bottom-Up-Structure-Construction(MP^1)
  Loop : $l = 1$ to L
  1. Composition: $\{P^l_\nu\} = \oplus(MP^{l-1}_{\mu,a}, MP^{l-1}_{\rho,b}, MP^{l-1}_{\tau,c})$ and $T^l_\nu = \{\mu^{l-1}, \rho^{l-1}, \tau^{l-1}\}$
  2. Clustering: $\{P_{\omega}{}^l_\nu\} = Clustering(\{P^l_\nu\})$
  3. Parsing: for each $P_{\omega}{}^l_\nu$, $\{MP^l_{\nu,a}\} = Parser(P_{\omega}{}^l_\nu, \{MP^{l-1}_\mu, MP^{l-1}_\rho, MP^{l-1}_\tau\})$
  4. Suspicious Coincidence (Pruning) : $\{P_{\omega}{}^l_\nu\}\{P\omega^l_\nu | Score(P\omega^l_\nu) > T_{pruning}\}$
  5. Competitive Exclusion: $\{(MP_{\omega}{}^l_\nu, CL_{\omega}{}^l_\nu)\}$ = $CompetitiveExclusion(\{P_{\omega}{}^l_\nu\}, \epsilon_{region})$ where ϵ_{region} is the size of the window W_{image} defined in regions which are covered by a set of images.
- $\tilde{P_\omega} = \arg\max_\nu Score(MP_{\omega}{}^L_\nu)$
- Top-Down-Structure-Extension
  Loop: $l = L$ to 2, for each node ν at level l within $\tilde{P_\omega}$
  - repeat
    1. $P_{\omega}{}^*_\rho = \arg\max_{P_{\omega}{}^{l-1}_\rho} Score(\tilde{P_\omega} \oplus P_{\omega}{}^{l-1}_\rho)$
    2. $\Delta = Score(\tilde{P_\omega} \oplus P_{\omega}{}^*_\rho) - Score(\tilde{P_\omega})$
    3. $\tilde{P_\omega} = \tilde{P_\omega} \oplus P_{\omega}{}^*_\rho$; $T^l_\nu = T^l_\nu \cup \rho^{l-1}$
  until $\Delta < T_{extension}$

Fig. 3. Bottom-up and Top-down Learning

in the concept vocabulary at level $l-1$). The clustering outputs a set of *concepts* at level l notated by $\{P^l_{\omega,\nu}\}$. Some of the parameters $\omega = (\mu, \sigma)$ of each concept are estimated from the second order moments (see above) and the remainder are inherited from the underlying concepts at level $l-1$.

3. Parsing. We parse the image dataset using the inference algorithm (see section (3.2)) for the set of concepts output by the clustering $\{P^l_{\omega,\nu}\}$. This gives a set of instances of the concepts in the dataset $\{MP^l_{\nu,a}\}$.

4. Pruning by Suspicious Coincidences. We remove concepts from $\{P^l_{\omega,\nu}\}$ if they do not have sufficient number of instances in the dataset (i.e. if there are instances of the concept in less than ninety percent of the images).

5. Competitive exclusion. We remove concepts from $\{P^l_{\omega,\nu}\}$ if their instances have significant overlap with instances of other concepts (and the other concepts have better scores).

The remaining concepts form the concept vocabulary $\{P^l_\omega\}$ at level l. Their instances in the dataset $\{MP^l\}$ have already been calculated (in the parsing step above). The process repeats itself until no new concepts are generated. In practice, this happens within 4-5 levels (see experimental section). The process is initialized for the leaf nodes. The vocabulary of concepts at level 1 $\{P^1_{\omega,\nu} : \nu = 1, ..., 4\}$ consists of four types of edgelets characterized by their orientations $\theta = m\pi/4$ for $m = 0, 1, 2, 3$. The size s is set to a default value s_1. We parse the training set of images to detect the instances of all the concepts (more precisely, we compute the set of points $(x, y) : E_d(\mathbf{\nabla} I(x, y), (x, y, m\pi/4, s_1) < T_1$, where T_1 is a threshold, for $m = 0, 1, 2, 3$). This gives the set of instances of each concept $\{P^1_{\omega,\nu}\}$.

4.2 The Top-Down Process

The top-down process fills in the missing parts of the hierarchy and also adds a dense representation at the lowest level (to enable segmentation of the image).

Missing parts of the hierarchy can occur for two reasons: (i) competitive exclusion may be too strict at eliminating proposals which only slightly overlap, (ii) shape variations at certain parts of the object may be big at small scales (hence rejected by suspicious coincidences) but are removed at larger scales. For example, there are big variations locally where the legs of a horse join the torso, which make it hard to detect small scale regularities by clustering. But there are large scale regularities between the legs and the torso which are found at higher levels in the bottom-up process. The top-down process can fill in the gaps at lower level by using the higher levels connections as "context" which allow the pruning criteria to be relaxed.

A greedy strategy is used to examine every node in the hierarchy. For each node ν^l at level l, we seek to add a substructure from the dictionary (the set of proposals obtained in the bottom-up processing) at the lower level $l-1$ as the new child of ν^l so that the extended structure fits the data best (locally). This extension operation for the node ν^l is repeated until the gain is less than some threshold. The extensions correspond to adding more energy terms defined in equation (4) and (5). In figure (5), one can observe that substructure are added into the hierarchy to capture the details of the head and hind leg.

5 Experimental Results

Performance Comparisons for segmentation and parsing. We applied our approach to learn hierarchical composition models for a number of different objects. We will first concentrate on the hierarchical model for the horse, learnt from data from the Weizmann database [3] which is divided to training (12 images without labeling) and testing (316 images with groundtruth) sets. These images cover many poses (standing, running, drinking, etc.), changes in viewing angles, different scales, textured body patterns and cluttered background. Our strategy to obtain segmentation, which is inspired by Grab-Cut [13] , is to obtain the parse by the inference algorithm on the HCMs and then segment object by graph-cut using the feature statistics inside the boundary as initializations (note that, unlike us, Grab-Cut requires initialization by a human). The comparisons using segmentation accuracy are shown in table 1. The methods in [14,15] are based on supervised learning. Only two methods [16,15] reported higher accuracy than ours: but (i) Obj Cut[16] is tested on only five images while our method is tested on more than 300 images; (ii) [15] assumes the position of object is given for both learning and testing (Note our method does not make this assumption). [6] manually designed the model (no learning) whose performance, i.e., about 91%, is similar to [14] and inferior to ours. [17] is not listed because they manually select the best among top results (all other methods output single result). In conclusion, our method achieves the comparable (if not the best) performance even though we use far less information (we only know that the object is present somewhere in the image, while the supervised methods know the precise location of the boundaries). In addition, our model simultaneously performs other tasks such as detection and parsing (labeling different parts of the horse). See figure 4

Table 1. Results on horse dataset. Columns 2 and 3 give the size of images taken for training and testing. Column 4 quantifies the segmentation accuracy. The last column shows the average time taken for one image. Note that [16] is tested on only 5 images and [15] assumes the position of the object is roughly given.

Method	Train	Test	Segmentation	Speed
Our method	12	316	93.3	16.9s
Ren [14]	172	172	91.0	–
Borenstein [18]	64	328	93.0	–
LOCUS [19]	20	200	93.1	–
OBJ CUT [16]	N/A	5	96.0	–
Levin [15]	N/A	N/A	95.0	–

Fig. 4. In the top two rows, Columns 1 ,3 and 5 show parse results of our method followed by segmentation results. The parse result is illustrated by the colored dots which correspond to the leaf nodes of the object. The correspondences are consistent for different images. Rows 3 and 4 show the parse results of our method and OBJ CUT (cropped from [16]) on 5 images used in [16] respectively. Note our method obtains more complete boundary.

for the typical segmentation and parsing results (we also cropped the results from [16] for comparison). Note that the color points indicate identities of object parts. Our method obtains more complete and stable boundary than [16] which learns object model from video where extra motion cues are used to make learning easier. The average time of our inference algorithm including both parsing and segmentation is 17 seconds.

Study the Hierarchy. The final hierarchy for horse is depicted in the left panel of figure 5. The mean position and orientation of edgelets of leaf nodes are depicted to sketch the model learnt by the unsupervised method. The colored

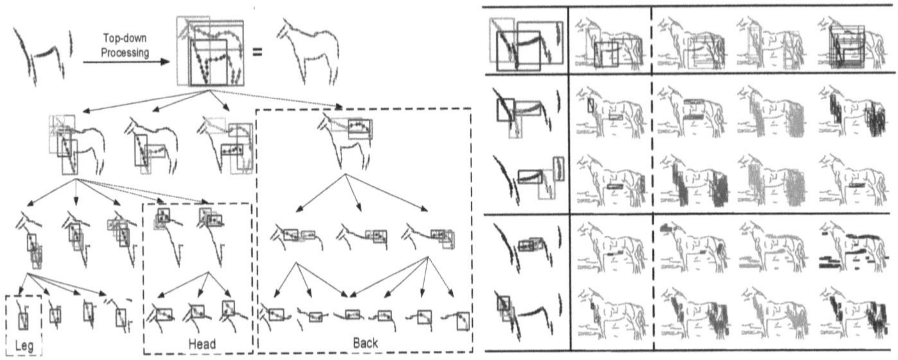

Fig. 5. Left: The hierarchy shows the learnt hierarchical model. The colored rectangles highlight the identities of the structures. All parts are obtained at the bottom-up state except that the rectangles in the dotted boxes show the parts of the model that were learnt in the top-down stage. Right: The rows illustrate structures at levels 4,3,3,2,2 (i.e. top row is level 4, next row is level 3,...). The first column gives the structure (with three children colour coded). The second column shows the structure detected on a specific image. The third, fourth, and fifth columns show the proposals for the sub-structure – colour coded.

rectangles in the dotted box highlight the parts filled in at the top-down stage. Note that the final skeleton of horse is nearly complete while the sketches obtained by the bottom-up processing miss several parts (head, leg and back). The responses of subparts of the hierarchy is depicted in the right panel of figure 5. One can see that the numbers of proposals decrease from low level to high level.

Complexity Analysis. Our experiments also show the relative effectiveness of our two principles as we form compositions at different levels of the hierarchy, see table (2). The reduction in compositions by *clustering* is fairly large (a reduction factor of 10) at level 1 but becomes negligible at higher levels as the compositions get more distinct. *Suspicious coincidences* causes major reductions in compositions with reduction factors ranging from 60 to 2000. This is because the vast number of compositions occur infrequently. *Competitive exclusion* has little effect at level 1 (a reduction factor of 5) but increases rapidly at higher levels to a reduction factor of 15 at level 4. This is because larger compositions are more likely to overlap and compete for the same niches in the images. Competitive exclusion principle is the main factor that causes the bottom-up process to stop at a specific level.

Analyze the Hierarchical Dictionary. It seems plausible that our algorithm will learn generic features (e.g., oriented straight lines, single curves, double edges, Y-junctions, T-junctions, etc.) at low-level of the hierarchy and more specific object features at the higher-levels. This is confirmed by observing the vocabularies that are extracted at different levels, see the left panel of figure (6).

Table 2. Columns from 2 to 5 show the numbers of proposals after composition (step 1), clusters after clustering (step 2), after pruning (step 4) and after competitive exclusion (step 5) respectively. The next column shows the time (seconds) taken for each level. Level 0 shows the results of the grouping of the edge points(360 degrees are divided into 4 angle ranges). all numbers are calculated over 12 real images in the training dataset.

L	Composition	Clusters	Suspicious Coincidence	Competitive Exclusion	Time
0				4	1s
1	167431	14684	262	48	117s
2	2034851	741662	995	116	254s
3	2135467	1012777	305	53	99s
4	236955	72620	30	2	9s

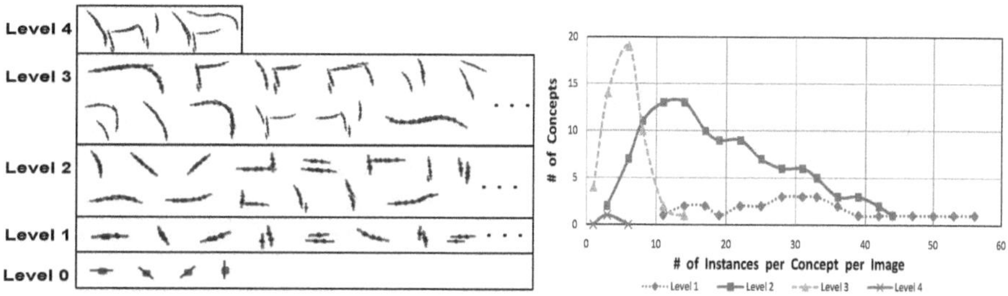

Fig. 6. Left: This figure shows elements of the vocabulary for horses at different levels (mean values only). Observe how the vocabulary contains "generic" shapes at low levels, but finds horse specific parts at the higher levels. Right: This figure shows the histogram of concepts at the different levels of the dictionary. A point in a curve quantifies the number of concepts which have a certain number of instances.

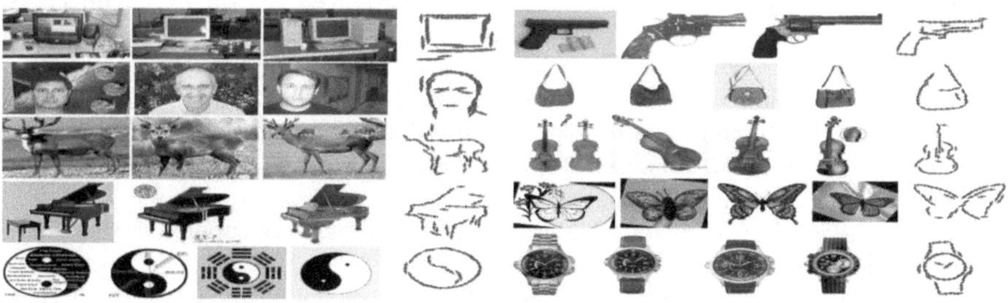

Fig. 7. This figure illustrates the generality of our approach. We show some typical training images which contain cluttered background, different shapes and deformations. The red-sketch images show the learnt models.

Recall that all these features are encoded by the same hierarchical composition principle. In the right panel of figure (6), we also plot the distribution of the concepts at the different levels of the dictionary. The peaks of levels 4, 3, 2 and

1 appear from right to left. Note that the concepts at level 1 have only one instance per image.

More objects. To demonstrate the generality of our approach, we apply it to learn models for a range of other objects collected from Caltech 101 [20], MIT LabelMe [21] and internet, see figure (7). These images cover different types of shapes (man-made structure and animal), cluttered background (monitor, face and deer) and rotation (violin). These models were learnt using a small amount of training data (12 images per object). They can be applied to parse images using the inference algorithm. This experiment shows the versatility of our approach while modeling different types of shapes of objects.

6 Conclusion

We described a new method for unsupervised structure learning of a hierarchical compositional model for deformable objects. The structure learning was performed by a bottom-up and top-down process. We tested our approach by using it to learn hierarchical models for horses and other objects (e.g. watches, purses, faces, grand pianos, violins, revolvers, butterflies). We evaluated the resulting models by comparing their performance to alternative methods.

Acknowledgements

We gratefully acknowledge support from the National Science Foundation with NSF grant number 0413214 and from the W.M. Keck Foundation.

References

1. Fergus, R., Perona, P., Zisserman, A.: Object class recognition by unsupervised scale-invariant learning. In: CVPR (2), pp. 264–271 (2003)
2. Zhu, L., Chen, Y., Yuille, A.L.: Unsupervised learning of a probabilistic grammar for object detection and parsing. In: NIPS, pp. 1617–1624 (2006)
3. Borenstein, E., Ullman, S.: Class-specific, top-down segmentation. In: Heyden, A., Sparr, G., Nielsen, M., Johansen, P. (eds.) ECCV 2002. LNCS, vol. 2351, pp. 109–124. Springer, Heidelberg (2002)
4. Fukushima, K.: Neocognitron: A hierarchical neural network capable of visual pattern recognition. Neural Networks 1, 119–130 (1988)
5. Jin, Y., Geman, S.: Context and hierarchy in a probabilistic image model. In: CVPR (2), pp. 2145–2152 (2006)
6. Chen, Y., Zhu, L., Lin, C., Yuille, A.L., Zhang, H.: Rapid inference on a novel and/or graph for object detection, segmentation and parsing. In: NIPS (2007)
7. Hinton, G.E., Osindero, S., Teh, Y.W.: A fast learning algorithm for deep belief nets. Neural Computation 18, 1527–1554 (2006)
8. Epshtein, B., Ullman, S.: Feature hierarchies for object classification. In: ICCV, pp. 220–227 (2005)
9. Serre, T., Wolf, L., Poggio, T.: Object recognition with features inspired by visual cortex. In: CVPR (2), pp. 994–1000 (2005)

10. Ahuja, N., Todorovic, S.: Learning the taxonomy and models of categories present in arbitrary image. In: ICCV (2007)
11. Fleuret, F., Geman, D.: Coarse-to-fine face detection. In: IJCV (2001)
12. Fidler, S., Leonardis, A.: Towards scalable representations of object categories: Learning a hierarchy of parts. In: CVPR (2007)
13. Rother, C., Kolmogorov, V., Blake, A.: "grabcut": interactive foreground extraction using iterated graph cuts. ACM Trans. Graph. 23, 309–314 (2004)
14. Ren, X., Fowlkes, C., Malik, J.: Cue integration for figure/ground labeling. In: NIPS (2005)
15. Levin, A., Weiss, Y.: Learning to combine bottom-up and top-down segmentation. In: Leonardis, A., Bischof, H., Pinz, A. (eds.) ECCV 2006. LNCS, vol. 3954, pp. 581–594. Springer, Heidelberg (2006)
16. Kumar, M.P., Torr, P.H.S., Zisserman, A.: Obj cut. In: CVPR (1), pp. 18–25 (2005)
17. Cour, T., Shi, J.: Recognizing objects by piecing together the segmentation puzzle. In: CVPR (2007)
18. Borenstein, E., Malik, J.: Shape guided object segmentation. In: CVPR (1), pp. 969–976 (2006)
19. Winn, J.M., Jojic, N.: Locus: Learning object classes with unsupervised segmentation. In: ICCV, pp. 756–763 (2005)
20. Fei-Fei, L., Fergus, R., Perona, P.: Learning generative visual models from few training examples: An incremental bayesian approach tested on 101 object categories. Comput. Vis. Image Underst. 106, 59–70 (2007)
21. Russell, B., Torralba, A., Murphy, K., Freeman, W.: Labelme: a database and web-based tool for image annotation. Technical Report (2005)

Contour Context Selection for Object Detection: A Set-to-Set Contour Matching Approach

Qihui Zhu[1], Liming Wang[2], Yang Wu[3], and Jianbo Shi[1]

[1] Department of Computer and Information Science, University of Pennsylvania
qihuizhu@seas.upenn.edu, jshi@cis.upenn.edu
[2] Department of Computer Science and Engineering, Fudan University
wanglm@fudan.edu.cn
[3] Institue of Artificial Intelligence and Robotics, Xi'an Jiaotong University
ywu@aiar.xjtu.edu.cn

Abstract. We introduce a shape detection framework called *Contour Context Selection* for detecting objects in cluttered images using only one exemplar. Shape based detection is invariant to changes of object appearance, and can reason with geometrical abstraction of the object. Our approach uses salient contours as integral tokens for shape matching. We seek a maximal, holistic matching of shapes, which checks shape features from a large spatial extent, as well as long-range contextual relationships among object parts. This amounts to finding the correct figure/ground contour labeling, and optimal correspondences between control points on/around contours. This removes *accidental alignments* and does not hallucinate objects in background clutter, without negative training examples. We formulate this task as a set-to-set contour matching problem. Naive methods would require searching over 'exponentially' many figure/ground contour labelings. We simplify this task by encoding the shape descriptor algebraically in a linear form of contour figure/ground variables. This allows us to use the reliable optimization technique of Linear Programming. We demonstrate our approach on the challenging task of detecting bottles, swans and other objects in cluttered images.

1 Introduction

We study the problem of object detection in natural images using shape. Visual objects can be represented at a variety of levels from signal (filter responses) to symbol (object parts). Our approach focuses on representation of the shape that is closer to the symbol level, which would allow abstract geometrical reasoning of the object. Shape description is invariant to color, texture, and brightness changes, which could enable significant reduction in the number of training examples, and increase of accuracy of the detection.

Object detection using shape alone is not an easy task. Most shape matching algorithms are susceptible to *accidental alignment*: hallucinating objects in background clutter. To avoid foreground (surface marking) and background clutter, shape descriptors are often computed within a window of limited spatial extent. Local window features are discriminative enough for detecting objects such as faces, cars and bicycles. However, for many simple objects, such as swans, mugs or bottles, local shape features are insufficient.

D. Forsyth, P. Torr, and A. Zisserman (Eds.): ECCV 2008, Part II, LNCS 5303, pp. 774–787, 2008.

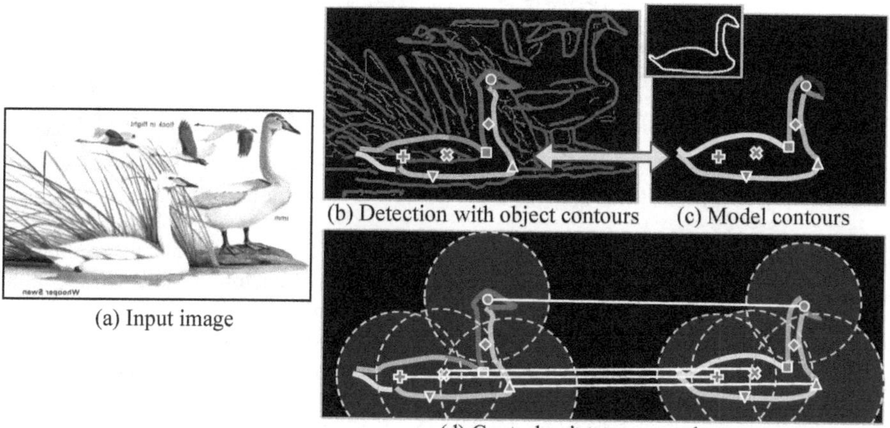

(a) Input image (b) Detection with object contours (c) Model contours (d) Control point correspondence

Fig. 1. Using a single line drawing object model shown in (c), we detect object instances in images with background clutter in (a) using only shape. Bottom-up contour grouping provides tokens of shape matching. Long salient contours in (b) provide distinctive shape descriptions, allowing both efficient and accurate matching. Image and model contours, shown by different colors in (b) and (c), do not have one-to-one correspondences. We formulate shape detection as a set-to-set matching task in (d) consisting of: (1) correspondences between control points, and (2) selection of contours that contribute contextual shape features to those control points, within a circular neighborhood.

To overcome this *accidental alignment* problem, we propose a shape detection approach called *Contour Context Selection* consisting of the following key ingredients:

1. We detect salient contours using bottom-up segmentation or contour grouping. Long contours are more distinctive, and maintaining contours as integral tokens for matching removes many false positives due to accidental alignment.
2. We break the model shape into its informative semantic parts, and explicitly check which subset of model shape parts is matched. Missing critical model parts can signal an accidental alignment between the image and model.
3. We seek holistic shape matching. We measure shape features from a large spatial extent, as well as long-range contextual relationships among object parts. Accidental alignments of holistic shape descriptors between image and model are unlikely.

Our *Contour Context Selection* reduces to finding a maximal, holistic matching between a *set* of image contours and a *set* of model parts. It searches over *figure/ground* labeling of the image and model contours, and *correspondences* between them. It is important to note that, in general, image contours and model contours do not correspond one-to-one. The holistic matching occurs only by considering a set of 'figure' contours together. To formulate this *set-to-set* matching task, we define control points sampled on and around the image and model contours. We compute shape features on the control points from the 'figure' contours within a large neighborhood (see Fig. 1). The task is to find the correct figure/ground contour labeling, such that there is an optimal one-to-one matching of the control points. This set-to-set matching potentially requires searching

(a) Accidental alignment (b) Missing critical parts

Fig. 2. Typical false positives can be traced to two causes: (1) Accidental alignment shown in (a). Our algorithm prunes it by exploiting contour integrity, *i.e.* requiring contours to be whole-in/whole-out. Contours violating this constraint is marked in white on the image. (2) Missing critical object parts indicates that the matching is a false positive. In (b), after removing the accidental alignment to the apple logo outline (marked in white), only the body can find possible matches and the neck of the swan is completely missing shown at the top-right corner of (b). Our approach rejects this type of detection by checking missing critical model contours after joint contour selection.

over exponentially many choices of figure/ground labeling. We simplify this task by encoding the shape descriptor algebraically in a linear form of contour selection variables, allowing to use the reliable optimization technique of Linear Programming.

This paper is organized as follows. Section 2 introduces the basic concept and formulation of Contour Context Selection. We present the computational solution for this framework using Linear Programming (LP) in Section 3. Section 4 describes related works and comparisons. Section 5 demonstrates our approach on the challenging task of detecting non-rectangular and wiry objects, followed by the conclusion in Section 6.

2 Shape Detection as a Set-to-Set Contour Matching Problem

Our goal is to detect objects in images using a *single* model and identify correspondences between the image and the model.

We use salient contours, extracted from bottom-up contour grouping, as tokens for image-model shape matching. Shape matching with contours instead of isolated edges has several advantages. Long salient contours are more distinctive, which leads to efficiency of the search as well as the accuracy of shape matching. Furthermore, by requiring the entire contour to match objects as a whole, we remove accidental alignment causing false postive detections (see Fig. 2 (a) for an example).

Using contour grouping as the starting point of shape matching carries risk as well. Contours could be mis-detected, or accidentally leak to background clutter. A good contour grouping algorithm is essential for shape matching. We utilize the approach in [1] which has demonstrated good performance in cluttered images detecting reliable contours. Furthermore, these contours groups are not disjoint, providing multiple hypotheses at places where contours can potentially leak to other objects (*e.g.* junctions).

To evaluate shape matching, one needs to measure the accuracy of alignment, and more importantly, determine *which* parts are aligned. For simple shapes, missing a small but critical object part can indicate a complete mismatch, see Fig. 2 (b). In this work, we manually divide the model into contours which corresponds to distinctive parts. Just as image contours, we require model contours to be matched as a whole.

The computational task of shape matching thus consists of parallel searches over image contours and model contours to obtain the maximal match of the image and model shapes. We cast the shape detection as the following problem:

Set-to-set contour matching. Given an image $\mathcal{I}$ and a model $\mathcal{M}$ represented by two sets of long salient contours:

- Image: $\mathcal{I} = \{C_1^I, C_2^I, \dots C_{|\mathcal{I}|}^I\}$, C_k^I is the kth contour;
- Model: $\mathcal{M} = \{C_1^M, C_2^M, \dots, C_{|\mathcal{M}|}^M\}$, C_l^M is the lth contour.

we would like to select the maximal contour subsets $\mathcal{I}^{sel} \subseteq \mathcal{I}$ and $\mathcal{M}^{sel} \subseteq \mathcal{M}$, such that object shapes composed by $\mathcal{I}^{sel}$ and $\mathcal{M}^{sel}$ match (see Fig. 1 for an image example).

Once this set-to-set matching is solved, to quantify shape matching we measure

- *which set of model contours are matched;*
- *how well the matched contours are aligned.*

The final classification cost function combines the following two terms:

$$C_{classification} = C_{config} \cdot C_{align} \tag{1}$$

where C_{config} evaluates the configuration of the matched model contours, and C_{align} measures quality of their alignment defined later in the following sections.

The main technical difficulty is that the image and model contours do not have one-to-one correspondence. Contours detected from bottom-up grouping and segmentation are different from the semantically meaningful contours in the model. However, as a whole they will have matches (see Fig. 1). Set-to-set contour matching bridges this semantic gap between the bottom-up grouping and the top-down model.

2.1 Solution for Set-to-Set Contour Matching

Our solution to the set-to-set matching problem includes three essential components:

Control point correspondence. While contour themselves do not correspond in one-to-one, their shape information can be evaluated at nearby *control points*, and those control points could have one-to-one correspondences. Suppose control points $\{p_1, p_2, \dots, p_m\}$ are sampled from the image and $\{q_1, q_2, \dots, q_n\}$ are sampled from the model. We define the correspondence matrix $(U^{cor})_{m \times n}$ from the image to the model as:

$$U_{ij}^{cor} = \begin{cases} 1, & \text{if } p_i \text{ matches } q_j \\ 0, & \text{otherwise.} \end{cases} \tag{2}$$

Note that these control points can be located anywhere in the image, not limited to contours. Computing dense point correspondences is unnecessary. Instead, rough matching of some control points is sufficient to select and match contour sets $\mathcal{I}^{sel}$ and $\mathcal{M}^{sel}$.

Feature representation: holistic shape features. The important question is, what will be the appropriate shape feature for matching these control points, and how to compute

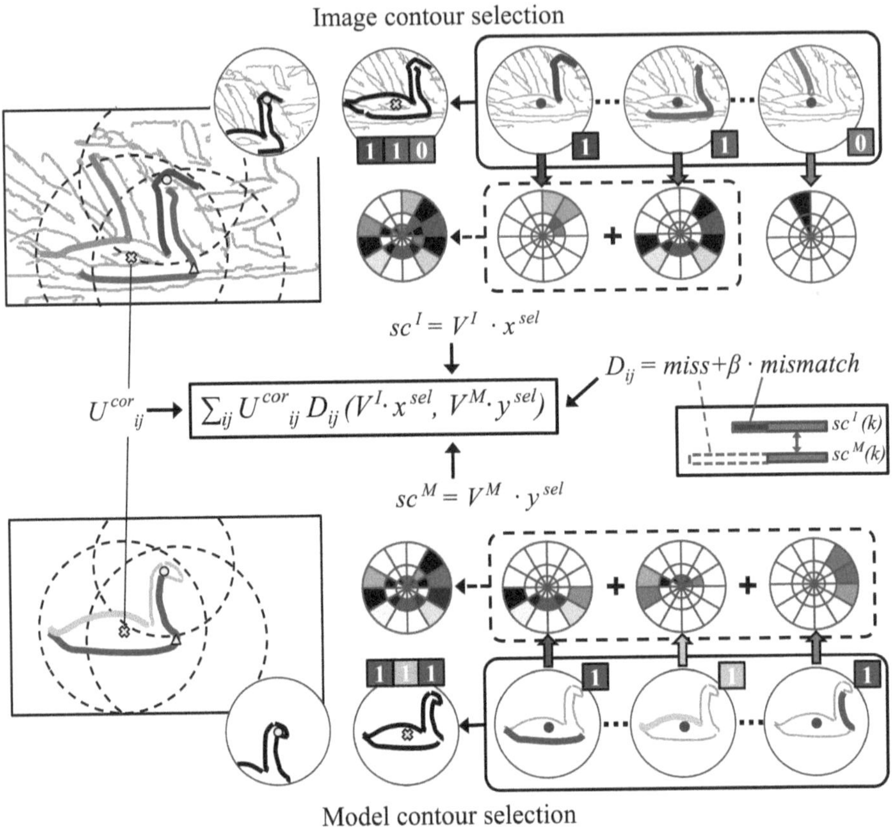

Fig. 3. Illustration of our computational solution for set-to-set contour matching on shape detection example from Fig. 1. The top and the bottom row shows the image and model contour candidate sets marked in gray. Each contour contributes its shape information to nearby *control points* in the form of Shape Context histogram, shown on the right. By selecting different contours (x^{sel}, y^{sel}), each control point can take on a set of possible Shape Context descriptions (sc^I, sc^M). With the correct contour selection in the image and model (marked by colors), there is a one-to-one correspondence U_{ij}^{cor} between (a subset of) image and model control points (marked by symbols). This is a computationally difficult search problem. The efficient algorithm we developed is based on an encoding of Shape Context description (which could take on exponentially many possible values) using linear algebraic formulation on the contour selection indicator: $sc^I = V^I \cdot x^{sel}$. This leads to the Linear Programming optimization solution.

shape dissimilarity D_{ij}. Comparing D_{ij} requires the feature to be common in the image and the model. Since there do not exist one-to-one correspondences between contours, the feature description is more appropriate on the contour set or global shape level rather than on the individual contour level. We propose a holistic shape representation at the control points covering not only nearby contours but also faraway contours (see Fig. 3).

The holistic shape representation immediately poses the problem of *figure/ground selection* since figure/ground segmentation is unknown and the shape feature is likely to include both foreground and background contours. Unknown segmentation introduces great difficulties to any shape features with a *fixed* context. A fixed context feature cannot adapt to the combinatorial possibilities of figure/ground labeling, each generating different contexts. Without the correct segmentation, background clutter and contours from other objects can corrupt the shape feature. Our strategy is to adjust the context of the holistic shape features during matching depending on the figure/ground selection. Therefore, we are able to select the right features and determine the figure/ground segmentation simultaneously.

Matching constraint: contour integrity. Contour selection implies that we restrict each contour to be an integral unit in matching. For each contour $C_k^I = \{p_1^{(k)}, p_2^{(k)}, \ldots, p_c^{(k)}\}$ where $p_i^{(k)}$'s are edge pixels, there are only two choices: either all the edge pixels $p_i^{(k)}$ participate in the matching, or none of them are included. Partially matched contours are not allowed. The same constraint is applied to model contours C_l^M as well. We introduce contour selection indicators $x^{sel} \in \{0,1\}^{|I|\times 1}$ in the *entire* test image and $y^{sel} \in \{0,1\}^{|M|\times 1}$ in the model defined as

$$\textbf{(Image contour selection)} \qquad x_\ell^{sel} = \begin{cases} 1 & \text{Contour } C_\ell^I \text{ is selected} \\ 0 & \text{otherwise} \end{cases}, \tag{3}$$

$$\textbf{(Model contour selection)} \qquad y_\ell^{sel} = \begin{cases} 1 & \text{Contour } C_\ell^M \text{ is selected} \\ 0 & \text{otherwise,} \end{cases} \tag{4}$$

The constraint of contour integrity makes matching robust to accidental alignment.

2.2 Context Sensitive Shape Features

Now we are ready to introduce the holistic shape representation called *Context Sensitive Shape Features* determined by the figure/ground selection of the contours induced by x^{sel} and y^{sel}. We choose Shape Context (SC) [2] as our basic shape feature descriptor. Measuring global shape requires the scope of SC to be large enough to cover the *entire* object. Define sc_i^I to be the vector of SC histogram centered at control point p_i, *i.e.* $sc_i^I(k) = \#$ of points in bin k. We introduce a contribution matrix V_i^I with size (#bin)×(#contour) to encode the contribution of each contour to each bin of sc_i^I:

$$V_i^I(k,l) = \#\text{ of points in bin } k \text{ from contour } C_l \tag{5}$$

Similar notations sc_j^M and V_j^M are defined for SC at control point q_j in the model.

The key observation is that shape features sc_i^I will be *different* depending on context x^{sel}, *i.e.* they are not fixed. Since each contour can have 2 choices, either selected or not selected, there exists 2^n possible contexts – exponential in the number of contours n. One advantage of histogram type of features such as Shape Context is that the exponentially many combinations of contexts can be written in a simple linear form:

$$sc_i^I(k) = \sum_l V_i^I(k,l) \cdot x_l^{sel} = (V^I \cdot x^{sel})_k \tag{6}$$

This allows us to cast the complex search as an optimization problem later.

Our goal is to find x^{sel} and y^{sel} such that they produce similar shape features: $V_i^I \cdot x^{sel} \approx V_j^M \cdot y^{sel}$. We evaluate and compare these two features by the context sensitive dissimilarity:

$$\textbf{(Context sensitivity)} \qquad D_{ij}(sc_i^I, sc_j^M) = D_{ij}(V_i^I \cdot x^{sel}, V_j^M \cdot y^{sel}) \tag{7}$$

The shape dissimilarity D_{ij} not only depends on the local attributes of p_i and q_j, but more importantly, on the context given by x^{sel} and y^{sel}. Matching object shapes boils down to minimizing D_{ij}, which is a combinatorial search problem on x^{sel} and y^{sel}.

2.3 Contour Context Selection Cost

Finding set-to-set contour matching finally becomes a joint search over correspondences U^{cor} and contour selection x^{sel}, y^{sel} by minimizing the following cost:

$$\min_{U^{cor}, x^{sel}, y^{sel}} C_{align}(U^{cor}, x^{sel}, y^{sel}) = \frac{1}{c} \sum_{i,j} U_{ij}^{cor} D_{ij}(V^I x^{sel}, V^M y^{sel}) \tag{8}$$

$$\text{s.t.} \quad U^{cor} \in \text{GeoSet} \tag{9}$$

where $c = \sum_{i,j} U_{ij}^{cor}$ is the number of control point correspondences. Correspondences U^{cor} from different object parts should have geometric consistency. We use a star model for checking global geometric consistency. Each correspondence (p_i, q_j) can predict an object center c_{ij}. For the correct set of correspondences, all the predicted centers should form a cluster, *i.e.* close to their average $center(U^{cor}) = \sum c_{ij} U_{ij}^{cor} w_{ij} / \sum U_{ij}^{cor} w_{ij}$, where w_{ij}'s are the weights on correspondences. Thus correspondences U^{cor} satisfying the geometric consistency constraint can be expressed as:

$$\text{GeoSet} = \{\|center(U^{cor}) - c_{ij} U_{ij}^{cor}\| \leq d_{max} \text{ if } U_{ij}^{cor} \neq 0\} \tag{10}$$

where d_{max} is the maximum distance allowed for deviation from the center.

What is the right matching cost $D_{ij}(V_i^I \cdot x^{sel}, V_j^M \cdot y^{sel})$? Recall that our problem is to search for the maximal 'common' subsets from the image and model contours such that their shapes are similar. This maximal condition on the contour subsets places additional requirement on the shape dissimilarity D_{ij}. A straightforward cost function, such as the L_1-norm: $D_{ij}(V_i^I \cdot x^{sel}, V_j^M \cdot y^{sel}) = \|V_i^I \cdot x^{sel} - V_j^M \cdot y^{sel}\|$, will simply result in the trivial solution which chooses empty sets from both sides (*i.e.* $x^{sel} = \mathbf{0}$, $y^{sel} = \mathbf{0}$). In fact all the norms as well as χ^2 distance suffer from the same problem.

We introduces the *joint selection cost* for D_{ij} which balances the maximal requirement on the match of contour sets and the quality of the match. We seek to match as many model contours as possible while the difference between image and model contours is small. Before describing the details, we first introduce a few variables. Set

- $sc_j^{\mathcal{MF}} = V_j^M y^{full}$ to be the shape context centered at model point q_j selecting the full model, where $y^{full} = \mathbf{1}_{|\mathcal{M}|\times 1}$ means selecting all model contours;
- $sc_i^I = V_i^I x^{sel}$ to be the shape context with selection x^{sel} on image at p_i;
- $sc_j^M = V_j^M y^{sel}$ to be the shape context with selection y^{sel} on model at q_j.

We use $sc_j^{\mathcal{MF}}(k)$, $sc_i^I(k)$, $sc_j^M(k)$ to denote the kth bin in the shape context.

Our joint selection cost consists of two terms: *miss* and *mismatch* (see Fig. 3). To match as many model contours as possible, the following difference between the number of matched points and that of full model points should be minimized:

$$\text{miss}_k^{(ij)} = sc_j^{\mathcal{MF}}(k) - \min(sc_i^I(k), sc_j^M(k)) \tag{11}$$

Here $\min(sc_i^I(k), sc_j^M(k))$ counts the number of matched contour points between the image and model in shape context bin k.

The above term $\text{miss}_k^{(ij)}$ alone does not measure how well the selected image contours match to the selected model contours. To ensure the matching quality, the amount of difference between the number of image and model contour points in all shape context bins needs to be minimized:

$$\text{mismatch}_k^{(ij)} = |sc_i^I(k) - sc_j^M(k)| \tag{12}$$

By combining Eq. (11) and Eq. (12), we have the following dissimilarity:

$$D_{ij} = \frac{\sum_k [\text{miss}_k^{(ij)} + \beta \cdot \text{mismatch}_k^{(ij)}]}{\sum_k sc_j^{\mathcal{MF}}(k)} \tag{13}$$

where $\beta > 1$ is a factor balancing the two types of costs. We use $\sum_k sc_j^{\mathcal{MF}}(k)$ to normalize the cost D_{ij} such that it is invariant to the number of contour points.

3 Computational Solution Via Linear Programming

Direct optimization of contour context selection cost function Eq. (8) is a hard combinatorial search problem. The shape dissimilarity $D_{ij}(V^I \cdot x^{sel}, V^M \cdot y^{sel})$ can only be evaluated given correspondences U^{cor}. However, finding the correct correspondences U^{cor} requires x^{sel} and y^{sel}. Therefore, the inference problem becomes circular. We approximate this joint optimization by breaking the loop into two steps: *single point figure/ground labeling* and *joint contour selection*. The first step focuses on finding reliable correspondences U^{cor} (maybe sparse) by matching image contours to the whole model. The second step selects contours simultaneously from both image contours labelled as figure and all the model contours being matched, based on the correspondences computed in the first step. In both steps, we optimize the cost function by relaxing it as an instance of Linear Programming (LP).

3.1 Single Point Figure/Ground Labeling

Our first step discovers all potential control point correspondences U_{ij} and computes the correspondent figure/ground labeling x^{sel} for them. We fix $y^{sel} = 1$ to encourage

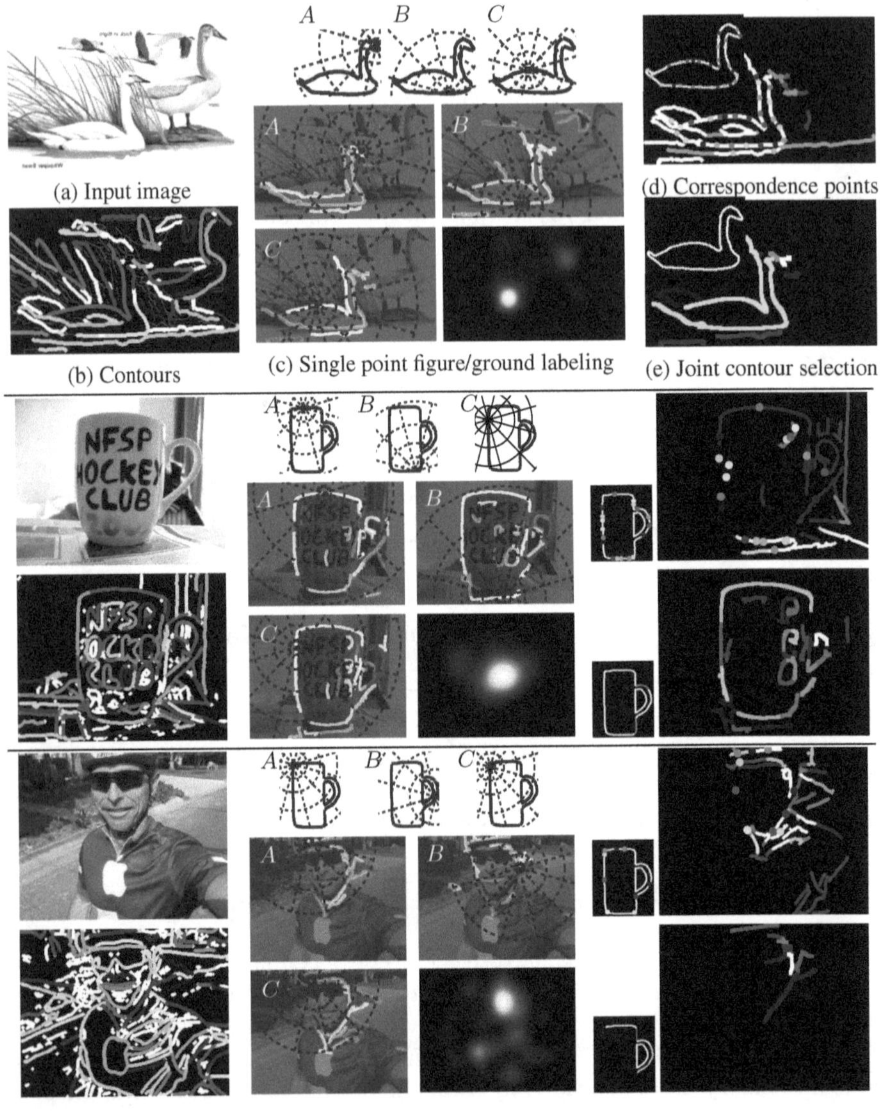

Fig. 4. Illustration of Contour Context Selection for shape detection. From the input image (a), we detect long salient contours shown in (b). For possible control point correspondences in (c), we select foreground contours whose global shape configuration most resembles to the model, with selection x^{sel} shown in gray scale (the brighter, the larger value of x^{sel}). Voting map for pruning geometrically inconsistent correspondences is shown at the right bottom corner of (c). (d) shows the consistent correspondences marked by different colors using the hypothesized correspondences. The optimal joint image-model contour selection is shown in (e). Note in the last example, model selection allow us to detect false match between the mug and the face.

matching to the full model as much as possible. Partial matches are undesired since the correspondences they produce are much less reliable. Therefore, only the $mismatch$ term in Eq. (13) is applied and hence the dissimilarity D_{ij} reduces to L_1-norm:

$$\min_{x^{sel}} \quad \|V_i^I \cdot x^{sel} - V_j^{\mathcal{M}} \cdot 1\|_1 \tag{14}$$

This cost will not collapse to zero because model contour selection is fixed ($y^{sel} = 1$).

Brute force approach of the above problem is formidable even for mid-size problems (20-30 contours). We compute an approximate solution by continuous relaxation on binary variable x^{sel} and add the constraint $0 \leq x^{sel} \leq 1$ thereafter. Since the norm in the cost function is L_1[1], this leads to an instance of Linear Programming (LP) which can be computed very efficiently.

Correspondences found from single point figure/ground labeling might not satisfy geometric consistency Eq. (10). Therefore, we enforce geometric consistency by pruning hypotheses of control point correspondences via a voting procedure [3]. Each image control point can predict an object center using the best match to model control points computed by Eq. (14). These predictions generate votes weighted by the shape dissimilarity, which accumulate to a voting map. We extract object centers from the local maxima and further backtrace the voting centers to identify consistent correspondences.

3.2 Joint Contour Selection

We have obtained the rough correspondences U^{cor} from the previous step. We optimize the contour selection cost Eq. (8) *w.r.t.* x^{sel}, y^{sel} to prune false positives and detect objects. The outcome includes both the matching cost C_{align} and model contours actually matched, indicated by y^{sel}. Both of them can be used to prune false positives. Note that it is not required to have a complete correspondence set U^{cor} since the cost Eq. (13) has been normalized by the number of correspondences.

LP can also be used to solve Eq. (8) for contour context selection by relaxing x^{sel} and y^{sel} to real value vectors. Eq. (13) and Eq. (8) translate to the following problem:

$$\min_{x^{sel}, y^{sel}} \sum_{U_{ij}^{cor}=1} \{\frac{1}{N_i} \sum_k [sc_i^{\mathcal{MF}}(k) - \min(sc_i^I(k), sc_j^M(k))] + \frac{\beta}{N_i} \|sc_i^I - sc_j^M\|_1\}$$

$$\text{s.t. } sc_i^I = V_i^I \cdot x^{sel}, \quad sc_j^M = V_j^M \cdot y^{sel}$$

where $N_i = \sum_k sc_i^{\mathcal{MF}}(k)$ is a normalization constant and $\min(x, y)$ computes the elementwise min of vectors x and y. The two terms in the summation are miss and mismatch in Eq. (13) respectively. The above problem can be relaxed to an instance of LP by adding slack variables $s_{ijk} \geq sc_i^I(k)$ and $s_{ijk} \geq sc_j^M(k)$ for $\min(sc_i^I(k), sc_j^M(k))$.

The selected model contours from joint contour selection form a shape configuration that are actually matched to image contours. Because the number of object model contours is typically very limited (usually 6 to 8), we can manually specify a dictionary

[1] Besides L_1, other distance functions such as L_2 and χ^2 for shape context can also be used. However, the relaxations will be computationally much more intensive.

of all possible configurations of true positives, *i.e.* setting C_{config} in Eq. (1) to be 0/1. Detection of model contours with bad configurations, e.g. missing critical parts, are rejected. This configuration checking together with the matching cost C_{align} can prune most of the false positives while preserving true positives.

4 Related Works and Discussion

Salient contours and their configurations are more distinctive than individual edge points for shape matching. Ferrari *et al.* ([4],[5]) represent objects by learning a codebook of Pairs of Adjacent Segments, which are consecutive roughly straight contour fragments. They achieve detection using a bag-of-words approach. Shotton *et al.* [6] learn a boosted contour-based shape features for object detection. These efforts utilize mostly short contour fragments, and therefore have to rely on many training examples to boost the discriminative power of shape features. In contrast, our work takes the advantage of contour grouping such as [1] to detect long salient contours, capturing more global geometric information of objects. More importantly, we constrain these long contours to act as a whole unit, *i.e.* they can either be entirely matched to an object, or entirely belong to the background. This characteristic makes shape matching more immune to accidental alignment to background clutter. Similar properties are exploited by grouping based verification approaches [7], and the recent work by Felzenswalb *et al.* [8].

From a broader perspective, our recognition framework is based on shape matching, which has a long history in vision. A large amount of research has been done on different levels of shape information. Early works [9,10] focused on silhouettes which are relatively simple for representing shape. Silhouette based approaches are limited to objects with a single closed contour without any interior edges with occlusions. Objects in real images are more complex, and may be embedded in heavy clutter. Efforts on dense matching of the edge points often focus on spatial configurations of key points, such as geometric hashing [11], decision tree [12] and Active Shape Models [13]. However, keypoints alone are insufficient to distinguish objects shapes in cluttered images [2].

Feature representation and shape similarity measurement are the key issues for matching. Shape Context [2] uses spatial distribution of edges points relative to a given point to describe shape. Inner Distance Shape Context (IDSC) refines it to account for articulated objects [14]. We build our basic shape feature representation on Shape Context, with contour as the unit. A much larger context window covering the whole object enables our approach to capture global shape configurations. We introduce a novel contour selection mechanism to extract global shape features against background clutter.

5 Experiments

We demonstrate our detection approach using only one hand-drawn model without negative training images, To evaluate our performance, we choose the challenging ETHZ Shape Classes [5] containing five diverse object categories with 255 images in total. Each image has one or more object instances. All categories have significant scale changes, illumination changes and intra-class variations. Moreover, many objects are

Table 1. Comparison of precision. We compare the precision of our approach to the precision in [5] at the same recall (lower recall in [5]). We convert the result of [5] reported in DR/FPPI into P/R since the number of images in each class is known. Our performance is significantly better than [5] in four out of five classes. The other class "Mugs" have some instances that are too small to be detected by contour grouping. Note that we did not use magnitude information which plays an important role in [5].

	Apple logos	Bottles	Giraffes	Swans	Mugs
Our precision/recall	**49.3%/86.4%**	**65.4%/92.7%**	**69.3%/70.3%**	**31.3%/93.9%**	25.7%/83.4%
Precision/recall in [5]	32.6%/86.4%	33.3%/92.7%	43.9%/70.3%	23.3%/93.9%	40.9%/83.4%

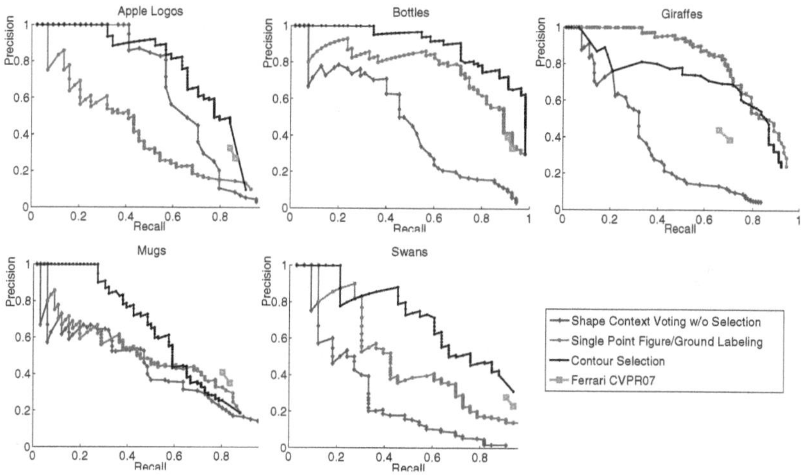

Fig. 5. Precision vs. recall curves on five classes of ETHZ Shape Classes. Our precisions on "Apple logos", "Bottles", "Giraffes" and "Swans" are considerably better than results in [5]: **49.3%/32.6%** (Apple logos), **65.4%/33.3%** (Bottles), **69.3%/43.9%** (Giraffes) and **31.3%/23.3%** (Swans). Also notice that we boost the performance by large amount compared to local shape context voting without contour selection.

surrounded by extensive background clutter and have interior contours. We have the same experimental setup as [5], using only a single hand-drawn model for each class and all 255 images as test set. To adapt to large scale variance, we generate multiple models by resizing the original one to 5 to 8 scales for each class.

We first use contour grouping proposed in [1] to generate long salient contours from images. Contours can have overlaps due to multiple possible groupings at junctions. Large window shape context for contour selection has 12 polar angles, 5 radial bins and 4 edge orientations. Moreover, blurring on bins [3] is used to increase the robustness of shape context to deformation and inner-class variations. This refinement can also be encoded into contribution matrices V^I, V^M as well. LPs arising from single point figure/ground labeling as well as joint contour selection are solved efficiently by using off-the-shelf toolbox SeDuMi. Single point figure/ground labeling for each hypothesized correspondence is computed within 0.2 sec. After selecting the figure contours,

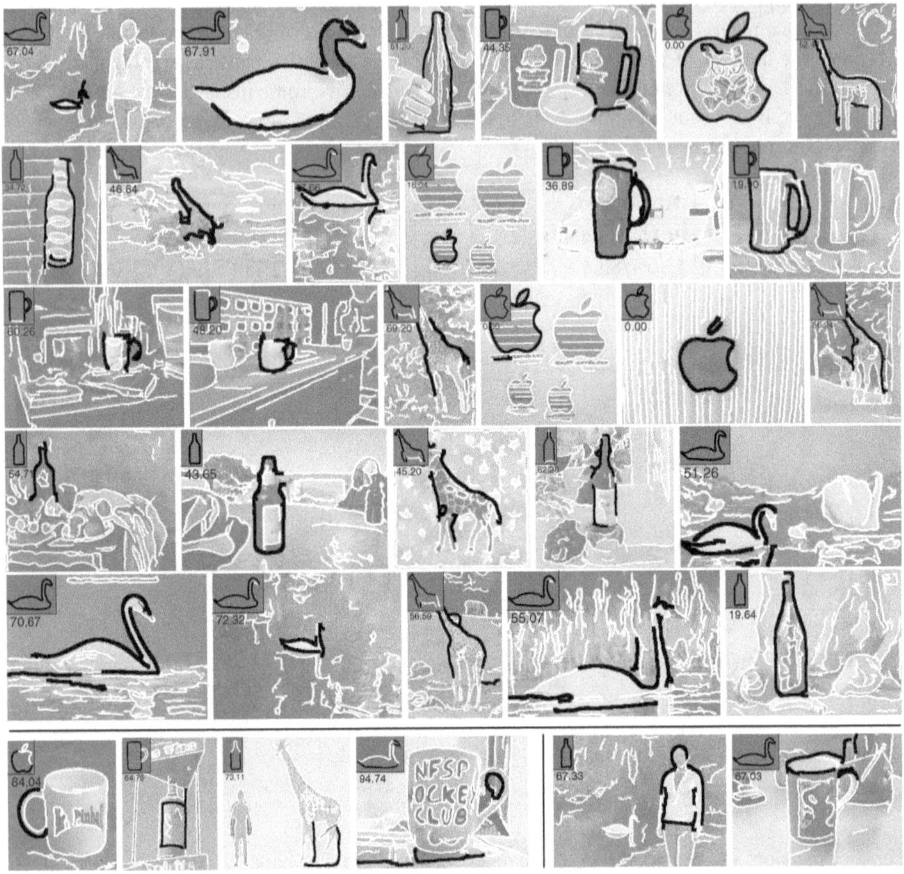

Fig. 6. Examples of contour context selection on model and image contours in ETHZ Shape Classes. The first five rows show detected objects from image with significant background clutter. In the last row, the first four cases are false positives successfully pruned by our algorithm due to bad configurations of selected model contours. The last two are failure cases. Each image only displays one detected object instance.

votes for object center weighted by shape dissimilarity are collected into a voting map. We extract local maximums in the voting map above certain threshold to generate object hypotheses. Since the correct object scale is unknown beforehand, voting is performed in a multiscale fashion, with non-maximum suppression on both position and scale.

Precision vs. recall (P/R) curve is used for quantitative evaluation. To compare with the results in [5] which is evaluated by detection rate (DR) vs. false positive per image (FPPI), we translate their results into P/R values. We choose P/R instead of DR/FPPI because DR/FPPI depends on the ratio of the number of positive and negative test images and hence is biased. Our final results on this dataset can be seen in Fig. 5. Results of the two steps of our framework are both evaluated. Single point figure/ground labeling only uses matching cost as the final evaluation for detection, while joint contour selection uses both matching cost and the detected shape configuration. Compared to the latest result in [5], our performance is considerably better on four classes out of five. We also compare voting using simple local shape context with our first step of contour selection. Contour selection greatly improves detection performances (see Fig. 5).

Our shape matching algorithm can reliably extract and select contours of object instances in test images, robust to background clutter and missing contours. Image results of detection with selected object and model contours are demonstrated in Fig. 6.

6 Conclusion

We introduce a novel shape based recognition framework called *Contour Context Selection*. We construct context sensitive shape features depending on selected contours and propose a method to search the best match. Joint selection on both image and model contours ensures detection to be robust to background clutter and accidental alignment. We are able to detect object in cluttered images using only one training example. Experiments on hard object detection task demonstrate promising results.

Acknowledgment. This work is partially supported by NSF grants: NSF-IIS-04-47953 (CAREER), NSF-IIS-03-33036(IDLP) and NSF-IIS004-31070. We would like to thank Alexander Toshev and Praveen Srinivasan for helpful discussions.

References

1. Zhu, Q., Song, G., Shi, J.: Untangling cycles for contour grouping. In: ICCV (2007)
2. Belongie, S., Malik, J., Puzicha, J.: Shape matching and object recognition using shape contexts. PAMI (2002)
3. Wang, L., Shi, J., Song, G., Shen, I.F.: Object detection combining recognition and segmentation. In: ACCV (2007)
4. Ferrari, V., Fevrier, L., Jurie, F., Schmid, C.: Groups of adjacent contour segments for object detection. PAMI (2007)
5. Ferrari, V., Jurie, F., Schmid, C.: Accurate object detection with deformable shape models learnt from images. In: CVPR (2007)
6. Shotton, J., Blake, A., Cipolla, R.: Contour-based learning for object detection. In: ICCV (2005)
7. Amir, A., Lindenbaum, M.: Grouping-based nonadditive verification. PAMI (February 1998)
8. Felzenszwalb, P.F., Schwartz, J.D.: Hierarchical matching of deformable shapes. In: CVPR (2007)
9. Zahn, C.T., Roskies, R.S.: Fourier descriptors for plane closed curves. IEEE Trans. of Computing (March 1972)
10. Gdalyahu, Y., Weinshall, D.: Flexible syntactic matching of curves and its application to automatic hierarchical classification of silhouettes. PAMI (1999)
11. Lamdan, Y., Schwartz, J.T., Wolfson, H.J.: Affine invariant model-based object recognition. IEEE Trans. Robotics and Automation 6, 578–589 (1990)
12. Amit, Y., Wilder, K.: Joint induction of shape features and tree classifiers. PAMI (1997)
13. Cootes, T.F., Taylor, C.J., Cooper, D.H., Graham, J.: Active shape models-their training and application. Computer Vision and Image Understanding 61(1), 38–59 (1995)
14. Ling, H., Jacobs, D.W.: Using the inner-distance for classification of articulated shapes. In: ICCV (2005)

Robust Object Tracking by Hierarchical Association of Detection Responses

Chang Huang, Bo Wu, and Ramakant Nevatia

University of Southern California, Los Angeles, CA 90089-0273, USA
{huangcha,bowu,nevatia}@usc.edu

Abstract. We present a detection-based three-level hierarchical association approach to robustly track multiple objects in crowded environments from a single camera. At the low level, reliable tracklets (i.e. short tracks for further analysis) are generated by linking detection responses based on conservative affinity constraints. At the middle level, these tracklets are further associated to form longer tracklets based on more complex affinity measures. The association is formulated as a MAP problem and solved by the Hungarian algorithm. At the high level, entries, exits and scene occluders are estimated using the already computed tracklets, which are used to refine the final trajectories. This approach is applied to the pedestrian class and evaluated on two challenging datasets. The experimental results show a great improvement in performance compared to previous methods.

1 Introduction

Tracking of objects is important for many computer vision applications. This is a relatively easy task when the objects are isolated and easily distinguished from the background. However, in complex and crowded environments, many objects are present that may have similar appearances, and occlude one another; also occlusions by other scene objects are common. We propose a method that can robustly track multiple objects under such challenging conditions.

Traditional feature-based tracking methods, such as those based on color [1], salient points [2] or motion blobs [3], do not have a discriminative model that distinguishes the object category of interest from others. Use of object detectors as discriminative models helps overcome this limitation. It also enables automatic initialization and termination of trajectories. However, though there has been significant improvement in object detection techniques, e.g. [4,5,6], the accuracy of the state-of-the-art object detectors is still far from perfect. Missed detections, false alarms and inaccurate responses are common (e.g. Fig.1). The tracking method must function with such failures, and also with the difficulties due to occlusions and appearance similarity among multiple objects.

In most existing association-based tracking methods, e.g. [6,7,8], an affinity score between detection responses (or tracklets) is computed once and fixed for all later processing. What we propose is a more flexible approach where associations are made in several levels and the affinity measure is refined at each level based on the knowledge obtained at the previous level. A scene model is also estimated from the tracklets and then used to generate the final object trajectories. This method is applied to track pedestrians in a number of videos; the experimental results show that our method achieves a large improvement in tracking accuracy without much increase in computational cost.

D. Forsyth, P. Torr, and A. Zisserman (Eds.): ECCV 2008, Part II, LNCS 5303, pp. 788–801, 2008.

Fig. 1. Misleading information provided by the pedestrian detector using the method of [6]

1.1 Related Work

The key issue of association based tracking is how to obtain correct associations robustly with noisy detection results. Multi-Hypothesis Tracking (MHT) [11], Joint Probabilistic Data Association Filters (JPDAFs) [12], and the particle filter based trackers [14,9] solve this problem by maintaining multiple hypotheses until enough evidence can be collected to resolve the ambiguity. Some other methods try to get better association by doing global analysis. For example, in [13], detection and tracking are coupled by Quadratic Boolean Programming that is solved by an EM-style algorithm. However, these approaches are limited for long-time association due to the combinatorial growth of hypothesis search space. Sampling methods such as MCMC [15] have also been employed to find the approximate solution in the high-dimensional hypothesis space.

The Hungarian algorithm [10] is another widely used approach for association. Wu and Nevatia [6] define an affinity measure based on position, size and color, and use the Hungarian algorithm to associate object hypotheses and detection responses at neighboring frames. Stauffer [7] first obtains tracklets by performing a conservative frame-to-frame correspondence, and then associates these tracklets by the Hungarian algorithm with an extended transition matrix that considers initialization and termination of each tracklet. He proposes a parametric source/sink model for the tracklet initialization and termination, and solves the coupled problem (estimating the source/sink model and associating tracklets) in an EM framework. Perera et al. [8] modify this extended transition matrix to maintain object identities even if their trajectories are merged or split. Kaucic et al. [17] extend Stauffer's method by introducing a segmentation based scene understanding module to estimate the locations of scene occluders. However, training the scene understanding module requires manually-labeling the background in a few frames of the video, which makes their approach not fully automatic. Singh et al.[16] use a Multiple Hypothesis Tracker to grow tracklets before associating them by the Hungarian algorithm. In these methods, the transition matrix for the Hungarian algorithm is computed only once and fixed for association. Hence, the errors in affinity computation caused by inaccurate detections are hard to be alleviated during the association process and are likely to be propagated to the higher level analysis.

1.2 Outline of our Approach

In our approach, object trajectories are obtained by progressively associating detection responses in a three-level hierarchical framework. As shown in Table.1, different models and methods are used to meet the requirements of association at different levels.

Table 1. Models and methods used in different levels of the hierarchical framework

	Motion	Appearance	Association	Scene model	Coordinates
Low level	N/A	Raw	Direct Link	N/A	Image
Middle level	Dynamic	Refined	Hungarian	General	Image
High level	Dynamic	Refined	Hungarian	Specific	Ground plane

First, at the low level, reliable tracklets are generated by linking detection responses in consecutive frames. A conservative two-threshold strategy is used to prevent "unsafe" associations until more evidence is collected to reduce the ambiguity at higher levels.

Second, at the middle level, the short tracklets obtained at the low level are iteratively associated into longer and longer tracklets. This is formulated as a MAP problem that considers not only initialization, termination and transition of tracklets but also hypotheses of tracklets being false alarms. In each round, positions and velocities of each input tracklet are estimated. This information helps refine the appearance model, and additionally provides a motion model to characterize the target. A modified transition matrix is computed and sent to the Hungarian algorithm to obtain optimal association.

Finally, at the high level, a scene structure model, including three maps for entries, exits and scene occluders, is estimated based on the tracklets provided by the middle level. Afterward, the long-range trajectory association is performed with the help of the scene knowledge based reasoning to reduce trajectory fragmentation and prevent possible identity switches.

In this framework, inaccuracies of detection responses, automatic initializations and terminations of trajectories, and occlusions by scene occluders are all taken into account. The main contributions of this paper include: 1) a novel three-level hierarchical framework to progressively associate detection responses, in which different methods and models are adopted to improve tracking robustness; 2) a modified transition matrix for the Hungarian algorithm to solve the association problem that considers not only initialization, termination and transition of tracklets but also false alarm hypotheses; 3) a novel Bayesian inference approach to automatically estimate a scene structure model as the high-level knowledge for the long-range trajectory association.

The rest of this paper is organized as follows: the proposed hierarchical association framework is elaborated in Section 2; experimental results on two challenging public video sets are shown in Section 3; some conclusions are made in Section 4.

2 Hierarchical Association of Detection Responses

We denote a detection response by $\mathbf{r}_i = (x_i, y_i, s_i, t_i, a_i)$, in which (x_i, y_i) is the position, s_i is the size, t_i is the occurrence frame index, and a_i is the color histogram; $T_k = \{\mathbf{r}_{k_i} | \forall i, t_{k_i} < t_{k_{i+1}}\}$ is an object trajectory/tracklet; $\mathcal{T} = \{T_k\}$ is the object trajectory/tracklet set; $\mathcal{T}^{\mathcal{L}}$, $\mathcal{T}^{\mathcal{M}}$ and $\mathcal{T}^{\mathcal{H}}$ are association results of the low level, the middle level and the high level respectively. We assume that a detection response can only belong to one tracklet/trajectory, which is formulized as a non-overlap constraint:

$$\forall \mathcal{E} \in \{\mathcal{L}, \mathcal{M}, \mathcal{H}\} \ \forall T_i, T_j \in \mathcal{T}^{\mathcal{E}}, \ \ T_i \cap T_j = \emptyset. \tag{1}$$

2.1 Low-Level Association

Denote by $\mathcal{R} = \{\mathbf{r}_i\}$ the set of all detection responses. The low-level association takes $\mathcal{R}$ as the input, and generates reliable tracklets by a simple and conservative method, *direct link*. Similar to that in [6], link probability between two responses is defined as the product of three affinities based on position, size and appearance:

$$P_{link}(\mathbf{r}_j|\mathbf{r}_i) = \begin{cases} A_{pos}(\mathbf{r}_j|\mathbf{r}_i)A_{size}(\mathbf{r}_j|\mathbf{r}_i)A_{appr}(\mathbf{r}_j|\mathbf{r}_i), & \text{if } t_j - t_i = 1 \\ 0, & \text{otherwise} \end{cases}. \tag{2}$$

Notice that the association can only happen between two consecutive frames. According to the non-overlap constraint in Equ.1, $(\mathbf{r}_i, \mathbf{r}_j)$ and $(\mathbf{r}_k, \mathbf{r}_l)$ are regarded as two *conflicting pairs* if $i = k$ or $j = l$. To prevent "unsafe" associations, two responses are linked if and only if their affinity is high enough and significantly higher than the affinity of any of their conflicting pairs:

$$\begin{aligned} &P_{link}(\mathbf{r}_i|\mathbf{r}_j) > \theta_1, \text{ and } \forall \mathbf{r}_k \in \mathcal{R} - \{\mathbf{r}_i, \mathbf{r}_j\}, \\ &\min\left[P_{link}(\mathbf{r}_i|\mathbf{r}_j) - P_{link}(\mathbf{r}_k|\mathbf{r}_j), P_{link}(\mathbf{r}_i|\mathbf{r}_j) - P_{link}(\mathbf{r}_i|\mathbf{r}_k)\right] > \theta_2, \end{aligned} \tag{3}$$

where θ_1 and θ_2 are the two thresholds.

Based on this *two-threshold strategy*, the low-level association can efficiently generate a set of reliable tracklets $\mathcal{T}^{\mathcal{L}} = \{T_k^L\}$. Any isolated detection response, which is not linked with any other one, is considered as a degenerate tracklet and also included in $\mathcal{T}^{\mathcal{L}}$. The low-level association does not resolve the ambiguity of conflicting pairs, as they can be figured out more effectively at higher levels.

2.2 Middle-Level Association

The middle level association is an iterative process: each round takes the tracklets generated in the previous round as the input and does further association. Take the first round, whose input is $\mathcal{T}^{\mathcal{L}}$, for example. We define a *tracklet association* as a set of tracklets: $S_k = \{T_{i_0}^L, T_{i_1}^L, \cdots, T_{i_{l_k}}^L\}$, where l_k is the number of tracklets in S_k. $T_k^M = \bigcup_{T_i^L \in S_k} T_i^L$ is the corresponding trajectory of S_k, and $\mathcal{S} = \{S_k\}$ is the *tracklet association set*.

The objective of the first round association can be formulated as a MAP problem:

$$\begin{aligned} \mathcal{S}^* &= \arg\max_{\mathcal{S}} P(\mathcal{S}|\mathcal{T}^{\mathcal{L}}) = \arg\max_{\mathcal{S}} P(\mathcal{T}^{\mathcal{L}}|\mathcal{S})P(\mathcal{S}) \\ &= \arg\max_{\mathcal{S}} \prod_{T_i^L \in \mathcal{T}^{\mathcal{L}}} P(T_i^L|\mathcal{S}) \prod_{S_k \in \mathcal{S}} P(S_k), \end{aligned} \tag{4}$$

assuming that the likelihoods of input tracklets are conditionally independent given $\mathcal{S}$, and the tracklet associations $\{S_k\}$ are independent of each other.

A Bernoulli distribution is used to model the probability of a detection response being a true detection or a false alarm. Let β be the precision of the detector, the likelihood of an input tracklet is defined as

$$P(T_i^L|\mathcal{S}) = \begin{cases} P_+(T_i^L) = \beta^{|T_i^L|}, & \text{if } \exists S_k \in \mathcal{S}, T_i^L \in S_k \\ P_-(T_i^L) = (1-\beta)^{|T_i^L|}, & \text{if } \forall S_k \in \mathcal{S}, T_i^L \notin S_k \end{cases} \tag{5}$$

where $|T_i^L|$ is the number of detection responses in T_i^L, and $P_+(T_i^L)$ and $P_-(T_i^L)$ are the likelihoods of T_i^L being a true detection and a false alarm respectively.

The tracklet association priors in Equ.4 are modeled as Markov Chains:

$$P(S_k) = P_{init}(T_{i_0}^L)P_{link}(T_{i_1}^L|T_{i_0}^L)\cdots P_{link}(T_{i_{l_k}}^L|T_{i_{l_k-1}}^L)P_{term}(T_{i_{l_k}}^L) \tag{6}$$

composed of an initialization term $P_{init}(T_{i_0}^L)$, a termination term $P_{term}(T_{i_{l_k}}^L)$ and a series of transition terms $P_{link}(T_{i_{i+1}}^L|T_{i_i}^L)$. Definitions of these terms will be given later in the subsection of implementation details.

Constrained by the non-overlap assumption in Equ.1, T_i^L cannot belong to more than one S_k. Thus, we rewrite Equ.4 by inserting $P_+(T_i^L)$ into its corresponding chain:

$$\mathcal{S}^* = \arg\max_{\mathcal{S}} \prod_{\forall S_k \in \mathcal{S}, T_i^L \notin S_k} P_-(T_i^L) \prod_{S_k \in \mathcal{S}} \Big[P_{init}(T_{i_0}^L)P_+(T_{i_0}^L)$$
$$P_{link}(T_{i_1}^L|T_{i_0}^L)\cdots P_{link}(T_{i_{l_k}}^L|T_{i_{l_k-1}}^L)P_+(T_{l_k}^L)P_{term}(T_{i_{l_k}}^L)\Big]. \tag{7}$$

This MAP formulation has a distinct property compared to the previous work [7,8]: it allows $\mathcal{S}^*$ to exclude some input tracklets, rejecting them as false alarms instead of receiving their initialization/termination penalties or transition terms by linking them.

Hungarian Algorithm. Supposing there are n input tracklets, we can transfer the MAP problem in Equ.7 into a standard assignment problem by defining a transition matrix:

$$\mathbf{C} = \left[\begin{array}{cccc|cccc} C_{11} & C_{12} & \cdots & C_{1n} & C_{1(n+1)} & -\infty & \cdots & -\infty \\ C_{21} & C_{22} & \cdots & C_{2n} & -\infty & C_{2(n+2)} & \cdots & -\infty \\ \cdots & \cdots & \cdots & \cdots & \cdots & \cdots & \cdots & \cdots \\ C_{n1} & C_{n2} & \cdots & C_{nn} & -\infty & -\infty & \cdots & C_{n(2n)} \\ \hline C_{(n+1)1} & -\infty & \cdots & -\infty & 0 & 0 & \cdots & 0 \\ -\infty & C_{(n+2)2} & \cdots & -\infty & 0 & 0 & \cdots & 0 \\ \cdots & \cdots & \cdots & \cdots & \cdots & \cdots & \cdots & \cdots \\ -\infty & -\infty & \cdots & C_{(2n)n} & 0 & 0 & \cdots & 0 \end{array}\right]_{2n\times 2n} \tag{8}$$

whose components are defined as

$$C_{ij} = \begin{cases} \ln P_-(T_i^L), & \text{if } i = j \le n \\ \ln P_{link}(T_j^L|T_i^L) + 0.5[\ln P_+(T_i^L) + \ln P_+(T_j^L)] & \text{if } i,j \le n \text{ and } i \ne j \\ \ln P_{init}(T_j^L) + 0.5\ln P_+(T_j^L), & \text{if } i = j+n \\ \ln P_{term}(T_i^L) + 0.5\ln P_+(T_i^L), & \text{if } i+n = j \\ 0, & \text{if } i > n \text{ and } j > n \\ -\infty, & \text{otherwise} \end{cases} \tag{9}$$

in which $\ln P_+(T_i^L)$ is divided into two halves that are integrated into the two neighboring transition terms respectively.

As stated before, the MAP formulation in this paper takes false alarm hypotheses into account. In particular, this is represented by the diagonal components of the transition matrix: each one is set to be the logarithmic likelihood of the tracklet being a false alarm, and the self-association of a tracklet is equivalent to rejecting it as a false alarm since it cannot be associated with any other tracklet, initialization or termination. Denoting $\Gamma^* = [\gamma^*_{ij}]_{2n \times 2n}$ as the optimal assignment matrix obtained by applying the Hungarian algorithm to the transition matrix $\mathbf{C}$, for each $\gamma^*_{ij} = 1$,

(1) if $i = j \leq n$, T_i^L is considered as a false alarm;
(2) if $i, j \leq n$ and $i \neq j$, link the tail of T_i^L to the head of T_j^L;
(3) if $i = j + n$, T_j^L is initialized as the head of the generated trajectory;
(4) if $i + n = j$, T_i^L is terminated as the tail of the generated trajectory.
In this way, we can compute $\mathcal{S}^*$ and its corresponding tracklet set $\mathcal{T}^{\mathcal{M}}$.

Implementation Details. The link probability between two tracklets is defined as the product of three components (appearance, motion and time):

$$P_{link}(T_j^L|T_i^L) = A_a(T_j^L|T_i^L)A_m(T_j^L|T_i^L)A_t(T_j^L|T_i^L). \tag{10}$$

To alleviate noises from inaccurate detections, for each input tracklet, a Kalman Filter is used to refine the positions and sizes of its detection responses and estimate their velocities. Color histograms of the detection responses are recomputed and integrated into a refined color histogram a_i^* for the tracklet by a RANSAC method.

The appearance affinity is defined by a Gaussian distribution:

$$A_a(T_j^L|T_i^L) = G(corr(a_i^*, a_j^*); 0, \sigma_c), \tag{11}$$

where $corr()$ calculates the correlation between a_i^* and a_j^*.

The motion affinity is defined as

$$A_m(T_j^L|T_i^L) = G\left(\mathbf{p}_i^{tail} + \mathbf{v}_i^{tail}\Delta t; \mathbf{p}_j^{head}, \mathbf{\Sigma}_{\Delta t}\right) G\left(\mathbf{p}_j^{head} - \mathbf{v}_j^{head}\Delta t; \mathbf{p}_i^{tail}, \mathbf{\Sigma}_{\Delta t}\right), \tag{12}$$

where Δt is the frame gap between the tail (i.e. the last detection response) of T_i^L and the head (i.e. the first detection response) of T_j^L; $\mathbf{p}_i^{head}$ (or $\mathbf{p}_i^{tail}$) and $\mathbf{v}_i^{head}$ (or $\mathbf{v}_i^{tail}$) are the refined position and estimated velocity of T_i^L at the head (or tail) (see Fig.2 for an illustration). The difference between the predicted position and the observed position is assumed to obey a Gaussian distribution.

The temporal affinity limits the maximum frame gap between two associated tracklets, and measures the probability of missed detections within the gap:

$$A_t(T_j^L|T_i^L) = \begin{cases} Z_\xi \alpha^{\Delta t - 1 - \omega}, & \text{if } \Delta t \in [1, \xi] \\ 0, & \text{otherwise} \end{cases}, \tag{13}$$

where α is the missed detection rate of the detector, ξ is an upper bound of frame gap, and Z_ξ is a normalization factor. Within the frame gap, ω is the number of frames in which the tracked object is occluded by other objects, and $\Delta t - 1 - \omega$ is the number of frames in which the tracked object is visible but missed by the detector. In practice, to compute ω, we interpolate detection responses within the frame gap (Fig.2) and check whether they are occluded by other objects by applying the occupancy map based occlusion reasoning method in [6] to $\mathcal{T}^{\mathcal{L}}$.

Initialization and termination probabilities of each tracklet are empirically set to be

$$P_{init}(T_i^L) = P_{term}(T_j^L) = Z_\xi \alpha^{\frac{1}{2}\xi}. \tag{14}$$

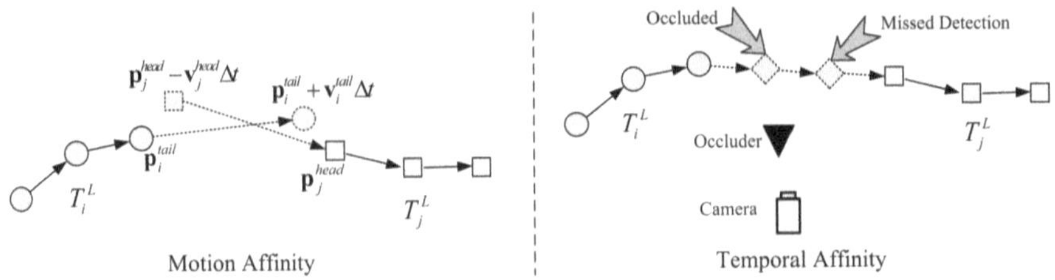

Fig. 2. Motion affinity and temporal affinity between two tracklets: the dashed circles and the dashed squares are predictions of T_i^L and T_j^L by their motion models, and the dashed diamonds are interpolated responses between the two tracklets

So far we have elaborated the first round at the middle level. In the following rounds, tracklets with longer frame gaps are associated by progressively increasing ξ.

2.3 High-Level Association

During the middle-level association, all tracklets have the same initialization/termination probabilities as there is no prior knowledge about entries and exits at that stage. This is equivalent to assuming uniform distributions of entries and exits in the scene. At the high level, an entry map and an exit map are inferred from $\mathcal{T}^{\mathcal{M}}$, which are used to specify the initialization/termination of each tracklet in the scene. In addition, a scene occluder map is also inferred from $\mathcal{T}^{\mathcal{M}}$ to revise the link probabilities. The three maps, as hidden variables, constitute a *scene structure model* in the high-level association. With a homography between the image plane and the ground plane, the scene structure model is estimated in the ground plane coordinates for better accuracy. We solve this coupled scene-estimation tracklet-association problem by an EM-like algorithm.

E-step. In the E-step, the probability distributions of entries, exits and scene occluders are calculated via Bayesian inference:

$$\frac{P\left(M_q(\overline{\mathbf{x}})=1|\overline{\mathbf{x}}\right)}{P\left(M_q(\overline{\mathbf{x}})=0|\overline{\mathbf{x}}\right)}=\frac{P\left(M_q(\overline{\mathbf{x}})=1\right)P\left(\overline{\mathbf{x}}|M_q(\overline{\mathbf{x}})=1\right)}{P\left(M_q(\overline{\mathbf{x}})=0\right)P\left(\overline{\mathbf{x}}|M_q(\overline{\mathbf{x}})=0\right)} \tag{15}$$

where $M_q(\overline{\mathbf{x}}) \in \{0,1\}$ is the indicator function for entries, exits or scene occluders ($q = \{en, ex, oc\}$) at position $\overline{\mathbf{x}}$ on the ground plane.

The positive and negative likelihoods are computed based on $\mathcal{T}^{\mathcal{M}}$. A tracklet T_k is used to propose a set of hypotheses for entries, exits and scene occluders as Fig.3.

The complete version of a tracklet T_k that includes missed detections, $\widetilde{T}_k = \{\mathbf{r}_{k_i}\}$, is obtained by filling the gaps between inconsecutive detection responses with interpolated ones. One observation is that an entry/exit is likely to be located close to the head/tail of a tracklet. Thus, two Gaussian hypotheses are proposed for the positive likelihoods of the entry and the exit respectively:

$$\begin{aligned} p_{en+}(\overline{\mathbf{x}}|T_k) &= G\left(\overline{\mathbf{x}};\, \overline{\mathbf{x}}_{en}, \sigma_p\right), \;\; \overline{\mathbf{x}}_{en} = \overline{\mathbf{p}}_k^{head} - \overline{\mathbf{v}}_k^{head}\Delta t_m \\ p_{ex+}(\overline{\mathbf{x}}|T_k) &= G\left(\overline{\mathbf{x}};\, \overline{\mathbf{x}}_{ex}, \sigma_p\right), \;\; \overline{\mathbf{x}}_{ex} = \overline{\mathbf{p}}_k^{tail} + \overline{\mathbf{v}}_k^{tail}\Delta t_m \end{aligned} \tag{16}$$

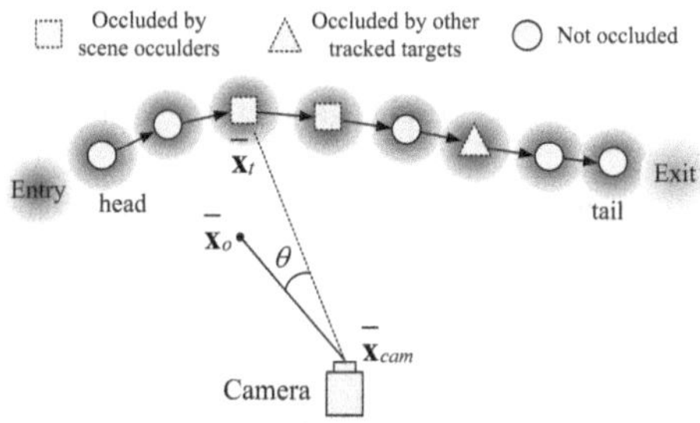

Fig. 3. Hypotheses for entries, exits and scene occluders proposed by a tracklet

where $\overline{\mathbf{p}}_k^{head}/\overline{\mathbf{p}}_k^{tail}$ and $\overline{\mathbf{v}}_k^{head}/\overline{\mathbf{v}}_k^{tail}$ are the estimated position and velocity at the head/tail of $\widetilde{T}_k^H$ by the Kalman Filter, and Δt_m is a short time span for predicting the positions of the entry and the exit. Another observation is that entries and exits are unlikely to be close to the passed region of a tracklet. Therefore, a mixture of Gaussian hypotheses is adopted to model their negative likelihoods:

$$p_{en-}(\overline{\mathbf{x}}|T_k) = p_{ex-}(\overline{\mathbf{x}}|T_k) = \frac{1}{|\widetilde{T}_k|} \sum_{\mathbf{r}_{k_i} \in \widetilde{T}_k} G\left(\overline{\mathbf{x}};\ \overline{\mathbf{x}}_{k_i}, \sigma_p\right) \tag{17}$$

where $\overline{\mathbf{x}}_{k_i}$ is the position of response $\mathbf{r}_{k_i}$.

Estimation of the scene occluder map focuses on the area between the camera and the tracklet. Given the camera position $\overline{\mathbf{x}}_{cam}$ on the ground plane, an angle-based Gaussian distribution is defined to model the occlusion relationship between a target at position $\overline{\mathbf{x}}_t$ and a scene occluder at position $\overline{\mathbf{x}}_o$:

$$\widetilde{G}_{angle}(\overline{\mathbf{x}}_o, \overline{\mathbf{x}}_t, \overline{\mathbf{x}}_{cam}) = Z_a G\left(angle(\overline{\mathbf{x}}_o - \overline{\mathbf{x}}_{cam}, \overline{\mathbf{x}}_t - \overline{\mathbf{x}}_{cam}); 0, \sigma_a\right) \tag{18}$$

where $angle()$ computes the view angle difference between the target and the occluder (i.e. θ shown in Fig.3), and Z_a is a normalization factor. The probability is maximized when the occluder lies on the line segment between the target and the camera.

An interpolated response of a tracklet is not detected by the detector. If it is not occluded by any other tracked target, it might be occluded by a scene occluder. Conversely, with high probability, there are no scene occluders lying between the camera and the detection responses. Based on these observations, a tracklet gives positive and negative likelihoods of scene occluders by

$$\begin{aligned} p_{oc+}(\overline{\mathbf{x}}|T_k) &= \frac{1}{|O_k|} \sum_{\mathbf{r}_{k_i} \in O_k} \widetilde{G}_{angle}\left(\overline{\mathbf{x}}, \overline{\mathbf{x}}_{k_i}, \overline{\mathbf{x}}_{cam}\right), \\ p_{oc-}(\overline{\mathbf{x}}|T_k) &= \frac{1}{|T_k|} \sum_{\mathbf{r}_{k_i} \in T_k} \widetilde{G}_{angle}\left(\overline{\mathbf{x}}, \overline{\mathbf{x}}_{k_i}, \overline{\mathbf{x}}_{cam}\right) \end{aligned} \tag{19}$$

where T_k is the original tracklet whose responses are all detected, and O_k is a subset of the complete tracklet $\widetilde{T}_k$, consisting of the interpolated responses that are not occluded by any other tracked target. Again the occlusion reasoning method in [6] is adopted here to decide the occlusion type of an interpolated response. Notice that interpolated responses that are occluded by other tracked targets make no contributions to the estimation of scene occluders.

Considering the hypotheses proposed by each tracklet as a set of i.i.d. samples extracted from the true likelihood distributions, we approximate the likelihoods in Equ.15 by the mixture of all hypotheses.

$$\begin{aligned} P\left(\overline{\mathbf{x}}|M_q(\overline{\mathbf{x}})=1\right) &= \frac{1}{|\mathcal{T}^{\mathcal{M}}|}\sum_{T_k\in\mathcal{T}^{\mathcal{M}}} p_{q+}(\overline{\mathbf{x}}|T_k), \\ P\left(\overline{\mathbf{x}}|M_q(\overline{\mathbf{x}})=0\right) &= \frac{1}{|\mathcal{T}^{\mathcal{M}}|}\sum_{T_k\in\mathcal{T}^{\mathcal{M}}} p_{q-}(\overline{\mathbf{x}}|T_k), \quad q=\{en,ex,oc\}. \end{aligned} \tag{20}$$

Eventually, the posterior probabilities in Equ.15 can be computed with their corresponding predefined prior probabilities.

M-step. In the M-step, the tracklets in $\mathcal{T}^{\mathcal{M}}$ are further associated to form even longer ones. Similar to the middle level, the association problem is formulated as a MAP problem and solved by the Hungarian algorithm. However, based on the scene structure model obtained from the E-step, the initialization and termination probabilities (Equ.14) of each tracklet are recomputed as

$$P_{init}(T_k) = Z_\xi \alpha^{\min(\frac{1}{2}\xi,\Delta t_{init})}, \quad P_{term}(T_k) = Z_\xi \alpha^{\min(\frac{1}{2}\xi,\Delta t_{term})}. \tag{21}$$

Δt_{init}(or Δt_{term}) is the frame number of missed detection part between the head (or tail) of T_k to the nearest entry (or exit):

$$\begin{aligned} \Delta t_{init} &= \inf\left\{\Delta t:\ P\left(M_{en}(\overline{\mathbf{x}})=1|\overline{\mathbf{x}}=\overline{\mathbf{x}}_k^{head}-\overline{\mathbf{v}}_k^{head}\Delta t\right)>0.5\right\}, \\ \Delta t_{term} &= \inf\left\{\Delta t:\ P\left(M_{ex}(\overline{\mathbf{x}})=1|\overline{\mathbf{x}}=\overline{\mathbf{x}}_k^{tail}+\overline{\mathbf{v}}_k^{tail}\Delta t\right)>0.5\right\}. \end{aligned} \tag{22}$$

where $\inf\{\}$ is the infimum of a set. Moreover, an interpolated response $\mathbf{r}_0$ at position $\overline{\mathbf{x}}_0$ is considered to be occluded by scene occluders if and only if

$$\max_{\alpha\in[0,1]}\left[P\left(M_{oc}(\overline{\mathbf{x}})=1|\overline{\mathbf{x}}=\alpha\overline{\mathbf{x}}_0+(1-\alpha)\overline{\mathbf{x}}_{cam}\right)\right]>0.5. \tag{23}$$

This is used to reivse the temporal affinity in Equ.13 by considering occlusions by scene occluders when counting the occluded frame number ω.

The scene structure model helps explain three important events for tracklet association: entering the scene, exiting the scene and being occluded by scene occluders. This greatly reduces the ambiguity of associating tracklets with long frame gaps. Compared to the parametric sources/sink model in [7] and the semi-automatic scene understanding module in [17], our approach is nonparametric and fully automatic. It directly calculates the posterior probability maps by Bayesian inference based on object tracklets.

Table.2 summarizes the overall algorithm of our hierarchical framework.

Table 2. The overall algorithm of hierarchical association framework

0) (Given): the detection response set $\mathcal{R}$, upper bounds of frame gap $\{\xi_i | \xi_i < \xi_{i+1}\}$ and the number of iterations in the middle-level association D

1) (Low-level association): obtain tracklet set $\mathcal{T}^{\mathcal{L}}$ by the direct link method.

2) (Middle-level association):

- Initialize: $\mathcal{T}^{\mathcal{M}} = \mathcal{T}^{\mathcal{L}}$.
- For $i = 1$ to D
  - ◇ (Affinity Revision): for each tracklet in $\mathcal{T}^{\mathcal{M}}$, obtain a motion model by Kalman filter and a refined appearance model by a RANSAC method; for each frame, compute an occupancy map according to $\mathcal{T}^{\mathcal{M}}$.
  - ◇ (Association): calculate the transition matrix $\mathbf{C}$ with ξ_i, obtain the optimal tracklet association set $\mathcal{S}^*$ and the corresponding $\mathcal{T}^*$. Set $\mathcal{T}^{\mathcal{M}} = \mathcal{T}^*$.

3) (High-level association):

- (E-step): estimate the scene structure model from $\mathcal{T}^{\mathcal{M}}$.
- (M-step): based on the scene structure model, repeat the iterative process in the middel-level association once with $\xi = \xi_{D+1}$ to obtain trajectory set $\mathcal{T}^{\mathcal{H}}$.

4) (Output): the complete trajectory set $\widetilde{\mathcal{T}}^{\mathcal{H}}$ by filling the frame gaps in $\mathcal{T}^{\mathcal{H}}$.

3 Experimental Results

We apply our hierarchical association framework to the multiple pedestrian tracking problem. In all experiments, the number of iterations D is set to be 3, and the upper bounds of frame gap ξ_i are set to be 8, 32, 128 for the middle level and 256 for the high level. We evaluate our method on two public video corpora: the CAVIAR set [19] and the i-LIDS AVSS AB set [20]. The CAVIAR set contains 26 videos captured in a corridor, and its ground truth contains 235 trajectories. The i-LIDS AVSS AB set contains three videos captured in a subway station, and its ground truth includes 135 trajectories. Both data sets are challenging due to heavy occlusions. We learn our pedestrian detector by the method in [6], and none of the videos in these two test sets are used for training.

Evaluation Metric. We adopt the metrics in the CLEAR evaluation [18] and use an automatic scoring software provided by the organizer. The three adopted metrics are:

(1) MOTA: Multiple Object Tracking Accuracy, calculated from the number of false alarms, missed detections, and identity switches;

(2) FGTIM: Fraction of Ground Truth Instances Missed; and

(3) FAPF: False Alarm Per Frame.

For the first score, higher is better; for the last two scores, lower is better.

For comparison, we also evaluate the method in [6] on the two sets. Table.3 lists the scores. Among them, "Wu & Nevatia's [6]" and "Our high level" are the final results of the two methods; "Our low level" and "Our middle level round i" ($i = \{1, 2, 3\}$) are the intermediate results obtained in our method, which are used to demonstrate the progressive improvement achieved by the hierarchical association framework. In addition, to show the advantage of the iterative process at the middle level, a simplified version of middle-level association with only one round, denoted as "Our middle level #", is evaluated. To show the benefit from the E-step at the high level, we evaluate a degenerate

Table 3. Evaluation results on CAVIAR and i-LIDS with CLEAR metric

	CAVIAR			i-LIDS		
	MOTA	FGTIM	FAPF	MOTA	FGTIM	FAPF
Wu & Nevatia's [6]	0.537	0.470	0.012	0.553	0.370	0.228
Our high level	0.800	0.200	0.025	0.684	0.290	0.137
Our low level	0.540	0.338	0.395	0.475	0.507	0.080
Our middle level round 1	0.627	0.332	0.141	0.490	0.507	0.042
Our middle level round 2	0.694	0.292	0.064	0.547	0.459	0.024
Our middle level round 3	0.759	0.235	0.032	0.640	0.358	0.059
Our middle level #	0.705	0.263	0.118	0.592	0.401	0.060
Our high level #	0.771	0.223	0.041	0.656	0.343	0.062

version of the high-level association (denoted as "Our high level #"), which replaces the estimated scene structure model with a general one. The homography between the image plane and the ground plane is obtained by manually labeling several points on the ground. This could also be easily derived from camera parameters.

CAVIAR Set. As shown in Table.3, from the low level to the high level, the MOTA score of our method is progressively increased, while the FGTIM score and the FAPF score are gradually decreased. The low-level association has the lowest detection rate and the highest false alarm rate since it just links detection responses in consecutive frames and doesn't take false alarm hypotheses into account. As the hierarchical association proceeds, the upper bound of frame gap ξ increases from 8 to 256. On one hand, it enables the algorithm to associate strong tracklets with long frame gaps, so that more and more missed detections can be recovered; on the other hand, it decreases the initialization and termination probabilities of each tracklet (see Equ.14), so that weak tracklets are likely to be rejected as false alarms. Compared to Wu & Nevatia's method [6], ours achieves a comparably low FAPF score, a much lower FGTIM score, and a much higher MOTA score. The first row (input detection responses) and the second row (tracking results) of Fig.4 demonstrate the improvements achieved by our method: 1) two false alarms in frame 133 are rejected by the proposed false alarm hypotheses in the MAP formulation; 2) three missed detections in frame 233 are recovered by associating long-gap tracklets; 3) an inaccurate response in frame 407 is corrected. The third row is the tracking result of another video in the CAVIAR set, and more example results are submitted as supplementary materials. The experimental results show that our method can overcome some deficiencies of the pedestrian detector as shown in Fig.1.

Compared to the iterative association approach ("Our middle level round 3") at the middle level, the simplified one ("Our middle level #") is inferior in terms of all three scores. Without the estimated specific scene structure model, the degenerate high-level association ("Our high level #") can not perform as well as the original one ("Our high level"). The comparisons justify the use of the iterative association approach and the specific scene structure model in our method.

i-LIDS Set. Similar to the results on the CAVIAR set, our method achieves big improvement for MOTA and FGTIM on the i-LIDS set. However, the FAPF score is not

Fig. 4. Tracking results of our method on the CAVIAR set: the first row and the second row are the input detection responses and the corresponding tracking results for a video; the third row is the tracking result for another video

monotonously reduced: it drops to the minimum at the second round of the middle level and rises afterward. We attribute this to the extremely heavy inter-occlusions in this data set (as shown in the second row of Fig.5): when a train stops at the station, many pedestrians rush in and out of the carriage in a short time. Tracklets of pedestrians in such crowded environments are highly fragmented and noisy. Associating these tracklets with long frame gaps is risky, since it is difficult to distinguish too many disordered targets with similar appearance based on color and motion. However, as the tradeoff between the detection rate and the false alarm rate, the MOTA score keeps increasing, and the final results of our method outperform those of Wu & Nevatia's [6] for all three scores. Again, the difference between the two modified methods ("Our middle level #" and "Our high level #") and their original versions ("Our middle level round 3" and "Our high level") justify the use of the iterative association approach and the specific scene structure model. Fig.5 shows some tracking results of our method on this set.

Scene Structure Models and Tracking Speed. In practice, the three probability maps in the scene structure model are implemented as discrete grids for efficiency. To visualize the estimated scene structure model, we binarize every probability map with a threshold of 0.5 as shown in Fig.6. It can be seen that our Bayesian inference approach in the E-step of the high level is capable of effectively reconstructing the crucial environment elements for the final association in the M-step. In particular, our method successfully locates the pillars as occluders in both scenes, which is important for the recovery of trajectories occluded by the pillar for a long time. The first row of Fig.5

Fig. 5. Tracking results of our method on i-LIDS set

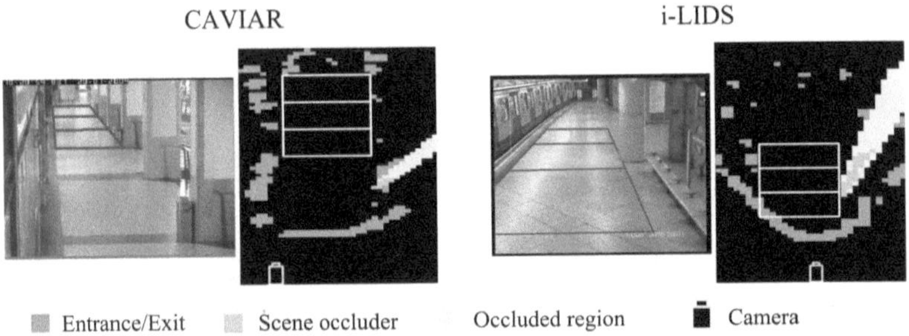

Fig. 6. Estimated scene structure models for CAVIAR and i-LIDS: the entry map and the exit map are both painted in red as they are mostly overlapped; the occluded region (in yellow) is inferred from the positions of camera (at the bottom) and scene occluders (in green); the three rectangles of each image indicate the ground plane coordinates.

shows an example from the i-LIDS set, in which the trajectory 15 is not broken although it is occluded by the pillar for more than 200 frames (i.e. 8 seconds).

For computational efficiency, a sliding window technique is used at the middle level to reduce the size of transition matrix for the Hungarian algorithm. Given the detection results, the speed of our method is about 50 FPS on a 3.0G Hz PC.

4 Conclusion

In this paper, we present a robust hierarchical association framework for the multiple object tracking problem. Experimental results on two challenging data sets show that the proposed method significantly improves the tracking performance by effectively associating tracklets with inaccurate detection responses and long-time occlusions. This hierarchical framework is a general approach, and other affinity measures or optimization methods can be easily integrated into this framework.

Acknowledgements. This research was funded, in part, by the U.S. Government VACE program.

References

1. Comaniciu, D., Ramesh, V., Meer, P.: The Variable Bandwidth Mean Shift and Data-Driven Scale Selection. In: ICCV (2001)
2. Tomasi, C., Kanade, T.: Detection and Tracking of Point Features. Technical Report CMU-CS-91-132, Carnegie Mellon University (1991)
3. Zhao, T., Nevatia, R.: Tracking Multiple Humans in Complex Situations. IEEE trans. on PAMI 26(9), 1208–1221 (2004)
4. Viola, P., Jones, M.: Rapid Object Detection using a Boosted Cascade of Simple Features. In: CVPR (2001)
5. Zhu, Q., Avidan, S., Yeh, M.C., Cheng, K.T.: Fast Human Detection Using a Cascade of Histograms of Oriented Gradients. In: CVPR (2006)
6. Wu, B., Nevatia, R.: Detection and Tracking of Multiple, Partially Occluded Humans by Bayesian Combination of Edgelet based Part Detectors. International Journal of Computer Vision (2007)
7. Stauffer, C.: Estimating tracking sources and sinks. In: IEEE Workshop on Event Mining in Video (2003)
8. Perera, A.G.A., Srinivas, C., Hoogs, A., Brooksby, G., Hu, W.: Multi-Object Tracking Through Simultaneous Long Occlusions and Split-Merge Conditions. In: CVPR (2006)
9. Li, Y., Ai, H., Yamashita, T., Lao, S., Kawade, M.: Tracking in Low Frame Rate Video: A Cascade Particle Filter with Discriminative Observers of Different Lifespans. In: CVPR (2007)
10. Kuhn, H.W.: The Hungarian method for the assignment problem. Naval Research Logistics Quarterly 2, 83–87 (1955)
11. Reid, D.: An algorithm for tracking multiple targets. IEEE Trans. Automatic Control 24, 843–854 (1979)
12. Fortmann, T., Shalom, Y.B., Scheffe, M.: Sonar tracking of multiple targets using joint probabilistic data association. IEEE J. Oceanic Engineering 8, 173–184 (1983)
13. Leibe, B., Schindler, K., Gool, L.V.: Coupled Detection and Trajectory Estimation for Multi-Object Tracking. In: ICCV (2007)
14. Okuma, K., Taleghani, A., Freitas, N.D., Little, J.J., Lowe, D.G.: A Boosted Particle Filter: Multitarget Detection and Tracking. In: ECCV (2004)
15. Yu, Q., Medioni, G., Cohen, I.: Multiple target tracking using spatio-temporal markov chain monte carlo data association. In: CVPR (2007)
16. Singh, V.K., Wu, B., Nevatia, R.: Pedestrian Tracking by Associating Tracklets using Detection Residuals. In: WMVC (2008)
17. Kaucic, R., Perera, A.G.A., Brooksby, G., Kaufhold, J., Hoogs, A.: A Unified Framework for Tracking through Occlusions and across Sensor Gaps. In: CVPR (2005)
18. CLEAREVAL, `http://www.clear-evaluation.org/`
19. CAVIAR, `http://homepages.inf.ed.ac.uk/rbf/CAVIAR/`
20. iLIDS, `http://www.elec.qmul.ac.uk/staffinfo/andrea/avss2007_d.html`

Improving the Agility of Keyframe-Based SLAM

Georg Klein and David Murray

Active Vision Laboratory, University of Oxford, UK
{gk,dwm}@robots.ox.ac.uk

Abstract. The ability to localise a camera moving in a previously unknown environment is desirable for a wide range of applications. In computer vision this problem is studied as monocular SLAM. Recent years have seen improvements to the usability and scalability of monocular SLAM systems to the point that they may soon find uses outside of laboratory conditions. However, the robustness of these systems to rapid camera motions (we refer to this quality as agility) still lags behind that of tracking systems which use known object models. In this paper we attempt to remedy this. We present two approaches to improving the agility of a keyframe-based SLAM system: Firstly, we add edge features to the map and exploit their resilience to motion blur to improve tracking under fast motion. Secondly, we implement a very simple inter-frame rotation estimator to aid tracking when the camera is rapidly panning – and demonstrate that this method also enables a trivially simple yet effective relocalisation method. Results show that a SLAM system combining points, edge features and motion initialisation allows highly agile tracking at a moderate increase in processing time.

1 Introduction

Real-time camera tracking or ego-motion recovery in known scenes is a well-studied problem in computer vision with immediate applications in robotics and augmented reality. Recent years have also seen an interest in the problem of tracking in unknown or only partially known scenes. This is a more difficult problem because the computer must not only estimate camera pose but also calculate a map of its environment. The prevalent approach is that of SLAM (Simultaneous Localisation and Mapping) which differs from recursive Structure-from-Motion (SfM) in that greater care is taken properly to correlate the emergent structure and camera pose.

A real-time, monocular SLAM implementation was demonstrated by Davison in 2003 ([1] is an up-to-date version). Davison's implementation is based on an Extended Kalman Filter which is updated every frame at $O(N^2)$ cost relative to map size, and this mandates the use of sparse and relatively small maps; further, tracking is based on small texture patches which are destroyed by motion blur; finally, the fact that every frame's measurements are integrated into the map requires a conservative approach to data association (i.e. small modelled accelerations and search ellipses). Consequently and subsequently there have been attempts to improve both the scalability and robustness of monocular SLAM. Chekhlov et al [2] improve the reliability of data association by employing the SIFT [3] descriptor, which allows for far bigger search ellipses and hence more agile tracking. Williams et al [4] also attach descriptors to map points, but do not

D. Forsyth, P. Torr, and A. Zisserman (Eds.): ECCV 2008, Part II, LNCS 5303, pp. 802–815, 2008.

use these during tracking, but instead to re-localise the camera in the event of a tracking failure. Eade and Drummond [5] employ a different statistical framework which allows denser maps, which improves tracking quality. Here, we start with our previous work [6] which de-couples the map-making process from frame-to-frame tracking, performing classical batch structure-from-motion estimation on a small number of *keyframes* as one process while tracking the camera relative to the map as another. This approach allows for more aggressive tracking, since data association errors are not automatically integrated into the map, and it is briefly described in Section 3.

Despite the advances made in monocular SLAM, we still find that rapid camera motions are generally not tracked. While the emergence of relocalisation methods has removed the need completely to re-start a SLAM system every time the camera is moved too quickly, it would be preferable if tracking did not fail in the first place. The problem with tracking rapid camera motions is that it produces large and unpredictable motion in the captured image: this large motion makes it difficult reliably to track individual features (data association) and it also produces motion blur. We use small commodity cameras with integration times approaching 30ms in indoor scenes, and so the extent of motion blur in the image can be very large (at times over 100 pixels). Most current monocular SLAM implementations use some form of interest point detector to achieve frame-rate operation (often Rosten and Drummond's FAST detector [7]), but even small amounts of motion blur will wipe the image clean of corner responses. Even if the interest point detector did fire, substantial motion blur will render useless the small texture templates typically used for feature matching. The problem can be somewhat alleviated by using features at multiple pyramid levels (as done in [6]) but high-level features are relatively sparse and even these are unreliable with heavy blur: Ultimately, tracking texture patches in the presence of blur is difficult.

In contrast to corner points, edges in the image are more resilient to motion blur. For a start, they are one-dimensional, and so locally is motion blur: One would therefore expect that even in a heavily blurred image, some edge features (those parallel to the local direction of blur) may remain intact. Further, even edges which are affected by blur can be tracked: Klein and Drummond [8] demonstrated that blurred edges can be used for tracking in real-time if an estimate for the magnitude of motion blur is known a priori. This motivates us to include intensity edges into the SLAM map. Monocular SLAM with edges has already been demonstrated by Smith et al [9] and Eade and Drummond [10], but neither of these systems attempted to exploit edges for agile tracking, as we do here. We adopt Eade and Drummond's concept of an *edgelet* – a very short, locally straight segment of what may be a longer, possibly curved, line – as our representation of intensity edges in the world. Section 4 describes the edgelet representation and our method for adding and optimising edgelets to our keyframe-based SLAM implementation, and Section 5 describes how these edgelets can be tracked robustly.

While one might expect the addition of edges to a SLAM system to improve tracking performance at high camera velocities, the same is not true for high acceleration. Rapid unmodelled accelerations cause problems for data association and can further lead to the use of an incorrect blur estimate for edge tracking – for example, when the camera rotates suddenly from rest. In [8] Klein and Drummond employed inertial sensors to provide a rotation prediction for each frame to work around this problem. Here we propose an alternative: In Section 6 we describe a procedure to estimate inter-frame rotation by full-frame direct minimisation. Finally, we show that this same method – together with the map's stored keyframes – allow a trivial relocalisation method to recover from tracking failure, and this is described in Section 7.

2 Related Work

A body of related work exists in the SfM literature for calibrated cameras. Viéville and Faugeras [11] developed an EKF that estimates the scene structure and the camera position using the measurements obtained from a system that tracked the positions of point and line features in the images. Faugeras et al [12] solved for camera motion and scene structure from three perspective images using an EKF to minimize an objective function based on the epipolar constraint. Related is Taylor and Kriegman's work [13] which proposed minimizing a cost function measuring the least squares distance between observed edge segments and the projection of reconstructed lines, and Bartoli and Sturm [14] discuss optimal line parametrisations for the the minimisation procedure. SfM using lines with uncalibrated cameras has also been demonstrated, leading to the exposition of Shashua [15] on projective tri-focal constraints.

As mentioned in the introduction, there have been previous demonstrations of monocular SLAM using edge features. The approaches of Smith et al [9] and Eade and Drummond [10] are very different, and this is in part due to differences in their underlying SLAM mechanisms: [9] employs an EKF which scales poorly with map size, and so this approach attempts to add large straight lines with two well-defined end-points to the map. By contrast [10] uses FastSLAM 2.0 and so can afford to insert large numbers of features into the map, and this is exploited by splitting large lines (straight or slightly curved) into small segments called edgelets, which are added to the map as independent units. Since the decoupled SLAM system used here scales easily to thousands of features we adopt a variant of the latter approach here.

3 Keyframe-Based SLAM Framework

This section provides a brief overview of the SLAM and maths framework [6] which is also used here. The SLAM system is split into two separate processing threads, which run simultaneously on a dual-core computer. One thread processes the 30Hz, 640×480 pixel, 8bpp grey-scale video input from an attached fire-i camera with a 2.1mm wide-angle lens: This is the tracking thread, which tracks the pose of the camera relative to the map, assuming that the map is fixed and certain. The second thread creates, expands and maintains the map from a small subset of these frames which are called keyframes. Features are added upon triangulation between two keyframes, and feature positions and keyframe poses are jointly optimised using bundle adjustment.

The map consists of a set of keyframes and 3D feature points. We define a world coordinate frame $\mathcal{W}$, a moving camera-centered coordinate frame $\mathcal{C}$, and each keyframe i also has an associated camera-centered coordinate frame $\mathcal{K}_i$. The jth point's position is stored as a 4-vector $p_{j\mathcal{W}} = (x_{j\mathcal{W}}\ y_{j\mathcal{W}}\ z_{j\mathcal{W}}\ 1)^T$ in the world coordinate frame. The ith keyframe stores its world-from-keyframe coordinate frame transformation $E_{\mathcal{W}\mathcal{K}_i}$, a three-octave image pyramid from the full 640×480-pixel image down to a 80×60-pixel scaled version, and a record of all point features measured in this keyframe.

Coordinate frame transformations are members of the Lie group SE(3), the set of 3D rigid-body transformations, and are expressed as 4×4 matrices with a rotation and a translation component

$$E_{\mathcal{AB}} = \begin{bmatrix} R_{\mathcal{AB}} & t_{\mathcal{AB}} \\ 0\,0\,0 & 1 \end{bmatrix}. \tag{1}$$

The subscript "$\mathcal{AB}$" is read as "coordinate frame $\mathcal{A}$ from coordinate frame $\mathcal{B}$", such that $E_{\mathcal{AC}} = E_{\mathcal{AB}}E_{\mathcal{BC}}$. Changes to coordinate frame transformations are achieved via another (small) transformation denoted M and minimally parametrised by a six-vector $\boldsymbol{\mu}$ via the exponential map

$$E'_{\mathcal{AB}} = ME_{\mathcal{AB}} = \exp(\boldsymbol{\mu})E_{\mathcal{AB}}, \tag{2}$$

where in this case the first and last three elements of $\boldsymbol{\mu}$ denote respectively translation and rotation about the x-, y-, and z- axes of coordinate frame $\mathcal{A}$. This representation allows for very easy differentiation of projection equations with respect to any change in coordinate frame transformations.

4 Adding Edgelets to Keyframe-Based SLAM

This section describes the method by which edgelet landmarks can be added to the SLAM system described above, so that both point feature and edgelets are tracked. We use [10]'s definition of an edgelet in an image: a locally straight, short segment of a large intensity discontinuity in the image.

4.1 Parametrisation

For each edgelet, two geometrical properties must be recorded: the position of the center of the edgelet and the direction of the edgelet. [10] store these five degrees of freedom in six numbers, using a position vector and a unit vector of direction. We find it convenient to overparametrise even further, storing each edgelet as a coordinate frame transformation $E_{\mathcal{WE}}$, where $\mathcal{E}$ is an edgelet-local coordinate frame. (With 12 internal numbers and 6 DOF this representation is far from minimal but it simplifies the 4-DOF differentiations needed later.) In the coordinate frame $\mathcal{E}$ the edgelet center is located at the origin and the edgelet direction is parallel to the z-axis, as illustrated in Fig. 1. To set the last degree of freedom, we further align the x-axis to point in the direction from dark to light in the frame in which it was first observed. While this information could be used to determine edgelet visibility, we ignore it in favour of the common method of inferring this direction from the edgelet direction, and recording visibility information in a different way (see Sect. 5.2).

4.2 Projecting and Finding Edgelets in Keyframes

To measure an edgelet in a keyframe, a good prior estimate of its position is first projected into the full-size image. The edgelet center is transformed to the keyframe's

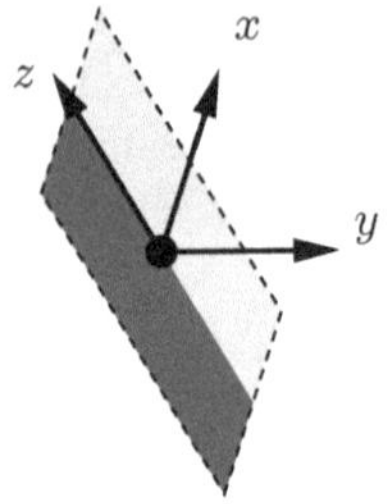

Fig. 1. The edgelet-centered 3D coordinate frame $\mathcal{E}$

camera-centered coordinate frame and then projected into pixel coordinates:

$$\boldsymbol{p}_{\mathcal{K}} = E_{\mathcal{K}\mathcal{E}} \left(0\,0\,0\,1\right)^T \tag{3}$$

$$\boldsymbol{c} = \begin{pmatrix} c_u \\ c_v \end{pmatrix} = \mathrm{CamProj}(\boldsymbol{p}_{\mathcal{K}}) \tag{4}$$

where the CamProj$(\cdot)$ function performs perspective division by z followed by the application of a calibrated camera model as in [6], which uses the radial distortion model of [16]. The edgelet direction $\boldsymbol{d}$ is obtained by differentiating the above with respect to motion along the z-axis in the $\mathcal{E}$-frame:

$$\boldsymbol{d} = \frac{\partial}{\partial \mu_3} \mathrm{CamProj}\left(E_{\mathcal{K}\mathcal{E}} M \left(0\,0\,0\,1\right)^T\right). \tag{5}$$

where μ_3 parametrises z-translation of the motion matrix M in the $\mathcal{E}$-frame. This is equivalent to projecting the rotated edgelet's z-axis into the image as done in [10].

Once projected, edgelets are measured directly in the full-size image, in a manner similar to most edge trackers [17,8,9]. The nearest locally-straight image edge is found by placing an edge of length 18 pixels in the image at the hypothesised position, populating five sample points along this edge, and then performing perpendicular searches at each sample point to find local gradient maxima of the correct polarity to sub-pixel accuracy. Data association is resolved by proximity (the perpendicular search range is short at only five pixels).

A straight line is fitted to the search results and the measurement accepted as valid only if the residual error is smaller than a tight threshold: this is to reject measurements of edgelets which are locally curved, or blurred, or confusingly close to other edges. The detected edgelet is parametrised as an infinite line $\boldsymbol{a} + \lambda\boldsymbol{b}$ as illustrated in Fig. 2.

4.3 Adding New Edgelets to the 3D Map

In both [9] and [10], new edges are added with initially unknown depth, and depth values calculated iteratively over a few frames of hopefully non-degenerate camera motion over which data association for the new edges must be maintained. This method is incompatible with keyframe-based SLAM for a number of reasons: 1. We do not want

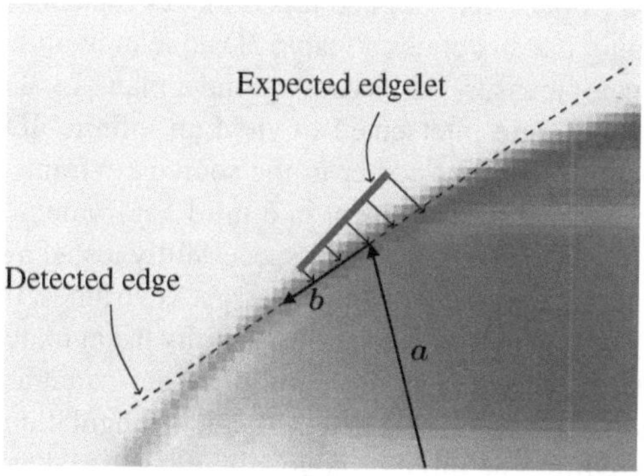

Fig. 2. Edgelet measurements in a keyframe

to impose the requirement that frame-to-frame motion should be smooth; 2. we want large quantities of features, and tracking large numbers of partially initialised features-in-making would place an unacceptably large computational burden on the frame-to-frame tracker; 3. it is not clear how to use the information from frame-to-frame depth estimation in subsequent bundle adjustments in a correct manner, short of just discarding it. By contrast, in [6] points are added by epipolar search and direct triangulation from two keyframes and we use an analogous approach here.

Every time a new keyframe is added to the map, a list of candidate edgelets which might be added to the map is generated. Where [10] suggest a fast edgelet extraction method which can be performed every frame while tracking, the map-making thread here is less bound to frame-rate constraints: We have time to perform a full Canny edge extraction [18] on the keyframe, with a modification to the edgel linker stage which breaks linked chains at points of high curvature. The result is a series of long, prominent, and reasonably straight edges in the image. (We also keep an image of all the maximal edgels, whether or not they were suppressed by the linker.) Candidate edgelets are formed by splitting these long edges into edgelet-length segments. Due to the use of Gaussian blur during the Canny extraction, edgel locations are not fully reliable and so the location of each candidates is refined by direct measurement in the keyframe as per Sect. 4.2, and candidates which are too close in the image to edgelets already existing in the map are rejected.

For each candidate edgelet, we perform an epipolar search in another near-by target keyframe to attempt to triangulate the edgelet. We do not search the entire epipolar line from zero to infinite depth, but rather restrict the search to a likely depth range based on the depth of features already found in the keyframe. The epipolar line is projected and rasterised into the target keyframe as a piecewise linear approximation (since the lenses used exhibit radial distortion, the epipolar line is not straight), and any maximal edgel encountered along the epipolar line is considered a potential match.

A single epipolar search can return a large number of matches in this way, and matches are culled by a number of heuristics. Often half of the matches can be rejected from light-to-dark polarity. Matches with gradients perpendicular to the epipolar line are further rejected. For any remaining match, an edgelet measurement in the target keyframe is attempted as per Sect. 4.2, and this removes matches which are curved or too short. Any remaining match defines a viable 3D edgelet hypothesis: The source and target keyframes' edgelet measurements each define a plane in 3D space in which the edgelet must lie, and these are intersected to yield an infinite 3D line. The edgelet's position along this is defined by its center in the source keyframe. Each hypothesis is now checked by attempting a measurement in a third keyframe; a new edgelet is only added to the map if exactly one hypothesis is successfully tested against this keyframe.

For any two keyframes chosen as source and target keyframes, there exists with high likelihood a set of edgelets which cannot be successfully triangulated due to the aperture problem (for example if the edgelets are parallel to the camera translation between the two views). Such incompatibility is detected, and triangulation of these edgelets is attempted again with a different, more orthogonally translated, keyframe.

4.4 Optimising the Map Using Bundle Adjustment on Points and Lines

Edgelets are added to the map with at least three measurements – one each from the source, target, and check keyframes – and more measurements are attempted in all old keyframes and all keyframes subsequently added. Whenever information is added to the map, the position of all edgelets, points and keyframes is optimised through bundle adjustment. To incorporate edgelets into the bundle adjustment requires the definition of an objective function, and the derivation of its differentials with respect to both keyframe pose parameters and the parameters of the edgelets.

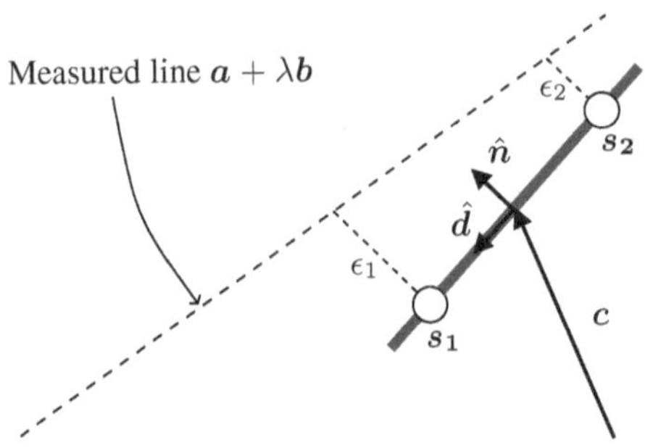

Fig. 3. Edgelet objective function measurements

Keyframe edgelet measurements provide two degrees of freedom, which can be expressed in a variety of ways. Here we use two perpendicular distances as illustrated in Fig. 3: Two virtual sample points $\boldsymbol{p}_1$ and $\boldsymbol{p}_2$ with projections $\boldsymbol{s}_1$ and $\boldsymbol{s}_2$ are initialised at a 5-pixel distance from the projected edgelet center, and the perpendicular distances from these sample points to the edge detected in the keyframe forms the error metric. Compared to a distance-plus-angle metric, this has the advantage that errors are expressed in pixel units, which makes them compatible with the M-Estimator also used for point features. The sample points are symmetrically distributed in space about the edgelet center. Represented in the $\mathcal{E}$-frame they have coordinates

$$\boldsymbol{p}_{1\mathcal{E}} = \begin{pmatrix} 0 \; 0 \; \frac{-5.0}{|\boldsymbol{d}|} \; 1 \end{pmatrix}^T, \quad \boldsymbol{p}_{2\mathcal{E}} = \begin{pmatrix} 0 \; 0 \; \frac{5.0}{|\boldsymbol{d}|} \; 1 \end{pmatrix}^T \tag{6}$$

and their projections in the image are $\boldsymbol{s}_n = \text{CamProj}(E_{\mathcal{K}\mathcal{E}}\boldsymbol{p}_{n\mathcal{E}})$. The error quantities ϵ are the edgelet-perpendicular distances from each sample point to the measured line

$$\epsilon_n = \hat{\boldsymbol{n}} \cdot \left(\boldsymbol{a} + \left(\frac{\hat{\boldsymbol{d}} \cdot \boldsymbol{s}_n - \hat{\boldsymbol{d}} \cdot \boldsymbol{a}}{\hat{\boldsymbol{d}} \cdot \hat{\boldsymbol{b}}} \right) \hat{\boldsymbol{b}} - \boldsymbol{s}_n \right) \tag{7}$$

where $\hat{\boldsymbol{n}}$ is perpendicular to $\boldsymbol{d}$ and a hat $\hat{\cdot}$ denotes a unit-normalized vector.

Differentials of the distance metric are obtained by considering the motion of points $\boldsymbol{p}_n$ in direction $\hat{\boldsymbol{n}}$ due to changes in keyframe pose and edgel parameters; for keyframe pose changes $\boldsymbol{\mu}_{\mathcal{K}}$ and edgelet motions $\boldsymbol{\mu}_{\mathcal{E}}$ they are

$$\frac{\partial \epsilon_n}{\partial \mu_{\mathcal{K}m}} = \hat{\boldsymbol{n}} \cdot \frac{\partial}{\partial \mu_{\mathcal{K}m}} \text{CamProj}\left(M E_{\mathcal{K}\mathcal{E}} \boldsymbol{p}_{i\mathcal{E}} \right), \; m = 1..6 \tag{8}$$

$$\frac{\partial \epsilon_n}{\partial \mu_{\mathcal{E}m}} = \hat{\boldsymbol{n}} \cdot \frac{\partial}{\partial \mu_{\mathcal{E}m}} \text{CamProj}\left(E_{\mathcal{K}\mathcal{E}} M \boldsymbol{p}_{i\mathcal{E}} \right), \; m = 1,2,4,5. \tag{9}$$

Only four motion parameters are considered for edgelet motion: those that have no influence on the reprojection error (translation along and rotation about the z-axis) are ignored.

5 Frame-Rate Tracking with Edgelets

5.1 Edgelet Search

The method described in Sect. 4.2 to find edgelets in keyframes is based on the premise that keyframes are not significantly corrupted by motion blur and a very good prior pose estimate exists: This allows edgelets to be found with high precision and using a very short search range. Such guarantees do not exist for frame-to-frame camera tracking, and edgelet tracking should then be fast, have a large search range, and be tolerant of motion blur. Further, while 2-DOF edgelet measurements are useful to determine edgelet orientation when building the map, the second degree of freedom is only marginally useful for constraining camera pose: For this task, a single-DOF measurement is sufficient. We therefore use a different method for measuring edgelets in the frame-rate tracker than we use for making the map.

Each edgelet is measured by performing a single perpendicular search from the edgelet center, providing a perpendicular distance measure ϵ. To provide some resilience to motion blur, we adopt the search method of Klein and Drummond [8]: Instead of searching for an intensity step in the image, the motion blur expected for each edgelet is calculated from a motion model, and the image is searched for an intensity ramp of the same length as the expected blur. The ramp search is performed by extracting a 1D intensity signal sampled along the edge normal, and cross-correlating with a single cycle of a sawtooth wave of period l. l is the edge-normal component of expected motion blur and is calculated by simply multiplying the expected camera motion γ_C during exposure by each edgelet's measurement Jacobian:

$$l = J\gamma_C \tag{10}$$

where J is the 1×6 Jacobian of an edgelet's perpendicular displacement with respect to changes in camera pose (with $\hat{n}$ calculated as before)

$$J_m = \frac{\partial \epsilon}{\partial \mu_{Cm}} = \hat{n} \cdot \frac{\partial}{\partial \mu_{Cm}} \text{CamProj}\left(M E_{C\mathcal{E}} \left(0\,0\,0\,1 \right)^T \right), \ m = 1...6. \qquad (11)$$

In contrast to Klein and Drummond [8] we search only for edge responses of the correct polarity. Further, in [8] each edgel search used the same maximum search length r_{search} and edge response threshold t_{min}. Here, we allow these parameters to vary with each edgelet, as described below.

5.2 Edgelet Visibility Table

Data association for edge tracking can be challenging because one intensity edge looks much like the next. Many systems [10,9] use only a 1-bit descriptor (the edge polarity) while [8] does not even use this. For SLAM systems without a pre-made CAD model to fall back on, the problem is exacerbated by the lack of prior occlusion information: The SLAM system will still look for an edge which has become hidden behind another, and likely lock onto the incorrect edge. One method of avoiding data association errors is to use a conservative motion model (and hence a small search range) but this is not completely effective, and would anyway be counterproductive here. In [5] a RANSAC step eliminates outliers and here we use an M-Estimator with a similar goal, but it would be preferable not to include outlier measurements in the first place.

We attempt to make edgelet measurements more reliable by building a per-keyframe database of visibility information. This is similar in spirit to the precomputed viewsphere originally employed by Harris [17] but we go beyond a binary visible/invisible flag, and our table also has an extra dimension of motion blur. For each edgelet, for each keyframe, and for each level of motion blur, we store binary visibility, max search length r_{search}, and expected edgelet response t_{min}. This table is built by the map-making thread. If an edgelet cannot be observed in a keyframe it is considered occluded from that view. On the other hand if an edgelet has been measured in a keyframe, the map-maker simulates the effects that various levels of motion blur would have on this edgelet, and records edge detection results for each case.

For each measured edgelet, the one-dimensional signal along the edgelet normal is extracted. To simulate motion blur, this is signal is convolved with box kernels of odd lengths from 3 to 65 pixels. For each blurred signal, the response to a cross-convolution with the appropriate sawtooth kernel is calculated. Each response is initially checked for presence of a maximum at the correct (edgelet-centered) location. Edges which are close to other edges will exhibit an off-center maximum with increasing blur length, and therefore the maximum level of blur under which such edgelets may be measured is recorded. Next, the edge response level t_{max} at the maximum is recorded, and we set $t_{min} = 0.5\, t_{max}$. Finally, the distance r_{near} to the nearest local maximum with response greater than t_{min} is recorded, and we set $r_{search} = 0.5\, r_{near}$.

6 Image-Based Motion Initialisation

This section describes a rapid method of estimating camera rotation between two frames. The method we use makes some simplifying assumptions that 1. the camera is purely rotating between frames, 2. either a telephoto lens is used or a spherical projection, 3. the exposure setting does not change much between frames and 4. there is no moving clutter. (None of these assumptions is valid in our application, but in practice only moving clutter becomes a problem.)

To estimate inter-frame rotation, we directly compare subsampled and blurred images of the two frames. Each incoming frame is already subsampled down to 80×60 pixels for tracking large-scale point features. We subsample this one octave further to 40×30 pixels and apply a Gaussian blur of $\sigma = 0.75$ pixels. The image thus formed for each new frame is next aligned to that of the previous frame so as to minimise the sum-squared-difference over the whole image. This is done by direct second-order minimisation [19]. We minimise over the three-parameter group SE(2) in image (pixel) space, allowing ten iterations for convergence. Finally, the resulting 3-DOF image-space transformation is converted to a best-fit 3D camera rotation by considering the motion of a few virtual sample points placed about the image, in a procedure similar to the unscented transform.

The blur parameter σ is a tuning variable: We find that increasing blur extends the convergence range, but also makes the method more susceptible to moving clutter.

7 Keyframe-Based Relocalisation

A failure-recovery system of some sort can be considered an essential requirement for any tracking system to be used outside of lab conditions. The underlying SLAM implementation described in [6] incorporates the recovery system of Williams et al [4], which performs relocalisation by training a fast classifier with map feature points. These map points can then quickly be detected when the system is lost, allowing pose recovery through the three-point-pose algorithm. Williams et al reported on an EKF-based SLAM system whose maps rarely exceeded 150 features, with a classifier storage requirement of 1.3 megabytes per map point. Here we operate with a map size of up to 15,000; scaling the classifier to this map size is clearly not practical. Even using only a fraction of the map points for relocalisation, the processing and memory requirements of the randomised list classifier become a concern.

Instead we exploit the fact that the SLAM system stores full keyframes, and relocalise directly from these. Instead of extracting some form of interest points and descriptors from keyframes and then matching a novel view against them, we find that keyframes are sufficiently densely distributed that the full image can be used as a descriptor: for each keyframe added to the map, we generate a sub-sampled 40×30 pixel image, apply a Gaussian blur of $\sigma = 2.5$ pixels, and finally subtract the mean image intensity. This zero-mean, heavily blurred image forms the keyframe's descriptor.

When tracking is lost, each incoming video frame is similarly subsampled, blurred, and mean-normalised. Next, the sum-squared-difference of this incoming image is compared to all keyframes. The camera pose is then set to the position of the keyframe with the lowest image difference. Finally, a camera rotation is estimated using the same procedure as in Sect. 6, but with extra degree of freedom to account for illumination variation: Minimisation is over the three SE(2) parameters and a mean intensity offset. Camera pose is rotated by the result of this alignment and normal tracking is re-started.

8 Implementation Notes

The motion initialisation algorithm is run at the beginning of every frame, and completely replaces the SLAM system's previous motion model. After this, points and edgelets are tracked jointly in two stages. In an initial coarse stage, only the highest-level point features and edgelets with a large (20+ pixels) expected search range are measured. After a pose update from these measurements further point features and edgelets are also measured in a fine stage. We currently employ no accurate estimate of measurement uncertainty: For points and edgelets in both tracking and bundle adjustment, measurements are assumed to have an independent single-pixel variance.

An important part of a relocalisation system is a mechanism to detect tracking failure. Such a mechanism must strike a balance between promptly detecting failure when it occurs and being lenient enough to permit tracking through difficult motions. Here we employ the same heuristic check as [6] – this is based on the ratio of point features expected to point features actually measured – however we suspend failure detection during rapid camera motion.

9 Results

9.1 Edgelets in the SLAM Map

Figure 4 shows a map of a desk generated by the system described above. The distribution of points to edgelets is fairly typical: Points outnumber edgelets by a factor of 5 to 1. This is partially explained by the fact that point features can exist at many scales,

Fig. 4. A 3D map of an office scene with 9400 point features, 1800 edgelets and 250 keyframes. Some structure is clearly visible, including a door-frame, cabinet, resting researcher and a desk with some computer screens.

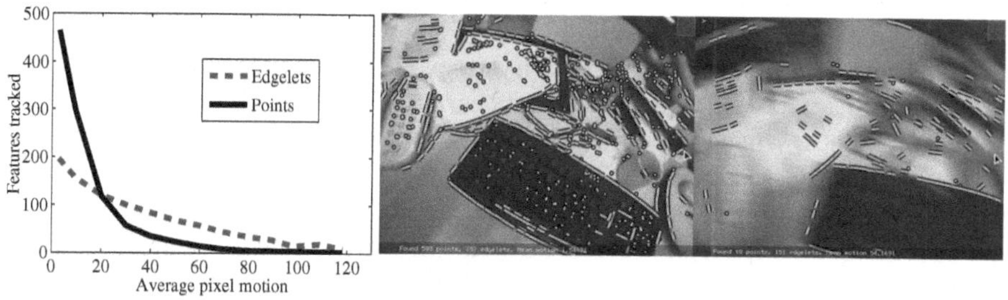

Fig. 5. The effect of image motion of number of features trackable. Left, average plots from 3500 frames. Center, in a stationary frame, 503 points and 307 edgelets are tracked. Right, with ~54 pixels of motion, 18 points and 151 edgelets are tracked.

and also that our method for adding edgelets to the map is comparatively conservative (for example, points are not checked against a third keyframe).

A slight change in user behaviour is required to properly populate the map with edges: due to the aperture problem, a baseline in two orientations must be provided to map all edges. In practice this means that exploration is now best performed with a sinusoidal or zig-zag camera motion, rather than a linear one.

9.2 Timings

On a 2.6 GHz dual-core computer, frame-to-frame processing for the illustrated map requires 12ms per frame when tracking only point features and 18ms when tracking points and edgelets. The frame-to-frame rotation initialisation is fast and adds only 0.5ms of processing time per frame. When adding a new keyframe to the map, performing edge extraction and adding edgelets can require up to 20ms per keyframe. This is twice as long as it takes to add points, but anyway pales compared to the cost of bundle adjustment, to which edgelets add a 25% extra processing burden. Edgelets are therefore individually more expensive than point features.

9.3 Agility

It is difficult to convey the behaviour of a real-time tracking system on paper, so we encourage the reader to refer to the attached results video which demonstrates the operation of the system. Subjectively, we note three changes in system behaviour: The first is an increased robustness to short and sharp camera pans, similar to a saccade of the eyes. Where the unmodified system would fail to track this and relocalise, the motion initialisation stage is able to track these sharp rotations with sufficient accuracy for the point tracker to converge at the end of the saccade. The addition of edge features is not necessary to support these motions. The second change is an improved ability to track fast linear translations perpendicular to the camera axis. For these motions, continual tracking is necessary and the addition of edge features beneficial. The final change in behaviour is negative: The motion initialisation method is susceptible to moving clutter in the image, such as a hand moved rapidly before the camera. Such moving clutter can cause tracking to fail, or (worse) shift tracking to a local minimum if observing repeated texture. These behaviours are illustrated in the accompanying video file.

Figure 5 illustrates the effect of rapid camera motion on tracking ability, sampled from 3500 frames. When the camera is moved slowly, point feature measurements outnumber edgelet measurements (as expected, since point features outnumber edgelets). As the camera is moved with increasing speed, causing motion blur, the number of point features measured drops rapidly below the number of available edgelet measurements.

9.4 Keyframe-Based Relocalisation

For a map with 250 keyframes, the new relocalisation method requires 1ms to find the best-match keyframe, and a further 0.5ms to calculate a rotation in this keyframe. This compares favourably with our previously used relocalisation method [4], which when used with 256 feature classes required up to 20ms for classification alone.

The methods differ in that our keyframe-based approach is not invariant to large rotation: While relocalisation is possible with a camera tilted up to circa 35°, full upside-down relocalisation from a novel view is not (but the utility of such a capability may be questioned). A further difference is that the keyframe-based method is more predictable in operation: Both methods occasionally fail to relocalise, and when this happens the new method guarantees relocalisation if the user simply moves the camera back to a 'known-good' location. By contrast the method of [4] may require a more random search pattern.

10 Conclusions and Further Work

This paper has demonstrated how the tracking agility of a monocular SLAM system can be improved. By combining and enhancing techniques from monocular SLAM, model-based 3D tracking and direct image-based tracking, we achieve an unprecedented level of robustness towards rapid camera motions. It is however important to distinguish between tracking agility and mapping agility: While we can track the camera during rapid motions exhibiting blur, such frames are not used for map-building. If anything, the rate at which a user can expand the map has slowed. The need for orthogonal baselines and three keyframes makes addition of edgelets to the map slower than that of points – incremental systems such as [10] can (given benign conditions) add new features more quickly. A combination of techniques may be an interesting area of future work.

Finally, the system makes no attempt to use edgelets beyond tracking (and bundle adjustment): No attempt is made to link them into longer constructs or use them to obtain occlusion information for points. While the addition of edgelets to the map makes this more visually useful to a human operator, future work should investigate how a computer can better exploit the rich geometric information available.

Acknowledgements

This work was supported by EPSRC grant GR/S97774/01.

References

1. Davison, A., Reid, I., Molton, N.D., Stasse, O.: MonoSLAM: Real-time single camera SLAM. IEEE Trans. Pattern Analysis and Machine Intelligence 29, 1052–1067 (2007)
2. Chekhlov, D., Pupilli, M., Mayol-Cuevas, W., Calway, A.: Real-time and robust monocular SLAM using predictive multi-resolution descriptors. In: Proc 2nd International Symposium on Visual Computing (November 2006)
3. Lowe, D.: Distinctive image features from scale-invariant keypoints. International Journal of Computer Vision 60(2), 91–100 (2004)
4. Williams, B., Klein, G., Reid, I.: Real-time SLAM relocalisation. In: Proc 11th IEEE International Conference on Computer Vision (ICCV 2007), Rio de Janeiro (October 2007)
5. Eade, E., Drummond, T.: Scalable monocular SLAM. In: Proc. IEEE Intl. Conference on Computer Vision and Pattern Recognition (CVPR 2006), New York, pp. 469–476 (2006)
6. Klein, G., Murray, D.: Parallel tracking and mapping for small AR workspaces. In: Proc Intl. Symposium on Mixed and Augmented Reality (ISMAR 2007), Nara (November 2007)
7. Rosten, E., Drummond, T.: Machine learning for high-speed corner detection. In: Leonardis, A., Bischof, H., Pinz, A. (eds.) ECCV 2006. LNCS, vol. 3951. Springer, Heidelberg (2006)
8. Klein, G., Drummond, T.: Tightly integrated sensor fusion for robust visual tracking. In: Proc. British Machine Vision Conference (BMVC 2002), Cardiff, pp. 787–796 (September 2002)
9. Smith, P., Reid, I., Davison, A.: Real-time monocular SLAM with straight lines. In: Proc British Machine Vision Conference (BMVC 2006), Edinburgh (September 2006)
10. Eade, E., Drummond, T.: Edge landmarks in monocular SLAM. In: Proc. British Machine Vision Conference (BMVC 2006), Edinburg (September 2006)
11. Viéville, T., Faugeras, O.D.: Feedforward recovery of motion and structure from a sequence of 2d-lines matches. In: Proc. 3rd Int. Conf. on Computer Vision, pp. 517–520 (1990)
12. Faugeras, O.D., Lustaman, F., Toscani, G.: Motion and structure from point and line matches. In: Proc. 1st Int. Conf. on Computer Vision, London, pp. 25–33 (1987)
13. Taylor, C.J., Kriegman, D.J.: Structure and motion from line segments in multiple images. IEEE Transactions on Pattern Analysis and Machine Intelligence 17(11), 1021–1032 (1995)
14. Bartoli, A., Sturm, P.: Structure from motion using lines: Representation, triangulation and bundle adjustment. Computer Vision and Image Understanding 100(3), 416–441 (2005)
15. Shashua, A.: Trilinearity in visual recognition by alignment. In: Proc. 3rd European Conf. on Computer Vision, Stockholm, May 1994, pp. 479–484. Springer, Heidelberg (1994)
16. Devernay, F., Faugeras, O.D.: Straight lines have to be straight. Machine Vision and Applications 13(1), 14–24 (2001)
17. Harris, C.: Tracking with rigid models. In: Blake, A. (ed.) Active Vision, MIT Press, Cambridge (1992)
18. Canny, J.: A computational approach to edge detection. IEEE Transactions on Pattern Analysis and Machine Intelligence (PAMI) 8(6), 679–698 (1986)
19. Benhimane, S., Malis, E.: Homography-based 2d visual tracking and servoing. Special Joint Issue on Robotics and Vision. Journal of Robotics Research 26(7), 661–676 (2007)

Articulated Multi-body Tracking under Egomotion

Stephan Gammeter[1], Andreas Ess[1], Tobias Jäggli[1], Konrad Schindler[1], Bastian Leibe[1,2], and Luc Van Gool[1,3]

[1] ETH Zürich
[2] RWTH Aachen
[3] KU Leuven, IBBT
{stephaga,aess,jaeggli,schindler,leibe}@vision.ee.ethz.ch

Abstract. In this paper, we address the problem of 3D articulated multi-person tracking in busy street scenes from a moving, human-level observer. In order to handle the complexity of multi-person interactions, we propose to pursue a two-stage strategy. A multi-body detection-based tracker first analyzes the scene and recovers individual pedestrian trajectories, bridging sensor gaps and resolving temporary occlusions. A specialized articulated tracker is then applied to each recovered pedestrian trajectory in parallel to estimate the tracked person's precise body pose over time. This articulated tracker is implemented in a Gaussian Process framework and operates on global pedestrian silhouettes using a learned statistical representation of human body dynamics. We interface the two tracking levels through a guided segmentation stage, which combines traditional bottom-up cues with top-down information from a human detector and the articulated tracker's shape prediction. We show the proposed approach's viability and demonstrate its performance for articulated multi-person tracking on several challenging video sequences of a busy inner-city scenario.

1 Introduction

Humans have truly amazing scene understanding capabilities. When walking in a busy street on our daily shopping tours, we are effortlessly aware of our surroundings; we can recognize other people visually, follow their tracks through crowded situations, and interpret their body poses from appearance cues. Computer vision is still a considerable way from this goal. While there has been a long history of research in articulated tracking (a good overview can be found in [8]), a vast majority of those papers focuses on recovering body poses of single persons in simpler environments [4,6,29,22,25,26] (notable exceptions include [2,15,17,20,21,31]). Although several approaches have demonstrated body pose estimation in static surveillance scenarios [15,31], none of those systems addresses the more challenging task of articulated tracking in unconstrained, busy street scenes, where many people overlap and partially occlude each other and where the camera itself may undergo egomotion.

This is by no means a coincidence. Articulated tracking under such conditions is extremely hard, and many factors contribute to this difficulty. Even when only tracking a single person, pose estimation and data association between frames contain significant challenges. Several state-of-the-art approaches build up articulated models from local parts in a bottom-up fashion [20,23]. Those approaches can easily get confused by the

D. Forsyth, P. Torr, and A. Zisserman (Eds.): ECCV 2008, Part II, LNCS 5303, pp. 816–830, 2008.

Fig. 1. An example for the articulated multi-person tracking scenarios considered in this paper. We address this task by first applying a robust multi-body tracker to handle the data association problem and identify individual tracks. An articulated tracker is then applied to each single-person track independently to infer precise body poses, which are in turn fed back to improve the observation model. As can be seen from our results, this procedure allows robust performance despite the presence of multiple people, temporary occlusions, scale changes, and camera motion.

abundance of human limbs that are visible in busy street scenes. Other approaches rely on global shape, which is difficult to extract in crowded situations due to clutter, overlap and occlusion, especially when the camera itself is moving [11].

When trying to track the articulations of multiple persons at the same time, additional difficulties arise from those persons' interactions. While algorithms that support multiple hypotheses can in principle deal with several people (see *e.g.* [20]), they typically do not explicitly distinguish between competing pose hypotheses for a single person in the image and different persons that are simultaneously visible. Also, relations between different subjects, such as temporary occlusion, cannot be modeled that way. A straightforward extension of a probabilistic inference algorithm to multiple subjects with occlusion reasoning requires a joint representation for the state space of multiple subjects [10], leading to an exponential increase in computational complexity.

In this paper, we propose an approach to overcome those difficulties in a system's context. The key insight behind this work is that it is not necessary to handle the complexity of multi-person interactions at the level of articulated tracking. Instead, we propose to carry out the global occlusion and multi-object reasoning on a coarser level and to only perform a more detailed articulated analysis on the output trajectories of the higher-level multi-body tracker (see Fig. 1). This allows us to also impart the articulated tracker with important information from trajectory analysis, such as a person's 3D walking direction, speed, and the knowledge when a trajectory is occluded. However, even a sophisticated multi-body tracker cannot solve the entire problem. Data association remains a challenging task: especially when multiple persons are walking close to each other, their limbs are often hard to distinguish. We address this issue by providing the articulated trackers with a guided segmentation that incorporates top-down knowledge from human detection. Together with a dynamic shape prediction from tracking, this observation model provides sufficiently precise measurements to support articulated multi-body tracking in very challenging street scenes.

In detail, this paper makes the following contributions. 1) We propose a combination of multi-body and articulated tracking for robust, multi-person 3D body pose estimation in inner-city scenes of realistic complexity. 2) We show how this general principle can be implemented in combination with a Gaussian Process (GP) articulated tracking approach based on global pedestrian silhouettes. This GP model incorporates learned prior knowledge of human shape and dynamics in order to capture the essence of human walking cycles. As regular GP training with sizable training sets is computationally very expensive, we propose several extensions to make learning tractable. 3) We augment the articulated tracker with a guided top-down/bottom-up segmentation procedure in order to reliably extract pedestrian silhouettes in busy scenes and under significant camera egomotion. 4) Finally, we demonstrate that our proposed system achieves robust performance in very challenging sequences.

The paper is structured as follows. The next section discusses related work. Section 3 gives an overview of our proposed system. The following three sections then present its different components in detail: the multi-body tracker (Sec. 4), our Gaussian Process articulated tracking approach (Sec. 5), and the guided top-down/bottom-up segmentation (Sec. 6). Finally, section 7 presents experimental results.

2 Related Work

The main challenges of 3D articulated tracking are the high-dimensional search spaces of body poses, multi-modal posterior distributions, and the fact that the images do not provide all the necessary information due to their 2D nature. Using multiple cameras and a controlled environment, ambiguities can be limited, and accurate 3D tracking results can be obtained [22,4,26]. We focus on realistic scenarios with noise and occlusions, where the scene is observed by a single camera or small-baseline stereo setup.

Many existing articulated tracking approaches can either be described as model-based generative top-down methods [25,6], or part-based bottom-up approaches [20,23] (see [8,18] for a comprehensive survey). The latter typically only allow for pairwise constraints between neighboring body parts in a graphical model of the human body. In order to infer 3D body poses from monocular or binocular image sequences, more powerful holistic prior models of possible 3D poses have been learned in [25,29].

More recently, several approaches have been proposed that learn the statistical properties of human body motion *and* the relationship between body poses and their image appearance. They rely on machine learning techniques such as kernel regressors or dimensionality reduction and can be divided into discriminative (*e.g.* [1,27]) and generative (*e.g.* [14,11,19]) methods. While discriminative approaches lead to more direct inference algorithms, they have to deal explicitly with ambiguities of the one-to-many discriminative mapping. Furthermore, they assume that the subject's 2D image location is known beforehand, which is not a trivial task for challenging multi-person scenarios such as the ones considered here. Generative approaches, on the other hand, suffer from the high dimensionality of the body pose space, which is a problem for both the learning and the generative tracking algorithms. Their performance can however be improved by

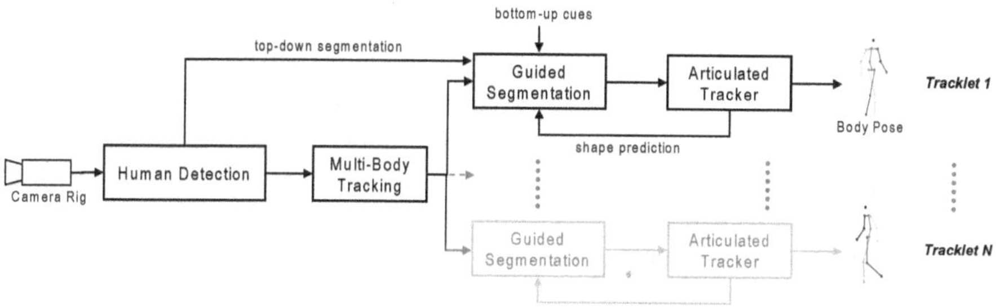

Fig. 2. Overview of the system

a suitable dimensionality reduction. [14,11] first learn such a low-dimensional pose representation and then model the mappings into the pose and appearance spaces, as well as the pose dynamics, using kernel regressors. [19] proposes an integrated formulation that obtains a dimensionality reduction in a Gaussian Process framework by estimating a low-dimensional latent space which simultaneously maps into the pose and appearance spaces. In this paper, we follow a similar line, but explicitly take into account also the dynamics, which prove to be very important for our application.

While most articulated tracking approaches consider only single persons, several methods have also been proposed for multi-person scenarios. In [17], multiple independent articulated trackers are initialized manually on different persons. [21] also automates this initialization stage by detecting stylized poses for 2D body pose estimation. Several approaches have demonstrated 3D body pose estimation in static surveillance scenarios [15,31]. Most directly related to our approach, [31] also applies a multi-object tracker to identify individual trajectories and estimate each tracked person's body poses over time. However, their tracking approach relies on background modeling, and their pose estimation process is relatively simple, based on a coarse discretization of the pose space. In very recent work, [2] propose an articulated pedestrian detector as basis for articulated multi-person tracking. While in this approach the articulation can help solve the data association problem, it is currently restricted to side-views and performs tracking only in 2D. In contrast, 3D articulated multi-body tracking from a moving, human-level perspective still remains an open issue.

3 System Overview

Fig. 2 shows the schematic layout of our multi-body articulated tracking system. A small-baseline (40cm) calibrated stereo rig mounted on a mobile platform captures two image streams and passes them on to a human detection module. Based on the obtained bounding boxes and rough stereo depth information, a multi-body tracker (Sec. 4) finds consistent object trajectories in 3D. Each trajectory is then passed to a single-person articulated tracker (Sec. 5), which estimates the person's 3D articulation based on a learned statistical representation. The estimation is made robust by a guided segmentation stage (Sec. 6) that combines the pedestrian detector's top-down segmentation with bottom-up image cues and a shape prediction inferred from the current state of the articulated tracker. This results, for every frame of the sequence, in one body pose estimate per tracked person, located in 3D world coordinates.

While in our approach, stereo-based depth computation supports the multi-body tracker and contributes to finding the subject's silhouette (see Sec.6) by setting it apart from the background, the accuracy of the depth information is limited by the small baseline between the cameras and does not allow for further disambiguation of the pose estimates (as would be possible in a true multi-camera setup [4,26,22]). The articulated pose estimation algorithm thus relies on image descriptors that are computed from the subject's silhouette. We do however take into account both image streams of the binocular sequences, which helps to alleviate problems that are caused by image noise; *i.e.* when one camera stream is temporarily corrupted by noise or occlusion, the algorithm can base its pose estimation on the second camera.

4 Multi-body Tracking

In order to reliably handle the complex interactions between multiple objects, we first address the task of tracking multiple pedestrians *without* taking into account the articulations, effectively factorizing the state space into independent "tracklets" for each visible pedestrian.

We adopt the tracking-by-detection approach from [7], which incorporates detection, stereo depth, and visual odometry to allow robust mobile tracking. The multi-body tracker itself is not the subject of this paper; we only give a high-level description of its functionality here and refer to [7] for details.

As input, the multi-person tracker takes two video streams recorded with our small-baseline stereo rig. A global world coordinate frame and ground-plane are recovered using structure-from-motion and stereo depth. Pedestrians are detected at each time-step with an ISM detector [16] and are placed in this global frame. Based on the space-time detections, an over-complete set of trajectories is tracked with independent Kalman filters. The best subset of this pool of hypotheses is selected through a global optimization procedure which enforces physical exclusion constraints, resulting in a consistent scene interpretation. The tracker is able to automatically initialize new tracks (usually, after about 5 detections) and to recover temporarily lost trajectories, thus enabling the system to track through occlusions.

The output of the tracker is a trajectory for each pedestrian in 3D world coordinates (including the person's 3D orientation, velocity, and bounding box), as well as the information when the person was occluded. As the articulated tracker is currently only trained on walking people, objects below and above a certain speed threshold are discarded. The unoccluded parts of each remaining trajectory (the "tracklets") can be processed independently by the subsequent articulated tracking module, which would otherwise become intractable. We want to point out, however, that data association between those tracklets still remains a challenging problem, as the limbs of adjacent persons may easily get confused. Section 6 therefore introduces a guided segmentation, which combines top-down information from the human detector with bottom-up image cues and which considerably improves the observation process.

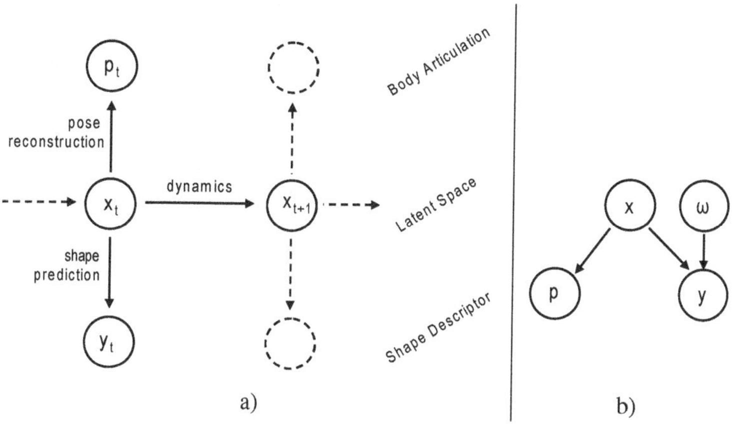

Fig. 3. Overview of the learned model. (a) Two slices of the temporal Markov chain. The arrows show the learned relationships between the variables. The low-dimensional pose representation ('Latent Space') is learned using LLE. (b) The prediction of the shape $\mathbf{y}$ depends on the low-dimensional body pose variable $\mathbf{x}$ and the orientation ω, while the body articulation $\mathbf{p}$ is only modeled as a function of $\mathbf{x}$.

5 3D Articulated Tracking

In this section, we present our articulated tracking approach. It operates on the output of the multi-body tracker and is provided with 3D trajectories and walking directions of the individual pedestrians, where many ambiguities and temporary occlusions are already resolved and accounted for by the previous stage.

The articulated tracking algorithm is based on a learned statistical model of body motions and their appearance. This model follows a generative approach to capture the relationship between body pose and image appearance and is conceptually similar to [19]; here we additionally learn a dynamical model and propose extensions that make learning tractable for sizable training sets.

The training data consists of corresponding pairs of body articulations and shape descriptors (silhouettes). The body articulations are represented as a list of spatial 3D body part locations from 20 joints of the human skeleton, as shown in Fig. 1. The matching shape descriptors are vectors computed from a detected person's bounding box, where each entry indicates whether a certain pixel lies on the foreground or on the background. We currently use bounding boxes with a resolution of 45×50 pixels and apply PCA dimensionality reduction on the resulting 2250-dimensional shape descriptor.

The tracking algorithm operates in a low-dimensional representation of the body poses that is obtained by applying Locally Linear Embedding (LLE, [24]) on the data set of body articulations. We then model the *reconstruction* of the original representation of the articulations, the *prediction* of the corresponding human shape in image space, and the temporal evolution (*dynamics*) of the body poses over time using Gaussian Process regression [12]. This model is illustrated in Figure 3.

Pose reconstruction. Gaussian Processes define probability distributions over functions and can be used to model the regression between two variables, in our case the reconstruction from the low-dimensional pose space $\mathbf{X}$ to the original articulation representation $\mathbf{P}$. Given a covariance function $k_{rec}(\mathbf{x}_i, \mathbf{x}_j)$ and a set of training pairs, a posterior *pdf* over expected reconstructions $\mathbf{p}^*$ can be computed for any point $\mathbf{x}^*$ in the low-dimensional pose space. Training a GP regression model entails finding good parameters β^{rec} of the covariance function (model selection). This can be done by maximizing the marginal likelihood of $\mathbf{P}$ with respect to the covariance parameters β^{rec}

$$P(\mathbf{P}|\mathbf{X}, \beta^{rec}) = (2\pi)^{-\frac{d_p N}{2}} |\mathbf{K}_{rec}|^{-\frac{d_p}{2}} exp(-\frac{1}{2} tr(\mathbf{K}_{rec}^{-1} \mathbf{P}\mathbf{P}^T)) . \tag{1}$$

Here, $\mathbf{X} \in \mathcal{R}^{N \times d_x}$ and $\mathbf{P} \in \mathcal{R}^{N \times d_p}$ are matrices containing the training data, N is the number of observations, and d_x and d_p are the respective dimensionalities of the appearance and pose data. The covariance matrix $\mathbf{K}_{rec} \in \mathcal{R}^{N \times N}$ is a function of the data $\mathbf{X}$ and the parameters β^{rec} of the covariance function, with elements $\mathbf{K}_{rec}^{i,j} = k_{rec}(\mathbf{x}_i, \mathbf{x}_j)$. We use standard squared exponential covariance functions with independent noise.

$$k_{rec}(\mathbf{x}_i, \mathbf{x}_j) = \beta_1^{rec} \exp\left(-\frac{\beta_2^{rec}}{2} ||\mathbf{x}_i - \mathbf{x}_j||^2\right) + \beta_3^{rec} \delta_{x_i, x_j} , \tag{2}$$

where $\beta^{rec} = \{\beta_1^{rec}, \beta_2^{rec}, \beta_3^{rec}\}$. The marginal likelihood (1) can then be optimized using numerical optimization methods such as scaled conjugate gradient.

Dynamics. In addition, our model is able to temporally predict future body poses according to a transition model $p(\mathbf{x}_{t+1}|\mathbf{x}_t)$. Similarly to (1), the marginal likelihood $P(\mathbf{X}|\beta^{dyn})$ is derived for the regression from $\mathbf{x}_t$ to $\mathbf{x}_{t+1}$ (see [30]), and optimized with respect to the parameters of the dynamics covariance function β^{dyn}.

Shape prediction. In contrast to the pose reconstruction, the shape prediction additionally depends on the orientation ω of the subject with respect to the observing camera (see Fig. 3b). In our training data, every body pose has a number of corresponding silhouettes, each viewed from a different angle. This results in NM training examples, where N is the number of poses and M the number of viewing directions in our training database. For the regression model, we thus have to optimize the marginal likelihood $P(\mathbf{Y}|\mathbf{\Omega}, \mathbf{X}, \beta^{app})$, where $\mathbf{\Omega}$ contains the viewing angles of the training shapes. Using a straightforward implementation, the complexity of the GP training algorithm scales with $(NM)^3$, since it involves the inversion of the covariance matrix $\mathbf{K}_{app} \in \mathcal{R}^{NM \times NM}$. This is impractical for the large datasets we use. We thus propose a covariance function that allows the covariance matrix to be written as a Kronecker tensor product, reducing the complexity to $O(N^3 + M^3)$ instead of the original $O((NM)^3)$. This can be done by defining the appearance covariance function as a product of a pose covariance function $k_{pose}(\mathbf{x}_i, \mathbf{x}_j)$ (*e.g.* squared exponential) and an orientation covariance function $k_{ori}(\omega_i, \omega_j)$,

$$k_{app}(\mathbf{x}_i, \omega_i; \mathbf{x}_j, \omega_j) = k_{pose}(\mathbf{x}_i, \mathbf{x}_j) k_{ori}(\omega_i, \omega_j) . \tag{3}$$

If for every pose $\mathbf{x} \in \mathbf{X} = \{\mathbf{x}_1 \ldots \mathbf{x}_N\}$ there are silhouettes for all possible viewing directions $\omega \in \mathbf{\Omega} = \{\omega_1 \ldots \omega_M\}$, then the appearance covariance matrix can be written as

$$\mathbf{K}_{app} = \mathbf{K}_{pose} \otimes \mathbf{K}_{ori} \tag{4}$$

Complexity can be further reduced by replacing the orientation covariance function with a delta function $k_{ori}(\omega_i, \omega_j) = \delta_{\omega_i,\omega_j}$. This makes sense during training of the GP regression, because the training samples only involve a number of discrete viewing directions $\omega \in \mathbf{\Omega}$. Once the regression parameters have been learned with this additional approximation, the orientation covariance function can then be replaced by one with a larger support (*e.g.* a Von Mises distribution), in order to allow for interpolations between the discrete viewing directions $\omega \in \mathbf{\Omega}$.

Learning the embedding. The marginal likelihood of the entire learned model can now be written as

$$P(\mathbf{P}, \mathbf{Y}, \mathbf{X}|\mathbf{\Omega}, \beta^{rec}, \beta^{app}, \beta^{dyn}) = P(\mathbf{P}|\mathbf{X}, \beta^{rec})P(\mathbf{Y}|\mathbf{\Omega}, \mathbf{X}, \beta^{app})P(\mathbf{X}|\beta^{dyn}). \tag{5}$$

Rather than just optimizing the regressors, as done here, (5) could be optimized with respect to the latent positions $\mathbf{X}$ as well, where the LLE coordinates serve as an initialization, similarly to [19] . This would lead to a multi-set extension of the Gaussian Process Latent Variable Model [12] with separate covariance functions for each of the mappings. However, our experiments suggest that this does not improve the tracking results; they thus do not justify the increase in the number of parameters to be optimized from less than ten to several thousand.

Articulated Tracking. The articulated tracking algorithm operates on the output trajectories of the multi-body pedestrian tracker of Sec. 4, which delivers 2D image locations, scales, and orientations of the tracked persons. Its observations are automatically estimated pedestrian silhouettes, obtained through the guided segmentation procedure of Sec. 6. A particle filter serves as an overall framework, where at time t the body pose hypotheses $\mathbf{x}_t^i$ are propagated in the low-dimensional pose space according to the learned dynamical model. For each particle $\mathbf{x}_t^i$, a shape $\mathbf{y}_t^i$ can be predicted by taking into account the 3D track orientation ω_t, estimated by the multi-body tracker. The particles are then weighted with their image likelihoods, obtained by comparing the predicted shape to the actually observed shape $\mathbf{y}_t^{obs}$,

$$w^i \propto p(\mathbf{y}_t^{obs}|\omega_t, \mathbf{x}_t^i) = \mathcal{N}(\mathbf{y}_t^{obs}; \mu_t^i, \mathbf{\Sigma}_t^i) , \tag{6}$$

where μ_t^i and $\mathbf{\Sigma}_t^i$ are the mean and covariance matrix of the predicted shape.

Finally, once the particle filter has been run on all images of a tracklet, a Viterbi algorithm extracts a smooth and consistent trajectory through the particle set (note that this can in practice be approximated with a fixed temporal look-ahead). Again, the transition costs between neighboring states are based on the learned dynamical model. In order to account for variations in the framerate of the sequence and the walking speed of the subjects, this step additionally chooses between different scaling factors of the predicted velocities, *i.e.* accelerated and slowed-down variants of the dynamical model.

6 Guided Adaptive Segmentation

As an interface between the multi-body tracker and the articulated tracker, we are using a set of automatically estimated figure-ground segmentations for each tracked person. In the majority of previous works [19,27,31], silhouettes are assumed to be available and are in practice often obtained using background modeling. Since we are dealing with a moving camera setup, we cannot use this option. Instead, we propose to obtain the segmentations by fusing top-down cues (from the detector and the articulated tracker) with bottom-up image information (from color and stereo depth). Keeping in line with previous work by several authors [3,5], the segmentation is formulated as an energy minimization problem with respect to the foreground/background labelling $C = \{c_0, c_1, \ldots\}$ of all pixels.

$$E(C) = \sum_i R(c_i) + \lambda \sum_{i,j \in \mathcal{N}} B(c_i, c_j) \,. \tag{7}$$

In the above equation, $R(c_i)$ denotes the region term for a pixel with index i, which has a label c_i (figure/ground). $R(c_i)$ is based on the top-down segmentation map $\mathbf{f}$ of the detector and the shape prediction map $\boldsymbol{\pi} = \sum_j w_t^j \mu_t^j$ of the articulated tracker, where w_t^j is the weight of sample j and μ_t^j is its predicted shape from (6).

$$R(c_i) = -\log(P_\pi(c_i)\; P_f(c_i)) \tag{8}$$

Here, P_π and P_f are the probabilities of a certain label given the segmentation maps $\boldsymbol{\pi}$ and $\mathbf{f}$ from the articulated tracker and from the detector respectively:

$$P_\pi(c_i) = \begin{cases} \pi_i & \text{if } c_i = 1 \\ 1 - \pi_i & \text{if } c_i = 0 \end{cases} \,. \tag{9}$$

The boundary term $B(c_i, c_j)$ encodes the belief that region boundaries typically coincide with intensity and depth discontinuities. It is defined on the 4-neighborhood $\mathcal{N}$ and penalizes neighboring pixels with different labels but similar colors $\mathcal{I}_i$ and depths $\mathcal{D}_i$

$$B(c_i, c_j) = e^{-\frac{|\mathcal{I}_i - \mathcal{I}_j|^2}{2\sigma_c^2}}\; e^{-\frac{|\mathcal{D}_i - \mathcal{D}_j|^2}{2\sigma_d{}^2}}\; \delta_{c_i \neq c_j} \,. \tag{10}$$

The resulting cost function can be minimized efficiently using standard graph-cut methods [3], yielding the binary foreground mask $\mathbf{y}_t^{obs}$. Together with the bounding box position and motion direction from the multi-body tracker, this mask serves as the input for inference in the articulated tracker.

By ways of the expected appearance $\boldsymbol{\pi}$, we can incorporate prior knowledge about human shapes from the articulated tracker into the segmentation task, which can effectively complete partial segmentation maps $\mathbf{f}$. We can furthermore bridge missing detections by feeding back the tracker's expectation and combining it with bottom-up information. This is demonstrated in the sequence shown in Fig. 4: at first, the segmentation works well, giving a good initialization to the particle filter. In later frames, the detector fails due to missing contrast, but the prediction, along with the depth map, is good enough to obtain a usable segmentation.

Fig. 4. Original images (top) and sample segmentations obtained using our guided segmentation (bottom). If no detection is available, the articulated tracker's prediction is used in conjunction with bottom-up cues (red silhouettes).

Of course, care has to be taken not to reinforce erroneous feedback, which might lead to hallucinated walking cycles. Therefore, the influence of π is kept low as long as a detection is present and its weight is only increased when the trajectory contains holes. Even in those cases, however, the boundary terms usually restrict the segmentation well enough.

7 Results

Training. For training the articulated tracker, we recorded motion capture data (at 30 Hz) of 6 different people walking at speeds between 3 and 6 km/h. The resulting data set consists of slightly more than 2000 different body poses, each represented by 20 joint locations (*i.e.* 60 dimensions). For every body pose, silhouettes were rendered for 36 different viewing directions. A three-dimensional LLE of the body pose data serves as the low-dimensional pose space for the GP regression model. The marginal likelihood was optimized with scaled conjugate gradients using the FITC sparse approximation with 200 inducing variables [28,13].

Results. We demonstrate our approach on 3 challenging video sequences showing real-world inner-city scenes. These videos were captured at about 13–14 fps from a mobile recording platform. Note that such a low framerate complicates the articulation reasoning considerably. Table 1 gives an overview over the sequences used in this paper. Video sequences are available on the following webpage: `http://www.vision.ee.ethz.ch/~stephaga/eccv2008/`

The first sequence (Fig. 5) was taken on a busy sidewalk. Even though the camera itself is standing still, traditional background subtraction would be difficult due to small

Fig. 5. Articulated multi-person tracking results for test sequence #1. The last row shows a 3D visualization of the estimated world state in the three images of the second row (This figure is best viewed in color).

camera shake, as well as passing trams and cars. While most people move sideways, they still occur at different depths and often have a slightly tilted trajectory, which we can account for by tracking directly in 3D. In the sequence's 454 frames, our system tracks 20 out of 23 people successfully with the multi-body tracker. One of the missed pedestrians runs too fast, and another one is at all times occluded by other persons. In addition to the 20 correct tracks, the system yields two additional tracks that contain errors due to wrongly estimated orientation or scale.

Counting each person individually, this amounts to a total track length of 932 frames, where a detection is available in 86% of the cases. We visually inspected all the resulting segmentations and found that 55% of these are well-defined (meaning the entire person is covered) in at least one camera. For the individual cameras, only 41% were well-defined. This underlines the usefulness of a stereo system in such real-world scenarios with frequent occlusions. While these numbers might seem low, we would like

Table 1. Sequences used for evaluating the proposed system. We report the number of (walking) persons, and the ones actually found by the multi-body tracker (MBT).

Seq.	# Frames	Pedestrians	Found by MBT
#1	454	23	20
#2	173	14	10
#3	242	21	17

Fig. 6. Articulated tracking results for test sequence #2. Note the robust articulation estimates of tracked pedestrians under scale changes and egomotion. (This figure is best viewed in color).

to note that the articulated tracker can also operate if only parts of the body are segmented correctly (most importantly, the legs). Based on these segmentations, the system tracks 74 walking cycles, 54 out of which were entirely correct. The remaining 20 cases mainly occur at the end of longer trajectories and are mostly due to multiple, consecutive bad segmentations or occlusions. Note, however, that the silhouettes generally did not contain enough information to unambiguously recover the arm positions, which additionally differed from our training examples since many people were carrying shopping bags or similar accessories. Example pose estimates of our system are shown in Fig. 5.

For the remaining sequences, we show qualitative results in Figs. 6 and 7. As the multi-body tracker takes care of the mapping between the world coordinate frame and

Fig. 7. Articulated tracking results for test sequence #3. This sequence shows a very challenging scenario with considerable egomotion and many pedestrians entering the visible scene at various distances and from different directions. (This figure is best viewed in color).

the local articulated trackers, we can apply our system to scenes captured under significant egomotion. Fig. 6 shows an example of such a case, where people enter the scene from several directions and undergo large scale changes. As the multi-body tracker restricts the sampling for orientations, we can still get acceptable results on such data. A more challenging case is shown in Fig. 7. Here, the system has to cope with more extreme scale changes and people moving in many different directions, while following one person through a busy pedestrian zone. In particular movement parallel to the viewing direction is highly ambiguous; still the articulation is identified in most cases.

8 Conclusion

We have presented a system for articulated multi-body articulated tracking from a moving platform. Our approach achieves good results in challenging real-world scenarios by factorizing the problem into separate tasks of multi-body tracking under occlusion and articulated body pose estimation for individual trajectories. This formulation allows the articulated tracker to benefit from trajectory-level information about the tracked person's speed and walking direction, which considerably simplifies inference and renders the problem tractable. We have further presented a way to implement this idea with an articulated tracker based on Gaussian Processes and have shown how the framework can be applied under egomotion with the help of a guided top-down/bottom-up segmentation module. Experimental results confirm the viability of our proposed approach.

Currently, our method is restricted to learned articulations from known actions such as walking and running; in contrast to bottom-up approaches it cannot recover arbitrary body poses. We are exploring ways to replaces the global body models by semi global ones to mitigate this issue. In addition, the results of our estimation could be used to learn specialized color models for different body parts, which then support more general pose recovery [21]. The method for pose inference from [2] could also be a promising extension, since it is based on local appearance and may enable a more direct interplay between detector and tracker. In addition, we will enlarge the training set of learned gait dynamics in order to more densely cover the range of different walking styles and will include gait modifications when carrying luggage items. Finally, we plan to extend the feedback from body pose estimation to the multi-body tracker in order to also improve the detection model by incorporating gait dynamics, similar to [9].

Acknowledgements. The authors thank Tobias Müller for developing the graph-cut-based segmentation. This project has been funded in parts by Toyota Motor Corporation/Toyota Motor Europe and the EU projects DIRAC (IST-027787) and HERMES (IST-027110).

References

1. Agarwal, A., Triggs, B.: Recovering 3d human pose from monocular images. PAMI 28(1), 44–58 (2006)
2. Andriluka, M., Roth, S., Schiele, B.: People-tracking-by-detection and people-detection-by-tracking. In: CVPR 2008 (2008)
3. Boykov, Y., Lea, G.: Graph cuts and efficient n-d image segmentation. IJCV 70(2) (2006)
4. Carranza, J., Theobalt, C., Magnor, M., Seidel, H.-P.: Free-viewpoint video of human actors. In: SIGGRAPH 2003 (2003)
5. Cremers, D., Rousson, M., Deriche, R.: A review of statistical approaches to level set segmentation: integrating color, texture, motion and shape. IJCV 72 (2007)
6. Deutscher, J., Blake, A., Reid, I.: Articulated body motion capture by annealed particle filtering. In: CVPR 2000 (2000)
7. Ess, A., Leibe, B., Schindler, K., van Gool, L.: A mobile vision system for robust multi-person tracking. In: CVPR 2008 (2008)
8. Forsyth, D., Arikan, O., Ikemoto, L., O'Brien, J., Ramanan, D.: Computational studies of human motion: Part 1. Foundations Trends Comp. Graphics Vision 1(2/3) (2006)
9. Giebel, J., Gavrila, D., Schnörr, C.: A bayesian framework for multi-cue 3d object tracking. In: Pajdla, T., Matas, J(G.) (eds.) ECCV 2004. LNCS, vol. 3021, Springer, Heidelberg (2004)
10. Hue, C., Cadre, J.-P.L., Prez, P.: Tracking multiple objects with particle filtering. IEEE Trans. on Aerospace and Electronic Systems 38(3) (2002)
11. Jaeggli, T., Koller-Meier, E., Van Gool, L.: Learning generative models for monocular body pose estimation. In: Yagi, Y., Kang, S.B., Kweon, I.S., Zha, H. (eds.) ACCV 2007, Part I. LNCS, vol. 4843, Springer, Heidelberg (2007)
12. Lawrence, N.: Probabilistic non-linear principal component analysis with gaussian process latent variable models. JMLR 6 (2005)

13. Lawrence, N.: Learning for larger datasets with the gaussian process latent variable model. In: Proc. Intern. Workshop on Artificial Intelligence and Statistics (2007)
14. Lee, C.-S., Elgammal, A.: Modeling view and posture manifolds for tracking. In: ICCV 2007 (2007)
15. Lee, M.-W., Nevatia, R.: Human pose tracking using multi-level structured models. In: Leonardis, A., Bischof, H., Pinz, A. (eds.) ECCV 2006. LNCS, vol. 3951, Springer, Heidelberg (2006)
16. Leibe, B., Leonardis, A., Schiele, B.: Robust Object Detection with Interleaved Categorization and Segmentation. IJCV 77(1-3) (2008)
17. Mitchelson, J., Hilton, A.: Simultaneous pose estimation of multiple people using multiple-view cues with hierarchical sampling. In: BMVC 2003 (2003)
18. Moeslund, T., Hilton, A., Krüger, V.: A survey of advances in vision-based human motion capture and analysis. CVIU 104(2) (2006)
19. Navaratnam, R., Fitzgibbon, A.W., Cipolla, R.: The joint manifold model for semi-supervised multi-valued regression. In: ICCV 2007 (2007)
20. Ramanan, D., Forsyth, D.: Finding and tracking people from the bottom up. In: CVPR 2003 (2003)
21. Ramanan, D., Forsyth, D., Zisserman, A.: Strike a pose: Tracking people by finding stylized poses. In: CVPR 2005 (2005)
22. Ren, L., Shaknarovich, G., Hodgins, J., Pfister, H., Viola, P.: Learning silhouette features for control of human motion. ACM Trans. Graphics 24(4) (2005)
23. Ren, X., Berg, A., Malik, J.: Recovering human body configurations using pairwise constraints between parts. In: ICCV 2005 (2005)
24. Roweis, S., Saul, L.: Nonlinear dimensionality reduction by locally linear embedding. Science 290(5500) (2000)
25. Sidenbladh, H., Black, M., Fleet, D.: Stochastic tracking of 3d human figures using 2d image motion. In: Vernon, D. (ed.) ECCV 2000. LNCS, vol. 1842, Springer, Heidelberg (2000)
26. Sigal, L., Bhatia, S., Roth, S., Black, M., Isard, M.: Tracking loose-limbed people. In: CVPR 2004 (2004)
27. Sminchisescu, C., Kanaujia, A., Li, Z., Metaxas, D.: Discriminative density propagation for 3d human motion estimation. In: CVPR 2005 (2005)
28. Snelson, E., Ghahramani, Z.: Sparse Gaussian processes using pseudo-inputs. In: NIPS 2006 (2006)
29. Urtasun, R., Fleet, D., Hertzmann, A., Fua, P.: Priors for people tracking from small training sets. In: ICCV 2005 (2005)
30. Wang, J., Fleet, D., Hertzmann, A.: Gaussian process dynamical models. In: NIPS 2006 (2006)
31. Zhao, T., Nevatia, R.: Tracking multiple humans in complex situations. PAMI 26(9) (2004)

Robust Real-Time Visual Tracking Using Pixel-Wise Posteriors*

Charles Bibby and Ian Reid

Active Vision Lab
Department of Engineering Science
University of Oxford
cbibby@robots.ox.ac.uk,ian@robots.ox.ac.uk

Abstract. We derive a probabilistic framework for robust, real-time, visual tracking of previously unseen objects from a moving camera. The tracking problem is handled using a bag-of-pixels representation and comprises a rigid registration between frames, a segmentation and online appearance learning. The registration compensates for rigid motion, segmentation models any residual shape deformation and the online appearance learning provides continual refinement of both the object and background appearance models. The key to the success of our method is the use of pixel-wise posteriors, as opposed to likelihoods. We demonstrate the superior performance of our tracker by comparing cost function statistics against those commonly used in the visual tracking literature. Our comparison method provides a way of summarising tracking performance using lots of data from a variety of different sequences.

1 Introduction

Fast and reliable visual tracking is a prerequisite for a vast number of applications in computer vision. Though it has been the subject of intense effort over the last two decades, it remains a difficult problem for a number of reasons. In particular, when tracking previously unseen objects, many of the constraints that give reliability to other tracking systems – such as strong prior information about shape, appearance or motion – are unavailable.

One technique that has shown considerable promise for its ability to perform tracking and segmentation within a unified framework is the use of an implicit contour, or level-set to represent the boundary of the target [1,2,3]. As well as handling topological changes seamlessly, tracking using level-sets can be couched in a fairly standard probabilistic formulation [4,5], and hence can leverage the power of Bayesian methods.

In this paper, we present a novel probabilistic framework for combined tracking and segmentation, which, as well as capturing all of the desirable properties of level-set based tracking, is very robust and runs in a few milliseconds on standard hardware. We base our framework on a generative model of image formation that represents the image as a bag-of-pixels [6]. The advantage of such

* We gratefully acknowledge the sponsorship of Servowatch Systems Ltd.

D. Forsyth, P. Torr, and A. Zisserman (Eds.): ECCV 2008, Part II, LNCS 5303, pp. 831–844, 2008.

a model – in common with other simpler density-based representations such as colour-histograms – is the degree of invariance to viewpoint this confers.

Like [4], we derive a probabilistic, region based, level-set framework, which comprises an optimal rigid registration, followed by a segmentation to re-segment the object and account for non-rigid deformations. Aside from issues of speed (which are not addressed in [4]) there are a number of key differences between [4] and our work, some of which stem from the generative model we use for image data (see Sect. 2). First, our derivation gives a probabilistic interpretation to the Heaviside step function used in most region based level-set methods [7,4]. Second, given this interpretation we propose a pixel-wise posterior term, as opposed to a likelihood, which allows us to marginalise out model parameters at a pixel level. As we show in Sect. 2, this derives naturally from our generative model, and is a subtle but absolutely crucial difference between our method and others e.g. [4,2,3], as our results show in Sect. 7. Third, in contrast to [7,4] and similar to [8,9] we assume a non-parametric distribution for image values as opposed to a single Gaussian (for an entire region). Finally, we introduce a prior on the embedding function which constrains it to be an approximate signed distance function. We show that this gives a clean probabilistic interpretation to the idea proposed by [10] and avoids the need for reinitialisation of the embedding function that is necessary in the majority of level-set based approaches.

Our work also bears some similarity to [11] who sought the rigid transformation that best aligns a fixed shape-kernel with image data using the Bhattacharyya coefficient. This work extended the pioneering work of this type [12,13] to handle translation+scale+rotation as opposed to translation only or translation+scale. In contrast to [11], however, we allow the shape to change online and propose a novel framework using pixel-wise posteriors, which removes the cost of building an empirical distribution and testing it with the Bhattacharyya coefficient. This has a second hidden benefit as it avoids the need to build a 'good' empirical distribution given limited data; we find in practice this gives a significant improvement over [12,13,11].

Unlike [4], [8,9] use a non-parametric distribution for image data. They derive contour flows based on both KL-divergence and the Bhattacharyya coefficient. Though they demonstrate that both are effective for tracking, they do not model rigid transformation parameters explicitly, they must recompute their non-parametric distributions at every iteration, and – as we show in Sect. 7 – objectives based on the Bhattacharyya coefficient are inferior to the one we propose.

Finally, it is worth mentioning template based tracking methods (see [14] for an excellent summary of past work). We include an ideal SSD cost in our comparative results (Sect. 7), which uses the correct template at each frame. Though this unfairly advantages the SSD method – since in reality the exact template is *never* available – it does suggest that in future there would be benefit in considering how spatial information can be incorporated. Thus although our method is currently based solely on non-parametric distributions we are currently investigating ways to augment it with spatial information.

The key problem with template tracking is how to adapt the template over time. Within our framework (and other similar work), because the segmentation is performed rapidly and reliably online, the appearance and shape models of the object can be updated over time without suffering from the significant problems of drift that plague other algorithms.

Our framework is general enough to be extended to various types of prior information and various imaging modalities, but in this paper we restrict ourselves to the problem of tracking the 2D projections of either 2D or 3D objects in ordinary colour video. In summary, the key benefits of our method are: (i) an extensible probabilistic framework; (ii) robustness - given by pixel-wise posteriors and marginalisation; (iii) real-time performance; (iv) excellent cost function characteristics; (v) no need to compute empirical distributions at every frame; (vi) online learning (i.e. adaption of appearance and shape characteristics); (vii) flexibility to track many different types of object and (viii) high invariance to view and appearance changes.

The remainder of this paper is organised as follows: Sect. 2 describes the representation of the object being tracked and derives a probabilistic framework from a simple generative model; Sect. 3 outlines the level-set segmentation; Sect. 4 shows the registration process; Sect. 5 describes our method for dealing with drift; Sect. 6 outlines the online learning process; Sect. 7 shows our results and Sect. 8 concludes with a summary and discussion.

2 The Generative Model

We represent the object being tracked by its shape $\mathbf{C}$, its location in the image $\mathbf{W}(\mathbf{x}, \mathbf{p})$ and two underlying appearance models: one for the foreground $P(\mathbf{y}|M_f)$ and one for the background $P(\mathbf{y}|M_b)$. Figure 1 illustrates this with a simple example.

Shape. is represented by the zero level-set $\mathbf{C} = \{\mathbf{x}|\mathbf{\Phi}(\mathbf{x}) = 0\}$ of an embedding function $\mathbf{\Phi}(\mathbf{x})$ [1,5]. The pixels $\mathbf{\Omega}$ in the object frame are segmented into two regions: one for the foreground $\mathbf{\Omega}_f$ and one for the background $\mathbf{\Omega}_b$.

Location. is described by a warp $\mathbf{W}(\mathbf{x}, \mathbf{p})$ that takes a pixel location $\mathbf{x}$ in the object frame and warps it into the image frame according to parameters $\mathbf{p}$. This warp must form a group [14]; however, this is acceptable as many common useful transformations in computer vision do form groups, for instance: translation, translation+scale, similarity transforms, affine transforms and homographies.

Appearance models. $P(\mathbf{y}|M_f)$ and $P(\mathbf{y}|M_b)$ are represented with YUV histograms using 32 bins per channel. The histograms are initialised either from a detection module or a user inputted initial bounding box. The pixels inside the bounding box are used to build the foreground model and the pixels from an inflated bounding box are used to build the background model. The two initial distributions are then used to produce a tentative segmentation, which is in turn used to rebuild the model. This procedure is iterated until the shape converges

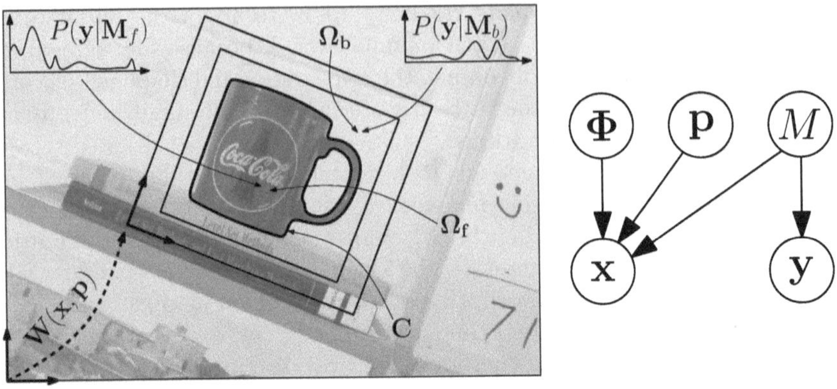

Fig. 1. (Left): Representation of object showing: the contour $\mathbf{C}$, the set of foreground pixels $\boldsymbol{\Omega}_f$, the set of background pixels $\boldsymbol{\Omega}_b$, the foreground model $P(\mathbf{y}|M_f)$, the background model $P(\mathbf{y}|M_b)$ and the warp $\mathbf{W}(\mathbf{x},\mathbf{p})$; (Right): Graphical representation of our generative model representing the image as a bag-of-pixels

(similar to [15]). Once tracking commences the appearance models and shape $\mathbf{C}$ are estimated (adapted) online, as described in Sect. 6. In summary, we use the following notation:

- $\mathbf{x}$: A pixel location in the object coordinate frame.
- $\mathbf{y}$: A pixel value (in our experiments this is a YUV value).
- $\mathbf{I}$: Image.
- $\mathbf{W}(\mathbf{x},\mathbf{p})$: Warp with parameters $\mathbf{p}$.
- $M = \{M_f, M_b\}$: Model parameter either foreground or background.
- $P(\mathbf{y}|M_f)$: Foreground model over pixel values $\mathbf{y}$.
- $P(\mathbf{y}|M_b)$: Background model over pixel values $\mathbf{y}$.
- $\mathbf{C}$: The contour that segments the foreground from background.
- $\boldsymbol{\Phi}(\mathbf{x})$: Shape kernel (in our case the level-set embedding function).
- $\boldsymbol{\Omega} = \{\boldsymbol{\Omega}_f, \boldsymbol{\Omega}_b\}$: Pixels in the object frame $[\{\mathbf{x}_0, \mathbf{y}_0\}, \ldots, \{\mathbf{x}_N, \mathbf{y}_N\}]$, which is partitioned into foreground pixels $\boldsymbol{\Omega}_f$ and background pixels $\boldsymbol{\Omega}_b$.
- $H_\epsilon(z)$: Smoothed Heaviside step function.
- $\delta_\epsilon(z)$: Smoothed Dirac delta function.

Figure 1 illustrates the simple generative model we use to represent the image formation process. This model treats the image as a bag-of-pixels [6] and can, given the model M, the shape $\boldsymbol{\Phi}$ and the location $\mathbf{p}$, be used to sample pixels $\{\mathbf{x},\mathbf{y}\}$. Although the resultant image would not look like the true foreground/background image to a human (the pixels would be jumbled up), the colour distributions corresponding to the foreground/background regions $\boldsymbol{\Omega}_f/\boldsymbol{\Omega}_b$ would match the models $P(\mathbf{y}|M_f)$ and $P(\mathbf{y}|M_b)$. It is this simplicity that gives more invariance to viewpoint and allows 3D objects to be tracked robustly without having to model their specific 3D structure. The joint distribution for a single pixel given by the model in Fig. 1 is:

$$P(\mathbf{x},\mathbf{y},\boldsymbol{\Phi},\mathbf{p},M) = P(\mathbf{x}|\boldsymbol{\Phi},\mathbf{p},M)P(\mathbf{y}|M)P(M)P(\boldsymbol{\Phi})P(\mathbf{p}). \tag{1}$$

We now divide (1) by $P(\mathbf{y}) = \sum_M P(\mathbf{y}|M)P(M)$ to give:

$$P(\mathbf{x}, \mathbf{\Phi}, \mathbf{p}, M|\mathbf{y}) = P(\mathbf{x}|\mathbf{\Phi}, \mathbf{p}, M)P(M|\mathbf{y})P(\mathbf{\Phi})P(\mathbf{p}), \tag{2}$$

where the term $P(M|\mathbf{y})$ is the pixel-wise posterior, of the models M, given a pixel value $\mathbf{y}$:

$$P(M_j|\mathbf{y}) = \frac{P(\mathbf{y}|M_j)P(M_j)}{\sum_{\{i=f,b\}} P(\mathbf{y}|M_i)P(M_i)} \qquad j = \{f, b\}. \tag{3}$$

Using this posterior is equivalent to applying Bayesian model selection to each individual pixel.[1] We now marginalise over the models M, yielding the pixel-wise posterior probability of the shape $\mathbf{\Phi}$ and the location $\mathbf{p}$ given a pixel $\{\mathbf{x}, \mathbf{y}\}$:

$$P(\mathbf{\Phi}, \mathbf{p}|\mathbf{x}, \mathbf{y}) = \frac{1}{P(\mathbf{x})} \sum_{\{i=f,b\}} \{P(\mathbf{x}|\mathbf{\Phi}, \mathbf{p}, M_i)P(M_i|\mathbf{y})\} P(\mathbf{\Phi})P(\mathbf{p}). \tag{4}$$

Note that the pixel-wise posterior and marginalisation are the subtle but crucial differences to the work in [4], which lacks the marginalisation step and uses a pixel-wise likelihood $P(\mathbf{y}|M)$. We show in Sect. 7 that our formulation yields a much better behaved objective. We consider two possible methods for fusing the pixel-wise posteriors: (i) a logarithmic opinion pool (LogOP):

$$P(\mathbf{\Phi}, \mathbf{p}|\mathbf{\Omega}) = \prod_{i=1}^{N} \left\{ \sum_M \{P(\mathbf{x}_i|\mathbf{\Phi}, \mathbf{p}, M)P(M|\mathbf{y}_i)\} \right\} P(\mathbf{\Phi})P(\mathbf{p}) \tag{5}$$

and (ii) a linear opinion pool (LinOP):

$$P(\mathbf{\Phi}, \mathbf{p}|\mathbf{\Omega}) = \sum_{i=1}^{N} \left\{ \sum_M \{P(\mathbf{x}_i|\mathbf{\Phi}, \mathbf{p}, M)P(M|\mathbf{y}_i)\} \right\} P(\mathbf{\Phi})P(\mathbf{p}). \tag{6}$$

The logarithmic opinion pool is normally the preferred choice and is most similar to previous work [4,5]; whereas the linear opinion pool is equivalent to marginalising over the pixel locations – this is allowed as our bag-of-pixels generative model treats pixel locations as a random variable. We continue our derivation assuming a logarithmic opinion pool for clarity, but also include results using a linear opinion pool for completeness. Note the term $\frac{1}{P(\mathbf{x})}$ has been dropped as it is constant for all pixel locations and we only seek to maximise $P(\mathbf{\Phi}, \mathbf{p}|\mathbf{\Omega})$.

3 Segmentation

The typical approach to region based segmentation methods is to take a product of the pixel-wise likelihood functions $\prod_{i=1}^{N} P(\mathbf{I}(x_i)|M_i)$, over pixel locations $\mathbf{x}_i$,

[1] It is also the distribution that would be computed in the E-step of an EM solution to the problem.

to get the overall likelihood $P(\mathbf{I}|M)$. This can then be expressed as a summation by taking logs and optimised using variational level-sets [1,5]. In contrast to these methods, our derivation leads to pixel-wise posteriors and marginalisation (5), a subtle but important difference.

For the remainder of this section, in order to simplify our expressions (and without loss of generality), we assume that the registration is correct and therefore $\mathbf{x}_i = \mathbf{W}(\mathbf{x}_i, \mathbf{p})$. We now specify the term $P(\mathbf{x}_i|\mathbf{\Phi}, \mathbf{p}, M)$ in (5) and the term $P(M)$ in (3) :

$$P(\mathbf{x}_i|\mathbf{\Phi}, \mathbf{p}, M_f) = \frac{H_\epsilon(\mathbf{\Phi}(\mathbf{x}_i))}{\eta_f} \qquad P(\mathbf{x}_i|\mathbf{\Phi}, \mathbf{p}, M_b) = \frac{1 - H_\epsilon(\mathbf{\Phi}(\mathbf{x}_i))}{\eta_b} \tag{7}$$

$$P(M_f) = \frac{\eta_f}{\eta} \qquad P(M_b) = \frac{\eta_b}{\eta}, \tag{8}$$

where

$$\eta = \eta_f + \eta_b, \quad \eta_f = \sum_{i=1}^{N} H_\epsilon(\mathbf{\Phi}(\mathbf{x}_i)), \quad \eta_b = \sum_{i=1}^{N} 1 - H_\epsilon(\mathbf{\Phi}(\mathbf{x}_i)). \tag{9}$$

Equation (7) represents normalised versions of the blurred Heaviside step functions used in typical region based level-set methods and can now be interpreted probabilistically as model specific spatial priors for a pixel location $\mathbf{x}$. Equation (8) represents the model priors, which are given by the ratio of the area of the model specific region to the total area of both models. Equation (9) contains the normalisation constants (note that $\eta = N$).

We now specify a geometric prior on $\mathbf{\Phi}$ that rewards a signed distance function:

$$P(\mathbf{\Phi}) = \prod_{i=1}^{N} \frac{1}{\sigma\sqrt{2\pi}} \exp -\frac{(|\nabla\mathbf{\Phi}(\mathbf{x}_i)| - 1)^2}{2\sigma^2}, \tag{10}$$

where σ specifies the relative weight of the prior. This gives a probabilistic interpretation to the work in [10]. Substituting (7), (8), (9) and (10) into (5) and taking logs, gives the following expression for the log posterior:

$$\log(P(\mathbf{\Phi}, \mathbf{p}|\mathbf{\Omega})) \propto \sum_{i=1}^{N} \left\{ \log\left(P(\mathbf{x}_i|\mathbf{\Phi}, \mathbf{p}, \mathbf{y}_i)\right) - \frac{(|\nabla\mathbf{\Phi}(\mathbf{x}_i)| - 1)^2}{2\sigma^2} \right\} + \\ N \log\left(\frac{1}{\sigma\sqrt{2\pi}}\right) + \log(P(\mathbf{p})), \tag{11}$$

where

$$P(\mathbf{x}_i|\mathbf{\Phi}, \mathbf{p}, \mathbf{y}_i) = H_\epsilon(\mathbf{\Phi}(\mathbf{x}_i))P_f + (1 - H_\epsilon(\mathbf{\Phi}(\mathbf{x}_i)))P_b$$

and

$$P_f = \frac{P(\mathbf{y}_i|M_f)}{\eta_f P(\mathbf{y}_i|M_f) + \eta_b P(\mathbf{y}_i|M_b)} \qquad P_b = \frac{P(\mathbf{y}_i|M_b)}{\eta_f P(\mathbf{y}_i|M_f) + \eta_b P(\mathbf{y}_i|M_b)}.$$

Given that we are about to optimise w.r.t to $\mathbf{\Phi}$ we can drop the last two terms in (11) and by calculus of variations [16] express the first variation (Gateaux derivative) of the functional as:

$$\frac{\partial P(\mathbf{\Phi}, \mathbf{p}|\mathbf{\Omega})}{\partial \mathbf{\Phi}} = \frac{\delta_\epsilon(\mathbf{\Phi})(P_f - P_b)}{P(\mathbf{x}|\mathbf{\Phi}, \mathbf{p}, \mathbf{y})} - \frac{1}{\sigma^2}\left[\nabla^2\mathbf{\Phi} - \text{div}\left(\frac{\nabla\mathbf{\Phi}}{|\nabla\mathbf{\Phi}|}\right)\right], \tag{12}$$

where ∇^2 is the Laplacian operator and $\delta_\epsilon(\mathbf{\Phi})$ is the derivative of the blurred Heaviside step function, i.e. a blurred Dirac delta function. Interestingly, $\delta_\epsilon(\mathbf{\Phi})$ can now be interpreted as a way of expressing uncertainty on the contour $\mathbf{C}$. If we were to use Gaussian uncertainty for the contour then the region based uncertainty would be expressed in terms of $\text{erf}(\mathbf{\Phi})$ instead of $H_\epsilon(\mathbf{\Phi})$ (we do not explore this any further in this paper). We seek $\frac{\partial P(\mathbf{\Phi},\mathbf{p}||\mathbf{\Omega})}{\partial \mathbf{\Phi}} = 0$ by carrying out steepest-ascent using the following gradient flow:

$$\frac{\partial P(\mathbf{\Phi}, \mathbf{p}|\mathbf{\Omega})}{\partial t} = \frac{\partial P(\mathbf{\Phi}, \mathbf{p}|\mathbf{\Omega})}{\partial \mathbf{\Phi}}. \tag{13}$$

In practice this is implemented using a simple numerical scheme on a discrete grid. All spatial derivatives are computed using central differences and the Laplacian uses a 3×3 spatial kernel. We use $\sigma = \sqrt{50}$ and a timestep $\tau = 1$ for all experiments. For stability $\frac{\tau}{\sigma^2} < 0.25$ must be satisfied (see [10] for details).

4 Tracking

It is possible to pose the tracking problem directly in a segmentation framework [8]. Instead, like [4] we model the frame-to-frame registration explicitly, by having the level-set in the object frame and introducing a warp $\mathbf{W}(\mathbf{x}, \mathbf{p})$ into (11). The main benefits of this approach are: (i) control over the interaction between registration (tracking) and segmentation (local shape deformation); (ii) by registering the embedding function first, fewer iterations are required to take account of shape changes (in fact we find one per frame is adequate for our sequences). We now drop any terms in (11) that are not a function of $\mathbf{p}$ in preparation for differentiation:

$$\log(P(\mathbf{\Phi}, \mathbf{p}|\mathbf{\Omega})) \propto \sum_{i=1}^{N}\Big\{\log\left(P(\mathbf{x}_i|\mathbf{\Phi}, \mathbf{p}, \mathbf{y}_i)\right)\Big\} + \log(P(\mathbf{p})) + \mathit{const}. \tag{14}$$

Introducing a warp $\mathbf{W}(\mathbf{x}_i, \Delta\mathbf{p})$ into (14) and dropping the prior term for brevity (we revisit this term in Sect. 5):

$$\log(P(\mathbf{\Phi}, \mathbf{p}|\mathbf{\Omega})) \propto \sum_{i=1}^{N}\Big\{\log\left(P(\mathbf{W}(\mathbf{x}_i, \Delta\mathbf{p})|\mathbf{\Phi}, \mathbf{p}, \mathbf{y}_i)\right)\Big\}, \tag{15}$$

where $\Delta\mathbf{p}$ represents an incremental warp of the shape kernel. There are many ways this expression could be optimised, the most similar work uses simple gradient ascent [4]. In contrast, we take advantage of the fact that all of the individual terms are probabilities, and therefore strictly positive. This allows us to write certain terms as squared square-roots and substitute in a first-order Taylor series approximation for each square-root, for example:

$$\left[\sqrt{H_\epsilon(\mathbf{\Phi}(\mathbf{W}(\mathbf{x}_i, \Delta\mathbf{p}))}\right]^2 \approx \left[\sqrt{H_\epsilon(\mathbf{\Phi}(\mathbf{x}_i))} + \frac{1}{2\sqrt{H_\epsilon(\mathbf{\Phi}(\mathbf{x}_i))}}\mathbf{J}\Delta\mathbf{p}\right]^2, \tag{16}$$

where:

$$\mathbf{J} = \frac{\partial H_\epsilon}{\partial \mathbf{\Phi}} \frac{\partial \mathbf{\Phi}}{\partial \mathbf{x}} \frac{\partial \mathbf{W}}{\partial \Delta\mathbf{p}} = \delta_\epsilon(\mathbf{\Phi}(\mathbf{x}_i)) \, \nabla\mathbf{\Phi}(\mathbf{x}_i) \, \frac{\partial \mathbf{W}}{\partial \Delta\mathbf{p}}.$$

Likewise we apply a similar expansion to $(1 - H_\epsilon(\mathbf{\Phi}(\mathbf{W}(\mathbf{x}_i, \Delta\mathbf{p})))$, allowing us then to optimise using Gauss-Newton[2]. This has the advantage that the Hessian itself is not required, rather, a first-order approximation of the Hessian is used. In consequence it is fast, and in our experience exhibits rapid and reliable convergence in our problem domain. It also avoids the issues highlighted in [17] of choosing the appropriate step size for gradient ascent. Excluding the full details for brevity we arrive at an expression for $\Delta\mathbf{p}$:

$$\Delta\mathbf{p} = \left[\sum_{i=1}^{N} \frac{1}{2P(\mathbf{x}_i|\mathbf{\Phi}, \mathbf{p}, \mathbf{y}_i)} \left(\frac{P_f}{H(\mathbf{\Phi}(\mathbf{x}_i))} + \frac{P_b}{(1 - H(\mathbf{\Phi}(\mathbf{x}_i)))}\right) \mathbf{J}^T\mathbf{J}\right]^{-1} \times$$
$$\sum_{i=1}^{N} \frac{(P_f - P_b)\mathbf{J}^T}{P(\mathbf{x}_i|\mathbf{\Phi}, \mathbf{p}, \mathbf{y}_i)}. \tag{17}$$

Equation (17) is then used to update the parameters $\mathbf{p}$ by composing $\mathbf{W}(\mathbf{x}_i, \mathbf{p})$ with $\mathbf{W}(\mathbf{x}_i, \Delta\mathbf{p})^{-1}$, analogous to inverse compositional tracking [14].

5 Drift Correction

Having the object represented by its location $\mathbf{p}$ and shape $\mathbf{\Phi}$ leaves an ambiguity where it is possible to explain rigid transformations of the shape either with $\mathbf{p}$ or $\mathbf{\Phi}$. Ideally, any rigid motion would be explained solely by $\mathbf{p}$; however, over time the shape $\mathbf{\Phi}$ slowly incorporates a rigid transformation. We define a prior on the location $P(\mathbf{p})$ which makes small corrections to keep the left/right and top/bottom borders[3] balanced and the minimum border distance equal to four pixels. This is implemented using a proportional controller that takes the four border distances as its input and outputs the prior $P(\mathbf{p})$.

[2] The Taylor expansion is poorly conditioned if $H_\epsilon(\mathbf{\Phi}(\mathbf{x}_i) = 0$; in practice this does not happen as the terms are never equal to zero.

[3] Smallest distances between the contour and the corresponding side of the foreground box.

6 Online Learning

Once registration and segmentation are completed both the foreground and background models are adapted online. This is achieved using linear opinion pools with variable learning rates α_i, $i = \{f, b\}$:

$$P_t(\mathbf{y}|M_i) = (1 - \alpha_i)P_{t-1}(\mathbf{y}|M_i) + \alpha_i P_t(\mathbf{y}|M_i), \qquad i = \{f, b\}. \tag{18}$$

In all experiments $\alpha_f = 0.02$ and $\alpha_b = 0.025$. For shape adaptation we control the evolution rate of the level-set using the timestep τ. Ideally these three parameters would change dynamically throughout the sequence to prevent learning occurring during times of confusion or if the object is lost; we intend to do this in future work.

7 Results

We have tested our system extensively on live video, as well as on a variety of recorded sequences which include objects that exhibit rapid and agile motion with significant motion blur, varying lighting, moving cameras, and cluttered and changing backgrounds. Figure 2 shows a qualitative evaluation of our method on three sequences. The first is a speedboat undergoing a 180° out-of-plane rotation – note how the shape is adapted online. The second is a person jumping around – note the motion blur and shape adaptation. Finally, the third is a hand being tracked from a head mounted camera past a challenging background which has a similar appearance to the object.

To perform a quantitative evaluation we have analysed the characteristics of the underlying cost function of our technique and compared this with competing alternatives, on a set of pre-recorded video sequences. Figure 3 shows still images taken mid-sequence from a subset of these sequences; the minimum length is 400 frames and the total number of frames is over 20,000. To facilitate visualisation of the results we use a 2D rigid transformation + scale, considering each of the four dimensions separately. The competing cost functions considered correspond to the following alternative methods of tracking: level-set methods based on likelihoods [4,2], mean-shift [12,13,11], inverse compositional [14] and distribution based tracking [8,9].

A good cost function has a single extremum at the true location. A poor one has multiple extrema and any local optimisation technique is liable to fall into one of these, which in practice is often the start of tracking failure. For each video frame and each dimension (translation in x and y, rotation and scale) we compute the objectives for the competing cost functions at 40 evenly spaced points over an interval centred at the true state. We then extract all local extrema from these objectives and examine how they are distributed across the interval. To summarise this information we compute a distribution for each dimension and each cost function, using our collection of over 20,000 frames. The ideal distribution would be a delta function centred on the true state; whereas a good distribution would be peaky around the true state and have low probability of local extrema within the region it will be required to converge from. A bad

Fig. 2. Qualitative evaluation: (top) a speedboat undergoing a 180° out-of-plane rotation illustrating shape adaptation; (middle) a person jumping around with significant motion blur; (bottom) a hand being tracked in front of a challenging background

Fig. 3. A selection of video frames from the data sets: (1st row) lifeboat, Coca-Cola mug, a face and a hand filmed from a head mounted camera; (2nd row) a hand using a mouse, a speedboat, a person from the caviar data set [18] and a tractor. The white indicates the current segmentation and the two black boxes indicate the object coordinate frame.

distribution would be relatively flat with high probability of local extrema over the entire space. The particular cost functions we consider are:

- **LogPWP:** Pixel-wise posteriors fused using a logarithmic opinion pool.
- **LinPWP:** Pixel-wise posteriors fused using a linear opinion pool.
- **LogLike:** Log likelihood, used in most level-set work [5,4,2].

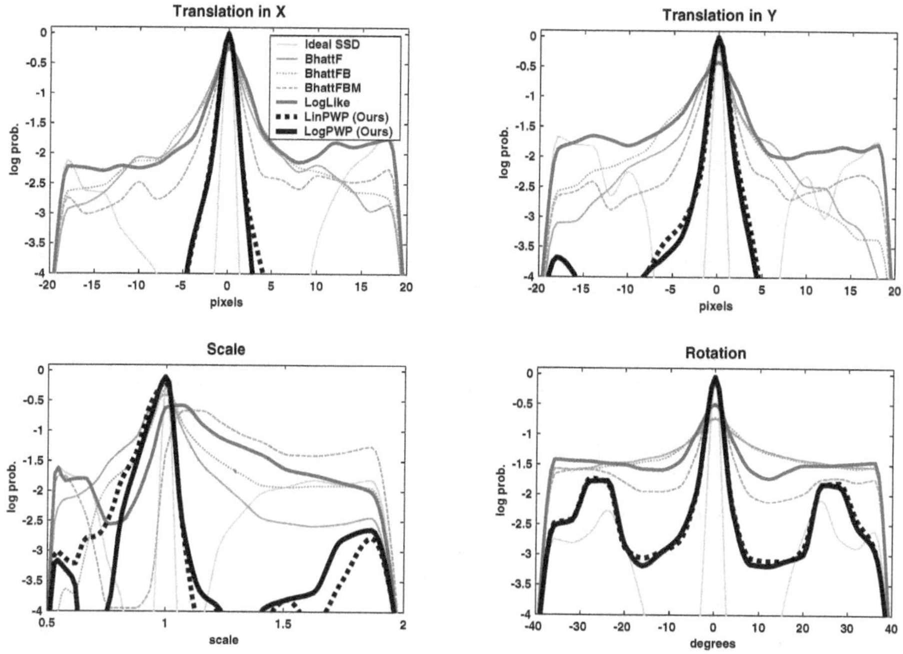

Fig. 4. Quantitative Analysis: Log probability distribution of extrema in the cost functions generated from 20,000 frames of real video data

- **BhattF:** Bhattacharyya coefficient:
  $B(\mathbf{\Omega}_f) = \sum_{j=1}^{V} \sqrt{P(\mathbf{y}_j|M_f)P(\mathbf{y}_j|\mathbf{\Omega}_f)}$, used by [12,13,11].
- **BhattFB:** Bhattacharyya coefficient with a background model:
  $B(\mathbf{\Omega}_f, \mathbf{\Omega}_b) = \sum_{j=1}^{V} \sqrt{P(\mathbf{y}_j|M_f)p(\mathbf{y}_j|\mathbf{\Omega}_f)} + \sum_{j=1}^{V} \sqrt{P(\mathbf{y}_j|M_b)P(\mathbf{y}_j|\mathbf{\Omega}_b)}$.
- **BhattFBM:** Bhattacharyya coefficient with a background mismatch:
  $B(\mathbf{\Omega}_f, \mathbf{\Omega}_b) = \sum_{j=1}^{V} \sqrt{P(\mathbf{y}|M_f)p(\mathbf{y};\mathbf{\Omega}_f)} - \sum_{j=1}^{V} \sqrt{P(\mathbf{y}_j|M_f)P(\mathbf{y}_j|\mathbf{\Omega}_b)}$, suggested by [9].
- **Ideal SSD:** Sum of squared pixel differences using the ideal template i.e. the template extracted at the current location $\mathbf{p}$. This is essentially what you would get if you had the perfect generative model giving the true pixel value at each pixel location including the noise. This of course is *never* going to be achievable but has been included as a useful benchmark and gives an indication of what effect incorporating texture may have.

Note: V is the number of pixel values i.e. $32 \times 32 \times 32$; $P(\mathbf{y}|\mathbf{\Omega}_i)$ $i = \{f, b\}$ is the empirical density built from the pixels $\mathbf{\Omega}_i$ and when computing Bhattacharyya coefficients we weight the contribution of each pixel according to our shape kernel, which is identical to Yilmaz's work [11].

Figure 4 shows distributions generated from over 20,000 real video frames for: translation in x, translation in y, scale and rotation.

- **Translation in x and y:** Our method has narrower distributions near the true state than all methods apart from ideal SSD and is significantly better than the log likelihood used by [4]. Unlike the other methods it also exhibits virtually no extrema outside a ± 5 pixel region – this means that our methods will converge to within ± 5 pixels of the true state from anywhere within the ± 20 pixel space we have evaluated.
- **Scale:** The Bhattacharyya method and Bhattacharyya with background mismatch both have poor localisation in scale, which is in agreement with the findings of many authors. The log likelihood also poorly localises scale compared with our posterior based methods.
- **Rotation:** All Bhattacharyya methods and the log likelihood are poor at correctly localising the rotation. The straight Bhattacharyya coefficient for example has more than a 1% chance of exhibiting extrema anywhere in the rotation space, at a 30Hz frame rate this corresponds to approximately 1 frame in every 3 seconds of video. It is worth noting that the side lobes (at approximately 25°) exhibited by our methods and ideal SSD are due to the self similarity corresponding to fingers in the hand sequences.

Experimentally we were unable to make the log likelihood successfully track several of our sequences, which is confirmed by its poor performance in Fig. 4. One possible explanation is that in other work [4,5,2], a single Gaussian parametric model is used. This implicitly enforces a smooth, unimodal distribution for the joint likelihood. Non-parametric representations do not exhibit these properties; however, they are better at describing complicated distributions and therefore desirable. The reason that our method can deal with these distributions is because of the normalising denominator in (3) and the marginalisation step in (4). These two steps prevent individual pixels from dominating the cost function hence making it smoother and well-behaved.

The work of [8] and its subsequent improvement [9] use distribution matching techniques to incorporate non-parametric distributions into a level-set framework. These methods, similar to the Bhattacharyya based methods, involve computing the empirical densities at every iteration of the optimisation, whereas our method avoids this extra cost. Not only is our method superior to these approaches in terms of cost functions (see Fig. 4), but it is computationally cheaper to evaluate as it does not require empirical distributions. This is a significant benefit because it not only reduces the cost per iteration, but avoids the issue of having to build 'good' distributions. One explanation for difference between the performance of these methods and ours, is that it is hard to build 'good' empirical distributions in real-time and most methods rely on simple histograms. Although this could be improved with Parzen or NP windowing techniques [19], it would almost certainly sacrifice real-time performance.

7.1 Timing

All terms in (17) include $\delta_\epsilon(\mathbf{\Phi}(\mathbf{x}_i))$ (blurred Dirac delta function). This means that an individual pixel's contribution to the optimisation diminishes the further from the contour it is. An efficient implementation, therefore, recognises this. Our implementation ignores pixels outside a narrow band and for an object size of 180×180 runs in 500μs on a P4 3.6GHz machine. On average the system runs at a frame rate of 85Hz for the complete algorithm and if shape and appearance learning are turned off (i.e. rigid registration only) it averages 230Hz.

8 Conclusions

We have proposed a novel probabilistic framework for robust, real-time, visual tracking of previously unseen objects from a moving camera. The key contribution of our method and reason for its superior performance compared with others is the use of pixel-wise posteriors as opposed to a product over pixel-wise likelihoods. In contrast to other methods [4,5], we solve the registration using Gauss Newton, which has significant practical benefits, namely: (i) the difficulty associated with step size selection is removed and (ii) reliable and fast convergence. We have demonstrated the benefits of our method both qualitatively and quantitatively with a thorough analysis of pixel-wise posteriors versus competing alternatives using over 20,000 video frames. Our results demonstrate that using pixel-wise posteriors provides excellent performance when incorporating non-parametric distributions into region based level-sets. It not only offers superior cost functions but avoids the need for computing empirical distributions [12,8,9,11] and is therefore faster.

In our ongoing research we are investigating in more detail the probabilistic interpretation of the blurred Heaviside step functions, and in particular the link with Gaussian uncertainty on contour location. Other work of future interest involves modifying the generative model to capture spatial information, and investigating how to incorporate multiple objects and online occlusion handling.

References

1. Osher, S., Paragios, N.: Geometric Level Set Methods in Imaging, Vision and Graphics. Springer, Secaucus, NJ, USA (2003)
2. Paragios, N., Deriche, R.: Geodesic active contours and level sets for the detection and tracking of moving objects. IEEE Transactions on Pattern Analysis and Machine Intelligence 22(3), 266–280 (2000)
3. Goldenberg, R., Kimmel, R., Rivlin, E., Rudzsky, M.: Fast geodesic active contours. IEEE Trans. on Image Processing 10(10), 1467–1475 (2001)
4. Cremers, D.: Dynamical statistical shape priors for level set based tracking. IEEE Transactions on Pattern Analysis and Machine Intelligence 28(8), 1262–1273 (2006)
5. Cremers, D., Rousson, M., Deriche, R.: A review of statistical approaches to level set segmentation: Integrating color, texture, motion and shape. International Journal of Computer Vision V72(2), 195–215 (2007)
6. Jebara, T.: Images as bags of pixels. In: Proc. 9th Int'l Conf. on Computer Vision, Nice (2003)
7. Chan, T., Vese, L.: Active contours without edges. IEEE Trans. Image Processing 10(2), 266–277 (2001)

8. Freedman, D., Zhang, T.: Active contours for tracking distributions. IEEE Transactions on Image Processing 13(4), 518–526 (2004)
9. Zhang, T., Freedman, D.: Improving performance of distribution tracking through background matching. IEEE Transactions on Pattern Analysis and Machine Intelligence 27(2), 282–287 (2005)
10. Li, C., Xu, C., Gui, C., Fox, M.D.: Level set evolution without re-initialization: A new variational formulation. In: Proc. 22nd IEEE Conf. on Computer Vision and Pattern Recognition, San Diego, California, vol. 1, pp. 430–436. IEEE Computer Society, Los Alamitos (2005)
11. Yilmaz, A.: Object tracking by asymmetric kernel mean shift with automatic scale and orientation selection. In: Proc. IEEE Conf. on Computer Vision and Pattern Recognition, Minneapolis, Minnesota (2007)
12. Comaniciu, D., Ramesh, V., Meer, P.: Real-time tracking of non-rigid objects using mean shift. In: Proc. 19th IEEE Conf. on Computer Vision and Pattern Recognition, Hilton Head Island, vol. 2, pp. 142–149 (2000)
13. Collins, R.T.: Mean-shift blob tracking through scale space. In: IEEE Conference on Computer Vision and Pattern Recognition, vol. 2, pp. 234–240 (2003)
14. Baker, S., Matthews, I.: Lukas-kanade 20 years on: A unifiying framework. International Journal of Computer Vision 69(3), 221–255 (2004)
15. Rother, C., Kolmogorov, V., Blake, A.: Grabcut: Interactive foreground extraction using iterated graph cuts. In: ACM Transactions on Graphics (SIGGRAPH 2004) (2004)
16. Evans, L.C.: Partial Differential Equations. AMS (2002)
17. Cremers, D., Osher, S.J., Soatto, S.: Kernel density estimation and intrinsic alignment for shape priors in level set segmentation. International Journal of Computer Vision 69(3), 335–351 (2006)
18. Fisher, R.: CAVIAR Test Case Scenarios, EC Funded IST 2001 37540. Online Book (October 2004)
19. Kadir, T., Brady, M.: Estimating statistics in arbitrary regions of interest. In: Proc. 16th British Machine Vision Conf., Oxford (2005)

Author Index

Lecture Notes in Computer Science

Sublibrary 6: Image Processing, Computer Vision, Pattern Recognition, and Graphics

For information about Vols. 1–4091
please contact your bookseller or Springer

Vol. 5303: D. Forsyth, P. Torr, A. Zisserman (Eds.), Computer Vision – ECCV 2008, Part II. XIX, 851 pages. 2008.

Vol. 5302: D. Forsyth, P. Torr, A. Zisserman (Eds.), Computer Vision – ECCV 2008, Part I. XXXVII, 801 pages. 2008.

Vol. 5259: J. Blanc-Talon, S. Bourennane, W. Philips, D. Popescu, P. Scheunders (Eds.), Advanced Concepts for Intelligent Vision Systems. XXIII, 1138 pages. 2008.

Vol. 5242: D. Metaxas, L. Axel, G. Fichtinger, G. Székely (Eds.), Medical Image Computing and Computer-Assisted Intervention – MICCAI 2008, Part II. LI, 1111 pages. 2008.

Vol. 5241: D. Metaxas, L. Axel, G. Fichtinger, G. Székely (Eds.), Medical Image Computing and Computer-Assisted Intervention – MICCAI 2008, Part I. LI, 1087 pages. 2008.

Vol. 5197: J. Ruiz-Shulcloper, W.G. Kropatsch (Eds.), Progress in Pattern Recognition, Image Analysis and Applications. XXI, 809 pages. 2008.

Vol. 5188: M. Sebillo, G. Vitiello, G. Schaefer (Eds.), Visual Information Systems. XIV, 340 pages. 2008.

Vol. 5166: A. Butz, B. Fisher, A. Krüger, P. Olivier, M. Christie (Eds.), Smart Graphics. XI, 277 pages. 2008.

Vol. 5158: S.N. Srihari, K. Franke (Eds.), Computational Forensics. IX, 229 pages. 2008.

Vol. 5128: T. Dohi, I. Sakuma, H. Liao (Eds.), Medical Imaging and Augmented Reality. XVI, 441 pages. 2008.

Vol. 5116: E.A. Krupinski (Ed.), Digital Mammography. XXVII, 769 pages. 2008.

Vol. 5112: A. Campilho, M.S. Kamel (Eds.), Image Analysis and Recognition. XXII, 1126 pages. 2008.

Vol. 5099: A. Elmoataz, O. Lezoray, F. Nouboud, D. Mammass (Eds.), Image and Signal Processing. XVI, 625 pages. 2008.

Vol. 5098: F.J. Perales, R.B. Fisher (Eds.), Articulated Motion and Deformable Objects. XIV, 458 pages. 2008.

Vol. 5096: G. Rigoll (Ed.), Pattern Recognition. XIII, 538 pages. 2008.

Vol. 5046: W. Liu, J. Lladós, J.-M. Ogier (Eds.), Graphics Recognition. IX, 336 pages. 2008.

Vol. 4992: D. Coeurjolly, I. Sivignon, L. Tougne, F. Dupont (Eds.), Discrete Geometry for Computer Imagery. XIII, 554 pages. 2008.

Vol. 4987: X. Gao, H. Müller, M.J. Loomes, R. Comley, S. Luo (Eds.), Medical Imaging and Informatics. XV, 388 pages. 2008.

Vol. 4958: V.E. Brimkov, R.P. Barneva, H.A. Hauptman (Eds.), Combinatorial Image Analysis. XVI, 446 pages. 2008.

Vol. 4931: G. Sommer, R. Klette (Eds.), Robot Vision. XI, 468 pages. 2008.

Vol. 4901: D. Zhang (Ed.), Medical Biometrics. XII, 324 pages. 2007.

Vol. 4889: A. Pasko, V. Adzhiev, P. Comninos (Eds.), Heterogeneous Objects Modelling and Applications. VII, 285 pages. 2008.

Vol. 4844: Y. Yagi, S.B. Kang, I.S. Kweon, H. Zha (Eds.), Computer Vision – ACCV 2007, Part II. XXVIII, 915 pages. 2007.

Vol. 4843: Y. Yagi, S.B. Kang, I.S. Kweon, H. Zha (Eds.), Computer Vision – ACCV 2007, Part I. XXVIII, 969 pages. 2007.

Vol. 4842: G. Bebis, R. Boyle, B. Parvin, D. Koracin, N. Paragios, S.-M. Tanveer, T. Ju, Z. Liu, S. Coquillart, C. Cruz-Neira, T. Müller, T. Malzbender (Eds.), Advances in Visual Computing, Part II. XXXIII, 827 pages. 2007.

Vol. 4841: G. Bebis, R. Boyle, B. Parvin, D. Koracin, N. Paragios, S.-M. Tanveer, T. Ju, Z. Liu, S. Coquillart, C. Cruz-Neira, T. Müller, T. Malzbender (Eds.), Advances in Visual Computing, Part I. XXXIII, 831 pages. 2007.

Vol. 4815: A. Ghosh, R.K. De, S.K. Pal (Eds.), Pattern Recognition and Machine Intelligence. XIX, 677 pages. 2007.

Vol. 4814: A. Elgammal, B. Rosenhahn, R. Klette (Eds.), Human Motion – Understanding, Modeling, Capture and Animation. X, 329 pages. 2007.

Vol. 4792: N. Ayache, S. Ourselin, A. Maeder (Eds.), Medical Image Computing and Computer-Assisted Intervention – MICCAI 2007, Part II. XLVI, 988 pages. 2007.

Vol. 4791: N. Ayache, S. Ourselin, A. Maeder (Eds.), Medical Image Computing and Computer-Assisted Intervention – MICCAI 2007, Part I. XLVI, 1012 pages. 2007.

Vol. 4781: G. Qiu, C. Leung, X.-Y. Xue, R. Laurini (Eds.), Advances in Visual Information Systems. XIII, 582 pages. 2007.

Vol. 4778: S.K. Zhou, W. Zhao, X. Tang, S. Gong (Eds.), Analysis and Modeling of Faces and Gestures. X, 305 pages. 2007.

Vol. 4768: D. Doermann, S. Jaeger (Eds.), Arabic and Chinese Handwriting Recognition. VIII, 279 pages. 2008.

Vol. 4756: L. Rueda, D. Mery, J. Kittler (Eds.), Progress in Pattern Recognition, Image Analysis and Applications. XXI, 989 pages. 2007.

Vol. 4738: A.C.R. Paiva, R. Prada, R.W. Picard (Eds.), Affective Computing and Intelligent Interaction. XVIII, 781 pages. 2007.

Vol. 4729: F. Mele, G. Ramella, S. Santillo, F. Ventriglia (Eds.), Advances in Brain, Vision, and Artificial Intelligence. XVI, 618 pages. 2007.

Vol. 4713: F.A. Hamprecht, C. Schnörr, B. Jähne (Eds.), Pattern Recognition. XIII, 560 pages. 2007.

Vol. 4679: A.L. Yuille, S.-C. Zhu, D. Cremers, Y. Wang (Eds.), Energy Minimization Methods in Computer Vision and Pattern Recognition. XII, 494 pages. 2007.

Vol. 4678: J. Blanc-Talon, W. Philips, D. Popescu, P. Scheunders (Eds.), Advanced Concepts for Intelligent Vision Systems. XXIII, 1100 pages. 2007.

Vol. 4673: W.G. Kropatsch, M. Kampel, A. Hanbury (Eds.), Computer Analysis of Images and Patterns. XX, 1006 pages. 2007.

Vol. 4642: S.-W. Lee, S.Z. Li (Eds.), Advances in Biometrics. XX, 1216 pages. 2007.

Vol. 4633: M.S. Kamel, A. Campilho (Eds.), Image Analysis and Recognition. XII, 1312 pages. 2007.

Vol. 4625: R. Stiefelhagen, R. Bowers, J.G. Fiscus (Eds.), Multimodal Technologies for Perception of Humans. XIII, 556 pages. 2008.

Vol. 4584: N. Karssemeijer, B. Lelieveldt (Eds.), Information Processing in Medical Imaging. XX, 777 pages. 2007.

Vol. 4569: A. Butz, B. Fisher, A. Krüger, P. Olivier, S. Owada (Eds.), Smart Graphics. IX, 237 pages. 2007.

Vol. 4538: F. Escolano, M. Vento (Eds.), Graph-Based Representations in Pattern Recognition. XII, 416 pages. 2007.

Vol. 4522: B.K. Ersbøll, K.S. Pedersen (Eds.), Image Analysis. XVIII, 989 pages. 2007.

Vol. 4485: F. Sgallari, A. Murli, N. Paragios (Eds.), Scale Space and Variational Methods in Computer Vision. XV, 931 pages. 2007.

Vol. 4478: J. Martí, J.M. Benedí, A.M. Mendonça, J. Serrat (Eds.), Pattern Recognition and Image Analysis, Part II. XXVII, 657 pages. 2007.

Vol. 4477: J. Martí, J.M. Benedí, A.M. Mendonça, J. Serrat (Eds.), Pattern Recognition and Image Analysis, Part I. XXVII, 625 pages. 2007.

Vol. 4472: M. Haindl, J. Kittler, F. Roli (Eds.), Multiple Classifier Systems. XI, 524 pages. 2007.

Vol. 4466: F.B. Sachse, G. Seemann (Eds.), Functional Imaging and Modeling of the Heart. XV, 486 pages. 2007.

Vol. 4418: A. Gagalowicz, W. Philips (Eds.), Computer Vision/Computer Graphics Collaboration Techniques. XV, 620 pages. 2007.

Vol. 4417: A. Kerren, A. Ebert, J. Meyer (Eds.), Human-Centered Visualization Environments. XIX, 403 pages. 2007.

Vol. 4391: Y. Stylianou, M. Faundez-Zanuy, A. Esposito (Eds.), Progress in Nonlinear Speech Processing. XII, 269 pages. 2007.

Vol. 4370: P.P. Lévy, B. Le Grand, F. Poulet, M. Soto, L. Darago, L. Toubiana, J.-F. Vibert (Eds.), Pixelization Paradigm. XV, 279 pages. 2007.

Vol. 4358: R. Vidal, A. Heyden, Y. Ma (Eds.), Dynamical Vision. IX, 329 pages. 2007.

Vol. 4338: P.K. Kalra, S. Peleg (Eds.), Computer Vision, Graphics and Image Processing. XV, 965 pages. 2006.

Vol. 4319: L.-W. Chang, W.-N. Lie (Eds.), Advances in Image and Video Technology. XXVI, 1347 pages. 2006.

Vol. 4292: G. Bebis, R. Boyle, B. Parvin, D. Koracin, P. Remagnino, A. Nefian, G. Meenakshisundaram, V. Pascucci, J. Zara, J. Molineros, H. Theisel, T. Malzbender (Eds.), Advances in Visual Computing, Part II. XXXII, 906 pages. 2006.

Vol. 4291: G. Bebis, R. Boyle, B. Parvin, D. Koracin, P. Remagnino, A. Nefian, G. Meenakshisundaram, V. Pascucci, J. Zara, J. Molineros, H. Theisel, T. Malzbender (Eds.), Advances in Visual Computing, Part I. XXXI, 916 pages. 2006.

Vol. 4245: A. Kuba, L.G. Nyúl, K. Palágyi (Eds.), Discrete Geometry for Computer Imagery. XIII, 688 pages. 2006.

Vol. 4241: R.R. Beichel, M. Sonka (Eds.), Computer Vision Approaches to Medical Image Analysis. XI, 262 pages. 2006.

Vol. 4225: J.F. Martínez-Trinidad, J.A. Carrasco Ochoa, J. Kittler (Eds.), Progress in Pattern Recognition, Image Analysis and Applications. XIX, 995 pages. 2006.

Vol. 4191: R. Larsen, M. Nielsen, J. Sporring (Eds.), Medical Image Computing and Computer-Assisted Intervention – MICCAI 2006, Part II. XXXVIII, 981 pages. 2006.

Vol. 4190: R. Larsen, M. Nielsen, J. Sporring (Eds.), Medical Image Computing and Computer-Assisted Intervention – MICCAI 2006, Part I. XXXVVIII, 949 pages. 2006.

Vol. 4179: J. Blanc-Talon, W. Philips, D. Popescu, P. Scheunders (Eds.), Advanced Concepts for Intelligent Vision Systems. XXIV, 1224 pages. 2006.

Vol. 4174: K. Franke, K.-R. Müller, B. Nickolay, R. Schäfer (Eds.), Pattern Recognition. XX, 773 pages. 2006.

Vol. 4170: J. Ponce, M. Hebert, C. Schmid, A. Zisserman (Eds.), Toward Category-Level Object Recognition. XI, 618 pages. 2006.

Vol. 4153: N. Zheng, X. Jiang, X. Lan (Eds.), Advances in Machine Vision, Image Processing, and Pattern Analysis. XIII, 506 pages. 2006.

Vol. 4142: A. Campilho, M.S. Kamel (Eds.), Image Analysis and Recognition, Part II. XXVII, 923 pages. 2006.

Vol. 4141: A. Campilho, M.S. Kamel (Eds.), Image Analysis and Recognition, Part I. XXVIII, 939 pages. 2006.

Vol. 4122: R. Stiefelhagen, J.S. Garofolo (Eds.), Multimodal Technologies for Perception of Humans. XII, 360 pages. 2007.

Vol. 4109: D.-Y. Yeung, J.T. Kwok, A. Fred, F. Roli, D. de Ridder (Eds.), Structural, Syntactic, and Statistical Pattern Recognition. XXI, 939 pages. 2006.

SPRINGER NATURE

GPSR Compliance

The European Union's (EU) General Product Safety Regulation (GPSR) is a set of rules that requires consumer products to be safe and our obligations to ensure this.

If you have any concerns about our products, you can contact us on ProductSafety@springernature.com

In case Publisher is established outside the EU, the EU authorized representative is:

Springer Nature Customer Service Center GmbH
Europaplatz 3
69115 Heidelberg, Germany

FSC
www.fsc.org
MIX
Papier | Fördert
gute Waldnutzung
FSC® C083411

Kolibri 360 GmbH
Ferdinand-Jühlke-Straße 7
99095 Erfurt, Deutschland
produktsicherheit@kolibri360.de